ACOUSTIC BEHAVIOUR OF ANIMALS

SOLE DISTRIBUTORS FOR THE UNITED STATES AND CANADA
AMERICAN ELSEVIER PUBLISHING COMPANY, INC.
52 VANDERBILT AVENUE, NEW YORK 17, N.Y.

SOLE DISTRIBUTORS FOR GREAT BRITAIN
ELSEVIER PUBLISHING COMPANY LIMITED
12B, RIPPLESIDE COMMERCIAL ESTATE
RIPPLE ROAD, BARKING, ESSEX

Library of Congress Catalog Card Number 61-8847

With 415 illustrations and 65 tables

ALL RIGHTS RESERVED
THIS BOOK OR ANY PART THEREOF MAY NOT BE REPRODUCED IN ANY FORM,
INCLUDING PHOTOSTATIC OR MICROFILM FORM,
WITHOUT WRITTEN PERMISSION FROM THE PUBLISHERS

ACOUSTIC BEHAVIOUR OF ANIMALS

EDITED BY

R.-G. BUSNEL

Laboratoire de Physiologie Acoustique,
Institut National de la Recherche Agronomique,
Centre National de Recherches Zootechniques,
Jouy-en-Josas, S.-et-O. (France)

FOR THE
INTERNATIONAL COMMITTEE OF BIO-ACOUSTICS

ELSEVIER PUBLISHING COMPANY
AMSTERDAM – LONDON – NEW YORK
1963

LIST OF CONTRIBUTORS

A. J. Andrieu, *Laboratoire de Physiologie Acoustique, Institut National de la Recherche Agronomique (I.N.R.A.), Centre National de Recherches Zootechniques (C.N R.Z.),* Jouy-en-Josas, S.-et-O. (France)

H. Autrum, *Zoologisches Laboratorium der Universität,* München (Deutschland)

W. F. Blair, *Department of Zoology, University of Texas,* Austin, Tex. (U.S.A.)

O. Brandt, *Laboratoire de Détection Sous-Marine, Marine Nationale,* Le Brusc, Var (France)

J. C. Bremond, *Laboratoire de Physiologie Acoustique, I.N.R.A., C.N.R.Z.,* Jouy-en-Josas, S.-et-O. (France)

W. B. Broughton, *Department of Zoology, Sir John Cass College,* London (England)

R. G. Busnel, *Laboratoire de Physiologie Acoustique, I.N.R.A., C.N.R.Z.,* Jouy-en-Josas, S.-et-O. (France)

P. Chauchard, *Laboratoire de Neurophysiologie de l'Excitabilité, École Pratique des Hautes-Études, Sorbonne,* Paris (France)

B. Dumortier, *Laboratoire de Physiologie Acoustique, I.N.R.A., C.N.R.Z.,* Jouy-en-Josas, S.-et-O. (France)

H. Frings, *Department of Zoology and Entomology, University of Hawaii,* Honolulu, Hawaii (U.S.A.)

F. Huber, *Zoophysiologisches Laboratorium der Universität,* Tübingen (Deutschland)

G. Kelemen, *Harvard Medical School,* Boston, Mass. (U.S.A).

A. Lehmann, *Laboratoire de Physiologie Acoustique, I.N.R.A., C.N.R.Z.,* Jouy-en-Josas, S.-et-O. (France)

D. Leston, *The Cottage, Green End,* Renhold, Bedford, formerly *Department of Zoology, University of Cambridge* (England)

P. Marler, *Department of Zoology, University of California,* Berkeley, Calif. (U.S.A.)

A. Moles, *Institut de Sociologie, Université de Strasbourg* (France)

J. M. Moulton, *Department of Biology, Bowdoin College,* Brunswick, Maine, *and Woods Hole Oceanographic Institution,* Woods Hole, Mass. (U.S.A.)

J. W. S. Pringle, *Department of Zoology, University of Oxford* (England)

O. Sotavalta, *Assistant Professor of Physiological Zoology, University of Oulu* (Finland)

G. Tembrock, *Zoologisches Laboratorium der Humboldt Universität,* Berlin (Deutschland)

A. E. Treat, *Department of Biology, The City College of New York,* N.Y. (U.S.A.)

B. Vallancien, *Laboratoire de Physiologie Otorhinolaryngologique, Faculté de Médecine,* Paris (France)

F. Vincent, *Laboratoire de Physiologie Acoustique, I.N.R.A., C.N.R.Z.,* Jouy-en-Josas, S.-et-O. (France)

N. I. Zhinkin, *Institute of Psychology,* Moscow (U.S.S.R.)

PREFACE

The idea of compiling this volume was first formed in 1956 at University Park State College, Pennsylvania, at the time of the meeting convened to form the International Committee of Bio-Acoustics (I.C.B.A.), which was organised on the initiative of our colleague and friend, Professor H. Frings.

Bio-Acoustics is a scientific discipline which has made enormous strides during the past decade and this book was designed to present the relevant data available at the time.

In spite of the generous collaboration of all concerned and of my own efforts, the intervals between the delivery of the various articles were unfortunately longer than had been hoped. For this reason, and also because of the constant influx of fresh results in recent years, the first articles to come in had to be supplemented with addenda.

On some points the book has already been outstripped by subsequent specialised studies which have added considerably to our knowledge in the last two years, such as those of Griffin, of Kellogg, of Tavolga and of Lilly in the U.S.A., those of Haskell and of Thorpe in Great Britain, and that of Tembrock in Germany. Nevertheless, I believe our book is original inasmuch as it covers a wide range of problems, with something of interest for every specialist. A further advantage is the diversity of the authors' scientific qualifications—including anatomists, zoologists, psycho-physiologists, entomologists, engineers, physicians, etc.—, thanks to which the phenomena can be compared in their different aspects and surveyed within a wide context.

The book shares the defects of every collective work, namely lack of unity, differences of style and in terminology; but the editor, wishing to preserve the original character of each article, has not attempted to eliminate these; notwithstanding, it is to be hoped that each contribution will, like the facets of a diamond, reflect the light and shed some brilliance upon the whole. I know that this book will only be up to date for a short time, but it will have the merit of presenting within its covers the facts that this new branch of biology has discovered up to the present.

Although the various authors wrote the articles in their own language, Elsevier Publishing Company asked us to publish them in English so that the book might be accessible to the widest possible professional public. The articles written in other languages had, therefore, to be translated, which accounts for a few awkward turns of phrase and sometimes a certain heaviness of style. For this I would crave my Anglo-Saxon readers' indulgence, begging them to remember that it is no easy matter to render a text, or even to correct one, in a foreign language.

I trust that my colleagues will approve of what I have done as editor and that they will pardon me for any mistakes or omissions I may have overlooked.

I offer my sincere thanks to all those who helped me to produce this book and to make its publication possible.

The fact should not be allowed to pass unmentioned, I feel, that all the contributors have kindly waived their copyright in favour of the I.C.B.A. for the purpose of establishing a fund to enable young research workers to undertake voyages of study.

R.-G. Busnel

AVANT-PROPOS

L'idée de cet ouvrage a pris naissance en 1956 à University Park, State College, Pennsylvania, lors de la réunion de fondation du Comité International de Bio-acoustique (I.C.B.A.) organisée à l'initiative de notre collègue et ami, le Prf. H. Frings.

La bio-acoustique connaît depuis 10 ans une croissance explosive, et, en principe, l'intention de ce livre était de faire le point des connaissances, propres à ce domaine.

Malgré les bonnes volontés et la redondance de mes efforts, je confesse que je n'ai pu obtenir avec tout le synchronisme souhaitable l'ensemble des manuscrits. De ce fait, et sous la pression des résultats qui n'ont cessé d'affluer ces toutes dernières années, les premiers articles ont dû être complétés par des addendas.

Sur certains points, le livre est déjà dépassé par des ouvrages spécialisés qui depuis 2 ans ont apporté des contributions importantes, que ce soient ceux de Griffin, de Kellogg, de Tavolga, de Lilly pour les U.S.A., ceux de Haskell et de Thorpe en Grande-Bretagne, ou celui de Tembrock en Allemagne. Néanmoins, je pense que notre volume tire son orginalité de ce qu'il couvre un large éventail de problèmes, permettant ainsi à chacun de pouvoir confronter sa spécialité. La diversité des origines scientifiques des auteurs: anatomistes, zoologistes, psychophysiologistes, entomologistes, ingénieurs, médecins, etc..., est également un gage d'intérêt en permettant d'obtenir une vision élargie et comparée des phénomènes.

Comme tout ouvrage collectif, ce livre en a les défauts inhérents: c'est-à-dire l'absence d'unité, les différences de style et d'expression, que l'éditeur n'a pas voulu corriger afin de garder à chacun son caractère original. Il faut espérer cependant que, tel un diamant, ses différentes facettes reflètent chacune la lumière et confèrent à l'ensemble un certain éclat. Je sais que ce livre ne sera d'actualité que pour peu de temps. Il aura eu le mérite de rassembler les données actuelles de cette nouvelle branche des sciences biologiques.

Malgré la diversité des moyens d'expression des auteurs, la Société d'Edition Elsevier nous a demandé de le publier en langue anglaise afin de lui assurer la plus grande diffusion possible. Tous les articles écrits en d'autres langues ont donc dû être traduits ce qui expliquera certaines gaucheries de tournures de phrases, et parfois les lourdeurs de style. Que les lecteurs anglo-saxons ne s'en offusquent pas, considérant qu'il est assez difficile de rédiger ou même de corriger un texte en langue étrangère.

Que mes collègues veuillent bien me donner quitus de mon rôle d'éditeur, et qu'ils soient assez indulgents pour les fautes, erreurs ou omissions que j'aurai laissé passer.

Que tous ceux qui m'ont aidé à réaliser ce livre, et à en rendre la publication possible, trouvent ici l'expression de mes sincères remerciements.

Il me paraît important de préciser ici que tous les co-auteurs ont bien voulu accepter de renoncer à leurs droits d'auteurs au profit de l'I.C.B.A., et ce, aux fins d'établir des bourses de voyages d'études pour de jeunes chercheurs.

R.-G. Busnel

INTRODUCTION

Advances in electronics and acoustics during the last twenty years have furnished the biologist with new and powerful tools for the production, analysis, and recording of sounds. These are particularly valuable to students of sensory physiology and animal behavior, making possible precise and critical studies in these areas. While today relatively few laboratories throughout the world are engaged in studies on animal sounds and acoustical communication, the future should witness a tremendous expansion of effort in these fields as the potential value of the newer acoustical devices is fully appreciated by biologists.

In view of the actual and potential increase in research on animal sounds and communication, an International Conference on Biological Acoustics was held, in April, 1956, at The Pennsylvania State University, U.S.A. Forty-five scientists from five countries attended to discuss the problems facing workers in bio-acoustics, particularly the problems of communication—not between the animals, but between the workers themselves.

As a result of this Conference, the International Committee on Biological Acoustics was formed, with two major objectives: (1) to arrange for an International Collection of Animal Phonography, for storage of recorded samples of sounds illustrating scientific articles acoustically, as pictures illustrate them visually; and (2) to support exchange of information about research in bio-acoustics among workers in the field. There are, at present, about 60 members of this Committee, representing fourteen countries.

This book represents the first large undertaking in furtherance of the second objective of this Committee. Comparative physiologists, animal behaviorists, sensory physiologists, neurophysiologists, ecologists, taxonomists, and evolutionists, all are finding studies of animal sounds valuable. In this book, members of the International Committee on Biological Acoustics summarize recent work in their fields of interest. The articles obviously cannot cover all the fields of bio-acoustics, nor all the active lines of research in each field. Instead they represent a selection from a much larger whole. Even today, another volume of the same size could be written without much duplication of what is here. Within a relatively few years, many volumes will be necessary to cover some of the fields here reviewed in single chapters. This volume is an indication of current and future paths, rather than a summation. It should be a source of critical information for the specialist, a review of current progress in selected areas of bio-acoustics for the interested non-specialist, and an inspiration for the beginner in the field. As such, it represents the first substantial realization of one of the aims of the International Committee on Biological Acoustics.

Department of Zoology and Entomology
University of Hawaii,
Honolulu, Hawaii (U.S.A.)

HUBERT FRINGS
General Secretary of the
International Committee
on Biological Acoustics

CONTENTS

List of Contributors IV
Preface V
Avant-propos VII
Introduction, by H. Frings IX

PART I. DEFINITIONS AND TECHNIQUES

Chapter 1. METHOD IN BIOACOUSTIC TERMINOLOGY
by W. B. Broughton

1. Summary 3
2. Problems and Principles 4
3. Review of Abused and Abusable Terms 6
    (a) Wave, wave-form, period, and frequency 6
    (b) Pulse 7
    (c) Click 12
    (d) Chirp 12
    (e) Phrase and sequence 12
    (f) Trill 13
4. The Modes and Levels of Analysis 14
5. Proposals 16
Synoptic Key to General Terminological Framework 22
Select References 24

Chapter 2. TECHNIQUES USED FOR THE PHYSICAL STUDY OF ACOUSTIC SIGNALS OF ANIMAL ORIGIN
by A. J. Andrieu

1. Introduction 25
2. Choice of the Microphone 25
    (a) Electrodynamic microphones 26
    (b) Ribbon microphones 26
    (c) Piezoelectric microphones 27
    (d) Electrostatic (or condenser) microphones 28
    (e) Contact microphones 29
    (f) Factors entering into the choice of a microphone 29
3. The Recording of the Sound 32
4. Methods of Storing Acoustic Signals 33
    (a) Magnetic recording 34
    (b) The electromechanical process 34
    (c) Photographic recording 34
5. Tape Recorders 35
    (a) Tape recorders fed by alternating current 35
    (b) Self-contained tape recorders 36
    (c) Rules to be observed in the use of tape recorders 36
    (d) Pre-amplifier 37
6. Measurement of Sonic and Ultrasonic Signals 37
    (a) Measurement of the sonic level 37
    (b) Oscillograms 38
    (c) The frequency spectra 39
    (d) Level recorder 44
    (e) Visual demonstration of a recording on a magnetic tape 44
7. Retransmission of the Sounds 44
References 47

Chapter 3. PRINCIPLES OF UNDERWATER ACOUSTICS
by O. Brandt

1. Generalities 48
2. Various Types of Transducers 50
3. Hydrophones 51
4. Transmitters 53

CHAPTER 4. EXAMPLES OF THE APPLICATION OF ELECTRO-ACOUSTIC TECHNIQUES TO THE MEASUREMENT OF CERTAIN BEHAVIOUR PATTERNS

by R. G. BUSNEL

1. Introduction 54
2. Detection, Recording and Measurement of Various Animal Activities 54
   (a) Actography of Drosophila 54
   (b) Actography of the rumination of a cow 55
   (c) Detection of the activities of burrowing animals 56
   (d) Study of the activity of a group of chicks in a brooder or hover 57
   (e) Study of the nocturnal activity of flocks of Corvidae in roosts 57
   (f) Actography of aquatic animals 58
   (g) Study of the activity of a bee-hive 58
   (h) Various vibratory movements 59
   (i) Study of a chirping insect 60
3. Analyses of the Sonic Signal Characterizing a Condition or Behaviour Pattern . . . 62
   (a) Studies on the specific variations of the signal 62
   (b) Study of the emission and reception behaviour between two or more individuals 63
   (c) Study of the phono-behaviour of response to a sonic decoy 64
References 65

PART II. GENERAL ASPECTS OF ANIMAL ACOUSTIC SIGNALS

CHAPTER 5. ON CERTAIN ASPECTS OF ANIMAL ACOUSTIC SIGNALS

by R. G. BUSNEL

1. Introduction 69
2. Quantitative Evolution of Acoustical Vocabularies 70
3. The Hierarchy of Signals in Behaviour: The Relative Importance of Acoustic Signals 70
4. The Particular Interest of Acoustic Signals 73
5. Preliminary Remarks on the Biophysical Relationship Between Signal and Behaviour 73
   (a) Connections between the intensity, the range and the density of population 73
   (b) Relationship between the duration of a signal and its range 74
   (c) The problem of the density of individual population in connection with the modes of expression of the signals 75
6. Acoustic Signals in Behaviour 76
   (a) Acoustic signals in sexual relationships 76
     (i) Meeting of the sexes: attracting call, 76 – (ii) Courtship song, 77 – (iii) Sexual competition, 79
   (b) Acoustic signals in family relationships 81
     (i) Relations between parents and young, 81 – (ii) Individual relationships, 82
   (c) Acoustic signals in connection with collective life 83
     (i) Co-ordination of the movements of the herd, 83 – (ii) Alarm behaviour, 83 – (iii) Experiments on alarm and distress calls, 85 – (iv) Food behaviour, 86
7. A Few Borderline Cases of Specificity and Interspecificity of Sound Signal Semantics 87
   (a) Specificity 87
   (b) Interspecificity 87
     (i) Mutual conditioning by geographical association, 88 – (ii) Non-conditioned interspecific response, 90 – (iii) Interspecificity resulting from signals having physical reactogenic characteristics common to several species, 91
   (c) Information given by certain signals 92
     (i) Work on birds, 93 – (ii) Work on Orthoptera, 95
   (d) Discussion 99
   (e) A work on fish 101
8. On some New Methods for Studying Hearing and Sound Behaviour 102
   (a) Phonokinesis Method 103
   (b) Phonoresponse Method 104
   (c) Phonotaxis Method 105
9. Review of our Knowledge of Animal Acoustic Behaviour 105
References 107

CHAPTER 6. ANIMAL LANGUAGE AND INFORMATION THEORY

by A. MOLES

1. Introduction 112
2. The Nature of Messages 112

CONTENTS

3. Redundancy and Complexity 116
4. Maximum Information and Redundancy 116
5. The Repertoire in Animal Languages 120
   (a) First method 121
   (b) Second method 124
   (c) Third method 125
6. Conclusion 130
References 131

CHAPTER 7. AN APPLICATION OF THE THEORY OF ALGORITHMS TO THE STUDY OF ANIMAL SPEECH: METHODS OF VOCAL INTERCOMMUNICATION BETWEEN MONKEYS
by N. I. ZHINKIN

1. Introduction 132
2. General Principles 136
   (a) Necessity of formalized approach to the objective interpretation of Monkey's signals 136
   (b) Substitution – Vertical and horizontal 137
   (c) Invariants 139
   (d) Some illustrations on the Russian language basis 140
   (e) Rules for forming a line of formal words 141
   (f) Founded alphabets hierarchy 144
   (g) Equality of unequal concrete words in the message 146
3. The sound Communicative System of Monkeys 150
   (a) General description 150
   (b) Some notes on registrating technique and methods of sound analysis 153
4. Discussion of Results 154
   (a) Review of human and monkey's signal sound spectra 154
   (b) Role of the pharynx in the phonation process of monkeys and human beings . . 162
   (c) The number of alphabets and algorithms of the monkey communication system 164
5. Conclusions and Hypothesis 169
   (a) Central control of the pharyngeal tube: organisation of acoustic transition processes necessary for formal words realisation (grammar) in the expanding message system 169
   (b) Philogenetic and ontogenetic formation of the system controlling sound communicative signals 171
6. Glossary of Terms Used in this Chapter 173
7. Notes on the Interpretation of the Figures 179
References 180

PART III. SPECIAL ASPECTS OF ANIMAL ACOUSTICS

CHAPTER 8. ACOUSTIC SIGNALS FOR AUTO-INFORMATION OR ECHOLOCATION
by F. VINCENT

1. Introduction 183
2. Echolocation and its Definition 183
   (a) History of the discovery of echolocation 183
   (b) Definition and principle 185
3. Echolocation in Chiroptera 185
   (a) Evidence of the emittor-receptor system of echolocation 185
   (b) Different types of echolocation in bats 186
   (c) Bats emitting high intensity signals 187
      (i) Vespertilionidae, 187 [1. Physical characteristics of the emissions, 187; 2. Production of these clicks in *Myotis* and *Eptesicus*, 189; 3. The auditory sense of Vespertilionidae and echolocation, 191; 4. The orientation of Vespertilionidae, 194; 5. Pursuit of preys, 196] – (ii) Rhinolophidae, 199 ([1. Behaviour of the animal during emissions, 199; 2. Physical characteristics of the emission of Rhinolophidae, 199; 3. Emission of the clicks, 200; 4. Echolocation in Rhinolophidae, 202] – (iii) Signals intermediate between the Vespertilionidae and the Rhinolophidae types: Asellia, 204 – (iv) Other families of bats in which principles of echolocation are akin to those of the Vespertilionidae and Rhinolophidae, 204 – (v) Conclusions, 208

    (d) Whispering bats 208
        (i) Phyllostomidae, 208 – (ii) Desmodontidae or vampire bats, 210 – (iii) Other whispering bats, 211 – (iv) Conclusions, 211
    (e) Orientation in Megachiroptera 211
        (i) Rousettus, 211 – (ii) Pteropus, 214
    (f) Conclusions 214
4. Other Terrestrial Mammals; the Rodents 215
5. Birds. 216
    (a) Steatornis 216
    (b) The swifts of "Bird's nests soup": Collocalia 217
    (c) Other birds 218
6. Insects 218
    (a) Prodenia 218
    (b) Gyrinus 219
7. Aquatic Vertebrates 219
    (a) Cetacea 219
        (i) Odontoceti, 220 – (ii) Mysticeti, 223
    (b) Fishes 223
        (i) Echo-fish, 224 – (ii) Cave dwelling species, 224 – (iii) Other fishes, 224
8. General Conclusions 225
References 225

CHAPTER 9. INHERITANCE AND LEARNING IN THE DEVELOPMENT OF ANIMAL VOCALIZATIONS
by P. MARLER

1. Inheritance as a Dominant Determinant of the Structure of Animal Vocalizations . 228
2. The Mechanisms of Genetic Control over the Development in Inherited Vocalizations 232
3. The Role of Learning from Conspecific Individuals in Development of Species-specific Vocalizations 234
    (a) Learned effects superimposed on inherited species-specific songs 234
    (b) Learning from conspecifics as a prerequisite for development of species-specific vocalizations 235
4. Learning from Other Species in the Development of Species-specific Vocalizations: Mimicry 238
5. Special Cases of Vocal Imitation of Alien Sounds by Captive Animals 239
6. Characteristics of the Learning Process in Vocalization Development 239
    (a) The disposition to learn 239
    (b) Control over which sounds are imitated 240
    (c) Sensitive periods 240
    (d) The nature of the learning process 241
References 241

CHAPTER 10. A STUDY OF THE AUDIOGENIC SEIZURE
by A. LEHMANN and R. G. BUSNEL

1. Definition and Description of the Phenomenon 244
    (a) Phases of the phenomenon 244
    (b) Animal species in which the phenomenon has been observed 245
2. Brief Historical Review 247
3. Physical Characteristics of the Inducing Acoustic Stimulus 248
4. Genetic Aspect of the Problem and its Relations to Senescence 248
5. Role of the Ear in the Seizure 250
6. Psychological Studies 251
7. Effect of Temperature 251
8. Relations Between the Seizure and the Time of Stimulation 251
9. Clinical Picture of the Physiological Repercussions of the Seizure 252
10. Experimental Studies on the Encephalic Centre 254
    (a) Surgical methods 254
    (b) Biochemical methods 255
    (c) Various drugs 257
    (d) Action of tranquillizers 258
    (e) Latent periods 260
11. Discussion 261
References 262

PART IV. EMISSION AND RECEPTION OF SOUNDS: MORPHOLOGICAL, PHYSIOLOGICAL AND PHYSICAL ASPECTS

A. Invertebrates

Chapter 11. MORPHOLOGY OF SOUND EMISSION APPARATUS IN ARTHROPODA
by B. Dumortier

1. Introduction	277
2. Sound Emission by Friction of Differentiated Regions	278
(a) Insects	279
(i) Cephalic pars stridens, 279 – (ii) Thoracic pars stridens, 281 – (iii) Abdominal pars stridens, 282 – (iv) Pars stridens on the legs, 290 – (v) Pars stridens on the elytra or the forewings, 293	
(b) Crustacea	305
(i) Stomatopoda, 305 – (ii) Decapoda, 305 [1. Penaeidae, 305 – 2. Alpheidae and Palaemonidae, 306 – 3. Palinura, 307 – 4. Anomura, 309 – 5. Brachyura, 310] – (iii) Isopoda, 315	
(c) Arachnida	315
(i) Araneida, 315 – (ii) Scorpionidaea, 319	
(d) Myriapoda	321
3. Sound Emission by Vibration of a Membrane Other than Wings	323
4. Sound Emission by Passage of a Fluid (Gas or Liquid) across an Orifice	323
(a) Passage by way of the mouth	323
(b) Passage through the spiracles or through glandular orifices	324
(i) Expulsion of liquids, 324 – (ii) Expulsion of air, 325	
(c) Passage through the anal region	325
5. Sound Emission by Percussion on the Substratum	325
(a) Coleoptera	326
(b) Psocoptera	326
(c) Isoptera	327
(d) Orthoptera	327
(e) Other examples	328
6. Sound Emission by Vibration of Appendages	328
7. Cases of Stridulation in which Non-differentiated Regions are Used	329
(a) Mandibular noises	329
(b) Wing noises	329
(i) Produced when at rest, 329 – (ii) Produced in flight, 330	
8. Sound Production by Larvae, Nymphs and Pupae	330
(a) Coleoptera	330
(b) Lepidoptera	331
(c) Orthoptera	333
(d) Other examples	333
9. Biological Problems Arising from Sound Emission Apparatus	334
(a) Adaptation and differentiation of apparatus	334
(b) Systematic distribution	335
(c) Origin of stridulatory apparatus	336
(d) Sound apparatus and sensory devices for perception	337
References	338

Chapter 12. THE PHYSICAL CHARACTERISTICS OF SOUND EMISSIONS IN ARTHROPODA
by B. Dumortier

1. Introduction	346
2. The Components of the Sound Emission	346
(a) Physical parameters	347
(i) Frequency, 347 – (ii) Intensity, 350 – (iii) Amplitude modulation and frequency modulation, 354 – (iv) Transients, 355	
(b) Biological parameters	356
(i) The pulses, 356 – (ii) The chirps, 357	
3. Analysis of the Mechanism of the Production of Sound Waves by the Stridulatory Apparatus	360
(a) Stridulatory mechanism	360

(b) Origin of sound waves produced by stridulatory apparatus 362
   (i) The case of Gryllodea, 362 – (ii) The case of Tettigonioidea, 364 – (iii) The case of Acrididae, 365 – (iv) Conclusion, 366
(c) Modification of the parameters under the action of temperature 366
   (i) Pulses and chirps, 366 – (ii) Frequency, 368
References 371

Chapter 13. THE FLIGHT-SOUNDS OF INSECTS
by O. Sotavalta

1. History 374
2. The Flight-Sound and its Recording Methods 374
3. Physiology of the Flight-Sound 377
4. Analysis of the Flight-Sound 383
5. Modifications of the Flight-Sound 386
6. Biological Significance of the Flight-Sound 386
References 389

Chapter 14. ACOUSTIC BEHAVIOUR OF HEMIPTERA
by D. Leston and J. W. S. Pringle

1. Introduction 391
2. Homoptera 392
   (a) Auchenorrhyncha 393
      (i) Anatomy, 393 – (ii) Physiology of sound production, 397 – (iii) Function of the songs, 400 – (iv) Sound reception, 400
   (b) Sternorrhyncha 401
3. Heteroptera 402
   (a) Pentatomomorpha 402
   (b) Cimicomorpha 404
   (c) Amphibicorisae 405
   (d) Hydrocorisae 405
   (e) Physical characteristics of the song 406
   (f) Functions of the song 407
   (g) Sound reception 408
4. The Evolution of Sound Production in Hemiptera 409
References 410

Chapter 15. ANATOMY AND PHYSIOLOGY OF SOUND RECEPTORS IN INVERTEBRATES
by H. Autrum

1. Comparative Anatomy of Sound Receptors 412
2. Physiology of the Sound Receptors 420
   (a) Frequency range of Invertebrate sound receptors 420
   (b) Absolute sound sensitivity 423
   (c) Principles of function of the sound-receiving apparatus 424
   (d) Orientation towards acoustic signals 428
References 431

Chapter 16. SOUND RECEPTION IN LEPIDOPTERA
by A. E. Treat

1. Historical Review 434
2. Morphology of the Tympanic Organs 435
3. Physiology 436
4. Ecological Significance 437
References 438

Chapter 17. THE ROLE OF THE CENTRAL NERVOUS SYSTEM IN ORTHOPTERA OF STRIDULATION DURING THE CO-ORDINATION AND CONTROL
by F. Huber

1. Introduction 440
2. Structure and Mechanism of the Stridulatory Apparatus 440
   (a) The stridulatory movements of the field cricket (*Gryllus campestris*) 441
   (b) The stridulatory movements of the acridid *Gomphocerus rufus* 444

3. The song patterns of *Gryllus campestris* and of *Gomphocerus rufus* 445
    (a) The songs of the field cricket. 445
    (b) The songs of *Gomphocerus rufus* 449
4. The Neurophysiological Basis of Sound Production. 450
    4A. The Function of the Thoracic Ganglion During the Stridulatory Movement . . 452
    (a) Anatomical remarks 452
    (b) The function of the isolated thoracic ganglion 453
    (c) The conduction of excitation from the brain to the thoracic ganglion 455
    (d) The nervous coupling of both halves of the thoracic ganglion 455
    (e) Local electrical stimulation of the neuropile in the thoracic ganglion 455
    (f) Discussion of results 457
    (g) Summary of the preceding results 459
    4B. The Function of the Brain During Sound Production 460
    (a) Anatomical remarks 460
    (b) Lesions in the brain. 462
    (c) Local injuries of the brain by punctures 464
    (d) Electrical stimulation in the protocerebrum of crickets 465
        (i) Methods, 465 – (ii) Results, 465 [1. Song patterns of the "normal song" type, 465 – 2. Song pattern of the "rivalry" song type, 467 – 3. Atypical song patterns, 469 – 4. Central inhibition of the stridulatory movement, 471 – 5. Competitive inhibition of running and singing, 471]
    (e) Localization of the points of stimulation releasing the stridulatory movement . 472
        (i) Method, 472 – (ii) Results, 473
    (f) Discussion of the stimulation experiments on the brain 473
        (i) Afferent activities of the antennae, 474 – (ii) Afferent activities of the cercus, 475 – (iii) Acoustical afferent activities, 476 – (iv) The role of the mushroom bodies during sound production, 476 – (v) The role of the central body during sound production, 477
5. Observations on a Reciprocal Action Between the Brain and the Thoracic Motor System 477
    (a) Relationship between the various elements of motion in courtship of *Gomphocerus rufus* 477
    (b) Relations between the motor systems in complex situations of behaviour . . . 478
    (c) Return of atypical songs to normal rhythms 479
6. The Peripheral Afferent Control of the Stridulatory Movement 479
    (a) The acoustical system 480
    (b) The system of proprioreceptors 480
    (c) Summary of Section 6. 483
7. Summary. 483
References 484

B. VERTEBRATES

CHAPTER 18. COMPARATIVE ANATOMY AND
PERFORMANCE OF THE VOCAL ORGAN IN VERTEBRATES
by G. KELEMEN

1. Lower Vertebrates 489
    (a) Fishes. 489
    (b) Amphibians 490
    (c) Reptiles. 490
2. Superior Vertebrates 491
    (a) Birds 491
    (b) Mammals 494
        (i) General view on the mammalian larynx, 494 – (ii) The laryngeal musculature, 498 – (iii) Primates, 505 – (iv) General considerations on mammals phonation, 512
References 518

CHAPTER 19. COMPARATIVE ANATOMY AND
PHYSIOLOGY OF THE AUDITORY ORGAN IN VERTEBRATES
by B. VALLANCIEN

1. Primary Phenomena 522
    (a) Elementary structures. 522
    (b) The nature of the excitation 525

CONTENTS

    (c) Mechanism of the transductor; nervous transmission of the stimulus 526
    (d) The chemical mediator 528
  2. Development of the Stato-acoustic Organ 529
    (a) General aspect of sensory structures 529
      (i) The maculae, 530 – (ii) The papillae, 530 – (iii) The auditory organ of Agnathes, 530
    (b) Evolution of the otic capsule 531
    (c) The organ of reception 534
    (d) The transmitting apparatus 535
      (i) Transmission in the middle ear, 535 [1. Aquatic vertebrates, 536 – 2. Reptiles, 541 – 3. Birds, 542] – (ii) Transmission in the inner ear, 546
  References 554

CHAPTER 20. EMISSION AND RECEPTION OF SOUNDS AT THE LEVEL OF THE CENTRAL NERVOUS SYSTEM IN VERTEBRATES
by P. CHAUCHARD

  1. Introduction: From Reflex to Mental 557
  2. The Elementary Centres of Audition and Speech 563
  3. The Apparatus Controlling Nervous Activity 566
  4. Instinctive Centres of Audition and Speech 568
  5. Audiophonative Mechanisms of the Cerebral Cortex 572
  6. Internal Speech and Human Thought 576
  References 580

PART V. ACOUSTIC BEHAVIOUR

A. INVERTEBRATES

CHAPTER 21. ETHOLOGICAL AND PHYSIOLOGICAL STUDY OF SOUND EMISSIONS IN ARTHROPODA
by B. DUMORTIER

  1. Sound Emission and its Significance in Behaviour 583
    (a) The place of the acoustic signal 583
    (b) Principal types of sound emissions 585
      (b-1) Emission ending with the creation of a situation which satisfies a need or a tendency 585
        (i) Calling song, 585 – (ii) Congregational song, 588 – (iii) Premating songs, 590 [1. Courtship song, 591 – 2. Agreement song, 594 – 3. Jumping sound, 596 – 4. Meaning and role of the premating songs, 596]
      (b-2) Emission associated with a "hostile" or defensive attitude and tending to put an end to a situation of constraint 599
        (i) Rival's song, 599 – (ii) Disturbance songs, 603 – (iii) Protest sound, 605 – (iv) Sound emission and stridulatory movements in the pre-imago stages, 608
    (c) The origin of the stridulatory movements 609
  2. Experimental Study of Acoustic Behaviour 610
    (a) Phonoresponse 610
      (a.1) Alternation 610
        (i) Specific alternation, 610 – (ii) Interspecific alternation, 612 – (iii) Alternation with artificial stimuli, 613 – (iv) The problem of learning, 616
      (a.2) Synchronism 617
    (b) Phonotaxis 618
      (i) Specific phonotaxis, 618 – (ii) Phonotaxis set off by artificial signals, 620 – (iii) Interspecific phonotaxis and "elements of specific recognition", 622 – (iv) Psychological aspect: phonotaxis and tropism, 625
  3. Song and Genetics 627
    (a) The song as a genetic feature 627
    (b) The hybrids 628
    (c) Song and reproductive isolation among species 631
    (d) Song and systematics 636
  4. Factors Determining the Song and Associated Behaviour Patterns 638
    (a) External factors 638
      (i) Physical factors, 638 – (ii) Biological factors, 639

(b) Internal factors 639
    (i) Observations on the male, 639 – (ii) Observations on the female, 640 [1. The cycle of sexual behaviour, 640 – 2. Factors determining the cycle, 641]
(c) Action of pharmacodynamic drugs 643
(d) Neurophysiology of acoustic behaviour 643
(e) Combined action of external and internal factors: daily rhythms of activity . . 644
    (i) Origin of the daily rhythm, 644 – (ii) Daily and annual rhythm, 646
Acknowledgements 646
References 649

B. Vertebrates

Chapter 22. ACOUSTIC BEHAVIOUR OF FISHES
by J. M. Moulton

1. Introduction 655
2. The Hearing of Fishes 656
3. The Definition of Hearing in Fishes 663
4. Uses of Sound in the Fisheries. 669
5. Sound Production by Fishes 671
6. The Relations of Sound to Fish Behaviour 680
7. Addendum 686
References 687

Chapter 23. ACOUSTIC BEHAVIOUR OF AMPHIBIA
by W. F. Blair

1. Introduction 694
2. The Caudata 694
3. The Acaudata. 695
   (a) Emission of sounds 695
      (i) Morphology of organs and mechanisms of function, 695 – (ii) Physical characters of emissions, 697 – (iii) Physiological characters, 703 – (iv) Behaviour and vocabulary, 703
   (b) Reception of sounds 705
      (i) Morphology of receptor apparatus, 705 – (ii) Behavioural responses, 706
References 706

Chapter 24. ACOUSTIC BEHAVIOUR OF BIRDS
by J. C. Bremond

1. Introduction 709
2. Classification of Sonic Emission 709
   (a) The problem of description. 709
   (b) The bird's repertoire of acoustic signals 711
3. Signals Emitted Under the Influence of Internal Inducing Factors and External Factors Instigating Behaviour. 711
   (a) General aspects. 711
      (i) Observations on the hormonal state as an internal motivating factor, 712 – (ii) Experimental proof, 712
   (b) Influence of external factors on the development of internal motivating factors 713
   (c) The song and its role in instigating acts of behaviour 714
   (d) Patterns of song 715
   (e) Song and territory 716
   (f) The song and its role during greeting and invitation ceremonies. 718
   (g) Acoustic signals other than song used during greeting and invitation ceremonies 719
   (h) Individual recognition. 720
      (i) Between adults, 720 – (ii) Between parents and their young, 720 – (iii) Recognition of the sex of a partner, 721
4. Emission of Acoustic Signals Induced by External Stimulation Factors 721
   (a) Call notes giving information on position and for co-ordination of activities . . 721
   (b) Calls related to environment 722
   (c) Calls accompanying feeding behaviour. 723
   (d) "Irrelevant" behaviour 724

5. Development of the Bird's Vocabulary with Relation to Age 725
   (a) Appearance of the first sounds 725
   (b) Distinction between innate and acquired signals 726
   (c) Subsong 728
   (d) Imitation 729
   (e) Psittacism 733
   (f) Dialects 734
6. Acoustic Signals – Their Role in Relation to Their Physical Properties 735
   (a) Range of frequencies emitted and comparison with Their auditory possibilities . 735
   (b) Recognition of the direction of a sonic source 738
   (c) Physical characteristics of sounds emitted during behaviour 738
   (d) Easily localized acoustic signals 740
   (e) Acoustic signals that may be situated with difficulty 741
   (f) Acoustic signals with a specific instigating value 742
   (g) Acoustic signals with a interspecific stimulus value 743
   (h) Does the concept of sonic environment exist in the bird? 744
7. Conclusion 745
References 746

CHAPTER 25. ACOUSTIC BEHAVIOUR OF MAMMALS
by G. TEMBROCK

1. Introduction 751
2. The Physical Characteristics of Sounds and the Physiological Factors in the Production of Sounds 752
3. Types of Vocalizations and Their Functions 761
   (a) The types of sound 761
       (i) Unvoiced noises, 761 – (ii) Voiced sounds, 762
   (b) Sounds in the service of pairing 768
       (i) Rival sounds, 768 – (ii) Pairing sounds, 768
   (c) Sounds in the service of rearing the young 769
       (i) Sounds of the adult, 769 – (ii) Sounds of the young, 770
   (d) Sounds and sociology 772
       (i) Alarm calls, 772 – (ii) Voice contact sounds, 773 – (iii) Group sounds, 774 – (iv) Sounds expressing a special mood, 774
   (e) Sound and protection 775
4. Sounds and Behaviour 775
   (a) Sounds and forms of behaviour 776
   (b) Sounds as stimuli 777
5. Problems of Homology Research 777
References 783

PART VI. APPENDIX

ADDENDA

Addendum to Chapter 8, by F. VINCENT 789
Addendum to Chapter 9, by P. MARLER 794
Addendum to Chapter 14, by D. LESTON and J. W. S PRINGLE 798
Addendum to Chapter 16, by A. E. TREAT 800
Addendum to Chapter 17, by F. HUBER 802
Addendum to Chapter 23, by W. F. BLAIR 803

SYSTEMATIC INDEX 805

GLOSSARIAL INDEX, by W. B. BROUGHTON 824

ALPHABETIC INDEX 911

PART I

DEFINITIONS AND TECHNIQUES

CHAPTER 1

METHOD IN BIO-ACOUSTIC TERMINOLOGY

by

W. B. BROUGHTON

1. Summary

1. The need for some standardization of bio-acoustic terminology is discussed. Some opinions have been tested by circulating a memorandum; the response is briefly analyzed.

2. A very large majority of terms are uncontroversial; these are not dealt with here, but in the Glossarial Index at the end of the book.

3. Abused and controversial terms are discussed. The practice of defining a term in a particular way "to hold for the purposes of this paper alone" is criticized as a question-begging procedure more often resorted to than justified.

4. Extensions of meaning, of concepts already defined in physical terms, must be no more than extensions; redefining by reference to biological criteria is not legitimate.

5. *Pulse*, in particular, probably the most difficult concept of all, is discussed in detail from the viewpoints of graphic representation, mode of production and effect on various types of ear. All these considerations demonstrate that the generally accepted physical connotation (a simple wave-train) is not only the sole legitimate but also the sole safe connotation. Certain special cases are reviewed.

6. Past usages of *chirp, phrase, sequence, trill* are discussed; a good many other terms receive incidental examination.

7. The available levels of analysis of animal sound are reviewed: the physical wave-form level (frequency, pulse properties etc.); the phonative act level (phoneme, syllable etc.); the level of subjective human audition (pitch, timbre, rhythm, melody, harmony, etc.); the level of the "relevant hearer's" audition (action potentials, and behaviour—combined with study of the emitter's behaviour).

8. Proposals are put forward based on the normal practical sequence of analysis. The author abandons his former specialized definition of *chirp* in terms of the motor act of emission, accepts the much more general and traditional concept (any *unitary sound*), and makes this the central concept in an empirical terminology for the logical first analysis of any given sound in terms of what the investigator himself hears.

9. To this are then grafted the data from progressive physical studies which, by filling in the gaps, complete and objectivize the empirical scheme into a rigorous system.

10. In some groups (*not including Man*), only one term is needed to convert this in turn into a scheme valid also for analysis by reference to the motor acts of emissions: for this, the German orthopterists' concept of *Silbe*, translated as *syllable*, is proposed; confusion with its human phonetic and etymological connotations (which differ from each other anyway) will not arise, since this particular scheme will not embrace human phonation.

11. Certain other animal groups, for various reasons which are mentioned, are not yet and may never be amenable to inclusion in a general scheme going beyond the empirical and physical levels (*e.g.* birds, Man). Workers on these must continue to use their own terminologies as now, but it is hoped that the principles put forward in this chapter may give at least some useful guidance.

12. A synoptic key to the general terminological framework of the subject of animal sound is given (pp. 22 and 23).

2. Problems and Principles

Two dangers face the student seeking to rationalize and codify a terminology that has grown up empirically and that is beginning to differentiate regionally or according to faculty or in other ways—as must always tend to happen. One danger is that of legislating prematurely and clumsily for hypothetical future requirements; the other is a too easy-going and long-sustained attitude of *laissez-faire* arising from the wish to let the mud settle before trying to penetrate the shadows of often obscure and chaotic usages. If the former danger must always be borne in mind, the latter is more insidious; while we wait for the mud to settle, divergence may be increasing, and we may be faced with the need to cure what we might have prevented. The first of the series of symposia in recent years on animal acoustics, held in Paris in 1954, made some attempt (Chavasse *et al.*, 1954) to meet this problem so far as it concerned French-, German-, and English-speaking workers in the limited field of grasshopper sounds. No decisions were attempted, largely for lack of time; and the present work affords an opportunity for a more radical stocktaking and for seeking by careful semantic prescription at least to avoid tendencies towards rough or even inadmissible usage, or to nip them in the bud. How far this can be combined with any attempt to standardize a general bio-acoustic terminology for all animals, is a much more difficult question.

With both these objectives in mind, a draft review of the problem was circulated in 1960 to a representative selection of those workers listed in 1957 by the International Committee for Biological Acoustics, and to a few other workers, totalling 36 in all, and sited widely over the world. The document was intended largely as an Aunt Sally to be potted at by its recipients; from the opinions expressed, it was hoped to get the whole problem in perspective and see just how much could be done at this stage. I had 17 replies of which 2 were oral: 7 expressed general agreement; another 5, more cautious approval, the caution being directed towards the first danger mentioned above, with, in one case, much discussion specifically on musical form; but of these 5, 3 expressly agreed with my strictures on the use of the word *pulse* in particular (*v. infra*), and to these can be added under this head a further 2 who were otherwise unclassifiable. Of the 3 remaining, 2 disagreed vigorously and the last, a mere courtesy acknowledgement, was unclassifiable.

These results would appear to be a statistical justification for going ahead with the draft substantially as it stood in 1960. Unfortunately for my task, but fortunately, I think, for the cause of better understanding, a few correspondents, notably M. Duijm and O. Sotavalta, made very long and careful critical analyses, running into many pages of close typing, for which I am extremely grateful—but which have thrown the whole of my ideas[*] back into the melting pot!

Let it first be said, however, that a very large number of terms in the repertoire of bio-acousticians are generally recognized to have precise and unambiguous meanings. These will be found either defined in the glossarial index at the back of this book, or defined in the text and referred to by page alone in the index. The present discussion seeks to deal rather with those terms whose meaning tends to vary with the author employing them and which, for this very obvious reason, stand in urgent need of codification. In this connexion, it is proper to examine the growing practice of using definitions expressly stated by their author to "hold only for the purposes of this paper". This may be justified where an author is carving a path through a jungle of blank ignorance and, for the purpose, using an unpreoccupied lay word, more or less provisionally, in his newly defined scientific connotation; but he has a clear responsibility, first, to see that the word is indeed unpreoccupied, that is, not already a precision scientific term, and second, to see that this applied meaning is related to the original meaning in a rational way. A hypothetical early explorer from Great Britain, seeing a tiger for the first time, could legitimately from his experience call it a great cat or something of the sort; but if he called it a great hedgehog, he would merely convince posterity that he was a great fool.

Whilst such misuses within a discipline are rare, they are all too easy in those polyglot buffer-state sciences, such as our own, where the frontiers of many disciplines meet. All the greater, then, is the author's responsibility to examine his terminology; and to eschew (even "for the purposes of this paper alone") usages that he recognizes to be improper or inexact. These are the considerations that lead me to discuss such terms, not particularly as an entomologist, a biologist, an acoustician or even a scientist, but rather as a layman and an amateur (*sensu proprio*) of English, recognizing the paramountcy of the exact word as the precision tool that has made scientific thought what it is (*v. infra*, Chapters 20 and 6, contributions of Chauchard and of Moles, on the significance of the word for the evolution of abstract thought).

The main sources of bio-acoustic terms are physics and its subdivisions (pulse, elementary wave-form, basic signal, principal frequency, etc.); phonetics (vowel, consonant, phoneme, syllable); poetry (syllable, line, stanza, canto = Silbe, Vers, Strophe, Lied respectively of the German orthopterists); music (staccato etc. of fish workers, note, phrase, song etc. of ornithologists, and especially the terminology of musical form, used by Sotavalta (1956) and perhaps by others); and lastly, the authors' own experience and imagination, upon which they have drawn like our hypothetical explorer (chirp, chirrup, stridulation, series, sequence, burst, volley and a host of onomatopoeisms). This last source has the advantage that, since the terms have no pre-existing exact scientific meaning, the author feels free to apply to them any meaning he chooses that is consonant in the roughest way with their everyday usage—which is nice for the author but can be bad for general understanding. Physical and phonetic sources have the diametrically opposite merit that the terms have usually very precise accepted scientific meanings—which the bio-acoustician must respect. If he limits any adaptation of them to clearly defined *natural extensions* of

---

* *Pulse* excepted, since the views I expressed appear to have overwhelming support in this respect (12 correspondents out of 17 who replied). It is also worth remembering that vigorous objection to a proposal usually stimulates a reply: the remaining 19 correspondents are therefore unlikely to include any with strong feelings against my interpretation of the word. Since the objectors are well worth convincing, however, I have argued the case below in some detail.

their existing meaning he will, at the cost of minor intellectual effort to himself, keep language fine and improve general understanding.

The poetic-musical sources have the advantage of greater elasticity than the physical, less looseness than the experience-imagination category; the danger in applying musical-form terminology seems to reside in its intrinsically artistic nature, assessments being made by appeal to the sensation of satisfaction and subjective completeness which, unobjectionable as a criterion in music itself (or at least the best that can be achieved) seems hardly rigorous enough for scientific application. Perhaps this is a counsel of despair; yet unless terms have a considerable measure of objectivity, they risk being used in different senses by different people, and this could lead to general ambiguity.

3. Review of Abused and Abusable Terms

(a) *Wave, wave-form, period and frequency*

Wave, wave-form, period* of a wave, and the various elementary ideas concerning frequency appear always to be used in senses very close to the original physical usage; and indeed the entities they represent are so near to pure physical entities that there seems very little risk in the future of misuse.

But it is probably worth while to note here the current tendency to restrict *frequency* to sine-waves or their derivatives, and to use the word *rate* for other periodicities; thus the old term *pulse repetition frequency* has largely given way to *pulse repetition rate*, with a considerable reduction of the scope for misunderstanding. *Basic signal, elementary vibration, elementary wave-form* (all three used in the sense of "sine-wave that would exist as a continuous wave in the absence of the observed modulation") are also unlikely to be misused (Fig. 1c).

It is when the concept of frequency is expanded to *principal frequency, dominant frequency* and analogous terms that the danger of ambiguity and misuse arises. Those who are familiar with harmonic analysis will know that when a *periodic quantity* (v. glossary) is not a pure sine-wave, it can be resolved by Fourier's Theorem into sinusoidal components** the first of which has the same frequency as the quantity itself, called the fundamental frequency; and the remainder of which, called harmonics, are whole multiples of this. Some musicians tend to call the first of these (*i.e.* the octave—2 × fundamental) the *first* harmonic; others, with physicists and engineers, call it the *second* (to agree with *two* times the fundamental) and regard the fundamental itself as the first. The latter is undoubtedly preferable.

It is a pity that *principal frequency* and *dominant frequency* are all too often used synonymously with fundamental frequency, for this is a waste of precision. W. Tavolga has urged (*in litt.*) that *dominant frequency* should always be reserved for the frequency of that harmonic which has the greatest amplitude, which need by no means be the lowest of the series. This is sound sense and makes full and proper use of the concept of dominance; and we heartily endorse it, the more so because we may have erred ourselves in this respect, at least in conversation.

---

* *Period* has a very distinct meaning in musical form, which is quite unlikely to be confused with the sense in which it is used in physics, if only because the two usages are unlikely to occur in the same context.

** Which, in a note, represent the constituent pure tones.

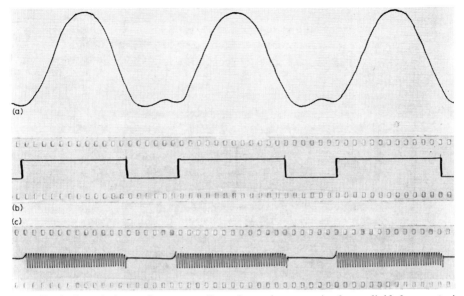

Fig. 1. Evolution of the modern conception of a pulse as a simple, undivided wave-train. (a) General aspect of the human ventricular pulse. (b) Comparable train of rectangular pulses ("rectangular waves"). (c) Continuous waves cut into pulses (simple wave-trains) by modulating with the wave-form of (b). The wave-form of (a) could be used to produce the same result.

(b) Pulse

Pulse is surely the most ill-used term ever taken over by the bio-acoustician. In extenuation let it be said at once that its precise meaning in physics has been debated for years, and that it was a physicist who first put biologists on the downward path of its misuse, by himself using it in no less than three radically differing senses: the venerable G. W. Pierce, in his charming and pioneering book (1948) uses it on p. 71 as equivalent to a cycle, *i.e.* a wave; on p. 26 as equivalent to one homogeneous assemblage of waves, *i.e.* a wave-train; and almost everywhere else in the book (*e.g.* p. 157, 181), as equivalent to assemblages of wave-trains (actually, assemblages of discrete "serrations"—*each* of the latter corresponding to one impact and thus representing (p. 100, by implication) a transient train of waves).

Again, the *right* of the physicist to the term may be questioned. I am indebted to Sir Julian Huxley (*in litt.*) for a hint about a certain propensity of physicists for misappropriating biological terms, giving them a precise physical meaning, and then complaining when the biologist fails to respect that meaning!—a hint which, intentionally or otherwise, led me to uncover half a column of letterpress in Van Nostrand's Encyclopaedia, on *pulse* in its medical connotations, followed by the phrase "also used in physics". Further researches in English and Latin dictionaries and elsewhere (for able help in which I am much indebted to Mr. K. D. C. Vernon, Librarian of the Royal Institution of Great Britain) show that the word in its medical usage dates back at least to Pliny, whereas I have so far been unable to trace its usage in physics back beyond the eighteenth century.

From historical considerations then, the biologist may properly growl over his bone at the snapping physicist, and apply any meaning he likes that is more or less consonant with the primary medical connotation of the word (heart-beat) (Fig. 1a). On the other hand, without rambling through eighteenth-century scientific publication, and without getting bemired in the physicists' arguments on interpretation of detail, we can concede that the physical concept of a pulse is a natural and legitimate extension of its medical counterpart (Fig. 1b and 1c). Its undisputed everyday usage in the sense of a simple wave-train, as in Pierce (1948) p. 26, and not a complex series of such wave-trains as on p. 157 and elsewhere, is a valuable contribution to precision of expression which we ought to be loth to sacrifice. Biologists who appear to have accepted this are Pringle (1954, p. 532), Leston and Pringle (Chapter 14, *infra*), Thorpe (1958, p. 546, and 1957, p. 170, and verbal comment thereon), Broughton (1952); bat workers, *e.g.* Möhres (1953, Impuls); all marine sound workers; and many others.

Nevertheless, though a pulse on paper in a biological vacuum (Fig. 1c) may be easy enough to grasp as a theoretical isolate, the minute we take into consideration the organism that is going to hear it, or that which produces it, we run into difficulties —difficulties of biological rather than physical origin. It is usual to regard a signal such as Fig. 2 as a somewhat irregular succession of somewhat irregular pulses, and

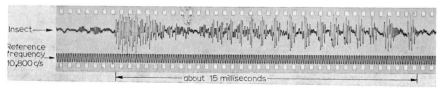

Fig. 2. Oscillogram of a complex pulse-train (probably one diplosyllable) from the courtship song of *Conocephalus dorsalis* (Latreille) (Orthoptera, Ensifera). The song is a long ripple of these syllables, arranged in two ways.

this indeed is what I myself have advocated. Before deciding the point, it is worth looking at this and some related wave-forms from the points of view of production and audition, to see how, if at all, this alters our opinion.

The four pure tones, C_{256}, E_{320}, G_{384}, C_{512} (Fig. 3a–d), when sounded simultaneously, produce the common chord of Middle C, that is, the wave-form of Fig. 3e. This is an exactly periodic quantity of (fundamental) frequency 64 c/s (the original four tones are harmonics 4, 5, 6 and 8 of this). This repetition rate is too fast for the human ear to hear as a rhythm pattern; instead it is heard as if it were a further pure tone of 64 c/s added to the original four, exactly, in fact, like the reference signal of 64 c/s shown in the figure. In other words the human ear interprets the fundamental frequency of the wave-form as a pure tone, as well as extracting from the wave-form the four real pure tones that make it up; this ear therefore acts as a harmonic analyzer.

As is well known from the work of Pumphrey and Rawdon-Smith (1939) and Pumphrey (1940), now confirmed by many others (though, more recently, questioned in its pitch-discrimination aspects by Horridge (1960)), a repetition rate of 64 periods per second is not too fast for the grasshopper ear to hear as a rhythm pattern; though 256, 320, 384, and 512 periods per second probably are. Moreover, this ear appears to possess no mechanism of either harmonic analysis or (*salve* Horridge) *pitch* (I do not say frequency) discrimination; therefore the wave-form of Fig. 3e, striking a grasshopper ear, will only produce a sensation of sound increasing and decreasing in intensity at a rate of 64 p.p.s. (with perhaps a subsidiary rhythm at 128 p.p.s.), that is, a response to the envelope of the wave-form—a *rectification* (*cf.* Fig. 1c and 1b). For this ear, in fact, a real series of discrete pulses—a sound of indifferent pitch, or

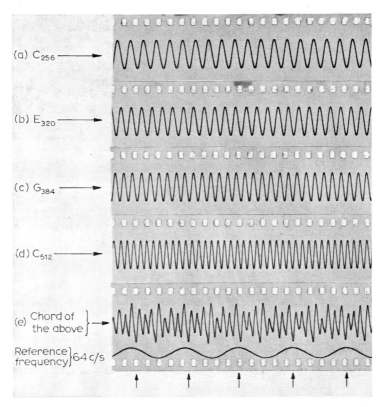

Fig. 3. Synthesis of the wave-form of the common chord of Middle C. Reference signal 64 c/s. Comparing (e) with the reference signal, we see that the chord is itself a periodic quantity repeating exactly every 64th of a sec. Very slight irregularities in the repeat pattern represent experimental inaccuracies due to using an improvised superposition technique for producing the chord.

no pitch, parcelled up into units repeated every 64th of a second (*e.g.* a rattle)—could mimic what to our ear is the common chord of C_{256}. Clearly then the mode of generation of a signal is not necessarily a reliable guide to its behavioural effect.

If now the common chord of Fig. 3e is replaced by the same chord in the scale of equal temperament (the piano-scale), in which the harmonics E and G are not quite exact multiples 5 and 6 of the fundamental of 64 c/s, the wave-form is of the type shown in Fig. 4, which is very nearly the periodic quantity of Fig. 3e but not quite; scrutiny shows that, in fact, nowhere is the wave-form exactly repeated—and this is not a matter of experimental error, but theoretical expectation. Fig. 4 is getting uncomfortably like the signals of some grasshoppers previously regarded as successions of ill-defined, impact-generated pulses.

The note of a musical instrument is rarely a pure tone such as those of Fig. 3a–d, but consists of a fundamental and several harmonics, so that even though one note is apparently struck, a chord in fact is produced. The wave-form of Fig. 5, from the work of D. C. Miller (1934), is that of a clarinet sounding C_{256}. It is instructive to compare this with Fig. 2, a grasshopper emission of the type generally regarded as pulsed;

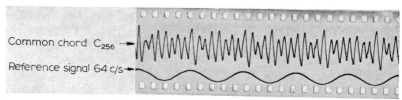

Fig. 4. The same common chord as in Fig. 3e, but in the scale of equal temperament. The wave-form is no longer exactly periodic, because the harmonics are no longer in simple ratio to the fundamental.

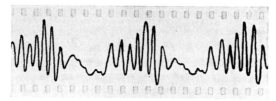

Fig. 5. Wave-form of C_{256}, played on a clarinet. (Redrawn from DAYTON C. MILLER, *The Science of Musical Sounds*, Macmillan Company, 1916 edition, by courtesy of the trustees and publisher.)

and to note that not only do the apparent pulses in that figure correspond fairly closely in general characteristics to the apparent pulses in the clarinet note, but also the spaces between carry similar small serrations, so that theoretically at least, this grasshopper signal could represent a note about 2 octaves above Middle C, and a series of harmonics of which about the 12th and 13th would be the most marked. This is not the place to develop these ideas beyond their relevance to the present context, which will presently be apparent; investigations now in progress will be reported elsewhere.

Lastly, preliminary experiments in another investigation now in progress strongly suggest that when a real pulsed electric signal* energizes a very small sound-emitting system, *i.e.* one whose dimensions are measured in millimetres at most, internal reflections inside the system are of tremendous importance. Under certain conditions the original pulses can be totally obliterated and even replaced by "false" secondary ones (which, arising as they do from echoes, presumably represent the resultant of the natural harmonics of the reverberation chamber of the system itself). In short, an essentially pulsatile signal has been replaced by an essentially harmonic (chordal) one, *even at the generating end*; but the new signal again *mimics* an irregularly pulsed one. And this could well happen in a grasshopper, where the emitting system (file and scraper) is of the right dimensions, and is indeed enclosed in a reverberation chamber made by the slightly raised tegmina. The wave-form of some acridids is substantially of this type (Fig. 10, *cf.* Fig. 11).

---

* i.e., one made by regularly switching on and off in some way, an otherwise continuous series of waves (Fig. 1 (c)).

The relevance of all this to our discussion is that since, on the one hand, a pulse-like wave-form may be generated either as a real (switched) pulse series or as a mere resultant of synchronous, blending pure tones; and since, on the other, it may (no matter how generated) be received in one type of ear as a real pulse and in others as a pure tone, the only safe test for applying the term *pulse* to it is, indeed, the physicist's test—what it looks like on paper, not the way it is generated nor what it sounds like to human or to any other ears.

It seems then that men started by giving the name *pulse* to a *biological* event, the heart-beat; we extended it to the wave-form representing that event on paper or cathode-ray tube, and so to other wave-forms of like character*; and lastly, to wave-trains whose envelopes were such wave-forms. Biologists took the first steps, physicists the later ones. For better or for worse, its adoption by physicists has given the term a connotation that is both general and precise. To dispute the rights of the matter, or to look askance because there is argument even among physicists on this or that ultimate detail of the concept, is a sterile business. The bio-acoustician using the term is not improvising it from his medical colleagues' vocabulary; he is taking over a ready-made precision term from the acoustic and other physical sciences, in which a pulse is universally accepted as a physical entity describable in principle in terms purely of the physical wave-forms composing it. And we have indicated very good reasons for regarding this as in practice the safest way of defining it. One biological event may *happen* to produce one pulse (*e.g.* the ventricular contraction itself); but to say: "I shall define as a pulse, the sound produced by one cycle of movement of the generating apparatus", is to risk the gall of finding at a later stage that this biological "pulse" consists of many physical pulses. And to say it, knowing full well that many physical pulses properly so called are indeed comprised in it, is plain disregard by one discipline of the useful work of another. In my draft, I named a minority of English and American entomologists who have done just this, and I sent the draft to them among others. Of them, only two replied—the two noted above as disagreeing. A book is not the place for polemics, therefore I mention no names here; I hope for the sake of general understanding, this much more detailed discussion will convince them that it is worth while and proper to fall into line with what appears to be an overwhelming majority of biologists in general. I also hope that the present discussion will allay any doubt in the minds of non-English-speaking workers concerning the proper use of the English language in this respect.

Before leaving the concept of a pulse as a wave-train, I am reminded by Dr. Tavolga of the need to distinguish in this respect between trains of sine-waves, square-waves and spikes. It is pre-eminently the simple train of sine-waves that has come to be called a pulse, by virtue, as we have seen, of its pulse-like envelope. But a square or rectangular wave (Fig. 1b) is itself one of the wave-forms that can properly be called a pulse—very frequently, indeed, it is used as the envelope of a train of sine-waves; hence it would be confusing to call a *train* of square waves a *pulse*; and in fact the term used is *pulse-train*. The same considerations apply to spikes, a train of which should be called a pulse-train, and not a pulse.

---

* The American Standard gives as a *basic* definition: "a variation of a quantity whose value is normally constant; this variation is characterized by a rise and a decay, and has a finite duration". This would describe equally well Figs. 1a and 1b.

(c) Click

I am indebted to Professor Dijkgraaf (*in litt.*) for drawing my attention to a cardinal error in my own conception of the term. So far mainly limited, in the biological context, to echolocation, it is finding its way to animal acoustics *via* echolocating animals, such as bats and oilbirds. It was originally used for a very short pulse of one or two waves only, separated by a much longer interval from its neighbour; but, in echolocation, it has come to be used for pulses of as many as 50 or 100 waves, still very short in duration, since the frequency is high, and still well separated from other clicks. Thus, the essence of the click is its short duration in time and its discreteness, rather than the number of cycles composing it (contrary to my former belief).

And once again, it is perhaps desirable to emphasize that, no more than a pulse (of which it is a special case), should a click be defined in terms of its generation or its reception.

(d) Chirp

In conformity with its lay connotation, the term chirp is usually employed for some kind of unitary sound. With this in mind, Broughton (1952) and Chavasse *et al.* (1954) tried to define it, for grasshoppers, as the sound-complex corresponding to a single movement of the apparatus (*i.e.* as equivalent to the *Silbe* of the German works—Faber (1929, 1932, etc.), Jacobs (1953, etc.)); but Pierce (1948), having mostly used "pulse" to denote this, had made "chirp" the next higher grade in his system, and spoke of "unipulse" and "multipulse" chirps. Alexander (1957) and Walker (1957), with analogous conceptions of a pulse, have followed his example. These three workers, in short, have used chirp as a multiple concept—at least potentially.

Bird workers, however, again use *chirp* with the general unitary connotation, though not in any exact or defined sense; moreover, here, some degree of qualitative description is implied (*e.g.* Thorpe, (1958), p. 555, describing the chaffinch's subsong as a long succession of chirps and rattles). Similarly, in marine sounds (Fish, 1954... "clicking and chirping, which are chewing sounds").

I must clearly withdraw my own definition, which equated *chirp* with a single movement, and thus with the German *Silbe*, because of its many other utilizations; at the same time, I have shown that *pulse* cannot properly be used for this either. Later in this chapter, we shall examine more specifically the problem of how far we can go towards finding another acceptable term for use in this context (*i.e.* equating to one movement) and of general application throughout the field of animal acoustics; meantime leaving *chirp to be retained as a useful empirical term for a sound which appears unitary to the human observer's unaided ear*. For this very happy thought, to which I shall return later, I am indebted to Dr. M. Duijm (*in litt.*), as indeed also for very many fertile ideas greatly contributing to the clearing up of my own mind on this whole subject of terminology.

(e) Phrase and sequence

I am again indebted to Dr. Duijm for reminding me of Baier's (1930) definition of *phrase*: combination of successive tones which is habitually of the same length, is species-specific, and has often a variety in pitch, arrangement, etc.; and to Dr. O. Sotavalta, for a careful elucidation of the concept in musical form, which may perhaps be paraphrased as: a characteristic unitary group of melodic figures.

The word has, however, been used much less definitively by later entomologists, *e.g.* Broughton (1952), who later (1953) dropped it for the more neutral term *sequence* (a discrete succession of strokes of the apparatus), Haskell (1957) (essentially similar); and many other ornithologists (discrete succession of notes forming—usually—a subdivision of a complete song). Whether this last is equivalent to the entomological usage is not known, since most sound elements in bird song cannot yet be equated with particular bodily events.

Faber (1953) used the German equivalent of *phrase*—entomologically again—with emphasis on the integral unity and invariably-reproduced characteristic form of the sound-series constituting it (Jacobs (1953) uses *Vers* for much the same idea, and Faber himself formerly used *Strophe* (1929, 1932)).

"Phrase" seems to be useful, well-established and accepted in bird-song terminology, but at most sporadic and not—as an English word—generally accepted in entomology. It would seem better to seek a different term in English to correspond to such concepts as Faber's, one which will better express the property of reproducible pattern—which is indeed better expressed by Faber's own former word *Strophe*, or Jacobs' *Vers*. We return to this later.

"Sequence", as both Duijm and Sotavalta point out (*in litt.*), is by itself *too* neutral to express this. Moreover, it has a widely different meaning in musical form, and is pre-occupied in bird work with yet another meaning—a succession of *songs* (each usually consisting of several *phrases*).

On the whole, therefore, both words should be probably dropped by entomologists so far as any exactly defined use is concerned—no great hardship, in view of the fewness of people affected. This would avoid false homology and leave their existing use in ornithology unambiguous; it would not, of course, prevent either being used in its normal lay sense outside ornithology, qualified or otherwise—and we shall refer to this later.

Other disciplines do not appear to use either word.

(f) Trill

Walker (1957) makes extensions of the very clear lay connotation of the word trill (a long and rapid succession of repeated sounds, resolvable as such by the human ear; a tremulous vibration) with precisely defined meanings. The extensions are natural, and the definitions fit snugly.

On the other hand, there is also the very well established usage of the term in music, where it is defined as a rapid alternation of two *different*, though (usually) *near*, notes; the lay use of *trill* accords not with this, but with the *tremolo* of the musician (rapid repetition of the *same* note—Fig. 7c, *cf.* 7a). With increasing knowledge of animal sounds, confusion could arise from these conflicting usages, and we may well consider, now, adapting our terms to conform to the musician's definitions, as being the most precise yet laid down. Walker's usages, like my own and many others', would then have to become tremolos rather than trills. Since it might be as well in the meantime to have some umbrella term for those cases where it is not yet known whether trill (s.s.) would be appropriate, I put forward, with some diffidence, the word *ripple*, which is fully descriptive but non-committal. See further in the section on **Proposals** (page 16).

4. The Modes and Levels of Analysis

In reviewing the sources of our terminology, one was left with the distinct impression of an *ad hoc* process resembling the pulling of rabbits from a conjurer's hat—or rather, from a row of some four or five adjacent hats. Small wonder, if in the past some of us finished up with the rabbit, the pigeon and the coloured flag in the wrong hats—and a terminology correspondingly confused.

We need to approach the problem from the opposite side: instead of studying available sources of terms that "might come in handy some day", we should first survey systematically the modes and levels in which a sound can be analyzed, decide which of them require, obligatorily, a set of terms, and which have the priority in this respect; and then find the terms in that order. We proceed now to arrange our ideas in this way.

1. We can log a sound *instrumentally* as a graphic wave-form, and measure and compare ultimate purely physical quantities such as frequency (fundamental and harmonics), pulse properties, and so on;

2. We can attack the *emission* of the sound, at two levels, *viz.*:

(*a*) That of *neuromotor physiology*, seeking to discover which motor act of the animal produces such-and-such a physical component of its objective wave-form.

It may or may not profit us to name such an emission element by reference to its corresponding motor act, but if we do, we are going to need terms in a *phonetic-poetic* category, such as *phoneme* (Prague school—other definitions have less relation to the motor act itself, v. glossary) or at least *phone* (vowel or consonant), *syllable, foot, line, verse* (or better, *stanza**, which is less ambiguous in translation) and *canto**;

(*b*) That of the accompanying *behaviour*, with which, however, we must usually take into account that of the "relevant hearer" too, 3(*b*) (ii) below;

3. We can study *reception* of the sound in two modes, and at two levels in the second of them, *viz.*:

(*a*) Its *impact upon our own unaided ears*, which will give us a set of *empirical* diagnostics by which to describe it, even before bringing it to objective instrumental study (pitch, timbre, rhythmicity or alternatively unitary quality; melody, and in principle harmony);

(*b*) Its *impact on the ears* for which it is "intended" (the "*relevant hearer*")—either the animal's own ears, or its fellows', or its synecological coactants', at the level:

(i) of the action potential and *neurosensory physiology*;

(ii) of the sign-stimulus and the *ethological and semantic study* of hearer and emitter alike.

The former will tell us which properties from any of the foregoing categories *can* be used for conveying information—or rather, which can not (leaving us to find out what purpose if any these do serve); the latter will tell us which of the potentially

---

* Dr. Sotavalta has gently twitted me for calling *stanza* and *canto* English terms. I would assure him and all possible critics on this score that, though of obvious Italian origin, their use is of very respectable antiquity in Britain, dating at least from the Elizabethan poets (*e.g.* Spenser's Faerie Queene).

utilizable properties in fact are used, and with what significance in the life of the animal and its coactants. Both will enable us to restrict our terminology.

Only when we can synthesize a copy of the signal out of its components, *accurate enough to produce substantially the same behaviour response*, can we feel satisfied with our analysis under all these heads. And to do it, we may have used a totally different succession of physiological (or artifical, pseudophysiological) events, such as mouth imitations of bird-and grasshopper-song, *appeaux*, synthetic voices and stridulators; and the signal we produce may to our own ears (3(a), above) sound nothing like the original.

The relationships between these various modes of analysis may be worth pursuing through a few examples. In human language, a word, with suitable intonation, *e.g.*, "Fire!", may trigger off a behaviour mechanism; or a sentence may be needed, especially if intonation (denoted above by the exclamation mark) is removed, and the message consists purely of *intelligence*, without *other information*, as telecommunication engineers put it.

Simpler from most points of view is the word "t...", expressive of disapproval. Oscillograms of a succession of such sounds (usually written, "tut, tut..."), with the intervals between deleted, are given in Fig. 6, which shows that some at least ap-

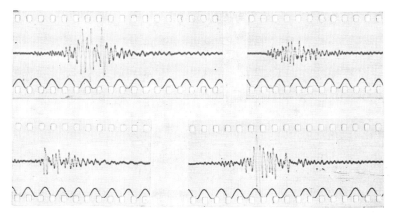

Fig. 6. Sequence of four human tongue-sounds ("t... t... t... t...") expressing disapproval. The second in particular *approaches* the wave-form of a simple pulse, but is in fact broken up rather indefinitely into three or more. Reference signal 800 c/s.

proach the wave-form of a simple pulse. Each member of the series is produced by a single set of motor acts, comparable with every other set; by fewer anatomical elements than a good many other words (crudely put: *tongue pressed on roof of mouth, suction applied, various depressors contracted until tongue jumps away*). Empirically, it is a single "chirp" (p. 12, and *infra*), linguistically a single syllable, phonetically a single phoneme (Prague sense); and it conveys its meaning by its "intelligence" alone, without "other information"—*e.g.* intonation of the voice. It can evoke a behaviour reaction, such as a scowl or a sly smile, in the human hearer.

The French word "tabac" is a little more complex. It consists of 2 unitary sounds (*ta* and *ba*) to the human hearer, but utilizes 3 phonemes (one twice); it requires at least two different sets of integrated motor acts (*ta* and *ba*, again); and it has two syllables; but it still only evokes one reaction in a given hearer, until intonation is used in addition by the utterer.

A *simple word* (*v.* glossary) such as "Fire!" or "Danger!" (with the! recognized as an integral part of it) can perhaps be reasonably equated to the call-note of a bird (in this case naturally the alarmcall). The song of the bird, on the other hand, is more like a sentence (grammatical sense) or perhaps even more, such as: "There is danger for any flat-footed male chaffinch with a face like the back of a bus that comes over my frontiers"; or more chastely: "I am a chaffinch; I have territory here. No other chaffinch is allowed, unless it shows female behaviour responses." Sotavalta points out that this particular song is also a sentence in the musical sense—constituting a single period, again in the musical sense.

On this basis, most animal sounds, other than courtship, are little more than words; courtships involving an acoustic ritual (or other behaviour sequences of such type) may, however, equate to sentences in this rough sense. To say this, however, is not, I hope, to imply that such rough relationships should be used as a basis for a new exact terminology.

5. Proposals

It seems, then, that the most promising approach to the whole problem is *via* the practical sequence of the analytical process itself. Since the emission has to be heard by the investigator before he even knows there is anything to record graphically, the first attempt at analysis is bound to be the provisional and empirical one of paragraph 3(*a*) of the preceding section—what does the sound sound like to him?— even though in some cases it may be very rapidly followed by visual analysis of the recorded waveform. Such an empirical first description accords well with the classical scientific method.

For a very large range of animal groups, the *audible ground rhythm* can be described in terms of the *chirp* (and its variants), and groupings of chirps. For chirp, Duijm (*in litt.*) has proposed essentially the following definition, which I feel will meet with general acceptance (I have expanded his wording slightly):

Chirp: the shortest unitary rhythm-element of a sound emission that can readily be distinguished as such by the unaided human ear.

This obviously presents no difficulty where an utterance is short and isolated (short chirp), or long but unmodulated (long chirp). But when it is both long and modulated, the modulation may be appreciated in any of three ways, which may lead to difficulties:

(i) *As a note*—in which case the whole utterance will still be a long chirp, of determinate pitch;

(ii) *As a tremulation*—in which case the utterance will become a *ripple* (Fig. 7) (defined as a chirp modulated at a rate below human fusion frequency). When frequency-modulated, the ripple is a *trill* or *shake*; when amplitude- or pulse-modulated, it is a *tremolo*;

(iii) *As a total separation of the sound* into disparate units—in which case it becomes a *first-order sequence* of *chirps*. (A second-order sequence will be an analogous succession of first-order sequences.)

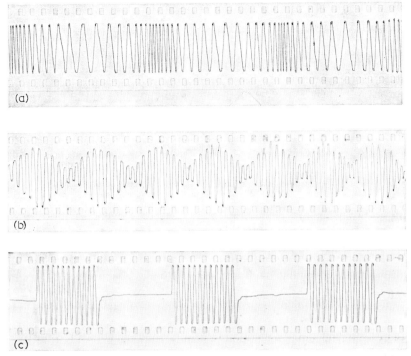

Fig. 7. Freehand drawings of three types of "ripple". (a) frequency-modulated = *trill* or *shake*. Note, however, that for clarity, a difference of pitch of about 2 octaves has been portrayed, which is far beyond the interval of the normal shake in music. Moreover, here, the frequency change is smooth, whereas in most musical instruments it is step-like. (b) amplitude-modulated = *tremolo* (or better, *celeste*). (c) pulse-modulated = *tremolo (sensu stricto)*.

Clearly, on the border-lines between (i) and (ii), and between (ii) and (iii), there will be difficulties in applying the definition, as always on all border-lines. These must be met *ad hoc*, by the exercise of horse sense. Other difficulties will arise from change of conditions: what is normally a ripple may be slowed down by a fall of temperature to a sequence of short chirps. Obviously here again, horse sense has to be used, and the term for the normal conditions retained, with reservations.

The audible ground rhythm is of course applied to some sort of *basic sound*. This may be recognized as:

(*a*) *A musical note* (fundamental and as yet undetermined harmonic pure tones, including the special case of the pure tone without harmonics);

(*b*) *A succession of different notes* (which can in principle have, independently of the ground rhythm, its own pattern or arrangement, the *melodic figure*—as in the extemporizations dear to the virtuoso of Edwardian soirées, taking a theme like Three Blind Mice or Baa Baa Black Sheep, and "beethovenizing", "chopinizing" or "handelizing" it by suitable adaptation of ground rhythm and harmony)*;

---

* This independence of melodic and rhythmic patterns is also admirably illustrated in the chanting of psalms, where the ground rhythm of the words has to be parcelled up in rather odd groups to make the very uneven prose fit the very regular pattern of the melody; in contrast to hymns, where word-rhythm and melodic figure fit rather closely.

(c) *A noise* (equivalent to a set of synchronous pure tones whose blending produces no regularity in the combined wave-form).

Type (b) is obviously *par excellence* the domain of bird (and lemur) song, where the melodic figure and its hierarchy (phrase, sentence period, etc.), if used, must replace for lack of an *independent* ground rhythm, the plain first, second and third order sequence etc. It looks as though the border-line difficulties of this musical-form terminology might be even greater than those of the rhythm concepts; its adoption would certainly require that workers had a good grasp of the principles of musical form, and a "feeling" for it. It is, however, for ornithologists themselves to arbitrate finally.

However that may be, there is an extremely interesting possibility in the adaptation of the concept of a *figure* to types (a) and (c) of basic sound, where by definition there is no *melodic* figure. This is to express the idea of a *patterned* succession of chirps by the term *figured sequence*, e.g. for a rhythm such as that of the spoken phrase: God save the Queen: ♩ ♩ ♩ ♪ ♪ ♩ ♩ This qualification expresses eloquently the concept of reproducible pattern implicit in Faber's *Strophe* or Jacobs' *Vers* already referred to, without being, as in their case, tied up in any way with motor events in the emitting animal.

Lastly, in this field of empirical analysis there is clearly still plenty of room for the onomatopoeisms and very apt lay terms like burst, string, volley, tattoo, that bring some colour to the drab landscape of scientific publication; and for terms in other categories of qualitative assessment, such as *sub-song* (orn.) and its synonyms (ent.).

We have said that the logical second stage of analysis is at the physical level. By then equating the physical time pattern of the emission to the ground rhythm of the empirical analysis, its frequency spectrum to the aural harmonic analysis, and its frequency variation in time to the melodic figure, common points of reference will be found: the *chirp*'s constitution in terms of *pulses*, or of pulse complexes (trains, or trains of trains); the relation of the aurally prominent harmonics to the dominant frequencies; and so on. But although we can for a particular emission *relate* the physical quantities to the biologically assessed empirical ones in this way, it must be stressed once again that these physical quantities (frequency, repetition rate, basic signal, elementary vibration, elementary wave-form, p. 6; harmonics, dominant and principal frequencies, p. 6; pulse, p. 7 ff. and Fig. 1; click, p. 12) must be *defined* by reference to the objectively measurable properties of the sound itself, and not by reference to any biological structure producing or receiving it. If this is kept clearly in mind, the actual wording of the definition is not so important, and can be less restrictive. Pulse itself may be defined then in some fairly unrestrictive way such as: *Pulse*: a unitary homogeneous parcel of sound waves of finite duration; a *simple* wave-train (*i.e.* one divisible into waves, but not into groupings of waves).

This filling of the gaps in the empirical analysis by means of the physical components converts the whole into a system which is fully objective and rigorous, but which as yet is without known *meaning* and without known relationship to the emitting mechanism. Stimulation of the "relevant hearer" with playback, with imitation, and with mutilations of the normal emission (with or without action-potential experiment to narrow the field), will elucidate the precise elements that transmit the meaning, *i.e.* the acoustic sign-stimuli, and enable us to label certain sections of the terminological framework with semantic data, *e.g.*: "this melodic figure heralds attack; that ground rhythm is a territorial announcement"—an essentially ethological terminology that needs no discussion here except in one important respect: insistence on these experimental tests' *preceding* the labelling of the sound-element with a behavioural tag.

But the very synthesis of these experimental emissions may, in some cases at least, depend on first completing our analysis in terms of motor mechanisms. It is difficult to believe that our present success (*e.g.* Lawrence, 1961) in synthesizing human speech could ever have been achieved without first disentangling the separate contributions of the breath, the vocal cords, the resonance chambers, the tongue, lips and all the other vocal machinery, to the global sonogram; and total success in response experiments on those complex sound-emitters, the Acrididae (Orth.), may well depend similarly on first resolving their sonogram into separate contributions by the stridulation-gear and the reverberation-chamber.

So correlation of motor events and sound components may have more importance than just that of completing the jig-saw. In this field particularly, the anarchic individualism of authors dismays the methodical mind; and yet, with such vast differences in emission-mechanisms in different animal groups, it is difficult to see how a common denominator can be found, as I myself discovered in trying to do it by widening the definitions of the German Orthopterists to fit all groups.

Thus, Jacobs (1953) defines *Silbe*, the lowest term of his motor hierarchy, as:

"Syllable": sound produced by one to-and-fro excursion of the femur. Going on from this, he defines a *diplosyllable* as a syllable in which both the *to-* and the *fro-* are acoustically effective, and contrasts it with a *disyllabic sound*, in which the femur makes 2 to-and-fro excursions. By implication, the syllable in which the excursion is effective in one direction only might properly be called a *haplosyllable*, though as far as I know, Jacobs does not say so expressly.

If now the basic definition of syllable be expanded to:

"Syllable": sound produced by one single cycle of contraction and relaxation of the operating muscles,

it still clearly applies to the grasshopper, one single cycle being all the contractions and relaxations between the moment when the apparatus leaves a certain position P and the moment when it takes it up again from the same direction.

It also clearly applies to any other repetitive signal, even (by natural extension) if this repetition is not quite exact. This even goes for the human voice in the simple case where it is agitatedly repeating, "tut, tut, tut, tut..." or "Fire! Fire!..." Nevertheless, with repetitive signals even as near to the Orthoptera as the cicadas (Homoptera), we meet what Pringle (*in litt.*) justly regards as a difficulty: each cycle of contraction and relaxation in some cicadas produces one *pulse*. And again, Tavolga (*in litt.*) points out that in some fish, each cycle of contraction and relaxation produces one *wave*! These are perhaps not insuperable problems, since to say that in the one, the "syllable" is *unipulsate*, and in the other, *monocyclic*, is not altogether without advantage as a terse description in both physical and physiological terms.

On the other hand, fitting bird sounds into such a scheme is impossible at present, for sheer lack of knowledge of motor correlations; and when they *are* known, they will presumably be of two categories—those of the melodic figure and those of the ground

rhythm. And the scheme fails utterly in a typically varied human utterance, since such a statement as:
"The quick brown fox jumps over the lazy dog"
becomes a single "syllable"! It is not only the word *syllable* itself that is objectionable here, but the whole idea of equating a simple to-and-fro movement, in one or two structures of one animal, to an enormously complex series of (admittedly differing) movements of a considerable number of structures of another.

It seems to me as to many others, therefore, that a fully general terminology *of motor-correlated units* is out of the question. It may be possible to adapt the orthopteran type of terminology to most stridulatory and analogous mechanisms, and this without even widening the Jacobs type of definition above; this still leaves it possible in such fields as Pringle's and Tavolga's to widen it in the way indicated, without lessening the precision of the orthopterists' connotations. But workers on birds, and other self-contained acoustic groups, must be left to work out their own salvation as functional studies progress; and in the present state of human phonetics, it seems best also to leave human utterances to their utterers.

So far as Orthoptera and their analogues are concerned, it is important to realize that, for our modified purposes, the German systems, excellent though they are, need not be taken over lock, stock and barrel. Our analytical procedure, as outlined above, can produce a description which is objective and complete in terms of human hearing and the physical sound-picture: one or more *sequences* of *chirps*, each with a known *pulse* composition, and each pulse with a known *wave* composition. All we need to do thereafter is to define *one* sound in terms of motor activity; it can then be fitted into the scheme at the appropriate point, and this will make the scheme complete in terms of the events of emission too. The obvious term is of course *Silbe*, as already reviewed, and there is no longer any objection to its being rendered by *syllable*, now that we have abandoned the idea of a general motor terminology applicable in the human context.

Thus, in *Chorthippus brunneus* (Thunberg), the unitary squirt (Fig. 8) of sound constituting the ordinary song, or that of the courtship song, is an unequivocal *chirp* in the sense here defined; observation shows that it is produced by a dozen or so to-and-fro movements, each of which is clearly apparent as a discrete train of many close-packed pulses in the oscillogram; in the stretched oscillogram (Fig. 10—high camera speed or slow-motion playback or both), the pulse limits are so indefinite as to suggest generation at least in part by internal reflection (p. 10). Thus, here it is the *chirp* that consists of several *syllables*, each of several *pulses*, each of one to several *waves*. And the chirp may well be described as *figured*, in recognition of its varied and characteristic syllable-pattern.

The rivalry song of *Chorthippus parallelus* (Zetterstedt), on the other hand, is a *figured, first-order sequence* of *chirps*. Each chirp is *disyllabic*, with a left and right *syllable* (and clearly *figured* too (Fig. 9). Each syllable consists of many clear-cut *pulses*, each of a very few waves (Fig. 11); and it is very likely that the syllable will also prove to be *figured*.

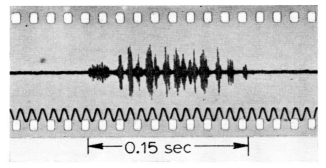

Fig. 8. *Chorthippus brunneus* (Thunberg) (Orthoptera, Acrididae). A single *chirp* from the courtship song. This example comprises fifteen or more syllables (or hemisyllables). Reference signal 100 c/s.

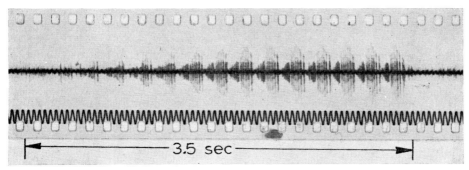

Fig. 9. *Chorthippus parallelus* (Zetterstedt) (Orth., Acrididae). A single *first-order sequence* from the rivalry song. It comprises 15 *disyllabic chirps*, each with one syllable produced by the left leg and one by the right. Reference signal 20 c/s.

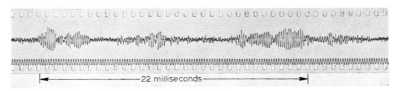

Fig. 10. *Ch. brunneus*. Part of a single syllable from the emission of Fig. 8. Note the rather indefinite intergrading of the pulses and the scarcity of transients; the song is sweet-toned (especially when slowed-down), and the pulses may be partly chordal. Reference signal 5,000 c/s.

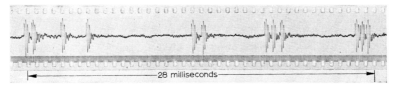

Fig. 11. *Ch. parallelus*. Part of a single syllable from Fig. 9. Here note the very disparate pulses, each beginning with a sharp transient (steep front), and thus, from theory alone, likely to represent one impact. Reference signal 10,000 c/s.

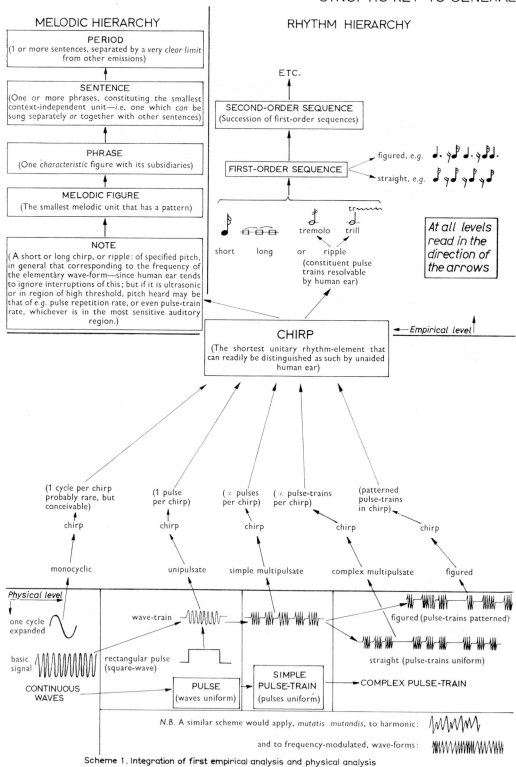

Scheme 1. Integration of first empirical analysis and physical analysis

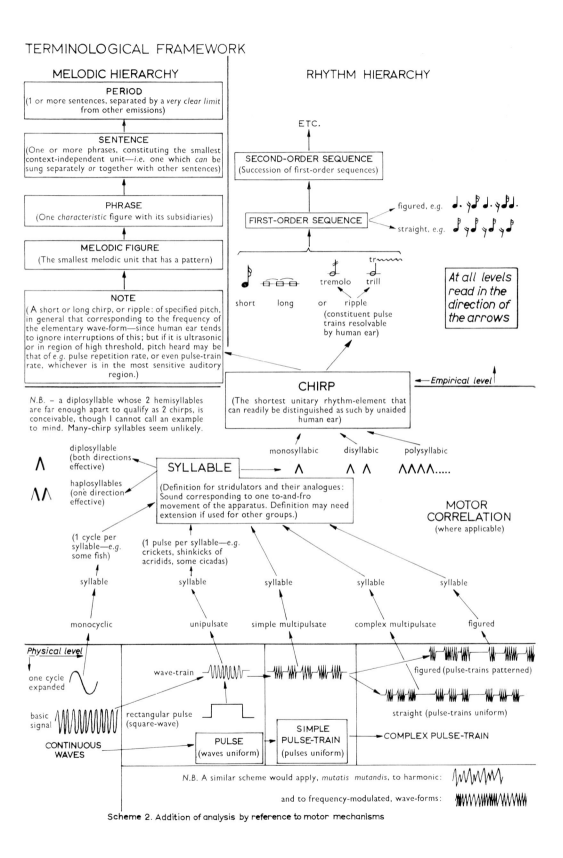

Scheme 2. Addition of analysis by reference to motor mechanisms

SELECT REFERENCES

Of the large number of works consulted (particularly in the fields of marine and insect sound), which it would be unprofitable to give in full, only a selection, illustrating particular problems discussed, are listed below. In the field of bird sound, I must admit to having adopted the lazy but economical expedient of consulting (orally or by letter) a few recognized authorities, rather than sifting the whole literature myself; I therefore owe and willingly give thanks, in particular to E. A. Armstrong, O. Sotavalta and W. H. Thorpe, who, however, must not be held responsible for my interpretation of their replies to my enquiries. I also owe grateful thanks to those who kindly replied to my circulated memorandum (referred to in the text).

ALEXANDER, R. D., 1957, Sound production and associated behaviour in insects, *Ohio J. Sci.*, *57*, (2), 101–113.
BAIER, L. J., 1930, Contribution to the physiology of the stridulation and hearing of insects, *Zool. Jb. Abt. allgem. Zool. Physiol. Tiere*, *47*, 151–248.
BROUGHTON, W. B., 1952, Gramophone studies of the stridulation of British grasshoppers, *J. S. W. Essex Tech. Coll.*, *3*, 170–180.
BROUGHTON, W. B., 1953, Further gramophone studies of grasshoppers and crickets, *Proc. R. ent. Soc. Lond. (C) 18*, 17–18.
CHAVASSE, P., R. G. BUSNEL, F. PASQUINELLY AND W. B. BROUGHTON, 1954, Propositions de définitions concernant l'acoustique appliquée aux insectes, *in* BUSNEL, R. G., *et al.*, 1955, L'acoustique des Orthoptères, *Ann. Epiphyt. Fasc. hors série, Paris*.
FABER, A., 1929, 1932, Lautäusserungen der Orthopteren I und II, *Z. Morph. Ökol. Tiere*, *13*, 745–803; *26*, 1–93.
FABER, A., 1953, Laut- und Gebärdensprache bei Insekten, *Mitt. Staatl. Mus. Naturk. Stuttgart*, No. 287.
FISH, M. P., 1954, Character and significance of sound production among fishes of Western N. Atlantic, *Bull. Bingham Oceanog. Colln.*, *14* (3), 1–109.
HASKELL, P. T., 1957, Stridulation and associated behaviour in certain Orthoptera, *Brit. J. Animal Beh.*, *5*, 139–148.
HORRIDGE, G. A., 1960, Pitch discrimination in Orthoptera, demonstrated by response of central auditory neurons, *Nature*, *185*, 623.
JACOBS, W., 1953, Verhaltensbiologische Studien an Feldheuschrecken, *Z. Tierpsychol.*, Beiheft I.
LAWRENCE, W., 1961, The use of artificial speech as a controlled stimulus for the study of perception, *Symp. Zool. Soc. Lond.*, *7*.
MILLER, D. C., *The Science of Musical Sounds*, New York, 1934.
MÖHRES, F. P., 1953, Über die Ultraschallorientierung der Hufeisennasen, *Z. vergl. Physiol.*, *34*, 547–588.
PIERCE, G. W., 1948, *The Songs of Insects*, Cambridge, Mass.
PRINGLE, J. W. S., 1954, Physiological analysis of cicada song, *J. exp. Biol.*, *31*, 525–560.
PUMPHREY, R. J. AND A. F. RAWDON-SMITH, 1939, "Frequency discrimination" in insects, *Nature*, *143*, 806.
PUMPHREY, R. J., 1940, Hearing in insects, *Biol. Rev.*, *15*, 107–132.
SOTAVALTA, O., 1956, Analysis of the song patterns of two Sprosser nightingales *Luscinia luscinia* (L), *Ann. Zool. Soc. "Vanamo"*, *17*, No. 4.
THORPE, W. H., 1957, Identification of Savi's, Grasshopper, and River Warblers by means of song, *British Birds*, *50*, 169–171.
THORPE, W. H., 1958, Learning of song patterns by birds, *Ibis*, *100*, 536–570.
WALKER, T. J., 1957, Specificity in the response of female tree crickets to calling songs of the males, *Ann. ent. Soc. Amer.*, *50*, 626–636.

CHAPTER 2

TECHNIQUES USED FOR THE PHYSICAL STUDY OF ACOUSTIC SIGNALS OF ANIMAL ORIGIN

by

A. J. ANDRIEU

1. Introduction

Various modern techniques and complex apparatus, whose characteristics will be examined, are indispensable for the study of acoustic signals in the animal world.

The acoustic signals are first transformed into electric signals which may be stored by a suitable recording method for subsequent measurements and analyses.

A microphone is of primary importance for any objective acoustic study, and the most delicate problem to be solved in animal acoustics is the recording of the sound which will provide a basis for any analytical study.

While special precautions are necessary in the study of each animal species, a certain number of general rules must be observed, whether the recording is carried out in the field or in the laboratory.

2. Choice of the Microphone

The same microphone is not suitable for all recordings. A series of classical microphones is used for terrestrial animals, while special microphones, or hydrophones, are used for marine animals.

The voltage collected must, in all cases, be proportional to the acoustic pressure provided by the signal. A large number of factors must be taken into account when choosing a microphone for a given recording:
1. The qualities of the sonic field:
   laboratory or fieldwork,
   frequency band to be recorded (depending on the species of animal),
   intensity of the signal, being a function of the distance between the transmitting animal (source) and the microphone.
2. Accuracy of the measurement:
   voltage output of the microphone,
   frequency distortion,
   non-linear distortion (response range),
   background noise.
3. Conditions of the recording:
   surrounding noise,
   temperature,
   humidity,

References p. 47

(These three factors are very variable in field work and influence the choice of material, the conditions for its use, and the value of the information collected.) mechanical vibrations, size and nature of the space (or the surroundings) where the recording is to be made.

Before discussing these different points, it appears desirable to state that the types of microphones available at present are based on different principles, a knowledge of which is indispensable for choosing the model to be used for a given study.

(a) Electrodynamic microphones

These are the most often used. Their working principle is as follows: a moving coil attached to a membrane placed in a magnetic field which is given by a magnet is set in motion by external sonic pressures. A current proportional to the speed of

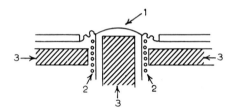

Fig. 12. Diagram showing principle of moving coil electrodynamic microphone in section (1) moving diaphragm, (2) moving coil, (3) permanent magnet.

displacement is then induced in the moving coil (Fig. 12). The response curve of such microphones varies greatly with the make. In general, a response above 10 kc/s presents a marked attenuation, due to the mass of the moving part. There are, however, very good electrodynamic microphones. Their response curve extends from 30 c/s to 15 kc/s at ± 5 dB. Their directivity is controlled by their dimensions and form and is, in general, not very efficient. The amount of harmonic distortion is low, and the background noise negligible. They withstand high levels of the order of 130–140 dB above 2×10^{-4} baryes (or dynes/cm^2). In all these points, their dynamic is very high.

They are robust, stable and practical. Since they have a low impedance, very long cables (several hundreds of meters) may be used for connecting them to the recording instruments.

They are well suited to the requirements of field work with most animals, with the exception of bats, certain rodents, birds, terrestrial Batrachia, and certain Invertebrates.

(b) Ribbon microphones

The working principle is similar to that of electrodynamic microphones: a very thin ribbon of a few microns in thickness replaces the moving coil and the membrane forming the diaphragm. The resonance frequency is generally very low, being less than 5 c/s.

Two methods of operation are used. In the so-called speed models, the sonic pressure is exerted on the two surfaces (front and back) of the ribbon. The vectorial sum of the pressures produces the effective electrical voltage. The frequency response varies according to whether the microphone is placed in a plane wave or a spherical wave field, since the speeds of vibration are different in each case.

In the so-called pressure models, only the front surface of the ribbon is used, as in the moving coil electrodynamic microphones.

In general, the response curve of ribbon microphones is linear within the range 30–15,000 c/s at ± 5 dB. The low frequencies are accentuated when these microphones are placed near the sonic source (up to one meter) due to the spherical form of the waves.

These microphones are stable, but bulky and fragile. They must not be used in the open air, for the ribbon may be deteriorated by air currents. Besides, even a slight wind causes quite a large amount of background. This type of microphone is therefore restricted to laboratory use with such species as have been indicated by field work with the electrodynamic microphone.

(c) *Piezoelectric microphones*

These microphones use the piezoelectric properties of certain crystals which produce electromotive forces under the action of external forces (pressure waves). Voltages developed are proportional to the torsions of the crystal in certain frequency limits (Fig. 13).

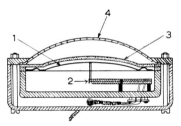

Fig. 13. Diagram of a piezoelectric microphone in section. (1) diaphragm, (2) crystalline piezoelectric element, (3) porgus metal plate, (4) perforated cover.

Five types of crystals are used, viz:. quartz, Rochelle salt (sodium-potassium tartrate), ammonium dihydrogen phosphate (ADP), lithium sulphate (LH) and barium titanate (ceramic) which is the latest substance to be used.

Quartz microphones are rarely used in air, since they are a thousand times less sensitive than those using Rochelle salt.

In general, two crystals are stuck together to form an arrangement of bimorphous elements which functions either by flexion or by torsion. This gives much greater efficiency and temperature and humidity have less influence on the mechanical and electrical constants.

The resonance frequency of a piezoelectric microphone depends on its size and the method of functioning (whether the crystal works by torsion or by flexion). If the resonance frequency is fairly high, linear characteristics may be obtained in certain wide frequency bands. The crystals may be used in two ways, either by direct action or by way of a diaphragm attached to the piezoelectric element. In the latter case, the sensitivity is increased, but the response curve is affected by resonances and is reduced at very high frequencies.

The sensitivity of the Rochelle salt microphone is equal to —60 dB based on the level 0, per barye at 1000 c/s. Microphones using ADP, lithium salt or barium titanate, have a very much lower sensitivity (—80 dB).

Piezoelectric microphones, except for those using barium titanate, are sensitive to humidity and temperature. The crystals must therefore be protected by a varnish or by a covering of oil (as it is the case for hydrophones).

The high impedance of this type of microphone does not permit the use of a linking cable more than a few meters in length, unless a pre-amplifier is connected to the microphone.

The characteristics of the various types of crystals used are given in Table 1[13].

Piezoelectric microphones are of interest, since they can be of very small dimensions (less than 1 cm³).

TABLE 1

COMPARATIVE TABLE SHOWING THE PROPERTIES OF THE VARIOUS CRYSTALS

Properties of the crystal	Type of crystal			
	Rochelle salt	A.D.P.	L.H.	Barium titanate
Type of cut with respect to crystal axes	X n Y	Z	Y	
Density (kg/m³)	1.77×10^3	1.79×10^3	2×10^3	5.5×10^3
Cut-off low frequency ($p/5$) at 25°C	0.1	12	1/1000	Very low
Maximum temperature at which it may be used (°C)	45	125	75	110
Maximum humidity at which it may be used (in %)	84	94	45	100
Dielectric constant K	200	15.3		1700
Piezoelectric modulus ($M/V \times 10^{-12}$)	165	24		78

Hydrophones, the microphones used for the study of marine animals, are piezoelectric microphones. They are discussed in Chapter 3 (p. 51).

(d) *Electrostatic (or condenser) microphones*

These microphones are now being used much more often since they surpass all other types in performance.

They are pressure microphones whose working principle is based on the variation of capacity of a flat condenser, one of the electrodes of which, forming the membrane of the microphone, is mobile. This membrane is very thin (from 10 to 15 microns) and is generally made of duralumin. Air is the dielectric and has a thickness of the same order of magnitude as the membrane. A continuous voltage is applied to the condenser terminals in series with a very high resistance R. Under the effect of an alternating sonic or ultrasonic pressure, there is a displacement of the mobile membrane and the appearance of an alternating voltage of the same frequency at the terminals of the resistance R (Fig. 14).

The output impedance of such microphones is very high, thus the first pre-amplifier stage is always attached very closely to the microphone. Nevertheless, there are recent models (capsule + pre-amplifier) with very much reduced dimensions (cylinders 2×12 cm) since subminiature parts are used.

The basic quality of these microphones is that they have a linear frequency response curve. It extends into the ultrasonic range. The background noise is generally low. The amount of distortion remains negligible up to levels of 130 dB above 2×10^{-4} baryes.

Their main disadvantage is their sensitivity to the moisture content of the atmosphere. However, certain recent models are capable of withstanding several days

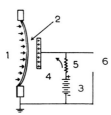

Fig. 14. Diagram showing principle of an electrostatic microphone. (1) sonic pressure transformed into movements of the diaphragm, (2) variable capacity, (3) battery, (4) direction of charging current, (5) resistance, R, (6) output. The charge is equal to the product of the capacity and the polarization tension.

of working in a humid atmosphere without difficulty. They are not sensitive to variations of temperature, 40 degrees of variation modifies the exit level by only 2 dB, with the exception of the resonance frequency which is, in general, very high.

The electrostatic microphone is the most suitable type for first-class recordings of sound. There remains merely the choice of the model to be used, bearing in mind the directivity diagram: omni-directional, cardioid, bi-directional (see p. 31).

It should be mentioned that electrostatic microphones represent the type most often used for the exact measurement of acoustic pressures. Their very small dimensions and the arrangement of the capsule allow a precise control of the sonic field, which even at high frequencies is very slightly disturbed by the microphone.

This type of microphone is used in studies on animals whose spectra are largely in the ultrasonic range: bats, rodents and numerous arthropods. Griffin[9] was the first to use them in his studies on the echo-location of bats.

(e) Contact microphones

These are electrodynamic or piezoelectric microphones specially adapted for placing against solid surfaces. They allow the measurement of both mechanical and acoustic vibrations. Their greatest application is in the laryngophone, a microphone which, when placed on the neck at the level of the vocal cords, transmits a clear vocal emission, in spite of a high surrounding sonic intensity. These microphones may be very small (1 cm³).

(f) Factors entering into the choice of a microphone

The principal characteristics and conditions of use of these various microphones are summarized in Tables 2 and 3. It is merely a matter of choice between the various types of microphones for recording and registering sound.

References p. 47

TABLE 2

CHARACTERISTICS OF VARIOUS TYPES OF MICROPHONES

Type of microphone	Output level	Impedance	Directional properties	Advantages	Disadvantages	Use	Counterindications
Electrodynamic	—50 to —65 dB	Low	Omni-directional Cardioid	Stable Robust Sensitive Not sensitive to wind	Limited range in high frequencies	Outdoors and indoors Possibility of using long connecting leads	
Ribbon	—60 to —70 dB	Low	Bi-directional Cardioid	True	Awkward Fragile Sensitive to wind	Inside	Outside
Piezoelectric 1. Rochelle salt 2. Ceramic	—80 to —60 dB	High	Omni-directional	Cheap	Not so true Affected by heat and humidity Not so true Short connecting leads	Indoors and outdoors	
Electrostatic	—60 to —50 dB at the output of the microphone pre-amplifier	Low	Omni-directional Cardioid Bi-directional	Very true Response curve which is very flat in high frequencies	Sensitive	Indoors and outdoors for high quality recording Transients of short rise-time Low, high and ultrasonic frequencies	Very high humidity

TABLE 3

CHOICE OF A MICROPHONE BASED ON CONDITIONS OF USE AND RESULTS SOUGHT

(The shaded squares refer to the conditions in which each type of microphone may be used)

Conditions of use and results	Electrodynamic (omni-directional)	Electrodynamic (cardioid)	Electrostatic (omni-directionnal)	Electrostatic (cardioid)	Ribbon (bi-directional)	Piezoelectric
Band covered 30–10,000 c/s	■	■	■	■	■	■
Band covered 30–15,000 c/s	■	■	■	■	■	
Band covered 30–50,000 c/s			■	■		
Dry weather	■	■	■	■	■	■
Damp weather	■	■			■	■
Heat	■	■	■	■	■	
Wind		■		■		
Distance Microphone–Animal: small	■	■	■	■	■	■
Distance Microphone–Animal: large		■		■	■	
Microphone placed at the centre of a group of animals	■		■		■	
Laboratory	■	■	■	■	■	■
Open air	■	■	■	■		
High quality	■	■	■	■	■	
Long lead from microphone to pre-amplifier (greater than 3 m)	■	■	■	■	■	

3. The Recording of the Sound

As this is a delicate operation a number of elementary rules have to be observed in order to obtain a satisfactory result.

It is advisable, in the laboratory or in the field, to work at the smallest distance compatible with the dynamics of the signal. Thus high sonic signal is obtained and the ratio signal/background noise is increased. Background noise[10] consists of noises natural to the premises—reverberation, resonance of the laboratory walls, surrounding noise, wind, noise originating from the apparatus, reflections and echoes. In the open air, about ten metres is considered to be a maximum distance. For this reason, when the animal cannot be approached, several microphones spaced from 15–20 metres apart may be switched on as required, depending on the position of the animal (particularly in the acoustic study of birds). When the animal was beyond approach, certain workers in Britain and America have used reflectors of either a parabolic form (Fig. 15), or a truncated cone with an opening of 30°, on the principle of

Fig. 15. Parabolic reflector used by D. J. Borror for recording bird songs. The author uses a portable tape recorder at the same time. The reflector has a diameter of 24 inches[3]. (Courtesy of Prof. Borror).

Fig. 16. Reflector used by C. R. Carpenter for recording the sound of the cries of gibbons. The microphone is in the centre. This device is a concentrator which uses the phenomenon of wave slippage[5].

wave slippage phenomena (Fig. 16)*. The microphone is at the focus of the reflector, thus obtained a concentration of the sonic level and a better insulation of the background noise. The results obtained should be interpreted with caution, particularly with paraboloid reflectors, since the resulting response curve may be be disturbed by reflection phenomena.

In the study of insects, especially those with very high frequencies in the acoustic spectrum (up to 100 kc/s), the microphone must be placed as close an possible to the

---

* See: OLSON, H. F. and I. WOLF, 1930, Sound concentration for microphones, *Journ. Ac. Soc. Am.*, (3) 410–417, and reference 20, from KELLOGG.

sonic source since there is a high absorption by the air for frequencies above 20 kc/s. This absorption is also dependent on temperature and humidity.

In laboratory work, a room insulated from external noise should be used, with the walls lined with an absorbent material to avoid any stray reflection. The microphone should be placed in such a way that transmission of the direct sound is not hindered by any obstacle.

In the same way in field work, the microphone should be placed as high as possible so as to avoid any reflection or absorption at ground level, particularly for high frequency recording.

Consideration of directivity in the use of microphones[1]

A microphone with a cardioid directivity diagram is the most suitable since two-thirds of the background noise and all sporadic noises, particularly reflections, are eliminated, and only the direct wave is recorded.

Thus relative insulation, which is the purpose behind the use of paraboloid reflectors, can be obtained.

The same applies in the field, particularly when the sonic source and the microphone are widely separated.

A microphone with an omni-directional directivity diagram is advisable when the microphone is close to the sonic source and when a group of animals is concerned.

When the microphone has to be placed between two sonic sources, it is advisable to use a microphone having a bi-directional directivity diagram.

The range of frequencies to be recorded governs the choice of microphone. The signal level, reverberation of the room, surrounding background noise, level at which the signal is emitted, the type of microphone and conditions of humidity and temperature are additional factors influencing the recording of the sound.

Conditions specific to the animal species under study and the environment are additional factors in the choice of a microphone.

4. Methods of Storing Acoustic Signals

Recorders based on various principles may be used for the temporal storing of acoustic signals that have been transformed into electrical signals by the microphone.

---

[1] *Definition of the directional diagram.* For a given frequency, representation of the response of an electro-acoustic transducer as a function of the direction of propagation of the wave in a fixed plane passing through the acoustic centre of the source.

Factor of directivity of a transducer. (a) For a transducer emitting sound at a given frequency, the ratio, in a free field, of the square of the acoustic pressure radiated, measured at a given point on the principal axis of the transducer, to the mean of the squares of the pressure on the surface of a sphere with the transducer as centre and passing through the above-mentioned point.

(b) For a transducer receiving sound at a given frequency, the ratio of the square of the electromotive force produced by the acoustic waves arriving in a direction parallel to the principal axis of the transducer, to the mean of the squares of the electromotive forces reaching the transmitter simultaneously from all directions produced by acoustic waves and of the same frequency and effective pressure as these waves, and reaching the transducer simultaneously from all directions.

Reference: Vocabulaire électrotechnique groupe 08, électroacoustique 1956, published by the S.N.I.R. (Syndicat National des Industries Radioélectriques), France.

References p. 47

(a) Magnetic recording

Since there has been much progress in this field in the last ten years, the tape recorder is now the basic working instrument for research on animal acoustics. The working principle is as follows[18]: a field producing a remanent induction proportional to the instantaneous value of the signal is applied to a ferromagnetic medium represented by the magnetic tape. An electromotive induction force is collected at the terminals of a suitable receiver in front of the magnetic tape when it is played back. The magnetic tape may be demagnetized by an intense magnetic field (Fig. 17).

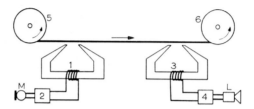

Fig. 17. Diagram showing principle of a tape recorder. (1) recording head, (2) recording amplifier, (3) playing back head, (4) playing back amplifier, (5) feed reel, (6) take-up reel, (M) microphone, (L) loud-speaker.

The system allows the immediate playing back of any recording and, theoretically, the indefinite use of the magnetic tape. The tape can run within a very wide range of speeds. Requiring no mechanical transformation, it is possible, to record currents varying between a very low frequency up to several Mc/s.

(b) The electromechanical process

This process had been used principally in the studies of mosquitoes by Kahn and Offenhauser[11], (Fig. 18). A groove engraved on a supple disc is modulated by the electrical signal. When playing back, a sapphire point follows the groove and excites an electrical transducer and the sound is immediately reproduced, but the disc cannot be used again. The quality of the recording is limited by the mechanical inertia of the parts which move when engraving and playing back. The possible frequency range recorded reaches 30–25 000 c/s, but the material is too awkward and large to allow this. This is beyond the range required in an animal acoustic laboratory.

However this method of recording on a disc pressed industrially in vinylite is at the moment the most successful for preserving documents of value.

(c) Photographic recording

The degree of opacity produced on a photo-sensitive surface is proportional to the signal and thus constitutes the recording. A photo-electric cell is used for playing back. This process has a drawback in that it is necessary to develop the film, and the frequency range is limited—between 30–10,000 c/s.

This is the process used for producing sound films for the cinema. It has been used formerly by Lutz[14] for studying the song of crickets, where a frequency band of 3–4000 c/s was certainly complied with (Movietone system).

5. Tape Recorders

The magnetic process, of all recording methods, appears to be the only one at present that merits retention for experimental studies in animal acoustics.

The choice of the tape recorder depends on the conditions of use and the electrical and mechanical performances required. A very high quality professional material is desirable in all cases.

Fig. 18. Recording on a disc, using a portable machine. W. Offenhauser, Cornell University Medical College, recording the sound of mosquitoes in flight[11]. (Courtesy of W. Offenhauser.)

Two types of tape recorders can be considered:
(a) Portable and fixed tape recorders fed by alternating current.
(b) Self-contained tape recorders fed by a battery or accumulator, which are particularly suitable for field work.

(a) Tape recorders fed by alternating current

This type of tape recorder may be used in the laboratory or in all instances where a source of alternating current is available. Depending on the running speed, the maximum performances which may, at present, be obtained with full-track recording on standard material of prefessional quality are:

$$\text{At 19 cm/sec } 40\text{--}15{,}000 \text{ c/s} \pm 2 \text{ dB}$$
$$\text{At 38 cm/sec } 30\text{--}20{,}000 \text{ c/s} \pm 2 \text{ dB}$$
$$\text{At 76 cm/sec } 30\text{--}25{,}000 \text{ c/s} \pm 2 \text{ dB}$$

The amount of distortion does not exceed 2%.

The possible maximum frequency for recording and playing back depends on the width of the clearance between the tape recorder heads and on the speed.

It is possible, with certain models, to reach 60,000 c/s at 76 cm/sec with ± 2 dB, and 100,000 c/s at 152 cm/sec with ± 5 dB.

The ratio signal/background noise is 60 dB for a signal recorded at the maximum level. This may reach 80 dB with certain tape recorders. The background noise is proportional to the band covered, *e.g.* a tape recorder covering the range 30–20,000 c/s has less background noise than a tape recorder covering the range 30–100,000 c/s. It would, therefore, be a mistake to choose a tape recorder covering the range 30–100,000 c/s for recording a low level signal covering the range 100–8000 c/s.

The tape recorder must be capable of covering the range 30–100,000 c/s for studying certain invertebrates, bats and marine mammals whose spectrum extends far into the ultrasonic range, and 30–25,000 c/s for studying birds; for other, mammals and Batracha, the band from 30–15,000 c/s is adequate.

Several studies on animal acoustics have been carried out from recordings made on semi-professional tape recorders, whose response curves are generally lower than the performances mentioned in the reviews. For this reason, the response curve should be verified before such models are used, by playing back standard alignment tapes specially recorded for this purpose.

(b) *Self-contained tape recorders*

These tape recorders which are fed by batteries are intended for reporting and, therefore, for outside work. They have a driving mechanism, either a clockwork motor, which limits the period of recording of a sequence, or an electric motor.

In general, these models have a running speed of 19 cm/sec and the frequency range is limited to 12 kc/s.

These recorders are perfectly suitable for outdoor recordings. They are practical and self-contained but, in general, their electrical performances are limited (low band covered, high background noise). Certain workers, in particular Möhres[16], have used special self-contained tape recorders which can be used successfully in the ultrasonic range. Still much smaller autonomous tape recorders are now produced, due to the progress of transistor technology. These often have better performances than the traditional types.

(c) *Rules to be observed in the use of tape recorders*

The following precautions should be taken in order to obtain the maximum results from the apparatus:

1. Regular cleaning of the magnetic heads with trichloroethylene on a piece of clean cotton to remove any possible deposit of magnetic oxide (particularly tape recorders which run at a high speed).

2. Magnetized objects or objects which may be magnetized should not be brought near to the magnetic heads. Anti-magnetic or copper scissors should be used at all times for editing tapes.

3. The magnetic heads should be demagnetized by means of a special apparatus.

4. Every tape recorder is regulated for use with one type of magnetic tape. If another type of magnetic tape is used the polarization should be regulated. However

all tapes may be played back on any recorder, if the recording curve of the tape and the playing back curve of the recorder used are aligned to the same standard.

Two international standards are in use at the moment: N.A.R.T.B. (New Association Recording Tape Broadcasting) and C.C.I.R. (Comité Consultatif International de Radiotélécommunication).

(d) Pre-amplifier

In general, professional tape recorders do not possess an input lead for the direct connection of microphones. A voltage pre-amplifier, inserted between the microphone and the tape recorder, is required in addition.

The input lead of professional tape recorders is, in general, intended for a signal of 0 dB (or 0.775 V). The preamplifier must, therefore, have an amplification of at least 80 dB, irrespective of the type of microphone. A wide response range is then covered (20–100,000 c/s). In this case the use of input and output transformers should be avoided. The same applies to the tape recorder, losses and phase rotations introduced by this system may cause distortion of the signal.

Magnetic recordings may be preserved for long periods under certain conditions of temperature ($15°$ at the most) and humidity with only very slight demagnetization. However, a gradual magnetization is produced during storage resulting in a pre-echo and post-echo affecting the high-level signals.

6. Measurement of Sonic and Ultrasonic Signals

The following measurements may be carried out on a sonic signal or series of signals once the acoustic signals have been recorded with the maximum fidelity: the duration, interval, form of the signal and its macroscopic composition. This information can be obtained with oscillograms. The frequency spectra are obtained with various types of frequency analysers and the levels may be measured by means of special registering devices.

(a) Measurement of the sonic level

Measurement of the level of the signal to be recorded is very useful, particularly when working in the field. For this a decibelmeter is employed. The working principle is as follows: a microphone supplies an electrical voltage proportional to the acoustic signal which is fed into an amplifier. The level of the signal, the background noise, or the noise coming from any other sonic source based on the reference level O equal to $2 \cdot 10^{-4}$ baryes may be obtained directly with a calibrated galvanometer. This level is generally expressed in decibels. Fig. 19 shows the relationships between the units of pressure (dynes/cm^2 or baryes), the acoustic energy in watts/cm^2 and decibels.

The response curve and the possibilities of a decibelmeter are a function of the frequency response curve of the microphone used. A decibelmeter is almost indispensable for recording in the field, especially when the sonic source is far away from the microphone. There is often a large amount of background noise and, under these conditions, it is useless to try to make a recording.

The information supplied by a decibelmeter should be deciphered carefully and the following taken into account:

(i) The indication obtained is exact in the case of a continuous noise (background noise or signal of quite a long duration) because the time constant of the amplifier and galvanometer system is negligible compared with the duration of the signal.

(ii) On the other hand, the indication obtained is always less than the true value when the level of a short signal having a rapid transient pulse is measured. The value of the peak may not be measured with this apparatus since its time constant and the inertia of the moving parts in the galvanometer limit it to a lower value. Only a mean

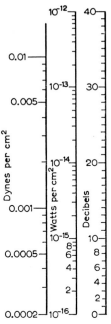

Fig. 19. Scale showing relationship between the different units used in electroacoustics: dynes/cm² or baryes, watts/cm² and decibels.

value of the level can be determined in practice. Cathode ray oscillography is used for measurements of the peak by means of an apparatus calibrated for voltage.

Great care must be taken in the interpretation of the results obtained with the decibelmeter and they must not be considered as absolute, particularly in the study of signals emitted by insects, although the conditions vary with the natural characteristics of the decibelmeter and the signal of the animal which is being studied.

It is always necessary to note the distance between the source and the decibelmeter in order to obtain a significant value for the measurement as the intensity decreases with the square of the distance. Recordings should be made as near to the source as possible, in order to eliminate the maximum amount of background noise.

(b) Oscillograms

A double beam oscilloscope with a bluish screen is preferable. Two spots are displaced vertically: one of them, which serves as a time basis, is excited by a sinusoid reference frequency and the other is excited by the signal originating either from the magnetic tape or directly from the microphone through a pre-amplifier. The screen may be photographed or filmed.

When the film is run at a constant high speed in front of the screen, a record of the amplitude of the signal is obtained as a function of the time (Fig. 20).

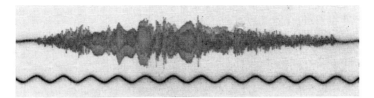

Fig. 20. Cathode oscillograph of a signal. The time basis is supplied by a sinusoid signal of 50 c/s. Example: Distress call of *Passer domesticus*.

The following information is obtained:
1. Duration of the signal and interval between each signal.
2. Form of the envelope of the signal.
3. Repetition of elementary signals or of a series of wave trains of the same signal.
4. Existence of harmonics and asymmetric components.
5. Measurement of the frequency of the signal obtained by slowing down the magnetic tape[4], or by passing the photographic paper very rapidly.

When it is desired to read recordings immediately, a second type of apparatus, the pen oscillograph, is used. This includes a device for unwinding the paper at a constant speed. An inked stylus traces a curve on the paper similar to the oscillograms obtained with a cathode ray tube. A second stylus fed by a current of sinusoid frequency serves as a reference and time basis. The inertia of the inscription systems restricts this type of apparatus to the study of low frequencies and the examination of rhythms in sequences of signals (Fig. 21).

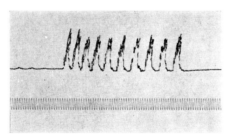

Fig. 21. Pen oscillograph with a 50 c/s reference signal given by a second oscillograph. Example: chattering of a *Pica pica*.

(c) *The frequency spectra*

Frequency spectra may be obtained by several models of analysers. The oldest one still in use is the electroacoustic spectrometer with filter bands (Siemens or Raytheon type). With this apparatus a signal may be analysed in a fraction of a second and a spectrum is shown immediately on the screen of a cathode ray tube. Signals which rapidly succeed one another (in time) may be decomposed into a series of frequency bands. The results may be filmed continuously. The analysis is carried out by means of twenty-seven filters of a third of an octave introduced successively by a commutator turning about twenty times per second. The image of the sonic spectrum appears on the screen in the form of light rays at the rate of one per frequency band. The height of the light rays is proportional to the voltages corresponding to the intensity of the components contained in each of the respective frequency bands (Fig. 22).

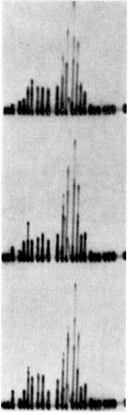

Fig. 22. Frequency study with the Siemens analyser. The abscissa gives the frequencies, the lines correspond to the different filters, the ordinate gives the amplitude of the components of each frequency band. Example: cinematographic recording of the noise of a bee-hive. The range of frequencies used here is from 40 to 16,000 c/s.

Depending on the type of apparatus, several ranges may be scanned, 10–700 c/s, 40–16,000 c/s, and 250–100,000 c/s.* This excellent apparatus has, however, two important disadvantages:

(i) Instead of an exact determination of frequencies, only the total energy present in frequency bands of a width of a third of an octave is obtained.

(ii) Twenty-seven filters have a relatively low selectivity. As a result, if two neighbouring filters are excited by voltages differing by a value greater than the selectivity of the filter (about 25 dB), the luminous ray corresponding to the weaker voltage cannot be obtained on the cathode screen. This ray becomes confused with a ray arising from excitation of bands adjoining the principal band corresponding to the higher voltage, and so there is no way of knowing whether sonic energy really

* See Chapters by Tembrock (p. 751) and Zhinkin (p. 132).

exists in this band. In the case where a component frequency of the spectrum is equidistant from the resonance of two neighbouring filters, two rays of equal amplitude are obtained in each of the bands, even though only one frequency exists.

This type of apparatus, which has the advantage of providing a spectrum of a signal very rapidly, has, for the reasons given, the inconvenience of uncertainty in the measurement and precise interpretation of the results.

Another type of analyser, using a different principle, is, therefore, preferable, particularly for animal acoustics.

The C.N.E.T.* frequency analyzer (Pimonow system)[17]. Analysis of a signal may be obtained with this apparatus within a time varying from a few seconds to several minutes. The working principle is as follows: the frequencies composing the signal are fed into a modulator with a continuously varying auxiliary frequency. Interferential frequencies $F_0 - F$ and $F^0 + F$ are obtained at the output. A highly selective quartz filter allows only the difference of the frequency to pass, after amplification. The variation of the auxiliary oscillator allows all the partial frequencies, whose voltages, after detection, are applied to the vertical deflection plates of a cathode ray tube, to pass through the filter successively. The horizontal scanning, corresponding to the axis of the frequencies, is adjusted mechanically to the variable condenser of the local oscillator.

Two selectivities are used, one of 3 c/s, the other of 50 c/s. Several minutes should be taken for analysis with the first; the second requires a shorter time. This time depends on the duration of the signal. Since the time constant of a filter increases with its selectivity, the time for analysis is proportional to the width of the range of analysis and to the selectivity (in practice to the square of the latter). A certain number of periods are therefore required to excite the filter. It is possible that the signal will be so short that it will not succeed in exciting the filter. In the study of transient phenomena the amplitudes of the components must therefore be interpreted with great caution. This apparatus does, however, provide the spectrum of frequencies with great precision (of the order of 5%) (Fig. 23).

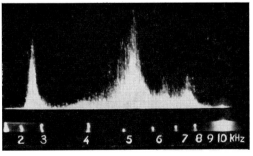

Fig. 23. Frequency study obtained by continuous playing back of a magnetic loop recorded with the C.N.E.T. frequency analyzer. Example: distress call of *Passer domesticus*.

The signal is recorded on a magnetic loop which is played continuously during an analysis. In order to obtain a complete analysis in a scanning time of, for example,

* Centre National d'Études des Télécommunications (France).

References p. 47

ten minutes, the magnetic loop must not have a duration greater than three seconds.

The frequency analyser (Sonagraph type)*. The frequency spectrum of a signal may be obtained with this apparatus as a function of time. A succession of instantaneous spectra is obtained giving a dynamic representation of the signal. The working principle is as follows: the signal or part of the signal lasting no longer than 2.4 sec is recorded on a magnetic drum. Reproduction takes place at a high speed. An interferential frequency system with a varying auxiliary frequency allows a mean frequency to be obtained in a similar manner to that of the analyser, C.N.E.T. type.

The recording is made on a special paper wound on a drum attached to the magnetic drum. The vertical displacement corresponds to the frequencies and the degree of blackening to their intensity (Figs. 24 and 25). We generally call this recording a "sonagram".

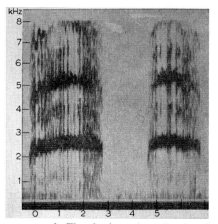

Fig. 24. Analysis using Kay sonagraph. The abscissa gives the time in 0.1 sec, the ordinate the frequencies. The intensity of blackness is proportional to the amplitude. Use of wide selectivity. Example: Distress call of *Passer domesticus*.

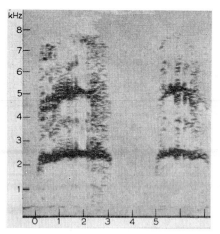

Fig. 25. Analysis using Kay sonagraph. Same analysis as in Fig. 24, but with narrow selectivity. Same example.

* From Kay Electric Company, U.S.A.

The band of frequencies which may be analysed is restricted to 100–8000 c/s. This range may be extended towards higher frequencies by slowing down the magnetic tape fed into the analyser or towards lower frequencies by speeding up the same.

With this apparatus (Fig. 26) the curve corresponding to the level of the signal

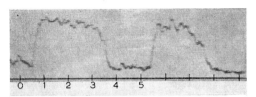

Fig. 26. Analysis using Kay sonagraph. Form of the envelope of the signal giving the relative levels. The abscissa gives the time in 0.1 sec. Same example as in Figs. 24 and 25.

may be obtained as a function of time. Instantaneous spectra may also be obtained by division of the signal. The amplitude of the resulting record is a function of the frequencies (Fig. 27).

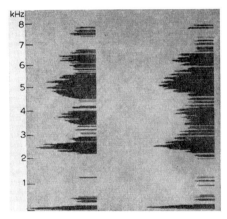

Fig. 27. Analysis using Kay sonagraph. Study of the relative instantaneous amplitude of different frequencies. Section at two different levels in time. The abscissa gives the amplitude of the different frequencies (1 mm = 1 db), the ordinate the frequencies. The same example as in Figs. 23, 24 and 25.

This apparatus is very useful for the biologist, since a quasi-three-dimensional representation of an acoustic animal signal may quickly be obtained. The information that may be obtained with this apparatus is much more important than that supplied by the other types of apparatus each being taken alone. However, since the recording paper has limited possibilities, certain reservations must be made concerning the exactitude of the amplitudes.

This is the apparatus that has been most often used by American and British biologists, in particular Borror et al.[2] and Thorpe et al.[19].

Other frequency analysers. Analysis of the frequencies of a complex sound may be made with other apparatus. The Panoramic Sonic Analyser is similar in principles to the C.N.E.T. analyser described above, but by its conception it is applicable more particularly to the study of continuous noises. The General Radio Analyser, although not an automatic analyser, is also similar.

References p. 47

(d) Level recorder (Brüel and Kjær or General Radio type)

Sonic levels may be measured by graphic recording on a paper with this apparatus. The paper may run at speeds varying from 0.03 to 100 mm/sec and from 5 to 190 cm/mm or hour on the Brüel and Kjær. The recording is logarithmic. Levels having variations of 75 dB in a band extending from 20 to 200,000 c/s may be recorded.

This type of apparatus can function continuously for several days, supplying useful information on the rhythms of the emission of signals, their durations and their levels.

This apparatus may be used either directly with the microphone serving for sound-recording (Fig. 28), or connected to the output of a tape recorder for analysis of a recording.

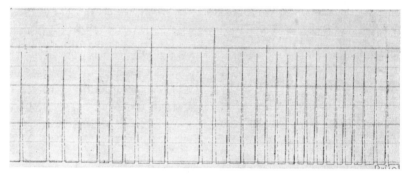

Fig. 28. Enumeration of the cries of *Ephippiger bitterensis* by level recording (Brüel and Kjær type). Speed of movement of paper 3 mm/sec.

(e) Visual demonstration of a recording on a magnetic tape

The signals may be visibly recorded on a tape. The tape is immersed in a viscous liquid (paraffin oil) containing an impalpable magnetic powder (iron carbonyl) in suspension. This technique has been used by Frings et al.[6].

7. Retransmission of the Sounds

Retransmission of the signals which have been recorded is necessary in the study of animal acoustics. A special apparatus and experimental precautions which will be described here are required.

Playback may be made with message repeaters on which is played a magnetic loop, contained in a charger, and lasting from 30 sec to a few minutes. These message repeaters were first used by Frings et al.[7] and later by Busnel et al. in their studies on the reactional behaviour of herring-gulls, starlings and crows[8].

Tape recorders do not contain incorporated power amplifiers. The retransmission chain consequently requires a power amplifier of 10–20 modulated watts, with a wide range (from 50 to 10,000 c/s) and a low background noise. The amplifier works either on the electric supply or on batteries. A converter is necessary except with the use of transistorized amplifiers which are highly recommended. Indeed the volume and weight of the energization is limited by the transistors which have a higher output than electronic valves.

Loud-speakers vary in type with the transmission desired. Compression horns are recommended for field-work since they have a more favourable output than membrane loud-speakers, as their frequency range does not generally exceed 10,000c/s. Compression speakers are usually equipped with an exponential horn (Fig. 29).

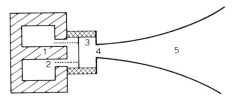

Fig. 29. Diagram showing principle of compression horn loud-speaker. (1) magnet, (2) moving coil, (3) compression horn, (4) mouthpiece, (5) exponential horn.

Bearing in mind their directivity, they may be connected in pairs at 180° or at 90° by using four of them. The emission becomes quasi-omni-directional by this method.

When playing back outdoors the following considerations must be kept in mind:

1. The loud-speakers have a variable radiation diagram which is a function of the frequency. Apart from the emission axis, there is an attenuation which is proportional to the frequency. There is no non-directive loudspeaker in the high frequencies.

2. It is advisable to place the loud-speaker at a certain level from the ground to avoid obstacles which can cause distortions of frequency, reflections and absorptions.

3. Absorption in the air increases in all cases with the frequency. For example, a decrease in the intensity of the emission of 50% is obtained at 220 m for 10,000 c/s, and at 55 m for 20,000 c/s.

4. Meteorological conditions also exert an important influence (temperature, humidity, wind).

All these factors influence the quality of a retransmission in free field and must be taken into consideration.

The loud-speaker is at present the weakest link in the playing back chain. For this reason, it is necessary to choose high quality material. The band and the range desired determine the choice. At frequencies under 4000 c/s it is possible to transmit correctly in a radius of 1 km with a power of 15 watts, if meteorological conditions are good.

First-class loud-speakers, either of the membrane, compression horn, or electrostatic type, or in the case of very high frequencies, of the ribbon type, may be used in the laboratory. A relatively wide and flat range (Fig. 30) may be obtained by combining several loud-speakers having different diameters. The Klein ionophone is used for the correct retransmission of high frequencies above 5000 c/s[12]. While normal loud-speakers have faults inherent in the use of a mechanical system for transforming electrical energy into acoustic energy, the ionophone is based on a radically different idea: the electroacoustical transformation is realised without mechanical means. Thus the response curve extends far into the ultrasonic range, up to several Mc/s.

The working principle is as follows: at atmospheric pressure certain bodies emit

References p. 47

ions at an elevated temperature. If a high-frequency current is applied between this source and an electrode, the ions are accelerated and cause an ionization of the air. If the high frequency is modulated by a low frequency signal, the ionization is pro-

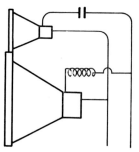

Fig. 30. Diagram showing wiring of two loud-speakers with a frequency separator. The loud-speaker with the large diameter transmits the low frequencies, the loud-speaker of small diameter the high frequencies. The inductance coil serves as a lowpass for the large diameter loud-speaker, the condenser as a highpass for the small diameter loud-speaker (compression horn).

portional to the latter and collision of the ions with the air molecules is proportional to the modulation. Sonic waves result (Fig. 31). This type of loud-speaker is recommended for the retransmission of very high frequencies and of short rise-time transients. It is also useful for supplying a pin-point sonic source which is for instance very useful in the acoustic study of animals for they can locate the emission more precisely.*

In order to obtain a very small sonic source, a compression speaker, the horn of which has been removed, may also be used. It is, in this case, impossible to transmit

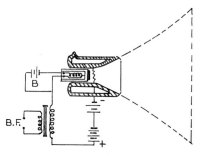

Fig. 31. Diagram showing principle of Klein ionophone.

frequencies below 1 kc/s correctly, the coupling with the air becoming very faulty, and damage to the membrane may follow.

When retransmitting in the open air, the extent of distortion does not appear to be a factor influencing the physiological reaction of the animals. On the other hand, the correct transmission of transients appears to be a necessary quality.

---

* Dukane Company, St. Charles, Illinois, U.S.A. See reference 21, on Ionovac.

REFERENCES

[1] BERGMANN, L., 1954, *Der Ultraschall*, Hirzel Verlag, Zürich.
[2] BORROR, D. J. and C. R. REESE, 1953, The analysis of birds' song by means of a vibralyser, *Wilson Bulletin*, *65*, (4) 271–276.
[3] BORROR, D. J., 1956, Capturing the melodies of birds, *Magnetic Film and Tape Recording*, *3*, (3) 22–25.
[4] BROUGHTON, W. B., 1954, Oscillographie économique du schéma de modulation par un appareil de coût et de fonctionnement bon marché, utilisable avec les appareils enregistreurs à vitesse variable, *L'Onde Electrique*, *324*, 204–211.
[5] CARPENTER, C. R., 1940, A field study in Siam of the behaviour and social relations of the Gibbon, *Comp. Psychol. Monographies*, *16*, (5) Ser. 84.
[6] FRINGS, H. and M. FRINGS, 1956, A simple method of producing visible patterns of tape recorded sound, *Nature*, *178*, No. 4528, 328–329.
[7] FRINGS, H., M. FRINGS, B. COX and L. PEISSNER, 1955, Recorded calls of Herring Gulls (*Larrus argentatus*) as repellents, *Science*, *121*, (3) 140, 340–341.
[8] FRINGS, H., M. FRINGS, J. JUMBER, R. G. BUSNEL, J. GIBAN and PH. GRAMET, 1958, Reactions of American and French species of Corvus and Larus to recorded communication signals tested reciprocally, *Ecology*, *39*, (1) 126–131.
[9] GRIFFIN, D. R., 1958, *Listening in the Dark*, Yale University Press, New Haven.
[10] GRIVET, P. and A. BLAQUIERE, 1958, *Le Bruit de Fond*, Masson, Paris.
[11] KAHN, M. C. and W. OFFENHAUSER, 1953, New Mosquito extermination plan uses Presto recordings to attract insects, *The Presto Recorder*, *V*, *5*, (5) 1–3.
[12] KLEIN, S., 1954, Ionophone ou haut-parleur ionique, *Colloque sur l'Acoustique des Orthoptères*, I.N.R.A., Paris, 46–49.
[13] LEHMANN, R., 1953, Les Microphones, Principes et caractéristiques, *Revue du Son*, *8*, 290–295.
[14] LUTZ, F. E., 1930, An analysis by Movietone of a Cricket's chirp (*Gryllus assimilis*), *Am. Museum Novitates*, No. 420, 1–14.
[15] MARLER, F. R., 1952, Variation in the song of the Chaffinch, *Ibis*, *94*, 458–472.
[16] MÖHRES, F. P., 1953, Jugendentwicklung des Orientierungsverhaltens bei Fledermäusen, *Naturwiss.*, *40*, (10) 298–299.
[17] PIMONOW, L., 1949, Analyseur de frequence à exploration rapide et automatique, *Annales des Télécommunications*, *4* (7) 241–255.
[18] SCHÜH, F. and N. MIKHNEWITCH, 1950, *L'Enregistrement Magnétique*, ed. GEAD, Paris.
[19] THORPE, W. H., 1954, The process of song learning in the Chaffinch as studied by means of the sound spectrograph, *Nature*, *173*, 465.
[20] KELLOGG, P. P., 1962, Capture Nature's sounds on tape, *Radio Electronics*, February, pp. 44–48.
[21] ANDRIEU, A. J., 1962, L'Ionovac, *Revue du Son*, No. 107–108, pp. 105–107.

REFERENCE BOOKS

BERNHART, J., 1949, *Traité de Prise de Son*, Eyrolles, Paris.
BUSNEL, R. G., 1954, *Colloque sur l'Acoustique des Orthoptères*, Institut National des Recherches Agronomiques, Paris.
CONTURIE, L., 1955, *Acoustique Appliquée*, Eyrolles, Paris.
DE GRAMONT, A., 1935, *Recherches sur le Quartz Piézoélectrique*. Editions de la Revue d'optique théorique et instrumentale, Paris.
GRIVET, P. and A. BLAQUIERE, 1958, *Le Bruit de Fond*, Masson, Paris.
PETZOLDT, E. G., 1957, *Elektro-Akustik*, Fachbuchverlag, Leipzig.
RICHARDSON, E. G., 1953, *Technical Aspects of Sound*, Elsevier Publishing Company, Amsterdam.
SCHÜH, F. and N. MIKHNEWITCH, 1950, *L'Enregistrement Magnétique*, Editions Gead, Paris.
WEILER, H., 1956, *Tape Recorders and Tape Recording*, Radio Magazines, Inc., Mineola, N. Y.

CHAPTER 3

PRINCIPLES OF UNDERWATER ACOUSTICS

by

O. BRANDT

1. Generalities

Water like air is a medium with a volumetric elasticity. The mode of sound propagation is identical in both cases and is carried on by means of compressional waves. The waves of transverse oscillation—such as those found in solid bodies—cannot exist here. The formation of waves at the surface of the water has nothing to do with sound waves.

As water is much less compressible than air, the amplitude of the oscillations is—with an equal intensity—60 times smaller, but the sound pressure is 60 times greater in water than in air. Therefore the sound pressure is predominant and the amplitude of the oscillations is very small.

This fact must be taken into account when dealing with sound and ultrasonic transmitting and receiving devices.

For instance an atmospheric loudspeaker with a very flexible diaphragm has a low efficiency in water because it is built to vibrate with a great amplitude and to work at low sound pressure.

The underwater transducers are therefore made much more rigid than the atmospheric transducers. Likewise, the hearing organs of animals are adapted in this manner.

In technical terms this is expressed by the matching of the acoustic impedance of the transducer to that of the medium.

The specific impedance is represented by the product $\varrho \cdot c$ where c is the speed of propagation and ϱ the density.

$$\varrho \cdot c_{water} = 1.5 \times 10^5 \qquad \varrho \cdot c_{air} = 41.5 \qquad (g\ cm^{-2}\ sec^{-1})$$

For a plane wave we find the following relations between intensity, amplitude, velocity of the oscillation, sound pressure and specific impedance:

$$I = \omega \cdot a \cdot p \qquad (1a)$$
$$I = v \cdot p \qquad (1b)$$
$$I = p^2/\varrho \cdot c \qquad (1c)$$
$$I = v^2 \cdot \varrho \cdot c \qquad (1d)$$

I = intensity
$\omega = 2\pi f$
f = frequency
λ = wave length
a = amplitude

v = oscillation velocity of particles
p = pressure
ϱ = density
$c = f \cdot \lambda$ = speed of propagation
$\varrho \cdot c$ = specific impedance of water.

If in these equations a is expressed in cm, p in microbar, v and c in cm/sec, the result must be multiplied by 10^{-7} in order to obtain I in W/cm².

The speed of propagation in water is approximately $c = 1500$ m/sec. The exact formula is:

$$c = 141{,}000 + 421t - 3.7t^2 + 110s + 0.018h \text{ (cm/sec)}$$

where t = the temperature in degrees centigrade
s = the salinity in g/kg
h = the depth in cm.

10^{-16} W/cm² has been chosen as the reference of the sound intensity. This intensity determines approximately the lowest limit for sound reception both in water and in air.

In fact, the noise caused by the thermodynamic movement in the audible frequency is of the same order.

The sound pressure at that limit is equal to 0.0002 microbar in air and to 0.012 microbar in water (see eqn. 1c). In order to keep the same reference, it has been agreed to express the level of sound pressure in decibels referred to 0.0002 microbar. Therefore:

$$N = 20 \log \frac{p}{0.0002} = 20 \log p + 74 \text{ (dB)} \qquad (2)$$

However in water the pressure of 0.0002 microbar does not correspond to anything and this reference is only a convention.

Indeed in underwater acoustics, it is more common to find the decibels referred to 1 microbar. This may lead to confusion if one forgets to add the reference level. For instance:

$$N = 20 \text{ dB (referred to 1 microbar)}$$

corresponds to:

$$N = 94 \text{ dB (referred to 0.0002 microbar)}$$

and indicates a pressure of $p = 10$ microbar.

The intensity of sound decreases with the distance from the source of sound. This decrease is due on the one hand to the geometrical divergence and on the other hand to a loss by damping. If p_r is the sound pressure at the distance of r meters and p_0 that at 1 m from the source, we have:

$$p_r = \frac{p_0}{r} \cdot 10^{-ar}$$

where a is the damping coefficient.
In sea water its value is:

$$a = 0.5 \times 10^{-10} \cdot f^{3/2} \qquad (f \text{ in c/s}; r \text{ in m})$$

In fresh water it is roughly 1/10th of that value. The absolute value of a depends very much on the impurities, the percentage of dissolved air or of air bubbles, the micro-organisms, etc.

We see that the geometrical divergence is more important with the audible frequencies, whereas damping is more important with the ultrasonic frequencies. The value p_0 is the level referred to a distance of 1 m from the source; it is called "transmission level".

We have already mentioned that the thermodynamic noise limits the sound reception. In practice, there is always in the sea, rivers or lakes, a fairly important background noise caused by the waves and their breaking, by the wind and other factors.

Thus, for instance, the background level in very smooth sea is still several tens of decibels higher than the thermodynamic noise. It is only above that already high background noise that the noise due to sources of a biological origine can be detected.

2. Various Types of Transducers

All the acoustics transducers used nowadays are of the electro-acoustic type.

Among the various types we shall mention:

The moving coil transducers

They use the same device as the ordinary loudspeakers, *i.e.* a coil placed in a magnetic field.

The magnetostrictive transducers

They are based on the magnetostrictive effect which appears for instance in nickel and in the ferrites. These materials are strained when subjected to the action of a magnetic field. In transmitting and receiving they work normally near their natural resonance frequency and therefore are used for the transmission of signals with a well defined frequency.

The piezoelectric transducers

Some natural crystals are endowed with piezoelectric properties. Some elements cut following certain crystallographic axes undergo a strain when they are subjected to an electrostatic field and vice-versa (Fig. 32). The piezoelectric elements used are mostly quartz, ammonium dihydrophosphate (ADP), Rochelle salt and lithium sulphate.

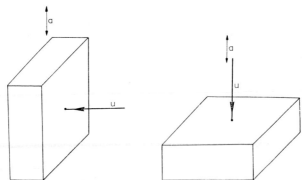

Fig. 32. Piezoelectric elements. Left: Transverse effect; the electric lines u are perpendicular to the axis of stress a. Right: Longitudinal effect; the electric lines u are parallel to the axis of stress a.

To these natural products has been added more recently the barium titanate which is a ceramic moulded under high pressure in the shape of plates, rods, or cylinders and baked at high firing temperatures. After firing it is polarized by means of a high tension voltage. It is the residual polarization which gives the piezoelectric characteristics. (Rochelle salt and barium titanate present rather the "electrostrictive" effect which in its exterior manifestations is identical with the piezoelectric effect.)

With piezoelectric transducers working at their resonant frequency one can transmit signals at very high power.

As receivers they are also used in the region below the resonant frequency where they cover a large frequency band with a constant sensitivity. They are, therefore, much in use as receiving and measuring hydrophones.

Because of its poor sensitivity quartz has become much less useful in this field. The Rochelle salt is very sensitive but has a high temperature coefficient.

The lithium sulphate has been used with some types of measuring hydrophones but its application has not been generalized.

On the other hand the ammonium dihydrophosphate (ADP) owing to its great sensitivity and its good stability has found a wide field of applications.

Barium titanate and recently lead zirconate are coming more and more in use.

3. Hydrophones

When the transducer is used for receiving it is called a hydrophone. The sensitivity is expressed in microvolts/microbar or decibels referred to 1 volt/microbar.

Fig. 33 represents a ammonium dihydrophosphate hydrophone (Model HP 40 of the Brusc Laboratory):

Fig. 33. Measuring Hydrophone. The sensitive element is made up of 4 plates of ammonium dihydrophosphate. The top casing contains a preamplifier-stage.

A set of four cristals is cemented between 2 half capsules of an insulating material. The element is situated behind a diaphragm made of rubber and filled with oil.

Between 150 and 20,000 c/s the level of sensitivity is about $S = -87$ dB (ref. to 1 volt/microbar, *i.e.* 45 microvolt/microbar).

We only mention this model but many similar models have been manufactured.

The sensitivity of this hydrophone in the audible band is the same in all directions except in the regions of the constructional components; it is therefore omni-directional.

To have a non-directional hydrophone one must respect the condition that its dimensions must be rather inferior to the wave-length.

If on the contrary a directional effect is required, the dimensions must be great compared to the wave-length. This is easy to achieve in the ultrasonic band, but leads to outsized dimensions in the audible band.

As a type of directional hydrophone we indicate the model illustrated in Fig. 34. It is a tube in barium titanate protected by a rubber envelope. In the horizontal plane the sensitivity is the same in all directions, *i.e.* there is no directivity. In the vertical plane however, the directivity is very pronounced.

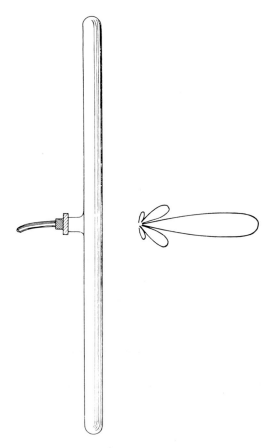

Fig. 34. Directional Hydrophone. The sensitive element is made up of a tube in barium titanate. Right: The directivity pattern.

Such transducers can be used to localize the source of sound or to eliminate a disturbing noise. Directivity patterns of transducers of different shapes and dimensions are dealt with in acoustics treatises.

We note that the directivity patterns are strictly identical in transmission and reception.

4. Transmitters

A model of an underwater loudspeaker is represented in Fig. 35. The coil (b) is in the field of a permanent magnet (m). The diaphragm (d) is elastically suspended. To allow it to withstand the hydrostatic pressure, a compensating device (c) is added.

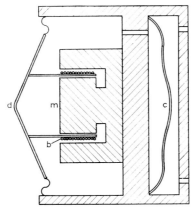

Fig. 35. Simplified model of an underwater loudspeaker.

For the transmission of pure frequency signals, the most varied devices are used, from underwater whistles to piezoelectric or magnetostrictive transmitters with an output of several tens of kilowatts.

CHAPTER 4

EXAMPLES OF THE APPLICATION OF ELECTRO-ACOUSTIC TECHNIQUES TO THE MEASUREMENT OF CERTAIN BEHAVIOUR PATTERNS

by

R. G. BUSNEL

1. Introduction

The electro-acoustic methods and techniques described in a previous Chapter by A. J. Andrieu (see p. 25) may be used with advantage in certain cases other than those essentially concerned with acoustic behaviour and to which they are more particularly applied.

The purpose of this chapter is to show, by a series of widely varied examples given here merely by way of illustration, what one may expect or obtain with the above-mentioned methods and apparatus.

In the first section, attention will be directed to results chosen from studies of animal activity, not primarily of acoustic interest, which may be obtained and measured by microphones of various types.

In the second section, examples will be given in which such physiological or ethological problems as hormonal state, relation of metabolism to temperature, territory, etc. are approached through the channel of measurable acoustic behaviour patterns.

2. Detection, Recording and Measurement of Various Animal Activities

Actography, with its various classical techniques, allows a faithful plotting of the total activities of an individual or of a group to be obtained in a certain number of cases. The use of certain acoustic methods enables interesting results to be obtained in this field, either from the preliminary magnetic recording, when working in the field, or from an oscillographic record collected directly from a microphonic receiver and an amplifier. The use of piezoelectric elements and of a recording bathymeter is the basis of the techniques used in most cases.

The following are a few concrete examples illustrating the experimentation and lay-outs used.

(a) *Actography of Drosophila*[6]

Under the action of certain gases or insecticide powders, *Drosophila*, which normally has a reduced activity at 20°C, becomes violently agitated, falls to the ground in convulsions, flies away, then falls again and dies. This action is inversely proportional, in its timing, to the rise of toxicity of the air per minute, therefore it is only a question of obtaining the chronoactographic recording of this phenomenon. An apparatus was constructed for detecting the shocks, falls and convulsions of the *Drosophila*.

A barium titanate piezoelectric bimetallic strip pick-up is used as the receiving device. It is connected by a taut wire to a thread made from a piece of nylon voile rendered semi-rigid by attaching it to a frame. This frame forms the base of a small cage into which the insects are introduced. When they are in a state of normal activity, they fly about or settle at the top of the cage; but when intoxication begins, they fall onto the thread and their falls and convulsions cause a series of damped oscillations. The bimetallic strip delivers an electric signal each time an insect strikes the thread. As the signal arising from the falling of a *Drosophila* is very much more intense than a noise of 100 dB, there is no interference from surrounding noises.

The electric signals provided by the bimetallic strip are amplified by two slightly selective stages, are then detected (grid detector) before passing to a power stage. The reading is obtained from a low inertia recording galvanometer, with a resistance of 4000 ohms, a natural frequency of 50 c/s, and a critical resistance of 600 ohms.

Either a single *Drosophila* or a group of, for example, twenty insects may be studied with a small cage of 50 to 100 cm³ capacity. Fig. 36 shows the record of an example of intoxication behaviour. It was likewise possible to connect the apparatus to an adding device for the impulses per unit time, which gives a total temporal report of the phenomenon in another form.

(b) Actography of the rumination of a cow

The actography of the masticatory movements of a cow in its stall is of interest

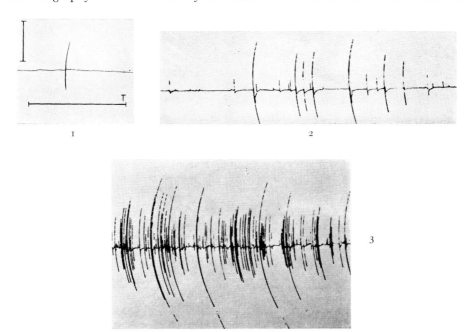

Fig. 36. Actogram of the intoxication of a group of 20 *Drosophila*. The vertical scale corresponds to 2 cm, thus indicating the width of deflection of the galvanometer pen for the shock of a single insect. Time basis (T): 30 sec, (1) Shock produced by one *Drosophila*, (2) Start of the intoxication, (3) Stage of intoxication immediately prior to death (Busnel and Pasquinelly[6]).

References p. 65

for the study of the physiology of digestion; at the same time an actography of the mooings of the animal may also be recorded, for example, in an analysis of their quantitative relationship to the phases of the rut.

It is difficult to isolate the animal from external noises, therefore a contact microphone of the laryngophone type with a piezoelectric element must be used.

The apparatus is placed on the animal's head, between the horns and at the level of the frontal bone. The receiver is protected by an enclosing leather head-phone attached to the head, since the animal often makes scraping and rubbing movements against the sides of the stall in which it is kept. The contact between the receiver and the head is assured by the pressure of the leather head-phone, supplemented by depilation of the skin and the application of a greasy ointment. The receiver is connected to a bathymetric recorder.*

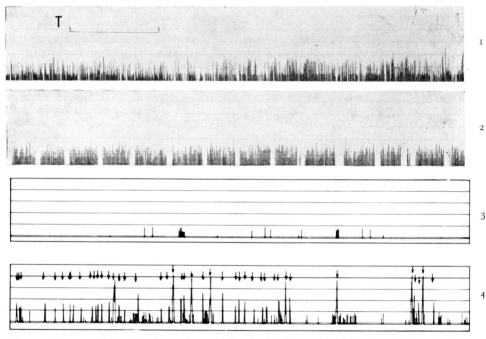

Fig. 37. Actogram of the behaviour of a cow. Time basis (T): 2 min, (1) Feeding, (2) Rumination, (3) Sleep, (4) Mooings with relation to an artificial oestrus (original, Brieu, Busnel and Signoret).

The records in Fig. 37 show the chronography of the various manifestations of the animal's activities: feeding, rumination, mooings.

(c) Detection of the activities of burrowing animals

The activity of burrowing animals which hollow out trees and grain is a source of vibrations and noises which may be demonstrated and measured by microphonic detection and recording. This procedure has been used for observing the behaviour of weevils, which are grain parasites[1], and of borers[10].

---

* It is possible to connect the piezodetector to an H.F. modulator and to transmit the signal by hertzian waves directly to the laboratory, the animal being free in the fields.

(d) Study of the activity of a group of chicks in a brooder or hover

An ordinary type of microphone is introduced into the brooder or hover containing the birds. It is connected, through an amplifier, to a recording bathymeter. The record obtained gives a faithful indication of the birds' activity, the breaks in the record corresponding to periods of sleep. In this case, the amplitude of the record bears a direct relationship to the sonic manifestations of behaviour, *i.e.* chirping and noises caused by movements.

This method makes it possible to analyze the periods of sleep and activity of the birds. The record is valid only if the birds are insulated from external noises which modify their behaviour and introduce stray noises into the record. Since the cries of chicks are of a fairly high frequency (3000–6000 c/s), a system of selective filters may be placed in the circuit to eliminate surrounding noises which are in most cases of low frequency.

The record in Fig. 38 is an example obtained on a group of ten or so chicks in a room insulated from external noises overnight, without special precautions being taken; a dynamic (Melodium type) microphone, placed 20 cm above a group of chicks distributed over 1 m², was used.

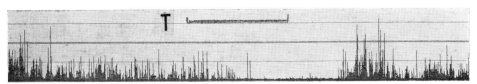

Fig. 38. Actography of a group of chicks. Time basis (T): 15 min. Periods of chirpings and cries, feeding and sleep (original).

(e) Study of the nocturnal activity of flocks of Corvidae in roosts

Observation of the sleeping habits of winter flocks of crows in roosts may be made entirely from a continuous magnetic sonic recording.

Bathymetric or simply oscillographic analysis enables the temporal analysis to be made from the recording. If the low frequencies are filtered for the analysis, the activities of jackdaws may be isolated from those of rooks, although they are mixed on the recorded tape. The progressive decrease in the intensity and number of the cries gives an excellent insight into the activity of the colony (Fig. 39).

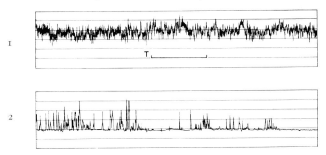

Fig. 39. Actogram of the sleeping behaviour in roosts of Corvidae (group of several thousands of birds in February, in Normandy). Time basis (T): 30 sec. (1) About 6.30 p.m., shortly after returning from flights, (2) About 7.45 p.m., all became still, the birds being perched on their places for the night (original, Busnel, Giban and Gramet).

References p. 65

With a sufficient intensity the recording range of the birds' cries in the open air during a winter-night covers a radius of 300–400 m, thus giving an analyzed area of activity of at least 2700 m².

Although, as far as we know, no other application has yet been undertaken of this technique of continuous recording followed by an analysis with filtering, it certainly could be applied successfully to studies of the diurnal or nocturnal activity of numerous species, for example the Batrachians. The selectivity of the analyzing filters will sometimes permit the isolation of the activity of one or other species amongst groups of different animal species.

(f) Actography of aquatic animals

The activity of certain aquatic animals may be studied using the same techniques, but with a hydrophone as the receiver, particularly the reaction time between the presentation of a prey and its capture. An example of this is given in the record in Fig. 40b of a fish *Rhombus maximus*, the first marked shock being due to a piece of food striking the water in the aquarium, the second corresponding to the noise made by the fish as it swallows it, followed by noises produced during mastication. The prey chosen in this case was a piece of crab. Fig. 40a shows the oscillograph record of the time a flat-fish *Solca solca* remained hiding in the sand. The sound of the rocking movements of the body of the fish were used for the analysis.

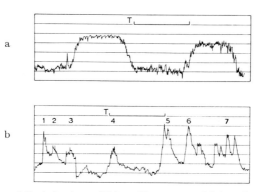

Fig. 40. Actography of the behaviour of fish. a. Movements of *Solca solca* hiding itself in the sand; each large variation in the record corresponds with total burial. b. Behaviour during the capture of prey by a *Rhombus maximus*. Time basis (T): 3 sec. (1) The prey is thrown into the water, (2) Violent movement of the fish as it turns round and makes for the prey, (3) The fish swallows the prey, (4)–(7) Masticatory noises produced by the vomerine teeth, (original).

Such procedures have been applied to the measurement of the motility of freely floating aquatic insects and animals[12].

(g) Study of the activity of a bee-hive

A microphone placed in a bee-hive with certain special precautions to prevent the bees from clogging it up with their propolis, enables an average noise to be collected, which, when recorded or analyzed, allows the activity of the colony to be measured. Special conditions may be shown depending on the intensity of the mean noise and its frequency, as revealed by spectrographic analysis. Apart from the permanent recording giving a temporal aspect of the activity, listening alone or recording at different moments may be used. An indication of the condition of the colony is given by the stethoscopic ausculation of the bee-hive by the response of the bees to the impact of the finger, through the tonal qualities and intensity of the sound. Apiarists have realized the importance of this tonal response of the bee-hive for a long time, and up to "28 variations in the speech of bees" have been described by this means, perhaps with a certain amount of exaggeration.

A rapid observation of the activity of the colony may be made by using the amplifier employed in deaf-aids, with microphones placed permanently in the hive. Analysis and control apparatus intended specially for this type of problem have recently become available[14].

Information and a visible pattern of certain phenomena may thus be transported from a distance due to the possibility of introducing a receiver for vibrations *in situ*, into a milieu which is not visible to the eye, such as a liquid medium, the interior of a social body (*e.g.* a bee-hive or a termitary), or a cave. Thus documents may be obtained which can be used for calculations on the relationships of the activity of the animals in the medium under study.

(h) Various vibratory movements

In the case of species-specific vibratory movements, if the animal or its support be connected to a receiver, all its movements become detectable and measurable in amplitude and timing alike, with greater sensitivity than that of the classical techniques of mechanical recording; for example the sexual tremulation of *Ephippiger* (Insect, Orthoptera), the pouncing movements of ant-lion larvae, and the vibration of the male *Schistocerca* reacting to the odour of the female[11].

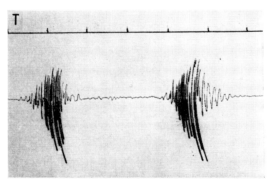

Fig. 41. Tremulation of a male *Ephippiger*. Time basis (T): 1 sec. (Busnel and Dumortier[7].)

(i) *Example*: The record in Fig. 41 is that of the tremors of a male *Ephippiger**. The following arrangement was used: the plant on which the animal is situated is

---

* We have also shown in the case of this insect that this tremulation corresponds to short-range information intended for the female regarding orientation, the male merely carrying out these movements when he has himself perceived the slight vibrations which are produced by an insect as it climbs the plant on which he has alighted. This tremulation is released, even though the male has no precise information concerning the visitor, who may be another male or an insect of another species.

References p. 65

connected by a wire to the piezoelectric element of a gramophone pick-up head, which is in turn connected through an amplifier to a recording galvanometer. The trace is recorded on a paper moving at a high speed, and so the detail of each oscillation of the insect may be obtained.

By the addition of a microphone to the experimental arrangement, likewise connected to an oscillograph (as shown in Figs. 42 and 43), both the sonic emissions and the tremors are recorded chronographically[7,8].

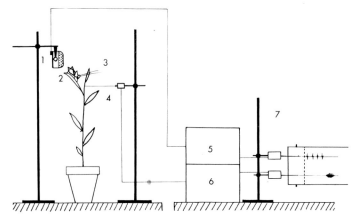

Fig. 42. Arrangement for graphic recording of acoustic and vibratory behaviour of a male *Ephippiger*. (1) Microphone, (2) Insect, (3) Plant acting as substratum, (4) Piezoelectric detector connected by a wire to the plant, (5)–(6) Amplifiers, (7) Galvanometer recording with pen on moving paper (Busnel and Dumortier[7]).

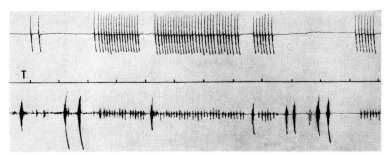

Fig. 43. Actography of the acoustic and vibratory behaviour of a male *Ephipiger bitterensis*. Upper record: Chirping, Lower record: Tremulation (it should be noted that when there is chirping, there is also natural movement of the animal's body transmitted by the plant on which it is situated (Busnel and Dumortier[7])). Time basis (T): 10 sec.

(ii) *Further example*: The study of the speed of very rapid movements such as the movements of a woodpecker's head while drumming. The sound is recorded with a microphone, and is analyzed either with an oscillograph or a bathymeter (Fig. 44).

(i) *Study of a chirping insect, permitting the analysis of a factor other than the sonic signal: Example of the insect Ephippiger (Orthoptera, Tettigonioidea)*

The behaviour of certain insects, in a large enough cages, may be natural enough to be virtually the same as behaviour in the free state. The possibility of insulating these insects from external noises permits a study of their sonic activities which are often associated with a large part of their sexual activities. An example taken from studies of *Ephippiger** will be given, chosen in this case to show how it was possible to illustrate from this study another factor entering into the behaviour pattern.

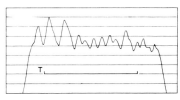

Fig. 44. Actography of the drumming behaviour of a wood-pecker. Time basis (T): 0.5 sec. Each peak in the record corresponds to a blow by the beak on the support used by the bird (original).

It may be noticed simply by the human ear that the male of this insect only chirps in the morning. What is the determining factor of this rhythm? To answer this question, it was necessary to carry out a continuous recording of the sonic activity of the animal, to specify the time-table of the stridulations. It is easy to obtain a record of the sonic activity by means of a permanent analysis from a galvanometer using a microphone or a preamplifier. It is even easier with a microphone having a high frequency selective filter system; this separates the emission of the insect from stray external sounds, since the stridulations have a band of wide amplitude between 8 and 12,000 c/s. This test was carried out using the same principle on another of the Tettigonioidea, *Barbitistes*, which emits practically only in the ultrasonic range, *i.e.* almost inaudible to the human ear. The record in Fig. 45 relating to this insect shows how the periods of song diminish and cease at certain hours.

Fig. 45. Actography of the periods of emission of acoustic signals of male *Barbitistes*. Time basis (T): 5 min. (1) Between 7 and 8 p.m., (2) Between 4 and 5 a.m., (3) Between 6 and 7 a.m. (original, Busnel and Dumortier).

The precise rhythm of the acoustic activity of these insects has been rendered visible and measured with this technique. A comparative measurement was however necessary for the study, *i.e.* the possibility of calculating the number of stridulations produced per unit time. This is a problem solved in principle by insertion in the circuit of a pulse counter recording the figure against a conventional time-base. This apparatus has been constructed and permits the total number of stridulations per unit time to be recorded (Fig. 46). From the results obtained in the case of the *Ephippiger* it may be concluded that the rhythm of the sonic emission is directly related to the light rhythm *via* the optical pathways (for full details, see p. 644 in Chapter 21 by B. Dumortier).

In studies concerning the rhythms of activity, techniques such as those which

References p. 65

have been used for *Ephippiger*, appear to be extremely valuable from the chronological result they provide, even when recourse to the summation of the sonic emissions is not necessary. Continuous recording at a very low speed and its analysis by reproduction at a high speed, to obtain a contraction of the time, may thus be visualized.

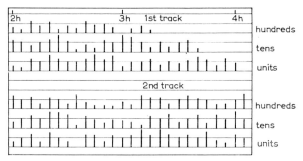

Fig. 46. Record of a two-track summation counter, used for the temporal notation of the chirping of *Ephippiger*. The summation is carried out every five minutes (original, Dumortier[9]).

The application of these methods to the observation of seasonal activities or vertical migrations, such as those of marine animals, for example *Synalpheus*, which have been observed to make violent noises with their chelae at certain periods, would certainly appear to provide important information on the ethology of these animals.

3. Analyses of the Sonic Signal Characterizing a Condition or Behaviour Pattern

(a) Studies on the specific variations of the signal

This is the true domain of acoustic methods. Visual reproduction and the physical characteristics of the parameters of the sonic signal make them as much a material for zoological study as morphological, anatomical or physiological features.

Apart from the collection of vocabularies, just as indispensable as the morphological collection, which it supplements admirably, the variations in the signal may also be of interest in specifying physiological or psychophysiological relationships. The following are a few examples of this:

(i) *Analysis of the influence of temperature on the emission behaviour.* An exact measurement of variations in activity associated with temperature is permitted in the case of poikilothermic species and more particularly for certain insects by a physical analysis of the recorded signal, in particular its oscillographic analysis. The best examples which may be given of this are probably that of the American Oecanthus, *Oecanthus niveus* (*Gryllidae*), the thermometer-cricket in which variations may be distinguished even by the human ear, and the crickets, which have recently been the subject of important analytical studies[2, 13]. In the case of *Ephippiger*, a systematic study undertaken with the oscillograph has enabled a correlation to be shown between the speed of opening and closing of the tegmina on the one hand and of temperature on the other (Fig. 47).

These analytical techniques also permit a precise study of the stridulatory movement and provide details of the signal corresponding to the movements of the stridulatory organs.

(ii) *Analysis of the evolution of the song relative to the hormonal condition in many birds*. Certain characteristics of the song are directly related to the hormonal state. These characteristics vary with the proportion of male or female hormone, and numerous earlier analytical acoustic studies based on listening with the human ear have shown this, but recording of the song and its transcription into a frequency spectrum and oscillogram allow this evolution to be measured (see Chapter 24, pp. 730, ff).

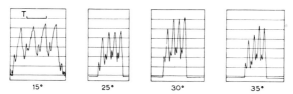

Fig. 47. Analysis of a temporal variation of an acoustic signal using the bathymeter. Relation between the temperature (15–25–30–35°C) and the duration of chirping in the case of *E. provincialis* male. Time basis (T): 0.1 sec (original. Busnel, Dumortier and Busnel).

(iii) *Miscellaneous studies*. Numerous examples of widely varying studies of the sonic signal, examined from various angles, may be found in different chapters of this book, *e.g.* genetic signals, learning, innate memory, specificity and interspecificity, phonetics (slowed-down or accelerated signals, the artificial recombining of the signals) autoinformation, etc.

(b) *Study of the emission and reception behaviour between two or more individuals*

These can be carried out from continuous recording of the sequence of activity. The following will serve as examples:

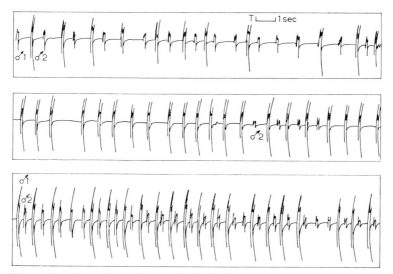

Fig. 48. Actography of a pattern of acoustic rivalry between 2 (1 and 2) *Ephippiger bitterensis* males. Time basis (T): 1 sec. Upper record: the two males reply to one another. After a stop by male No. 2, resumption of rhythm. Middle record: appearance of the signal of male No. 2 which occurs exactly between the signals of No. 1. Lower record: alternation between the signals of the 2 insects is well spaced—when one stops, the other speeds up slightly, then resumes its place (Busnel, Dumortier and Busnel[4]).

References p. 65

Sexual rivalry. In *Ephippiger*, if one male chirps and a second male enters into competition, this is shown by a modification of the rhythm of the emissions of the two males (record shown in Fig. 48). This can be proved experimentally by rendering one of the insects deaf by piercing the tympanal organs at the level of the front legs, when it will be found that the rhythm of the emission does not change[4].

With certain *Chorthippus* (Insecta-*Orthoptera-Acrididae*), the rivalry behaviour between males is very clear, the sexual calling signal is replaced by a second shorter signal and the emission rhythm changes. A very rapid exchange of signals takes place between two males in competition with one another (Fig. 49).

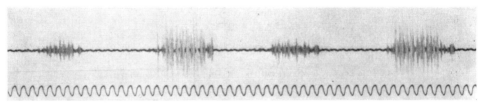

Fig. 49. Actography of the rivalry behavior between two male *Chorthippus brunneus* Charp. Time basis (T): 50 c.s. (Busnel and Loher[5]).

Oscillographic analysis gives a very accurate account of this behaviour. It hardly need be stressed that the times measured comprise the detection of the signal by the peripheral receiving organs, the transmission of the information to the central nervous system, its psychophysiological integration by the centres and then the emission of a motor order to the emitting organs. All this sequence of internal nervous activity is therefore chronicled in the receiving insect as from its reception of the stimulus up to the setting-off of its phono-response, which is the final term of this whole series of physiological activities[5].

Another example (Fig. 50) refers to the exchange of nocturnal signals between a male and female owl in the course of sexual display, when the male was in a cage. The recording was made on a magnetic tape and its temporal analysis on a recording galvanometer.

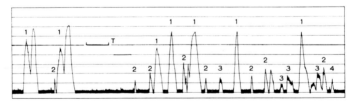

Fig. 50. Actography of a nocturnal exchange of call and reply signals of male and female *Strix aluco* in the sexual behaviour. Time basis (T): 4 sec. (1) Male calling in cage, (2) Female, (3) Another male, (4) Another female. Recording kindly made available by Rocher (original analysis).

(c) *Study of the phono-behaviour of response to a sonic decoy*

In the case of certain species it is possible to evoke phono-responses from an animal by a stimulus of the same physical nature, *i.e.* sonic, which may be compared to a decoy, since it will be indistinguishable from the natural signal of the species. A special study of this question has been carried out in the laboratory with certain

Orthoptera and the tree-frog *Hyla arborea*. Oscillography of the sonic recording of the stimulus and the phono-response give a chronographic visible record of the periods, and therefore of the phenomenon. This is independent of the study of the stimulus itself which is carried out by classical methods of analysis. Fig. 51 shows an example of this type, obtained from *Hyla*[3].

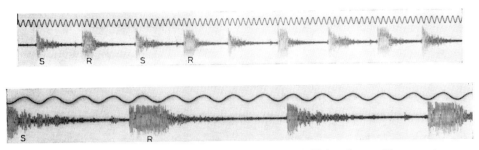

Fig. 51. Actography of the phono-response behaviour of a male *Hyla arborea* with one metronome signal. Time basis (T): 50 c/s. (S) metronome signal, (R) croaking response by frog (Busnel and Dumortier[3]).

The importance of these methods for research on the homophonies or dysphonies of artificial signals is so obvious that it is not necessary to stress it any further.

REFERENCES

[1] ADAMS, R. E., J. E. WOLFF, M. MILNER and J. A. SHELLENBERGER, 1954, Detection of internal insect infestation by sound amplification, *Cereal Chem.*, 31, 271–276.
[2] ALEXANDER, R. D., 1956, A comparative study of sound production in Insects with special references to the singing Orthoptera and Cicadidae of the Eastern United States, *Thesis*, Ohio State University, Columbus.
[3] BUSNEL, R.G. and B. DUMORTIER, 1955, Phonoréactions d'*Hyla arborea* à des signaux acoustiques artificiels, *Bull. Soc. Zool. France*, 80, 60–69.
[4] BUSNEL, R. G., B. DUMORTIER and M. C. BUSNEL, 1956, Recherches sur le comportement acoustique des Ephippigères, *Bull. Biol.*, 90, 221–185.
[5] BUSNEL, R. G. and W. LOHER, 1954, Sur l'étude du temps de la réponse au stimulus acoustique artificiel chez les Chorthippus et la rapidité de l'intégration du stimulus, *Compt. Rend. Soc. Biol.*, 148, 862.
[6] BUSNEL, R. G. and F. PASQUINELLY, 1954, Nouvelle application de la technique électroacoustique d'enregistrement de mouvements physiologiques de très faible amplitude à l'actographie de la Drosophile intoxiquée par des insecticides, *Compt. rend. Soc. Biol.*, 148 1587.
[7] BUSNEL, R. G., F. PASQUINELLY and B. DUMORTIER, 1955, La trémulation du corps et la transmission aux supports des vibrations en résultant comme moyen d'information à courte portée des Ephippigères mâles et femelles, *Bull. Soc. Zool. France*, 80, 18–22.
[8] BUSNEL, R. G. and F. PASQUINELLY, 1953, Détection et enregistrement sans inertie de mouvements de faible amplitude par une technique électro-acoustique, *J. physiol.* (*Paris*), 45, 61–66.
[9] DUMORTIER, B., 1957, Facteurs externes contrôlant le rythme des périodes de chant chez Ephippiger ephippiger mâle, *Compt. rend. Acad. Sci.*, 244, 2315–2318.
[10] ETZ, N., 1951, Schiffswurm-Detektor, *Umschau*, 51, 19.
[11] LOHER, W., 1958, An olfactory response of immature adults of the Desert Locust, *Nature*, 181, 1280.
[12] MULLER, H., 1954, Ueber ein neues Verfahren zur Aktivitätsregistrierung, *Z. Biol.*, 107 (4) 298–307.
[13] WALKER, T. J., 1957, Specificity in the response of female tree crickets to calling songs of the males, *Ann. Entom. Soc. USA*, 50, 626–636.
[14] WOOD, E. F., 1956, Queen piping, *Bee World*, 37 (10) 185–195; (11) 216–219.

PART II

GENERAL ASPECTS OF ANIMAL ACOUSTIC SIGNALS

CHAPTER 5

ON CERTAIN ASPECTS OF ANIMAL ACOUSTIC SIGNALS

by

R. G. BUSNEL

> "Il n'y a Beste ne Oiseau
> Qu'en son jargon ne chante ou crye".
> *Rondel de Charles d'Orléans* (1391–1465)

1. Introduction

The study of animal behaviour brings out the fact that one part of animal activity includes reactions of individuals to informative signals coming from other individuals. The knowledge of animal signals has been dealt with scientifically for only a relatively short period of time, for experiments are necessary before the term "signal" can be properly used, as it is only applicable if an observable, and preferably measurable, reaction is caused in a responding individual.

Signals are generally defined according to the physico-chemical nature of the stimulus or the sensory properties of the organs which act as detector-receptors. They are chemical (smell), visual (sight), electrical (galvanic sensibility) tactile and kinaesthetic (vibratory sense) and acoustical (hearing).

Animal acoustical signals seem to be of particular interest, for, thanks to modern techniques, they are easily recorded, analysed and reproduced. They are certainly not so elaborate as certain kinaesthetic signals, such as the information conveyed by bees in their dance. Because of this, the term "language" is not warranted for the use of acoustical messages by animals. Indeed, bee dances convey complex associative information of a mathematical kind, in a symbolic form which is characteristic of language. Acoustical messages express, from a semantic point of view, only one piece of information at a time, although they contain, within themselves, other underlying information, characteristic of the physical quality of the message, which may be called intrinsic information: the presence of an individual of the species, position in space (direction, distance, height), the sex of the animal emitting the signal and possibly individualization.

Phonetic animal expressions convey, almost without exception, subjective situations and aspirations. They are affective sounds which seldom tend to become objective designations or denominations. They express the idea of immediate time only, of a present situation or one which will occur in the immediate future. They cannot express abstract ideas which are unconnected with organic behaviour.

After a general examination of the problems raised by these studies, certain aspects of which will be intentionally omitted since they are treated in other chapters of this book, several examples will be given of experimental work dealing more particularly with "animal acoustical information".*

2. Quantitative Evolution of Acoustical Vocabularies

The number of signals which make up a repertory varies with the species, but the higher one goes in the Animal Kingdom the greater their number. Few documents are reliable on this subject. Moreover this idea would be more interesting if a complete table could be composed of all types of signals and not merely of the acoustical ones.

Table 4 gives quantitative values for certain vocabularies which have now been tabulated for different animal species.

This table shows that the number of acoustical signals is relatively small for invertebrates and lower vertebrates, at least as far as our present knowledge goes, in spite of highly specialized emitting organs, particularly those of certain insects.

The data in the table also demonstrate the extensive use of acoustical signals in the communication of marine fishes and mammals. In this typically homogeneous medium the propagation of sound in the sound channel should be greatly facilitated by the speed with which sound waves travel in a liquid medium, the density of which is equal to 1. Because of the difficulty of studying these animals our present knowledge is limited. It is very probable, when one considers the highly developed cortex of marine mammals and the enormous development of the auditory area that they use numerous signals.

The greatest number of signals has so far been tabulated for birds and higher mammals.

3. The Hierarchy of Signals in Behaviour: The Relative Importance of Acoustical Signals

The behaviour of an animal of a given species must be considered as forming a whole and, although it is convenient to split it up to study this or that particular point, the results obtained should always be fitted into the general framework and the different parts should be connected together in their natural sequence.

In this connection animal signals can be fully understood only when knowledge of the different signals is integrated. But it so happens that up to the present time acoustical signals are those which are the best known and for which we are beginning to get numerical values.

In general signals of different kinds may be grouped, or emitted and used in an hierarchical order, which very often depends on the transmitting channel and the distance between the source and receiver. The use of one signal or another also depends on the particular behaviour during which they are emitted.

In certain types of behaviour, acoustical signals obviously play a preponderant role, as illustrated in the following examples:

Brückner's Experiment on the Hen and her Chicks (1933)

This relates to the maternal behaviour of a hen with her brood of chicks which is released only by acoustical signals: the cheeping of the chicks. If a chick is isolated

---

* *Studies in Communication*, published by Secker and Warburg, London, 1955, may be profitably consulted in this connection.

TABLE 4

A FEW EXAMPLES OF THE NUMBER OF ACOUSTICAL SIGNALS IDENTIFIED AT PRESENT WHICH MAKE UP THE REPERTORIES OF SELECTED ANIMALS

INVERTEBRATES		Insects		
		Diptera	1	Mosquito *Culex* (Roth, 1948[118])
		Orthoptera	2	*Ephippiger* (Busnel, 1953[20])
			3	Field cricket, *Gryllus* (Regen, 1913[115])
			3	Tree cricket, *Oecanthus* (Busnel, 1953[20])
			5	Acridian, *Locusta migratoria* (Busnel, 1953[20])
			6	Grasshopper, *Chorthippus* (Faber, 1932[45])
		Homoptera	1–6	Cicada – Cicadellidae (Pringle, 1954[113]; Ossiannilsson, 1949[107])
		Hymenoptera	10–12	Bees, *Apis mellifica* (Hansson, 1945[65])
			26	Bees, *Apis mellifica* (Martin, 1880[91])*
		Lepidoptera	1	Moth *Androsa* (Arctiidae) (Peter, 1911[112])
VERTEBRATES				
	Lower	Fishes	1	Sea horse, *Hippocampus* (Fish, 1953[50])
			2	Sea robin, *Prionotus* (Fish, 1948[49])
			3	Eel *Anguilla rostrata* (Fish, 1948[49])
		Anura	2	Frog *Rana pipiens* (Noble and Aranson, 1942[105])
		Reptiles	1–4	Alligator, *Alligator mississipiensis* (Beach, 1944[5])
	Higher	Birds	5	Chick of the domestic hen (Collias and Joos, 1953[38])
			6	Herring gull, *Larus argentatus* (Frings and Coll., 1954[53,55])
			9	Jackdaw, *Corvus monedula* (Lorenz, 1931[85])
			12	Grey lag goose, *Anser anser* (Lorenz, 1938[86,87])
			12	Domestic canary, *Serinus canarius* (Poulson 1959[144])
			13–15	Blackbird, *Turdus merula* (Messmer, 1956[93])
			17	Black capped chickadee, *Parus atricapillus* (Odum, 1942[106])**
			20	Domestic hen and cock (Schjelderup-Ebbe, 1922[124])
			20	European Chaffinch, *Fringilla coelebs* (Marler, 1952[88–89])
			24	American song sparrow, *Melospiza melodia* (Nice, 1947[104])**
			25	Whitethroat, *Sylvia c. communis*, (Sauer, 1954[119])
MAMMALS		Sea	3	Bottlenosed dolphin, *Tursiops truncatus* (Bride and Hebb, 1948[19])
			5	Pilot whale, *Globicephala melaena* (Kritzler, 1952[82])
			5–7	White porpoise, *Delphinapterus leucas* (Schevill, 1949[121])
		Land	4	Bats, Vespertilionidae, Rhinolophidae, Chiroptera (Griffin, 1941[63], 1958[64], Möhres, 1953[97])
			5–8	Bovidae, domestic and wild cow (Schloeth, 1956[128])
			10	Prairie dog, *Cynomys ludovicianos* (King, 1955[79])
			23	Pig, *Sus scrofa* (Grauvogl, 1958[62])
			12–15	Gibbon, *Hylobates lar* (Boutan, 1913[15]; Carpenter, 1940[37])
			23–32	Chimpanzee, *Pan troglodytes* (Yearkes and Learned, 1925[143]; Kohts, 1935[80])
			36	Fox, *Vulpes vulpes* (Tembrock, 1957[136])

* Old observations should be verified.
** In these figures the authors have not discriminated between signals and songs.

References p. 107

under a bell-jar the hen no longer hears the signals, and although she is able to see her chick, she no longer reacts, and thus abandons it. In this behaviour, there is no visual signal. Only the acoustical signal is able to set off the protective reaction between the young and the mother (Fig. 52).

Fig. 52. Brückner's experiment.

In Orthoptera of the genus Ephippiger

In this species the sexually mature females[42] are attracted at a distance by the acoustical signal of the males. If the sound emission of the male is artificially stopped when the female approaches within an inch or two, and if a second insect emits signals at a distance X, where no visual stimulus is possible, the female will imamediately leave the first male to join the second, although she receives visual information from the first male. A signal given with a Galton whistle will attract the female to the experimenter even if she is near to mute male insects when the whistle is blown, since visual perception has hardly any value as a signal with this species. This may also be shown by covering the eyes or destroying them; the blinded female will find the male without any apparent difficulty[30].)

There may exist a time hierarchy in the use of the perceptive sensorial elements of the receptors depending on the range of the different types of signals. This may vary in different kinds of behaviour, as illustrated by the case of the dog. The visual signals, the attitude of the ears and the tail, can be perceived at 30–40 m. The acoustical signal, barking, has a very great range (several hundred yards). Finally, chemical information comes into play in differing degrees at a distance of a few meters. During the sexual period, the female's chemical signal not only has a greater range than the acoustical signal, but also remains effective longer as is characteristic of most chemical signals (conveys information of the past).

One of the important aspects of animal behaviour is the existence of particular kinds of directed behaviour in the course of which only one kind of action appears, or during which the receptivity (motivation) only comes into play in connection with a stimulus of well-defined nature and origin. The previous examples have illustrated this in the field of the sound message.

It may be useful to recall that not all the noises or sounds emitted by animals are to be considered as signals, at least until this has been proved. For example, the cry of death's head hawk-moth (*Acherontia atropos*), although well structured, has never yet been connected with a behaviour pattern. The noises made by the special device of shrimps of the family *Alpheidae*, though intense and produced by highly specialized organs, have still not been defined psychologically and it would perhaps be bold, at the present time, to ascribe to them the value of a signal. On the other hand, noises resulting from an activity, for example the flight-tone of the female mosquito, have the value of a signal for the ♂ which are attracted although it does not seem, at first sight, to be a special flight with a particular rhythm (as would be the case for some of the wing movements of the male fruit fly (*Ceratitis capitata*), the noise of which guides the female). Finally, the noise emitted by the activity of an animal may be considered as a signal for an animal of another species if there is some relationship in the behaviour of the second species with regard to the first (the noise made by a moving animal for the animal which preys on it). If the noise made by the prey is specific, there is no doubt that it has the value of a signal for the predator which hears it.

4. The Particular Interest of Acoustical Signals

Acoustical signals have particular properties on account of the physical nature of their medium, the sound wave, and seem to be well adapted to animal needs.

(1) Their range: which may sometimes attain several miles (Table 5).

(2) Their value in absence of visibility: at night; in the forest or undergrowth areas; in very deep or muddy water.

(3) Their centre: that is to say an easily locatable source, both as regards direction and distance.

(4) Their lack of permanence: (*verba volens*) as against the chemical signal which impregnates the marked zones for a long period of time. Sound signals may be considered as instantaneous.

(5) Loss of information due to background noises in the transmission channel (air of water), seems to be of less importance than for the other types of signals.

(6) The quantity and the nature of the information that they transmit per unit time may be relatively greater than with the other types of signals (codification).

5. Preliminary Remarks on the Biophysical Relationship between Signal and Behaviour

From a statistical point of view, the probabilities of individuals' meeting (male and female, for example) depend on their density per surface unit, on their total mobility, and on the richness and qualities of the means of information they have at their disposal enabling them to increase their individual field of radiation and their perceptive universe.

The emission of sound is only one part of the total repertoire signals, but its possibilities of emitting at great distance and with a large amount of information per time unit makes it well adapted for the purpose of communication.

It is possible to conceive a logical connection between the physical characteristic of the message and the density of individuals.

(a) *Connections between the intensity, the range and the density of population*

The range of a signal is in direct proportion to its intensity, other physical conditions of the medium and the signal being equal. As a result, the reception areas

TABLE 5
EXAMPLES OF THE RELATIVE RANGE OF PERCEPTION OF SIGNALS**

Radius of a few receptive fields

Visual		Acoustical		Olfactory	
Species	Distance	Species	Distance	Species	Distance
Fiddler Crab*	20 m	Mosquito (Ins)	3 m	Necrophorus (Ins)	100 m
		Oecanthus (Ins)	30 m		
Spider*	20 cm	Locust (Ins)*	38 m	Saturnia (Ins)*	3000 m
Glow-worm	3–5 m	Conocephalus (Ins)	3 m	Skunk	30–100 m
Dragonfly larva*	130 cm	Chorthippus (Ins)	10–12 m	Goat	200–500 m
Firefly	10–30 m	Cicada (Ins)	500–800 m	Wolf-dog	1500–2000 m
Blenny (fish)*	40–125 cm	Batrachia	100–500 m	Horse*	4400 m
Archer (fish)*	50 cm	Pigeon	500 m		
Batrachia	10–50 cm	Jay	500 m		
Rat*	50 cm–3 m	Crow	500–1000 m		
Giraffe*	700 m	Nightingale	150 m		
Vulture	3–7000 m	Crocodile	1500 m		
Pigeon	500–1000 m	Lion	2500 m		
		Sperm-whale	1500 m		
		Marmot*	100–350 m		
		Lyre bird*	250 m		
		Long-tailed grouse*	2000 m		
		Wood grouse*	100 m		
		Black woodpecker*	600–800 m		
		Greater spotted woodpecker*	140–200 m		
		Argus*	1000 m		

* The values marked with an asterisk are taken from the literature on the subject and are more often the result of "observations" than "experiments".

** The quantitative data of this table are only approximate, figures measured experimentally being rather rare. They should not be considered as absolute values for, according to circumstances (background noise in the sound channel), they may fall to zero. The figures given in the table should be understood as upper limits. It is easy to understand the difficulties of obtaining quantitative information, as the experiments have generally to be carried out in natural surroundings, from free animals, from perception reactions of minimum intensity, coming from a signal of maximum intensity at the source. It should equally be borne in mind that the intensity of a signal, and consequently its range, can vary with the meaning of the message.

These figures express the distance at which a highly receptive animal can still react to a signal given with maximum intensity by the emitter, usually both belonging to the same species. For example, for the dog, it is the maximum distance at which a male can still be attracted by the chemical signal of a female in heat; for the nightingale, it is the maximum distance from the centre of the territory of a male, at which another male can approach, during the sexual period, without giving rise to attack behaviour from the possessor of the territory; for a starving Batrachian, it is the maximum distance at which its prey no longer causes a capture reaction.

covered are smaller for signals of low intensity. The density of the individual population per surface unit, which is the inter-individual distance corresponding to the field of reception, will come into play in the following sense: if the density of the animal population is high, signals of low intensity will suffice for the information to be received, while if the density is low, the emission intensity must be greater to attain the same aim.

(b) Relationship between the duration of a signal and its range

There are species in which only sound signals cause an oriented reaction in the receivers. If the sound signal is of short duration, the receivers, if not far away, can reach the source without any appreciable error, since the time necessary for the journey is short. But, if the receivers are at a great distance, the signal, if brief, must be frequently repeated, unless the receiver has a "directional acoustical memory"*; otherwise the signal must be repeated practically without stopping. In all cases the emitter must be at "a fixed point", the word "fixed" having a variable temporel significance for different individuals, conditions of the moment and species.

This fixed area, the place from which the call is emitted, can be expected in species living in low density populations but has up to now only be studied in a few cases. We shall give it the name of "call territory" (this term corresponds to other "territorial" ideas, based on other criteria, especially the need for a certain feeding area). This "call territory" is always relatively small. It would seem that more precise studies of this idea would enable us to establish a very close relationship between the area of the "call territory" and the average angle of error in the orientation of attracted receivers.

(c) The problem of the density of individual population in connection with the modes of expression of the signals

(1) *High density*. At small inter-individual distances a particular signal may be perceived by several individuals at once. If this signal has a call value, several receivers may arrive at the same time, thus creating a complicated social situation. Several other means of information which then come into play may lead to various solutions having an isolation value: choice, refusal, rivalry, battles, etc. Examples of signals which come into use at short range include the play of the antennae, olfactory stimuli and postures. The first signal used may be re-emitted, but it will be modified in intensity, by a change of rhythm or by a variation in duration, and it then acquires a new specific meaning for the given situation. Indeed social life necessitates an increase in the richness of signal repertoires.

(2) *Low density*. On the other hand, if the population density is low, and even very low, the connections between the emitter, the receiver and the signal may be interpreted in a different way. As a matter of fact, the distance between the emitter and the receiver will play an important role in the selection of the receivers attracted. Since for the same distance, the animal most receptive to the call will reach the emitter in the shortest possible time, being physiologically prepared in advance to react to the information. On arrival, and contact having been established, a more specialized acoustic signal at short range is not usually given out. The main role of the signal has been a call from a distance and a guide for localisation. If, however, the social evolution is more complex, additional releasing mechanisms may come into play for short distances, with complementary action. This brings us back to the first case. Thus, to a certain degree, the density seems to determine the nature of the signals and of the inter-individual relationship (sexual or social).

(3) *Another aspect*. If the abundance of individuals reaches a certain point, and if signals are emitted, the message will not have a really discriminating value unless there is a meaning, *i.e.*, a psychically valid codification. Moreover, it will not be

---

* We call "directional acoustical memory" the receiver's memory for the direction of origin of a perceived sound signal. It may be observed by following the receiver's journey to reach the emitter if no other stimuli are received and if the receiver is able to avoid the obstacles and yet as a rule, reach the emitter.

References p. 107

necessary to have very long emissions; on the contrary, brief emissions will enable a greater quantity of information to be sent out in the minimum of time, and possibly a rapid reciprocal exchange of signals. According to the degree of evolution, we shall find a richness of acoustic repertory which enables more and more complex situations to be expressed. The highest stage is human vocabulary, where complexity is associated with individual culture, which may be thought of, to a certain extent, as evidence of advanced evolution.

6. Acoustic Signals in Behaviour

General Classification of Sound Emissions in Connection with Behaviour

Sounds emissions having the value of signals may be classified in connection with the different kinds of behaviour to which they are related. They correspond to the following individual or collective relationships:

(A) Sexual relationships: attracting call; courtship; competition.

(B) Family relationships: between young and parents and reciprocally; individual recognition.

(C) Relationships with community life: group activities; alarm behaviour; food behaviour.

(D) Sound ranging—Echolocation: Only scattered examples will be given here concerning these studies, for certain of these points will be dealt with in detail in other chapters of the book.

(a) Acoustic signals in sexual relationships

(i) *Meeting of the sexes (attracting call)*

Among the signals used by the different animal species to facilitate meeting and sexual recognition, the sound signal has an essential place, at least for species which are able to send signals.

For invertebrates, a large part of the known sound signals have value mainly as sexual attractants and transmit to the receiver both the spatial location and the sexual state of the emitter. The larvae or immature stages are usually unable to send out sounds. This applies also to numerous amphibians and to birds. Bird calls may be, in certain species, very particular signals (the tapping of woodpeckers, the bleating of snipes, the rattle of nightjars). This type of sound, therefore, has the value of a social release-mechanism, for its aim is generally to lead, in a directional way, one of the two sexes, usually the female, to the other. For certain *Diptera* (mosquitoes), it is the male who travels, attracted by the tone of the female's flight, which thus takes on the value of a sexual signal. Experimental proofs of sexual phonotaxis have been obtained for a certain number of species, by using either the signal itself of the males or females, or a retransmission of the recorded signal. The oldest experiment in this field was made by Regen (1913[115]), who attracted the females of the field cricket by emitting the male's song through a telephone receiver, working as a loudspeaker.

This experiment has been repeated, with various technical modifications, on several other animal species:

Insects: Mosquito[75,118], *Ephippiger*[25,42,43], *Acrididae*[21,22,23,68,84]. Anura: Toads and Frogs[92]. Fishes:[101,132,133]. Birds, Mammals:[135,136].

For Ephippiger, where the female shows a remarkable phonotaxis to the male's signal, independent of visual perception or stimuli, the intensity of the signal is one of the important features of orientation and choice. If the resonant membrane of the male's acoustical organ is destroyed (which will modify the intensity of the signal more than its frequency range), the sound emitted will still be capable of attracting a sensitive female. If, however, a second male suddenly emits a signal, when the female is but a few centimetres from the first male, she will change her direction and will go toward the second male. Here the volume of the sound determines the female's phonotaxic behaviour.*

A variation of this type of signal has been described[107] in the exchange of sound signals of certain Homoptera (*Doratura stylata*). The male emits two different signals and the female a third. If we call the male's signals A and C and the female's B, the sequence of the exchange of signals leading to copulation is as follows:

(1) Male: A.A.A.

(2) A female located in the male's field of sound will, if receptive, reply by B.B.B., which is a reply and call signal.

(3) The male hears B.B.B. and stops emitting.

(4) The male emits C.C.C. in reply to the female's B.B.B. (reply signal).

(5) The male stops emitting C.C.C. and goes towards the female, which continues to emit B.B.B. until the two insects meet.

In this case, there is no male territory, but a female one, and she does not move till the male arrives.

The fact that hunters use bird-calls which emit male or female signals (a horn which imitates the cry of the roe in rut, net-traps using male gurnets, which attract the fish of the other sex by their signals), shows the strength of the taxic effect produced on the receiver of the opposite sex when it is in the proper hormonal condition (motivation).

For animals which have no sexual dimorphism, sound signals, whether associated or not with other signals, may have, according to the species, a discrimination value. This seems to be the case for certain birds, such as the house wren *Troglodytes aedon* and *Thryomanes bewicki*[94].

(ii) *Courtship song*

For numerous species of vertebrates and for some invertebrates, sexual approach is preceded by a ritual of attitudes, corresponding to a progressive exchange of various signals, which take place over a certain period of time, sometimes during several days. This ritual, taken as a whole, is normally called "courtship" or "nuptial display" (fish-birds). Sound signals often play an important role here.

These nuptial displays have been interpreted in many different ways, some examples of which are given below:

Although they delay copulation, they would seem to have a hormonal role, for they enable genital products to be either emitted or matured. On the other hand, their specificity prevents interspecific copulations and limits the repetition of individual copulations. Finally, they play a role in natural selection by allowing a better choice to be made. These different points of view are certainly valid, but they relate to particular species and cannot be generalised.

---

* M. C. and R. G. Busnel, unpublished results.

By way of example, we may cite the displays of several birds, and in particular the pigeon. The male is sexually mature before the female and the nuptial display is spread out over several days, the stimulation provided by the display of the male gradually bringing the female to a receptive hormonal state.

The nuptial displays of certain large bustards is accompanied by the inflation of the air-sac under the tongue, which forms a sound-box and increases the voice power (*Choriotis australis, Otis tarda*). The air-sac of this latter species stretches down to the ground, along with the skin and feathers which cover it and this makes it look like an enormous beard. In certain species (birds, mammals), acoustical signals may be produced during calls or courtship, along with beating of wings. Examples are the snapping of the rectrices of humming-birds, who fly at a fixed point in front of the female, the beating of the wings of the lapwing, the snapping of the beak of storks and owls.

For certain herons, the attraction of the female by the male is brought about by gestural signals. But, at short range, the male calms the apprehensions of the female about coming into its territory by attitudes and sound signals to which she responds. These sound signals seem to be essential, for if the male is rendered deaf by blocking up their ears, the nuptial display cannot develop. The courting signals of the cricket seem to aim at bringing the female to a quiescent state, which stops her continual moving about and allows copulation to take place[67].

The female of the gobiid-fish, *Bathygobius soporator*, reacts to the signal of the male, transmitted through a loudspeaker, by increasing her breathing rhythm and her activity of movement. This motor activity is not oriented, however. A visual signal must be allied to the sound signal before a directional reaction is manifested. The visual signal alone does not bring about any modification of behaviour[132,133].

The same observation would apply to a fresh-water fish, *Aplodinotus grunniens* Raf. found in Winnebago Lake (Wisconsin, U.S.A.). In this species, only the mature males emit sounds, and only during the reproduction time (pulsed sounds of 200 msec, at 5 sec interval, frequency between 250 and 400 c/s)[145].

For the Mammals, we have some interesting examples:

Pressure applied to the back of an oestrous sow evokes a characteristic reflex. The animal is completely immobile and actively resists pressure from behind. The back is arched, and the ears and tail are elevated. This reaction, well known to stock breeders as the immobility reflex, is a certain sign of oestrus.

But, only 42 % of all oestrous sows respond to dorsal pressure in the absence of a boar, while all do when he is present; it may therefore be concluded that the boar emits stimuli which contribute to the release of the immobilization reflex.

In the presence of an oestrous sow, and particularly when she avoids him or cannot be reached by him, the boar emits a series of characteristic grumblings that one can interpret as a courting song. It consists of a series of brief, low frequency sounds which succeed one another in a rapid (6 to 8 per second) and regular rhythm. The intensity of the sound is 85 to 95 dB at one meter and is emitted for periods of 3 to 4 seconds separated by inspirations. When this signal is broadcast by a loud-speaker, no boar being present, about half of the 58 % of pressure-negative oestrous sows react to the immobilization test.

Thus, even though the immobilization of the oestrous animal can be obtained by kinesthetic stimulus alone, the courting signal of the male, that sometimes sets up a positive phonotaxis, facilitates the releasement of the reflex; this is probably only an inhibition to the usual reaction of flight. Other signals of the boar have been found that have also an important role, especially the chemical signal of the preputial glands. In the presence of the scent from these glands, positive reactions were obtained in all of the group of pressure-negative and sound-negative oestrous sows[146].

For the mare, the seighing that the stallion emits during the course of the courtship, releases the opening of the vulva and the outburst causes the erection of the clitoris.

Courtship songs, which are mainly emitted by males, are almost always of weak intensity in comparison with call signals. They chiefly occur in species where sexual competition is relatively strong, and may intervene in selection. This phase of behaviour may sometimes be necessary for the isolation of the species and for natural selection.

(iii) Sexual competition

Rivalry signals: Examples of sexual rivalry expressed by acoustical signals may be those studied in two Orthoptera: *Ephippiger* and *Chorthippus*.

If a male *Ephippiger* stridulates and a second male enters into acoustical competition with him, this is expressed by an alternation in the rhythm of emission of both males.

Demonstration of acoustical rivalry: Males which have been made deaf by cutting their tympanic organs no longer modify their rhythm.

For certain species of *Chorthippus*, rivalry behaviour between two males is very clear: the sexual call signal is replaced by a second signal, which is shorter and has a changed rhythm. There is an extremely rapid exchange of signals between the two males. The time lag between the two signals changes from 1.6 seconds (rhythm of the call signal) to 0.57 seconds (rhythm of the rivalry signal).[84]

For *Chorthippus brunneus*, the stimulation is so strong that there is even a sequel of acoustical activity after the stimulation provided by the situation, an after-effect which is expressed by a series of sound movement which gradually die away[23].

Similar phenomena are found in Batrachians and certain fish[101]. The study of "choirs" heard in natural surroundings shows, indeed, that for certain species, this apparent harmonisation is really made up of individual signals from members of the same sex with an identical sexual state, who give out more rapid rhythms than if they were alone. These signal exchanges form the essential part of this rivalry behaviour.

Acoustical rivalry followed by attacks on territory which is limited acoustically: Although no quantitative studies have been made on this point, two kinds of observation may be given:

Birds.—Certain small passerine birds have territories, the area of which varies during the course of the year according to the bird's activities. These territories are generally related to food and there is probably a relationship between area and quantity of food. But in spring, or during the mating season, the male's territory takes on a sexual importance and, at this time especially, the male birds defend their territories against the other males of the species by their songs (robins, nightingales), the essential role of which seems to be that of indicating and marking out their territories and attracting the females. In addition, other stimuli come into play, these even may lead to battles as can be illustrated with sound recordings retransmitted through hidden loudspeakers. Male nightingales and robins even answer in such cases to their own signals. The bird reacts violently to the recorded song, it flies swiftly and aggressively towards the source of sound, even alighting on it, and if a stuffed bird is placed on the loudspeaker, it will attack it.

Some birds, such as the Argus pheasant, and the Tetrao (capercaillie), have a special territory (arena) on which the male bird calls and displays. There appears to be a direct relationship between the distance separating these territories and the intensity of the bird's call. An acoustical rivalry has been observed between males who reply to the calls of their neighbours and it may be possible that, in the case of the Argus pheasant, for example, intensity of the male's cry is related to its sexual maturity. Acoustical rivalry in display and possession of territory are commonly observed in birds.

The common tern, *Sterna hirundo*, for instance, when it possesses a territory, begins by uttering a special cry of growing intensity at the sight of a trespassing neighbour before adopting a threatening attitude, which is the second and highly stereotyped step in defense. No precise measurement of this acoustic threat has been made[108].

In the case of some birds, it may be that certain songs correspond to an attitude of rivalry, which might be a substitute for real sexual competition (eg. grouse).

Mammals.—Hunters have often noticed that the sex-call of the stag, the "belling", rapidly changes into a "troat" if, within its hearing and on its territory they imitate a stag's signal with a hunting horn, a pierced seashell or lamp-glass.

The answering signal of the male sounds enraged and the animal is immediately attracted and comes to the spot to attack its rival. This sexual rivalry and tendency to defend its territory, which lead to the phonotaxis of the male, are the basis of the practice, current in some hunts, of using lures, in which the hunters imitate the male signal during the rutting season and the stags answer and some to the spot. This technique is used in stag, reindeer and moose hunting.

In the Canadian Rockies, the noise of the earliest train whistles aroused this reaction of rivalry and attack, at different time of the year, bringing a number of male moose charging the trains; so that the whistles had to be changed. (Original contribution by Mr. Bernard, Scientific Attaché, French Embassy, Ottawa, 1956.)

Addendum

French explorers between 1634 and 1761, noted that Indians of the Menomini tribe, belonging to the Algonquin nation, living near Lake Michigan, U.S.A., used wooden whistles to attract the does, by imitating the cry of the fawn. They noticed, too, that this way of hunting was dangerous because it also attracted wild cats and wolves. (Documents in the Alanson Skinner collection, 1910–1913, No. 50/9843, case 22, Indian Collections of the New York Natural History Museum.)

These Indians also used, for other purposes, a wooden flute which they called "the lovers' flageolet". In the William Jones Collection 1903 in the same Museum, case 23, the following inscriptions appear beside them:

"Flageolet played chiefly by men in courting, the belief being that the sound has magic power to influence women." (Does this refer to a conditioned reflex?)

"Such an instrument, when reputed as a successful aid in courtship, may be rented out by its owner, who receives good payment."

(b) Acoustic signals in family relationships

(i) Relations between parents and young

Especially in the case of the higher vertebrates, sound signals emitted by the young transmit information to the parents, indicating particular psychological or physiological states induced by hunger, fear or desertion. The parents have innate reactions to these signals, which vary considerably according to the species and situations.

Thus the cries of young crocodiles in their eggs, before hatching, attract the adult females, who come to help them break out of their shells.

Experiments carried out, mostly on birds, emphasise the important part played by sound signals of parents in the behaviour of the young.

Newborn blackbirds (*Turdus merula*) brought up in separate nests[93] isolated from all vibrations, respond to acoustical stimuli of pure frequency, produced by a generator (300 to 10,000 c/s) by opening their beaks wide, as they would to parents bringing food. Sounds between 800 and 1000 c/s are very effective and the same phonokinesis can be obtained by human whistling. Subsequent experiments showed that, after the third day, the birds have learned (are conditioned to) a given signal, and no longer respond to different ones.

It is according to the same kind of "learning" conditions that young of one species, brought up by parents of another species, respond, at least partially, to the signals of the foster-parents. For example, chickens hatched and brought up by a turkey respond to the turkey's signals, although they are different from those of the hen. However, no observation has been made on the reciprocity of the danger reaction of mother to young in these cases, since the turkey's reaction, like the hen's, is to gather her brood together, huddled under her tail. This would be an interesting experiment and, though it is highly probable that the foster-mother understands the signals of the young of the other species, there is no proof that the young would show a reaction, which is generally considered innate, to the danger signal of the foster-mother's species. Young pheasants, brought up by a hen, react instinctively to the danger signal of their foster-mother, but these species are genetically closely related.

The goslings which follow Lorenz when he calls them as can be seen in his well-known films certainly react to a foster-parent, but in this case the foster-father imitates the parental cry of the species of the young and does not try to impose his own signals on the goslings.

In the case of a young crow, vocal imitation of the same kind produces phono-responses[61]*.

There is very often a definite taxis of the parents towards the young emitting a distress signal, the acoustical signal being, for some birds, more effective at certain stages than any other means of communication (Brückner's experiment on the hen). Cries of young dolphins have been recorded[123] and the reactions of the mother towards its young described. A young sperm-whale, wounded by the stem-post of the survey ship *Calypso*, uttered cries which caused the gathering of the whole school around it (Cousteau's film "Le Monde du Silence", 1956). Hunters of African crocodiles some-

---

* For a definition of this term see p. 104.

times make a young crocodile vocalise by biting its tail; these cries attract adults in the neighborhood.

The reactions of the young to parent's warning signals vary with the species: they may, for instance, freeze to the ground or flee for shelter. If the scientific truth has been respected in a film of Walt Disney, a good example can be seen when young bears, at an acoustic signal of the mother, immediately climb trees and hide in the forks of branches. In the case of communication between mother and young, in wild cattle, two different acoustical signals for recognition at a distance have been observed (one for the mother and one for the young) and two other acoustical signals for close communication[128].

Chicks of the domestic hen react to two signals of the mother, the food and danger calls.

In both cases, the chicks gather around the mother, but in the case of the danger signal, there is the additional reaction of huddling under her. This signal is given by the mother when she perceives danger in the air, especially sparrow hawks and buzzards.

In the case of wild Gallinaceous birds, the ancestors of the domestic hen, natives of southern Asia (Viet-Nam), the reactions of the chicks to the mother's danger signal are different. Instead of gathering under the mother as in the case of the domestic hen, they scatter in all directions, freeze and become invisible, while the mother flies off and diverts the enemy (Darwin, confirmed by Boutan[16]).

(ii) Individual relationships

Sound signals, especially in birds, also serve probably for individual recognition. Although as yet there are no precise experiments, numerous observations have been made on the subject, especially on what is called the "welcoming to the nest" reactions, with appeasement ritual—for example, clapper of the storks' beaks. ("The stork, that ever-welcome traveller, perched on his slender legs, whose beak clapper like castanets." Petronius: *Satyricon*).

It is probable that some forms of individual recognition are related to different acoustical signals characteristic of the individual, although associated with numerous other aspects, but there are practically no experimental results on the purely "personal" character of animal signals.

In the case of the oven-bird (*Furnarius rufus*, Passeriform), it is observed that, as soon as it sees its companion appear, the bird in the nest emits loud and regular cries or sometimes a trill. When the approaching bird utters isolated notes, the first bird answers in a different rhythm, with rapid triplets. Both birds sing rhythmically, the former with a lively beating of the wings to a rapid succession of notes and the latter by a more measured rhythm, in harmony with its slower song.

A water-bird, a Common Screamer (*Chauna torquata*—Anseriform) has also been observed to sing antiphonally. When one of the mates approaches, it calls to the other. The one in the nest rises and answers and both birds sing together for about 15 seconds, the male on a lower note than the female.

Similarly synchronised male–female duets have been observed in South America, in the case of the Wood Quail of Guyana (*Odontophorus gujanensis marmoratus*) and the Wood Rail of Cayenne, *(Arimides cajanea)* both birds facing each other and singing together, male and female taking quite unlike parts. Antiphonal singing has been observed also for many types of owls (*Strigidae*), for the cardinal, the lyre bird and many others.

Communal singing, when more than two birds aggregate to sing, seems to be a bivalent social behavior sexually stimulatory and intimidatory. Examples are seen in the Spur Winged Lapwing of La Plata *(Belonopterus chilensis)* and in the Oystercatcher (*Haematopus ostralegus*) piping parties, in which two to a dozen birds may participate. When two males have piped together to one female they often do not attack each other.

According to Möhres[97,98] individual recognition between mother and young in bats, who have their young grouped in colonies, is acoustical, each young having a special call. He experimented by marking and scenting mothers and young, without altering the results in individual recognition.

(c) *Acoustic signals in connection with collective life*

(i) *Co-ordination of the movements of the herd*

(Signals of collective information value, possible indication of despots or leaders.) One of the most interesting example of the effect of an acoustical signal on a group was given by Lorenz on the Grey-Lag Goose which is described in details in the article of Bremond, p. 721[86,87]. Although this is one of the best studied occurrences of coordination, it is very probable that there are numerous signals of this type in the case of other birds and also in mammals, where it appears that there is, at least in certain cases, co-ordination of movements in the group, for instance in monkeys (*Cynocephalus* and others), wolves, horses, cattle and elephants.

(ii) *Alarm behaviour*

Despite the absence of experimental findings, numerous observations have shown that sound signals play a great part in alarm behaviour and that, in the higher vertebrates, very different signals can give rise to types of behaviour which are probably innate, specific and directly connected with a particular signal. Moreover, alarm signals can, in some cases, be preceded by warning signals, putting individuals or the group on their guard.

Examples: (a) *Mammals*[79]: the black-tailed prairie dog (*Cynomys ludovicianus*) is the prey of many larger animals. It has a special sound signal to indicate the approach of enemies of the colony (eagles or screech-owls) consisting of a series of rapid barks. When the danger is past, a sentry, by means of another signal, announces the "all clear" and all the members of the colony leave their holes. They have another sound to signal the approach of a coyote, a cry which has no collective value but varies in frequency and intensity according to the movements of the coyote. Prairie dogs also bark at the approach of antelope, bison and man, to warn their colony. When they have young, the barks are more frequent and it seems that there is a social training period. Sound signals in young reared in laboratories appear to be different.

These sound signals are one of the most characteristic features of prairie dog grouped in cities. Observations of reactions produced by the signals can be divided into: a warning bark, danger at the approach of a bird of prey, defensive bark, signal marking territory, muffled cry, dispute, derision (?), fear, battle warning, gnashing of teeth (competition?).

References p. 107

The warning bark varies in intensity and frequency with the stimulus and may last more than one hour. The collective result of the signal is a warning to the community. The animals stop what they are doing, sit up and look for the potential danger. If this is visible, they flee towards their holes; if not, after a few moments, they resume their activities.

The warning for birds of prey is given on a shriller, more sustained note and members of the colony immediately flee to their holes, without any preliminary reaction of attention. Hence this signal is much more effective.

The females, as well as the males, defend their territories. The signal marking the territory is very distinctive and the animal, in emitting it, sits up on its hind legs. The signal is long, in two phases. It is used very often inside the colony and always causes reactions among those who hear it and who answer in turn.

The same kind of observations have been made concerning the Siberian marmot or tarbagan (*Marmota sibirica*). When an enemy enters the territory of the colony, the animals flee to their holes, sit up on their hind legs at the entrance and whistle ceaselessly in response to each other. If the intruder is a man, the radius of the circle of whistling tarbagans is between 100 and 350 metres; if it is a dog, the circle is smaller and the whole colony stands up and whistles.

One of the most interesting types of behaviour is that of certain field mice, living mostly under dead leaves, which, when they move, make a noise which the others take no notice of. But if the leaves rustle for any other reason, the whole colony rushes into their holes. It has been discovered that, before moving, each mouse utters an ultrasonic warning cry, the purpose of which is to show the harmless nature of the rustling which will follow and prevents the alarm reaction of the other mice (Lorenz and Schleidt, quoted by Pumphrey[114]).

The distinction between the "potential danger" signal and the "real danger" signal as two types of signals has also been made on the basis of observations on the kangaroo (*Macropus benetti*[16]).

The male kangaroo lives with a harem of females. It has two signals, a kind of whistle, which have a meaning for the whole group. The first is a look-out signal which the male, who is on the watch, utters as soon as he perceives a real or possible danger. If the danger is proved real, a second signal, different from the first, results in the flight of the group.

(b) Birds: Two kinds of alarm calls are frequently noticed among birds, one answering to a surprise attack and the other to a bird of prey observed at a potentially dangerous distance. Reactions to these two danger calls may be entirely different, the former leading the birds to hide, crouch or similar emergency reaction, the latter leading to either a group attack or a flight into carefully chosen hiding.

The signals which produce such reactions have accordingly been given the appropriate name of "social releaser", suggested by Tinbergen.

They may even be extended to other groups, having then an interspecific value, a conception which will be examined later.

Some species of birds organise real attacks against birds of prey. These attacks are made in cooperation by the whole group and the method of attack is the following. An individual reacts to the bird of prey with an alarm signal. The response of the others to the danger is to unite with the first bird in a joint attack. The survival value of the process of perception and signalling involved in spotting the bird of prey lies in the fact that the discovery of the latter by the group depends on the speed of reaction of the most circumspect and the most alert individual. The survival value of such an alarm, signal is therefore evident.

In the case of the buzzard, the male, who is hovering, gives the signal to the female, who is sitting on her nest.

If certain encampments of gregarious birds are attacked, they fly together, wheel and utter cries of distress and fury (seagulls, crows).

Some birds, such as the St. Martin's buzzard, the black-headed gull and the shrike, even attack man. Sparrows seldom attack a cat, but they gather above it in a bush, ceaselessly uttering their alarm signal.

Such collective defence can in certain species, be of help to a wounded bird by driving off the predator. This is signalled to the group by a special cry or a clacking of the beak. Such is the case with the Corvidae and the Terns.

Some birds can be deceived by their own danger cry. Davis[39] noted that a particular cry of *Crotophaga ani* (smooth-billed ani) gave the alarm to a whole colony. An American mocking-bird, having once incorporated this cry in its song, involuntarily caused panic among neighbouring Anis.

When the young are born, numerous cries ensure the protection of the nest against approaching danger.

Wrynecks and tits, who nest in holes, hiss like snakes when they are attacked in the nest. This behaviour was studied in *Parus carolinensis* (a tit), where the whistling appears to be similar to that of the snake *Agkistrodon mokasen*.

(iii) Experiments on alarm and distress calls

Birds: It was shown experimentally (by playing back to the birds their recorded calls) that sound signals of distress or alarm had really a value of social releaser at the group level. Initial experiments were carried out by Frings and Jumber (1954)[53,55], in the USA on starlings and seagulls, followed by the work of Busnel, Giban and Gramet in France (1955)[27] on crows, undertaken with the purpose of driving the birds away, either for the protection of crops or airfields or landing grounds, or even to ensure the cleanliness of small towns where starlings nest.

The cries used are the so-called cries of "distress", obtained by holding the bird by its feet or uttered when attacked by a bird of prey. Signals of alarm recorded in the course of earlier observations on natural behaviour may also be used.

Records of these signals can be broadcasted at sufficient intensity to cover a considerable distance. This is done by day or by night according to the species. The results obtained are satisfactory. As regards herring-gulls, for instance, with signals broadcast at a intensity of 75 dB* (at a distance of 1 m), the result was conclusive within a radius of 800 m. The birds stay outside the area for an average time of 30–45 minutes after each one minute broadcast that consists of a signal repeated five times with 7 seconds' silence between each signal. In the course of two-day experiments, it was found that broadcasts reduced to 10 or 20 seconds were sufficient to cause flight.

Results of the same kind but even more marked were obtained by night with starlings and Corvidae living in roosts (up to 20,000 birds in the case of starlings and 5–10,000 in the case of Corvidae).

---

* In this paper, intensity is defined on the basis of 0 dB = 10^{-16} Watt/cm^2.

References p. 107

Cries of distress were broadcasted at night near the roosts. These lasted on an average one minute, and were sometimes repeated several times in the course of the night. The birds abandoned the roost almost instantaneously and collectively.

Three or four nights' broadcasting has, in some American towns, kept starlings away for several weeks and even up to one year. With Corvidae, in France, one single broadcast sometimes has an effect lasting from a few days to a few months.

In these different experiments, it was observed that reactions to the signals of distress and alarm were specific (starlings and herring-gulls), while several species of Corvidae particularly in France, reacted to the signal of any related species.

As regard the Corvidae, if the broadcasts took place in broad daylight, the distress call, instead of causing a flight reaction, attracted the birds to the loudspeaker. Birds can be called from great distances (sometimes several kilometres) by such a procedure**.

It is to be noted that in all these field experiments, the quality of the broadcast was not very good, since considerable amplification was used, (with amplifiers of 20–30 W, generally speaking, giving 90–110 dB at 1 m from the loudspeaker). This did not result in a very faithful rendering of the signal. But, in spite of distortion and the increase in background noises, the reactions were very good. It was possible to prove the relative semantic value of different signals by such distinctive taxic reactions***.

(iv) Food behaviour

In addition to parent–young reactions in regard to food requirements, Frings et al.[55–58] experimented on a specific sound signal relating to food among herring-gulls. This signal has a marked attractive power for other members of the species within range of sound. When recorded and retransmitted, this signal attracted birds from afar. They flocked around the loudspeaker, where they waited for 10 to 15 minutes before flying off again. Such signals, when used on the seashore from powerful loudspeakers, can attract herring-gulls from 3 to 5 km.

The joint use of alarm and distress signals and call for food (when they exist for the same species) can be used as an effective means of protecting certain areas from damages caused by birds.

For certain carnivorous birds, especially nocturnal birds, the noises or cries of their prey act as a signal. A hunting technique based on this idea uses the attraction of owls, foxes and screech owls by the recorded cries of mice and rabbits, broadcast through loudspeakers. Certain carnivorous fish in African rivers (Senegal, Niger) are attracted by the noise made by herbivorous fish when eating, and the native fishermen use a frictional noise-making device producing the same type of sound. This is the "cotio-cotio" of the Senegalese, which is also found in the Sudan, Nigeria and as far east as Lake Tchad[33].

The most curious case of acoustical signals in relation to food behaviour is that of an African bird, the Honey-guide, which will be discussed on p. 724.

** Experiments carried out in Canada in 1954[129] to induce flight in ducks by means of sounds of great intensity from sirens giving comparatively pure tone were entirely negative. On the other hand, in West Africa, the Africans make a successful use of violent noises uninterrupted during the night to chase away the Quelea from the colonies where these small birds gather in millions[32].

*** See Colloque sur la "Protection Acoustique des Cultures" edited by R. G. BUSNEL and J. GIBAN, *Ann. Ephiphyties*, Paris 1960, 245 p.

Among insects, a flight reaction caused by an acoustical signal was the subject of experiments on the owlet moth and moths of the Phalaenae families[141]. These moths, as soon as they hear certain acoustical signals, generally supersonic, drop to the ground if they are in flight or flatten themselves against the ground. It was suggested that this was a flight behaviour to enable them to escape from bats. Without extending this interpretation, until further experimental checks are made, we might note that these moths can hear ultrasonic signals of a frequency range which overlaps that of various bats, reaching 250 kcs.[139,140].

7. A few Borderline Cases of Specificity and Inter-specificity of Sound Signal Semantics

(a) Specificity

The sound signal, in numerous species, appears to have a very specific release value. In some races, or even geographical sub-races, with which preliminary experiments have been carried out, the sound signal would even seem to be so specific that a very significant value might almost be attributed to local dialects. Frings' use of the distress signal of starlings to chase this gregarious bird from the roosts in certain towns of the United States and of herring-gulls' calls led certain European experimental stations to use copies of these American signals for the same purpose. Such was the case in Holland and in England. But, contrary to all expectations, these signals of American origin did not cause the expected reaction, *i.e.* desertion of the birds' usual concentration areas. From a systematic point of view, however, these were of the same zoological species, *Sturnus vulgaris* and *Larus argentatus*. Dutch scientists[66], working this time with recorded distress calls of birds captured locally in the Netherlands, obtained the desired reaction. This counter-experiment, confirming the specificity of the signal, would be worth following up by analysis of the signals, in view of discovering their physical differences and perhaps by this means explaining the different results. For the human ear no appreciable variations can be distinguished. In tests on jackdaws, in some regions in the south of France, the same failures resulted but it has not yet been possible to show by broadcasting signals recorded with local birds that the latter produced the expected reactions either.

It is more easily understandable that, as regards different species, the signals of one species have no informative value, and the same authors have pointed out other species of sparrows, habitually associated with starlings or Corvidae in colonies or feeding-grounds, which do not react to the distress signals of the latter.

(b) Interspecificity

On the contrary, for other species of birds, it has been shown that the signals of one species could be perceived as such, that is to say that their semantics, being integrated, produce phonoreactions of the same kind in one or several other species.

These phenomena may have several causes.

(1) A mutual conditioning brought about by fairly close geographical association which is found in different populations.

(2) Non conditioned interspecific response.

(3) Existence in the signal of common reactogenic characteristics to several species.

References p. 107

The experimental studies of interspecificity of sound signals is very recent, although observations on wild animals carried out by hunters have supplied some data on this subject.

The oldest known example of the interspecific value of acoustical signals is that represented in a mural fresco in a Theban necropolis of the 18th dynasty, the tomb of Menna, where a hunter holding the feet of birds of one species (cormorant) is depicted; birds of another species (Pintail type ducks) are attracted, towards the source by the cries of the former, near enough for the hunter to throw his boomerang (Fig. 53).

Fig. 53. Fresco from the tomb of Menna (Drawing from a Hassia plate taken from "Peinture Egyptienne", in the "Arts du Monde" serie, Hachette, Paris).

Numerous examples can be given of the interspecificity of alarm and distress signals among birds—the alarm call of the jay or the magpie, which makes sparrows take to the air and flee, and the distress call of the former two species, which attracts Corvidae.

Head erected, some penguins, utter certain calls which can be discerned as signals by others and provoke vocal reactions in neighbouring colonies of other species of sea birds, such as the braying of *Sphenisens* the trumpeting of *Apterrodytes* or the song of *Pygoscelis*[117] (a group phenomenon called "sonic panurgism"). The same phenomenon can also be observed in a colony where different species are grouped, in a hen-house for instance.

(*i*) *Mutual conditioning by geographical association*

In the course of preliminary experiments on the interspecificity of acoustical signals with *Corvidae* carried out in France, it was found that a positive phonotaxis could be obtained with Corvidae: (the jackdaw, *Corvus monedula*, or the rook, *C. frugilegus*, or the carrion crow, *C. corone*) with the aid of acoustical signals from one of the two other species, and this even when the signals were physically different. These reactions show a semantic interspecificity in the signal, and this conclusion has been further strengthened by a study of the reaction of the three species to the calls of the American crow, *C. brachyrhynchos*, of North-East American origin. Tests have also been carried out on the reaction of this latter species to signals from French Corvidae[31,58,59].

The signals of the American crow were the "assembly calls" uttered by birds having spotted a bird of prey or a cat. The French signals were the "distress calls" of a bird held in the hand. The two signals are unmistakably different from the point of view of physical acoustic and cannot be confused by a human ear. Retransmissions made in the two countries with an identical message-repeater and the same loud-speakers were carried out outdoors on groups of birds at distances ranging from 200 to 1200 meters from the transmitter.

In France, the experiments took place in winter, at close range, with mixed groups of Corvidae on the feeding ground, the transmissions lasting from 1 to 2 minutes. The reaction was as follows:

Shortly after noticing the transmission, the birds flew off, wheeled and then, forming a fairly compact group high up, came towards the origin of the signal. Then, uttering sharp cries, the birds circled in a flock above the source of the sound. After this, little by little, the group became looser and looser and the birds scattered. These reactions lasted from two to five minutes.

However this reaction is variable, according to outside factors and to the biological status of the group. Its intensity can be measured: $+$ = positive; $-$ = negative; o = null and intermediates (Table 6).

TABLE 6

RESULTS OF TESTS CARRIED OUT ON REACTION OF THE FRENCH SPECIES TO THE ASSEMBLY CALL OF AMERICAN CROW AS COMPARED TO THE DISTRESS CALL OF JACKDAW

Notation of reaction	Assembly call of American crow		Distress call of jackdaw	
	Number of tests	Reaction %	Number of tests	Reaction %
	52		50	
$+++;++$ or $+$		31		40
$+-$		23		16
$-$		11		20
o		35		24

The results obtained show that the groups of French Corvidae reacted to the assembly call of the American crow as to a call of distress of the jackdaw. Details of the notation show, however, that the American call is slightly less effective in provoking a positive phonotaxis than the French one.

In America, "calling tests" (that is experiments carried out when the birds were not, at first, in sight), gave the following results:

(1) With the call proper to the species:

(a) Maine, in summer; out of 30 tests, 27 gave positive results, *i.e.* attraction of groups of 2 to 30 individuals, nearly up to the source of sound, in 1 to 5 minutes;

(b) Pennsylvania, in winter; out of 50 experiments, 44 gave positive results *i.e.*: attraction of groups of 5 to 200 individuals.

(2) In both these regions, at the same seasons, the signals of the French species seemed to be entirely ineffective. When, however, the same experiments were repeated

References p. 107

in Pennsylvania, not in winter but at the beginning of June, with the same signal of the jackdaw, it was found that the American crow reacted as if to the signals of its own species, whereas the same experiments, resumed a few days later in Maine, confirmed the negative character of the reactions already observed in summer.

The analysis of results obtained in the United States points to a different reactivity in birds according to the season and the place and has led to the following interpretation, based on the migration of populations. It is believed that the population which breeds in Maine migrates in winter, particularly to Pennsylvania, where it replaces the population of the same species which, at this time of the year, has left for the Southern States (Florida). This latter population then associates with other Corvidae, whilst the population of Maine remains practically isolated.

The isolation of the population from Maine and, on the other hand, the association of that from Pennsylvania with other Corvidae might explain the differences in reactions to signals of the French species. This hypothesis of association and of interspecificity of acoustical signals is supported by the fact that the herring gull (*Larus argentatus*), which inhabits Maine in the same surroundings as the *C. brachyrhynchos*, reacts to the calls of the latter, and also to the signals of the French species.

Similarly in France, in the course of experiments in Normandy, where the carrion crow consorts with the herring gull on the beaches and neighbouring localities, the carrion crow seems to react by a positive phonotaxis to the food calling of the signal of the herring gull.

The absence of reaction of the *C. brachyrhynchos* population of the North-East of the USA to signals of the French species does, therefore, give a well-defined semantic value to the specific signal. It leads to the conclusion that, with the segregation of populations, certain acoustical signals may have either an interspecific value or a specific value for different species or sub-species of birds that associate. The comprehension of the semantics of the message is probably due to an acoustical conditioning, helped by the natural association of several species in the same biotope*.

It is of interest to quote another series of examples taken from research work on positive phonotaxic reactions of French Corvidae to acoustical signals of jay, starling, dog and rabbit. The results are given in Table 7, with reactions measured and recorded as they are in Table 6, the values are expressed in percentages of experiments. They can be compared with the values given in the case of various sounds and imitations which are annexed to Table 6.

(ii) Non-conditioned interspecific response

In the food relationship of prey to predator, the sound signals or the noises of the prey can produce a positive phonotaxis in the predator receiver, which is utilised as a method of capture. Apart from invertebrates, these reactions do not appear to be innate, (*e.g.* squeaks of mice drawing the attention of a fox, barking of a dog attracting crocodiles etc.).

---

* It should also be pointed out that, with French Corvidae, the distress call of the jay and the starling cause the same positive phonotaxis as that observed in the case of their own signals. On the other hand, the effect on starlings of distress calls given by Corvidae is only a non-directive agitation. Finally, in most of the experiments, it was noticed that domestic cows and horses were soon attracted by distress calls from crows, jays and starlings. It has not yet been possible to discover the basis of this phonotaxis.

TABLE 7

REACTION OF FRENCH CORVIDAE TO RECORDED DISTRESS SIGNALS EMITTED BY DIFFERENT SPECIES

Reaction	Basic distress signal of jackdaw (in %)	Distress signal (in %)			Barking of dog (in %)
		Jay	Starling	Rabbit	
+ + + +; + + +; + +	60	68	42	5	0
+	25	23	26.5	20	0
+ —	5	4.5	5	15	27
— or 0	10	4.5	26.5	60	73

Reaction	Imitation of crow's call made by man (in %)	Use of a lure imitating distress signal of hare (in %)	Hand rattle (in %)
+ + + +; + + +; + +	18	45.5	0
+	37	36.5	0
+ —	18	9	0
— or 0	27	9	100

The example of symbiotic relationship which presently may be considered as unique and which is one of the most important to mention in relation to this aspect of the interspecificity of acoustical signals is that of the African bird *Indicator* or honey-guide. By means of its optical, and more particularly its acoustical signals, this bird produces a directed phonotaxis of species such as the rattel and the baboon. This will be developed on p. 724.

(iii) Interspecificity resulting from signals having physical reactogenic characteristics common to several species

In this case, signals given by one species and perceived by another produce in the receiver a null or negative phonoreaction (phonoresponse or phonokinesis or phonotaxis and sometimes both). Such phenomena have been described, particularly among invertebrates and some lower forms of vertebrates.

It may be thought that, some of these signals have a similarity of physical characteristics corresponding to those of the signal as a whole, while others, even though they seem very diverse more probably contain a physical reactogenic character common to all. It is also possible to assume a slight discriminatory capacity in the central nervous system of the receiving agent, possibly linked with a strong momentary motivation, since these experiments have, for the most part, been carried out in connection with sexual behaviour. It seems that this problem is important and, if it has been borne out by observations in regard to "natural" signals, *i.e.* signals given by an animal of another species, only the use of physical techniques strictly within the acoustical field has enabled comparison between one or more physical and reactogenic factors common to several signals, to be made.

Experimental observations of natural signals have shown phonoresponses between different species of Orthoptera, Acrididae (*Chorthippus*) among themselves or with Tettigoniidae (*Ephippiger*), also between Tettigoniidae, whose signals are sometimes physically very different[29,84,142].

It has even been possible to obtain, in the laboratory, different types of phono-

References p. 107

taxis, including the meeting of individuals of the opposite sex and of different species. In some cases they have resulted, failing other discriminatory signals, in attempts at copulation. In other cases, after the phonotaxic phase, specific differentiation signals come into play. Although the results are only experimental, obviously the problem of the segregation of the species and that of the isolating mechanism arises. It is probable that, for some animal populations which are akin from a morphological point of view (species and sub-species), the fact of being in strictly defined geographical territories contributes, with the genetic factor, to the isolation and integrity of their specific maintenance.

For some species, even very similar, this problem has not been solved. Indeed, as a result of the work of Jacobs[72] and more particularly Perdeck[111], who produced hybrids of *Chorthippus brunneus* and *Chorthippus biguttulus*, the parents of which are capable of reacting to the signals of each of the species*, it was shown that neither the female *brunneus* nor the female *biguttulus* would react to the song of the male hybrid but that, on the other hand, the female hybrid reacted to the signals of the parents as well as to the hybrid males.

This shows, in any case, that for some at least of the species in question, the acoustical signals have an essential reactogenic value in their behaviour and that there is, moreover, a common reactogenic character in these different signals, capable of producing these interactions. It is probably the marked physical resemblance between the sexual noises of the two species of fish, *Bathygobius soporator* and *Chasmodes bosquianus*, which induces the former to have oriented reactions to the experimental transmission of signals given by the latter[132,134].

Observations of artificial signals, naturally led to research into the reactogenic factor common to these different signals of various animals. Successive analyses give the impression that, among other factors, the existence of a system of groups of impulses, or of impulses that can be described as transitory (the rapid setting-up of a wavelength of a certain strength), could produce these interspecific reactions in various species.

In certain species the deformation of the natural signal by successive copying or the synthesis of similar signals (lures) confirm this freedom of reaction to a non-specific signal, provided that it has the necessary physical characteristics (in particular intensity and rhythm)[21, 23, 25, 84].

It has still to be decided whether these artificial signals do, in fact, produce, in comparatively primitive animal forms, reactions which might be considered as "supra-normal". In all cases this absence of close specificity of signals raises the question of the discrimination of reception, which seems very slight, at least in the cases studied.

But the validity of these points of view will be confirmed only if it can be shown that the animals concerned do not distinguish the natural signal from the synthetic signal, when these are used at the same time, with the same strength and in conditions in which if the animal made a choice it would take on a truly functional value.

(c) Information given by certain signals

Researches in this sphere have been undertaken for some years, with the object

---

* This author came across hybrids of these two species in the natural state in the Netherlands.

of determining what is the essential reactogenic character of signals. Two quite detailed, because little known, examples of this will be given, one relating to birds' signals and the other to Orthoptera.

Before enlarging upon various aspects of this, it might be well to cite one example which well illustrates this notion of "information". It is the case of a visual signal, studied on birds by Lorenz and Tinbergen—a figure-target launched into the sky. It has vaguely the form of a bird with extended wings and, according to which side one looks at it, it appears to have a long neck and a short tail or a long tail and a short neck. When made to fly in the direction in which it looks like a "long-necked" bird, it causes no reactions on young Gallinaceous birds on the ground; if launched in the opposite direction it causes an immediate reaction of fright.

Therefore, it is not to the shape as such that the birds react since it is the same in both cases, but to shape as it is recognized through the direction of movement.

(i) Work on birds[35]

Following the same general idea, that is an inversion of the message but in this case of an acoustic nature, quantitative studies on the reactions of birds to acoustical signals of distress, according to whether they were emitted in the correct sequence or backwards, were made with three species of Corvidae: crow, jackdaw, and rook. No distinction was made as to whether they were the signals of the same species or those of jay or starling (see the chapter on Interspecificity, p. 87). Table 8 clearly

TABLE 8

POSITIVE PHONOTAXY OF CROWS TO A DISTRESS SIGNAL EMITTED NORMALLY OR TEMPORARILY INVERSED (100 TESTS)

Codification of reaction	Number of positive reactions to the basic signal of					
	Jackdaw		Jay		Starling	
	normal	inversed	normal	inversed	normal	inversed
+++; +++; +++ ..	60	24	68	40	42	45
+ ...	25	14	23	25	26,5	25
+— ...	5	24	4,5	30	5	10
— or 0 ...	10	38	4,5	5	26,5	20

shows the slight reactogenic sensitivity of these species to the temporal aspect of the message. This is obviously bound up with the physical structure of the signal, which all have some similar characteristics.

Another experiment was carried out with a double goal: (1) To see if each call of a series (in this case the alarm signal of a jackdaw) was equivalent in its effect on the bird's behavior; and (2) If there was any redundance in the bird's signal. At first a series of 40 successive natural calls was used in their original content. This series (called C 2) lasted 1 mn and was tested twenty times, as can be seen in Table 9. This serie C 2 was then cut in two parts containing an unequal number of calls (23 for C 3 and 15 for C 10) each was repeated so that the end signal lasted 1 mn as the original C 2. Then these two sequences were again cut: C 3 in two, giving C 5 which was made up of 10 calls, and C 7 of 12 calls. C 10 was divided in three: C 13 made up of 12 calls, C 9 of 1 and C 15 of 2. Each sequence was as before repeated

References p. 107

TABLE 9

SUMMARY OF RESULTS OBTAINED ON THE POSITIVE PHONOTAXY OF THE CROWS TO THE JACKDAW DISTRESS CALL DIVERSELY CUT UP BEFORE THE EMISSION (see Fig. 54)

Reference No. of the signal	No. of original calls in each signal	No. of repetitions to obtain 1 mn	Summary of results (%)	
			From very positive to mildly positive	From indifferent to negative
C 2	40	2	85	15
C 3	23	3	90	10
C 10	15	6	60	40
C 15	10	5	20	80
C 7	12	7	55	45
C 9	1	41	9	91
C 13	12	8	83	17
C 15	2	40	64	36

as many times as necessary to last 1 minute. Fig. 54 illustrates the cuts made in the original signal.

C 3 and C 13 have thus proved to be as reactive for the birds as the original call C 2 but while the most reactogenic factor obtains as much as 90% positive phonotaxie, the less reactogenic ones have only 9%, therefore different parts of a signal have quite different semantic meanings.

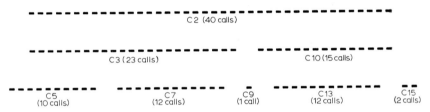

Fig. 54. Different experimental cuts in a distress call to test the reactogenic power of small parts of a sequence (see Table 9). C 2 was cut in two, C 3 and C 10. C 3 was again cut in two, C 5 and C 7, and C 10 was cut in three, C 9, C 13 and C 15.

To further the comprehension of these factors a part of C 13 was used to determine the value of the intervals of silence in a sequence. The one used was $1/3$ of C 13 which had obtained 63.5% positive reactions when emitted without any interval between the sequences (each lasting 4.5 sec). When each sequence was separated by 4.5 or 2.5 sec of silence the positive results fell to 36.5 and 30% respectively (see Table 10 and Fig. 55).

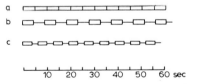

Fig. 55. Different intervals between sequences of the distress call of the jackdaw, tested for reactogenic power.

It appears certain, therefore, that information is associated more with some parts of the calls of a sequence than with other and that the interval separating the calls has also a semantical value.

TABLE 10

IMPORTANCE OF THE INTERVAL BETWEEN SEQUENCES OF AN DISTRESS CALL FOR ITS REACTOGENIC POWER ON CROWS

Nature of the reconstruction	Results (in %)	
	Negative results	Positive results
a. Transmission without interval	63.5	36.5
b. Interval of 4.5 seconds between each sequence of 4.5 sec	36.5	63.5
c. Interval of 2.5 seconds between each sequence of 4.5 sec	30	70

Of course, these results are, as yet neither final nor complete and they are only given here as an indication of an analytical method which seems interesting. The same writers adopted the same procedure with Orthoptera and with a relatively more immediate success, but it is quite certain that the more developed behavior of birds is more complex to define than that of insects.

(ii) Work on Orthoptera

These researches bear on a series of experiments undertaken on a small field grasshopper, *Chorthippus brunneus*, in order to determine the influence of the relationship $\Delta i/\Delta t$ of a signal evoking a phonoresponse[36].

Some phonotaxic or phonokinetic* reactions of Orthoptera, male or female, can be released by artificial acoustical signals, emitted by whistles or frequency generators[30,84]. These studies led to the conclusion that, within certain limits, frequency was not the major reactogenic factor for the receiving insect. The latter was more sensitive to the wave-form of the signal and particularly to the transient at the beginning or end of the sound**. We define a transient as a quick rise from intensity zero to maximum intensity or reversed. The effectiveness of these transients holds true if the artificial signal is at least 30 dB above the natural intensity of the insect's emission.

It has been shown that square waves, or short clicks, in the limits of indicated intensities, are sufficient to cause the reaction, whereas signals of the same intensity, but in which this parameter increases very gradually and slowly, are ineffective.

It is therefore the suddenness of change in intensity which produces the best results. In other words it is a question of the time (speed) of the setting up of the signal, or its return from the intensity reached to zero. (Fig. 56).

---

* For a definition of these terms see pp. 103, 104 and 105.
** This characteristic of the wave form of the signal, which is connected with the well-known "Gestalt" theory of psychology, seems indeed to recur in other types of acoustical signals, and especially those used in echolocation (see Chap. 8). In fact, the frequency of these signals is very variable according to the species: 10 to 160 kc/s in certain marine mammals, 10 to 120 kc/s for bats, 7,000 c/s for *Steatornis* and 4,000 c/s for *Callocalia*. They are effective for all these animal types but they have in common the same wave form, *i.e.* short clicks, essentially of a transient type.

References p. 107

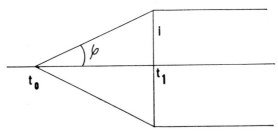

Fig. 56. Diagram of the slope of a signal, corresponding to the timed installation of constant amplitude.

Analysis of these signals leads, in fact, to the measurement of the relationship $\Delta i/\Delta t$, the angle ψ varying according to the value of Δi and of Δt. It can be expressed as: $\psi = 1\sqrt{i^2 + t^2}$.

The researches have, therefore, been carried out by using signals in which the angle ψ varied.

The experiments were carried out on male *Chorthippus brunneus*, with signals in which $\psi = 90°$, with frequencies between 50 and 20,000 c/s, and a sound level around 80 dB. The insects reacted by chirps which were supposed to be phonoresponses*** when taking place within a very short time (0.7 to one second) after the stimulus. These phonoresponses then constituted the test of psychophysiological sensitivity.

Since no particularly preferential frequency was found, an arbitrarily chosen signal at 10,000 c/s was used with intensities varying from 65 to 90 dB at the insect's level.

The insects were isolated in a plastic cylinder with a net base to prevent any absorption or reflection of sound on the walls, the cylinder being placed on the axis of the loudspeaker. All the laboratory tests were carried out on individually tested specimens, at an average temperature of 25°, the ground noise being about 45 dB. The signals were transmitted by a variable slope sound generator.

It is possible with this generator to obtain symmetrical or asymmetrical signals simply by adjusting knobs. The time necessary for the sound to obtain its maximum amplitude was from 2 msec to 1 sec, and of course the value of the angle ψ varies with this time of onset and with the intensity of the signal. The length of the signal used during these experiments was from 0.1 to 2 sec. (Fig. 57.)

Characteristics of the insect's reaction

The insect does not react to the stimulus when engaged in certain biological activities, such as cleaning the antennae, the palps or the eyes, defecation, or movement from place to place.

When not occupied in one of these activities, its reaction to the stimulus consists of a normal sound emission which might be considered as having the value of a signal for another insect of the same species. In this animal, the reaction, from a neuromuscular point of view, corresponds to frictional movements of the hind legs against the elytra; it is not comparable to a series of uncoordinated movements which

---

*** For a definition of phonoresponse see p. 104.

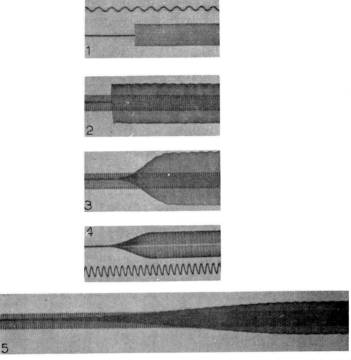

Fig. 57. Various oscillograms setting out the variants to t_1, as a function of i (variable between the examples given), at the time of the setting up of a signal of 10,000 c/s. The overall duration of the signal varied between 0.1 and 1 second.

    1 to t_1:   0   Basic time: 200 c/s
    2 to t_1:   2 msec   Basic time: 450 c/s
    3 to t_1: 60 msec   Basic time: 450 c/s
    4 to t_1: 130 msec   Basic time: 50 c/s
    5 to t_1: 300 msec   Basic time: 450 c/s

(From Busnel and Loher[36].)

incidently involve the production of a sound. It is, however, important to note that the stimulus may cause several successive phonoresponses, apparently uncontrolled in their time sequence (on first analysis) and which follow without any rhythm in the course of the few seconds after application of the stimulus. It is possible to count up to 8 of them. In the experiments, only one positive response was counted, whatever the number of emissions produced by the insect as the result of one stimulus. (In the acoustical rivalry behavior between two males, similar phenomena have been noted.)

With symmetrical signals the following results we found:

Variant: Δt (These signals have an identical Δt value at the beginning and end of the signal.) As an example, for a Δi value of 80 dB, the results shown in Table 11 are obtained.

A quick analysis of Tables 11 to 13 shows:

1. (a) Δt has a great influence on the percentage of positive phonoresponse (86% at or around 0.1 msec—0% at or above 300 msec); (b) The shorter Δt the better the result, since at 2 msec already the percentage of positive results has fallen from 86 to 40% in Table 11 and from 71 to 34% in Table 12.

References p. 107

TABLE 11

Δt value in msec	No. of stimuli	% of positive responses
< 0.1	345	86
2	964	40
20	281	48
60	178	36
100	155	16
130	61	15
220	95	9.5
300	30	0

An other example, in which $\Delta i = 70$ and 90 dB, is shown in Table 12.

TABLE 12

Δt value in msec	No. of stimuli	% of responses	Δi
< 0.1	231	71	
2	301	34	70 dB
20	174	17	
60	29	3	
2	297	42	
20	112	64	
60	106	51	90 dB
100	132	30	
130	51	2	

Variant: Δi. In this example, the factor $\Delta t = 2$ msec.

TABLE 13

db intensity	No. of stimuli	% of positive responses
65	40	0
70	301	34
80	964	40
85	241	45
90	297	42
95	55	20
100	45	9.5

2. If one takes 30 to 40% of positive response as reference it can be seen that:
for 70 dB it is reached at a Δt value of 2 msec (34%),
for 80 dB it is reached at a Δt value of 60 msec (36%),
for 90 dB it is reached at a Δt value of 100 msec (30%).
Therefore, a variation of Δi influences the percentage of positive responses.

The statistical study of the results gives the following values:

A χ^2 test was applied to these figures.

Is there a significant difference for $t = 2$ milliseconds ? At 70 dB and 90 dB χ^2: 3.293— P significance: .10 (non significance .05). For $\Delta t = 20$ milliseconds at 70 and 90 dB χ^2: 71.49. Very significant difference for all values of P. For $\Delta t = 60$ milliseconds, the test is useless. The difference (70 dB = 1/29) 3% with (90 dB = 54/100) 51% is, therefore, significant.

The analytical results of Table 12 taken as an example are condensed in Table 14.

TABLE 14

	ratio + Total		% +		T^2	Conclusion*
	70 dB	90 dB	70 dB	90 dB		
$\Delta t = 2$ ms	103/301	123/297	34	42	** at P = .10	NS
$\Delta t = 20$	26/174	72/112	17	64	*** at P = .01	S
$\Delta t = 60$	1/29	54/106	3	51	*** at P = .01	S

* NS = not significant. S = significant.

For the values of Δt between 2 and 60 msec has the intensity an influence upon the % of positive responses? In other words, is a better result obtained with 90 dB than with 70 dB. For $\Delta t = 2$ msec—no influence on the intensity (χ^2:NS); for $\Delta t = 20$ msec—distinct influence on the intensity (χ^2:S); and for $\Delta t = 60$ msec—distinct influence on the intensity (χ^2:S). In other words, in these two cases, better results were obtained at 90 than at 70 dB.

These results obtained with symmetrical signals, the beginning and end of which had the same slope of attack, led the way to researches on asymmetrical signals having a single abrupt variation, to see if this was sufficient to cause the reaction. In this case either the beginning or the end of the signal had a slope of attack at or above 220 msec, which has been proved to have no reactional value.

The experimental results obtained in these tests with a frequency of 10,000 c/s and an intensity of 80 db are shown in Table 15.

TABLE 15

Slope at beginning of stimulus Δt	Slope at end of stimulus Δt	No. of signals	% of positive responses
2 msec	220 msec	59	42
2	300	34	41
2	340	24	42
2	500	35	66
500	2	24	21
220	2	12	17

On examining this table it can be shown that the percentages of responses obtained with an asymmetrical signal (a single transient), starting abruptly, are of the same order as those obtained with a symmetrical signal (two transients). However, when the abrupt variation is at the end of the signal, the percentage of response, is considerably less although not completely absent.

(d) Discussion

The results obtained by these experiments define the limits within which the wave form of a signal $\Delta i/\Delta t$ which is equal to the tangent ψ is effective. They also confirm the fact that in the examples and conditions given above, and even with high intensities, a sudden rise in intensity is much more effective than a gradual one. In other words the smaller the angle ψ the poorer the reaction.

The acoustical stimulus produces in the animal used for the experiment an overall psychomotor reaction which is the sum and the result of neuro-muscular excitations, central in origin but governed by peripheral reception. Invertebrates, in this case *Ch. brunneus* (and also *Ephippiger*[30]) can be considered as organisms with less self-control, *i.e.* with less regulating inhibitory mechanisms and hence a lesser degree of freedom than the higher vertebrates. This is confirmed by the fact that their reactions are of the forced or tropistic type, of which there are few or no examples among the higher vertebrates (in particular in Homeotherms). This is probably the origin of their gross responses to stimuli. In higher vertebrates, reactions are confined to the neuro-muscular axes.

These total reactions to acoustical stimuli are comparable with the results obtained on neuro-muscular preparations, using electric, mechanical or chemical stimuli. It might be worth while to summarise them briefly here.

The excitation, to be effective, requires a variation of the intensity of the stimulus, and it is most important that this variation should be abrupt. If it is a case of the closing of an electric circuit, the current must start abruptly, *i.e.* reach its full strength in a very short time. If the setting up of the current is gradual, its effectiveness is lessened or disappears.

Below a certain intensity (liminal intensity) or above a certain intensity, variable according to the species studied, the reaction cannot be noticed any more. These are called the upper or lower threshold of excitation. It is possible that the upper threshold is due more to a too great distortion of the signals when the intensity is too high, or to a nociceptive action than to a true psychophysiological threshold.

Within the limits of time and of intensities indicated in this type of experiment, it appears that the mechanisms which come into play correspond to those which are subject to the general laws of neuro-muscular excitability.

It might be interesting to point out that by the way it reacts, it seems that for insects the stimulus as described above is perceived as having an information value. This is only effective as in the experiments just mentioned, if it corresponds to the terms of the laws of neuro-muscular excitability. Although the case may be different for other animals or in other types of experiments. In our work we have found that the effective part of this kind of signal which provokes a phonobehavior in insects can be defined in terms of the relationship $\Delta i/\Delta t$. The best results were here obtained with the biggest Δi for the smallest Δt.*

In support of these results, as well as others which are not enumerated here, but which can be found in various publications[24,26], it is worth while recalling that human speech is exclusively a sequence of transients insuring the transmission of the semantic content of the message as brought to light by the theory of information. Carl Stumpf (1926) showed that a sound with constant pitch and intensity loses part of its character if the transient at the beginning of the sound is cut out by some process or another. Recording on tape-recorders prove this, and all that is necessary is to cut off the portion in question with a pair of scissors. There then remains, after a sudden attack, the "body" of the sound producing object, the characteristics of which are no longer variable in time. This author showed that instrumental sounds, thus amputated, become very difficult to distinguish from one another: the sound of a flute can be mistaken for a tuning-fork, a cornet for a trumpet, a clarinet for an oboe and a

---

* These results are now confirmed electrophysiologically, by AUTRUM (1960), Phasische und tonische Antworten vom Tympanalorgan von *Tettigonia viridissima*, *Acustica*, *10*, No. 5-6, 339-348.

bassoon for a cello; the distinction even becomes impossible between qualities of sound as different as those of a cornet and a violin, a hunting-horn and a flute.

As Winckel[147] wrote in a book, little known to bio-acousticians, "Klangwelt unter der Lupe" (Berlin, 1952), tone is not only determined by the appearance of the spectra of partials in the stationary state (formantic structure), but an important part is played by the transient of attack and extinction of the sound vibrations.

(e) A work on fish

Similarly it is worth while drawing attention to some experiments with a fish, the gobiid fish, *Bathygobius soporator*, of which Tavolga made an interesting study in 1958[133, 134]. This fish has a taxic reaction to sounds during the sexual period. This applies both to males and females. The author, after having obtained the reactions in the laboratory, confirmed them in the sea.

The natural sound signal of the male has an average of 225 msec, and is uttered in series with intervals of about 100 msec. The frequency of maximum amplitude of this grunting is between 110 and 150 c/s. This fish reacts in the same way to artificial sounds consisting of impulses obtained with a sinusoidal wave of 200 c/s, the duration of the impulses being 75 msec. On the basis of this test and by varying the physical characteristics of the signal, the following results were obtained:

Duration of the artificial signal: Normal reactions are obtained for durations between 75 and 150 msec, *i.e.* 1/3 to 2/3 of the mean duration of the natural signal. Yet a positive reaction can be obtained with a duration of 25 msec or less, if the rhythm of transmission is varied. Above 500 msec, the signal becomes ineffective.

Frequency of artificial signal: From 100 to 300 c/s, a positive effect is obtained, which can, moreover, be apparent up to 500 c/s. Beyond this the fish no longer responds. It reacts therefore to wavelengths between 2.80 and 14 metres.

Rhythm of artificial signal: This is the number of signals per unit of time. It can be varied by modifying the time between two impulses, which in the natural signal is 100 msec. A positive response is always obtained between 0.1 and 10 seconds. Shorter intervals of from 0.05 to 0.01 sec had variable effects, depending upon the pulse duration. With a duration of 50 or more msec, there were few or no positive responses at these short intervals. The short intervals, however, were effective at a pulse duration of less than 25 msec.

Groups of pulses can be adjusted and spaced so as to produce a positive response; *e.g.* pulses of 10 msec each, in groups of ten, with 10 msec intervals between them will elicit a positive response. Spikeform signals can be used as well as sine and square waves. The component and harmonic frequencies become immaterial in these conditions.

The effectiveness of a signal, in this fish for example, is not impaired by a wide range of variations in the physical parameters of the signal. If for the sake of discussion one assigns the mean frequency of 200 c/s to the natural signal (200 c/s being the most effective frequency for artificial signals) knowing that the mean duration of a natural grunt is of 225 msec, it is easy to calculate that this sound will be made up of 45 oscillations, while only 5 to 15 will make up the most efficient artificial signal (frequency 200 c/s duration 25–75 msec). This raises the problem of the tonal perception in the fish, studied in the classic conditioning methods which have not been up to the present very concerned with the variations of thresholds in terms of what

Moles calls "the thickness of the present". This term is defined as the minimum duration during which 2 consecutive physical events appear psychologically to be simultaneous. It is this which can be approximately determined here by these studies of the variation in the duration of the signal.

The factor which appears to be the most important is that of rhythm, which lowers the thresholds of perception, since a response can be obtained with groups of impulses of 10 msec, representing only 2 oscillations. The interference of various parameters with the information imparted by the signal at the level of the receiver is more complex here than in the case of the insect, and this reaction might lead to interesting studies on the whole of the psycho-physiology of audition and integration of the signal for the fish.

8. On Some New Methods for Studying Hearing and Sound Behaviour

Apart from man who can bear witness to his feelings, the exploration of acoustical sensitivity of animals can only be done by approximative methods; "The reality of perception cannot be ensured by the existence of sensory organs but of behaviour" (Buytendijk).

Electrophysiology and its methods of studying cochlear microphonic potentials, action potentials of the acoustic nerve or encephalic acoustic areas, have added considerably to the sum of our knowledge on animal hearing at the level of the individual receiver, considered as a physiological mechanism. The technique of using permanent micro-electrodes, coupled with the method of conditioned reflexes, has greatly enlarged the problem and it cannot be doubted that this sphere will in the coming years show promising results.

It seems necessary to point out that, apart from these methods, which may all spring from a homogeneous mode of thinking, such as is offered by modern electrophysiology, with all the techniques which make this line of study an end in itself, there are other methods producing results which at the same time cut across and complete them, because they bring in other reactional conceptions.

Although it is not new, it seems necessary to emphasize the importance of the method of conditioned reflexes in the study of animal hearing, where it is most often based on association with an alimentary reflex*. It is useful to recall, for instance, the results obtained by von Frisch on the minnow *Phoxinus laevis* (1936)[61]; Brand and Kellog (1939) on birds[18]; Kriszat (1940) on the mole[81]; Schevill and Lawrence[122] (1953) on the dolphin; and Reinert (1957) on the elephant[116].

By this method thresholds and curves of sensitivity can be studied. The response of the animal really represents the whole sequence: reception of the signal, transmission to the CNS, integration, association with the reflex and motor response that have been learnt. (One example is that of the response of the decorticated dog of Pavlov and Babkine[110] to inverted sound sequences, an experiment too often forgotten.

---

* An interesting application of conditioning to a sound signal was carried out in a Piggery (Animal husbandry department) of the Zootechnical Research Institute at Jouy-en-Josas, (France). The pigs are generally disturbed by maintenance work and the arrival of visitors, for they associate man with the distribution of food. To prevent this disturbing without refusing visitors, a sound signal, a bell, was associated with the distribution of food. The animals are now perfectly quiet during the day whoever enters thus being in the best conditions for fattening. Only at the sound signal are they awakened and they can be seen to get up together and grunt loudly while waiting for their rations.

This result really complicated the problem of cortical projection, as defined by Tunturi and Bremer.)

We shall see that the methods of phonokinesis, phonoresponses and phonotaxis are somewhat different in spirit.

(a) Phonokinesis method

The following definition might be proposed: Reflex motor reaction, innate and unoriented, caused by an acoustic stimulus which has not necessarily the value of a signal in terms of behavior (of animal in a free state). It can be observed in an animal *in toto* or in separate organs.

Examples: I. On an awakened animal. (a) Preyer's reflex, on the ear of mammals: sharp movements of the auricle, and a start at short sounds. Very well observed in the dog, the guinea-pig (studied particularly by Schleidt on numerous rodents[125,126]). (b) Rapid shiver of the vibrissae which form the whiskers of some rodents (rat, mouse, guinea-pig, etc.). (c) Start of the whole body, movement of wings, flight of some insects (Orthoptera, caterpillars, butterflies, etc.), either free on the ground or removed from the ground to eliminate the reflex inhibition caused by contact of the tarsi (Fig. 58)[1,26,95,96,140,141]. (d) Temporary stopping or modification of the rhythm of the respiratory movements of the vocal organs of Anura and of certain lizards,

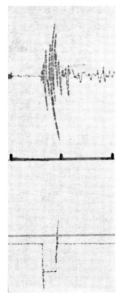

Fig. R-7. Phonokinesis of a male *Nomadacris* (Orthoptera) to a sound of 500 c/s, 110 dB. Above, the reaction of the insect recorded on a galvanometer, below the duration of the signal, in the middle, scale 1/sec (from M. C. Busnel, B. Dumortier and R. G. Busnel[34]).

(*Lacerta viridis*) with movements of the eyes. (e) Opening of the beak in the young blackbird (experiments of Messner on *Turdus merula*)[86].

II. On isolated organs: In some species (insects, mammals) separate organs or fragments of tissue (head, abdomen, thorax, muscles) may react to periodic vibratory stimuli, even when the tympanic receivers are not in the circuit (work of Mrs. Busnel

References p. 107

on insect organs; of Nassonov and Ravdonik[102, 103] on isolated spinal ganglions of rabbits and isolated striated muscles of frogs). This points not only to a muscular contraction of the frog's muscle at between 2,000 and 3,500 c/s, but above all to an increase in the rate of fixation of a vital due.

Phonokinetics enables the study of curves of sensations perceived and integrated, generating reflex motor responses, to be made. As such, and especially on the animal *in toto*, they present a great interest, as they appear to contribute more to the analysis of an overall psycho-physiological reaction than to the measurement of a potential of which no more can be seen than the neuronic reception at the level studied.

(b) *Phonoresponse method*

The phonoresponse is a form of phonokinesis in which the motor response of an organism *in toto* is an emission of sound. The stimulus may or may not have the value of a signal in the natural behaviour of the receiver, but its response has this value for other members of the species which hear it. It can only be obtained with animals equipped for sound production. They must be free, awake and in certain particular physiological states. Studies have been made on grasshoppers[23,29], tree frogs (*Hyla arborea*)[28] crows[27, 31, 35, 54, 56, 58] and turkeys[127]. It has been noticed that the American alligator[5] responds with a roar to sounds produced by a tuning-fork, a vibrating blade, a cello or even a French horn, starting from 57 c/s up to 341 c/s, the best responses being to 57 c/s and its first harmonic. The stimulus can also be a sound produced by an audio-oscillator, the beats of a metronome and so forth.

Phonoresponses enable very precise measurements of the time of integration of the stimulus to be made, according to each psychophysiological state. Recordings

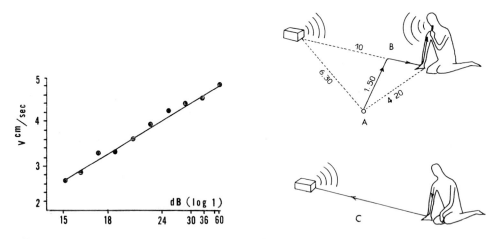

Fig. 59. Speed of movement of *Ephippiger*, female towards the male, in terms of the distance from the transmitter, and hence in terms of the intensity of sound for the receiver (average of 10 experiments). In the abscissa, the logarithm of the intensity expressed in dB. In the ordinate the speed of the insect in progress, in cm/sec (from Busnel and Dumortier[28]).

Fig. 60. Experiment on *Ephippiger*: A. The female is motionless at the beginning of the experiment. B. The female is attracted by the signal of a Galton whistle, in spite of the presence of signals from the male. C. When the whistle stops, the female leaves the experimenter and rejoins the cage of the males (from Busnel and Dumortier[28]).

of the stimulus and response can be made on the same apparatus (tape-recorder). Oscillograph analysis gives very precise time measurements.

(c) *Phonotaxis method*

Phonotaxis is a directed motor reaction of an organism with an acoustical stimulus having the value of a signal for the receiver.

Acoustical stimuli can be tested on animals which, in their natural behavior, move in the direction of a source of sound. Regen[115] as related page 76 was the first to attract an insect with sounds (positive phonotaxis).

(1) Natural sound signals of the species played back, or even modified similar signals (lures) can be used as the stimulus.

(2) Artificial ones (for certain primitive forms: invertebrates and lower vertebrates) can also be used.

This method enables a study to be made of several important psychophysiological phenomena, at the integrating and reactional level. Thus the sensation perceived may be analyzed. For instance:

(a) The study of the psycho-physiological receptivity of the animal: example, reaction of arrival at the loudspeaker of a male robin on hearing its own signal in its own territory; male mosquitoes attracted by a tuning-fork; and alligator attracted by the sound of the French horn or the noise of a bar being struck. Beach's experiments on the alligator[5], is an aquarium demonstrated that, in addition to the phono-response, the animal turned and moved towards the source of sound, taking up an aggressive attitude. In correlation with these experimental observations, it may be added that in Africa, some crocodile hunters use these phonotaxis and that the reptiles approach the signal, sometimes from a distance of one kilometer. Another example is found in fishes (*Protopterae* of Senegal) attracted by the sound of a drum.

(b) The directional acoustic memory.

(c) The duration of the after-effect of the sound stimulus on the directed motor reaction.

(d) The speed of movement in relation to increase in intensity of the stimulus, which grows progressively as the animal approaches the source (Weber-Fechner Law). Example: *Ephippiger* (Fig. 59[30]).

(e) The study of the reactogenic value of a signal in relation to another (hierarchy). Since a signal can be modified physically, the specific sensitivity of the animal to the signal and its variants can be progressively tested. Example: competition between the artificial and the natural signal (Fig. 60).

(f) The specificity of a signal or its interspecificity. Examples of animals attracted by signals coming from another species: Magpie, jay crow, cormorant duck, etc.

Experimental difficulties in studies on phonotaxis

In most cases the experiments can only be carried out on free animals in the natural state. Since very rapid conditioning occurs in animals experimented on, in captivity, the results thus obtained should be treated with caution.

9. Review of our Knowledge of Animal Acoustic Behaviour

In concluding this chapter, a short review of these types of studies may be useful. This will include their historical developments and the major lines of progress which have been achieved and which will henceforth constitute a homogeneous chapter in comparative psychophysiology, such as was probably foreseen by Landois[83] when he wrote his book, "Tierstimmen" in 1874.

"Man has a great power of speech, which is to a large measure vain and false. The animals have little, but that little is useful and true, and a small and sure thing is better than a great lie." (*dixit* Leonardo da Vinci).

This conception of "animal speech" is very old and man has always been intrigued by these "languages" and has wanted to have access to them, as is related in the myths of Orpheus, of King Solomon, the sphepherd Opien and works as ancient as those of Aristotle.

Man undoubtedly succeeds in interpreting some of the songs or cries of animals. He sometimes imitates them with sufficient precision to deceive and attract them. The old technique of decoying birds is probably as old as the first civilisations. Already in a 16th century book, the French surgeon Ambroise Paré showed the interest man took in "animal language", quoting 35 different words characterising the "speech" of 35 animal species. In the course of the last hundred years, special importance has been attached to the study of these problems. As an example, up to 1960, more than 1,752 publications appeared on insects alone[57]. On one of them, the *Acherontia atropos*, the death's head hawk moth, about 60 were counted from the time of Réaumur to 1939.

If, between 1920 and 1945, important new developments were achieved, it is especially since this time that remarkable strides have been made in this sphere.

It is difficult to make a choice of what can be considered the "most outstanding" work of the contemporary period, and this choice could be disputed. It does nevertheless seem that 10 groups of results would be especially worth recording since, in various ways, they have enlarged the limits of bio-acoustics and opened the way to numerous researches. The following are suggested:

(1) Researches of electro-physiologists and ethologists on the hearing of insects: Demonstration of the perception of ultra-sounds by grasshoppers (Auger and Fessard, 1928[1]).

Work of Autrum[2,3,4] on the hearing of grasshoppers, showing their perception up to 100 kc/s, starting with amplitudes equal to 25 times the diameter of the hydrogen atom.

First ethological observations on the acoustical behavior of grasshoppers by Faber (1928–1953)[45–48].

(2) Demonstration of the hearing of fishes by K. Von Frisch, 1936[60].

(3) Work of Griffin and Galambos (1941)[63,64], Dijkgraaf (1943)[40,41], Möhres (1953–)[97,98]: Echo location of bats by transmission of ultra-sounds.

(4) Study of the transmission of sounds by sea creatures (1942–1945): Work of the American and Japanese Admiralties during the Second World War, followed by researches developed by Mrs. Fish on fishes[49,51]. Demonstration of the use of ultra-sonic signals in echo-location by sea mammals: McBride (1948)[19], Schevill (1949)[121,123] and Kellogg (1953)[76,78].

(5) Study of the behaviour of honey-birds, parasitic birds of Africa by Friedmann (1955)[52].

(6) Study of the pseudo-language of monkeys: work by Yerkes[143], Carpenter[37], Jinkin[73].

(7) Studies of innate or acquired signals of birds, undertaken partly by Thorpe[137, 138] and Marler[88, 90] and by Prof. Koehler's School (Sauer[119], Messmer[93]).

(8) Studies of regional variations in acoustical signals of Anura: Blair[8–12]; of birds: Stadler[130, 131] and Borror[13, 15].

(9) The attraction of mosquitoes to low frequency sounds: work of Kahn and Offenhauser[75] and Roth[118] and the results of Frings[53–56] et al. on the movement reactions of starlings and gulls.

(10) The acoustical reactions of the cricket after destruction of specialised cerebral zones by Huber[69, 70].

However, it is important to point out that a large part of these studies was only made possible by a parallel development of electro-acoustical apparatus. Special mention must be made of the perfecting of tape-recorders, by Von Braunmuhl and Weber, in Germany, between 1941 and 1945, from the original idea of the Frenchman Paul Janet who, in 1887, made a report to the Academy of Science containing the description of the principle of the method. The first tape-recorder was presented in Paris at the Universal Exhibition of 1900 by the Dane, Valdemar Poulsen.

To these must be added the achievements of the first panoramic analysers of frequency in France by Pimonow (C.N.E.T.) (1949) and the development and application of the analyser of the type called "Visible Speech" (the Sonagraph of the Kay Electric Co., U.S.A.,) to animal acoustical signals (Borror[13], 1953).

REFERENCES

For authors working on insects, only a few articles have been cited since they can be found in the specialized bibliography (see FRINGS, reference 57)

[1] AUGER, D. and A. FESSARD, 1928, Observation sur l'excitabilité de l'organe tympanique du criquet, *Compt. rend. Soc. Biol.*, 99, No. 23, 400–401.
[2] AUTRUM, H., 1940, Über Lautäusserungen und Schallwahrnehmung bei Arthropoden II. Das Richtungshören von Locusta und Versuch einer Hörtheorie für Tympanalorgane vom Locustidentyp, *Z. vergleich. Physiol.*, 28, 326–352.
[3] AUTRUM, H., 1941, Über Gehör und Erschütterungssinn bei Locustiden, *Z. vergleich. Physiol.*, 28, 580–637.
[4] AUTRUM, H., 1961, Der Einfluss des Verhältnisses i/t eines künstlichen akustischen Signals auf das Verhalten in toto und auf das peripherische Rezeptorennervensystem von einigen Orthopterenarten. Phasische und tonische Antworten vom Tympanalorgan von *Tettigonia viridissima*, *Acustica*, 10, 339–348.
[5] BEACH, F. A., 1944, Responses of captive alligators to auditory stimulation, *Am. Naturalist*, 78, 481–505.
[6] BEATTY, R. T., 1932, *Hearing in Man and Animals*, G. Bell and Sons Ltd., London.
[7] BEBEE, W., 1922, *A Monograph on the Pheasants*, London, Chap. 4.
[8] BLAIR, F. W., 1955, A differentiation of mating call in spadefoots, genus *Scaphiopus*, *Texas J. Sci.*, 7, No. 2, 183–188.
[9] BLAIR, F. W., 1956, Call difference as an isolation mechanism in southwestern toads (genus *Bufo*), *Texas J. Sci.*, 8, No. 1, 87–106.
[10] BLAIR, W. F., 1956, Mating call and possible stage of speciation of the Great Basin Spadefoot, *Texas J. Sci.*, 8, No. 2, 236–238.
[11] BLAIR, W. F., 1957, Structure of the call and relationships of *Bufo microscaphus* Cope, *Copeia*, No. 3, 208–212.
[12] BLAIR, W. F., 1958, Mating call in the speciation of Anuran Amphibians, *Am. Naturalist*, 92, No. 862, 27–57.
[13] BORROR, D. J. and C. R. REESE, 1953, The analysis of bird song by means of a vibralyzer, *Wilson Bull.*, 65, No. 4, 271–276.
[14] BORROR, D. J. and C. R. REESE, 1954, Analytical studies of henslow's sparrow songs, *Wilson Bull.*, 66, No. 4, 243–252.
[15] BORROR, D. J. and C. R. REESE, 1956, Vocal gymnastics in wood thrush songs, *Ohio J. Sci.*, 46, No. 3, 177–182.
[16] BOUTAN, L., 1913, Le pseudo-langage; observations effectuées sur un Anthropoïde, le Gibbon, *Actes Soc. Linn. (Bordeaux)*, 67, 5–76.

[17] BRANDT, A. R. and P. P. KELLOGG, 1939, The range of hearing of canaries, *Science*, *XC*, New Series, No. 2237, 354.
[18] BRANDT, A. R. and P. P. KELLOGG, 1939, Auditory responses of starlings, English sparrows, and domestic pigeons, *Wilson Bull.*, *51*, No. 1, 38–41.
[19] McBRIDE, A. F. and D. HEBB, 1948, Behavior of the captive bottle-nosed dolphin *Tursiops truncatus*, *J. Comp. and Physiol. Psychol.*, *41*, 111–123.
[20] BUSNEL, M. C., 1953, Contribution à l'étude des émissions acoustiques des Orthoptères. I. Recherches sur les spectres de fréquence, et sur les intensités, *Ann. Epiphyties*, No. *3*, 333–421.
[21] BUSNEL, R. G. and W. LOHER, 1954, Sur l'étude du temps de la réponse au stimulus acoustique artificiel chez les Chorthippus, et la rapidité de l'intégration du stimulus, *Compt. Rend. Soc. Biol.*, *148*, 862–865.
[22] BUSNEL, R. G. and W. LOHER, 1954, Mémoire acoustique directionnelle du mâle de *Chorthippus biguttulus* L. (Acrididae), *Compt. Rend. Soc. Biol.*, *148*, 993–995.
[23] BUSNEL, R. G. and W. LOHER, 1951, Recherches sur le comportement de divers mâles d'Acridiens à des signaux acoustiques artificiels, *Ann. Sci. Nat. Zool.*, *16*, 11e série, 271–281.
[24] BUSNEL, R. G., 1955, Mise en évidence d'un caractère physique réactogène essentiel de signaux acoustiques synthétiques déclenchant les phonotropismes dans le règne animal, *Compt. Rend. Acad. Sci.*, *240*, 1477–1479.
[25] BUSNEL, R. G., B. DUMORTIER and F. PASQUINELLY, 1955, Phonotaxie de femelle d'Ephippiger (Orthoptère) à des signaux acoustiques synthétiques, *Compt. Rend. Soc. Biol.*, *149*, 11–13.
[26] BUSNEL, R. G., 1955, Probabilité du rôle prédominant d'un des caractères physiques des signaux acoustiques artificiels dans le déclenchement de phonotaxies dans le règne animal, *J. Physiol.*, *47*, 123–128.
[27] BUSNEL, R. G., J. GIBAN, PH. GRAMET and F. PASQUINELLY, 1955, Observations préliminaires de la phonotaxie négative des Corbeaux à des signaux acoustiques naturels, *Compt. Rend. Acad. Sci.*, *241*, 1846–1849.
[28] BUSNEL, R. G. and B. DUMORTIER, 1955, Phonoréactions de mâle d'*Hyla arborea* à des signaux acoustiques artificiels, *Bull. Soc. Zool. France*, *LXXX*, No. 1, 66–69.
[29] BUSNEL, R. G., M. C. BUSNEL and B. DUMORTIER, 1956, Relations acoustiques interspécifiques chez les Ephippigères (Orthoptères, Tettigoniidae), *Ann. Epiphyties*, *3*, 451–469.
[30] BUSNEL, R. G., B. DUMORTIER and M. C. BUSNEL, 1956, Recherches sur le comportement acoustique des Ephippigères (Orthoptères, Tettigoniidae), *Bull. Biol.*, *France et Belg.*, *90*, 3, 219–286.
[31] BUSNEL, R. G., J. GIBAN, PH. GRAMET, H. FRINGS, M. FRINGS and J. JUMBER, 1957, Interspécificité des signaux acoustiques ayant une valeur sémantique pour des Corvidés européens et nord-américains, *Compt. Rend. Acad. Sci.*, *245*, 105–108.
[32] BUSNEL, R. G. and P. GROSMAIRE, 1958, Enquête auprès des populations du fleuve Sénégal sur leur méthode acoustique de lutte traditionnelle contre le Quéléa, *Bull. I.F.A.N.*, *20*, Sér. A No. 2, 623–633.
[33] BUSNEL, R. G., 1959, Etude d'un appeau acoustique pour la pêche utilisé au Sénégal et au Niger, *Bull. I.F.A.N.*, *21*, Sér. A, No. 1, 346–360.
[34] BUSNEL, M. C., B. DUMORTIER and R. G. BUSNEL, 1959, Recherche sur la phonocinèse de certains insectes, *Bull. Soc. Zool. France*, *84*, No. 5–6, 357–370.
[35] BUSNEL, R. G. and J. GIBAN, 1960, Colloque sur la protection acoustique des cultures, *I.N.R.A.*, Paris.
[36] BUSNEL, R. G. and W. LOHER, 1961, Influence du rapport i/t d'un signal acoustique artificiel sur les comportements *in toto* et sur le système nerveux récepteur périphérique de certains Orthoptera. Déclenchement de phonoréponses chez *Chorthippus brunneus* (Thunb.) *Acridinae*, *Acustica*, *11*, 2, 65–70.
[37] CARPENTER, C. R., 1940, A field study in Siam of the behavior and social relations of the Gibbon (*Hylobates lar*), *Comp. Psychol. Monogr.*, *16*, No. 5, 1–212.
[38] COLLIAS, N. and M. JOOS, 1953, The spectrographic analysis of sound signals of the domestic fowl, *Behaviour*, *5*, 175–187.
[39] DAVIS, D. E., 1942, The phylogeny of social nesting habits in the Crotophaginae. *Quart. Rev. Biol.*, *17*, 115–134.
[40] DIJKGRAAF, S., 1943, Over een merkwaardige functie van den gehoorzin bij vleermuizen. *Verslag Ned. Akad. Wetensch. Afd Natuurkunde*, *52*, 622–627.
[41] DIJKGRAAF, S., 1946, Die Sinneswelt der Fledermäuse, *Experientia*, *2*, 438–448.
[42] DUIJM, M. and T. VAN OYEN, 1948, Sprinkhanen, *De Levende Natuur*, *51*, 1–7.
[43] DUIJM, M. and T. VAN OYEN, 1948, Het tsjirpen van de zadelsprinkhaan, *De Levende Natuur*, *51*, 81–87.
[44] FABER, A., 1929, Die Lautäusserungen der Orthopteren (I), *Z. Morphol. Ökol. Tiere*, *13*, 3/4, 745–803.

⁴⁵ FABER, A., 1932, Die Lautäusserungen der Orthopteren (II), *Z. Morphol. Ökol. Tiere*, 26, 1–93.
⁴⁶ FABER, A., 1936, Die Laut- und Bewegungsäusserungen der Oedipodinen, *Z. Wiss. Zool.*, 149, 1–85.
⁴⁷ FABER, A., 1953, Laut- und Gebärdensprache bei Insekten. Orthoptera (Geradflüger) (I), *Staatl. Museum Naturk. Stuttgart.*
⁴⁸ FABER, A., 1954, Laut- und Gebärdensprache bei Insekten. Orthoptera (Geradflüger) (I), *Z. Tierpsychol.*, 11, No. 3, 522–524.
⁴⁹ FISH, M. P., 1948, Sonic fishes of the Pacific, *Pacific Ocean; Biol. Project Techn. Rept.*, 2, 1–144.
⁵⁰ FISH, M. P., 1953, The production of underwater sound by the northern sea-horse *Hippocampus hudsonius*, *Copeia*, No. 2, 98–99.
⁵¹ FISH, W. R., 1953, A method for the objective study of bird song and its application to the analysis of bewick wren song, *Condor*, 55, No. 5, 250–257.
⁵² FRIEDMANN, H., 1955, The honey-guide, *U.S. Natl. Museum Bull.*, 208, Smithsonian Institution, Washington, D.C., 279 pp.
⁵³ FRINGS, H., M. FRINGS, B. COX and L. PEISSNER, 1954, Auditory and visual communication in the herring gull, *Larus argentatus*, *Anat. Record*, 120, No. 3, 734.
⁵⁴ FRINGS, H. and J. JUMBER, 1954, Preliminary studies on the use of a specific sound to repel starlings (*Sturnus vulgaris*) from objectionable roosts, *Science*, 119, No. 3088, 318–319.
⁵⁵ FRINGS, H., M. FRINGS, B. COX and L. PEISSNER, 1955, Auditory and visual mechanisms in food-finding behavior of the herring gull, *Wilson Bull.*, 67, No. 3, 155–170.
⁵⁶ FRINGS, H., M. FRINGS, B. COX and L. PEISSNER, 1955, Recorded calls of herring gulls, (*Larus argentatus*) as repellents and attractants, *Science*, 121, No. 3140, 340–341.
⁵⁷ FRINGS, M. and H. FRINGS, 1960, *Sound Production and Sound Reception by Insects*, a bibliography, Pennsylvania State University Press, 108 pp.
⁵⁸ FRINGS, H. and M. FRINGS, 1957, Recorded calls of the eastern crow as attractants and repellents, *J. Wildlife Management*, 21, No. 1, 91.
⁵⁹ FRINGS, H., M. FRINGS, J. JUMBER, R. G. BUSNEL, J. GIBAN and PH. GRAMET, 1958, Reactions of American and French species of *Corvus* and *Larus* to recorded communication signals tested reciprocally, *Ecology*, 39, 1, 126–131.
⁶⁰ FRISCH, K. VON, 1936, Über den Gehörsinn der Fische, *Biol. Rev.*, 11, 210–246.
⁶¹ GRAMET, PH., 1956, Etude du phonocomportement d'un couple de Freux (*Corvus frugilegus*) lors du nourrissage de la femelle pendant la couvaison et des jeunes après l'éclosion, *Bull. Soc. Zool. Franc.*, 81, No. 2–3, 126–131.
⁶² GRAUVOGL, A., 1958, Über das Verhalten des Hausschweines unter besonderer Berücksichtigung des Fortpflanzungsverhaltens, *Thesis*, Freie Universität, Berlin.
⁶³ GRIFFIN, D. R. and R. GALAMBOS, 1941, The sensory basis of obstacle avoidance by flying bats, *J. Exptl. Zool.*, 86, No. 3, 481–506.
⁶⁴ GRIFFIN, D. R., 1958, *Listening in the Dark*, New Haven, Yale University Press.
⁶⁵ HANSSON, A., 1945, Lauterzeugung und Lautauffassungsvermögen der Bienen. *Opuscula entomol., Suppl. VI*, 1–124.
⁶⁶ HARDENBERG, J. D. F., 1960, Colloque sur la Protection acoustique des Cultures contre les Oiseaux, *Ann. Epiphyties, Fasc. Hors Sér.*, I.N.R.A. Paris.
⁶⁷ HASKELL, P. T., 1953, The stridulation behavior of the domestic cricket, *Brit. J. Animal Behaviour*, 120–121.
⁶⁸ HASKELL, P. T., 1957, Stridulation and associated behaviour in certain Orthoptera. 1. Analysis of the stridulation of and behaviour between males, *Brit. J. Animal Behaviour*, V, No. 4, 139–148.
⁶⁹ HUBER, F., 1952, Verhaltensstudien am Männchen der Feldgrille (*Gryllus campestris* L.) nach Eingriffen am Zentralnervensystem, *Verhandl. Deut. Zool. Ges. Freiburg*, 138–149.
⁷⁰ HUBER, F., 1955, Sitz und Bedeutung nervöser Zentren für Instinkthandlungen beim Männchen von *Gryllus campestris* L., *Z. Tierpsychol.*, 12, No. 1, 12–48.
⁷¹ HUBER, F., 1956, Heuschrecken- und Grillenlaute und ihre Bedeutung, *Naturwissenschaften*, 43, No. 14, 317–321.
⁷² JACOBS, W., 1950, Vergleichende Verhaltensstudien an Feldheuschrecken, *Z. Tierpsychol.*, 7, No. 2, 169–216.
⁷³ JINKIN, N. I., 1958, *Mechanism of Language* (in Russian), Edition of the Academy of Pedagogical Sciences, Institute of Psychology, (Moscow), 371 pp.
⁷⁴ JUMBER, J. F., 1956, Roosting behavior of the starling in Central Pennsylvania, *Auk.*, 73, 411–426.
⁷⁵ KAHN, M. C. and W. OFFENHAUSER, 1949, The first field tests of recorded mosquito sounds used for mosquito destruction, *Am. J. Trop. Med.*, 29, No. 5, 811–825.
⁷⁶ KELLOGG, W. N., 1953, Ultrasonic hearing in the porpoise, *Tursiops truncatus, J. Comp. and Physiol. Psychol.*, 46, No. 6, 446–450.

[77] KELLOGG, W. N., R. KOHLER and H. N. MORRIS, 1953, Porpoise sounds as sonar signals, *Science*, *117*, No. 3036, 239–243.
[78] KELLOGG, W. N., 1958, Auditory perception of submerged objects by porpoises, *J. Acoust. Soc. Am.*, *31*, No. 1, 1–6.
[79] KING, J. A., 1955, Social behavior, social organization and population dynamics in a black-tailed prairie dog "Town" in the black hills of South Dakota, contribution from the laboratory of vertebrate biology, No. 67, University of Michigan, Ann. Arbor, 1 vol., 123 pp.
[80] KOHTS, N., 1935, *Infant Ape and Human Child*, Scientific memoirs, Museum Darwinianum, Moscow (in Russian, English summary), 580 pp., 145 plates.
[81] KRISZAT, G., 1940, Untersuchungen zur Sinnesphysiologie, Biologie und Umwelt des Maulwurfs (*Talpa europaea*), *Z. Morphol. Oekol. Tiere*, *36*, 446–511.
[82] KRITZLER, H., 1952, Observations on pilot whale in captivity, *J. Mammal.*, *33*, 321–334.
[83] LANDOIS, H., 1874, *Tierstimmen*, Freiburg.
[84] LOHER, W., 1957, Untersuchungen über den Aufbau und die Entstehung der Gesänge einiger Feldheuschreckenarten und den Einfluss von Lautzeichen auf das akustische Verhalten, *Z. vergleich. Physiol.*, *39*, 313–357.
[85] LORENZ, K., 1931, Beiträge zur Ethologie sozialer Corviden. *J. Ornithol.*, *79*, 1, 67–127.
[86] LORENZ, K. and N. TINBERGEN, 1938, Taxis und Instinkthandlung in der Eirollbewegung der Graugans, *Z. Tierpsychol.*, *2*, 1–29.
[87] LORENZ, K., 1941, Vergleichende Bewegungsstudien an Anatinen, *J. Ornithol.*, *3*, 19–29.
[88] MARLER, P. R., 1952, Variation in the song of the chaffinch (*Fringilla coelebs*), *Ibis*, *94*, 458–472.
[89] MARLER, P. R., 1956, The voice of the chaffinch and its function as a language, *Ibis*, *98*, No. 2, 231–261.
[90] MARLER, P. R., 1957, Specific distinctiveness in the communication signals of birds, *Behaviour* (*Netherlands*), *11*, No. 1, 13–39.
[91] MARTIN, D. R., 1880, *Les Abeilles*, Enchiridion apicole ou Manuel d'Apiculture rationnelle, Gounouilhon, Editeur, Bordeaux.
[92] MARTOF, B. S. and E. F. THOMPSON, 1958, Reproductive behavior of the chorus frog *Pseudacris nigrita*, *Behaviour*, *13*, 243–258.
[93] MESSMER, E. and I. MESSMER, 1956, Die Entwicklung der Lautäusserungen und einiger Verhaltensweissen der Amsel (*Turdus merula merula* L.) unter natürlichen Bedingungen und nach Einzelaufzucht in schalldichten Räumen, *Z. Tierpsychol.*, *13*, 3, 341–441.
[94] MILLER, L., 1952, Auditory recognition of predators, *Condor*, *54*, No. 2, 89–92.
[95] MINNICH, D. E., 1925, The reactions of the larvae of *Vanessa antiopa* to sounds, *J. Exptl. Zool.*, *42*, 443–469.
[96] MINNICH, D. E., 1936, The responses of caterpillars to sounds, *J. Exptl. Zool.*, *72*, 439–453.
[97] MÖHRES, F. P., 1953, Über die Ultraschallorientierung der Hufeisennasen, (Chiroptera-rhinolophinae), *Z. vergleich. Physiol.*, *34*, 547–588.
[98] MÖHRES, F. P., 1953, Jugendentwicklung des Orientierungsverhaltens bei Fledermäusen. *Naturwissenschaften*, *40*, No. 10, 298–299.
[99] MOLES, A., 1952, *Physique et Technique du Bruit*, Dunod, Paris, 150 pp.
[100] MOLES, A., 1958, *Théorie de l'Information et Perception Esthétique*, Flammarion, Paris, 221 pp.
[101] MOULTON, J. M., 1956, Influencing the calling of sea robins (*Prionotus* spp.) with sound, *Biol. Bull.*, *111*, No. 3, 393–398.
[102] NASSONOV, D. N. and K. S. RAVDONICK, 1947, Reakcjia izolirovannykh popieretsh nopolusatykh nyshe liagushke slyshymyie rvuki, *Fiziol. Zhur., S.S.S.R.*, *33*, 25.
[103] NASSONOV, D. N. and K. S. RAVDONICK, 1950, Neposredstvennoié vlianié slyshymykh svoukov na nervnyé kletky izolirovannykh spino-mozgovykh ovzlov krolika, *Doklady Akad. Nauk S.S.S.R.*, *71*, 985–987.
[104] NICE, M. M., 1947, Studies on the life history of the song-sparrow, *Trans. Linnaeus Soc.*, *6*, 1–328.
[105] NOBLE, K. G. and L. R. ARANSON, 1942, The sexual behavior of Anura. The normal mating pattern of *Rana pipiens*, *Bull. Am. Museum Nat. Hist.*, *80*, 127–142.
[106] ODUM, E. P., 1942, Annual cycle of the black-capped chickadee (III), *Auk.*, *59*, 499–531.
[107] OSSIANNILSSON, F., 1949, Insect drummers, *Opuscula Entomol.*, Suppl. X, 1–45.
[108] PALMER, R. S., 1941, A behavior study of the common stern, *Proc. Boston Soc. Nat. Hist.*, *42*, 1–119.
[109] PARE, A., 1954, *Animaux, Monstres et Prodiges*, Edition du Club français du livre, Paris.
[110] PAVLOV, I., 1912, Résultats d'expériences sur l'extirpation de régions diverses de l'écorce par la méthode des réflexes conditionnels, *Trav. Soc. Méd. Russes St. Petersbourg*, in *Oeuvres Choisies*, Editions en langues étrangères, Moscow (1954) (in French,) Chap. VI, pp. 305–327.
[111] PERDECK, A. C., 1958, The isolating value of specific song patterns in two sibling species of grasshoppers, *Behaviour*, *12*, 1–75.
[112] PETER, K., 1911, Über einen Schmetterling mit Schallapparat *Endrosa* (*Setina*) *aurita* var. *ramosa*, *Greifswald Mitt. Nat. Ver.*, *42*, 24–31.

[113] PRINGLE, J. W. S., 1954, A physiological analysis of cicada song, *J. Exptl. Biol.*, *31*, No. 4 525–560.
[114] PUMPHREY, R. J., 1955, Rapports entre la réception des sons et le comportement, Colloque Acoustique Orthoptères, *Ann. Epiphyties, Fasc. Hors Sér.*, p. 320–337.
[115] REGEN, J., 1913, Über die Anlockung des Weibchens von *Gryllus campestris* L. durch telephonisch übertragene Stridulationslaute des Männchens, *Arch. Physiol. Menschen u. Tiere*, *155*, 193–200.
[116] REINERT, J., 1957, Akustische Dressurversuche an einem indischen Elefanten, *Z. Tierspychol.*, *14*, No. 1, 110–126.
[117] ROBERTS, B. B., 1940, The breeding behaviour of penguins with special reference to *Pygoscelis papua* (Forster), *British Graham Land Expedition (1943–1947), Sci. Reps.*, *1*, 195–254.
[118] ROTH, L. M., 1948, A study of mosquito behaviour—An experimental laboratory study of the sexual behaviour of *Aedes aegypti* (Linnaeus), *Am. Midland Naturalist*, *40*, No. 2, 265–352.
[119] SAUER, F., Die Entwicklung der Lautäusserungen vom Ei ab schalldicht gehaltener Dorngrasmücken (*Sylvia C. communis*, Latham) im Vergleich mit später isolierten und mit wildlebenden Artgenossen, *Z. Tierpsychol.*, *11*, No. 1, 10–93.
[120] SCHALLER, F. and C. TIMM, 1949, Das Hörvermögen der Nachtschmetterlinge, *Z. vergleich. Physiol.*, *32*, 468–481.
[121] SCHEVILL, W. E. and B. LAWRENCE, 1949, Underwater listening to the white porpoise, (*Delphinapterus leucas*), *Science*, *109*, 143–144.
[122] SCHEVILL, W. E. and B. LAWRENCE, 1953, High frequency auditory response of a bottlenosed Porpoise, *Tursiops truncatus* Montagu, *J. Acoust. Soc. Am.*, *25*, No. 5, 1016–1017.
[123] SCHEVILL, W. E. and B. LAWRENCE, 1956, Food-finding by a captive porpoise (*Tursiops truncatus*), *Breviora*, *53*, 15.
[124] SCHJELDERUP-EBBE, T., 1922, Zur Sozialpsychologie der Vögel, *Z. Psychol.*, *88*, 225.
[125] SCHLEIDT, W. M., 1951, Töne hoher Frequenz bei Mäusen, *Experientia*, *7*, 65.
[126] SCHLEIDT, N. M., 1952, Reaktionen auf Töne hoher Frequenz bei Nagern, *Naturwissenschaften*, *39*, No. 3, 69–70.
[127] SCHLEIDT, N. M., 1954, Untersuchungen über die Auslösung des Kollerns beim Truthahn (*Meleagris gallopavo*), *Z. Tierpsychol.*, *11*, No. 3, 417–435.
[128] SCHLOETH, R., 1956, Quelques moyens d'inter-communication des taureaux de Camargue, *Terre et Vie*, *103*, No. 2, 83–93.
[129] SHAW, E. A. G. and G. J. THIESSEN, 1954, High-intensity sound as a repellent for mallard ducks and other water fowl, *J. Acoust. Soc. Am.*, *26*, No. 1, 141.
[130] STADLER, H., 1930, Le dialecte des oiseaux, *Alauda, Suppl. to No. 1*, 66 pp.
[131] STADLER, H., 1934, Der Vogel kann transponieren, *Ornithol. Monatsschr.*, *59*, No. 1/2, 1–9.
[132] TAVOLGA, W. B., 1954, Reproductive behavior in the gobiid fish, *Bathygobius soporator*, *Bull. Am. Museum Nat. Hist.*, *104*, 427–461.
[133] TAVOLGA, W. N., 1956, Visual, chemical and sound stimuli as cues in the sex discriminatory behavior of the gobiid fish, *Bathygobius soporator*, *Zoologica*, *41*, 49–64.
[134] TAVOLGA, W. N., 1958, The significance of underwater sounds produced by males of the gobiid fish, *Bathygobius soporator*, *Physiol. Zool.*, *31*, No. 4, 259–271.
[135] TEMBROCK, G., 1957, Zur Ethologie des Rotfuchses (*Vulpes vulpes* L.), unter besonderer Berücksichtigung der Fortpflanzung, *Zool. Garten (NF)*, *23*, 4/6, 289–532.
[136] TEMBROCK, G., 1958, Lautentwicklung beim Fuchs sichtbar gemacht, *Umschau*, *18*, 566–568.
[137] THORPE, W. H., 1954, The process of song learning in the chaffinch as studies by means of the sound spectrograph, *Nature*, *173*, No. 4402, 465–469.
[138] THORPE, W. H., 1958, Further studies on the process of song learning in the chaffinch (*Fringilla coelebs gengleri*), *Nature*, *182*, No. 4635, 554–557.
[139] TREAT, A. E., 1955, The response to sound in certain Lepidoptera, *Ann. Entomol. Soc. Am.*, *48*, No. 4, 272–284.
[140] TREAT, A. E., 1956, The reaction time of noctuid moths to ultrasonic stimulation, *J. N.Y. Entomol. Soc.*, *64*, 165–171.
[141] TURNER, C. H., and E. SCHWARZ, 1914, An experimental study of the auditory powers of the giant silkworm moth (Saturniidae), *Biol. Bull.*, *27*, 325–332.
[142] WEIH, A. S. V., 1951, Untersuchungen über das Wechselsingen (Anaphonie) und über das angeborene Lautschema einiger Feldheuschrecken, *Z. Tierpsychol.*, *8*, No. 1, 1–41.
[143] YERKES, R. M. and B. W. LEARNED, 1925, *Chimpanzee Intelligence and its Vocal Expressions*, Williams and Wilkins Company, Baltimore.
[144] POULSON, H., 1959, Song learning in the domestic canary, *Z. Tierpsychol.*, *16*, No. 2, 173–178.
[145] SCHNEIDER, H., and A. D. HASLER, 1960, Laute und Lauterzeugung beim Süsswassertrommler *Aplodinotus grunniens* Rafinesque (Sciaenidae, Pisces), *Z. Tierpsychol.*, *17*, 499–517.
[146] SIGNORET, J.-P., F. DU MESNIL DU BUISSON and R. G. BUSNEL, 1960, Rôle d'un signal acoustique de verrat dans le comportement réactionnel de la truie en oestrus, *C. R. Acad. Sci.*, *250*, 1355–1357.
[147] WINCKEL, F., 1952, *Die Klangwelt unter der Lupe*, Max Hessel Verlag, Berlin.

CHAPTER 6

ANIMAL LANGUAGE AND INFORMATION THEORY

by

A. MOLES

1. Introduction

The scope of this paper is of a strictly rationalist character: an organism or individual is objectively defined by its external behaviour, which can be epitomized in a statistical study. The behaviour to be explained is that which is the most probable, and exceptional behaviour is *a priori* left aside as requiring an individual study which may be regarded as of secondary importance.

The organisms that interest us here react to messages which reach them from the outside world (Umwelt). The Umwelt sends messages to the organism and the organism undergoes modifications of its behaviour which are a function of these messages (*cf. infra*). Cybernetics is the science of the reactions of the individual and the mechanisms behind them, and Communication Theory is the science concerned with the study of the messages.

Cybernetics accepts as axiomatic that it can make no distinction in principle between artificial and biological organisms, and that it must consider them, to the extent that they exhibit the same behaviour, as equivalent. This assumption is the basis of the *theory of models*, which looks like playing an important rôle in the animal physiology of the future.

The object of this introduction to communication theory is to sketch out its main lines: the concept of information considered as a quantity, the concepts of repertoire and code, then the concept of acoustic quantification and the translation of the acoustic signal into a finite repertoire, finally, the application of these ideas to animal communication in particular. A final question: is the idea of "animal language" statistically valid?

2. The Nature of Messages

The messages received by an organism are of considerable diversity. They are received by the sense organs and include *inter alia*, in animals, those of the olfactory, tactile, visual, acoustic, thermal and other types. Physiology indicates that the criteria for classifying these messages must rely upon the sense organs that they stimulate, and one of the purposes of animal psychology is to make a particular study of the different receptors that in cybernetics are called *affectors*.

One of the first observations made by information theory is that there exist general laws—of near-numerical status—and which are valid for all the receptor systems: eye, ear, smell etc. Though it is easy in higher organisms, such as Man in particular, to distinguish clearly between cybernetics and information theory, this is much harder with lower animals. In a general way, we shall arbitrarily call the impacts of the outside world upon the organism, that do not destroy it but that do evoke a reaction, of whatever magnitude, *messages*.

We shall make the assumption—which is at the root of many practical difficulties in animal psychophysiology—that it is possible to detect all the reactions of an organism, *i.e.* that we possess a very extensive account of its psychophysiological functioning, and that we have knowledge not only of its external functions (eating, walking, perspiring, calling, etc.) but also of the physicochemical variations of the internal medium, the electroencephalogram, the cardiogram and so on. This assumption broadens the field of action of the theory of behaviour and takes into account a certain number of silent, non-gestural reactions which command attention. Actually, the assumption is purely one of principle, for the lower the organism is in the animal scale, the more difficult it becomes to analyze its internal reactions. Nevertheless it is essential to note that, for very general reasons related to the complexity of the systems, these reactions must be statistically more limited in number, so that the uncertainty in the knowledge of them is offset by the reduction of their diversity.

From the experimental point of view, a message will be considered as a modification of the physical medium, of the external environment, which is transmitted to the organism, *i.e.* which reaches its boundary and produces some variation of the internal environment.

For instance: modifications of the temperature of the skin—whether by convection or under the influence of radiation—, expressed as a function of time, constitute a *thermal message*; continuous modifications of hydrostatic pressure on the external surface of a fish, and more particularly on its acoustic organs, build a particular *tactile message* expressing modifications of the state of the external environment; variations of the atmospheric pressure of sufficient rapidity and received by appropriate receptors, constitute an *acoustic message* and so on...

All these messages, from the experimental point of view, can be recorded with appropriate detectors; microphones, photoelectric cells, enable measurement of the stimuli and recording of their values on a tape. The whole of these elements is the "scientific portrayal of the outside world" of the organism (Umwelt), made at the place where the latter is.

In practice, and especially in animal psychophysiology, it is useful to distinguish simple messages (such as auditory or visual or other messages belonging to a single category of physical phenomena and making use of a single channel of transmission) from multiple messages which arise by superposition in the integrative nervous system: a fire, for example, is mediated for a human individual or an animal organism by *olfactory, visual, thermal* and *acoustic* messages. It is proper to study them separately and make a synthesis of them later.

It would appear that a phenomenology of animal perception must take into account the complexity of animal behaviour, and the corresponding complexity demanded in natural messages: the organism systematically uses the consistency of the outside world as it knows it (either innately or from its "apprenticeship in the world") in order to supplement one message by another, and in order to connect itself with the outside world. This is the notion that information theory uses under the name of *redundancy*, which will be further developed later.

Most of the messages we are accustomed to consider take place during a certain time: they are sequences of stimuli, and information theory has dealt with them first,

References p. 131

e.g. the sound message radiating in the course of time is a temporal signal. But the theory of messages is also valid for spatial messages, *e.g.* the thermal distribution of water temperature in the different layers of a lake is a special message, the distribution of sounds in space is another. Many visual messages are spatial messages, especially in higher organisms.

Consequently, we can classify messages according to their origin from the physical world outside or specifically from another organism, though to distinguish between the two is not always very easy. The distinction is, however, clear enough when the two "organisms" belong to the same species, and we shall concern ourselves here with messages between two individuals, the source or emitter (E), and the receiver (R), although it may often be also applicable in the case of signals between different species (interspecific signals) (Fig. 61).

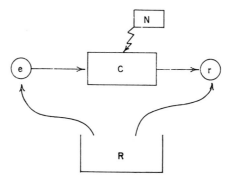

Fig. 61. The transmission channel between two individuals of the same species.
e   emitter   R   repertoire common to emitter and receiver
r   receiver   C   physical channel (acoustical, optical, etc.)
N   noise source

These inter-individual messages across the spatio-temporal framework (essentially across space, *i.e.* from one point to another; but sometimes across time, *i.e.* from one epoch to another) constitute what is properly termed *communication*, which is distinguished from perception as follows: communication may be defined as a particular case of the perception of messages between two organisms. We shall call these two organisms "individuals".

Languages are particular messages. They imply on the one hand the existence of at least two individuals of the same species, one being the source and the other the receiver; and on the other hand, the existence of a *repertoire* (*sign-set*) common to these two individuals. They come about through the intermediation of signals. A secret language is one whose repertoire is known only to a limited number of individuals as compared with the whole population of individuals in the species. "Foreign" languages (*i.e.* of other species) are secret messages.

The communication of intelligible messages may be represented by the scheme of Fig. 61. This implies that a message transmitted in a channel from the source to the receiver is composed of elements drawn from a repertoire common to the receiver and the source.

From the anthropomorphic point of view, the elements of the repertoire are often called *symbols* or *words*, and the repertoire itself, when classified in coherent order, is termed a *dictionary*.

A language may employ an extremely simple repertoire which may even be limited to *one* symbol, one sign, whose presence or absence constitutes two possibilities and whose successive combinations of presence and absence theoretically allow the expression of an unlimited number of messages. In fact, the simpler the repertoire the longer the message, and the most advanced organisms tend to use the most extensive repertoires, thus increasing the density of information. In practice, when a greater variety of symbols is available, the source can in a given time provide a much greater number of different expressive signals. Another simple example is the "all-or-nothing" law, valid for messages sent from one neuron to another, or for the Morse code. Such messages comprise only two elements: switch off or on, stimulated or not stimulated, etc. One of these alternatives is called a unit of information or "bit".

Animal signals always comprise a multiplicity of distinct items of information, and the theoretical scheme of communication by alternatives, reducing the repertoire to two symbols, is no more than an ideal case, only approximated by human signals such as for instance Morse code.

It is convenient to distinguish messages with one element from repeated temporal messages: an electric lamp can be either on or off: there is only one message, a single alternative, and the information concerning the lighting of the room is reduced to one bit. The firefly emits a single signal only which it repeats indefinitely; the only alternatives are emission and darkness.

But the same lamp can be used to communicate or send a new message every second; interest then is no longer focussed on the state of illumination of the room, but on the modes of transmission of the individual who may switch off and on three times, then leave off for a certain number of seconds, and so on. Every second, every unit of time (time quantum) that the message is renewed, another message is sent; and the totality of these constitutes a global message which is much more complex. The length of this latter will be the number of temporal quanta that have been used, here a certain number of seconds. Each second there were two available possibilities, consequently one bit per second was transmitted. Information theory shows that it is possible to encode any message, however complex, into a series of binary messages. These being the simplest imaginable, the elementary alternative constitutes a good unit of measurement for the quantity that the message can transmit. We shall see later, through examples, just how any message can be encoded into a binary repertoire.

One of the essential results arising from communication theory is that, whatever the nature of a message, whatever its channel, whatever the use made of it, every message carried represents a certain quantity transferred from source to receiver. It is this quantity that is designated by the scarcely appropriate word "information". (If this latter had any connection with what is usually called information, in the everyday sense, it is only in the abstract and etymological connotation, signifying more or less "instruction to put in form".)

In fact, information is, *before all else*, the *complexity* of a particular form; the information provided by the page of a journal, by a drawing, by a sensory pattern, is a measure of the complexity of that message, that pattern.

References p. 131

In animal psychology, we measure by a certain "quantity of information" the complexity of the stimuli to which the receiver organism is subjected; it is in this way that we measure the complexity of the shapes that have to be recognized by a rat situated in a maze or in front of a latch-trap, to get its food. The quantity of information will measure the complexity of the multiple messages received by an animal at one and the same time through his thermal, acoustic, visual and other senses.

There are then as many values of information as of types of message. In the case of exploration of the outside world by a single sense, we get a measure of the complexity of the signal received by this channel; but if we examine a multiple message comprising two or three channels (optical, acoustic and so on) the complexity of the outside environment is increased, and this is why we must emphasize that information is nothing more than a measure of the quantity of complexity conveyed. We have to realize that the transmission of messages from one point to another, is actually the *increasing of complexity at the receiving end as a function of a certain complexity created at the emitting end*.

3. Redundancy and Complexity

When a given communication channel is used, its physical nature determines the repertoire from which the transmitter must draw his symbol. The symbols "on" and "off" will represent a repertoire of two signs in a printed text, there is a system of 27 signs (letters of the alphabet + interval); in a French text a system of some 30,000 supersigns—30,000 dictionary entries—and so on (further examples of acoustic messages will be found in Table 4, p. 71, essay by Busnel[2]).

An inter-individual message consists of a group of these signs or symbols, assembled by the source at a given point where it augments the complexity of its immediate surroundings; it is transmitted through the channel to another point of the universe, namely the Umwelt of the receiver; this complexity of the environment affects the receiver by modifying his subsequent behaviour.

4. Maximum Information and Redundancy

Starting from the foregoing data, we can investigate what message, regarded as a complex system, would possess the maximal complexity or, in other words, would carry the maximum information. Although such an investigation is of practical interest only when applied to systems which effectively exploit maximum information—that is to say, artificial organisms—it is nevertheless of great theoretical interest, for it enables us to define the actual limits of the understanding of a message and the rôle of what we call *redundancy* in bringing order out of the chaos of the natural surroundings.

We have then at our disposal a repertoire, and we have to draw from it a certain number of elements to be assembled into a sequence. This sequence constitutes the message.

If the length is constant, it is easy to see that the more uniformly the chosen elements are used (*i.e.* the more perfectly we exploit the total resources of the repertoire) the more complex and original is the message that can be carried: this amounts to saying that all elements would have a virtually equal probability of being chosen. Because of this, there is no linkage, no internal rigidity in the message and it is impossible to predict any given element from that which precedes it; a mathematician would say that there was no autocorrelation in such a message. Nothing is foreseeable, everything is novel, hence the maximum of information is carried. The most original message conveys the most information, and the word "originality" appears synonymous with "complexity". Thus information, as a measure of complexity, increases when we use the totality of elements in the repertoire in an equiprobable manner.

In the case of the simplest binary repertoire, the maximum of information will be carried if the *a priori* probability of using the yesses or the noes is the same for each, or in other words if nothing, strictly speaking, is known about which can be expected next. This is the case in tossing a coin; the latter has neither awareness nor memory, and it is impossible to know beforehand which "signal" will be recorded, heads or tails. The same holds with messages containing more elements in their repertoire. It can be shown that the content of the language will be a maximum when all the words of the language, all the symbols, are equally probable, when, in short, each is used without taking account of what preceded it; in such a case, each word would carry an integrally novel idea.

The moment we apply this doctrine to an animal language, that of Man for example, we discover its essentially theoretical nature: it is plainly impossible to make sense by drawing words at random from a dictionary and stringing them one

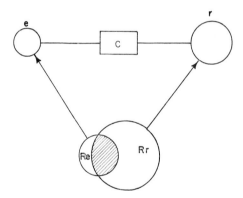

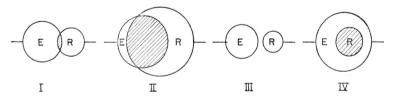

Fig. 62. The various situations of the emitter-receiver couple are determined by the degree of commonness in their repertoires of symbols.
    I   scarcely overlapping repertoires
    II nearly common repertoires
    III two completely exclusive repertoires
    IV the repertoire of the emitter exceeds largely that of the receiver

References p. 131

after the other. If we did, we should get a completely incomprehensible language, since every word would be deprived of its connection with every other.

It is here that the idea of *intelligibility* comes in, and this is the second major concept that information theory has recently delineated.

But what is this intelligibility? Intelligibility is essentially the perceivability of the internal organization and structure of the message, and indeed specifically of the connection between words that follow one another. We know that in the human language, nouns follow articles, that this letter follows that, that verb follows noun, that in general each sentence has a verb, and so on... We know a lot of things *a priori* from the simple fact of the continuity of the world on our scale. Outside the purely linguistic sphere, we also know that we are talking about coherent things, that nature does not jump, that where there is smoke there is fire, and *vice versa*; in short, that there just are *no isolated, totally unforeseeable events*—except those very special cases which mathematicians have taken great pains to discover: cointossing for example. So the structure of the message is bound up with its intelligibility; and the more structure it has the more intelligible it is, but the less information it can carry. Conversely, the less information it carries, the more intelligibly it carries it. This brings us up against one of the fundamental dialectics of the theory of messages: intelligibility and originality are very exactly opposed. We have to place ourselves between these two dialectic poles if we are to achieve real messages—to the detriment of the "information efficiency" of the message *sensu stricto*.

Why then throw away all this energy, why dissipate thus the efficiency of the messages?—For it is the transmitter who has to pay in every case for the wastage in lost time, lost symbols, unnecessary effort. Is this waste really unnecessary? Is there not perhaps a fundamental reason for such profligacy?

On examination, two reasons appear—the first the most essential. The concept of intelligibility itself is bound up with that of predictability, of banality. Now understanding is the *perception of structures* coupled with apprehension and deduction. Here we are concerned with the simple experimental and objective viewpoint; whether this perception be directed towards getting a mere general line or towards a more elaborate, more abstract understanding does not concern us here—this is a problem of cybernetics that we shall purely and simply set aside.

What is really important is to see what this word "apprehension" means. If the message is too original, if it consists of words out of order, deprived of arrangement, grammar, syntax or logic, these words will convey perhaps a maximum of information; but only provided they can be analyzed, that is, decoded into an immediately assimilable language (as in a telegram in which articles, verbs, and words eliminated as superfluous by the sender are replaced by the recipient). Either way, the message itself, as it stands, is unusable; it is not grasped, because it has no internal structure. We only apprehend *structures*, and the function of the message is to be perceived. In order to be perceived, then, it must have a structure; it has therefore to suffer some wastage of elements, since we have just seen that all structure implies foreseeability and must herefore be in direct contradiction with the maximum information. We now see very quickly that the idea of information itself is actually only a mensurational concept. The measure is not the thing itself, it has nothing vital nor essential about it; what clearly interests us much more is intelligibility, understanding, and therefore structure.

How is a structure or a pattern achieved? A structure is precisely defined in psychology as being *an assemblage of elements which are perceptibly not the result of chance*. It is exactly the previous definition in reverse.

But there is another reason for realizing that the message cannot have the maximum information content. This reason hangs upon the concept of *noise*, of the spontaneous destruction of the message. The message is a pattern, this pattern is transmitted through the external medium to the recipient, the receiver; it is subjected to the perpetual fluctuations of this medium, to a degradation which results from natural conditions. This constant degradation is what theoreticians call *noise*. The word has an extraordinarily general connotation. It is customary to think of it as an acoustic term, but it is no such thing. It means in reality any perturbation, any interference, any involuntary act, any system that is outside a message and interferes with it or is superposed on it. Noise is the message of nature, in contradistinction to the message of the transmitter, and appears as randomness compared to the artificial order that the transmitter sought to establish. Noise, then, more or less mutilates, distorts, deforms or destroys the pattern. The very existence of a pattern or form is, according to recent experiments in psychology, something which *resists* distortion: we know that a square inclined, transformed, deformed, remains a square for the Gestalt psychologist. It was formerly possible to think that Gestalt psychology was more or less in opposition to behaviourist psychology; but now it would seem simply a matter of their being separate departments of an extended objectivist psychology. In the field of animal acoustic behaviour, the study of *dummies* is indeed the application of the idea of "form" to the signal: the "dummy" retains the effective pattern of the signal over and above its incidental variations (lures).

We have thus found two essential reasons why the message of maximal content, using all the elements of the repertoire in equiprobable fashion, would be in practice unusable with biological receivers. On the one hand such a message would not be intelligible, not assimilable as it stands, by the receiver; on the other hand, it would be *fragile*, and may be destroyed by noise, since each element is unforeseeable and unconnected to its neighbours and therefore, if destroyed, impossible to reconstruct by reference to the context. The information then appears to represent a measure of the complexity of the patterns constituting the message.

Here appears a quantity opposed to information, and more important in practical application: this is *redundancy*, the percentage of "superfluity" in the message. Since certain of the latter's elements are deducible from their antecedents, at least approximately, and since drawing one element from the repertoire influences the probability of one or other of the succeeding elements, we can say that this partial predictability is reflected, certainly in a lessening of content, but above all in an increase of intelligibility, because of the statistical linkages of elements with one another. The relative percentage of what is regarded as superfluous is expressed by the redundancy, which represents to some degree a complement to the relative information. This redundancy would be maximal when the information was zero—if for example the same symbols were constantly repeated. It is useful, however, to observe clearly in this connection that although repetition automatically augments redundancy, the two ideas are quite distinct and there are many other ways of increasing the redundancy of a message, *viz.*: all those which increase the *a priori* internal connections between the elements.

References p. 131

Reverting to what we have observed earlier, we find then that redundancy protects the communication channel against noise and against external disorder. In the dialectic order/disorder, order is the message, disorder is the noise.

We are thus led to transfer our attention from the concept of information to that of redundancy. From the very nature of the problem, we are concerned with probabilities of occurrence, hence with measuring complexity or banality. Redundancy is a measure of the relative banality of the message, whatever its length; information is a measure of the maximal complexity that a message of the same length could have if all its symbols were equiprobable.

Actually, since all the symbols are not equiprobable, we are going to have to determine:

1. What is the extent of the repertoire common to source and receiver (Fig. 62);
2. What are the probabilities of occurrence of the various symbols of the repertoire in this language (whether it be human, animal or mechanical).

Since we are here more especially interested in animal language, we shall proceed to examine its repertoire and indeed the possibility of its establishment.

5. The Repertoire in Animal Languages

To start with, we shall limit ourselves to those messages *via* the acoustic channel which constitute spoken languages in the lay sense of the word, leaving aside gestural, visual etc. languages, to which the same considerations would, however, apply: the study would be conducted in the same way and further study would encompass the whole assemblage of channels so as to reconstruct the multiple messages in the environment of the receiver. The establishment of the repertoire is the equivalent of studying a dictionary relating to an unknown language; it has to take as basis, as elements, as "letters" of the dictionary, not this or that phonetic symbol (of a too anthropomorphic character) but the acoustic elements direct. After that, the probability distribution in the common repertoire has to be worked out: this will be a statistical study of the sound elements—a *technology of language*.

With animals, it is obvious that questions of method will arise straight away: for example, the possibility of applying statistics to a very small number of elements or even the possibility of reasoning statistically at all in order to extract laws of organization from curves plotted from very few known coordinates.

The establishment of the repertoire can be carried out according to three methods: the first related to the *receiver*, the second related to the *source*, the third related to their reciprocal *situations*.

1. It is quite clear that there is no command of words unless and insofar as the words are understood. If a language consists of a variety of elements, there is only a possibility of real expression if the receiver carries in his memory system a sufficient variety of these elements. Investigation, therefore, of the variety of the elements perceivable by the receiver must lead to a certain maximal repertoire of elements that will give an indication of the *subtlety* or *refinement* of perception. It will give for instance in the case of the higher animals, an idea of the possibility of establishing a conditioned reflex in those animals for the perception of more or less complex messages. In short, the maximum of possible elements is the maximum of elements that are able to be differentiated in the perceptual field.

2. The second way relates to the source. This will examine which are the various elements that the source is capable of producing, and of emitting in *distinguishable* or *voluntary* fashion.

3. Lastly, there is the third possible method, which may be called phenomenological, consisting of examining what, in the situation of the animal, are the different categories of message that it has the opportunity to express, that it expresses effectively, and that it is *capable* of using. At this point we find a maximal field of expressible situations; this assemblage of expressible situations is also an estimate of the vocabulary.

It is now clear that a concordance of the three methods will give extremely strong evidence of the extent of this vocabulary, and this point will be examined later.

In similar manner, we can seek to define what would be the possible level of redundancy, to what degree the animal makes use of the concept of redundancy. This can be done (having first defined the repertoire of messages) by investigating what, from the noise point of view, are the situations in which it is placed and what is the minimum of completely perceptible messages. The binary logarithm of the number of repetitions, or of the number of times that the actual signal contains this minimum signal, gives an idea of the redundancy.

(a) First method (by reference to receiver)

Studying a given species of animal, we have to determine with great care its threshold of audibility, its normal auditory field; and we have to start from the —likely enough—hypothesis that when signals are too weak, being in practice too often submerged by the natural background noise, the animal will not be able to make much use of them for communication. In short, we must concentrate attention on signals of fair intensity. We must determine the natural background noise in the acoustic frequency range that is in fact used for hearing in the species under consideration. Thus we have to take into account a range of frequencies and a range of sound-levels.

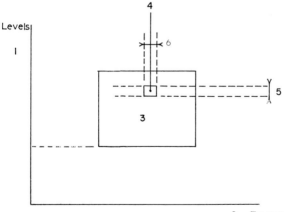

Fig. 63. The sound channel has two dimensions, the "frequencies" coordinate and the "levels" coordinate. The psycho-physiological "repertoire" is determined by a set of differential thresholds for frequency and level discrimination.

1 axis of level
2 axis of frequency
3 audible area
4 element of repertoire
5 differential threshold for frequency
6 differential threshold for level

References p. 131

In these two dimensions of sound, we must then investigate *differential thresholds*: relative differences in the character of the sound, whether of pitch or intensity, which are susceptible of being perceived systematically and unambiguously by the animal (Fig. 63). Now it is quite evident that such an investigation, simple enough in methodology, will be extremely laborious in experimental practice, since the only correct way of carrying it out is by using conditioned reflexes. Other methods, based on phonotactic or phonokinetic reactions, can in practice, however, give very substantially useful indications.

Thus we get a sound repertoire, certainly valid for certain individuals: for Man comes into this category. But the extent to which the "signal" properly so called makes use of this repertoire is not established; we know for instance that the lower animals make use of a signal that is very little differentiated in pitch and intensity.

Actually of course, this procedure only gives two dimensions of sound, those of the sound-chart, and it has to provide the plot of the sound chart from which the elements are to be chosen. What constitutes the actual repertoire is the *sound event* itself, in the evolution, as a function of time, of the elements of this plot that are used each instant. This sound event is in effect an evolving system, and we at once perceive that in animal language the part that evolves in time plays a much more important rôle than the static part, that is, the spectrum, a mere combination of simple differential elements of which intensity, varying as it does so much with distance, can only have extremely rough thresholds: and absolute pitch, which is merely the mean position of the spectrum.

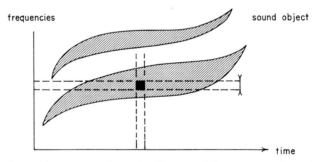

Fig. 64. The sound event is represented on the diagram of frequency against time (sonagram) and can be divided into elementary thresholds occurring during slices of time ("thickness of the moment" of psychologists), roughly corresponding to Gabor's "logons".

Now it is known today—particularly since the work of Busnel and his collaborators—that in animal language, as in human, the consonants, here defined as transients, have much greater value than the vowels; they carry much more information, the vowels serving exclusively, it would seem, as a *support* for the "consonants". The vowel sound gives the timbre; this is the utilization of a greater or less number of acoustic quanta, and it seems likely that this timbre is nothing more or less than a—very approximate—indication of species, sex and age of the emitting individual: an identity-sign. From the semantic point of view, what counts is the consonant, the transient part of the phenomenon, exactly indeed as in human language, but here perhaps with a crucial importance, at least in the lower groups. The vowel sound itself is only the support of a temporal form: the support of the transient phenomenon.

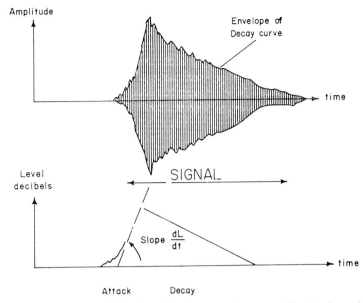

Fig. 65. Shape of signal (above) and envelope. The abruptness is measured by the slope of the curve of level against time.

As for the structure of these transient phenomena, it even appears probable, according to some recent German work (Winckel)[11] on the psychology of time, that these may be determined in both man and animals by the delays in the functioning of the synapses between neurons, and that the concept of the duration of the present (taking into account the minimum delay for perception and for separation of apparent simultaneities) may be, in mammals as in man, of the order of a sixteenth of a second. At any rate, in other species it can but be greater (snails 1 sec), which simplifies the problem: we have only to cut our sound signals into slices certainly thicker than 1/16 sec, but in no case less.

If the method that starts from the receiving subject and determines the variety of signals that he can receive as distinct entities, is peculiarly cumbersome, due to the application of conditioned reflex techniques, there are nevertheless several possible approximations. Two are worth noting here:

One is the method of *lures**—that is by fabricating signals more or less similar to the normal signals emitted by the animals, and in all cases pragmatically defined by the fact that the animals react to the *lure* exactly as they do to the signal of another individual. It appears then that all the information contained in the animal message is contained in the *lure*, since the reaction is the same.

The other method consists in applying one of the most general procedures in the aesthetics of perception, namely *modification*, the systematic mutilation of the signal, starting from an actual signal recorded for instance on tape, and deforming it progressively. There comes a moment when the receiver individual no longer reacts to this signal, or reacts differently; we have then reached a threshold of

---

* Artificial gadgets for imitating bird and animal noises.

References p. 131

differentiation indicated by the measure of variation to which we have subjected the signal.

This method has been applied by Busnel to the modification of the temporal form of the signal and its initial transient. It is possible, for example, to abstract the initial transient and curtail the vowel sound supporting it to the point where there is no longer any reaction. Or again, the steepness of this attack signal (which is measured as the derivative of level with respect to time dL/dt) can be deformed and attenuated so as to determine the threshold of the animals to transients: the moment when they no longer react. (See further remark on p. 131.)

(b) Second method (by reference to source)

The second method of determining the maximum repertoire at the disposal of an animal species starts with the source. Working always in the sound channel of the sound-producing organs, it seeks to determine the *maximal diversity* of signals that the animal is capable of emitting. It is closely bound up with neuro-physiology; in effect, the purpose is to determine the various distinguishable signals that are possible to the source. These signals must necessarily differ acoustically from one another, but above all must differ in that each corresponds to that well-defined but unique muscular process which produces it. This is tantamount to saying that the articulation patterns of the various vowels, of the various consonants, and more generally of the various sound-events—the various phonemes of the language or its various words—must be distinct from one another. This leads us to ask neurophysiology to indicate what, in the nerve ganglia, are the various possibilities of differentiation, what number of combinations and *what* combinations are followed by differing muscular contractions.

In certain Orthoptera, for example, the sound-producing organs are in the nature of files rubbing on membranes. We may ask what are the various speeds with which the files can be moved, what the various rhythms with which they can repeat their movement, what the displacements to which the animal can subject the file relative to the membrane and so on...

In principle, it is for the physiologists concerned to answer these questions. But there is anyhow a plainly different way of facing the problem, one whose results are tolerably difficult to forecast in the present state of affairs. We know in all cases, that animal organisms as we know them only use a part of the differentiation-potential available to them. Do they restrict their vocabulary thus merely for reasons of redundancy, founded on the adaptation of the species to situations in an environment permeated by noise (where too great a subtlety of vocabulary could do more harm than good)?—or is it because there is a loss between the organization of their nervous system and their physiological mechanisms? Therein lies the problem to be solved.

Busnel has shown that the richness of the vocabulary is generally bound up with the spatial density of the social field relative to the range of the signals. In species of low density, a simple signal suffices to attract the nearest individual, and at the closer distance, no ambiguity exists as to destination, since emitter and receiver are alone. In species of high density with numerous individuals in range of a given signal, a selection has to be made between them by the caller: accordingly he enriches his vocabulary. These results agree with the work of Zipf.

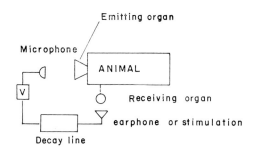

Fig. 66. The Lee effect is obtained when the organism is stimulated by its own messages delayed in time.

What is the exact relationship between the repertoires obtainable by examination of the source and of the receiver as previously outlined? It would seem from the most recent work done on human phonation and audition that there is a far closer correlation between the two than was suspected a few years ago, and that a feedback mechanism is involved, audition operating as a constant control on phonation. In other words, since reception by an individual is a control upon his own emission, there can be no emission without reception; this is the case with deaf mutes and from it we can deduce that the result of differentiations of emission cannot be very different from that with which differentiations of reception provide us. What happens when an animal is deaf from birth or artificially made so? Therein lies an experimental method of approach to this problem (Schwarzkopf). We can, again, use the method of delayed speech (Lee effect), which arrives with a timelag at the ears of the source, and thus falsifies his cybernetic servo-mechanism circuit.

(c) *Third method (by reference to situations)*

Lastly, the third method is essentially one of *situation*. This is the one that has been practised by naturalists and observers of the habits of insects, which can furnish a considerable documentation on the various categories of messages capable of being expressed. This, furthermore, leads indirectly to the problem of sentences, for the message is a succession of elements (symbols) drawn from a repertoire. In evolved languages, this succession of symbols expresses, in its very continuity, ideas—that is, developing patterns; it assembles words. There seems to be no such thing as "sentences" in lower animals; the capacity to assemble ideas marks a sort of threshold in the intellectual development of species, below which we find just words one after another, strung out at relatively large distances and not assembled into sentences; these generally are repetitions pure and simple. It is apparent at once that apart from all other considerations, one essential cause of this repetitive tendency is noise. Possibly problems of sociometric diffusion also come in, but this point needs to be further clarified by biology.

In certain of the most highly evolved species of grasshoppers (Acrididae), it is known that there exists a basic vocabulary comprising about six different signals of which the dictionary is roughly as follows:

Signal I: It is fine, life is good;
Signal II: I would like to make love;
Signal III: You are trespassing on my territory;
Signal IV: She's mine (of the female of course);

References p. 131

Signal V: Oh, how nice it would be to make love!
Signal VI: How nice to have made love!

The prairie dog, on the other hand, seems to possess a much more extensive vocabulary, of some thirty differentiated signals which he can employ in relatively rapid succession.

Now let us imagine (purely arbitrarily, obviously) an extremely intelligent "grasshopper", "wishing" to transmit a message consisting of the six signals we mentioned above, each during about one second, in some specified order, such as:

III IV I VI V II IV III III II I V IV II VI etc. ...,

and let us further suppose that the six are equiprobable, *i.e.* equally frequent in the life of the animal. This message would transmit a maximal information measured by the binary logarithm of 1/6, say about 15. This message of 15 bits, taking into account the duration of the signals, each of which lasts at least 1 sec (interval included) would thus take 6 sec. The density of information, or quantity per second, would thus be 2.5.

In fact of course, the grasshopper which uttered such a message might perhaps be extremely intelligent and full of imagination, but not terribly logical within itself—since it would be saying in succession: "Ah! life is good!", and "I'm dying to make love", which is somewhat contradictory; then, after that again, "Ah! how good to have made love!" which is flatly contradictory. There is then a certain mutual exclusiveness within the words of this rudimentary vocabulary, and already a certain minimum of redundancy appearing, since the appearance of the signal "I want to make love" renders extremely improbable the appearance in the next second of the signal "How nice to have made love"—the one relating as it does to an act performed, the other to an act one wishes to perform. There could thus be some incoherence here, but we must suppose our grasshopper something of a surrealist, with a poetic urge to tread upon the corns of logic.

We see from this paradoxical example that this message, both informative and surrealist, is quite simply impossible, and this not for physiological but for logical reasons, primarily for the receiver: the nearby grasshopper, however intelligent, receiving this message from its congener, would probably be completely flummoxed by it and would be incapable of *assimilating* the flood of contradictions that reached it. In addition, such a message is inconsistent also through taking no account of the consistency and coherence of nature, and consequently rejecting the natural world. Now animals, even more closely than Man, are integrated into the natural world: only Man can write surrealist poetry...

We approach the natural message when we examine the expression of a particular sentiment sufficient to fill the whole "mentality" (in the cybernetic sense) of the animal with this single sentiment, with one sensation and one expression. The animal repeats this expression many times. If for instance he is in a state of war with another who is on the point of invading his territory, he chants: "You're trespassing", and repeats it for a relatively considerable time, several seconds, indeed specifically for the time the stimulus persists. This message makes a communication, partakes of the nature of a veritable language in every accepted sense of the word... By repeating its signal, the animal lessens its originality for the receiver; if the receiver has a slow wit, receiving the signal several times repeated, say for 10 sec, the total rate of information transmitted will drop progressively from its initial value.

Returning to our fictitious grasshopper, we recall that in choosing from six different signs it provides in each signal an information of 2.5 bits. If it utters ten signals, it provides ten times as much. The density of information which it provides is 2.5 bits/sec if all the signals are different and all used with equal frequency.

Let us now suppose that our grasshopper, instead of six signs, emits only one, which it repeats ten times; this gives an information of 2.5 bits in 10 sec, that is only 0.25 bits/sec. In the 10 sec it could have provided $2.5 \times 10 = 25$ bits, so the redundancy is $(25 - 2.5)/25 = 90\%$. Thus it establishes a redundancy of 90%.

All this rests on the arbitrary (and false) hypothesis that the grasshopper could use equally all the six signals in its vocabulary. In fact, for reasons some of which we have already met, it does not use them in equiprobable fashion, but some very often, others very seldom. It is obvious that the rarely used ones convey more information than the very common ones; the unexpected gives us much more information than what we already expected. *Signals we expect do not change the structure of our universe*, in particular they do not change the complexity of the world we live in, since we have already assimilated it. On the other hand, what is novel adds a detail to the picture, adds something to our vision of the Umwelt. It is more informative and profitable to make use of rare signals than of common ones. But if the individual used rare signals systematically often, they would *ipso facto* become no longer rare in the long run. There we leave the matter; it is a problem for the statistical morphology of vocabularies.

Since symbols can in practice never be equiprobable, some are much less frequent than others and have consequently more information value than the commoner ones.

This shows that in principle *we always lose* in this exchange; if we gain in rate of information by symbols, what we lose on the common symbols exceeds what we gain on the rare ones. The sum total diminishes, and the optimum is the very special case we have just discussed in which all the symbols are equiprobable. It is impossible to consider this as anything but a theoretical limiting case. On the other hand, an essential difficulty in languages which comprise only a few signs is to decide their relative frequencies of utilization; this will need long observations and remains a problem of the future, implying long magnetic recordings on the linguistic habits of grasshoppers, crows or other animals, and analysis by frequency of occurrence of the different words in the language. We shall in this way get a vocabulary qualified by coefficients of probability—coefficients of frequency—from which we shall be able to determine the numerical rate of information.

We may sometimes wonder whether, with such a great simplicity of signs, this operation is worth carrying out. Recent work in linguistics has shown that it was indeed very important, for it led to penetration of the statistical structure of the language—which is the second problem already referred to above about language. The fundamental studies on this question were made by the American psychologist Zipf. If we construct a histogram of frequencies: if in fact we study the relative frequency of the "words", give them a number, arrange them in diminishing order of frequency and plot these ranks as abscissae (or rather their logarithms); and if as ordinates we plot the number of times each of the words of the various ranks has been used, we get a curve which naturally decreases (by construction, since it was obtained by the supposition that the rank should decrease from the most frequent to the least frequent), but which, for languages with a large number of symbols, shows great regularity, reflecting the statistical structure of the language (Fig. 67).

Granted, where only six or even thirty symbols are involved, such a curve scarcely has any meaning, but it is worth noting that the term language need not necessarily be restricted to sound language nor even mimetic language: there are many others. Not to mention the systems of transmission of vibrations through the integument of two grasshoppers at opposite ends of a branch, one communication-channel that appears from all we now know to comprise quite a large number of perfectly distinct and explicit words, is the *olfactory* channel. Dogs, cats and higher mammals appear to have a sense of smell as subtle from the quantitative point of view as that of human beings.

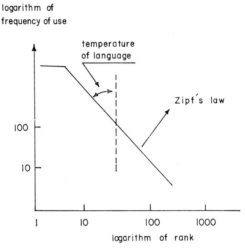

Fig. 67. Curve of log utilization frequency against log rank of the elements of the repertoire (Zipf's law points out the regularity of this curve).

From the qualitative point of view, we have certain knowledge that half a score of perfumiers each make five or six distinct perfumes that a reasonably experienced chemist is perfectly capable of distinguishing, if he has learned the language of scents. It is likely that mammals possess an olfactory vocabulary if not as big, at least as varied, constituting veritable messages. These messages come either from the outside world (these we leave aside) or from another individual; and we could speculate indefinitely on the way dogs investigate the scent of this or that bitch, or this or that street lamp. Whichever way we look at it, in this field there are certainly frequencies which are very different from one another. There are alien bitches, voluptuous bitches, pedigree bitches and untouchable lovelies.

Incidentally, to put the problem of animal language better in relation to human language, it is interesting to place on record the reason why, in the olfactory field, individuals cannot make up a "sentence". To make a sentence is to put together a series of words (or scents, or elementary odours) in a complex that presents a pattern, a structure; it is to assemble them. Now, from the simple fact of the nasal cavities, of the volume of air breathed and so on, olfactory perception is a slow perception, each "word" needing several seconds and the passage from one word to another requiring several seconds, since the olfactory channels have to be swept clear. We cannot then make olfactory sentences, for the condition of perception of a temporal structure is to grasp, and apprehend through the memory, a whole assemblage of connected words. If our instantaneous memory is too brief, we do not manage to capture in one group the relations between the successive words, and we can only speak in monosyllables. This is perhaps why no serious effort has been made up to now in Man to construct olfactory sentences; the situation will probably only change when we know how to excite the olfactory sense directly by electrical means.

This digression at any rate shows how we can reason about other forms of animal language such as those with a small number of symbols and an extremely slow pace. In the acoustic field, or the optical, more familiar still, upon which the efforts of physiologists have up to now been spent, and with a language comprising a greater number of symbols, it is likely that we can construct curves of frequency distributions. One essential interest bound up with these frequency curves and this statistical study of language is due to the theoretical interpretation given by Zipf. He remarks that, in every communication problem, a constant dialectic equilibrium is established between two contradictory tendencies of the source and receiver individuals. One tendency is the social tendency which demands a message as precise and complex as possible, and consequently a language as varied as possible, since the social body comprises a large number of individuals; the other is the tendency to laziness, to least effort, and thus to brevity, all of these tendencies being in the domain of the source. Between these two tendencies is established, in a given social context, a dynamic equilibrium[12].

It is quite clear that considerations of this kind can only be applied to sufficiently evolved animals, and this leads to a final question of a general character: what is the exact rôle that information theory can play in animal languages? It appears in the light of the considerations presented above that even the framework of a particular optical or acoustic channel is for animals too narrow a framework. One of the main difficulties is that *information theory has an essentially statistical character*, yet how can language statistics be done with too small a number of elements?

It seems that information theory furnishes a fairly elaborate theoretical model leading at any rate to a possible animal linguistics. But in order to be effectively applied, it requires that the sense of the word "language" be broadened, and the question arises whether we should not do better to apply the term to a multiple system of communications with the outside world. The individual is connected to the surrounding world and to his congeners not by a channel exclusively auditory or visual, but by the whole gamut of sensory channels, and this in an "occasional" way. More exactly, let us say that such an animal will utilize simultaneously his smell, hearing and lastly touch, to communicate with his congeners; and he will dispose of a repertoire of complex symbols containing the totality, the whole lot of the various storable images that he possesses.

We know for instance that birds recognize a certain range of avian types, much as aviators recognize the marks and nationalities of aircraft from their profiles, with their morphological indications extremely reduced.

Is it possible to extend the theory of languages and of information to this multiple connection with the outside world? The question stands unanswered, for the repertoire in particular now becomes a fluctuating and ill-defined thing for

References p. 131

which it is difficult to calculate the probabilities; the chief obstacle being that the repertoire now depends on the situation. Some bird, out of song range, salutes another bird which clearly cannot use the sound channel; it automatically therefore loses a certain number of pages from its global dictionary. The language then depends on the situation; we have to make the problem of the situation and the problem of languages come into the probability theory of perception. Depending on the distance of the perceptual field and the physicochemical range of the signal, the stimuli unfold in a certain hierarchy; and from this moment we see that the statistics of language open out more widely into the statistics of situations, that is to say into a quantitative study of the acts of animal life—opening up interesting perspectives of a statistical ethology.

6. Conclusion

If this essay has dwelt upon examples taken from the field of animal *acoustics*, it is because for technical reasons this type of signal can be more readily logged, transcribed, analyzed, than all the others, the numerical quantitative study of which remains embryonic.

It remains, from amongst the various ideas brought out by information theory, to recall the following:

1. From the conceptual point of view, the concepts of the *channel*, the *common repertoire*, *redundancy* and the *hierarchy* of repertoires;

2. It has emerged clearly from this exposition that the notion of redundancy is infinitely more important than that of information properly so called. *Information is the measure of the complexity of the patterns*; redundancy is the measure of their relative banality and especially of their intelligibility, of their standardized character;

3. Again, we must recall in a more exact terminology, the concept of Zipf (law of least effort) which opposes *exactitude* against *brevity*;

4. Lastly, we recall the dialectic concept order/disorder; opposition of noise to message, the message representing order relative to the two organisms that communicate, opposing the disorder of the surrounding world.

Among the possible applications of information theory to animal signalling, we point out:

5. A setting in order (ranking) of patterns (Gestalts, formes) consisting of complex stimuli, according to the measure of their complexity;

6. A ranking of animal series according to the complexity of their synaptic systems;

7. The beginnings of a theory of artificial animal models—a direction extremely fertile in new vistas on behaviour;

8. A clarification of the concepts of relations between organisms, which unites some very disparate notions and some related sciences in the same architectural structure: that of communication theory.

From the philosophical point of view, it is interesting to note that, despite appearances, there is here no properly finalistic aspect. Cybernetics achieves a kind of operational finalism in the sense that it objectively discerns in problems, oppositions between the "intentions" which are only valid for the organism (be it machine, living thing or human being). These oppositions are all centred on the individual, and on the grander scale this individual finalism melts away. There is no finalism except that of the individual. Put differently, if we take up the position of the individual himself we perceive the development of the message as an "intention"; but if we place ourselves outside the individual, we perceive it as a "fact". This remark leads in some degree to the position of Husserl; it furnishes an objective interpretation of the finalistic vocabulary.

Addendum to page 124, section (a)

Since this paper was written, Busnel and Moles[13] have made studies on signals, conveying a seemingly large amount of information, among very different kinds of societies. The first example is taken from shepherds of the Pyrenean mountains who, like those of the Canary Islands, communicate by means of a whistled speech. This whistled language is a modulated carrier whose inflexion in pitch and transients conveys the semantic content. The second is the underwater communication of porpoises. The shape and pattern of their emission is very similar to the whistled speech quoted above, and seems also to convey a notable amount of information.

REFERENCES

[1] BENVENISTE, E., 1952, Communication animale et langage humain, *Diogène*, Nov., pp. 1–8.
[2] BUSNEL, R. G., Signaux acoustiques chez les animaux, *Colloque: Communications et langages*, 1959, 22–25 January. Gauthiers-Villars Ed., Paris, in press.
[3] CHERRY, C., 1957, *On Human Communication*, Wiley, London–New York.
[4] KAINZ, F., 1954, 1956, *Psychologie der Sprache*, Enke Verlag, Stuttgart.
[5] MILLER, G., 1957, *Language and Communication*, McGraw-Hill.
[6] MOLES, A., 1957, Technologie du langage parlé et du langage écrit, *Ere Atomique*, 7, 5, 39, 52.
[7] MOLES, A., 1958, *Théorie de l'Information et Perception Esthétique*, Flammarion, Paris.
[8] SHANNON, C., and W. WEAVER, 1949, *Mathematical Theory of Communications*, University of Illinois Press.
[9] ZIPF, G. K., 1936, *Psychobiology of Language*, Houghton Mufflin Co., New York.
[10] HALDANE, J. B. S., 1955, Communication in Biology, *Studies in Communication*, Secker and Warburg, London, p. 29–43.
[11] WINCKEL, F., 1953, Die Klangwelt unter der Lupe, Max Hesses Verlag, Berlin.
[12] See ref. 2, articles by MOLES, VALLANCIEN and FRY.
[13] BUSNEL, R. G., and A. MOLES, 1962, Sur une langue sifflée pyrénéenne, *Logos*, in press.

CHAPTER 7

AN APPLICATION OF THE THEORY OF ALGORITHMS TO THE STUDY OF ANIMAL SPEECH*

METHODS OF VOCAL INTERCOMMUNICATION BETWEEN MONKEYS

by

N. I. ZHINKIN

1. Introduction

Many eminent explorers have tackled the subject of the sounds produced by monkeys on the supposition that "language" was the heritage of man and of him alone. The fault lies not with the experts. It is just that the concept of language and the concomitant concept of thought remain insufficiently defined even to this day. An impression appeared to be forming that monkeys had a kind of language and yet, most probably, there was no such thing. This is why attempts had been made at staging *negative* experiments. The question was put in the form "how could a monkey be made to do something or other which even a child could do with such ease?"

Starting with this conception, Yerkes and Learned[17], tried to teach a chimpanzee to say *ba-ba* to a banana and *na-na* to the process of eating it. Their attempts were a failure. Yet Kellog and Kellog[7], found that their chimpanzee correctly carried out 58 verbal orders of the type "kiss", or "show me your nose", or "stop", etc. This surprised nobody since everyone knows that, with skilful training, even a dog can be made to obey a man no less faithfully.

Yet even this negative approach has brought to light a number of facts which are worth pointing out. In 1949, while working in the nursery for monkeys at Sukhumi, Tikh[14] made many attempts to transform some gregarious signal sounds of the baboons into sounds with another meaning, such as food. Not one of these attempts was a success. Yet no sooner had she selected, instead of a gregarious sound, one of the kind known as "organic", such as groaning, or hiccupping, or munching, than there followed, comparatively easily, the vocal reaction *kh* to the sight of a nut and a triple *kh-kh-kh* to that of a lump of sugar. Similar results were obtained in the L.G. Voronin Laboratory by Pankratov[16]. Here the gregarious communicative sounds *gm-gm* and *tse-tse* of the *Papio anubis* were being transformed into a food reflex. In the end this was successfully accomplished with a few of the monkeys, although only with great difficulty; a failure of the higher nervous activity occurred with the majority of the anubis, there arose a neurosis, the conditioned reflex very quickly disappeared and a negative attitude to the whole experiment followed.

Have these facts anything to tell us? They seem in a way, yet they leave many more questions entirely without answer. In the end, through this traditional approach to the study of the sounds made by certain animals, we arrive at a somewhat trite

---

* For a glossary of terms, definitions and symbols used in this Chapter, see p. 173.

conclusion that an animal's "language" cannot be transformed into a human one, although an animal diligently carries out certain orders given by a human. This indeed was the conclusion arrived at, in his day, by Sarris[12], who had studied the behaviour of dogs. "A dog," he writes, "is capable only of what conforms to its canine nature (...dem Hund gemäss ist)." All this vividly recalls the wisdom of a tradition which is widely spread among the tribes of the Far North. There was a time, we are told, when dogs freely talked with men. But on one occasion man had so cruelly offended the dogs that they resolved never to speak to any man again. So the dogs still know everything but they keep their silence.

Despite the simple charm of these truly profound observations, the approach to the study of the "language" of an animal will probably have to be somewhat different. I wish to give the reader advance notice of what he may find after reading a few pages of this article. Much, very much indeed, depends not only on facts but on the *approach* to their collection and treatment. The approach used in the present article is that of constructive algebra. Sounds made by monkeys are regarded as sequences of elements which have been defined in a certain way, and rules for the construction of these sequences are explained. With this sufficiently formalised and indisputably objective approach we shall be able to describe, using identical terms, both human speech and the calls of monkeys, likewise the chirp of a grasshopper, the ultra-sound of a gnat, etc. In this manner we free ourselves from an open or tacit admission that vocal communication is peculiar to man and to him alone. Nor shall we be faced with the question whether, to put it bluntly, a monkey is capable of "conversing" like a human. It would be more interesting to find out how vocal communication between the monkeys themselves is carried on. If the sequences of sounds are not random, then they are capable of control to the extent to which entropy is destroyed. It remains to be discovered, under what conditions the control of the order of the elements in a sequence will constitute a communicative system.

The object of this article is the approach to the study of communicative systems. Monkeys merely afford a welcome and most convenient means of achieving this object. The vocal apparatus of a monkey is very similar to that of man but their communicative systems are very different. A comparison of this kind gives rise to quite an interesting outlook.

I shall now briefly outline some of the problems suggested. It does not follow that they will be specifically discussed in this article but then every approach is interesting only in proportion to the promise it holds out.

It is a point of dispute among linguists, whether it is possible to describe a language (la langue) without recourse to the term meaning. This problem has recently acquired a not unimportant practical significance in connection with informational machines, because it is still not quite clear, what kind of meanings they are capable of dealing with. When a linguist insists that he has described, say, the system of phonemes of a language he does not know, somebody may still doubt whether he, the linguist, having lived during a certain time among people using this language, has completely failed to guess the meaning of one single word. To put it more simply, is it necessary to be a partner in a communicative system in order to differentiate between verbal meanings? Now suppose that we possess a description of the communicative system of monkeys. Nobody is likely suspect that this description was

References p. 180

made possible only because the discoverer had reached a mutual understanding with the monkeys.

For the purpose of formulating this problem, it is necessary to treat the so called meanings as objectively as the components of a sequence which is capable of control. Generally speaking, an objective definition of meaning is quite simple. If, within a chain of signals $\alpha_1 \alpha_2 \alpha_3 \ldots \alpha_n$, certain parameters α_n coincide, within given tolerance limits, with the parameters α_1, i.e. α_1 and α_n are equivalent $(\alpha_1 \sim \alpha_n)$, then we say that the signal α_n has preserved the meaning of α_1. Now a signal may have a complicated structure; it will then become apparent that the criterion of equivalence will differ for the different layers of this complex structure*.

If follows from this that the old-fashioned yet widely used term *meaning of words* will become definite only after the structure of words has been investigated. An irresistible urge to employ this term before it has been more precisely defined is readily understandable, yet it is capable of resulting in the same kind of misunderstandings as the terms language, thought and certain other terms are apt to produce.

The problem of the "transformation of words used by monkeys" has been stressed earlier in this foreword. It is the same problem of meaning. Note that the experiments were designed to investigate the transformation of the meanings of signals. Whatever the outcome of these experiments, the theory of the problem remains incomplete. If the same problem is approached in the constructive manner, it is found that one can distinguish two types of communicative system, i.e. (a) a system of non-expanding messages and (b) a system of expanding messages (p. 148, 149). Their comparison makes it perfectly clear that sequences of the first system cannot be transformed into sequences of the second because the first system possesses, for this purpose, neither corresponding alphabets nor algorithms. It would be interesting to investigate the effect of this circumstance on the meanings of signals in these distinct systems.

The problem of meanings must, in all probability, be approached from various points. Here is another one. Without entering into any discussion, let us reckon that the modern translating machine translates not words and, certainly, not thoughts, but syntagmas. Now a translator still tries to grasp the theme or the situation which is related in the text. And yet, not only for the purposes of control, but at all times, a translator acts exactly like a machine (it would have been more accurate to have put it the other way round, but this does not change matters) simply because, if he fails to take account of the structure of the syntagmas, he will translate nothing. We can thus assume that a translator's chain of coded transitions in parts coincides and in part diverges from that of a machine. The constructive approach affords means of detecting both the coincidences and the divergences**. Monkeys are a help to a certain extent in this case also, for their "language", being devoid of syntagmas, is intelligible even without translation, as we would expect from theoretical considerations. But of special interest is the fact that the sounds which are made by monkeys do nevertheless convey meanings***.

---

* See Glossary, para. 27.
** The problem reduces itself to a selection of concrete words (p. 147).
*** Without entering into any further discussion of this problem, we shall note that speech (la parole) without meaning is only senseless muttering. Two assertions are then possible.
(i) A language (la langue) which is manifested (represented) in action, i.e. as speech, requires no translation. The ability to learn one's native language and the so called direct method of teaching

The list of questions which, in the constructive approach, are framed in a new manner, could be considerably extended. We shall add to it only one more.

One of the cardinal objects of study both for anthropologists and for the students of monkeys is the problem of anthropogenesis. In theory, this problem is reduced to that of reconstruction of successive structures. A certain structure is assumed to be basic. The task lies in establishing the consecutive replacements of elements which lead to a series of adjacent structures A, B, C, D. The elements here considered are those which are concerned with the transition from a communicative system of non-expanding messages (A) to one of expanding messages (B). An analysis of sound sequences of both systems shows that they are similar in the structure of certain components (vowels) yet quite distinct in the capacity to initiate acoustic transitional processes. The ordinary fact that these transitional processes are formed in the pharynx acquires a key position in the approach here offered. An indication may be furnished by the relative position of the epiglottis. This is placed high in the vocal apparatus of the system A and low in the system B. Hence the conclusion that the period during which the epiglottis descends from its high to its low position is the period of the change from the system A to the system B. The summary formula (that is one which does not enumerate the elements one by one) of the transition from system A to system B will be the substitution or exchange of two resonance chambers (the oral and the pharyngeal) for the twin oro-pharyngeal resonance chamber. The exact time of the transition can also be calculated with reasonable accuracy. Since the remains of the ancestors of man are only certain bones, and no cartilages are preserved, we must employ a method of correlation in which, by using available parts, we determine the probable form and position of other parts of the same organ*.

It is possible that this method, when used in coordination with other methods now employed, will be of use in the solution of the question of the period of the appearance of human speech.

If we now attempt to formulate very briefly the essence of the approach adopted in this article then we can put it this way: a communicating signal is complex, therefore its study must start with the simplest elements which cannot be further resolved and must proceed to their synthesis. This postulate is as simple as it is obvious. And yet, when a language is theoretically studied, complex and integrated components are often discussed, while elements are neglected. The journey which begins at the elementary level may be long, yet its end is most rewarding.

I wish to add that the reader could usefully refer to the "Glossary of terms employed in the present article" contained in the appendix, because he will find in it certain additional explanations of the fundamental concepts.

---

a foreign language are both based upon this fact. In the long run a description of the language of monkeys amounts to the same.

(ii) A language can be described without the use of the term *meaning*, if a certain passage of a text of some definite length has been previously translated into (a) the same language (equivalent paraphrase), the text being the manifestation of the language in speech, or, (b) some other living language, this living language being the manifestation in speech.

* We assume of course that the osteological remains have already been chronologically placed.

References p. 180

2. General Principles

(a) Necessity of formalized approach to the objective interpretation of monkey's signals

This essay is an approach to the study of vocal means of intercommunication employed by monkeys. It is based upon recordings of calls uttered by baboons, *Papio hamadryas*, when living under near-normal conditions.

Monkey calls have long attracted the attention of students of these animals. The method hitherto commonly used for the description of these sounds was based upon hearing. The movement of the fundamental tone was estimated by ear and recorded in musical notation[8,17].

No scientific phonetic transcription could, obviously, be employed, simply because no such transcription is intended for recording animal calls. Investigators, naturally, bear in mind that a non-phonetic transcription of monkey calls, if based on hearing, merely recalls the sounds of human speech. This circumstance introduces so substantial an element of subjectivity that the results obtained, apart from being scanty, lack in reliability.

Yet it cannot be denied that even aural investigations have their uses, if confined within some definite and scientifically objective limits. They are indispensable for any preliminary description.

Firsov of Leningrad[3] employed oscillograph recording of monkey sounds and thus eliminated any danger due to uncertainty in subjective estimation. Yet no overall record of any complex sound, unless resolved into spectra, contributes much of value towards any conclusion that could be drawn concerning a system of communication by sound. Material thus obtained can be compared with similar material only as regards the contour of the oscillogram and the frequency of the fundamental tone, and this is manifestly insufficient. This is why the present writer applied, to the study of sounds uttered by monkeys the methods of spectral analysis. The spectrum is the final, or, if you wish, the initial element of any acoustic phenomenon. Figuratively speaking, it is the microscopic analysis of the sound. The study of spectra, while yielding a wealth of data, sets, at the same time, a number of puzzles which can only be resolved through the study of the instrument producing the sound and of the system controlling any intercommunication maintained by this means.*

To solve the first of these puzzles, we employed X-ray observations of the epilaryngeal tube of the baboon in course of phonation. Attempts at the solution of the other puzzle, that of the control of the acoustic apparatus, have met with more serious obstacles. Any data obtained through the analysis of spectra or through the observation of the modulation of the vocal apparatus can be regarded as facts, but no judgement on the subject of the control of the signals forming a communication can be treated as anything but an hypothesis based upon these, unfortunately not always complete, facts. Yet a bare recital of facts would take us only the shortest distance on the road towards the truth.

The danger of subjectivity is particularly grave in any discussion of the problem of the control of intercommunication by sound, the more so that, in our case, the behaviour studied is that of animals. This is why, before any attempt is made to present the facts themselves, an agreement must be arrived at concerning the method

---

* See detailed explanations on page 179.

of their interpretation. It remains a moot point whether it is at all possible to talk of monkey sounds as of means of intercommunication and to regard them as signals possessing some definite meaning. Are we permitted to surmise that there exists a peculiar monkey language and, in such a case, to what, in general, could the term "language" be applied, what is meant by "a separate sound", "a word", "a syllable", "a combination of sounds", etc.? No answer to any of these questions is, as yet, sufficiently definite even in the matter of human speech; they are especially difficult in the case of monkeys, the more so as an undefinable similarity in these phenomena, as manifested in a monkey and in man, is bound up with an equally vague dissimilarity. Yet if we refuse to answer any questions like these, then any interpretation of either spectra or of X-ray records, loses all perspective.

The way out of this difficult situation may be found in such a description of the facts, as would be completely objective and formalised to such an extent, as to fit not only any sounds made by monkeys or any language spoken by man, but also other signals, or groups of such signals, their selections or combinations, including communicated messages. It is an established fact that, far from obliterating any specific differences, formalisation reveals their existence.

The author has arrived at the conclusion that the theory of algorithms provides such a formal approach. This theory is, at the present time, in the process of evolution as a constituent branch of higher algebra*.

Sections 2 (b) to (g) of this essay are in no way to be regarded as a discussion of mathematical ideas but merely as an attempt by the author briefly to explain and then apply some of them. Its object is to arrive at an agreement concerning the application of these mathematical ideas in the course of a discussion of the facts recorded in section 4 (c).

(b) Substitution: vertical and horizontal

Suppose that a number of symbols, to be called "letters", no two of which are identical, constitute an alphabet. In this manner, the list $\left\{\begin{matrix} a, b, \\ c, d \end{matrix}\right\}$, constitutes an alphabet. The term "letter" will be applied only to a continuously described and complete transcription of it, no breaking up of letters into elements is allowed. In other words, a letter is one single element. Any meaning may be attached to a letter selected from the alphabet, but this meaning must always be quite definite and must permanently remain attached to it. If this condition has been satisfied, then no two objects or phenomena denoted by two distinct letters are alike. Thus one letter could denote a closing relay, another one—an opening relay. Again, following the example set by Gamow[4], who attempted to find a code of inherited information, letters could be used to denote the bases of the polynucleotid chain of the chromosome (adenine, thymine, guanine, cytosine). Letters may denote sounds which can be distinguished by the human ear or by an apparatus which has been specially designed for this purpose.

Sound spectra can also be denoted by letters. Even some definite part of a spectrum, which differs from other parts of the same spectrum or from the remaining spectra, can be so denoted. Not only can a sound be exchanged for a letter, but a

---
* In the Soviet Union this subject is being studied by, *e.g.* Makarov, Novikov and their pupils[9,11].

References p. 180

letter can also be exchanged for a sound (as in reading aloud). In such a case, a sound and a letter are considered interchangeable signs. A case where a sign of one material origin (as an exchange of a letter for a sound, or a sound for a letter for a sound, or a sound for a letter, or an electric signal for a sound) will be called a vertical substitution. Since the elements exchanged in a vertical substitution are signs of different material origin, it should be considered that some of these signs belong to some alphabet A and others to some alphabet B. Rules exist for the transition from alphabet A to alphabet B. These rules are usually referred to as translation of one code into another.

When one or more letters have been selected from an alphabet, they can be arranged in a line. A conjunction of this type is called a word. Thus *a, acd, aaaa, acdcc*, are words formed from some definite alphabet.

Suppose the alphabet $\left\{ \begin{array}{l} a, b, \\ c, d \end{array} \right\}$ contains the word *acdadad* (1). Then the words *dad, da, ad, ac, cd*, are words of the same alphabet, and one can say that they form part of the composition of the word (1) and that each of the words *dad, ad, da*, has two entries, whereas the words *cd*, and *ac* have only one.

Now write down the formulae *dad—ac; cad—b*, where the symbol — is not a part of the alphabet but denotes the operation of substituting the word on the left of it by the one on the right. Then, with the help of the first formula, two other words may be formed from (1), namely *acacad* (2) and *acdaac* (3). Similarly, using the second formula, (2) can be transformed to *acab* (4). The words (1) and (2), (1) and (3), (2) and (4), are known as adjacent words. Adjacent words are equivalent. The words (2), (3) are not adjacent words. Any exchange of one word for another one in a horizontal transcription of some sequence of words consisting of letters of one and the same alphabet will be called a horizontal substitution, as distinct from a vertical substitution, where the words are from different alphabets A and B.

If the rules had permitted other substitutions which could be applied to the words (2), (3) then two chains of adjacent words could be formed, each of which would terminate in a word for which no substitution from the original list would apply. We shall call such words final and we shall call an algorithm the set of rules in accordance with which initial words are transformed into final ones*.

Once the meaning of the letters has been determined, then it may be possible, in the case of some particular field of study, to establish, as a result of observation or experiment, either certain algorithms, or certains chains of adjacent words, or certain final and initial words, or, lastly, the entire alphabet or some part thereof. The problem before us is to discover the entire system used for transforming initial into final words. Thus, in the case of the exploration of the polynucleotid chain, an alphabet of four letters is known to exist; these letters denote the bases referred to above. A number of final words, *i.e.* the sequence of aminoacids contained in the various specific forms of albumen, is also partially known. It is required to find the initial sequence of words and the algorithm by means of which they are transformed into final words.

Where the history of some human language is studied, we begin with a certain known set of final words and algorithms in use at the present time, also with a

---

* This definition of an algorithm may not be mathematically precise, but it will be sufficient for the discussion of our problems.

considerably smaller number of final words and algorithms which were in use in the course of its earlier stages of development. We require to reconstruct the nearest alphabet for which no record exists, the basic words and such algorithms used for their transformation, as would lead to the final words of the modern language.

In the case of monkey calls, we must reckon with the fact that neither alphabet, nor initial words, nor algorithms are known and that the recordings of final words are insufficiently accurate. Yet a model evolved for the study of more complicated structures helps to understand the simpler ones too. At the same time, a comparative study of simple and complex structures helps to understand the latter more fully.

(c) *Invariants*

In theory, a word may have any number of entries, because one can always add to a line, on the right, either one and the same sequence of letters, or even just one letter. If it so happens that, after many substitutions had been made, the adjacent word so obtained fails to coincide with any final word known from experience, this does not yet mean that some one basic word is not equivalent to a given final word, because not all possibilities of sequences of a chain of equivalent words have as yet been exhausted, even where the alphabet consists, for instance, of only four letters. But suppose the basic word is restricted; we can add no other word on its right. Even then, in a number of cases, no solution to the problem, by examining every combination of letters, is practically possible. This, for instance, is the case of the polynucleotid chain, where the alphabet has four letters. As Gamow remarks*, if a computing machine had been set, in the days of the Roman Empire, to check one million variations per second, and worked, without interruption, twenty-four hours a day, year in year out, then the work of computation would be incomplete even today. But even if the initial word happens to be so short that the number of possible sequences of letters is easily established, then the problem still remains of finding the algorithm used in the transformation of one word into another. Finally, in the case where it can be shown that an initial word coincides with a final one, *i.e.* where there is no transforming algorithm, then, the conclusion, in its comparative aspect, is still of scientific value, especially in investigations of communicative systems.

There is no need to calculate all the possible sequences of initial words in order to solve the problem of the existence, in any given system, of any algorithm for the transformation of words. The method of finding *invariants* is sufficient for this purpose. By comparing such final words as are known to us, we can discover the properties which remain unchanged for some definite group of final words. Such properties are called invariant. It is easy to discover an algorithm of substitution for a group of invariant words, if the constancy of the entry of certain words is noted, and to find a chain of adjacent words. But if two or three groups of words prove to be invariant as regards some feature, then the algorithm used to transform one group into another can be found. Finally, we may discover such invariants, as are peculiar to every word of a given system, and, accordingly, the structure of the initial words will be established. In this case, the set of final words of the given system can be regarded as being regulated by a definite number of algorithms which transform initial words into final ones.

---

* Ref. 4, p. 214 of the Russian translation.

References p. 180

(d) *Some illustrations* (*on Russian language basis*)

With the object of revealing the fundamental features of the model in which we are interested, let us now examine some of the simplest cases of communication by speech, as exemplified by the Russian language.

We shall use the Latin alphabet, instead of the Russian, and form, with its help, the word *stol* = K. We shall, further, introduce the symbol ∧ which is not a letter but denotes a null-word, *i.e.* a word which contains no letters of the alphabet. We shall suppose that there exists the following algorithm of substitutions to be applied six times in succession (column I).

COLUMN I*	COLUMN II	
1. K — ∧	1. stol·∧	(∧·a table)
2. ∧ — a	2. stol·a	(of·a table)
3. ∧ — u	3. stol·u	(to·a table)
4. ∧ — ∧	4. stol·∧	(∧·a table)
5. ∧ — om	5. stol·om	(by with·a table)
6. ∧ — e	6. stol·e	(about·a table)

With the help of this algorithm we obtain six words each of which is composed of two words. We shall place between them a point, which is not a letter of the alphabet, and record them in Column II. Every word in Column II occurs in the Russian language (it is an example of the declension of a Russian noun); we can therefore conclude that the algorithm in question correctly transforms the word *stol·*∧. It is easily seen that the chain of equivalent words so obtained has two invariant properties: (1) there are six, and only six, adjacent words and (2) all words to the left of the point are identical as to letter sequence.

It appears that the same system contains other words, such as *dom, les, vagon, kolos*, etc., which, as shown in Column III, are subject to the same algorithm of substitutions.

COLUMN III		
1. dom·∧	1. les·∧	1. vagon·∧
2. dom·a	2. les·a	2. vagon·a
3. dom·u	3. les·u	3. vagon·u
4. dom·∧	4. les·∧	4. vagon·∧
5. dom·om	5. les·om	5. vagon·om
6. dom·e	6. les·e	6. vagon·e etc.

We conclude that there exists yet a third invariant feature, namely that all words to the right of the point are identical as entries into an adjacent word bearing a definite number, for a certain group of words each of which lies to the left of the same point. For this reason, adjacent words of this group can be represented as in Column IV.

COLUMN IV	COLUMN V
1. S·∧	1. sten·a
2. S·a	2. sten·i
3. S·u	3. sten·e
4. S·∧	4. sten·u
5. S·om	5. sten·oj
6. S·e	6. sten·e

* The symbol — denotes the operation of a horizontal substitution (*cf.* p. 176, para. 20).

Here S denotes any one of the group of such words which, when placed to the left of the point, accept the entries to the right of it, as shown in the column. The words S contain various sequences of letters. This means that the algorithm which selects letters for a composition into a line has not yet been determined. This is why the adjacent words $S \cdot \wedge$, $S \cdot a$, etc. can be written down only in column form. This also means that these adjacent words are *not final*. It is as yet unknown which of the words of the group $S \cdot \wedge$ will be selected at any given moment.

While examining other words of the same language, we readily distinguish such groups of words as *stena*, in which the chain of six adjacent words is transformed by means of another algorithm, namely $a-a$, $a-t$, $a-e$, $a-u$, $a-oj$, $a-e^*$.

This group of words contains, for example, the words *muha* (a fly), *komanda* (a command), *gora* (a mountain), etc. In all other respects the invariance of the $stol \cdot \wedge$ group of words is retained, therefore the $stol \cdot \wedge$ group can be denoted by the symbol $S_1 \cdot \wedge$ and the $sten \cdot a$ group by $S_2 \cdot a$. The suffixes 1, 2, here stand for the number of the group, they are not letters of the alphabet. The symbol S_0 will then denote the class of words which are invariant with respect to the groups of words $S_1, S_2, S_3, \ldots$, as regards the features already referred to. (They are nouns.)

This examination can be easily extended to discover words of other classes which, under the influence of special algorithms, give rise to chains of adjacent words. These classes comprise, for instance, adjectives P, verbs C, pronouns M, numerals L, etc., as well as words to which no algorithm for transforming into adjacent words applies. An example of the latter class are the adverbs N. In the case of N, the invariant property is the non-existence of any algorithm for the transformation into adjacent words. Algorithms exist for the transference of words from one class to another and from group to sub-group. Example: $kolos \cdot \wedge$ (class S_0, group S_1, sub-group S_1^1), $kolos \cdot ik \wedge$ (class S_0, group S_1, sub-group S_1^2), $kolos \cdot its' a^{**}$ (class C). Words for which no algorithm for forming any adjacent words exists, (*e.g.* adverbs), are themselves transformed from other classes by means of definite algorithms. Thus the adverb $beg \cdot om$ (at the run) was transformed, by the application of the algorithm shown in Column I, in exactly the same manner as the fifth adjacent word $stol \cdot om$ in Column II. This is equivalent to the formula $S_1^1 \cdot om$. For our purpose it is important to note that there exists a class of words which neither possess algorithms for the formation of adjacent words, nor are transformed by any algorithm from words of any other class. Such are the interjections *ach*, *oj*, *o*, *a*, *ura*, etc. We shall denote them by the symbol $\bar{A}$ meaning words without algorithm.

(e) *Rules of forming a line of formal words*

Setting aside the class of words which is not subject to any algorithm, we shall first draw some preliminary conclusions and then proceed with further considerations.

We can show that various algorithms exist, in the language under our review, which prescribe any entries to be made to a word, both on its right and its left, as well as in the middle, yet, in every case, the number of entries is limited and itself algorithmically determined. We are thus enabled to introduce two new concepts

---

* The symbol t here represents a letter which is peculiar to the Russian language and is written ы.

** The symbol s' denotes a softened s and is a letter of the alphabet.

References p. 180

every entry made either in front of, or following, the point will be called an *incomplete word*, and the formation as a whole—a *complete word**. It will be easily seen that there can be no algorithm by means of which, within some given synchronous system, incomplete words could be formed from an alphabet of letters. But there are algorithms for entering incomplete words for the purpose of forming complete words. This means that *incomplete words are initial words* and a complete word—an algorithmic transformation of incomplete words. We must therefore prepare a special list of incomplete words with the help of which, and using certain algorithms, we can carry out the operation of selecting incomplete words which are suitable for the formation of complete words. We shall call this list the *Alphabet of Incomplete Words*. In this manner, on account of a shortage of algorithms, we are forced to introduce, in addition to the alphabet of letters, a new alphabet of a second order which is superimposed upon the first one. This alphabet drastically curtails the colossal number of possible sequences formed from the 39 letters of the Russian alphabet, when it comes to the formation of complete words containing, sometimes, more than ten elements. There is, for instance, no word *auiooj* in the Russian language, etc. Since the number of entries of incomplete words is limited and since every complete word is of final length, therefore there exists, between every pair of complete words, a gap which we shall denote by the symbol & which forms no part of either the alphabet of incomplete words of the given language. The formula $P \cdot oj \& S_1^1 \cdot \wedge$ stands for a combination of two complete words, the first of which is an invariant of the class P, the other an invariant of the class S_1^1. (Example: *bol'sch·oj stol·* $\wedge$.)**. The algorithm for the formation of combinations of this kind, known as syntagmas, can also be presented in the form of an implication, e.g. $(S_1^1 \cdot \wedge) \rightarrow (P \cdot oj)$.

The implication is interpreted as meaning "if a complete word belongs to class S_0 and group S_1^1, then the other complete word of class P, which is entered in the same line, will have, on its right, the entry of the incomplete word *oj*. Similarly, the formula $P \cdot aja \& S_2 \cdot a$ (as, for example, *bol'sch·aja sten·a*) is subject to the algorithm in the implication $(S_2 \cdot a) \rightarrow (P \cdot aja)$ and is interpreted in a similar manner.

We have now obtained a recording of two complete words arranged not in a column but in a line, *i.e.* we have arrived at a combination of complete words. Yet a combination of this kind must be regarded as an *incomplete syntagma*, because a further application of the same algorithm will once again lead to a recording, in column form, of six pairs of adjacent complete words.

COLUMN VI

1. $\wedge \rightarrow oj$	1. *stol·*$\wedge$ & *bol'sch·oj*
2. $a \rightarrow ogo$	2. *stol·a* & *bol'sch·ogo*
3. $u \rightarrow omu$	3. *stol·u* & *bol'sch·omu*
4. $\wedge \rightarrow oj$	4. *stol·*$\wedge$ & *bol'sch·oj*
5. $om \rightarrow im$	5. *stol·om* & *bol'sch·im*
6. $e \rightarrow om$	6. *stol·e* & *bol'sch·om*

This is how the algorithm operates. Suppose, for instance, the fifth adjacent word of the class S_0 (*stol·om*) to be selected, then the fifth, and only the fifth, adjacent word of class P (*bol'sch·im*) can be put in the same line with it. Since the invariant

---

* An incomplete word is usually known as a "morpheme", a complete word as simply a "word". Neither term is convenient for our purpose.

** The ' following the letter *l* denotes a softened *l*; *l'* is a letter of the alphabet.

group of adjacent words comprises a variety of elements and since the algorithm points only to the number of the adjacent word and to the class of words, then the sequence of letters in the word bearing some selected number remains indefinite. This means that syntagmas of the type $P \cdot oj \& S_1^1 \cdot \wedge$ or $P \cdot aja \& S_2 \cdot a$ are not final words.

As can be seen, a syntagmatic algorithm transforms a columnar recording of adjacent words into a linear recording. It is as if, in accordance with a definite rule, it "scooped up" adjacent words from different columns and arranged them in a line. Only such elements can find themselves arranged in a line, as are compatible, according to some algorithm. It is obvious, too, that the checking for compatibility must be carried out word for word in each pair. The syntagmatic algorithm may therefore be referred to as the *algorithm of compatibility of two words arranged in a line*.

If we wish to extend the linear recording, we must know for which particular word a compatible one is being selected. No continuation of any linear recording is otherwise possible. Thus suppose we have selected the third adjacent word of the class S_1^1, namely $S_1^1 \cdot u$ (*stol·u*; *dom·u*) and obtained the syntagma *bol'sch·omu stol·u*; *bol'sch·omu dom·u*. But we could have selected the fifth adjacent word *stol·om*; *dom·om* instead. We would then have obtained the syntagma *bol'sch·im stol·om*; *bol'sch·im dom·om*. Since the selection of the first compatible word is indefinite, the problem has more than one solution. This is precisely why syntagmas of this type are incomplete.

If any linear recording is to be continued, we must find the *algorithm of the complete syntagma* which determines the number of the first adjacent word with which another adjacent word is compatible. For every word thus selected there can be assigned, in accordance with the algorithm of incomplete syntagmas, places for any entry of other adjacent words taken from columns which correspond to one or other class, group and sub-group of invariant words.

An algorithm for the transformation of words into a complete syntagma can be formulated in the following manner. The first word of a complete syntagma may be either a first adjacent word of class S_0, or an equivalent first adjacent word of another class, or a word which has no adjacent words. The second word of a complete syntagma may be taken from various classes of words but it must always be equivalent to the first word of the complete syntagma.

To enable us to apply this algorithm, we must introduce the idea of a linear equivalence of words of various classes*. Assume the existence of the syntagma $C \cdot it \& S_1^1 \cdot \wedge$ (*sto·it stol·* $\wedge$). The word $S_1^1 \cdot \wedge$ satisfies the requirements of the algorithm referred to. (It is the first adjacent word of the class S_0.) It is, therefore, the first word of a complete syntagma. Table 16 contains an algorithm of linear equivalence of three classes of words: C, M and S_0. It shows that each of the three adjacent words of class C has a corresponding adjacent word of class M. Yet only the first adjacent word ($S_1^1 \cdot \wedge$) of the class S_0 corresponds with the third adjacent words of the classes C and M. No other correspondence exists**.

---

* We have so far discussed only the equivalence of words of the same class, group, sub-group, and the transformation of words of one class into another.
** This correspondence is established simply by examining a passage of some length written in the given language. It then appears that one comes across only the syntagma *stol·* $\wedge$ & *sto·it*; never *stol·* $\wedge$ & *sto·ju* or *stol·* $\wedge$ & *sto·isch*.

References p. 180

TABLE 16

Class C	Class M	Class S_0
1. $C \cdot iu$ (sto·ju) (stand·∧)	1. ja (I)	0. —
2. $C \cdot isch$ (sto·isch) (stand·∧)	2. tt (thou)	0. —
3. $C \cdot it$ (sto·it) (stand·s)	3. { on (he) / ona (she) / ono (it) }	1. $S_1^1 \cdot \wedge$ (stol·∧) (table·∧)
		2. $S_1 \cdot a$ (stol·a)
		3. . . . . . .
		4. . . . . . .

It follows that the words $S_1^1 \cdot \wedge$; on; $C \cdot it$ are linearly equivalent. We shall denote this by the symbol $\sim$.

Thus, in accordance with the algorithm of the complete syntagma, any separate entry of the word $C \cdot it$ to two such words is admissible in a line and is sufficient for the formation of a complete syntagma. We shall use the symbol $\&$ for the combination of words to form a complete syntagma. Thus $S_1^1 \cdot \wedge \& C \cdot it$; $on \& C \cdot it$ are examples of complete syntagmas.

An entry of an incomplete syntagma may be made to the word $S_1^1 \cdot \wedge$, as, for instance, $(P \cdot oj \& S_1^1 \cdot \wedge) \& C \cdot it - bol'sch \cdot oj\ stol \cdot \wedge\ sto \cdot it$. With the help of algorithms of incomplete syntagmas an entry may be made to any word of a complete syntagma.

A complete syntagma may consist of one complete word. This can be proved by checking lines for linear equivalence. The first adjacent word of the class M, namely ja, has, in addition to the adjacent words shown in Table 16, yet another chain of adjacent words in all respects analogous with the six adjacent words of the class S_0, namely the words ja, men'a, mn'e, men'a, mnoj, mn'e. Therefore $ja \sim S_1^1 \cdot \wedge$ and hence $ja \& C \cdot ju - ja \& sto \cdot ju$ is a complete syntagma. Here the elements of the combination of a complete syntagma are incomplete words.

If a further investigation were undertaken into the linear equivalence of series in the classes of words, other types of combinations of words to form a complete syntagma could be discovered, but such an investigation would lead us away from our subject.

Thus the entire set of algorithms for the entry of complete words is just a list of rules for the selection of complete words from columns of the class of equivalent words, so as to form a line of words. This statement is theoretically correct for any human language, since the sequence of elements of a line of words cannot be arbitrary. Otherwise, in a message in a certain language, an arbitrary order of elements would be taken by a recipient for a disorder. Sets of algorithms for different languages will, naturally, be different. For instance, there will be some difference in the case of analytic and of synthetic languages. In analytic languages (*e.g.* English), as distinct from synthetic (*e.g.* Russian), there will be a smaller quantity of numbers in a column of adjacent words, or some numbers will be filled either by incomplete words of a special class (prepositions) or by null-words which will correspond to the related number of a non-null-word in a line of words (the place of the word in a syntagma).

(f) Founded alphabets hierarchy

To sum up, algorithms of complete and incomplete syntagmas transform adjacent words of various classes from a column into a line. As soon as a complete syntagma has been obtained, the flow, into the combination, of further entries of words may be interrupted at will and the same operation (that of the selection of complete and incomplete syntagmas, with the help of the same algorithms) repeated. We shall introduce the symbol $\mathbf{I}$ to denote the repetition of the application of an algorithm. We can write $S_1^1 \cdot \wedge \& C \cdot it \mathbf{I} M_3 \& P \cdot oj$, (for example, $stol \cdot \wedge sto \cdot it \mathbf{I} on\,bol'sch \cdot oj$)*. We now possess every algorithm used to add to a line, on its right, still further entries of complete syntagmas. Yet the word so obtained (in accordance with out definition, everything written in one line is a word), however long, cannot be regarded as final, because it has not been made clear, in accordance with which rule any invariant complete word may be chosen for entry into a line from the variety of possible claimants. Thus, instead of the word: *The table· ∧ stand·s* $\mathbf{I}$ *It is big*, the word *The flower·s grow· ∧* $\mathbf{I}$ *They are bright* would equally suit the formula. The solution of problem would depend upon the existence of an algorithm for the selection of concrete words, yet no such algorithm exists in the language under review. Moreover, even with the use of every algorithm known to this language, we can still obtain lines containing incompatible words, *e.g. The table stands. The table does not stand*. Both words are algorithmically correctly constructed, yet they are obviously incompatible, when contained in the same line.

Just as, on p. 142, we were forced, owing to an absence of algorithms, to prepare an alphabet of incomplete words, and only with their help were enabled to discover the algorithms for the transformation of incomplete words into complete words, so now we must prepare an alphabet of complete (concrete) words, thus settle the question of the possible existence of an algorithm of compatibility of complete words placed in the same line and decide to which system this algorithm belongs.

In section 2(b) it was explained that a letter of the alphabet was merely a symbol which differs from any other symbol. As soon as we have selected one or more letters and arranged them in a line, then each of them or their sequence became a word of this alphabet. We can now attach to a word any meaning we please. We have thus obtained words of various types, *e.g.* incomplete and complete words, words which are incomplete syntagmas and others which are complete syntagmas. The question arises of the type of words to which a meaning may be attached, and if so, then what particular meanings.

Suppose that every letter of our alphabet A exactly corresponds with a letter of the alphabet of some actually existing language, say French or English. At this instant we have made the first vertical substitution and, using this arbitrary algorithm, have obtained an equivalent word. Any passage in the selected language may now be transcribed in our alphabet which has now been strictly limited. We shall now carry out a second vertical substitution · using a definite code, we shall transform each letter into an electric signal. The result is another word in a new alphabet (see para. 21 of the Glossary). These electric signals may now be transmitted along a line of communication, provided that the words retain their equivalence, *i.e.* so long as they do not lose any vertical substitutions which are prescribed for them and remain adjacent words throughout their entire journey. At the output end of the line, a final word is decoded and thus reconverted to the initial word, *i.e.* to a letter of the alphabet belonging to the selected language. The entire passage, as accepted at the input end of the line, is now completely equivalent to the passage received at its

---

* M_3 here denotes the third adjacent word of class M namely *on*, in Table 16.

References p. 180

output end. We may say that every word was converted into itself by means of algorithms invented for the purpose of coding; this is how the transmission was effected.

The two important points to note are: (a) only one of the embedding alphabets (see para. 27 of the Glossary) was necessary, *i.e.* the funding alphabet of letters of some particular language, (b) all adjacent words arranged themselves directly into a line; there was no arrangement in columns or any selection of adjacent words from the embedded alphabet of in complete words. It follows that for certain systems only an initial, embedding alphabet of letters and a limited number of rules for coding are sufficient, because the algorithms of complete words, those of incomplete words and the algorithms of both kinds of syntagmas were already applied to the text and reflected in the sequence of letters before its transmission along the line.

A machinal translation from one language into another presents a different picture. In this case, besides an alphabet of letters, we must take into account also an alphabet of incomplete words, because we must discover the algorithms for the equivalent transformation of incomplete words into complete words in two different languages. A translating machine transforms the input of final words of one language into an output of final words of a different language. Yet, here too, no alphabet of complete words is necessary, because the selection of complete words has already been made in the text at the input end and has been reflected in the sequence of the letters of the incomplete words.

If we now turn to the natural mechanism of communication either of man or of an animal, then we are inevitably faced with the need to account not only for an alphabet of letters and of incomplete words but also for an alphabet of complete concrete words, because only then are the invariant series of words limited and the final word, *i.e.* an exact sequence of letters, produced.

(g) *Equality of unequal concrete words in the message*

Thus when we exchanged some abstract alphabet $A = \left\{ \begin{array}{c} a, b, \\ c, d \end{array} \right\}$ for the alphabet of letters of a certain language we made a vertical substitution with the meaning: $a \vdash$ a letter of that language (where $\vdash$ denotes a vertical substitution); $b \vdash$ another letter of the same language, etc. When the letter K of that language was exchanged for the electric signal e.s. ($K \vdash$ e.s.), then e.s. became the meaning of K.

When, in a translating machine, we have exchanged the alphabet of such incomplete words of a given language, as can be transformed into complete words of the same language, for an alphabet of such incomplete words of another language, as can be transformed into complete words of this other language, then we have likewise made a vertical substitution with the meaning $Q_1 \vdash Q_2$ (where Q_1, Q_2 are the two languages).

The question now arises whether any vertical substitution is made and, if so, then what meaning it thereby acquires, when, within a system of communication a selection of complete concrete words is made.

It was pointed out that the language we have earlier schematically described has no algorithm for the selection of concrete words from a group of invariant words (*i.e.* there is no rule for determining whether the word *cow* and not the word *wall*, or the word *Alfred* and not the word *Victor* should be selected). It is this property of the language that makes it into a means for transmitting a communication (message) or endows it with the function of maintaining communications.

For suppose, to begin with, that the recipient of a message does not possess all the alphabets and algorithms of the given language. Then the probability of obtaining, at the receiving end of any information concerning the word order within the limits of invariant groups is less than unity. It means that no matter what selection of concrete words was made at the source of the message, these words will, at the receiving end fail to arrange themselves in a line, and the message will be misunderstood.

Now suppose that both the sender and the receiver possess all the alphabets and a complete list of algorithms of some language. Then the probability of receiving the information concerning the word order within the limits of invariant groups is unity, which corresponds to zero information. But the probability of receiving any concrete word is less than unity, therefore, the selection of a certain concrete word at the source of the message will bring to the receiver (in the absence of any noise) an amount of information which is equal to $-\log_2 p$, where p denotes the probability of the appearance at the receiving and of this concrete word.

The weight attached to the selection of concrete words in the human intercommunication by speech can be roughly determined. Since the set of algorithms transforming initial words into final ones is the same for the set of invariants of complete words of all classes, then certain letters, of the limited number contained in some alphabet, will regularly repeat themselves in the next, being subject to the strict algorithm of the composition of complete words. This means that the probability of the occurrence of a certain sequence of letters could be predicted.

In many languages such a superfluity in the sequence of letters actually constitutes about 50%. Consequently about 50% of the remaining uncertainty in the sequence of letters in a text is due to the selection of concrete words.

Thus a complete syntagma (let us denote it by Σ) contains no message until the selection of concrete words has taken place. We shall call such a syntagma—a formal word.

Only after the selection of concrete words has a syntagma become a message ($\mathcal{M}$). We must explain the nature of the ratio $\Sigma/\mathcal{M}$. We shall then know what is meant by the meaning of a concrete word. We shall take for granted that may $\mathcal{M}$ consist of several elements. The question of the ratio $\Sigma/\mathcal{M}$ is then reduced to the problem of so matching the concrete words in Σ that the elements in $\mathcal{M}$ become compatible.

Suppose that we are watching a bonfire. A sequence of gradual changes in the state of the burning material takes place, until a black spot and charred grass remain where the fire had been before. We shall call the series of states, which the burning material had assumed, a real series. We shall denote each of these states by a letter of an arbitrarily chosen alphabet. Then each letter, or any definite sequence of them, will be a complete concrete word which will be equivalent (in accordance with our agreement) to the corresponding state of the real series.

In this case each complete word is compatible with the preceding one, because the order of states in any real series cannot be incompatible. Therefore a selection of letters, with the purpose of obtaining a line of compatible concrete words, is possible if we introduce a vertical substitution of the order of the elements of a real series by concrete words.

An important difference exists between an objective series and its equivalent series of concrete words. We could call the initial state the input and the final state the output. Here a state is transformed into an adjacent one in accordance with some kind of algorithm. Yet the entire chain of an objective series is infinite or, in any case, incapable of a complete survey. Therefore both its beginning and end are arbitrarily defined in terms of a linear interval. On the other hand, a series of concrete words is invariably some finite and cyclic system. The same alphabet must exist and the same algorithms must function at the output end as at the input. Otherwise no message can be received. There is no sense in transmitting a message which cannot be received.

If a series of complete words exactly corresponds with an objective series of states, then the number of complete words or the volume of the vocabulary is equal to the number of elements contained in the objective series. In this case the message can contain nothing more than what is contained in the lexicon. Such a system may be called a system of *non-expanding* messages. One, two, or more of its N words may be selected to denote some definite objective situation.

Another system is also possible containing, besides an algorithm of vertical objective substitution, also an algorithm for horizontally substituting or exchanging one concrete word for another. This becomes possible as a result of operations with words.

Let the complete syntagma Σ convey the message $\mathcal{M}$: *A birch is a tree*. We shall divide Σ and $\mathcal{M}$ simultaneously into two parts in such a way that the final* vertical substitutions, so denoted in M, are equal and the concrete words in Σ unequal. If we mark the place of division of Σ and $\mathcal{M}$ by the same line | we shall get: *a birch* | *is a tree*.

Final vertical substitutions in $\mathcal{M}$ will be denoted by the Greek letters B and Δ. Let us postulate the following logical conditions: $\forall$ is a quantifier of generality denoting "all" and $\exists$ a quantifier of existence denoting "there exist some (certain)".

The symbols $\forall$ and $\exists$ are not letters of the alphabet A from the elements of which the words in Σ are composed. Let the concrete word *birch* $= b$ and the concrete word *is a tree* $= t$. Then we have $b \neq t$, $\exists \Delta = \forall B$ which means that there exist such (some) Δ that are equal to all B. (*All birches are trees.*)

Thus the equality of two objects conveyed in $\mathcal{M}$ which we observe under definite logical conditions permits the combination in Σ of unequal concrete words. We shall call the property acquired, in this case, by concrete words their lexical meaning, and the final formation obtained when such lexical meanings are combined —a concept (Π). Lexical meanings were obviously formed as a result of combinations of final vertical substitutions for two concrete words. This can be denoted by the sign $\longleftrightarrow$. Thus $b \longleftrightarrow t$ (or, in a general form, $\Sigma/\mathcal{M} = \Pi$). Π, as we can see, is a null-word in the alphabet A from the elements of which Σ is formed, because neither of the letters b or t occurs in Π**. We can now draw a conclusion which is even more important to our thesis. Under definite conditions one concrete word may be substituted for another concrete

---

* By a final vertical substitution we understand one after which no other vertical substitution is possible.

** This is obvious, because a concept of something is formed by the juxtaposition of words, or, as we say, in the context. Concepts are to be found not within the words themselves but between them, as it were.

word, because they were equated in their lexical meanings through the equality of the significances in $\mathscr{M}$. Thus the word *a birch* may, under certain conditions, be substituted for the word *a tree*. An obvious example is the concrete word *Peter* when used instead of *son* and the word *pine* when used instead of *tree*. In order to make such an exchange effective, we require a special message concerning the alteration in the meaning of the words contained in the message. Under such a system, every word of a lexicon may have a directive added to it concerning its exchange with another word with a fresh meaning which has been previously defined as an objective substitution for a certain word. Since an objective substitution can only be introduced into complete concrete words forming part of a complete syntagma (line), then the algorithm of the exchange of words through the mechanism of a complete syntagma transforms only concrete words, without affecting complete, incomplete or formal ones. This means that every algorithm which regulates adjacent words in a column is preserved and every complete formal word is correctly transformed. In the outcome, if the line of words has been correctly constructed, the number of concrete words in an alphabet will be less than the number of possible objective substitutions. To put it differently, the same concrete words may be used to communicate different things. We shall refer to such a communicative system as one of *expanding* messages. Let the line of words *acadblo* be recorded in this system and suppose that these words are given a definite objective meaning. An entry, in this line, of a new word, such as *aril*, may alter the objective substitution of the preceding words. For instance, the addition, to the complete Russian words for *steep slope*, of the words *of the mountain* or *of the instep*, will change the objective meaning of this incomplete syntagma (steep slope of a mountain, steep slope of the instep of a foot). Such a change in the meaning may cause a loss in the equivalence of the input and output of a communicative system. This occurs not infrequently in the phenomenon of mutual misunderstanding.

This effect can be eliminated, if an algorithm for the equivalent exchange of words can be found. In that case it becomes possible to find the final compatible complete words, without having, every time, to check the correspondence of a word with an objective substitution. It will be sufficient to carry out such a check only in the case of one initial word and then, if the algorithms are known which transform the initial words into the equivalent adjacent ones, the final word will consist of entries of compatible words and will correspond with the objective series. An algorithm for the exchange of words is a message concerning a message. One message becomes an objective substitution for another.

Suppose an alphabet of complete words contains such words as A, B (having any desired objective substitution) and such symbols of operation for the exchange of words (none being a word or a letter of the alphabet) as $^-$ (sign of negation, *e.g* $\bar{A}$ means Non-A); $\vee$ (sign of disjunction—either...or); $\rightarrow$ (sign of implication—if...then); & (sign of combination—both...and); () (brackets). Then if we denote equivalence by the sign $\sim$, the word $A \rightarrow B$ may be exchanged for the word $\bar{A} \vee B$, *i.e.* $(A \rightarrow B) \sim (\bar{A} \vee B)$, which follows only from the algorithm of operations with words, irrespective of any *objective* substitution. In the same manner, $A \rightarrow B \sim \overline{A \& \bar{B}}$; and $A \vee B \sim (\bar{A} \rightarrow B) \& (\bar{B} \rightarrow A)$; and $A \sim \bar{\bar{A}}$; and $\overline{A \vee B} \sim \bar{A} \& \bar{B}$, etc.

It is evident that the final line will contain indications concerning compatible

References p. 180

and incompatible words. Thus $\overline{A \& B}$ shows that A and $\overline{B}$ are incompatible. Such algorithms for the exchange of words are called logical algorithms and their domain that of meta-language. The set of logical algorithms does not form part of the set of algorithms for the transformation of letters into incomplete and complete words and into the words of a syntagma. These are two distinct links of a communicative system. Algorithms for the transformation of letters are algorithms governing the correctness of a message, they ensure the equivalence of the input and output of the communicative system. Logical algorithms are algorithms of correctness and incorrectness. They ensure the continued existence of an objective substitution, once it has been adopted.

If both sets of algorithms are employed, then final concrete words having a definite sequence of letters can be obtained from any given alphabet of letters. *This is the ultimate aim of any communicative system.*

We have given a general sketch of the most complicated communicative system. An actual, living communicative system, *e.g.* one used by man, fails, at times, to function correctly and accurately. The analysis of failures in the operation of algorithms is of great interest to psychoneuropathologists. Of no less interest is the process of building the mechanism while the use of algorithms is being acquired during the process of ontogeny. But the same conception of a communicative system can be applied also to the process of formulating the problems of phylogeny. An analysis of vocal communicative signals used by monkeys offers some useful material in this connection.

3. The Sound Communicative System of Monkeys

(a) General description

The Academy of Medical Sciences maintains, at Sukhumi, on the Black Sea shore of the Caucasus, a nursery for monkeys. It houses a number of species, of these the baboons (*Papio hamadryas*) are best suited for our investigation. The chests of the males of these large monkeys of canine appearance and with narrow noses are covered with thick silvery hair. A small family group of these animals was imported from Abyssinia. The conditions under which they live in Sukhumi closely resemble their native habitat. Their breeding habits are normal. The herd now numbers some tens of specimens who freely roam a sort of spacious aviary with an area of not less than one hectare.

Although the baboons belong to the lower apes, they use a sufficiently wide variety of communicative calls. The reason is that the animals are tied by rather complicated herd relations and ruled by a strong leader. Monkey calls of one sort or another are heard all day long almost without interruption; they die down only with the onset of sleep. Prof. Tikh of the University of Leningrad, who has studied the gregarious behaviour of the baboons for several years, in connection with the problem of anthropogeny, reckons that they can utter up to 18 different signal calls[14]. If their several versions are also taken into account, then, in her opinion, this number may rise to 40. Our present investigation was limited to an acoustic analysis of only a few of the more widely used calls. The object in view was to reveal the general structure of this peculiar communicative system.

There is a widely held opinion that animal calls, including monkey calls, possess no objective meaning. They are commonly looked upon as merely signs of emotional states. Such a view cannot be regarded as being well founded, at all events not in the case of the baboons. Even a simple observation, made without instruments of any sort, shows that the calls made by the baboons possess a sufficiently definite meaning as signals carrying a message. Using the term which we have introduced in section 2 (g), we may say that the call is regulated by an algorithm of objective vertical substitution. It is true that a series of varying states of expression evolves simultaneously with the call, but the call is a signal concerning a situation, not a mere utterance given to this state. This conclusion follows as a result of even the simplest experiments carried out in the enclosure here described.

If somebody, unseen by the monkeys, takes his station behind a large stone not far from the herd and suddenly shows to the animals a long stick carrying a net used for catching them, then loud calls immediately resound. These are usually denoted, in a conventional manner, by *ak, ak, ak**. We shall, at a later stage, present an acoustic analysis of these sounds. The call is repeated by other members of the herd, while every animal turns in the direction of the object which has so suddenly appeared. In the intervals between the calls, some of the individuals, mainly males and, in particular, the leader, make a characteristic gesture of scratching the ground with a fore-paw. The call is repeated even by those baboons who, to begin with, never saw the object. To a human ear, the *ak, ak, ak* call is the same with all baboons, the only difference being in the fundamental tone; the pitch is lower and the sound itself louder in the case of the leader.

If a human imitates the *ak, ak, ak* call, many baboons repeat it. Yet the reaction of the herd is less vigorous in the absence of the objective situation which has given rise to the calls. A similar but more highly pitched *ak, ak, ak* call is uttered also by the macaque (*Macaca*), a monkey of the genus of narrow-nosed, sub-family of Cercopithecinae. In Sukhumi, the macaque kept in large cages placed at a higher level than the enclosure of the baboons. Any calls made by the macaque are heard by the baboons. They produce no response from the latter. If a monkey of a different species is carried past a cage occupied by the macaque, there is silence, but if the monkey happens to be itself a macaque, the inmates of the cage meet it with loud calls of *ak, ak, ak*.

We thus have reason to believe that the *ak, ak, ak* call of the baboons possesses an objective meaning as a signal. It is uttered in any new and unexpected situation and is a message concerning danger. Prof. Tikh justly calls it "the danger signal". We must suppose that these animals are equipped with a mechanism for the reception and transmission of messages. It is evident, too, that they communicate not their emotional states but changes in an objective situation. On the input side, the message is in the ever unchangeable vocal form of *ak, ak, ak*. Individual animals transform the message they have received each in his own way. Some scratch the ground with a fore-paw, the reaction of other animals is to turn their heads, as if taking bearings, mothers lift their young onto their backs, without invariably repeating the call. If one of the monkeys was engaged in "searching" another one (a common practice among monkeys of cleaning each other's skin and fur), then this activity stops as soon as the danger call has been heard. Thus the information conveyed by this call

---

* Monkey calls will be described by Latin letters, for instance a letter *a* in *ak* must be read as in "bath", a letter *i* in *ee* must be read as in "keep", a letter *au* in *au* must be read as in "now".

References p. 180

passes through the most different channels and is transformed in a different way in the case of each member of the herd.

As a rule, the *ak, ak, ak* call is repeated several times but, according to Prof. Tikh, the baboons have a version of the *ak* call which is uttered only once. This shrill sound with a very high fundamental tone signifies extreme danger. On hearing it, the frightened herd takes to flight.

The existence of a communicative significance in monkey calls becomes more certain if a comparison is made between a number of calls and the corresponding objective situations.

If one or more individuals are removed from the herd and carried away in a cage, there follows a long period during which calls of *au, au, au, au* are heard from both sides. There is no difficulty in discovering the communicative significance of these sounds. Members of a herd keep, as a rule, close together; they do not stray outside the limits of some definite spatial contact. The interval of this contact varies with different age groups. If a mother loses her young out of her sight, she utters the *au* call. The more mature and fully grown baboons may lag behind the herd at a greater distance. The *au* calls then come from these stragglers and the herd responds with the same sound. Thus the *au* call signifies loss of contact with the herd and the reaction produced is that of search by the herd of the lost member and *vice versa*.

While the *ak, ak, ak* calls follow each other rapidly in bunches, lengthy pauses are noticeable between the *au, au* calls, in the course of which any reply calls from the straying animal may be heard. They are a sort of roll call, a sort of "check" on the presence of the complete herd or family group. Both as regards the situation and the character of the call, there is a strong similarity between it and the calls exchanged between members of a company of men roaming the forest. The difference which the ear can distinguish is only in the stress which the baboons put on the *a* in *au* and the humans, usually, on the *u*, and the fact that the human *u* is protracted. The baboons use also an *ou* call in the same situation. The latter call has, presumably, no distinct communicative significance. Some individuals use the *au*, others the *ou* call. Sometimes one and the same baboon produces both calls indiscriminately.

The influence of the leader of a herd of baboons is very considerable. If two monkeys fight so viciously that one of them, probably the weaker, utters a shrill squeal, then the leader interferes, catches the stronger one and gives it such a thrashing that the squeal now comes from this one. The intensity of the squeal constantly changes; it has a high pitch and resembles an *ee* sound in its timbre. The situation which invariably gives rise to these sounds leads us to believe that its communicative significance is that of an appeal to the leader. This is obvious from his behaviour. On hearing the squeal he, as a rule, interferes in the fight. His power is enhanced by the fact that the squeals of the animal he has punished remain unanswered. Except for the leader, no member of the herd interferes in a fight.

Mutual relations between members of a herd of baboons are rather varied. Besides the special relations between the fully grown on the one hand and the young and adolescent on the other, various contacts are formed between the fully grown animals. These are apparent both from gestures and from sounds. The more typical forms of gesture are "searchings", huggings and "presentations". In the last case, one of the monkeys (irrespective of sex) turns its hindquarters towards another who either responds to the gesture or turns away from the first one. Before the huggings and searchings begin, in their course and, frequently, quite independently of them, a soft and rather complicated signal of mutual satisfaction can be heard. Very roughly it could be represented by *hon* in which the *h* is something like an aspirate which is formed within the stricture when the rear part of the back of the tongue is raised with only a small amount of tension. The *on* consists of a clearly audible *o* with a nasal resonance. The *hon* sound is undoubtedly a sign of some physiological state but, at the same time, it is a message concerning the presence of this state, because it may be either accepted or rejected by the partner. Only certain quite definite pairs of monkeys exchange these sounds.

In addition to the sounds we have already described, we have succeeded in detecting a muted, voiceless sound which is unlike any sound produced by humans. It is produced by a clearly visible pressing together of the lips and by some sort of complicated movement of the tip of the tongue. This whisper is of so low an intensity that it has eluded every attempt at recording on magnetic tape, since the microphone must still be placed at a certain distance from the animals. As to its communicative significance, the sound is best described as being orientative, yet concerning only an individual animal, not the entire herd. It is uttered when an individual becomes aware of a change in the situation, as, for example, when a mother approaches her young or when one monkey is close to another. The sound is accompanied by looking round.

It is thus possible to establish quite accurately the existence, among the baboons, of at least seven different signal calls, the two versions of *ak*, the *au*, the *ou*, the *hon*, the squeal (*ee*) and the lingual and labial whisper which we shall denote by *ptpt*. These are the most common sounds used by the baboons and can be heard without difficulty, with the exception of the high pitched version of the *ak*. The list is incomplete, yet it is sufficient to raise the question of the structure of the vocal system of intercommunication between the baboons.

One ought, first of all, to note that the sounds here described can be classified in accordance with their intensity. Every version of the *ak*, *au* and *ou* calls is always very intense. The level of intensity varies, as a rule, within the limits of 90 to 110 db., the *ak* being louder than the *au* or *ou*. The *hon* sound is always considerably less intense, it reaches a level of only 40–50 db. The softest sound is the whispered *ptpt*. These distinctions are very marked from the point of view of the acoustic character of monkey calls. The point is that not only does the situation which forms the subject of the message determine the intensity of the call, but that the phonic system of the baboon which is responsible for the production of these calls is incapable of regulating their intensity in a continuous manner. The sounds are either very loud or very soft; there is no transition from one degree of loudness to the next. This peculiarity to which explorers, as a rule, pay little attention, is a key and a good indication to the solution of the problem of algorithms of the sequence of sounds of the communicative system of monkeys.

(b) Some notes on registrating technique and methods of sound analysis

Sounds uttered by baboons living in herds in normal conditions were recorded on magnetic tape which, like the microphone used for the purpose, could register frequencies of up to 10,000 c/s. The magnetic recordings were analysed by means of a spectrometer. The terms of our investigation required no greater accuracy in the

References p. 180

analysis of the frequency components of the spectrum than is necessary for the definitions of formant regions which constituted the object of the study are contained within a sufficiently wide frequency band of the spectrum. It was more important to estimate the level of the sound. A spectrometer was therefore used which enabled to obtain, from a continuous sound, a series of spectra taken at intervals of 20 msec[*].

The impulse to be investigated was simultaneously fed into 32 filters and the result of the analysis was recorded in a band oscillograph. A record which is obtained for every 20 msec will be called a frame. Since the bands of resonance of two consecutive filters overlap by up to 10 %, an allowance was made, during the examination of the summary spectrum, for the total duration of the sound and for the frequency of the appearance, in a given filter, of a given harmonic. The diagrams which appear further in this paper show summary spectra. The abscissae contain the numbers of the filters, while the ordinates are the linear amplitudes of the frequency plotted in arbitrary units of length. The limiting frequency of the first filter is 62 c/s, the sensitivity band 80 c/s, the resonance frequency 102 c/s. In future we shall indicate only the number of the filter whose resonance frequency is shown in Table 17.

TABLE 17

Filter No.	Resonance frequency	Filter No.	Resonance frequency	Filter No.	Resonance frequency	Filter No.	Resonance frequency
1	102	9	1 119	17	2 942	25	6 277
2	193	10	1 292	18	3 552	26	6 905
3	300	11	1 983	19	3 582	27	7 596
4	415	12	1 687	20	3 937	28	8 355
5	536	13	1 902	21	4 322	29	9 191
6	666	14	2 132	22	4 737	30	10 110
7	806	15	2 382	23	5 187	31	11 121
8	956	16	2 652	24	5 706	32	12 233

4. Discussion of Results

(a) Review of human and monkey's signal sound spectra

Let us first examine the group of sounds of the *ak* type. Those frequencies in the spectrum which have relatively wide amplitudes we shall call formant frequencies. Fig. 68 shows four spectra of the *ak* sound as uttered by different baboons. It is obvious that the formant frequency of the 13th filter appears in each of them. In addition, there appears, with an even higher amplitude, the frequency of either the 8th, or the 9th, or the 10th filter. If these spectra are compared with the spectra of the other calls uttered by baboons of the same herd, c.f. Figs. 72, 76, 77 and 78, one is forced to conclude that the baboons possess an invariant group of *ak* sounds, each invariably accompanying a definite situation and each differing from every sound of the other invariant sound groups.

Fig. 69 shows the spectra of two imitations of the *ak* call made by a human

---

[*] This spectrometer was made in the Laboratory for Experimental Phonetics and Psychology of Speech, in the First Government Pedagogical Institute of Foreign languages in Moscow. (Prof. V. A. Artemov is in charge of this laboratory.)
It is more or less similar to the Siemens spectrometer described by Andrieu on p. 25 of this book.

voice. Here too the formant frequencies of the 13th and 9th filters are prominent. Since the baboons respond to the imitated *ak* sound, we must conclude that the information contained in its message is transmitted through these formant characteristics of the spectrum. All formant information may be regarded as being objective,

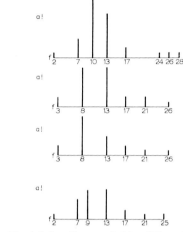

Fig. 68. *ak* (*a!*) calls recorded by different baboons.

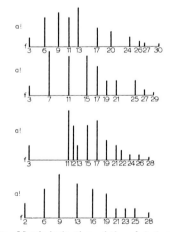

Fig. 69. Man's imitation of the *ak* (*a!*) call.

situational vertical substitutions produced through the operation of the algorithm of the exchange of a uniform spectrum of a casual noise for a non-uniform, selective spectrum. The appearance, within the uniform spectrum of a casual noise, of any particular frequency is random. The algorithm consists in that this spectrum is replaced by a non-uniform and selective one, with the separation of certain frequencies in such a way that a definite non-uniform spectrum is always followed by an equivalent objective situation. A column of two adjacent words thus makes its appearance. The selection which has been carried out in the spectrum corresponds with the selection of one situation from all those capable of identification in the system under discussion.

References p. 180

Fig. 70 shows the spectrum of the *a* sound of the *ak* syllable, as pronounced by a human. It is clear that the formants arrange themselves in the 9th and 11th filters but the common envelope of the spectrum is similar to those of the spectra in Figs. 68 and 69. The proximity of the formant zones and the similarity of the envelopes testify to the fact that the *a* sound, both in the case of man and of baboon, belongs to a common and related zone of spectra. Experiments carried out by ear only are reliable in this case, they lead to the same facts as established by more objective methods. It should be said that, in general, vowels pronounced separately (without any other sounds of a word) by people of various nationalities are recognised as being similar. This became evident from the investigations which were carried out by Artemov and Zimnija in connection with ten languages (Russian, English, French, Czech, Bulgarian, Albanian, Armenian, Georgian, Spanish and Lettish)[2].

Fig. 70. Man's normal pronunciation of the *a* in the *ak* syllable.

Although there is a similarity in the formants, the eye is struck by the sharp difference in the level of the human and the simian *a*.

The difference appears in respect of the three parameters—time, strength and attack. In no record of the *ak* call obtained by us from baboons does the duration exceed 120 msec. A human imitation of the same call lasts from 160 to 240 msec. In normal speech, the duration is 240 msec.

Differences in strength are even more marked. Normal speech does not rise above 75–80 dB in level. The *ak* call of the baboon invariably exceeds 100 dB. It is a very powerful call. But the greatest difference is apparent in the attack of the sound and in its ending. Like any vowels, the spoken *a* is smoothly modulated in loudness. The amplitudes of the harmonics do not begin to grow before 60 msec (in the fourth frame) and the frequency bands have, even then, not yet been stabilised. Everybody who has studied the spectra of speech dynamically is familiar with this fact. In the *ak* call of the baboon, the amplitudes reach their maximum in the first 20 msec (in the first frame). This is a case of a very abrupt attack of a sound. The ending of the sound is most significant. Spectrum analysis shows that there is no *k* of any kind at the end of a baboon's *ak*. When recording the *ak* sound, as pronounced by man, the oscillograph shows a gap after the *a*, where it is joined to the *k* followed by a spectrum containing a large number of components of small amplitude (a period of explosive resonance).

An oscillographic record of a baboon's *ak* shows neither gap at the junction nor explosive resonance. Here the *a* stops instantly. This phenomenon cannot be explained in any other way but by a laryngeal union. It can be quite definitely asserted that the baboons have *no consonants preceded or followed by vowels* and that the *a* vowel is produced by momentarily closing the vocal sphincter (a very abrupt attack) and, in the end, again momentarily opening it (abrupt ending)*. Therefore any addition of a *k* to the aurally recorded *ak* is wrong.

---

* Checked ending.

In future the "danger" call of the baboon and the macacus will be denoted by
a!, where ! means an abrupt ending to the vowel. Abrupt (but not explosive) attacks
on vowels are known in human languages, for example, in German. Abrupt endings
are met with in the case of some sounds in certain Caucasian, African, Burmese
and Red-Indian languages. When the a! call of a baboon is imitated the achievement
of the abrupt ending presents the greatest difficulty. This is why, with increased
power, the imitated sound is of longer duration. This lack of abruptness on the one
hand and the absence of a visual image on the other presumably reduce the vigour
of a baboon's reaction to an a! call, when it is imitated by a human voice. The result
is that there is no complete equivalence between the input and the output of the
communicative system between monkey and man.

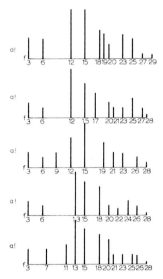

Fig. 71. a! calls recorded by different macaques.

Fig. 71 shows five spectra of the a! call as uttered by the macaque. It will
be seen that the formants are distributed either between the 12th and the 15th, or
the 13th and 15th filters. In addition, there is a noticeable increase in energy in the
right (high-frequency) part of the spectrum, where we meet with frequencies with
large amplitudes in filters Nos. 17, 18, 20, 23, 35. And yet, a common envelope of
the spectra is similar to the envelope of any a sound. The duration of the sound is
between 280 and 320 msec, its strength up to 100 dB. There is an abrupt ending.
Since baboons fail to respond to calls uttered by macaques (and *vice versa*), we must
conclude that the differential displacement of the formant frequencies of the spectrum
from the 9–13 to the 12–15 filters and an addition of high-frequency components is
distinguished in both systems.

The a sound has an abrupt ending only in the a_4 (danger) signal. In the search
call *au* there is a smooth transition from the a to the u sound. Fig. 72 shows six
spectra of the *au* call, as uttered by different baboons. In each of these sound-combinations the spectra of the a and the u are shown separately. The duration of the
entire sound is from 320 and 360 msec. If each spectrogram is examined frame by

References p. 180

frame, the acoustic process of transition from the *a* to the *u* becomes apparent. As will be made clear in a later part of this paper, this transition is effected only by changing the volumes of the oral resonance chamber. An examination of the spectra (*cf.* Figs. 68 and 72) shows that the general picture of the *a* is preserved. In each

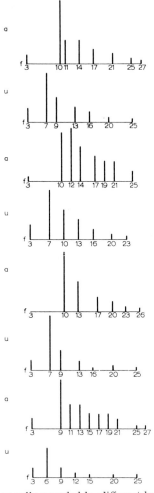

Fig. 72. *au* calls recorded by different baboons.

case there is a large amplitude of the formant of the 10th (and 9th) filter and a smaller amplitude of the formant of the 13th (or 12th or 14th) filter. It would seem, from the spectroscopic components of the *a* both in the *a!* and in the *au*, that the *a* could be regarded as an invariant component. Yet such an inference is premature. All that could be said is that, although *a!* and *au* are different sound-combinations, yet whether the *a* of the *a!* possesses the same distinguishing qualities as the *a* of the *au* is as yet unknown. The formants of the *au* are slightly displaced to the right, which can probably be accounted for by the raising of the fundamental tone at the *a* and then dropping it at the *u*. The spectrum of the *u* is entirely stable. In three cases there is a formant of the 7th filter and in one case, where the *u* was short (120 msec) and of low intensity, this formant has moved to the sixth filter. A second and weaker formant appears in the 9th and 10th filters.

Fig. 73 shows the spectrum of the English diphthong *au* and Fig. 74 that of the corresponding Lettish diphthong with a falling intonation. Both were pronounced by the same woman's voice*. A considerable similarity between the diphthongs of the human speech and the *au* call of the baboons is obvious. In both cases the greatest amplitude of the formant in the first element of the diphthong lies in the 9th and

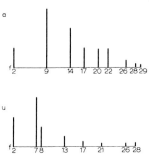

Fig. 73. The English *au* diphthong.

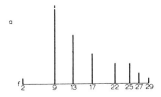

Fig. 74. The Lettish *au* diphthong.

10th filters, the second formant belongs to the 13th and 14th filters. The second element of the diphthong preserves the formants of the 6th and 7th filters. The envelopes of the two spectra are also similar. The difference lies in that, to the spectrum of the *a* of the human diphthong, a low frequency from the 2nd filter has been added. This addition can be regarded as an influence, on the *a*, of the low-frequency spectrum of the *u* of the second element of the diphthong. In the sounds produced by the baboons this influence is much less felt and becomes evident only in a small amplitude from the 3rd filter. The transition is shorter and more complicated in the human diphthongs than in the *au* calls of the baboons. Even those nations in whose language there are no diphthongs (like the Russians) pronounce the *au* as one syllable. The

---

* Both these spectra were recorded by the same apparatus and were kindly placed at my disposal by Mrs. A. A. Neuland of Riga. They were made in connection with her work (not yet published) on the comparison of English and Lettish diphthongs.

diphthong of a baboon is quite distinct. While it cannot be regarded as consisting of two syllables, yet it is difficult to regard it as only one.

The dynamics of this phenomenon are represented in Fig. 75 where the levels of the *a!* and the *au* calls are plotted on a logarithmic scale. In the case of the *a!*, the intensity is seen to rise and fall sharply. In the *au* diphthong, there is a noticeably prolonged acoustic process which is well reproduced in the spectrogram (in three

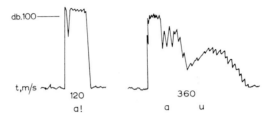

Fig. 75. The curve of the dynamic level of the *a!* and *au* calls as recorded by a baboon.

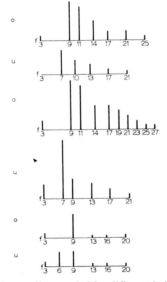

Fig. 76. *ou* calls recorded by different baboons.

frames). The process is *exclusively* confined to the oral resonance chamber and is generated by stretching the lips far forward. As we shall see later, in man, every transitional modulation, be it inside a syllable, or between syllables, takes place in the pharyngeal region. This explains the semi-diphthongal dynamics of the *au* of a baboon.

Fig. 76 shows three spectra of a baboon's call which, to the human ear, sounds like *ou*. It is obvious that here the second element (the *u*) coincides in every respect with the same element of the *au* call. There is no marked difference between the spectra of the *a* and of the *o*. The maximum amplitude of the 10th filter frequency has (in the *a*) moved only to the 9th filter (in the *o*). The corresponding shift in the human *o* sound is more pronounced (filters 7–9). We are led to believe that the *a* and *o* sounds of the baboon are only slightly differentiated. In some cases we get a typical *a*, while other records show something approaching *o*. As already pointed out, neither is the communicative significance of the *au* and the *ou* calls capable of objective separation.

Now it is important to note that the *hon* sound, with its nasal resonance, shows a very characteristic spectrum. All seven available spectra of this sound are almost

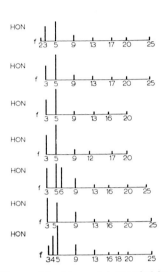

Fig. 77. The *hon* sound recorded by different baboons.

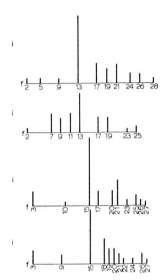

Fig. 78. The *ee* squeal of a female baboon.

identical (see Fig. 77). Formants from the 3rd and 5th filters are invariably present in this very weak sound with prevalent low frequencies. The shifting of the spectrum to the left is due to the powerful nasal resonance. We must therefore note that, when combined with the *u*, the *o* hardly differs from the *a*, whereas the *hon*-combination differs sharply from all the others.

The spectrum of the loud squeal uttered by a female baboon presents a special interest (see Fig. 78). Frequencies of various amplitudes are scattered almost all over the spectrum. The clearly predominant formant comes from the 13th or 15th filter. The interest of this sound lies not so much in its slight resemblance to the human *ee* in which the component frequencies are similarly widely scattered over the spectrum, as in its duration. If the duration of the *a!* never exceeds 120 msec, the squeal lasts 700 msec or even longer. And yet it forms only one syllable. While it lasts it varies appreciably both in its dynamic level and in the structure of its spectrum. Yet, despite this variability, no chain of syllables is ever obtained. In other words, the baboons possess no syllabic conjunction and, consequently, no algorithm for the transformation of sounds into a syllabic line. This monkey has a monosyllabic communicative system. Acoustic observations by themselves are, however, insufficient for a complete answer to the question. We must turn to the examination of the apparatus used by the animals to produce phonation and resonance.

References p. 180

(b) Role of the pharynx in the phonation process of monkeys and human beings

Disregarding such differences in the structure of the human and of the simian larynx as are of little importance to our thesis, let us turn our attention to the relationship between those parts of the mouth and the throat (pharynx) which cause resonance to occur (Fig. 79). The epilarynx of a baboon is placed very high,

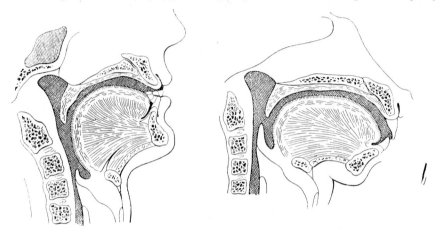

Fig. 79. Left: Larynx of Man, after J. Sobotta, *Atlas der deskriptiven Anatomie des Menschen*, Abt. 1–3, 3rd Edition, Lehmann, Munich, 1919–20. – Right: Larynx of a baboon, after W. L. H. Duckworth, *Handbook of Morphology and Anthropology*, Cambridge, 1915.

so that only a very narrow slit remains between the mouth and the pharynx. The two resonance chambers are separated. The epilarynx of man has descended much lower. The regions of the mouth and the pharynx have been united to form one combined oro-pharyngeal resonance chamber. It is interesting that the epilarynx of a newly born child is also placed very high; it descends towards the period of speech formation.

We do not here propose to produce any factual evidence to show the part played by the pharyngeal resonance chamber in human speech. This will be found in other papers by the same author[18, 20] where he describes his study of the epilaryngeal tube by filming X-ray records. The fundamental facts which are essential to the development of the present thesis reduce themselves to the following.

In man, the meso- and hypopharynx are modulated in the process of speech. They assume varying volumes and shapes for each sound element of a syllable. During this period the sounds are fused into a syllable, *i.e.* an intra-syllabic acoustic process of transition is at work. In addition, the pharyngeal tube modulates with every syllable, thus forming, within the dynamics of acoustics, a syllabic curve of loudness, and joins them into a chain of syllables of unequal power. If there are no pharyngeal modulations, then no chain of syllables can be formed within the structure of the apparatus of speech. The apparatus of speech is a pneumatic system of tubes of various diameters. It is controlled by muscles which, in turn, are controlled by nervous impulses. The various sounds of speech comprise some acoustically very powerful vowels, like *a* or *o*, and some very low-powered, noisy consonants, like *p* or *t*. If the acoustic power of these sounds remained unequal, then, if they occurred in the same syllable or even in comparatively closely following syllables, a powerful sound would overshadow a low-powered one. The pharynx could be regarded as a control system of the return communication which, at the output end of the apparatus of speech, regulates the final mechanical effect which is equivalent to the acoustic result and informs the central control of the requirement of air to be supplied by the respiratory system. The result is that powerful sounds are reduced and low-powered ones amplified, the overshadowing effect is eliminated and it becomes possible to produce a line of distinguishable sounds. This, briefly, happens when man speaks.

The study of the functioning of the pharyngeal tube of a monkey presents far greater difficulties. With present-day techniques it is obviously impossible to obtain any X-ray records of this tube while it emits sounds under natural conditions. Difficulties arise even if such records are made in a special laboratory. The animal has to be held by assistants and either laid on the table or pressed against the screen. It assumes a state of complete inhibition and, as a rule, makes no sounds of any kind. Nevertheless, after a series of attempts, we succeeded, in the end, in obtaining a number of X-ray records of the epilaryngeal tube of a baboon while the animal was uttering the powerful *ee* squeal of the type already described. These records show that *the pharyngeal tube of the baboon does not modulate when the uttered sound is strengthened or weakened*.

This is a crucial point in the interpretation of the spectograms previously discussed and in the problem of the existence of a communicative system between monkeys. The absence of any transitions from vowels to consonants (and *vice versa*) in any calls made by baboons now becomes clear. The resonance apparatus of the baboon has no regulator of air pressure; no low-powered consonant can, therefore, be combined with any powerful vowel. Instead of an *ak* there is an *a!*, i.e. the oral *k* is replaced by a guttural stroke. Air pressure is increased, before it reaches the resonance chamber, without special control but by simply momentarily stopping the respiratory system. It was pointed out that, in addition to the very loud *a!* call, there was a peculiar muted signal which we had agreed to denote by *ptpt*. The distinctive nature of these two sounds clearly shows the impossibility of combining sonorous and mute sounds in this system. It is possible that the laryngeal *!* and the muted muttering of the *ptpt* type are genetic predecessors of the human consonants which, at a later epoch, when the oro-pharyngeal resonance chamber was formed and the epilarynx lowered, were included in the human syllabic chain.

I reckon that the other species of monkeys are equally incapable of combining a vowel with a non-sonorous consonant. The calls of the capuchin monkey which we have so far, not yet spectroscopically, examined, are of interest in this connection. They are intermittent sounds with a very high fundamental tone and recall the trill of a song bird. Firsov[3] records one of these sounds by the letters *ikrr* and produces an oscillogram which shows that the sound is, in fact, intermittent but that, after an interruption, there is another sonorous sound, not a muted *k*. The oscillogram shows, in all probability, the combination *i!rr*, i.e. an *i* followed by a guttural stroke and a tremulous and sonorous *r*. A sonorous consonant is acoustically the most powerful of consonants, while *i* is the weakest of all the vowels. A combination of this sort, which includes an interruption effected by means of a laryngeal stroke (or without it), is fully within the power of the phonic apparatus of the monkey.

The smoothness of the transition from the *a* to the *u* in the diphthongeal *au* and *ou* calls of the baboon is effected not by means of any modulation of the pharyngeal

References p. 180

tube but by altering the volumes of the oral resonance chamber. This is clearly visible to the eye. During the transition to the *u*, the baboon stretches his lips far out.

The absence of any pharyngeal modulations results also in that, in the case of the baboon, all amplitudes of the spectral frequencies reach their maximum from the first instant, while the pharyngeal modulations of man damp the beginning and the end of the syllable, thus forming a smooth syllabic curve of loudness.

An examination of the pharyngeal tube with X-rays thus leads to a sufficiently clear explanation of the spectra of sounds produced by monkeys.

(c) The number of alphabets and algorithms of the monkey communication system

Our next problem is to determine, from the analysed sounds of the baboons the number of alphabets and the algorithms of the communicative system used by monkeys.

We shall reckon that the following elements make up the sounds which a baboon can utter: *a, !, o, u, h, n, t, p, ee*. The *a!* signal occurs in two versions, one with a high tone and another with a lower one. It thus appears that we have to reckon with at least ten elementary components of sound. Let us suppose that these elements can be arranged in all possible permutations of three at a time *(e.g. hon)*. Then it is possible to make $10^3 = 1000$ permutations of incomplete words. If each of these permutations is reckoned as a unit, *i.e.* if each of them is regarded as a complete word, and, having prepared an alphabet of these words, the permutations are selected only two at a time, then there will be 10^6 final messages. In actual practice we have found only seven final messages which possess a definite meaning as signals. If we consider such variations as we have failed to take into account, then, according to other investigators, the total number of signals used by the baboon is still not more than forty. Consequently there exists some sort of algorithm by means of which only certain definite permutations are selected from all the possible ones.

It is the same with a human language, where not all possible permutations of the initial alphabet of elements are used. Thus, in the Russian language, the combination *tkpstr*, etc. is impossible. In section 2 (e) (p. 142) we have established that the list of admissible sequences of the alphabet of letters used in a human language is determined by the alphabet of incomplete words. The question arises what kind of alphabet is employed in the communicative system of monkeys?

To enable us to answer this question, we must first agree on the meaning we shall attach to the letters *a, b, c, d*, ... of the initial alphabet. Since we are discussing an acoustic system of sounds, the letters must denote sounds. We must lay down exactly what, in a complex sound, corresponds to a letter of the alphabet defined as an element which is distinct from any other element. It has earlier been remarked that we mean by a letter its entire configuration, independently of the elements which make it up. The entire object of acoustic analysis lies in the separation of the elements of a sound and in the determination of whether these elements are invariant or not, in other words, whether there is an algorithm governing the combination of elements. Therefore a letter must be used to denote not the entire sound but only a certain element of it.

We shall regard a spectrum of a random noise, when recorded during a certain period of its duration, to be incapable of resolution into elements, because the probability of the appearance of any one frequency band is equal to the probability of the appearance of any other (paras. 17, 18 of the Glossary). It would then appear that the required element which ought to be denoted by a letter shall be any non-random spectrum containing a selection of a frequency band or of amplitudes. Yet even in this manner there would be no end to the search for the elements of a spectrum. For suppose that one of the spectra is selective, *i.e.* contains a limited number of components (a tonal spectrum) and another spectrum is continuous, *i.e.* with a large number of components and variations in amplitude in accordance with a known law (a noise spectrum). These two spectra will differ from each other. Yet even the first, the tonal, spectrum will differ from other tonal spectra as regards the zone of distribution of frequencies, *e.g.* on the left, in the middle, slightly to the right, etc. The same is possible also in the case of a continuous spectrum, if account is taken of the envelope of the amplitudes. Both the selective and the continuous spectrum can show, in addition, different degrees of concentration of the sound energy. In some the energy may be concentrated in this or some other frequency band, in another case the energy is diffused over the entire spectrum. Degrees of concentration of sound energy are also possible. One ought further to add that even when their spectra have similar elements, one complex sound may differ from another in its fundamental tone, its attack and its ending. Therefore, in order to solve the problem of the comparison of the elements of a spectrum (for the purpose of denoting them by letters), we must define the terms of comparison first. Such terms are the *characteristics* of a sound which we have enumerated above (see paras. 8, 13, 15 of the Glossary).

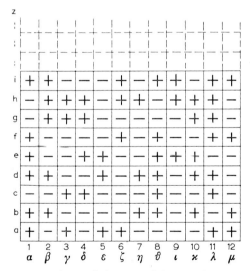

Fig. 80. A grid of binary comparisons of elements of the sound spectrum of speech distinguishable by man.

The result is a grid of one–one comparisons in which one spectrum differs from another in one characteristic and, in another characteristic, the first also finds a distinguishable pair, etc. An abridged scheme of such a grid is shown in Fig. 80, where the characteristics are numbered and the letters *a, b, c, d, ..., z* of the Latin

References p. 180

alphabet represent the conjunctive products of the distinguishable elements of the spectrum which can be split in pairs. These then are the required elements of the sound spectrum which cannot be further resolved and which we shall denote by letters of the Latin alphabet.

The term distinguishable elements means distinguishable in the given system. Once the distinctness of given particular elements is found, then the given particular system of these elements is also defined. In the preceding pages the letters $a, b, c, d,\ldots$ denoted *any* distinguishable elements whether they were sounds, or relays, or anything else.

Jacobson has elaborated a theory of binary distinguishable features of sounds, as applied to the human languages[5,6]. This conception opens the way to a successful search for the algorithms which govern the structure of the sounds of human speech. We shall briefly define them in a very general form.

This author establishes 12 pairs of features of sound in juxtaposition. He calls them distinctive features.

1. Vocalic—Non-Vocalic
2. Consonantal—Non-Consonantal
3. Interrupted—Continuant
4. Checked—Unchecked
5. Strident—Mellow
6. Voiced—Voiceless
7. Compact—Diffuse
8. Grave—Acute
9. Flat—Plain
10. Sharp—Plain
11. Tense—Lax
12. Nasal—Oral

These names are not always adhered to in the terminology of the spectra but they can be fully reduced to equivalence with spectral elements, because the acoustic series is equivalent to the series of volumes of the resonance chambers.

If we denote each of these pairs by letters of the Greek alphabet we obtain a column of twelve disjunctions.

1. $\alpha \vee \bar{\alpha}$
2. $\beta \vee \bar{\beta}$
3. $\gamma \vee \bar{\gamma}$
4. $\delta \vee \bar{\delta}$
5. $\varepsilon \vee \bar{\varepsilon}$
6. $\varsigma \vee \bar{\varsigma}$
7. $\eta \vee \bar{\eta}$
8. $\theta \vee \bar{\theta}$
9. $\iota \vee \bar{\iota}$
10. $\kappa \vee \bar{\kappa}$
11. $\lambda \vee \bar{\lambda}$
12. $\mu \vee \bar{\mu}$

By selecting one letter from each disjunction, we can construct a line of combinations of letters, for instance

$$\bar{\alpha} \& \beta \& \bar{\gamma} \& \bar{\delta} \& \bar{\varepsilon} \& \bar{\varsigma} \& \bar{\eta} \& \theta \& \bar{\iota} \& \bar{\kappa} \& \lambda \& \bar{\mu}$$

is a combination which gives a complete definition of the sound t (hard) of the given language.

Thus the definition of a given particular sound t is obtained by selecting each of the parameters of the spectrum of those distinguishable in the given language. The algorithm used for the selection of these parameters transforms a column of disjunctive features of a spectrum into a line of combination. One line of combinations corresponds to the sound a, another line to the sound b, etc. Each line is formed from the features $\alpha, \beta, \gamma, \delta$, etc. which refer to the objective series of the spectrum and are therefore none other than a vertical substitution and the objective meanings of the letters $a, b, c, d, \ldots, z$ (cf. the grid in Fig. 80). Therefore $a, b, c, d, \ldots$ are sounds but they are now distinguishable in accordance with a definite algorithm, *i.e.* they are distinguishable invariants.

As we have pointed out in section 2 (e) (p. 142), the sounds a, b, c, d form combinations of incomplete words (*sten, stol, a, oj, ik, om*, etc.). But there the objective meaning of the sounds $a, b, c, d, \ldots$ or, which is the same thing, the spectral substitution, may change. Thus, in the first adjacent incomplete word *stol·*∧, the third letter from the left is *o*, yet in the second adjacent incomplete word *stal·á* (spelt *stolá* but pronounced almost like *stalá*) the third sound has changed to *a*. This means that the second word ceases to be adjacent and equivalent to the first, and it looks as if the entire algorithm of the column of incomplete words for the class S_0 has lost its power. This phenomenon is due to the transference of the stress of the complete word *stól·∧* from the middle to the end *stal·á*. Since this phenomenon is invariant (*vól—valá, kól—kalá*, etc.), then there exists an algorithm which governs the exchange of one objective meaning for another in an incomplete word ($o-a$), or a rule for the change of the formant frequencies of the spectrum. In the Russian language such a rule actually comes into force in the circumstances we have just mentioned. The additional algorithm preserves the equivalence of the incomplete words.

Every time a complete word is formed from sounds, a change of the formant frequencies can take place in one or another element of the combination of sounds of the complete word. This depends upon the syllabic dynamic level of the complete word, its sound composition and its place within the complete or incomplete syntagma. Therefore, in addition to the above mentioned algorithm for the transformation of the spectrum into sounds $a, b, c, d, \ldots, z$, there exist supplementary algorithms for the exchange of formant frequencies. Continuing the terminology which we have already adopted, we shall call the sounds $a, b, c, d, \ldots, z$ incomplete sounds (they are usually referred to as phonemes), and the sounds which make up complete words (with the help of supplementary algorithms)—complete sounds or syllables and shall denote them by the capital letters of the Latin alphabet A, B, C, D. Only $a, b, c, d, \ldots$ go to form the alphabet of sounds, because A, B, C, D are obtained from the first set in accordance with a definite algorithm (see para. 14 of the Glossary). The series $\alpha, \beta, \gamma, \delta, \ldots$ refers to the series of vertical spectral substitutions.

There are thus three alphabets in a human language: an alphabet of incomplete sounds (phonemes), one of incomplete words (morphemes) and one of complete words (a lexicon)*. The transition from one alphabet to another is effected by means of algorithms which transform incomplete sounds into complete words or messages, and the meaning of an incomplete sound is determined by a spectral substitution.

In order to discover whether, in a given system, there exists an alphabet of incomplete sounds, it is sufficient to compare any two such complete words in which all complete sounds are equal each to each, except for one pair. Then the sound combinations of the two complete words will differ in this, and only this, that there is preserved an invariant distinction between two distinguishable complete sounds. This then is the invariant of an incomplete sound. Suppose that there exist the complete words *DOM* and *DAM*, then the complete sound *o* contains the spectral substitution of the incomplete sound *o*, whereas the complete sound *A* contains the incomplete sound *a*. This system contains at least two incomplete sounds a, o, each being invariant because it always retains the constant function of discriminating between a pair of words (any words, whether complete or incomplete). The continued existence of the function of discrimination is effected by a spectral substitution and by an algorithm for the exchange of this substitution by a substitution of a spectrum of a different type (under definite conditions). In accordance with this algorithm, an incomplete sound is transformed into an equivalent complete sound.

---

* To that, the alphabet of distinctive features should be added (*cf.* Glossary, para. 8, 13 and 17).

Let us apply this criterion to the communicative system used by the baboons. Here we meet the complete words $AU-OU$. The U is the same in each combination. May we, on the basis of this fact, draw the conclusion that in the system of sounds, as used by the baboons, there exist the complete sounds a, u? Such a conclusion is possible only in the case where the objective, signalling substitutions of the combinations $AU-OU$ are different, because the communicative system transmits not a difference between sounds but, through this difference, a difference between objective, signalling meanings equivalent to them. We have earlier established that the AU is indistinguishable from the OU, as regards the situation denoted. Thus a comparison of AU and OU gives no basis for any conclusion concerning the existence, in the communicative system of the baboons, of any alphabet of incomplete sounds. We arrive at the same result if we compare any other sound combinations which we have made the subject of our study. Thus $A!$ and AU cannot be compared, because their final elements are different. This is equally true of OU and HON. The $PTPT$ combination cannot be compared with any other. A comparison, as to their tonality feature, is possible only between the high pitched version of the $A!$ and the low $A!$. But the common feature a here exists only within the combination containing $!$ and the $!$, by itself, has nothing one could compare it with. Since the a is inseparable from the $!$, they cannot constitute distinguishable elements of an alphabet. We shall arrive at the same conclusion if we note that, within the possible combinations A_4, OU, AU, HON, there are no permutations of complete sounds such as UA, UO, or combinations like AO, OA, AH, HAN, etc.

The communicative system of the baboons has, therefore, *no alphabet of incomplete sounds (phonemes)*. Such a conclusion can be based upon and explained (in a general form, probably, for all species of monkeys) by established facts. Monkeys possess no means of controlling the apparatus which converts incomplete sounds into complete ones. Incomplete sounds form the statics of the sound system, because they arise before and without the application of the supplementary algorithm for the change of the formant meanings of $a, b, c, d, \ldots$ Complete sounds, on the other hand, are the dynamics of the system, because they are transformed in the supplementary algorithm of syllabic modulations. The absence of any control of the resonance volumes of the pharyngeal tube prevents, in this pneumatic mechanism, any application of an algorithm for the transformation of the statics into syllabic dynamics. Only incomplete dynamics remain here, *i.e.* the invariable formation, with the uncontrollable energy of the entire respiratory system, of one and only one syllable. Let us recall that every sound combination uttered by a baboon is either very strong or very weak.

Since there is no controlling algorithm for the formation of a syllabic chain, therefore no permutation of the syllabic elements $a, b, c, d, \ldots$ is possible either. Consequently no alphabet of incomplete words can be formed. The recorded combinations $A!$, HON, AU, OU at once become complete words; they are not transformed from some incomplete words by means of any definite algorithms. This, in turn, leaves no possibility of extending a line of a message by means of incomplete or complete syntagmas.

The communicative system of the monkeys contains, therefore, merely an alphabet of complete words and two algorithms for objective substitutions.

We can state the first algorithm, in the form of substitutions in the following way:

$A!_1 \vdash$ Spectrum No. 1*
$A!_2 \vdash$ Spectrum No. 2
$OU = AU \vdash$ Spectrum No. 3
$HON \vdash$ Spectrum No. 4
$PTPT \vdash$ Spectrum No. 5
$EE \vdash$ Spectrum No. 6

etc., if we can find any complete words or any of their versions and their corresponding spectra. This algorithm establishes the rule for the selection, from a random noise, of a spectrum of a definite composition. The result is that any one sound combination can be distinguished from any other.

The second algorithm can be stated as follows:

Spectrum No. 1 ⊢ Situation** No. 1 (Danger)
Spectrum No. 2 ⊢ Situation** No. 2 (Extreme danger)
Spectrum No. 3 ⊢ Situation** No. 3 (Loss of a member of the herd)
Spectrum No. 4 ⊢ Situation** No. 4 (Mutual satisfaction)
Spectrum No. 5 ⊢ Situation** No. 5 (Orientation)
Spectrum No. 6 ⊢ Situation** No. 6 (Appeal to the leader)

It will be seen that the second algorithm transforms the first one and gives the rule for distinguishing between situations. Apart from the second algorithm, the first one has no meaning. Like any other communicative system, that of the monkeys consists of two links of a vertical substitution. In the first, a symbol acquires a material and objective meaning (in our case it is the selection of the sound spectrum), in the second link, this symbol replaces a situation and becomes a signal of a situation. The alphabet of complete words and these two algorithms are equivalent at the input and output ends of every communicative system used by monkeys, when in herds.

Since no combination of complete words is possible in this system, then *a complete word = a message and a complete word is a final word*. This means that the number of messages is equal to the number of complete words. If we reckon that there are 6 situations and that they are all equally probable, then the quantity of information which is communicated every time is $\log_2 6 = 2.58496$. Similarly, if there are 40 situations and 40 complete words, the value is 5.32193. We have earlier called this system of selecting messages from a fixed list of complete words a system of *non-expanding* messages. It is, at the same time, unilateral. On hearing the $A!$ signal, a baboon either repeats it or alters his behaviour (*e.g.* scratches the ground with his paw, or flees). The response can contain nothing that would explain the situation more fully, because the lexicon contains no other corresponding complete word or algorithm for the exchange of words. The line of the message is closed at its right end.

5. Conclusions and Hypothesis

(a) Central control of the pharyngeal tube: organisation of acoustic transition processes necessary for formal words realisation (grammar) in the expanding message system

On the basis of the material here set out and of the conception which seems to emerge, it is possible to make three observations of a general character which cannot, however, be regarded as being other than hypothetical. In many quarters they may possibly be regarded as highly controversial.

---

* The spectra are here numbered irrespective of the numbering of the figures which appear earlier in this paper.
** A situation is a final vertical substitution.

A. First observation. The ultimate conclusion seems, at first sight, to be contradictory and yet more significant than a mere terminological distinction. In the communicative system of monkeys, initial words coincide with final words. This means that there is no algorithm of any kind that would transform an alphabet of letters into words. Neither is there any alphabet of letters, there are only (complete) words of limited length.

Could such a formation be called a communicative system, could it, further, be called a language? For a language is invariably, and with good reason, regarded as a set of rules which provide a means of communication, *i.e.* a set of algorithms for the construction of words. No language is possible without it, nor is it at all clear whether any communication would also be possible.

If stated in this way, the question remains purely terminological. Suppose we do not call the sounds made by monkeys a language. The importance lies elsewhere, namely in the equal necessity of either finding the algorithm of a phenomenon or proving its non-existence or impossibility. In the latter case it would be necessary to prepare a table or a list of corresponding objects, so that the required object may be selected from it. Note that, in a human language, there is no algorithm for filling incomplete words with letters. Thus arises a need for preparing an alphabet of incomplete and complete words. In section 2 (d) (p. 141) it was pointed out that, in a human language, there exists a class of words ($\bar{A}$) which are not transformed by means of algorithms from other words and which have no algorithms for the formation of adjacent words. They are interjections. In a human language there is not even any strict algorithm for the selection of complete words. Only after some person or another has selected a sufficiently large number of complete words is it possible to estimate, with an error of a magnitude as yet undetermined, a certain measure of probability of the forthcoming selection of words.

In consequence, in what we have decided to call a language, we ought to be able to find not only such domains as are subject to the rule of algorithms but also other domains which are not so ruled. It is, therefore, all the more necessary to examine and compare the various communicative systems. It will then be found that a communicative system may not contain any algorithm for the transformation of a number of symbols but that there will invariably exist two algorithms of vertical substitution for the transformation of messages. This is what constitutes a communicative system.

B. Second observation. There is a quite generally held opinion that the signal calls of the monkeys and of the other animals are unconditioned, inborn reflexes. This is contrary to the evidence derived from the observation of birds. Imitation and an ability to learn play a part in the formation of their sound sequences. There is also the established fact that the sounds of a young monkey are distinct from those uttered by a fully grown animal. Considered by itself, this fact does not yet point to an ability to learn sounds, because it is possible to assume that an unconditioned reflex begins to function only at a certain stage of ontogeny. A more substantial argument in favour of the conventional nature of the signal calls used by monkeys is to be found in the information derived from an analysis of this communicative system. If we admit that a signal call is a selection from a certain number of possible situations, then we also admit the existence of individual experience in the act of assimilating these situations, their retention within memory and a mechanism effecting a selection in a particular case. A sufficiently precise answer to this problem is possible only after a study of signal calls in use during ontogeny and after a more thorough classification of the situations which form the subject of a signal. It is possible that both the calls of the baboons and the situations which they signal differ, in some way, from herd to herd, in accordance with the conditions in which they live and the territory they roam. Acoustic analysis will, naturally, prove to be much more fruitful in this field of enquiry than simple recording by ear.

C. *The third observation* must be made with even greater caution. The notion which has been formed, by the successors of Pavlov, concerning a second signalling system, amounts to that, with man, a word is a signal of signals. The reaction to a light seen or a sound heard forms the first signal, while the replacement of these first signals by a word forms the signal of signals.

It is the same with the baboons; there is a reaction not only to a sound, such as AU, but to a situation signalled by this sound. This is the meaning of the two algorithms of vertical substitution here discussed. The baboon receives a message concerning a change in the situation and proceeds to search for the lost member of the herd. It would be more correct to argue that the distinction between the signalling systems lies not in the presence of a vertical objective substitution, or in its absence, but in the fact that one of them belongs to the class of *non-expanding* messages, the other to the class of *expanding messages*. This is the correct interpretation of Pavlov's famous statement that the word of a fully grown person is connected with *every* external and internal excitation, signals them *all* and replaces them *all*. The communicative system of the monkeys has no algorithm for the replacement of words. It is a fixed system.

The establishment of these facts is interesting as an approach to the objective definition of what is known as "a concept". Earlier there has been given the formula $\Sigma/\mathcal{M} = \Pi$. Though there are $\mathcal{M}$ (messages) in the so-called "monkey language", it has no Σ (syntagmas); consequently, there is no ratio $\Sigma/\mathcal{M} = \Pi$. Only where there is in Σ a combination of two concrete words, such a simultaneous division of Σ and $\mathcal{M}$ into two parts is possible, and the lexical meanings of the concrete words are equated in $\mathcal{M}$. In other words, the concept forming mechanism exists in the entire system of grammatical divisions, beginning with the division into incomplete and complete sounds, passing to incomplete and complete words and ending with the division into incomplete and complete syntagmas. Yet, as in the case of monkeys, a transmission of messages is possible even without concepts.

Finally we should draw the following rather paradoxical principal conlusion: the human ability to form concepts arose with the appearance of the mechanism for the control of the pharyngeal tube. Without modulations of the pharynx no acoustic transition processes can be formed and no horizontal permutation of the elements of the alphabet of sounds $\begin{Bmatrix} a, b, \\ c, d \end{Bmatrix}$ is possible. Consequently no grammar of any sort is possible. But since vertical substitutions are possible, transmission of messages is also possible.

(b) *Philogenetic and ontogenetic formation of the system controlling sound communicative signals*

In Chapter IV of his widely known book, entitled *Cybernetics and Society*, Norbert

Wiener remarks upon the limited nature of animal sounds. The roar of one lion is hardly distinguishable from that of another. This is certainly true; lions roar, dogs bark and cats mew. But even the acoustic apparatus of man can do no more than produce a chain of syllables which we call speech. Man uses sound-syllables not only to talk but to sing, laugh (*cha-cha*; *chi-chi*; *che-che*; etc.), cry, groan, shout, etc. Perhaps only the whistling of a tune contains no speech sounds, although the inhabitants of Gomera (one of the Canary Islands) have learned to modulate the Spanish vowels even in a whistle. Their whistling speech provides them with means of communication over long distances, across ravines, woods and cliffs.

The entire difference between the signal calls of animals and those of man lies only in the fact that, in speech, the sequence of syllables is governed, by a definite number of algorithms, in such a way that the sound elements are broken down into invariant classes, groups and sub-groups. The set of these algorithms, which are acquired through learning, constitutes the phenotype. Speaking generally, the effective acoustic apparatus of a monkey could also produce a chain of syllables which would be like human speech almost in every respect. Note that the resonator region of the mouth of even the dog-like baboon can modulate such types of vowel as *a*, *o*, *u*, *ee*, and that the pharyngeal tube is capable of restriction and expansion and the epiglottis of rising and falling, as, for instance, in swallowing. But monkeys have no programme of algorithms to control this effective apparatus in such a way as to obtain a chain of syllables formed with the aid of repeated selections from invariant groups of sound elements. In these circumstances there is no need for any training, so important to man, of the pharyngeal tube to modulate at every speech sound, a fact which has, as yet, been insufficiently appreciated. This, probably, is the reason why, during ontogeny, the teaching of signal calls (*i.e.* the establishment of equivalence between signal and situation) proceeds, in monkeys, so quickly that the observer simply fails to notice it or forms only a vague idea of what exactly he must watch for.

In the long run it must be admitted that the very ability of a human brain to assimilate the phenotype which corresponds to some language or another is typical of his genus and depends upon the structure of the controlling system. It is possible that, at some future date, an investigation of this structure, in its every aspect, may lead to its improvement; our present task is merely to describe it and to construct its model.

6. Glossary of Terms Used in this Chapter

N.B. 1. References to paragraphs of this glossary are contained in round brackets. The same kind of bracket is used also to refer to a page of the article itself, where this is necessary. Square brackets contain references to items in the accompanying bibliography. The first figure within these brackets gives the number of the item listed in the bibliography, the second number shows the relevant page.

N.B. 2. The terms here listed are used in the article not with the purpose of research into mathematical ideas or problems, they are rather regarded merely as means of describing facts which were empirically established within the domain of the comparative study of certain problems connected with the means of communication used by humans and by monkeys.

1. *Algorithm.* An algorithm in an alphabet A (para. 3) is an ordered set of rules, or a list of directives which determine the process of the successive transformation of abstract words (para. 25) in A. Any word P in A may be a *basic* word (para. 27). The term algorithm is used in the article also to describe the rules for the transition from one alphabet to another one (para. 21, also 24).

The definition of an algorithm here given as an object of a mathematical theory is not regarded as being completely accurate. This applies, above all, to the precision with which each directive is given. The step accomplished in the transition from one directive to another may vary. If, besides direct steps, integrated steps are taken, the latter will cause, in the word which is being transformed, changes of any conceivable significance. Markov[9], introduces the concept of a *normal* algorithm and, by resolving into simple rules of a standard type, finds such methods of regulating the directives issued by an algorithm that their simplicity and clarity are beyond any doubt. But for the purpose of our article no detailed iterated transcription of any algorithm is necessary in every case.

2. *An algorithm transforms a word into another one.* This phrase is equivalent to the statement that any algorithm $\mathscr{A}$, when applied to a basic word P, transforms it, by means of a substitution (para. 20) into some word Q. The equation $\mathscr{A}(P) = Q$ is read "algorithm $\mathscr{A}$ transforms P into Q" [*ref.* [9]].

3. *Alphabet. An Abstract Alphabet.* An abstract and non-null alphabet is a finite set of abstract letters $\left\{ \begin{array}{l} a, b, \\ c, d \end{array} \right\} = A$ (para. 10).

4. *Alphabet. A Null Alphabet* contains no letters $\left\{ \quad \right\}$. Any non-null alphabet may be regarded as an expansion of a null alphabet (para. 17).

5. *Communicative System.* In this article, a communicative system is looked upon as consisting of two independent (in the sense of the reception and the treatment of information) divisions A and B[*] each of which is equipped with two links, *e.g.* reception of information $a_1;a_2$ and transmission of information $b_1;b_2$. $A(a_1,b_1)$; $B(a_2,b_2)$.

If both A and B contain a central control (with a return link) possessing similar or partly similar alphabets and algorithms for the transformation (by means of a final vertical substitution) (paras. 12, 19) of the words P,R... which they have received into words Q,K... for transmission, then A–μ–B is a communicative system. The symbol μ is used to denote that the two divisions (A and B) are equal or partly equal in the transmission of the message M. The formula A–μ–B is not a word in the algorithms of the system A and B.

6. *Concept.* A concept Π is a combination within a message $\mathscr{M}$ of two lexical meanings of concrete words (para. 11) which combination was effected in accordance with a logical algorithm. This can be expressed by the formula

$$Б \longleftrightarrow \mathscr{D} = \Pi \qquad (1)$$

where $Б$ and $\mathscr{D}$ are concrete words within a concrete alphabet of the given language. The sign $\longleftrightarrow$ is not a letter of this alphabet but denotes a combination of final vertical substitutions (para. 19) to the words $Б$ and $\mathscr{D}$ of the message $\mathscr{M}$. Subsequent entries may be added to the first pair of lexical meanings *i.e.* lexical meanings of concrete complete words may be exchanged (see p. 148). A concept may be written in the general form

$$\Sigma / \mathscr{M} = \Pi \qquad (2)$$

where the sign / is not a letter of the alphabet of the given concrete language. It signifies the relationship of the complete syntagma (para. 22) to the message $\mathscr{M}$ (para. 12). It follows from these definitions that a concept is a null-word.

---

[*] The divisions A and B are of course complementary. Components of any system are complementary if each implies the existence of the other. Thus right implies left and *vice versa*. In the case of communicative systems the partners are complementary.

$$\Pi = \wedge \qquad (3)$$

and that a complete syntagma can always be found in which Π is realised in the various letters of some concrete alphabet C. A concept is thus on the border of two calculi—the calculus of syntagmas and the calculus of predicates (para. 27).

7. *Entry.* Any word (para. 25) contained within another word consitutes an entry. Thus the word K is an entry in R if R = MNKO.

8. *Formant.* A formant is a word of the type S″ of the alphabet Φ (paras. 13, 17) which, upon the application of a vertical substitution (para. 21), gives rise to a strictly defined word of the type d (para. 13) of the alphabet Sd. and enters as an element into the embedded alphabet of the phenomenes Θ.

$$S'' \to d$$
$$S'' \leftarrow d$$

9. *Invariants.* Deductive variants is the name given to two or more chains (columns) of words which possess the same properties as a result of a transformation of words into these chains by means of a definite algorithm. Thus the words $stol \cdot \wedge$; $dom \cdot \wedge$; $les \cdot \wedge$; $vagon \cdot \wedge$ (see p. 140) are invariants.

10. *Letter.* A graphic symbol which is distinct from other like symbols (without being dismembered) is called a letter. One occurrence of a letter coincides with another occurrence of the same letter with a precision which is tantamount to an identity [ref. [9], p. 8]. We are thus enabled to speak of *abstract* letters. Concrete letters, irrespective of their numbers, if identical with a given letter, are regarded as representations of *one and the same* abstract letter.

11. *Lexical Meaning.* The function acquired by a concrete word (para. 29) at the moment when, under certain logical conditions, words which are unlike in the sequence of their letters are transformed into like words as regards their relations to like objects contained in a message $\mathscr{M}$. This is achieved by applying a final vertical substitution (para. 19).

When a similar transformation takes place in respect of an incomplete word (para. 27), we may speak of the grammatical meaning of an incomplete word. Grammatical meanings are not discussed in the present article.

12. *Message.* A message $\mathscr{M}$ is a word which has been transformed in the complete syntagma Σ (para. 22), as in a final formal word (para. 27), into a final concrete word K_0 (para. 29) by selecting from a lexicon (para. 28) the concrete words k_1; k_2; k_3; ... k_n and applying a final vertical substitution (para. 19), while certain logical conditions (quantifiers) are fulfilled (see p. 148).

Logical quantifiers are absent in any system in which

$$k_1 = N = P = \Sigma$$
$$k_2 = N = P = \Sigma$$
$$\dots\dots\dots\dots\dots$$
$$k_n = N = P = \Sigma \quad (\text{para. 27})$$
$$\mathscr{A}(k_1) = k_1; \mathscr{A}(k_2) = k_2; \dots \mathscr{A}(k_n) = k_n$$

i.e. in which messages are distinguished from each other only by distinct vertical substitutions to the words $k_1, k_2, \dots k_n$.

13. *Phoneme or an incomplete speech sound.* This name is given to a word in the alphabet

$$\text{Sd.} = \begin{Bmatrix} a, b, c, \\ d, e, f, \\ g, h, i, \\ \dots \end{Bmatrix} \quad (\text{para. 15})$$

Thus $akrho = d$ is a phoneme.

Since the words of the alphabet Sd. are, in certain circumstances, (para. 8), recoded in vertical substitutions (para. 21) into words of the alphabet Φ (para. 17), *i.e.*

$$\text{Sd.} = \begin{Bmatrix} a, b, c, \\ d, e, f, \\ g, h, i, \\ \dots \end{Bmatrix} \vdash \begin{Bmatrix} \alpha, \beta, \\ \gamma, \delta, \\ \varepsilon, \zeta, \\ \eta, \vartheta, \end{Bmatrix} = \Phi,$$

then the word d may be written in the alphabet Φ, *i.e.* in words of a non-random spectrum (pp. 166, 167).

The set of words of the type d (phonemes) constitutes the alphabet

$$\Theta = \begin{Bmatrix} a, b, \\ c, d, \end{Bmatrix}$$

APPLICATION OF THE THEORY OF ALGORITHMS

After the transformation of phonemes into syllables (para. 14), incomplete words (para. 27) and an embedded alphabet of incomplete words are formed from the alphabet Θ. In the embedded alphabet these words are algorithmically transformed into complete syntagmas.

14. *Sound. A complete speech-sound or syllable* is a word-continuum which arises as a result of a transformation, by means of a special algorithm s, of one or more phonemes regarded as separate elements of the alphabet

$$\Theta = \begin{Bmatrix} a, b, \\ c, d, \\ e, .. \end{Bmatrix} \quad \text{(para. 13)}$$

The following general description will explain the working of this algorithm. Suppose there are two basic words which are sound sequences of the phonemes bc and cb. If each of these sequences remained discrete, then, in accordance with the definitions of an abstract letter and of an abstract word (paras. 10, 25), b of the first sequence would have been equal to b of the second sequence, i.e., $b = b$ and, likewise, $c = c$. But if the sequences bc and cb are continuous, then $b \neq b$ and $c \neq c$. For a continuous sound sequence could be recorded in the form

$$(bc) \quad bbbbc \cdot bcccc \quad (1)$$

$$(cb) \quad ccccb \cdot cbbbb \quad (2)$$

if we agree to record the process of acoustic transition in this manner. Formulae (1) and (2) show that equal entries of words between connecting points (where a point is not a letter of the alphabet Θ) and the correlates of the words of the words of the discrete sequences bc, cb are unequal

$$bbbbc \neq cbbbb \quad (3)$$

$$ccccb \neq bcccc \quad (4)$$

Thus the algorithm s transforms

the word bc into the word $bbbbc \cdot bcccc$ and $\quad (5)$

the word cb into the word $ccccb \cdot cbbbb$ $\quad (6)$

A *complete speech-sound* or *syllable* can, accordingly, be called the generator of phonemes which translates their discrete sequence into a continuous one (continuum). This transition takes place in the *output* link of a communicative system b_1, b_2 (para. 5), namely in the modulations of the pharyngeal tube when they give rise to a process of acoustic transition. It provides the means for making various algorithmic entries of words chosen from an alphabet of phonemes (para. 13). It is evident that in the substitutions

$$bc \longrightarrow bbbbc \cdot bcccc$$

$$cb \longrightarrow ccccb \cdot cbbbb$$

the basic exchangeable words bc, cb retain their equality $b = b$, $c = c$, in every strictly algorithmic transformation of words placed to the right of —, no matter how radical it may be.

That this equality is preserved is provided for in the *receiving* link of the communicative system a_1, a_2 (para. 5) by the application of the algorithm with the converse substitution

$$bbbbc \cdot bcccc \longrightarrow bc \quad (7)$$

$$ccccb \cdot cbbbb \longrightarrow cb \quad (8)$$

It is this converse algorithm which provides, during the reception of a continuous sequence of syllables, for the restoration of the discrete words of the alphabet

$$\Theta = \begin{Bmatrix} a, b, \\ c, d, \\ e, .. \end{Bmatrix}$$

in the form of functional invariants (see pp. 166–167).

15. *Sound. Distinctive features of a sound* are its qualitative elements (pitch, power, duration, quality, etc.) which are *distinguished by the ear* but which are not independent individuals within it. The set of these elements may be formed into an alphabet

$$Sd. = \begin{Bmatrix} a, b, c, \\ d, e, f, \\ g, h, i, \\ \ldots \end{Bmatrix} \quad \text{(paras. 8, 13, 14)}$$

16. *Spectrum.* A scale of distribution of the energy of a sound in frequency bands is known as its spectrum.

17. *Spectrum.* A non-random spectrum is an expansion of the null-alphabet of a random spectrum (para. 18) according to some definite algorithm. Thus suppose we write down the alphabet of a random spectrum in null-words

$$\begin{Bmatrix} \wedge \wedge \wedge \wedge \wedge \wedge \\ \wedge \wedge \wedge \wedge \wedge \wedge \end{Bmatrix}$$

Assume the existence of the alphabet

$$\varPhi = \begin{Bmatrix} \alpha, \beta, \gamma, \\ 1 \; 2 \; 3 \\ \delta, \varepsilon, \zeta, \\ 4 \; 5 \; 6 \end{Bmatrix}$$

in which all letters which correspond to frequency bands of definite width are distributed in strict numerical order in such a way that small numbers correspond to low frequency bands and vice versa. We shall write down a certain scheme of elementary substitutions (para. 20)

$$\wedge - \alpha; \; \wedge - \delta; \; \wedge - \lambda$$

If we apply this algorithm to the alphabet of the random spectrum, we obtain the word $\alpha\delta\lambda = S''$ of the alphabet $\varPhi$, where S'' is one of the representatives of a non-random spectrum*.

18. *Spectrum.* A *random spectrum* is one in which the appearance of any frequency band is equally probable. A random spectrum can, accordingly, be regarded as a null-alphabet.

19. *Substitution.* A *final vertical substitution* is one after which no other vertical substitution (para. 21) can be made in the same direction. Thus any object which constitutes the subject of a message $\mathcal{M}$ is a final vertical substitution in a concrete complete word contained within a complete syntagma (paras. 11, 12, 22).

20. *Substitution.* A *horizontal substitution* is an elementary step performed in an algorithm and consists of two words formed from letters of the same alphabet. On the left is placed a word instead of the entry of which a substitution is effected; on the right the word which is substituted in accordance with the general formula

$$A - B$$

The — in this formula is not a letter of the alphabet of the algorithm; it shows only that the word B replaces the word A.

In the calculation of syntagmas (paras. 6, 27, 28), all permissible algorithmic substitutions are horizontal, because, they take place in the words of the basic (funding) alphabet A (para. 27).

21. *Substitution.* A *vertical substitution*. Signs of different material origin are met with in concrete, actually existing communicative systems or in special technical devices used to exchange messages. They are actual letters, or sounds, or signals carried by nerves, or electric, electromagnetic or flag signals, or Braille, or the sign language used by the deaf and dumb, etc. All signs of this sort may be regarded as *constructive objects* or words contained in certain alphabets (para. 24).

A *vertical substitution* is one in which a word, *i.e.* a sign of one material nature, is transformed into a word which is a sign of another material nature in accordance with the general formula

$$A \vdash \alpha \; **$$

where A and α are words belonging to different alphabets and $\vdash$ belongs to neither. A vertical substitution may be directed either to the right or to the left. Thus

$$A \vdash \alpha;$$
$$A \dashv \alpha;$$

or, which is the same thing,

$$A \vdash \alpha;$$
$$\alpha \vdash A.$$

Two types of chain of vertical substitution are possible:

$$(i) \; A \vdash \alpha \vdash Б \vdash \triangledown$$

---

* If a non-random spectrum were to be defined more precisely, it would be necessary to consider, apart from ordinal numbers, also the quantifier value of the sound energy of the frequency bands. This could be done by attaching numerical suffixes to the words.

** In the explanation which follows formulae are written horizontally.

where A, α, Б, $\triangledown$ are words taken from different alphabets and

(ii) $A \vdash \alpha_0\alpha_1\alpha_2\alpha_3\ldots\alpha_n \vdash A$

where $\alpha_0\alpha_1\alpha_2\alpha_3\ldots\alpha_n$ is a word which belongs to the alphabet $\{\alpha\}$. In this case the vertical substitution transforms the word A into itself. In this manner the transmission of A is made possible by the use of signals which are words of the alphabet $\{\alpha\}$ (see p. 145). The rule governing the transition from A to α or from α to A is a translation of one code into another one.

22. *Syntagma*. A complete syntagma Σ is a combination, in accordance with a definite algorithm $\mathscr{B}$ (para. 27), of either two incomplete words, or of two complete words, or of two incomplete syntagmas (para. 23), and their arrangement in a line. After the application of the closing formula of the algorithm $\mathscr{B}$, a renewal of the syntagmatic (grammatical) process is permitted (see p. 145).

23. *Syntagma*. An incomplete syntagma σ is a combination, in accordance with a definite algorithm $\mathscr{C}$, of two complete words (para. 26) selected from columns of adjacent words (para. 9). The algorithm $\mathscr{C}$, like algorithm $\mathscr{A}$, contains no closing formula (para. 27) and, like $\mathscr{B}$, increases the number of entries into N.

24. *Theory of Algorithms*. During the last 25–30 years, this theory has engaged the attention of the mathematicians Turing, Post, Church, Kleene and, in the Soviet Union, Markov, Novikov and their school. The new ideas which they propounded and the new problems which they set were, from the middle thirties onwards, developed into the *constructive trend* in mathematics[13].

The words of any alphabet A are regarded as *constructively definable* objects. Words considered in this manner are not only those that actually exist but also those which could potentially be formed as the result of an application of definite rules to the transcription of, or the exchange or addition of some letter of the alphabet under consideration, to the outcome of some previous operation on a word. The following two premisses are accepted:

(i) "Only constructive objects which admit of being expressed (coded) in words of certain alphabets are regarded as objects of our study;

(ii) In the course of the examination of these objects the abstraction of their potential existence is admitted, but the application of the abstraction of actual infinity is excluded" [ref.[13], p. 228 and ref.[9], p. 15].

The idea of the abstraction of actual infinity consists in this that the process of word-building is thought of not only as being capable of unlimited continuation but also as completed, *i.e.* the words of the alphabet A are regarded as simultaneously existing. It is therefore possible to formulate, for instance, the existence theorem of a real number which satisfies certain conditions without, at the same time, any indication as to the method by which such a number could be calculated [ref.[10], p. 316].

While admitting the abstraction of potential existence, the theory of algorithms poses the problem of determining a method for calculating a constructive object (*e.g.* a real number). This abstraction is obtained from the real boundaries of our constructive potentialities. This allows us to talk of words of any length whatsoever as being capable of existence. The existence of these objects is potential [ref.[9], p. 15].

The theory of algorithms applies not only to objects and problems of classical mathematics but also to logic in the form of the calculation of predicates and of statements. It is only natural to apply this theory to the calculation of syntagmas, *i.e.* to grammatical calculus and to the investigations of various communicative systems and arrangements regarded as constructive objects.

25. *Word*. An abstract word is a sequence of any number of abstract letters following each other in a line. The sign for a null-word (para. 30) or other signs which do not belong to the alphabet of letters in this sequence may be inserted between these abstract letters. A combination of two words of the alphabet A is a word of the alphabet A. Words which are equal as regards the sequence of letters are regarded as *one* abstract word.

26. *Word. Adjacent words*. Two or more words N, P, R, Q, ... of which each consecutive word has been formed, by means of a permitted elementary substitution, from an initial N so as to constitute a deductive chain (column) of words (para. 9) are said to be adjacent words. (See also p. 140.)

27. *Word. Basic word, initial word, incomplete word, final word*. These concepts are best explained by comparison. In accordance with the definition of an algorithm (para. 1), any word P in A may become a *basic* word on which an algorithm in the same alphabet could operate. Two types of algorithm exist. We denote them $\mathscr{A}$ and $\mathscr{B}$ and define them as follows:

(a) Let the algorithm $\mathscr{A}$, when applied to the *basic* word P, $\mathscr{A}(P)$, *reduce* the number of entries in P in such a way that the process is cut short at some word N. Then the word N is an *initial* word in A. If $\mathscr{A}(N)$ takes place, *i.e.* if P = N, then the algorithm $\mathscr{A}$ transforms N into itself. For the sake of uniformity in the terminology here used, *initial* words were also referred to, in the article, as *incomplete* words (see para. 28 for complete words). An alphabet of incomplete words of the alphabet of letters A (see p. 142) has to be compiled for the purpose of the application of subsequent grammatical algorithms. The relation of the alphabet of letters A to the alphabet of incomplete words can be described as that of *embedding*. We shall use the term mutually embedding to describe two alphabets in which the second, embedded, alphabet enumerates the initial words composed of letters of the first, embedding, alphabet.

(b) We shall denote $\mathscr{B}$ the other algorithm which *increases* the number of entries to an initial word N, so that the process ends only with a complete syntagma Σ (para. 22). Then Σ which, in this article, is referred to as a formal word, will, at the same time, be a *final formal word*. At the moment when, through the inclusion of concrete words (para. 29), Σ is transformed into $\mathscr{M}$ (para. 12), a *final concrete word* is formed in the alphabet of letters of the given language.

It follows from what has just been stated that the definition in para. 1 of this Glossary could, where it applies to the domain of grammar, be made more precise as follows: an algorithm in the abstract alphabet A is an ordered set of rules or a list of directives which govern the process of finding and successively transforming initial (incomplete) words N in the embedding alphabet of letters A into a final formal word Σ. The application of the set of these algorithms is the province of the grammatical calculus*.

The distinction between the algorithms $\mathscr{A}$ and $\mathscr{B}$ corresponds with what Markov [ref. [2], p. 56] calls simple and closing formulae of substitutions by an algorithm of the alphabet A. When a simple formula is applied, the process of the transformation of a word ceases every time its continuation becomes impossible. When a closing formula is applied, the process stops the moment this formula is applied. The process must continue in all other cases. Thus, during the formation of the word Σ, the process ends after the application of a closing formula, the sign I is written (see p. 145) and we are once more allowed to refer to the lists of directives contained in $\mathscr{C}$ and $\mathscr{B}$.

If, within a certain system, the following equations always hold

$$P = N = K = \Sigma$$

i.e. the basic, initial and final words coincide, then the algorithms $\mathscr{C}$ and $\mathscr{B}$ are absent and the algorithm $\mathscr{A}$ always transforms every word of the alphabet A into itself and only into itself. This does not preclude the use of an algorithm with a vertical substitution (paras. 1, 21). As pointed out on p. 169, this is the actual scheme whereby words are transformed in the communicative system used by monkeys.

28. *Word*. A *complete word* is one which may enter as an element (*i.e.* as an irreducible whole) into various (as to structure, *i.e.* methods of combining the words) incomplete syntagmas (para. 23). Complete words comprise those that were transformed by the algorithm $\mathscr{A}$ (para. 27) either from incomplete (initial) words into some word Q or into the basic word P = N, (which was transformed by $\mathscr{A}$ into itself).

A list of the complete concrete words is known as a lexicon of the given language. When calculating syntagmas (paras. 6, 27), there is no need to resort to a lexicon of concrete words, or to lexical or grammatical meanings (para. 11). In this calculus, complete words are written as words in the alphabet of abstract letters A, and the combinations of both complete or incomplete words, as well as of incomplete or complete syntagmas are denoted by special signs which do not belong to the alphabet of these words. But a lexicon providing lexical and grammatical meanings is needed when selecting concrete words in Σ (para. 22).

29. *Word*. A *concrete and complete word* is a transcription of a sequence of concrete letters in some alphabet of a definite language (English, Russian, etc.). It fulfills the following conditions:
   (i) This word is a complete word in accordance with the definition in para. 28,
   (ii) it has been identified in passages written (or spoken) in this language.
Concrete words which consist of the same sequence of letters are *different* concrete words, no matter how many times they are repeated, *i.e.* the number of these words is equal to the frequency of their occurrence.

30. *Word*. A *null-word* is one which contains no letters. The sign $\wedge$ is used to denote such a word. This sign is not a letter of the alphabet of words in which it occurs.

---

* The grammatical calculus has two divisions:
   (i) The calculation of adjacent words in classes, groups and subgroups of incomplete words N, by applying the algorithm $\mathscr{A}$ to a certain number of basic words P of the language under investigation (analytic process).
   (ii) The calculation of syntagmas, by applying, to the results of the operation of the algorithm $\mathscr{A}$, of the algorithms $\mathscr{C}$ and $\mathscr{B}$ (synthetic process).

CONVENTIONAL SIGNS USED IN THE ARTICLE AND IN THE GLOSSARY

—	Horizontal substitution.
⊢	Vertical substitution.
·	Combination of incomplete words.
&	Conjunction, combination of complete words in an incomplete syntagma.
&	Conjunction, combination of words in a complete syntagma.
∧	Null-word.
¯	Negation (placed on top of a letter).
∨	Disjunction, a combination of words through "either—or".
→	Implication, a combination of words through "if—then".
↔	Combination of two final vertical substitutions.
∀	A quantifier of generality, "all".
∃	A quantifier of existence, "there exist some (certain)."
~	Equivalence.
$\mathscr{A}\text{-}\mu\text{-}\mathscr{B}$	Equality or partial equality in the transmission of messages between the two divisions of a communicative system.
Σ	Complete syntagma.
$\mathscr{M}$	Message.
Π	Concept.

7. Notes on the Interpretation of the Figures

The registration of the signals was carried out at the Sukhumi Nursery (Caucasus). The microphone of MD-55 type and the tape recorder of MAG-8 type with a frequency range from 50 to 10,000 c/s were applied. Acoustical analysis was carried out in the Laboratory of Phonetics in the Moscow Institute of Foreign Languages, with the assistance of I. A. Zimnyaya.

(1) *The first group of signals* (Fig. 68, p. 155). Four cries *a!* (*ak*) of hamadryads of different sex and age. The cries have signal meaning of danger at the appearance of unfamiliar phenomena. Spectrum intensity is about 100 dB (at one meter distance). The second element of the cry *a!* (*ak*) is guttural explosion "!". The first element is similar to man's type. Formant frequencies are 950, 1120 and 1900. The intensity curve is shown in Fig. 75, on p. 160. Sharp rise and fall of the curve are of importance.

(2) *The second group of signals* (Fig. 69, p. 155). Man's imitation of hamadryads' cries (with imitator's fundamental frequency of 150 c/s). Four cries *a!* (*ak*) are presented. Formant frequencies are 800, 1120 and 1690–1900. In comparison to the spectrum in Fig. 68 the number of harmonics is increased.

(3) *The third group of signals* (Fig. 70, p. 156). Man's natural pronunciation of the *a* sound in the *ak* syllable. Formant frequencies are 1120–1480. The form of the intensity curve is similar to that of the spectrum in Fig. 68 (a hamadryad's cry "*a!*").

(4) *The fourth group of signals* (Fig. 71, p. 157). Five cries *a!* (*ak*) of macaques of different sex and age. The cries have signal meaning of danger, of tentative reflex (orienting on any kind of alarm). In comparison to the spectrum in Fig. 68 of hamadryads' cries the whole formant structure is moved to the right. Hamadryads do not react to these cries.

(5) *The fifth group of signals* (Fig. 72, p. 158). Four cries *au* of hamadryads. The cries have signal meaning of mutual searching when being behind the herd. The spectrum curve of the *a* sound in *au* differs from that of the *a* sound in "*a!*". Compare Fig. 72 to Fig. 68. The intensity level curve is shown in Fig. 75 (p. 160).

(6) *The sixth group of signals* (Fig. 76, p. 160). Three cries *ou* of hamadryads of different sex and age. The signal meaning of the cries does not differ from that of *au* cries.

(7) *The seventh group of signals* (Fig. 77, p. 161). Seven cries *hon* of hamadryads of different sex and age. The cries have signal meaning of mutual satisfaction. All the seven spectra of *hon* cries are similar. Their intensity level is about 40 dB (at one meter distance). An *o* sound in *ou* greatly differs from the *o* sound (nasal) in *hon*.

(8) *The eighth group of signals* (Fig. 78, p. 161). Four cries of *i* type squeal. The duration of every squeal is about 400–700 msec. The formant structure is shifting all the time to the right and the left but the formant frequencies 1900–2380 are constant.

REFERENCES

[1] Artemov, V. A., 1956, *Experimental Phonetics*, published by Literature in Foreign Languages, Moscow, p. 130 (in Russian).
[2] Artemov, V. A. and I. A. Zimnija, 1958, *Phonematic Spectra and their Employment in Mechanical Translation*, one of a number of *Theses* presented to the Conference on Mechanical Translation, Moscow, May 15–21, 1958, pp. 31–33, publications of the Ministry of Higher Education of the USSR (in Russian).
[3] Firsov, L. A., 1954, Oscillographic investigation of the vocal reactions in monkeys, in *Physiol. J. USSR*, *40*, (1) 18, 19, published by the Academy of Sciences of the USSR (in Russian).
[4] Gamov, G., A. Rich and M. Icas, 1956, Problems of the Transmission of Information from the Nucleic Acids to the Albumens; Problems of Biophysics, in *Advances in Biological and Medical Physics*, Vol. IV, New York.
[5] Jakobson, Roman, C. Gunnar Fant and Morris Halle, 1955, *Preliminaries to Speech Analysis (The Distinctive Features and their Correlates)*, Acoust. Laboratory Massachusetts Inst. of Technology, 2 pr., p. 40.
[6] Jakobson, Roman and Morris Halle, 1955, *Fundamentals of Language*, pp. 29–31.
[7] Kellog, W. N. and L. A. Kellog, 1933, *The Ape and Child*, New York–London, pp. 275–306.
[8] Ladigina-Kots, N. N., 1935, *The Child of a Chimpanzee and the Human Child in their games, instincts, emotions and expressive gestures*, published by the State Darwin Museum, Moscow, pp. 241–248 (in Russian).
[9] Markov, A. A., 1954, The theory of algorithms, *Transactions of the Steklov Mathematical Institute*, *42*, publications of the Academy of Sciences of the USSR, Moscow–Leningrad (in Russian).
[10] Markov, A. A., 1958, Concerning constructive functions, *Transactions of the Steklov Mathematical Institute*, 52, 315–348 (in Russian).
[11] Novikov, P. S., 1955, Concerning the algorithmic insolubility of the problem of the identity of words in the theory of groups, *Transactions of the Steklov Mathematical Institute*, 44 (in Russian).
[12] Sarris, E. G., 1931, *Sind wir berichtigt vom Wortverständnis des Hundes zu sprechen?*, Leipzig, p. 129.
[13] Shanin, N. A., 1958, Concerning the constructive understanding of mathematical judgements, *Transactions of the Steklov Mathematical Institute*, 52, 226–311 (in Russian).
[14] Tikh, N. A., 1950, The Gregarious Life of Monkeys in the Light of the Problem of Anthropogenesis, *Thesis*, Leningrad, pp. 26–32 (in Russian).
[15] Wiener, Norbert, 1954, *Cybernetics*, Chap. IV, Wiley, New York.
[16] Voronin, A. G., 1952, *An Analysis and a Synthesis of Complex Stimuli in the Case of the Higher Animals*, Medgiz, p. 73 (in Russian).
[17] Yerkes, R. M. and B. M. Learned, 1925, *Chimpanzee Intelligence and its Vocal Expressions*, Baltimore, pp. 53–56.
[18] Zhinkin, N., 1956, New data concerning the functioning of the motor speech analyser, when in combination with an auditory one, *News of the Academy of the Pedagogical Sciences of the RSFSR*, No. 81, Moscow, p. 207 ff. (in Russian).
[19] Zhinkin, N., 1957, The paradox of speech respiration, in *Physiol. J. USSR*, *52*, (2) 145, published by the Academy of Sciences of the USSR, Moscow–Leningrad (in Russian).
[20] Zhinkin, N., 1958, *The Mechanisms of Speech*, p. 219 ff., published by the Academy of Pedagogical Sciences of the RSFSR (in Russian).
[21] Zhinkin, N., 1959, Towards the study of the mechanism of speech. In *The Science of Psychology in the USSR*, Vol. I, pp. 470–487, published by the Academy of Pedagogical Sciences of the RSFSR (in Russian).

PART III

SPECIAL ASPECTS OF ANIMAL ACOUSTICS

CHAPTER 8

ACOUSTIC SIGNALS FOR AUTO-INFORMATION OR ECHOLOCATION

by

F. VINCENT

1. Introduction*

Griffin's work on the echolocation of bats forms the basis of the major part of our knowledge of this highly specialised acoustic behaviour. Echolocation was discovered subsequently in a diversity of other animals ranging from certain birds to marine mammals.

To a very large extent these researches lent fresh impetus to world-wide work on animal acoustics as a whole by arousing interest in a new field of enquiry, the wealth of potentialities and possibilities of which Griffin had thus revealed.

Griffin published a detailed account of the question and of his results in an excellent book of 413 pages called "Listening in the Dark"[26], to which the reader is referred as a fundamental work on this problem (1958).

Within the limits of this book, we have only attempted to give a general view of this problem in a short synthesis.

2. Echolocation and its Definition

(a) *History of the discovery of echolocation*

Echolocation was discovered in bats, those fliers by night whose amazing ability to avoid obstacles when moving rapidly and with precision in the dark had for long puzzled many students.

The first of these, Spallanzani, performed experiments on these animals from 1794 onwards. He found that blinded bats did not detract from the precision of their flight, even if the room in which he was experimenting contained numerous obstacles[78]. At Spallanzani's request, Vassalli, Rossi and Senebier[75] repeated these experiments and came to the same conclusion. All four inferred from these results that bats possessed a highly developed sense of touch, probably located at the level of the wings, and compared it to that possessed by some blind people**.

These first observations led Spallazani to carry out other experiments, which produced other results. He found that bats were able to direct their flight in a tunnel with angles and curves quite well without touching the walls even when their wings had been carefully coated with varnish. Thus he was forced to assume that the sense of touch did not account for the phenomenon.

---

* For references in parentheses, see the Addendum to this Chapter.
** For blind men, see Addendum.

References p. 225

He next eliminated the senses of taste and smell by removing the tongue and obstructing the nostrils of his experimental animals. This made no difference to their flight at all.

When, however, he covered the animals' heads completely, thus depriving them of vision, hearing, smell and taste at the same time, he found that they collided with the obstacles. He was unable to obtain conclusive results by plugging the bats' ears with wax.

This author and his collaborators, therefore, came to the conclusion that bats are endowed with a sixth sense.

At about the same period, Jurine[38] repeated Spallanzani's previous experiments with blind bats; but, when he had introduced various waxy substances into the animals' ears, they were no longer able to avoid the obstacles. From that moment the part played by hearing in the orientation of these mammals was established.

At that time, Cuvier[8] refused to admit that these conclusions were well founded; nor would he acknowledge the quality of the experiments without, however, trying to verify any of them.

Cuvier's prestige was so high among the scientific minds of the age that the nineteenth century brought nothing new in this domain. It was not until 1900 that Rollinat and Trouessart, going over Spallanzani and Jurine's experiments, obtained certain proof that hearing plays a dominant part in this phenomenon, the other senses being merely subsidiary[68]. Moreover, when they blocked the nose and mouth of one individual bat, its flight became less well-directed.

In 1908, Hahn[32] performed the same experiments with great care upon North American species. He obtained the same results, but failed to recognise the part played by the auditory sense, as such, in this phenomenon. He postulated a sixth sense of "direction" after a whole series of experiments on conditioning bats in captivity*.

It was Hartridge, 1920[33], who suggested a possible kinship between the system by which bats orientate themselves and the principle promulgated by Langevin and used for the acoustic detection of submarines in the 1914–1918 war[48]. His hypothesis introduced the intervention of ultrasonic sounds, and not infrasonic, the wavelengths of which are far too long to be of any use on small objects (for 15 c/s, 26 m).

Taking up this hypothesis again, Pierce and Griffin[63], with the aid of adequate apparatus, were the first (in 1938) to detect and demonstrate the emission of ultrasonic sounds by bats.

Finally, Griffin and Galambos (1940), scrupulously repeating all previous experiments, came to the following conclusions:

1. Nothing incapacitates a bat from avoiding obstacles except loss of hearing and obturation of the mouth[27].

2. Bats in flight emit ultrasonic sounds of a maximum energy of 50,000 c/s. Their ears are sensitive to this type of emission[21].

---

* In 1912, the *Titanic* sank after hitting an iceberg. Maxim, an engineer, suggested that calamities of this kind could be avoided by emitting infrasounds and picking up their echo. He was convinced that bats oriented themselves by hearing the echo from their wingbeats (roughly 15 c/s).

3. Bats hear the echo of the ultrasonic sounds they emit through the mouth and orientate themselves by integration of their variations.

All these data were confirmed independently by Dijkgraaf in Holland as from 1943[10].

Since 1949, Möhres[52], studying species different from those used by the American researchers (Rhinolophidae and *Rousettus*), has shown that there is great diversity in the systems of emission utilised by different families of bats.

(b) Definition and principle

In 1944, Griffin[22] suggested the term "echolocation" to describe a system of self-information comprising, in one and the same animal, an organ for emitting an acoustic signal and an organ for receiving the echo of that signal, something like a feed-back physiology. The fundamental principle of echolocation is as follows:

The animal emits acoustic waves which are reflected when they impinge upon an obstacle. The animal hears the acoustic energy that has returned from the object and is able to judge the distance of the obstacle and, if the emitted beam sweeps a certain sector, he can also discern the outline of the obstacle in that sector.

3. Echolocation in Chiroptera

Different types of echolocation have been described as associated with various groups of bats, often linked to the anatomy and specific habits of each family or sub-family.

(a) Evidence of the emittor-receptor system of echolocation

Once Hartridge had propounded the hypothesis that Chiroptera orientate themselves by echolocation (1920), Griffin and Galambos verified it by experiments, applying the following methods to the Vespertilionidae, chiefly to *Myotis lucifugus*.

Sixteen wires of 1.2 mm diameter were hung vertically at 33 cm intervals in a single plane across the middle of a large and completely darkened laboratory. The experiment consisted of calculating the percentage of flights successfully accomplished by a single individual (flying between the obstacles without hitting them) in relation to the total number of flight tests.

These different percentages were compared from one experiment to another. As allowance also had to be made for chance, it was calculated that the percentage of successful flights due to mere chance was between 30 and 35, according to the results of Hahn who threw into a flight room a dummy "bat" and counted the numbers of barriers hit (see Table 18).

The bat always scores many more misses with its ears plugged and mouth muffled for the experiment than when intact or deprived of sight. This places the evidence in support of the importance of those two sensory organs beyond all doubt.

On the other hand, blinded bats have a slight tendency to fly with greater accuracy, as though the visual information it normally receives impairs the acuity of its auditory perception.

Timm[80] reports a highly interesting experiment, in which he succeeded in inversing the information received by the ears of Vespertilionidae by introducing tubes into their ears. These tubes were bent so that the one of the left ear opened on the right side of the head and vice versa. Then the bats flew in the direction opposite

TABLE 18

Number of bats used	Experimental treatment	Experimentals		Controls	
		Number of flights	Average % misses*	Number of flights	Average % misses*
28	Both eyes covered (controls untreated)	2016	76	3201	70
12	Both ears covered (controls untreated)	1047	35	1297	66
9	Ears and eyes covered (controls with only the eyes covered)	654	31	832	75
8	Closed glass tubes in ears (controls with the same tubes in ears but open)	580	36	636	66
12	Both ears covered (controls with one ear covered)	853	29	560	38
6	Eyes and one ear covered (controls with only the eyes covered)	390	41	590	70
7	Mouth covered (controls with eyes covered or intact)	549	35	442	62

* Flights in which the bats skimmed the walls of the room or avoided the wires by turning away were not included in the scores. Further, in a room so beset with a network of obstacles, a certain percentage of errors in all these cases was unavoidable.

to that which they should have followed. His report, however, was insufficiently detailed and Möhres[53], who tried to repeat his experiments (1953), was unable to set them up satisfactorily.

To discover what part receptors on the wings might play in this orientation, Griffin and Galambos carried out a series of experiments which provided the following results:

	% of flight without hits
Intact bats	81
Bats with the lower side of the wings coated with collodion. (Flight is mechanically impeded by this treatment.)	62
Bats with lower side of wings coated with collodion, eyes and ears covered	47
Bats with eyes and ears covered (wings untreated with collodion do not facilitate flight*)	33

At this stage the authors themselves point out (Galambos[16], 1942) that they have reached virtually the same point as Spallanzani and Jurine in 1798, but with considerable numerical precision. Besides this, these results could be related to those obtained first by Pierce and Griffin, then by Griffin and Galambos, associated with the production of ultrasonic sounds by bats in flight; they proved to be short pulses of the order of 5 to 10 milliseconds produced in bursts within a purely ultrasonic frequency spectrum, with a maximum energy at 50 kc/s. These noises were emitted even if the animals were deafened—in which case their orientation was very poor. If their mouths were carefully obturated, no signal at all was recorded and their orientation was also very poor. Thus the whole of Hartridge's hypothesis was confirmed.

(b) *Different types of echolocation in bats*

During the last fifteen years, Griffin in the U.S.A. and Möhres in Germany have

* Otherwise Dijkgraaf[11] reports experiments in which nerves supplying the wings were cut—with no effect on the orientation power.

adduced evidence to show that there are various modalities of echolocation among Chiroptera. These different types will now be considered one by one in association with the most familiar species.

Our subjects can be divided into three main groups, *viz.*

1. Bats whose orientation is assisted by clicks of high intensity (attaining to 20 to 100 dynes/cm^2–100 to 114 dB*, measured at 10 cm from the mouth) with a frequency range of 10 to 100 kc/s. This group contains insectivorous and partly insectovorous bats.

2. "Whispering" bats, whose emitted sounds are difficult to detect (intensity at the level of the mouth being less than 5 dynes/cm^2–88 dB) and cover a scale of very high frequencies. These are predominantly fruit-eating and carnivorous bats. Apparently, therefore, the echolocation of moving prey requires far greater energy than that utilised for the detection of static prey like fruit and sleeping animals.

3. Bats directed by hearing, sight and smell.

(c) *Bats emitting high-intensity signals*

(i) *Vespertilionidae*

1. *Physical characteristics of the emissions.* The sounds emitted by Vespertilionidae** for echolocation are series of ultrasonic pulses, or clicks, each lasting at most 10 msec.

In the case of *Myotis lucifugus* the duration of the click varies from 1 to 5 msec. Hence the wavelength, in air, of an average click of 2 msec will be 68.4 cm. If the object is about 137 cm in front of the bat, the latter will receive two distinct sounds, *i.e.* the initial sound between 0 and 2 msec and the echo between 8 and 10 msec. This interval facilitates hearing the second sound, which is very attenuated. And in the case of a click of 1 msec, the echo of an object 35 cm away will still be heard with an interval of more than 1 msec between the two.

Only a few of these clicks are emitted when bats are at rest, and twenty or so a second when they are active on the ground or are preparing for flight. As they begin to fly, the rhythm increases, easily reaching 50 per second when they are flying close to small obstacles or when they are about to land. Just as they pass between the obstacles the rhythm decreases, reaching again its normal value in flight (20 to 30 per second). There is no such drop in rhythm when they collide with an obstacle.

Each click is accompanied by a very faint audible sound (Dijkgraaf's "Ticklaut" and "the audible click" of Griffin and Galambos). During accelerated emission, these faint sounds more often than not merge and produce a slight "buzzing" (or "Ratterlaut").

It is interesting to note, by the way, that Chiroptera are capable of emitting quite audible sounds unassociated with orientation by echolocation. There is, besides, a whole series of emitted sounds intermediate between the audible cries and the echolocating clicks, especially those emitted by bats still drowsy after their daytime sleep ("Tageschlaflethargie"), during which their body temperature is lowered. These

---

* 0.0002 dynes/cm^2 = 0 dB.
** The numerical values reported here concern *Myotis*, but other Vespertilionidae seem to have quite similar characteristics.

References p. 225

emissions have a very strong audible component. Griffin has noticed that the more dominant the distinctly audible component of these signals is, the less useful they are for avoiding small obstacles, and it would seem that the best adaptation to echolocation consists in reducing the low frequencies and increasing the ultrasonic components.

The frequencies of the middle of the click of *Myotis lucifugus* reach 109 decibels or 60 dynes/cm² at 8 cm from the microphone. This is a very great intensity, but it is compensated for by the brevity of each click. All the same, the clamour assailing the ears of one bat among a whole flight of animals of its kind is almost unimaginable.

Pierce's analyzer showed a maximum energy of 50 kc/s, a figure that can be computed from the oscillograph. But an heterodyne analyzer likewise shows considerable energy at other ultrasonic frequencies. It will, in fact, be seen on closer study, even with the oscillograph, that the frequency is almost never constant. Always very high at the beginning of the click, it drops progressively by an octave towards the end of the pulse. Acoustically, this appears to be typical of the clicks of Vespertilionidae and kindred families. This modulated frequency occurs whatever the activity of the animal may be. For all that, no two clicks are exactly alike, even in the emissions of a single bat. Thus the initial frequency of each pulse may vary from 72 to 120 kc/s*. Nevertheless, the highest frequency always occurs at the beginning of the click. In the event of overlapping (*e.g.* when obstacles are as close as 20 cm and less), this drop in frequency within the span of the click may enable the animal to differentiate between the emitted pulse and its echo.

This modulated frequency accounts for the rather vast spectrum obtained for this type of emission. It should be borne in mind, moreover, that the extremities of each click may contain very much higher and very much lower frequencies (from 20 kc/s to 153 kc/s) within a fleeting instant. Their energies are very low compared with those of the rest of the spectrum and, as far as the very high frequencies are concerned, they may probably be considered as harmonics of frequencies lower than 100 kc/s.

By reducing the ultrasonic part of a click and amplifying its low-frequency part, oscillographs of the two parts can be obtained on the same record. It will then be seen that the low-frequency component has a spectrum between 3 and 10 kc/s (and after having cut the cricothyroid muscle of a bat, the animal emits sounds from 8 to 12 kc/s) with a sound pressure level of 0.03 dyne/cm², or 44 dB (the animal being very near the microphone). This appears 0.1 to 0.2 msec before the main pulse (see Fig. 81).

It may well be, indeed, that this cry continues throughout the pulse but remains invisible on account of the great difference in pressure between the audible and ultrasonic parts.

Finally, the greater the skill of a bat at avoiding obstacles, the fainter is the audible component of the sounds it emits. In the same way, this low-frequency part of the pulse is more important in young bats than in adults (learning, perhaps).

The inference would seem to be that low frequencies are undesirable for the accomplishment of perfect echolocation.

Griffin[26] showed that the intensity of the emitted wave is not distributed

---

* The highest measurable frequency may vary due to signal-to-noise level.

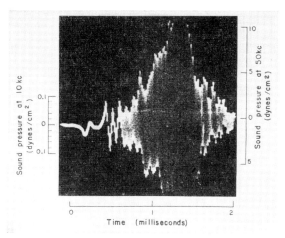

Fig. 81. Clicks emitted by *Myotis lucifugus*. On the ordinate axis are two scales of intensity on both sides of the figure; one for the audible low frequency component and the other one for the frequencies of the rest of the click. On the abcissa: time (after Griffin[26]).

uniformly in all directions. There is a decided maximum in front. Nevertheless, variations are observed according to the frequencies emitted.

For his part, Möhres[53] measured the intensities emitted by a Vespertilionid, *viz. Nyctalus noctula* Schreb, and obtained evidence of a maximum intensity in the axis of the animal. But the lateral attenuation was not, in his view, as marked; even with the microphone set up in the extreme lateral position, he recorded surprisingly intense pulses.

2. *Production of these clicks in* Myotis *and* Eptesicus. The photographs of flying bats show that the animals' mouths are always open (Figs. 82 and 86). Careful examination by Griffin of a bat's throat while emitting ultrasonic sounds with its mouth wide open revealed a transitory flutter at the back of the tongue at the very moment at which the analyzer registered a click. The source of the emission was thus located in the respiratory tract.

Fig. 82. *Myotis lucifugus* flying, the mouth open, characteristic for the specific behavior (after Vesey-Fitzgerald[84]).

Anatomical studies have shown that bats have a remarkably large larynx of highly specialised structure[14].

When, on the other hand, the mouth of *Myotis* is completely sealed, some individuals manage none the less to orientate themselves. This led to the conjecture that the nostrils might serve as the passage for the emission. Yet, if the nostrils of these animals are obturated, they emit their signals through the mouth and their orientation is quite normal. This is one of the main characteristics of the echolocation of the Vespertilionid. It may be noted that *Plecotus*, the particular behaviour of which will be dealt with later, is able to emit signals either through the nostrils or the mouth.

Griffin succeeded in introducing small glass cannulae into the trachea of Vespertilionidae under anesthesia, so that they were able to breathe without using their larynx. No further sounds were recorded after these animals had recovered.

It is evident that the voluminous muscular masses of these animals' larynx make it a far more important organ, relative to the size of the trachea, than in the case of Man.

Its internal anatomy is also specialised: *Eptesicus fuscus*, for example, has two pairs of very thin membranes, about 6 to 8 microns thick, which rest on the walls of the larynx above a small, shallow cavity (0.25 mm), elongated in shape (roughly 2 mm) (Fig. 83). The cricothyroid muscles attached to the two principal cartilages of the larynx (thyroid and cricoid) are able to stretch these membranes, something like the skin on a drum.

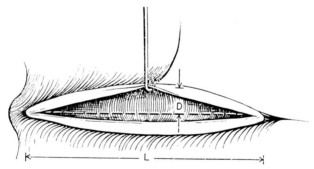

Fig. 83. Scheme of the membranes which cover one side of the cavity of the larynx of *Eptesicus fuscus*, the upper membrane was held up artificially (after Griffin[26]).

If these membranes are cut, the only sounds obtained are very low and very faint.

Man needs the inferior laryngeal nerves for phonation. Their suppression in bats makes no apparent difference to the sounds emitted by the animal. In point of fact in Man it is the arytaenoids and the muscles and tissues associated with them which move backwards and forwards in phonation. In bats the two arytaenoids are welded together and are partly ossified, making the whole system very rigid. This lead to the assumption that it had no part in the emissions.

If the superior laryngeal nerves (which innervate the cricothyroid muscles) are cut, only the frequency of the clicks seems affected; it diminishes notably. Moreover, it becomes virtually constant (10 kc/s, *i.e.* of the same order as the audible component of the normal clicks). It would seem, therefore, that the effect of these muscles is to lower the frequency by one octave or more during the short pulse. Hence one may assume that, during the emission, the muscles relax progressively, the frequency being proportional to the tension of the membranes.

When the two roots of the hypoglossis have been cut, clicks of long duration (10 msec) are sometimes obtained, while other clicks are perfectly normal (2.5 msec). When all three pairs of nerves have been cut, the bat can still emit short clicks normally for a considerable time. The mechanisms which release and stop the clicks within 1 or 2 msec are still unknown, but it is reasonable to suppose that the control is exercised at the level of the epiglottis.

On the other hand, if one calculates the resonance frequency of the membranes, one finds 45 kc/s, which agree with the figures found experimentally.

Lastly, Novick[60] placed electrodes in the upper laryngeal nerve and cricothyroid. For several milliseconds before the emission of the click he observed intense activity, which diminished rapidly directly after the initial burst of sound. From this it may be inferred that the contraction of the cricothyroid begins before the click is emitted, and that this muscle might be completely inactive during the emission itself, thus bringing about the modulated frequency to which reference has been made.

Möhres[53] studied the production of clicks by the European *Myotis* and concluded: "these are crackling sounds ('Sprenglaute') produced by the intermittent accumulation of the flow of respiratory air, and the abrupt opening of the obturating system. This genesis accounts for the form of the pulse and the drop in frequency during the span of a pulse: the vocal cords* are at first stretched by the increase in accumulated pressure and thus produce the octave of the fundamental vibration associated with the drop in pressure. There are no other clues to the origin of the Vespertilionid type of pulses".

If that is so, however, in view of the pulsed but continuous character of the emission, it is difficult to understand how a bat can breathe without ceasing to emit**. One would have to fall back on inspiration similarly pulsed, unless the expired air is not the generator of the clicks and breathing takes place independently of the emission. In that event the inspired air might pass through the nasal passages to the lungs without necessarily interfering with the vibrating structures. This interesting problem should be studied but is still now an unsolved one.

This controversy is reminiscent of that between the upholders of the neuro-muscular theory and the champions of the neurochronaxic theory in respect to phonation in Man. Pending the results of new experiments, it would be prudent to assume that these two hypotheses are complementary. (See Addendum.)

3. *The auditory sense of Vespertilionidae and echolocation.* To complete their system of auto-information, Chiroptera must undoubtedly have very sensitive hearing, enabling them to detect the very faint echoes of their emitted sounds and to distinguish them from other noises in their surroundings more often than not of greater

---

* The 6–8 micron membranes and the vocal cords may or may not be homologous but there is now no evidence on this point.
** Emission of orientation sounds is often at a steady rate of 30–50 per second for a time equal to several respiratory cycles. Normal ventilation rates are on the order of 5–8 per second, and possibly increased somewhat during flight. In the case of the very high repetition rates (150–250 per second) it is a little bit different for they occur only in bursts and there may be one expiration per burst.

References p. 225

intensity than their own. It is known that a bat is completely disorientated when its ears are sealed.

At first sight the temptation to link the evident morphological complexity of the external ear of Microchiroptera with their astonishing auditory prowess is irresistible. Quite often, the size of these organs is greater than that of the head itself (Fig. 84). Many of them, moreover, have folds and several other important refinements. One of these is the tragus, a small straight column that rises from the base of the ear and is situated exactly in the path of the sound waves at the entrance to the tympanic canal. It is important to note that the tragus and the two halves of the pinna which it delimits have widths of the same order of magnitude as the lengths of the waves of the echolocating clicks.

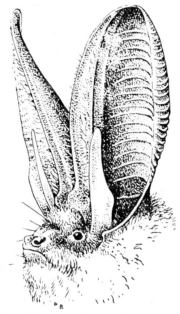

Fig. 84. Head of the long-eared bat, *Plecotus auritus* which shows the particularly complex structure of the external ear (after Bourliere, in Grasse[20]).

The purpose of the pinna is probably to focus sounds. This focussing varying with the frequency, the start of the clicks will be far better directed than the remainder of the pulse, which is emitted at a lower frequency.

The cochlear organ of a bat is similar to that of other mammals, with the following important differences:

(1) The part of the cochlea nearest the middle ear, which is concerned with high frequencies, is more developed in bats.

(2) The fenestra rotunda (which is not really round in bats) is in contact with the fluid of the inner ear in a manner quite different from what we are accustomed to find in other mammals: it is situated, not at the end of the cochlea, but at 1 mm after its first coil.

(3) The basilar membrane which, in other mammals is always less wide and slightly more rigid at the end near the internal ear, is in this case so additionally

thickened that the portion which is apparently free to vibrate in response to sound is narrower than normal.

In order to form some idea of the auditory sensitivity of bats, Galambos[17] recorded and measured the microphonic potentials from the ears of members of four species of Vespertilionidae, viz. *Myotis lucifugus, M. koenii, Pipistrellus subflavus* and *Eptesicus fuscus* (Fig. 85)*. He obtained cochlear potentials from frequencies as low as

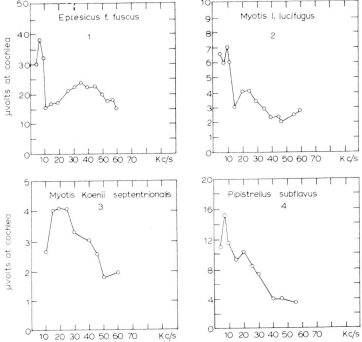

Fig. 85. Curves of the response of cochlear potentials of 4 species of bats, elicited by sounds of 5 to 60 kc/s (after Galambos[17]).

30 c/s, noting a maximum amplitude at 10 kc/s (except in *Myotis koenii*, where it occurred between 20 and 30 kc/s), then a less pronounced maximum, between 30 and 50 kc/s, in *Eptesicus* (which corresponds to the frequency of emission). The apparatus used did not enable him to experiment at frequencies higher than 98 kc/s, at which, nevertheless, he still obtained some important potentials (a guinea-pig is sensitive only up to 52 kc/s). No records are obtained if the experimental animal's ears are obturated.

A penetrating study has shown[64] that bats possess a specialised brain related to their mode of life; the tactile and auditory areas in particular, are highly developed**.

Recently Pye([22]) established a theory that, though not verified experimentally, offers a simple explanation of certain aspects of echolocation.

Let us suppose that the ear of the bat is a non-linear receptor, a distorted apparatus. A call and its echo perceived simultaneously (as with an object brought close enough to the ear so there is an overlap of the initial wave by the return wave) will form harmonics of higher and lower frequencies than those of the two funda-

---

* See Addendum. ** See note p. 223.

References p. 225

mentals. The first order tone difference will be a lower frequency than either the call or the echo, and it will be audible as a beat note. (The other tones and harmonics appear to have weak enough amplitudes or high enough frequencies so that the author does not consider them pertinent to this problem.)

In the case of the Vespertilionidae, with the target at a very short distance, at the time of the overlap the echoes are multiplied, and a discrimination based on their intervals in time would be so complex that it seems hardly possible. By listening to the beats instead of to the echo, Vespertilionidae can obtain accurate information. The fall of frequency of calls is not linear, and the beats will not have a constant frequency. They will be, however, very characteristic of the echo delay. The varied frequencies of the beats of the multiple echoes will be easily differentiated by the two ears for objects not in the median plane. By a comparative frequency analysis, at the level of the cochlea or of the central cervical auditory apparatus, the bat can know the distance and direction of the obstacle. This would explain the habit of the Vespertilionidae in turning his whole head in the course of a flight (the ears having synchronized movements). The animal would seek thus to equalize the frequencies of the beats received by his two ears to give him immediately the direction of the target. This winning hypothesis is confirmed by the loss of the ability for echolocation by the Vespertilionidae when one ear is blocked.

Finally, Dijkgraff[11] and Möhres[53] have shown that the maximum distance to which the pulse-echo overlaps occur, corresponds to the maximum range at which accurate echolocation is utilized with the best output for each type (50 cm for a 3 msec call).

4. *The orientation of Vespertilionidae.* Griffin and his collaborators used various tests to find out more about bats' performance and to discover the limits of their skill.

Steel wires, 1.2 mm in diameter, were suspended vertically and spaced 33 cm apart; the laboratory was thus divided into two parts by this barrier. After several thousand trial flights with 129 *Myotis lucifugus* in this room, the results gave an average of 35% hits and 65% misses. By selecting especially adept individuals, Curtis[7] found 82% misses.

It is well to bear in mind that the lengths of the sound waves emitted by bats are within the range of 3 to 10 mm, and that the size of the obstacles is therefore, approximately one third of the shortest wave emitted. Now, the total energy contained in the trains of reflected waves decreases considerably when the size of the reflecting object is less than the length of the wave. The detection by bats of objects of this size thus becomes an interesting subject of study. Looking at Sgonina's works[76] we can see that when the diameter of the obstacles was reduced to 0.3 mm, most of the bats failed to avoid them except by chance. But *Myotis natteri* scored 55% misses and when it was confronted with obstacles 0.8 mm in diameter it scored 72% misses. Similarly, the following results were obtained with a *Plecotus auritus*:

Size of objects (in mm)	Per cent flights without hits
0.8	44
0.9	67
1.0	79
1.3	88
1.5	88
1.7	91

These results, however were obtained with one individual only of the two species. Curtis took up this question with Griffin[26]. All the tests were carried out with steel obstacles aligned and spaced at 33 cm, the diameter increasing from 0.07 to 4.8 mm. The species tested was *Myotis lucifugus*. The results seem to show that the obstacle of 3 mm corresponded to a critical diameter. The length of a wave of that size would correspond to 100 kc/s which is above the range of auditory sensitivity of almost all known animals. Hence bats are able to detect the faint echoes of objects smaller in size than one wave length. Admittedly, it might be assumed that an emission of sounds of smaller wave length than anything discovered so far exists, but it seems highly improbable.

Rawson and Griffin[26] studied the influence of size of the spaces between the obstacles (vertical metallic rods 1.2 cm in diameter) upon the orientation of *Myotis lucifugus* with the following results:

Spacing (in cm)	Per cent flights without hits
11	32
31	75
55	95

The minimal wing-spread of bats being 6 cm, the percentage of misses due to chance should be zero in the case of the obstacles spaced at 11 cm. Unlike birds, however, the two wings of these animals are entirely independent and they have no difficulty, therefore, in folding one more than the other, which enables them to pass through passages far narrower than anything birds of the same wing-spread could get through. When the obstacles are arranged horizontally, the results are better still, because the animal is able to pass with its wings fully extended.

Möhres and Oettingen-Spielberg[58], then Griffin and Rawson[26], kept one individual of the same species for several weeks in succession in a vast room divided by a highly reflective partition, across roughly two-thirds of the width of the room, thus leaving a passage through which the bat made it a habit of flying in order to avoid the obstacle. After this apprenticeship, the partition was shifted to the other side when the animal was not in the room, so that the old passage was blocked and a new one made at the other end. When the bat was again released in the room, it could be heard very clearly emitting its clicks in a calm rhythm (approximately 20 a second); yet it flew straight at the old opening and hurtled against the partition (in spite of the fact that it reflected strongly). It returned and started all over again colliding with the partition.

As the emission of its clicks was normal, its ears were evidently giving it the necessary signals for obstacle avoidance. It would appear, therefore, that after a certain amount of habituation, the animal pays less and less attention to the echoes it hears and pursues its flight mechanically without, however, ceasing to click.

First Hahn[32] and then Möhres and Oettingen-Spielberg[58] obtained evidence of another peculiarity of the sense of direction in bats. After conditioning a bat to find its food put in a cage placed in a certain manner in a room, the experimentators rotated the cage through an angle of 90° or 180° to see if it was able to find the door in its new position. The bat had great difficulty in foiling this trick. It evidently

References p. 225

orientated itself by the echoes received from the room as a whole, or at any rate a considerable portion of it, rather than by those of the cage itself. The bat required as long to learn the location of the feeding place after the cage had been related as it had in the first period of training. It seems probable that orientation was not lacking in this case, but that it was based on the aggregate echoes of the room. This problem is undoubtedly related to those associated with the orientation of migrating bats.

While on this subject it will be interesting to note the experiments performed by H. C. Mueller[59] (personal communication relating to experiments carried out in 1958–1959). Intact and blind individuals were carried 5 miles from their habitat and then released. The return of both lots was observed and it was found that the time taken was about the same. The author believes that bats can also utilise echo location for long-distance travel, if not directly, at least so as to learn to know a large territory perfectly by overflying it in all directions and using their echolocation. But some recent experiences lead to doubt that bat homing can be explained by simple recognition of topographic features*.

It may be said in conclusion that echolocation is not an automatic phenomenon but requires sustained vigilance; this would account for the evident fatigue of bats. This subjected to numerous consecutive tests.

5. *Pursuit of preys.* Since bats depend so much on their system of echolocation for finding their way in the dark, it was tempting to speculate whether their search for food is likewise based on this principle. Indeed, most pursuits take place at twilight or at night. Spallanzani was the first to experiment in this way: he captured 52 individuals and, after blinding them, released them. Four days later, he returned early in the morning to the place of capture and recovered four of the animals he had blinded. He killed them and found as many insects in their stomachs as in those of intact animals.

Sight, therefore, apparently plays but a small part, if any, in the pursuit of prey**. It was then suggested that bats orientate themselves by the buzzing of insects[30,82]. Plausible as this may sometimes appear***, however, many of the insects captured by bats emit no sound whatever. It is interesting to note[82] that some nocturnal moths are sensitive to ultrasonic sounds, particularly between 15 and 60 kc/s, and this may enable them to detect the approach of predatory bats and, sometimes, to evade them.

The utilisation of echolocation for capturing insects is not an improbable hypothesis, as the captured prey are larger than the smallest obstacles which bats are able to avoid. Moreover, if they were guided by the noises of their quarry, they would have to be almost silent themselves and to hold their emission; whereas if they pursue by echolocation, they would continue to emit their clicks in order to locate accurately.

From an analysis made by Gould[18] of the contents of bats' stomachs it has been calculated that captures vary, all according to the size of the insects, from 67 to 5000 an hour (Fig. 86). A reasonable average is 500 an hour, or one insect every 7 seconds! Griffin recorded the emissions of bats flying in natural surroundings.

---

* See note in Addendum.
** See p. 185, section (a).
*** Experiences of masking the flight sounds showed that this hypothesis cannot apply in the case of *Drosophila*[30].

He watched both, the animal hunt down an insect and capture it, and a cathode ray oscillograph. The latter recorded the clicks that he could associate visually with the capture by the variations in the rhythms of the emissions*. Cruising clicks had a duration of 10 to 15 msec. Hence bats, when flying at more than several metres from the ground (or the walls of a flight chamber) emit wave trains 3 to 5 metres long. Furthermore, the frequency is in the neighbourhood of 30 kc/s and practically constant throughout one click. Nevertheless, slight modulations of frequencies can be observed as well as different frequencies from one pulse to another (30 kc/s to 70 kc/s). According to the distance from the ground, the rhythm of emission may vary from 4 to 10 per second. It is reasonable to suppose that these cruising clicks enable the bat to locate the ground, trees and other objects of the scene at a certain distance.

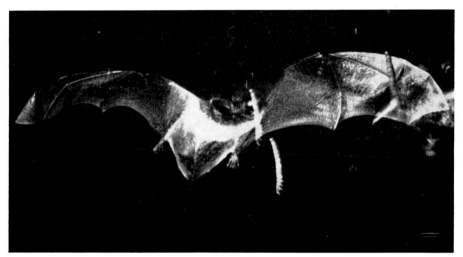

Fig. 86. A *Myotis* that catches an insect larve in flying (by courtesy of Webster[30] and *Endeavour*).

The clicks of capture, on the other hand, are very similar to those recorded in the laboratory. The transition from one to the other is gradual, effected by reducing the duration of each click and, at the same time, that of the interval between two clicks. There is, moreover, a reduction in the sound pressure of each click and a lowering in the internal frequency from one pulse to another[30] (Fig. 87).

Previously recorded series of clicks emitted by flying bats can be played back with a loud-speaker, but it has not been possible to obtain conclusive results on account of technical difficulties. For example when a bat comes too close to the loudspeaker it quickly becomes aware of the difference between that emission and the reflection of clicks by an insect. Griffin succeeded in diverting several bats temporarily towards his loud-speaker the emission of which might have been mistaken for those of insect prey. (See Addendum on *Plecotus*.)

Thus the clicks are emitted in slow rhythm when the animal is flying several metres above the ground. This prevents it from confusing the echoes from different

---

* These variations are elicited by the detection of the insect. The distance from the insect at which detection occurs is generally of about 50 cm for *Drosophila* and it can occasionally be as much as one metre with the same or with mosquitos *(Culex)*[30].

References p. 225

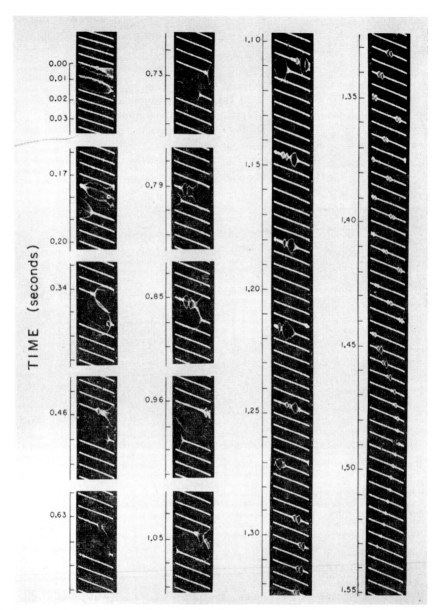

Fig. 87. Unceasing oscillogram of the clicks emitted by *Eptesicus fuscus* in the wild. During the first second the bat flew at a level of 10 or 15 m, between 1 and 1.25 sec it dived on a quarry and the rhythm of the clicks increases considerably (after Griffin[26]). (See also Fig. 96 with the same behaviour observed on *Delphinus delphis*.)

clicks. Conversely, the fact that the bat is concentrating on the echoes coming from the insect accounts for the accelerated rhythm while it is hunting its quarry. The interval between two clicks then becomes shorter, allowing the echo of each click to be perceived before the emission of the next. If, however, the intervals are further reduced, the animal is confronted with a fresh problem, *viz.*, the risk of overlapping between the click and its echo. This is circumvented by shortening the duration of each click, in such fashion that the length of the corresponding wave train, in air, is smaller than the distance between the bat and its quarry.

As regards the wave-lengths used, we can estimate that too high frequencies (50 to 70 kc/s) would be inefficient in cruising clicks, as they would be absorbed by the air for distances of several metres, whereas at 20 to 30 kc/s attenuation by the air is reduced to a minimum. High frequencies are more useful during the pursuit of insects, because the echoes of remoter objects can thus be eliminated; moreover, their short wave length is more effectively reflected by small objects.

(ii) The Rhinolophidae

These are animals of the Old World, clearly distinguishable from the Vespertilionidae by a highly developed nasal appendage, a nasal cavity which widens at the top, very mobile ears having no tragus, an immensely large cochlea and very specialised larynx.

Purely from the acoustic standpoint, they differ from the Vespertilionidae by their manner of emitting sounds and by the signals themselves.

Two European species have been studied more especially by Möhres and his collaborators[53] since 1949, *viz. Rhinolophus ferrum equinum* Schreb and *Rhinolophus hipposideros* Bechst.

1. Behaviour of the animal during emissions. The mouth is closed, or almost so. The nostrils, at the centre of the horsesoe-shaped nasal apparatus, sserve as emitting canals while the free "crests" of the horseshoe vibrate slightly from time to time. The head is inclined during flight, another point of difference as compared with Vespertilionidae.

If one covers a Rhinolophid's mouth without interfering with its nostrils, it orientates itself normally. But if its nostrils are sealed, there is an end to its astonishing performance and it is no longer able to avoid obstacles. At the same time, the movements of the ears, characteristic of orientated flight, are suppressed. It has to be noticed that Rhinolophidae have great difficulty to breathe through the mouth.

2. Physical characteristics of the emissions of Rhinolophidae. Though consisting of clicks, they are nevertheless very different from the type of emissions characteristic of the Vespertilionidae.

Möhres took 168 measurements of the clicks emitted by *Rhinolophus ferrum equinum*. Their average duration is 65.3 msec, (extremes 24.2 and 138.8 msec) when the animal is held in the hand (Fig. 88). 25 measurements during flight averaged 90 msec. In every way these figures are considerably higher than those obtained with Vespertilionidae. The lesser Rhinolophid's clicks have a comparable duration to that of the greater Rhinolophid.

It has been shown[61] that other species emit shorter pulses and that a variation of a few kc/s may be found within an isolated click of these species.

As a plot of the pulses against time suggested, the cadence of emission is slow, varying from 4 per second at rest, to 5 or 6 per second in flight. Emission and respiration are synchronised, though each respiration need not necessarily produce an emission. This also applies to flight. Here we have further distinguishing features between Rhinolophidae and Vespertilionidae.

References p. 225

Owing to the long duration of these pulses, the oscillographic picture has a "ribboned" aspect, the amplitude curve being gently rounded, often flat during the major part of the emission. In this case, therefore, the very rapid variations in amplitude of the Vespertilionidae are lacking; yet, with due allowance for the time scale, there is still some degree of transitoriness.

Fig. 88. Click of *Rhinolophus ferrum equinum* (after Möhres[54]).

It had not been possible to make any sound pressure measurements. But, by comparing the minimum distances necessary to detect the emissions of clicks with a microphone between *Myotis* and *Rhinolophus*, Möhres was able to show that the latter produce far greater intensities, and that these can be amplified by the animals (the expiration is also correlatively more energetic).

The frequency is constant, or nearly so, in Rhinolophidae, 80 kc/s in *Rh. ferrum equinum*, 100 kc/s in *Rh. hipposideros*, except in the very end of each pulse[61]. Perhaps these relatively high frequencies are associated, as in *Carollia*, with emission through the nostrils.

Furthermore, during emissions of high intensity, a buzzing audible at 60 cm accompanies the clicks.

3. *Emission of the clicks*. The parallelism—both in intensity and time—between respiration and emission on the one hand, and the junction between larynx and nasal cavities on the other, suggests that the larynx may produce the pulses.

In view of the purity and regularity of the frequencies of the clicks emitted by the larynx of Rhinolophidae, a study of that organ in bats was bound to prove very fruitful. It revealed some characters typical of Chiroptera, but far more accentuated than in other members of that order: the larynx is highly developed, largely ossified and supported by powerful muscles[14].

Fig. 89 clearly shows the direct connection between the larynx and the nasal cavities in Rhinolophidae.

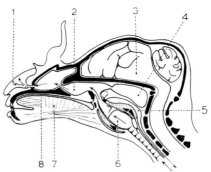

Fig. 89. Schematic sagittal section of the head of a Rhinolophe: 1. Horse shoe; 2. Nostrils; 3. Brain; 4. Pharyngeal bag; 5. Oesophagus; 6. Larynx with its blocking system; 7. Tongue; 8. Buccal cavity (after Möhres[53]).

When a Rhinolophid is prevented from breathing through its nostrils, the larynx falls back and a connection is established with the mouth; and, though this allows the animal to breathe and emit a cry of anger, it debars it from emitting ultrasonic pulses. On abrupt removal of the obstruction from the nostrils, the larynx reverts to its normal position and the animal is again able to emit its clicks. No satisfactory explanation has yet been suggested for the dependence of the emissions upon the high position of the larynx.

This specialisation enables Rhinolophidae to capture a quarry, masticate it and even to swallow it without ceasing to emit ultrasonic clicks.

Rhinolophidae have very poorly developed olfactory organs. They will willingly eat insects with foetid glands, like *Periplaneta*, which *Myotis* disdain. The nostrils are situated at the centre of the nose leaf (the horseshoe) which consists of three parts, the lancet, mounted on the muzzle and provided with cavities; the horseshoe, descending to the upper lip and surrounding the nostrils; the median crest, connecting the lancet to the horseshoe and widening towards the base into a saddle. The whole of this appendage is simply a fold of tegument, supported in the middle by a cartilage, and lacking tactile corpuscles, but surrounded by a double crown of very sensitive vibrissae. This appendage can be seen to vibrate slightly, especially at the level of the free margins of the horseshoe, when the animal's interest is aroused by a particular interest. These vibrations accompany an intensive ultrasonic emission, doubled by the buzzing mentioned above.

The emission is not in the least affected by suppression of the lancet, median crest and saddle. The horseshoe seems to serve some purpose, as this organ probably has a guiding influence on the ultrasonic sounds issuing from the nostrils; the emission is at 80 kc/s, *i.e.* with a wave length of 4.25 mm. Now the distance between the two nostrils being half a wave length allows the maximum reinforcement of the interfering beams. In front of the nostrils, therefore, the beams cancel each other out and this affords protection to the ears. The ratio is maintained in the lesser horseshoe bat: distance between the nostrils 1.6 mm, for a wave length of 3.4 mm.

The directivity is thereby improved and the intensity increased.

Möhres measured at several angles the drop in intensity received as a function of the distance between the microphone and the subject. The resulting curves are very much flatter than the curve representing the decline in intensity as a function of the square of the distance (Fig. 90). This is due to the position of the nostrils

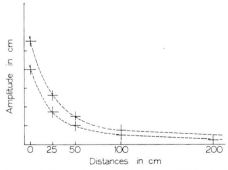

Fig. 90. Curves of the ultrasonic intensity emitted by two Rhinolophes, measured in front of the animal and in his axe, in relation to the distance (after Möhres[53]).

References p. 225

in a kind of funnel canalising the beams, especially as regards measurements taken in front of the animal, and at an angle of 22.5°.

The interference hypothesis is confirmed by the fact that the intensity curve received facing the subject is situated well above that collected at an angle of 22.5°, a fact which does not apply to the Vespertilionidae.

4. *Echolocation in Rhinolophidae.* We shall here try to mark down some of the laws governing the function of echolocation in the horseshoe bats, and to define certain limits.

The beam of ultrasonic sounds extends to 8 to 9 metres in Rhinolophidae, whereas the maximum among the Vespertilionidae is 1 metre*. It is, thus, more difficult to capture a Rhinolophid than a Vespertilionid, as the former takes evasive action when the hunter is several metres away from it. This early detection would appear to be due to the long range of this bat's emissions. The lengthening of the range is associated with increased respiratory movements, and with reinforcement of the beam (by narrowing of the outlet and modification of the profile of the horseshoe).

Möhres carried out a series of experiments for measuring the ability of bats to discover a 20 × 20 cm aperture in the wall of a cage. (Tests with the door of the cage closed enabled him to eliminate the possibility of locating being taken from the wall as a whole.) An opening of this kind can be detected by a horseshoe bat at 6.40 metres. No indication is given, however, of the dimensions of the bars of the cage which, judging by the results obtained with *Myotis*, must undoubtedly have influenced the precision of the method. Furthermore, this distance of 6.40 metres may not be the limit, as experiments at greater distances could not be made in that particular chamber.

Whereas Vespertilionidae, clinging by their four claws to vertical surfaces, usually head downwards, can only probe their surroundings by lateral movements of the head, Rhinolophidae preferably attach themselves to vaults by the posterior claws and are therefore able to make rapid rotations in all directions. The head, trunk, spinal column, and knees are well-adapted to these movements, which allow a sweep of 180° in the vertical plane and 360° in the horizontal.

The shape and movements of the pinnae are of interest. Horseshoe bats' ears have no tragus. A lower fold of the exterior border (antitragus) surrounds the base of the ear and protects the ears from returning ultrasonic sounds. The antitragus and the interior margin are separated by a groove which widens towards the base into a second orifice. Thus Rhinolophid has two funnels: the smaller one in front, having little mobility; the larger (limited by the pavilion of the ear and the antitragus) being directed laterally and upwards is very mobile.

The movements of these ears are highly complicated and very rapid: lateral rotations of the pinna, slow movements, quick movements forwards and backwards. Both ears perform the first two movements together in the same direction and, with the free movements of the head, provide the bat with the ideal means of scanning its environment. More rarely the ears perform these movements in opposite directions. All these movements may be superposed or mixed together.

In the detection of objects the Rhinolophidae, unlike the Vespertilionidae, orientate themselves very well with one ear or both.

---

* These are not absolute values, but the threshold of detection for a given sound-detecting apparatus.

Other experiments were attempted with one of the two funnels of horseshoe bats' ears stopped in turn. In both cases obstacle avoidance is good, though slight deterioration in keenness of perception is to be observed when the upper orifices are plugged. The animal then aims a little higher than ordinarily when making for its perch, for instance. Mistakes of the same order are to be seen if the pinnae are blocked or pressed down backwards. It would seem, therefore, that the lower orifices provide only a rough perception of obstacles, whereas the upper ones afford more precise information, sharpened by the mobility of the ears.

Studies of the physical properties of the clicks and of the monaural hearing of Rhinolophidae lead us to believe that these animals do not perceive objects by the shifted echoes which reach their ears. More particularly, it would seem to be incompatible with the duration of each click (there being no frequency modulation* in these clicks, Griffin's hypothesis in regard to Vespertilionidae does not hold in this case). Besides, the head movements of the animal which is scanning space, bring a multitude of echoes from various directions and distances; moreover, the formation of the echo depends upon the moment when the object enters the beam. All these features of the emission are against the hypothesis of perception of the shifting of echoes in time. Nor would the hypothesis appear to be valid from the point of view of reception, since the animal's monaural hearing is perfect. By the same token, monaural hearing refutes the hypothesis of perception of a difference in phase.

The possibility remains, therefore, of a perception of differences of intensity. This hypothesis, which, according to Möhres, is the only valid one, rests on several facts. First of all, the higher the frequency of an acoustic wave is, the more linear is its propagation. There would thus appear to be the phenomenon of an acoustic "shadow" which the movements of the ears might reveal[43]. On the other hand, the bat's receptor must be adapted to the maximal intensity of the reflected ultrasonic signals, and in this case we are confronted with the necessity of a pure sound, without modulation, which is the actual fact. Good regulation cannot be achieved with a modulated frequency any more than with variations in intensity.

The movements of the ears probably enable the animal to judge accurately the distance of objects, with the help of its two pairs of auditory orifices, one of which is immobile in relation to the head.

Thus Möhres and co-workers consider the differences in intensity between the echoes to be responsible for horseshoe bats' proficiency at echolocation. Griffin suggests a different solution to this delicate problem. The frequency of Rhinolophidae's clicks being practically constant, there would be a slight difference in frequency between the outgoing wave and the echo on account of the movement of the bat relative to the reflecting obstacle. This is the Doppler-Fizeau effect in which a sound appears to have a slightly higher frequency to a listener moving towards the source, and vice versa. The amplitude of the phenomenon depends upon the bat's speed. It is accentuated in the case of an echo, because it will be perceptible both on the outgoing and return trip and the two effects will be additive. Most of the small bats do not fly faster than 1 or 2 % of the velocity of sound, so no more than a 2 to 4 % change in frequency could result from the Doppler-Fizeau effect. This magnitude of change is easily detectable by the human ear.

---

* Novick[61] has shown that there is a slight drop in frequency at the very end of the *Rhinotophus'* pulse.

References p. 225

Let it be said that, in the case of the Vespertilionidae, which modulate the frequencies by an octave within each click this hypothesis is not likely to be true.

The theory of the beat-notes can be called on here regarding the subject of the frequency variations due to the Doppler effect. The frequency of the beats is thus proportional to the relative speed of the bat proceeding to the target. Monaural hearing permits as good an echolocation as does binaural hearing, as the differences of the frequency beats lost by the two ears is slight. This lets one suppose that the detection is made rather owing to the movement of the ears than owing to the binaural hearing (difference of phase, shifts in time).

For very short distances, one can consider an information of the Vespertilionidae type, made so the end of each call presents a fall in frequency comparable to that seen in a click of Vespertilionidae; furthermore the rapid, unsynchronized movements of the two ears can vary the Doppler effect. Finally, let us remember that Möhres[53] showed that the Rhinolophidae practiced echolocation with the greatest efficiency at a distance not exceeding 6.4 m. That corresponds to the maximum distance for which the click-echo overlap occurs in this species.

(iii) Signals intermediate between the Vespertilionidae and Rhinolophidae types: Asellia

Asellia tridens (Hipposideridae of Africa) has been studied by Möhres and Kulzer (1955)[46, 57]. It is able to emit in a few different ways: (a) clicks of the type of those of Vespertilionidae (a few msec—45 to 60 kc/s—modulation of frequency) for detecting relatively distant objects*; (b) clicks of the Rhinolophidae's type (longer, though not reaching Rhinolophidae's 90 msec—120 kc/s—no modulation) for detecting objects at close range; (c) composite clicks of both types (a long initial pulse, without modulation, of 120 kc/s; a short concluding modulated part, of 45 to 60 kc/s); (d) clicks having all the gradations between the foregoing when the animal is approaching or withdrawing from the object. The intensity of these clicks is lower than those of Rhinolophidae; *Asellia*'s nose lead is also smaller.

Faced with so much complexity of emission, an investigator may ask himself which hypothesis is plausible to account for this bat's perception of objects. It is difficult, as yet to be sure.

(iv) Other families of bats in which the principles of echolocation are akin to those of the Vespertilionidae and Rhinolophidae

These are families of bats whose echolocating emissions resemble those of the Vespertilionidae, though they have not been studied as thoroughly as the latter.

1. The *Molossidae* (free-tailed bats): Griffin and Nocivk[28] studied some bats of this family (*Tadarida* more especially) and heard them generate clicks like those of Vespertilionidae, except that the frequency was slightly lower. These two families are very similar in their hunting behaviour.

The other insectivorous bats utilise types of acoustic orientation which are all adapted to their particular problems.

2. *Plecotus*: Though a member of the Vespertilionidae, the kind of echolocation used by this bat is slightly different from that employed by *Myotis* and has more possibilities. *Plecotus* is remarkable for its enormous ears and its leisurely flight.

---

* Though we have seen (p. 202) that the Rhinolophidae's emissions were more adapted to the detection of rather distant objects than those of Vespertilionidae.

The animal often hovers in front of an obstacle before passing round it. Yet it is equally capable of seizing insects on leaves as of swallowing them during flight. If they are guided in the former case by echolocation, they must be able to receive meaningful echoes not only from small objects but also from the insects clinging to leaves, which must themselves send back significant echoes. Although we have no experimental studies to go upon, it is tempting to speculate on the part played by this bat's peculiarly developed external ears.

The echolocating emissions of *Plecotus rafinesquii* are very similar to those of the other Vespertilionidae, *viz.* short clicks, with a progressive drop in frequency which, however, is slightly lower than that observed in *Myotis*. The intensity is lower. It differs in no way from *Myotis* in its ability to avoid obstacles but, while the latter always flies with its mouth at least partly open, emitting its signals through this passage, *Plecotus* often flies with its mouth firmly closed. Indeed, this species can emit its clicks just as well through the nostrils as through the mouth, in which characteristic it bears some resemblance to the Rhinolophidae (see Addendum).

3. Piscivorous Bats: Certain specialisations go along with the fish-eating habit of some bats, such as enlargement of the hind claws to form sharp prehensile hooks by which the animals can pluck fish near the surface. These bats are found in several families, *viz:*

(a) Vespertilionidae with *Myotis macrotarsus* and *Myotis (Rickettia) pilosa*, of Asia, which undoubtedly feed on fish.

(b) *Pizonyx vivesi* and *Noctilio leporinus* (Noctilionidae) of Central and South America are the best known. *Pizonyx*, whose habitat is the coast, emits frequency-modulated clicks like the Vespertilionidae in captivity. *Noctilio* is not wholly piscivorous. Like *Pizonyx*, it hunts both freshwater and marine fish, but it also eats insects. A closely allied bat, *Dirias albiventer minor*, is not equipped with large hind claws and, although living with *Noctilio*, it eats insects only.

Griffin found that *Dirias* emits frequency-modulated pulses (with some exceptions) of high intensity.

It has recently been observed[4,5] that *Noctilio* held in captivity capture their prey with their hind claws and immediately afterwards take it between their teeth.

In these experiments, and those made by Griffin in natural conditions, no audible sound was emitted, but ultrasonic clicks like those of the Vespertilionidae were recorded: an average duration of 5 msec, drop in frequency 38 to 28 kc/s and 20 to 60 dynes/cm^2 intensity at a distance of 50 cm from the microphone (100 to 109 dB).

While skimming the surface of the water, before and during the capture of their quarry, the animals emit short frequency-modulated pulses at a high repetition rate (60 per sec), like those generated by one of the Vespertilionidae when approaching a pond or other still water to drink, rippling the surface of the water and dipping the lower jaw into it for the purpose, no doubt, of scooping up a drop or two. This has often been seen done by captive *Myotis*. Since the control of flight by echolocation is presumably similar in a fishing *Noctilio* and a drinking *Myotis*, it is not surprising that the short, frequency-modulated pulses should be emitted at high repetition rates in the two cases.

The other pulses emitted by *Noctilio* are different in character. They are longer (8 to 15 msec), gradually increasing in amplitude to a practically constant level before dropping progressively at the end of the pulse. In other cases there is a long initial portion when the frequency is high but the amplitude low; then, during the last 2 or 3 msec, the amplitude increases sharply, forming a peak. The frequency of these longer pulses likewise declines progressively from start to finish, but the rate of frequency change is not always the same and tends towards a maximum in the peak, when there is one. Typical values for the initial frequency are 45 to 64 kc/s, while at the end they are of the order of 25 to 40 kc/s. It might be inferred from these variations in the pulses that they correspond to different kinds of information elicited, but we have no established evidence in support of this supposition.

Does echolocation serve to the detection of fish? The animals usually fish at night; hence vision would not seem to be involved, though experiments have not yet been undertaken to verify this. Besides, Griffin and Bloedel[26] have never seen ripples on the surface of the water, possibly betraying the presence of fish just beneath the surface to the bats. It would seem more reasonable to suppose that fish close to the surface are detected by echolocation. There are, however, certain physical considerations for and against this hypothesis and the matter has not yet been definitely resolved.

Most of the energy of a waves train will be reflected rather than refracted by the surface of water, viz. 9.9% reflected as against 0.1% refracted. Starting from this 0.1%, how much of the fish's echo would be left after the second trip through the surface?

Let us not forget, however, that an *Eptesicus* hunting an insect 1 cm in diameter can detect it by echolocation at 2 metres' distance (the signal, therefore, making a round-trip of 4 metres). *Noctilio* are far closer still to their quarry (10 to 20 cm at most, i.e. one-tenth the distance of *Eptesicus* from its prey) and fly slowly. But at 2 metres in air, the echo would contain only 1/10,000 of the energy received by a 20 cm distant prey (varying inversely as the square of the distance). On the other hand, in water, the double passage through the air–water dioptre corresponds to a lowering of the residual energy by 10^{-6}. There is, therefore, a ratio of 100/1 in favour of the insectivorous bat, between the two types of energy received by an *Eptesicus* at 2 m from its prey in the air, and a *Noctilio* at 20 cm from its fish. But the sound pressure level of *Noctilio*'s cry is 4 to 5 times higher than that of the Vespertilionidae's, which corresponds to 16 to 25 times greater energy. Though not yet fully resolved, the problem is already beginning to look simpler.

It should be added that fish having swim bladders, like minnows—one of bat's favourite preys—are very good reflectors of sound.

One important question still outstanding is the long duration of the pulses compared to the time required by acoustic waves to make the round-trip between the bat's mouth and the fish's swim bladder. *Noctilio*'s shortest pulses last 3 msec (often from 5 to 15 msec), whereas the wave travelling 20 cm takes 1.2 msec there and back. But the results of experiments performed by Möhres with Rhinolophidae have shown that this is not incompatible with perfect echolocation. Nor is the perception of the Doppler-Fizeau effect impossible.

Another difficulty arises from the fact that most of the acoustic energy generated will be reflected by the surface of the water, which will contribute in no small measure to masking the echo from the fish. Yet if the emitted sound were pointed in a very narrow beam just ahead of the animal upon a smooth surface of calm water, the angle of reflection would equal the angle of incidence, with the result that little would reach the bat's ears. Nothing is known about the directivity of *Noctilio*'s emitted sounds, except that there is no marked focussing of the emissions of these animals when held in the hand in a laboratory. They have, moreover, slightly pendulous lips and it may be that they use them for concentrating acoustic energy into a point.

Let it be said in conclusion that, if it be difficult to prove at the moment that bats resort to echolocation for the detection of fish, it would be premature to state categorically that it is out of question.

4. *Emballonuridae* (sac-winged bats). A few species have been studied by Bloedel, Griffin and Novick[26, 28] and by Möhres and Kulzer[54].

Saccopteryx bilineata and *Rhynchiscus naso* emit pulses of sound when they are crawling on the ground or flying. Those of *S. bilineata* are often audible, sometimes at several metres' distance. Most of their energy is concentrated between 14 and 25 kc/s. The duration of the pulses varies from 2 to 15 msec.

Rhynchiscus naso is very "talkative", but its pulses are inaudible. These are very small bats (weighing 3 to 4 g) and their emissions very much resemble those of the Vespertilionidae in intensity, duration (5 msec an average) and frequency (35 to 103 kc/s).

In Panama, one *Rhynchiscus* emitted clicks incessantly of 90 msec, scarcely without any interruption between them. This was apparently a different type of signal, emitted at a pure frequency ranging from 21 to 30 kc/s. This type of pulse, very similar to that of Rhinolophidae, though of lower frequency, puts *Rhynchiscus* in a class apart, being capable of emitting two very different types of signals, one closely resembling that of the Vespertilionidae and the other very like those emitted by the Rhinolophidae.

These clicks are emitted with frequency changes less pronounced than in Vespertilionidae and some pulses are constant in frequency.

Taphozous orientates itself by short pulses of relatively intense sound of the Vespertilionid type, emitted when about to take off and in flight. The average duration of these pulses is 4.2 msec. They are emitted at a repetition rate of 60 to 160 per second, accompanied by a slight audible noise (Dijkgraaf's "Ticklaute"). The frequency spectrum of emission ranges from 25 to 80 kc/s with several harmonics in each click. The sounds are emitted from the animal's mouth. Its eyes are well developed, but we do not know to what extent they are involved in orientation. Thus we have here a family resembling the Vespertilionid type, but with certain characters peculiar to itself (very little or no frequency modulation) and varying from species to species (more especially the two kinds of emission by *Rhynchiscus*).

5. *Rhinopomatidae* (mouse-tailed bats). *Rhinopoma microphyllum* was studied by Möhres and Kulzer[54]. Its acoustic habits are allied to those of *Taphozous*, i.e. it emits short pulses (an average of 4 msec) which become intense at the take-off and during flight, at a repetition rate of 130 to 300 per second. The frequency spectrum is narrower (50 to 60 kc/s). The animal emits its pulses through the mouth. Its eyes are well developed.

6. *Chilonycterinae* (sub-family of the Phyllostomidae). These bats have no nose leaf, which differentiates them from the other Phyllostomidae. They are insectivorous. Their emissions have the same intensity levels as those of the Vespertilionidae, each pulse lasting 1.5 to 29 msec, varying with the species. The frequency is pure, without any modulation (average 23 kc/s), with one or several harmonics.

References p. 225

This, then, is a family closely allied to the Emballonuridae. As with them, the animal was studied held in the hand or crawling in a cage.

7. *Nycteridae*. *Nycteris thebaica*[54] has a nasal appendage from which the nostrils open, through which, in turn, the emitted sounds issue (thanks to an anatomical apparatus analogous to that possessed by the Rhinolophidae). The nostrils have a distance of half a wave-length (1.8–2 mm). The ears of this species have very little mobility—unlike those of the Rhinolophidae. They are large and have a tragus. These animals have a pronounced, very sure and very accurate sense of orientation. The emissions range from 125 to 260 pulses per second (each of 1.9 msec duration), the frequency being pretty well localised around 88 kc/s. There does not appear to be any audible component. This is a species resembling the Rhinolophid type far more closely than any of those previously mentioned.

(v) Conclusions

This first group of bats includes only the Microchiroptera, hunters of moving quarry (insects or fish). They all have in common the fact that the pulses of their acoustic emissions are strong in a range of frequencies below 100 kc/s. Two main types of emission have been thoroughly studied by American and German authors, *viz*. the Vespertilionid type with its short frequency-modulated pulses emitted by mouth in a very narrow beam, and the Rhinolophid type with its much longer, pure-frequency clicks emitted through the nostrils, with a wider beam.

In addition to these two types of emissions, there are others, much less studied, that vary considerably. *Plecotus* and the Chilonycteridae have some of the possibilities of the Rhinolophidae, and some of the possibilities of the Vespertilionidae while *Asellia* and *Rhynchiscus* have most characteristics of both types.

Often the emissions are highly individualistic, comparable neither with the Vespertilionidae nor the Rhinolophidae (the largely audible clicks of *Saccopteryx*; the high repetition rate of *Taphozous* and *Rhinopoma*).

It will take many years of laborous experimentating with all the species of bats to clarify this whole field. The bases on which the amazing acuity of perception displayed by these animals is founded are still veiled from us. All we have are hypotheses, interesting admittedly, but not yet experimentally verified, concerning the Doppler-Fizeau effect, the perception of shifted echoes in time, and the perception of differences in intensity.

(d) Whispering bats

At present we know of two main families belonging to this group of bats—the Phyllostomidae which are vegetarians, and the Desmodontidae or vampire bats—which, though they differ in diet, have similar acoustic behaviour. All animals of this group either are frugivorous or hunt sleeping animals. The emitted pulses of these bats are of low intensity and generally of very high frequencies, which may be related to their feeding on non-moving prey.

(i) The Phyllostomidae*

Inhabiting Central and South America, these bats were studied by Griffin and Novick[28]. They have a spear-shaped nose leaf.

---

* The Chilonycterinae, which are members of this family but are insectivorous and have no nose leaf, have entirely different acoustic habits, as described above.

1. In 1953 Griffin took recordings of *Carollia perspicillata azteca* in Panama. The nose leaf of this species is triangular, while the ears are very mobile and independent, a feature which they have in common with the Rhinolophidae. It was impossible to pick up any sound at all with the usual equipment. When blindfolded, the animals found their way about with great dexterity, better, in fact, than the Vespertilionidae. When their ears were covered, however, they were completely incapacitated.

On returning to Cambridge, Novick succeeded in picking up Dijkgraaf's "Ticklaute". After cutting the superior laryngeal nerves (which, in the case of *Myotis*, lowered the frequency), he recorded distinct sounds of 10 to 15 kc/s and, at a lower level, ultrasonic components. From this he inferred that the normal emissions of *Carollia* are of far-higher frequencies than those of the insectivorous bats.

With the aid of more sensitive equipment, and by placing the microphone only a few centimetres or decimetres from this bat, he was able to pick up sounds, which could be described as whispers rather than cries. It is not known how important this whispering is to the exceptional adaptness of these animals at echolocation. Probably more contributory is their ability to turn completely round in a cage little wider than their own wing-spread (which is of the order of 30 cm), whereas *Myotis* (wing-spread 25 cm) cannot do this. The percentage of misses (that is of successful flight without touching a wire) scored by *Myotis* in obstacle tests with wires spaced 30 cm was 36 %. Here, for comparison are the scores of the two genera:

Diameter of obstacles (in mm)	Per cent misses	
	Myotis	Carollia
0.070	36	—
0.175	—	56
0.260	52	—
0.275	—	65
0.350	72	—

The sound pressure of *Carollia*'s pulses at less than 33 cm from the microphone is on an average 3.6 dynes/cm^2 or 85 dB. Under the same conditions, the Vespertilionidae emit pulses of 100 dynes/cm^2 pressure (or 114 dB). The average duration of these clicks is 1.4 msec (extremes 0.9 and 2.3). They are emitted just as well through the mouth as the nostrils. Components can be demonstrated at almost any frequency from 15 to 128 kc/s, but in flight they are rarely below 60 kc/s (sounds emitted by the mouth are more richly endowed with frequencies below 60 kc/s). There is no modulation of frequency, but there are often complex mixtures and variables of frequencies, with the sudden appearance and disappearance of multiple components. On the other hand, pulses are sometimes obtained with almost constant frequency. Complex pulses could be said to have a fundamental frequency, F, plus its harmonics $2F$, $3F$, $4F$, etc.; but the higher harmonics may be much greater in amplitude than the lower ones and the fundamental may be lost in the noise level of the apparatus. This greatly complicates the interpretation of the oscillograms and leaves us guessing as to how *Carollia* manage to sort out the echoes they receive.

2. Other Phyllostomidae studied are fruit-eating bats (*Uroderma bilobatum*, *Artibeus jamaicensis palmarum*), nectar feeders (*Glossophaga soricina leachii*, *Loncho-

phylla robusta), or fruit eating but predatory on occasion (*Phyllostomus hastatus panamensis, Macrophyllum macrophyllum, Lonchorhina aurita*). They all emit sounds similar to *Carollia's*, that consist essentially of faint clicks (sound pressure less than 10 dynes/cm² or 94 dB with the microphone 5 cm from the mouth) of short duration (less than 5 msec, or even less than 1 msec in the case of the nectar-feeding bats).

There is no homogeneity where the frequencies are concerned. These range, from species to species, from 11 to 128 kc/s:

Genera	Emitted frequencies in kc/s
Lonchorhina	11 to 14
Macrophyllum	20 to 30
Phyllostomus	30 to 50
Uroderma	60 to 90
Artibeus	75 to 90
Lonchophylla	65 to 119
Glossophaga	73 to 128

In these spectra there is a fundamental frequency and harmonics of considerable amplitude. In *Artibeus* there is, at times, even an amplitude modulation of one third of the predominant frequency as in *Carollia*.

The following are the results obtained with the animals tested:

Diameter of obstacles (in mm)	Carollia	Glossophaga	Artibeus
1.050	—	80 % (34 flights)	67 % (39 flights)
0.600	—	68 % (91 flights)	85 % (47 flights)
0.275	65 % (243 flights)	75 % (314 flights)	85 % (68 flights)
0.175	56 % (816 flights)	89 % (560 flights)	72 % (54 flights)
Spacing of targets	30 cm	30 cm (60 cm for 0.175)	60 cm
Wing-spread	30 cm	25 cm	45 cm

When bats are hunted in Panama with mist nets made of fine linen thread, *Myotis* and *Rhynchiscus* are very rarely caught, whereas *Carollia, Glossophaga, Artibeus* and *Phyllostomus* very often are. The latter two manage to escape, but not to avoid the nets. This would seem flatly to contradict the results of the obstacle avoidance tests tabulated above and at the moment we do not know how to account for these facts. This apparent contradiction can be supposed to lie in a greater utilisation of spatial memory by Phyllostomidae. Indeed they are caught with nets in places with which they are probably quite familiar.

(ii) Desmodontidae or Vampire bats

These bats feed on blood only. It is interesting to note that dogs, which are sensitive to ultrasonic sounds, are not bitten by vampire bats. They are awakened by the pulsed sounds emitted by the approaching bats.

To study these bats the same apparatus is needed as is used for *Carollia* on account of their faint pulses (2 dynes/cm² intensity, or 80 dB). Their band of frequencies commonly ranges from 60 to 75 kc/s and, more rarely, from 15 to 92 kc/s. Harmonics are often prominent and, in this way, considerable energy is emitted above 100 kc/s. The average duration of the pulses is 2.2 msec.

(iii) Other whispering bats

These are certain members of the Nycteridae* and Megadermidae[61]. The duration of their pulses is less than 1 msec, and they are of high frequency, with harmonics sometimes exceeding 100 kc/s. The intensity of the pulses is very low. The flying habits of these animals are similar to those of *Carollia*.

(iv) Conclusions

We have, then, been considering a group of bats endowed by specialisation for detecting stationary objects (fruit or sleeping animals), all of whose emitted pulses are very faint (less than 10 dynes/cm^2, or 94 dB, at a few centimeters from the microphone), with a high and complex spectrum in which harmonics are liable to mask the fundamental sound. This spectrum varies within each pulse, the duration of which does not exceed 10 msec. The ears of these species appear to play a predominant part in their duration. Experiments demonstrate that these animals have extraordinary capabilities that they do not utilize under natural conditions.

As a whole, this group is less well known than that of the high-intensity emitting bats and a closer study of it will no doubt bring forth many surprises.

(e) The Orientation of the Megachiroptera

These animals are vegetarians. Unlike the Microchiroptera they have very well developed nasal apparatus and eyes. Because of their small ears their common name is "flying foxes" ("flying dogs" as the Germans call them).

Möhres and Kulzer have studied[57] two species of the genus *Pteropus* viz. *Pteropus giganteus* (wing-spread 1.3 m) and *P. poliocephalus* (far smaller), as well as *Rousettus aegypticus* (wing-spread 60 cm). Novick has recently studied other members of the Megachiroptera in the Philippines, Ceylon and Central Africa[61].

(i) Rousettus

A great many of these little bats live entirely in the dark in natural or artificial caves.

Rousettus aegypticus (the "tomb bat") was studied in captivity. It was found to emit clearly audible clicking sounds of low intensity when taking off and landing. The clicks are more intense during flight, and sound like a buzzing. Although often co-ordinated with the wing-beats, these metallic clicks are independent of them.

At the approach of objects or other individuals, the repetition rate of the emissions is increased as is the intensity. This depends on the light conditions: the darker the cave, the louder the clicks. When deprived of all means of visual orientation, these bats still orientate themselves perfectly, even in the presence of complicated obstacles. They easily perceive a network made of 4 mm diameter wires (Möhres does not give the size of the mesh), or single wires stretched horizontally, from which they dangle without difficulty. Under those same conditions, they become

---

* Other members of the Nycteridae were dealt with on p. 208.

completely helpless if their ears are stopped, but recover their skill the moment a glimmer of light appears.

All this points to a dual means of orientation in these animals, auditory and optical, either alternating or complementary to each other, depending on light conditions. Möhres has, furthermore, demonstrated the outstanding importance of their olfactory sense in the search for food.

The analysis of the echolocating clicks emitted by *Rousettus*, having a very similar shape, shows them to be closely allied to the Vespertilionid type. The frequency spectrum ranges from 6.5 to 100 kc/s (the audible component having a maximum at about 7 kc/s). There is no frequency modulation within these clicks, which have an average duration of 5.5 msec. The repetition rate is 6.9 per sec, varying during flight (Fig. 91).

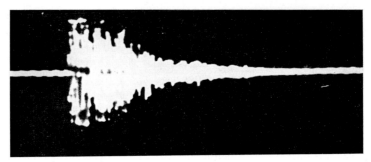

Fig. 91. Oscillogram of a click of *Rousettus aegypticus* (after Möhres[55]).

According to measurements of range, the intensity is in the neighbourhood of that of the Vespertilionidae, like *Nyctalus noctula* and *Lasiurus borealis*. It increases approximately sixfold as the light dims from twilight to absolute darkness*.

Rousettus flies with its mouth closed, or almost so, or habit which is markedly different from the Vespertilionidae. An anatomical connection was found between the larynx and the nasal cavity, similar to that observed in the Rhinolophidae. The sounds were, therefore, thought to be emitted through the nostrils.

Experiments performed by Kulzer (1956)[46] showed that cutting the laryngeal nerves—one by one or simultaneously—and blocking the larynx and the nostrils in no way prevents *Rousettus* from emitting their signals; nor is their orientation impaired. Kulzer then studied the buccal cavity and found that the tongue makes rapid movements while the sounds are being emitted. It makes no difference if the individual nerves or all the branches of the lingual nerve are cut. But as soon as the two hypoglossal nerves are cut, the production of orientating signals ceases. This result was verified by applying local anaesthetics to various portions of the tongue. As a matter of fact, the same thing was found in *R. amplexicaudatus* and *R. seminudus*[60].

Hence it is by these clicking movements of the tongue that *Rousettus* produce their echolocating sounds (Fig. 92).

---

* This was found in birds as well (*cf.* p. 217).

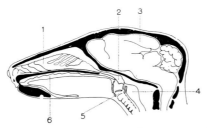

Fig. 92. Schematic sagittal section of the head of a *Roussettus:* 1. Nasal cavity; 2. Nostrils; 3. Brain; 4. Oesophagus; 5. Larynx; 6. Buccal cavity (after Kulzer[46]).

This was tested[29] on a male of 75 cm wing-spread, in a flight chamber of 10 × 4 × 2.5 m, in the same way as with *Myotis*. Table 19 presents a few of the results obtained with the obstacles spaced 53 cm apart compared with those obtained in the case of *Myotis*[7] with 30 cm spacing of vertical obstacles.

TABLE 19

COMPARISON OF THE OBSTACLE AVOIDANCE SCORES OF A *Rousettus aegypticus* WITH THOSE OF *Myotis l. lucifugus* (CURTIS[7]). THE WIRES OR OTHER CYLINDRICAL OBSTACLES WERE ARRANGED VERTICALLY AND SPACED 53 cm APART FOR *Rousettus* AND 30 cm APART FOR *Myotis*.

	Diameter of obstacle (in mm)	Myotis l. lucifugus		Rousettus aegypticus	
		No. trials	% misses	No. trials	% misses
Cardboard tubes	25	—	—	109	76
Rubber tubing	19	—	—	161	78
Rubber tubing	13	—	—	100	77
Rubber tubing	6	—	—	50	80
Metal rods	4.76	140	85	—	—
Insulated metal wires	3	—	—	442	85
Bare metal wire	1.5	—	—	200	77
Bare metal wire	1.21	3820	82	—	—
Bare metal wire	1.07	—	—	280	68
Bare metal wire	0.68	480	77	—	—
Bare metal wire	0.65	—	—	225	58
Bare metal wire	0.46	—	—	134	45
Bare metal wire	0.35	660	72	—	—
Bare metal wire	0.28	—	—	50	18
Bare metal wire	0.26	660	52	—	—
Bare metal wire	0.12	530	38	—	—
Bare metal wire	0.07	460	36	—	—

We find, therefore, that the effective critical size is in the neighbourhood of 0.3 mm, whereas it is about 0.1 mm for *Myotis*.

Some interesting studies of masking have been made. Twenty loudspeakers were used in an attempt to disorientate the bat with partly "white" noises. Whereas it scored 79 % success in missing obstacles 3 mm in diameter in the dark, its score rose to 90 % in the light and with a masking noise. In the dark and with a "white" noise in the range of 0 to 25 kc/s, the bat was completely disorientated. Still in the dark, but with masking noise including frequences above 15 kc/s, its score was 27 %.

Thus the acuity of this species' system of orientation was tested very thoroughly by this method. It appears that the frequencies within the range of 15 to 25 kc/s play a predominant part in dazing the bat.

References p. 225

Conclusions. The Megachiroptera are distinguished from other animals by their varied means of orientation, *i.e.* vision, hearing and smell. Though of unequal acuity, the bat utilizes the three senses according to circumstances and its own needs.

The other Megachiroptera are not as well endowed as *Rousettus*.

(ii) *Pteropus*

The importance of vision to the orientation of these animals is to be inferred from their large eyes, the pupils of which vary in diameter with the intensity of the light. The movements of their ears, on the other hand, although not as rapid as those of the Rhinolophidae, suggest acoustic orientation.

With their ears stopped, *Pteropus* lose none of their orientating ability as long as there is a glimmer of light, however dim. (This has also been found in owls, see p. 216.) In complete darkness their flight is abruptly interrupted. When they were forced to fly in the dark they hurtled into the walls and obstacles. Hence these animals are incapable of echolocation. Apparently the movements of their ears serve for the locating of strange noises, to which their hearing is very sensitive.

The same phenomena are found in the *Cynopterus* of the Philippines, which feed on nectar, and in *Eidolon, Ptenochirus, Lissonycteris, Eonycteris* and *Macroglossus*[26].

Thus we have here a group of bats not well adapted to nocturnal orientation. These animals are, indeed, less inclined to seek the dark in nature than are *Rousettus*.

(f) *Conclusions*

At the end of this rapid survey of current knowledge on the acoustic habits of Chiroptera, the physical properties of the echo locating pulses have been shown, and sometimes even the modalities of their genesis. Certain definite facts on the limits of this system of orientation have also come to light. These data, however, vary so much from one species to another, indeed sometimes within one and the same species (*e.g. Asellia*), that the matter becomes quite bewildering, especially when we consider that the order contains a great many species of which only a few have been studied.

This study has produced a clear-cut division between the Microchiroptera, which orientate themselves by echolocation, and the Megachiroptera, whose orientation is essentially visual and, to a minor extent, olfactory, or even auditory, as in the genus *Rousettus* which represents an exception in this sub-order. It is tempting to associate this division to that on the adaptation to flight, which is far less evolved in the Megachiroptera. It might be postulated, therefore, that the latter are descended from the Insectivorous and were the ancestors of the more specialised Microchiroptera.

Palaeontological studies, however, have shown that the most ancient Chiroptera known were clearly Microchiroptera. No Megachiroptera fossils prior to the Oligocene have been found so far. It would therefore be more reasonable to regard the Megachiroptera as the descendants of the Microchiroptera.

It might then be conjectured that *Rousettus* either represents a transitional type from the Microchiroptera to the Megachiroptera, or else stands for a Megachiroptera that was specialised for echolocation long after the division occurred between the two groups. This second hypothesis derives from the different way in which the pulses are emitted between *Rousettus* and the Microchiroptera.

Within the sub-order Microchiroptera itself, a diphyletic origin had already been tentatively attributed to the Vespertilionidae and to the Rhinolophidae because of the numerous differences existing between these two families[31]. Now the study of the acoustic habits of members of these families tends to support this hypothesis.

The classification attempted in the foregoing review will probably be changed with further increases in our knowledge.

We know virtually nothing about the perception of echoes by bats and the information thus conveyed to them. No hypothesis has yet been confirmed, even respecting the animals best known (Vespertilionidae and Rhinolophidae). The numerous varieties of emitting pulses suggest that there is no one over-all theory adequate to account for the perception of all bats. Although several suggestions have been advanced here, none has been experimentally verified.

Another major problem which has hitherto defied solution is how bats congregated in large communities manage not to confuse their own emitted pulses and the echoes returned with those of their nearby congeners. Similarly, if we are to assume that each individual is protected from its own emissions—sounds of such intensity as to probably deafen them very rapidly otherwise—how then do bats avoid this danger from the emissions of their neighbours?

We are likewise ignorant about the existence of an apprenticeship of echolocation by young bats.

4. Other Terrestrial Mammals: The Rodents

Ultrasonic components of up to 54 kc/s have been recorded in the squeals of guinea-pigs *Cavia cobaya*, whose whistles are within the audible range[1].

The squeals of laboratory rats, *Rattus norvegicus*, cover frequencies from 19 to 29 kc/s; their snuffling rise to 80 kc/s. This species furthermore emits pure frequencies between 21.5 and 28 kc/s having no audible component. These particular sounds last from 1 to 2.5 seconds and present a rapid modulation of frequency of 2 kc/s. They are emitted when the rats are quiet and are possibly related to the animals' thoracic movements.

As it was thought that all these facts might be associated with the utilisation of an ultrasonic system of autoinformation to probe the surroundings, studies on the rat were undertaken with the aid of a simple maze[69], in which ten blinded rats were tested. The obstacles used were different for each test. All the rats learned to choose the correct paths, seven of them doing so 18 times in 20. As the barriers were painted black, radiated energy was ruled out. If they were inclined at an angle of 45° to the direction in which the rats were moving, the scores fell, only to rise again to the above percentage when the barriers were replaced at 90°. Probably the sound at an angle of 45° was not reflected towards the rat, but to the side.

In experiments with rats deprived of the use of their ears these animals were hopelessly disorientated.

The experiments were then repeated with instruments installed to record the noises made by the animal in the maze. Although some ultrasonic signals were recorded, they were very rarely emitted and would not seem to be related to the rat's means of orientation. However, the animals produce almost ceaseless noises of many kinds: they sniff, sneeze, scrape the floor, and their teeth chatter. All these noises, including that of their scampering feet, are easily audible to a listener at close quarters. In certain cases, however, no sound at all could be perceived despite the fact that the animals were following the prescribed route. Hence either they orientated themselves correctly by mere chance, or else they did so by emitting inaudible sounds.

These experiments were further pursued by this method. With metal barriers set perpendicular to the path of the animals, they scored 82 % successful runs, this score dropping to 56 % when the barriers were inclined at 45°. With barriers of fabric at 90° the percentage of successful runs dropped to 52 %. Individuals with impaired hearing, but not completely deafened, scored only 57 % with metal barriers at 90°.

An interesting point is the important part played by the floor[77] and the walls of the maze[83], as well as by outside noises in such experiments.

This work has certain affiliations with experiments[49] in which rats, reared in the dark, were found able to jump from one platform to another at the first attempt in the light. It is possible for these animals to orientate themselves by their auditory sense, as was demonstrated with a hamster and a dormouse[39], and this would explain why blind rats managed to jump between two platforms 15 to 20 cm apart[69].

Many points remain to be cleared up, but it was of interest to mention rodents here. Although echolocation has not been as rigorously investigated as in bats, the role of hearing in the orientation of rats has been clearly demonstrated.

5. Birds

The orientation of nocturnal birds aroused the curiosity of Spallanzani, who studied both owls and bats. He noticed that the former were hopelessly disorientated in complete darkness. More recent studies, especially those made by Curtis[7], showed that *Tyto alba* (a member of the Strigidae family) was able to see quite well in what seemed complete darkness to the human eye, provided there was just enough light, however dim, for its specially adapted eye to perceive. This system of orientation does not involve hearing at all. The same apparatus used to record bats' emitted pulses showed that the flight of this bird is partly silent. It has since been discovered that some of these nocturnal birds are optically sensitive to infra-red radiated by the environment. These birds do not roost in caves hundreds of feet below the ground, impenetrable to the faintest ray of light, as do bats. This, however, does not apply to all nocturnal birds, or birds inhabiting caves, as will be seen when we come to consider at least two species.

(a) *Steatornis*

In 1799, Alexander von Humboldt and Aimé Bonpland visited a great cavern at Caripe in Venezuela[35] inhabited by Guacharo*, the only known nocturnal bird to feed on fruit. These birds, assembled by thousands, make a terrific noise in the dark portions of the cavern, where their nests are sometimes as much as 20 metres from the ground. This species is *Steatornis caripensis* (the oilbird of Caripe) of the Steatornithidae family, related to the Caprimulgidae, a subclass of the Carinatae.

They nest, at certain hours in any event, in total darkness, which does not prevent them from flying in company with the bats, emitting audible clicks as they fly.

Griffin[24] went to study them in their habitat, taking with him the apparatus he had used to such good effect in his studies of bats. He exposed a film for 9 minutes

---

* In Spanish: "One who cries and laments".

in the caves and then compared the developed film with an unexposed one. He proved thus that no light reached the birds, yet they circled noisily. A microphone was held up in the vicinity of the ceiling in order to be near the birds when they flew out of the cave at night to search for food. The sounds were different from those heard previously. They were clicks with frequencies ranging from 6.1 to 8.75 kc/s and with pulse duration from 1 to 1.5 msec. These clicks are not emitted regularly, but in "bursts". The sound pressure could not be measured, but the clicks were readily audible at a distance of 200 metres.

By stretching a net across the lowest part of the entrance to the cavern, four of these birds were caught alive; most of the others, however, had detected the net and flew above it. Griffin noticed that the emission of clicks stopped as soon as a beam of light reached the bird; hence acoustic orientation was apparently only utilised in the dark. The captured birds were studied in a flight chamber, in which they flew about perfectly well, with and without illumination; but in the dark, none was able to avoid frequent collisions with an electric light flex suspended from the ceiling (these birds had had no food for more than 24 hours and were therefore not in their normal condition). Tests were then carried out with each of the three birds, one at a time, with their ears plugged. In each case the animal flew straight into the first wall it came to, but immediately recovered its previous skill in avoiding collisions once the plugs were removed. With the room lights on, these animals orientated themselves very well, even with their ears stopped.

Hence, like the bat *Rousettus*, these birds display a double system of orientation, which we may call audiovisual.

As yet, however, we know nothing about the minimum size of detectable objects (the wave length of the sounds emitted by *Steatornis* is approximately 5 cm), nor about the auditory sensitivity of this bird. Does the limited range of its hearing force the oilbird to use longer wave lengths than the insectivorous bats, or do these low frequencies suffice because it is concerned with larger objects?

(b) *The swifts of "Bird's Nest Soup"*

It is the *Collocalia* of China (of the Apodidae family) which constructs the nests relished as a delicacy in that country. These birds feed on insects, which they hunt chiefly by day. One species living in southern India and Ceylon shares dark caves with bats. Their nests are often built deep in long, twisting caverns where total darkness prevails. Several bird watchers have observed the habits of these swifts, which proved to be very similar to those of *Steatornis*, particularly as regards their acoustic behaviour. Novick[61] carried out experiments with *Collocalia brevirostris unicolor* in Ceylon in 1957, using the apparatus that had served for the study of bats. This species nests in caves from 10 to 15 metres deep into which a faint glimmer of light penetrates, to some portions of the cave at any rate. When flying into this retreat, these birds emit very sharp clicks, the duration of which is similar to that of the clicks generated by *Steatornis* (Fig. 93). The clicks become more intense when the animal is fling towards the darkest parts of the cave. The rate of clicking reaches a maximum while hovering in front of the nest before landing. On the other hand, the bird virtually stops clicking as it flies towards the light.

If their ears are stopped or their eyes covered, these birds fly more slowly, emitting clicks continuously. If both sensory organs are put out of action simultaneous-

References p. 225

ly, the birds refuse to fly and, if they are forced to do so, bump into obstructions. Novick then repeated these experiments in a dark chamber. When free, the birds flew around with ease and did not collide with the walls. With their ears covered, their flight became laboured and they often collided.

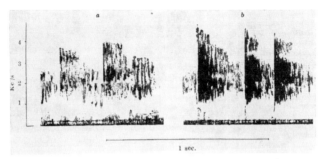

Fig. 93. *Collocalia*; sonagram of the echolocation clicks. a = with filter; b = without filter (after Medway[51]).

An analysis of the signals showed that the frequency band was of the order of 3 to 4 kc/s, the duration of the pulses being approximately 100 msec. The pattern of the signal is comparable to a click[51].

Hence the only two species in which echolocation has been detected orientate themselves equally well by vision, depending on the lighting conditions. This double system of orientation links these birds with the Egyptian *Rousettus* bat.

(c) Other birds

There are perhaps other birds which utilise echolocation:

The Swifts of Europe, for example, which sometimes fly and hunt all night[47].

The common nighthawk detects its quarry from considerable heights; while hunting insects at night, it emits sounds composed of a series of short pulses.

6. Insects

Our knowledge of echolocation in this branch of the animal kingdom is limited to the findings of Roeder and Treat[67] in *Prodenia* and to those of Eggers (cf. Griffin[23]) in *Gyrinus* in 1953.

(a) Prodenia*

This moth and several other nocturnal moths display a tendency to take evasive action or simulate death when exposed to ultrasonic stimulii. The tympanic nerve of *Prodenia eridania* is sensitive to frequencies from 3 to 240 kc/s, mainly between 15 and 60 kc/s. The destruction of these tympanic organs suppresses such reaction[70,82]. Now many bats are insectivorous and detect their quarry by ultrasonic echolocation, and the pulses emitted by bats have been demonstrated to provoke this type of reflex in these insects, which might be compared to an evasive or defensive reaction[82].

---

* See Treat's article, Chap. 16.

It was Hinton[34], however, who suggested that some of these nocturnal insects used echolocation. When one of these moths was suspended, in flight, at a fixed point, and a preparation of the tympanic nerve of another individual was brought within 22 centimetres of the suspended moth, recordings were obtained from the preparation of a series of short pulses corresponding exactly to the wing-beats of the fluttering animal. An analysis of the sounds produced by the flying moth reveals, in addition to the fundamental frequency of the wing-beats, which is too low to stimulate the tympanic organ, a pulse with a frequency close to 15 kc/s (with a higher component) which is emitted at an exact moment of the wing-beat. The irresistible inference from this datum was that these insects are sensitive not only to the emitted pulses of their fellows, but also to their own echoes. Unfortunately, these experiments have hitherto only been performed upon a single species, but complementary studies will probably be made to define the limits of the use of this echolocation. Roeder and Treat([24]) carried out further experiments on the reception of bats' cries by the moths.

(b) *Gyrinus*

The eyes of some of these fresh-water Coleoptera were destroyed; however, they were still able to swim about on the surface of the water in an aquarium without bumping into the glass walls. When their antennae were cut off, they could no longer do this. These antennae are rather elaborate, containing inflated structures and a pedicel which has a fringe of sensory hairs floating on the surface of the water. It is this pedicel which enables *Gyrinus* to avoid collisions, as it detects the waves generated by its swimming movements and reflected from obstacles. However, dust particles and the thin film of oily substances usually covering stagnant water seem to play some part in this type of orientation, for the beetle's obstacle avoidance in distilled water is far less skilful.

Wave motion is clearly involved, but in this case vertical waves are undoubtedly more important than longitudinal waves. Similar types of orientation occur in fish, due largely to the lateral line organs. True, the tactile sense is called upon rather than hearing, but there is a certain analogy between this system and an auto information one. See also Addendum on crustacean echolocation.

7. Aquatic Vertebrates

As light does not penetrate deeply in the ocean, visual orientation is limited to that minority of marine animals which live in the upper layers, and to those deep-sea animals endowed with vision and a system of luminous organs. After his discovery of echolocation in bats, Griffin began to speculate on the existence of some sort of acoustic orientation in at least a few of these marine animals. The suggestion seems all the more plausible in that some fish, Crustaceans and Cetaceans have been known from remotest Antiquity to emit sounds.

(a) *Cetacea*

From the point of view of acoustics, the best known of all marine animals are the Cetaceans. The Odontoceti have been the subject of observation or experimentation. Aristotle mentions their emissions, which have also been reported by mariners and fishermen[45]. But it is the use of submarine echolocating apparatus employed during the last war which has made us realise how "talkative" and "noisy" these animals are.

References p. 225

Since then, various species of Odontoceti chiefly have been vigorously studied by Dreher, Engels, Kellogg et al.[42], Lilly and Miller[15],[16], McBride and Kritzler[6,44], Norris et al.[21], Schevill and Lawrence[73], Wood[86], and Worthington[87] in the U.S.A.; Tomilin[81] in the U.S.S.R.; Fraser and Purves[6] in England; Dudok van Heel[13], Sedee[74], Reysenbach de Haan[65] in the Netherlands; Vincent[85] in France.

(i) Odontoceti

In addition to their remarkable phonetic abilities, these animals have excellent hearing, whereas their peripheral and central sensory systems of sight and smell are rather poor. Hence their use of echolocation seemed probable enough, since many of these animals often move about in turbid water, in which it would seem impossible to see anything, and use to swim in schools at night as well as in daytime.

Since 1840, when sperm whale – *Physeter catodon* – fishing began, Bennett[3] and Davis[9] reported hearing "creaking" noises emitted by these animals. More recently, one of the last of the whale hunters of New Bedford told Schevill that he heard a number of sound pulses while he was waiting for the sperm-whales to rise. Gilmore[6 bis] speaks of echolocation-like sounds which could be "calls for help" in the same species.

Wood[86], using submarine listening apparatus, heard *Tursiops truncatos* emit rasping and grating sounds when a strange object was introduced into these animals' tank, or when they were probing their surroundings. *Stenella*, in the same tank, did not do this.

The acoustic emissions of *Tursiops* have been studied[72] as they approach their quarry. They consist of a series of creaks, or more precisely clicks, emitted at repetition rates varying from less than 10 to more than 400 per second, the slowest ones sounding rather like knocks and the quickest ones like grunts or groans.

Sounds of the same type have been heard coming from the pilot whale (*Globicephala*)[44], though they are not associated with any particular behaviour. Other species, such as Beluga, have acoustic habits closely related to those of *Tursiops* and

Delphinus delphis[85]

All authors noticed that the repetition rate of the emissions increases when the animals are hunting and capturing their quarry. In such cases the sounds have been suggestive of combined mewing and rasping sounds[86], or just mewing[85].

At the moment we have few analyses of these signals, though some of their physical properties have been established in the cases of *Tursiops truncatus*, *Physeter cathodon* and *Delphinus delphis*. They consist of very frequent series of short, sudden and strong clicks which strike the ear as being of variable frequency (Fig. 94).

The spectra are always complex, ranging from 6 to 200 kc/s in *Tursiops* and from 4 to 15 kc/s in *Delphinus* (the band above that value could not be investigated with the apparatus used) (Fig. 95). In point of fact, these signals have a spectrum of theoretically infinite frequencies, comparable to that of a white noise, as essentially it is composed of short lessened impulses.

The clicks are always of very short duration: in *Tursiops* it is sometimes no more than 1 msec; in *Delphinus* the average duration is 20 msec and sometimes much shorter.

The repetition rate varies from species to species. In *Delphinus* it may increase from 5 to more than 100 clicks per second. The rhythm is not a steady one, but is liable to be accelerated, especially while the animal is feeding (Fig. 96). *Physeter* has a normal repetition rate of 2 to 5 pulses a second in series which may include as many as 73 clicks. The groans heard[87] seem to be series of very short clicks with very narrow intervals between them.

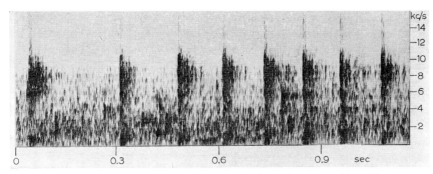

Fig. 94. Sonagram of the echolocation clicks of *Delphinus delphis*[85].

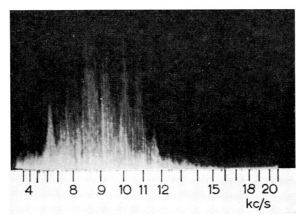

Fig. 95. Frequencies spectrum of echolocation clicks of *Delphinus delphis* corresponding to Fig. 94.

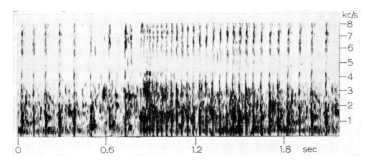

Fig. 96. Sonagram of some fast "cracking" of *Delphinus delphis* emitted during a meal (locating the prey and catching it)[85].

References p. 225

All the physical properties of these signals have the characteristics that have been recognised as necessary for echolocation.

Since 1947[6] the following facts have been recorded. A *Tursiops* does not break the nets put out to catch small fish but leaps over them. To capture them it is necessary to use nets of 25 cm^2 mesh. When one of these animals was being netted in this way, others leapt over the nets to avoid them. These catches are, moreover, made at night and in turbid water. Hence, in conditions when there can be no question of optical perception, these animals are capable of discerning the small meshes of fishing nets and of avoiding them. They even "know" that the margin of the net is the boundary of their free field.

The Odontoceti have extraordinary acuity of hearing. They easily discover the exact place in a tank where one is rippling the water with the tips of the fingers. It might be supposed, therefore, that the impact of a fish on the surface of the water would be enough to direct the animals to their prey. But if one plunges a fish into the water and pulls it out again very quickly, the disturbance of the surface alone does not entice the Cetacean. Now, at night the animals are perfectly capable of finding their way to a fish in turbid water. The inference is that vision plays no part at all in their orientation.

These early observations have recently been repeated upon two individuals of *Tursiops truncatus*, 2.40 metres long, and Kellog[41] conducted a number of very elegant experiments in order to study the various aspects of these problems.

A fish plunged into the tank is immediately discovered and captured, without any hesitation by Dolphins. Even if the Cetaceans are silent before the experiment, the impact of the fish upon the water elicits a shower of clicks which continues till well after the capture.

The effects of various objects producing a similar disturbance of the surface of the water to that caused by a fish were studied. Like the fish, they provoked a veritable torrent of emitted sounds.

One way of rippling the surface, without throwing any objects into the water, was to let drops of water rain down on the surface. The dolphins immediately emitted their clicks but, when they received no echo, they soon stopped.

Targets silently slipped into the water evoked no emissions. It was not until a series of occasional exploratory clicks were emitted that these objects were detected.

The immediate reaction to the impact of an object on the surface of the water is the emission of exploratory clicks, which is only continued if echoes are perceived. Furthermore, these two dolphins appeared to be perfectly able to discriminate between two quarries, such as a *Mugil cephalus* and a *Leiostomus xanthurus*.

Under the same experimental conditions, a metal maze forming corridors 2.6 metres long and 2.6 metres wide was lowered into the tank. In twenty minutes the two animals had only four collisions. The other tests failed to produce any collision. The metal was replaced by a net of 12.5 cm mesh. Some free openings were let into it, while others had a transparent "door" of plexiglass. The dolphins skilfully avoided both the threads and the plexiglass, passing only through the free openings.

These experiments with Dolphins eliminated vision as a source of orientation (experiments made at night and in very muddy water). The olfactory sense in these animals is known to be very poor, and the rapid reactions ruled out taste as the orienting sense. That these dolphins emit sounds adapted to the circumstances indicates that they orient themselves by a type of echolocation comparable to the general principle defined in the case of bats.

A fairly large number of recent researches[62,65] have shown that Cetaceans have virtually no olfactory nerve centres at all and a very limited visual area, but that, by contrast, the acoustic areas and fibres are highly developed. The acoustic midbrain colliculi are larger than the optic pair in bats and, more strikingly, in Cetaceans([3]). The Odontoceti, more especially, have well developed auditory organs of great sensitivity, a fact which is borne out by the many observations of seafarers who report that the transmission of ultrasonic pulses by echo-sounders frightens these animals away[15]. Indeed, in the aquarium this sensitivity was found to exceed all expectations, *Tursiops truncatus* reacting well to frequencies between 100 c/s and 80,000 c/s[40]. Apparently these porpoises, or at any rate *Phocaena phocaena*, can locate the direction of the source of sound in water with properties very similar to those of the human ear in air, for wavelengths of the same order of magnitude[23].

It is now possible to affirm, on the basis of careful experiments and up-to-date observations of the emissions, hearing and habits of the Odontoceti, that these animals are guided by a system of auto-information. (See Addendum.)

(ii) Mysticeti

The few observations made respecting the sounds emitted by these animals and their auditory ability cannot be said to represent a study of the problem of echolocation.

Among the Rorquals (Balaenidae), emissions have been reported[25] from *Megaptera boops* (*Rorqual longimane*, humpback whale). These are loud, low-pitched sounds. Of the true whales, sounds have been heard from *Eubalena* (the right-whale) as well as from five species of Mysticeti reported by Tomilin[81], though no details were given of the conditions under which the emissions were forthcoming, nor of the characteristics of the sounds produced. This is not surprising in view of the difficulties encountered at sea.

It is known from other sources that the sensitivity of the Rorquals (*Balaenoptera musculus* and *B. physalus*) to the ultrasonic signals of Asdic is similar to that of the Delphinidae. They frighten the whales away, a reaction which the Norwegians have eagerly utilised for their own advantage when hunting these huge animals with what they call a "whale-startling device"*.

In conclusion there are too few detailed facts available at present to serve as evidence of the practice of echolocation by the Mysticeti. However, this function may well exist in these animals since their brain although not exhibiting a development of the acoustic centres comparable to that of the Odontoceti[62], is clearly more "acoustic" than "visual" or "olfactory".

(b) Fishes

Little is known at present about the acoustic orientation of fishes. The lateral line of these animals, which is highly sensitive to vibrations, would seem to provide them with a kinaesthetic rather than an acoustic sensitivity, although it is difficult to differentiate these two vibratory domains, and the distinction between them is

---

* See *The Norwegian Whaling Gazette*, 1 (1953) 23.

quite arbitrary. Nevertheless, it would appear possible, not to say probable, that deep-sea[2] or cave-fishes[50] species practise echolocation in some degree.

(i) The "Echo Fish"

A study of sound recordings made North of Puerto Rico enabled Griffin[26] to detect several series of short notes (300 to 1,500 msec) of approximately 500 c/s, sometimes reaching to 75 to 80 dB above the background noise level (1 to 2 dynes/cm^2). Each call of every series was followed at a fairly constant interval by a faint repetition, almost completely masked by the background noise. Knowing the depth and the interval between the call and its "echo", Griffin located the source as being between 1,250 and 3,925 metres from the surface*. It hardly seems possible for a single animal to produce two types of sound alternately with such regularity. Less plausible still would be the assumption that a second individual could answer the first with that same regularity. The most likely explanation is, therefore, that the second sound is an echo reflected by the ocean bed**. The depth and the physical properties of the emitted sounds suggest that a fish is responsible for them. It may well be that the animal uses such echoes for its orientation, but this is no more than a hypothesis receiving some support from the known fact that fish are endowed with auditory sensitivity[12].

(ii) The cave-dwelling species

Although generally blind, these fishes are perfectly capable of orientating themselves.

A *Stygicola dentatus*, Brotulidae inhabiting caves in Cuba[26], was observed three times in one and a half hours to pass through a hole in a rocky partition without bumping into the rock. Another avoided a stalagmite every time it passed.

Similarly, *Amblyopsis speloca* (Amblyopsidae)[2, 26], of the North American caves, avoids colliding with the bottom or sides of the ponds in which it lives. This animal has a swim bladder which can be used to emit sounds, and a particularly well-developed lateral line. It is by no means impossible that echolocation may be practised by certain fishes, but we have no experimental evidence that it is. In any event it has not been discovered whether they emit audible or ultrasonic sounds.

(iii) Other fishes

The observations which Benl[2bis] made on two *Xenomystus nigri* (Notopteridae) must finally be noticed. These fresh water fishes living in Africa emit isolated or grouped gruntings, by day or night, especially at sunrise. Their main emissions are during courting parades, in which the "attacking" fishes always give the signal. The Botiens (*B. horae*) give out signals of the same type by rapidly opening and closing the mouth. They only rarely emit when isolated.

Can one, in such cases, talk of echolocation, or do those signals have a psychological meaning in the behavior of the fish? It is too early to conclude on this point, but it would be interesting to compare this behaviour to that of porpoises.

---

* According to the possible distance of the source from the hydrophone.
** The hypothesis of an acoustic channel would necessitate the presence of multiple echoes. Furthermore, such a channel is usually rather unstable and would not have enabled recordings to be made of the same phenomenon at different times.

8. General Conclusions

The system of echolocation corresponding to acoustic auto-information has been demonstrated by Griffin and his school with remarkable precision and admirable insight.

This type of signal is an exceptional kind as compared with the signals serving for communication between one individual and another, and it is more widespread than was initially suspected. It is probable that in years to come numerous species will be found to be endowed with the ability to use it. The acquired knowledge demonstrates some very individualistic properties of the organs of phonation and hearing, and highly specialised nervous systems.

The extension of the process to groups as different as terrestrial and marine animals and the variants of the signal in frequencies suggest that the modalities vary more or less with the species. It may perhaps be premature to attempt a detailed analysis of the mechanisms of perception, but it seems probable in the light of electrophysiology that perception of the returning waves will usually be associated with the form and the rhythm. There can be no doubt that this field of bio-acoustics is still a fertile one, and that much strenuous work will greatly enrich our future knowledge.

Acknowledgement

I wish to extend my thanks to Prof. D. R. Griffin for his constructive criticisms, in reading over the manuscript of this chapter.

[For Addendum to this Chapter, see p. 789]

REFERENCES

[1] ANDERSON, J. W., 1954, The production of ultrasonic sounds by laboratory rats and other animals, *Science*, *119*, 808–809.
[2] BACKUS, R. H., 1958, Sound production by marine animals, *J. Underwater Acoust.*, *8*, 2, 191–202.
[2 bis] BENL, G., 1959, Lautäusserungen beim Afrikanischen Messerfisch und bei Botien, *D.A.T.Z.*, *12*, 4, 108–111.
[3] BENNETT, F. D., 1840, *Narrative of a Whaling Voyage around the Globe*, London, 2, 206.
[4] BLOEDEL, P., 1955, Observations on the life histories of Panama bats, *J. Mammal.*, *36*, 232–235.
[5] BLOEDEL, P., 1955, Hunting methods of fish-eating bats, particularly *Noctilio leporinus*, *J. Mammal.*, *36*, 390–399.
[6] McBRIDE, A. F., 1956, Evidence for echolocation by cetaceans, *Deep-Sea Research*, *3*, 153–154.
[7] CURTIS, W. E., 1952, Quantitative studies of echolocation in bats (*Myotis l. lucifugus*), studies of vision of bats (*Myotis l. lucifugus* and *Eptesicus f. fuscus*) and quantitative studies of vision of owls (*Tyto alba practincola*), thesis, Cornell University, Ithaca (N.Y.).
[8] CUVIER, G., 1795, Conjectures sur le sixième sens qu'on a cru remarquer dans les Chauves-Souris, lues à la Société d'Histoire Naturelle, le 17 Ventôse, *Magasin encyclopédique*, 6, 297–301.
[9] DAVIS, W. M., 1874, *Nimrod of the Sea*, pp. 236–355, New York.
[10] DIJKGRAAF, S., 1943, Over een merkwaardige functie van den gehoorzin bij vleermuizen, *Koninkl. Ned. Akad. Wetensch. Verslag Gewone vergader. Afd. Nat.*, *52*, 622–627.
[11] DIJKGRAAF, S., 1946, Die Sinneswelt der Fledermäuse, *Experientia*, *II*, No. 11.
[12] DIJKGRAAF, S., 1947, Über die Reizung des Ferntastsinnes bei Fischen und Amphibien, *Experientia*, *3*, 206–208.
[13] DUDOK VAN HEEL, W. H., 1959, Audio-direction finding in the porpoise (*Phocaena phocaena*), *Nature*, *4667*, 1063.
[14] ELIAS, H., 1907, Zur Anatomie des Kehlkopfes der Mikrochiropteren. *Morphol. Jahrb.*, *37*, 70–118.
[15] FRASER, F. C., 1947, Sound emitted by dolphins (*Delphinapterus leucas*), *Nature*, *160*, 759.

[16] GALAMBOS, R., 1942, The avoidance of obstacles by flying bats: Spallanzani's ideas (1794) and later theories, *Isis, 34,* 132–140.
[17] GALAMBOS, R., 1942, Cochlear potentials from bats by supersonic sounds, *J. Acoust. Soc. Am., 14,* No. 1, 41–49.
[18] GOULD, E., 1955, The feeding efficiency of insectivorous bats, *J. Mammal., 36,* 399–407.
[19] GOULD, J. and C. T. MORGAN, 1941, Hearing in the rat at high frequencies, *Science, 94,* 168.
[20] GRASSE, P. R., 1955, *Traité de Zoologie,* XVII, 2, Masson et Cie.
[21] GRIFFIN, D. R., 1941, The sensory basis of obstacle avoidance by flying bats, *J. Exptl. Zool., 86,* 481–506.
[22] GRIFFIN, D. R., 1944, Echolocation by blind men, bats and radar, *Science, 100,* 589–590.
[23] GRIFFIN, D. R., 1953, Sensory physiology and the orientation of animals, *Am. Scientist, 41,* 209–244.
[24] GRIFFIN, D. R., 1953, Acoustic orientation in the oil bird, *Steatornis, Proc. Natl. Acad. Sci., 39,* 884–893.
[25] GRIFFIN, D. R., 1955, Hearing and acoustic orientation in marine animals, *Deep-Sea Research,* suppl. to *3,* 406–417.
[26] GRIFFIN, D. R., 1958, *Listening in the Dark,* Yale University Press. 413 p.p.
[27] GRIFFIN, D. R. and R. GALAMBOS, 1940, Obstacle avoidance by flying bats, *Anat. Rec., 78,* 95.
[28] GRIFFIN, D. R. and A. NOVICK, Acoustic orientation of neotropical bats, *J. exptl. Zool., 130,* 251–300.
[29] GRIFFIN, D. R., A. NOVICK and M. KORNFIELD, 1958, The sensitivity of echolocation in the fruit bat *Rousettus, Biol. Bull., 115,* No. 1, 107–113.
[30] GRIFFIN, D. R., F. A. WEBSTER and C. R. MICHAEL, 1960, The echolocation of flying insects by bats, *J. Animal Behavior, 8,* 141–154.
[31] GROSSER, O., 1900, Zur Anatomie der Nasenhöhle und des Rachens der einheimischen Chiropteren, *Morphol. Jahrb., 29,* 1–77.
[32] HAHN, W. L., 1908, Some habits and sensory adaptations of cave-inhabiting bats, *Biol. Bull., 15,* 135–193.
[33] HARTRIDGE, H., 1920, The avoidance of objects by bats in their flight, *J. Physiol., 54,* 54–57.
[34] HINTON, H. E., 1955, Sound-producing organs in the Lepidoptera, *Proc. Roy. Entomolog. Soc., London, Ser. C., 20,* 5–6.
[35] HUMBOLDT, A. VON and A. BONPLAND, 1824–25, *Voyages aux régions équinoxiales du Nouveau Continent fait en 1799–1804,* Paris, F. Schoeller.
[36] IKEDA, YOSHINDO and T. YOKOTE, 1939, Über einige, teils bisher noch unbekannte Eigentümlichkeiten in der Schnecke einer Art von Fledermäusen (*Rhinolophus Nippon* Temminck) *Nagasaki Igakki Zassi, 17,* No. 4, 1041–1060.
[37] IWATA, N., 1924, Über das Labyrinth der Fledermaus mit besonderer Berücksichtigung des statischen Apparates, *Aitchi J. Exptl. Med., Nagoya, 1,* 41–173.
[38] JURINE, L., 1798, Experiences sur les Chauves-Souris qu'on a privées de la vue, *J. de Phys., 46,* 145–148.
[39] KAHMANN, H. and K. OSTERMANN, 1951, Wahrnehmen und Hervorbringen hoher Töne bei kleinen Säugetieren, *Experientia, 7,* 268–269.
[40] KELLOGG, W. N., 1953, Ultrasonic hearing in the porpoise *Tursiops truncatus, J. Comp. and Physiol. Psychol., 46,* 446–450.
[41] KELLOGG, W. N., 1958, Echo-ranging in the porpoise, *Science, 128,* 982–988.
[42] KELLOGG, W. N., R. KOHLER and H. N. MORRIS, 1953, Porpoise sounds as sonar signals, *Science, 117,* 239–243.
[43] KLENSCH, H., 1949, Die Lokalisation des Schalles im Raum, *Naturwiss., 36,* 145.
[44] KRITZLER, H., 1952, Observations on pilot whale in captivity, *J. Mammal., 33,* 321–334.
[45] KULLENBERG, B., 1947, Sound emitted by dolphins, *Nature, 160,* 648.
[46] KULZER, E., 1956, Flughunde erzeugen Orientierungslaute durch Zungenschlag, *Naturwiss., 43,* 117–118.
[47] LACK, D., 1956, *Swifts in a Tower,* London, Methuen, 239 pp.
[48] LANGEVIN, P. and C. CHILOWSKY, 1916, *Procédés et appareils pour la production de signaux sous-marins dirigés et pour la localisation à distance d'obstacles sous-marins,* French Patent No. 502, 913.
[49] LASHLEY, K. S. and J. T. RUSSELL, 1934, The mechanism of vision: XI. A preliminary test of innate organization, *J. Genet. Psychol., 45,* 136–144.
[50] LOWENSTEIN, O., 1957, The acoustico-lateralis system, in M. E. BROWN (Edit.) *Physiology of Fishes,* Academic Press, New York, Vol. 2, pp. 155–186.
[51] MEDWAY, L., 1959, Echolocation among *Collocalia, Nature, 184,* 1352–1353.
[52] MÖHRES, F. P., 1950, Aus dem Leben unserer Fledermäuse, *Kosmos, 46,* 291–295.
[53] MÖHRES, F. P., 1953, Über die ultraschallorientierung der Hufeisennasen (Chiroptera-Rhinolophinae), *Z. vergleich. Physiol., 34,* 547–588.

[54] MÖHRES, F. P., 1955, Untersuchungen über die Ultraschallorientierung von vier afrikanischen Fledermäusen Familien, *Verh. Deut. Zool. Gesell.*, 59–65.
[55] MÖHRES, F. P., 1956, Über die Orientierung der Flughunde, *Z. vergl. Physiol.*, *38*, 1–29.
[56] MÖHRES, F. P. and E. KULZER, 1955, Ein neuer, kombinierter Typ der Ultraschallorientierung bei Fledermäusen, *Naturwiss.*, *42*, 131–132.
[57] MÖHRES, F. P., and E. KULZER, 1957, *Megaderma*—ein konvergenter Zwischentyp der Ultraschallpeilung bei Fledermäusen, *Naturwiss.*, *44*, 21–22.
[58] MÖHRES, F. P. and T. OETTINGEN-SPIELBERG, 1949, Versuche über die Nahorientierung und das Heimfindevermögen der Fledermäuse. *Verh. Deut. Zool.*, 248–252.
[59] MUELLER, H. C. and J. T. EMLEN, 1957, Homing in bats, *Science*, *126*, 307–308.
[60] NOVICK, A., 1955, Laryngeal muscles of the bat and production of ultrasonic sounds, *Am. J. Physiol.*, *183*, 648.
[61] NOVICK, A., 1957, Orientation in paleotropical bats: I. Microchiroptera, *Federation Proc.*, *16*, 95–96.
[62] OGAWA, T. and S. ARIFUKU, 1948, On the acoustic system in the cetacean brains, *Sci. Rept. Whales Research Inst.*, *2*, 1–20.
[63] PIERCE, G. W., and D. R. GRIFFIN, 1938, Experimental determination of supersonic notes emitted by bats, *J. Mammal.*, *19*, 454–455.
[64] POLYAK, S., 1926, The connections of the acoustic nerve, *J. Anat.*, *60*, 465–469.
[65] REYSENBACH DE HAAN, F. W., 1956, De ceti auditu, over de gehoorzin bij de walvissen. Thesis, Utrecht, and *Acta Oto-Laryngologica*, suppl. *134*, (1957).
[66] RILEY, D. A. and M. R. ROSENZWEIG, 1957, Echolocation in rats, *J. comp. and Physiol. Psychol.*, *50*, No. 4, 323–328.
[67] ROEDER, K. D. and A. E. TREAT, 1957, Ultrasonic reception by the tympanic organ of noctuid moths, *J. exptl. Zool.*, *134*, 127–158.
[68] ROLLINAT, R. and E. TROUESSART, 1900, Sur le sens de la direction chez les Chiroptères, *Compt. Rend. Soc. Biol.*, 600–607.
[69] ROSENZWEIG, M. R., D. A. RILEY and D. KRECH, 1955, Evidence for echolocation in the rat, *Science*, *121*, 600.
[70] SCHALLER, F. and C. TIMM, 1950, Das Hörvermögen der Nachtschmetterlinge, *Z. vergleich. Physiol.*, *32*, 468–481.
[71] SCHEVILL, W. E., 1953, Auditory response of a bottle-nosed porpoise, *Tursiops truncatus*, to frequencies above 100 kc/s, *J. exptl. Zool.*, *124*, 147–165.
[72] SCHEVILL, W. E., 1956, Food-finding by a captive porpoise (*Tursiops truncatus*), *Breviora-Mus. comp. Zool., Harvard Univ.*, *53*, 1–15.
[73] SCHEVILL, W. E. and B. LAWRENCE, 1949, Underwater listening to the white porpoise (*Delphinapterus leucas*), *Science*, *109*, 143–144.
[74] SEDEE, G. A., 1957, Ph.D. Thesis, Utrecht.
[75] SENEBIER, J., 1807, *Rapports de l'air avec les êtres organisés* (tirés des journaux d'observations et d'expériences de Lazare Spallanzani, avec quelques mémoires de l'éditeur sur ces matières), three volumes, Geneva (especially Vol. 2).
[76] SGONINA, K., 1935, Der spallanzionische Fledermausversuch, *Zool. Anzeiger*, *109*, 325–327.
[77] SHEPARD, J. F., 1929, An unexpected cue in maze learning, *Psychol. Bull.*, *26*, 164–165.
[78] SPALLANZANI, L., 1932, *Opere di Lazzaro Spallanzani*, 5 vols, Milan, Ulrico Hoepli. (Letters describing experiments with bats are in Vol. 3.)
[79] SMITH, P. F., 1954, Further measurements of the sound-scattering properties of several marine organisms, *Deep-Sea Research*, *2*, 71–79.
[80] TIMM, C., 1950, Aus der Sinneswelt der Fledermäuse, *Umschau Wiss. u. Tech.*, *50*, 204.
[81] TOMILIN, A. G., 1955, On behavior and sonic signalling in cetaceans (in Russian), *Akad. Nauk SSSR, Trudy Inst. Okean.*, *18*, 28–47.
[82] TREAT, A. E., 1955, The response to sound in certain Lepidoptera, *Ann. Entomol. Am. 48*, 272–284.
[83] TSANG, Y. C., 1934, The functions of the visual areas of the cerebral cortex of the rat in the learning and retention of the maze, *Comp. Psychol. Monogr.*, *10*, 56.
[84] VESEY-FITZGERALD, B., 1947, Les fonctions sensorielles des Chauves-souris, *Endeavour*, *VI*, No. 21, 36–41 (French Edition).
[85] VINCENT, F., 1960, Etudes préliminaires des émissions acoustiques de *Delphinus delphis* en captivité, *Bull. Inst. Oceanog.*, No. 1172.
[86] WOOD JR., F. G., 1954, Underwater sound production and concurrent behavior of captive porpoises *Tursiops truncatus* and *Stenella plagiodon*, *Bull. Mar. Sci. Gulf and Caribb.*, *3*, 120–133.
[87] WORTHINGTON, L. V. and W. E. SCHEVILL, 1957, Underwater sounds heard from sperm whales, *Nature*, *180*, 291–292.

CHAPTER 9

INHERITANCE AND LEARNING IN THE DEVELOPMENT OF ANIMAL VOCALIZATIONS

by

P. MARLER

One of the most distinctive biological attributes of man is his language, and one of the characteristics of this language which was long thought to separate him from animals is the fact that the vocalizations which comprise it are learned. We now know that man is not unique in this respect since some birds, at least, learn the species-specific songs and calls from individuals of the previous generation. The processes underlying the development of animal vocalizations have thus attained special interest, and it is the aim of this chapter to review our present knowledge in this field.

The common assumption that there is more or less strict genetic control over the form of most animal vocalizations is based on sound observation, and the evidence bearing on this aspect will be considered first. As a corollary, the neglected subject of vocal inventiveness will receive attention, and finally we shall deal with the role of learning from sounds produced by others.

1. Inheritance as a Dominant Determinant of the Structure of Animal Vocalizations

If an animal raised in isolation develops the normal species-specific vocalizations, as heard in the natural state, we may assume that genetic factors play a predominant role in their development. The precise method by which genetic control is achieved need not at present concern us, as long as the "self-differentiating" character[10] is established. If the vocalizations of inter-specific hybrids have been examined, we can approach closer to understanding the type of genetic control which is operating.

Orthoptera. Among lower animals the assumption that genetic control prevails is so widespread that studies of animals raised in isolation are uncommon. In the Orthoptera we must rely instead upon hybrid studies, of which there are several.

An early example of the relevance of behavior to taxonomy was Fulton's study of song in hybrids of two subspecies of *Nemobius fasciatus*, *N.f. tinnulus*, and *N.f. fasciatus*[11]. The song of the F1 generation was intermediate between those of the two parents in several respects, including tempo. Some of the F2 generation also had intermediate songs, others were like *tinnulus*, none were like *fasciatus*. Although several genes are evidently involved, genetic control of song is clearly established, since, although they were apparently exposed to parent songs, the F1 hybrids developed an intermediate song type.

F1 hybrids between *Chorthippus brunneus* and *C. biguttulus* also have songs which are clearly distinguishable from those of both parents, and intermediate in several respects, particularly in the duration of the notes, which is much shorter then in *biguttulus*, but longer than in *brunneus* (Fig. 97)[43].

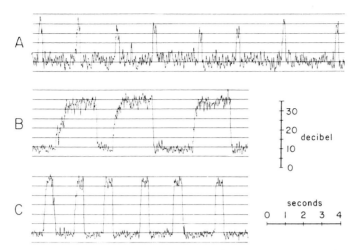

Fig. 97. Sound-level analyses of one phrase of the song of *Chorthippus brunneus* (A), *C. biguttulus* (B), and the hybrid between them (C). Note the intermediate nature of the hybrid song. (From Perdeck, 1957[43].)

A rather different aspect of vocalization was studied in hybrids between *Gryllus bimaculatus* and *G. campestris*[17]. These two species behave differently at copulation. The male *G. bimaculatus* gives an average of 4.4 courtship calls before mounting, by repeated raising of the elytrae from rest to the stridulation position. *G. campestris* males raise the elytrae once, without producing a sound. The F1 males raised the elytrae and produced an average of 1.4 calls before mounting. A careful quantitative study of the F2 generation and of back-crosses provides good evidence for monofactorial inheritance of these characteristics.

These experiments establish beyond doubt a predominance of genetic control in the development of Orthopteran vocalizations, though it must be admitted that no one has actually set out to induce the learning of new sounds by insects. A number of other studies of pre-imaginal Orthopteran behavior seem to support the fact of genetic control without being expressly directed at this problem[19, 24, 65]. A point which emerges from these and other studies is the close coordination between the form of a vocalization and the structure of the sound-producing organ. Since the morphology of hybrids tends to be intermediate between that of the two parent species we might expect an effect on sound production. However, it is clear that heritable neuromuscular coordinations also play an important role, as evinced by the performance by larval forms of appropriate movements for the production of species-specific sounds, before the morphological structures necessary for sound production have developed[19, 65].

Anura. The situation in Orthoptera and Anura appears to be rather similar. Again, no studies have been found concerned with the development of vocalization in isolated individuals, and we must once more rely on hybrid studies. Blair has analyzed the calls of many North American Anurans, including several hybrids[3, 4, 5]. In each case the voice proved to be intermediate between the two parents. In an overlap zone between *Microhyla olivacea*, and *M. carolinensis*, eight presumed hybrids were found, intermediate between the two parents in duration and mid-point frequency of their calls[3]. In a similar study of toads, a presumed natural hybrid between *Bufo americanus* and *B. woodhousei* had calls intermediate in every respect between those of the parent species (Fig. 98)[4]. Artificial hybrids between two more distantly related species, *Bufo woodhousei* and *B. valliceps* behaved differently, producing either rudimentary or imperfect calls of one parent—*B. valliceps*. Here the mechanism of genetic control cannot be elucidated, but in the two former examples the role of inheritance in voice determination is clear. In all cases the hybrids were exposed to calls of one or both parents in the natural state.

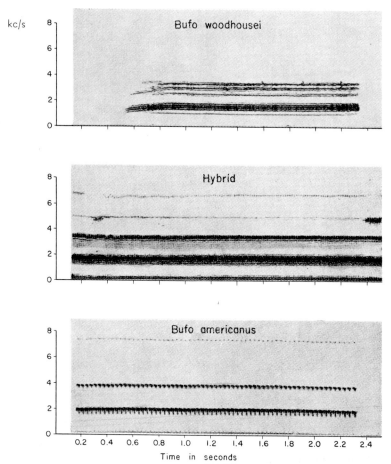

Fig. 98. Sonagrams of the calls of *Bufo woodhousei*, *B. americanus* and a presumed natural hybrid between them. Note the intermediate trill rate and side band structure in the hybrid's song. (From Blair, 1956[4].)

Aves–Calls. The great volume of data on vocal development in birds is derived from isolation experiments. To the best of the writer's knowledge, there is no careful study of song in a hybrid between parents whose song is predominantly inherited, a serious omission. Since the mode of control of the song and of calls is often different, these are dealt with separately in the following paragraphs. The evidence consistently indicates that with a few exceptions the calls of both passerine and non-passerine birds are inherited. This is true, for example of *Gallus domesticus*, as can readily be shown by raising chicks in an incubator[50]. The same kind of experiment is much more difficult to perform with small nidicolous passerines, but has been accomplished with *Sylvia communis*[48] and *Turdus merula*[34]. In both cases the calls developed normally, all 25 calls of the former, for example, matching closely with those of wild members of the species. In another approach, birds have been deafened centrally by extirpation of the lagena at an age of several days. *Turdus merula* operated upon in this way, developed calls similar to those of wild birds[34].

Many more species of nidicolous passerines have been taken from the nest at various times between hatching and fledging, and raised by hand, in varying degrees of isolation from their own and other species. It seems unlikely that experience of species-specific sounds during the pre-isolation period has a direct effect on the subsequent development of calls, though the possibility cannot be entirely ruled out. The calls of small passerines raised in this way almost always develop normally. This has been demonstrated in *Sturnella magna* and *S. neglecta*[20] and *Muscicapa hypoleuca*[9]. A wild hybrid between *Muscicapa hypoleuca* and *M. albicollis*, whose calls are mainly very similar[14], developed an intermediate form of the one call which is distinctly different in the two species[25]. The calls of hand-raised *Melospiza melodia* are normal[18], and the same is true of *Carpodacus mexicanus*[31, 36] and of all but one of the calls of *Fringilla coelebs*[29]. Heinroth, on the basis of wide experience of hand-raising nestlings in this way, suggests that calls in general, of passerines and non-passerines alike, share this same dependence on inheritance as the dominant factor controlling their development[15]. Although we shall see that there are a few exceptions most of the experiments conducted since his time have supported that generalization.

Aves–Songs. By similar experimental methods the development of song has been studied. Stadler suggests that the songs of non-passerines are inherited[58] and the small amount of evidence available certainly supports this. Incubator-raised domestic cockerels crow normally[50]. *Streptopelia risoria* and *Geopelia* sp., incubated and raised by another species developed the species-specific pattern of cooing[7,8]. A hand-raised *Cuculus canorus* gave a normal "cuckoo".

Many passerines have been studied. Incubator-raised *Sylvia communis*[48] and *Turdus merula*[34], and deafened *T. merula*[39], all developed species-specific patterns of song. In hand-raised nestlings, normal song development has been demonstrated in *Hirundo rustica*[45] and *Riparia riparia*[15], *Troglodytes troglodytes*[15], *Certhia brachydactila*[15], *Anthus trivalis*[44], *Quiscalus quiscula*[21], *Oriolus oriolus*[45], *Turdus merula*[13, 22], *T. ericetorum*, and *T. viscivorus*[15], *Phylloscopus collybita*[15,58] and *P. sibilatrix*[44], *Sylvia communis*[48], *Locustella naevia*[15], *Emberiza schoeniclus*[44,64], *E. calandra*[64], and *Pyrrhula pyrrhula*[16].

In addition to this abundance of examples of inherited songs it is interesting to note the transformation of the song of certain birds during domestication. As a result of direct selection for voice characters, the song of the roller canary is now very different from that of wild *Serinus canarius*[30]. And the distinctive "tours" by which the quality of the song is assessed develop normally in males raised by a non-singing female, in a sound-proof room[35]. Around the turn of the century a special type of domestic pigeon, the trumpeter pigeon, was bred for an unusual type of cooing, though this habit was subsequently neglected by breeders and lost[23]. And finally, there is an ornamental breed of cockerel in the orient, bred for a very attenuated crow, and matched in crowing contests in much the same way as the roller canary in Europe and North America.

Mammalia. The information available on voice development in mammals other than man is very sparse indeed, even in the primates. While it seems likely that the species-specific sounds of sub-primate mammals, and even of such primates as *Hylobates* and *Anthropopithecus* are inherited, very few studies have been explicitly directed at this problem. Seitz has raised a number of small carnivores in isolation, and found for example, that 6 calls and their variations all develop normally in *Nyctereutes procyonoides*[56]. Two domestic cats raised in isolation developed normal calls[66]. Among the Lagomorpha, the variety of calls of *Ochotona princeps* are either inherited or are learned during the first month of life[57]. Perhaps most interesting is the observation by Boutan on *Hylobates leucogenys* that all 13 of the species' calls develop normally in a hand-raised animal[6], though it is not clear how familiar he was with the voice of wild individuals. This problem recurs in primate studies. Hand-raised chimpanzees for example, may develop up to 32 different vocalizations[68], but we have no information on how these compare with calls of wild animals. There seems to be no information on development of voice in such outstanding vocalizers as Insectivora, Chiroptera, Pinnipedia or Cetacea.

In spite of the lack of information on many groups, it already seems highly probable that the great majority of animal vocalizations are inherited. This appears to be true from the Orthoptera, through Anura and birds, up to mammals and even including at least one primate. Species abnormalities of voice in individuals raised out of hearing of their own kind are very much the exception. The extent of this inherited vocabulary may range from a few sounds in Orthoptera and Anura up to as many as 20–30 distinct vocalizations in some birds and mammals. However, although the development of such sounds is more or less normal in isolated animals, we should not exclude the possibility that changes through learning could well be superimposed, if the situation permitted this. In certain birds with inherited species-specific songs such elaboration by learning is known to take place and may be more widespread among other groups than is at present appreciated.

2. The Mechanisms of Genetic Control over the Development in Inherited Vocalizations

In many cases genetic control over voice appears to be mediated by effects on morphological characteristics which are directly involved in sound production, such as the number of pegs on the stridulating organs of Orthoptera, the dimensions of resonators such as elytra, or in Anura, the form of the throat sacs. Even here, however, it is necessary to postulate some control over the neuro-muscular coordinations underlying the movements of sound production, and in birds and mammals this probably becomes the dominant factor. The variations in structure of the vocal apparatus certainly have effects on voice, but there is no evidence that the marked differences in song of closely related birds, for example, can be directly ascribed to an obvious structural difference in the syrinx, though it must be admitted that few such studies have been made. The differences seem to be related mainly to variations in patterns of neuro-muscular activity.

The two most obvious ways in which such control could operate are either by an inherited pattern of motor input to the sound producing organs, or by an inherited auditory "template" to which the animal would match the sounds produced, as a result of hearing its own voice. At present we cannot choose between these two. Some experiments with deafened birds suggest that auditory feed-back may not be involved[34]. On the other hand, the vocalizations of *Pyrrhula pyrrhula* deafened as adults may show some change[18,51]. The song and certain calls remain normal, but the social call may become higher pitched with a more squeaky tone, and with more "syllables". However, we may note that according to Heinroth[16] and Nicolai[42], the social call is the only vocalization of this species which is not inherited. Thus it may be that inherited vocalizations are not dependent on auditory feed-back either for development or maintenance, while the reverse may be true in both cases for learned vocalizations. Much more information is needed on this point.

There is little evidence on the nature of the genetic mechanisms involved in animal vocalizations. One study suggests mono-factorial inheritance of a song difference in two species of *Gryllus*[17], while other orthopteran song differences may be affected by several genes. There is no information on higher animals, although the susceptibility of bird song to continuous selection during domestication suggests the possibility of polygenic control.

A question arises concerning the degree of precision of genetic control over the form of inherited vocalizations. Certain birds with supposedly inherited songs show a considerable variation between individuals in the same population. Sometimes this variation seems to be a result of learning which is superimposed, as for example in *Turdus merula*[31] and *Pyrrhula pyrrhula*[42] (see below). Sometimes it appears to be a result of variable expressivity in the genetic mechanism. Analysis of the structure of testosterone-induced crowing in several inbred lines of male white leghorn chicks reveals a general pattern common to all, together with much individual variation which is not related to individual parentage[42]. Since there is no evidence for learning in this case, it may be more reasonable to ascribe the individual differences in crowing to variable expressivity of the genetic factors which are concerned.

The role of improvisation in development of bird song

Certain species of birds, deprived of the opportunity to learn sounds from others, may still develop a surprisingly large and varied repertoire of song themes, implying an important degree of "improvisation". A deaf *Turdus merula*, for example, developed three song motifs which were recombined in different ways in exactly the same way as in wild birds[34]. In the same study wild birds enlarging the repertoire by learning from others, would occasionally add entirely new themes which did not seem to be copied from another bird. There is a suggestion of the same process in *Mimus polyglottos*[22], and it may also intrude in species with simpler songs. At its simplest it may just involve the temporal rearrangement of sounds which are themselves inherited as in *Sylvia communis*[48], but at times there is a suggestion of the development of new tonal units as perhaps in the very variable songs of *Emberiza schoeniclus*, whose song is thought to be inherited.

Again the limits of the extent of this improvisation are probably set by inheritance. Once more, we may regard it as manifestation of variable expressivity of the

References p. 241

genetic mechanism, though the limits are much broader than in the examples considered in the previous section.

3. The Role of Learning from Conspecific Individuals in Development of Species-specific Vocalizations

It is evident from the foregoing survey that the great majority of the species-specific vocalizations of animals are inherited. The known exceptions to this rule are found among the birds, where learning sometimes plays an important role in voice development. The simplest case to be considered first is learning from conspecific individuals, as a means of elaborating vocalizations, which would nevertheless develop in species-specific fashion without this learning.

(a) *Learned effects superimposed on inherited species-specific songs*

Some of the birds with an inherited normal song nevertheless modify it by learning in nature. Perhaps the best example is *Turdus merula*[34], which develops the species-specific song when raised in an incubator, yet learns many song themes in nature from other blackbirds. Messmer states that "even in their first spring, the young blackbirds try to imitate the motifs of adults partly or wholly. When the singing of adult birds dies down in midsummer the learning of some of the young vanishes as well. Throughout the winter the single learned motifs are clearly discernible in the juvenile song. Most young of the third hatching who, having been born after adult birds had stopped singing in midsummer, had no opportunity to learn previously, learn their first motifs in early spring from neighboring songsters. The adult bird uses up to eight motifs partly learned from other birds in the neighborhood, partly own inventions."

The inherited song of *Turdus ericetorum* also seems to be elaborated by learning, particularly since imitations of other species are not infrequent in the song of wild song thrushes[67]. *Pyrrhula pyrrhula* evidently develops the species-specific song when raised in isolation, but Nicolai has demonstrated a marked influence on song through learning from other bullfinches, in birds raised in semi-natural conditions in open-air aviaries[42].

In this species the young male normally learns his song from his father, and the imitations may be rather precise. If the father's song is abnormal, either from special training or through having been fostered by another species, the son still learns it, even though there may be other adults giving normal song within hearing. Finally the role of learning in development of championship song in roller canaries is widely acknowledged by breeders[58], even though normal song develops in isolated birds[35].

Thus the possession of an inherited song by no means precludes the possibility of the intervention of learning in normal song development. This is important in the interpretation of experiments in which a bird has been raised away from its own species but in company with other species. The learning of alien songs by such "isolated" birds has been regarded as implying the absence of inheritance of the species-specific song. While this may be the case, it is also possible that the inherited song has been lost in the process of imitating other species.

(b) Learning from conspecifics as a prerequisite for development of species-specific vocalizations

In a few birds there is evidence for abnormalities in the calls of individuals raised in isolation. The social call of *Fringilla coelebs* is sometimes abnormal in isolated birds and may be changed by learning from other chaffinches in nature[29]. The same is true of the social call of *Pyrrhula pyrrhula*[42], which in some respects has taken over the function served by song in other finches[41]. It has been reported that 17 of the 20 calls of *Parus major* depend on learning from other great tits for normal development[47], but details of this study are not available. It may be that these birds suffered from a deficiency of health or environment, for another worker reported that calls of this species are inherited[59]. As Nicolai has pointed out it is essential that birds be in perfect health if differences between their behavior and that of wild birds are to be meaningfully interpreted. In the same way, the appropriate physical environment is important. The abnormal song of a hand-raised caged *Alauda arvensis*, or *Anthus pratensis*, for example, may be partly a result of inability to perform the soaring flight which normally accompanies full song[27].

The best authenticated case of learning as a necessary prerequisite for normal development of species-specific song is *Fringilla coelebs*. A number of workers have raised chaffinches by hand and are agreed on the greater simplicity of the song which develops, as compared with wild birds[15,44,58,61,64]. The presence of other species does not seem to affect the form of this abnormal song, the learning capacity being employed, with two exceptions[59,62], only in imitation of other chaffinches[61,64].

The typical song of wild *Fringilla coelebs* in Britain is from 2–2.5 seconds in duration, divided by Thorpe into three phrases[61,64]. Phrase 1 consists of from 4–14 notes, with one or more stepwise changes in mean frequency. Phrase 2 is usually distinct, with 2–8 notes, usually lower in frequency than phrase 1. The song terminates with phrase 3, consisting of phrase 3a, with from 1–5 notes, together with a more or less complex terminal flourish 3b.

The songs of chaffinches taken from the nest at an age of about 5 days and raised alone in sound-proof rooms lack almost all of what we can regard as normal features of chaffinch song. There is only a slight trace of the division into two phrases, and the terminal flourish. The mean frequency of notes tends to be lower than normal, and the stepwise frequency changes are lacking. While the song has about the normal length and number of notes, the latter have an abnormally uniform tonal quality, the first and last notes often being almost identical in structure (Fig. 99).

It is clear from the various experiments which Thorpe has conducted, and from field studies[28], that the differences between this song of isolated birds and normal chaffinch song are a result of learning processes which occur in nature in two phases. Young chaffinches are fledged from April to June. They come into full song in the following spring, although more or less elaborate subsong may be heard from young birds in the fall[28]. If young wild males are captured in September and isolated thenceforth, the song which develops in the following spring is slightly less elaborate than the normal, but is divided into three phrases with a terminal flourish[64] (Fig. 99). Thus experience during normal fledgeling and early juvenile life, when adults are still singing, leads to the subsequent appearance of some characteristics of the normal song, learned before the bird itself is able to produce any kind of full song.

References p. 241

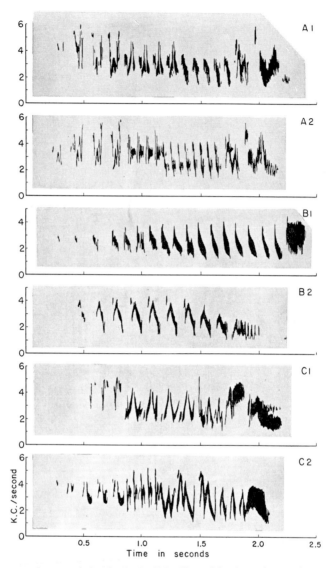

Fig. 99. Sonagrams of songs of the chaffinch, *Fringilla coelebs*. A1 and 2 are the songs of two wild, adult males; B1 and 2 are songs of two males raised in isolation from an age of about 5 days; C1 and 2 are songs of two males caught and isolated at an age of about 4–5 months. Note how birds in the last group have developed many of the elaborations of normal song. (From Thorpe, 1958[64].)

The second phase of learning takes place in the following spring, in competitive singing with other members of the group. At this stage details are added to the song, particularly to the end phrase, and wild birds have been recorded at this season imitating the songs of different neighbors. Once this is completed the song remains virtually unchanged for the rest of the spring and in subsequent seasons. The cumulative result of these two stages of learning is the characteristic, species-specific song of *Fringilla coelebs*, quite distinct in many respects from the inherited song of this species.

Less detailed studies of *Emberiza citrinella* suggests a similar influence of learning on song development. Birds raised away from their own species develop songs with a normal introduction, but lacking the distinctive end phrase of normal songs of this species[15, 58, 64]. Evidently there are no complications of learning from other species by experimental birds. The existence of "dialects" in yellow bunting song, which are characterized especially by the form of the endphrase is another parallel with *Fringilla coelebs*.

Melospiza melodia also develops abnormal song when raised in isolation[40]. Three individuals taken from the nest at about 5 days of age and raised by hand, developed songs which "in form, length, timing and even the number of songs were typical of the species, but not the quality". The birds imitated each other, and Nice records a similar process among young wild song sparrows, "repeating exactly the songs of their adult rivals" during the establishment of territory. Furthermore, she was able to show the absence of any detailed relationship between the songs of young males and those of their fathers or grandfathers. Thus it is clear that learning plays an important role in song development, in wild song sparrows, and it is probable that species-specific song cannot develop without it.

Carduelis cannabina raised by hand in isolation from the age of a week, developed songs shorter than normal, with fewer notes, uttered in a much slower rhythm[46]. The pitch and the timbre seemed to be innate, but the rhythm and melody were not. Only one phrase, the "crowing" was completely normal. Captive males of this species are known to imitate songs of other species, and it seems reasonable to suppose that this capacity is used in nature in learning the species-specific song from conspecific adults, although there is no direct evidence on this point.

Two workers have reported abnormal song in hand-raised *Muscicapa hypoleuca*[9, 15], and another has field observations of males learning the song of *Muscicapa albicollis*, in a part of their range where they are sympatric and occasionally form mixed pairs[25]. A young *M. hypoleuca* hatching from an egg placed in a nest of *M. albicollis* developed *albicollis* song in the following year. Thus *M. hypoleuca* clearly has the capacity to modify the song by learning, though it is difficult to be certain on the basis of present evidence if this learning is necessary for species-specific development.

Lanyon has studied another interesting species pair, *Sturnella magna* and *S. neglecta*[20]. Both of these, isolated from their own species at the time of fledging, failed to develop the species-specific song, but acquired songs from other species which they could hear. A *neglecta* male for example, learned songs of *Contopus virens* and *Geothylpis trichas*, and males of both species copied *Agelaius phoeniceus* songs. It became clear that each species is capable of learning the other's song. This learning capacity is undoubtedly exercised in the wild state in learning from conspecifics, but to what extent it is a necessary precursor for development of species-specific song is still uncertain until birds are raised in complete isolation from other species. Lanyon has demonstrated that the difference in size of the repertoire—twice as large in *magna* as in *neglecta*—is inherited, for the same difference appears in his hand-raised birds.

While not revealing much about the mode of development of species-specific song, the experiments of Scott demonstrate how widespread is the tendency for captive birds raised away from their own species to copy songs of other birds within hearing. One species, *Icterus galbula* was raised away from all other birds, and the abnormal song suggests that learning is necessary for normal song development in this species[52]. Birds in a second brood learned the abnormal songs of the first brood. Many species developed unusual songs through imitation of other species, including *Dolychonyx oryzivorus* and *Agelaius phoeniceus*[54], *Pheucticus ludovicianus*[55], and others studied in less detail such as *Turdus migratorius, Sialia sialis, Hyocichla mustelina, Dumetella carolinensis, Toxostoma rufum, Icteria virens, Icterus spurius, Molothrus ater, Quiscalus quiscula, Richmondena cardinalis*[53]. In some cases, at least there is evidence that completely isolated individuals can nevertheless develop the species-specific song, as in the last two species mentioned[21,60]. Heinroth[15] and Poulsen[44] also recorded abnormal songs of isolated *Luscinia megarhyncos, Anthus pratensis,* and *Sylvia curruca* and other species on which further study is needed.

While some of the above experiments have not been conducted under the most strictly controlled conditions, there is a sufficient bulk of evidence to leave us in no doubt that some species of birds need experience of the vocalizations of older members of the species to develop the species-specific song. So far as can be ascertained, birds are unique in this respect, though it is possible that careful study of higher mammals might reveal something similar.

4. Learning from other Species in the Development of Species-specific Vocalizations: Mimicry

The songs of certain species of birds include in nature a proportion of imitations of other species. At times the incidence of imitation is high enough that we can speak of it as a normal element in the species-specific song. In *Turdus ericetorum* for example, "imitations of call-notes and simple phrases of other species are not uncommon" in the song[67], and the same could be said of *Toxostoma rufum*[49]. The song of *Acrocephalus palustris* is "mimetic to a remarkable extent"[67], and in *Sturnus vulgaris*, "imitations of other birds and animals and even artificial sounds are often incorporated"[67]. *Mimus polyglottus* is another classic example, and much of the song is made up of imitations of other birds, and not only these, but of other kinds of animal life and of inanimate sounds. These are interpolated into the middle of the song, and sometimes songs are largely made up of imitations[49]. *Dumetella carolinensis* is somewhat less accomplished in this respect. To judge from the literature, some of the most accomplished mimics are the bower birds. *Chlamydera maculata* for example, shows "remarkably faithful reproduction of... many other species, as well as the barking of dogs, the noise of cattle breaking through scrub, the noise of a maul striking a splitter's wedge, of sheep walking through fallen dead branches, of Emus crashing through twanging fence wires, and of odd camp sounds". Marshall gives many examples of precise imitations by this and other species of *Ptilonorhynchidae*[33].

There are many problems posed by these mimics which, in the absence of detailed studies, we cannot answer. If the imitations are actually part of the species-specific song, then learning is obviously necessary for its development. However, there is a strong suggestion that the vocal inventiveness of an isolated *Mimus polyglottus* for example, suffices for the development of much of the species-specific quality of normal song, at least to the human ear[22].

It may be asked whether we can really speak of species-specificity in the songs of mimics. However, in many cases the imitations are colored by modifications which add a species-specific quality. While in *Mimus polyglottus*, "individual notes, calls or phrases of another's song may be produced with great fidelity, the mimic does not go so far as to reproduce the method of the song, its time relations, or its structures"[1]. *Sturnus vulgaris* adds its peculiar shrill or squeaky quality, either to the imitations, or the other notes interposed between them. The song of *Acrocephalus palustris*, which is full of imitations, includes a characteristic, nasal call, "never absent from the song if of any duration"[67]. And we should note a tendency for observers to overestimate the proportion of imitations in the songs of mimics. Miller suggests that on the average imitations only comprise ten percent of most mockingbird song[37].

5. Special Cases of Vocal Imitation of Alien Sounds by Captive Animals

The capacity of certain birds and mammals to imitate alien sounds, and especially human speech, has received much attention from psychologists. The work has been well reviewed by Bierens de Haan[2], Mowrer[38] and Thorpe[63], and since, to the writer's knowledge, there have been no recent studies in this direction, it will only be briefly mentioned here.

The custom of teaching birds to imitate tunes on musical instruments was common in Europe in the eighteenth century and special scores were published for this purpose[13]. Many species of song bird were used, particularly *Pyrrhula pyrrhula*, which is very accomplished in this respect. The ability to learn human speech is also well known, particularly in parrots and there are well documented cases of large vocabularies[2,38,63].

No sub-human mammal has been found which approaches these birds in its capacity for imitating human speech. Some dogs are said to imitate a few words. After very exhaustive experimentation, Furness was able to induce an orangutan to say a few words[12]. Similar attempts with a chimpanzee failed[12,69]. Yerkes' conclusions is that "despite the possession of a vocal mechanism that closely resembles the human, and a tendency to produce sounds which vary greatly in quality and intensity, the chimpanzee has surprisingly little tendency to reproduce others of the sounds which it hears than those characteristic of the species"[69]. Thus the capacity to imitate is clearly subject to control by inheritance, and it is no accident that the best ability to imitate alien sounds is found in the one group where learning plays a significant role in normal, species-specific development, namely the birds.

6. Characteristics of the Learning Process in Vocalization Development

(a) *The disposition to learn*

There are several aspects of the learning processes involved in the development of vocalization worthy of special mention. The first of these is the control of inheritance over the "disposition to learn"[63]. We have seen how some bird songs, like the majority of animal sounds, are inherited. Others are learned, but the disposition to modify song by learning from others is not restricted to these examples, and may also impose effects on inherited songs. Thus genetic control over the readiness to learn is, to some extent distinct from control over development of species-specific vocalization as such.

(b) Control over which sounds are imitated

In some birds there is also control over what sounds the individual will imitate, and there seem to be several mechanisms for this. The first possibility is limitation imposed by the structure of the vocal organs. We know so little about these at present that the effects cannot be assessed, but the broad capabilities of those birds which will imitate alien sounds, not accompanied by any obvious structural specializations of the syrinx, suggest that syringeal structure does not control the details of what can and cannot be imitated.

In *Fringilla coelebs* there seems to be an inherited restriction of imitation to sounds of a certain pattern. By playing songs of various kinds on an endless tape to hand-raised birds in sound-proof rooms, Thorpe[64] has been able to induce; first, normal imitation of a typical chaffinch song; second, a degree of imitation of a reversed chaffinch song, and of a song of *Anthus trivialis* which "is probably tonally nearer to the chaffinch than that of any other British bird"; third, a failure to imitate various artificial songs with abnormal tonal quality. Thorpe's conclusion is that there is an inborn restriction of imitation to sounds of a certain duration and more especially with a certain tonal quality, which might be learned through experience of its own voice.

It is important to note that in the chaffinch, the control of the sounds to be imitated is imposed directly. In *Pyrrhula pyrrhula* Nicolai has found another mechanism[42]. This species only learns sounds given by a bird (or in special circumstances a mammal) with which an emotional bond is established in youth. In nature the young male becomes attached in this way to his father and learns his song and calls. If a man is adopted as the parent figure, the young male will learn human sounds. In this way a male abnormal both in song and in his social call may result. If this bird now breeds, his sons will learn his abnormal song and call, even though they may be able to hear normal sounds from other adult males at the same time. In this way Nicolai has been able to perpetuate abnormal songs for several generations, with no reversal to the species-specific type. Thus, here there is no direct control over what sounds will be imitated, although in nature the restriction of learning to sounds of the father will achieve the same effect.

(c) Sensitive periods

The characteristic mentioned above is shared with other cases of what Lorenz has called imprinting[26]. Another property is the restriction of learning to a certain period. While in some birds such as *Turdus merula*[34] and *Mimus polyglottus*[22], song learning may continue throughout life, in others such as *Fringilla coelebs*[64] and *Carduelis cannabina*[46], it is restricted to a certain sensitive period. In *Fringilla coelebs* this period lasts in nature for about the first ten months of life, is "very indefinite at onset, but culminating in a short period of intense experience and rapid learning brought suddenly to a close"[64]. The physiological mechanism which terminates the learning period is unknown. Subsequent to this time the song remains constant for life, with certain rare exceptions among experimental birds[64]. In *Sturnella neglecta* the "critical period of song learning appeared to fall between the start of subsong, at about four weeks of age, and the first winter"[14].

(d) The nature of the learning process

Thorpe has pointed out that the processes of vocal learning do not fit neatly into any of the usual categories of learning[63], and Mowrer emphasizes the same point, both with regard to learning of alien sounds by birds and also with reference to human speech[38,29]. In particular, the nature of the reward, or reinforcement, is in some doubt. In common with the other cases of "imprinting" the performance of the act of vocalizing appears to be, in a sense, self rewarding. Mowrer suggests that in some cases parental sounds are imitated because they "sound good", through previous connotations acquired during parental care. This is consistent with the need for a human experimenter to establish an intimate social relationship with such a bird as a parrot, before it can be taught to speak. At least one case of song learning in nature, in *Pyrrhula pyrrhula* would fit into this scheme[41]. However, other examples are less consistent, since learning may take place without any other social contact than that provided by the sound itself. A chaffinch isolated in a sound-proof room will learn songs from a loudspeaker[64]. The occurrence of latent learning, with sounds first imitated after a considerable lapse of time since they were last heard, is also difficult to fit into such a theory. Although auditory feedback is clearly concerned in reinforcement, there may be more than one method by which it can be involved, either in matching vocalization to any sounds which the animal hears and can produce, as in some natural mimics; in matching with sounds having acquired rewarding connotations, as Mowrer suggests; in matching sounds produced by a particular individual with whom a special relationship has been established; or by matching sounds for which the animal has some kind of inherited auditory recognition without a corresponding inherited motor coordination, as seems to be true in *Fringilla coelebs*. It will be difficult in any given case, to study the learning process in sufficiently quantitative terms to decide which type of reinforcement is effective in a given case, and we should be prepared for the possibility that more than one may operate in the same animal.

[For Addendum to this Chapter, see p. 794]

REFERENCES

1 ALLARD, H. A., 1939, Vocal mimicry of starling and mockingbird, *Science*, 90, 370–371.
2 BIERENS DE HAAN, J. A., 1929, Animal language in relation to that of men, *Biol. Rev.*, 4, 249–268.
3 BLAIR, W. F., 1955, Mating call and stage of speciation in the *Microhyla olivacea M. carolinensis* complex, *Evolution*, 9, 469–480.
4 BLAIR, W. F., 1956, The mating calls of hybrid toads, *Tex. J. Sci.*, 8, 350–355.
5 BLAIR, W. F., 1958, Mating vall in the speciation of Anuran amphibians, *Am. Nat.*, 92, 27–51.
6 BOUTAN, L., 1913, Le Pseudo-language, Observations effectuées sur un anthropoide: le Gibbon (*Hylobates leucogenys*-Ogilby), *Act. Soc. Linn. Bordeaux*, 47, 5–81.
7 CRAIG, W., 1908, The voice of pigeons regarded as a means of social control, *Am. J. Sociol.*, 14, 66–100.
8 CRAIG, W., 1941, Male doves raised in isolation, *J. Anim. Beh.*, 4, 121–133.
9 CURIO, E., 1959, Verhaltensstudien am Trauerschnäpper, *Z. f. Tierpsychol.*, Suppl. 3, 1–118.
10 EWER, R. F., 1957, Ethological concepts, *Science*, 126, 599–603.
11 FULTON, B. B., 1933, Inheritance of song in hybrids of two subspecies of *Nemobius fasciatus* (Orthoptera), *Ann. Entom. Soc. Am.*, 26, 368–376.

[12] Furness, W. H., 1916, Observations on the mentality of chimpanzees and orangutans, *Proc. Amer. Phil. Soc. Philadelphia*, *55*, 240–246.
[13] Godman, S., 1955, The bird fancyer's delight, *Ibis*, *97*, 240–246.
[14] Haartman, L. von, and H. Löhrl, 1956, Die Lautäusserungen des Trauer- und Halsbandfliegenschnäppers, *Muscicapa h. hypoleuca* (Pall.) und *M. a. albicollis* Temminck, *Ornis Fenn.*, *17*, 85–97.
[15] Heinroth, O., 1924, Lautäusserungen der Vögel, *J. Ornith.*, *72*, 223–244.
[16] Heinroth, O. and M., 1924–1926, *Die Vögel Mitteleuropas*, Berlin.
[17] Hörmann-Heck, S. von, 1957, Untersuchungen über den Erbgang einiger Verhaltensweisen bei Grillenbastarden, *Gryllus campestris* L. and *G. bimaculatus* De Greer, *Z. f. Tierpsychol.*, *14*, 137–183.
[18] Hüchtker R. and J. Schwartzkopff, 1958, Soziale Verhaltensweisen bei hörenden und gehörlosen Dompfaffen (*Pyrrhula pyrrhula* L.), *Experientia*, *15*, 106.
[19] Jacobs, W., 1953, Verhaltensbiologische Studien an Feldheuschrecken, *Z. f. Tierpsychol.*, Suppl. 1.
[20] Lanyon, W. E., 1957, The comparative biology of the meadowlarks (*Sturnella*) in Wisconsin., *Publ. Nuttall. Ornith. Club*, *1*, 1–67.
[21] Laskey, A. R., 1937, Notes on the song of immature birds, *Migrant*, *8*, 67–68.
[22] Laskey, A. R., 1944, A mockingbird acquires his song repertoire, *Auk.*, *61*, 211–219.
[23] Levi, H., 1951, *The Pigeon*, Columbia S. C.
[24] Loher, W., 1957, Untersuchungen über den Aufbau und die Entstehung der Gesänge einiger Feldheuschreckenarten und den Einfluss von Lautzeichnen auf das akustische Verhalten, *Z. vergl. Physiol.*, *39*, 313–356.
[25] Löhrl, H., 1955, Beziehungen zwischen Halsband- und Trauerfliegen-schnäpper (*Muscicapa albicollis* and *M. hypoleuca*) in demselben Brutgebiet, *Proc. 11th Int. Ornith. Congr.*, 1954, Basel, 333–336.
[26] Lorenz, K., 1935, Der Kumpan in der Umwelt des Vogels, *J. Ornith.*, *83*, 137–213; 289–413.
[27] Marler, P., 1956, The voice of the chaffinch, *New Biology*, *20*, 70–87.
[28] Marler, P., 1956, Behaviour of the chaffinch, *Behaviour*, Suppl. 5, 1–178.
[29] Marler, P., 1956, The voice of the chaffinch and its function as a language, *Ibis*, *98*, 231–261.
[30] Marler, P., 1959, Developments in the study of animal communication, in *Developments since Darwin*, ed. by P. Bell, Cambridge, England.
[31] Marler, P., unpublished, The voice of hand-raised house finches, *Carpodacus mexicanus*.
[32] Marler, P., M. Kreith and E. Willis, 1962, An analysis of testosterone-induced crowing in young domestic cockerels, *Animal Behaviour 10*, 48–54.
[33] Marshall, A. J., 1954, *Bower birds*, Oxford.
[34] Messmer, F. and E., 1956, Die Entwicklung der Lautäusserungen und einiger Verhaltensweisen der Amsel (*Turdus merula merula* L.) unter natürlichen Bedingungen und nach Einzelaufzucht in schalldichten Raümen, *Z. f. Tierpsychol.*, *13*, 341–441.
[35] Metfessel, M., 1945, Roller canary song produced without learning from external sources, *Science*, *81*, 470.
[36] Miller, L., 1921, The biography of Nip and Tuck, *Condor*, *23*, 41–47.
[37] Miller, L., 1938, The singing of the mockingbird, *Condor.*, *40*, 216–219.
[38] Mowrer, O. H., 1950, Psychology of talking birds, in *Learning theory and personality dynamics*, New York.
[39] Mowrer, O. H., 1958, Hearing and speaking: an analysis of language learning, *J. Speech and Hearing Dis.*, *23*, 143–152.
[40] Nice, M. M., 1943, Studies in the life history of the song sparrow, II Behavior of the song sparrow and other passerines, *Trans. Linn. Soc. N. Y.*, *6*, 1–328.
[41] Nicolai, J., 1956, Zur Biologie und Ethologie des Gimpels (*Pyrrhula pyrrhula* L.), *Z. f. Tierpsychol.*, *13*, 93–132.
[42] Nicolai, J., 1959, Familientradition in der Gesangentwicklung des Gimpels (*Pyrrhula pyrrhula* L.), *J. Ornith.*, *100*, 39–46.
[43] Perdeck, A. C., 1957, The isolating value of specific song patterns in two sibling species of grasshoppers (*Chorthippus brunneus* Thunb. and *C. biguttulus* L.), *Behaviour*, *12*, 1–75.
[44] Poulsen, H., 1951, Inheritance and learning in the song of the chaffinch (*Fringilla coelebs* L.), *Behaviour*, *3*, 216–228.
[45] Poulsen, H., quoted by W. H. Thorpe, 1951, *Ibis*, *93*, 252–296.
[46] Poulsen, H., 1954, On the song of the linnet (*Carduelis cannabina* L.), *Dansk. Ornith. Foren. Tidsskr.*, *48*, 32–37.
[47] Promproff, A. N., and E. V. Lukina, 1945, Conditioned reflectory differentiation of calls in Passeres and its biological value, *Compt. rend. Acad. Sci. USSR*, *46*, 382–384.

[48] SAUER, F., 1954, Die Entwicklung der Lautäusserungen vom Ei ab schalldicht gehaltener Dorngrassmücken (Sylvia c. *communis* Latham) im Vergleich mit später isolierten und mit wildlebenden Artgenossen, *Z. f. Tierpsychol.*, *11*, 10–92.
[49] SAUNDERS, A. A. 1935, *A guide to bird songs*, New York.
[50] SCHJELDERUP-EBBE, T., 1923, Weitere Beiträge zur Sozial- und Individualpsychologie des Haushuhns, *Z. Psychol.*, *92*, 60–87.
[51] SCHWARTZKOPFF, J., 1949, Über Sitz und Leistung von Gehör und Vibrationssinn bei Vögeln, *Z. vergl. Physiol.*, *31*, 527–608.
[52] SCOTT, W. E. D., 1901, Data on song in birds. Observations on the song of Baltimore Orioles in captivity, *Science*, *14*, 522–526.
[53] SCOTT, W. E. D., 1902, Data on song in birds: the acquisition of new songs, *Science*, *15*, 178–181.
[54] SCOTT, W. E. D., 1904, The inheritance of song in passerine birds. Remarks and observations on the song of hand-raised bobolinks and red-winged blackbirds (*Dolichonyx oryzivorus* and *Agelaius phoeniceus*), *Science*, *19*, 154.
[55] SCOTT, W. E. D., 1904, Remarks on the development of song in the red-breasted grosbeak, *Zamelodia ludoviciana* (Linn.) and the meadowlark, *Sturnella magna* (Linn.), *Science*, *19*, 957–959.
[56] SEITZ, A., 1955, Untersuchungen über angeborene Verhaltensweisen bei Caniden, III Beobachtungen an Marderhunden (*Nyctereutes procyonoides* Gray), *Z. f. Tierpsychol.*, *12*, 463–489.
[57] SEVERAID, J. H., 1958, The natural history of the pikas (Mammalian genus *Ochotona*), Diss. Ph. D. Univ. Calif., Berkeley.
[58] STADLER, H., 1929, Vogelstimmenforschung als Wissenschaft, *Proc. 6th Int. Ornith. Congr.*, 1926, Kopenhagen 338–357.
[59] STRESEMANN, E., 1947, Baron von Pernau, pioneer student of bird behavior, *Auk.*, *64*, 35–52.
[60] SUTTON, G. M., quoted in M. M. NICE, 1943, *Proc. Linn. Soc. N. Y.*, *6*, 143.
[61] THORPE, W. H., 1954, The process of song learning in the chaffinch as studied by means of the sound spectrograph, *Nature*, *173*, 465–467.
[62] THORPE, W. H., 1955, Comments on "the bird fancyer's delight": together with notes on imitation in the subsong of the chaffinch, *Ibis*, *97*, 247–251.
[63] THORPE, W. H., 1956, *Learning and instinct in animals*, London.
[64] THORPE, W. H., 1958, The learning of song patterns by birds with especial reference to the song of the chaffinch, *Fringilla coelebs*, *Ibis*, *100*, 535–570.
[65] WEIH, A. S., 1951, Untersuchungen über das Wechselsingen (Anaphonie) und über das angeborene Lautschema einiger Feldheuschrecken, *Z. f. Tierpsychol.*, *8*, 1–41.
[66] WEISS, G., 1952, Beobachtungen an zwei isoliert aufgezogenen Hauskatzen, *Z. f. Tierpsychol.*, *9*, 451–462.
[67] WITHERBY, H. F., F. C. R. JOURDAIN, N. F. TICEHURST and B. W. TUCKER, 1944, *The handbook of British birds*, London.
[68] YERKES, R. M., and B. W. LEARNED, 1925, *Chimpanzee intelligence and its vocal expression*, Baltimore.
[69] YERKES, R. M., and A. W. YERKES, 1929, *The great apes*, New Haven, Conn.

CHAPTER 10

A STUDY OF THE AUDIOGENIC SEIZURE

by

A. LEHMANN and R. G. BUSNEL

1. Definition and Description of the Phenomenon

The audiogenic seizure is a disturbance in behaviour consisting of a set of psychomotor reactions which may be induced in certain animals, particularly rodents, by an acoustic stimulus of relatively high intensity.

A description of the seizure as it occurs in the mouse is given below.

(a) *Phases of the phenomenon*

When the animal is placed in an enclosed space it shows reactions of exploration, and then remains more or less immobile. The following phases are observed after acoustic stimulation.

Phase 1. Immediately after the acoustic emission, the animal gives a jump (reaction of attention), then moves for a while in a normal manner around the cage, with its ears pricked up. This is the so-called latent period which may last for a fraction of a second up to one or two minutes.

Phase 2. The animal suddenly begins to run in all directions, within the cage whatever its size may be[238]. This desperate running is extremely rapid and violent. It lasts for several seconds, and persists even after the stimulus has ceased. It may stop before the end of the signal. This period, known as the "running fit", is characteristic; it does not occur in any other convulsive seizures, such as those induced by C.N.S. stimulants of the strychnine type, picrotoxin, amphetamine or oxygen under pressure.

Phase 3. (a) Phase 2 stops as abruptly as it started. This may be the only manifestation observed, the animal immediately resuming its normal behaviour.

(b) In most cases, however, the animal reaches the convulsion stage. It falls violently onto its side, its whole body being seized with clonic convulsions. This convulsive stage lasts for a few seconds and may have two results: either, in the case of a "clonic" seizure only, the animal gets up again and resumes its normal activity.

(c) Or, in certain animals, the clonic seizure is followed by spectacular manifestations. The animal, which is lying on its side having convulsions, ceases to move, then the back and front feet are slowly and completely extended and the body becomes stiff and insensible. The eyes are shut and the respiratory and cardiac rhythms are modified. This is called the "clonic-tonic" phase. It lasts for a few seconds (Fig. 100).

Phase 4. There are now two possibilities: (a) so-called "recovery": at the end of phase 3c the animal, which is still stiff and on its side, micturates, then slowly opens its eyes, which gradually exhibit exophthalmia; its body relaxes, is agitated by slight clonic convulsions and respiration resumes its normal rhythm. The animal finally gets up and resumes normal behaviour. This phase lasts from 10 to 15 seconds.

(b) The animal suddenly becomes limp because it has died from anoxia. It is sometimes possible to avoid death if thoracic massage is applied in time.

Note. Phase 3 and 4 are, to a certain extent, independent of the duration of the

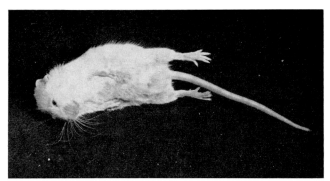

Fig. 100. Mouse during the clonic-tonic seizure.

stimulus, *i.e.* once phase 2 has started, the cycle of the seizure takes place and it is impossible to stop it, even if the stimulus is suppressed.

The phenomenon is summarized in Fig. 102 and in the records in Fig. 101. Fig. 103 shows an experimental lay-out used for tests of this type, and Figs. 104 and 105 show examples of actual actographic records[148].

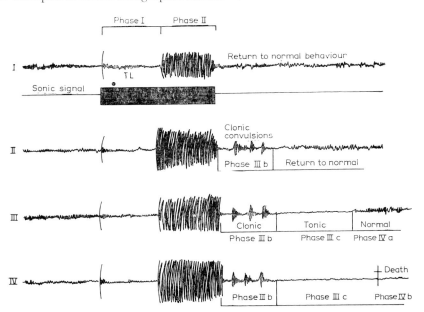

Fig. 101. Oscillographic records of the various phases of the audiogenic seizure in the mouse. Time scale: 2 sec = 1 cm.

(b) *Animal species in which the phenomenon has been observed*

The phenomenon has always been observed in special strains of the following species: rat, mouse, rabbit, chicken, dog (according to Ginsburg[119]) and goat (Lush).

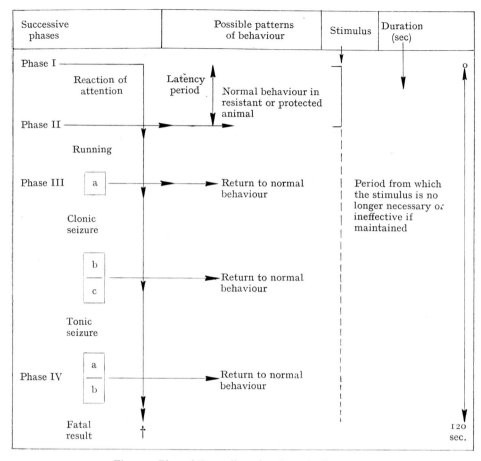

Fig. 102. Plan of the audiogenic seizure in the mouse.

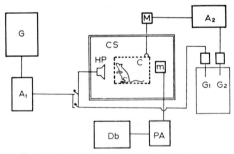

Fig. 103. Drawing showing the experimental arrangement.

A_1 and A_2 = Amplifiers
C   = Cage for animal
CS   = Sound-proof room
Db   = Decibelmeter
G   = Frequency generator
G_1 and G_2 = Recording fixed pen galvano-
           meters moving on an uncoiling
           sheet

HP = Electrostatic loud-speaker
M = Piezoelectric microphone picking up
     mechanical vibrations from the ani-
     mal's cage
m = Decibelmeter microphone
PA = Pre-amplifier

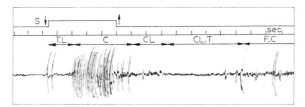

Fig. 104. Actogram of a clonic-tonic seizure.

Upper Record	= Acoustic signal	C	= Running phase
Middle Record	= Time scale in seconds	CL	= Clonic phase
Lower Record	= Actogram	CL.T	= Clonic-tonic phase
T.L.	= Latent period	F.C.	= End of seizure

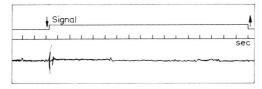

Fig. 105. Actogram of the behaviour of a resistant mouse or one protected by a tranquillizer.

Generally speaking, the seizure takes place in these different species according to the scheme given above. There are various modifications in certain strains or species, particularly in the so-called "running fit", which, in certain strains of rats, for example, is accompanied by violent leaps, and, sometimes, following frequently repeated tests, by uncoordinated movements of different parts of the body (myoclonus), to which Soviet authors have given the name "tics"[171, 175].

2. Brief Historical Review

The first mention of the phenomenon dates from 1914; it was described in the rat by Donaldson[72] as a reaction to the noise made by a bunch of keys. Vassiliev[324] observed it in mice in 1924, and Nachtsheim[262] in the rabbit in 1941.

There have been numerous studies on this phenomenon in various American schools of psychology and physiology, especially since 1937, on the rat by Bevan[18-24], Farris and Yeakel[73-78], Finger[79-83], Griffiths[128-139], Maier[193-219], Morgan[250-255] and Patton[266-284] and since 1947 on the mouse, by Chance[55-59], Frings[85-98], Fuller[99-105], Ginsburg[111-119] and Vicari[327-336]. There are at the moment about 360 publications which are almost entirely of American origin.

Krushinski[170-181] has been studying this phenomenon on the rat since 1949 in the U.S.S.R.

In France there are only two groups of work to be mentioned, one on the rat by Cain, Mercier and Canac (1948-1953)[49-54, 228-237], the other on the mouse by Busnel and Lehmann (1953-1959)[47, 147, 148, 184-188].

The phenomenon was first called reflex epilepsy due to sound[324], then audiogenic seizure[251], audioepileptic seizure[311], and sonogenic convulsions[244], the latter being the exact name from the etymological point of view. "Audiogenic seizure" is the term generally used.

References p. 262

The reason for the importance given to this phenomenon by research workers, is due to its resemblance to human epilepsy and the various pathological states which may accompany it (keratitis, nystagmus, catatonic states, myoclonus), leading sometimes to death by cerebral haemorrhages. In addition, it has been used during the past few years as one of the pharmacodynamic techniques for the study of "tranquillizers" and monoaminoxidase inhibitors and in this respect it may be useful for testing chemical series of these compounds. Finally, it is interesting to stress that certain problems of the biochemistry of the brain relating to the disturbances in behaviour caused by epilepsy may be approached by studies on the audiogenic seizure and research work on these lines promises interesting developments.

3. Physical Characteristics of the Inducing Acoustic Stimulus

Complex sounds may be used, for example those produced by jingling a bunch of keys, a Galton Whistle, an electric bell, a siren, or pure tones of a known frequency produced by a generator and transmitted by a loudspeaker. The necessary intensities are between 90 and 134 dB. According to different authors[9, 10, 70, 122, 125, 250, 254, 259, 292] the frequency band extends from 4 to 80 kc/s, depending on the species.

The effective intensities may vary with the frequency of the sounds. Moreover, the percentage of seizures at the different frequencies in the rat and mouse seems to follow the audiogram curves of these animals, to a certain extent[92, 107].

The latent period and the intensity of a seizure caused by a stimulation resulting from a continuous or non-continuous sound are similar, providing the intervals do not exceed $\frac{1}{2}$–1 second[104, 107].

Nevertheless, although stimuli repeated every 2 minutes can cause successive seizures, a continuous sound of even much longer duration does not cause a succession of seizures.

Krushinski et al.[175, 178, 181] have likewise used stimulations alternating in intensity and duration under the following conditions:

Loud, continuous ringing of a bell (white noise) for 90 sec, followed by an alternation every 10 sec for 15 min (i.e. silence – loud noise, silence—weak noise) and then, from the 16th to the 19th min, silence. The cycle is then concluded by 90 sec of loud noise.

In these tests the authors obtained a special convulsive behaviour which develops throughout the stimulation, causing a parabiotic state in the animal, in the pavlovian sense of the term* frequently ending in a fatal cerebral haemorrhage (Fig. 106).

4. Genetic Aspect of the Problem and its Relations to Senescence

Only certain strains or sub-strains of animals show this psychomotor reaction. Its genetic aspect has been studied particularly in the mouse. The following list shows the strains in this species amongst which susceptible sub-strains are found.

D B A	A/jax
C 57	*Mus musculus*
Swiss albinos	*Peromyscus*
C E	BUE/Wi
C_3H	

---

* For this use of the term, see: *The Central Nervous System and Behaviour*, edited by M. A. B. Brazier, sponsored by Josiah Macy Jr Foundation, New York, 1958, p. 159.

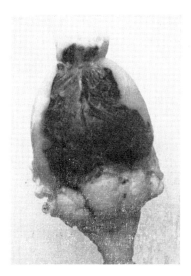
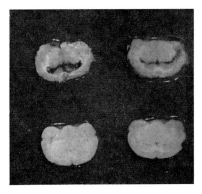
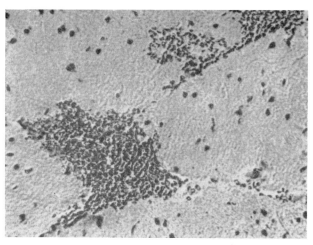

Fig. 106. Cerebral haemorrhages. (1) Brain seen from above, (2) Transversal sections of the brain, top: haemorrhages; bottom: normal brain, (3) Microphotograph of a section of the cortex showing extravasations of blood. (After Krushinski[175,181].)

They are almost all kept in the U.S.A., at the Roscoe B. Jackson Memorial Laboratory at Bar-Harbor, Maine.

At first it was thought that the susceptibility of the mouse was due to a single dominant gene[112, 144, 331, 340]. It now appears that there are several more or less dominant genes[87, 100, 113].

A recent study has been made on the rat showing that the genetic origin of the seizure is complex, and that it depends on several genes[181], as in the mouse.

In the case of the rabbit, the sensitive sub-strains are Vienna white and EcAp[11, 262].

The character or characters are not related to sex[74, 201]. However, certain agents, such as glutamic acid, have an action on the seizure varying with the sex of the animals[105]. Moreover male rats are more likely to die from cerebral haemorrhages than female rats.

Frings has established three sub-trains in the mouse by extensive breeding and narrow selection, one of which is resistant, the second undergoing clonic seizures only, and the third undergoing clonic-tonic seizures[90, 95, 96]. For the past eight years, carrying out similar breedings from these strains in our laboratory with 5–6 generations a year, we have found that too strict an inbreeding brings about such an increase of the sensitivity that all the animals die during the seizure. By maintaining a relative consanguinity, and keeping only the most sensitive animals, strains of mice are obtained of which 95% regularly undergo clonic-tonic seizures and whose initial latent period of 10–30 sec becomes 0.5–8 sec.

This special hereditary aspect of the phenomenon is therefore important; it indicates that it is probably caused by modifications of physicochemical factors.

It is interesting to note that it has been found, only quite recently, that the so-called "resistant" mice, using Frings'[94] terminology, have, in fact, only a reduced sensitivity to the sonic stimulus. Frings[85] usually used sounds of 10–12 kc/s, with an intensity of 100 dB, and Anthony[9] brought on the seizure in 40% of these so-called resistant mice by using 134 dB at 40 kc/s.

Influence of age

In the rat, as in the mouse, the age factor is important. The number of reactions reaches a maximum in the rat at 3 weeks and becomes practically nil at the 20th month[73, 201].

In the mouse, the seizure can only be brought on in animals having a minimum age of 12–15 days, which is undoubtedly related to a certain level of development of the endocrine, enzymatic and neuro-muscular systems. The period of maximum sensitivity is approximately at the 30th day. It then decreases from the 50th day to about 1 year[34, 41, 93, 96, 331]. As their sensitivity decreases, the animals which had clonic-tonic seizures have clonic seizures, then cease to react. At the same time their latent period increases progressively[93].

5. Role of the Ear in the Seizure

An animal whose outer ears are closed does not undergo seizures. In the rat it has been shown that sensitivity is increased when it has a purulent otitis media, which is perhaps related to a lowering of the auditory threshold of pain[167, 224, 276, 281, 284]. This likewise partly explains the inequality and frequent instability of responses in breeds of rats, even in those which have been selected.

Otitis does not occur in the mouse[88, 242], nevertheless 95% of the animals are found to react.

An attempt has been made to determine the role played by the vestibular function in the rat by injecting streptomycin. Seizures occur nevertheless, and so it would seem that excitation at the level of the vestibule is not directly related to the reaction[83].

6. Psychological Studies

American psychologists have attempted to compare the audiogenic seizure with the behaviour patterns observed in situations of conflict, and have interpreted the convulsive phase as being a situation of escape, which does not explain very much and hardly advances the problem[193–196, 203–217]. They have likewise observed that the seizures do not substantially modify the responses of the animal to a certain number of tests, such as brilliance, maze and classical patterns of learning[57, 305, 306, 309, 316].

All the tests relating to a conditioning of association (conditioned reflex) of the seizure to a visual or electrical stimulus, have been negative[31, 64, 123, 304].

Tests have been carried out to establish a relation between the sensitivity to the seizure and the emotionality of the animal[25, 78, 131, 146, 190]. The results obtained have been quite variable. On the other hand, when the mouse can be placed in a hut (non-soundproof) the number of seizures decreases[57].

External motivations, or a strict restraint, decrease the frequency of the seizures and can even abolish them completely[39, 40, 52, 53, 92, 138, 238].

Also, the fact that mice are reared or tested single or in a group can lead to a modification in the frequency of the seizures[168, 291]. The phenomenon may therefore be considered as relatively delicate, and so quite rigorous conditions are necessary for its comprehensive study.

7. Effect of Temperature

According to BURES[44], the animals no longer react to the sonic stimulation if their body temperature is reduced to 25 °C.

8. Relations between the Seizure and the Time of Stimulation

Laboratory animals are generally studied during the day, and observations have not often been carried out at night. Particularly in the case of rodents, numerous activities are nocturnal and relatively recent tests carried out at night have revealed new aspects of certain problems.

This has been the case for the audiogenic seizure in the mouse. The greatest susceptibility of this animal is nocturnal, between 8 and 10 p.m., when the rectal temperature is highest and the amount of blood eosinophils is lowest[142].

If the natural light cycle is reversed by means of artificial illumination the susceptibility of the animals displaced. Maximum susceptibility is reached several hours after illumination, corresponding to the normal diurnal physiological rhythm, it is therefore changed in phase[143].

References p. 262

9. Clinical Picture of the Physiological Repercussions of the Seizure

Numerous analyses have been carried out in this field. The clinical picture which results is as follows:

According to certain authors[192, 227, 342] *the arterial pressure* increases in the rat.

Stechenko on the contrary has found a significant fall in the arterial pressure[318]. This fall is all the more significant when the initial level of the pressure is high and the predisposition to cerebral haemorrhages seems greater in animals with a higher initial arterial pressure.

The cardiac rhythm is greatly disturbed. Initially accelerated, it falls, then increases again. These alterations in rhythm are accompanied by a modification of the amplitude. During the period of coma and tetany in the rat, the rhythm becomes very slow and the E.C.G. is modified; the QRS waves possibly disappear completely, but the P wave always remains; this indicates interference with transmission of the impulse from the auricle to the ventricle[120, 138, 189, 221, 222, 343].

Blood. The time of coagulation is reduced from 114 to 35 seconds after the seizure and the CO_2 in the blood also decreases in a significant manner from 14,5 to 8.1 millimols/litre. On the other hand, the quantity of phosphorus and potassium in the plasma does not change appreciably.

Sensitive rats normally have a slightly higher amount of serum proteins than resistant rats, while the quantity of glucose is almost the same. However, if the quantitative variations of these substances are studied for a month, much greater variations are found in sensitive rats than in rats which are resistant to the seizure[63, 182]. Electrophoresis of the plasma of sensitive and resistant mice has shown that sensitive strains exhibit a value for, and a distribution of the protidic fractions differing in frailty and quality from those of resistant strains of mice. In sensitive strains, the albumin fraction is higher and the γ-globulin fraction is much reduced (50 %). An increase in the α-globulins and a division into α-1 and α-2 is found in blood removed immediately after the seizure, as in all cases of stress or shock. However, the almost complete breakdown of the γ-globulins half an hour after the seizure[186] (Figs. 107 and 108) is specific to the audiogenic seizure.

A slight eosinopenia sets in three hours after the seizure in mice, and recovery occurs within 24 hours. It should be noted that the resistant animals submitted to the sonic stimulation have a similar, but less accentuated, reaction[6, 7].

The pH of the blood seems to be particularly important. Acidosis and a decrease of the total CO_2 occurs after the seizure in the rat. The pH drops from 7.40 to 6.96[63]. Alkalosis produced artificially by sodium bicarbonate injection, increases the number of seizures, while acidosis produced by acetazolamide injection or by inhaling CO_2, reduces the seizures[119, 245, 246], possibly even leading to their total suppression.

Glands. These various changes seem to be the indication of a profound disturbance of the adrenalino-sympathetic function which appears to be confirmed by the fact that the adrenals are hypertrophied at the level of the fasciculated soudanophilic zone after the mice have been exposed to noise[7]. Nevertheless adrenalectomy in the rat, as in the mouse, has given few appreciable results[105, 134, 136], and the injection of ACTH has no influence on the susceptibility of the animals[160, 161]. On the other hand, parathyroidectomy sensitizes the animal[172].

The smooth musculature of the sphincters is affected in the rat as in the mouse.

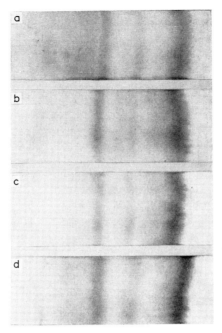

Fig. 107. Electrophorograms of the blood of mice during audiogenic seizures. (a) Normal mouse (b) ¼ h after clonic seizure, (c) Idem after ½ h, (d) Mouse with acquired resistance.

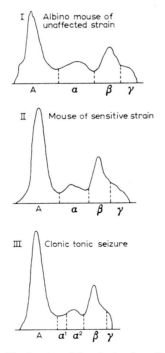

Fig. 108. Planimetry of the electrophoretic results.

At the end of the clonic-tonic phase of the seizure, there is a relaxing of the sphincters with micturition and very often defecation.

Precise measurements may be made of the dilation of the pupil by measuring its diameter with a slot lamp provided with a micrometer eyepiece. The diameter of the pupil is between 0.5 and 1 mm in sensitive mice before sonic stimulation. Immediately after stimulation, when the animal is in the tonic phase of the seizure, the pupil dilates and its diameter reaches 3–3.5 mm. This dilatation is quite transitory and the diameter of the pupil returns to about 1.5 mm in 30–40 sec., *i.e.*, the time necessary for the animal to resume its normal behaviour. After a few minutes, the diameter is again the same as it was before the seizure*.

Central nervous system. According to certain authors the electro-encephalogram appears to be modified as in the case of electroshock or epilepsy, but it is not certain that these differences are not due to artefacts[43, 138, 189], although other authors do not find the modifications[53, 302]. Semiokhina[302] has noted a synchronisation rhythm in the sub-cortical zone only, during the "running" phase, and slow waves of large amplitude in the cortical and sub-cortical zones during the clonic seizure. She observed a decrease in amplitude and an increase of frequency in both zones during the tonic seizure.

Injections of a small amount of curare or cutting the glossopharyngeal nerve reduce the sensitivity of the animals, although cutting the spinal nerve appears to cause death during the seizure[33, 189].

Histological and cytological structures. It has not been possible to find any cytological relation between the fine cerebral structures and the susceptibility of the different strains, or to find any histological changes attributable to the seizures.

10. Experimental Studies on the Encephalic Centres

As the central origin of the phenomenon seems definite, numerous surgical, biochemical or nutritional tests have been undertaken on the brain.

(a) Surgical methods

Removal of the cerebral lobes does not modify susceptibility or the latent period of the seizure; but the number of seizures may decrease or increase, according to different authors. Decortication or cutting the medulla at the cervical level seems to have no effect[14, 20, 54, 191, 338].

Tests carried out by Koenig (1957) show that interruption of the intercerebral pathway of the auditory canals by bilateral electrocoagulation at the level of the geniculate bodies only causes a very slight attenuation of the frequency and intensity of the seizures (Fig. 109), indicating that the cortex does not have a significant role in bringing on the seizure[169].

According to Kotelar (personal communication) removal of 90% of the cortex does not modify the nature of the seizure in very excitable rats (in the sense given by the author); on the other hand, this destruction renders normally less reactive animals, sensitive. Neither does section of the corpus callosum alter the reaction. In short, the complete removal of one hemisphere (right or left) does not modify the seizure, the

---

* A. LEHMANN, J. MAWAS and R. G. BUSNEL, unpublished studies.

movements of the animal remain bilateral thus showing the sub-cortical origin of the phenomenon.

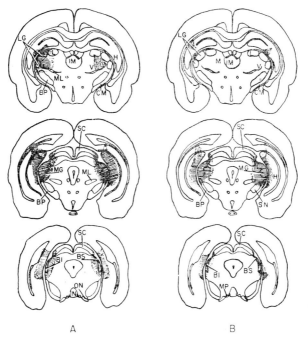

Fig. 109. Transverse sections of the brain of a rat showing three levels of massive (A) or slight (B) destruction undergone by two animals and causing modifications in the intensity and frequency of the seizures. The shaded sections represent primary areas of destruction and the dotted sections areas of slippage. (After Koenig[169].)

BI = Brachium of inferior colliculus
BP = Basis pedunculi
BS = Brachium of superior colliculus
CM = Commissure of Meynert
H = Hippocampus
IM = Intermediate Mass
IN = Interpeduncular nucleus
L = Lateral thalamic nucleus
LG = Lateral geniculate nucleus
M = Medial thalamic nucleus
MG = Medial geniculate nucleus
ML = Medial lemniscus
MP = Mamillary peduncle
ON = Oculomotor nerve root
SC = Superior colliculus
SN = Substantia nigra
V = Ventral thalamic nucleus

(b) Biochemical methods

1. *Direct.* (a) Hydration of the cerebral tissue, either by intraperitoneal injections of isotonic glucose or by intraventricular serum injection, protects the animal, although these treatments increase epileptic seizures and decrease the threshold of convulsions due to cardiazol and to electroshocks[236].

(b) Application of hydrate of alumina cream to the motor and acoustic areas, as well as to the colliculus, renders 33% of the resistant rats sensitive to the audiogenic seizure [303] (Fig. 110).

(c) An increase of glycogen in the brain has been found in the mouse after the seizure[60].

(d) There is a central decrease and a peripheral increase of free acetylcholine during the seizure in the mice of strain DBA[112].

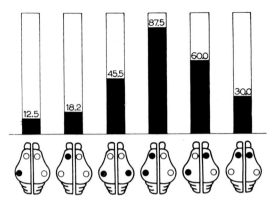

Fig. 110. Audiogenic epilepsy evoked in rats after application of alumina-cream to the cerebral cortex. Top: percentage of audiogenic epileptic seizures, bottom: diagram of position of application of the aluminacream. (After Servit and Sterc[303].)

(e) Sensitive mice of strain DBA, compared with the resistant animals of strain C_{57}, show a decrease in the turnover of high-energy phosphates (P∼) of the brain during their period of sensitivity to the seizure, and therefore a disturbance of the cerebral metabolism, which disappears with the loss of susceptibility. As this biochemical modification vanishes along with age in the DBA strain, it is possibly related to a genetic retardation in the maturity of an enzyme system of the C.N.S., whose nature and localisation are at present unknown[2].

2. *Indirect*. (a) The vitamins, aneurine and pyridoxine, which play a particularly large part in neuro-vegetative mechanisms, are of importance. A deficiency in these vitamins causes a higher sensitivity, while an excess can give protection to the animals[132, 267, 278, 287].

(b) A deficiency in amino acids[19], and of certain trace elements, or of metals, such as magnesium, coupled with certain enzymes, renders the animal more sensitive[133, 183, 273, 277].

(c) Glutamic acid, whose role in the tricarboxylic cycle is well known, has a protective effect against the seizure; it prevents a fatal outcome when added in excess to the diet. The antimetabolites of this amino acid inhibit its protective effect, thus suggesting an enzymatic action[105, 114, 115, 116, 117, 150].

(d) Calcium[179] appears to play an important part in the pathological manifestations following the audiogenic seizure. Limbs of females after lactation and of young animals during the period of growth are frequently fractured during convulsions in the strains of rats studied in the U.S.S.R.; this does not occur if the animals are given a supplementary diet consisting of powered shellfish, which is rich in calcium. The animals are also protected from spinal haemorrhages, which cause paralysis and death after seizures, since these haemorrhages are caused by a fault in certain vertebrae due to decalcification (Fig. 111).

Calcium carbonate and lactate avoid these accidents, although less efficiently than an equal amount of powered shellfish. Even though parathyroidectomy aids the onset of the seizure, it nevertheless guards against spinal haemorrhages.

Recent experiments* have shown that calcium lactate in amounts of 600 mg/kg

---

* R. G. BUSNEL and A. LEHMANN, unpublished work.

corresponding to 39 mg/kg calcium, and calcium gluconate in amounts of 600 mg/kg corresponding to 55.8 mg/kg calcium, satisfactorily protect the mouse against the audiogenic seizure. This protection, which starts 30–60 minutes after the injection, lasts for more than 24 hours and ceases at the end of 48 hours. This action may be due to the well-known sedative effect of calcium on the encephalic centres.

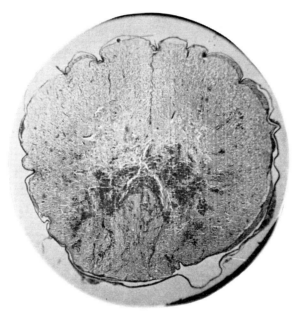

Fig. 111. Effusion of blood in the spinal cord at the end of convulsive seizures due to a sonic stimulation. (After Krushinski and Molodkina[179].)

(c) Various drugs

A large number of drugs promote or inhibit the audiogenic seizure. The following is a list of these substances, which vary in their action with the dose, the species of animals studied, and, sometimes, even with different authors.

In general, it may be stated that, apart from certain exceptions, the convulsants and parasympathicomimetics promote the seizure, while the anti-convulsants, sympathicolytics, sympathicomimetics and sedatives (barbiturates or others) have an inhibiting action on the onset of the seizure or on the violence of the convulsions[3, 12, 35, 48, 50, 51, 56, 62, 68, 71, 86, 97, 111, 112, 127, 135, 158, 159, 163, 166, 176, 199, 228–235, 237, 295, 312, 313, 322, 335, 244]*.

---

* Most of these tests were carried out on the rat and sometimes the results obtained were less significant. Discrepancies in the results may have been due to the choice of the animal, since rats were generally not chosen from selected breeds, but susceptible or resistant animals were chosen from a random batch. The absence of homogeneity of reactions may explain the discordant results of the various authors.

In addition, since a high proportion of rats suffer from otitis, this illness may play a part both in the initial susceptibility of the animals and in its reaction to drugs. Also certain conclusions have been drawn after experimenting on only a small number of animals.

It is certainly possible to carry out more precise experiments with the mouse, since it does not suffer from otitis and since continuous strains exist in which all the animals are susceptible or resistant to a stimulation of well-determined frequency and intensity.

References p. 262

Promoting substances	Inhibiting substances
Caffeine	Ephedrine
Nicotine	Atropine
Eserine	Lactic acid
Dimethylaminoethanol	Succinic acid
Malic acid	Dextrose
Sodium pyruvate	Tridione
α-Ketoglutaric acid	Diphenylhydantoin
α-Methylglutamic acid	Cocaine
Strychnine	Pseudococaine
Metrazol	Procaine
Picrotoxin	Papaverine
Camphor	Ethyl alcohol
Thyroidine	Radioactive I and P
Nicotinic acid hydrazide	Phenobarbital
Isonicotinic acid hydrazide	Sodium Bromide
Anticholinesterases	Benzedrine

(d) Action of tranquillizers

Many of these compounds have been studied with regard to their action on the audiogenic seizure in the mouse[47, 147, 148, 291], and in the rat[294]. Most of them have a protective action against the seizure, without an apparent modification of the normal behaviour of the animal at the effective doses, *i.e.* without a hypnotic action[47]. The following are examples of such substances:

Phenergan
Diparcol
Methyl-3-pentyne-1-ol-3 carbamate
Meprobamate
Benactyzine

However, the mechanism of the action of these products is still not very well understood. Electro-encephalography of the various animals submitted to the action of these drugs appears to show that their action is probably situated at the sub-cortical, hypothalamic or reticulated substance levels. But there is still no certain experimental proof of their mode of action[30].

Apart from the use of animals susceptible to the audiogenic seizure as a means of testing this class of drugs, these studies would not perhaps have helped to advance the problem, if the introduction of a new drug, reserpine, had not led to new developments. It has been established[23, 135, 319] that reserpine increases the intensity of convulsions due to the audiogenic seizure, causing a very high proportion of deaths from the seizure. When reserpine is injected into mice usually subject to clonic seizures, tonic seizures are produced, which become fatal 24 or 48 hours after the injection. Reserpine injected into mice usually subject to non-fatal clonic-tonic seizures, causes death during the seizure in a large number of these animals during the first test carried out 1 hour after the injection and in others during successive tests. Some results obtained are summarized in Tables 20 and 21, showing the effect of reserpine on strains of mice with different susceptibilities to the audiogenic seizure[185]. The doses used in these two cases were very small (1 mg/kg i.p.).

It is known that this drug causes release and breakdown of the biogenic amines in the encephalic centres, in particular noradrenaline and 5-HT, which are normal

TABLE 20

STRAIN USUALLY SUBJECT TO LESS SEVERE CLONIC SEIZURES

Dose of reserpine in mg/kg	Number of mice	Reactions after									
		1 h					24 h			48 h	
		×	=	C	O	†	×	=	†	=	†
0.5	10	10					3	7	5	—	—
1	8	6		2			2	6	6	2	—
2	6	5			1			6	6	—	—
3	12	10			2			12	6	6	6

TABLE 21

STRAIN USUALLY SUBJECT TO CLONIC-TONIC SEIZURES

Dose of reserpine in mg/kg	Number of mice	Reactions after							
		1 h			24 h			48 h	
		×	=	†	×	=	†	=	†
0.5	10		10	5	5	5			
1	10		10	7	3	3			
3	12	4	8	6		6		6	6
4	10		10	10					
5	12		12	12					

Abbreviations used: = clonic-tonic seizure; × clonic seizure; C running (partial protection); O absence of seizure (total protection); † death during seizure.

N.B. The resistant mice stand doses of 10 mg/kg whithout a resultant sensitivity to the sonic stimulus. This effect of reserpine lasts for 3–4 days.

metabolic constituents of the brain[16, 289, 307]. Modification in the concentration of these substances begins 30–60 minutes after the administration of reserpine and lasts for several days, even though, as the injection of reserpine-C^{14} has revealed[321], all the reserpine has disappeared from the brain. The action of this drug on the seizure therefore, may be correlated with the concentrations of the amines in the brain tissues.

Pletscher[290] showed that synthetic products derived from benzoquinolizine caused the same falls of 5-HT and noradrenaline as reserpine in the encephalic centres. Injections of benzoquinolizine derivatives or of reserpine into mice resulted in an identical action on the audiogenic seizure[188], thus confirming the hypothesis that the mode of action of these products on the seizure may be due, in both cases, to the decrease of 5-HT and noradrenaline in the brain. Subsequent tests have confirmed the important part played by these amines, and in particular 5-HT on the intensity of the convulsion phenomena of the seizure. Injecting mice with 100 mg/kg iproniazid, a monoaminoxidase inhibitor, for three days, resulted in an accumulation of 5-HT and noradrenaline in the brain[323]. In this case, as the Table 22 shows, a high proportion of mice were protected, this proportion being very much more significant in mice over 100 days old than in younger animals. In unprotected mice 50 % undergo clonic seizures only; 80 % of the mice are protected when a single injection of 100 mg/kg iproniazid is given, followed by an injection of 120 mg/kg of the precursor of 5-HT, 5-hydroxytryptophan. The remainder are subject only to clonic seizures. The latter effect lasts for 2 h, corresponding to the time during which 5-hydroxytryptophane causes an increase in the amount of 5-HT in the mouse's brain[28].

Lysergic acid diethyl amide[153] causes an accumulation of 5-HT in the homogenates of the brain *in vitro*, similar to that caused by iproniazid. Amphetamine[220] enters into competition with the monoaminoxidase which destroys 5-HT. It is possible that the protection obtained by these products against the audiogenic seizure is related to their effect on 5-HT metabolism. However, the injection of 5-HT whether or not preceded by an injection of iproniazid, has a slight effect only. This is possibly because it is destroyed before reaching the nervous centres, or because it cannot pass the blood-brain barrier. Modification of the concentrations of noradrenaline and 5-HT is prevented by the injection of iproniazid or other monoaminoxidase inhibitors together with reserpine. These amines are then maintained at their normal values in the brain. Under these circumstances the usual action of reserpine (death during the seizure) is reversed and the mice do not die after the sonic stimulation. This test can be used to study the potential value of the monoaminoxidase inhibitors. No protection is obtained by associating lysergic acid diethyl amide and reserpine, but the increase in the severity of the convulsions usually associated with reserpine is sometimes inhibited. This association between the concentrations of noradrenaline and 5-HT, and the violence of the convulsions due to the sonic stimulus is the only reliable evidence available at present concerning the intimate biochemical mechanism of the audiogenic seizure. Nevertheless these amines do not seem to bear a direct relation to the onset of the seizure, for their reduction by reserpine does not lead to the production of seizures in animals of the resistant strain.

TABLE 22

PROTECTION OF MICE AGAINST THE AUDIOGENIC SEIZURE AFTER ONE OR SEVERAL INJECTIONS OF 100 mg/kg IPRONIAZID[188]

	Age of the animals	
	Less than 100 days	More than 100 days
Test carried out 1 h after the 1st injection	20% (24)*	20% (20)
Test carried out 1 h after the 2nd injection (at 24 h intervals)	45% (20)	61% (18)
Test carried out 1 h after the 3rd injection (at 24 h intervals)	43% (31)	66% (64)

* Number of animals.

(e) *Latent period*

It has been shown* that there is an association between the occurrence of seizures and unidentified neuro-humoural agents. The existence of a latent period between the start of the sonic stimulus and the onset of the seizure also supports this hypothesis. By submitting the animals to an acoustic stimulus every five minutes, the latent period increases in proportion to the number of repetitions; the latent period has been found to increase from 1 to 15 seconds after 10-12 consecutive tests in the same animal. The severity of the seizures decreases progressively, and then the animal becomes insensible for from one to several hours. Repeated tests have the same effect

---

* A. LEHMANN, 1960, *Compt. rend. Soc. Biol.*, *154*, 57-61.

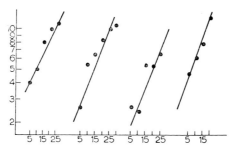

Fig. 112. Graph showing the evolution of the latent period in the course of successive stimuli (at intervals of 5 minutes) for 4 different animals. Abscissa: arithmetic time scale of stimuli, Ordinate: logarithmic scale of latent period in seconds. Each record corresponds to a single mouse.

as age in causing a progressive suppression of the seizure. The results of these tests are given in Fig. 112.

11. Discussion

From these studies, it is possible at the moment to formulate the following hypotheses concerning the audiogenic seizure.

It appears to be established that the seizure originates from a peripheral excitation which spreads through the auditory apparatus to the central nervous system. The sub-cortical nervous centres are reached either directly or by way of the reticulated substance; the excitation is then transmitted to the motor centres, thus determining the convulsion phase of the seizure.

There are strong reasons for thinking that the onset of this phenomenon depends on the genetic pattern of the animals. The cause is certainly due to numerous mechanisms; the existence of a latent period, increased by the repetition of the seizures, suggests the intervention of a chemical mediator. The rapidity of the phenomenon as well as the slight non-significant variations in the constitution of the blood, suggest that a modification occurs in the central nervous system.

Disturbance of the biochemical equilibrium of the brain at the sub-cortical level following peripheral excitation may be due to a modification in the lability of a protein-coenzyme bond, to a deficiency of one of the products necessary for an enzymatic reaction, to a modification of pH, or to a combination of several of these phenomena.

The following facts may be emphasized in favour of this hypothesis:

Deficient diets, particularly those deficient in substances which may be considered as cofactors of specific enzymatic reactions in the glycolytic and tricarboxylic acid cycles increase the severity of seizures. The antimetabolites of glutamic acid prevent its protective effect. The decrease in the noradrenaline and serotonin content of the brain following injections of reserpine or similar drugs, increases the violence of the convulsions and the number of deaths, but is not sufficient to bring on seizures in resistant mice. Although these changes in the concentrations of indole and catecholamines increase the severity of the convulsions, they are not sufficient to provoke them. Conversely, an excess of glutamic acid has a protective influence or reduces the intensity of the convulsions and the risks of mortality during the seizure. The same result is obtained when there is an increase in the amount of 5-HT in the central nervous system, brought about artificially by injection of either iproniazid, 5-hydr-

References p. 262

oxytryptophan, LSD 25, or d-amphetamine. Weakening or disappearance of the seizure also occurs with age, under the influence of iproniazid (the action of which is accentuated in the case of old mice), or during repetition of the seizures. Protection of the animal may therefore be associated with a progressive accumulation of one or more active substances in the process that leads to the convulsions; a decrease in these substances would then increase the number or violence of the seizures.

This hypothesis is in accordance with that put forward by Abood and Gerard[2]. These authors showed that the seizure is associated with a disturbance in the phosphorylatic system of the brain. In actual fact the specific activity of high energy phosphates is greatly diminished in sensitive mice, solely during their period of susceptibility. The relation to the age of the animal suggests that this aspect of the phenomenon may be related to the speed of maturity of an enzymatic system that has not yet been identified. This system may be responsible for the liberation of a chemical mediator. Other supporting evidence in favour of the hypothesis of an enzymatic action is the pH effect of the blood on the reaction, since CO_2 inhalation and acetazolamide injection act as a protection against the seizure, while alkalosis increases the sensibility of the animals.

Calcium probably exerts its protective influence through a slightly different mechanism: the diffusion of the substance responsible for the excitation is perhaps prevented, due to a modification of the cellular polarisation and therefore of the permeability of the membranes.

The influence of barbiturates and tranquillizers in suppressing the onset of the seizures may be brought about by their action on either the encephalic centres or on the reticulated substance, and the subsequent inhibition or blockage of these pathways would prevent the peripheral excitation from reaching the motor centres.

The convulsant phase of the audiogenic seizure, particularly in its external manifestations, is, in many aspects, similar to other convulsive phenomena (epilepsy, electroshock, chemical convulsivants etc.). The audiogenic seizure differs, however, in that it is the only convulsion phenomenon which originates from a purely sensorial excitation and which exhibits a specific type of first phase known as the "running fit".

REFERENCES

[1] Abood, L. G. and R. W. Gerard, 1952, Phosphorus metabolism and function in the nervous system, *Federation. Proc., 11*, 3.
[2] Abood, L. G. and R. W. Gerard, 1955, A phosphorylation defect in the brain of mice susceptible to audiogenic seizure, *Biochemistry of the developing nervous system*, Academic Press. Inc., New York, p. 467–472.
[3] Adams, J. S. and W. J. Griffiths Jr., 1948, The effect of tridione on audiogenic fits in albino rats, *J. Comp. Physiol. Psychol., 41*, 319–326.
[4] Allen, C. H. and I. Rudnik, 1947, Sound kills mouse, *Sci. News Letters, 1*, 274.
[5] Anthony, A., 1955, Harmful effect of sound in mice, *Anat. Rec., 122*, 431.
[6] Anthony, A., 1955, Effects of noise on eosinophil levels of audiogenic seizure-susceptible and seizure-resistant mice, *J. Acoust. Soc. Am., 27*, 1150–1153.
[7] Anthony, A. and E. Ackermann, 1955, Effect of noise on the blood eosinophil levels and adrenals of mice, *J. Acoust. Soc. Am., 27*, 1144–1149.
[8] Anthony, A., 1956, Changes in adrenals and other organs following exposure of hairless mice to intense sound, *J. Acoust. Soc. Am., 28*, 270–274.
[9] Anthony, A. 1959, Biological effect of noise in vertebrate animals, *W.A.D.C. Technical Report 57-647; ASTIA Document* No. AD 142.078, p. 78–81.
[10] Antonitis, J. J., 1954, Intensity of white noise and frequency of convulsive reactions in DBA/I mice, *Science, 120*, 139–140.

[11] ANTONITIS, J. J., D. D. CRARY, P. B. SAWIN and C. COHEN, 1954, Sound induced seizures in rabbits, *J. Heredity*, *45*, 278–284.
[12] ARNOLD, M. B., 1944, Experimental factors in experimental neurosis, *J. Exp. Psychol.*, *34*, 257–281.
[13] AUER, E. T. and K. U. SMITH, 1940, Characteristics of epileptoid convulsive reactions produced in rats by auditory stimulation, *J. Comp. Psychol.*, *30*, 255–59.
[14] BEACH, F. A. and T. H. WEAVER, 1943, Noise induced seizures in the rat and their modification by cerebral injury, *J. Comp. Neurol.*, *79*, 379–391.
[15] BEACH, F. A., 1951, Body chemistry and perception, in R. R. BLAKE and G. U. RAMSEY, *Perception and approach to personality*, Ronald, New York, Chap. 3.
[16] BEIN, H. J., 1956, The pharmacology of Rauwolfia, *Pharmacol. Rev.*, *8*, 435–483.
[17] BENEDICT, L. L., 1951, Sound induced seizures in rats reared on diets with high sugar content, *Thesis*, Univ. of New Mexico, unpublished.
[18] BEVAN, J. M., 1951, Effects of stress upon certain physiological mechanisms and behaviour of the albino rat, *Thesis*, Duke University, unpublished.
[19] BEVAN, W., JR., J. S. HARD and U. S. SEAL, 1951, Sound induced seizures in rats fed in amino-acid deficient diet, *J. Comp. Physiol. Psychol.*, *44*, 327–330.
[20] BEVAN, W., JR. and E. L. HUNT, 1953, Proprioceptive inflow and susceptibility to experimentally induced seizures, *J. Comp. Physiol. Psychol.*, *46*, 218–224.
[21] BEVAN, W., 1955, Sound precipitated convulsions 1947 to 1954, *Psychol. Bull.*, *52*, 473–503.
[22] BEVAN, W., M. GRODSKY and G. BOSTELMANN, 1956, Taming and susceptibility to audiogenic convulsions, *Science*, *124*, 74–75.
[23] BEVAN, W. and R. MCCHINN, 1957, Sound induced convulsions in rats treated with reserpine, *J. Comp. Physiol. Psychol.*, *50*, 311–314.
[24] BEVAN, W., E. L. HUNT and R. MCCHINN, 1957, Sound induced convulsions in rats subjected to cerebellar damage, *J. Comp. Physiol. Psychol.*, *50*, 307–310.
[25] BILLINGSLEA, F. Y., 1941, The relationship between emotionality and various other salients of behaviour in the rat, *J. Comp. Psychol.*, *31*, 69–77.
[26] BINGHAM, W. E. and W. J. GRIFFITHS, JR., 1952, The effect of different environments during infancy on adult behavior in the rat, *J. Comp. Physiol. Psychol.*, *45*, 307–312.
[27] BITTERMANN, M. E. and C. J. WARDEN, 1943, The inhibition of convulsive seizures in the white rat by the use of electric shocks. *J. Comp. Psychol.*, *35*, 133–137.
[28] BOGDANSKI, D. F., H. WEISSBACH and S. UDENFRIEND, 1958, Pharmacological studies with the serotonin precursor 5-hydroxytryptophan. *J. Pharmacol. Exp. Therap.*, *122*, 182–194.
[29] BOGOYAVLENSKAYA, N. V., 1957, The system of blood coagulation and haemorrhage into the brain in conditions of nervous trauma, *Bull. Exp. Biol. Med.*, *9*, 52–56. (in Russian, summary in English).
[30] BOVET, D., V. G. LONGO and B. SILVESTRINI, 1957, Les méthodes d'investigations électrophysiologiques dans l'étude des médicaments tranquillisants. Contribution à la pharmacologie de la formation réticulaire, *Psychotropic Drugs*, Elsevier, Amsterdam, pp. 193–206.
[31] BRADY, J. V., W. C. STEBBINS and R. GALAMBOS, 1953, The effect of audiogenic convulsions on a conditioned emotional response, *J. Comp. Physiol. Psychol.*, *46*, 363–367.
[32] BRADY, J. V., 1956, Assessment of drug effects on emotional behavior, *Science*, *123*, 1033–1034.
[33] BRDAR, P. and P. MOYER, 1949, The relationship of the glossopharyngeal and vagus nerves to audiogenic seizures, *Persona*, *1*, (?) 15–17
[34] BRUCE, W. C., 1952, Behavior associated with age and audiogenic seizure in an inbred strain of mice, *Proc. Iowa Acad. Sci.*, *59*, 367–372.
[35] BUCHEL, L., A. DEBAY, J. LEVY and O. TANGUY, 1959, Sensibilisation des rats à la crise convulsive audiogène, Application à l'étude des substances protectrices, *J. physiol. (Paris)*, *51*, 421–422.
[36] BUREŠ, J., 1952, Pathophysiology of the reflex epileptic seizure, *Thesis*, Univ. of Prague.
[37] BUREŠ, J., 1953, Contribution to the analysis of reflex influence on paroxysmal susceptibility, *Cesk. Fysiol.*, *2*, 14–27 (in Czech).
[38] BUREŠ J., 1953, External inhibition on reflex epilepsy in the rat and the mouse, *Cesk. Fysiol.*, *2*, 28–37 (in Czech).
[39] BUREŠ, J., 1953, Beitrag zur Analyse der reflektorischen Beeinflussung der paroxysmalen Bereitschaft, *Cesk. Fysiol.*, *II*, 12–28 (in Russian, summary in German).
[40] BUREŠ, J., 1953, Äussere Hemmung der Reflexepilepsie bei Ratten und Mäusen, *Cesk. Fysiol.*, *II*, 29–40 (in Russian, summary in German).
[41] BUREŠ, J., 1953, Susceptibility to convulsions in reflex epilepsy in the ontogenesis of rats and mice, *Cesk. Fysiol.*, *II*, 274–282.
[42] BUREŠ, J., 1953, Beeinflussung der Krampfbereitschaft durch Ausschaltung der Analysatoren bei Ratten und Mäusen, *Cesk. Fysiol.*, *II*, 284–293 (in Russian, summary in German).

[43] Bures, J., 1953, Experiments on the electrophysiological analysis of the generalisation of an epileptic fit, *Cesk. Fysiol.*, *II*, 347–356.
[44] Bures, J., 1953, Einfluss totaler Herabsetzung der Körpertemperatur auf die Krampfbereitschaft zur reflexiven Epilepsie bei Ratten und Mäusen, *Cesk. Fysiol.*, *II*, 357–362 (in Russian, summary in German).
[45] Bures, J. and O. Buresova, 1956, The influencing of reflex acoustic epilepsy and reflex inhibition ("animal hypnosis") by spreading EEG depression, *Physiol. Bohemosloven*, *5*, 395–400.
[46] Bures, J., 1956, On the problem of the reflex influencing of susceptibility to convulsions experimentally and clinically, *Casopis. Lekarn. Cesk.*, *45*, 393–397 (in Czech, summary in English).
[47] Busnel, R. G., A. Lehmann and M. C. Busnel, 1958, Étude de la crise audiogène de la souris comme test psychopharmacologique: son application aux substances du type "tranquilliseur", *Path. Biologie*, (9–10) 749–762.
[48] Busquet, H. and Ch. Vischniac, 1950, Identité d'action du principe sympathicomimétique du genêt et de l'adrénaline sur la perméabilité capillaire et sur la crise audiogène, *Compt. rend. Soc. Biol.*, *144*, 398–400.
[49] Cain, J., 1952, La crise audiogène du rat albinos, Univ. of Aix, Marseille, France. Thesis.
[50] Cain, J. and J. Mercier, 1948, Influence de certains médicaments anticonvulsivants sur la crise audiogène du rat albinos, *Compt. rend. Soc. Biol.*, *142*, 688–691.
[51] Cain, J. and J. Mercier, 1948, Influence exercée par le tridione (triméthyloxazolidinedione) sur les crises audiogènes du rat albinos, *Compt. rend. Soc. Biol.*, *142*, 993–994.
[52] Cain, J. and R. Naquet, 1950, Influence de la contention et des circonstances expérimentales sur la crise audiogène du rat, *Compt. rend. Soc. Biol.*, *144*, 1192–1193.
[53] Cain, J., J. Mercier and J. Corriol, 1951, Sur l'électroencéphalogramme du rat albinos soumis à la crise audiogène, *Compt. rend. Soc. Biol.*, *145*, 915–916.
[54] Canac, F., V. Gavreau and J. Cain, 1948, Recherche sur les crises audiogènes provoquées par le son chez le rat, *C.R.S.I.M.–C.N.R.S.*, note No. 158, Marseille.
[55] Chance, M. R. A., 1947, Factors influencing the toxicity of sympathomimetic amines to solitary mice, *J. Pharmacol. Exp. Therap.*, *89*, 289–296.
[56] Chance, M. R. A., 1948, Aggression, a component of post-epileptic automatism in *Peromyscus*, *Nature*, *161*, 101–102.
[57] Chance, M. R. A., 1954, The suppression of audiogenic hyperexcitement by learning in *Peromyscus maniculatus*, *Brit. J. Beh.*, *2*, 31–35.
[58] Chance, M. R. A. and D. C. Yaxley, 1949, New aspects of the behaviour of *Peromyscus* under audiogenic hyper-excitement, *Behaviour*, *2*, 96–105.
[59] Chance, M. R. A. and D. C. Yaxley, 1950, Central nervous function and changes in brain metabolite concentration. I. Glycogen and lactate in convulsing mice, *J. Exp. Biol.*, *27*, 311–323.
[60] Chance, M. R. A., 1953, Rat behavior, *Nature*, *172*, 488–489.
[61] Clementi, A., 1929, Stricnizzione della sfera corticale uditiva id epilepsia sperimentale da stimuli acustici, *Arch. Fisiol.*, *27*, 388.
[62] Cohen, L. M. and H. W. Karn, 1943, The anticonvulsivant action of dilantin on sound induced seizures in the rat, *J. Comp. Psychol.*, *35*, 307–312.
[63] Cole, W. H., E. H. Yeakel and E. J. Farris, 1942, A preliminary study of changes in the blood of grey Norway rats following audiogenic seizures, *Anat. Rec.*, *84*, 524–525.
[64] Colman, K. W., 1952, Conditioning and audiogenic seizure, *Psi Chi Newsletter*, 13–14.
[65] Corn, M., D. Lester and L. A. Greenberg, 1955, Inhibition effects of certain drugs on audiogenic seizure in the rat, *J. Pharmacol. Exp. Therap.*, *113*, 58–63.
[66] Critchley, 1937, Musicogenic epilepsy, *Brain*, *60*, 13–27.
[67] Davis, E. W., W. S. McCulloch and E. Roseman, 1944, Rapid changes in the O_2 tension of cerebral cortex during induced convulsions, *Am. J. Psychiat.*, *100*, 825–829.
[68] Dember, W. N., P. Ellen and A. B. Kristofferson, 1953, The effect of alcohol on seizure behavior in rats, *Quart. J. Stud. Alcohol.*, *14*, 390–394.
[69] Dice, L. R., 1935, Inheritance of waltzing and of epilepsy in mice of the genus *Peromyscus*, *J. Mammal.*, *16*, 25–35.
[70] Dice, L. R. and E. Barto, 1952, Ability of mice of the genus *Peromyscus* to hear ultrasonic sounds, *Science*, *116*, 110–111.
[71] Dobrokhotova, L. P., 1957, Influence of methylthiouracile on haemorrhagic shock states, their evolution under the influence of nervous trauma, *Doklady Acad. Sci. SSSR*, *114*, 1320–1321 (in Russian).
[72] Donaldson, H. H., 1924, *The rat*, Mem. Wist. Inst., Philadelphia, No. 6, 2nd ed.
[73] Farris, E. J. and E. H. Yeakel, 1942, The effect of age upon susceptibility to audiogenic seizures in albino rats, *J. Comp. Psychol.*, *33*, 249–251.
[74] Farris, E. J. and E. H. Yeakel, 1942, Sex and increasing age as factors in frequency of audiogenic seizures in albino rats, *J. Comp. psychol.*, *34*, 75–78.

[75] FARRIS, E. J. and E. H. YEAKEL, 1942, Effects of a neurophysiological stimulus on the breeding of albino rats, *Anat. Rec., 84*, 454.
[76] FARRIS, E. J. and E. H. YEAKEL, 1943, Susceptibility of albino and grey Norway rats to audiogenic seizures, *J. Comp. Psychol., 35*, 73–80.
[77] FARRIS, E. J. and E. H. YEAKEL, 1944, Breeding and rearing of young by albino rats subjected to auditory stimulation, *Anat. Rec., 89*, 325–330.
[78] FARRIS, E. J. and E. H. YEAKEL, 1945, Emotional behavior of grey Norway and Wistar albino rats, *J. Comp. Psychol., 38*, 109–118.
[79] FINGER, F. W. and H. SCHLOSSBERG, 1941, The effect of audiogenic seizure on general activity of the white rat, *Am. J. Psychol., 54*, 518–527.
[80] FINGER, F. W., 1942, Factors influencing audiogenic seizure in the rat: repeated stimulation and deprivation of food and drink, *Am. J. Psychol., 55*, 68–76.
[81] FINGER, F. W., 1943, Factors influencing audiogenic seizures in the rat: Heredity and age, *J. Comp. Psychol., 35*, 227–232.
[82] FINGER, F. W., 1947, Convulsive behavior in the rat, *Psychol. Bull., 44*, 201–248.
[83] FINGER, F. W., R. C. BICE and W. F. DAY, 1952, Audiogenic seizure and streptomycine induced vestibular dysfunction, *J. Comp. Physiol. Psychol., 45*, 163–169.
[84] FORSTER, F. M. and L. MADOW, 1950, Experimental sensory induced seizures, *Am. J. Physiol., 161*, 430–434.
[85] FRINGS, H. and M. FRINGS, 1950, Recherches sur l'action des vibrations soniques et ultrasoniques sur les rongeurs, *Les conférences du Palais de la Découverte*, Série B, No. 26.
[86] FRINGS, H. and J. E. O'TOUSA, 1950, Toxicity to mice of chlordane vapor and solutions administered cutaneously, *Science, 111*, 658–660.
[87] FRINGS, H., M. FRINGS and A. KIVERT, 1951, Behavior patterns of the laboratory mouse under auditory stress, *J. Mammal., 32*, 60–76.
[88] FRINGS, H. and M. FRINGS, 1951, Otitis media and audiogenic seizure in mice, *Science, 113*, 688–90.
[89] FRINGS, H. and I. SENKOVITS, 1951, Destruction of the pinnae of white mice by high intensity airborne sound, *J. Cell. Comp. Physiol., 37*, 267–281.
[90] FRINGS, H. and M. FRINGS, 1952, The development of strains of albino mice with predictable susceptibility to audiogenic seizures, *Anat. Rec., 113*, 550–551.
[91] FRINGS, H. and M. FRINGS, 1952, Audiogenic seizure in the laboratory mouse, *J. Mammal., 33*, 80–87.
[92] FRINGS, H. and M. FRINGS, 1952, Acoustical determinants of audiogenic seizures in laboratory mice, *J. Acoust. Soc. Amer., 24*, 163–169.
[93] FRINGS, H. and M. FRINGS, 1952, Latent periods of audiogenic seizures in mice, *J. Mammal., 33*, 487–491.
[94] FRINGS, H., M. FRINGS, J. L. FULLER, B. E. GINSBURG, S. ROSS and E. M. VICARI, 1952, Standardization of nomenclature describing audiogenic seizure in mice, *Behaviour, 4*, 157–160.
[95] FRINGS, H. and M. FRINGS, 1953, Development of strains of albino mice with predictable susceptibilities to audiogenic seizures, *Science, 117*, 283–284.
[96] FRINGS, H. and M. FRINGS, 1953, The production of stocks of albino mice with predictable susceptibilities to audiogenic seizures, *Behaviour, 5*, 305–319.
[97] FRINGS, H. and A. KIVERT, 1953, Nicotine facilitation of audiogenic seizures in laboratory mice, *J. Mammal., 34*, 391–393.
[98] FRINGS, H., M. FRINGS and M. HAMILTON, 1956, Experiment with albino mice from stocks selected for predictable susceptibilites to audiogenic seizures, *Behaviour, 9*, 44–52
[99] FULLER, J. L., 1949, 1950, Genetic control of audiogenic seizure in hybrid between DBA subline 2 and C 57 black subline 6, *Rec. Genet. Soc. Am., 18*, 86–87, *Genetics, 35*, 106–107.
[100] FULLER, J. L., C. EASLER and M. E. SMITH, 1950, Inheritance of audiogenic seizure susceptibility in the mouse, *Genetics, 35*, 622–632.
[101] FULLER, J. L. and E. WILLIAMS, 1951, Gene controlled time constants in convulsive behaviour, *Proc. Natl. Acad. Sci, 37*, 349–356.
[102] FULLER, J. L. and A. RAPPAPORT, 1952, The effect of wetting on sound induced convulsions in mice, *J. Comp. Physiol. Psychol., 45*, 246–249.
[103] FULLER, J. L. and M. E. SMITH, 1952, Kinetics of the mass excitation of the mammalian nervous system by sound, *Anat. Rec., 113*, 547.
[104] FULLER, J. L. and M. E. SMITH, 1953, Kinetics of sound induced convulsions in some inbred mouse strains, *Am. J. Physiol., 172*, 661–670.
[105] FULLER, J. L. and B. E. GINSBURG, 1954, Effect of adrenalectomy on the anticonvulsivant action of glutamic acid in mice, *Am. J. Physiol., 176*, 367–370.
[106] GALAMBOS, R. and H. DAVIS, 1943, The response of single auditory nerve fibres to acoustic stimulation, *J. Neurophysiol., 6*, 639–658.

[107] GALAMBOS, R. and C. T. MORGAN, 1943, The production of audiogenic seizures by interrupted tones, *J. Exp. Psychol.*, *32*, 435–442.
[108] GILBERT, P. F. and G. C. V. GAWAIN, 1950, Sonic and ultrasonic effects on maze learning and retention in the albino rat, *U.S.A.F. Techn. Report*, No. 6030, Wright Patterson Air Force Base.
[109] GILBERT, P. F. and G. C. V. GAWAIN, 1950, Sonic and ultrasonic effects on maze learning and retention in the albino rat, *Thesis*, Pennsylvania State College, Psychology Publication No. 2314.
[110] GILBERT, P. F. and G. C. V. GAWAIN, 1954, Sonic effects of maze learning and retention in the albino rat, *Denison Univ. Bull. J. Sci. Lab.*, *43*, 61–65.
[111] GINSBURG, B. E. and E. HUTH, 1947, Some aspect of the physiology of gene controlled audiogenic seizures in inbred strains of mice, *Genetics*, *32*, 87.
[112] GINSBURG, B. E. and R. B. HOVDA, 1947, On the physiology of gene controlled audiogenic seizures in inbred strains of mice. *Anat. Rec.*, *99*, 621–622.
[113] GINSBURG, B. E., D. MILLER and M. J. ZAMIS, 1949, 1950, On the mode of inheritance of susceptibility to sound induced seizures in the house mouse, *Rec. Gen. Soc. Am.*, *18*, 89; *Genetics*, *35*, 109.
[114] GINSBURG, B. E., S. ROSS, M. J. ZAMIS and A. PERKINS, 1950, An assay method for the behavioral effect of L-glutamic acid, *Science*, *112*, 12–13.
[115] GINSBURG, B. E., S. ROSS, M. J. ZAMIS and A. PERKINS, 1951, Some effect of L(+) glutamic acid on sound induced seizures in mice, *J. Comp. Physiol. Psychol.*, *44*, 134–141.
[116] GINSBURG, B. E. and E. ROBERTS, 1951, Glutamic acid and central nervous system activity, *Anat. Rec.*, *111*, 492–493.
[117] GINSBURG, B. E. and J. L. FULLER, 1954, A comparison of chemical and mechanical alterations of seizure patterns in mice, *J. Comp. Physiol. Psychol.*, *47*, (4), 344–348.
[118] GINSBURG, B. E., 1954, Genetics and the physiology of the nervous system. Genetics and the inheritance of integrated neurological and psychiatric patterns, *Proc. Assoc. Res. in nervous and mental disease*, *33*, 39–56.
[119] GLASER, N. M., 1941, Autonomic charges associated with abnormal behaviour in the rat. I. Analysis of changes in heart rate occuring as a result of responses in auditory stimulation. II. The effect of metrazol upon heart rate, *Thesis*, Univ. of Michigan.
[120] GOLDBERG, D., 1952, Audiogenic seizures and related behavior in the albino rat, *Thesis*, Emory Univ., unpublished.
[121] GOLUB, L. M. and C. T. MORGAN, 1945, Patterns of electrogenic seizures in rats. Their relation to stimulus intensity and to audiogenic seizures, *J. Comp. Psychol.*, *38*, 239–245.
[122] GOODSELL, J. S., 1955, Properties of audiogenic seizures in mice and effect of anticonvulsant drugs, *Fed. Proc.*, *14*, 345.
[123] GOODSON, F. E. and M. H. MARX, 1953, Increased resistance to audiogenic seizures in rats trained on an instrumental wheel-turning response, *J. Comp. Physiol. Psychol.*, *46*, 225–30.
[124] GOUSSELNIKOVA, K. G., 1959, Data on the mechanism of epileptic seizures caused by sound in rats, *Nauchn. Dokl. Vysshei Shkoly, Biol. Nauki* (1), 69–73 (in Russian).
[125] GRAEL, L. J. MC., 1949, Pure tone and audiogenic seizures in mice, *Thesis*, Penna State College Penn.
[126] GREENBERG, D. M., M. D. D. BOELTER and B. W. KNOPF, 1942, Factors concerned in the development of tetany by the rat, *Am. J. Physiol.*, *137*, 459–467.
[127] GREENBERG, L. A. and D. LESTER, 1953, The effect of alcohol on audiogenic seizures in rats, *Quart. J. Stud. Alcohol*, *14*, 385–389.
[128] GRIFFITHS, W. J. JR., 1942, The effect of dilantin on convulsive seizures in the white rat, *J. Comp. Psychol.*, *33*, 291–296.
[129] GRIFFITHS, W. J. JR., 1942, Transmission of convulsions in the white rat, *J. Comp. Psychol.*, *34*, 263–279.
[130] GRIFFITHS, W. J. JR., 1942, The production of convulsions in the white rat, *Comp. Psychol. Monographies*, *17*, 1–29.
[131] GRIFFITHS, W. J. JR., 1944, Absence of audiogenic seizures in wild Norway and Alexandrine rats, *Science*, *99*, 62–63.
[132] GRIFFITHS, W. J. JR., 1945, The effect of thiamine hydrochloride on the incidence of audiogenic seizures among selectively bred albino rats, *J. Comp. Psychol.*, *38*, 65–68.
[133] GRIFFITHS, W. J. JR., 1947, Audiogenic fits produced by magnesium deficiency in tame domestic Norway rats and wild Norway and Alexandrine rats, *Am. J. Physiol.*, *149*, 135–141.
[134] GRIFFITHS, W. J. JR., 1948, The effects of adrenalectomy on the incidence of audiogenic seizures in rats, *Am. Psychologist*, *3*, 332–333.
[135] GRIFFITHS, W. J. JR. and M. COHEN, 1949, Effects of morphine sulphate and diaminon hydrochloride on incidence of audiogenic seizures in albino rats, *J. Comp. Physiol.*, *42*, 391–397.
[136] GRIFFITHS, W. J. JR., 1949, Effect of adrenalectomy on incidence of audiogenic seizures among domestic and wild rats, *J. Comp. Physiol. Psychol.*, *42*, 303–312.

[137] GRIFFITHS, W. J. JR., 1950, Self-selection of diet in relation to audiogenic seizures in rats, *Science*, *112*, 786–787.
[138] GRIFFITHS, W. J. JR., 1953, The influence of behavioral factors on the incidence of audiogenic seizures in rats, *J. Comp. Physiol. Psychol.*, *46*, 150–152.
[139] GRIFFITHS, W. J. JR. and W. F. STRINGER, 1952, The effect of intense stimulation experienced during infancy on adult behavior in the rat, *J. Comp. Physiol. Psychol.*, *45*, 301–306.
[140] GRÜNEBERG, H., 1947, *Animal genetics and Medecine*, Paul B. Hoeber Inc., New York, xii 296 pp.
[141] HALBERG, F., M. B. VISSCHER and J. J. BITTNER, 1953, Eosinophile rythm in mice: range of occurence; effect of illumination, feeding and adrenalectomy, *Am. J. Physiol.*, *174*, 109–122.
[142] HALBERG, F., J. J. BITTNER and R. J. GULLY, 1955, Twenty four hour susceptibility to audiogenic convulsions in several stocks of mice, *Federation Proc.*, *14*, 67–68.
[143] HALBERG, F., E. JACOBSEN, G. WADSWORTH and J. J. BITTNER, 1958, Audiogenic abnormality spectra, twenty four hour periodicity and lighting, *Science*, *128*, 657–658.
[144] HALL, C. S., 1947, Genetic differences in fatal audiogenic seizures between two inbred strains of house mice, *J. Heredity*, *38*, 2–6.
[145] HALL, C. S., 1951, The genetics of behavior, in S. S. STEVENS (Ed), *Handbook of experimental psychology*, Wiley, New York, pp. 309–310; 317–319.
[146] HALL, C. S. and P. H. WHITEMAN, 1951, The effects of infantile stimulation upon later emotional stability in the mouse, *J. Comp. Physiol. Psychol.*, *44*, 61–66.
[147] HALPERN, B. N. and A. LEHMANN, 1956, Action protectrice du carbamate de methyl-3-pentyne-1-ol-3 (CMP) contre la crise convulsive audiogène, *Compt. rend. Soc. Biol.*, *150*, 1863–1866.
[148] HALPERN, B. N. and A. LEHMANN, 1957, Bases expérimentales de l'action thérapeutique d'une nouvelle médication sédative et antianxieuse le carbamate de méthyl-3-pentyne-1-ol-3, *Presse médicale*, (27), 622–625.
[149] HAMBURGH, M. and E. VICARI, 1958, 1960, Physiological mechanisms underlying susceptibility to audiogenic seizures in mice, *Anat. Rec.*, *132*, 450; *J. Neuropath. and Exp. Neur.*, *19*, 461–472.
[150] HAMILTON, H. C. and E. B. MAHER, 1947, The effects of glutamic acid on the behavior of the white rat, *J. Comp. Physiol. Psychol.*, *40*, 463–468.
[151] HARLOW, H. F., 1949, Physiological psychology, in V. E. HALL (Ed.), *Ann. Rev. Physiol.*, 284–286.
[152] HARRIMAN, A. E. and A. M. BRIAN, 1956, Learned inhibition of "sound-induced seizures" in the rat, *Am. J. Psychol.*, *69*, 100–102.
[153] HOAGLAND, H., 1957, A revue of biochemical changes induced in vivo by lysergic acid diethylamide and similar drugs, *Ann. N.Y. Acad. Sci.*, *66*, 445–458.
[154] HOFELD, J. R., 1948, Changes in metabolic rate during audiogenic seizures in the rat, *Am. Psychologist*, *4*, 359.
[155] HUMPHREY, G. and F. MARCUSE, 1939, New methods of obtaining neurotic behavior in the rat, *Am. J. Psychol.*, *52*, 616–619.
[156] HUMPHREY, G. and F. MARCUSE, 1941, Factors influencing the susceptibility of albino rats to convulsive attacks under intense auditory stimulation, *J. Comp. Psychol.*, *32*, 285–306.
[157] HUMPHREY, G., 1941, Experiments on the physiological mechanism of noise induced seizures in the albino rat, *Bull. Can. Psychol. Assoc.*, *1*, 39–41.
[158] HUMPHREY, G., 1942, Experiments on the physiological mechanism of noise induced seizures in the albino rat. The action of parasympathetic drugs, *J. Comp. Psychol.*, *33*, 316–323.
[159] HUMPHREY, G., 1942, Experiment on the physiological mechanism of noise induced seizures in the albino rat: The site of action of the parasympathetic drugs, *J. Comp. Psychol.*, *33*, 325–341.
[160] HURDER, W. P. and A. F. SANDERS, 1953, Audiogenic seizure and the adrenal cortex, *Science*, *117*, 324–326.
[161] HURDER, W. P. and A. F. SANDERS, 1954, A correction and additional observation, *Science*, *119*, 476–477.
[162] JAMES, W. T. and C. BOYLES, 1950, The effect of seizure in rats on food and water intake, *Am. J. Psychol.*, *63*, 284–286.
[163] JENNEY, E. H. and C. C. PFEIFFER, 1958, The convulsant effect of hydrazides and the antidotal effect of anticonvulsants and metabolites, *J. Pharmacol. Exp. Therap.*, *122*, 110–123.
[164] JENSEN, G. D. and E. STAINBROOK, 1949, The effect of electrogenic convulsions on the oestrus cycle and weight of rats, *J. Comp. Physiol. Psychol.*, *42*, 502–505.
[165] JONES, M. R., 1944, Some observations on the effect of phenobarbital on emotional responses and air-induced seizures, *J. Comp. Psychol.*, *37*, 159–163.
[166] KARN, H. W., C. H. LODOWSKY and R. A. PATTON, 1941, The effect of metrazol on the susceptibility of rats to sound induced seizures, *J. Comp. Psychol.*, *32*, 563–567.
[167] KENSHALO, D. R. and K. D. KRYTER, 1949, Middle ear infection and sound induced seizures in rats, *J. Comp. Physiol. Psychol.*, *42*, 328–331.

168 KING, J. T., Y. CHIUNG OUH LEE and M. B. VISSCHER, 1955, Single versus multiple cage occupancy and convulsions frequency in C3H mice, *Proc. Soc. Exp. Biol. Med.*, 88, 661–663.
169 KOENIG, E., 1957, The effect of auditory pathway interruption on the incidence of sound induced seizures in rats, *J. Comp. Neurol.*, 108, 383–392.
170 KRUSHINSKI, L. V., 1949, New data on the study of experimental epilepsy and on its basic physiological mechanism, *Uspekhi Sovremennoĭ Biol.*, 28, (4) 108–133 (in Russian).
171 KRUSHINSKI, L. V. and L. N. MOLODKINA, 1949, Paralysis caused by effusions of blood in the C.N.S. after experimental epileptic seizures in the rat, *Dokl. Ac. Sc. SSSR*, 66, (2) 289–292 (in Russian).
172 KRUSHINSKI, L. V., L. N. MOLODKINA and I. A. KITSOVSKAIA, 1950, The role of the parathyroid gland on the onset of experimental epileptic seizures, *Bull. Exp. Biol. Med.*, 31, (8) 136–140 (in Russian).
173 KRUSHINSKI, L. V., D. A. FLESS and L. N. MOLODKINA, 1950, Physiological analysis of the basic process of experimental reflex epilepsy, *J. Obshch. Biol.*, 11, (2) 104–119 (in Russian).
174 KRUSHINSKI, L. V., D. A. FLESS and L. N. MOLODKINA, 1952, Study of the after-limited inhibition by a sound-excitation method, *Bull. Exp. Biol. Med.*, 33, (4) 13–16 (in Russian).
175 KRUSHINSKI, L. V., L. P. PUSHKARSKAIA and L. N. MOLODKINA, 1953, Experimental study of the effusions of blood in the brain under the influence of nervous trauma, *Vesti. Mosc. Univ. Ser. Physiol. Math. et Estestv. Sci.*, (12) 25–44 (in Russian).
176 KRUSHINSKI, L. V., M. I. A. CEREISKI, L. P. PUSHKARSKAIA and G. I. FEDOROVA, 1955, Experimental study of new anti-epileptic preparations, *Zh. Vysshei Nervnoi Deyatel'nosti*, 5, (6) 892–900 (in Russian).
177 KRUSHINSKI, L. V., 1954, Studies of the connection between excitation and inhibition by a method of normal and pathological sonic excitations, *Uspekhi Sovremennoĭ Biol.*, 37, 74–93 (in Russian).
178 KRUSHINSKI, L. V. and D. A. FLESS, 1958, Role of the restoration of the after-limited inhibition in pathological nervous activity, *Trudi Vsesoyouz Obshch. Fiziol.*, 4, 41–47 (in Russian).
179 KRUSHINSKI, L. V. and L. N. MOLODKINA, 1957, Haemorrhages in the spinal cord following experimental epileptic seizures, *Uspekhi Sovremennoi Biol.*, 44, 220–231 (in Russian).
180 KRUSHINSKI, L. V., V. A. KORZHOV and L. N. MOLODKINA, 1958, The influence of electroshock on the pathological conditions caused in rats by a sound stimulus, *Zh. Vysshei Nervnoi Deyatel' nosti*, 8 (1), 95–102 (in Russian, summary in English).
181 KRUSHINSKI, L. V., 1959, Genetic investigations in experimental pathophysiology of higher nervous activity, *Bull. Moskov. Obshch. Biol.*, 64, 105–117. (in Russian, summary in English).
182 LACEY, O. L., 1945, The dependance of behavior disorder in the rat upon blood composition. Audiogenic seizure as a function of blood composition, *J. Comp. Psychol.*, 38, 257–270.
183 LAZOVIK, A. D. and R. A. PATTON, 1947, The relative effectiveness of auditory stimulation and motivational stress in precipitating convulsions associated with magnesium deficiency, *J. Comp. Physiol. Psychol.*, 40, 191–202.
184 LEHMANN, A., 1956, Analyse chronoactographique de la crise audiogène de la souris, *Compt. rend. Soc. Biol.*, 150, 1860–1863.
185 LEHMANN, A., B. N. HALPERN and R. G. BUSNEL, 1957, Recherches sur la crise audiogène de la souris: ses modifications sous l'influence de la réserpine et du carbamate de méthyl-3-pentyne-1-ol-3, *J. physiol. Paris*, 49, 265–268.
186 LEHMANN, A., A. DRILHON and R. G. BUSNEL, 1957, Etude électrophorétique des globulines de la souris dans leur rapport avec la crise audiogène, *Compt. rend. Soc. Biol.*, 151, 1090–1094.
187 LEHMANN, A., B. N. HALPERN and R. G. BUSNEL, 1957, Etude sur la crise audiogène de la souris d'un antagonisme entre le carbamate de méthyl-3-pentyne-1-ol-3 et la réserpine, *Compt. rend. Soc. Biol.*, 151, 1094–1096.
188 LEHMANN, A. and M. C. BUSNEL, 1958, Antagonismes entre la réserpine, la tétrabenzine, le carbamate de méthylpentynol et leur association avec l'iproniazide étudiés avec le test de la crise audiogène de la souris, *Neuropsychopharmacology*, Elsevier, pp. 348–351.
189 LINDSLEY, D. B., F. W. FINGER and CH. E. HENRY, 1942, Some physiological aspects of audiogenic seizures in rats, *J. Neurophysiol.*, 5, 185–198.
190 LINDZEY, G., 1951, Emotionality and audiogenic seizure susceptibility in five inbred strains of mice, *J. Comp. Physiol. Psychol.*, 44, 389–393.
191 LONGHURST, J. U., 1948, Effects of brain injury to the rat on seizures produced during auditory stimulation, Thesis, Univ. of Michigan, unpublished.
192 MC CANN, S. M., H. B. ROTHBALLER, E. H. YEAKEL and H. A. SHENKIN, 1948, Adrenalectomy and blood pressure of rats subjected to auditory stimulation, *Am. J. Physiol.*, 155, 128–131.
MC GRAEL, L. J., see ref. 125.
193 MAIER, N. R. F., 1939, *Studies of abnormal behavior in the rat*, The neurotic pattern and an analysis of the situation which produces it, Harper, New York.

194 Maier, N. R. F. and N. M. Glaser, 1940, Studies of abnormal behavior in the rat. II. A comparison of some convulsions producing situations, *Comp. Psychol. Monogaphies*, *16*, 1–30.
195 Maier, N. R. F., 1940, Studies of abnormal behavior in the rat. IV. Abortive behavior and its relation to the neurotic attack, *J. Exp. Psychol.*, *27*, 369–393.
196 Maier, N. R. F. and N. M. Glaser, 1940, Studies of abnormal behavior in the rat. V. The inheritance of the "neurotic pattern", *J. Comp. Psychol.*, *30*, 413–418.
197 Maier, N. R. F. and J. Sacks, 1941, Studies of abnormal behavior in the rat. VI. Patterns of convulsive reactions to metrazol, *J. Comp. Psychol.*, *32*, 489–502.
198 Maier, N. R. F. and J. B. Klee, 1941, Studies of abnormal behavior in the rat. VII. The permanent nature of abnormal fixations and their relation to convulsive tendencies, *J. Exp. Psychol.*, *29*, 384–389.
199 Maier, N. R. F., J. Sacks and N. M. Glaser, 1941, Studies of abnormal behavior in the rat. VIII. The influence of metrazol on seizures occurring during auditory stimulation, *J. Comp. Psychol.*, *32*, 379–388.
200 Maier, N. R. F. and N. M. Glaser, 1942, Studies of abnormal behavior in the rat. IX. Factors which influence the occurence of seizures during auditory stimulation, *J. Comp. Psychol.*, *34*, 11–21.
201 Maier, N. R. F. and N. M. Glaser, 1942, Studies of abnormal behavior in the rat. X. The influence of age and sex on the susceptibility to seizures during auditory stimulation, *J. Comp. Psychol.*, *34*, 23–38.
202 Maier, N. R. F. and J. Sacks, 1942, Studies of abnormal behavior in the rat. XI. Factors that influence the type of reaction to metrazol, *J. Comp. Psychol.*, *34*, 331–340.
203 Maier, N. R. F. and J. B. Klee, 1943, Studies of abnormal behavior in the rat. XII. The pattern of punishment and its relation to abnormal fixations, *J. Exp. Psychol.*, *32*, 377–398.
204 Maier, N. R. F. and S. Wapner, 1943, Studies of abnormal behavior in the rat. XIII. The effects of punishment for seizures on seizure-frequency during auditory stimulation, *J. Comp. Psychol.*, *35*, 247–248.
205 Maier, N. R. F., 1943, Studies of abnormal behavior in the rat. XIV. Strain differences in the inheritance of susceptibility to convulsions, *J. Comp. Psychol.*, *35*, 327–335.
206 Maier, N. R. F. and S. Wapner, 1944, Studies of abnormal behavior in the rat. XV. The influence of maze behavior on seizures occurring during auditory stimulation and the effect of seizures on maze performance, *J. Comp. Psychol.*, *37*, 23–34.
207 Maier, N. R. F. and S. Wapner, 1944, Studies of abnormal behavior in the rat. XVI. A case of generalized inhibitory neurosis, *J. Comp. Psychol.*, *37*, 151–158.
208 Maier, N. R. F. and J. B. Klee, 1945, Studies of abnormal behavior in the rat. XVII. Guidance versus trial and error in the alteration of habits and fixations, *J. Psychol.*, *19*, 133–163.
209 Maier, N. R. F. and W. Parker, 1945, Studies of abnormal behavior in the rat. XVIII. Analysis of stomachs of rats repeatedly exposed to auditory stimulation, *J. Comp. Psychol.*, *38*, 335–341.
210 Maier, N. R. F. and R. S. Feldman, 1946, Studies of abnormal behavior in the rat. XIX. Water spray as a means of inducing seizures, *J. Comp. Psychol.*, *39*, 275–286.
211 Maier, N. R. F., R. S. Feldman and J. U. Longhurst, 1947, Studies of abnormal behavior in the rat. XX. Change in seizure patterns with repeated testing, *J. Comp. Physiol. Psychol.*, *40*, 73–86.
212 Maier, N R. F. and J. U. Longhurst, 1947, Studies of abnormal behavior in the rat. XXI. Conflict and "audiogenic" seizures, *J. Comp. Physiol. Psychol.*, *40*, 397–412.
213 Maier, N. R. F. and R. S. Feldman, 1948, Studies of abnormal behavior in the rat. XXII. Strength of fixation and duration of frustration, *J. Comp. Physiol. Psychol.*, *41*, 348–363.
214 Maier, N. R. F. and P. Ellen, 1952, Studies of abnormal behavior in the rat. The prophylactic effect of "guidance" in reducing rigid behavior, *J. Abnormal Social Psychol.*, *47*, 109–116.
215 Maier, N. R. F., 1948, Experimentally induced abnormal behavior, *Scientific Monthly*, *67*, 210–216.
216 Maier, N. R. F., 1949, *Frustration*, The study of behavior without a goal, McGraw-Hill, New York, pp. 134–140.
217 Maier, N. R. F. and P. Ellen, 1951, Can the anxiety-reduction theory explain abnormal fixations, *Psychol. Rev.*, *58*, 435–445.
218 Maier, N. R. F., J. U. Longhurst and P. Ellen, 1951, Effects of lactose in the diet on seizure behavior of male and female rats, *J. Comp. Physiol. Psychol.*, *44*, 501–506.
219 Maier, N. R. F. and P. Ellen, 1951, The effects of lactose in the diet on frustration susceptibility in rats, *J. Comp. Physiol. Psychol.*, *44*, 551–556.
220 Mann, P. J. G. and J. H. Quastel, 1940, Benzedrine and brain metabolism, *Biochem. J.*, *34*, 414–431.
221 Marcuse, F. L. and A. U. Moore, 1941, Heart rate and respiration preceding and following audiogenic seizures in the white rat, *Proc. Soc. Exp. Biol. Med. N.Y.*, *48*, 201–202.

[222] MARCUSE, F. L. and A. U. MOORE, 1943, Heart rate in the comatose state of audiogenic seizures, *J. Exp. Psychol.*, *32*, 518–521.
[223] MARTIN, R. F. and C. S. HALL, 1941, Emotional behavior in the rat: The incidence of behavior derangements resulting from air blast stimulation in emotional and non-emotional strains of rats, *J. Comp. Psychol.*, *32*, 191–204.
[224] MARX, M. H. and M. JURKO, 1950, The relationship between the audiogenic seizure and middle ear disease, *J. Genet. Psychol.*, *76*, 221–239.
[225] MARX, M. H. and R. M. CHAMBERS, 1952, Incidence of audiogenic seizures following experimental induction of middle ear disorder, *J. Comp. Physiol. Psychol.*, *45*, 239–245.
[226] MARX, M. H. and W. J. VAN SPANKEREN, 1952, Control of the audiogenic seizure by the rat, *J. Comp. Physiol. Psychol.*, *45*, 170–180.
[227] MEDOFF, H. S. and A. M. BONGIOVANNI, 1945, Blood pressure in rats subjected to audiogenic stimulation, *Am. J. Physiol.*, *143*, 300–305.
[228] MERCIER, J., 1949, Note préliminaire sur l'influence exercée par quelques médicaments sympathicomimétiques sur la crise audiogène du rat albinos, *Compt. rend. Soc. Biol.*, *143*, 1125–1127.
[229] MERCIER, J., 1949, Influence de la caféine et de la théophylline sur la crise de comportement dite crise audiogène du rat albinos, *J. physiol. (Paris)*, *41*, 229A–231A.
[230] MERCIER, J., 1949, Influence du camphre et de la picrotoxine sur la crise audiogène du rat albinos, *J. physiol. (Paris)*, *41*, 231A–233A.
[231] MERCIER, J., 1950, Influence exercée sur la crise audiogène du rat albinos par quelques alcaloïdes naturels ou synthétiques dérivés de l'opium, *J. physiol. (Paris)*, *42*, 683–686.
[232] MERCIER, J., 1950, Influence exercée par la cocaïne, la pseudocaïne droite et la procaïne sur la crise audiogène du rat albinos, *J. physiol. (Paris)*, *42*, 679–683.
[233] MERCIER, J., 1950, Influence du gardénal, du métabromogardenal et du bromure de sodium sur la crise audiogène, *Compt. rend. Soc. Biol.*, *144*, 1174–1176.
[234] MERCIER, J., 1950, Influence des substitutions dans les noyaux de l'acide barbiturique et de l'hydantoïne, *Compt. rend. Soc. Biol.*, *144*, 1177–1178.
[235] MERCIER, J., 1951, Influence exercée par deux sympathicolytiques sur la crise audiogène, *Compt. rend. Soc. Biol.*, *145*, 906–908.
[236] MERCIER, J. and L. GARNIER, 1951, Influence exercée par l'hydratation du tissu cérébral sur la crise audiogène, *Compt. rend. Soc. Biol.*, *145*, 1199–1201.
[237] MERCIER, J., 1953, Etude physiologique et pharmacodynamique des crises épileptiques expérimentales. *Thesis*, Univ. of Aix, Marseille, France.
[238] MICHELS, K. M. and W. BEVAN JR., 1952, Audiogenic seizure in rats as a function of the volume of the test chamber, *J. Genet. Psychol.*, *81*, 185–191.
[239] MILLER, D. S., 1957, Influence of safe radiation levels on susceptibility to audiogenic seizures in mice (FDBA/1 C 57 132/6), *Anat. Rec.*, *128*, 589–590.
[240] MILLER, D. S. and M. Z. POTAS, 1956, Coincidence of low-level radiation and increased susceptibility to audiogenic seizures in mice, *Anat. Rec.*, *125*, 649–650.
[241] MILLER, D. S. and M. Z. POTAS, 1956, The influence of castration on susceptibility to audiogenic seizures in DBA mice, *Anat. Rec.*, *124*, 336.
[242] MILLER, D. S. and M. J. ZAMIS, 1949, Independent occurrence of audiogenic seizures and middle ear infections in inbred mice, *Anat. Rec.*, *105*, 556–557.
[243] MILLER, D. S., B. E. GINSBURG and M. Z. POTAS, 1952, Inheritance of seizure susceptibility in the house mouse (*Mus musculus*), *Genetics*, *37*, 605–606.
[244] MIRSKY, I. A., S. ELGART and C. D. ARING, 1943, Sonogenic convulsions in rats and mice. I. Control studies, *J. Comp. Psychol.*, *35*, 249–253.
[245] MITCHELL, W. G. and E. OGDEN, 1954, Influence of blood pH on susceptibility of rats to audiogenic seizures, *Am. J. Physiol.*, *179*, 225–228.
[246] MITCHELL, W. G. and R. C. GRUBBS, 1956, Inhibition of audiogenic seizures by carbon dioxyde. *Science*, *123*, 223–224.
[247] MITCHELL, W. G. and F. A. HITCHCOCK, 1956, Influence of decrease ambiant oxygen on susceptibility of rats to audiogenic seizures, *Am. J. Physiol.*, *187*, 571–572.
[248] MOLODKINA, L. N., 1956, Physiological analysis of experimental motor neurosis obtained by a method of sonic excitation, *Thesis*, Moscow (in Russian).
[249] MOORE, W. T., B. MOORE, J. B. NASH and G. A. EMERSON, 1952, Effect of maze running and sonic stimulation on voluntary alcohol intake of albino rats, *Texas Reports Biol. and Med.*, *10*, 59–65.
[250] MORGAN, C. T. and J. GOULD, 1941, Acoustical determinants of the "neurotic pattern" in rats, *Psychol. Rec.*, *4*, 258–268.
[251] MORGAN, C. T. and H. WALDMANN, 1941, Conflict and audiogenic seizures, *J. Comp. Psychol.*, *31*, 1–11.
[252] MORGAN, C. T., 1941, The latency of audiogenic seizures, *J. Comp. Psychol.*, *32*, 267–284.

253 MORGAN, C. T. and J. D. MORGAN, 1939, Auditory induction of an abnormal pattern of behavior in rats, *J. Comp. Psychol.*, *27*, 505–508.
254 MORGAN, C. T. and R. GALAMBOS, 1942, The production of audiogenic seizures by tones of low frequency, *Am. J. Psychol.*, *55*, 555–559.
255 MORGAN, C. T. and E. STELLAR, 1950, *Physiological Psychology*, McGraw-Hill, New York, pp. 523–527.
256 MORIN, G. and J. CAIN, 1947, Sur un comportement anormal provoqué par certains bruits sur le rat, *Compt. rend. Soc. Biol,141*, 1245–1246.
257 MORIN, G. and J. CAIN, 1947, Sur la signification des convulsions et des attitudes catatoniques provoquées par le son chez le rat, *Compt. rend. Soc. Biol.*, *141*, 1247–1248.
258 MORIN, G. and J. CAIN, 1949, Une forme connue d'épilepsie expérimentale: l'épilepsie audiogène, *La Médecine*, *30*, 12–16.
259 MORIN, G., F. CANAC, J. CAIN and G. GAVREAU, 1948, Obtention des crises audiogènes du rat par des sons définis, *Compt. rend. Soc. Biol.*, *142*, 359–360.
260 MORTON, J. and E. B. HALE, 1953, *The effect of high intensity sound on maze learning in albino rat*. Report submitted to Bio-Acoustic Unit. Aeromedical Lab, Washington D. C. Wright Patterson, A.F.B., Ohio.
261 MUNN, N. C., 1950, *Handbook of psychological research on the rat*, Houghton-Mifflin, Boston, pp. 420–439
262 NACHTSHEIM, H., 1939, 1941, Krampfbereitschaft und Genotypus. I. Die Epilepsie der Weissen Wiener Kaninchen, *Z. Mensch. Vererbungs und Konst. E.*, *22*, 791–810; II. Weitere Untersuchungen zur Epilepsie der Weissen Wiener Kaninchen, *ibid*, *L*, *25*, 229–244.
263 NOLAN, C. Y., 1952, Stimulational seizures without pseudopregnancy in white rats, *J. Comp. Physiol. Psychol.*, *45*, 183–187.
264 NORKINA, L. N., 1950, Influence of strong stimuli on the higher nervous activity of animals, *Fiziol. Zhur. SSSR*, *36*, 524–529 (in Russian).
265 NORTON, H. W., 1954, On audiogenic seizure and adrenal cortex. *Science*, *119*, 475–476.
266 PATTON, R. A. and H. W. KARN, 1941, Abnormal behavior in rats subjected to repeated auditory stimulation, *J. Comp. Psychol.*, *31*, 43–46.
267 PATTON, R. A., 1941, The effect of vitamins on convulsive seizures in rats subjected to auditory stimulation, *J. Comp. Psychol.*, *31*, 215–221.
268 PATTON, R. A., H. W. KARN and C. G. KING, 1941, Studies on the nutritional basis of abnormal behavior in albino rats. I. The effect of vit. B1 and vit. B complex deficiency on convulsive seizures, *J. Comp. Psychol.*, *32*, 543–50.
269 PATTON, R. A., H. W. KARN and C. G. KING, 1942, Studies on the nutritional basis of abnormal behavior in albino rats. II. Further analysis of the effect of inanition and vit. B1 on convulsive seizures, *J. Comp. Psychol.*, *33*, 253–258.
270 PATTON, R. A., H. W. KARN and C. G. KING, 1942, Studies on the nutritional basis of abnormal behavior in albino rats. III. The effect of different levels of vit. B1 intake on convulsive seizures. The effect of others vitamins in the B-complex and mineral supplements on convulsive seizures, *J. Comp. Psychol.*, *34*, 85–89.
271 PATTON, R. A., H. W. KARN and H. E. LONGNECKER, 1944, Studies on the nutritional basis of abnormal behavior in albino rats. IV. Convulsive seizures associated with pyridoxine deficiency, *J. Biol. Chem.*, *152*, 181–191.
272 PATTON, R. A. and H. E. LONGNECKER, 1945, Studies on the nutritional basis of abnormal behavior in the rat. V. The effect of pyridoxine deficiency upon sound induced magnesium tetany, *J. Comp. Psychol.*, *38*, 319–334.
273 PATTON, R. A. and A. D. LAZOVIK, 1946, Sensory preconditionning on the convulsions associated with magnesium deficiency in the rat, *J. Comp. Psychol.*, *39*, 265–273.
274 PATTON, R. A., 1946, The effect of rice polish concentrate on the incidence of sound induced convulsive seizures in young albino rats, *Am. Psychologist*, *1*, 275.
275 PATTON, R. A., 1946, Nutritional deficiency and sound induced convulsive seizures in the rat, *Ment. Health Bull.*, *24*, 16–20.
276 PATTON, R. A., 1947, The incidence of middle ear infection in albino rats susceptible to sound induced seizures, *Am. Psychologist*, *2*, 320.
277 PATTON, R. A., 1947, Sound induced convulsions in the hamster associated with magnesium deficiency, *J. Comp. Physiol. Psychol.*, *40*, 283–289.
278 PATTON, R. A., 1947, Vitamin B-complex concentrates and the incidence of sound induced seizures in young albino rats maintained on purified diets, *J. Comp. Physiol. Psychol.*, *40*, 323–332.
279 PATTON, R. A., 1947, The experimental approach to convulsive seizures, *Ment. Health Bull.*, *25*, 15.
280 PATTON, R. A., 1947, Maternal nutritional deficiency and the incidence of sound induced convulsions in young albino rats, *Am. Psychopath. Assoc.*, *36*, 204–209.

[281] PATTON, R. A., 1947, Purulent otitis media in albino rats susceptible to sound induced seizures, *J. Psychol.*, *24*, 313–317.
[282] PATTON, R. A., R. W. RUSSEL and J. F. PIERCE, 1949, The effects of auditory stimulation on the electro-convulsive threshold, *Am. Psychologist*, *4*, 233.
[283] PATTON, R. A., 1951, Abnormal behavior in animals, in C. P. STONE (Ed.), *Comparative psychology*, (3rd ed.) Prentice Hall, New York, pp. 479–498.
[284] PATTON, R. A. and L. M. SABARENKO, 1951, Further observations on the incidence of middle ear infection and sensitivity to sound induced convulsive seizures in young albino rats, unpublished manuscript, Univ. of Pittsburgh, cited in C. P. STONE (Ed.), *Comparative Psychology*, (3rd ed.), Prentice Hall, New York, p. 492.
[285] PENNINGTON, L. A., 1941, The effect of cortical destructions upon response to tones, *J. Comp. Neurol.*, *74*, 169–191.
[286] PFEIFFER, C. C., E. H. JENNEY, W. GALLACHER, R. P. SMITH, W. R. BEVAN, K. F. KILLAM and E. K. W. BLACKMORE, 1957, Stimulant effect of dimethylaminoethanol, *Science*, *126*, 610–611.
[287] PILGRIM, F. J. and R. A. PATTON, 1948, Reversibility of sound induced convulsions associated with pyridoxin deficiency, *Am. Psychologist*, *3*, 359.
[288] PILGRIM, F. J. and R. A. PATTON, 1949, Production and reversal of sensitivity to sound induced convulsions associated with a pyridoxin deficiency, *J. Comp. Physiol. Psychol.*, *42*, 422–426.
[289] PLETSCHER, A., A. P. SHORE and B. B. BRODIE, 1956, Serotonin as a mediator of reserpine action in brain, *J. Pharmacol. Exp. Therap.*, *116*, 84–89.
[290] PLETSCHER, A., 1957, Release of 5-Hydroxytryptamine by benzoquinolizine derivatives with sedative action, *Science*, *126*, 507.
[291] PLOTNIKOFF, N. P. and M. D. GREEN, 1957, Bioassay of potential ataraxic agents against audiogenic seizures in mice, *J. Pharmacol. Exp. Therap.*, *119*, 294–298.
[292] POLAND, R. G., E. A. HELSTROM and R. T. DAVIS, 1952, Running speed of rats as a function of auditory stimuli, *Proc. South Dakota Acad. Sci.*, *31*, 117–181.
[293] PORTMANN, M. and C. PORTMANN, 1952, Action comparative des sons intenses sur les systèmes de fibres nerveuses afférentes et efférentes de la cochlée, *Compt. rend. Soc. Biol.*, *146*, 1110–1111.
[294] QUADBECK, G., 1956, Die Auswertung anticonvulsiver Verbindungen am audiogenen Krampf der Ratte. *Naunyn-Schmiedeberg's Arch. exp. Pathol. Pharmakol. (Germany)*, *228*, 178–182.
[295] QUADBECK, G. und E. ROHM, 1956, Konstitutionsspezifität der anticonvulsiver Wirkung von 5-p-chlorophenyl-5-methyl-5-hydantoin beim audiogenen Krampf der Ratte, *Arzneimittel-Forsch.*, *6*, 531–533.
[296] QUADBECK, G. und G. D. SARTORI, 1957, Über den Einfluss von Pyridoxin und Pyridoxal-5 phosphat auf den Thiosemicarbazid-Krampf der Ratte, *Naunyn-Schmiedeberg's Arch. exp. Pathol. Pharmakol.*, *230*, 457–461.
[297] RABE, P. L., 1952, The cumulative frustration effect in the audiogenic seizure syndrome of DBA mice, *J. Genet. Psychol.*, *81*, 3–17.
[298] REESE, H. H., 1948, The relation of music to diseases of the brain, *Occup. Therap. and Rehabil.*, *1*, 12–18.
[299] REISS, B. F., 1948, The theorical basis of convulsive therapy in relation to animal experimentation, *J. Personality*, *17*, 9–15.
[300] RUSSEL, R. W., 1954, Comparative psychology, in C. P. STONE (Ed.), *Ann. Rev. Psychol.*, *5*, 234–237.
[301] SCOTT, J. P., 1950, The use as test material of inbred strains of mice having high frequency of audiogenic seizures, *Science*, *111*, 583.
[302] SEMIOKHINA, A. F., 1958, Electrophysiological study of the acoustic and motor analysers in a model of experimental motor neurosis, *Zh. Vysshei Nervnoi Deyatel'nosti*, *8*, (2), 278–285 (in Russian, summary in English).
[303] SERVIT, ZD. and J. STERC, 1958, Audiogenic epileptic seizures evoked in rats by artificial epileptogenic focus, *Nature*, *181*, 1475–1476.
[304] SHAFER, J. N. and D. R. MEYER, 1953, Effects of intense light stimulation on sound induced seizures in rats, *J. Comp. Physiol. Psychol.*, *46*, 305–306.
[305] SHAW, W. A., A. J. UTECHT and E. A. FULLANGER, 1953, The effect of auditory stimulations upon the immediate retention of a previously learned maze behavior in the albino rat, *J. Comp. Physiol. Psychol.*, *46*, 212–215.
[306] SHOHL, J., 1951, Effects of exposure to sound and discrimination performance in the rat, *Psychol. Monographies*, *65*, 1–19.
[307] SHORE, P. A. and B. B. BRODIE, 1957, Influence of various drugs on serotonin and norepinephrin in the brain, *Psychotropic Drugs*, Elsevier, Amsterdam, 423–427.
[308] SIEGEL, P. S. and O. LACEY, 1946, A further observation of electrically induced "audiogenic seizures" in the rat, *J. Comp. Psychol.*, *39*, 319–320.
[309] SISK, H. L., 1942, The effect of experimentally induced audiogenic seizures upon relearning in the white rat, *J. Psychol.*, *14*, 85–88.

310 Sisk, H. L., 1944, Maze learning ability and its relation to experimental audiogenic seizures in the rat, *J. Gen. Psychol.*, *30*, 89–91.
311 Smith, K. V., 1941, Quantitative analysis of the pattern of activity in audioepileptic seizure in rats, *J. Comp. Psychol.*, *32*, 311–328.
312 Smith, R. P. and W. Bevan, 1957, Maze performance, emotionality, audiogenic seizure susceptibility in rats and mice treated with 2-dimethylaminoethanol, *Proc. Exp. Biol. Med.*, *96*, 382–385.
313 Snee, T. J., C. F. Terrence and M. E. Crowly, 1942, Drugs facilitation of the audiogenic seizure, *J. Psychol.*, *13*, 223–227.
314 Stainbrook, E. J., 1942, A note on induced convulsions in the rat, *J. Psychol.*, *13*, 337–342.
315 Stainbrook, E. J. and H. de Jong, 1943, Symptoms of experimental catatonia in the audiogenic and electroshock reactions of rats, *J. Comp. Psychol.*, *36*, 75–78.
316 Stainbrook, E. J., 1947, The experimental induction of acute animal behavior disorders as a method in psychosomatic research, *Psychosomatic Med.*, *9*, 256–259.
317 Stainbrook, E. J., 1948, Experimentally induced convulsive reactions of laboratory rats. II. A comparative study of post convulsive maze behavior, *J. Gen. Psychol.*, *39*, 191–210.
318 Stechenko, A. P., 1959, Physiological study of haemorrhagic shock states, *Nauchn. Dokl. Vysshei Shkoly, Biol. Nauki*, (1), 74–78 (in Russian).
319 Swinyard, E. A. and G. B. Fink, 1959, Maximal audiogenic and electroshock seizures in mice and their alteration by psychopharmacologic drugs, *Federation Proc.*, *18*, 449.
320 Trent, S. E., 1956, Peripheral sensory inhibition of pain with a parietal lobe lesion. With a note on audiogenic fits, *J. Nerv. Mental Dis.*, *123*, 356–364.
321 Tsien, W. H., E. B. Sigg, H. Sheppard, A. J. Plummer and J. A. Schneider, 1956, Uptake of reserpine C_{14} by various areas of the cat brain, *Federation Proc.*, *15*, 493.
322 Turchioe, R., 1945, The effect of coramine on the facilitation of the audiogenic seizure, *J. Comp. Psychol.*, *38*, 103–107.
323 Udenfriend, S., D. F. Bogdanski and H. Weissbach, 1957, *Biochemistry and Metabolism of serotonin as it relates to the nervous system*, in Metabolism of the nervous system, D. Richter Ed, Pergamon Press, London, pp. 566–575.
324 Vassiliev, Y. A., 1934, On the nature of Parfenov's reaction, *Russ. Fysiol. J.*, *6*, 80 (in Russian).
325 Vassilieva, V. M., 1958, EGG study of motor cortical regions in white rats during epileptiform reactions, *Zh. Vysshei Nervnoi Deyatel'nosti*, *8*, 602–610 (in Russian).
326 Vassilieva, V. M., 1955, Study of the electrical activity of the brain in rats during the reflex epileptic seizure, *Thesis*, Moscow (in Russian).
327 Vicari, E. M., 1947, Etablissement of differences in susceptibility to audiogenic seizures of 5 endocrinic types of mice, *Anat. Rec.*, *97*, 407.
328 Vicari, E. M., 1948, A study of the genetic constitution of fatal and non-fatal audiogenic seizures in seven strains of mice. *Genetics, a*, *33*, 128–129.
329 Vicari, E. M., 1948, Age threshold of gene controlled sonogenic convulsions in mice modified by endocrine action and associated with physiologic threshold, *Genetics, b*, *33*, 632.
330 Vicari, E. M., 1949, Thiouracil prevents death which normally follows convulsions in sonogenic mice, *Anat. Rec.*, *103*, 594.
331 Vicari, E. M., 1950, Genetic aspects of the age threshold of susceptibility to sound induced convulsions in mice, *Genetics*, *35*, 137.
332 Vicari, E. M., 1951, Fatal convulsive seizure in the DBA mouse strain, *J. Psychol.*, *32*, 79–97.
333 Vicari, E. M., 1951, Effect of 6-n-prophylthiouracil on lethal seizures in mice, *Proc. Soc. Exp. Biol.*, *78*, 744–756.
334 Vicari, E. M., 1952, Hereditary fatal convulsions prevented by steroid hormones, *Genetics*, *37*, 634.
335 Vicari, E. M., A. Tracy and A. Jongbloed, 1952, Effect of epinephrine, glucose and certain steroids on fatal convulsive seizures in mice, *Proc. Soc. Exp. Biol. Med.*, *80*, 47–50.
336 Vicari, E. M., 1957, Audiogenic seizures and the A/jax mouse. *J. Psychol.*, *43*, 111–116.
337 Watson, M. L., 1939, The inheritance of epilepsy and waltzing in *Peromyscus*, *Contr. Lab. of vertebrate genetic. Univ. Mich.*, *1*, (11) pp. 1–24.
338 Weiner, H. M. and C. T. Morgan, 1945, Effect of cortical lesion upon audiogenic seizures, *J. Comp. Psychol.*, *38*, 199–208.
339 Williams, R. J., L. J. Beny and E. J. Beerstecker, 1949, Individual metabolic patterns alcoholism and genetrophic diseases, *Proc. Natl. Acad. Sci.*, *35*, 265–271.
340 Witt, G. and C. S. Hall, 1947, The genetics of audiogenic seizures in the house mouse, *Am. Psychologist*, *2*, 324.
341 Witt, G. and C. S. Hall, 1949, The genetics of audiogenic seizure in the house mice, *J. Comp. Physiol. Psychol.*, *42*, 58–63.
342 Yeakel, E. H., H. A. Shenkin, A. B. Rothballer and S. M. McCann, 1948, Blood pressure of rats subjected to auditory stimulation, *Am. J. Physiol.*, *155*, 118–127.

[343] YOSHII, N. and H. SASAKI, 1951, Electrocardiographic study of audiogenic seizure in the rat, *Med. J. Okasa Univ.*, 2, 121-131.
[344] ZJUZIN, I. K. and T. S. ZAICKINA, 1955, Influence of radioactive isotopes on convulsive seizures artificially induced in animals, *Zh. Nevropathol. Psikhiatr.*, 55, 343-344 (in Russian).
[345] ANONYMOUS, 1947, Death rings the bell, *Science Illustrated*, 2, 33.

ADDENDUM

ALLEGRANZA, A., 1956, Effect of anti-convulsivant and neuroplegic drugs on experimental epilepsy induced with ultrasonics, *Rev. neurol.*, 94, 395-399.
ANTHONY, A. and B. MARKS, 1959, Noise-induced convulsions in mice, *Experientia*, 15 (8), 320-322.
BEVAN, W. and R. McCHINN, 1956, Audiogenic convulsions in male rats before and after castration and during replacement therapy, *Physiol. Zool.*, 29, 309-313.
D'AMOUR, F. E. and A. B. SHAKLEE, 1955, Effect of audiogenic seizures on adrenal weight, *Am. J. Physiol.*, 183, 269-271.
DOBROKHOTOVA, L. P., 1958, The effect of hyperthyroidization on the functional condition of the nervous system in development of shock haemorrhagic conditions in animals under the effect of nervous trauma, *Problemy Endokrinol. i Gormonoterap.*, 4, (3) 12-21 (in Russian, summary in English).
GOUSSELNIKOVA, K. G. and N. L. KRUSHINSKAIA, 1958, Bioelectric changes in certain parts of the Cerebellum and of the cortical motor centers during epilepsy caused by sound stimulants, *Nauchn. Dokl. Vysshei Shkoly, Biol. Nauki*, (2), 78-82 (in Russian).
GOUSSELNIKOVA, K. G., 1959, Study on the bioelectrical activity of the olfactory lobes of rat's brain during epileptic seizures, *Problems of Epilepsy*, Moscow, 270-275 (in Russian).
KOCH, R., 1958, Audiogene sowie durch Pentomethylentetrazol und Strychnin hervorgerufene Krampfe bei adrenalektomierten Ratten, *Arzneimittel-Forsch.*, 8, 90-94.
KRUSHINSKI, L. V. and L. P. DOBROKHOTOVA, 1957, The influence of the thyroid gland on the mortality rate in shock haemorrhagic conditions caused by strong sound stimulants. *Bull. Exp. Biol. Med.*, (8), 46-49 (in Russian, summary in English).
KRUSHINSKI, L. V., D. A. FLESS and N. V. DUBROVINSKAYA, 1957, Peculiarities in the development of parabiotic stages during pathologic states of the brain provoked by sound stimulants in the rat. Summary of contributions to the Conference in memory of N. E. VEDENSKY on the 35th anniversary of his death. Vologda, 21-24 Oct., 1957, pp. 221-227 (in Russian).
KRUSHINSKI, L. V., L. N. MOLODKINA and N. A. LEVITINA, 1959, Time and conditions of restoring and exhausted inhibitory process under the action of acoustic stimuli, *Zh. Vysshei Nervnoi Deyatel'nosti*, 9, 566-572 (in Russian, summary in English).
KRUSHINSKI, L. V., 1959, Investigations on the physiological mechanism of reflex epileptic seizures, *Problems of Epilepsy*, Moscow, pp. 245-259 (in Russian).
LEHMANN, A and R. G. BUSNEL, 1959, Etude sur la crise audiogène, *Archives des Sciences Physiologiques*, 13, 193-225.
LEVY, G. W. and W. BEVAN, 1958, A failure to find social facilitation of audiogenic seizures in the rat, *Animal Behaviour*, 6, 43-44.
LUSH, J. L., 1930, Nervous goats, *J. Hered.*, 21, 242-247.
MOLODKINA, L. N. and B. I. KOTELAR, 1957, Influence of promedol on excitability and on the state of the inhibition process of the central nervous system. Summary of contributions to the Conference in memory of N. E. VEDENSKY on the 35th anniversary of his death. Vologda, 21-24 Oct., 1957 p. 86-87 (in Russian).
MOLODKINA, L. N., 1959, New investigations in the field of nervous activity, *Priroda*, (4), 97-99 (in Russian).
NELLHAUS, G., 1958, Experimental Epilepsy in Rabbits. Development of a Strain Susceptible to Audiogenic Seizures, *Amer. J. Physiol.*, 193, (3), 567-572.
NOVAKOVA, V., 1958, Changes in higher nervous activity following experimental epileptic seizures in rats, *Physiol. Bohemosloven*, 7, (1), 102-108.
PAVLOVA, E. B., 1957, Effect of a strong sound stimulus on higher nervous activity in the rat, *Zh. Vysshei Nervnoi Deyatel'nosti*, 7, 754-764 (in Russian).
PROKOPETZ, I. M., 1958, Experimental study of the role of the functional cataleptic state in recovery and defence, *Nauch. Dokl. Vysshei Chkoly Biol. Naouki*, (3), 84-89 (in Russian).
SEMIOKHINA, A. F., 1959, Electrophysiological study of the acoustic and motor analysers during the reflex epileptic seizures, *Problems of Epilepsy*, Moscow, pp. 259-269 (in Russian).
WERBOFF, J. and J. CORCORAN, 1959, Sex differences in audiogenic seizures susceptibility, *Anat. Rec.*, 134, 3, 652-653.
WILSON, C., 1959, Drug antagonism and audiogenic seizures in mice, *Brit. J. Pharm.*, 14, 415-419.

PART IV

EMISSION AND RECEPTION OF SOUNDS:
MORPHOLOGICAL, PHYSIOLOGICAL AND PHYSICAL ASPECTS

A – INVERTEBRATES (Chapters 11–17)
B – VERTEBRATES (Chapters 18–20)

CHAPTER 11

MORPHOLOGY OF SOUND EMISSION APPARATUS IN ARTHROPODA

by

BERNARD DUMORTIER

1. Introduction

Entomological literature has been enriched during the last hundred years by a considerable number of works on the sound emissions of insects, thus giving us today a fairly precise knowledge of the phenomenon. Nevertheless, although the publications dealing with the morphology and operation of the sound apparatus of particular insects are numerous, attempts to correlate these scattered data are more unusual.

However, one can cite in this connection the general reviews of Swinton (1881)[233], Prochnow (1908)[207], and Weiss (1914)[254], as well as the publications of Gahan (1900)[98], Düdich (1920–1922)[73-75], and Meixner (1934)[178] on Coleoptera in general; those of Arrow (1904)[11] on lamellicorn beetles, and those of Marcu (1930–1939)[173-175] on various other groups. As regards the Orthoptera which (at least for European and North American forms) represent the order most extensively studied from the acoustic point of view, the most general work is that of Pierce (1948)[200]. Kevan (1955)[152] has completed the picture by drawing up a list of the Orthoptera that use "unorthodox" methods. Studies on Homoptera have been chiefly carried out by Pringle. Ossiannilsson (1949)[190] published a very thorough monograph on the auchenorrhynchous Homoptera of northern Europe. The important contribution of Alexander (1956)[4] on Orthoptera and Cicadidae of the Eastern United States should also be mentioned.

It will be noticed that these works deal only with insects. General surveys of stridulation in other Arthropoda are rarely found: Berland (1932)[27] on Arachnida; Balss (1921 and 1956) ([18-19]) on Crustacea; Guinot-Dumortier and Dumortier (1960)[111] on crabs in particular.

As to syntheses giving a general picture of stridulation among Arthropoda as a whole, the only ones that can really be mentioned are those of Landois (1874)[159], among more recent studies the publication of Scharrer (1931)[218], and especially the "Tierstimmen" of Tembrock (1959)[234]. The last work, more concerned with vertebrates than with arthropods, indicates the present status of acoustics in the animal kingdom.

In this chapter we do not intend to carry out an exhaustive study of the problem, but rather to give primary attention to recent discoveries while reviewing the essential features of the knowledge acquired over a long period. Still we must make a few reservations in this preamble as to the value of certain early observations to which we have had to refer for want of recent data.

Two broad types of processes are responsible for sound emissions in the animal kingdom:
   1. the vibration of elastic structures in the breathing tube, and

References p. 338

2. the friction of rigid parts often provided with a special surface.

The sounds produced by vertebrates come within the first category, while most invertebrates* use methods belonging to the second.

Although terrestrial vertebrates — birds on the one hand with their syrinx, mammals on the other hand with their larynx — show a consistent arrangement of the vocal apparatus throughout their respective lines, the opposite is true in the invertebrates, where there is the greatest disparity in the morphology of the sound organs. Paradoxically we shall see that certain processes used by insects (see pp. 323 and 325) are very close to those which characterize the terrestrial vertebrates.

Sound emissions of invertebrates may have five different origins:
1. friction of differentiated parts,
2. vibration of membranes,
3. expulsion of a fluid (gas or liquid),
4. shocks to the substrate,
5. vibration of appendages.

To these five categories a sixth should be added comprising animals in which stridulation is related to the use of non-differentiated regions.

We shall consider in turn each of these six categories, dealing separately with insects, crustaceans, and arachnids. We shall indicate whenever possible whether stridulation has in fact been heard, as well as the circumstances in which it is produced.

At the end of the description of methods of sound emission in adults, a special paragraph will be devoted to sound emissions in pre-imaginal forms (see p. 330).

A more detailed study on the stridulation of Hemipteroida has been given in Chapter 14 by D. Leston and J. W. S. Pringle. Consequently, insects of this order will be studied in less detail in the following section.

2. Sound Emission by Friction of Differentiated Regions

This method is ordinarily termed "stridulation". Though etymologically the term stridulation may be applied to every emission of grating or piercing sounds, no matter how they are produced, we shall employ it more especially for those sounds which originate through the rubbing of two specialised surfaces against one another.

The stridulatory apparatus is always composed of two parts. The first, named "*pars stridens*" (or again file or strigil) is a part of the animal's body on which surfaces have developed, the composition and size of which are extremely variable (hairs, spines, tubercles, teeth, ridges, ribs, etc.).

The second part is the "*plectrum*" (scraper or strigilator). Here again the morphology differs very widely according to the animals in question. The plectrum is sometimes a protrusion with a sharp edge, or a tooth, or a line of denticulations; at times, also, it is composed of the tapering edge of an appendage (elytron) or of a joint (femur), etc. Stridulation is produced by the rubbing of the plectrum on the pars stridens (or *vice versa*).

Actually, the distinction between these two components of the stridulatory

---

* This actually refers only to arthropods: when other invertebrates produce "noises" (rapid snapping of the valves of the pecten, tapping of the pedicellariae in sea-urchins, etc.), these are only the consequence of some other activity (taking in food, locomotion), and there is no reason to believe that they play a special role in behaviour.

apparatus is generally convenient but somewhat artificial. Indeed, at times it becomes impossible. It is frequently the case with crustaceans and arachnids, where the two parts of the apparatus may have equally large or even identical protrusions. In addition, although the pars stridens is usually the place where the sound waves seem actually originate under the effect of the mechanical excitation by the plectrum, certain structures suggest that the reverse is equally possible. In this case, the part which corresponds morphologically to the pars stridens sets the plectrum in vibration.

To these reservations we shall add another: certain authors' descriptions are not accurate enough to distinguish with absolute certainty which is the plectrum and which is the pars stridens.

However, we shall use this terminology to make the description easier, though we are fully aware of the element of approximation which it inevitably introduces.

(a) Insects

(i) Cephalic pars stridens

Antenno-antennary method

Only a few examples of this method seem to exist. It was discovered in female *Pulchriphyllium crurifolium* (Serv.) and in *Phyllium athanysus* Westw. (Phasmoptera, Phylliidae)[122, 152] (Fig. 113).

The apparatus, situated on the third segment of each antenna, is composed of a striated line (plectrum), and a row of small tubercles (pars stridens). The stridulation is produced when the two joints rub against each other.

Buccal method (maxillo-mandibular)*

A stridulatory apparatus making use of this method has been described in the genus *Cylindracheta* Kirby (Orthoptera, Tridactyloidea) (Fig. 114). At the proximal

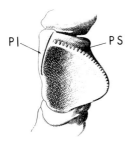
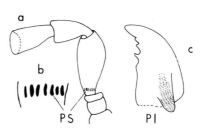
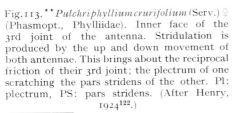

Fig. 113.** *Pulchriphyllium crurifolium* (Serv.) ♀ (Phasmopt., Phylliidae). Inner face of the 3rd joint of the antenna. Stridulation is produced by the up and down movement of both antennae. This brings about the reciprocal friction of their 3rd joint; the plectrum of one scratching the pars stridens of the other. Pl: plectrum, PS: pars stridens. (After Henry, 1924[122].)

Fig. 114. *Cylindracheta arenivaga* Tind. (Orth., Cylindrachetidae). (a) right maxillary palpus showing the pars stridens, (b) enlarged pars stridens, (c) left mandible with a rough knob (= plectrum).
Pl: plectrum, PS: pars stridens. (After Tindale, 1928[237], modified by Kevan, 1955[152].)

---

* The noises produced by the mandibles will be dealt with in connection with stridulations made in the absence of a differentiated apparatus (see p. 329). A maxillo-mandibulary apparatus is found in larvae of several Coleoptera (see p. 330).

** The drawings of this chapter (except those related to Crustacea) were done by Mr. J. Rebière.

References p. 338

part of its 3rd joint, the maxillary palp has a rasp against which a rugged protrusion of the mandible rubs[152, 237].

Rostro-tarsal method

Species of *Corixa* Geoff. (Heteroptera, Corixoidea) scratch the transversal ridges of the clypeus with the point of the tarsus[135] (Fig. 115). A similar method has been pointed out in the related genus *Sigara* Fabr.[38]. These insects have moreover other methods of stridulation (see p. 292).

Cranio-prothoracic method

This method appears to be represented only in the Coleoptera where numerous examples are known[12, 73, 98].

In different genera belonging to the families Languriidae, Nitidulidae, Endomychidae, Hispidae, etc. the vertex has one or two sagittal striated lines, which rub against the sharp edge of the pronotum through a vertical movement of the head (Fig. 116, see also p. 304).

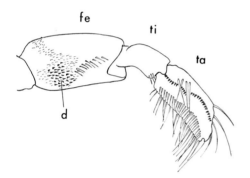
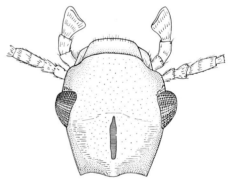

Fig. 115. *Corixa striata* L. (Hempipt., Corixidae). Foreleg of the male showing the denticulated area of the femur. This area is scratched by the tarsus of the opposite leg. d: denticulated area, fe: femur, ta: tarsus, ti: tibia. (After v. Mitis, 1936.*)

Fig. 116. *Enoplopus velikensis* Piller (Col., Tenebrionidae). Dorsal view of the head showing the pars stridens on the vertex. (After Dudich, 1920[73].)

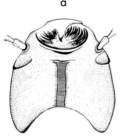
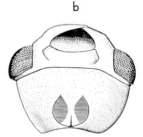

Fig. 117. Two examples of stridulatory apparatus on the gula. Ventral view of the head of (a) *Scolytus destructor* Ol. (Col., Ipidae), (b) *Priobium castaneum* Sturm (Col. Anobiidae). (After Gahan, 1900[98].)

---

* *Z. Morphol. u. Ökol. Tiere, 30.*

The position of the pars stridens is reversed in various Tenebrionidae, Chrysomelidae, and Scolytidae, where it is situated on the gula. A sharp fold of the prosternum serves as plectrum (Fig. 117).

(ii) Thoracic pars stridens

Prosterno-rostral method

Various genera of the families Reduviidae and Phymatidae (Heteroptera) have a longitudinal channel on the proternum, transversely striated, which the animal scratches with the point of its rostrum[257] (Fig. 118).

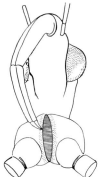
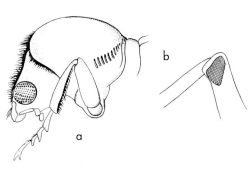

Fig. 118. *Coranus subapterus* (De Geer) (Hémipt., Reduviidae). Ventral view of the head and prosternum, showing the tip of the rostrum scratching the intercoxal pars stridens. (After Poisson, 1951[203].)

Fig. 119. *Phonapate nitidipennis* Waterh. ♀ (Col., Bostrychidae). Pronoto-femoral apparatus, (a) anterior region of the insect showing the ribs on the prothorax, (b) inner side of the foreleg showing the striated area on the femur. (After Gahan, 1900[98].)

Prosterno-mesosternal method

The genus *Serica* Mc Leay (Coleoptera, Melolonthinae) has a stridulatory apparatus formed by a striated protrusion of the prosternum against which the edge of the mesosternum rubs. However, in *S. brunnea* L., the stridulation occurs when an apophysis of the metasternum moves against a striation situated in the intercoxal area of the prosternum[257].

Pronoto-femoral method

In the genus *Phonapate* Lesne (Coleoptera, Bostrychidae), the female has a striated area on the inside of the distal end of the front femur which rubs against a row of 6 or 7 ribs on the side of the prothorax (Fig. 119)[12].

Mesonoto-pronotal method

This is the characteristic method of Coleoptera Cerambycidae. The front of the mesonotum dips downwards, and its inflated median part has a very fine striation over which the tapered hind edge of the pronotum passes (Fig. 120). There is a similar apparatus in the genus *Clythra* Laich (Coleoptera, Chrysomelidae)[12].

References p. 338

Metathoracic methods

In the Lepidoptera of the genus *Endrosa* Hb. and *Euprepia* O. (Arctiidae), the episternum is considerably hypertrophied in a voluminous bulb, smooth or striated. This structure undoubtedly plays a part in the noise made by the male when flying, but the exact mechanism of this emission is not known[31,195,196]. *Rhodogastria bubo* Walker and *R. lupia* Druce (Arctiidae) have a convex oval formation on the metapleurites which is striated and scrapes against the prothorax[51] (Fig. 121). These two insects also exude a frothy substance. In the case of *R. bubo* this expulsion occurs at the same time as the sound emission (see p. 325).

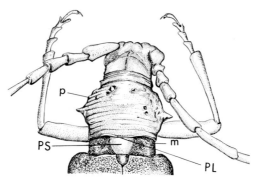

Fig. 120. *Ergastes faber* L. (Col., Cerambycidae). The head and the pronotum are bent downwards to show the median swelling of the mesonotum (pars stridens). m: mesonotum, p: pronotum Pl: plectrum, PS: pars stridens (original).

(iii) Abdominal pars stridens

Abdomino-pronotal method

This form of stridulation appears to exist only in *Cyphoderris monstrosus* Uhl. male and female (Orthoptera, Prophalangopsoidea). It consists of a striated area on the sides of the first abdominal tergite which is rubbed by the postero-lateral edge of the pronotum fitted with small teeth (Fig. 122). This archaic insect also possesses, although in a rudimentary stage, the classic elytral apparatus of the Ensifera[7].

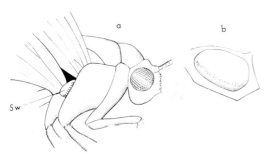

Fig. 121. *Rhodogastria lupia* Druce (Lépid., Arctiidae). (a) schematic view of the right anterior part showing the striated metapleural swelling (sw), (b) enlarged view of the swelling. (After Carpenter, 1938[51].)

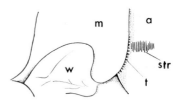

Fig. 122. *Cyphoderris monstrosus* Uhler ♀ (Orth., Prophalangopsidae). Stridulatory apparatus (left side). a: 1st abdominal tergite, m: metanotum, str: striae of the 1st abdominal tergite, t: teeth of the metanotum, w: atrophied hind wing. (After Ander, 1938[7].)

Abdominal method

Under this heading we group the mode of stridulation used by ants. The majority of the lower families (Poneridae, Dorylidae, primitive Myrmicidae) have a stridulatory apparatus consisting of a transverse striation on the first tergite of the gaster, which is scratched by the posterior dorsal face of the last petiole segment ending in a sharp blade[93, 139, 140, 224] (Fig. 123). (This method is abdominal since the last segment of the petiole and the first segment of the gaster are in fact segments 3 and 4 of the abdomen.)

In *Pogonomyrmex marcusi* (Myrmicidae) there is a double stridulatory apparatus situated at the juncture of the petiole and the gaster, one ventral and the other dorsal[177] (Fig. 124).

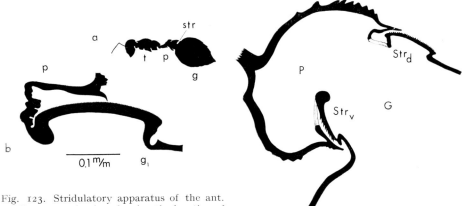

Fig. 123. Stridulatory apparatus of the ant. (a) diagrammatic view showing the location of the apparatus, (b) diagrammatic sagittal section of the apparatus of *Myrmica rubra laevinodis* Nyl.
g: gaster, g_1: first tergite of the gaster, p: petiole, str: stridulatory apparatus, t: thorax, ((b) after Janet, 1894[140].)

Fig. 124. *Pogonomyrmex marcusi* (Hym., Myrmicidae). Schematic section showing the two stridulatory devices. G: gaster, P: petiole, str_d: dorsal stridulatory device, str_v: ventral stridulatory device. (Simplified after Marcus and Marcus, 1951[177].)

Dorymyrmex emmaericaellus (Kusn) has a stridulatory apparatus located differently from the classic type: the pars stridens on the petiole is rubbed by a laminated expansion of the thorax[177].

It seems certain that the stridulations of a number of species are effective and audible to man (*e.g.* the large tropical Poneridae[28, 60]). But among smaller types, the fact that the movements of the gaster against the petiole are not accompanied by any audible sound perceptible to our ears, often leads to the apparently erroneous belief that the stridulation may be ultra-sonic (see p. 350).

In some species of Mutillidae (Hymenoptera), it has been noticed that females produced stridulations (*Mutilla barbara brutia* Petagna, *M. europaea* L.). The posterior edge of a segment scratches the anterior edge of the following one which is striated.

Another example of abdominal stridulation is supplied by the Lepidoptera Lymantriidae, *Lymantria monacha* L. and *Stilpnotia salicis* L. The males of these have a cavity on the 3rd urite, one part of which is rugged, and the other striated. Through the action of a particular muscle these two areas can rub against each other[31, 157].

Arcte caerulea Guénée (Noctuidae) has a broad sagittal crest, transversally ribbed, on one of the last abdominal tergites. This formation may have some connection with the clicking noise the insect makes in flight[109]. It can be seen in the 4 species of the genus (Fig. 125).

Abdomino-tibial method

A few rare examples in Heteroptera, Homoptera Aphididae, and some Orthoptera Acrididae illustrate this method. *Nabis flavomarginatus* Schltz. (Heteroptera, Reduvioidea) rubs its hind tibiae on the sides of the 9th abdominal tergite, which plays the role of pars stridens.

In the genera *Tetyra* Fabr. and *Pachychoris* Burm. (Heteroptera, Pentatomidae) the friction is produced against the 5th and 6th abdominal sternites[203].

The same principle is found in the Homoptera, Aphididae, with the plant-lice of the genera *Toxoptera* Kock, *Longiunguis* V. D. Goot and *Aphis* L. The noise of a fairly large group of *T. aurantii* (Fonsc.), is audible eighteen inches[261] away*. This sound emission may result from the rubbing of the short hairs of the hind tibiae, against lines of raised and serrated cuticle on the 5th and 6th abdominal sternites[77] (Fig. 126).

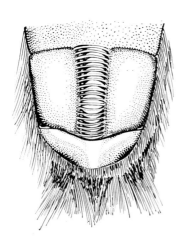
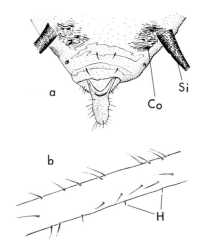

Fig. 125. *Arcte caerulea* Guenee ♂ (Lepid., Noctuidae). Posterior part of the abdomen seen from above showing the ribbed crest (hairs and scales partially removed). (After Gravely, 1915[109].)

Fig. 126. *Toxoptera aurantii* (Fonsc.) (Homopt., Aphididae). (a) ventral end of the abdomen, (b) part of hind tibia highly enlarged, showing the modified hairs.
co: cuticular ornamentation (pars stridens), h: stridulatory hairs, si: siphunculus. (After Eastop, 1952[77].)

An ornamentation of clusters of vertical flutings on the 4th to 8th abdominal tergites is seen in the genus *Charora* Sauss. (Orthoptera, Acrididae). This pars stridens is rubbed by the robust spines of the hind tibiae[29, 124] (Fig. 127). The same kind of apparatus is found in the genus *Egnatioides* Vosseler[124].

---

* In this species, the sound is emitted in a synchronized way by all the individuals grouped on a leaf. It is accompanied by a rhythmical raising and lowering movement of the abdomen.

Abdomino-femoral method

Chopard (1938)[55] and especially Kevan (1954)[152] have reviewed this method very thoroughly in the Orthoptera, which, with the Coleoptera, are without doubt the only insects to use it. Numerous examples are found in the Orthoptera, Ensifera (Gryllacridoidea), and Caelifera (Acrididae, Pamphagidae and Pneumoridae).

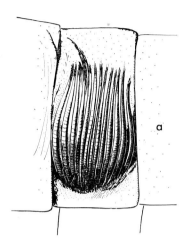 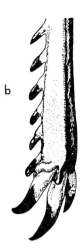

Fig. 127. *Charora crassinervosa* Sauss. ♂ (Orth., Acrididae). (a) sixth abdominal tergite (left) showing the stridulatory flutings, (b) hind tibia with strong spines. (After Henry, 1942[124].)

In Gryllacridoidea, abdomino-femoral stridulation appears commonly in both sexes, and begins in the final nymphal stages[7,8]. The apparatus consists of one or several rows of granules on the 2nd and 3rd abdominal segments, against which rubs a longitudinal protrusion or a rough surface of the femur (Fig. 128). The stridulation of the New Zealand species *Deinacrida megacephala* Buller has been heard[131].

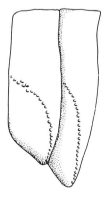 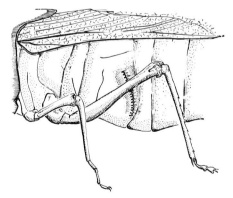

Fig. 128. *Ametroïdes kilonotensis* (Sjöst.) ♀ (Orth., Gryllacridoidea). Second and third abdominal segment (left) with a row a stridulatory tubercles rubbed by a crest of the hind femur. (After Kevan, 1955[152].)

Fig. 129. *Physemacris variolosa* L. (Orth., Pneumoridae). Stridulatory apparatus (original).

In Pneumoridae (Acridoidea), the pars stridens is much more distinct. Actually, it stands out on the side of the 2nd abdominal tergite, in the form of a semi-lunar projection covered with large radial ribs. Friction is produced by the hind femora, which are armed with rows of denticles (Fig. 129).

The stridulation of the males Pneumoridae has been long known, since Thunberg discovered it in 1795[236]. It is encountered in the genera *Pneumora* Thunb., *Bullacris* Roberts, *Physemacris* Roberts. The abdomen of these insects, considerably dilated, probably serves as a sound box[152].

Some Acrididae also practise the abdomino-femoral method. In *Xyronotus aztecus* Sauss., (Fig. 130), the pars stridens is fairly similar to that of *Pneumora*. On the other hand, *Phonogaster cariniventris* Henry offers a special arrangement. Two parallel ridges run the length of tergites 3, 4 and 5. Friction is ensured by a line of denticulations on the femur[124].

The stridulatory apparatus of the Pamphagidae, Batrachotetriginae and Pamphaginae (Acridoidea), is based on the same principle as that of the Pneumoridae which it resembles. The noise of the South-African species *Trachypterella andersoni* (St.) is quite intense[216].

Some examples of this method are found among Coleoptera, where it has been reported in the genus *Lomaptera* Gory & Perch. (Cetoninae). The stridulatory area on which the femur rubs is situated on the 2nd and 3rd abdomina sternites[223].

In the Heteroceridae, which live in colonies in hollowed-out galleries in the sand of beaches, the stridulatory apparatus is formed by a curved projecting line situated on the first abdominal sternite across which the femur of the hind legs passes. An abdomino-femoral apparatus has also been described in Hemiptera, Aradidae[241] (Fig. 131).

Abdomino-elytral method

This method is practised by a great number of Coleoptera, but this is not neces-

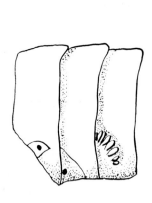

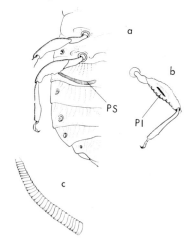

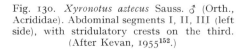

Fig. 130. *Xyronotus aztecus* Sauss. ♂ (Orth., Acrididae). Abdominal segments I, II, III (left side), with stridulatory crests on the third. (After Kevan, 1955[152].)

Fig. 131. *Pictinus* sp. (Héteropt., Aradidae). (a) ventral view showing the pars stridens (PS) on the second abdominal segment, (b) hind leg with plectrum (Pl) on the femur, (c) detail of the pars stridens. (After Usinger, 1954[241].)

sarily a sign of systematic connections, since it appears among Hydrophilidae, Chrysomelidae, Tenebrionidae, Scrabaeidae (numerous examples), Carabidae, Curculionidae, Silphidae and probably some others.

The functional principle is simple: one of the abdominal tergites offers a single- or double-striated surface (pars stridens) on which a differentiated or non-differentiated part of the elytra (plectrum) scratches, during a forward and backward movement of the abdomen[98].

Among Hydrophilidae, an abdomino-elytral apparatus is found in the genus *Spercheus* Kugel. In *Spercheus emarginatus* Schall., it consists of a pair of fluted swellings, borne by the sides of the first abdominal segment. At the same level the inside edge of the elytra presents an area covered with tiny protuberances. The apparatus would be put into action by a pendular movement of the abdomen. Males and females sing in both air and water. The stridulation is produced when the animal is disturbed, and when it is moving about[40,96].

It has been reported that, during mating, male *Hydrophilus piceus* L. produces an almost uninterrupted stridulation caused by the friction of a slightly roughened area of the elytra against the more or less rough edge of the 2nd abdominal tergite. In *Berosus* Leach, the apparatus would be the same, but the 2nd tergite would show a more developed specialised growth in the form of transverse crests[35].

The Chrysomelidae also offer several examples. The apparatus of *Lema trilineata* Ol. (male and female) is formed by a triangular striated surface of the pygidium, against which the distal edge of the elytra passes[253] (Fig. 132).

The apparatus in the genus *Crioceris* Geoffr. is closely related (Fig. 133). In the genus *Clytra*, Darwin[70] recognised an analogous apparatus, but this observation is contested by Gahan[98] who attributes the stridulation to a mesonoto-pronotal method (see p. 281).

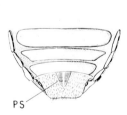
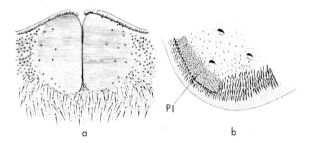

Fig. 132. *Lema trilineata* Ol. (Col., Chrysomelidae). Dorsal view of the abdomen, with pars stridens (PS) on the pygidium. (After Walker, 1899[253].)

Fig. 133. *Crioceris duodecimpunctata* L. (Col., Chrysomelidae). (a) pygidial pars stridens, (b) underside of the elytron with corrugated area (plectrum: Pl). (After Dingler, 1932, in Meixner, 1934[178].)

In Tenebrionidae, some cases are also known. *Heliophilus cribratostriatus* (Mulsant) has three or four striated areas on the apical edge of the elytra near the sutural angle, which rub against a file on the pygidium. The same formation is found in *Olocrates gibbus* F. and male *O. abbreviatus* Ol. (Fig. 134). The insect stridulates if it is turned on its back, or generally speaking, when it is disturbed. It does not seem that this stridulation has any effect on the female[211].

The most numerous examples certainly seem to occur in the family Scarabaeidae, where Arrow[11] noted, in the sub-family Dynastinae alone, about twenty genera performing the abdomino-elytral stridulation. In male *Oryctes nasicornis* L. the propygidium is very finely grained in its median part (Fig. 135). The stridulation of the related species* *O.rhinoceros* L. has been heard[109]. This differentiation of the pygidial or propygidial region is most marked in the genus *Camelonotus* Fairm. (Dynastinae), where the pars stridens is "one of the best-developed to be found among insects"[11].

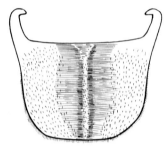 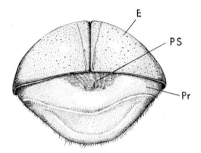

Fig. 134. *Olocrates abbreviatus* Ol. ♂ (Col., Tenebrionidae). Pygidium. (After Remy, 1935[211].)

Fig. 135. *Oryctes nasicornis* L. ♂ (Col., Scarabaeidae). Back view showing the pars stridens on the propygidium. E: elytron, Pr: propygidium, PS: pars stridens. (Original).

Sound emission has been reported in several species of *Geotrupes*. It is caused by the friction of the elytra against some surfaces covered with tiny spines on the abdominal tergites, this method being in addition to the typical coxo-metasternal method of the genus[246] (see p. 291).

We shall cite two examples in the family Carabidae: the genera *Elaphrus* Fabr. and *Blethisa* Bon. in which the apparatus is almost the same[98]. The one part consists of a file situated on each side of the propygidium, striated in lines parallel to the axis of the body (Fig. 136). As for the other part, the epipleural fold of the elytra is itself striated longitudinally. The orientation of the striation of the two parts does not allow friction to be effected by a backwards and forwards movement of the abdomen, as happens in most cases. It necessitates, rather, a lateral movement like that already reported in *Spercheus* (Hydrophilidae) (see p. 287).

Some Curculionidae, as Landois[159] has already observed, also possess a stridulatory apparatus. In the genera *Cryptorrhynchus* Kl., *Camptorrhinus* Schonh., *Gasterocercus* Lap & Brullé and *Ectatorrhinus* Lacord., the pars stridens is pygidial in the female (with plectrum on the elytra), while in the male the arrangement is the reverse: an elytral pars stridens and a pygidial plectrum[98] (see p. 293).

The Silphidae of the genus *Necrophorus* F. have two transversely striated bands on either side of the median line on the 5th abdominal tergite on which passes the apical edge of the elytron which does not show any particular differentiation, (Fig. 137).

Finally, the stridulation which can be heard from the Scarabaeidae, *Polyphylla fullo* L. and *Anoxia villosa* F. when they are disturbed, may originate from the rubbing

---

* These species are very remarkable since the larva and pupa stridulate also (see pp. 330, 331).

of tergites 3, 4, 5 and 6 covered with very small rough spots, on the outer edge of the elytron[20,21].

Outside Coleoptera, examples of the abdomino-elytral method are much rarer. However, we can cite *Archiblatta hoeveni* Voll. (Dictoypt., Blattidae), which possesses a rudimentary organ on the first abdominal segment, rubbed by the lower face of the elytra[55,148].

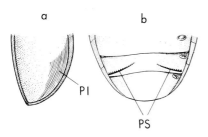

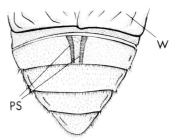

Fig. 136. *Blethisa* sp. (Col., Carabidae). (a) underside of the left elytron with plectrum (Pl) (b) pars stridens (PS) on the propygidium. (After Gahan, 1900[98].)

Fig. 137. *Necrophorus vespillo* L. (Col., Silphidae). Dorsal view of the abdomen (elytra removed) PS: pars stridens, W: wing. (After v. Frankenberg, 1936.)*

Abdomino-alary method

This method, quite infrequent, appears only to occur in the Coleoptera Heteroptera and Diptera. Fabre[87], in his "Souvenirs Entomologiques" had made mention of a sound emission—though quite weak—in *Bolbelasmus gallicus* Muls. (Geotrupidae) and Arrow[11] followed this observation with a description of an apparatus consisting of surfaces covered with spinules on the last abdominal tergites, in contact with which there is a spinulous region situated right against the costal vein of the wing. According to this author this apparatus is found equally in *B. unicornis* Schrank, and *B. bocchus* Er. In *Passalus cornutus* Fabr. (Passalidae), the fifth tergite has two oval surfaces covered with very small spines. The movement of the abdomen makes these regions

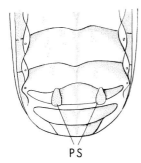

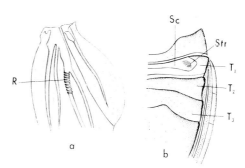

Fig. 138. *Passalus cornutus* Fabr. (Col., Passalidae). Spiny area of the 5th abdominal tergite (pars stridens: PS). (After Babb, 1901[16].)

Fig. 139. *Tessaratoma papillosa* Thunb. (Homopt., Pentatomidae). (a) ribbed wing-vein, (b) dorsal view showing the spatulate sclerite with stridulatory area.
R: ribs of the vein, Sc: sclerite, Str: stridulatory area, T_1, T_2, T_3: abdominal tergites. (After Kershaw, 1907[151].)

* *Natur u. Volk*, 66, 1.

rub against the wings (Fig. 138). The removal of the elytra does not suppress the sound emission, which, in a general way, takes place when the animal is disturbed[16].

In the genus *Proculus* Kaup. (Passalidae), the wings, considerably atrophied, are covered with tiny conical teeth on their lower side. Opposite the wings the abdomen has some differentiated zones where the friction is effected[11].

The stridulation heard in the South-American Dynastinae of the genus *Phileurus* Latr., apparently must be produced by the passage of a chitinous wing plate across the two tergites anterior to the propygidium[11].

As for the Hemiptera, a case of abdomino-alary stridulation is known in *Tessaratoma papillosa* Thunb. (Pentatomidae). At the level of the first abdominal tergite is found a sclerite broadened out into a spatula shape at its two extremities, which are also provided with a finely striated convex surface. This structure can move at an angle of 35°, which brings it into contact with a wing-vein covered with transversal ribs[151] (Fig. 139).

Except for the Coleoptera and Hemiptera, this process of sound emission is not very widespread. Mention might, however, be made of Diptera of the genus *Dacus* Fabr. (Trypetidae) (Fig. 140). In male *D. tryoni* Frogg. the sides of the 3rd abdominal tergite have a row of some twenty bristles against which the wing rubs; the wing itself is not provided with conspicuous stridluatory structures[187]. Sound emissions are short (80–350 msec)[184]. An identical apparatus exists in *D. oleae* Gmel.[90].

(iv) *Pars stridens on the legs**

Coxo-prosternal method

In the imagos and nymphs of *Ranatra quadridentata* Stal (Heteroptera, Nepidae) which stridulate just as well in water as in air, the fore coxa have a file that comes into

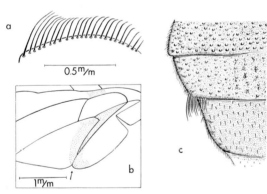
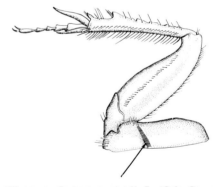

Fig. 140. *Dacus tryoni* Frogg. ♂ (Dipt., Trypetidae). (a) right abdominal comb (frontal view), (b) posterior part of the left wing (the arrow indicates the region of contact with the abdominal comb), (c) *D. oleae* Gmel., abdomen of the male with the comb on the posterior margin of the 3rd tergite. ((a) and (b) after Monro, 1953[184], (c) after Féron, 1960[90].)

Fig. 141. *Geotrupes vernalis* L. (Col., Geotrupidae). Hind leg showing the pars stridens (arrow) on the coxa. (Original.)

---

* A roughened nipple-like protuberance and a disc on the interior face of the hind coxa (Pearman's coxal organ) has been discovered in certain Psocoptera. This may be a stridulatory apparatus[17,193]. Several species also produce sounds by percussion on the substratum (see p. 326).

contact with another file situated on the inner surface of the cephalic margin of the lateral plate of the coxal cavity. To stridulate the insect moves its front legs back and forth either together or independently[238, 239].

Coxo-metasternal method

This is a typical method of the Coleoptera Scarabaeoidea. Rarely, it is found in the Orthoptera.

In the genus *Geotrupes* Latr. (Geotrupidae) the coxa of the hind leg has an oblique line of microscopic striae on its upper face. These scratch against the back edge of the coxal cavity (Fig. 141).

In *Heliocopris* Hope and *Synapsis* Bates (Scarabaeidae), there is the same apparatus. It is found again in various other genera of the same family such as *Taurocerastes* Phil., *Orphnus* Mc. Leay, *Hybalus* Brullé, *Aegidium* Westw., *Aegidinus* Arrow, *Frickius* Germ., *Idiostoma* Arrow, and others[11]. In *Plagithmysus* Motsch (Cerambycidae) it is found in conjunction with a coxo-mesosternal apparatus. In addition this insect possesses an elytro-femoral apparatus (see p. 296) and the classic mesonoto-pronotal system of the family (see p. 281).

Apart from Coleoptera, only the family Phylophoridae (Orthoptera, Tettigonioidea), not provided with an elytral apparatus, has a coxo-metasternal mechanism on the third pair of legs[47,152].

Coxo-femoral method

A stridulatory apparatus has been described in a few male Thysanoptera belonging to the genera *Diceratothrips* Bagnall, *Sporothrips* Hood and *Anactinothrips* Bagnall. The pars stridens is situated on the outer face of the fore coxa and consists of an alignment of small crests over which a thin flange of the basal extremity of the femur passes[129] (Fig. 142).

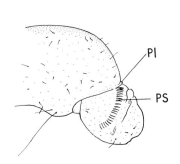

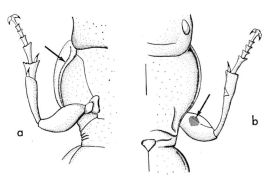

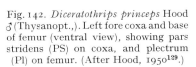

Fig. 142. *Diceratothrips princeps* Hood ♂ (Thysanopt.,). Left fore coxa and base of femur (ventral view), showing pars stridens (PS) on coxa, and plectrum (Pl) on femur. (After Hood, 1950[129].)

Fig. 143. *Siagona fuscipes* Bon., (Col., Siagonidae). (a) ventral view showing the denticulated line of the prothoracic epipleuron (arrow), (b) dorsal view with striated area on the forefemur (arrow). (After Bedel and Francois, 1897[25].)

Femoral methods

Several Coleoptera and Heteroptera have a pars stridens on the femur which scrapes against another part of the body, varying according to the case. For instance, the stridulation is femoro-prothoracic in some species of the genus *Siagona* Latr. (Coleoptera, Caraboidea). The prothoracic epipleuron has a denticulated line throughout its length over which passes the fore femur which bears a striation parallel to its axis[25] (Fig. 143).

In the Coleoptera Rutelinae of the genera *Macraspis* Mc. Lean and *Lagochile* Hoffm., the hind femur has near its apex a striated crest. Movements of the appendage place this crest in contact with an abdominal area bearing a series of well marked grooves[11,98,188].

The Heteroptera also reveal a few well-known examples of femoral stridulation: species of *Corixa* and *Sigara* (Corixidae) use several processes of stridulation (we have already mentioned the bucco-tarsal method) (see p. 280). The fore femur has a denticulated area which rubs against the tip of the tarsus of the opposite leg (the friction might also occur with the sharp edge of the clypeus) (Fig. 115).

The American genus *Buenoa* Kirk. (Notonectidae) has three stridulatory mechanisms, in two of which the fore femora are used: on the one hand there is a swelling covered with bristles at the distal end of the joint which comes into contact with a sharp-cornered area of the head (not in Fig. 144); on the other hand, a striated area in the median position rubs against an enlarged bristle situated at the base of the fore coxa[132, 203] (Fig. 144).

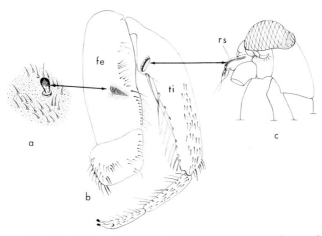

Fig. 144. *Buenoa* sp. (Hemipt., Notonectidae). Two stridulatory devices. a–b: coxal bristle-striated area of the fore femur, b–c: tibial comb-rostral spur.
a: bristle of the fore coxa, b: foreleg, c: head and pronotum. fe: femur, ti: tibia, rs: rostral spur. (After Poisson, 1951[202].)

The males of *Velia* (Veliidae), and more precisely *Stridulivelia* Hung., have a stridulatory area at the rear end of the hind femur which rubs against the prominent lateral margin (connexivum) of the abdomen[203].

Tibio-buccal method

An example in Heteroptera is supplied by the genus *Buenoa* which has just been mentioned. The apparatus consists of a tibial comb rubbing against the 3rd segment of the rostrum which is extended into a spur (Fig. 144) (this apparatus is also found in the closely related genus *Anisops* Spinola).

(v) *Pars stridens on the elytra or the forewings*

Elytro-abdominal method

This method is fairly widespread among Coleoptera. Thus, species of *Trox* Fabr. (Trogidae) have a pars stridens on the inner face of the elytra along the sutural edge. A double transversal projection situated on the propygidium acts as a plectrum[225]. An apparatus of the same nature was described a long time ago in *Hygrobia hermanni* F. (Hygrobiidae)[220]. Examples of this method are also known among Curculionidae (Fig. 145): in *Cryptorrhynchus lapathi* L., the male has a triangular area along the sutural edge that rubs against a keel on the rear abdominal tergite[98,174,176]. This apparatus is also found in various species of the genera *Acalles*

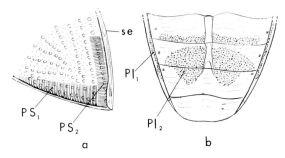

Fig. 145. *Dorytomus longimanus* Forst., (Col., Curculionidae). (a) inner face of the elytron showing the double pars stridens (PS_1, PS_2), (b) dorsal view of the abdomen with both corresponding plectra (Pl_1, Pl_2), se: sutural edge. (After Dudich, 1920[73].)

Schonh., *Sibinia* Germ.[235], *Pissodes* Germ.[260], *Lepyrus* Germ.[154]. We have already described (see p. 288) the abdomino-elytral apparatus of some other female Curculionidae. In the males of these species the apparatus is reversed, with the pars stridens being situated on the elytra[98].

An apparatus of the elytro-abdominal type has likewise developed in the family Scarabaeidae: in the genera *Ochodaeus* Serv. (Fig. 146) (where stridulation has been heard)[130] and *Ligyrus* Burm. In the genus *Copris* Geoffr. the elytra are provided with two protuberances embedded in a cavity of the propygidium. This apparatus helps to

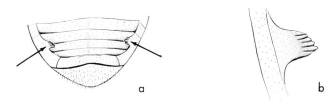

Fig. 146. *Ochodaeus maculipennis* Arrow (Col., Scarabaeidae). (a) abdomen seen dorsally (elytra removed) with stridulatory ridged knobs (arrows), (b) stridulatory knob enlarged. (After Arrow, 1904[11].)

keep the elytra completely closed, but it also acts as a stridulatory apparatus. Each elytral protuberance is striated like the walls of the cavity where it moves, and a backwards movement of the abdomen makes them work against each other[11].

Various methods used by Lepidoptera

Rather than classifying them in terms of their location, we shall group under this heading the apparatus found in the Lepidoptera, (most often the Noctuidae), which bring into play the forewings. Observations on these go back a long way and, as regards some, seem a little doubtful.

In *Aegocera tripartita* Kirby and *A. mahdi* Pagenst. (Noctuidae), the male has a deeply striated area without scales on the costa of the forewings. It must be the long spines of the fore tarsus, which during flight rub against this striation of the wing. The sound produced is a series of clicks[31,113] (Fig. 147).

In the Australian genus *Hecatesia* Boisduval (Noctuidae), the apparatus is quite similar. There is also, however, a grooved fold on the costal margin which may contact the tarsus of the middle legs, the spines of which can rub against the striation of the wing during flight[113]. However, in *H. fenestrata* Boisduval and in *H. thyridion* Feisthamel, it is thought that the striated area may be rubbed by the end of the thick capitate antenna[79].

Scharrer[218], in his review of the stridulatory apparatus of Arthopoda, has pointed out in agreement with Jordan[145], the case of the male *Musurgina laeta* Jordan (Noctuidae). Here, the thickened cubital vein of the forewing is folded transversally. The striated hind tarsus could pass over this vein, giving rise to stridulation (Fig. 148).

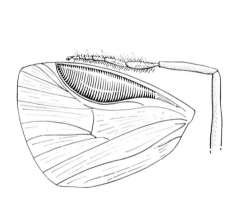
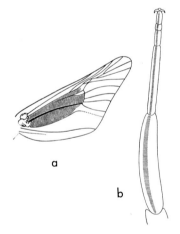

Fig. 147. *Aegocera tripartita* Kirby (Lepid., Noctuidae). Striated area on the costa of the forewing, and foreleg with spiny tarsus. (After Hampson, 1892[113].)

Fig. 148. *Musurgina laeta* Jordan ♂ (Lepid., Noctuidae). (a) forewing with striated area, (b) hind tarsus transversally ridged. (After Jordan, 1921[145].)

In various other Noctuidae, the wing has a marked concavity or convexity, frequently striated, against which one of the legs scrapes. This is the case, for example, with *Thecophora fovea* Tr. (hind wing) (Fig. 149), *Pemphigostola synemonistis* Strand (Fig. 150), *Platagarista tetraplura* Meyr, etc.[31,114,249]. We probably can refer the ticking of *Heliochelus paradoxus* Grote (Noctuidae), audible at 20 ft., to the passing of tarsi on the forewings[118].

Darwin[70] had already drawn attention to the clicking noises made during flight by various species of the American genus *Ageronia* Hubner (Nymphalidae). It seem that the sound is only produced when one individual is pursuing another. *A. arethusa* Cram. has a hollow pyriform organ at the base of the forewing, with two club-shaped expansions. Opposite this organ, the thorax bears two solid hooks which strike against the wing expansions when in movement[113] (Fig. 151).

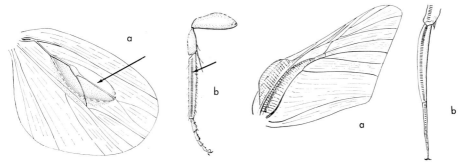

Fig. 149. *Thecophora fovea* Tr., (Lepid., Noctuidae). (a) left hind wing (ventral face) showing the stridulatory swelling (arrow), (b) hind leg with striation (arrow) on the first joint of the tarsus. (After Hannemann, 1956[114].)

Fig. 150. *Pemphigostola synemonistis* Strand (Lepid., Noctuidae). (a) swelling of the forewing (dorsal view, (b) striation of the tarsus of the middle leg. (After Viette, 1955[249].)

We shall give other examples of stridulation among Lepidoptera in the paragraph dealing with sound emissions in absence of differentiated apparatus (see p. 329).

Elytro-femoral method

These methods have been typically developed in Orthoptera, Acridoidea, and more particularly in the family Acrididae where the morphology of the apparatus is very constant. In a more erratic manner, elytro-femoral stridulation is found in a number of Coleoptera.

 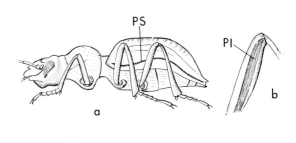

Fig. 151. *Ageronia arethusa* Cram. (Lepid., Nympalidae). Stridulatory apparatus (see explanation in the text). (After Hampson, 1892[113].)

Fig. 152. *Cacicus americanus* Sol. (Col., Tenebrionidae). (a) pars stridens (PS) made of a tuberculated line on the side of the elytron, (b) striation on the hind femur (plectrum: Pl). (After Gahan, 1900[98].)

We shall consider Coleoptera first. In the genus *Geniates* Kirby (Scarabaeidae) the distal end of the hind femur has a very finely striated area which rubs against a grooved ridge surrounding the outside edge of the elytron[188].

In a South American species, *Chiasognathus granthi* Steph. (Lucanidae), the periphery of the elytron is swollen and covered with a chain of tubercles. Opposite this structure the hind femur has sharp longitudinal crests which serve as plectrum. The stridulation of this insect was noticed by Darwin[70].

Species of the genus *Plagithmysus* Motsch, (Cerambycidae) have an elytro-femoral formation, consisting of a stridulating file on the periphery of the elytron which scratches the hind femur[98]. This formation is in addition to the recognised apparatus of the family (see p. 281) and the two coxo-sternal devices (see p. 291).

The genera *Ctenoscelis* Serv. (Cerambycidae), *Oxychila* Dej. (Cicindelidae) and *Cacicus* Sol. (Tenebrionidae), also have this kind of stridulatory apparatus. In the latter, however, the elytral denticulation is on a curved line some distance above the elytral margin. The hind femur is striated longitudinally (Fig. 152). The stridulation of *C. americanus* Sol. has been heard,[98]

In *Prionus coriarius* L. (Cerambycidae), the hind femur has a sharp crest which passes over the external margin of the elytron. No differentiation of the elytron is visible[22].

In male and female *Graphopterus variegatus* Fabr. (Masoreidae), a ground beetle of the desert regions of Africa which runs across the sand stridulating, the pars stridens is composed of a denticulation of the outer edge of the elytron as in the preceding cases, but this projection is also present on the edge of the abdominal sternites (Fig. 153). The hind femur has simple rounded crests. The same is true of *G. serrator* Forsk.[201].

Elytro-femoral stridulation is particularly characteristic of the Orthoptera Caelifera of the super-family Acridoidea*. The mechanism is simple: the hind femur has a row of small tubercles on a crest which runs longitudinally along its inner face (Figs. 154, 155). During stridulation the leg is pressed against the body and

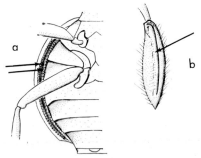

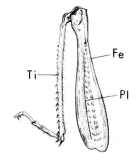

Fig. 153. *Graphopterus variegatus* Fabr., (Col., Masoreidae). (a) ventral view showing the double pars stridens (arrows), (b) hind femur with keel acting as a plectrum (arrow). (After Pocock, 1902[201].)

Fig. 154. *Chorthippus jucundus* (Fisch.) ♂ (Orth., Acrididae). Inner face of the hind leg showing the denticulated plectrum along the femur. Fe: femur, Pl: plectrum, Ti: tibia (original).

---

* In Tridactyloidea, the second super-family of the sub-order Caelifera, there are other devices: maxillo-mandibulary in *Cylindracheta* (Cylindrachetidae) (see p. 279) and alary-elytral in the Tridactylidae (see p. 305).

exerts a pivoting movement which causes the femoral denticulation to rub against a vein of the elytron (vena radialis media) (see p. 365) (Fig. 156 and 221).

In Oedipodinae, this device is reversed. The denticulation is on the elytron (vena intercalata) (Fig. 157), whereas the femur has only a crest without spines. The movement of the leg is, however, the same.

In acridids the stridulatory apparatus is only well developed in the male sex and is often absent in the female, or it is merely an irregular line of hairs of rudimentary spines on the femur. However, regardless of statements to the contrary, stridulation is observed among female Acridinae[137, 208]. Thus the females of *Chorthippus brunneus* Thunb. stridulate as loudly as the males. In this species the stridulation of the female is part of the behaviour (see p. 594), as in the case of *Stenobothrus lineatus* (Panz.) and *Omocestus viridulus* (L.). Stridulation has also been observed in the females of *Myrmeleotettix maculatus* (Thunb.), *Gomphocerus rufus* (L.), etc.

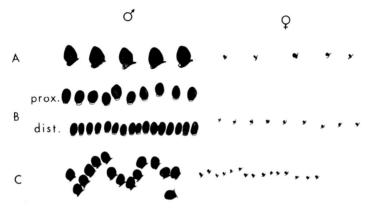

Fig. 155. Detail of femoral teeth in some Orthoptera, Acridinae ♂ and ♀. (A) *Chorthippus brunneus* Thunb., (B) *Myrmeleotettix maculatus* (Thunb.) (C) *Gomphocerus rufus* (L.) (After Jacobs, 1953[137].)

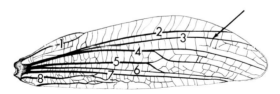

Fig. 156. Elytron of Acridinae: *Chorthippus montanus* (Charp.) ♂ showing the principal veins (after Brunner's nomenclature). The femoral plectrum rubs against the vena radialis media (arrow) (After Faber, 1929.)*

1: vena mediastina	4: v. radialis posterior	7: v. dividens
2: v. radialis anterior	5: v. ulnaris anterior	8: v. plicata
3: v. radialis media	6: v. ulnaris posterior	

In *Chorthippus parellelus* (Zett.), the female carries out stridulatory movements without sound emission because of the shortness of the elytra. We shall come back to this (see p. 594) and to the mechanical problems of the emission (see p. 365).

---

* *Zool. Anz., 81*, 1.

References p. 338

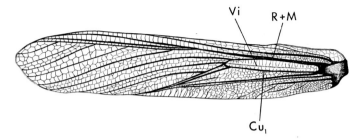

Fig. 157. Elytron of Oedipodinae: *Locusta migratoria* L. showing the location of the pars stridens on the vena intercalata. Cu_1: first cubitus, R + M: base of radius and media, Vi: vena intercalata. (Original.)

Finally it should be pointed out that numerous Acridoidea do not use their elytra, but various formations situated on the sides of the abdomen (see pp. 285 and 286). Nevertheless, the movement of the femur which produces the sound emission is very much the same as in the case of forms with the recognised elytro-femoral apparatus.

Elytro-tibial method

This method seems to be found only among Acrididae and Pamphagidae (Orthoptera). Kevan[152] made a detailed study of it, and thinks that this method is derived—at least among the Acrididae—from the elytro-femoral method. *Stethophyma grossum* (L.) (Acrididae) produces a clicking noise by striking its tibiae against its elytra[82,85]. *Nomadacris septemfasciata* (Serv.) (Acrididae) has a peculiar behaviour during mating: the insect makes its hind tibiae vibrate with their spines striking against the elytra[137]. In *Psectrocnemus longiceps* I. Bol. (Acrididae) the hind tibial points are striated and rub against the radial vein of the elytron (Fig. 158). This insect has in addition the typical femoral denticulation. In the genus *Lamarckiana* Kirby (Pamphaginae) and in the *Porthetini*, the hind tibia has a distal spur which comes into contact with the modified subcostal area of the elytron[43,55,243].

Fig. 158. *Psectrocnemus longiceps* Bol. (Orth., Acrididae). Hind tibia showing the striated spines. (After Henry, 1940[123].)

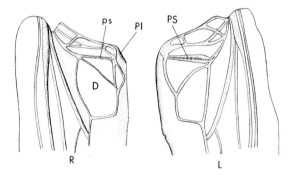

Fig. 159. *Conocephalus maculatus* (Le Guil.) (Orth., Conocephalidae). Elytra (inner face). The pars stridens (more developed on the left elytron) appears on the cubital posterior vein. The plectrum is on the right elytron which bears also a drum of thin and transparent cuticle (non stippled). The corresponding region in the left elytron is made of normal cuticle. Legend see Fig. 163. (After Roy-Noel, 1954[213].)

Elytro-elytral method

The interaction of the two elytra for stridulation appears to be used only among present-day insects, in the Ensifera Orthoptera belonging to the super-families Tettigonioidea and Gryllodea* (see pp. 362-365). In Prophalangopsoidea a stridulatory apparatus of a primitive type exists on the elytra of the male, and in the genus *Cyphoderris* Uhl., it is associated with an abdomino-pronotal device (see p. 282). In Gryllacridoidea, elytral stridulation is completely non-existent but, on the other hand, a characteristic abdomino-femoral method has developed in this group (see p. 285).

(a) *Tettigonioidea*. The stridulatory apparatus is situated in the forward area of the elytron (cubito-anal area) (Fig. 159). When the elytron is atrophied, this area undergoes no change (Ephippigeridae). The left elytron, which covers the right when at rest, has a thick and denticulated line on the under surface of its very enlarged cubital vein. This constitutes the pars stridens (Fig. 160). When the elytra open or close, this ridge is rubbed by the inside edge (plectrum) of the right elytron. The latter has generally a pars stridens which is well developed though smaller and less chitinised than that of the left elytron. This pars stridens is obviously not functional. On the same right elytron, a smooth oval area bounded by cubital and cubital posterior veins has been developed.** This structure bears the name of drum or disc; (it is also termed—most unfortunately—tympanum). Some authors have attributed to it a role in the strengthening of sounds (see p. 366).

0.5 m/m

Fig. 160. *Ephippiger ephippiger* (Fiebig) ♂ (Orth., Ephippigeridae). Longitudinal section of the pars stridens showing the prismatic shape of the teeth. (Original.)

Females in exceptional cases only reveal a stridulatory apparatus of the same type as the males (the case of Ephippigeridae, see below). However, it appears that simple and only slightly evolved devices are not rare. In some fifty species of North American Tettigoniidae, structures of a stridulatory nature have been found in the females[97] (Fig. 161). In these, the plectrum is represented by a curvature towards the base of the inner side of the left elytron in the neighbourhood of the anal area. The veins of the upper face of the right elytron are covered with small spines which are scratched by the plectrum when the elytra are in movement (genera *Scudderia, Orchelium, Microcentrum, Neoconocephalus*). The arrangement is different in *Pterophylla camellifolia* (Fabr.): the vein forming the inner edge of the right elytron is rounded

---

* In *Schistocerca gregaria* Forskål (Acridoidea), however, an elytro-elytral stridulation has been observed (see p. 329).
** This is generally the case, but in some families both elytra have a drum (ex. Pseudophyllidae, genus *Pterophylla*; Ephippigeridae, genus *Ephippiger*).

References p. 338

above and covered with small granules. The veining of the lower face of the two elytra forms a projecting network of numerous small keels.

In some species the stridulation of the female has in fact been heard: *Microcentrum rhombifolium* (Sauss.), *Scudderia curvicauda* (De Geer), *S. texensis* (Sauss. & Pictet), *S. furcata* Brunner, *Amblycorypha rotundifolia* (Scud.)[5,97] (see p. 595).

The male apparatus as we have described it, exists in all families of the Tettigonioidea, apart from the Phyllophoridae where it is of the coxo-metasternal type (see p. 291). In two other families, however, it has some fairly noteworthy peculiarities.

In the Meconemidae *Cyrtaspis scutata* Charp, the male has very atrophied discoidal elytra (about 3 mm) concealed under the pronotum. These two elytra are similar and have a crescent-shaped file of eight teeth on their lower face, and a hairy area on their upper face in the same region. Stridulation results not from the passage of the inside edge of the right elytron over the file of the left elytron, but from the passage of this file over the hairs which cover the back of the other elytron[165] (Fig. 162).

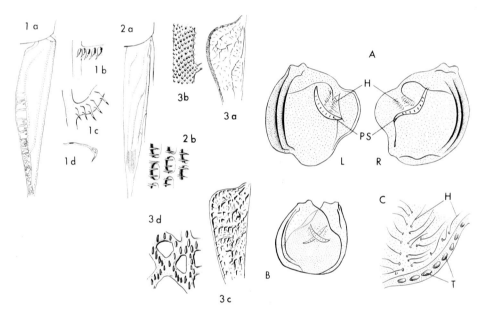

Fig. 161. Stridulating organs of females Tettigoniidae. (1) *Scudderia curvicauda* (De Geer). a: right elytron, b, c: different types of spines on the veins, d: cross section of the margin of the left elytron acting as a plectrum on the spiny areas of the right elytron. (2) *Neoconocephalus triops* (L.) a: right elytron, b: enlarged part of the anal area (3) *Pterophylla camellifolia* (Fahr.) a: right elytron, b: vein (enlarged) on inner margin, c: left elytron (ventral view), d: enlarged part. (After Fulton, 1933[97].)

Fig. 162. *Cyrtaspis scutata* Charp. ♂ (Orth., Meconemidae). (A) Draft of elytra (dorsal view). The pars stridens (on the inner face) is seen by transparency. On the upper face, hairy area, (B) Elytra in stridulatory position, (C) Enlarged portion of the right elytron. H: hairs, PS: pars stridens, T: teeth of the pars stridens. (After Lienhart, 1921[165].)

The case of *Meconema thalassina* (De Geer) belonging to the same family will be discussed later (see pp. 328 and 329).

In Ephippigeridae, the elytra are very reduced in size and are partially or totally concealed under the pronotum (Fig. 163). The male possesses, as is general, two pars stridens, that of the left elytron being, however, stronger than that of the right elytron. But the species of the genus *Ephippiger* Berthold have two pecularities: as already pointed out, the presence of a drum on the two elytra (and not on the right elytron alone), and the existence in the female of a functional stridulatory apparatus. It should, moreover, be noticed that the position of the file is the opposite to that in the male: it is on the upper surface of the right elytron, with the result that the right file which takes on the stridulatory function (see p. 595) is more developed than the left one[107].

A study of the venation and the development of the sound apparatus in the genera *Tettigonia* L., *Conocephalus* Thunb. and *Decticus* Serv., has shown that the stridulatory region is already well marked in first instar nymphs and that the teeth of the pars stridens appear a little before the imaginal moult[213].

A very exceptional case of elytro-elytral stridulatory apparatus was discovered

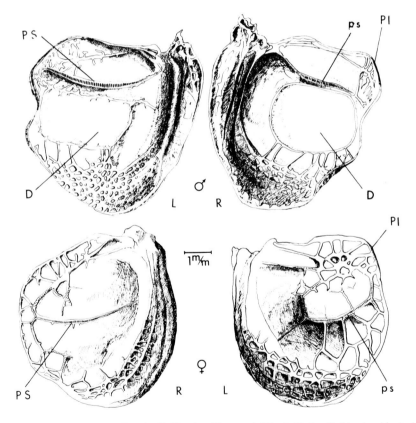

Fig. 163. Elytra of male and female *Ephippiger bitterensis* Finot (Orth., Ephippigeridae). Above: male's elytra (ventral view). The functional pars stridens (PS) is on the left elytron (underside); the pars stridens of the right elytron is useless (ps). Below: female's elytra (dorsal view). The functional pars stridens is on the right elytron (upper side). The location of the plectrum is also reversed, male: right elytron, female: left elytron. D: drum, Pl: plectrum, PS: functional pars stridens, ps: unfunctional pars stridens. (Original.)

a Diptera fossil in Baltic amber—*Eohelea stridulans* which has a series of 15 parallel ridges at the distal end of each wing[198].

(b) *Gryllodea*. The stridulatory apparatus occupies a larger part of the elytron than in the case of the Tettigonioidea, but its structure and principle are fundamentally the same: the pars stridens on the under surface of an elytron (lower cubital vein[213]), the plectrum on the antero-internal edge of the other elytron (Figs. 164–166). The females, except for the Gryllotalpidae[23, 24, 173] are normally deficient in any stridulatory apparatus. There are, however, two differences:

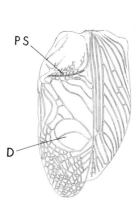

Fig. 164. *Gryllus bimaculatus* De Geer ♂ (Orth., Gryllidae). Right elytron. D: drum, PS: pars stridens. (After Stärk, 1958.)*

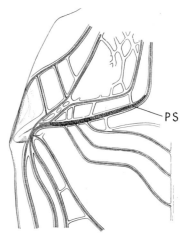

Fig. 165. *Acheta domesticus* (L.) (Orth., Gryllidae). Right elytron. Enlarged view of the stridulatory region. PS: pars stridens. (After Roy-Noel, 1954[213].)

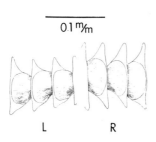

Fig. 166. *Gryllus campestris* L. (Orth., Gryllidae). Highly enlarged view of the teeth of the left and right elytron. These lamellar teeth are quite different from those of Tettigonioidea. (After Stärk, 1958.)*

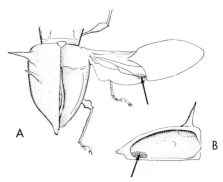

Fig. 167. *Amphisternus* sp. (Col., Endomychidae). (A) dorsal view of the insect; the left wing is closed and covered by the elytron, the right elytron is removed and the wing is spread out showing the stridulatory area (arrow), (B) inner side of the right elytron with the complementary stridulatory area (arrow). (After Arrow, 1924[12].)

* *Zool. Jb.*, 77, 9.

1. The two elytra are generally similar, each one having a file and a drum. The drum is in a more distal position than in the Tettigonioidea. It is, however, absent in the Gryllotalpidae[53].

2. The right elytron usually covers the left (see p. 360).

The study of the development of the elytra of the male and of the female in *Gryllus bimaculatus* De Geer[222], has shown that dimorphism only appeared at the 5th instar.

Alary-elytral method

Some Coleoptera show instances of this method which we are placing here although it is difficult to know which is the plectrum and which the pars stridens.

TABLE 23

QUANTITATIVE DATA ON THE PARS STRIDENS OF SOME INSECTS ♂

Species	Situation of the P.S.	Nature of the elements composing the P.S.	Number of the elements	Length of the P.S. (mm)	Author
ORTHOPTERA					
Acheta domesticus (L.)	right elytron	laminated expansions	133	1.9	Pierce, 1948[200]
G. assimilis Fabr.	right elytron	laminated expansions	142	3.9	Pierce, 1948[200]
Nemobius fasciatus (De Geer)	right elytron	laminated expansions	118	0.99	Pierce, 1948[200]
Oecanthus niveus (De Geer)	right elytron	laminated expansions	39	2	Pierce, 1948[200]
O. quadripunctatus Beutenmüller	right elytron	laminated expansions	49	1.38	Pierce, 1948[200]
Amblycorypha rotundifolia (Scudder)	left elytron	crests	92	2.4	Pierce, 1948[200]
Pterophylla camellifolia (Fabr.)	left elytron	crests	67	4	Pierce, 1948[200]
Conocephalus brevipennis (Scudder)	left eyltron	crests	35	1.18	Pierce, 1948[200]
Ephippiger bitterensis Finot	left elytron	crests	40–55	—	Pasquinelly & Busnel, 1955*
Chloealtis conspersa Harris	hind femur	tubercles	104	4.8	Pierce, 1948[200]
Chorthippus curtipennis Harris	hind femur	tubercles	121	4.45	Pierce, 1948[200]
C. apricarius (L.)	hind femur	tubercles	156	4.0	Jacobs, 1953[a137]
C. biguttulus (L.)	hind femur	tubercles	100	3.3	Jacobs, 1953[a137]
C. brunneus Thunb.	hind femur	tubercles	63	2.9	Jacobs, 1953[a137]
C. parallelus (Zett.)	hind femur	tubercles	95	3.9	Jacobs, 1953[a137]
Stenobothrus lineatus (Panz.)	hind femur	tubercles	388	6.6	Jacobs, 1953[a137]
S. stigmaticus (Ramb.)	hind femur	tubercles	100	2.8	Jacobs, 1953[a137]
Omocestus rufipes (Zett.)	hind femur	tubercles	117	3.7	Jacobs, 1953[a137]
Gomphocerus rufus (L.)	hind femur	tubercles	185	3.7	Jacobs, 1953[a137]

* *Colloque sur l'Acoustique des Orthoptères*, 1955, fascicule hors série des *Ann. des Epiphyties*, I.N.R.A. Paris, p. 146.

TABLE 23 (continued)

Species	Situation of the P.S.	Nature of the elements composing the P.S.	Number of the elements	Length of the P.S. (mm)	Author
Arphia sulphurea (Fabr.)	elytron	tubercles	175	6	Pierce, 1948[200]
Griotettrix verruculatus (Kirby)	elytron	tubercles	249	8.7	Pierce, 1948[200]
Locusta migratoria gallica Rem.					
♂ phase *solitaria*	elytron	tubercles	175	—	Busnel, 1953[*]
♂ phase *gregaria*	elytron	tubercles	125	—	Roerich and Moutous, 1955[**]
COLEOPTERA					
Crioceris merdigera L.	pygidium	striae	120–130	0.5	Landois, 1874[159]
Cerambyx cerdo L.	mesonotum	striae	238	3.4	Landois, 1874[159]
Geotrupes stercorarius L.	hind coxa	striae	84	0.36	Landois, 1874[159]
Serica brunnea L.	prosternum	striae	180	0.27	Landois, 1874[159]
Necrophorus vespillo L.	5th abdominal tergite	striae	126–140	1.95	Landois, 1874[159]
Elaphrus riparius L.	propygidium	teeth	21	0.45	Landois, 1874[159]
Enoplopus velikensis Piller	vertex	striae	135–150	0.75	Dudich, 1920[73]
Hispa testacea L.	vertex	striae	50–60	0.35	Dudich, 1920[73]
Ceutorrhynchini	elytron (distal end)	striae	70–100	0.3–0.6	Dudich, 1920[73]
HYMENOPTERA (Formicoidea) ☿					
Neomyrma rubida Latr.	1st tergite of the gaster	striae	180	—	Raigner, 1933[***]
Crematogaster scutellaris Ol	1st tergite of the gaster	striae	160	—	Raigner, 1933
Myrmica sulcinodis Nyl.	1st tergite of the gaster	striae	80	—	Raigner, 1933
Solenopsis fugax Latr.	1st tergite of the gaster	striae	30	—	Raigner, 1933

[a] Average values for quantities between 10 and 53.

In Erotylidae and Endomychidae, tropical mycetophagous insects, there is, in addition to a cranio-prothoracic apparatus (see p. 280), another device that is considered stridulatory[12]. It consists of a granulated area of the wing situated at its distal end when it is folded. Opposite it, the elytron has a rugose region where the friction occurs (Fig. 167). A similar structure is also found among Dytiscidae. It has been noticed that in the case of *Cybister aegyptiacus* Peyron[56], stridulation occurs when the insect is on the ground. The elytra are partially open and the wings vibrate rapidly. This somewhat faint stridulation may last one minute.

---

[*] *Ann. des Epiphyties*, 3, 333
[**] *Colloque sur l'Acoustique des Orthoptères*, 1955, fascicule hors série des *Ann. des Epiphyties*, I.N.R.A., Paris, p. 91.
[***] *Brotéria*, 2, 51.

Among Orthoptera, the alary-elytral method seems to be fairly common in the Tridactylidae[152]. Here, a plectrum situated at the base of the radial vein of the wing, acts on an elytral pars stridens[53]. The case of female *Anoedopoda lamellata* (L.) (Tettigonioidea, Mecopodidae) seems more exceptional: hairs of the subcostal area of the wing rub against the inner face of the elytron, the subcostal vein of which is very prominent[152,251].

(b) Crustacea

The particular morphology of Crustacea (Fig. 174) has led to the development of the stridulatory apparatus in positions where it is not found in insects. If there are mechanisms and devices peculiar to Crustacea (Alpheidae, Palinuridae), the majority are nevertheless of the regular pars stridens-plectrum type. Thus it seemed to us preferable to link together all animals belonging to that class, adopting the systematic order, rather than separating them into different groups in terms of the principle of their apparatus functioning as was done in the case of insects.

Stridulatory devices have been found only in Malacostraca; here they have been observed in nearly 50 genera, mostly belonging to the order Decapoda. Among these, crabs are specially well represented, since they alone account for more than 30 genera which include stridulating species[111].

There are only a very few studies on the stridulation of Crustacea. Almost all that can be mentioned are the comparatively old works of Aurivillius (1893)[15], Ortmann (1901)[189], and Alcock (1900–1902)[1,2], as well as the summary reviews of Hansen (1921)[115], Balss (1921, 1956)[18,19], Scharrer (1931)[218], Schmitt (1931)[221], and Tembrock (1959)[234]. More recently Guinot-Dumortier and Dumortier (1960)[111] have made a study confined to Brachyura.

(i) Stomatopoda

In *Squilla empusa* Say, *S. mantis* L. and *Lysiosquilla excavatrix* Brooks, a stridulation, sometimes vigorous, is produced by the uropods passing over the lower surface of the telson[18,36,101]. This stridulation has been observed in disturbed animals.

Gonodactylus chiragra L. is another mantis-shrimp which produces a sharp clicking sound with its raptorial claws (pmx_2). The dactyl of this appendage fits into a groove of the propodus like the blade of a pocket-knife in its handle. At the proximal end of the propodus there is an erectile spine. It is the impact of the tip of the dactyl against this spine, when the claw is abruptly opened, that should cause the noise[2]. It is the same in other species: *G. oerstedii* Hansen[144], *G. glabrous* Brooks and *G. demani* Henderson[149].

Actually this noise is probably only a consequence of the mechanical process of the stretching of the appendage.

(ii) Decapoda

(1) *Penaeidae*

This family gives a few examples of a rather special type of stridulatory apparatus. It is made of a row of small crests on the sides of the cephalothorax (branchial region) against which rubs the antero-lateral edge of the first segment of the abdomen.

In *Penaeopsis stridulans* (Wood Mason) there are 5–12 stridulating crests, in *Metapenaeopsis barbatus* (De Haan) (= *Parapenaeus akayebi* Rathbun), about 20

(Fig. 168), and in *M. acclivis* (Rathbun) (= *P. acclivis* Rathbun) from 13–18, *M. durus* Kubo has 28–35[18, 158].

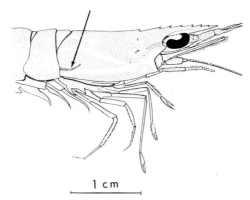

Fig. 168. *Metapenaeopsis barbatus* (de Haan) (Decapoda, Penaeidae). Stridulating crest of the cephalothorax (arrow). (After Kubo, 1949[158].)

(2) Alpheidae (and Palaemonidae)

It was in 1843 that Krauss[156] first observed the way snapping shrimps produce a characteristic noise with their large chela. Subsequently numerous writers tackled this problem, but it seems that the exact origin of this noise has not yet been proved. In the genera *Alpheus* Fabr.*, *Synalpheus* Bate (Alpheidae), as well as in various species of the genera *Pontonia* Latreille, *Typton* Costa, and *Coralliocaris* Stimpson (Palaemonidae)**, the large chela shows a peculiar structure: the dactyl (movable finger) has a strong tooth, often very bulky, on its inner face. When the claw is closed, this tooth fits into a cavity on the upper face of the fixed finger of the propodus (pollex) like a piston in a cylinder (Fig. 169). The noise produced by the animal

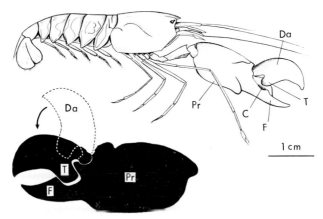

Fig. 169. *Alpheus* sp. (Decapoda, Alpheidae). C: Cavity of the fixed finger, Da: dactyl, F: fixed finger, Pr: propodus, T: tooth of the dactyl (original).

* By a resolution of the Commission of Zoological Nomenclature, the generic name *Alpheus* is now valid in place of *Crangon* used by some American authors.

** In the genus *Amphibetaeus* Coutière (Alpheidae), the same apparatus is found, but is seems that the animal does not use it for sound production[62].

is a very brief clap (½ to 1 msec) of considerable intensity (124 db at 1 m, in *Synalpheus lockingtoni* Coutière[144]). The detonation of *A. strenuus* Dana has been compared to the noise made by striking with full force the rim of a glass dish with a wooden ruler[62]. Various hypotheses have been suggested as to the way in which this noise is produced. Some thought that its origin was the sudden extraction of the dactyl from the cavity of the propodus, in the way a cork is drawn from a bottle[262]. Others, on the other hand[37], agreed that it would result from the impact of the two fingers when the claw closed. Volz (1938)[250], who made a thorough study of this type of behaviour among the Alpheidae, thinks that the noise arises from the movement of the tooth of the movable finger against a protuberance on the inside surface of the fixed finger, which, in addition, is used to slow down the closing of the claw.

Finally, Coutière (1899)[62] proposed another explanation, that the noise arose from impact of the tooth of the movable finger against the water in the cavity of the fixed finger. As the water cannot freely escape, it offers a certain resistance and acts as a hard body when the percussion occurs. However, in *A. malleodigitus* Bate, the fixed finger has disappeared and consequently the movable finger beats the air, but there is nonetheless a vigorous noise. The exact means by which this noise is produced is, therefore, still to be discovered.

It is paradoxical to observe that while the Alpheidae are far and away the noisiest of Crustacea, this noise seems to have no biological significance, and to be only the consequence of another activity. In fact, the sudden thrust of the tooth of the dactyl into the cavity of the propodus results in spurting a fairly strong current of water forwards to raise the mud 2 cm from the claw, and may even break the glass jar in which the animal is enclosed[147]. This would represent a defensive or offensive type of behaviour to cause an enemy to retreat or to forbid entry of another individual into the cavity in which the animal lives[19, 250], or again to stun and then to capture a prey passing near[147].

These Crustacea were the subject of serious preoccupation to the American Navy at the beginning of the Japanese-American War. The ships' listening devices revealed in some areas an intense noise, similar to the crackling of a burning bush or to the rattling of coal rolling down a metal chute. It was thought to be some new device used by the enemy and it remained a complete mystery until the end of 1942. It was then that the biological laboratories of the Navy finally discovered that all this din was the work of Alpheidae. It was exactly a century previously that Krauss[156] had published the first work on this subject...

Alpheidae are found in large numbers at depths not exceeding 30 fathoms, when the bottom is rocky or covered with coral, shells, sponges, calcareous seaweed, etc. in which these creatures take refuge. In these biotopes the noise is continual, and the intensity level in the region of 3–10 kc reaches about 60 db[144]. A diurnal snapping rhythm has been observed, characterised by an increase of 2–5 db during the night[143].

Alpheidae are found all over the world and the 11° C winter surface isotherm marks the approximate northern and southern limits of their continuous range[143].

(3) *Palinura: Palinuridae*

The spiny lobsters seem certainly to be the first Crustacea known to stridulate, for Athenaeus[14] mentioned the noise they produce in the 3rd century of our era. The first scientific descriptions were made by Leach[164] in 1815 on *Palinurus vulgaris* Latr., by Möbius[183] in 1867 and subsequently by a number of other writers[72, 103, 105, 119, 150, 167, 185, 189].

The basal joint of the antenna (basicerite) bears on its inner face a process. Its enlarged and concave end fits exactly over a longitudinal swelling running along the side of the rostrum. On the lower surface of the antennary process there is a striated oval membrane (stridulatory membrane); when the antenna moves from front to back, the stridulatory membrane glides along the longitudinal swelling of the rostrum like a hand on the rail of a staircase. The thrust of the stridulatory membrane in this movement, combined with the presence on the rostral swelling of microscopic teeth pointing forward[185], is the basis for the grinding noise produced by the lobster. Moreover, the antennary process has a small knob on the side of the stridulatory membrane, which inserts itself in a groove running along the entire length of the swelling. This device guides the apparatus and keeps it in contact throughout its course.

This apparatus is found in the genera *Palinurus* Weber, *Panulirus* White and *Linuparus* White. Möbius (1867)[183] very aptly compared the noise the creature makes to that of a boot rubbing against a table leg.

In *P. argus* (Latreille) two different types of emission have been observed: a slow rattle and a rasp[185].

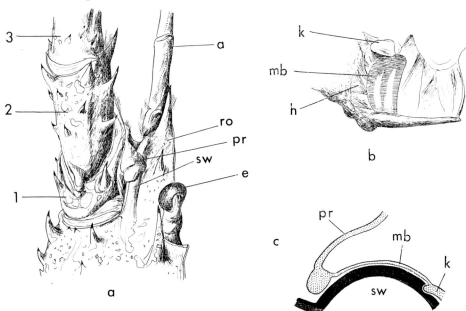

Fig. 170. *Palinurus vulgaris* Latr. (Decapoda, Palinuridae). (a) half anterior part of the cephalic region, (b) innerside of the stridulatory process, (c) schematic section of the stridulatory process. a: antennula, e: eye, h: hairs, k: knob, mb: stridulatory membrane, pr: stridulatory process of the basicerite, ro: rostrum, sw: swelling of the rostrum, 1, 2, 3: joints of the antenna. (Original.)

Various observations have shown that the animals frequently waved their antennae without necessarily emitting sound; the sound emission corresponds therefore to a particular movement of the appendage[72, 167, 185].

The rasp was heard in *P. vulgaris* during a fight between two individuals. It is more usually produced when the creature is captured and is accompanied by violent abdominal contractions. The slow rattle is observed when the animals are sufficiently numerous, and to an observer it does not seem to be connected with any well-defined activity.

(4) *Anomura*

1. *Thalassinidea.* Two stridulating species are known in the family Thalassinidae: *Thalassina anomola* (Herbst) and *Gebia issaeffi* Balss. The former lives in the Philippines and deep burrows in the beaches. The noise, similar to that of a cork being drawn from a bottle, is made by the creature in its burrow. It may be obtained by passing the front leg over the spines on the side of the carapace[194]. No differentiated stridulatory apparatus is visible.

In *Gebia issaeffi* the dactyl of the legs of the first pair has two fluted crests, one on the upper edge, the other on the inner edge. The noise might be produced by the mutual friction of these two limbs[18] (Fig. 171).

2. *Paguridea.* Two different types of stridulatory apparatus have been developed in this tribe, one in the family Coenobitidae, the other in Paguridae.

To the first belongs *Coenobita rugosus* H. Milne Edwards which has a row of ribs on the upper exterior area of the claw of the cheliped. A denticulate crest extends over the inner face of the dactyls of the 2nd and 3rd pairs of walking legs, parallel to the axis of the joint[125, 189]. This apparatus only exists on the left side (Fig. 172). The noise produced is probably due to the passing of the teeth of the dactyl over the ribs of the palm. "When shut up in a large tin box, these animals continually give out a low chirping sound, though it is not possible to discover how they do this"[30]. An apparatus of the same kind is found in *C. perlatus* H. Milne Edwards.

Fig. 171. *Gebia issaeffi* Balss (Decapoda, Thalassinidae). Inner face of the dactyle of the first walking leg showing the fluted crest. (After Balss, 1921[18].)

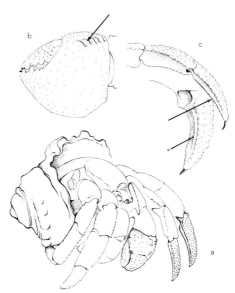

Fig. 172. *Coenobita rugosus* H. Milne Edwards (Decapoda, Coenobitidae). (a) feature of the animal in its shell, (b) enlarged view of the left chela showing the ribs (arrow), (c) first and second walking legs with the denticulate crest (arrows). (Original.)

In the family Paguridae, the seven species of the genus *Trizopagurus* Forest, are provided with stridulatory structures on the inner face of the palm of the cheliped, in the extension of the dactyl. They consist of rows of thick, horned regularly-spaced bristles[94] (Fig. 173). Less well developed bristles also extend to the dactyl. Stridulation (never observed) may be the result of mutual friction of the two claws, both of which have this apparatus*.

This type of apparatus and its working somewhat recall what has been observed among some mygales (see p. 316). The supposition that it might have a stridulatory role was first made in 1888 in regard to the species *Clibanarius strigimanus* (White) (= *Trizopagurus strigimanus* (White).[121]

(5) *Brachyura*

Among the crabs (Fig. 174), the processes of sound emission can be grouped into two main categories.

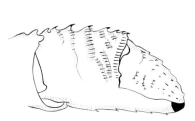

Fig. 173. *Trizopagurus tenebrarum* (Alcock) (Decapoda, Paguridae). Stridulating bristles of the cheliped. (After Forest, 1952[94].)

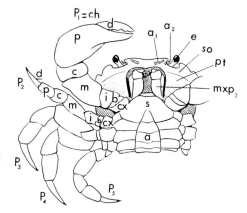

Fig. 174. External morphology of a crab, ventral view. (After Guinot-Dumortier and Dumortier, 1960[111].)

a: abdomen
a_1: antennula
a_2: antenna
b: basis
c: carpus
cx: coxa
d: dactyl
e: eye
i: ischium
m: merus
mxp_3: third maxilliped
p: propodus
P_1 = ch: cheliped
P_2, P_3, P_4, P_5: ambulatory legs
pt: pterygostomian region
s: sternum
so: sub orbital region

(i) Friction of appendages against the cephalothorax,
(ii) Mutual friction of appendages.

Friction of appendages against the cephalothorax

In most cases there is an action of the cheliped on the cephalothorax. In male and female *Ovalipes ocellatus* (Herbst) (Portunidae), the pterygostomian region bears a pars stridens formed of a curved crest provided with about fifty flutings. The plectrum is a double row of ribs on the merus of the cheliped[209] (Fig. 175). A very similar arrangement is found in male and female *Ommatocarcinus macgillivrayi* White (Goneplacidae).

---

* This stridulatory apparatus owes its name to the genus *Trizopagurus* (τρίζειν = to grate) which includes the species formerly grouped in the genera *Pagurus*, *Aniculus* and *Clibanarius*.

The Xanthidae *Pseudozius bouvieri* (A. Milne Edwards) has a similar structure in both sexes, with the plectrum on the carpus of the cheliped.

Fig. 175. *Ovalipes ocellatus* (Herbst) (Decapoda, Portunidae). Stridulatory apparatus. Ch: cheliped, Pl: plectrum, PS: pars stridens. (After Guinot-Dumortier and Dumortier, 1960[111].)

In the catometopous crabs of the genera *Acmaeopleura* Stimpson (Fig. 176), *Macrophthalmus* Desmarest, *Metaplax* H. Milne Edwards, etc. the pars stridens is in a sub-orbital position. It is formed of granules, more or less numerous, and the plectrum consists of a smooth crest on the merus of the cheliped. In the species *Helice tridens* De Haan, the sub-orbital crest is particularly well developed (Fig. 177).

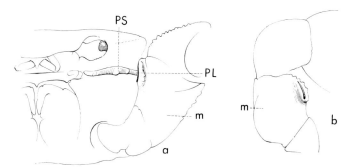

Fig. 176. *Acmaeopleura parvula* Stimpson (Decapoda, Grapsidae). (a) front view of the crab showing the pars stridens in a sub orbital position and the plectrum on the merus of the cheliped, (b) another aspect of the plectrum.
m: merus of the cheliped, Pl: plectrum, PS: pars stridens.

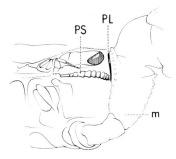

Fig. 177. *Helice tridens* De Haan (Decapoda, Grapsidae). Another example of the same type of stridulatory apparatus. (Both, Fig. 176 and 177 after Guinot-Dumortier and Dumortier, 1960[111].)

The arrangement can, however, be reversed: pars stridens on the cheliped, and plectrum on the cephalothorax. Thus in various species of the genus *Matuta* Weber

(Oxystomata, Calappidae) the propodus of the cheliped has a deeply striated area on its inner face, which plays against the tubercles of the pterygostomian region (Fig. 178). Stridulation has been heard in these animals[2, 189]. The same structure is also found in the genus *Acanthocarpus* Stimpson (Calappidae). In *Menippe* De Haan (Xanthidae) the pars stridens, which occupies the same position as in the two preceding genera, resembles a portion of a finger print (Fig. 179); it rubs against the protuberances of the orbital border. Stridulation of these species has been heard[65]. The pars stridens is also on the cheliped in the Goneplacidae of the genera *Hexapus* De Haan, *Lambdophallus* Alcock and *Hexaplax* Doflein.

The process of stridulation is very different in certain freshwater crabs of the genus *Potamon* Savigny (Brachyrhyncha)[15].

In *Potamon africanum* (A. Milne Edwards) the coxa of the walking legs p_2 and p_3 has on its very convex upper face an area of wiry bristles, short and swollen with a hard point at their end. Opposite the spiny area of the coxa, a cluster of a few thick, dark bristles, inflated at their ends, hangs down on the free edge of the branchiostegite (Fig. 180). A backward movement of the legs makes the coxal bristles scrape against the large bristles of the branchiostegite. The result is a quite intense chirping noise.

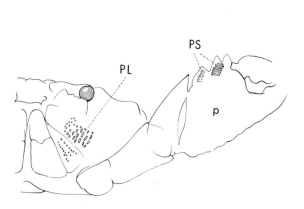
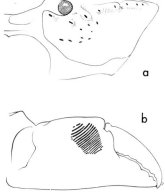

Fig. 178. *Matuta lunaris* (Forskål) (Decapoda, Calappidae). Double pars stridens in the palm and granulous plectrum in the pterygostomian region. p: palm (inner face of the propodus of the cheliped), Pl: plectrum, PS: pars stridens. (After Guinot-Dumortier and Dumortier, 1960[111].)

Fig. 179. *Menippe obtusa* Stimpson (Decapoda, Calappidae). (a) protuberances of the orbital border (plectrum), (b) finger print pars stridens on the inner face of the palm. (After Guinot-Dumortier and Dumortier, 1960[111].)

Various authors[15, 221] have had the opportunity to hear stridulation of walking *Dotilla* Stimpson (Scopimerinae), gregarious crabs living on beaches. Probably, the noise is produced by the passage of the merus of the walking legs across the pleural region of the carapace, which is covered with short bristles (Fig. 181).

Mutual friction of appendages

The modes of stridulation that we are grouping here are much more heterogeneous.

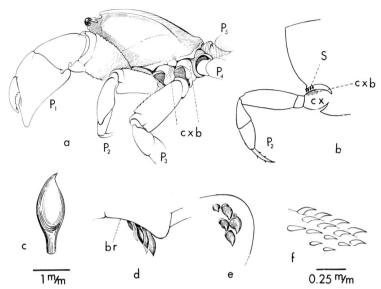

Fig. 180. *Potamon africanum* (A. Milne Edwards) (Decapoda, Potamonidae). (a) lateral view of the animal showing on P_2 and P_3 the areas of coxal bristles, (b) diagrammatic transversal section of the animal at the level of a stridulatory region, (c) enlarged inflated bristle of the branchiostegite, (d) and (e) a cluster of bristles of the branchiostegite, (f) area of coxal bristles. br: branchiostegite, cx: coxa, cxb: coxal bristles, P_1: cheliped, P_2–P_5: walking legs, S: bristles of the branchiostegite. (After Guinot-Dumortier and Dumortier, 1960[111].)

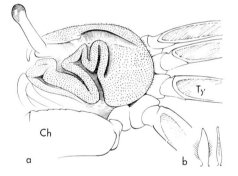

Fig. 181. *Dotilla fenestrata* Hilgendorf (Decapoda, Ocypodidae) (a) left side of the animal showing the convolutions and the rough aspect of the carapace. On the walking legs, one can see a flattening of the merus called "tympanum". (b) enlarged front and side view of one of the minute bristles of the carapace. Ch: cheliped, Ty: tympanum. (After Guinot-Dumortier and Dumortier, 1960[111].)

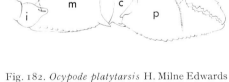

Fig. 182. *Ocypode platytarsis* H. Milne Edwards (Décapoda, Ocypodidae) large cheliped with the stridulatory apparatus. c: carpus, i: ischium, m: merus, p: propodus, Pl: plectrum, PS: pars stridens. (After Guinot-Dumortier and Dumortier, 1960[111].)

The method used by males and females of the genus *Ocypode* Weber (Ocypodidae)* is certainly unique among arthropods. The apparatus is uneven. It only exists on the

---

* It seems that all species stridulate, except *O. cordimana* Desmarest.

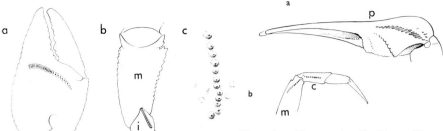

Fig. 183. *Ocypode cursor* (L.) (a) pars stridens, (b) plectrum, (c) enlarged view of the pars stridens of *O. nobilii* De Man. i: ischium, m: merus. (After Guinot-Dumortier and Dumortier, 1960[111].)

Fig. 184. *Uca musica* Rathbun (Decapoda, Ocypodidae). (a) inner side of the large cheliped showing the tubercles on the propodus, (b) the row of granules on the merus and carpus of the first walking leg. c: carpus, m: merus, p: propodus. (After Rathbun, 1914, *U.S. Nat. Mus.*, 47, 117.)

larger of the two chelipeds. The propodus has on its inner face, at the base of the two fingers, a rasp which extends at right-angles to the axis of the appendage. The ischium is supplied with a smooth crest (plectrum) parallel to the axis of the joint, against which the rasp rubs when the cheliped is folded (Fig. 182 and 183). Stridulation has been heard in various species by several writers[2,9,64,69,111,189,221].

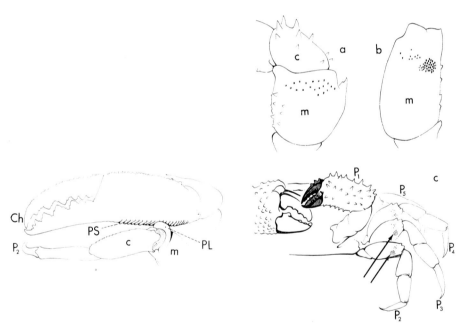

Fig. 185. *Ovalipes punctatus* De Haan (Decapoda, Portunidae). The supposed stridulatory apparatus (compare with Fig. 175). c: carpus, Ch: cheliped, m: merus, P_2: first walking leg, Pl: plectrum, PS: pars stridens. (After Guinot-Dumortier and Dumortier, 1960[111].)

Fig. 186. *Globopilumnus stridulans* Monod (Decapoda, Xanthidae). (a) external face of the merus of the cheliped with a granulous area, (b) internal face of the merus of the first walking leg with another granulous area, (c) front view of the animal showing the stridulatory areas (arrows) of the internal face of the merus of walking legs P_2, P_3, P_4, c: carpus, m: merus, P_1: cheliped, P_2–P_5: walking legs. (After Guinot-Dumortier and Dumortier, 1960[111].)

Also among Ocypodidae, there is the genus *Uca* Leach in which some species (*U. musica* Rathbun, *U. terpsichores* Crane) have structures considered to be stridulatory. The inner face of the propodus of the large cheliped has a series of elongated tubercles on its lower edge. When the appendage is flexed, this crest comes into contact with a row of granules on the merus and the carpus of the first walking leg on the same side[64] (Fig. 184) (see also p. 328).

The apparatus of *Ovalipes punctatus* De Haan (Portunidae) is of the same type. We mention it particularly for its interest when compared with that of another species of the same genus, *O. ocellatus* (see p. 310). The sound organ consists of a row of large ribs on the lower edge of the palm of the cheliped, which can scrape against a horned rim partially surrounding the distal end of the merus of the first pair of walking legs (Fig. 185).

Mutual friction of the appendages is a process used by both male and female in species of the genus *Globopilumnus* Balss (Xanthidae). The regions opposite the merus of the cheliped and the walking legs have granular areas which give rise to stridulation when they pass against one another (Fig. 186).

A last method is represented by *Sesarma* Say (Grapsidae) where friction of one cheliped against the other has been observed. The dactyls here have a row of large tubercles[240].

(iii) Isopoda

In the genus *Androniscus* Verhoeff (Oniscoidea) an apparatus, presumably stridulatory, has been observed. It constitutes, moreover, one of the characteristics of the genus[218, 231, 247]. The base of the ischium of the 7th pair of thoracic limbs has two rows of minute scales. Opposite these scales the carpus and propodus of the 6th pair of legs are supplied with similar projections and scaly formations. It is not known whether these animals use this device.

Although, apart from Malacostraca, there are no known species with stridulatory apparatus, it has been noticed that various species of the genus *Balanus* Da Costa (Cirripedia, Thoracica), produce a crackling sound which, because of the density of the population, gives a continuous noise constituting an important component of the underwater noises of biological origin. This crackling consists of clicks lasting about one millisecond produced at the rate of about ten a minute[45].

(c) Arachnida

Stridulatory apparatus has been observed in numerous species of both Araneida and Scorpionidea which we shall examine in turn.

(i) Araneida

Berland (1932)[27] made a summary of the various devices found in this class. The stridulatory regions may occupy five main positions:
- chelicerae,
- chelicera-pedipalp,
- pedipalp-p_1,
- coxa-abdomen,
- abdomen-cephalothorax.

References p. 338

The chelicera-chelicera type

This type is found in various mygales where the inner face of the basal joint of the chelicerae has an area of club-shaped bristles. The friction of the two appendages against each other produces a clearly audible metallic noise. This is the case, for instance, in the genera *Selenogyrus* Pocock (Fig. 187), *Euphrictus* Hirst (Theraphosidae)[128].

The chelicera-pedipalp type

The basal joint of the chelicera bears a long striated area on its exterior face against which rubs a row of teeth or a simple tubercle of the femur of the pedipalp. This device is found in male and female *Sicarius hahni* (Karsch) (Sicariidae) (Fig. 188), *Stridulattus stridulans* Petrunk. (Salticidae), in the genera *Leptiphantes* Menge (Linyphiidae) and *Diguetia* Simon (Sicariidae).

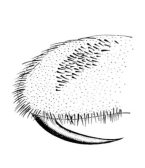
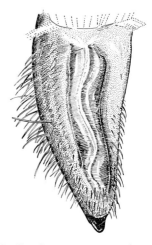

Fig. 187. *Selenogyrus aureus* Poc. (Araneida, Teraphosidae). Inner face of the chelicera showing the stridulatory bristles. (After Hirst, 1908[128].)

Fig. 188. *Sicarius hahni* (Karsch) (Araneida, Sicariidae). Outer face of the chelicera showing the striated area. (After Millot, 1949[180].)

An apparatus of a similar type is present in the mygale *Chilobrachys stridulans* (Wood-Mason) (Theraphosidae) and other species of the genus, but striation of the chelicerae and teeth of the pedipalp are replaced by bristles[102] (Fig. 189). In *Phrictus crassipes* L. K. (= *Selenocosmia crassipes* (L. K.)) (Theraphosidae) these bristles are on the basal joint of the pedipalp, and the frictional area of the chelicera is covered with spines[180,228,229] (Fig. 190). The same is the case in the genus *Musagetes* Pocock (Theraphosidae).

The pedipalp-p_1 type

In the genus *Hysterocrates* Simon (Theraphosidae) the coxa of the pedipalp has spines on its rear surface. On its front surface the coxa of p_1 has vibrant bristles.

The coxa-abdomen type

In some Erigonidae the male has developed an apparatus which seems to be peculiar to this family. A spine on the distal postero-medial face of coxa IV scrapes

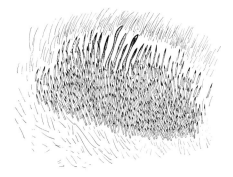

Fig. 189. *Chilobrachys samarae* Giltay (Araneida, Teraphosidae). Stridulatory bristles of the pedipalp coxa. (After Giltay, 1935[102].)

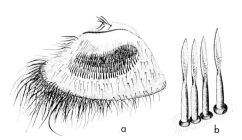

Fig. 190. *Selenocosmia crassipes* (L. K.) (Araneida, Theraphosidae). (a) inner face of the tarsal joint of the pedipalp showing the stridulatory bristles, (b) four of these bristles mostly enlarged. (After Spencer, 1896[229].)

against a striation of the lung book cover. This apparatus was first observed in a very small species (1.5 mm), *Entelecara broccha* L. Koch[49], and afterwards found in other species of this genus. This device is also a characteristic for the genus.

The same structure characterises the male *Gongylidiellum vivum* O.P. Camb.[110] (Fig. 191), *G. mursidum* Simon and various species of the genus *Erigone* Audoin.

The abdomen-cephalothorax type

This kind of apparatus is typical for the family Theridiidae. It consists of either a denticulation or a series of crests on the front side of the abdomen that rubs against a striation of the rear surface of the cephalothorax. In the male *Theridium ovatum* Cl., the back of the cephalothorax has two striated areas on the two sides of the sagittal plane. A flange in the form of a crescent, spans the pedicel at the level of its abdominal insertion; each side has three superposed crests which come into contact with cephalo-thoracic striated areas[110] (Fig. 192). There is a similar apparatus in *Steatoda castanea* Cl. and in *S. bipunctata* L. where it represents a male characteristic which only appears at the last moulting. Stridulation seems to be produced only at the time of reproduction when the male is in search of the female[179].

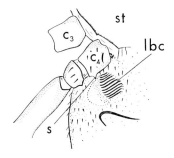

Fig. 191. *Gongylidiellum vivum* O. P. Camb. ♂ (Araneida, Erigonidae). Stridulatory apparatus. c_3: coxa$_3$, c_4: coxa$_4$, lbc: striated lung book cover, s: spine of the coxa$_4$, st: sternum. (After Guibé, 1943[110].)

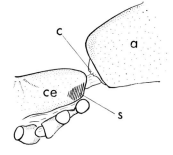

Fig. 192. *Theridium ovatum* Clerk ♂ (Araneida, Theridiidae). Stridulatory apparatus. a: abdomen, c: crests of the flange of the pedicel, ce: cephalothorax, s: striae of the cephalothorax basal angle. (After Guibé, 1943[110].)

References p. 338

A similar device is found in the genera *Asagena* Sundevall, *Lithyphantes* Thorell, *Argyrodes* Simon (Theridiidae) and in the genus *Cambridgea* L. Koch (Agelinidae)[197].

It will be noticed that this abdomino-cephalothoracic device looks like that found in the Crustacea Decapoda of the genus *Penaeopsis* (see p. 305)*.

Other examples

Apart from Araneida and Scorpionidea which will be dealt with later, two orders of Arachnida include stridulating species.

In the order Solifugae, the inner face of the chelicerae, which are considerably developed appendages, shows various ornamentations. Stridulation is due to the passing of spiny bristles of one chelicera over ridges of the other[181] (Fig. 198).

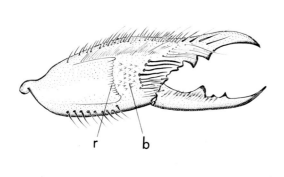

Fig. 193. *Galeodes araneoides* (Pally.) ♀ (Solifugae, Galeodidae). Inner side of the left chelicera. b: spiny bristles, r: ridges. (After Roewer, 1934, in Millot and Vachon, 1949[181].)

Fig. 194. *Musicodamon atlanteus* Fage (Amblypygi, Tarantulidae). Club-shaped bristles of the inner side of the chelicera. (After Fage, 1939[88].)

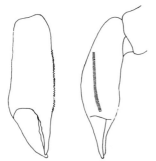
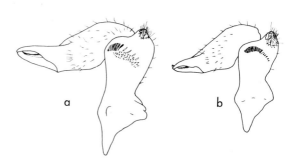

Fig. 195. *Cryptobunus silvicolus* (Opilionidea, Triaenonychidae). Second cheliceral segment, seen from the front and the inner side, showing the row of stridulatory ridges. (After Lawrence, 1937[162].)

Fig. 196. (a) *Nemastoma dentipalpes* Auss., (b) *N. argenteo lunulatum* Simon (Opilionidea, Nemastomatidae). Inner side of the right chelicera of the male showing the stridulatory ribs on the basal segment. (After Juberthie, 1957[146].)

---

* In some other spiders the sound emission is produced by percussion on the substratum (see p. 328).

In *Musicodamon atlanteus* Fage, belonging to the order Amblypygi, a stridulatory apparatus has been found that is very much akin to that of the mygales. It is formed of club-shaped bristles on the inner face of the chelicerae[88] (Fig. 194).

In harvest-spiders several examples of stridulatory apparatus are known. In the South-African genera *Biacumontia* Lawrence, *Lispomontia* Lawrence, *Lawrencella* Lawrence, *Cryptobunus* Lawrence, of the sub-order Laniatores, a rasp consisting of small parallel ribs extends throughout the inner face of the 2nd joint of each chelicera[162] (Fig. 195). In the Palpatores *Nemastoma dentipalpes* Auss., and in male and female *N. argenteo lunulatum* Simon (Nemastomatidae), the basal joint of the chelicera has a curved line of small ribs[146] (Fig. 196).

(ii) *Scorpionidea*

Pocock[202] was undoubtedly among the first to study the possibilities of stridulation of certain scorpions. Following his work other examples have been found which may be grouped in four categories in accordance with the method used.

The pedipalp-walking leg type

In *Heterometrus swammerdami* Simon (Scorpionidae), the coxa of the pedipalp (the appendage bearing the claw) has an area of stiff, curved bristles which comes into contact with a rasp situated on the coxa of the first pair of walking legs (the "keyboard" of Pocock, Fig. 197). This arrangement is reversed in *Pandinus imperator* (C. L. Koch) (Scorpionidae)[3,255].

The chelicera-cephalothorax type

This method of emission recalls that found among the mygales (see p. 316)

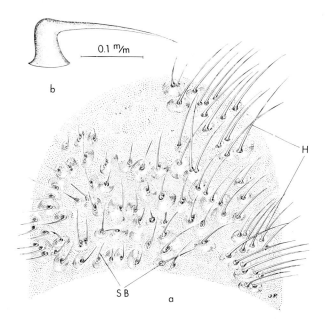

Fig. 197. *Heterometrus costimanus* Simon (Scorpionidea, Scorpionidae). (a) stridulatory region of the coxa of the pedipalp; the area of right-angle curved bristles is bounded to the right with long hairs; (b) enlarged stridulatory bristle. H: hairs, SB: stridulatory bristles. (Original.)

because the stridulatory structures appear on the same appendage. But in fact the working mechanism is different. In ten or so species of the genus *Opisthophthalmus* (Scorpionidae): *O. latimanus* C. L. Koch, *O. wahlbergi* (Thor.), *O. glabrifrons* (Pet.), etc.[33,255]; the inner face of the basal joint of the chelicera has two areas of bristles: an upper area of small bristles bent back at the base and curved at right-angles (the *scaphotrix* of Pavlowski), and a lower area formed of a few large bristles flattened out in raquet form (the *trichocopae* of Pavlowski). The animal stridulates by an alternating to and fro movement of its chelicerae, but the sound is not produced by a mutual rubbing of the two chelicerae. It comes from the passing of the scaphotrix against an interior sagittal carina of the cephalothorax. This carina is striated (Fig. 198).

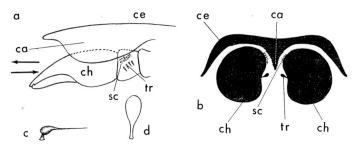

Fig. 198. *Opisthophthalmus latimanus* C. L. Koch (Scorpionidea, Scorpionidae). (a) diagrammatic sagittal section at the level of the carina, showing the right chelicera with its two kinds of bristles (sc and tr). The arows indicate the stridulatory movement of the chelicera. (b) transversal section at the level of the basal joint of the chelicera showing the bristles and the striated carina, (c) an enlarged scaphotrix (stridulatory bristle), (d) an enlarged trichocopa.
ca: carina, ce: cephalothorax, ch: chelicera, sc: scaphotrix, tr: trichocopa. (After Vachon, original.)

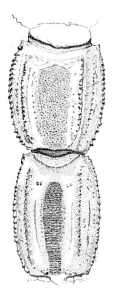

Fig. 199. *Parabuthus laevifrons* Simon (Scorpionidea, Buthidae). First and second caudal segments showing the granulous areas scratched by the sting. (Original.)

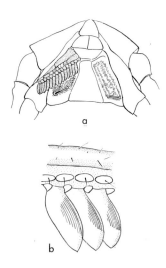

Fig. 200. *Rhopalurus borelli* Pocock (Scorpionidea, Buthidae). (a) ventral view (left pecten removed) showing the rough stridulatory area of the third sternite, (b) enlarged view of three striated pectinal teeth. (After Pocock in Werner, 1934[255].)

The trichocopae (the role of which appears to be one of chemoreception) might also come into play, but they would only have a small part in the sound emission[3]. Stridulation behaviour is easily started in both sexes by a touch or an air puff. The sound is a low, soft chirping[244].

The tail-sting type

In *Parabuthus planicauda* (Poc.), *P. flavidus* (Poc.) (Buthidae) and other species of the genus, stridulation has been observed[3, 161, 255]. It is produced by the passing of the point of the sting over the first two segments of the tail (metasoma) (Fig. 199). In *P. brachystylus* Lawrence, the functional area of the sting is striated. The same mechanism is found in some other species of the genus *Androctonus* (Buthidae)[182].

The pectine-sternite type

This method of stridulation is one of the first to have been described. In 1828 Burchell discovered it in *Rhopalurus borelli* Poc. (Buthidae)[202, 255]. In this species the pectinal teeth are striated inside and rub against two granulous areas of the 3rd mesosomal sternite (Fig. 200).

(d) Myriapoda

Myriapoda have not often been quoted in studies on sound emission. However, Scharrer (1931)[218] and Tembrock (1959)[234] mentioned stridulatory apparatus in Sphaerotheridae. According to Saussure and Zehntner (1902)[217] "African and Madagascan Sphaerotheridae are eminently stridulatory animals". In the male, the sound apparatus, which appears on the copulatory appendages, is variably shaped (Fig. 201). For instance:

The first copulatory leg (22nd pair) has carinae which may be rubbed by a tubercle situated on the last ambulatory leg (21st pair) (Fig. 201 g, h, i);

A pygidial tubercled swelling might also come into contact with the striae of the 2nd joint of the second copulatory leg (23rd pair) etc. (Fig. 201a,b)[32,217,248]. Moreover, several of these devices can be found in the same animal.

In the female, the sub-anal plate is striated and a tooth of the 2nd joint of the last pair of legs scrapes against this striation (Fig. 202).

The Malay species *Sphaeropoeus volzi* Carl has a nocturnal stridulation which can be heard from far away and which resembles the call of the midwife-toad[48]. In the African genus *Alipes* Imhoff (Scolopendridae), the tarsus of the last pair of appendages is enlarged into a granulated patella. A small expansion, or a rugged area of the tibia, would be capable of rubbing against the tarsal patella[61, 100]. The rapid vibration of these appendages produces a rustling sound[163].

Although stridulating myriapods appear to be fairly rare, they do provide the most surprising examples.

In the scutigeromorph centipede, *Scutigera decipiens* (Verhoeff), a leg detached from the body by autonomy, is agitated by vigorous rhythmical contractions giving rise to a loud creaking sound. On the ventral face of the femur, in its basal part, a small transverse slit with a row of hooks can be observed. This may perhaps be a stridulatory apparatus, but it is not known which area of the appendage this projection can rub against. Identical structures are found on the other legs but stridulation only occurs in this species with the autotomised legs[10]. On the other hand, in other species the sound is also produced by the legs *in situ*. It has been pointed out, moreover, that the pill-millipede *Arthrosphaera aurocineta* Pocock vibrated very distinctly when it was picked up, and that this vibration was accompanied by a squeaking noise[109].

In the scolopendromorph centipede *Rhysida nuda togoensis* Kräp., the anal legs are easily autotomised and then, for some forty seconds, show a spontaneous activity accompanied by a faint creaking sound, audible at 2 or 3 ft. However, there is no discernible apparatus which appears to correspond to this emission. The suggestion has been put forward that the sound might be the result of the stresses set up in the

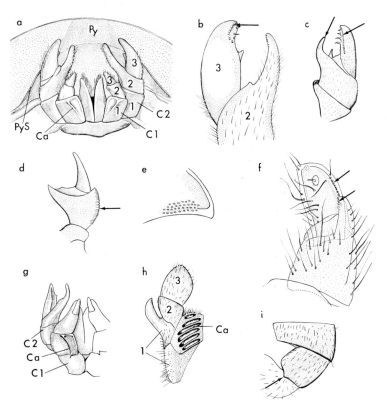

Fig. 201. Examples of presumed stridulatory devices in Myriapoda, Sphaerotheridae ♂. (a) ventral view of the pygidial region showing the pygidial swelling and the two pairs of copulatory legs; on the first, the basal joint bears a carina, (b) distal part of the second copulatory leg, the deniticulation (arrow) of the third joint (movable finger) may rub against a pygidial swelling (see a), (c) second copulatory leg: the sharp tip of second joint may scratch a row of tubercles along the movable finger (arrows), (d) a striated edge (arrow) of the second copulatory leg rubs a granulous area of the pygidium, (e) granulous area of the pygidium, (f) first copulatory leg showing the possibility of scratching between tuberculate areas of the first and the second joint (arrows). This method is like the one represented in (c), (g, h) two types of carinae on the first joint of the first copulatory leg (see also a), (i) last ambulatory leg showing the tooth (arrow) scratching the carinae of the first copulatory leg (see a, g, h).
C_1: first pair of copulatory leg (22nd pair of legs), C_2: second pair of copulatory legs (23rd pair of legs), Ca: carina, Py: pygidium, PyS: pygidial swelling, 1, 2, 3: 1st, 2nd and 3rd joint of the copulatory legs. *Species*: (a, b) *Sphaerotherium coquerelianum*, (c) *S. anomalum*, (d, e) *Bournellum retusum*, (f) *Borneopoeus costatus*, (g) *Sphaerotherium campanulatum*, (h, i) *Sphaeropoeus musicus*.
(a, b, c, d, e, g, h, i after Saussure and Zehntner, 1902[217], f after Verhoeff, 1928[248].)

cuticle when the limb is flexed and straightened. These stresses are caused by large apodemes attached to the proximal end of one segment and lying within the retractor muscle of the next[57, 58].

Fig. 202. Presumed stridulatory devices in Myriapoda, Sphaerotheridae ♀ *(Sphaerotherium acteon)*. (a) sub anal plate showing the crests on which scratches a rough area (arrow) of the last pair of ambulatory legs, (b). (After Saussure and Zehntner, 1902[217].)

3. Sound Emission by Vibration of a Membrane (other than Wings)

The Homoptera represent almost the only arthropods in which stridulation is caused by the vibration of a membrane. These insects and this method are specially dealt with by D. Leston and J. W. S. Pringle in Chapter 14, and so we shall only mention them here in passing.

One single example of the use of this method apart from the Homoptera, seems to be that of the moth *Hylophila prasinana* L. (Noctuidae) which emits a sharp stridulation audible at 10 ft. The insect has on the metasternum a half-moon shaped opening covered with a folded membrane on which muscles adhere. The contraction of these muscles can cause a vibration of the membrane[259]. This observation, made rather long ago, is worth checking.

4. Sound Emission by Passage of a Fluid (gas or liquid) across an Orifice

It seems that sound emissions produced by expulsion of fluids is found only among the insects. Several methods are used.

(a) Passage by way of the mouth

The only example is that of the "Death's Head Moth", *Acherontia atropos* L. (Lepidoptera, Sphingidae*), the cry of which, noticed for the first time by Reaumur in 1734[210], has been the subject of nearly 70 publications. Most of these have erroneous observations as their basis. The first valid explanation was given in 1828 by Passerini[192] but the only complete and profound study is due to Prell (1920)[206].

The method which the moth uses is unique among the arthropods. The cry consists of two very short sequences, repeated rapidly: one sequence low-pitched and one sequence high-pitched.

The low part of the cry is due to the dilation of the pharyngeal cavity which produces a breath of air across the proboscis (Fig. 203). This rising column of air causes the epipharynx, a rigid crescent-shaped clapper which closes the buccal orifice, to vibrate (Fig. 204a). The high sequence is caused by the expulsion of air by the proboscis which acts like a whistle while the epipharynx is kept erect (Fig. 204b).

---

* A comparable emission has been noticed in *Smerinthus populi* L. (Sphingidae)[63] and in *Arctia caja* L. (Arctiidae)[169], but these observations will have to be checked.

Recently, the hypotheses of Prell have been confirmed by some methods of acoustical analysis which have permitted the study of the physical structure of the cry as well (see Table 24, p. 348 and Fig. 216, E, p. 351)[44].

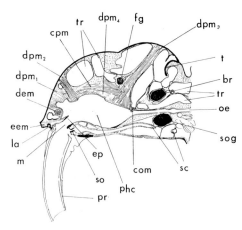

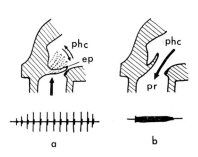

Fig. 203. *Acherontia atropos* L. (Lépid., Sphingidae). Sagittal section of the head showing the different structures playing a role in the sound emission. (After Prell, 1920[206].)

Fig. 204. Working of the epipharynx of *Acherontia atropos* L. during the sound emission. (a) low pitched sequence, (b) high pitched sequence, (big arrows indicate the way of the air current). Below: schematic drawing of oscillographic analysis of each sequence. ep: epipharynx, phc: pharyngeal cavity, pr: proboscis. (After Busnel and Dumortier, 1959[44].)

br: brain; com: constrictor muscle of oesophagus; cpm: constrictor muscle of pharynx; dem: depressor muscle of epipharynx; dpm_1 and dpm_2: anterior dilator muscles of pharynx; dmp_3: medio-posterior dilator muscle of pharynx; dpm_4: lateral dilator muscle of pharynx; eem: elevator muscle of epipharynx; ep: epipharynx; fg: frontal ganglion; la: labium; m: mouth; oe: oesophagus; phc: pharyngeal cavity; pr: proboscis; sc: salivary canal; so: salivary orifice; sob: suboesophageal ganglion; t: tentorium; tr: trachea.

(b) *Passage through the spiracles or through glandular orifices*

Older writers often ascribed the humming of Diptera and Hymenoptera to a vibration of lamellae situated at the opening of the spiracles, caused by the passage of air breathed in or out[42, 159]. These ideas have since been almost abandoned. However, it is probable that these insects can emit some sounds with an origin other than the vibration of their wings.

On the other hand, some more recent observations have shown that the expulsion of air by the spiracles in certain insects might be accompanied by a sound emission. Similarly the production of foam at the level of certain glandular orifices or of actual spiracles, may be accompanied by a noise.

(i) *Expulsion of liquids*

This occurs in Orthoptera and Lepidoptera.

Among the first, *Taeniopoda picticornis* (Wlk.) (Acridoidea) ejects foam through the mesothoracic spiracles, a phenomenon which is accompanied by a sound comparable to that produced by crumpling paper[55, 76, 152, 170]. In the genus *Dictyophorus* Thunb. (Pyrgomorphidae), the emission of foam gives rise to a whistling sound[50, 51].

Among Lepidoptera, several examples are known[51]. *Dysschema tiresias* Cram. (Arctiidae) emits with a chirping noise a foamy matter which flows out through two pairs of orifices situated on the top and sides of the thorax.

The same phenomenon is observed in *Composia fidelissima* Herr.-Schäf., *Rhodogastria bubo* Walker, *R. lupia* Druce (Arctiidae) and in female *Erasmia sanguiflua* Drury (Zygaenidae).

It should be mentioned that, despite of the fact that this emission of foam is accompanied by a sound, it is in no way certain whether it means anything. It is possible that the noise is only a consequence of the production of foamy liquid.

(ii) Expulsion of air

A humming different from stridulation (see p. 304) has been noted among some Coleoptera Dytiscidae. The humming may have as origin an expulsion of air by the first pair of spiracles[256]. The same phenomenon is observed in the larva of *Cybister confusus* Sharp (Dytiscidae) (see p. 331).

In 1956, Woods[263] explained one of the many noises of the bee-hives, the "Queen piping" (see Table 24, p. 349). This sound, which is heard distinctly at 20 ft., is produced by the newly hatched queen, when another queen is already present in the hive. The releasing stimulus of the sound emission is olfactory in nature (the smell of the first queen).

Simultaneously with the piping, the wings vibrate laterally in a kind of scissors motion, keeping close to the body. The rate of the vibration of the wings corresponds to the frequency of the sound, but it is not the wings themselves which are responsible for the sound emission.

The nerve centers which control the opening of the valves of the spiracles are connected to those which control the wing muscles. Therefore at the same time that the wings vibrate, the valves are worked at the same rhythm by an alternating movement of shutting and opening. The column of air which circulates in the tracheal trunks is then broken up according to the frequency of the vibration, which causes the production of a sound wave of identical frequency*.

(c) Passage through the anal region

The detonations which various species of the genera *Aptinus* Bon. and *Brachynus* Weber (Coleoptera, Brachynidae) produce have been known for a long time[159, 254]. The anal glands of these insects are highly developed and their secretion is collected by two contractile pouches which emerge by two canals above the anus. The liquid produced in these glands contains, among other things, nitric acid and a lipid. Ejected outside the body, it explodes and volatilises into a whitish vapour, phosphorescent in the dark, which has the strength to blister the skin. In certain tropical forms, the vapour seems to be formed mostly of an oxide of nitrogen which stains the skin yellow[254].

5. Sound Emission by Percussion on the Substratum

Production of sounds by percussion on the substratum, although most often not requiring any specialized structure, is of certain importance from the biological point of view. One can even consider that this is a more highly evolved method than stridulation since the same result (sound emission) is attained without recourse to a specialized apparatus, hence much more simply.

---

* It is then a real *vocal* apparatus, the functioning principle of which is identical to that which Husson proposed for the vocal chords of man[133,134].

(a) Coleoptera

The most famous case is that of the Death watches, *Anobium pertinax* L., *A. striatum* Oliv., etc., *Xestobium tesselatum* Oliv. (Anobiidae), which have been studied for a very long time[68, 78, 159, 160, 258].

In the female *Xestobium tesselatum*, the percussion is effected in two stages: first the insect rises up on its forelegs, its abdomen sloping downwards, then all the body swings towards the front, under thrust from the hind legs, with the middle legs serving as a pivot. Then the head can knock the substratum (Fig. 205). In this way the insect emits salvoes of 7 to 8 percussions per second (a little less if the temperature is low)[99]. Some older observations made on *Anobium* are in agreement (see pp. 586 and 616).

Another case has been described in *Nothorrhina muricata* Dalm. (Cerambycidae). This insect takes up a position between two ridges of bark and makes its body vibrate rapidly, striking the walls of the crevice in which it is lying. The noise thus produced is a faint purring[86].

A slightly different procedure is used by the mycetophagous Tenebrionidae *Bolitotherus cornutus* (Panzer). During the preliminaries of copulation, the male rubs its abdominal sternites against two tubercles on the thorax of the female. The rasping sound thus made is heard 6–8 ft. away. It lasts one to two minutes[166].

(b) Psocoptera*

The males of different species make a rapid crepitation by percussion of a protuberance of the subgenital plate on the substratum. Like the Anobiidae, these insects have been named Death-watches, *Horloge de la Mort* or *Totenuhr*.

In *Atropos pulsatorium* L. (Fig. 206) the rate of percussion is 5–6 blows per second. The ticking may sometimes go on for a minute. It is shorter (3 sec) in *Lepinotus inquilinus* Heyd. These insects, very small in size (2 mm for *A. pulsatorium*), are

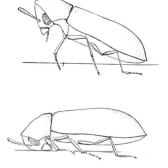

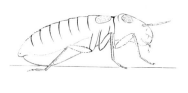

Fig. 205. *Xestobium tesselatum* Oliv. ♀ (Col., Anobiidae). Tapping movement with the head. (After Gahan, 1918[99].)

Fig. 206. *Atropos pulsatorium* L. (Psoc. Trogiidae). Tapping movement with the abdomen. (After Pearman, 1928[193].)

---

* A coxal organ presumed to be stridulatory has likewise been described in the Psoci[193] (see foot-note p. 290).

without doubt among the smallest to produce sounds. But the role of this activity is unknown[17,112,193].

(c) Isoptera

Some quite old observations[39,95,106,153,155], not precise as to species or genus, note a tapping behaviour in different termites. In a Brazilian species, a break in a gallery makes some black-headed termites (soldiers?) run up. These bang the ground rapidly with their heads and assemble yellow-headed individuals (workers?) who get to work on repairing the structure. The sound produced is like the noise of sand falling on a sheet of paper[106]

Various other observations, despite their vagueness, agree that the tapping produced by the soldiers is an assembly signal for the workers (see p. 590). More recently, the snapping by soldiers which is set off by a disturbance has been noted in various species of the genera *Syntermes* Holmg., *Cornitermes* Wasm., *Armitermes* Wasm., *Rhinotermes* Hagen, *Bellicositermes* Emerson, *Capritermes* Wasm., etc. The percussive region is the gula or the mandibles, according to the genera. In *Capritermes*, the snapping is performed with such a force that the animal is thrown 1 or 2 cm into the air by the recoil. According to the kind of substratum, the noise is heard from 5–20 ft. away[80]. Actually this question of the distance the sound carries is almost certainly unimportant, for the "message" is probably received by a mechano-receptor through vibrations transmitted by the substrate[81].

A loud chirping produced by the snapping of thousands of soldiers has often been heard as a response to a disturbance by a blow on the nest[108].

It seems that the termites (more especially workers) use another type of vibratory emission in addition. As a matter of fact, the animals are often seen jerking themselves back and forth rapidly while pivoting upon their legs. The same behaviour has been observed in a termitophile Staphylinidae (*Termitogaster emersoni* Mann). The vibration make a worker approach at once and start to lick the beetle all over the body[80]. It is possible that workers behave in the same way among themselves.

(d) Orthoptera

Various Oedipodinae (Acrididae) male and female, bang the ground with the tarsi of their hind legs in certain phases of their sexual behaviour. Faber (1936)[83] has made a very detailed study of this display.

In male *Oedipoda coerulescens* (L.), the percussion rate is of the order of 12 per second; for the female the rhythm is a little slower. Some similar observations have been made on *Bryodema tuberculata* (Fabr.). This tapping may be accompanied by a stridulation (by passage of the femur over the elytron) in male and female *Psophus stridulus* (L.). In the male, this behaviour is a reaction to the presence of another male (rivalry), and it is generally preceded by antennary contacts. In the female it represents an attitude of refusal towards the male.

It seems that this percussion on the substratum may be understood as a derivative of a stridulatory movement. The noise produced, which evidently depends on the nature of the substratum, may act as a signal for the partner or the rival. But it is also possible that the actual signal is visual (movement of the legs), and that the blow on the ground is only incidental.

References p. 338

It is interesting to note that in response to the same situation, different stages of passages between this tapping and the real stridulation are found in Oedipodinae. In *Celes variabilis* (Pall.), there is vibration of the legs without raps or stridulation. In various *Oedipoda* and in *Bryodema tuberculata*, there is percussion on the substratum. This percussion is accompanied by a stridulation in *Psophus stridulus*. Finally, the same movement is purely stridulatory in *Locusta migratoria* L. (*solitaria* phase)[83]. This tapping has been likewise observed in the American species *Encoptolophus sordidus* Burm.[200].

The percussion of the abdomen appears less frequent, but it occurs in *Aiolopus strepens* (Latr.)[85], and for a long time it was thought responsible for the noise produced by *Meconema thalassina* (De Geer) (Tettigonioidea, Meconemidae). However, some recent observations[67] seem to indicate that the abdomen does not enter into contact with the substratum. The elytral stridulation of this insect will be dealt with later (see p. 329).

In all the examples which have just been cited, the signal (if it really is a signal) may be received acoustically through the air, but it may just as well be by mechanoreception of the vibrations transmitted by the substrate.

(e) Other examples

Among Hymenoptera, the percussion on the ground is found in some ants (*Dendromyrmex*, Formicinae)[106]. In some species, a remarkable regularity of percussion rhythm is shown[92]*.

Tapping has likewise been observed in various species of Perlidae (Plecoptera). The males of these species have a plate on the 9th abdominal sternite, with which the percussion appears to be made on the substrate. In *Dinoceras cephalotes* Curtis, the noise consists of three or four successive quite intense raps. These series of raps are repeated three or four times in *Chloroperla grammatica* Scop.[33, 172].

Among spiders, some examples of tapping have also been noticed: *Lycosa chelata* (Müller)[54], *Tarentula pulverulenta* (Cl.)[34] (Lycosidae), male *Pisaura mirabilis* (Cl.) (Pisauridae)[205]. During the reproductive period, females *Pardosa lugubris* Walck. (Lycosidae), rap rapidly with the end of their abdomen on the dry leaves in which they live. The noise, lasting barely a second, may have a sexual significance[54].

6. Sound Emission by Vibration of Appendages

The most numerous cases refer to the vibration of wings, but this question of flight sound is the subject of Chapter 13 and will not be dealt with here.

Other examples are very rare. Mention can hardly be made of the vibration of the antennae in the genus *Phyllomorpha* Lap. (Heteroptera, Coreidae)[203] and that of the large cheliped of the fiddler crabs (*Uca*). In *Uca pugilator* (Bosc)[41, 71] a rapid movement of the appendage folded against the body has been observed (at a rate of about 10 per second). The sound produced is a low humming, but it could not be due to the rubbing of the claw against the carapace nor to impact with the ground. This emission is noticed at night, and it arises from the gathering of individuals around or inside their burrows (see p. 586).

---

* It is known that a well differentiated stridulatory apparatus is found in certain genera. (See pp. 283 and 350.)

7. Cases of Stridulation in which Non-differentiated Regions are Used

It is certain that the vibration of the antennae in *Phyllomorpha* or the ticking of *Anobium* represents emissions of noises, the origin of which are organs or regions not specifically adapted to this use. Nevertheless, they have no place in this sixth category which includes solely noise soriginating in the *friction* of appendages or of two areas with only an accessory stridulatory role. In any case, they have none of the usual differentiations for the function of sound emission.

(a) Mandibular noises

A certain number of examples are known among Orthoptera, Acridoidea[116,152]. *Oedaleonotus fuscipes* Scud. produces a clicking with the end of its mandibles[245]. In the European Catantopidae, *Calliptamus italicus* (L.), Faber (1949)[84] noticed the production of mandibular noises connected with stridulatory movements of the hind legs without sound emission. In these two acridids the noise seems to have a sexual significance.

More recently a new example of this type of emission has been discovered in the American species *Paratylotropidia brunneri* Scud.[6]. The South-African Gryllacrididae *Henicus monstrosus* makes a rasping noise in the same way[226]. Other cases are also known: *Mesembria dubia* Wlk. (Acrididae)[124], *Locusta migrata* L. (Acrididae). Snapping of the mandibles is also observed in various caterpillars (see p. 331).

(b) Wing noises

(i) Produced when at rest

Kevan (1955)[152] made a detailed study of some Orthopteroidea that are capable of producing humming noises. It is well known that the adoption of a fighting attitude in the praying mantis is accompanied by a sort of whistling caused by the spreading of the wings. Certain stick-insects (Palophinae) when disturbed emit a sharp whistle which has the same origin.

The genus *Phymateus* Thunb. (Acridoidea, Pyrgomorphidae) shows an aposematic coloration of the wings, which when spread out, give rise to an audible rustling[52].

In *Meconema thalassina* (De Geer) (Meconemidae) there is an elytral stridulation without a clearly defined stridulatory apparatus. (see also p. 328)[46].

It has also been observed[168] that in *Schistocerca gregaria* Forskål (Acrididae) there are two stridulations very closely linked with sexual behaviour, one produced by the elytra alone, the other by the vibration of wings and elytra. Although no differentiated apparatus has been observed, the duration of these emissions, lasting as long as 6 sec, and the manner in which they are produced, must qualify them as real stridulations and not just as noise accompanying another activity (spreading of the wings, for instance). It is a unique example of an elytro-elytral stridulation among Acridoidea.

In Lepidoptera numerous observations relate the buzzing noise made by various species of Nympalidae when the forewings slowly pass over the hind wings (*Euvanessa antiopa* L., *Aglais urticae* L., *A. polychloros* L., *Vanessa io* L.). This behaviour seems more frequent in individuals aroused from their hibernation[79,214,230,232].

After fertilisation (presence of the sphragis), the female *Parnassius mnemosyne* L. (Papilionidae) scrapes the external face of its hind wings, with the spines of the tarsi

References p. 338

and the spurs of the hind tibiae. The resultant noise, resembles that of a needle scratching a parchment, and may be heard 12 ft. away[31,142].

(ii) Produced in flight

Let us recall that we are considering only noises other than those originating in the simple vibratory movement of the wings (see Chapter 13). The chattering noises made in flight by Orthoptera, Oedipodinae, have often been pointed out. It seems that this noise is a peculiarity of the male, and a sexual significance is attributed to it. This chattering may come from the opening and closing of the hind wings, where in some species the veins are especially thick[136].

Flight noises are also found among other Orthoptera apart from Oedipodinae[55,85,242]. *Orphulella speciosa* (Scud.) (Truxalinae) beats its wings during flight and thus emits a series of rapid clicks[200]. Male *Stenobothrus rubicundus* (Germ.) (Acridinae) produce in flight a rustling noise.

8. Sound Production by Larvae, Nymphs and Pupae

As we pointed out at the beginning of this survey all that we have hitherto described refers to adults, but numerous immature forms are endowed with a stridulatory apparatus, often well differentiated, which sometimes is not found in the imago*. Stridulation in the pre-imago instars is only observed in insects. We shall study it on a systematic basis, rather than according to the morphology of the apparatus.

(a) Coleoptera

Stridulation in larvae is principally found among Scarabaeoidea. In Melolonthinae, Rutelinae, Dynastinae and Cetoninae, the maxillary stipes has spines rubbing against a striated or granulated plate of the sternal face of the mandible[11,109,141] (Fig. 207).

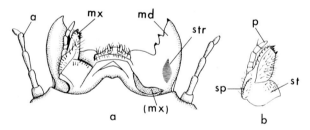

Fig. 207. *Oryctes rhinoceros* L. (Col., Scarabaeidae). Stridulatory apparatus of the larva. (a) ventral view of mouth parts (left maxilla removed) showing the striated area of the mandible, (b) tergal face of the left maxilla with the row of spines on the stipes.
a: antenna, md: mandible, mx: right maxilla, (mx): basis of the left maxilla removed, p: maxillary palpus, sp: stridulatory spines of the stipes, st: stipes, str: striated area of the sternal face of the mandible. (After Gravely, 1915[109].)

In Geotrupidae and Passalidae the third pair of legs has small sharp protuberances which scrapes a striated area of the mesocoxa or the trochanter (Figs. 208 and 209). In the genera *Lucanus* Scop., *Dorcus* McLeay, *Platycerus* Geoffr. and *Sinodendron* Hellw.

---

* In the Holometabola the imago and the larva both sometimes have a stridulatory device but the location is hardly ever the same.

(Lucanidae) the device is reversed. The coxa of the mesothoracic legs is granulated, while a striated crest appears on the well-developed trochanter of the hind legs[11].

A noise that can be produced either in the air or in the water has also been observed in the larva of *Cybister confusus* Sharp. (Dytiscidae). In the water the emission is accompanied by a release of bubbles from the right mesothoracic stigmata[186].

A case of pupal stridulation is known among Coleoptera. It is that of *Oryctes rhinoceros* L. (Dynastinae). The apparatus is made up of gin-traps* of the abdominal tergites 1 to 6. The sound is produced when the pupa is disturbed[109].

It will be noticed that the site of the stridulatory apparatus is not the same in the larva as in the adult (Rutelinae, Dynastinae) and that some families have stridulating species only in larval instars (Melolonthinae, Lucanidae).**

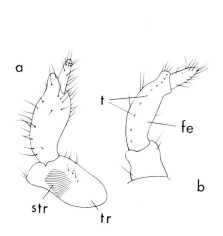

Fig. 208. *Geotrupes sylvaticus* L. (Col., Geotrupidae). Stridulatory apparatus of the larva. (a) mesothoracic leg with striation on the trochanter; (b) metathoracic leg with a row of teeth along the femur.
fe: femur, str: striated area, t: teeth, tr: trochanter. (After Lengerken, 1927.)***

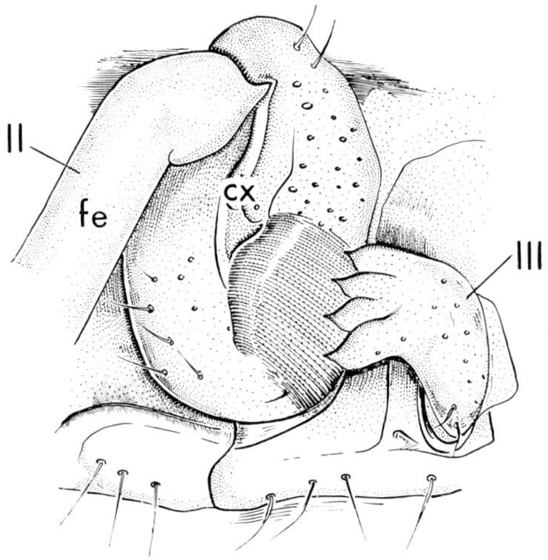

Fig. 209. Larval stridulatory apparatus of Coleoptera, Passalidae. cx: coxa of the mesofthoracic leg with its striated area, f: femur of the mesothoracic leg, II: mesothoracic leg, III: atrophied metathoracic leg used as a plectrum to scratch the striated area of the mesocoxa. (After Sharp, 1899, in Meixner, 1934[178].)

(b) Lepidoptera

Mandibular crackling is fairly common among the caterpillars of Sphingidae and Saturnidae when disturbed. The head movements observed in *Rhodia fugax* Butler (Saturnidae) and in *Sphecodina abbotti* Swainson (Sphingidae)[191], as an accompaniment to the production of noise, are undoubtedly only a reaction to excitation, and the

---

* Hinton (1948, 1955)[126,127] classed under this name, devices probably with a defensive role, which are found in the pupae of certain Lepidoptera and Coleoptera. They are formed at the junction of two segments by an invagination, the edges of which are sclerotized or denticulated. The gin-traps are normally open, but the upward curve of the abdomen causes them to close like jaws capable of seizing or wounding a small insect.
** With one exception: *Chiasognathus granthi* (see p. 296).
*** *Biologie der Tiere Deutschlands*, IV (24), part 40.

origin of the stridulation is probably mandibular. It is not impossible that the same origin is true for the genera *Dicranura* Bsd. and *Cerura* Schrank (Notodontidae). These caterpillars emit a loud scraping sound when they are disturbed, by swinging the foremost part of the body from one side to the other, thereby pressing the mouth-parts against the surface of the leaf[89]. In the genus *Drepana* Schrank (Drepanidae), the anal prolegs have disappeared and have been replaced by two small chitinous teeth. The caterpillar produces a scrapping sound by rubbing these teeth against the leaf it lies on[89].

Pupae, seldom noted for their stridulation[31, 204, 212, 219] have just as many examples of quite remarkable apparatus. Hinton (1948, 1955)[126, 127] counted more than 80 species capable of producing noises during their pupal stage. Four methods are used.

Among several Hesperiidae[199] and Lycaenidae[26], the pupa responds to mechanical stimuli by swift strokes against the substratum to which it adheres.

The most common method, since it is found in about ten families, is the mutual friction of one or several pairs of abdominal segments. The apparatus is generally situated on the dorsal region. It consists of two projections made up of tubercles or coarse transverse ridges (Fig. 210). One is on the front edge of a segment, the other on the rear edge of the preceding segment. Contractions of the abdomen make these formations rub against each other.

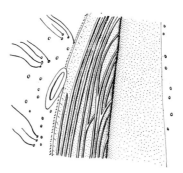

Fig. 210. Pupal stridulatory apparatus of *Lymantria viola* Swinhoe, (Lepid., Lymantriidae). Stridulatory surfaces of the fourth and fifth abdominal segments. (After Hinton, 1948[126].)

This type of apparatus is found in Hesperiidae, Lycaenidae, Papilionidae, Lymantriidae, Saturnidae, and so forth. In several Sphingidae the stridulatory apparatus is lateral, and in some cases (*Acherontia atropos* L., for instance), it seems derived from a gin-trap. (See foot-note* p. 331.)

A third process is used by the hesperid *Gangara thyrsis* F.[26]. On each side of its sagittal axis, the 5th abdominal segment has a sharp transverse ridge. The friction of the very long and striated proboscis against this double projection produces a hissing noise.

Finally, Hinton[126] recognised in 26 species of Sarrothripinae and Careinae (Noctuidae) the possibility of stridulation by the passage of an abdominal projection over the inner wall of the cocoon. This projection appears on the tergal surface of the 10th segment of the abdomen in the form of a series of longitudinal crests variable in number (80 in *Blenina metascia* Hamps., 6 to 8 in *Symitha molalella* Walk., (Sarrothripinae)). It is in the sub-family Sarrothripinae that this method has reached its highest development. Opposite the abdominal crest of the pupa, the inner wall of the cocoon has another row of crests in its lower part (Fig. 211). There is no relationship between the number of the crests of the cocoon (which varies from one cocoon to another) and that of the abdominal crests. In *Plotheia decrescens* Walk., for example, 32 abdominal crests and 7 in the cocoon have been counted, while in *Eligma narcissus* Cr., 20 abdominal crests and 8 to 18 in the cocoon have been noted[91, 120].

Supposing that this apparatus is fundamentally stridulatory, it could be considered the most extraordinary that we know of, since the caterpillar which will have only the plectrum (abdominal crests) during its pupation, builds itself the pars stridens.

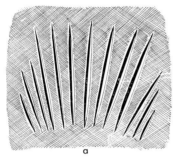 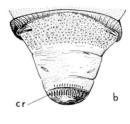

Fig. 211. Pupal stridulatory apparatus of *Eligma hypsoides* Walk (Lepid., Noctuidae). (a) first part of the apparatus made of crests on the inner surface of the cocoon, (b) end of abdomen of the pupa showing the row of crests (cr) on the last segment (second part of the apparatus). (After Hinton, 1948[126].)

(c) Orthoptera

Crackling noises of mandibular origin have been noticed in the nymphal instars of the genera *Podisma* Latr. and *Calliptamus* Serv. (Catantopidae)[138]. They are also observed among adults (see p. 329). Apart from these sound emissions, properly speaking, it has often been noted that immature acridids perform stridulatory movements with their hind legs. These are silent, as there is no appropriate apparatus present. For instance, among the nymphs of *Locusta pardalina* Walk, in the three phases (*gregaria, solitaria, transiens*) a "stridulation" without production of sound, has been observed. In half of these observations the stridulatory movement seemed to have been caused by the presence of another individual[227].

(d) Other examples

In Odonata, a well developed apparatus has been detected in *Epiophlebia superstes* Selys (Anisozygoptera), which is a relic of an extinct group of the Cretaceous period. During its larval life, this species has a dorso-ventrally striated area on the lateral edge of the abdominal tergites 3 to 7. A movement of the abdomen causes this striation to rub against the hind femora. The sound produced is strident and the animal only makes it when disturbed[13].

This is the only known example in Odonata, but it may be wondered whether the presence of a stridulatory apparatus in this archaic species does not indicate that such an apparatus existed in some forms that have now disappeared.

References p. 338

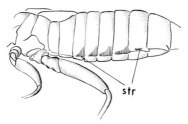

Fig. 212. *Epiophlebia superstes* Selys (Odonata, Anisozygoptera). Stridulatory apparatus of the larva. str: striated areas of abdominal segments 3–7. (After Asahina, 1939[13].)

The production of a noise audible at several yards has been noticed in the pupa of the Hymenoptera *Phytodictus polyzonias* Forst (Ichneumonoidea), a parasite of the caterpillars of *Tortrix*[171].

Vespa crabro L. larva produces a rasping sound by rubbing its cephalic region against the walls of the cell where it lives. This behaviour only observed in hungry larva, could mean a "food begging call"[104].

9. Biological Problems Arising from Sound Emission Apparatus

(a) *Adaptation and differentiation of apparatus*

The observation of apparatus working by friction (stridulatory s. str.) shows that they are always superimposed on regions or appendages, the fundamental role of which is not stridulation (legs, wings, elytra, etc.). The use of such apparatus is therefore bound up with the functioning of the parts on which they are situated*, but the movement which gives rise to the stridulation can be really specialised: a movement of elytra in Orthoptera *Ensifera*, of the hind legs in *Caelifera*, of the claw in Ocypod crabs, of the tail or the chelicerae in certain Scorpions. This movement may also be trifling and visible in non-stridulant forms: among Coleoptera, a movement of the head in the species with an occipital pars stridens, of the legs against the elytra (*Graphopterus*), a movement of the mandibles in Orthoptera and caterpillars, etc.

It is difficult to appreciate the more or less considerable adaptation of the apparatus to the function since criteria are lacking. In fact, as soon as a device is capable of producing a noise which becomes an integral part of the animal's behaviour, it fulfills its role and the degree of differentiation adds nothing. The granulated area of the pygidium against which the end of the elytra rubs in *Oryctes* is a very rough structure. On the other hand the double striated crest of *Necrophorus* which is used in the same manner, indicates a much more advanced device. But if these two beetles use their apparatus with the same efficiency, the greater delicacy and the higher degree of perfection of the latter are of no apparant advantage to the insect.

However, if we forget for a moment the character of functional finality and limit our attention to the morphological aspect, it is undoubtedly possible to recognise degrees of differentiation.

---

* The sound instrument of the Cicadidae with its special musculation and innervation which give it an autonomous character, is on the other hand a real *organ* of emission. Meanwhile, in many Tettigonioidea, elytra and often wings are so atrophied that they can no longer be used for flight. In these cases since the only role of the organ is now stridulatory, there has been a real transformation of its function.

For instance, this differentiation is important, in the Ocypod crabs since the two parts of the apparatus are isolated in the middle of two joints. They do not, therefore, arise from the modification of an already existing structure, as is the case in a number of arthropods where there is simply a striation or a denticulation on a crest or a vein which is also found in unstridulating forms.

It must also be frankly admitted that devices such as those of certain mygales or scorpions, point to a fairly simple evolutionary invention, since the stridulatory regions are areas simply covered with bristles.

There are lastly, and at the lowest level of the scale, those in which the sound emission results from regions without any morphological peculiarity.

Undoubtedly a place apart should be given to the pupal apparatus of certain Noctuidae since half of it is really fabricated by the caterpillar (see p. 333).

If the structure of the sound devices is quite variable, their location is just as diverse. The preceding pages have shown that there are hardly any parts of the body or appendages where they have not appeared. But whatever its shape and wherever it is found, the pars stridens-plectrum unit shows in nearly all cases, a particularly precise arrangement. The plectrum is situated where its mechanical efficiency is greatest, and the projection of the pars stridens is placed in such a way that there is maximum friction. Faced with such a mutual adaptation of the two parts, each formed independently of the other, it is possible to speak of *coaptation* in the sense that Cuénot (1926)[66] defined the term.

Although very often, the stridulatory devices have only been discovered by means of examination of preserved specimens, the use made of these devices is nevertheless probably effective in a large number of cases for the reason we have just explained. In addition, apart from secondary sexual ornamentation or hypertelic formations, evolution seldom causes the appearance of a complex structure if it is of no use to the species, and selection does not maintain it.

(b) Systematic distribution

Considered from the systematic aspect, a survey of stridulating arthropods enables us to recognise 3 categories. First of all there are the typically stridulatory groups such as Orthoptera and Homoptera. Opposed to these, is a second category in which the stridulation is much more accidental, such as Coleoptera* or Crustacea Decapoda. Then lastly there are those in which the possession of a stridulatory device is unusual (Lepidoptera, Diptera)

In the first case the evolutionary drive directed towards the perfecting or the acquisition of new structures or mechanisms useful to the species has been in evidence throughout the whole order. All the super-families in Orthoptera have representative cases of sound emission apparatus; families without them are rare. The same can be said for Homoptera (*Auchenorrhyncha* in particular). Concomitantly, considerable homogeneity governs the constitution of the apparatus. This homogeneity is still evident throughout the series of auchenorrhynchous Homoptera where a tymbal apparatus is found[190]. Among Orthoptera, elytro-elytral stridulation is typical of the super-families Tettigonioidea and Gryllodea, whereas the femoro-elytral method (with its derivatives) characterises Acridoidea. On the other hand

---

* The number of stridulating genera in this order is, however, not negligible, since it reaches at least 380 without including those belonging to the family Cerambycidae[75].

this constancy in the principle and the structure of the apparatus maintained among Orthoptera up to the level of the super-family, disappears in the order Coleoptera, giving way to a multiplicity of forms and localisations. Only exceptionally is there a type that is characteristic of a familly (Geotrupidae); most often various devices are found together (*e.g.* among Scarabaeidae), and correspondingly the same method is found in widely separated families: abdomino-elytral stridulation in Tenebrionidae, Chrysomelidae, Curculionidae, Silphidae, Hydrophilidae, Scarabaeidae, and Carabidae; elytro-femoral stridulation in Tenebrionidae, Curculionidae, Lucanidae, Masoreidae and Cicindelidae.

In Crustacea Decapoda, stridulatory devices have appeared in a very erratic way and show considerable polymorphism. More successful, however, is the case of Brachyura where stridulating genera represent more than half the examples of stridulating Crustacea known at present; but the same may be said of them as with Coleoptera. Even among crabs there is the rather astonishing case of the genus *Ovalipes* in which the apparatus of the species *ocellatus* is strikingly different from that of the species *punctatus* (see pp. 310 and 314). Dispersion among the crabs has, however, been canalised into only two out of the four tribes which they constitute, the Brachygnatha and the Oxystomata[111].

The picture is quite similar for Arachnida. While among Araneida, the family Theridiidae has the typical abdomino-cephalothoracic system, that of the mygales Theraphosidae is as well represented by an apparatus of the chelicera-chelicera type, as by a chelicera-pedipalp or pedipalp-p_1 type.

We have come across many examples in which the same apparatus is found within an order in families very remote from each other. It is not without interest to point out that this can be the case between different classes. Thus in *Cyphoderris monstrosus* Uhl. (Orthoptera Prophalangopsidae), among the Theridiidae spiders and in the Crustacea Decapoda of the genus *Penaeopsis*, the pars stridens-plectrum unit is localised on the front part of the abdomen and the rear part of the region directly anterior (cephalothorax for *Penaeopsis* and Theridiidae, metanotum for *Cyphoderris*). (See pp. 282, 305 and 317.) This is a quite interesting case of morphological and functional convergence.

To summarize, if an exception is made of the two fundamentally stridulatory groups, Orthoptera and Homoptera, the distribution of the different forms of structure of the stridulatory apparatus never squares with the systematic divisions.

(c) Origin of stridulatory apparatus

There can obviously be no question of explaining, except by hypotheses, how stridulatory devices originate. Has there been a progressive development of a projection on the two places where a mutual friction naturally works, or on the other hand, from a more neo-darwinist angle, is a mutation or a series of mutations a starting point? It is sufficiently clear that the first "explanation" is purely verbal and that the second is incapable, without recourse to an inadmissible finality, of throwing light upon a coapted structure such as the pars stridens-plectrum unit.

However, without going as far as to offer a solution, one idea may shed some light on the problem. It is possible that, in certain cases at least, the apparatus we now call stridulatory, originally had a different role: that of immobilising at rest mobile appendages. Elytro-elytral or abdomino-elytral formations, for instance, may very well have served initially to keep the elytra closed and pressed against the abdomen. Arrow (1904)[11] detected several of these closing devices in beetles of the genus *Copris*, some of which were in addition used for stridulation (see p. 293).

If we are in the dark as to the morphogenesis of stridulatory apparatus, the characteristic forms under which they make their appearance as well as their distribution are easier to understand. The polymorphism and the multiplicity of localisation within the family, the absence of transitional forms between one type and another, and the impossibility of recognizing a phylogenetic connexion in these erratic appearances, entitle us to think that, at least in Crustacea, Arachnida, and as regards insects, in Coleoptera, stridulation is a late acquisition. It is a response of the specific potentialities of each group to an urge, a tendency, a "need" experienced by the group*. On the other hand the homogeneity found among Orthoptera and Homoptera leaves room for the supposition of more ancient development (or at least potentiality of development) in the evolutionary process.

The study of fossil Prophalangopsidae gives an idea of what the evolution of the stridulatory apparatus has been. In the Upper Palaeozoic, the elytron has a primitive and variable venation, and a stridulatory apparatus is generally lacking (on the other hand, a tibial tympanal organ is already present). Later, the apparatus develops and takes nearly the whole forewing of the male. Finally, in the Upper Jurassic, one finds forms in which the apparatus is restricted to the cubito anal area. In existing Tettigonioidea it is still found in this location[264].

(d) Sound apparatus and sensory devices for perception

It might be considered that to a certain extent, the presence of a sound emission apparatus and especially its use, are indications of the existence in the species of a system of perception capable of detecting one of the physical phenomena brought into being by the working of the apparatus of other individuals of the species.

In opposition to other devices adapted to a specified role (the fossorial forelegs of the mole-cricket, the raptorial legs of mantis-shrimps or praying-mantis, apparatus for connecting the wings in Hymenoptera and certain Lepidoptera, etc.), stridulatory devices seem to have no utility for the animal *itself*. Their existence and their use are only understood if one considers at the same time the existence and the use of a sensory equipment that can inform a second individual of the activity of the former and possibly determine an appropriate type of behaviour.

The Chapter 21 will show that sound emission must in most cases be considered as a means of information and communication, and the very idea of information implies the presence of at least two elements: a transmitter device and a receiver device.

This being so, and without excessive finality, it is difficult to see how evolution could have created the first instrument so many times with such diversity and

---

* What seems to bear evidence as to the strength of this urge is the fact that the same result (sound emission) is obtained in different ways in closely connected forms, either in the same family (elytro-femoral method of the Oedipodinae and femoro-elytral method of the Acridinae) or in the same genus (case of the crabs *Ovalipes ocellatus* and *O. punctatus*) or even within the species (elytro-abdominal method in the male of certain Curculionidae and abdomino-elytral in the female, characteristic apparatus of the male Tettigoniidae, and atypical structures in the female).

precision, without having worked out the second, since the efficiency of the one is entirely conditioned by the presence of the other*.

In fact, although theoretically the receiver *must* exist**, in practice there are still many cases in which it remains to be discovered. This gap is perhaps due to the fact that there is a tendency to consider more particularly the sound aspect of stridulation and to search for a receiver capable of perceiving that part of the phenomenon directly transmitted by air (or by water in the case of aquatic arthropods). In point of fact, a not insignificant fraction of the energy passes through the substratum (either acoustic energy or energy resulting from the vibration of the body at the time of stridulation)[117]. The possible role of a number of mechano-receivers as systems of perception should not, therefore, be neglected. Among Crustacea[59] and Arachnida[215, 252] it even seems that only vibrations transmitted by the substratum are perceived.

---

* Notice, however, that the contrary is not true, and that an auditory receiver can be part of the sensory equipment of an animal without either it or other members of its species making use of acoustic messages. The use of such a receiver is nevertheless certain, for it ensures a knowledge of the sound aspect of the environment thus completing the picture given by the other senses. Palaeontology recently confirmed this view since there are not yet any stridulatory apparatus in the oldest Prophalangopsidae fossils but one already finds a tympanal organ on the fore-tibiae[264].

** However, in Gryllacrididae, that possess abdominal stridulatory apparatus, the tympanal organs do not exist.

REFERENCES

[1] ALCOCK, A., 1900, Materials for a carcinological fauna of India No. 6; The Brachyura Catometopa or Grapsoidea, *J. Asiat. Soc. Bengal (Calcutta)*, 69, 279.
[2] ALCOCK, A., 1902, *A Naturalist in Indian Seas*, London. 328 pp.
[3] ALEXANDER, A. J., 1958, On the stridulation of Scorpions, *Behaviour*, *12*, 339.
[4] ALEXANDER, R. D., 1956, A comparative study of sound production in Insects, with special reference to the singing Orthoptera and Cicadidae of the eastern United States, *Thesis*, Ohio State University. 529 pp.
[5] ALEXANDER, R. D., 1960, Sound communication in Orthoptera and Cicadidae, in W. E. LANYON and W. N. TAVOLGA, Animal Sounds and Communication; Am. Inst. of Biol. Sci., p. 38, Washington, *Publication, No. 7*.
[6] ALEXANDER, R. D., 1960, Communicative mandible-snapping in Acrididae, *Science*, *132*, 152.
[7] ANDER, K., 1938, Ein abdominales Stridulationsorgan bei *Cyphoderris*, und über die systematische Einteilung der Ensiferen (Saltatoria), *Opusc. Entom.*, *3*, 32.
[8] ANDER, K., 1939, Vergleichend-anatomische und phylogenetische Studien über die Ensifera (Saltatoria), *Opusc. Entom.*, *3*, Suppl. II, VIII.
[9] ANDERSON, A. R., 1894, Note on the sound produced by the Ocypode Crab, *Ocypoda ceratophthalma*, *J. Asiat. Soc. Bengal (Ca'cutta)*, *63*, 138.
[10] ANNANDALE, N., J. COGGIN BROWN and F. H. GRAVELY, 1913, The Limestone Caves of Burma and the Malay Peninsula, Myriapoda, *J. Asiat. Soc. Bengal (Calcutta)* (N.S.), 9, 415.
[11] ARROW, G. J., 1904, Sound production in the Lamellicorn Beetles, *Trans. Entom. Soc. (London)*, 709.
[12] ARROW, G. J., 1924, Vocal organs in the Coleopterous families Dytiscidae, Erotylidae and Endomychidae, *Trans. Entom. Soc. (London)*, 134.
[13] ASAHINA, S., 1939, Tonerzeugung bei Epiophlebia-Larven (Odonata, Anisozygoptera), *Zool Anz.*, *126*, 323.
[14] ATHENAEUS, III. cent. A. D., *Le Banquet des Savants*, translated by LEFEBURE DE VILLEBRUNE, Lamy Ed. Paris, 1789–91.
[15] AURIVILLIUS, C. W. S., 1893, Die Beziehungen der Sinnesorgane amphibischer Dekapoden zur Lebensweise und Atmung, *Nova acta reg. Soc. Sci. (Upsala)*, (ser. 3), *16*, 1.
[16] BABB, G. F., 1901, On the stridulation of *Passalus cornutus* Fabr., *Entom. News*, *12*, 279.
[17] BADONNEL, A., 1951, Les Psocoptères, in GRASSÉ, *Traité de Zoologie*, (Masson Ed. Paris), Vol. 10, p. 1301.

[18] BALSS, H., 1921, Über Stridulationsorgane bei Dekapoden Crustaceen, *Naturwiss. Wochschr.*, *20*, 697.
[19] BALLS, H., 1956, Decapoda, in BRONNS, *Klassen und Ordnungen des Tierreichs*, Vol. 5 (1), Akademische Verlagsgesellschaft, Leipzig, p. 1369.
[20] BAUDRIMONT, A., 1923, Sur la "musique" du Hanneton du Pin, *Proc. Soc. Linn. Bordeaux*, *75*, 174.
[21] BAUDRIMONT, A., 1923, Sur le bruissement d'*Anoxia villosa*, *Proc. Soc. Linn. Bordeaux*, *75*, 180.
[22] BAUDRIMONT, A., 1923, Sur le Prione tanneur, sa façon de protester, *Proc. Soc. Linn. Bordeaux*, *75*, 181.
[23] BAUMGARTNER, W. J., 1905, Observations on some peculiar habits of the molecricket, *Science*, *21*, 885.
[24] BAUMGARTNER, W. J., 1910, Observations on the Gryllidae. *Kansas Univ. Sci. Bull.*, *5*, 309.
[25] BEDEL, L. and PH. FRANCOIS, 1897, Sur l'appareil stridulatoire des *Siagona* Latr., *Bull. Soc. Entom. France*, 38.
[26] BELL, T. R. D., 1918–25, The common Butterflies of the Plains of India, *J. Bombay Nat. Hist. Soc.*, 25.
[27] BERLAND, L., 1932, *Les Arachnides*, Lechevalier Ed. Paris. 485 pp.
[28] BERLAND, L., 1951, Les Hyménoptères, in GRASSÉ, *Traité de Zoologie*, Masson Ed. Paris, Vol. 10 (1), p. 813; Vol. 10 (2), pp. 1002, 1006, 1042, 1058.
[29] BOLIVAR, I., 1899, Orthoptères du voyage de M. Martinez Escalera dans l'Asie Mineure, *Ann. Soc. Entom. Belgique*, *54*, 583.
[30] BORRADAILE, L. A., 1901, Land Crustaceans, *Fauna and Geography of Maldive and Laccadive archipelagoes*, *I*, 1.
[31] BOURGOGNE, J., 1951, Lépidoptères, in GRASSÉ, *Traité de Zoologie*, Masson Ed. Paris, Vol. 10, p. 174.
[32] BOURNE, C. G., 1885, On the anatomy of *Sphaerotherium*, *Linn. Soc. J. Zool.*, *19*, 161.
[33] BRIGGS, C. A., 1897, A curious habit in certain ♂ Perlidae, *Entom. Month. Mag.*, (2nd ser.), *8*, 207.
[34] BRISTOWE, W. S. and G. H. LOCKET, 1926, The courtship of British lycosid spiders, *Proc. Zool. Soc. (London)*, 317.
[35] BROCHER, F., 1911–12, L'appareil stridulatoire de l'*Hydrophilus piceus*, et celui de *Berosus aericeps*, *Ann. de Biologie Lacustre*, *5*, 215.
[36] BROOKS, W. K., 1886, Report on the Stomatopoda collected by H.M.S. Challenger during the years 1873–76, *Report of H.M.S. Challenger Zool.*, *45*, 1.
[37] BROOKS, W. K. and F. H. HERRICK, 1892, The embryology and metamorphosis of the Macrura, *Mem. Nat. Acad. Sci. (Washington)*, *5*, 321.
[38] BRUYANT, CH., 1894, Sur un Hémiptère aquatique stridulant *Sigara minutissima* L., *Compt. rend. Acad. Sci.*, *118*, 299.
[39] BUGNION, E., 1910–1917, Le bruissement des Termites, *Bull. Soc. Entom. Suisse*, *12*, 125.
[40] BUHK, F., 1910, Stridulationsapparat bei *Spercheus emarginatus* Schall, *Z. Wiss. Insektenbiol.*, *6*, 342.
[41] BURKENROAD, M. D., 1947, Production of sound by the fiddler crab *Uca pugilator* (Bosc), with remarks on its nocturnal and mating behavior, *Ecology*, *28*, 458.
[42] BURMEISTER, H., 1833, Des sons que produisent certains Insectes, *Rev. entomol.*, *1*, 161.
[43] BURTT, E. D., 1946, Observations on East African *Pamphaginae* (Orthoptera) with particular reference to stridulation, *Proc. Roy. Entom. Soc. (London)*, *A*, *21*, 51.
[44] BUSNEL, R. G. and B. DUMORTIER, 1960, Vérification par des méthodes d'analyse acoustique des hypothèses sur l'origine du cri du Sphinx *Acherontia atropos* L., *Bull. Soc. Entom. France*, *64*, 44.
[45] BUSNEL, R. G., and A. DZIEDZIC, 1962, Rythme du bruit de fond de la mer à proximité des côtes et relations avec l'activité acoustique des populations d'un Cirripède fixé immergé, *Cahiers Océanographiques*, No. 5, 293.
[46] CAPPE DE BAILLON, P., 1921, Note sur le mécanisme de la stridulation chez *Meconema varium* Fabr. (Phasgonuridae), *Ann. Soc. Entom. France*, *90*, 69.
[47] CARL, J., 1906, L'organe stridulateur des *Phyllophorae*, *Arch. Sci. Phys. Nat. (Genève)*, *23*, 406.
[48] CARL, J., 1906, Diplopoden aus der malayischen Archipel (Reise von Dr. Walter Volz), *Zool. Jb.*, *24*, 227.
[49] CARPENTER, G. H., 1898, The smallest of stridulating spiders, *Nat. Sci.*, *12*, 319.
[50] CARPENTER, G. D. H., 1913, Notes on the struggle for existence in tropical Africa, *Bedrock*, *(London)*, *2*, 358.
[51] CARPENTER, G. D. H., 1938, Audible emission of defensive froth by insects, *Proc. Zool. Soc. (London)*, *108*, 243.
[52] CARPENTER, G. D. H., 1946, The relative edibility and behaviour of some aposematic grasshoppers, *Entom. Month. Mag.*, *82*, 1.

[53] CARPENTIER, F., 1936, Le thorax et ses appendices chez les vrais et chez les faux Gryllotalpides, *Mém. Mus. Hist. Natl. Belg.*, *4*, (2) 1.
[54] CHOPARD, L., 1934, Sur les bruits produits par certaines araignées, *Bull. Soc. Zool. France*, 59, 132.
[55] CHOPARD, L., 1938, *La Biologie des Orthoptères*, Lechevalier Ed. Paris. 541 pp.
[56] CLAINPANAIN, R. P. J., 1917, Le chant du Dytique bordé (*Cybister aegyptiacus*), *Bull. soc. roy. entom. Egypte*, 125.
[57] CLOUDSLEY-THOMPSON, J. L., 1958, *Spiders, Scorpions, Centipedes and Mites*, Pergamon Press. 228 pp.
[58] CLOUDSLEY-THOMPSON, J. L., 1961, A new sound-producing mechanism in Centipedes, *Entom. Month. Mag.*, *96*, 110.
[59] COHEN, M. J., 1955, Function of receptors in the statocyst of the Lobster *Homarus americanus*, *J. Physiol.*, *130*, 9.
[60] COLLART, A., 1925, Quelques observations sur les Fourmis *Megaponera*, *Rev. Zool. Afric.* (Brussels), *13*, 26.
[61] COOK, D. F., 1896–97, The species of Alipes, *Brandtia*, *17*, 71.
[62] COUTIÈRE, H., 1899, Les Alpheidae, morphologie externe et interne, formes larvaires, bionomie, *Thesis*, Masson Ed. Paris. 560 pp.
[63] COWL, E. M., 1901, Stridulation of *Smerinthus populi*, *Entom. Rec. & J. of Var.*, *13*, 164.
[64] CRANE, J., 1941, Eastern Pacific Expedition of the N.Y. Zoological Society, XXVI, Crabs of the Genus *Uca* from the West Coast of Central America, *Zoologica*, *26*, 145.
[65] CRANE, J., 1947, Intertidal Brachygnathous Crabs from the West Coast of tropical America, with special reference to ecology, *Zoologica*, *32*, 69.
[66] CUENOT, L., 1926, Les coaptations, *La Science moderne*, 3ème année, 39.
[67] CURRIE, P. W. E., 1953, The "drumming" of *Meconema thalassinum Entom. Rec. & J. of Var.*, 65, 93.
[68] DALE, J., 1834, On the ticking of Anobium, *Mag. Nat. Hist.*, (ser. *1*), 7, 472.
[69] DANA, J. D., 1852, Crustacea; U.S. exploring expedition during the years 1838, 1839, 1840, 1841, 1842, Part 1, 13, VIII + 685, Atlas 13, p. 1–27.
[70] DARWIN, C., 1894, *The Descent of Man and Selection in Relation to Sex*, 2nd ed., London.
[71] DEMBOWSKI, J., 1925, On the speech of the fiddler crab *Uca pugilator*, *Trav. Inst. Nencki*, *3*, 1.
[72] DIJKGRAAF, S., 1955, Lautzeugung und Schallwahrnehmung bei der Languste (*Palinurus vulgaris*), *Experientia*, *11*, 330.
[73] DUDICH, E., 1920, Über den Stridulationsapparat einiger Käfer, *Entom. Blätter*, *16*, 146.
[74] DUDICH, E., 1921, Beiträge zur Kenntnis der Stridulationsorgane der Käfer, *Entom. Blätter*, *17*, 136, 145.
[75] DUDICH, E., 1922, Beiträge zur Kenntnis der Stridulationsorgane der Käfer, *Entom. Blätter*, *18*, 1.
[76] DUNCAN, C. D., 1924, Spiracles as sound producing organs, *Pan-Pacific Entom.*, *1*, 42.
[77] EASTOP, V. F., 1952, A sound production in the Aphididae and the generic position of the species possessing it, *Entomologist*, *85*, 57.
[78] EDMONDS, R., 1834, The death-watch; the ticking of *Anobium*, *Mag. Nat. Hist.*, 7, 468.
[79] EDWARDS, H., 1889, Notes on noises made by Lepidoptera, *Insect Life*, 2, 11.
[80] EMERSON, A. E., 1930, Communication among Termites, *Proc. 4th Intern. Congr. Entom., Ithaca*, I, 722.
[81] EMERSON, A. E. and R. C. SIMPSON, 1929, Apparatus for the detection of substratum communication among termites, *Science*, 69, 648.
[82] FABER, A., 1928, Die Bestimmung der deutschen Geradflüger nach ihren Lautäusserungen, *Z. wiss. Ins. Biol.*, *23*, 209.
[83] FABER, A., 1936, Die Laut- und Bewegungsäusserungen der Oedipodinen, *Z. wiss. Zool.*, *149*, 1.
[84] FABER, A., 1949, Eine bisher unbekannte Art der Lauterzeugung europäischer Orthopteren: Mandibellaut von *Calliptamus italicus* (L.), *Z. Naturforsch.*, *46*, 367.
[85] FABER, A., 1953, *Laut- und Gebärdensprache bei Insekten: Orthoptera*, Teil I, Vergleichende Darstellung von Ausdrucksformen als Zeitgestalten und ihre Funktionen, Stuttgart. 198 pp.
[86] FABER, A., 1953, Eine unbekannte Art der Lauterzeugung bei Käfern: das Trommeln von *Nothorrhina muricata*, *Jh. Ver. vaterl. Naturk. Württemberg*, *108*, 71.
[87] FABRE, J. H., 1900, 1907, *Souvenirs entomologiques*, 6th series, p. 195; 10th series, p. 160. Final edition (1922 and 1924) Delagrave Ed., Paris.
[88] FAGE, L., 1939, Sur une Phryne du Sud Marocain pourvue d'un appareil stridulant, *Musicodamon atlanteus* n. gen. n. sp., *Bull. Soc. Zool. France*, *64*, 100.
[89] FEDERLEY, H., 1905, Sound produced by Lepidopterous larvae, *J.N.Y. Entom. Soc.*, *13*, 109.
[90] FÉRON, M., 1960, L'appel sonore du mâle dans le comportement sexuel de *Dacus oleae* Gmel. (Dipt. Trypetidae), *Bull. Soc. Entom. France*, 65, 139.

[91] Fletcher, T. B., 1919, Second Hundred Notes on Indian Insects, *Bull. Agric. Research Inst. Pusa*, 89. 102 pp.
[92] Forbes, H. O., 1881, Sound producing Ants, *Nature*, 24, 102.
[93] Forel, A., 1886, Etudes myrmécologiques en 1886, *Ann. Soc. Entom. Belgique*, 30, 131.
[94] Forest, J., 1952, Contribution à la révision des Crustacés Paguridae. I. Le genre *Trizopagurus*, *Mém. Mus. Hist. Nat.*, (Ser. A) V, 1.
[95] Fotheringham, I., 1881, Sound producing Ants, *Nature*, 25, 55.
[96] Frankenberg, G. von, 1940, Ein brutpflegender und musizierender Wasserkäfer (*Spercheus emarginatus*), *Natur u. Volk*, 70, 1.
[97] Fulton, B. B., 1933, Stridulating organs of female Tettigoniidae (Orthop.), *Entom. News*, 44, 270.
[98] Gahan, C. J., 1900, Stridulating organs in Coleoptera, *Trans. Entom. Soc. (London)*, 433.
[99] Gahan, C. J., 1918, The death-watch: notes and observations, *Entomologist*, 51, 153.
[100] Gerstaeker, C. E., 1854, Über eine neue Myriapoden und Isopoden Gattung, *Stett. entom. Z.*, 15, 312.
[101] Giesbrecht, W., 1910, *Fauna and Flora des Golfes von Neapel*, 33, Monogr. Stomatopoda, Berlin, p. 201.
[102] Giltay, L., 1935, Liste des Arachnides d'Extrême-Orient et des Indes Orientales recueillis, en 1932, par S.A.R. le Prince Léopold de Belgique, *Bull. Mus. Roy. Hist. Natl. Belg.*, 11, 1.
[103] McGinitie, G. E. and N. McGinitie, 1949, *Natural History of Marine Animals*, Mc. Graw-Hill Book Co., Inc., N.Y.
[104] Gontarski, H., 1941, Lautäusserungen bei Larven der Hornisse (*Vespa crabro*), *Natur u. Volk*, 71, 291.
[105] Goode, G. B., 1878, The voices of crustaceans, *Proc. U.S. Natl. Mus.*, 1, 7.
[106] Gounelle, E., 1900, Sur les bruits produits par deux espèces américaines de Fourmis et de Termites, *Bull. Soc. Entom. France*, 168.
[107] Goureau, 1837, Essai sur la stridulation des insectes, *Ann. Soc. Entom. France*, 6, 31.
[108] Grassé, P. P., 1949, Ordre des Isoptères, in Grassé, *Traité de Zoologie*, Masson Ed. Paris, Vol. 9, p. 408.
[109] Gravely, F. H., 1915, Notes on the habits of Indian Insects, Myriapods and Arachnids, *Rec. Indian Mus.*, 11, 483.
[110] Guibé, J., 1943, Présence d'un appareil stridulatoire chez le mâle de deux espèces d'Araignées *Theridium ovatum* Cl. et *Gongylidiellum vivum* O. P. Camb., *Bull. Soc. Zool. France*, 68, 65.
[111] Guinot-Dumortier, D. and B. Dumortier, 1960, La stridulation chez les Crabes, *Crustaceana*, 1, 117.
[112] Haller, G., 1874, Über einige bis jetzt weniger bekannten Tonapparate der Insekten, *Der zool. Garten*, Frankfurt a. M., p. 106.
[113] Hampson, G. F., 1892, On stridulation in certain Lepidoptera, and on the distortion of hind wings in the males of certain Ommatophorinae, *Proc. Zool. Soc. (London)*, 188.
[114] Hannemann, H. I., 1956, Über ptero-tarsale Stridulation und einige andere Arten der Lauterzeugung bei Lepidopteren, *Deut. Entom. Z.*, 3, 14.
[115] Hansen, H. J., 1921, *Studies on Arthropoda*, I, On stridulation in Crustacea Decapoda, Gyldendalske Boghandel, Copenhagen, p. 56.
[116] Harz, K., 1957, Beobachtung von Mandibellauten bei Angehörigen der Acridinae, *Nachrichtenblatt der Bayerischen Entom.*, 6, Sept. 1957.
[117] Haskell, P. T., 1955, Vibration of the substrate and stridulation in a Grasshopper, *Nature*, 175, 639.
[118] Hebard, M., 1922, The stridulation of a North-American Noctuid *Heliocheilus paradoxus* Grote, *Entom. News*, 33, 244.
[119] Heldt, H., 1929, Rapport sur la Langouste vulgaire (*Palinurus vulgaris* Latreille), *Rapports et Procès-Verbaux des réunions de la Commission Internationale pour l'exploration scientifique de la Mer Méditerranée*, 4, 113.
[120] Hemmingsen, A. M., 1947, A chrysalis stridulating by means of instrument on inside of cocoon, *Entomologiske Meddelelser*, 25, 165.
[121] Henderson, J. R., 1888, Report on the Anomura collected by H. M. S. Challenger, *Report of H. M. S. Challenger Zool.*, 27, 1.
[122] Henry, G. M., 1924, Stridulation in the Leaf-Insect, *Spol. zeylan.*, 12, 217.
[123] Henry, G. M., 1940, New and little known South Indian Acrididae, *Trans. Roy. Entom. Soc. (London)*, 90, 497.
[124] Henry, G. M., 1942, Three remarkable stridulatory mechanisms in Acrididae, *Proc. Roy. Entom. Soc. (London)*, A, 17, 59.
[125] Hilgendorf, F., 1869, Crustaceen, *Baron Carl Claus von der Decken's Reisen in Ost Afrika in den Jahren 1859-1865*, 3, 67.
[126] Hinton, H. E., 1948, Sound production in Lepidopterous pupae, *Entomologist*, 81, 254.

[127] HINTON, H. E., 1955, Protective devices of endopterygote pupae, *Trans. Soc. Brit. Entom.*, 12, 49.
[128] HIRST, A. S., 1908, On a new type of stridulating-organ in mygalomorph spiders, with the description of a new genus belonging to the sub-order, *Ann. & Mag. Nat. Hist.*, II, (8th ser.), 401.
[129] HOOD, J. D., 1950, Thrips that "talk", *Proc. Entom. Soc. (Washington)*, 52, 42.
[130] HORN, G. H., 1876, Revision of the United States species *Ochodaeus* and other genera, *Trans. Am. Entom. Soc.*, 5, 177.
[131] HUDSON, G. V., 1919, The sound producing organ of *Deinacrida megacephala*, *Entom. Month. Mag.*, 5, 232.
[132] HUNGERFORD, H. B., 1924, Stridulation of *Buenoa limnocastris* Hung. and systematic notes on the *Buenoa* of the Douglas Lake region of Michigan, with the description of a new form, *Ann. Entom. Soc. Am.*, 17, 223.
[133] HUSSON, R., 1956, Données expérimentales concernant les deux premiers registres de la voix humaine. *J. physiol. (Paris)*, 48, 573.
[134] HUSSON, R., 1957, La vibration des cordes vocales de l'Homme sans courant d'air et les rôles d'une pression sous glottique éventuelle, *J. Physiol. (Paris)*, 49, 217.
[135] HSÜ, F., 1937, Structure de l'appareil soit-disant phonateur de *Corixa*, *Ann. Soc. Sci. Bruxelles*, 57, 128.
[136] ISELY, F. B., 1936, Flight stridulation in American Acridians (Orthop., Acrididae), *Entom. News*, 47, 199.
[137] JACOBS, W., 1953, Verhaltensbiologische Studien an Feldheuschrecken, *Z. f. Tierpsychol.*, Beih. 1. 228 pp.
[138] JACOBS, W., 1957, Einige Probleme der Verhaltensforschung bei Insekten, insbesondere Orthopteren, *Experientia*, 13, 97.
[139] JANET, C., 1893, Note sur la production des sons chez les Fourmis et sur les organes qui les produisent, *Ann. Soc. Entom. France*, 62, 159.
[140] JANET, C., 1894, Sur l'appareil de stridulation de *Myrmica rubra*, *Ann. Soc. Entom. France*, 63, 109.
[141] JEANNEL, R., 1949, Ordre des Coléoptères, in GRASSÉ, *Traité de Zoologie*, Masson Ed. Paris, Vol. 9, p. 771.
[142] JOBLING, B., 1936, On the stridulation of females of *Parnassius mnemosyne* L., *Proc. Roy. Entom. Soc. (London) A*, II, 66.
[143] JOHNSON, M. W., 1948, Sound as a tool in marine ecology, from data on biological noises and the deep scattering layer, *J. Marine Research*, 7, 443.
[144] JOHNSON, M. W., F. A. EVEREST and R. W. YOUNG, 1947, The rôle of snapping shrimp (*Crangon & Synalpheus*) in the production of underwater noise in the sea, *Biol. Bull.*, 93, 122.
[145] JORDAN, K., 1921, On the replacement of a lost vein in connection with a stridulating organ in a new Agaristid Moth from Madagascar, with description of two new genera, *Nov. Zool.*, 28, 68.
[146] JUBERTHIE, C., 1957, Présence d'organes de stridulation chez deux *Nemastomatidae* (Opilions), *Bull. du Mus.*, 2ème sér., 29, 210.
[147] KAESTNER, A., 1959, *Lehrbuch der Speziellen Zoologie*, Vol. 1 (4), Jena, p. 659–979.
[148] KARNY, H., 1924, Beiträge zur malayischen Orthopterenfauna, V, Bemerkungen über einige Blattoiden, *Treubia*, 5, 3.
[149] KEMP, B. A., 1913, An account of the Crustacea Stomatopoda of the Indo-Pacific region, *Mem. Indian Mus.*, 4, 1.
[150] KENT, W. S., 1877, Sound producing arthropods, *Nature*, 17, 11.
[151] KERSHAW, J. C., 1907, Life history of *Tessaratoma papillosa*, *Trans. Entom. Soc. (London)*, 253.
[152] KEVAN, D. K. Mc. E., 1955, Methodes inhabituelles de production de son chez les Orthoptères, *Colloque sur l'Acoustique des Orthoptères*, Fascicule hors série des Ann. des Epiphyties, I.N.R.A, Paris, p. 103.
[153] KIRBY and SPENCE, 1817, *Introduction to Entomology*, London, Vol. II, p. 375.
[154] KLEINE, R., 1918, Der Stridulationsapparat der Gattung *Lepyrus* Germar, *Entom. Blätter*, 14, 257.
[155] KÖNIG, J. G., 1779, Naturgeschichte der sogenannten weissen Ameisen, *Berliner Ges naturf. Freunde*, 4, 24.
[156] KRAUSS, F., 1843, *Die Südafrikanischen Crustaceen*, Eine Zusammenstellung aller bekannten Malacostraca, Bemerkungen über deren Lebensweise und geographische Verbreitung, nebst Beschreibung und Abbildung mehrerer neuer Arten, p. 1–68.
[157] KRÜGER, P., 1913, Über das Stridulationsorgan und die Stridulationstöne der Nonne *Lymantria monacha* L., *Zool. Anz.*, 41, 505.
[158] KUBO, I., 1949, Studies on the Penaeids of japanese and its adjacent waters, *J. Tokyo College of Fisheries*, 36, 1.
[159] LANDOIS, H., 1874, *Thierstimmen*, Herder'sche Verlagshandlung Freiburg im Breisgau. 229 pp.
[160] LATREILLE, P., 1800, Rapport généraux des travaux, *Soc. Philom. (Paris)*, 4, 67.

[161] LAWRENCE, R. F., 1927, Contribution to a knowledge of the fauna of S. W. Africa, *Ann. S. African Mus.*, *25*, 217.
[162] LAWRENCE, R. F., 1937, A stridulating organ in harvest-spider, *Ann. & Mag. Nat. Hist.*, *20*, (10th ser.), 364.
[163] LAWRENCE, R. F., 1953, *The biology of the cryptic fauna of forests*, Cape Town, Balkema.
[164] LEACH, W. E., 1815, *Malacostraca Podophthalmata Britanniae*, or descriptions of such British species of the linnean genus *Cancer* as have their eyes elevated on footstalks. 124 pp.
[165] LIENHART, R., 1921, Le mécanisme de la stridulation chez *Cyrtaspis scutata* Charp. (Meconemidae), *Ann. Soc. Entom. France*, *90*, 156.
[166] LILES, M., 1956, A study of the life history of the forked fungus beetle *Bolitotherus cornutus* (Panzer) (Coleopt. Tenebrionidae), *Ohio J. Sci.*, *56*, 329.
[167] LINDBERG, R. G., 1955, Growth, population dynamics, and field behavior in the spiny lobster, *Panulirus interruptus* (Randall), *Univ. Calif. Publ. Zool.*, *59*, 157.
[168] LOHER, W., 1959, Contribution to the study of the sexual behavior of *Schistocerca gregaria* Forskål (Orthoptera: Acrididae), *Proc. Roy. Entom. Soc. (London)*, *A*, *34*, 49.
[169] LOVETT, E., 1881, Stridulation in *Arctia caja*, *Entomologist*, *14*, 178.
[170] LUTZ, F. E., 1924, Insect Sounds, *Bull. Am. Mus. Nat. Hist.*, *50*, 333.
[171] LYLE, G. T., 1911, Stridulation in the pupa of an Ichneumonid, *Entomologist*, *44*, 404.
[172] MACNAMARA, CH., 1926, The "drumming" of Stoneflies (Plecoptera), *Can. Entom.*, *58*, 53.
[173] MALENOTTI, E., 1926, I canti delle Gryllotalpe, *Atti dell'Accad. Agric. Sc & Lettr. (Verona)*, *3*, (5).
[174] MARCU, O., 1931–32, Zur Kenntnis der Stridulationsorgane einiger Curculioniden, *Verh. Mitt. Siebenburg ver. Naturw. (Hermannstadt)*, *81–82*, 66.
[175] MARCU, O., 1934, Die Stridulationsorgane einiger Scarabaeidae, *Bull. Fac. Sti. Cernauti*, *8*, 71.
[176] MARCU, O., 1944–47, Die Stridulationsorgane der Gattungen *Acallorneuma mainardi* und *bohemanius* Schultze (Curculionidae), *Ann. Sci. de l'Univ. de Jassy*, *30*, 14.
[177] MARCUS, H. and E. E. MARCUS, 1951, Los nidos y los organos de stridulacion y de equilibrio de *Pogonomyrmex marcusi* y de *Dorymyrmex emmaericellus* Kusn., *Folia Universitaria*, *5*, 117, Cochabamba, Bolivia.
[178] MEIXNER, J., 1934, Coleoptera, in KUKENTHAL, *Handbuch der Zoologie*, Vol. 4 (2) W. de Gruyter, Berlin, p. 1040.
[179] MEYER, E., 1928, Neue Sinnesbiologische Beobachtungen an Spinnen, *Z. Morphol. u. Ökol. Tiere*, *12*, 1.
[180] MILLOT, J., 1949, Les Aranéides, in GRASSÉ, *Traité de Zoologie*, Masson Ed. Paris, Vol. 6, p. 610.
[181] MILLOT, J. and M. VACHON, 1949, Les Solifuges, in GRASSÉ, *Traité de Zoologie*, Masson Ed. Paris, Vol. 6, p. 482.
[182] MILLOT, J. and M. VACHON, 1949, Les Scorpions, in GRASSÉ, *Traité de Zoologie*, Masson Ed. Ed. Paris, Vol. 6, p. 386.
[183] MÖBIUS, K., 1867, Über die Entstehung der Töne, welche *Palinurus vulgaris* mit den äusseren Fühlern hervorbringt, *Arch. f. Naturgeschichte*, *33*, 73.
[184] MONRO, J., 1953, Stridulation in the Queensland fruit fly *Dacus (Strumeta) tryoni* Frogg, *Austr. J. Sci.*, *16*, 60.
[185] MOULTON, J. M., 1957, Sound production in the spiny Lobster, *Panulirus argus* (Latreille), *Biol. Bull.*, *113*, 286.
[186] MUKERJI, D., 1929, Sound production by a larva of *Cybister* (Dytiscidae), *J. Bombay Nat. Hist. Soc.*, *33*, 653.
[187] MYERS, K., 1952, Oviposition and mating behaviour of the Queensland fruitfly (*Dacus tryoni* Frogg. and the Solanum fruitfly (*Dacus cacuminatus* Hering), *Austr. J. Sci. Research*, *B*, *5*, 264.
[188] OHAUS, 1903, Verzeichnis der von Herrn Richard Haensch in Ecuador gesammelten Ruteliden, *Berlin Entom. Z.*, *48*, 215.
[189] ORTMAN, A. E., 1901, *Crustacea, Malacostraca* in BRONNS, *Klassen und Ordnungen des Tierreichs*, Vol. 5, Part 2 (2), Akademische Verlagsgesellschaft, Leipzig, p. 1245–1249.
[190] OSSIANNILSSON, F., 1949, *Insect Drummers*, *Opuscula Entomologica*, Suppl. X, Berlingska Boktryckeriet, Lund. 145 pp.
[191] PACKARD, A. S., 1904, Sound produced by a japanese saturnian caterpillar, *J. N.Y. Entom. Soc.*, *12*, 92.
[192] PASSERINI, C., 1828, Note sur le cri du Sphinx à tête de Mort, *Ann. Sci. Nat.*, *13*, 332.
[193] PEARMAN, J. V., 1928, On sound production in the Pscocoptera and on a presumed stridulatory organ, *Entom. Month. Mag.*, *64*, 179.
[194] PEARSE, A. S., 1911, On the habits of *Thalassina anomala* Herbst, *Philippine J. Sci.*, *6*, 213.
[195] PETER, K., 1910, Über einen Schmetterling mit Schallapparat, *Endrosa (Setina) aurita* var. *ramosa*, *Mitt. Naturw. Ver. Neuvorpommern u. Rügen*, *42*, 24.
[196] PETER, K., 1912, Versuche über das Hörvermögen eines Schmetterlings (*Endrosa* var. *ramosa*), *Biol. Zentr.*, *32*, 724.
[197] PETRUNKEVITCH, A., 1930, The spiders of Porto-Rico, *Trans. Conn. Acad.*, *30*, 181.

[198] Petrunkevitch, A., 1956, *Eohelea stridulans*, a striking example of paramorphism in a Baltic amber gnat, *Science*, *123*, 675.
[199] Piepers, M. C. and P. C. T. Snellen, 1910, *The Rhopalocera of Java, Hesperiidae*, The Hague. 60 pp.
[200] Pierce, G. W., 1948, *The songs of Insects*, with related material on the production, detection and measurement of sonic and supersonic vibrations, Harvard Univ. Press, Cambridge, Mass. vii + 329 pp.
[201] Pocock, R. I., 1902, The stridulating organ in the Egyptian Beetle *Graphopterus variegatus* F. (Masoreidae), *Ann. Mag. Nat. Hist.*, *10*, 154.
[202] Pocock, R. I., 1904, On a new stridulation organ in Scorpions discovered by W. J. Burchell in Brazil in 1828, *Ann. Mag. Nat. Hist.*, *7*, 62.
[203] Poisson, R., 1951, Ordre des Hétéroptères, in Grassé, *Traité de Zoologie*, Masson Ed. Paris, Vol. 10 (2), p. 1657.
[204] Prell, H., 1913, Über zirpende Schmetterlingspuppen, *Biol. Zentr.*, *33*, 496.
[205] Prell, H., 1917, Über trommelnde Spinnen, *Zool. Anz.*, *48*, 61.
[206] Prell, H., 1920, Die Stimme des Totenkopfes (*A. atropos*), *Zool. Jb. Abt. Syst. Geog. Biol. Tiere*, *42*, 235.
[207] Prochnow, O., 1907–08, Die Lautapparate der Insekten, *Int. Entom. Z. Guben*, *1* (1907), 133.
[208] Ragge, D. R., 1955, Le problème de la stridulation des femelles Acridinae (Orthoptera, Acrididae) *Colloque sur l'Acoustique des Orthoptères*, Fascicule hors série des *Annales des Epiphyties* I.N.R.A., Paris, p. 171.
[209] Rathbun, M. J., 1930, The Cancroid Crabs of America, *U.S. Natl. Mus. Bull.*, *152*. 609 pp.
[210] Reaumur, (R. Ferchault de), 1734, *Mémoires pour servir à l'Histoire des Insectes*, T. 1, 7ème mémoire, p. 294; T. 2, 7ème mémoire, p. 289.
[211] Remy, P., 1935, L'appareil stridulant du Coléoptère Ténébrionide *Olocrates abbreviatus*, *Ann. Sci. Nat. Zool.*, *18*, 7.
[212] Romanoff, N. M., 1892, *Mémoires sur les Lépidoptères*, Vol. VI, St. Petersburg, p. 232.
[213] Roy-Noel, J., 1954, Contribution à l'étude de l'appareil musical des Ensifères, *Ann. Sci. Nat. Zool.*, 11ème sér., 65.
[214] Salisbury, E. J., 1940, Stridulation in *Nymphalis Io*, *Entom. Month. Mag.*, *76*, 117.
[215] Salpeter, M. M. and C. Walcott, 1960, An electron microscopical study of a vibration receptor in the Spider, *Exp. Neurol.*, *2*, 232.
[216] Saussure, H. de, 1888, Prodromus Ædipodiorum Insectorum ex ordine Orthopterorum, *Mém. Soc. Phys. Hist. Nat. Genève*, *28*, (9). 254 pp.
[217] Saussure, H. de and L. Zehntner, 1902, *Myriapodes de Madagascar*, Imprimerie Nationale Ed., Paris, 356 pp..
[218] Scharrer, E., 1931, Stimm- und Musikapparate bei Tieren und ihre Funktionsweise, in *Handbuch der normalen und pathologischen Physiologie*, *15*, 1223.
[219] Schild, F. G., 1877, Miscellen. Fünf Entwicklungsstände gleichzeitig, Zirpende Insektenpuppen, Darwinistische Erwägungen, *Stett. Entom. Z.*, *38*, 85.
[220] Schmidt, W. L., 1840, Über die Töne welche *Pelobius hermanni* hören lässt, *Stett. Entom. Z.*, *1*, 10.
[221] Schmitt, W. L., 1931, *Crustaceans*, Smithsonian Scientific series, p. 87.
[222] Sellier, R., 1952, La différenciation de l'appareil sonore élytral des mâles de Gryllides, *Compt. rend. Acad. Sci.*, *234*, 1639.
[223] Sharp, D., 1874, Note on the existence of stridulating organs in the genus *Lomaptera*, *Entom. Month. Mag.*, *11*. 136.
[224] Sharp, D., 1893, On stridulation in ants, *Trans. Entom. Soc. (London)*, 199.
[225] Sharp, D., 1897, On the stridulatory organs of Trox, *Entom. Month. Mag.*, *33*, 206.
[226] Skaife, S. H., 1953, *African Insect life*, Longmans Green & Co., Cape Town–London–New York. viii + 387 pp.
[227] Smit, C. J. B. and A. L. Reyneke, 1940, Do nymphs of Acrididae stridulate?, *J. Entom. Soc. S. Africa*, *3*, 72.
[228] Spencer, B., 1895, The presence of a stridulating organ in a spider, *Nature*, *51*, 438.
[229] Spencer, B., 1896, On the presence and structure of a stridulating organ in *Phlogius* (*Phrictus*) *crassipes*, *Rep. Horn Exp. Central Austr. Zool.*, *2*, 412.
[230] Stainton, M. T., 1888, The noise or sound produced by butterflies of genus *Vanessa*, *Entom. Month. Mag.*, *25*, 225.
[231] Strouhal, H., 1929, Bemerkungen zu einigen *Androniscus*-Arten (Isop. terr.), *Zool. Anz.*, *85*, 69.
[232] Swinton, A. H., 1876–77, On stridulation in the genus *Vanessa*, *Entom. Month. Mag.*, *13*, 169.
[233] Swinton, A. H., 1881, *Insect variety, its propagation and distribution*, London. 326 pp.
[234] Tembrock, G., 1959, *Tierstimmen*, Ziemsen Verlag, Wittenberg Lutherstadt. 285 pp.
[235] Tempere, G., 1937, La stridulation chez *Acalles* et *Sibinia* (Curculionidae), *Miscellanea Entomologica*, *38*, 127.

[236] THUNBERG, C. P., 1795, *Travels in Europe, Africa and Asia performed between the years 1770 and 1779*, London (1st English ed. 1793).
[237] TINDALE, N. B., 1928, Australasian Molecrickets of the family Gryllotalpidae, *Rec. S. Austr. Mus.*, *4*, 1.
[238] TORRE BUENO, J. R. DE LA, 1903, Notes on the stridulation and habits of *Ranatra quadridentata*, *Can. Entom.*, *35*, 235.
[239] TORRE BUEONO, J. R. DE LA, 1905, Notes on the stridulation and habits of *Ranatra quadridentata*, *Can. Entom.*, *37*, 85.
[240] TWEEDIE, M. W. F., 1954, Notes on Grapsoid Crabs from the Raffles Museum *Bull. Raffles Mus.*, *25*, 118.
[241] USINGER, R. L., 1954, A new genus of Aradidae from the Belgian Congo with notes on stridulatory mechanisms in the family, *Ann. Mus. Roy. Congo Belge*, Tervuren, Miscellanea Zoologica H. Schouteden, *I*, 540.
[242] UVAROV, B. P., 1928, *Locusts and Grasshoppers*, a handbook for their study and control, Vol. 14, London.
[243] UVAROV, B. P., 1943, The tribe Thrinchini of the subfamily Pamphaginae and the interrelations of the Acridid subfamilies, *Trans. Roy. Entom. Soc. (London)*, *93*, 1.
[244] VACHON, M., B. DUMORTIER and R. G. BUSNEL, 1958, Enregistrement de stridulations d'un Scorpion sud-africain, *Bull. Soc. Zool. France*, *83*, 253.
[245] VARLEY, G. C., 1939, Stridulation in *Œdaleonotus fuscipes*, *Proc. Roy. Entom. Soc. (London)*, C *4*, 10.
[246] VERHOEFF, K. W., 1902, Über die zusammengesetzte Zirpvorrichtung von *Geotrupes*, *Sitzber. Ges. Naturforsch. Freunde, Berlin*, p. 149.
[247] VERHOEFF, K. W., 1908, Über Isopoden *Androniscus* nov. gen., *Zool. Anz.*, *33*, 129.
[248] VERHOEFF, K. W., 1928, Klasse Diplopoda, in BRONNS, *Klassen und Ordnungen des Tierreichs*, Vol. 5 (2), Akademische Verlagsgesellschaft, Leipzig, p. 197.
[249] VIETTE, P., 1955, Position systématique et appareil stridulant de *Pemphigostola synemonistis* Strand, de Madagascar, *Bull. Soc. Entom. France*, *60*, 176.
[250] VOLZ, P., 1938, Studien über das "Knallen" der Alpheiden, nach Untersuchungen an *Alpheus dentipes* Gueren und *Syhalpheus laevimannus* (Heller). *Z. Morphol. u. Ökol. Tiere*, *34*, 272.
[251] VOSSELER, J., 1907, Einige Beobachtungen an ostafrikanischen Orthopteren, *Deut. Entom. Z.*, 241.
[252] C. WALCOTT and W. G. VAN DER KLOOT, 1959, The physiology of the spider vibration receptor, *J. Exp. Zool.*, *141*, 191.
[253] WALKER, C. M., 1899, The sound producing organs of *Lema trilineata*, *Entom. News*, *10*, 58.
[254] WEISS, O., 1914, Die Erzeugung von Geräuschen und Tonen, in WINTERSTEIN, *Handbuch der vergleichenden Physiologie*, Bd. 3, 1 Hälfte, I. Teil, p. 249.
[255] WERNER, F., 1934, Scorpions, Pedipalpi, in BRONNS, *Klassen und Ordnungen des Tierreichs*, Vol. 5 (4), Part 1, Akademische Verlagsgesellschaft, Leipzig.
[256] WESENBURG-LUND, 1911–12, Biologische Studien über Dytisciden, Biol. Suppl. V, Ser. V, zur *Inter. Nat. Rev. ges. Hydro, Bio, Hydrog.*
[257] WESTRING, N., 1847, Bidrag till historien om insekternes stridulations-organer, *Natuurhist. Tidsskr.*, (2nd ser.), *2*, 334.
[258] WESTWOOD, J. O., 1834, Note about the ticking of *Anobium*, *Mag. Nat. Hist.*, *7*, 470.
[259] WHITE, F. B., 1872, Note on the sound made by *Hylophila prasinana*, *Scottish Nat.*, *I*, 213.
[260] WICHMANN, H., 1912, Beitrag zur Kenntnis des Stridulationsapparates der Borkenkäfer, *Entom. Blätter*, 8.
[261] WILLIAMS, C. B., 1922, Co-ordinated rhythm in insects; with a record of sound production in an Aphid, *Entomologist*, *55*, 173.
[262] WOOD MASON, J., 1878, Stridulating Crustaceans, *Nature*, *18*, 53.
[263] WOODS, E. F., 1956, Queen piping *Bee World*, *37*, 185; 216.
[264] ZEUNER, F., 1934, Phylogenesis of the stridulating organ of locusts, *Nature*, *134*, 460.

For a more detailed bibliography on Insects sounds, see M. FRINGS and H. FRINGS, 1960, *Sound Production and Sound Perception by Insects, A Bibliography*, The Pennsylvania State University Press, 108 pp.

CHAPTER 12

THE PHYSICAL CHARACTERISTICS OF SOUND EMISSIONS IN ARTHROPODA

by

B. DUMORTIER

1. Introduction

The song of Insects has attracted the attention of man for a very long time, but only in the last twenty years thorough studies have been undertaken on this subject. The studies have followed the same course of development as the physical equipment upon which they depend completely. As a matter of fact, if the musical transposition of calls (the only method formerly used) gives an approximate idea of the rhythm, it is very inaccurate in regard to frequency, since only the dominant frequency (to the ear) can be picked up, and even then, only if it is within the restricted range of the musical scale. Moreover it is certain that only individuals endowed with particularly good and well-trained hearing can use this method.

Kreidl and Regen[40] in 1905 were the pioneers of the physical study of the song of insects. They determined the frequency spectrum of the field cricket's song with the aid of a stroboscope and a gramophone. Lutz (1924)[46] was probably one of the first to begin to use electro-acoustic equipment for the same purpose, but we are especially indebted to the American physicist Pierce[55] for having developed this branch of research since 1936, both in the field of analysis, in the strict sense, and in the study of the mechanics of emissions. Mention must also be made of important data of Busnel and Chavasse[23] (1951) and since 1951 of Busnel and co-workers, of Broughton[15, 16, 17, 18], Loher[41, 42], Pringle[57], and Alexander[1, 3].

Today* these investigations can be carried out with increased thoroughness, thanks to the continual improvement of apparatus (Fig. 213): microphones, amplifiers, and tape-recorders with a broad band and a low distortion factor, frequency analysers etc.**, and the time is probably not far off when the precision of the equipment will exceed our power of interpretation.

We shall study in turn the phsysical nature of acoustical emissions, and the mechanisms by which they are produced.

2. The Components of the Sound Emission

Among the parameters to be considered in the sound signal of animals, some are inherent in the physical nature of sound or the structure of the stridulatory apparatus;

---

* It is interesting to note that the French entomologist Goureau[31] proved the need for a physical apparatus for the study of insect song. He wrote as early as 1837 on certain Orthoptera, the stridulation of which seemed inaudible to him: "I believe that we could hear their songs if our ears were sensitive enough to perceive them. But this remarkable existence of insensitive noise and silent sounds cannot be proved by experiments at the moment. To acquire this sort of proof one would need an instrument amplifying sound and producing for the ear the same effect that the microscope produces for the eyes."

** A critical study of this apparatus will be found in Chapter 2.

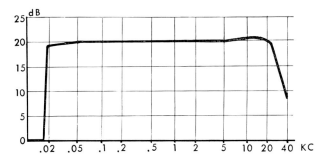

Fig. 213. Typical frequency characteristics of the Brüel and Kjær condenser microphone type 4134, showing the response in a diffuse sound field. All recordings used for frequency analysis (see Fig. 215–217) were made with this microphone.

such are frequency, intensity, modulation, and transients. Others concern more the manner in which the animal uses its apparatus: duration of the stridulation, speed of displacement of the vibrating parts, and differences in the spacing of stridulatory movements. We shall deal with *physical parameters* first and *biological parameters* second.

(a) *Physical parameters*

(i) *Frequency*

Up to 100 kc/s the acoustic spectrum is completely covered by the different sonic and ultra-sonic emissions of arthropods: the flight tones in low frequencies; the songs of crickets, entirely localised in the audio-range; the songs of Acridids, with a wider range of frequencies including ultra-sonic components; the stridulations of Tettigoniids, of some Coleoptera, and the noises produced by certain Crustacea, penetrating very considerably into the ultra-sonic scale (Fig. 214).

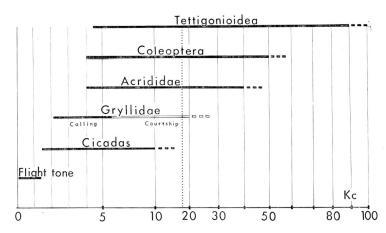

Fig. 214. Spectral characteristics of insect sounds. The three points which end the lines that give the extent of the spectra, mean that the limit given here, is imposed by the microphone. Obviously, higher frequencies can exist in the signal. The dotted line indicates the limit between sonic and ultra-sonic range for the human ear.

References p. 371

TABLE 24

FREQUENCIES IN THE ACOUSTIC EMISSIONS OF SOME ARTHROPODS

(In regard to the Orthoptera, the reference is for the calling song, unless otherwise indicated, Figs. 215–217.)

Notes: (a) Superposed figures represent dominant frequencies. (b) Two figures joined by a hyphen indicate a continuity of the spectrum between these two limits.

INSECTS

Order and Family	Species	Frequency (c/s)	Author
ORTHOPTERA			
Gryllidae	*Acheta assimilis* Fabr.		
	calling song	4,900	Pierce, 1948 (i)[55]
	courtship	17,000	
	G. campestris L.	5,000	Busnel unpubl.
	Acheta domesticus (L.)	4,500	Busnel, 1953[21]
	Miogryllus verticalis (Serv.)	6,000 (24°C)	Alexander, 1956[1]
	Nemobius carolinus Scud.	5,430	Pierce, 1948[55]
	N. allardi Alex. & Thom.	7,740	Pierce, 1948[55]
		7,500–8,000 (27°C)	Alexander and Thomas, 1959[5]
	N. palustris Blatchley	6,000 (29°C)	Pierce, 1948[55]
Oecanthidae	*Oecanthus nigricornis* Walk	3,700 (24°C)	Walker, 1958[70]
	O. niveus (De Geer)	2,600 (24°C)	Walker, 1958[70]
	O. latipennis Riley	2,000 (18°C)	Alexander, 1960[3]
	O. exclamationis Davis	2,200 (23°C)	Alexander, 1960[3]
	O. pellucens (Scop.)	3,100 (28°C)	Busnel, Dumortier, unpubl.
	Neoxabea bipunctata (De Geer)	3,000 (21°C)	Alexander, 1960[3]
Trigonidiidae	*Anaxipha exigua* (Say)	7,000 (21°C)	Alexander, 1960[3]
Eneopteridae	*Orocharis saltator* Uhler	3,500 (17.5°C)	Alexander, 1960[3]
Gryllotalpidae	*Gryllotalpa gryllotalpa* (L.)	3,200 (25°C)	Busnel, 1953[21]
	G. hexadactyla Perty	2,000 (21°C)	Alexander, 1960[3]
Phanaeropteridae	*Amblycorypha oblongifolia* (De Geer)	9,160	Pierce, 1948[55]
	Scuderia curvicauda (De Geer)	11,666	Pierce, 1948[55]
	Barbitistes fischeri (Yers.)	12,000–90,000	Dumortier, unpubl. (ii)
Pseudophyllidae	*Pterophylla camellifolia* (Fabr.)	18,000–63,000	Pierce, 1948[55]
Conocephalidae	*Neoconocephalus ensiger* (Harris)	13,700	Pierce, 1948[55]
	Neoconocephalus ensiger (Harris)	4,500–50,000	Borror, 1954[13]
	Homorocoryphus nitidulus (Scop.)	10,000–16,000	Busnel, 1953 (ii)[21]
	Conocephalus brevipennis (Scud.)	34,000 47,000	Pierce, 1948[55]
	C. spartinae (Fox)	31,000 40,000	Pierce, 1948[55]
	Orchelimum vulgare Harris	7,700 16,000 27,000	Pierce, 1948[55]
Tettigoniidae	*Atlanticus testaceous* (Scud.)	14,960	Pierce, 1948[55]
	Decticus verrucivorus (L.)	5,000–100,000	Busnel and Chavasse, 1951 (ii)[23]
	Tettigonia viridissima L.	7,000–100,000	Busnel, 1953 (ii)[21]
Ephippigeridae	*Ephippiger ephippiger* (Fiebig)	6,000–88,000	Busnel and Chavasse, 1951 (ii)[23]
Acrididae	*Locusta migratoria* L.	3,000–18,000	Busnel and Chavasse, 1951 (ii)[23]

TABLE 24 (continued)

Order and Family	Species	Frequency (c/s)	Author
	Chloealtis conspersa Harris	7,300	Pierce, 1948[55]
	Chorthippus curtipennis Harris	8,150	Pierce, 1948[55]
	C. biguttulus (L.)	4,000–50,000	Loher, unpubl. (ii)
	Stenobothrus lineatus (Panz.)	5,000–16,000	Dumortier, unpubl. (iii)
HOMOPTERA			
Cicadidae	*Tibicen canicularis* Harris	7,400	Pierce, 1948[55]
		5,000–7,200	Alexander, 1960[3]
	Okanagana rimosa (Say)	8,100	Pierce, 1948[55]
		8,000–10,000	Alexander, 1960[3]
	Platypleura octoguttata (Fabr.)	5,700	Pringle, 1954[57]
LEPIDOPTERA			
Sphingidae	*Acherontia atropos* L.		
	low-pitched call	5,000–15,000	Busnel and Dumortier,
	high-pitched call	3,500–20,000	1959[24]
COLEOPTERA			
Geotrupidae	*Geotrupes* sp.	2,000–40,000	Autrum, 1936[8]
Silphidae	*Necrophorus* sp.	–28,000	Autrum, 1936[8]
	Necrophorus vespilloides Herbst	3,000–30,000	Dumortier, unpubl.
Cerambycidae	*Cerambyx cerdo* L.	2,560–8,625	Tembrock, 1960[67]
	id.	4,500 > 20,000	Dumortier, unpubl. (iii)
Passalidae	*Passalus cornutus* Fabr.	4,000–5,000	Park, 1937[52]
HYMENOPTERA			
Apidae	*Apis mellifica* L.		
	Queen a–virgin	180	Woods, 1956[72]
	b–just before copulation	350–380	
DIPTERA			
Trypetidae	*Dacus tryoni* Frogg.	3,000	Monro, 1953[49]
Culicidae	*Anopheles gambiae* ♀	420*	Kahn and Offenhauser, 1949[38]
	Aëdes aegypti L. ♀	600*	Kahn and Offenhauser, 1949[38]
	id.	450–600*	Wishart and Riordan, 1959[71]
ARACHNIDA			
ARANEIDA			
Theridiidae	*Steatoda castanea* Cl.	435	Meyer, 1928[48]
	Steatoda bipunctata L.	325	Meyer, 1928[48]
CRUSTACEA			
DECAPODA			
Alpheidae	*Alpheus* sp.	3,000–50,000	Johnson, Everest and Young, 1947[37]
Palinuridae	*Palinurus argus* (Latr.)	6,000–8,000	Moulton, 1957[50]

* The fundamental frequency depends on the age of the insect and if it is engorged or not[61].

(i) The analyser used by Pierce was able to detect chiefly the dominant frequency and that is the figure given here. This obviously does not mean that the emission of the insect under consideration is in all cases confined to this frequency.

(ii) Recordings made for the audio range with an electrodynamic microphone sensitive up to 15 kc/s ($\pm$ 5 dB) and for the ultra-sonic range with a quartz microphone sensitive from 15–100 kc/s ($\pm$ 10 dB); Pimonow analyser.

(iii) With an electrostatic microphone sensitive up to 30 kc/s.

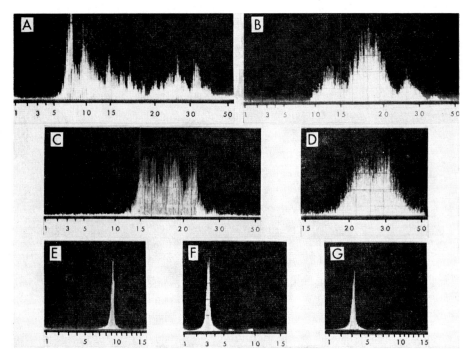

Fig. 215. Spectrograms of the song of some Tettigoniids and Gryllids. (A) *Tettigonia cantans* Fuessly, (B) *Decticus verrucivorus* (L.), (C) *Metrioptera bicolor* (Phil.), (D) *Metrioptera brachyptera* (L.), (E) *Nemobius fasciatus* (De Geer,) (F) *Oecanthus pellucens* (Scop.) (G) *Oecanthus quadripunctatus* Beuten. All spectrogramms except E and G made from recordings with Bruël and Kjær condenser microphone (see frequency characteristics on Fig. 213) and Ampex tape-recorder model 307, 76 cm/sec, ± 3 dB, 200–40,000 c/s; panoramic spectrum analyzer LEA, Pimonow system, model AF 10 S. Frequency in kc/s. (E and G from recordings kindly communicated by Prof. R. D. Alexander.)

The performances of the "transmitters" are often superior to those of the receivers (microphones) that we use for their analysis. This is the problem in conducting and drawing conclusions from experiments on frequency spectra*. Table 24 may be subject to change with the advent of improved techniques.

While the number of species in which the sound spectrum is known is somewhat limited, we can make a few remarks concerning these data:

1. The presence of ultra-sonic frequencies is very common[28,54,62].

2. The breadth of the spectrum is extremely variable; for example it covers a band of 90 kc/s in *Decticus verrucivorus*, whereas in *Oecanthus* it is limited to about 500 c/s.

3. Among the Orthoptera, spectra types appear generally to be superfamily characteristics (Fig. 217): a narrow spectrum in the Gryllodea, a wide spread in the Acridoidea with beginning in the medium of the audio range, and a very wide spread in the Tettigonioidea, with beginning in far audio range. These spectral characteristics are obviously associated with the morphology of apparatus and their method of working.

Here we must tackle the problem of the stridulation of ants. In fact, according

---

* The origin of many erroneous results with sound analysis may be attributed to the use of a microphone that cuts frequencies above 15 kc/s (the case with most electro-dynamic microphones).

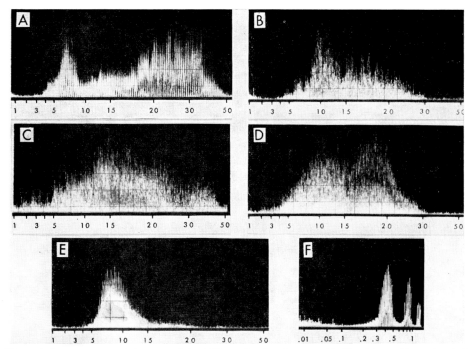

Fig. 216. Spectrograms of the song of Acridids and other insects (A) *Chorthippus biguttulus* (L.) (calling song), (B) *Chorthippus brunneus* Thunb. (rival's song), (C) *Aromia moschata* L. (Col., Cerambycidae), (D) *Necrophorus vespilloides* Herbst (Col., Silphidae), (E) *Acherontia atropos* L. (Lépid., Sphingidae), (F) *Culex* sp. (Dipt., Culicidae). Recordings made in the same conditions as those of Fig. 215, frequency in kc/s.

to a common opinion, the frequency spectrum of these insects' stridulation generally would be in the ultra-sonic range. As a matter of fact, certain authors had drawn this conclusion in studying non stridulating species[43, 44]. Actually many species stridulate audibly: *Myrmica ruginodis* Nyl.[66], *Monomorium salomonis* L., *Tetramorium caespitum* L., several species of the genus *Messor*[63], etc... Raignier (1931)[59] gives a rather long list of species, in which the stridulation has been heard.

The fact that in some cases the stridulatory movement does not give rise to any audible sound, proves in no way that the emission is ultra-sonic. Therefore, one may suppose that the weak intensity of vibrations, together with the minute emitting surface, hinders any air transmission[9]. The radiation of the vibratory energy through the substratum (because of the less important discontinuity between body and soil than between body and air) is however possible[34]. It is the same with the transmission by close contact of individuals. As a matter of fact, one can perceive an emission if the ant is placed on the sensitive surface of the microphone, but nothing, if the insect is a few centimetres away from it.

(ii) Intensity

The accurate study of the sound intensity of arthropods' emission involves as many difficulties, if not more, as does that of frequency. Added to the difficulty generally encountered in detecting all the frequencies of the signal (which means, in consequence, a measuring of intensity only in the fraction picked up by the microphone,

References p. 371

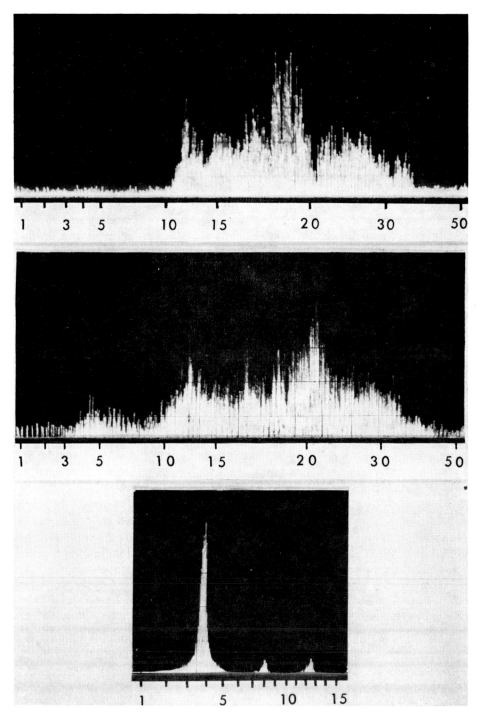

Fig. 217. Frequency characteristics of Tettigoniids, Acridids and Gryllids. From top to bottom: *Ephippiger ephippiger* (Fiebig), *Arcyptera fusca* (Pallas), *Gryllus campestris* L. (see harmonics 1 and 2). Recordings made in the same conditions as those of the Fig. 215, frequency in kc/s.

there is the fact that most apparatus (apart from cathode ray oscillographs) have much too long a time constant to make it possible to integrate sound emissions that are often less than a hundredth of a second long. Moreover, measurements taken in the field are often rendered inaccurate on the one hand by the absorption caused by vegetation, and on the other, by considerable thermic variations in the first few centimetres above the soil (microclimate). This results in the formation of layers of air at different temperatures where considerable acoustic refractions are produced[64]. Caution must again be payed in interpreting measurements taken in rooms where no special provision is made to reduce the reverberation time. In addition, the insect is not comparable to a pinpoint radiating spherical waves, and there is no doubt that its emissions are partially focused; hence all measurements of intensity have to take into account the direction the animal is facing.

It seems to us quite illusory to consider as valid any measurements which do not allow a margin of error at least equal to ± 3 dB.

In Table 25 are some examples from Orthoptera, Acrididae (with an approximation reckoned at 15 %).

TABLE 25

INTENSITY IN dB OF SOME ACRIDIDS; INSECTS AT 30 cm FROM THE MICROPHONE
(According to Haskell, 1955[32].)

Species	Measurements taken											
	in the laboratory specimen No.						in the field specimen No.					
	1	2	3	4	5	6	A	B	C	D	E	F
Chorthippus parallelus (Zett.)	36	37	35	32	37	34	32	31	32	30	34	32
C. brunneus Thunb.	44	47	49	45	46	47	38	40	40	46	43	44
Stenobothrus lineatus (Panz.)	35	31	33	31	33	30	37	—	31	—	30	32
Omocestus viridulus (L.)	—	—	—	—	—	—	36	32	35	37	34	36

It will be noted that the figures of measurements taken in the field are lower than those carried out in the laboratory. The dispersion and absorption of sound waves due to vegetation and microclimate seem to be the cause of this difference.

The calling song of *Acheta domesticus* (L.) measured in the laboratory at 60 cm from the insect reaches about 47 dB. The same song in *Oecanthus pellucens* is about 60 dB. at 10 cm[112].

The following values have been found[21, 41] among various Orthoptera Acridoidea placed at 10 cm in the axis of the microphone (the approximation of the data is comparable to that previously given).

Locusta migratoria L. (*solitaria*) ♂, disturbance
    (warning) song during mating   35 dB
    calling song   30 dB
Locusta migratoria L. (*solitaria*) ♀, stridulation
    indicating refusal to mate (warning song)   35 dB
Chorthippus biguttulus (L.), calling song from 38 to 62 dB
C. jucundus (Fisch.), calling song from 38 to 54 dB
C. brunneus Thunb., calling song from 35 to 47 dB

The Tettigonioidea reach very considerable intensities. The stridulation of *Tettigonia viridissima* L. and that of *Ephippiger terrestris* (Yersin) give an average value of 90–95 dB at 1 m., which corresponds to the roar of a lion at a few metres distance... The crests of these signals are still higher, attaining even 110 dB. Considerable as they are, these values are undoubtedly not exceptional in this superfamily, but the fact that the greater part of the energy liberated is within frequency ranges to which the human ear is not sensitive or only slightly so, gives us an erroneous impression.

Very few measurements have been taken in other groups. We may, however, point out that the intensity of the cry of the moth *Acherontia atropos* L. is about 65 dB at 5 cm distance[24], and that the flight tone of the mosquito *Aëdes aegypti* L. ♀ has an intensity of 40 dB at 15 mm distance[71]. The congregational song of two Cicadidae, *Magicicada septemdecim* (L.) and *M. cassinii* (Fisher) reaches 94 dB for the former and 74 dB for the latter[1]. The scorpion *Opisthophthalmus latimanus* C. L. Koch produces a rather faint stridulation which, measured at a few centimetres distance, is barely more than 50 dB[69]. On the other hand the snapping of certain Alpheid shrimps is extremely intense and rises sometimes above 120 dB at 1 m[37].

(iii) Amplitude modulation and frequency modulation

The variation in terms of time of the individual parameters which make up the signal is called modulation. This variation may affect the amplitude or the frequency. In the first case, one has amplitude modulation (A.M.), in the second, frequency modulation (F.M.) (Fig. 218).

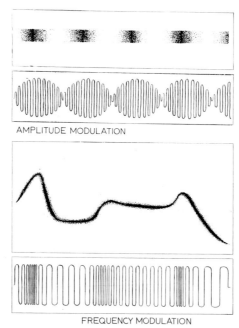

Fig. 218. Schematic representation of amplitude and frequency modulation by means of spectrographic and oscillographic analysis. On spectrograms: time in abscissae, frequency in ordinate. (Kay sound spectrograph).

The A.M. appears quite frequently in the sound emissions of arthropods where the intensity of the signal undergoes more or less regular fluctuations. It is linked with the cycle of movement of the emission apparatus (*e.g.* passing of the femur across the elytron in the Acridinae, wingstroke in Gryllidae).

The phenomenon of A.M. is frequently pushed to its extreme limits, since the amplitude can decrease to nothing after having attained its maximum, and there are then two cycles of amplitude separated by a period of silence (as in *Oecanthus*). The modulating frequency then corresponds to the pulse rate (see p. 356 and Fig. 223).

As for the F.M., represented by a deviation of frequency in the course of the signal, it is only possible to study it in emissions with a narrow spectral band. There is a F.M. in the case where, for example, between the beginning and the end of emission the frequency of the signal takes on different values f_1, f_2, f_3, f_2, etc... (see example p. 740, Fig. 395).

Actually, this instability is difficult to detect. That is why it seems to us that the concept of F.M. does not exactly relate to the acoustics of arthropods. On the other hand it is observed frequently in birds. This difference derives from the structure of the apparatus of emission, which allows the birds quick and extensive tonal variations, but only enables insects to change rhythm or intensity during the signal.

Although it is easier, on the other hand, to transfer concept of A.M. to the bioacoustic field, as we have shown earlier, it should be noted that:

(i) the carrier frequency is not necessarily a pure frequency

(ii) it is the carrier frequency which is heard, while the modulation frequency does not provide any tonal impression*,**.

Here verifications may be made by ear, oscillograph or frequency analyser. There is a considerable probability that for insects A.M. is one of the most important factors in the specific recognition of the signal (see p. 624).

(iv) Transients

This is the name of the "phenomena which take place in a system owing to a sudden change of conditions and which persist for a relatively short time after the change has occurred" (British Standards Institution 205/1575). A transient is thus produced at the beginning or at the end of a stable periodic phenomenon.

This definition is obviously qualitative ("sudden change"); that is why Pimonow (1957)[56], in trying to express the phenomenon quantitatively, admitted that, for the human ear, the vibratory phenomenon had a transitory character during 165 periods, no matter what the frequency***. Nothing, of course, justifies us in considering this evaluation as valid for receivers other than the human ear. Busnel, who was the first to draw attention to the role of transients in the releasing of phonoresponses and phonotaxis (see pp. 615 and 620) defined them as swift variations of considerable amplitude. As emissions within a stable vibratory phenomenon are rare, transients thus appear very frequently in biological acoustics.

It is probable that in Tettigonioidea, the numerous transients which analysis

---

* However, it has been demonstrated that when a pure tone frequency F (carrier frequency) is modulated in amplitude at a frequency f (modulating frequency) two lateral frequencies appear which enclose the carrier frequency, one $F-f$, the other $F+f$, giving the spectrum $F-f, F, F+f$. The amplitude of the lateral frequencies is then a variable of the rate of modulation. In some Orthoptera Gryllidae or Oecanthidae, the carrier frequency of which is relatively pure, the modulating frequency fairly high (from 50 to 100 c/s) and the rate of modulation 100%, it should theoretically be possible to trace the presence of these two lateral frequencies in the spectrum.

** Actually, when the wingstroke rate is high enough, and the insect in close proximity to the ear, it is possible to perceive a noise superimposed to the sound of the stridulation itself. This noise may not be an objective acoustical phenomenon, but rather a distortion which originates in our middle ear, due to the considerable intensity of the sound.

*** In the band of frequencies where the threshold of differential perception is: $\Delta F'/F = 3.10^3$.

reveals (sudden impacts of the teeth on the plectrum) are partly responsible for the large proportion of ultra-sonic frequencies observable in the spectra of these insects. In fact, it is known that a transient phenomenon, as the analysis in the Fourier's series shows, gives rise to a whole group of frequencies, wider as the phenomenon becomes shorter (Fig. 219).

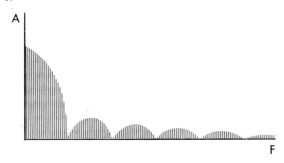

Fig. 219. Theoretical frequency spectrum of a transient obtained by the analysis in Fourier's series. F: frequency, A: amplitude.

(b) *Biological parameters*

These characterise the form and distribution of sound emissions in time. To make the definition of these parameters easier a special terminology has been devised, but it must be recognised that at present there is still a great confusion over its usage*.

Almost all English and American authors use the term *"pulse"* for the sound emission produced by an entire movement of the stridulatory apparatus and the term *"chirp"* for a regular group of *pulses*, separated before and after by a silence.

It was proposed at the Seminar on the Acoustics of Orthoptera, which was held at the "Laboratoire de Physiologie Acoustique", Jouy-en-Josas, France in 1954, to give the name "chirp" to the "sound-complex corresponding to a single movement of the stridulatory apparatus", and to confine the term "pulse" to a small group of homogeneous oscillations** within a "chirp"[25]. These proposals do not seem to have been adhered to, and we shall conform here to the present usage, with the sole intention of not increasing confusion.

We shall, then, call:

Pulse: the emission resulting from a passing of the plectrum over the pars stridens. This is the elementary stridulatory movement, the complete cycle of the apparatus, beginning and ending at the position of rest (Figs. 220 and 221).

Chirp: a series of pulses (= a repetition of the elementary stridulatory movement) distinct or linked together, each series being separated at the beginning and end by a distinguishable time space.

Pulse rate: (pulse repetition frequency—P.R.F.): "the number of pulses occurring per second, when they are produced in a reasonably regular manner"[33].

We shall avoid the use of terms of note, verse, sequence, phrase, strophe, syllable, since authors use them differently, and more often than not, these terms cannot be clearly defined except in particular cases.

---

* For a critical examination of this terminology see Chap. 1.
** This definition evidently fitted in much more with that of the electro-acousticians from whom the term has been borrowed. It is a pity that its meaning has been modified in passing into the hands of bioacousticians.

(i) *The pulses*

The duration of the pulse is extremely variable (Fig. 222), but it does not seem linked to a particular type of stridulatory apparatus, since among Acridinae, the pulse of *Stenobothrus lineatus* (Panz.) lasts 1 sec while that of *Chorthippus brunneus* Thunb.* or *Chloealtis conspersa* Harris has a duration of about 6 to 8 msec.

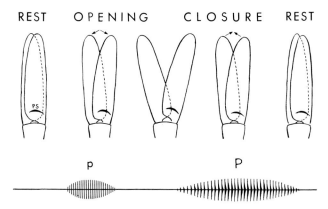

Fig. 220. Schematic analysis of a stridulatory movement in a Tettigoniid with oscillographic representation of the following sound emission. P: main pulse produced at the closure of the elytra, p: pulse sometimes produced at the opening of the elytra (see text), PS: pars stridens.

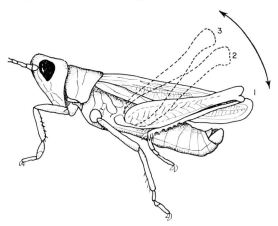

Fig. 221. Typical stridulatory movement of an Acridid, *Chorthippus albomarginatus* (De Geer): 1 (rest) 2, 3, 2, 1. (After Brown, 1955[20].)

In Tettigonioidea, the spans are just as important: *Gampsocleis glabra* (Herbst) (Tettigoniidae) completes a passage of the file in 5 msec while this same movement is done in 60 to 80 msec by *Decticus albifrons* (F.) (Tettigoniidae) and in 630 msec by *Uromenus rugosicollis* (Serv.). *Microcentrum rhombifolium* (Sauss.) (Phanaeropteridae) seems to have one of the slowest movements, since in its ticking song, the closing of the elytra is produced in 4 to 5 sec**.

---

* See first footnote p. 360.
** In this species, Alexander calls "pulse" each of the separate "ticks" (one tothstrike) produced during the closing of the elytra.

References p. 371

358   B. DUMORTIER

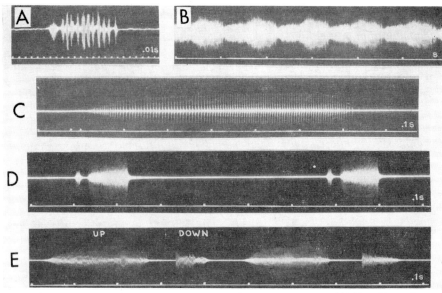

Fig. 222. Oscillographic study of the pulse. Length of the pulse (see text) (A) *Chorthippus brunneus* Thunb. (timing mark 1/100 sec), (B) *Stenobothrus lineatus* (Panz.) (timing mark 1 sec), (C) *Uromenus rugosicollis* (Serv.) (timing mark 1/10 sec). Examples of double pulse: (D) *Ephippiger ephippiger* (Fiebig) (timing mark 1/10 sec), (E) *Cerambyx cerdo* L. (timing mark 1/10 sec), stridulation by an up (first pulse) and down (second pulse) movement of the thorax.

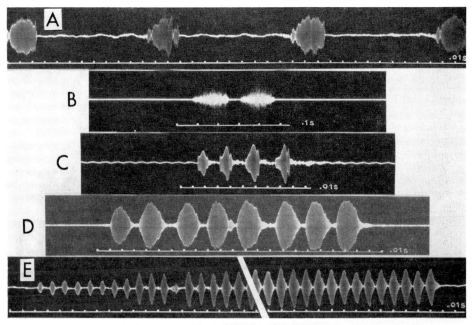

Fig. 223. Oscillographic study of the time-distribution of pulses. (A) one pulsed chirp of *Nemobius tinnulus* Fulton (timing mark 1/100 sec), (B) bipulsed chirp of *Pterophylla camellifolia* (Fabr.) from North of the Appalachian Mountains (timing mark 1/10 sec), (C) *Gryllus campestris* L. (timing mark 1/100 sec), (D) *Oecanthus niveus* (DeGeer) (timing mark 1/100 sec), (E) *Oecanthus angustipennis* Fitch., beginning and end of the trill (3–4 sec) (timing mark 1/100 sec). (A, B, D, E from recordings kindly communicated by Prof. R. D. Alexander.)

(*i*) *The chirps*

The time-distribution of pulses occurs in different ways (Fig. 223):

(i) *Regularly-spaced pulses*. The pulses are not grouped, but are produced in slow series. This is not a frequent case among Insects, but it occurs in Ephippigeridae: *Ephippiger bitterensis* Finot, *E. ephippiger* (Fiebig), in Gryllidae: *Nemobius tinnulus* Fulton.

The chirp which in this case is the same as the pulse, is called "one pulsed"*.

(ii) *Grouped pulses*. Within a group (= chirp), each pulse may be separated from the following and preceding ones by a very short space (distinct-pulses). This is the case in the majority of Gryllidae, Oecanthidae, Gryllotalpidae, some Acrididae and numerous Tettigonioidea. On the other hand the pulses may not be separated (linked-pulses) as in certain other Acrididae, in many Cicadidae, in the scorpion *Opisthophthalmus latimanus*, etc.

The number of pulses making up the chirp, *i.e.* the duration of the chirp, likewise shows a great diversity: *Gryllus campestris* L. (Gryllidae): chirp of 3 pulses; *Acheta pennsylvanicus* (Burm.) (Gryllidae): chirp of 3, 4 or 5 pulses; *Nemobius fasciatus* (De Geer) (Gryllidae): chirp of 7 or 8 pulses; *Oecanthus niveus* (De Geer): chirp of 8 pulses; *O. pellucens* (Scop.) (Oecanthidae): chirp of 15–20 pulses[22]; *Gryllotalpa gryllotalpa* (L.) (Gryllotalpidae): chirp of about 20 pulses.

In other species, on the other hand, the pulses are emitted without interruption, with a generally high pulse rate, for several seconds, indeed several minutes. Such a mode of emission is called "trill".* It is the case for example in the American Gryllids *Nemobius melodius* Thomas & Alexander (pulse rate = 24 at 18°C)[2,68], *Oecanthus nigricornis* Walker (pulse rate = 65 at 28°C), *O. quadripunctatus* Beutenmuller (pulse rate = 37.7 at 23°C)[70], *Neoxabea bipunctata* (De Geer) (pulse rate = 100 at 23°C)[1], and in the European genus *Pteronemobius*. Continuous stridulation is also found in the Tettigonioidea, such as *Neoconocephalus robustus* (Scud.) (pulse rate = 158)[1], *Homorocoryphus nitidulus* (Scop.) and many other Conocephalidae.

Not all stridulations can be included in this simple classification. Actually certain Orthoptera have an irregular song. Thus in *Scudderia septentrionalis* (Serv.) (Phaneropteridae) chirps with about a dozen pulses are observed, separated by numerous very short clicks. The song of *Amblycorypha oblongifolia* (De Geer) (Phaneropteridae) is composed of four pulses of increasing duration (8, 13, 21, 38 msec). In the Conocephalidae, *Conocephalus brevipennis* (Scud.) and *C. fasciatus* (De Geer) fairly numerous clicks fill the space between two successive chirps, similarly in *Orchelimum concinnum* Scudder and *O. vulgare* Harris[1] In the genus *Decticus* Serv. (Tettigoniidae), the song begins with some isolated clicks, which are emitted at an increasingly quick rate, and finally become a continuous emission. The song of *Amblycorypha uhleri* Stal (Phaneropteridae) seems to have the most complexity, since in the course of a single sequence, which lasts for 40 to 50 sec, the pulses are emitted at 3 different rhythms making up "phrases" with very characteristic structures[1,3].

We should note that this stridulatory activity, which is not sporadic but generally well spread over the 24 hours (see p. 644), represents a considerable expenditure of energy. In fact, though very few studies have been made in this direction,

---

* It is obviously impossible to distinguish strictly between "regularly spaced pulses" and "trill", the difference being only in the pulse rate. We may admit that the trill is a song of long duration in which the pulses are not perceptible individually.

an hour's courtship stridulation has been observed in *Stenobothrus lineatus* (Panz.) (Acrididae), interrupted only by rare pauses of a few seconds. In *Gryllus campestris* L. (Gryllidae), 42,000 stridulations have been counted in 4 h 20 min. *Lilioceris lilli* Scop. (Coleopt. Chrysomelidae), when held between the fingers reached the considerable figure of 13,665 stridulations in half an hour[10]. An average of 30,000 stridulations per 24 h has been found in *Ephippiger ephippiger* (Fiebig) (Orthopt. Ephippigeridae).

3. Analysis of the Mechanism of the Production of Sound Waves by the Stridulatory Apparatus

(a) Stridulatory mechanism

Starting from its position of rest the stridulatory appendage, leg, elytron, chele, chelicera, etc., necessarily produces a forward and a backward movement Figs. 220 and 221). As a general rule, among the Orthoptera, Ensifera in particular, the emission only occurs during the closing movement[55,60].

In a few cases, however, an emission is observed in both directions (double pulse). This appears particularly clearly in the Tettigonioidea of the European genus *Ephippiger* (Ephippigeridae) and in those of the American genera *Orchelimum* (Conocephalidae) and *Atlanticus* (Tettigoniidae)[1] (Figs. 220, 222, D,E, and 225).

In the Caelifera and especially in the Acridinae, the movement is often more complex. The femur on reaching the upper position makes a few rapid oscillations of slight amplitude before lowering. Such is the case, with *Chorthippus jucundus* (Fischer) *C. brunneus* Thunb., and *C. biguttulus* (L.)[42].*

We have already seen (see p. 299) that among the *Ensifera* the corresponding position of the elytra is a superfamily characteristic: the right elytron covering the left R/L in Gryllodea, the left elytron covering the right in Tettigonioidea (L/R). Since the two elytra have a *pars stridens***, it is theoretically possible for stridulation to occur when the elytra are in the inverted position. Various experiments to this end have been carried out on Gryllodea[30,39,45,60,65].

The proportion of individual insects with naturally inverted elytra is very small: 1/2000 in *Gryllotalpa gryllotalpa* (L.) and *Acheta domesticus* (L.)[30]. The figures are appreciably different for *Acheta assimilis* Fabr., *A. pennsylvanicus* (Burm.) and *A. rubens* Scud. since among these species 4–10 % of the individual insects examined (50 for each species) are left-winged[60]. If the position of the elytra in the adult *G. bimaculatus* De Geer is inverted, it is unusual for the animal not to re-establish the normal position in a few minutes. If, on the other hand, the experiment is made with newly emerged adults, the majority retains the inverted position and stridulates. It has even been observed that if subsequently the elytra are put back into the normal position (R/L) the insect attempts to return to the L/R-position in which it was put at the outset[65]. The song of L/R-insects is generally weaker than that of normal ones, and they are also incapable of emitting their fight-song (same observation with *G. campestris* L.[39]). This doubtless arises from the fact that the elytra, once they are inverted, cannot be raised sufficiently to take the convenient position for emitting this type of song. These experiments, even if they do not explain the origin and the reason for a double pars stridens, do at least

---

* These oscillations, which must be considered as pulses, can be seen only in slow motion picture or on a high speed oscillogram. They pass easily unnoticed, so the structure of the song is misinterpreted by some authors[33].

** In the Gryllodea the two pars stridens are generally identical, but in the Tettigonioidea the pars stridens of the right elytron (the lower one) is less robust.

TABLE 26

BIOLOGICAL PARAMETERS OF THE CALLING SONG OF SOME ORTHOPTERA

(The sign — means that the indication which should appear here was not given by the author quoted.)

Species	Temperature (°C)	Pulse Length of time (msec)	Pulse rate	Type*	Chirp Number of pulses composing the chirp	Chirp Length of time (sec)	Author
Acrididae							
Chorthippus brunneus Thunb.	—	8**	60–80	——	12	0.2	Loher and Broughton, 1955[42]
Stenobothrus lineatus (Panz.)	—	1000**	1	— — —	—	up to 30	Haskell, 1957[33]
Omocestus viridulus (L.)	—	40**	12–14	— — —	up to 680	up to 60	Loher and Broughton, 1955[42]
Chloealtis conspersa Harris	—	6.7	73**	— — —	8–9	0.1	Pierce, 1948[55]
Gryllidae							
Anaxipha exigua (Say)	23	15**	36**	— — —	trill	—	Alexander, 1956[1]
Gryllus assimilis Fabr.	—	8–15	35**	— — —	3–5	0.15 (5 pulses)	Pierce, 1948[55]
Acheta domesticus (L.)	23.5	20	26	— — —	2–3	0.075 (2 pulses)	Dumortier, Busnel, unpubl.
Nemobius confusus Blatchley	18	12**	46	— — —	39	0.8**	Alexander, 1956[1]
N. tinnulus Fulton	29	30	9	▬			Alexander and Thomas, 1959[5]
Oecanthidae							
Oecanthus niveus (De Geer)	23.5	12**	50	— — —	8	0.16**	Walker, 1957[70]
O. quadripunctatus Beutenmuller	23.5	15**	38	— — —	trill	—	Walker, 1957[70]
O. argentinus Saussure	29	—	50	— — —	trill	—	Alexander, 1956[1]
O. pellucens Scop.	—	30	36	— — —	18	0.5	Busnel, 1955[22]
Gryllotalpidae							
Gryllotalpa gryllotalpa (L.)	—	4.7	75	— — —	19	0.24	Busnel, 1955[22]
G. hexadacyla Perty	21	8**	65	— — —	8–19	0.15* (12 pulses)	Alexander, 1956, 1960[1,3]
Ephippigeridae							
Ephippiger provincialis (Yers.)	24	95	8	— — —	3–6	0.48 (4 pulses)	Dumortier, Busnel, unpubl.
Uromenus rugosicollis (Serv.)	—	630		▬			Dumortier, Busnel, unpubl.

* ——— linked pulses; - - - - distinct pulses; ▮ regularly spaced pulses.
** Values deduced from the information given in the work of the author referred to.

show that the animal by its plasticity can adapt itself to unusual conditions in order to exhibit a form of behaviour arising from an intense motivation. On this subject it should be noted that in a North American species, *Acheta veletis* Alex. & Big., the destruction of the plectrum of the left elytron, causing the abolition of stridulation in the normal position (R/L), brought about a spontaneous movement to the inverted position (L/R) in which the song again becomes possible[60].

(b) Origin of sound waves produced by stridulatory apparatus

Our knowledge in this field is almost solely limited to Orthoptera and Hemiptera (refer to Chapter 14 for this order). We shall, therefore only consider the first here*.

Knowing the structure of the apparatus and the way in which the insect uses it, the problem is to find the physical laws which explain the results of the analysis (frequency spectrum and oscillogram).

Unfortunately, the conclusion is rapidly reached that no single law rules the emission procedures. They are so complicated that the search for an equation formula, which is always found with well defined physical phenomena, has to be given up. Thus we have to proceed by means of approximations and hypotheses.

(i) The case of Gryllodea

This is probably the least complex case, since the frequency emitted is almost pure (see Table 24 and Fig. 215) as is shown by the analyses made by various writers on a certain number of Gryllidae and Oecanthidae.

By studying the film of a stridulating *Acheta pennsylvanicus* (Burm.), frame by frame, Pierce (1948)[55] established that only 47 % of the length of the pars stridens was covered by the plectrum in the course of a pulse. Knowing that the pars stridens of this insect is made up of 142 teeth, this means, then, that 67 teeth are used in a single stridulatory movement.

Now, it is shown by study of the oscillograms** of this species, that the average number of oscillations in a pulse is also 67.

From these results and from those obtained before by Kreidl and Regen (1905)[40] and by Lutz and Hicks (1930)[47], we can conclude that each oscillation corresponds to the impact of the plectrum against a tooth of the pars stridens. "As each tooth of the file passes over the scraper, a thrust in one direction is given to the file, and an opposite thrust to the scraper. These thrusts are periodic forces with a frequency determined by the number of impacts per second of the file teeth with the scraper"[55] (Fig. 224). This number depends on the speed of the stridulatory movement, the number of teeth and their spacing.

This explanation of the phenomenon given by Pierce seems an extremely probable one. But from then on two hypotheses are possible:

(i) either the toothstrike rate corresponds to the natural frequency of all or part of the elytron***.

---

* See pp. 323 and 325 for the explanation of the functioning of the vocal system of *Acherontia atropos* L. (Lepidoptera) and of the queen bee.
** Actually, Pierce calculates the number of sound waves by multiplying the frequency of the signal by the duration of the stridulatory movement.
*** The comparison between the file toothstrike rate and the sound frequency is possible, since in both cases it is a question of the determination of the number of times a certain phenomenon is reproduced in one second. In the first case this phenomenon is the impact of the teeth against the plectrum, in the second it is a cycle of the periodic movement by which the sound is represented.

(ii) or, on the other hand, the elytron works in forced vibrations at frequencies different than its own frequency.

In the first hypothesis, the resonator (elytron) receives an impulse (tooth/plectrum impact) for each oscillation. The direction and timing of this impulse are such as to increase the oscillatory movement. The unit is then comparable in its functioning principle to various sustained systems: pendulum, radio-electric oscillator, sustained pitch-fork[53].

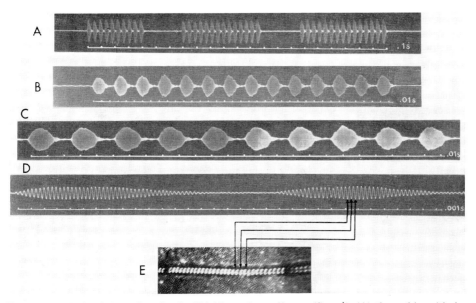

Fig. 224. Analysis of the pulse of a Gryllid (*Oecanthus pellucens* (Scop.)). (A) three chirps (timing mark 1/10 sec), (B) and (C) one chirp filmed at two different speeds (timing mark 1/100 sec), the structure of the pulses becomes more and more conspicuous, (D) two pulses filmed at a very high speed (timing mark 1/1000 sec), showing the sound waves (here about 3,300 c/s), (E) a microphotograph of the pars stridens, each toothstrike gives rise to one oscillation (see text). (Modelled after Busnel, 1955[22].)

But if this theory is held, it implies that the resonance of the elytron may be produced not on one frequency, but on a band of several kc/s in width. Actually it is established that in certain species the emitted frequency varies within quite wide limits according to the speed at which the pars stridens sweeps across the plectrum, hence in the final analysis, in accordance with the file toothstrike rate (see p. 366). Thus in *Nemobius carolinus* Scudder (Gryllidae) the emission is produced on a frequency within the range 4,100 and 6,400 c/s. In *Orocharis saltator* Uhler (Eneopteridae) this band goes from 3,800 to 5,800 c/s and from 2,000 to about 3,200 in *Oecanthus niveus* (De Geer) (Oecanthidae)[1,3] (Fig. 230).

This amounts to saying that each toothstrike rate will stimulate one of the resonance frequencies of the elytron.

The second hypothesis gives no role to the resonance frequency and introduces a comparison of the sound apparatus of Gryllodea with an electrodynamic loud speaker. In this apparatus, the membrane (cardboard cone) is submitted to forced vibrations which are transmitted to it by an exciter, the moving coil. This membrane has a natural frequency, but in fact, works at frequencies which are distinctly higher. The elytron may be compared with the membrane, and the pars-stridens-plectrum system to the moving coil (exciter). If we suppose that the natural frequency of the unit is lower than the vibration frequency of the exciter (*i.e.* toothstrike rate), the alternating movements of approach and separation of the elytra, will cause an acoustic phenomenon which will follow exactly the excitation frequency whatever it may be. Any change in this frequency will be reproduced by an identical change in the sound frequency emitted. This range of frequencies (see Fig. 320) may be compared with the bandwidth of a loud speaker.

The variation of frequency in accordance with the speed of the stridulatory movement may be explained equally well by either hypothesis. Only actual measurement of the natural frequency of the apparatus will allow a choice to be made.

(ii) The case of Tettigonioidea

Here the spectrum is much more complex than in Gryllodea, at times covering nearly a hundred kc/s (see Table 24). Hence we may suppose that the acoustic principles capable of explaining this type of emission are somewhat different. The thorough study[53] which has been made on *Ephippiger**, leads to some conclusions which may certainly be app ied to numerous other Tettigonioidea.

The oscillographic analysis reveals a succession of impulses with vertical attack-edge and an exponential decay. Each wavetrain has the regular form of the series of damped waves of a resonator excited by an initial impact. Only by the impact of the

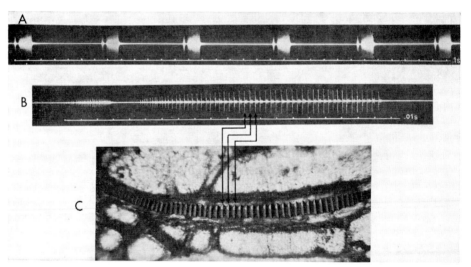

Fig. 225. Analysis of the pulse of a Tettigoniid (*Ephippiger ephippiger* Fiebig). (A) six double pulses (timing mark 1/10 sec), (B) one double pulse filmed at high speed (timing mark 1/100 sec), (C) a microphotograph of the pars stridens, each toothstrike gives rise to a damped wavetrain (see text). (Modelled after Pasquinelly and Busnel, 1955[53].)

---

* In the species of this genus the pulse is double, the opening and closing of the elytra causing a sound emission (see p. 360). However the stridulation of closing is much longer and intenser than that of opening.

plectrum against teeth of the pars stridens can each of these impulses be initiated. Teeth and impulses are, moreover, equal in number (Fig. 225).

The width of the spectrum leads us to think that actually it is not a resonator which is excited by each impact, but a number of resonant systems. The shifting of the plectrum's point of percussion along the pars stridens during the stridulatory movement, must put these resonant systems into action successively. In *Ephippiger* a certain lowering of the frequency at the end of stridulation is actually established. The emission bands of each of the systems in question, when put side by side, may evidently cover quite a broad area, but to explain the presence of the high ultrasonic frequencies (up to 100 kc/s), it is certainly necessary to take account of transients produced at each impact against a tooth. It is known that the analysis of such phenomena reveals a theoretically infinite spectrum (see p. 355 and Fig. 219).

Although everything seems to point to the pars-stridens-plectrum unit as the source of the acoustic phenomenon, we should note that some experiments on *Ephippiger bitterensis* in which the pars stridens was totally destroyed (burnt by thermocautery with perforation of the elytron (Busnel, unpubl.)), have shown that though the operation produced a fall in intensity of twenty decibels, the principal components of the spectrum only underwent little variations.

(iii) The case of Acrididae*

Very few studies have been made on the mechanism of stridulation of Acrididae, and it is principally to Jacobs (1953)[36] and Loher (1957)[41] that we owe the most thorough observations and experiments for understanding these emission processes.

The frequency spectrum of Acrididae is somewhat akin to that of Tettigonioidea, since it extends at times, into the ultrasonic range. However, the fine structure of the pulse, as shown by the oscillograph, does not show in any certain way anything comparable with the impulses which the impact of the rasp-teeth against the plectrum produce in the Tettigonioidea.

By progressive amputations of the elytra, Loher (1957)[41] established that the intensity of the signal could be profoundly affected by these operations. However, if the frequency spectrum showed a tendency to extend towards the high frequencies as the area of the elytra became smaller, the dominant frequencies remained almost stable. The destruction of the femoral teeth only affects the frequencies a little, but makes the intensity of the signal considerably lower.

Hence, it seems that the teeth of the pars stridens do not play a role on their own. The important thing apparently is rather the whole of the relief they compose, and the simple crest on which they are set, is sufficient to produce an emission in which the frequencies are practically the same[41]. Moreover, the female *Chorthippus brunneus* Thunb. which at times stridulate as vigorously as the male, actually have only hairs or a rudimentary denticulation as femoral pars stridens.[36,58].

Sound emission in Acridids seems then to be much more like a rubbing than a succession of percussions. As for the width of the spectrum, it is possibly due to the displacement of the point of contact between the pars stridens and the plectrum during a passage of the femur. In this way the resonance of various portions of the elytron is produced. It has been observed in fact that in animals which have several stridulations,

---

* In fact we shall only consider the single sub-family of Acridinae.

each one of which corresponds to the sweeping of a different region of the elytra, the tonality varied according to the position of the femur (case of *Eremogryllus hamadae* Krauss[19,20]).

(iv) Conclusion

It must be agreed that all these explanations are very hypothetical. The intensity of the signal, often great, and the spectrum, from the pure frequency to the range several kc/s in width, both remain facts, the bases of which are almost a complete mystery. However, the cutting of the elytra in such a way as to leave only the plectrum and the pars stridens [experiments made on *Pterophylla camellifolia* (Fabr.)[1], *Gampsocleis glabra* (Herbst) (Tettigonioidea) (Busnel, unpubl.), various *Chorthippus* (Acridoidea)[41], and *Gryllus campestris* L. (Gryllodea) (Dumortier, unpubl.)] considerably modifies (lessens) the intensity, while the frequency is much less affected. This leaves room for the supposition that the "frequency generator" is doubtless represented by the pars-stridens-plectrum system. The rest of the elytron mostly playing the role of an acoustic coupler, enabling the air transmission of the feeble vibratory energy produced at the level of this "generator". The efficiency of the coupling being in proportion to the surface which is involved in it, the probability is that almost all the elytron is used (and not simply a formation such as the drums, as has often been claimed without experimental verification).

However, the fact that total destruction of the pars stridens (*Ephippiger, Chorthippus*) weakens, but does not suppress the sound production, seems to show that the sensitivity of the resonating assembly is such that a deformed excitation remains sufficient, though it has nothing in common with normal excitation, except the region where it is operated and the nature of the movement which produces it.

*(c) Modification of the parameters under the action of temperature**

(i) Pulses and chirps

In 1882 Brooks[14] discovered that the chirping rate of *Oecanthus niveus* (De Geer) (Orthopt. Oecanthidae) varied in accordance with the temperature. Some years later, Dolbear (1897)[26] proposed an empirical formula for this same insect, relating the temperature expressed in degrees Fahrenheit to the number of chirps emitted per minute:

$$T = \frac{N}{4} + 40^{**}$$

where T = temperature in °F
N = number of chirps in 1 min.

A little later, Bessey and Bessey (1898)[12] gave another formula, valid between 60° and 80° F (= 15.5 and 26.5° C):

$$T = 60 + \frac{N-92}{47}^{***}$$

---

* See also p. 638.
** Conversion into °C: $T = \frac{N}{7.2} + 4.4$
*** Conversion into °C: $T = 15.5 + \frac{N-92}{84.6}$

Edes, in 1899,[27] took up the preceding observations. He verified the formula of Dolbear and found a margin of error as regards degree.

A rise in temperature not only increases the chirping rate, but also the pulse rate. In the Gryllidae *Nemobius allardi* Alex. & Thom., Pierce (1948)[55] thus finds the equation:

$$T = 5N - 3^*$$

where N = number of pulses in 1 sec.

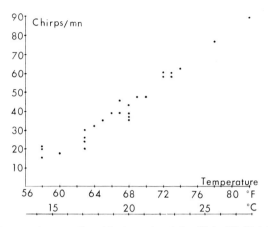

Fig. 226. Effect of temperature on the chirping rate of the Katydid *Cirtophyllus perspicillatus*. (After the data of Hayward, 1901[35].)

This acceleration of the stridulatory movements resulting from every rise in temperature appears to be general, and Figs. 226–228 give the form of the phenomenon for Gryllodea and Tettigonioidea (it seems that no measurements of this kind have been made in Acridoidea). As far as can be ascertained, it is a question of linear functions. However, in *Neoconocephalus ensiger* (Harris), the relationship temperature–chirping rate is expressed by an exponential curve**[29].

Parallel with this increase in their number per unit of time, there is a diminution of duration of the acoustic complexes considered (chirps or pulses). For instance, in *Ephippiger provincialis* (Yers.) (Orthopt. Tettigonioidea) endemic species of the South of France, the duration of a chirp of 4 pulses, which is 900 msec at 14° C, falls to 350 at 40° C, in following a curve of hyperbolic shape (Dumortier, Busnel, unpubl. Fig.229).

This relation between the temperature and the activity of the muscles is a result of the law of Van 't Hoff which governs the kinetics of most chemical systems involved in the functioning of muscle.

It is of interest to note that, even in vertebrates, temperature may have an effect on sound emissions. In various Anura, it has been established that the number

---

* Conversion into °C: $T = 2.77 N - 19.4$
** In species with a complex stridulation, the temperature does not act necessarily in the same way upon the different parts of the song. For instance, at 25°C, the song of *Orchelimum agile* De Geer (Tettigoniidae) is composed of two equal parts: a series of about 35 distinct pulses (8–10 sec) and a trill (8–10 sec). As the temperature falls, the pulses are less numerous (they even can disappear) and the length of the trill increases up to 30 sec[6].

References p. 371

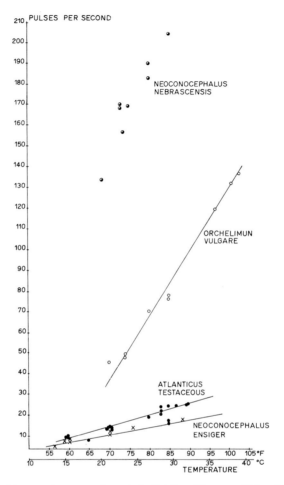

Fig. 227. Effect of temperature on pulse rate in Tettigoniid songs. (After Alexander, 1956[1].)

of breeding calls per unit of time increases with the temperature of the water, and that in the same relation, their length is in inverse ratio to that value[11,73].

Besides temperature, other atmospheric factors may have an influence on the parameters characterizing stridulation. Hence the chirping rate of *O. niveus* rises if a draught produced by a fan is directed on the insect[7].

The air disturbance either sets off a kinesthetic excitation, or it works by modifying the hydration or the respiratory rhythm of the insect. We cannot tell, but perhaps this observation explains the differences in stridulation rhythms which are sometimes observed according to whether the animals are on low branches or near the top of a tree.

(ii) Frequency

We have already noted that in the Gryllodea the speed of the plectrum's passage over the teeth of the pars stridens, and consequently the frequency, rise with temperature.

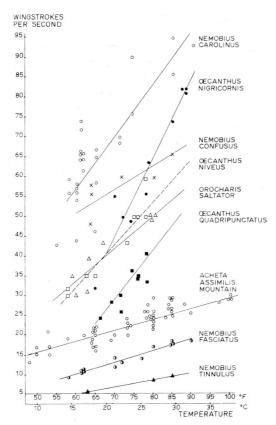

Fig. 228. Effect of temperature on rate of wingstroke in cricket songs. (After Alexander, 1956[1].)

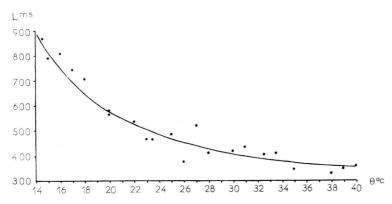

Fig. 229. Effect of temperature on the length of the chirp (four pulses) of *Ephippiger provincialis* (Yers.) L: length of the chirp in msec, θ: temperature in °C. (After Dumortier and Busnel, unpubl.)

Fig. 230 gives the results obtained by Alexander (1956)[1]. In the European Gryllid *Oecanthus pellucens* (Scop.), the frequency, which is approximately 2,500 c/s at 4.5° C goes to 3,300 for 33° C (Busnel, Dumortier, unpubl.). In *Gryllus campestris* L., on the other hand, the frequency stays practically constant at all temperatures (Busnel, unpubl.).

In Tettigonioidea and Acridoidea, the spectrum of which covers a very wide band at times, it does not seem possible to distinguish a frequency slippage.

Outside the Orthoptera very little research has been done to find out the variation in frequency in relation to temperature. However, it has been noted that in male *Doratura stylata* (Boh.) (Homoptera, Jassoidea), the frequency changes from 121 c/s at 25° C, to 153 c/s at 41° C[51].

We have already indicated that in various Anura the length of the cry diminishes when the temperature is raised. In *Bombina variegata variegata* (L.) the frequency is equally sensitive to heat changes, since from 550 c/s at 17° C, it rises to 650 c/s at 26° C[73]. However, it is probable that the explanation of this phenomenon is quite different from that given for the insects, since the functioning of the sound apparatus in vertebrates is based on a totally different principle.

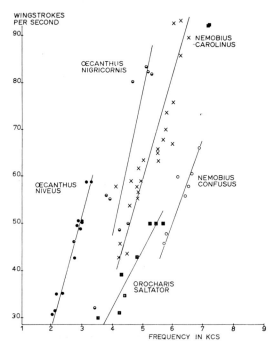

Fig. 230. Change in frequency in cricket songs with change in wingstroke rate. (After Alexander, 1956[1].)

REFERENCES

[1] ALEXANDER, R. D., 1956, A comparative study of sound production in Insects, with special reference to the singing Orthoptera and Cicadidae of the eastern United States, *Thesis*, Ohio State University. 529 pp.
[2] ALEXANDER, R. D., 1957, The song relationship of four species of ground Crickets (Orthopt., Gryllidae, *Nemobius*), *Ohio J. Sci.*, 57, 153.
[3] ALEXANDER, R. D., 1960, Sound communication in Orthoptera and Cicadidae, in Animal Sounds and Communication, proceedings of the AIBS meetings, Bloomington (Indiana), 1958, edited by W. E. LANYON and W. N. TAVOLGA, 38–92, *Amer. Inst. of Biol. Sci.*, Washington.
[4] ALEXANDER, R. D. and T. E. MOORE, 1958, Studies on the acoustical behavior of 17-years Cicadas, *Ohio J. Sci.*, 58, 107.
[5] ALEXANDER, R. D. and E. S. THOMAS, 1959, Systematic and behavioural studies on the crickets of the *Nemobius fasciatus* group, *Ann. Entom. Soc. Am.*, 52, 591.
[6] ALLARD, H. A., 1929, The last meadow Katydid; a study of its musical reactions to light and temperature, *Trans. Am. Entom. Soc.*, 55, 155.
[7] ALLARD, H. A., 1930, Changing in the chirp-rate of the snowy tree cricket (*Oecanthus niveus*) with air currents, *Science*, 72, 347.
[8] AUTRUM, H., 1936, Über Lautäusserungen und Schallwahrnehmungen bei Arthropoden, *Z. vergl. Physiol.*, 23, 332.
[9] AUTRUM, H., 1936, Die Stridulation und das Hören der Ameisen, *Sitzber. Ges. Naturforsch. Freunde*, Berlin, p. 210.
[10] BAIER, L. J., 1930, Contribution to the physiology of the stridulation and hearing of Insects, *Zool. Jb.*, 47, 151.
[11] BELLIS, E. D., 1957, The effects of temperature on Salientian breeding calls, *Copeia*, 2, 85.
[12] BESSEY, C. A. and E. A. BESSEY, 1898, Further notes on thermometer crickets, *Am. Naturalist*, 32, 263.
[13] BORROR, D. J., 1954, Audiospectrographic analysis of the song of the cone-headed grasshopper *Neoconocephalus ensiger* (Harris), *Ohio J. Sci.*, 54, 297.
[14] BROOKS, M. W., 1882, Influence of temperature on the chirp of the cricket, *Popular Sci. Month.*, 20, 268.
[15] BROUGHTON, W. B., 1952, Gramophone studies of the stridulation of British grasshoppers, *J. South-West Essex Techn. Coll.*, 3, 170.
[16] BROUGHTON, W. B., 1953, Further gramophone studies of grasshoppers and crickets, *Proc. Roy. Entom. Soc. (London)*, C, 18, 17.
[17] BROUGHTON, W. B., 1954, Oscillographie économique de schéma de modulation par un appareil de coût et de fonctionnement bon marché utilisable avec les appareils enregistreurs à vitesse variable, *L'onde Electrique*, 324, 204.
[18] BROUGHTON, W. B., 1955, L'analyse de l'émission acoustique des Orthoptères à partir d'un enregistrement sur disque reproduit à des vitesses ralentis, *Colloque sur l'Acoustique des Orthoptères*, Fascicule hors série des *Ann. des Epiphyties*, I.N.R.A., Paris, p. 82.
[19] BROWN, E. S., 1951, The stridulation of *Eremogryllus hammadae* Krauss (Orth., Acrididae), *Proc. Roy. Entom. Soc. (London)* A, 26, 89.
[20] BROWN, E. S., 1955, Mécanismes du comportement dans les émissions sonores chez les Orthoptères, *Colloque sur l'Acoustique des Orthoptères*, Fascicule hors série des *Ann. des Epiphyties*, I.N.R.A., Paris, p. 168.
[21] BUSNEL, M. C., 1953, Contribution à l'étude des émissions acoustiques des Orthoptères; 1er memoire: recherches sur les spectres de fréquence et sur les intensités, *Ann. des Epiphyties*, 3, 333.
[22] BUSNEL, M. C., 1955, Étude des chants et du comportement acoustique d'*Oecanthus pellucens* (Scop.), *Colloque sur l'Acoustique des Orthoptères*, Fascicule hors série des *Ann. des Epiphyties*, I.N.R.A., Paris, p. 175.
[23] BUSNEL, R. G. and P. CHAVASSE, 1951 Recherches sur les émissions sonores et ultrasonores d'Orthoptères nuisibles à l'agriculture: Étude des fréquences, *Supplemento al vol. VII, Ser. IX del Nuovo Cimento*, (Bologna), p. 1.
[24] BUSNEL, R. G. and B. DUMORTIER, 1959, Vérification par des méthodes d'analyse acoustique des hypothèses sur l'origine du cri du Sphinx *Acherontia atropos* L., *Bull. Soc. Entom. France*, 64, 44.
[25] CHAVASSE, P., R. G. BUSNEL, F. PASQUINELLY and W. B. BROUGHTON, 1955, Propositions de définitions concernant l'Acoustique appliquée aux Insectes, *Colloque sur l'Acoustique des Orthoptères*, Fascicule hors série des *Ann. des Epiphyties*, I.N.R.A., Paris, p. 21.
[26] DOLBEAR, A. E., 1897, The Cricket as a thermometer, *Am. Naturalist*, 31, 970.
[27] EDES, R. T., 1899, Relation of the chirping of the tree-cricket (*Oecanthus niveus*) to temperature, *Am. Naturalist*, 33, 935.

[28] EYRING, P., 1946, Jungle's acoustics, *J.A.S.A., 18*, 257.
[29] FRINGS, H. and M. FRINGS, 1957, The effect of temperature on chirp-rate of male cone-headed grasshoppers, *Neoconocephalus ensiger, J. Exp. Zool., 134*, 411.
[30] GOLDA, H. and W. LUDWIG, 1958, Rechts-links Fragen bei der Flügellage und dem Zirpen von *Gryllotalpa vulgaris* L., *Zool. Anz., 161*, 1.
[31] GOUREAU, 1837, Note sur les sons insensibles produits par les Insectes, *Ann. Soc. Entom. France, 6*, 407.
[32] HASKELL, P. T., 1955, Intensité sonore des stridulations de quelques Orthoptères britanniques, *Colloque sur l'Acoustique des Orthoptères*, Fascicule hors série des *Ann. des Epiphyties*, I.N.R.A., Paris, p. 154.
[33] HASKELL, P. T., 1957, Stridulation and associated behaviour in certain Orthoptera; 1. Analysis of the stridulation of, and behaviour between, males, *Brit. J. Animal Beh., 5*, 139.
[34] HASKINS, C. P. and E. V. ENZMANN, 1938, Studies on certain sociological and physiological features in the Formicidae, *Ann. N.Y. Acad. Sci., 37*, 97.
[35] HAYWARD, R., 1901, The Katydid's call in relation to temperature, *Psyche, 9*, 179.
[36] JACOBS, W., 1953, Verhaltensbiologische Studien an Feldheuschrecken, *Z. f. Tierpsychol.*, Beih. 1 228 pp.
[37] JOHNSON, M. W., F. A. EVEREST and R. W. YOUNG, 1947, The role of snapping shrimp (*Crangon* and *Synalpheus*) in the production of underwater noise in the sea, *Biol. Bull., 93*, 122.
[38] KAHN, M. C. and W. OFFENHAUSER JR., 1949, The identification of certain West African mosquitoes by sound, *Am. J. Tropical Med., 29*, 827.
[39] KEILBACH, R., 1935, Über asymmetrische Flügellage bei Insekten und ihre Beziehungen zu anderen Asymmetrien, *Z. Morphol. u. Ökol. Tiere, 29*, 1.
[40] KREIDL, A. and J. REGEN, 1905, Physiologische Untersuchungen über Tierstimmen; Stridulation von *Gryllus campestris* (I. Mitteilung), *Sitzber. Akad. Wiss. Wien, Math. Nat. kl.*, Abb. *III, 114*, 57.
[41] LOHER, W., 1957, Untersuchungen über den Aufbau und die Entstehung der Gesänge einiger Feldheuschreckenarten und den Einfluss von Lautzeichen auf das akustische Verhalten, *Z. vergl. Physiol., 39*, 313.
[42] LOHER, W. and B. BROUGHTON, 1955, Études sur le comportement acoustique de *Chorthippus bicolor* (Charp.) avec quelques notes comparatives sur des espèces voisines (Acrididae), *Colloque sur l'Acoustique des Orthoptères*, Fascicule hors série des *Ann. des Epiphyties*, I.N.R.A., Paris, p. 248.
[43] LUBBOCK, J., 1881, Observations on bees, ants and wasps, *Nature, 23*, 255.
[44] LUBBOCK, J., 1913, *Ants, Bees and Wasps*; A record of observations on the habits of the social Hymenoptera, Appleton & Co., N.Y. xix + 448 pp.
[45] LUTZ, F. E., 1906, The tegminal position in *Gryllus*, *Can. Entom., 38*, 207.
[46] LUTZ, F. E., 1924, Insect Sounds, *Bull. Am. Mus. Nat. Hist., 50*, 333.
[47] LUTZ, F. E. and W. R. HICKS, 1930, An analysis by Movietone of a cricket's chirp (*Gryllus assimilis*), *Am. Mus. Novitates, 420*, 1.
[48] MEYER, E., 1928, Neue sinnesbiologische Beobachtungen an Spinnen, *Z. Morphol. u. Ökol. Tiere, 12*, 1.
[49] MONRO, J., 1953, Stridulation in the Queensland fruit fly *Dacus* (*Strumeta*)*tryoni* Frogg., *Austr. J. Sci., 16*, 60.
[50] MOULTON, J. M., 1957, Sound production in the spiny lobster *Panulirus argus* (Latr.), *Biol. Bull., 113*, 286.
[51] OSSIANNILSSON, F., 1949, *Insect Drummers, Opuscula Entomologica*, supplementum X, Lund, Berlingska Boktryckeriet. vi + 145 pp.
[52] PARK, O., 1937, Studies in nocturnal activities. Further analysis of activity in the beetle, *Passalus cornutus*, and description of audio-frequency recording apparatus, *J. Animal Ecology, 6*, 239.
[53] PASQUINELLY, F. and M. C. BUSNEL, 1955, Études préliminaires sur les mécanismes de la production des sons par les Orthoptères, *Colloque sur l'Acoustique des Orthoptères*, Fascicule hors série des *Ann. des Epiphyties*, I.N.R.A., Paris, p. 145.
[54] PIELMEIER, W. H., 1946, Supersonic insects, *J.A.S.A., 17*, 337.
[55] PIERCE, G. W., 1948, *The Songs of Insects*; with related material on the production, detection and measurement of sonic and supersonic vibrations, Harvard, Cambridge, Mass.
[56] PIMONOW, L., 1957, Audition et analyse des vibrations en régime transitoire, *Rev. Laryngol*, 11-12, 972.
[57] PRINGLE, J. W. S., 1954, A physiological analysis of Cicada song, *J. Exp. Biol., 31*, 525.
[58] RAGGE, D. R., 1955, Le problème de la stridulation des femelles Acridinae (Orthoptera, Acrididae), *Colloque sur l'Acoustique des Orthoptères*, Fascicule hors série des *Ann. des Epiphyties*, I.N.R.A., Paris, p. 171.
[59] RAIGNIER, A., 1933, Introduction critique à l'étude phonique et psychologique de la stridulation des fourmis, *Brotéria*, Revista de Sciencias Naturaes, Lisbon, *2*, 51.

[60] RAKSHPAL, R., 1960, Sound producing organs and mechanism of sound production in field crickets of the genus *Acheta* F. (Orthopt., Gryllidae), *Canad. J. Zool.*, *38*, 499.
[61] ROTH, L. M., 1948, A study of Mosquito behavior. An experimental laboratory study of the sexual behavior of *Aëdes aegypti* L., *Am. Midland Naturalist*, *40*, 265.
[62] SABY, J. S. and H. A. THORPE, 1946, Ambiant ultrasonic noises of the jungle, *J.A.S.A.*, *18*, 271.
[63] SANTSCHI, F., 1909, Sur un moyen très simple d'entendre les sons de très petits insectes, *Bull. Soc. Entom. France*, *18*, 310.
[64] SCHIELLING, H. K., N. P. GIVENS, W. L. NYBORG, W. A. PIELMEIER and H. A. THORPE, 1947, Ultrasonic propagation in open air, *J.A.S.A.*, *19*, 222.
[65] STÄRK, A., 1958, Untersuchungen am Lautorgan einiger Grillen und Laubheuschreckenarten, zugleich ein Beitrag zum Rechts-Links-Problem, *Zool. Jb.*, *77*, 9.
[66] SWINTON, A., 1877, Note on the stridulation of *Myrmica ruginodis* and other Hymenoptera, *Entom. Month. Mag.*, *14*, 187.
[67] TEMBROCK, G., 1960, Stridulation und Tagesperiodik bei *Cerambyx cerdo* L. *Zool. Beitr.*, N.F., *5*, 419.
[68] THOMAS, E. S. and R. D. ALEXANDER, 1957, *Nemobius melodius*, a new species of cricket from Ohio, *Ohio J. Sci.*, *57*, 148.
[69] VACHON, M., B. DUMORTIER and R. G. BUSNEL, 1958, Enregistrement des stridulations d'un Scorpion sud-africain, *Bull. Soc. Zool. France*, *83*, 253.
[70] WALKER, T. J., 1958, Specificity in the response of female tree crickets (Orth. Gryllidae, Oecanthinae) to calling songs of the males, *Ann. Entom. Soc. Am.*, *50*, 626.
[71] WISHART, G. and D. F. RIORDAN, 1959, Flight responses to various sounds by adult males of *Aëdes aegypti* L. (Diptera: Culicidae), *Can. Entom.*, *91*, 181.
[72] WOODS, E. F., 1956, Queen piping, *Bee World*, *37*, 185; 216.
[73] ZWEIFEL, R. G., 1959, Effect of temperature on call of the Frog *Bombina variegata*, *Copeia*, *4*, 322.

For a more detailed bibliography on Insect sounds, see M. FRINGS and H. FRINGS, 1960, *Sound Production and Sound Perception by Insects, A Bibliography*, The Pennsylvania State University Press, 108 pp.

CHAPTER 13

THE FLIGHT-SOUNDS OF INSECTS

by

O. SOTAVALTA

1. History

The buzz of insects has drawn the attention of man certainly since prehistoric times. Many phenomena of nature were, in early history, reviewed by Aristotle who also discussed the possible origin of the hum emitted by insects. Later following his ideas, numerous scholars of natural history, even up to fairly modern times, have described a variety of organs in the insect body which were thought to produce this sound with the aid of a never-ending stream of "internal air" flowing through the internal canals of the body and its numerous orifices, a stream strong enough to make the sound well audible at a distance of several yards. Even at the present time, some stray textbooks still maintain the view that the insects buzz by pressing air through their tracheal stigmata.

The association of the buzz with the motion of the wings was first noticed by Albertus Magnus (d. 1280), and later explained by the 17th century scientists Camerarius[4] and Goedart[12]. They considered it to be a frictional sound, similar to the whizz produced by a thin rod moving through the air. The first correct interpretation of the origin of sound in physical vibratory phenomena, which was given by Mersenne in 1636 and re-demonstrated by Stancari in 1706, was not applied to insect flight-sounds until 1832, when it was mentioned in an unusual, popular book on *Music of Nature*[9], although earlier in 1799 there appeared a short notice with a similar suggestion by an unknown observer[1].

In the 1860s this explanation of the origin of flight-sound began to get a foothold in scientific literature[28], and the classical author of flight-sound observations was Landois who in 1867 published his observations[20], and later enlarged his work into *Thierstimmen*[21]. Since the publication of this monograph, only stray data on flight-sounds have appeared in the literature, and not until 1947 was an extensive list of pitch determinations of flight-sounds published[44]. Apart from this, several authors have approached the problem of wingbeat frequency from a "non-acoustical" point of view, the earliest of which was Marey[25, 26], the inventor of the kymograph. Also Magnan who made decisive improvements to high-speed film photography has published an extensive list of wingbeat frequency determinations in insects[23].

2. The Flight-sound and its Recording Methods

The flight-sound is the direct acoustical consequence of the wing vibration, and its pitch has a direct relationship to the beating frequency of the wings. Therefore two categories of methods have been used to determine the physical characteristics of these related problems, depending on whether the vibratory phenomena themselves, the origin, or the sound, the consequence, are to be analyzed. The first category includes visual observation as a subjective method, and the application of the kymograph, chronophotography, the stroboscope and photocell techniques as objective methods, while the other category includes acoustic observation as a subjective method and the application of the cathode-ray oscillograph and various "spectrographic" and automatic sound analyzers as objective methods. The subjective methods naturally require strictly critical conditions and generally can only be used by experienced observers with suitable qualifications. The use of the acoustic method by which the majority of pitch determinations have been made, assumes that the observer has had musical training, and an inexperienced observer will unconsciously encounter certain difficulties including the psychological "soprano-tenor-error", *i.e.*, the equating of standard "soprano" and "tenor" tones which are actually one octave apart from each other in spite of the fact that their notation is identical. This has produced a systematic error of one octave even in the material of Landois himself, and has caused much disagreement between the wingbeat frequency data obtained by various other methods and authors.

Most of the objective methods require the insect to be attached to a support and this increases the chances of obtaining "unnatural" results—although they do not always occur. High-speed photography can be used with minimum interference to the natural behavior of the insect[23], and it is also possible to register oscillographically the flight-sound of an insect in free flight[15, 44]. The simultaneous use of two objective methods to register both the visual wingbeat frequency stroboscopically and the acoustic flight-tone oscillographically has shown excellent agreement between the results obtained from both techniques, and this proves definitely that the fundamental frequency of the flight-sound is the same as the frequency of wingbeat[53].

The pitch of the flight-sound of insects with the lowest wingbeat frequencies is inaudible; the audible frequencies range from about 16 c/s to more than 1000 c/s, corresponding to the range between the note C of the 32-ft organ pipe and the note c two octaves above middle c or c^1 of the piano (*i.e.*, from C_2 to c^3 according to the European notation). By special measures this range can be extended by another octave, to c^4 (2200 c/s, three octaves above middle c)[47]. It is frequently imagined that the flight-sound pitch of small mosquitoes is so high that it approaches the upper limit of hearing (20,000 c/s) or even exceeds it, but this, however, is a gross exaggeration. Fig. 231 gives a survey of the flight-sound pitch of some common insects. The flight-tones of Lepidoptera and most Coleoptera correspond to the bottom octaves of the piano, and the flight-tones of Hymenoptera and Diptera to the middle octaves.

The flight-tone is generally specific for different species of insects, and also relatively constant. While in many cases its pitch in related species does not differ essentially, in other cases the species and even races of the same species can be distinguished by their flight-tone[39, 44, 46]. The normal variation of the flight-tone is generally within the range of one or two tones; in flies this range can be as large as one octave. As well as the normal variation, for which there is no apparent reason, the flight-tone may also vary in a number of cases where an obvious reason is responsible for it.

The wingbeat frequency shows some relationship to certain morphological characteristics of insects. Small insects have, as a rule, a high, and large insects a

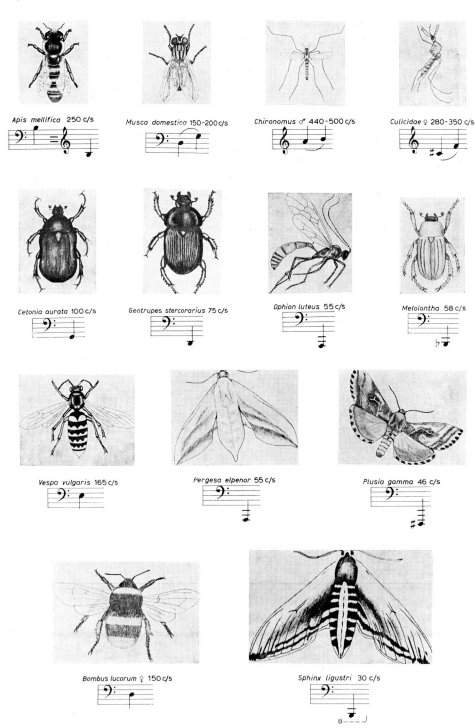

Fig. 231. The flight-sound pitch of some common insects.

low wingbeat frequency, and insects with proportionally short and/or light wings a higher wingbeat frequency than those with long and/or heavy wings. Also a small wingbeat amplitude is associated with a higher frequency than a large wingbeat amplitude. Several attempts have been made to work out by statistical or physical methods some laws relating the above characteristics[2, 24, 38, 44, 45]. For the present, the following formula seems to agree best with data obtained from random investigations of insects[45,50]:

$$n = k \cdot m^{0.50} \cdot I^{-0.33} \cdot f_0^{-0.67}$$

where n = wingbeat frequency, m = mass of the insect, I = moment of inertia of the wings, and f_0 = half-amplitude of wingbeat. When the cgs-system is used for the dimensions, the average value of k is about 25.5–26 (log k = 1.41). This relationship has been deduced from the formula relating the wing inertia and wingbeat frequency in wing-loading alteration experiments (see p. 378). It is apparent that several secondary factors are involved, so that it is difficult to improve this formula by statistical means. It seems to be valid only when the mass of thorax muscles is proportional to the mass of the insect, as assumed in the deduction of the formula.

3. Physiology of the Flight-sound

The dense rhythm of the wingbeat which is the origin of the flight-sound is generated by the action of the flight muscles. In insects with a direct flight mechanism, such as Odonata and Neuroptera, the depressing and elevating muscles of the wings are contracted alternatively; in insects with an indirect flight mechanism which include most of the other orders such as Diptera, Hymenoptera, Coleoptera, Lepidoptera and Hemiptera, the alternating opposing contractions take place between the dorsoventral and longitudinal thoracic muscles which act upon the wings by means of a lever system formed by the dorsal wall or roof (*scutum*) of the thorax and by arrangements in the wing joints[6]. In four-winged insects of this type the fore and hind wings are connected together by a hook mechanism (in Hymenoptera) or by the so-called frenulum (most Lepidoptera), or the action of fore and hind wings is in some other way linked together (part of Lepidoptera).

The physiology of the flight muscles which have to maintain such a rhythm shows some unique features. Each portion of the muscles is innervated by a single nerve fibre[35]. In insects with a low wingbeat frequency each stroke is separately controlled by a nerve impulse (synchronous type), while insects with high wingbeat frequency maintain a continuous contraction rhythm which is myogenic without receiving separate nerve impulses per stroke (asynchronous type)[35, 41]. This unique myogenic system which also appears in the tymbal muscle of the sound organs of the cicadas (Hemiptera Homoptera)[36] and which is capable of contraction with an extremely high frequency, has been discovered by Pringle. According to his present hypothesis[35–37], the contracting muscle stretches its antagonist and thus creates a tension in it, and either this tension or the sudden release of it induces a contraction in the antagonist which then acts retroactively upon the first muscle, thus completing the cycle. The nerve impulses control only the rhythm as a whole, its start, end, and "modulation".

The wingbeat frequency of an individual insect is very easily affected by a change in the mechanical load on the wing, and experiments show that the wing inertia is

References p. 389

one of the most important factors which affect the dynamics of the wing motion[44, 45]. When the wings were loaded with collodion or "sub-loaded" by cutting off pieces, it was found that the wingbeat frequency was approximately inversely proportional to the cube root of the moment of inertia of the wings (Fig. 232). In contrast to this, altering the air pressure or density has a varying effect on the wingbeat frequency, obvious only in insects of small size and/or low wingbeat frequency, while insects with large size and/or high wingbeat frequency are apparently unaffected by it.

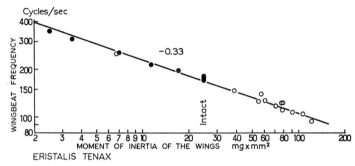

Fig. 232. The effect of wing inertia on the wingbeat frequency in a specimen of *Eristalis tenax*. ○, collodion loading; ●, "sub-loading" by mutilation. (Redrawn from Sotavalta[45].)

The wing cycle consists of a complex movement which cannot be expressed exactly by any simple physical function. If it is assumed that the asymmetries in the various directions of this movement are positive and negative and thus cancel each other to a certain extent, a simple harmonic rotary motion gives a rough approximation. Further assuming that there is a complete damping in each single downward and upward stroke, the power necessary to oscillate the wings would be, if n = wingbeat frequency, I = moment of inertia of the wings, and f_0 = half-amplitude of the wingbeat[45]:

$$P_i = 4 \cdot \pi^2 \cdot n^3 \cdot I \cdot f_0^2$$

whence, P_i and f_0 remaining constant, altering I changes n:

$$n \propto I^{-0.33}$$

which shows exact agreement with experimental results. Since the aerodynamic power output is apparently in certain cases only a fraction of P_i[45], it is understandable that the effect of air pressure or density changes on the wingbeat frequency may appear to be nonexistent, as shown above. Later investigation has paid attention to torque, click mechanism and other possible means of preserving wasted energy[37, 52], but this does not seem to be convincing since it provides no alternative explanation for the above cases, and also has been mainly based on observations of flight that in many respects differs from the average type of flight which occurs in insects.

On the basis of the evidence that the power input remains constant when the wing load is altered[48], the above experiments also indicate that the influence of the reflex relating the wing inertia to the operation of the flight motor system is insufficient to produce any effective regulation of power output[37], and also strongly suggest that, at least in larger insects with dimensions similar to bees and large flies, the efficiency of the wing motor system is very low[45]. This means that the majority of the energy expended is wasted in keeping the wings themselves in motion.

Attempts have recently been made to correlate the wingbeat frequency to the specific resonance frequency of a mechanical oscillator system possessing the dimensions of the wings and adjacent structures[55, 56, 57]. However, since these have so far been performed partly with dragonflies (*Odonata*), which are insects that do not

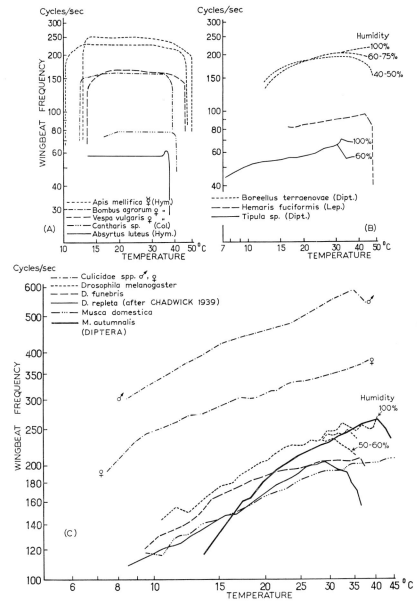

Fig. 233. The wingbeat frequency as a function of environmental temperature in certain insects. (Redrawn from Sotavalta[44].)

represent the average type of wing dimensions and flight energetics, or partly give rather incoherent results, it is best for the time being to wait for further investigations in this problem without taking a definite standpoint to these recent hypotheses. It is true that the author of one of these papers[56] has succeeded in conforming the dimensions of a great variety of insects and birds of all sizes under a common plan, and therefore, by analogy or otherwise, he has apparently closely approached the essence of the oscillating mechanism that lies in the background of the function of the wings.

Other factors that affect the wingbeat frequency and thus the pitch of the flight-tone are mainly physiological. It has been found that in small Diptera (*Drosophila*, mosquitoes) in particular, but also in certain larger flies (*Musca*, *Boreellus*), craneflies (*Tipula*) and in the moth *Hemaris* the temperature of the environment has a definite effect on the wingbeat frequency (Fig. 233 B–C), while in larger Hymenoptera (*Apis*, *Vespa*, etc.) and also in the beetle *Cantharis*, the frequency is not affected by the temperature (Fig. 233 A)[5, 10, 44, 46]. It should however be noted that in the former group the relative humidity can play an important role at temperatures higher than 30°C. In certain mosquitoes the flight-tone pitch is such a sensitive indicator of the environmental temperature that it is, at least in theory, possible to train oneself to determine the air temperature from the hum that mosquito swarms emit when flying in calm air[44].

However, the matter is not quite so simple. The effect of the temperature of environment appears to be indirect, through the internal temperature of the thorax, which is dependent upon the balance between the heat produced by the flight muscles and the heat dissipated by the body, which partly depends on the environmental temperature[49]. This balance naturally varies in different insects and under different circumstances. Fig. 234 shows the correlation between the wingbeat frequency and the temperature conditions in the honeybee (*Apis*) and in the bumblebee (*Bombus*) which are insects with effective thoracic muscles and a small relative surface area of the body, and in *Polistes*, an insect with less effective thoracic muscles and a large relative surface area.

The facts known about the thoracic temperature of insects only relate to species having the same size as bees, bumblebees, wasps, large flies and moths, since a temperature-measuring device small enough for mosquitoes, fruit-flies (*Drosophila*), and other small insects has not yet been constructed. Therefore as far as small insects are concerned we can make only certain suggestive inferences. Since the heat dissipation is proportional to the surface area of the insect (l^2) and the heat produced by the thoracic muscles is proportional to the insect's volume (l^3), the ratio heat dissipation : heat production will show considerably higher values in small insects than in large ones[10, 49]. Test observations on insects with largely varying weights indicate that the energy input appears to vary with the 1.4th power of the weight[48], and if the energy output (heat production) is proportional to the input, in two insects with 0.1:1 ratio of their linear dimensions there is a 167:1 ratio of heat dissipation : heat production in the smaller insect as compared to 1:1 ratio in the larger insect, under otherwise similar circumstances (wingbeat frequency, amplitude, etc.). It is therefore doubtful whether small insects in general are able to maintain any significant temperature difference between the thorax and the environment, and thus the effect of the environmental temperature is much more *direct* in small insects.

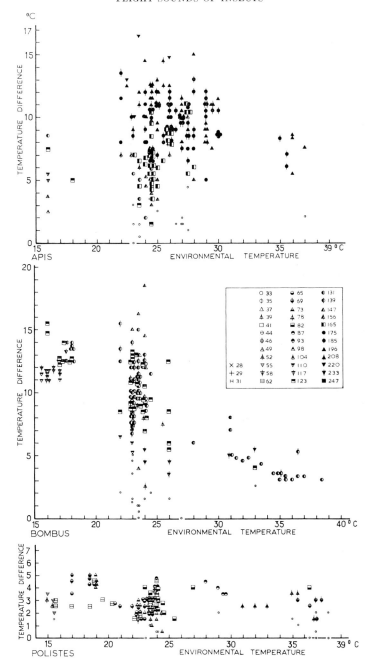

Fig. 234. The relationship between the environmental temperature, the temperature difference between the thorax and the environment, and the wingbeat frequency in *Apis*, *Bombus* and *Polistes*, and key to the symbols indicating wingbeat frequency. Small circles indicate temperature difference at rest. (From Sotavalta[49].)

Later survey of the material where oxygen consumption data exist[58] shows that the carbohydrate and fat consumers among flying animals are well comparable to

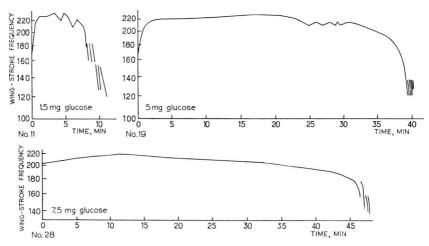

Fig. 235. Wingbeat frequency curves showing the onset of fatigue in three experiments on honeybees compared with the different amounts of food taken. (From Sotavalta[48].)

each other, and that their oxygen uptake per hour varies with about 1.2 power of the weight of the animal (insect, hummingbird), throughout a range of 1:10,000 weight variation. This ratio corresponds to a 33:1 ratio (instead of 167:1) of heat dissipation: heat production of a small animal as compared to a 1:1 ratio of a geometrically similar large animal having 10 times as large a linear dimension as the former.

Several large insects are able to raise their thoracic temperature before flight by an accelerating wing whirring of small amplitude, which can be heard as a soft sound of rising pitch. By this and other warming up methods, a rise of as much as 10°C/min can be obtained, the final thoracic temperature being as much as 15–20°C higher than the environmental temperature (cf. Fig. 234)[49]. Without pre-flight warming, such insects can generally produce only a faint flapping of the wings, and thus normal flight activity requires a minimum level of the thoracic temperature. If the temperature of the environment is well over 30°C, flight can start without preliminary warming-up.

Depending on the store of food immediately available during flight, fatigue occurs sooner or later if an insect flies without interruption. The first sign of this is a drop in the wingbeat frequency and in the sound-pitch which at the beginning is very slight but which increases relatively rapidly[5, 48]. Fig. 235 shows a typical wingbeat frequency pattern for the honeybee (*Apis mellifica*).

The insect is also able to modify the pitch of its flight-sound within a certain range "by its own free will", and insects such as angry bees generally emit a buzz higher than the normal[15]. The flight-tone is modified to a certain extent by altering the amplitude of the wingbeat; a decreased amplitude means an increased frequency and vice versa. This seems to be related to the ability of the insect to maneuver its flight by altering its flight speed or probably by shifting the power output between propulsion and sustainment of the flight[44, 45]. This modulation of the flight-tone readily appears in Diptera, and can vary in this case within as large a range as one octave, as already has been mentioned.

4. Analysis of the Flight-sound

The wing motion is not sinusoidal in form, but this does not exclude the possibility that the complex sound-wave generated can be divided into harmonic components having sinusoidal form. Since in the rotary movement of the wing there are no separate independent cyclic movements involved, such as occur in the elastic vibration of bells, rods, etc., the existence of possible inharmonic partials cannot have any essential significance in the flight-sound. The flight-sound is in any case a complex tone, and harmonic analysis therefore will give a completely satisfactory picture of its physical components. However by splitting up the complex tone into its partials the true intensity (loudness) relationships of the partials as they are subjectively heard, are not generally given. The relation between the intensity of the different partials is not yet completely understood, because the subjective loudness of the tones consists, in addition to the objective intensity, of many other factors among which the most important are the so-called combination tones and the acoustic sensitivity.

Fig. 236 shows oscillogram samples of the flight-tone of *Apis mellifica*, *Bombus lucorum*, *Chironomus* sp. and *Forcipomyia* sp.[44, 47]. The fundamental frequency of the samples varies between 198–235, about 130, 600–680 and 800–950 c/s, respectively. A Mader harmonic analysis of the 25 periods of the *Bombus* sample (Fig. 236 B) gives

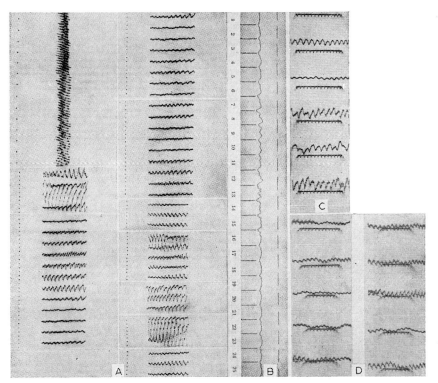

Fig. 236. Oscillograms of flight-sounds of insects in free flight. (A) *Apis mellifica*; (B) *Bombus lucorum*; (C) *Chironomus* sp.; (D) *Forcipomyia* sp. (From Sotavalta[44, 47].)

the values for the absolute amplitude of the first 9 partials as shown in Table 27[44]. The unit has only a relative value.

The intensity of a tone is dependent on the square of the vibration amplitude and frequency when the vibrating mass is the same. We can therefore obtain another

TABLE 27

Partial No.						Period No.							
	1	2	3	4	5	6	7	8	9	10	11	12	13
1	0.45	1.30	1.37	1.69	1.45	1.45	2.26	2.50	0.77	0.95	1.48	3.16	1.80
2	0.28	0.54	0.64	0.67	0.81	1.11	1.61	2.08	1.78	1.35	1.39	1.40	2.18
3	0.29	0.33	0.35	0.45	0.41	0.43	0.71	0.98	0.32	0.18	0.42	0.06	0.40
4	0.17	0.18	0.08	0.21	0.01	0.14	0.15	0.26	0.26	0.17	0.08	0.57	0.43
5	0.09	0.09	0.01	0.07	0.03	0.10	0.14	0.29	0.12	0.14	0.27	0.35	1.01
6	0.09	0.08	0.02	0.09	0.00	0.01	0.04	0.15	0.08	0.09	0.10	0.26	1.18
7	0.06	0 04	0.08	0.07	0.09	0.09	0.10	0.06	0.02	0.10	0.25	0.16	0.21
8	0.07	0.09	0.00	0.06	0.06	0.08	0.07	0.17	0.07	0.05	0.09	0.19	0.30
9	0.02	0.06	0.04	0.07	0.00	0.04	0.06	0.11	0.07	0.11	0.05	0.04	0.21
	14	15	16	17	18	19	20	21	22	23	24	25	
1	1.14	2.04	1.64	1.77	1.32	1.30	1.03	1.29	0.89	1.16	0.87	1.07	
2	1.48	1.72	0.60	0.32	0.39	0.33	0.29	0.38	0.31	0.48	0.31	0.34	
3	1.15	0.75	0.68	0.31	0.45	0.51	0.29	0.33	0.43	0.44	0.38	0.36	
4	0.50	0.47	0.43	0.31	0.36	0.26	0.22	0.20	0.26	0.24	0.18	0.12	
5	0.43	0.21	0.22	0.16	0.16	0.14	0.14	0.09	0.10	0.11	0.07	0.07	
6	0.13	0.10	0.13	0.09	0.12	0.09	0.10	0.03	0.10	0.04	0.08	0.07	
7	0.14	0.06	0.04	0.04	0.06	0.08	0.04	0.06	0.06	0.08	0.08	0.08	
8	0.17	0.07	0.13	0.01	0.10	0.05	0.07	0.05	0.05	0.05	0.04	0.06	
9	0.09	0.03	0.06	0.02	0.03	0.05	0.06	0.04	0.02	0.06	0.05	0.07	

TABLE 28

	1	2	3	4	5	6	7	8	9	10	11	12	13
1	1	1	1	1	1	1	1	1	1	1	1	1	1
2	0.55	0.69	0.87	0.63	**1.26**	**2.37**	**2.07**	**2.77**	**21.41**	**8.09**	**3.56**	0.78	5.89
3	**1.36**	0.58	0.58	0.64	0.72	0.80	0.91	1.38	1.55	0.32	0.73	0.00	0.45
4	0.82	0.31	0.05	0.25	0.00	0.15	0.07	0.17	1.83	0.51	0.05	0.52	0.92
5	0.36	0.12	0.00	0.04	0.01	0.12	0.10	0.34	0.61	0.54	0.84	0.31	7.91
6	0.52	0.14	0.01	0.10	0.00	0.00	0.01	0.13	0.39	0.32	0.17	0.24	**15.54**
7	0.32	0.05	0.16	0.08	0.19	0.19	0.10	0.03	0.03	0.54	1.41	0.13	0.67
8	0.29	0.31	0.00	0.08	0.11	0.19	0.06	0.30	0.52	0.18	0.24	0.23	1.79
9	0.05	0.17	0.07	0.14	0.00	0.06	0.06	0.16	0.68	1.09	0.09	0.01	1.11
	14	15	16	17	18	19	20	21	22	23	24	25	
1	1	1	1	1	1	1	1	1	1	1	1	1	
2	6.75	**2.84**	0.53	0.13	0.35	0.26	0.32	0.35	0.48	0.68	0.50	0.39	
3	**9.16**	1.21	**1.54**	0.28	1.05	**1.38**	0.71	0.59	**2.11**	**1.29**	**1.72**	**1.00**	
4	3.08	0.85	1.10	0.49	**1.19**	0.64	0.72	0.38	1.37	0.68	0.69	0.20	
5	3.56	0.26	0.45	0.20	0.37	0.29	0.46	0.12	0.32	0.22	0.16	0.10	
6	0.45	0.09	0.23	0.09	0.30	0.17	0.34	0.02	0.46	0.04	0.30	0.15	
7	0.74	0.04	0.03	0.03	0.10	0.18	0.08	0.11	0.23	0.23	0.41	0.26	
8	2.41	0.07	0.40	0.00	0.37	0.09	0.29	0.10	0.20	0.12	0.13	0.20	
9	0.51	0.02	0.11	0.01	0.04	0.12	0.27	0.08	0.04	0.21	0.26	0.34	

The loudest partial has been printed in boldface.

table by squaring the numbers of Table 27 and multiplying the results by the square of the ordinal of the partial in question. Taking the intensity of the fundamental in each period as unity, we obtain Table 28.

Since the combination tones and the acoustic sensitivity of the ear further distort the picture, the former by increasing the fundamental and lower partials unequally, the latter similarly increasing the loudness of the higher partials, this table can only give a suggestive description of the relative loudness of the partials.

The patterns of the sound-wave at various directions in space given out by the stationary flying insect differ considerably (Fig. 237)[53]. From oscillograms obtained in this way it is possible to proceed "backwards" to a certain degree and work out the details of the wing position and inclination in respect to the oscillogram, provided that this is correctly interpreted and attention is paid to the nature of the sound-wave and interference. In spite of the fact that the sound-wave and the airflow have the same origin, it is certainly unwise to equate the sound-wave with the airflow generated by the wings as they move through the air. The sound-wave proceeds spherically through the air at a speed of 330 m/sec irrespective of the speed of the beating wing, while the speed of the airflow is determined by the acceleration and speed of the wing, and its direction by the resultant of the acting forces that the wings develop while doing work against the air. In *Drosophila* with a wing length (r) of 2 mm, a wingbeat frequency (n) of 180 c/s and a stroke-angle of 120°, the linear speed of the wing-tip during a half-stroke is, as a rough approximation:

$$2n \times \frac{2\pi r}{3} = 360 \times \frac{2 \times 3.14 \times 0.2}{3} = 150 \text{ cm/sec}$$

which is 1/220 of the sound-wave speed. The difference in pattern of the sound-wave at various directions in space has a *direct* significance only when describing the effect of interference in the propagating sound-waves.

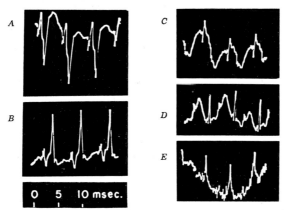

Fig. 237. The variation of form of the flight-sound pattern recorded at various space directions from *Drosophila funebris* in fastened flight. (A) anterior; (B) posterior; (C) dorsal; (D) ventral; (E) at right side. (From Williams and Galambos[53], reprinted in Chadwick[6].)

The intensity of the sound produced by a swarm of insects depends on various factors: the intensity of the sound of a single member of the swarm, the number of insects in the swarm, the distance of the swarm from the observer, the phase of the wingbeat cycle of the members of the swarm, and their wingbeat amplitude and

References p. 389

frequency, the latter also depending upon the auditory threshold of the observer's ear[14]. The insects in the swarm are assumed to fly with an uniform wingbeat frequency and amplitude, but naturally with different phases of the wingbeat cycle at a certain instant. Statistical compensation eliminates the effects of those insects which happen to be in opposite phases of the cycle, thus causing the "efficacious" number of insects to vary from time to time. The significance of the statistical compensation is largest when the number of insects in the swarm is highest. The problem is very complicated when absolute values of intensity are concerned, but is simplified considerably if a single insect is taken as unity and the calculations are made in relative terms. The intensity of the sound generally increases proportionally to the logarithm of the number of the members of the swarm (according to Fechner's Law), and decreases inversely proportionally to the square of the distance between the swarm and the observer. A swarm of 20 insects therefore emits the same intensity of sound at a distance of 2.27 m as a swarm of 200 insects at 7.18 m, as a swarm of 2000 insects at 22.7 m and as a single insect at a distance of 1 m^{14}.

5. Modification of the Flight-sound

Certain insects belonging to the orders Hymenoptera and Diptera are able to produce a shrill and high sound when they are not flying. A "squeal" is emitted by bumblebees when collecting nectar or pollen from a flower, and by bumblebees, Syrphid flies and mosquitoes when held between fingers or forceps so that the wing movements are impeded. Similar sounds are also produced by a tired bumblebee unwilling to fly, and by a Syrphid fly resting on a leaf or twig between flights. Furthermore, the "dialogue" of the queen bees in the hive also uses this kind of sounds. The pitch of these sounds is approximately one octave higher than the pitch of the usual flight-tone of the species in question[15, 21, 44]. During this sound-production the wings and the thorax are held in a state of fine tremor and in a "singing" Syrphid nothing other than the wing-bases are vibrating. The same muscles that by their normal activity produce the wing-beats and the normal flight-sound, are also responsible for this "squeal", and they are apparently able to maintain a rapid rhythm of partial contractions almost without any visible motion. It is as impossible for this sound, as for the normal flight-tone, to be produced by blowing air through the tracheal stigmata or by the halteres, although this is frequently believed. The latter, moreover, are absent in Hymenoptera, and can be cut off in Diptera without producing any effect on the "squeals". Certain Hymenoptera are able to perform the pre-flight warming up by some kind of direct oxidative process in the thoracic muscles without producing any audible sound or visible contraction rhythm at all[49, 51], but if such an insect is disturbed during this process, it emits a regular "squeal" which shows a rising pitch which is correlated to the thoracic temperature[49]. Since the "singing" tone emitted by a resting Syrphid also rises in pitch, it is apparently by this means that the insect raises the thoracic temperature to the required level for normal flight. Moreover, similar observations have been recorded in Dytiscid beetles during the preparatory process for flight[13, 16].

6. Biological Significance of the Flight-sound

It has been much discussed whether the flight-sounds of insects play any role in the biological relationships between these insects and other insects and animals.

Much research has been done on the auditive powers of various insects, and this has led to conclusions which vary from case to case. Simple experiments show that honeybees cannot hear during flight[8] and are not affected by the hum of their fellows, and this has also been confirmed by more extensive experimental investigations[15]. However, bees have a keen sense of vibration and can perceive sound-waves which are propagated through a solid support under their legs[15]. On the other hand, male mosquitoes have a considerable sense of hearing in their antennae[27, 40, 42], and can use this ability to find their mate by her hum, as well as to keep together in all-male swarms. The swarm can be gathered together by sounding a tone of appropriate pitch, a fact that was known already to Landois, and its members can be similarly induced into copulatory reflexes when no female is present[42]. It is also easy to see the vibration of the antennae evoked as a response to a tone in an active male mosquito fixed under a microscope. Buzzing telephone wires, arc lamps and various sound-producing electrical equipment have been observed similarly to attract mosquito swarms, while loud discontinuous sounds and noises such as an engine whistling or the oar clatter from a boat, affect the swarm to the extent of putting it into disorder and even finally scattering it[7, 18, 30, 32]. The use of the method of sound attraction has also been tried for the destruction of large numbers of male mosquitoes by "electrocution" in heavily mosquito-infested areas[19, 31], but apart from the initial experiments the results seem to have been rather meagre.

The above phenomena are confirmed by recent investigations[54] showing that male mosquitoes of *Aëdes aegypti* are easily attracted by sounds having the frequency of 300–700 c/s, and that maximum attraction was obtained with the sound intensity of 68–85 dB at one inch distance. When the intensity exceeded 100 dB, copulatory reflexes were observed. The frequency of the fundamental is the only significant part of the sound, since when the fundamental was eliminated by filtering, the plain harmonics had no effect. The sounds of other mosquito species, when brought to the appropriate frequency range, evoked a similar response. The mosquitoes can also locate an attractive sound in the presence of considerable background noise, but this ability breaks down if the background is too loud (over 90 dB). The sound receptors seem also to adapt themselves to only one frequency at the same time, since if two simultaneous sounds of different frequencies within the appropriate frequency range had the same intensity, the mosquitoes failed to respond.

Certain male Braconid wasps emit a characteristic type of sound with their wings during the pre-copulatory rites when they are in the immediate neighborhood of the female (Fig. 238)[44], but it is not yet determined whether this has any acoustical significance for the female. A similar behavior appears in Syrphid males, which "court" by flying with a high-pitched hum straight above a female on the flower, and in certain bees (*Megachile, Anthophora, Anthidium*) that behave likewise[29].

Fig. 238. Sound-pattern of two "courting" Braconid males. (From Sotavalta[44].)

The triungulin larva of the Meloid beetle *Hornia minutipennis* which lives parasitically in the nests of wild bees, is induced to great motile activity and clasping reflexes by the flight-tone of its prospective host as it approaches the flower where the triungulins are waiting, and thus clear auditory powers are shown[17]. Since other Meloid beetles have similar life histories, it is most likely that they behave also in a similar manner.

Several authors have recorded observations on the role that the flight-sound of certain insects can play in their ecological relationships to other animals. Gad-flies, horse-flies, bot-flies and warble-flies are said to cause a negative response in cows and other animals when flying around them and trying to bite or deposit their eggs[3, 21, 43]; certain Lepidopterous caterpillars show a similar negative response to the parasitic flies that pursue them[44]. Flying nocturnal insects cause a positive response in bats that catch their prey in flight[22] and that are also otherwise attracted by whizzing sounds. A further case of the possible ecological significance of the flight-sound may be the audio mimicry, which can be an adjunct to color mimicry[11]. In some, although definitely not in all, of the well-known cases of color mimicry when a defenseless insect resembles in its appearance an aculeate or other wasp, the flight-tones are also close to each other in pitch. This is the case for certain Syrphid and Bombyliid flies as well as for Aegeriid moths. Perhaps the commonest example is the drone fly (*Eristalis tenax*), the appearance of which in flight much resembles a honeybee, and which also has a flight-tone pitch only a few tones lower than the bee's hum.

A curious controversy occurred about the significance of the flight-tone associated with the appearance of "trumpeters" in bumblebee colonies. Already Camerarius and Goedart observed the existence of bumblebee specimens which buzzed loudly at or on their nest early in the morning[4, 12], and they considered this phenomenon to be a regular morning "reveille" calling other members of the nest to work. This was once more observed by a later author[34], before being sentenced, by a word from the great Réaumur, to the realm of nonsense for more than a hundred years. However, the phenomenon was rediscovered by Angus about a hundred years ago[33], and it was then given another interpretation: that of ventilatory function, commonly known in honeybees. Later observers were inclined to again revive the "reveille" idea of the old authors and the phenomenon became a matter of controversy and polemics, until it was definitely proved that the phenomenon really served for the ventilation of an overheated nest[33].

The flight-sound of insects is primarily an involuntary consequence of the wingbeat frequency which is rather delicately balanced with regard to the mechanical and physiological system of the flight movements. Any alteration which occurs in the factors involved in the system, is therefore sensitively indicated by the wingbeat frequency and the flight-sound. Thus to a certain extent the investigation of the flight-sound is a way of quantitative approach to the study of insect flight. The above observations show that in certain cases the flight-sound also can have a biological significance either for the insect itself and for the individuals within the circle of its own species, or for the ecological system of which the producer of the sound is an inseparable part.

REFERENCES

[1] ANON, (S. R.), 1799, On the vibration of the wings of a fly. *J. of nat. Philos., Chem. and Arts*, ed. by NICHOLSON, *3*, 35-36.
[2] ATTILA, U., 1947, Betrachtung des Flügelschlags bei Insekten an Hand eines physikalischen Modells, *Acta entom. fenn*, *5*, 1-9.
[3] BURMEISTER, 1832, *Handbuch der Entomologie*, I, 696 pp.
[4] CAMERARIUS, J. R., 1627, Apum industria, justicia, cura, munia. Apum murmur unde. *Syllog. memorab. med. et mirab.*, *9*, 1-117.
[5] CHADWICK, L. E., 1939, Some factors which affect the rate of movement of the wings in Drosophila, *Physiol. Zoöl.*, *12*, 151-160.
[6] CHADWICK, L. E., 1953, The motion of the wings, *Insect Physiology*, ed. by K. D. ROEDER, pp. 577-614.
[7] CHILD, C. M., 1894, Ein bisher wenig beachtetes antennales Sinnesorgan der Insekten, mit besonderer Berücksichtigung der Culiciden und Chironomiden. *Z. wiss. Zool.*, *58*, 475-528.
[8] FRISCH, K. v., 1923, Über die "Sprache" der Bienen. *Zool. Jb. Physiol.*, *40*, 1-186.
[9] GARDINER, W., 1832, *Music of Nature*, London, 530 pp.
[10] GAUL, A. T., 1951, A relation between temperature and wing beats, *Bull. Brooklyn Entom. Soc.*, *46*, 131-133.
[11] GAUL, A. T., 1952, Audio mimicry: an adjunct to color mimicry, *Psyche*, *59*, 82-83.
[12] GOEDART (GOEDARTIUS), J., 1667, *Metamorphoseos et historiae naturalis II; De insectis*. Medioburgi, p. 186 ff.; 1685, *De insectis in methodum redactus; cum notularum additione*, Londini, 356 + 45 pp.
[13] GRIFFINI, A., 1896, Observations sur le vol de quelques dytiscides et sur les phénomènes qui le précèdent, *Arch. ital. biol.*, *25*, 326-331.
[14] GUYE, C. E., 1937, Sur le bruit que produit un essaim d'insectes bourdonnants, *Arch. sci. phys. nat.*, *19*, 53-70.
[15] HANSSON, Å., 1945, Lauterzeugung und Lautauffassungsvermögen der Bienen, *Opusc. entom.*, Suppl. 6, 1-124.
[16] HIRSCH, J., 1904, Die Lautäusserungen der Käfer, *Soc. entom.*, *19*, 82-83, 89-91, 97.
[17] HOCKING, B., 1949, *Hornia minutipennis* Riley: a new record and some notes on behaviour, *Can. Entomol.*, *81* (3), 1-6.
[18] HOWARD, L. O., 1901, *Mosquitoes, how they live*, London-New York, p. 15.
[19] KAHN, M. C. and W. OFFENHAUSER JR., 1949, The first field tests of recorded mosquito sounds used for mosquito destruction, *Am. J. Trop. Med.*, *29*, 811-825.
[20] LANDOIS, H., 1867, Die Ton- und Stimmapparate der Insekten in anatomisch-physiologischer und akustischer Beziehung, *Z. wiss. Zool.*, *17*, 105-184.
[21] LANDOIS, H., 1874, *Thierstimmen*, Freiburg i. Br., 229 pp.
[22] LANE, F. W., 1941, Flight in nature, *Flight*, *40*, 316-318.
[23] MAGNAN, A., 1934, *La locomoticn chez les animaux, Le vol des insectes*, Paris, 186 pp.
[24] MAGNAN, A. and A. SAINTE-LAGUË, 1933, Le vol au point fixe, *Act. sci. ind.*, *60*; *Exp. morph. dyn.*, *4*, 31 pp.
[25] MAREY, E. J., 1868, Détermination expérimentale du mouvement des ailes des insectes pendant le vol, *Compt. rend. Acad. Sci.*, *67*, 1341-1345.
[26] MAREY, E. J., 1869, Mémoires sur le vol des insectes et des oiseaux I, *Ann. sci. nat. Zool.*, *5* (12), 49-150.
[27] MAYER, A. M., 1874, Researches on acoustics. Experiments on the supposed auditory apparatus of the *Culex* mosquito, *Am. J. Sci.*, *8*, 81-103, reprinted in *Am. Nat.*, *8*, 577-592; *Ann. nat. hist.*, *15* (1875) 349-364.
[28] MÜHLHÄUSER, F. A., 1866, Ueber das Fliegen der Insecten, *Jber. Pollichia*, *22-24*, 37-42.
[29] MÜLLER, H., 1872, Anwendung der Darwin'schen Lehre auf Bienen, *Verh. naturh. Ver. Rheinl.*, *29*, 1-96.
[30] NUTTALL, G. H. F. and A. E. SHIPLEY, 1902, Studies in relation to malaria II, *J. Hyg.*, *2*, 58-84.
[31] OFFENHAUSER, W. H. and M. C. KAHN, 1949, The sounds of disease-carrying mosquitoes, *J. Acous. Soc. Am.*, *21*, 259-263.
[32] OSTEN-SACKEN, 1861, Entomologische Notizen, *Entom. Ztg. Stettin*, *22*, 51-55.
[33] PLATH, O. E., 1923, Observations on the so-called trumpeter in bumblebee colonies, *Psyche*, *30*, 146-154.
[34] PLUCHE, N. A. DE, 1735/50, *Le spectacle de la nature I*, Amsterdam-Paris, 560 pp.
[35] PRINGLE, J. W. S., 1949, The excitation and contraction of the flight muscles of insects, *J. Physiol.*, *108*, 226-232.
[36] PRINGLE, J. W. S., 1954, The mechanism of the myogenic rhythm of certain insect striated muscles, *J. Physiol.*, *124*, 269-291.

[37] PRINGLE, J. W. S., 1957, *Insect flight*, Cambridge Univ. Press, 133 pp. (Also further literature refs.)
[38] PROCHNOW, O., 1907, *Die Lautapparate der Insekten*, Guben, 175 pp. (*Int. entom. Z.*)
[39] REED, S. C., C. M. WILLIAMS and L. E. CHADWICK, 1942, Frequency of wing-beat as a character for separating species, races and geographic varieties of Drosophila, *Genetics*, *27*, 349–361.
[40] RISLER, H., 1953, Das Gehörorgan der Männchen von *Anopheles stephensi* Liston (Culicidae), *Zool. Jb. Anat.*, *73*, 165–186.
[41] ROEDER, K. D., 1951, Movements of the thorax and potential changes in the thoracic muscles of insects during flight, *Biol. Bull. Woods Hole*, *100*, 95–106.
[42] ROTH, L. M., 1948, A study of mosquito behavior. An experimental laboratory study of the sexual behavior of *Aedes aegypti* (L.), *Am. Midland Nat.*, *40*, 265–352.
[43] RUDOW, 1896, Die Töne, welche Insekten hervorbringen, *Ins. Börse*, pp. 79–81.
[44] SOTAVALTA, O., 1947, The flight-tone (wing-stroke frequency) of insects, *Acta entom. fenn.*, *4*, 1–117. (Also further refs.)
[45] SOTAVALTA, O., 1952, The essential factor regulating the wing-stroke frequency of insects in wing mutilation and loading experiments and in experiments at subatmospheric pressure, *Ann. Zool. Soc. Vanamo*, *15* (2), 1–67.
[46] SOTAVALTA, O., 1952, On the difference and variation of the wing-stroke frequency in wing mutants of *Drosophila melanogaster* Mg. (Dipt., Drosophilidae), *Ann. entom. fenn.*, *18*, 57–64.
[47] SOTAVALTA, O., 1953, Recordings of high wing-stroke and thoracic vibration frequency in some midges, *Biol. Bull. Woods Hole*, *104*, 439–444.
[48] SOTAVALTA, O., 1954, On the fuel consumption of the honeybee (*Apis mellifica* L.) in flight experiments, *Ann. Zool. Soc. Vanamo*, *16* (5), 1–27.
[49] SOTAVALTA, O., 1954, On the thoracic temperature of insects in flight, *Ann. Zool. Soc. Vanamo*, *16* (8), 1–22.
[50] SOTAVALTA, O., 1954, The effect of wing inertia on the wing-stroke frequency of moths, dragonflies and cockroach, *Ann. entom. fenn*, *20*, 93–101.
[51] WEIS-FOGH, T., 1953, The ventilatory mechanism during flight of insects in relation to the call for oxygen, *16th Int. Zool. Congr.*, Copenhagen. (Abstract.)
[52] WEIS-FOGH, T. and M. JENSEN, 1956, Biology and physics of locust flight I, Basic principles in insect flight, A critical review. *Phil. Trans. Roy. Soc. London B*, *239*, 415–458.
[53] WILLIAMS, C. M. and R. GALAMBOS, 1950, Oscilloscopic and stroboscopic analysis of the flight sounds of Drosophila, *Biol. Bull. Woods Hole*, *99*, 300–307.
[54] WISHART, G. and D. F. RIORDAN, 1959, Flight responses to various sounds by adult males of *Aedes aegypti* (L.) (Diptera: Culicidae), *Can. Entomol.*, *91*, 181–191.
[55] DANZER, A., 1956, Der Flugapparat der Dipteren als Resonanzsystem, *Z. vergl. Physiol.*, *38*, 259–283.
[56] GREENEWALT, C. H., 1960, The wings of insects and birds as mechanical oscillators, *Proc. Amer. Philos. Soc.*, *104*, 605–611.
[57] RUSSENBERGER, H. and M. RUSSENBERGER, 1960, Bau und Wirkungsweise des Flugapparates von Libellen, *Mitt. Naturf. Ges. Schaffhausen*, *27*, 1–88.
[58] SOTAVALTA, O. and E. LAULAJAINEN, 1961, On the sugar consumption of the drone fly (*Eristalis tenax* L.) in flight experiments, *Ann. Acad. Sci. Fenn. A IV*, *53*, 1–25.

CHAPTER 14

ACOUSTICAL BEHAVIOUR OF HEMIPTERA

by

D. LESTON AND J. W. S. PRINGLE

1. Introduction

The Hemiptera are commonly supposed to have arisen in the lower Permian from a Psocopteran-like ancestor with a unique feeding mechanism, confined to a fluid diet[5,11]. Despite this restriction the type proliferated, and Hemiptera now show as wide a diversity of form and occupy as wide a variety of habitats as any insect order. It is this proliferation, successful whether measured by survival time, number of species or number of individuals, that compels any study of hemipteran activity to be viewed in an evolutionary context. The widespread but fragmented occurrence of stridulation, the many different stridulatory mechanisms, the various functions subserved by sound production—these no less than the more obvious differences between shield bug and aphid, waterbug and cicada, demonstrate the extraordinary plasticity of the hemipter.

Modern instrumental analysis has revealed two distinct forms of song within the terrestrial Heteroptera, which are associated with two classes of behaviour. The songs were categorized by Leston[26] as extraspecific communication and intraspecific communication respectively, and this dichotomy is now shown to be of wider application. In this chapter the meaning of the terms is as follows. Intraspecific communication involves songs such as courtship, aggregational or mating calls, often different in the two sexes, usually absent in larvae, exhibiting a regular pulse repetition frequency or a precise pulse pattern. Extra-specific communication involves songs such as warning or defensive calls, usually similar in both sexes of adult and sometimes present in the larva, not exhibiting any form of regularity in pulse repetition. The former are manifested either as the result of internal drives or in response to external stimuli, the latter only in response to external stimuli.

In addition to this categorization of songs by function and sound structure, which has wide applicability among the insects and parallels Marler's[29] classification of bird songs, it is possible within the Hemiptera to adopt an alternative classification based on the physical mechanism of sound production. With few exceptions sound production (stridulation) in insects makes use of the elastic properties of the exoskeleton. By special structural modifications this elasticity allows an otherwise slow movement to be transformed into a rapid click movement (or a series of clicks) containing frequencies high enough to excite sound vibrations. In the more usual method, strigilation, one part of the hard exoskeleton, the plectrum, is rubbed over another, the strigil, and the clicks or pulses are generated either by friction or by the action of the ridges of the strigil. In the other method, apparently peculiar to the Hemiptera, the clicks are generated by the sudden distortion of a doubly-curved cuticular surface from one configuration to another, the energy being supplied by a muscle or muscles inserted directly on the cuticle near the sound-producing region or tymbal[30, 35, 39]. Both these basically different mechanisms are found within the Hemiptera, and each appears to have arisen more than once in evolution. Care is therefore needed in the use of apparent homologies in taxonomic arguments.

In all hemipteran songs the sound is highly modulated and is usually broken up into discrete pulses, interspersed by periods of silence, though the high repetition frequency of the pulses masks this sound structure from the human ear and makes it necessary to use an oscilloscope or other recording instrument in order to detect the true nature of the song. Nothing has so far been described within this order resembling the song of the Gryllidae, in which the frequency of contact of the teeth of the strigil is the fundamental frequency of the sound[36]. The fundamental frequency of the sound in hemipteran song is the resonant frequency of the excited cuticular structure, and the pulse or modulation frequency is that of the rate of contact of the teeth of the strigil with the plectral surface on another part of the cuticle or, in the second method of sound generation, of the buckling of the tymbal. Full spectral analyses have been published only for a few cicadas. Audiospectrograms[3] of the sounds of *Magicicada septemdecim* and *M. cassinii* show that the former emits a reasonably pure tone at about 1500 c/s, while the latter's song contains frequencies ranging from 400 to 9000 c/s; frequencies from 4000 to 7000 c/s have also been found in the songs of nine species of large cicada from Ceylon[35]. The abdominal resonator of the male cicada is probably responsible for the narrow-band emission of many of these insects, and both oscillograms and descriptions of the subjective impression of many other songs suggest a wider frequency spectrum in most Hemiptera.

From the rather limited evidence available, the pulse repetition frequency or the pulse pattern in intraspecific songs appears to be remarkably constant within the species, and to vary from a few per second in some of the smaller Homoptera[31], Pentatomidae[21], Cydnidae, Lygaeidae and Piesmidae[19] up to 500–600 per second in some cicadas[35].

Organs for sound reception, where known, will be described in the separate sections. It is a general feature of sound reception in insects[38] that the information relayed to the central nervous system up the sensory nerves relates to the pulse patterns of the received sound and not to the fundamental frequency. This makes it understandable that the specific character of intraspecific communication in insects resides in the pulse pattern. The fundamental sound frequency acts as a carrier wave for this pulse information, and the receptor may be more or less sharply tuned to this fundamental so as to achieve greater sensitivity and selectivity. In many of the smaller Hemiptera with less highly evolved mechanisms than those of the cicadas there may be little evidence of a sound carrier wave, and in these cases any organ sensitive to vibrations is a potential receptor for the song; this consideration probably accounts for the failure so far to locate the receptor in some of the smaller species.

2. Homoptera

Recent Homoptera are conveniently divided into Auchenorrhyncha (cicadas, membracids, cicadellids, cercopids, fulgorids, etc.) and Sternorrhyncha (psyllids, aleurodids, aphids and coccids), though some authorities would deny the later monophyleticity. In addition there is the family Peloridiidae, usually ranking as the separate series Coleorrhyncha[34]. Within the suborder sound production has been described, or implied on anatomical grounds, in most of the Auchenorrhyncha[31], and in one aphid genus[10] and certain psyllids[32] among the Sternorrhyncha.

(a) Auchenorrhyncha

(i) Anatomy

Until recently it was supposed that cicadas alone among Homoptera were capable of sound production, but in the last ten years it has become apparent that intraspecific communication is the rule throughout the Auchenorrhyncha and that there is a true homology in the anatomy of the apparatus for sound production throughout the series[31, 33]. In addition to the numerous Eurymelidae, Cicadellidae, Membracidae, Cercopidae, Cicadidae and Fulgoroidea, there are a number of relict families some species of which have survived with little change from early Mesozoic times[6, 12]. These, particularly Tettigarctidae and Aetalionidae, give clues to the probable nature of the primitive mechanism.

The organs for sound production are located in the 1st abdominal segment, the lateral part of the tergum of which is modified to form the striated tymbal (Fig. 239). Details of the muscles and nerves of this region in the male of *Tettigarcta tomentosa*

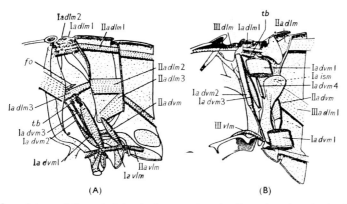

Fig. 239. Musculature of the 1st abdominal segment and adjacent regions in Auchenorrhyncha. (A) *Neophilaenus campestris* (Fall.) ♂ (Cercopidae), (B) *Aphrodes bicinctus* (Schrnk.) ♂ (Euscelidae, celidae).

III = metathorax
Ia, IIa, IIIa = 1st, 2nd, 3rd abdominal segments
dlm = dorsal longitudinal muscle
dvm = dorso-ventral muscle

ism = intersegmental muscle
vlm = ventral longitudinal muscle
fo = internal strengthening of tergal plate
tb = striated tymbal

are shown in Fig. 240. Of the various dorsoventral muscles present in the 1st abdominal segment, one (muscle $Ia\ dvm_1$ of Ossiannilsson[31]) appears to be the functional tymbal muscle whose contraction buckles the tymbal and supplies energy for sound production. One, two, or three other dorsoventral muscles may occur in association with the tymbal and inserted on its anterior edge. In *Tettigarcta* and in cicadas there is only one complex accessory muscle, which is labelled "tensor muscle" in Figs. 240

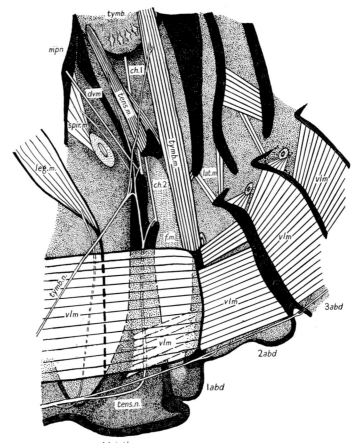

Metathorax

Fig. 240. *Tettigarcta tomentosa*, ♂. Details of muscles and nerves in the ventral part of the metathorax and abdomen (from Pringle[37]).

ch. 1, ch. 2 = chordotonal organs 1 and 2
f.m. = folded membrane
mpn = metapostnotum
tymb. = striated tymbal
1abd, 2abd, 3abd = abdominal segments
dvm = dorso-ventral muscle
lat.m. = lateral abdominal muscles
leg.m. = metathoracic leg muscle
spir.m. = spiracular muscle
tens.m. = tensor muscle
tymb.m. = tymbal muscle
vlm = ventral longitudinal muscles
tens.n. = tensor nerve
tymb.n. = tymbal nerve

Fig. 241. The muscles and nerves concerned in sound production and reception in *Platypleura capitata* (Oliv.) (Cicadidae). The tymbal muscle has been cut across near its base and the abdominal, auditory and leg nerves are cut near the ganglion (from Pringle[35]).

chit.V = base of chitinous V
ch. 1, ch. 2 = chordotonal organs 1 and 2
f.m. = folded membranes
lat. = lateral arm of chitinous V
III leg = metathoracic leg
d.m. = dorsal muscles
det.m. = detensor tympani muscle
dors.abd.m. = dorsal longitudinal abdominal muscles
dvm. = dorso-ventral muscle
leg m = metathoracic leg muscles
m.int.mes. = musculus intersegmentalis mesothoracicus
spir.m. = spiracular muscle
tens.m. = tensor muscle
vent.abd.m. = ventral longitudinal abdominal muscles
vent.m. = ventral longitudinal muscle of metathorax
w.m. = metathoracic direct wing muscles
aud.n. = auditory nerve
tens.n. = tensor nerve
tymb.n. = tymbal nerve

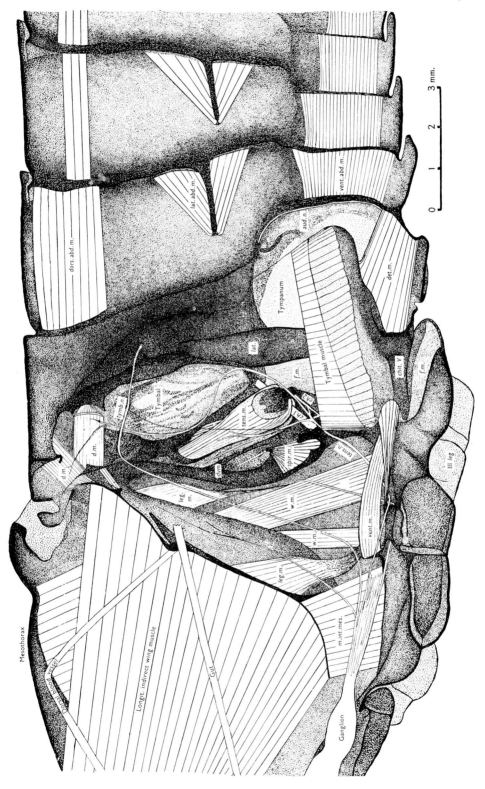

and 241 on the view that its tonic contraction controls the stress in the tymbal and so modifies the sound produced; this has been verified experimentally in cicadas[35]. In various Cercopoidea, Cicadelloidea and Fulgoroidea a dorsoventral muscle of the 2nd abdominal segment and various dorsal and ventral longitudinal muscles of both the 1st and 2nd abdominal segments appear to be enlarged and may also be functionally associated with the mechanism of sound production[31], and in a few cases where muscle $Ia\ dvm_1$ is absent it is possible that some of these may have taken over its function. In nearly all the Homoptera the essential apparatus is present in both sexes, though it is often better developed in the male. The female call has been described in a number of Cicadelloidea and Cercopidae and occasionally (for example *Paropia scanica*) the female has the better-developed musculature. In *Aetalion reticulatum* (Aetalionidae) a tymbal and associated muscles are found in all nymphal instars except the first[12]. It is only in the Cicadidae that the female lacks the means of sound production and that sexual dimorphism is fully evolved.

In the male cicada muscle $Ia\ dvm_1$ (the tymbal muscle) is very large, and the anatomy of the anterior abdominal segments becomes highly modified to accommodate it (Fig. 241). The 1st abdominal sternum, which bears the ventral insertion of the tymbal muscle, is greatly strengthened and develops lateral apophyses which form the "chitinous V" of earlier authors[30]; in the mid-ventral line two other apophyses meet and fuse in the higher cicadas to enclose a ventral canal containing the abdominal nerves (Fig. 242). The tymbal is also more elaborately folded and is surrounded by a thickened rim which isolates it mechanically from the surrounding cuticle; this and

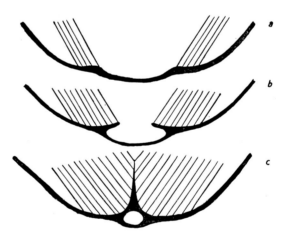

Fig. 242. Evolution of the ventral attachment of the tymbal muscle in cicadas. (a) primitive condition, as in *Tettigarcta*, (b) intermediate condition found in some cicadas, (c) condition in *Platypleura* (from Pringle[37]).

the presence of the large abdominal air sacs contribute to the great efficiency of the organ for sound production and result in a volume of sound which is very much greater than in any other type of insect.

An accessory stridulating apparatus of the strigil and plectrum type is found in the cicada subfamily Tettigadinae[30]. It consists in both sexes of a striate area on the lateral angle of the mesonotum, which is said to make contact with the posterior edge of the tegmen.

(ii) Physiology of sound production

The tymbal method of sound production yields a song which consists essentially of double pulses, the resonant vibrations of the cuticular structure at the IN and at the OUT click of the tymbal. This double-pulse structure becomes apparent in oscillograph records (Fig. 243A) made in the laboratory, but may be masked in field recordings by echoes and by inequality in the intensity of sound production produced by the two tymbal movements. Thus in *Platypleura capitata* (Cicadidae) contraction of the tensor muscle increases the intensity of sound produced by the IN click; the apparently single pulses of the natural song of some cicadas (Fig. 243B) are to be interpreted as IN clicks of the tymbal strained by tonic contraction of the tensor muscle to such an extent that the OUT click is undetectable[35].

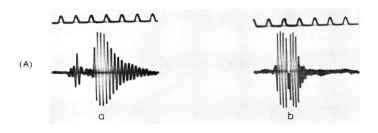

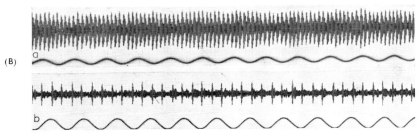

Fig. 243. (A) Oscillograms of the double pulse of sound produced by the IN–OUT movements of the tymbal in cicadas. (a) *Platypleura capitata*, (b) *P. octoguttata*. Time marker, 1 msec. (B) Oscillograms of the aggregational or common song of cicadas. (a) *Platypleura capitata*, (b) *Terpnosia ransonetti*, Time marker, 50 c/s (from Pringle[35]).

The number of cycles of sound vibration produced at each click of the tymbal depends on the damping of the cuticular or tracheal resonator, and is naturally much greater in cicadas which have an abdominal air sac than in the smaller Auchenorrhyncha which lack this structure. In many of the Cercopids and Cicadellids the damping is so high that not more than one or two rapidly decrementing cycles of sound vibration are produced per tymbal click, so that the oscillogram of the song is little more than a series of pulse transients. This may be of little significance to the insect since the specific information is carried by the pulse pattern, but the distance

References p. 410

over which communication is effective must be much less in these cases, since the sound receptor cannot be sharply tuned to a particular carrier frequency.

The apparatus for sound production in cicadas consists of three functional components; the tymbal and tymbal muscle generating the sound, the tensor muscle straining the tymbal and increasing the sound intensity and the remaining accessory musculature altering the shape of the abdomen and controlling the resonance. In the genera *Platypleura*[35] and *Meimuna*[16] the rhythm of contraction of the tymbal muscle is myogenic; that is, it is determined by the mechanical loading of the muscle and not by the time of arrival of motor nerve impulses. In these insects the tensor muscle, by altering the loading of the tymbal muscle, also controls the pulse repetition frequency. In certain other genera, including *Magicicada*[8], *Graptopsaltria*, *Tanna* and *Oncotympana*[15] the rhythm of contraction of the tymbal muscle is neurogenic; that is, it is determined by the time of arrival of motor nerve impulses. Information is urgently needed* on the distribution of this physiological property in the family, since it is an essential component in any understanding of the mechanism of patterning of the song. The elaborate nature of this patterning in some species of cicada is illustrated in Fig. 245, which shows the aggregational songs of nine species from Ceylon. A complex patterning is also found in the songs of some of the smaller Homoptera.

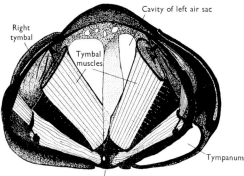

Fig. 244. Anterior view of the isolated abdomen of a male cicada; the tymbal muscles are easily exposed for direct stimulation (from Pringle[35]).

The central nervous organisation for sound production in cicadas is located in special "sound-production lobes" of the composite thoracic ganglion[7]. A recent investigation[17] has been made of the neural mechanism of activity in the sound reflex of the neurogenic species, *Graptopsaltria nigrofuscata*. Sound is evoked by stimulation of hair sensilla on the body surface and can also be produced by electrical excitation of caudal nerves. In this species the two tymbal muscles, left and right, are excited alternatively at 100 contractions/sec, giving a sound pulse repetition frequency of 200/sec. The rhythm appears to originate in an internuncial neuron which fires at 200/sec and drives the motor neurons on each side by alternate impulses.

---

* A simple experiment consists in stimulating the isolated abdomen (Fig. 244) with about 4 V of alternating current (50 or 60 c/s) from a bell transformer. If the sound produced is heard to have a frequency of 50 or 60 c/s, the mechanism is neurogenic; if a higher frequency is heard the mechanism is myogenic. The second author would be glad to learn the results of any experiments carried out on cicadas which have been identified or of which a specimen can be forwarded.

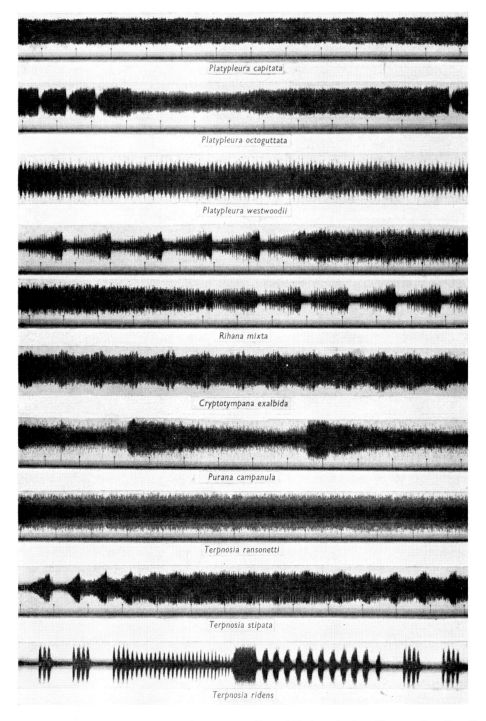

Fig. 245. Oscillograms from magnetic tape recording of Ceylon cicadas. Time marker (for all records) 0.5 sec (from Pringle[35]).

(iii) Function of the songs

The attempt has been made[33] to fit the types of song heard in the smaller Auchenorrhyncha into the elaborate scheme of classification proposed by Faber[13] for Orthoptera. The "common song", which is given by males of most species, would include the "aggregational song" of cicadas[35]. It may occur when the male is isolated, but can also be induced by the insect hearing the songs of other males. In cicadas it undoubtedly has an aggregational function, since the whole local population of one species may be found singing together in a small area; this may have a protective function, as vertebrate predators cannot locate aurally any individual when several are singing together. The tymbals of the nymphs of Aetalionidae are functional organs of sound production[12] and may have the same aggregational function in these gregarious nymphs, which live above ground as do those of the Membracidae, Biturridae and Eurymelidae.

Distinct "calls of courtship" have been recorded from *Platypleura octoguttata*[35] and from several American cicadas[1,3] and have been heard from a few cicadellids[31]. They have a different pulse rhythm from the aggregational song and are emitted by the male only when in close proximity to the female, or by the female in species where this sex sings.

In *P. octoguttata* the intensity of sound is very much less than in the aggregational song, but the pulse repetition frequency is higher and rises with increasing sexual excitement. "Calls of pairing" emitted during the act of copulation have been heard in certain cercopids.

In additional to these intraspecific calls nearly all Auchenorrhyncha emit "calls of distress" when roughly treated; the phenomenon is not reported in Fulgoroidea and Typhlocybidae. Cicadas can produce a very loud raucous noise when handled or when captured by a predator, and the effect might well be to secure the insect's release; in the smaller Auchenorrhyncha the intensity of sound is hardly enough to be of value in this way. These calls are always without pattern in the pulses, and may perhaps be a secondary function of a mechanism primarily evolved for intraspecific communication.

(iv) Sound reception

The location of the organ of sound reception is known with certainty only in the cicadas, which possess a well-developed tympanum on the ventrolateral wall of the 2nd abdominal segment (Fig. 241). This is richly supplied with chordotonal sensilla[45] innervated by a separate auditory nerve, and the sensilla respond to the song of the species[35]. Since in the males the tympanum is physically coupled to the same tracheal resonator as the sound-producing tymbal, there is probably an automatic tuning of the receptor to the correct sound carrier frequency; the tympanal organ of some Japanese cicadas responds maximally to the carrier frequency of the song of their own species[22]. The tympanum is saved from damage during stridulation by a detensor muscle which bends its rim and allows the delicate membrane to go slack[35]. The female cicada also has a well-developed tympanum.

Chordotonal organs lying in the same position as those of cicadas occur in *Tettigarcta*[37], but without the tympanal membrane (Fig. 246). The song of *Tettigarcta* has not been heard but if, like the other Auchenorrhyncha which lack the tracheal resonator, it consists essentially of a sequence of pulse transients, an adequate receptor would be provided by such a group of chordotonal sensilla sensitive to vibration. It seems likely that a homologous group of chordotonal sensilla in the 2nd abdominal segment may be present throughout the Auchenorrhyncha, and that the more efficient communication system of the cicadas has evolved from the primitive condition by the addition of the tracheal resonator and the tympanal membrane. In this way the elaborate central nervous organisation which must be required for the generation and perception of the pulse patterns could undergo a continuous process of evolutionary elaboration.

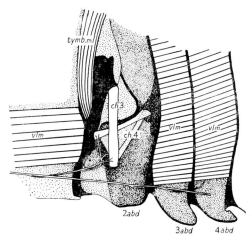

Fig. 246. The organ of sound reception in *Tettigarcta tomentosa*. Ventral part of the base of the abdomen, with the ventral longitudinal muscle of the second abdominal segment removed. Lettering as in Fig. 240; ch. 3, ch. 4 = chordotonal organs 3 and 4 (from Pringle[37]).

(b) Sternorrhyncha

There have been only isolated observations of sound production in these families. When colonies of the aphid *Toxoptera aurantii* (= *coffeae* auctt.) were disturbed they produced a "distinct scraping sound, audible with a big colony as much as eighteen inches away from the leaf". An irregular patch of ridged cuticle had earlier been reported on the abdomen of this species. A review of the situation from a systematist's standpoint[10] led to the conclusion that *Toxoptera* should be re-defined as a stridulating genus to include *T. aurantii*, *T. citricidus* (= *taversi* auctt.) and *T. odinae*; although the mechanics of strigilation have not been observed it is suggested that the rugose cuticle functions as a strigil and that certain irregularly scattered and somewhat feeble hairs function as plectra. However the sound is made, it seems that the colony emits an extraspecific "call of distress", but at what type of enemy has not been determined.

Livia juncorum, *Trioza acutipennis* and *T. nigricornis* are psyllids that have been heard producing one-second bursts of sound, apparently by vibrating the wings[32]. Both sexes call, and available evidence suggests that at least some Psyllidae emit intraspecific songs of low energy.

Nothing is known regarding the evolutionary history of the Sternorrhyncha other than that the psyllids are probably the nearest of the four recent families to the prototype Sternorrhyncha and that the family dates back to the Permian[6]. It seems best to regard stridulation by *Toxoptera* as an isolated phenomenon and to suspend judgement on the psyllids until more genera have been investigated.

3. Heteroptera

The major classification of Dufour, as recently extended[27], provides a convenient background on which to study sound production. The terrestrial Heteroptera (Geocorisae) are divided on morphological criteria into Cimicomorpha and Pentatomomorpha; the first attempt to use songs as sytematic characters at high level resulted in confirmation of the validity of the two series[26]. Both the so-called watersurface bugs (Amphibicorisae) and waterbugs (Hydrocorisae) arose from landbugs, probably independently from cimicomorphan ancestors.

(a) Pentatomomorpha

The discovery of the diversity of stridulating mechanisms and the widespread occurrence of sound production in this series is so recent that any survey must necessarily be incomplete: field or laboratory observations on the thousands of species of tropical Pentatomoidea and Coreoidea (includes Lygaeoidea of most authors) are wanting. Unlike the Auchenorrhyncha, where one method of stridulation occurs throughout, the Pentatomomorpha have a great number of methods of sound production and, as the investigated sample is largely Palaearctic, many more probably await discovery[26].

Table 29 shows the known instances of stridulation; almost any two parts of the sclerotized integument which can be brought together with relative movement can give rise, it appears, to a stridulating mechanism; however, some methods have more evolutionary potential than others—for example the wing strigil, with an adjacent enclosed air space (subhemelytral, subalar, or abdominal air sacs). Some strigil and plectrum organs involve very little modification of musculature, but those involving wing strigils require a great deal of alteration to the muscles when the plectrum is a striate dorsal plaque, the lima, as in *Piesma* (Fig. 247). Here terga I and II are fused and the dorsal longitudinal muscles hypertrophied; contraction of the muscles deforms the terga and drags the attached lima across the strigil. Modification of the anterior abdominal terga occurs also in Cydnidae[25]; the cydnid and piesmid sound-producing components are good examples of convergence but in the former the strigil is part of the 1st vannal vein, in the latter of the cubitus. Sexual dimorphism may affect the strigil, as in *Piesma*, where the female shows reduction, or the plectrum, as in *Aethus flavicornis*, where the female has a reduced lima[21].

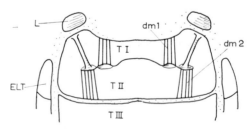

Fig. 247. *Piesma quadrata*, ♂. 1st and 2nd abdominal terga from below, to show the muscles and lima. T.I, T.II, T.III = abdominal terga, dm 1, dm 2 = dorso-longitudinal muscles, L = lima, ELT = externo-lateral tergite.

TABLE 29

SOUND PRODUCTION IN PENTATOMOMORPHA

(After Leston[26], with additions)

Group	Genera	Sex etc.*	Strigil	Plectrum	Remarks
ARADIDAE					
Calisiinae	Aradacanthia	m f	ridge, meta-pleuron	ridge, middle femora	Sounds not reported
Mezirinae	Artabanus Rossius	m f l	tergum IV	teeth, hind tibiae	ditto
	Illibius	m	tergum III	tubercle, hind femora	ditto
	Pictinus Stelgidocoris Psectrocoris	m f	tergum III	tubercle, hind femora	ditto
	Strigocoris	m f	tergum IV	knife-edge, hind femora	ditto
TESSARATOMIDAE					
Tessaratominae	? all	m f	wing, vein IV	lima, tergum I	Sounds heard
Piezosterninae	? all	m ?f	unknown		ditto
SCUTELLERIDAE					
Tetyrinae	all	m f l	sterna V–VII	teeth, hind tibiae	Sounds not reported
Scutellerinae:					
Sphaerocorini	all	m	pygophoral	conjunctival appendage	Probably a grasping organ
ACANTHOSOMIDAE	Elasmucha	m	unknown		Sound heard
CYDNIDAE					
Sehirinae	all	m f	wing, vein IV	lima or ridge tergum I	ditto
Cydninae	all	m f	ditto	ditto	ditto
Corimelaeninae	all	m f	ditto	ditto	ditto
PENTATOMIDAE					
Mecideinae	all	m f	sterna II–IV	tubercles, hind femora	Sounds not reported
Pentatominae	? all	m f	—	—	Tymbal, terga I–II, Sounds reported, Females sing in a few genera only
Amyoteinae	? all	m	—	—	Tymbal, terga I–II, Sounds reported
COREIDAE					
Coreinae	Centrocoris	m	unknown		Sounds heard
	Phyllomorpha	m	ditto		ditto
	Spathocera	m	ditto		ditto
LARGIDAE					
Larginae	Arhaphe	m f	margin of hemelytron	lima, hind femora	Sounds not reported
LYGAEIDAE					
Orsillinae	Kleidocerys	m f	wing, accessory vein	ridge, postnotum 3	Sounds heard
PIESMIDAE	? all	m ?f	wing, vein Cu	lima, pre-1st tergite	ditto

* m, male; f, female; l, larva.

That the majority of the strigils have evolved independently seems an inescapable conclusion; a recent survey of aradid systematics[44] puts the stridulating genera within Mezirinae almost randomly within the subfamily whilst the isolated group of sound producing coreids and the presence of a strigil in *Araphe*[24] but not related genera of Largidae all suggest parallel evolution of strigils and plectra.

Sound production in a large number of genera of Pentatomidae *sensu stricto* has been thought to involve the 1st and 2nd abdominal terga[25], but the essential modification of the sclerites and musculature, enabling them to function as tymbals, has been missed.

In *Sciocoris* (Fig. 248), *Mecistorhinus*, *Agonoscelis*, *Carpocoris* and *Dictyotus* (all distantly related genera of Pentatominae) and in *Glypsus* and *Stiretrus* (Amyoteinae) abdominal terga I and II are fused and the suture line is deflected backwards bilaterally by a large apodeme. The 1st dorsal longitudinal muscle is a large flat sheet, well seen in *Sciocoris*, originating on the posterior margin of the third thoracic phragma and inserted on the inter-tergal apodeme. A similar muscle, the tergal longitudinal, (TL1), which is apparently that now under discussion, has also been described from the pentatomine *Nezara viridula* but without note of its functional significance[28].

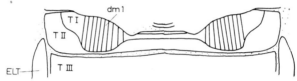

Fig. 248. *Sciocoris cursitans*, ♂. 1st and 2nd abdominal terga from below, to show the tymbal muscle (dm 1). Lettering as in Fig. 247.

The cuticle above the apodeme, the lateral areas of the fused 1st and 2nd abdominal terga, shows a varying degree of differentiation from that of the remainder of these two terga and from that of terga III to VII: it is often vaguely striate and there is no doubt that when pulled by TL1 it deforms sufficiently to produce sounds and therefore falls within the definition of a tymbal implied in the introduction to this present review.

Sound production probably occurs throughout Pentatominae and Amyoteinae, save for cases of secondary loss. In *Carpocoris* and *Palomena* both sexes emit calls: in the other genera calls have only been reported so far from males. There is sexual dimorphism both in the sculpturing of the tymbal area of the cuticle and in the size of the tymbal apodeme (hence probably in the size of the muscle).

(b) Cimicomorpha

The reduvioid prosternal furrow is cross-striate and played upon by the tip of the rostrum; this stridulation was reported in the eighteenth century and was used in delimiting Reduviidae by all the classic systematists; until recently, however, little has been known of its function except that it provides "un moyen passif de défense".

Reduvioidea comprise the Elasmodemidae, Phymatidae and Reduviidae, the latter with some 30 subfamilies; almost all species, in males, females and larvae, have an identical form of stridulating apparatus varying intraspecifically in length of prosternal furrow, in number of striations within the furrow and in the spacing of the striations[18, 26]. The Triatominae, which, unlike the vast majority of Reduviodea, prey on vertebrates rather than on insects, have apparently lost the strigil in many instances. In *Pirates* the anterior margin of the prosternum is in contact with the underside of the head, a modification of the usual mode of stridulation[34].

The minute schizopterid *Ptenidiophyes* has a series of pegs in a groove of the hemelytra in macropterous males; this could be a funtioning strigil[26]. In both other cases where sound-producing organs have been described, Nabidae and Saldidae, doubt is cast on this function for the alleged strigils by the negative result of prolonged attempts to hear any sounds[26].

(c) Amphibicorisae

The only sound production reported in this group covers isolated instances in the probably polyphyletic family Veliidae. *Microvelia diluta*, "stridulates when irritated, producing a shrill scraping sound which is perceptible for some yards"[9]; this bug is 2.5 mm long. In the subgenus *Stridulovelia*, a plectrum on the front of the posterior femora plays on a series of small teeth situated on the antero-lateral margins of the abdomen[34]. All that can be said at present is that Veliidae would repay experimental investigation.

(d) Hydrocorisae

Stridulation under water has long been known to occur amongst insects but the physical problems involved have not yet been investigated and must await development of a small underwater microphone. Nearly all the described types of hydrophone are far too cumbersome for use in small aquaria.

The first account of stridulation by a waterbug was due to observations by two English amateurs[4]; for a time it was thought that the abdominal "strigil" of Corixidae was a sound producing organ. Later workers disallowed this function to the abdominal strigil and claimed that the palae, a set of teeth on the front tibio-tarsus of male corixids, were sound producing. Many Corixidae produce one or two distinct calls by rubbing a part of the inner surface of the front femora, covered by a patch of more or less modified setae, against the sharp edge of the head[46]. However, the degree of modification of the general hair cover of the male femur need be very slight, as, for example, in *Sigara distincta*[40].

So far sound production by Corixinae has been reported in the west European genera *Callicorixa*, *Corixa*, *Arctocorisa*, *Sigara*[42] and the Nearctic *Palmacorixa*. Of the other genera in which it has been suggested on morphological evidence, *Hesperocorixa* (British species) have given negative results in experiments.

In Micronectinae, a further subfamily of Corixinae, loud sounds are produced apparently by the abdominal "strigil"[46]. Thus there are two known and accepted methods of sound production in Corixidae but the possibility that the palae are also in fact sound producing is not yet excluded; sounds, too faint to record satisfactorily, have been heard from *Sigara distincta* accompanied by movement of the palae. Among Notonectidae *Notonecta* is known to sing (Notonectinae) whilst *Buenoa* and *Anisops* (Anisopinae) have a combination of strigils and plectra[34, 42]. In male *Buenoa limnocastoris*[20] it is claimed that:

1. The distal end of the front femur plays on the angular side of the head (this in females also);

2. A comb on the front tibia proximally plays on the striate lateral face of the third antennal segment,

3. A spur on the coxa of the front leg plays on a striate area on the inner face of the front femur.

Other families which are known to include stridulating members are Naucoridae, Pleidae and Nepidae[34]. However, no survey can be offered of stridulating methods in Hydrocorisae at present; the known instances are confined to the better known members of the European and North American faunas, suggesting that a world survey would reveal the presence of many different types of strigil-and-plectrum organs. In fact, the situation shows stridulation by waterbugs to be as widespread in its occurrence and as diverse in its methods as in the terrestrial Pentatomomorpha.

(e) Physical characteristics of the song

Instrumental recordings of the songs of strigilating Heteroptera have been made only for the pentatomomorphan genera *Kleidocerys*, *Piesma* and *Sehirus* and the reduviid *Coranus subapterus*[19]. In all cases the sounds were of the pulsed type expected from a strigil-and-plectrum mechanism. Where two or more distinctive calls have been recorded from the same species, the difference has been in the pulse repetition frequency (Fig. 249); a difference in the frequency of pulses was also noted between two species of *Kleidocerys*[19]. The calls of water bugs are also interrupted (pulse-modulated)[41]. All the calls of Pentatomomorpha are of low intensity, being audible to the human ear only at a distance of a few centimetres. The same is true of most

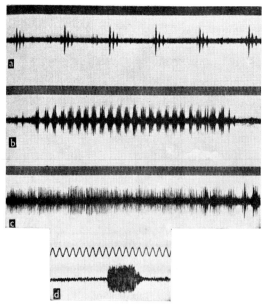

Fig. 249. Stridulation of *Sehirus bicolor;* (a) "normal" song of the male, (b) "courtship" song of the male, (c) third song of the male (d) fourth song of the species, possibly of female. Timing wave in all records, 50 c/s (from Haskell[19]).

Hydrocorisae but *Sigara dorsalis* and *S. striata* are more powerful and *Micronecta poweri*, a bug less than 2 mm long, has a range for the human ear of 7 m in air. All these calls are of the intraspecific type.

The reduviid *Coranus subapterus* has a call irregular in waveform and frequency, with no suggestion of a regular pulse repetition[19]: the bursts of song vary within and between individuals and are therefore of a characteristic extraspecific type.

The tymbal organ of Pentatominae and Amyoteinae produces sound frequencies in the range 120–125/sec in *Aelia acuminata* and 150–200/sec in *Carpocoris*, with a pulse repetition frequency of 9–12/sec[21]. It is difficult to see how these low sound frequencies can be produced by a resonant mechanism of these small dimensions, and the oscillograms (Fig. 250) suggest rather that each pulse is a double IN–OUT click of the tymbal, as in cercopids and jassids.

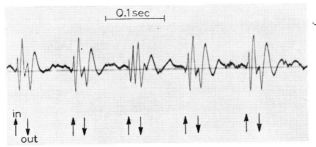

Fig. 250. Oscillogram of the song of *Aelia acuminata*, (Pentatominae), showing suggested interpretation as a series of double pulses. Time marker, 0.1 sec. (Oscillogram from Jordan[21]).

(f) Function of the songs

When *Coranus subapterus* is touched on the head or thorax it stridulates; the same response can be elicited from an isolated head and thorax[19]. All the many observations on Reduvioidea, together with the facility of the larvae to produce sounds when touched, show that stridulation in the group is defensive in function (Faber's "calls of distress"). The sudden loud outbursts of a large reduviid when picked up by a trained observer often causes him to drop it: there is no reason to doubt that stridulation would sometimes scare off a reptilian, amphibian or avian predator. In the case of *Coranus* there is reason to think the recipient of sounds may often be a lycosid spider.

Elsewhere in Heteroptera the calls are intraspecific and include what is probably in most species a "common song". *Buenoa limnocastoris* males sing by day or night and are anaphonic[20]. The song changes abruptly at a distance of about a centimeter from a female to a second call, which can be categorised as a "call of courtship". If the female escapes without copulating then "common song" recommences; whilst emitting this a male will attempt to follow any other individual in its vicinity. Two calls are given out by *Sigara dorsalis* and *S. striata*; they cause the females to swim rapidly in small circles, which is effective since the males try to copulate with any moving object[41, 46]. Observations on *S. dorsalis* show that "common song" may occur spontaneously or be triggered off by the "common song" of another individual of the same species. In addition, it can be elicited by a low frequency noise: in aquaria it may be started by the vibration, perhaps carried to the water through floor and

References p. 410

bench, of a passing heavy vehicle, by striking the bench heavily or by scraping on the walls of the aquarium (unpublished): an individual gives out one burst of song in response to the single stimulus but continues anaphonically if another individual takes up the cry. With a single male it was found that it would call once at a definite interval (about 2.6 sec) after each stimulus and that repeating the stimulus up to four or five times—but seldom more—would each time give rise to a response.

That "common song" causes heterosexual aggregation is apparent from experiments with *Kleidocerys* in small arenas; observations on *S. dorsalis* also suggests that the call is not to be separated from a rivalry or territorial function. A male calls and is answered by a second male; one approaches the other, still singing; usually one of the two is nudged away by its colleague and takes rapid evasive action.

The intensity of the "courtship song" is generally much less than that of the "common song". The change to "courtship song" has been noted in Corixidae and Notonectidae amongst the water bugs[46] and in Cydnidae, Pentatomidae and Piesmidae amongst the terrestrial Geocorisae[19]: it would seem that the sequence common song to courtship song, so well established in terrestrial Orthoptera and Homoptera, is of equal value to aquatic insects. In the pentatomid *Carpocoris pudicus*[21] there is a female call in addition to the two usual male songs; it is perhaps an "acceptance song", releasing male copulation behaviour, but may have, too, an orientation function. The female call of *Sehirus bicolor* may be of this type[19].

In *S. bicolor* male "courtship song" led to a third type of male call, apparently by elimination of the intervals between the bursts of pulses of the former song form. No function was observed to tie up with Call 3 but it is apparently an essential step between courtship and copulation, again perhaps with an orientation or positional function.

Some workers report that Cydnidae will also call when touched or picked up. If so it is likely that a "call of distress" has been evolved upon a pre-existing intraspecific repertory, as among certain Homoptera.

(g) Sound reception

The auditory receptors of water-bugs (Corixidae, Notonectidae) has long been assumed to be the tympanum and associated scoloparia near the wing base on the mesothorax[23,41]. This is in contact externally with the air bubble carried by the insect and is well situated physically to be a sound receptor. This mesothoracic organ is, however, merely one of a series of two-celled chordotonal organs situated also in the metathorax and 1st abdominal segments; these and a one-celled chordotonal organ on the 1st and 2nd abdominal segments are present in many Amphibicorisae and Hydrocorisae, including some which have not been heard to produce sounds[23]. A hydrostatic or a general vibration sense therefore cannot be excluded, but the greater development and taut membrane of the mesothoracic organ of *Corixa* suggests that this is a true auditory receptor.

In landbugs the organs known as Tullgren's trichobothria are probably hearing organs[43]. They occur only in the Pentatomomorpha and have been recognized throughout the group except in Aradoidea: males and females have groups of two or three trichobothria ranged regularly, serially and bilaterally on the abdominal sterna (usually on III to VII)—the arrangements provide major taxonomic characters. Each trichobothrium is apparently multicellular and is pivoted basally within a shallow pit; it could act as a receptor for sound in the same manner as the fine cercal hairs of the cockroach[38]. It is significant that the great group of intraspecific-calling landbugs coincides, save for Aradoidea, with the group bearing trichobothria[26]; as for Aradoidea, their cuticle bears masses of pegs, setae and tubercles of all kinds—many, from the animals' way of life, probably tactile organs—sufficient to prevent detection, so far, of any auditory organs.

4. The Evolution of Sound Production in Hemiptera

There are perhaps two general problems of insect acoustics that can be better discussed in relation to the Hemiptera than to any other order. Put in the form of questions these are: (1) why are stridulatory mechanisms so widespread in this particular order, and (2) how does an entirely new type of stridulatory mechanism, the tymbal, arise in the course of evolution?

In any discussion of the first question, it is important to bear in mind the two fundamentally distinct functions of sound production, extra-specific and intraspecific communication. It does not necessarily follow that the same biological situation will favour the evolution of the two types of behaviour, and indeed some insect orders appear to show one type only. That both functions for sound production are found in the Hemiptera is a fact demanding explanation.

The order has probably been characterized since the late Palaeozoic by its method of feeding by sucking juices from plants. Such a mode of life demands a cryptic appearance and a passive, rather than an active, defence against predators. It is not difficult to see how, in such a situation, the chance production of sounds by the movement of one part of the body over another could have developed on the one hand into a deterrent to predators, and on the other into a means of ensuring the meeting of the sexes. Special considerations obviously apply to those groups which have secondarily adopted a different way of life, for example the Hydrocorisae and the predatory Cimicomorpha.

The strigil-and-plectrum method of sound production has clearly proved to be the easiest to evolve in insects with their hard but elastic exoskeleton. It had earlier been suggested that the tymbal method arose in a quite different manner as a development of movements made by the insects during actual copulation[37]: a certain minimum amount of movement in the cuticle is necessary in order to produce a sound click, and it is difficult to see how, except by actual contact between the sexes, small movements could initially be effective enough to ensure their preservation and development in evolution. However, the discovery of a tymbal mechanism among the Pentatominae suggests another and perhaps more probable evolution. In the Pentatomomorpha there are families in which the movement by the dorso-longitudinal muscles of the 1st abdominal tergum operates a plectrum (the lima) which works on a wing strigil; it is possible that a mechanism of this sort might evolve into a tymbal mechanism by accentuation of the movement of the 1st abdominal tergum and the loss of the wing strigil. At present we know too little of either the anatomy or the physiology of these insects to decide whether or not this is a likely evolutionary development, but it does at least offer hope of relating the two very different mechanisms of sound production found in the order.

For Addendum to this Chapter, see p. 798.

References p. 410

REFERENCES

[1] ALEXANDER, R. D., 1956, A comparative study of sound production in insects, with special reference to the singing Orthoptera and Cicadidae of the eastern United States, Thesis, Ohio State University.
[2] ALEXANDER, R. D., 1957, Sound production and associated behaviour in insects, *Ohio J. Sci.*, 57, 101–113.
[3] ALEXANDER, R. D. and T. E. MOORE, 1958, Studies on the acoustical behaviour of seventeen-year cicadas, *Ohio J. Sci.*, 58, 107–127.
[4] BALL, R., 1845, On noises produced by one of the Notonectidae, *Report Brit. Assoc. Adv. Sci.*, 64–65.
[5] BEKKER-MIGDISOVA, E. E., 1940, Fossil Permian cicadas of the family Prosbolidae from the Sojana River, *Trudy Paleont. Inst.*, 11 (2), 5–79 (in Russian).
[6] BEKKER-MIGDISOVA, E., 1949, Mesozoic Homoptera from central Asia, *Trudy Paleont. Inst.*, 22, 1–68 (in Russian).
[7] BINET, A., 1894, Contributions á l'étude du système nerveux sous-intestinal des insectes, *J. Anat. (Paris)*, 30, 449.
[8] BOETTIGER, E. G., 1957, (personal communication).
[9] DISTANT, W. L., 1910, *The Fauna of British India, Rhynchota*, Vol. 5, Taylor and Francis, London, p. 140.
[10] EASTOP, V. F., 1952, A sound production mechanism in the Aphididae and the generic position of the species possessing it, *Entomologist*, 85, 57–61.
[11] EVANS, J. W., 1956, Palaeozoic and Mesozoic Hemiptera (Insecta), *Austr. J. Zool.*, 4, 165–258.
[12] EVANS, J. W., 1957, Some aspects of the morphology and interrelationships of extinct and recent Homoptera, *Trans. Roy. Entom. Soc. London*, 109, 279–294.
[13] FABER, A., 1929, Die Lautäusserungen der Orthopteren, I, *Z. Morphol. Ökol. Tiere*, 13, 745–803.
[14] FABER, A., 1932, Die Lautäusserungen der Orthopteren, II, *Z. Morphol. Ökol. Tiere*, 26, 1–93.
[15] HAGIWARA, S., 1956, Neuro-muscular mechanism of sound production in the cicada, *Physiol. Comp. Oecol.*, 4, 142–153.
[16] HAGIWARA, S., H. UCHIYAMA and A. WATANABE, 1954, The mechanism of sound production in certain cicadas with special reference to the myogenic rhythm of insect muscles, *Bull. Tokyo Med. Dent. Univ.*, 1, 113–124.
[17] HAGIWARA, S. and A. WATANABE, 1956, Discharges in motoneurons of cicada, *J. Cell. Comp. Physiol.*, 47, 415–428.
[18] HASE, A., 1933, Über die Lauterzeugung sowie deren mutmassliche Bedeutung bei der Wanze Panstrongylus, Hemiptera U. Familie Triatomidae, *Biol. Zentr.*, 53, 607–614.
[19] HASKELL, P. T., 1957, Stridulation and its analysis in certain Geocorisae (Hemiptera, Heteroptera), *Proc. Zool. Soc. London*, 129, 351–358.
[20] HUNGERFORD, H. B., 1924, Stridulation of Buenoa limnocastoris Hungerford and systematic notes on the Buenoa of Douglas Lake region of Michigan, with the description of a new form, *Ann. Entom. Soc. Amer.*, 17, 223–226.
[21] JORDAN, K. H. C., 1958, Lautäusserungen bei den Hemipteren-Familien der Cydnidae, Pentatomidae und Acanthosomidae, *Zool. Anz.*, 161, 130–144.
[22] KATSUKI, Y. and N. SUGA, 1958, Electrophysiological studies on hearing in common insects in Japan, *Proc. Japan. Acad.*, 34, 633–638.
[23] LARSÉN, O., 1957, Truncale Scolopalorgane in den pterothorakalen und den beiden ersten abdominalen Segmenten der aquatilen Heteropteren, *Acta Univ. Lundensis, (Lund)*, [2] 53, 1–68.
[24] LATTIN, J. D., 1958, A stridulatory mechanism in Arhaphe cicindeloides Walker (Hemiptera: Heteroptera: Heteroptera: Pyrrhocoridae), *Pan-Pacific Entom.*, 34, 217–219.
[25] LESTON, D., 1954, Strigils and stridulation in Pentatomoidea (Hem.): some new data and a review, *Entom. mon. Mag.*, 90, 49–56.
[26] LESTON, D., 1957, The stridulatory mechanisms in terrestrial species of Hemiptera Heteropteram, *Proc. Zool. Soc. London*, 128, 369–386.
[27] LESTON, D., J. G. PENDERGAST and T. R. E. SOUTHWOOD, 1954, Classification of the terrestrial Heteroptera (Geocorisae), *Nature*, 174, 91–92.
[28] MALOUF, N. S. R., 1932, The skeletal motor mechanism of the thorax of the stink bug, Nezara viridula L., *Bull. soc. roy. entom. Egypt*, 16, 161–203.
[29] MARLER, P., 1956, The voice of the chaffinch and its function as a language, *Ibis*, 98, 231–261.
MITIS, H. VON, see ref. 46.
[30] MYERS, J. G., 1929, *Insect singers*, Routledge, London.
[31] OSSIANNILSSON, F., 1949, Insect drummers, *Opusc. Entom.*, (Suppl.) 10, 1–146.
[32] OSSIANNILSSON, F., 1950, Sound-production in psyllids (Hem. Hom.), *Opusc. Entom.*, 15, 202.
[33] OSSIANNILSSON, F., 1953, On the music of some European leafhoppers and its relation to courtship, *Trans. IXth. Int. Congr. Entom.*, 2, 139–141.

[34] POISSON, R., 1951, Ordre des Hétéroptères, in Grassé, P.-P., *Traité de zoologie (Paris)*, *10*, (2), 1657–1803.
[35] PRINGLE, J. W. S., 1954, A physiological analysis of cicada song, *J. Exp. Biol.*, *31*, 525–560.
[36] PRINGLE, J. W. S., 1956, The physiology of insect song, *Acta. physiol. pharmacol. neerl.*, *5*, 88–97.
[37] PRINGLE, J. W. S., 1957, The structure and evolution of the organs of sound-production in cicadas, *Proc. Linn. Soc. London*, *167*, 144–159.
[38] PUMPHREY, R. J., 1940, Hearing in insects, *Biol. Rev.*, *15*, 107–132.
[39] RÉAUMUR, R. A. F. DE, 1740, *Mémoires pour servir à l'Histoire des Insectes*, *5*, 145–206 (Paris).
[40] REICHENBACH-KLINKE, H., 1949, Klammerapparate und Schrillfelder der Ruderwanzen (Hem., Corixidae), *Entomon*, *1*, 161–163.
[41] SCHALLER, F., 1951, Lauterzeugung und Hörvermögen von Corixa (Callicorixa) striata L., *Z. vergl. Physiol.*, *33*, 476–486.
[42] SOUTHWOOD, T. R. E. and D. LESTON, 1959, *Land and water bugs of the British Isles*, Warne, London.
[43] TULLGREN, A., 1918, Zur Morphologie und Systematik der Hemipteren, I, *Entom. Tidskr.*, *39*, 115–133.
[44] USINGER, R. L. and R. MATSUDA, 1959, *Classification of the Aradidae (Hemiptera-Heteroptera)*, British Museum (N.H.), London.
[45] VOGEL, R., 1923, Über ein tympanales Sinnesorgan, das mutmassliche Hörorgan der Singzikaden, *Z. ges. Anat. Entw.-gesch.*, *67*, 190–000.
[46] VON MITIS, H., 1936, Zur Biologie der Corixiden, Stridulation, *Z. Morphol. Ökol. Tiere*, *30*, 479–495.

CHAPTER 15

ANATOMY AND PHYSIOLOGY OF SOUND RECEPTORS IN INVERTEBRATES

by

H. AUTRUM

1. Comparative Anatomy of Sound Receptors

There are three basic types of sound receptors in arthropods, classified according to their anatomical structure:

1. Movable *hairs*, supplied with sensory cells, articulating on the body surface.

2. *Scolopophorous organs*, including a wide range of differentiated types of stimulus conductors.

3. Slits in the chitin, covered by thin chitin membranes, and provided with sensory cells: *lyriform organs* of spiders.

1. *Hair sensillae* are found in insects and spiders on the body surface, the extremities and appendages (for example the anal cerci, Figs. 251, 266C). They are

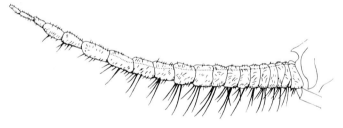

Fig. 251. *Periplaneta americana*. Anal cercus of the cockroach, side view. Note the long, sound-sensitive hairs on the ventral side. (From Autrum.[5])

flexibly articulated in the surrounding chitin; at the bases are dendrites from one or several sensory cells (Fig. 266C). They attain their highest state of differentiation on the anal cerci of insects, and on the extremities of many spiders (*trichobothriae*). Hair sensillae react to sound waves. When situated on the articular membranes of insects, they can also perceive vibrations of the extremities (see p. 422).

2. The *scolopophorous organs* contain sensory cells (bipolar neurons) with a characteristic structure: they possess apical bodies (scolopidia; Fig. 252). There are almost always accessory cells in addition to these sensory cells: a sheath cell encases the dendrite of the sensory cell, and an attachment cell mediates the suspension of the sensory cell on the secondary structures of the sensory organ, or on the body wall. The scolops has a complex ultramicroscopical structure[29].

The number of sensory cells and the anatomical differentiation of the structures for stimulus reception and stimulus conduction is extremely varied. The scolopophorous organs can thus be adapted to many specialized functions.

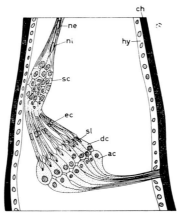

Fig. 252. *Formica sanguinea* ♀. Scheme of the subgenual organ in the leg of *Formica*. (From Weber[64].)

ac = accessory cells
ch = chitin cuticle
dc = cap cells
ec = enveloping cell
hy = hypodermis

ne = nerve
nl = nucleus of a neurilemma cell
sc = sense cells
sl = scolopidia

(a) In the simplest case, one finds vibration-sensitive scolopophorous organs stretched between two points of the body wall (probably in *Dytiscus*[35]). There are no accessory structures.

(b) Simple *chordotonal organs* are situated between tibia and tarsus in Diptera, Coleoptera and Hemiptera, and provide for reception of vibrations from the ground[11].

(c) The *subgenual organs* are groups of sensory cells, which in many insects are located in the legs at the proximal end of the tibia, near the knee. Their anatomical structure has been described by Eggers[24] and Debaisieux[20,21]. They consist of a very thin veil of sensory cells affixed to each other in a fan-like shape. This veil is spanned or hung in the hemolymph of the leg, obliquely to its longitudinal axis (Figs. 252, 260). Subgenual organs are missing in Diptera and Coleoptera. The subgenual organs are receptors for ground vibration[4,6,11,57].

(d) *Johnston's Organ** is a highly differentiated scolopophorous organ, which is situated between the second and third antennal segments of insects (for anatomical description see Eggers[24] and Risler[50,51]) and consists of numerous scolopophorous cells and accessory cells. The degree of differentiation of this organ varies between different orders of insects. It is relatively simple in some Orthoptera, and most highly developed in *Culicidae* and *Chironomidae*, as well as in *Gyrinus*. In the *Culicidae*, Johnston's Organ has been demonstrated to have an auditory function in males. The sensory cells here are situated in the second antennal segment (pedicellus); on one side they are suspended inside the spherical hollow of the enlarged pedi-

---

* Johnston's Organ has several functions; it does not function exclusively for hearing[13].

cellus, and on the other side suspended on the prong-shaped processes from the circular plate at the base of the next (third) antennal segment, which at the same time is the initial segment of the hairy flagellar portion of the antenna (Fig. 253).

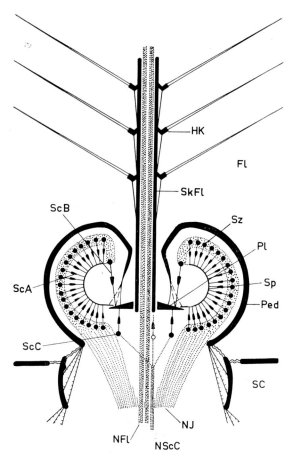

Fig. 253. Scheme of the antenna of *Culicidae*, with Johnston's Organ. Nerves stippled. (From Risler[51].)

Fl	= flagellum	SC	= scapus
HK	= hair crown of the flagellum	ScA	= outer ring of scolopidia of Johnston's Organ
NFl	= flagellar nerve		
NJ	= nerve complex of the Johnston's Organ	ScB	= inner ring of scolopidia
		ScC	= single scolopidium
NScC	= nerve of the three single scolopidia	SkFl	= inner skeleton of flagellum
Ped	= pedicellus	Sp	= chitin prongs (insertion of scolopidia)
Pl	= circular chitin plate	Sz	= nuclei of sensory cells

The plate is firmly connected with the base of the first flagellar segment (third antennal segment). Chitinous prongs insert on the plate. The sensory cells are arranged in one or two rings; there are in addition three single scolopophorous cells situated between the scapus and the first flagellar segment. The flagellum is relatively long, and carries a varying number of hairs. This flagellum is a sound-receiving organ, the movements of which are detected by Johnston's Organ (Fig. 253). In the male *Culex*, the whole organ is more strongly developed than in the female.

(e) The highest degree of differentiation of scolopophorous organs is found in the *tympanal organs*, which are provided with tympanic membranes. They range between quite simple organs with only one (*Plea*) or two sensory cells (the metathoracic tympanal organs of the Lepidoptera; *Corixa*) and highly complicated structures with several hundred (about 1500 in the *Cicadidae*) sensory cells and accessory structures. The sensory cells are either immediately affixed to a very thin, movable chitinous membrane (tympanum) or they lie in the vicinity of the tympana without any direct connection.

Around the tympanum there is often a thickened chitin frame. A trachea or a tracheal air bladder lies touching the inner side of the tympanum.

On the basis of their position in the insect body, one can distinguish between:

(i) *Pteral* tympanal organs on the wings: in many Lepidoptera, on the base of the first pair of wings, for example in *Satyridae*.

(ii) *Truncal* tympanal organs, on the thorax or on the first or second abdominal segment. They lie on the mesothorax in *Cryptocerata* (Heteroptera); these are small tympana with only one (*Plea*) or two (*Corixa*) sensory cells. In many Lepidoptera they lie on the metathorax (for example *Notodontidae, Lymantriinae, Noctuidae* (Fig. 254), *Arctiinae*). Here there are only two sensory cells, but in addition very

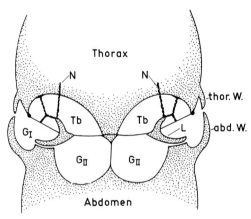

Fig 254. *Catocala*. Schematic horizontal section through both tympanal organs of a Noctuid. (From Weber[64].)

abd.W. = abdominal ridge
G I = tympanal pit
G II = opposite tympanal pit
L = lamella
N = tympanal nerve
Tb = tracheal vesicle
thor.W. = thoracal ridge

complex accessory structures whose function is still unknown (Fig. 255). A sensory cell lies on one of these structures (the stirrup or Bügel, Fig. 255); it cannot be excited acoustically and is probably a proprioceptor[52]. The tympanal organs lie on the first abdominal segment in many Lepidoptera (*Geometridae, Pyralididae*; four sensory cells), and also in the *Locustidae* (Fig. 256). In these groups, the tympanal organ is already highly specialized. Two muscle bundles insert on the

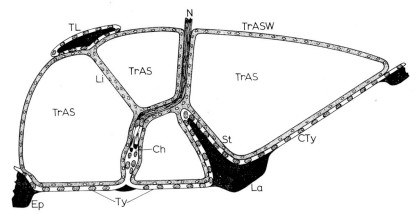

Fig. 255. *Catocala*. Left thoracal tympanal organ of a Noctuid (enlargement of the left organ of Fig. 254. (From Weber[64].)

Ch	= chordotonal organ	N	= nerve
CTy	= counter tympanum	St	= stirrup
Ep	= epaulette	TL	= tensor ledge
La	= lamella	TrAS	= tracheal air sacs
Li	= ligament	TrASW	= tracheal air sac wall
		Ty	= tympanum

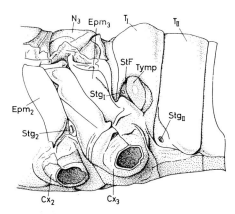

Fig. 256. *Mecosthetus grossus* ♀. Side view of the mesometathorax (2, 3) and base of the abdomen. (From Weber[64].)

Cx	= coxa	N	= notum	T	= tergum
Epm	= epimerum	Stg	= stigma	Tymp	= tympanum

tympanum, but their function is unknown. The tympanum itself consists of a layer of epidermal chitin, touched on its inner side by a large tracheal air sac (Fig. 257): this outer tracheal air sac borders on two more medially situated air sacs, one anterior and one posterior. The innermost tracheal air sacs of the right and left sides border on each other in the midline of the body, so that there is an air cushion communicating between the left and right tympani. The tympanum possesses a total of four small *tympanal bodies*, sclerotizations which have developed through folding and thickening of the chitin. The sensory cells (scolopidia) insert on these tympanal bodies, via cap cells. The cap cells encapsulate the distal ends of the sensory cells, on one end; and

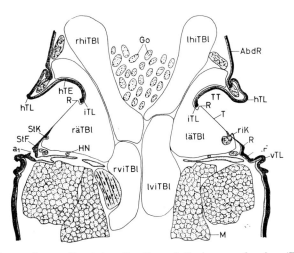

Fig. 257. *Oedipoda coerulescens*. Frontal section through the tympanal region. (From Weber[64].)

a_1	= first abdominal stigma	StK	= peduncular body
AbdR	= abdominal ring	T	= tympanum
Go	= gonads	TbI	= tracheal vesicle
HN	= auditory nerve	hTE	= posterior tympanal framework
riK	= groove-shaped body	hTL	= posterior tympanal ridge
M	= muscles	iTL	= internal tympanal ridge
R	= rims	vTL	= anterior tympanal ridge
Stf	= stigmen field	TT	= tympanal pouch

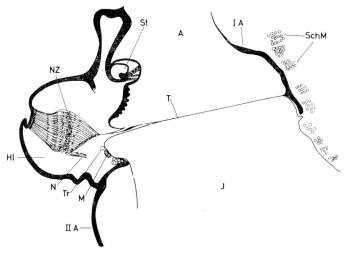

Fig. 258. *Cicadetta coriaria* ♂. Frontal section through the base of the abdomen, with left auditory capsule, tympanum and adjacent organs. (From Weber[64].)

A	= external air	NZ	= nerve cells of the scolopophorous organ
IA, IIA	= anterior and posterior edge of the second abdominal sternum	SchM	= acoustic plate muscle
Hl	= hemolymph	St	= stigma
J	= internal air	T	= tympanum
M	= muscle	Tr	= trachea
N	= nerve		

References p. 431

on the other end they extend fibrous processes distalwards, which anastomose with the epidermis of the tympanal bodies. The tympanal nerve leads from the sensory cells to the first abdominal ganglion, which is fused with the third thoracic ganglion.

In the *Cicadidae*, the tympana occur on the second abdominal segment, in the depths of tympanal cavities, which are covered by double folds of skin, leaving only a narrow open slit (Fig. 258). Again a large air sac, derived from tracheae, borders on the inner surface of the tympanum. The sensory cells (up to 1500) lie in a bundle, which is connected with the tympanum via cap cells. A ligament provides for the suspension and tension of the sensory organ. Tensor muscles are also present in the Cicadidae (Fig. 259). The air bladder (tympanal vesicle) again borders on an adjacent air vessel, so that a countertympanum is formed (which, however, is not innervated by sensory cells). The right and left air bladders of the counter tympana border on each other in the midline of the body.

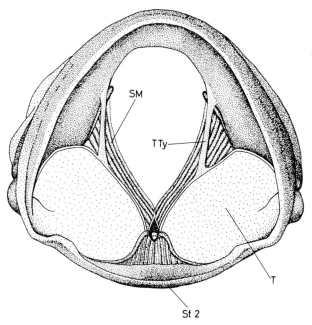

Fig. 259. *Huechys incarnata* A.u.S. ♂ (from Java). Enlarged about 20×. The tympanum, with tensor and acoustic muscle, seen from behind. (From Vogel[62].)

SM = acoustic muscle T = tympanum
St_2 = second sternum TTy = tensor tympani

(iii) *Pedal* tympanal organs occur in the tibia of the first pair of legs in the *Tettigoniidae* (less differentiated organs without tympana can also be found in the second and third pairs of legs). Each tympanal organ is provided with two tympana in the *Tettigoniidae* (Fig. 260); in the *Grylloidea* there is usually (though not always) only one tympanum present (Fig. 261A, B). The tympana lie free, or else are sunk deep into chitin pockets, formed by skin folds. A large trachea goes through the lumen of the tibia and splits into anterior and posterior branches in the region of the tympana; further distal, the two branches reunite (Fig. 260). The outer walls of these tracheal branches border here on the thin cuticle of the tibial wall, thus forming the tympana. The walls of the tracheal branches touch each other in the median plane, so that a thin membrane is formed between them. The trachea does not completely fill the lumen of the tibia; room is left dorsally for a hemolymph canal, and ventrally for a muscle strand. The sensory cells of the tympanal organs lie in a long row in the hemolymph canal. They insert, therefore, not on the tympanum, as in the truncal organs, but on the dorsal wall of the anterior tracheal branch. The

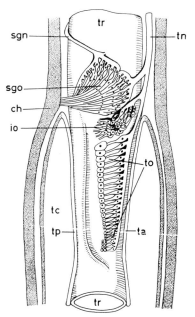

Fig. 260. *Decticus verrucivorus*. Auditory organ in the foreleg of a grasshopper. (From Autrum[9].)

ch = chitin cuticle
io = intermedial organ
sgn = nerve of subgenal organ
sgo = subgenual organ
ta = anterior tympanum

tc = tympanic cavity
tn = tympanal nerve
to = tympanal organ
tp = posterior tympanum
tr = trachea

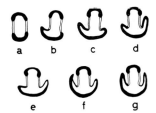
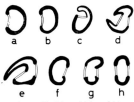

Fig. 261. Schematic cross-section through the forelegs of various *Tettigoniidae* (A) and *Grylloidea* (B). (From Weber[64].)

(A) a   = *Phylloptera*
    b,c = *Phaneroptera* sp.
    d   = *Haaria*
    e   = *Phaneroptera* sp.
    f   = *Phylloptera* sp.
    g   = *Decticus*

(B) a = *Endacusta*
    b = *Cophus*
    c = *Gryllotalpa*
    d = *Orocharis*
    e = *Platydactylus*
    f = *Gryllus toltecus*
    g = *Gryllus campestris*
    h = *Oecanthus*

References p. 431

sensory cells are bent to form an angle. Their free distal ends lie in the hemolymph canal, covered by the cap cells (p. 416), which face into the canal. The length of the scolopidia decreases distally. In the vicinity of the sensory cells of the tympanal organ are the *subgenual organ* (a vibration receptor) and the sensory cells of the *intermedial organ* (Fig. 260), whose function is unknown.

3. In *spiders*, the sensory slits, combined into lyriform organs near the joints of the legs, serve to detect vibrations. The sensory cells are situated below slits in the chitin membrane, 5–160 μ long and 2–3 μ wide, and are covered on the outside by a thin membrane. The slit widens to form a canal in its center; the dendrite of a subepidermal sensory cell extends into this canal, to insert at the outer covering membrane. Several such slits, of varying lengths, lie parallel to each other like the strings of a lyre[36, 39a, 40, 63].

2. Physiology of the Sound Receptors

(a) Frequency range of Invertebrate sound receptors

(i) Stimulation by air-borne sound. *Trichobothriae* and *antennal sound receptors* respond only to low frequencies: hairs on the anal cerci of *Periplaneta* respond between 50–400 c/s, synchronously with the frequency of the sound; up to 800 c/s with partial synchronization; and over 800 c/s asynchronously. Such asynchronous responses can be recorded up to 3000 c/s, but only when very high stimulating intensities are employed. The synchronization up to 800 c/s was derived from the entire cercal nerve; probably a rotation of activity of the individual receptors is occurring here.

The *Johnston's Organs* at the base of the antennae respond only to a narrow range of frequencies. For instance, in *Anopheles subpictus*, the frequency range detected extends between 150–550 c/s, and has a maximal sensitivity at 380 c/s; according to Tischner[58] this maximum is due to a resonance of the antennal flagellum. The upper and lower limits are partly determined by the characteristics of the receptors; for if one causes the antennae to vibrate by applying a vibrating rod, there are no action potentials below 100 or above 500 c/s, in the Johnston's Organ of the male *Aedes aegypti*[38]. The frequency of the female's flight tone lies near 380 c/s; the males (whose flight tone frequency is about 550 c/s) are attracted towards the females by their flight tone[53, 58]. Males of *Pentaneura aspera* (Diptera; Tendipedidae) are attracted by tones between 125 and 250 c/s[27]; small intensities are sufficient for these reactions, which are sharply confined to a certain frequency range (at 125 c/s, 13–18 dB; at 250 c/s, 3–8 dB above the background noise). Alarm reactions can be provoked by loud sounds (over 80 dB) between 80 and 800 c/s.

The frequencies which stimulate the *tympanal organs* have been measured only for continuous tones. Thus only one component of sound has been measured of the two sound components essential for (at least certain) tympanal organs (see p. 427). All tympanal organs react to clicks of sufficiently steep onset. In *Tettigonia*, the electrical response of the tympanal nerve to constant sound differs from that for steeply modulated sound events[10].

Corixa reacts to tones between 2000 and 40,000 c/s[54]. In 38 species of Lepidoptera, Schaller and Timm[55, 56] found maximal reactions between 40,000 and 80,000 c/s. According to Haskell and Belton[33] *Notodontidae* and *Arctiidae* react to frequencies between 3000 and 20,000 c/s (as determined electrophysiologically). In *Prodenia evidania* the tympanal nerve reacts to sounds between 3000 and 240,000 c/s, with a maximal sensitivity between 15,000 and 60,000 c/s[52].

In all *Tettigoniidae*, the range of hearing extends well into the region of ultrasound. Regen[48], using the Galton whistle, found that *Thamnotrizon apterus* perceives tones up to 25,000 c/s. Wever and Bray[67] obtained reactions up to 45,000 c/s in *Amblycorypha oblongifolia* and *Pterophylla camellifolia*. Autrum[3,4] showed that in *Tettigonia cantans*, *T. viridissima*, and *Decticus verrucivorus* the region of hearing extends at least up to 90,000 c/s, and probably beyond. The maximal sensitivity occurs between 10,000 and 60,000 c/s[69] (Fig. 262).

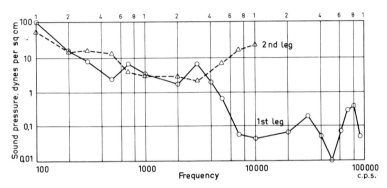

Fig. 262. *Conocephalus strictus*. Thresholds from the first and second legs of one side of a specimen of *C. strictus*. (From Wever and Vernon[69].)

Very high frequencies can be detected with the tympanal organ also in many Acrididae: in *Locusta migratoria migratorioides*[45], *Arphia sulfurea*[66], *Paroxya atlantica*[68,69], the region of maximal sensitivity lies between 3000 and 15,000 c/s (Fig. 263); in *Anacridium aegyptum*[1] it lies above 20,000 c/s, in *Sphingonotus coeruleans* above 37,000 c/s.

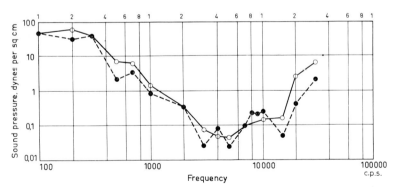

Fig. 263. *Paroxya atlantica paroxyoides*. Auditory thresholds for the two tympanal organs of a grasshopper. The curves show the minimum sound pressure required to produce impulses in the tympanal nerve. (From Wever and Vernon[69].)

As a rule, however, the region of maximal auditory sensitivity lies at lower frequencies in the Acrididae than in the Tettigoniidae. The curve for auditory threshold in *Locusta migratoria* and in *Arphia* already begins to flatten out at 5000

References p. 431

to 8000 c/s, whereas the sensitivity in *Tettigonia* and *Decticus* is still increasing above 10,000 c/s. The maximal sensitivity in *Locusta migratoria* lies near 15,000 c/s[14], in *Tettigonia* and *Decticus* it certainly occurs at much higher frequencies[69].

The region of maximal sensitivity is lowest in Grylloidea[69] (Fig. 264) and Cicadidae. Wever and Bray[67] found no responses of the tympanal nerve between 6000 and 11,000 c/s in *Gryllus assimilis*. Also in *Gryllus* (*Acheta*) *campestris* there is hearing only below 8000 c/s. The lower auditory limits, however, occur at relatively lower frequencies (about 300 c/s), whereas in Tettigoniidae and Acrididae it is near 800–1000 c/s. The maximal sensitivity of Cicadidae is about 1300 c/s (*Tanna japonensis*) and around 3000–4000 c/s in other Japanese cicadas[37]. Cicadas are not able to hear tones above 20,000 c/s.

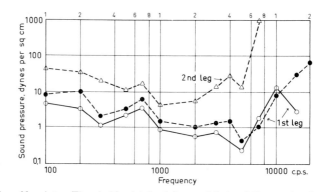

Fig. 264. *Gryllus abbreviatus*. Thresholds obtained from crickets of the species *G. abbreviatus*. The two lower curves represent thresholds obtained from the forelegs, and the upper curve those obtained from the second leg. (From Wever and Vernon[69].)

In numerous Tettigoniidae and Grylloidea, there occur *rudimentary scolopophorous organs* in the second and third pairs of legs, which are comparable to the tympanal organs but have no tympanum. Their auditory sensitivity is lower than that of the tympanal organs, and they do not react to frequencies above 10,000 c/s[4,69].

(*ii*) *Vibration receptors*. In insects, two main groups can be distinguished with respect to sensitivity towards vibration stimuli from the ground (Fig. 265 and Table 30)[11]: in the one group, the action potentials obtained from the extremities upon vibration of the ground consist in asynchronous impulses. The absolute vibration sensitivity may be very high; at optimal frequency it lies between 0.004 and 57 mμ. The optimal excitation frequency is between 1000 and 3000 c/s. All Orthoptera, Hymenoptera and Lepidoptera so far investigated belong to this group (Table 30, Group 1a, 1b). The vibration-sensitive organs are true subgenual organs, situated in the tibia.

In the second group, the action potential derived from the nerve of the extremity consists of synchronous or of irregular impulses, with a much lower frequency than those of the first group (Table 30, Group 2). The sensitivity in the second group is much lower than that in the first; the optimal frequencies lie between 200 and 400 c/s, and the threshold amplitudes between 1300 and 6600 mμ. No action potentials can be observed in the second group at higher frequencies. The Hemiptera, Diptera and Coleoptera belong to this group. The vibration receptors are chordotonal organs, located between tibia and tarsus, or hair sensillae on the articular membranes. Subgenual organs are missing in this group.

TABLE 30

VIBRATION RECEPTION IN INSECTS

	Insect	Optimal frequency (c/s)	Threshold amplitude at optima frequency (mμ)	at 400 c/s (mμ)
1. (a)	Periplaneta americana	1400	0.004	0.72
	Locusta cantans	2000	0.068	0.64
	Tachycines asynamorus	1000	0.1	2.2
	Liogryllus campestris	1500	0.1	1.6
	Pyrameis atalanta	3000	0.29	26
	Satyridae	2000	0.4	2.0
	Acrididae	3000	0.9	12.0
(b)	Apis mellifica	2500	13	400
	Andrena nitida	1000	18	1490
	Camponotus sp.	2000	19	177*
	Bombus soroeensis	1500	32	270
	Vespa crabro	1500	36	1950*
	Agrotis sp.	1200	57	630
2.	Rhodnius prolixus	400	1300	
	Carabus hortensis	300	5500	
	Eristalis sp.	200	6590	

* Threshold amplitude at 300 c/s.

Liesenfeld[39] found the lower limits of vibration reception in the spider, *Zygiella x-notata*, to be about 10 c/s, as determined in behavioral experiments. Since the resonance frequency of the radial threads of the net (where the lurking spider places her legs) is about 5.2 c/s, the resonance oscillations of the net are thus not perceived. Walcott and Van der Kloot[63] claim that the vibration receptors of the spider *Achaearanea tepidariorum* (lyriform organs on the tarsus, near its articulation with the metatarsus) respond to sound frequencies between 20 and 45,000 c/s. In their experiments, however, vibrations of the electrodes or of the ground, as a result of the sound vibrations, have not been excluded. The sensitivity curves published by these authors suggest that the sound had caused resonance vibrations in the apparatus, rather than exciting the vibration receptors directly. The vibration threshold of the lyriform organs of *Achaearanea* is near 25 Å, when vibrations of about 200 c/s are applied directly. It is claimed that the individual receptors of the lyriform organs are tuned to certain frequencies, so as to make frequency differentiation possible. Walcott and Van der Kloot[63] could not, however, furnish convincing proof for this claim.

(b) *Absolute sound sensitivity*

Absolute sound sensitivities have been determined for the tympanal organs of *Tettigonia*[3,6], for the antennal sound receptors of *Aedes aegypti*[38] and for the subgenual organs of *Periplaneta*[6,7]. Under optimal stimulus conditions, the threshold power which must be applied to the sensory organs was as follows:

Tympanal organ of *Tettigonia*	4.4×10^{-17} watt
Antenna of *Aedes*	10^{-13}–10^{-14} watt
Subgenual organ of *Periplaneta*	5.9×10^{-17} watt
Ear (human)	8×10^{-18}–4×10^{-17} watt

The smallest absolute amplitudes which still act as a stimulus are about 10^{-10} cm for the subgenual organ of *Periplaneta* and are thus at the borderline of what is physically possible[3,6]. For physical reasons, the organs could not be more sensitive than they in fact are. The threshold amplitude for sound reception in *Aedes* is about 0.1–0.2 μ, computed on the basis of an antennal length of 1 mm[38].

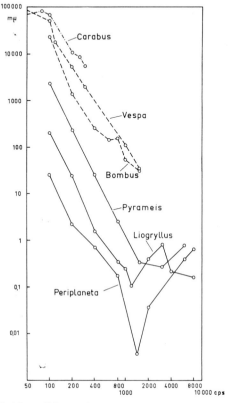

Fig. 265. Vibration thresholds at different frequencies for a few characteristic species of different sensitivities. Abscissa: vibration frequency. Ordinate: threshold amplitude in mμ (= 10^{-7} cm). (From Autrum and Schneider[11].)

(c) *Principles of function of the sound-receiving apparatus*

(i) The typical *vibration receptors* (subgenual organs of insects, lyriform organs of spiders) are unable to deliver information to the central nervous system about vibration frequency. They lack the ability for frequency analysis. Only the very simple chordotonal organs in the legs of flies, Coleoptera and Hemiptera are an exception to this rule[57]: they respond to very low frequencies, in the range from 50 to 300 c/s, with spikes which are synchronous with the vibrations.

Since the excitation of the subgenual organs is not synchronous with stimulus frequency, the periodic stimulus must be transformed into a constant, aperiodic one (rectified). Autrum[5] has proposed a hypothesis to explain the mechanism of this transformation in the subgenual organ: rotating eddies are said to arise in the blood canal above and below the subgenual organ, when the tibia is set into vibration. The pressure of these eddies then excites the subgenual organ. In models, the existence of such eddies can be demonstrated.

(ii) The hairs, antennae, and tympanal organs of the arthropods, being sound receptors, are receptors of displacement, physically speaking (Fig. 266); they react to pressure *gradients*, or to periodic displacement of air molecules. Essentially this ability depends on the condition that their dimensions are small in comparison with sound waves, so that no periodic pressure differences can develop on the surface of the hair or tympanum[2].

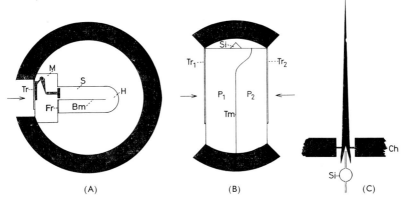

Fig. 266. Schematic representation of the various auditory receptors in the animal kingdom. (A) *Pressure receptors*: the mammalian ear. The tympanum is accessible to sound waves on only one side. The head acts as a firm casing. The apparatus of the middle ear (hammer, anvil, stirrup) inserts on the tympanum, and conducts movements of the tympanum to the inner ear (cochlea, with Organ of Corti). Bm, basilar membrane; Fr, round window; H, helicotrema; M, middle ear; S, cochlea; Tr, tympanum. (B) *Displacement receptors*: The tympanal organs of the grasshoppers (e.g. Tettigoniidae) as *pressure-gradient receptors*. The sound pressure affects the two opposing tympana Tr_1 and Tr_2, which are coupled with the tracheal membrane Tm through the air cushions P_1 and P_2 of the tracheal air cavities. The difference between the pressures acting on Tr_1 and Tr_2 is the pressure which acts on Tm. Si, sensory cells of the Crista acustica. The tympana in A and B consist of two parts: a rigid part (drawn double in the diagram) which turns like a door about an axis in its upper edge; and a soft-skinned part (drawn with single lines in diagram) which make possible the movement of the rigid portion. (C) *Displacement receptor*: mobile hairs of the arthropods. The hair is movably articulated in the chitin armor (Ch); a sensory cell inserts on its basis. (From Autrum[5].)

Since pressure gradients and the displacement of air molecules always have a certain directional characteristic, displacement receptors should in principle be capable of furnishing information about the direction from which the sound waves come, or the direction in which they proceed. If one rotates a receiver, the response of which depends on the direction of the sound, within the field of sound, its reactions should show maxima and minima; it then has a directional characteristic.

Correlation between the response of sound receptors and their orientation within the field of sound has been demonstrated for the tympanal organs of Orthoptera by Autrum[3], Pumphrey[43] and Autrum, Schwartzkopff and Swoboda[12].

The sound-sensitive hairs of the insects and spiders and the insect antennae reacting to sound are also actually displacement receptors. Whether they have a directional characteristic or not depends on the construction of the stimulus-perceiving structures (hair, antennae) and their movability with respect to the body (kind of insertion), as well as on the anatomical position of the sensory cells. Hairs which are

References p. 431

equally flexible in all directions will of course be set into movement equally by sounds from any direction parallel to their plane of insertion. And if the sensory cells insert in a fashion radially symmetric to the longitudinal axis of the hair, then such a hair need not have any directional characteristic, in a sensory-physiological sense. This means that *a priori* they need not furnish the central nervous system with any information about the direction of displacement within the sound wave. These problems have not been investigated for the sensory hairs of insects and spiders.

(iii) Tischner[58] has developed a hypothesis for the *Johnston's Organs* of male *Culicidae*. These organs lie at the basis of the long, hairy flagellar portion of the antennae, which is set into motion by the field of sound. This organ is also a displacement receptor[2,5]. Movement of the air molecules in the field of sound, in a direction parallel to the length of the flagellum, causes a movement of the basal plate (Pl, Fig. 253), parallel to its own plane. Thus through a movement towards the base of the antenna, the distal sensory cells of the outer sensory cell (Sc A) ring are stretched and through a counterdirected movement the proximal sensory cells are stretched.* A sinusoidal tone must thus effect an action potential (and thereby an excitation) with a frequency twice that of the stimulating frequency; this has indeed been observed. In addition, the sensory cells of the inner ring (Sc B) respond with the same frequency as the stimulating frequency, the basic frequency. If, however, a sound component occurs which is perpendicular to the longitudinal axis of the flagellum, then the plate (pl) will be tilted. Then the sensory cells of the inner ring (ScB) will also respond with double the stimulating frequency. This tendency for the basic frequency to disappear becomes increasingly pronounced as the displacement component parallel to the flagellum becomes smaller. Since in *Anopheles* the antennae have a sharp resonance frequency near 380 c/s, the directional information received by the relative proportions of the basic frequency and the first harmonic is unambiguous. These views can be supported electrophysiologically: from Johnston's Organ, one obtains varying proportions of the respective potentials (the basic frequency and its first harmonic), depending on the direction of the field of sound.

(iv) The tympanal organs of insects are also displacement receptors[3,5,8,43] (Fig. 266). The wavelengths of the stimulating sound approach the dimensions of the tympanum only at much higher frequencies, in the ultrasound region. One does not find any relationship between the stimulating sound frequency and the response of the animals (or of the tympanal nerves), by means of either behavioral or electrophysiological analysis. The frequency of the impulses in the nerves or in the central nervous system is asynchronous within the entire spectrum of sound leading to excitation of the tympanal organs. The tympanal organs are unable to distinguish between sinusoidal tones of different frequencies, nor are they able to analyze the components in a mixed sound.

Katsuki and Suga[37], in leading off from individual ganglion cells in the prothoracal ganglion of *Gampsocleis buergeri* (Tettigoniidae), find two kinds of ganglion cells: one is sensitive mainly to higher frequencies, the other to lower frequencies. The assumption of these authors that this finding offers a possibility for frequency

---

* For the mechanoreceptors of the Johnston's Organ, stretching—*not* compression or shortening—is the stimulus leading to excitation[61].

differentiation does not seem to be quite conclusively proved: The tympanal nerve in Tettigoniidae contains afferent fibers both from the tympanal organ and from the subgenual organ, whose frequency characteristics are quite different. It is also quite possible that the two types of neurons represent tonic and phasic units, so that one type responds to transients, the other to constant sound, as made probable by Autrum[10] for *Tettigonia* (see below).

Tympanal organs are very sensitive to amplitude modulation of sound[47]; especially for very steep and sudden modulation, that is, toward *transients*[15]. *Tettigonia* reacts (1) to constant sound (with asynchronous impulses in the tympanal nerve; tonic response) and (2) to transients consisting of single impulses or of sequences of few impulses (phasic response). The reactions probably differ in their basic nature, and stem from different receptors of the tympanal organ. This is suggested since (1) their latency periods are different—the electrical response to transients in the tympanal nerve of *Tettigonia* appears after about 2 msec, the response to non-modulated sound after about 4 msec; (2) the impulses arising as a reaction to the constant (non-modulated) sound appear during the refractory period of the impulses elicited by transients[10]. The steeper the transients, the larger is the electrical response in the tympanic nerve. A single cycle of a sound wave within the audible spectrum is sufficient to elicit an impulse; in *Tettigonia* this means one cycle of sound between 10,000 and 100,000 c/s. In contrast, the reaction to non-modulated sound in *Tettigonia* (as measured by the nerve impulse) requires a tone lasting at least 1 msec. The tympanal organs react to modulation frequencies or frequencies of transients up to 300 or even 400 per second, with sufficiently high stimulus intensities. Naturally, no lower frequency limit exists. Upon application of 80 sufficiently steep modulations (transients) per second, the reactions of the tympanal nerve are almost as large as upon application of a single, equally steep modulation.

It seems apparent, therefore, that the tympanal organs of Orthoptera react to transients as well as to non-modulated sounds, and with seemingly different receptors of the tympanal organ. Thus the following parameters are available for the recognition of species-specific songs: (1) the rhythm of the song (frequency of the modulations or of the transients); (2) the steepness of the transients; and (3) the duration of the individual sound impulses. Sufficient variables are available, therefore, to label and identify the songs of species members, including the behavioral variations.

Volleys of nerve impulses, synchronous with the sound impulses, arise in the tympanal organs of male and female cicadas, when the acoustic organs are stimulated by sounds produced by other cicadas[41,42]. Thus the tympanal organs of cicadas also respond to pulse-modulations in song. In this species, the tympanum is tightened by a tensor muscle (Fig. 259) during emission of the animal's own song, so that it is protected against the high sound intensity produced by the animal itself. In four kinds of Acrididae, Haskell[30,31,32] also showed that the tympanal organs respond to the songs from other individuals with volleys of nerve impulses synchronous with the pulse frequency. The flight sound of *Schistocerca gregaria* likewise produces impulse volleys in the tympanal nerves of members of the same species; these volleys are synchronous with the wing beat frequency[32].

The thoracic tympanal organs of Noctuidae are not able to differentiate frequencies[52]. The only two sensory cells which insert at the tympanum have different thresholds; they respond to clicks and constant sounds in the region between 3000 and 240,000 c/s. Clicks and short sound pulses elicit after-impulses in the nerve fibers; there is a fast, but incomplete adaptation for constant sounds, as measured by the number of impulses in the nerve fibers. In the tympanal nerve of *Prodenia evidania* there is a receptor unit, the impulses of which cannot be influenced by acoustical stimulation; the corresponding sensory cell does not lie at the tympanum, but on the so-called "stirrup" or Bügel (Fig. 255). This sensory cell is probably a proprioceptive organ which responds to deformations of the chitinous tympanal frame. Roeder and Treat[52] have proposed two hypotheses, as yet not experimentally tested, to explain the role of this acoustically unresponsive unit: (1) in flight, the tympanal organ might be deformed with the rhythm of the flying movements, through the muscles which insert on it. This would lead to changes in ultrasound sensitivity, synchronous with the wingbeat frequency. Thus the detection of sound pulses coming from the wing movements might be suppressed. (2) Alternatively, the tympanal frames and tympanal muscles might neutralize deformations of the tympanum which would result from active wing movements; thus a constant sensitivity for ultrasound could be maintained. The non-acoustic unit would be, in both cases, the feedback mechanism for the suggested functions. There are no experimental studies of the significance of the accessory apparatus (tympanal muscles, etc.).

(d) *Orientation toward acoustic signals*

(i) Acoustic signals can release non-oriented reactions in Invertebrates. For example, many insects and other arthropods respond to loud noise with fright or alarm reactions. Numerous Noctuids and Geometrids accelerate, and deviate from their previous direction of flight, when hit by ultrasound; or they stop flying and fall to the ground. Crawling Lepidoptera take flight, sitting ones run away[55]. In their normal surroundings, ultrasound is emitted by their principal natural enemies, the bats, which they can often avoid with these escape reactions. Reaction times are dependent on temperature and sound intensity; the reactions provide successful escape maneuvers for these Lepidoptera[60]. Non-oriented reactions have been described for many insects[15,18,27,58]. The swarming males of *Culicidae*[53] and of *Ceratopogonidae*[23] erect the hairs of the antennal flagellum, when hit by sound to which they are sensitive. Ant lions (*Myrmeleontidae*) react to the vibrations produced by their prey; the receptors are probably located in or on the first thoracal segment[49]. *Palinurus vulgaris* also reacts to sound with non-oriented antennal movements[22].

(ii) Displacement receptors are dependent on the vectorial magnitude of displacement of air molecules in the field of sound; in principal they can be used to determine the direction of sound.

Male *Culicidae* are attracted by the flight sound of the females (Roth[53], including summary of the older literature; see also Tischner[58,59]). For the antennae of *Culicidae*, it has been shown[38,58,59] that action potentials, arising on the Johnston's Organ, contain the basic frequency (resonance frequency) and the first harmonic. The relation of the amplitude of the first harmonic to the amplitude of the basic frequency (the distortion factor) varies with the direction which the incoming sound takes with respect to the longitudinal axis of the antennal flagellum. When the sound is parallel to the flagellum (angle of sound 0°) the distortion factor is 0; only the basic frequency is found in the action potential. The distortion factor increases to about 250 % when the direction of sound is perpendicular to the antenna; it is almost independent of the amplitude of antennal movement, and independent of the stimulus frequency. Thus the distortion factor is an indicator of the direction of the sound source, relative to the animal. According to this, one antenna should suffice for direction-finding; this has been confirmed by behavioral experiments[53]: males with only one antenna were easily able to find females, which they detected by means of their flight sound.

(*iii*) The abdominal and tibial tympanal organs can localize the direction of sound. Equal sounds will be heard with different intensity depending on the direction from which they come (Autrum[3], for *Tettigonia viridissima*; Autrum, Schwartzkopff and Swoboda[12], Pumphrey[43] for *Locusta migratoria*) (Fig. 267). There have as yet been no behavioral experiments to show how this is utilized by the Acrididae for orientation with respect to the sound source. Autrum[3] has proposed a hypothesis for the orientation towards sound source in *Tettigonia* and *Decticus*, using the tibial tympanal organs: orientation would depend on a collaboration of right and left tympanal organs. If the animal crawls, the movement of the legs causes a sound source to be heard with greater or smaller intensity, according to its position relative to the plane of symmetry of the tympanal organs.

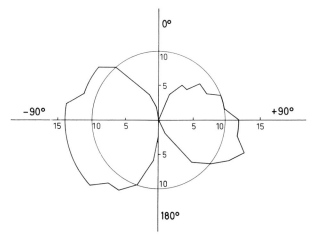

Fig. 267. *Locusta*. Sensitivity of an isolated tympanic organ plotted on polar coordinates as a function of the direction of incidence of the test stimulus. Sensitivity (log reciprocal of threshold amplitude) is plotted radially and the minimum sensitivity is arbitrarily taken to be zero. The line 0–180° lies in the sagittal plane of the animal, and for angles of positive or negative sign the test stimulus is incident on the external or internal aspect of the tympanic organ respectively. (From Pumphrey[43].)

Let us suppose that the sound comes from a direction which makes the angle x with the longitudinal axis of the body; the grasshopper begins by moving the leg situated on the side corresponding to the sound source, turning it in such a way that the intensity clearly decreases. The critical zone in the directional diagram (Figs. 269, 270) reveals the position of the sound source. The leg facing the sound source cannot assume all possible angles toward the source; normally, a leg can be turned forward until the femur makes an angle of 40° with the longitudinal axis of the body. But the movement is stopped before the insect reaches an angle of $40° + x$, at which the sound source penetrates the critical zone of the tympanic organ, where the intensity of the perceived tone diminishes suddenly. In turn, the symmetrical

References p. 431

leg, further removed from the sound source, follows the direction of rotation forward. The combination of these two movements permits the animal to turn to the side whence the sound comes. The angle x is decreased with each step and, finally, the insect advances directly toward the sound source.

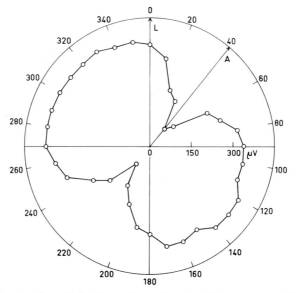

Fig. 268. *Calliphora erythrocephala*. Dependence of potential size on sound direction. The head of the animal is directed upwards in the plane of the sketch. The potential was led off from the right antenna, after the left had been removed. Spontaneous activity 70 μV, stimulus 150 c/s, 80 dB. (Original.)

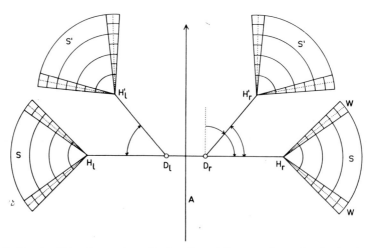

Fig. 269. *Tettigonia*. This scheme shows the location of the area of poor hearing(s) during movement of the leg. A, longitudinal axis of the grasshopper; D_l, D_r, left and right axes of rotation, respectively, of the legs on the body; H_l, H_r, location of the left and right tympanal organs, respectively, as the leg makes its maximal backwards rotation. H'_l, H'_r, location of left and right tympanal organs as the leg makes its maximal forwards rotation; W, tangent to point of inflection. (From Autrum[5].)

This hypothesis still awaits experimental verification. According to Regen[48], the females of *Gryllus campestris* are able to advance in the direction of the male even when the tympanic organ of one side is damaged. However, whereas the walk of insects provided with both their tympanic organs is a rather imposing, skilful performance, the females bearing these lesions orient themselves toward the sound source with much less surety. Therefore, these observations certainly do not positively contradict the hypothesis.

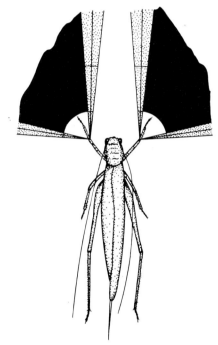

Fig. 270. *Tettigonia cantans*. The range of reduced sensitivity of the tympanic organs is shown in black. Sound coming from this direction is perceived less strongly than that originating from other directions. When a sound source is outside of the range shown in black and enters the stippled area, the loudness increases strongly. (From Autrum[8].)

REFERENCES

[1] AUGER, D. and A. A. FESSARD, 1928, Observations sur l'excitabilité de l'organe tympanique du criquet, *Compt. rend. soc. biol.*, 99, 400–401.
[2] AUTRUM, H., 1936, Über Lautäusserungen und Schallwahrnehmung bei Arthropoden. I. Untersuchungen an Ameisen. Eine allgemeine Theorie der Schallwahrnehmung bei Arthropoden, *Z. vergleich. Physiol.*, 23, 332–373.
[3] AUTRUM, H., 1940, Über Lautäusserungen und Schallwahrnehmung bei Arthropoden. II. Das Richtungshören von Locusta und Versuch einer Hörtheorie für Tympanalorgane vom Locustidentyp, *Z. vergleich. Physiol.*, 28, 326–352.
[4] AUTRUM, H., 1941, Über Gehör und Erschütterungssinn bei Locustiden, *Z. vergleich. Physiol.*, 28, 580–637.
[5] AUTRUM, H., 1942, Schallempfang bei Tier und Mensch, *Naturwiss.*, 30, 69–85.
[6] AUTRUM, H., 1943, Über kleinste Reize bei Sinnesorganen, *Biol. Zentr.*, 63, 209–236.
[7] AUTRUM, H., 1948, Über Energie- und Zeitgrenzen der Sinnesempfindungen, *Naturwiss.*, 35, 361–369.

[8] AUTRUM, H., 1955, Analyse physiologique de la réception des sons chez les Orthoptères. *Ann. inst. natl. recherche agron.*, *Sér. C, Ann. epiphyt.*, fascicule spécial: L'acoustique des Orthoptères, *6*, 338–355.
[9] AUTRUM, H., 1959, Nonphotic receptors in lower forms. *Handbook Physiol.-Neurophysiol.*, Am. Physiol. Soc., Vol. I, Chapter 16, p. 369–385.
[10] AUTRUM, H., 1960, Phasische und tonische Antworten vom Tympanalorgan von *Tettigonia viridissima*, *Acustica*, *10*, 339–348.
[11] AUTRUM, H. and W. SCHNEIDER, 1948, Vergleichende Untersuchungen über den Erschütterungssinn der Insekten, *Z. vergleich. Physiol.*, *31*, 77–88.
[12] AUTRUM, H., J. SCHWARTZKOPFF and H. SWOBODA, 1961, Der Einfluss der Schallrichtung auf die Tympanalpotentiale von *Locusta migratoria* L., *Biol. Zentr.*, *80*, 385–402.
[13] BÄSSLER, U., 1958, Versuche zur Orientierung der Stechmücken: Die Schwarmbildung und die Bedeutung des Johnstonschen Organs, *Z. vergleich. Physiol.*, *41*, 300–330.
[14] BUSNEL, M. C. and R. G. BUSNEL, 1956, Sur une phonocinèse de certains Acridiens à des signaux acoustiques synthétiques, *Compt. rend.*, *242*, 292–293.
[15] BUSNEL, R. G., 1956, Étude de l'un des caractères physiques essentiels des signaux acoustiques réactogènes artificiels sur les Orthoptères et d'autres groupes d'insectes, *Insectes sociaux*, *3*, 11–16.
[16] BUSNEL, R. G. and W. LOHER, 1953, Recherches sur le comportement de divers Acridoides mâles soumis à des stimuli acoustiques artificiels, *Compt. rend.*, *237*, 1557–1559.
[17] BUSNEL, R. G. and W. LOHER, 1954, Sur l'étude du temps de la réponse au stimulus acoustique artificiel chez les Chorthippus et la rapidité de l'intégration du stimulus, *Compt. rend. soc. biol.*, *148*, 862.
[18] BUSNEL, R. G., W. LOHER and F. PASQUINELLY, 1954, Recherches sur les signaux acoustiques synthétiques réactogènes pour divers Acrididae ♂, *Compt. rend. soc. biol.*, *148*, 1987.
[19] CHRYSANTHUS, F., 1953, Hearing and stridulation in spiders, *Tijdschr. Entomol.*, *96*, 57–83.
[20] DEBAISIEUX, P., 1935, Organes scolopidiaux des pattes d'insectes. I. Lépidoptères et Trichoptères, *Cellule rec. cytol. histol.*, *44*, 273–314.
[21] DEBAISIEUX, P., 1938, Organes scolopidiaux des pattes d'insectes. II. *Cellule*, *47*, 79–202.
[22] DIJKGRAAF, S., 1955, Lauterzeugung und Schallwahrnehmung bei der Languste (*Palinurus vulgaris*), *Experientia*, *11*, 330.
[23] DOWNES, J. A., 1955, Observation on the swarming flight and mating of Culicoides (Diptera: Ceratopogonidae), *Trans. Roy. Entomol. Soc. (London)*, *106*, 213–236.
[24] EGGERS, F., 1928, Die stiftführenden Sinnesorgane; Morphologie und Physiologie der chordotonalen und der tympanalen Sinnesapparate der Insekten, *Zool. Bausteine*, *2*, 353.
[25] FRINGS, H. and M. FRINGS, 1956, Reactions to sounds by the wood nymph butterfly *Cercyonis pegala*, *Ann. Entomol. Soc. Am.*, *49*, 611–617.
[26] FRINGS, H. and M. FRINGS, 1958, Uses of sounds by insects, *Ann. Rev. Entomol.*, *3*, 87–106.
[27] FRINGS, H. and M. FRINGS, 1959, Reactions of swarms of Pentaneura aspera (Diptera: Tendipedidae) to sound, *Ann. Entomol. Soc. Am.*, *52*, 728–733.
[28] FRINGS, M., 1955, Bibliographie générale sur l'acoustique des insectes. *Ann. inst. natl. recherche agron.*, *Sér. C, Ann. epiphyt.*, fascicule spécial: L'acoustique des Orthoptères, *6*, 401–446.
[29] GRAY, E. G. and R. J. PUMPHREY, 1958, Ultra-structure of the insect ear, *Nature*, *181*, 618.
[30] HASKELL, P. T., 1956, Hearing in certain Orthoptera. I. Physiology of sound receptors, *J. exptl. Biol.*, *33*, 756–766.
[31] HASKELL, P. T., 1956, Hearing in certain Orthoptera. II. The nature of the response of certain receptors to natural and imitation stridulation. *J. exptl. Biol.*, *33*, 767–776.
[32] HASKELL, P. T., 1957, The influence of flight noise on behaviour in the desert locust *Schistocerca gregaria* (Forsk.), *J. Insect Physiol.*, *1*, 52–75.
[33] HASKELL, P. T. and P. BELTON, 1956, Electrical responses of certain lepidopterous tympanal organs, *Nature*, *177*, 139–140.
[34] HORRIDGE, G. A., 1960, Pitch discrimination in Orthoptera (Insecta) demonstrated by responses of central auditory neurones, *Nature*, *185*, 623–624.
[35] HUGHES, G. M., 1952, Abdominal mechanoreceptors in Dytiscus and Locusta, *Nature*, *170*, 531.
[36] KASTON, B. J., 1935, The slit sense organs of spiders, *J. Morphol.*, *58*, 189–209.
[37] KATSUKI, Y. and N. SUGA, 1958, Electrophysiological studies on hearing in common insects in Japan, *Proc. Japan Acad.*, *34*, 633–638.
[38] KEPPLER, E., 1958, Über das Richtungshören von Stechmücken, *Z. Naturforsch.*, *Pr.b, 13*, 280–284.
[39] LIESENFELD, F. J., 1956, Untersuchungen am Netz und über den Erschütterungssinn von *Zygiella x-notata* (Cl.) (Araneidae), *Z. vergleich. Physiol.*, *38*, 563–592.
[39a] LIESENFELD, F. J., 1961, Über Leistung und Sitz des Erschütterungssinnes von Netzspinnen, *Biol. Zentr.*, *80*, 465–475.
[40] MCINDOO, N. E., 1911, The lyriform organs and tactile hairs of Araneads, *Proc. Acad. Nat. Sci. Phila.*, *63*, 375–418.

[41] Pringle, J. W. S., 1953, Physiology of sound in Cicadas, *Nature*, *172*, 248–249.
[42] Pringle, J. W. S., 1954, The mechanism of the myogenic rhythm of certain insect striated muscles, *J. Physiol. (London)*, *124*, 269–291.
[43] Pumphrey, R. J., 1940, Hearing in insects, *Biol. Revs. Cambridge Phil. Soc.*, *15*, 107–132.
[44] Pumphrey, R. J. and A. F. Rawdon-Smith, 1936a, Hearing in insects: the nature of the response of certain receptors to auditory stimuli, *Proc. Roy. Soc. (London) B*, *121*, 18–27.
[45] Pumphrey, R. J. and A. F. Rawdon-Smith, 1936b, Sensitivity of insects to sound. *Nature*, *137*, 990.
[46] Pumphrey, R. J. and A. F. Rawdon-Smith, 1936c, Synchronized action potentials in the cercal nerve of the cockroach (*Periplaneta americana*) in response to auditory stimuli, *J. Physiol. (London)*, *87*, 4P–5P.
[47] Pumphrey, R. J. and A. F. Rawdon-Smith, 1939, "Frequency discrimination" in insects: a new theory, *Nature*, *143*, 806.
[48] Regen, J., 1926, Über die Beeinflussung der Stridulation von *Thamnotrizon apterus* Fab. Männchen durch künstlich erzeugte Töne und verschiedenartige Geräusche, *Sitz. ber. Akad. Wiss. Wien, Math.-Naturw. Kl., Abt. I*, *135*, 329–368.
[49] Richard, G., 1952, Contribution à l'étude de la biologie des fourmilions, *Bull. soc. zool. France*, *77*, 252–263.
[50] Risler, H., 1953, Das Gehörorgan der Männchen von *Anopheles stephensi*. *Zool. Jahrb., Abt. Anat. u. Ontog. Tiere*, *73*, 165–186.
[51] Risler, H., 1955, Das Gehörorgan der Männchen von *Culex pipiens* L., *Aedes aegypti* L. und *Anopheles stephensi* Liston, eine vergleichend morphologische Untersuchung, *Zool. Jahrb., Abt. Anat. u. Ontog. Tiere*, *74*, 478–490.
[52] Roeder, K. D. and A. E. Treat, 1957, Ultrasonic reception by the tympanic organ of noctuid moths. *J. exptl. Zool.*, *134*, 127–158.
[53] Roth, L. M., 1948, A study of mosquito behavior. An experimental laboratory study of the sexual behavior of *Aedes aegypti* (L.). *Am. Midland Naturalist*, *40*, 265–352.
[54] Schaller, F., 1951, Lauterzeugung und Hörvermögen von *Corixa* (*Callicorixa*) *striata* L., *Z. vergleich. Physiol.*, *33*, 476–486.
[55] Schaller, F. and C. Timm, 1949, Schallreaktionen bei Nachtfaltern, *Experientia*, *5*, 162.
[56] Schaller, F. and C. Timm, 1950, Das Hörvermögen der Nachtschmetterlinge, *Z. vergleich. Physiol.*, *32*, 468–481.
[57] Schneider, W., 1950, Über den Erschütterungssinn von Käfern und Fliegen, *Z. vergleich. Physiol.*, *32*, 287–302.
[58] Tischner, H., 1953, Über den Gehörsinn von Stechmücken, *Acustica*, *3*, 335–343.
[59] Tischner, H. and A. Schief, 1954, Fluggeräusch und Schallwahrnehmung bei *Aedes aegypti* L. (Culicidae). *Verhandl. deut. zool. Ges. Tübingen*, *51*, 453–460.
[60] Treat A. E., 1956, The reaction time of noctuid moths to ultrasonic stimulation, *J.N.Y. Entomol. Soc.*, *64*, 165–171.
[61] Uchiyama, H. and Y. Katsuki, 1956, Recording of action potentials from the antennal nerve of locusts by means of micro-electrodes, *Physiol. comparata et Oecol.*, *4*, 154–163.
[62] Vogel, R., 1923, Über ein tympanales Sinnesorgan, das mutmassliche Hörorgan der Singzikaden, *Z. Anat. Entwicklungsgeschichte*, *67*, 190–231.
[63] Walcott, Ch. and W. G. van der Kloot, 1959, The physiology of the spider vibration receptor, *J. exptl. Zool.*, *141*, 191–244.
[64] Weber, H., 1933, *Lehrbuch der Entomologie*, Fischer, Jena.
[65] Wellington, W. G., 1946, Some reactions of muscoid Diptera to changes in atmospheric pressure, *Can. J. Research, D*, *24*, 105–117.
[66] Wever, E. G., 1935, A study of hearing in the sulfur-winged grasshopper (*Arphia sulphurea*). *J. Comp. Psychol.*, *20*, 17–20.
[67] Wever, E. G. and Ch. W. Bray, 1933, A new method for the study of hearing in insects, *J. Cellular Comp. Physiol.*, *4*, 79–93.
[68] Wever, E. G. and J. A. Vernon, 1957, The auditory sensitivity of the atlantic grasshopper, *Proc. Natl. Acad. Sci. U.S.*, *43*, 346–348.
[69] Wever, E. G. and J. A. Vernon, 1959, The auditory sensitivity of Orthoptera, *Proc. Natl. Acad. Sci. U.S.*, *45*, 413–419.

CHAPTER 16

SOUND RECEPTION IN LEPIDOPTERA

by

ASHER E. TREAT

1. Historical Review

Comparatively few adult Lepidoptera make sounds that can be heard by man[11]. Perhaps it is mainly on this account that a sense of hearing in these insects was long doubted despite many claims for its existence[29]. To Charles Bonnet (1780) is ascribed the first published statement that lepidopterous larvae respond to sounds, but not until the present century were the sensory structures identified as integumentary hairs[1, 20, 21]. A sense of hearing has never been claimed for the pupae, though a few of these are known to produce sounds[14]. In 1790 Gabriel Bonsdorff of Åbo in Finland wrote of certain "Phalaenae" that "during the night, scarcely a voice could be raised than they would turn round very swiftly, and the antennae appear to be, as it were, convulsed". Kirby and Spence in 1826 contributed a similar observation, and shared Bonsdorff's supposition that a sense of hearing resided in the antennae. Romanes noted in 1876 that certain moths would cease to vibrate their wings when high pitched tones were sounded, which moved F. Buchanan White (1877)[34] to suggest, perhaps for the first time, that moths might avoid insect-eating bats through an ability to perceive their high-pitched squeaks.

In 1832 De Villiers[6] described swollen cuticular and membranous organs found beneath the hind wing bases of the sound-producing lithosiid *Chelonia pudica*. Similar structures, regarded as sound-producing organs, were soon discovered in many arctiid and related moths[8]. The true tympanic organs of these insects were still overlooked, however, even when Peter (1912)[23] gave evidence that resting females of the alpine lithosiid *Endrosa aurita ramosa* F. E. S. respond by synchronized and audible wing vibrations to sounds produced by the males in flight. Meanwhile Swinton (1877)[30] had given the first careful description of a lepidopterous tympanic organ in a noctuid (*Catocala* sp). Swinton surmised that this structure had an auditory function, and believed it homologous to the tympanic organs of the Acridiidae. Although Swinton's work was cited by Minot[22] in 1882, and although Jordan in 1905 called attention to "the sensory organ situated at the base of the abdomen" in the Geometridae, most entomologists took little notice of the tympanic organs until they were rediscovered many years later by Eggers (1911)[7] and by von Kennel (1912)[16]. In an admirable series of detailed studies, these authors (particularly the former) laid the foundation for all subsequent work. Their papers are too well known to require extensive citation; the most important references are given in von Kennel and Eggers (1933)[16]. Eggers' pupils, Heitmann (1934)[13] and Gohrbandt (1939)[10] greatly extended the knowledge of the tympanic organs, and these structures have become the basis for important revisions in concepts of moth phylogeny[18, 24, 28].

Although Eggers' experiments confirmed Swinton's idea of an auditory function for the tympanic organs, it remained for Schaller and Timm (1950)[27] to demonstrate that the most effective stimuli were those in the ultrasonic range of vibrational frequency. These authors also reported observations suggesting that the ultrasonic cries of bats could elicit behavioral effects similar to those observed in moths stimulated with artificially generated ultrasonics. Treat (1955)[31] studied a variety of behavioral responses by kymographic recording, and by the same method estimated the reaction times of two species of noctuids. Electrical responses from tympanic nerves of arctiids and notodontids were first recorded by Haskell and Belton (1956)[12], while Roeder and Treat (1957)[25] used electrical methods in analyzing the physiological characteristics of the noctuid tympanum. The latter authors reported tympanic nerve responses to flying bats, and in one instance to the wing beats of a moth in stationary flight. They also give evidence that a stretch-sensitive nervous element in addition to the scolopophorous sensilla occurs in the tympana of these moths.

True tympanic organs are not found in the Rhopalocera, the Sphingidae, Saturnoidea, or most of the microlepidoptera. Nevertheless, responses to acoustic stimuli, particularly to sounds of relatively low frequency and high intensity, have often been reported in atympanous species or in tympanotomized moths. The sensory structures in these insects have never been identified. Turner's results (1914)[32] with saturniid moths may be referable to the scolopophorous organs of the wing bases[19, 33] or to those of the legs[2, 5]. Where these have been eliminated, as in some of the experiments of Frings (1956, 1957)[9] displacement-sensitive hairs may be involved. The sense organs at the bases of the wing veins are in some Lepidoptera provided with tympanum-like membranes, but since they have not been proved to have an auditory function they will not be discussed here. The subgenual organs and acoustic hairs are also omitted from discussion.

2. Morphology of the Tympanic Organs

A comparison of the tympanic organs of different Lepidoptera suggests a diversity of evolutionary origins. Among the abdominal tympana it is possible to recognize several distinct types. These were classified by Sick (1935)[28] as (1) the geometrid and (2) the pyralid type, both occurring in the first segment, (3) the drepanid-cymatophorid type and (4) the uraniid-epiplemid types, both in the second segment, and (5) the little-known axiid type in the seventh segment*. Kiriakoff (1956)[18] has greatly expanded this classification. Recently a new type of abdominal tympanum, somewhat

---

* Forbes' figure of the "supposed tympanum" of *Cimelia* Led. (= *Axia* Hbn.) occurs on Plate III of his paper "The classification of the Thyatiridae", *Ann. Entom. Soc. America*, 29, (1936) 779–803. Only the skeletal parts are shown, and the text (p. 789) contains only a passing reference without description. Sick's mention of Forbes' work must have resulted from a then unpublished communication. The axiid tympanum does not appear to have been studied subsequently.

resembling that of the Pyralididae, has been discovered in a cossid moth*. The metathoracic tympana of the noctuoid families are less diversified, but even these are regarded by Kiriakoff as di- or triphyletic in origin.

A characteristic feature of a true tympanic organ is a tracheal air sac some portion of which borders upon a thin external cuticular membrane capable of being set in motion by the vibrational displacement of air particles. In most instances the scolopophorous sensilla (2 in the noctuoid and uraniid-epiplemid types, 4 in the others thus far studied) are enclosed in a cylindrical sleeve of tracheal epithelium attached at one end to the tympanic membrane. In the drepanid-cymatophorid tympana, however, the sensilla are spread between two layers of tracheal epithelium partially dividing a double internal chamber which borders elsewhere upon the exterior. Various accessory cavities and membranes, both internal and external, are often provided. Though sometimes described as resonators, these might more appropriately be deemed merely reflectors, since their dimensions are invariably small in comparison with the wave lengths to which the organs respond. These accessory structures are not innervated.

3. Physiology

In the moths most intensively studied (various noctuidae) the maximum sensitivity of the tympanic organs is for vibrational frequencies between 15 and 60 kc/s. In this range, even under unnatural experimental conditions, sound pressures as low as 0.01 dyne/cm^2 are adequate to excite the tympanic nerve, and it seems likely that considerably lower thresholds may prevail in nature. At higher intensities, responses have been recorded to pure tone frequencies as low as 3 and as high as 200 kc/s or higher, though it is possible that experimental artefacts were involved in the higher range since microphones adequate for monitoring these stimulus tones were not available. The two scolopes mediate responses of different intensity thresholds, but there is no evidence of frequency discrimination. The responses from the two units are distinguishable by their different spike heights. Adaptation to continuous pure tones is rapid but incomplete, spike frequencies falling from initial values as high as 1000 per second to about half that rate in 0.1 second and slowly declining thereafter. Brief after-discharge is observed following clicks and short tone pulses. An irregular "resting" discharge is usually noted.

In various types of abdominal tympanic organs, muscles have been described whose contraction is supposed to alter tension relationships. That these muscles are capable of changing the sensitivity of the organs is a plausible but untested conjecture. No such muscles are known in the noctuoid families, and although in these

---

* I am indebted to Dr. Harry K. Clench of the Carnegie Museum, Pittsburgh 13, Pennsylvania, for information regarding the tympana and "prototympana" found in certain of the Cossidae. Although referred to in print by Kiriakoff (1956)[18], CLENCH, *Mitt. Münchner Entom. Ges.*, *47*, 122–142, 1957, and others, these organs have not yet been thoroughly studied or described, and it is not known whether their function is auditory. The clearest examples are said to be found in the rare specimens of *Dudgeonea* Hampson, from Madagascar, Africa, India, New Guinea, and Australia. Tympanum-like structures have also been noted in some of the Psychidae (Clench, personal communication), in the Glyphipterygidae (Kiriakoff, *op. cit.*), and in a saturniid (GOHRBANDT, *Z. wiss. Zool.*, *Abt. A*, *149*, 537–600, 1936). Indeed Kiriakoff has proposed the hypothesis that "Tout groupe de Lépidoptères possède, ou a possédé, effectivement ou potentiellement des organes tympanaux."

moths a stretch-sensitive neurone closely associated with the acoustic sensilla contributes a continuous rhythmic discharge to the impulse pattern, its activity has not been found to have any relation to the acoustic response. Abdominal tympanic organs have as yet been little studied electrophysiologically. Those of *Tetracis crocallata* Gn. (Geometridae) show complex impulse patterns such as would be expected from the four scolopes of their sensilla (Treat and Roeder[1959]).

Behavioral responses to sound vary widely both in moths of different species and in a single insect at different times. Resting noctuids exposed to ultrasonic stimulation will sometimes instantly take flight. Often, however, they show only twitching of the antennae or slight postural changes. Repeated stimuli may arouse more vigorous responses such as running movements culminating in flight, or they may evoke no response at all. Some species when aroused appear to be especially reactive; others, particularly among the arctiids and amatids, momentarily cease activity at each stimulus. The character and intensity of response seems related to periodic variations in physiological condition and to the concurrent pattern of stimulation via other sensory channels.

During free flight the approach of a bat or a brief artificial sound with strong ultrasonic components often elicits a rapid swerving or diving movement. Among strong, high fliers such as many noctuids these movements as a rule merely alter the flight path in a seemingly erratic way. Weaker or lower fliers often dive into the grass and do not immediately resume flight. These reactions are best observed in dim light. Close to a strong light source the flight path is often too irregular for responses to sound to be recognized, although many moths while fluttering about a brightly lighted screen will show vigorous darting and whirling movements at the sound of a galton whistle or the jingling of keys or coins. The response decays in intensity, however, and after a few repetitions the stimulus becomes ineffective.

Little is known of the central nervous pathways concerned with acoustic behavior in these insects. Both thoracic and abdominal tympanic organs are innervated from the posterior part of the pterothoracic ganglion. Motor responses are greatly reduced though not always abolished by decapitation.

4. Ecological Significance

Despite known acoustic factors in the courtship of a few kinds of moths, it seems unreasonable on the basis of present evidence to suppose that these species represent the rule rather than the exception. Similarly tenuous is the suggestion that moths with tympanic organs may make use of echolocation. Both electrophysiological and behavioral evidence indicate that the acoustic senses of these insects are adapted to the detection of the ultrasonic cries of their predators the insectivorous bats. A systematic study of the interactions of bats and moths under field conditions is now in progress. It remains to be discovered whether the tympanic organs serve their possessors in the avoidance of other enemies, or in other contexts as yet unsuspected.

[For Addendum to this Chapter, see p. 800]

REFERENCES*

[1] ABBOTT, C. E., 1927, The reaction of *Datana* larvae to sounds, *Psyche*, *34*, 129–133.
[2] AUTRUM, H. and W. SCHNEIDER, 1948, Vergleichende Untersuchungen über der Erschütterungssinn der Insekten, *Z. vergl. Physiol.*, *31*, 77–88.
[3] BONNET, C., 1779, *Oeuvres d'histoire naturelle et de philosophie*, Tome I, Fauche, Neuchatel, p. 273; 286–287 (not, in this edition, Tome II, p. 36-37, as sometimes cited).
[4] BONSDORF(F) G. 1790, *Fabrica, usus, et differentiae antennarum in insectis*, Diss., Aboae, Finland, Original not seen; passage quoted from translation by J. SHARP, 1833, *Field Naturalist*, *1*, 292–299.
[5] DEBAISIEUX, P., 1935, Organes scolopidiaux des pattes d'insectes, I, Lépidoptères et trichoptères, *Cellule*, *44*, 271–314.
[6] DE VILLIERS, 1832, Observations sur l'Ecaille pudique de Godart, genre *Eyprepria* d'Ochs, *Ann. Soc. Entom. France*, [1] *1*, 203–204.
[7] EGGERS, F., 1911, Über das thoracale Tympanal-Organ der Noctuiden, *Sitzber. naturf. Ges. Univ. Jurjew (Dorpat)*, *20*, 139–145.
[8] FORBES, W. T. M. and J. FRANCLEMONT, 1957, The striated band (Lepidoptera, chiefly Arctiidae), *Lepid. News*, *11*, 147–150.
[9] FRINGS, H. and M., 1956, Reactions to sound by the wood nymph butterfly *Cercyonis pegala*, *Ann. Entom. Soc. Am.*, *49*, 611–617.
FRINGS, H. and M., 1957, Duplex nature of reception of simple sounds in the scape moth *Ctenucha virginica*, *Science*, *126*, 24.
[10] GOHRBANDT, I., 1939, Ein neuer Typus des Tympanalorgans der Syntomiden, *Zool. Anz.*, *126*, 107–116.
[11] HANNEMAN, H. J., 1956, Über ptero-tarsale Stridulation und einige ander Arten der Lauterzeugung bei Lepidopteren, *Deut. Entom. Z.*, *N.F. 3*: 14–27.
[12] HASKELL, P. T. and P. BELTON, 1956, Electrical responses of certain lepidopterous tympanal organs, *Nature*, *177*, 139-140.
[13] HEITMANN, H., 1934, Die Tympanalorgane flügunfähiger Lepidopteren und die Korrelation in der Ausbildung der Flügel und der Tympanalorgane, *Zool. Jb., Anat., 59*, 135–200.
[14] HINTON, H. E., 1948, Sound production in lepidopterous pupae, *Entomologist*, *81*, 254–269.
[15] JORDAN, K., 1905, Note on a peculiar secondary sexual character found among Geometridae at the sensory organ situated at the base of the abdomen, *Nov. Zool.*, *12*, 506–508.
[16] KENNEL, J. VON, 1912, Über Tympanal-organe im abdomen der Spanner und Zünsler, *Zool. Anz.*, *39*, 163–170.
KENNEL, J. VON and F. EGGERS, 1933, Die abdominalen Tympanalorgane der Lepidopteren, *Zool. Jb., Anat., 57*, 1–104.
[17] KIRBY, W. and W. SPENCE, 1826, *An Introduction to Entomology*, Vol. 4, London, p. 241.
[18] KIRIAKOFF, S. G., 1956, Sur l'origine et l'évolution des organes tympanaux phalénoïdes (Lépidoptères), *Bull. ann. soc. roy. entom. Belgique*, *92*, XI-XII, 289–300.
[19] LE CERF, F., 1926, Contributions à l'étude des organes sensoriels des Lépidoptères, *Encyc. Entom. sér. B, III, Lepidoptera*, *1*, 131–158.
[20] MINNICH, D. E., 1925, The reactions of the larvae of *Vanessa antiopa* to sounds, *J. Exp. Zool.* *42*, 443–469.
[21] MINNICH, D. E., 1936, The responses of caterpillars to sounds, *J. Exp. Zool.*, *72*, 439–453.
MINNICH, D. E., 1937, The reactions of fragments of the larvae of *Aglais antiopa* Linn. to sounds, *Bull. Mt. Desert Island Biol. Lab.*, p. 19–20.
[22] MINOT, C. S., 1882, Comparative morphology of the ear IV, *Am. J. Otology*, *4*, 89–168.
[23] PETER, K., 1912, Versuche über das Hörvermögen eines Schmetterlings (*Endrosa v. ramosa*), *Biol. Zentr.*, *32*, 724–731.
[24] RICHARDS, A. G., 1933, Comparative skeletal morphology of the noctuid tympanum, *Entomologia Americana*, *13* (n.s.) (1) p. 1–44.
[25] ROEDER, K. D. and A. E. TREAT, 1957, Ultrasonic reception by the tympanic organ of noctuid moths, *J. Exp. Zool.*, *134*, 127–158.
[26] ROMANES, G. J., 1876, Sense of hearing in birds and insects, *Nature*, *15*, 177.
[27] SCHALLER, F. and C. TIMM, 1950, Das Hörvermögen der Nachtschmetterlinge, *Z. vergl. Physiol.*, *32*, 468–481.
[28] SICK, H., 1935, Die Bedeutung der Tympanalorgane der Lepidopteren für die Systematik, *Verh. Deut. Zool. Ges.*, *37*, *Zool. Anz. Suppl.*, *8*, 131–135.
SICK, H., 1937, Die Tympanalorgane der Uraniden und Epiplemiden, *Zool. Jb., Anat. Gent. Tiere*, *63*, 351–398.

* References include key papers only. For short reviews and bibliographies the reader should consult these and FRINGS, M. and H., 1960, *Sound Production and Sound Reception by Insects, A Bibliography*, Pennsylvania State Univ. Press, University Park, Pennsylvania, 108 pp.

[29] STOBBE, R., 1911, Über das abdominale Sinnesorgan und über den Gehörsinn der Lepidopteren mit besonderer Berücksichtigung der Noctuiden, *Sitzber. Ges. naturf. Freunde Berlin*, p. 93–105.
[30] SWINTON, A. H., 1877, On an organ of hearing in insects with special reference to Lepidoptera, *Entom. mont. Mag.*, *14*, 121–126.
[31] TREAT, A. E., 1955, The response to sound in certain Lepidoptera, *Ann. Entom. Soc. Am.*, *48*, 272–284.
TREAT, A. E., 1956, The reaction time of noctuid moths to ultrasonic stimulation, *J. N.Y. Entom. Soc.*, *64*, 165–171.
TREAT, A. E. and K. D. ROEDER, 1959, A nervous element of unknown function in the tympanic organs of moths, *J. Insect Physiol.*, *3*, 262–270.
[32] TURNER, C. H., 1914, An experimental study of the auditory powers of the giant silkworm moths (Saturniidae), *Biol. Bull.*, *27*, 325–332.
TURNER, C. H. and E. SCHWARZ, 1914, Auditory powers of the *Catocala* moths; an experimental field study, *Biol. Bull.*, *27*, 275–293.
[33] VOGEL, R., 1912, Über die Chordotonal-organe in der Wurzel der Schmetterlingsflügel, *Z. Wiss. Zool.*, *100*, 210–244.
[34] WHITE, F. B., 1877, (Untitled comment on a letter of G. J. Romanes), *Nature*, *15*, 293.

CHAPTER 17

THE ROLE OF THE CENTRAL NERVOUS SYSTEM IN ORTHOPTERA DURING THE CO-ORDINATION AND CONTROL OF STRIDULATION*

by

F. HUBER

1. Introduction

Among invertebrates the insects have a highly organized and frequently very complex behaviour. It is expressed in movements of the body, the appendages, in certain postures and in the positioning of the whole insect in space. From the physiological point of view the behaviour can be analysed right down to the functions of single muscles and skeletal elements; but normally several muscles co-operate and develop a movement pattern which is organized in space and time. The co-operation between various muscles and parts of the body is performed by a nervous process, the *co-ordination*.

In the last ten years more and more literature on insects has appeared, describing the functional organization of neural mechanisms which co-ordinate and control the modes of behaviour. The behaviour of stridulating Orthoptera is a very suitable subject for neurophysiological analysis since the number of behaviour patterns is limited and surveyable. It is the purpose of this work to show the results of experiments on the participation of different ganglia and their particular areas in the central nervous system (CNS) during the process of stridulation, to localize the nervous apparatus and to demonstrate its function. For the experiments two representatives of the order Saltatoria were used: *Gryllus campestris* L. (Gryllidae) and *Gomphocerus rufus* L. (Acrididae).

2. Structure and Mechanism of the Stridulatory Apparatus

The members of the order Saltatoria have developed sound-producing organs on various parts of the body. Moreover, the representatives of the families Tettigoniidae, Gryllidae and Acrididae each possess different mechanisms of sound production. The sound patterns and their biological significance have been described ([42-47, 77-83, 17-36, 1-5]).

The songs have a communicative value and they control the relationship between individuals and between sexes during the period of reproduction. The terms "courtship"–"rivalry"– or "calling song" characterize the respective biological functions. Each song has a double meaning: for the stridulating insect it is an expression of a

---

* This investigation was supported by a grant from Die Deutsche Forschungsgemeinschaft.

certain mood (Stimmung), *i.e.* a reaction towards exogenous stimuli in a given endogenous situation. In the soundreceiving insect the song is the releaser of movements aimed at changing or ending this situation.

In this paper only the two "orthodox" forms of sound production in Orthoptera will be described: the stridulatory movements of the elytra (Tettigoniidae, Gryllidae) and of the hindfemora (Acrididae). The two stridulatory apparatus work on the same principle: a sclerotized row of pegs or lamellae (the file) rubs in a certain rhythmical way against a smooth, sclerotized ridge (the scraper) and causes a vibration of parts of the body.

(a) *The stridulatory movement of the field cricket Gryllus campestris*

In the male of *Gryllus campestris* parts of the elytra are modified as sound-producing organs. Each elytrum bears on its under side a file; it corresponds morphologically to the post-cubital vein[110,111]. The scraper in crickets represents the posterior elytral margin between the end of the file and the region of the anal field. During stridulation both the elytra are raised and their dorsal parts slightly tilted towards each other. Then the file of the right tegmen rubs against the scraper of the left. The elytra are held in different positions depending on the type of song and the positions are as characteristic as the sound patterns produced. In the case of the "calling" and "rivalry" songs the tegmina are held steeply (40–45°); during the "courtship" songs they are only slightly raised (15–20°), but their dorsal parts tilt more steeply towards eachother. The elytral position is important for determining the intensity of the various songs. The stridulatory movement is the result of a collaboration between many muscles and skeletal parts in the second thoracic segment. The muscles receive nerve impulses from different motor neurons of the second thoracic ganglion via various nerve branches (Table 31, Figs. 271, 272). Commands from the CNS must contain information concerning the angle of elytral position and the tension, as well as about speed, rhythm, duration, and intensity with which the elytra are rubbed against each other.

TABLE 31

Gryllus campestris, STRIDULATORY MUSCLES AND
EFFERENT NERVES RESPONSIBLE FOR THE ELYTRAL MOVEMENT

Phase	Muscles (Fig. 271)	Nerves (Fig. 272)
1. *Extension* of the elytra; control of the stridulatory position	depressor of the tergal arm (4) = protractor of the elytrum	n.lateralis 2
	depressor of the prescutum and scutum (6, 7, 8, 9) = elevator of the elytrum	n.laterales 2, 3, 5
	depressor of the basalar sclerite (10, 11) = protractor of the elytrum	n.lateralis 2
2. *Stridulation*		
(a) inward movement of the elytra	dorsal longitudinal muscles (1, 2, 3) = depressor of elytrum	n.anterior 2b
	deflector of elytrum (5, 12) = adductors	n.lateralis 5
	deflector of elytrum (13)	n.lateralis 5
(b) extension of the elytra	muscles of phase 1 (4, 8, 10, 11) = abductors	n.laterales 2, 3, 5

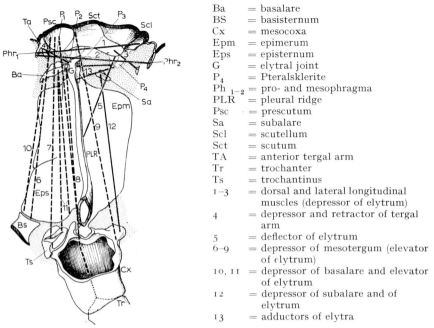

Ba	= basalare
BS	= basisternum
Cx	= mesocoxa
Epm	= epimerum
Eps	= episternum
G	= elytral joint
P_4	= Pteralsklerite
Ph_{1-2}	= pro- and mesophragma
PLR	= pleural ridge
Psc	= prescutum
Sa	= subalare
Scl	= scutellum
Sct	= scutum
TA	= anterior tergal arm
Tr	= trochanter
Ts	= trochantinus
1–3	= dorsal and lateral longitudinal muscles (depressor of elytrum)
4	= depressor and retractor of tergal arm
5	= deflector of elytrum
6–9	= depressor of mesotergum (elevator of elytrum)
10, 11	= depressor of basalare and elevator of elytrum
12	= depressor of subalare and of elytrum
13	= adductors of elytra

Fig. 271. *Gryllus campestris*. Top view of the right half of the 2nd thoracic segment, with parts of the exoskeleton and the stridulatory muscles. Skeleton white, membrane dotted, elevator muscles dotted lines, depressor muscles solid lines.
The muscles 7–8–9–11–12 also move the middleleg (Huber[72]).

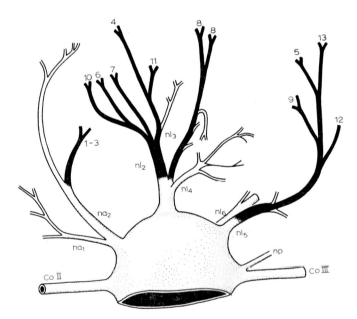

Fig. 272. *Gryllus campestris*. Right half of 2nd thoracic ganglion, seen dorsally, with lateral nerve branches. The numbers 1–13 indicate the muscles (Fig. 271), which are innervated by the various nerves. Co II and III = connectives, na = nervus anterior, nl_{2-6} = n.laterales, np = n.posterior.

The following description of the elytra is only based on a morphological investigation of the mesothoracic exo- and endoskeleton and its muscles. Work is in progress on selective elimination of single efferent nerves, and on recording of muscle activity during stridulation (see[135,136]).

Phase 1: The muscles (6, 7, 8, 9) attached to the prescutum and scutum pull down the anterior part of the mesotergal plate. The membranes between the scutum and the elytra tighten and raise the tegmen from the resting position. At the same time the muscles (4, 8, 10, 11) attached to the tergal arm and the basalar sclerite contract, the tergal arm is pulled down ventrally and the sclerite approaches the pleural body wall. Thus the pulling force which is directed towards the middle of the second thoracic segment causes the elevation and extension of the elytra. In addition, the structure of the wing articulation and the distribution of the axillary sclerites produce a torsion of the elytrum along its longitudinal axis, so that the dorsal parts of the elytra are tilted towards each other. All the active muscles of the left and right half of the thoracic segment contract synchronously. It follows that the elytra perform the same movement simultaneously, although the left tegmen lags behind a little because in the resting position it is covered by the anal field of the right one. The angle of incidence of the left tegmen remains smaller and this difference is very important for the sliding to and fro of the elytra.

Phase 2: Stridulation begins: the file of the right tegmen and the scraper of the left are rubbed against each other at nearly the same amplitude, *i.e.* the relevant muscles function alternately. The first group of muscles involved comprises the wing depressor (12) which lowers the elytrum by pulling the subalare into the ventral direction; it is assisted by the posterior tergopleural muscle (5). The most important muscles for the stridulatory movement seem to be 13, which are inserted on the pleural nob of the wing articulation, and attached to a sclerotized area on the base of the anal field of each tegmen. When they contract, they draw the elytrum inwards, and when they relax, the elytra return into the initial position due to the contraction of the elevator and promotor muscles (4, 7, 8, 9, 10, 11).

Although the male cricket possesses a morphologically perfect stridulatory apparatus on each tegmen, only the file of the right elytrum and the scraper of the left are normally used for stridulation. It is, however, possible experimentally to get males to stridulate with the complimentary parts of the second apparatus, *i.e.* from left to right (Linksgeiger) ([111]; Huber[140]). Their songs will be discussed later.

In *Gryllus campestris* it is still not known whether the sound is produced during the inward or outward movement. Voss[116] concluded from morphological studies in *Acheta domesticus* that sound is only produced during the inward phase and Pierce[91] arrived at the same result in *Gryllus assimilis* by analysing the sound and recording the elytral movements photographically. Frings and Frings[48] also provides data showing that the tettigoniid *Neoconocephalus ensiger* stridulates during the closing phase of the elytral movement. Pasquinelli and M. C. Busnel[90] investigated the stridulation of *Ephippiger bitterensis* and found that sound is produced during closing and opening of the short tegmina. Each impact of a single lamella of the file against the scraper causes a strongly damped oscillation which dies away before the next lamella touches the ridge. The number of sound units corresponds so accurately to the number of lamellae that one can predict defects of the file from the changes in the sequence of sound units. *Ephippiger* belongs therefore to that group of Orthoptera in which certain parts of the song pattern are determined by the structure of the stridulatory apparatus. In crickets each movement of the elytra causes a short and nearly pure tone impulse (Fig. 273); the oscillation is only slightly damped. The main or carrier frequency of *Gryllus campestris* ranges from 3.0 to 5.8 kc/s with harmonics at 9 and 17 kc/s (Fig. 274)[87,72,4]. For the various songs the crickets do not rub the full length of the file against the scraper, so that the chirps can be of different lengths (Fig. 273)[91,72,4].

References p. 484

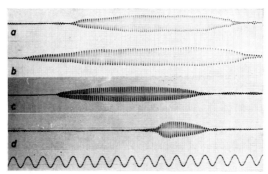

Fig. 273. Oscillograms of *G. campestris* sounds emitted during one elytral movement (= chirp). Chirp of (a) "calling" song, (b) "rivalry" song, (c) cantus fortior of "courtship" song, and main chirp of cantus mitior of "courtship" song. Time mark 600 c/s (Huber[72]).

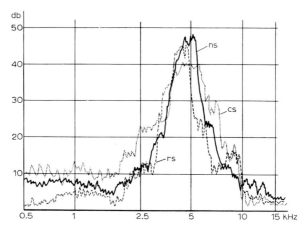

Fig. 274. Frequency spectra of 3 *G. campestris*-songs (ns = "calling" song, rs = "rivalry" song cs = "courtship" song). Ordinate = intensity in dB, abscisse = frequency in kc/s (Huber[72]).

Pasquinelli and M. C. Busnel[90] have shown that the number of elementary waves in the chirps of *Oecanthus pellucens* corresponds to the number of lamellae. According to their hypothesis the impact of each lamella against the scraper induces a single oscillation. In the field cricket it has been shown that the elytra produce a main frequency of 4–5 kc/s regardless of how fast or with how many lamellae the file rubs against the scraper[87]. But the question remains as to whether the structure and number of the lamellae merely determine duration and intensity of a chirp, or whether they are responsible together with the speed of the movement for the carrier frequency (see also[4]).

(b) *The stridulatory movement of the acridid Gomphocerus rufus*

Many Acrididae produce sound by rubbing the hindfemur against a pronounced sclerotized elytral vein of the same side. According to the species the femora are moved in either a syndromic or an antidromic way. The file of the male and female of *G. rufus* consists of a row of pegs on the inner side of each femur; the thickened vena radialis media of the tegmen serves as scraper (*Stenobothrus*-type after Jacobs).

Operations on the stridulatory apparatus of *Chorthippus biguttulus* by Loher[86] who removed parts of the elytrum or took off the pegs of the file, only changed the intensity of the song, not its pattern. Thus the rhythmical element in the songs of Acrididae is an image of muscular activity and an expression of the central nervous process of co-ordination. The ultimate command to stridulate comes from the 3rd thoracic ganglion which innervates all the muscles necessary for stridulation (Table 32, Figs. 275, 276).

TABLE 32

Gomphocerus rufus, MUSCLES AND NERVES INVOLVED IN THE STRIDULATORY MOVEMENT

Phase	Muscles (Fig. 275)	Nerves (Fig. 276)
1. Elevation of hindfemora	Elevator of trochanter (1, 2, 3)	n.lateralis 4
2. Moving the femur away from the elytrum	Anterior rotator of coxa (7)	n.lateralis 5
	protractors and abductors of coxa (11, 12)	n.laterales 2, 3
3. Depression of the femur	Depressor of trochanter (4, 5, 6)	n.laterales 3, 6
4. Pressing the femur against the elytrum	Posterior rotator of coxa (7)	n.lateralis 5
	Adductors and retractors of coxa (8, 9, 10)	n.lateralis 5

The stridulatory movement of the Acrididae represents a special form of femoral movement. Origin and insertion of the metathoracic stridulatory muscles ensure a rotation of the leg at the monocondylic pleurocoxal joint, and an up-and-down movement of the femur at the dicondylic coxotrochanteral joint. In principle the femur can produce sound in both phases. Morphological studies on the position of pegs[79,81] and cinematographical records[86] suggest, however, that the elytrum is only touched by the femur during the downstroke. Here, too, the motion can be subdivided:

Phase 1: Before stridulation the insect presses the hindtibiae closely against the hindfemora and lifts the legs off the ground (Fig. 277), so that the body is resting on the fore- and middlelegs. Then each femur is brought upwards by the contraction of the elevator muscles of the trochanter, which is represented in the jumping leg by a narrow chitinous ring tightly connected with the femur. During elevation the femur is deflected outwards, *i.e.* away from the elytrum, because of contraction of the anterior rotator and the abductors of the coxa as they turn the coxa outwards.

Phase 2: The femora are lowered by the contraction of the depressor muscles which are inserted over the same tendon on the ventral edge of the trochanter and originate partly in the body (4) and partly in the coxa (5, 6). At the same time the posterior rotators and the adductors of the coxa contract and press the coxa and the femur against the elytrum.

3. The Song Patterns of *Gryllus campestris* and of *Gomphocerus rufus*

Hitherto studies on the songs of Orthoptera have been mainly concerned with elucidating their biological significance and with their utilisation for clarifying relationships amongst primitive taxonomic groups[42–47,77–83,1–5]. They have also been concerned with the analysis of sensory processes[17–36,86,56–59,118,4]. However, there are no studies especially aimed at employing the song pattern as a tool for analysing the central nervous co-ordinations on which they are based.

(a) The songs of the field cricket

Rhythm is the main characteristic of cricket songs. In the case of representatives

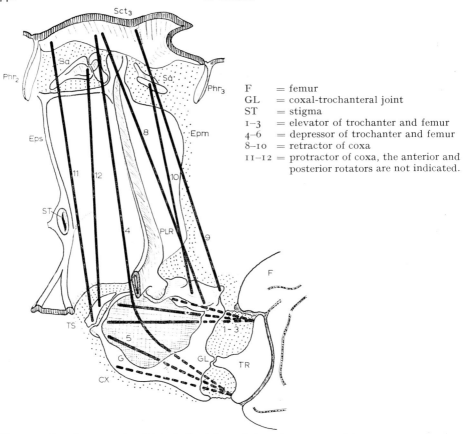

Fig. 275. *Gomphocerus rufus*. Right half of the 3rd thoracic segment with the exoskeleton and stridulatory muscles (= solid lines). See Fig. 271 for further explanation of symbols.

F = femur
GL = coxal-trochanteral joint
ST = stigma
1–3 = elevator of trochanter and femur
4–6 = depressor of trochanter and femur
8–10 = retractor of coxa
11–12 = protractor of coxa, the anterior and posterior rotators are not indicated.

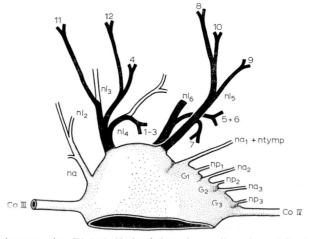

Fig. 276. *Gomphocerus rufus*. Right half of 3rd thoracic ganglion, the 3 abdominal ganglian fused with it (G_{1-3}), and the lateral nerve branches. Co III and Co 4 = connectives; for explanation of symbols see Fig. 272. na_{1-3} = n. anteriores, np_{1-3} = n.posteriores of the 3 abdominal ganglia, ntymp = n.tympani, numbers 1–12 indicate the muscles (Fig. 275), which are supplied from the nerves.

Fig. 277. Male of *G. rufus* in stridulatory position. The hindlegs are lifted off the ground and the tibiae pressed against the femora.

of the genus Oecanthus, Walker[118] has shown that the females can interprete the rhythm. Similar results were obtained by the analysis of songs of the field- and housecricket[87, 17, 47, 72]. Here the sounds are characterized by 3 parameters: carrier frequency, amplitude, and modulation of amplitude (rhythm). The frequency spectra of the different songs of a species are nearly identical, in other words, a female is unable to "understand" the various songs of its own species on the basis of their frequencies. Moreover, it has been shown that the tympanic organs probably cannot discriminate between pure tones of different frequencies[8-12], but that they are very sensitive to variations of amplitude[100-103]. Recently, in *Schistocerca gregaria*, *Locusta migratoria* and *Acheta domesticus* neurons have been found which respond differently to pure tones according to the frequency used. This seems to prove that the Orthoptera also can discriminate pure tone frequencies[61, 84, 112]. The sound intensity may vary, but for the field cricket no exact measurements of this are yet available. The decisive criterium for placing a song in a certain category is therefore the innate stridulatory rhythm.

The basic element of the gryllid song is the chirp (Silbe), that is, the sound produced during a single movement of the elytra. The various songs are distinguished from each other by the number of chirps, the chirp rate in the sequence and by the sequence rate. In a sequence the intensity may rise and sink in a regular or irregular way.

"*Calling*" song. In the field cricket the sequences which are perceived by the human ear as single trills consist of 4 or less frequently 3 chirps, and they follow each other at approximately constant intervals (Fig. 278a). The intensity increases from the first to the last chirp and the intervals range from 18 to 25 msec. The rhythm of the "calling" song is understood as a modulation of the carrier frequence of 5 kc/s with 30 to 40 c/s. The sequence rate is influenced by temperature[118, 48, 5]. The field cricket normally stridulates at the rate of 3–5 sequences per second (25–30° C)

References p. 484

and it has been found that the sequence rate increases when the male has not copulated for a long time. The stridulatory rhythm is therefore closely related to the physiological state of the insect, *i.e.* its mood—Stimmung— (see Section 4, p. 467).

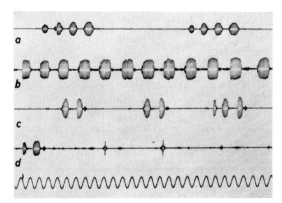

Fig. 278. Song rhythms of *G. campestris* (a) 2 sequences of the "calling" song, (b) part of a sequence of the "rivalry" song, (c) 3 sequences of the c.fortior, (d) transition from c.submitior to c.mitior of the "courtship" song, with main and interchirps. Time unit 50 c/s (Huber[72]).

"*Rivalry*" *song*. The change from the "calling" song to the "rivalry" song is accompanied by an increase in the number of chirps per sequence (Fig. 278 (b)) and thus the sequence length is altered. Frequently the periodical modulation of intensity disappears as well, but chirp length and chirp interval remain nearly constant. When the males are fighting the sequences become longer; after flight of the opponent the song decreases and readjusts itself to the pattern of the "calling" song.

"*Courtship*" *song*. Much more striking is the modification of the song pattern during the "courtship" song. Faber[47] distinguishes 3 phases: Cantus fortior, c.submitior and c.mitior (Figs. 278 (c) and (d)). In c.fortior the number of chirps per sequence is reduced, and owing to the excitement of the male, the succession of sequences normally increases. The change from "calling" song to c.fortior is vague, but from c.submitior to c.mitior it is very abrupt. Here the elytral position changes as well and there appears a new sound element which had evidently been suppressed before; we shall call it the "interchirp rhythm" (Zwischensilbenrhythmus). The oscillograms of Fig. 279 exclude the possibility that this new element is a former chirp rhythm which has diminished in intensity. They represent a "calling" song and the transitional songs which lead into courtship. The singer was a male which had been brought into contact with a female for the first time 6 days after ecdysis. Shortly after antennal contact (which may release courtship or fight[60,66]), interchirps of low amplitude appear between the 4 main chirps (Fig. 279(a). They are produced by the vibratory movements of the elytra and show somewhat wider frequency spectra with maxima ranging from 17 to 19 kc/s[87]. The various phases of the "courtship" song contain numerous interchirps which form the leading sound element in the c.mitior (Figs. 279(b)–(d)).

The "courtship" song is thus distinguished from the other two patterns by a shortening of the sequences, a change in sound intensity and by the appearance of a new element, the interchirp. Recently it has been shown that the interchirps of

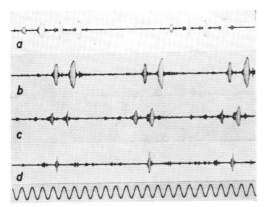

Fig. 279. Song rhythms of *G. campestris* (a) 2 sequences of the "calling" song leading to the "courtship" song, with 4 mainchirps and some interchirps, (b)–(d) sequences of c.fortior, submitior and mitior of the "courtship" song with main and interchirps. Time unit 50 c/s.

the "courtship" song are produced not only by elytral friction, but also by an impact of the costal field of the elytra against the body (Huber, unpubl.). This is in accordance with the high proportion of ultrasonic frequencies found in the interchirp[87].

When these results on the structure of song patterns are translated into the mode of operation of the nervous system, they suggest that the relevant neural mechanisms are working in a rhythmical way; that song patterns are caused by several different rhythmical processes in the CNS. The first process determines the repetition frequency of the elytral movements. A second process controls the sequence rate and simultaneously modulates the first process. The third rhythmical action is concerned with the periodical change of intensity, especially during the "calling song". This is obviously due to an increase of the amplitude of the elytral movement causing a prolongation of the chirp and a stronger stroke of the file against the scraper. It would mean that the stridulatory muscles were contracting more intensively near the end of a sequence than at the beginning. In the "rivalry" song particularly the central rhythm controlling the structure of the sequence is changed, and in the "courtship" song yet a further factor is altered, namely the elytral position.

(b) The songs of Gomphocerus rufus

For *G. rufus* Faber[46] and Jacobs[79] describe two main stridulations: the "calling" song and the "courtship" song. As both of these songs, as well as the "responding" song between male and female, are produced in the same way (by the same sound producing component), only the stridulatory movements of the male's "calling" song are dealt with here.

A single sequence lasts 3.5–6 sec in medium sunshine[46,79]. During stridulation the hindfemora carry out a double movement. They strike up and down 28–34 times with a high amplitude and during this phase they also perform a vibrating movement of low amplitude and high frequency. The result is a sound with frequency portions of 2–15 kc/s and a song pattern which contains 4–6 intensive impulses/sec. The identically constructed stridulatory apparatus of *G. rufus* co-operate nearly synchronously in all songs. The sequence rate of the "calling" songs depends on the temperature and the mood of the male.

References p. 484

A sound analysis of *G. rufus* has been made (Huber, Loher, unpubl.). Some information about the nervous control of the stridulatory movements can be reported here. The double movement of the hindfemora means that there must be two rhythmical processes which are directed towards the same effector system. They are responsible for the up and down stroke and for the vibrating movement. It is still not known, if there is a division of labour between the elevators and depressors of the trochanter. Either certain muscles are responsible for the strokes of high amplitude and low frequency, and others for the vibrations, or one and the same muscle contracts in stages. Both hindfemora always swing in the same phase, therefore: they must be centrally coupled. Even more complex is the action of the CNS in the "courtship" song of this species. Here the stridulatory movement is only one element out of a chronologically fixed program of motion[46, 79, 67].

4. The Neurophysiological Basis of Sound Production

The morphological study of the stridulatory apparatus has shown that nerves of the 2nd thoracic ganglion in gryllids and of the 3rd thoracic ganglion in acridids supply the stridulatory muscles. The two ganglia represent the final control system of the singing movement. But it is also well known from behaviour studies that different types of songs are released by optical, acoustical or mechanical stimuli. The sense organs involved are connected to different parts of the CNS and we may therefore assume that several parts of the CNS collaborate during sound production.

A large number of operations on the nerve cord of *Gr. campestris* and *G. rufus* revealed those parts of the CNS which guarantee the normal functioning of the stridulatory movement. The various operations are summarized in Table 33.

Gryllus campestris: In a male cricket only the following parts of the CNS are necessary for co-ordination of the whole repertory of songs: the brain, the thoracic ganglion and one connection between them, and the lateral nerve branches supplying stridulatory muscles (Fig. 280(a)). In addition, to release the "rivalry" and "courtship" songs an antennal stimulation is required, whose repetition rate and intensity are important[66].

If the nerve cord is interrupted anywhere between the 2nd thoracic and the last abdominal ganglion, the "courtship" song does not normally occur, although it can be experimentally evoked[66]. After this operation the male responds to the antennal contact of a female as if copulation had already finished. The male lowers the antennae sidewards and vibrates, while watching each movement of the female. Zippelius[134] called this kind of behaviour after-courtship (Nachbalz), because it occurs regularly after copulation and after transference of the spermatophore and makes possible a migration of the sperms into the receptaculum of the female[66] (see also[4]).

A cricket male always behaves as if it had no spermatophore for the first few weeks after severance of the abdominal nerve cord. How is the presence of the spermatophore normally signalled? The gonads and those glands which provide the material for the spermatophore seem to have no influence upon the acoustical behaviour, as has been proved by extirpations. The "courtship" song, however, is suppressed after cutting the nerves of the last abdominal ganglion supplying the spermatophore sac and the accessory glands. We may assume therefore that sense organs—probably proprioreceptors in the region of the copulatory apparatus—participate in the control of stridulation. Although this operation eliminates nervous control, it does not influence the co-ordination of the stridulatory movement. Crickets can sing and court if the afferent nerves to the last abdominal ganglion are unilaterally undamaged, or if the connective is only interrupted on one side. But even under those circumstances courtship and after-courtship of the male alternate frequently in the presence of a female[66].

TABLE 33

OPERATIONS ON THE CNS OF *Gryllus campestris* AND *Gomphocerus rufus* AIMED AT LOCATING THE NERVOUS STRUCTURE NECESSARY FOR THE STRIDULATORY MOVEMENT

Place of operation	Stridulations of							
	Gryllus				Gomphocerus			
	N_1	gG	RG	WG	gG	WG	AG	N_2
1. Unilateral severance of the abdominal nervous system	8	+	+	(+)	+	+	+	18
2. Bilateral severance of the abdominal nervous system	7	(+)	+	—	+	+	+	11
3. Co III severed unilaterally	5	+	+	+	+	+	+	3
Co II severed bilaterally	7	+	+	+	—	—	—	7
4. Co II severed unilaterally	5	+	+	+	+	+	+	4
Co II severed bilaterally	10	—	—	—	—	—	—	6
5. Co I severed unilaterally	8	+	+	+	+	+	+	4
Co I severed bilaterally	4	—	—	—	—	—	—	11
6. Sko severed unilaterally	3	(+)	(+)	(+)	(+)	(+)	(+)	5
Sko severed bilaterally	3	—	—	—	—	—	—	3
7. Co III severed at the right and Co II at the left side					—	—	—	3
8. Co III severed at the left and Co I at the right side	3	—	—	—	—	—	—	2
Co II and Co I severed at the right side	2	+	+	+	+	+	+	4
9. Thoracic ganglion III halved					—	—	—	3
Thoracic ganglion II halved	3	—	—	—	+	+	+	2
Thoracic ganglion I halved	2	+	+	+	+	+	+	3

gG = "calling" song; RG = "rivalry" song; WG = "courtship" song; AG = "responding" song; N_1 = number of operated crickets, and N_2 = number of operated grasshoppers; Ko = connective; Sko = circumoesophageal-connective of the CNS. The numbers are explained in Fig. 280, the symbol (+) in the text.

The male suppresses the "calling" song after copulation until the new spermatophore has been pressed into the sac and has hardened[65,66]. One would expect, therefore, that the "calling" song preceding courtship would be suppressed as well as the courtship song after the abdominal nerve cord is cut. This is not the case; the operated males do sing, though distinctly less frequently (Table 33; (+)). This result makes it likely that inhibition of the "calling" and "courtship" songs after copulation is induced by the change of excitation of the sense organs in the region of the copulatory apparatus. The renewal of the calling song after sectioning of the nerve cord or after removal of the accessory glands necessary for the periodical formation of the spermatophore must be due to factors now unknown.

Unilateral interruption of the oesophageal connectives in gryllids and acridids (Table 33, (+)) induces the insect to move in narrow circles, turning in the direction of the undamaged side[113,64,65,66,106,107]. This circus movement makes it difficult to

References p. 484

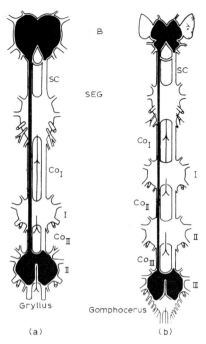

Fig. 280. Diagram of the CNS of *G. campestris* and *G. rufus* with structures important for sound production (black). B = brain, SEG = suboesophageal ganglion, I–III = thoracic ganglia, Co = connective, Sc = circumoesophageal connective. The drawn out lines demonstrate the nerve tract between brain and thoracic ganglion. (a) Huber[72].

readjust the posture for the fighting and courtship behaviour, and its is observed that sound production now occurs only very occasionally (Table 33 (+)).

Gomphocerus rufus: Sound production in acridids also depends on an intact brain and one nerve connection to the third thoracic ganglion and the innervation of the hindfemora (Fig. 280(b)). Severance of the two acoustical nerves leading from the tympanic organ in the first abdominal segment to the first abdominal ganglion, eliminates the "responding" songs of males and females. Furthermore, the release of courtship behaviour and its characteristic song requires impulses to be transmitted from the eyes and antennae to the brain.

The compound eyes and the tympanic organs enable the male to localize the female and to recognize an individual of the same species[79]. Antennae and eyes lead the male into the right position in front of the female, where courtship begins. The interruption of the abdominal nerve cord in the male does not affect courtship and mounting; in the female oviposition is, of course, disturbed (Huber, unpubl.).

In conclusion we find in gryllids and acridids that two nervous structures, brain and a thoracic ganglion, co-operate during sound production. They are actively controlled by different sense organs. The questions and experiments arising from this chapter will be described in detail in the following section.

4A. The Function of the Thoracic Ganglia during the Stridulatory Movement

(a) Anatomical remarks

The ganglia of the nervous system in Orthoptera are surrounded by a double sheath, an outer layer, the neurolemma, which is optically homogenous, and an inner layer called the perineurium. The cell bodies of motor and interneurons are situated in groups near the periphery of each ganglion. Their fibres, the neurites and dendrites, enter the central region of the ganglion, the neuropile, and ramify into numerous branches. Here the nerve cells form synapses within and between the halves of a ganglion. The terminations of the sensory fibres are in close contact with the motor dendrites and the terminations of the connective fibres. The neuropile is also the origin of the large efferent nerves and here the ganglion receives the nerve impulses from other parts of the CNS.

So far the best description of the distribution of ganglion cells and fibres in any thoracic ganglion has been given by Zawarzin[132, 133] for the ganglia of Aeschna larvae and by Power[91a] for Drosophila. The anatomical structure of the ganglia in gryllids and acridids is not yet clear. The connective fibres entering the ganglion may traverse it dorsally (the dorsal group of fibres) and ventrally (the ventral group of fibres) and give off branches to the neuropile (Fig. 281). Directly beneath the dorsal connective fibres lies the region of the dorsal roots, *i.e.* the motor neurons and their fibres. Between them and the ventral roots representing the incoming sensory fibres lies the basal neuropile. The cell groups of the thoracic ganglia consist mainly of motor neurons and to a smaller extent of interneurons. In contrast to nearly all the motor neurons and many of the sensory fibres, these interneurons give off branches in both halves of the ganglion. They therefore constitute an important link in the coupling of functionally equal elements of both halves of the ganglion (Fig. 282).

This histological investigation has led to the conclusion that the fibres of the motor neurons and the branches of the ventral roots are primarily limited to the same half of the ganglion. The connection between ganglion halves is mainly due to the fibres of interneurons and the terminations of the connective fibres. Further studies are required to complete this picture.

(b) The function of the isolated thoracic ganglion

In order to explain the function of the thoracic ganglion for sound production in gryllids and acridids the connective fibres in front and behind the ganglion were cut. The isolated ganglion was capable of sending efferent nerve impulses to the muscles via the lateral nerve branches and could receive afferent impulses from the segmental sense organs. After the operation, neither the gryllids nor the acridids were able to stridulate (Table 33). The change of behaviour observed is reported in a later Section (p. 478).

Once the thoracic ganglion had been isolated, the cricket slightly raises the elytra. This may be due to a change of the tonus in the elevator and depressor muscles after the elimination of the superior ganglia (see[104, 106–107, 66, 70]). After mechanical stimulation of the sensory hairs at the base of the elytral articulation, the male reacts either with cleaning movements of the elytra, or less frequently with a short soundless "stridulatory movement". These movements do not last much longer than the duration of the stimulus, the elytra being moved 3–10 times against each other. Thus an inadequate stimulus releases a co-ordinated movement of the elytra for only a short time, and their position does not correspond to that in any of the described songs of crickets.

References p. 484

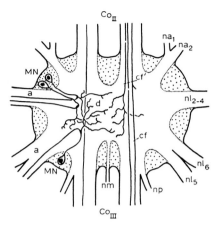

Fig. 281. 2nd thoracic ganglion of *G. campestris* in longitudinal section with the lateral nerves na, nl, and np (Fig. 272) and the connectives (Co II and III). Central white area = neuropile, dotted region = places of ganglion cell bodies, MN = motor neurons with dendrites (d) and axon (a), cf = connective fibre without, and cf' = with ramification in the neuropile, nm = n. medianus. Semidiagrammatic reconstruction of sections with fibre impregnation.

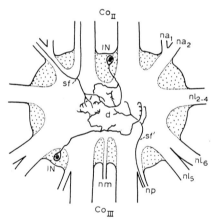

Fig. 282. 2nd thoracic ganglion of *G. campestris*. IN = interneuron with dendrites (d) in both halves of the ganglion, sf and sf' = sensory fibres of sense organs from the same segment. For other symbols see Fig. 280.

If the third thoracic ganglion of an acridid is isolated, the insect reacts to mechanical or chemical stimulation of the coxa, the hindfemur and particularly of the distal region of the leg with a strong kicking of the stimulated extremity. A stronger stimulation may even lead to a co-ordinated jump, with the hindlegs moved synchronously, but a stridulatory movement of the hindfemur can no longer be released. Occasionally, however, the insect raises both legs and vibrates spontaneously, which means that the third thoracic ganglion is still able to co-ordinate a basic form of the stridulatory movement. During walking the hindlegs are dragged behind for several days after the operation, owing to the loss of tonus in the muscles. Even later the original walking rhythm does not reappear completely (see also[74,75]).

(c) The conduction of excitation from the brain to the thoracic ganglion

Stridulation can take place in both species if only one pathway between the brain and the thoracic nervous system is intact. So the supply of nervous excitation along one conducting tract is sufficient to provoke a rhythmical and co-ordinated activity. At the beginning of stridulation the wing or leg movement of the operated side shows a certain retardation. The cricket raises the elytra more slowly, but during the stridulation it is impossible to judge from the movements or from the sound pattern whether or not a nerve tract is interrupted. Insects of the same species understand the sound signal produced. In *Gomphocerus* the beginning of stridulation is marked by a short phase shift in the homolateral leg movement. The femur of the operated side lags behind the other femur during the first second of stridulation, but then they move together. This deviation in sound production has apparently no behavioural significance, for female crickets can be attracted as before and grasshoppers respond promptly to these songs.

In a further series of experiments the connectives were cut unilaterally between the different ganglia, either on the left or on the right side, and also bilaterally. Table 33 shows that impulses coming from the brain are transported to the connective fibres, which traverse the preceding ganglia without decussating (operations 7, 8, 9).

(d) The nervous coupling of both halves of the thoracic ganglion

During stridulation a cricket raises the elytra simultaneously and rubs them against each other. The hindlegs of a grasshopper are also coupled to synchronously perform the same rhythmical movements. Stridulation can still take place—apart from a phase shift at the beginning of the song—when the connectives are unilaterally cut, or after stimulation of the sense organs of one side. It is concluded therefore that the fibres of each connective influence the whole population of motor neurons either directly or through mediation of thoracic interneurons. Otherwise synchronous movement of the extremities would not be ensured. Anatomically both circuits are possible. A direct communication between the motor neurons has been demonstrated in Aeschna larvae[133]. Electrophysiological investigations on crustaceans[121–123] and insects[53,119] have shown that impulses from interneurons of one side can stimulate the motor neurons in both halves of the ganglion. In crustaceans the neurons of the opposite side react with a longer latency. If the 2nd thoracic ganglion in crickets or 3rd thoracic ganglion in grasshoppers is halved, males cannot start or maintain a stridulatory movement, although both halves of the ganglion are connected to the brain. Mechanical stimulation of the sensory hairs on the wing cannot release either stridulation or flight. Grasshoppers can still perform defensive movements under the given conditions, but only the pro- and mesothoracic legs are co-ordinated.

(e) Local electrical stimulation of the neurophile in the thoracic ganglion

In Orthoptera, Auger and Fessard[6,7] and Hughes[73] have exposed various thoracic ganglia from the ventral side and stimulated then in the longitudinal and transverse direction with direct current. In this way they could release, amongst other things, rhythmical leg- and wing movements. Consequently it is shown that the nervous elements of the thoracic ganglia can induce a sequence of rhythmical and co-ordinated movements. Similar experiments have recently been carried out in field- and house crickets. After isolation of the 2nd thoracic ganglion the insects were fixed on a

References p. 484

holder, with the ventral side upwards, so that they could move their middle legs and elytra freely. Then a tungsten electrode of 15 μ, coated with lacquer except for the tip, was inserted through the sheath of the ganglion, and a platinum wire was put into the dorsal hemolymph sac of the prothorax. Fig. 283 shows the second thoracic ganglion with the points of stimulation in the central neuropile and Table 34 summarizes the responses.

TABLE 34

POINTS OF STIMULATION AND MOVEMENTS OF THE MIDDLELEGS AND ELYTRA DURING ELECTRICAL STIMULATION OF THE ISOLATED 2ND GANGLION IN THE FIELD CRICKET

F	I	N	Released movements	Points of stimulation
15	30	15	The middlelegs are drawn towards the body and tremble	Area of the connective fibres II, left
15	24	11	The right middleleg twitches, the elytra vibrate	Neuropile in the area of the origin of the n.laterales 2 and 4, right
15	10	12	Twitching of the stretched left middleleg and vibration of the elytra	Neuropile in the region of origin of the n.laterales 5 and 6, left central neuropile
15	18	7	Stridulatory movements of the elytra, independent of the stimulus frequency	Central neuropile

F = impulse frequency; impulse form was rectangular, duration 1 msec; I = strength of stimulus in μamp; N = number of experiments per insect. Stimulation was carried out with double threshold current.

In the first 3 experiments the movements of the middle legs and the elytra were synchronised with the rhythm of stimulation, i.e. their frequency changed with that. This result indicates that the electrical stimulus probably has an effect near the motor neurons or near their fibres in the central neuropile. After a short latent period (0.5–1 sec) the middlelegs and elytra always moved in the same pattern showing that in every case the same groups of muscles had been stimulated. At a frequency of 30–45 impulses/sec a summation of these single movements occurred. Leg and elytrum then remained in a certain position or beat in a natural rhythm. This fusion frequency is in good accordance with the results of Pringle[92–93] and Hoyle[62–63] who specify that in various thoracic muscles of insects a stimulation frequency of 20–40 impulses is required to produce complete tetanus. As a single point of stimulation induces either both middlelegs or both elytra to move synchronously, the stimulus must excite some structure in the neuropile which participates in the coupling of both halves.

The last experiment of Table 34 is of particular interest here. In the thoracic ganglion there are regions which, after stimulation, induce a stridulatory movement independent of the stimulus frequency. After a short latent period the male lifted the elytra and stridulated continuously throughout the stimulation. But the soft sound did not fit into any of the known gryllid songs. The stimulus must have affected a structure which excited the stridulatory muscles of both sides in a rhythmical way. Furthermore, the experiment demonstrated that the thoracic system co-ordinates the elytral movement, but not in a specific song rhythm.

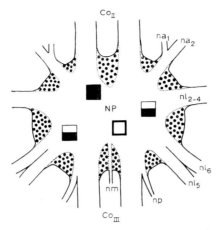

Fig. 283. 2nd thoracic ganglion of *G. campestris* with stimulation points in the central neuropile (NP). ■ = stimulation point for movements of middleleg at the stimulus frequency, ▆ = stimulation point for movements of middlelegs and elytra at the stimulus frequency, □ = stimulation point for co-ordinated and rhythmical elytral movement (basic form of stridulation), independent of the stimulus frequency. (Huber[72]).

(*f*) *Discussion of results*

The problem discussed in this chapter is the function of the thoracic ganglia during stridulation. Without the supply of cerebral nerve impulses these ganglia are unable either to begin or to maintain a normal song rhythm. But they do possess all the mechanisms and circuits necessary for distributing the excitation to both halves of the ganglion and ensuring the co-operation of the functionally equivalent neurons.

In the following section some observations on stridulatory movements in operated insects are described and the possibilities of co-operation between the thoracic neurons discussed.

(*i*) *Muscle tonus and stridulatory movement*. After bilateral severance of the connectives in front of the 2nd thoracic ganglion gryllids raise their elytra slightly and keep them in that position for several days[66]. By comparison, acridids lower the femora and stretch the tibiae and tarsi if the connectives are cut in front of the 3rd ganglion. Therefore the legs drag during walking. After unilateral severance of the connectives the insects can soon use the elytrum or hindfemur of the operated side for stridulation. These observations are explained in the following way: In various representatives of Orthoptera it has been found that the tonus of the thoracic and abdominal muscles is controlled by the headganglia[13, 104, 66, 70]. The controlling impulses are mainly transmitted along the connective fibres to the same side of the body. Since each movement is dependent on a posture which requires a certain muscle tension, it is evident that changes in the muscle tonus would not only influence the posture of a resting insect, but also its movements. The tonic asymmetry is therefore sufficient explanation for both locomotory changes and the phase shift seen during the synchronous stridulatory movement.

(*ii*) *Control of tonus*. Muscle tension in insects is determined by the frequency of discharge of the motor neurons[62, 96-97], which are themselves under the control of higher centres. The brain and suboesophageal ganglion (SEG) frequently work as antagonists: elimination of the brain in Mantis causes an increased contraction of the lateral and dorsal neck muscles[104]; in crickets the adductors of the femora and tibiae are more strongly contracted and head and body are raised and lifted off the ground, thus indicating the removal of central inhibition. When the SEG is extirpated, the leg muscles relax, indicating a lowered excitation of the motor neurons.

The thoracic and abdominal motor neurons constantly send impulses to the muscles, as was shown by Weiant[119] in the case of the metathoracic neurons of the roach. These motor neurons discharge spontaneously. They increase their frequency after elimination of the head ganglia (removal of inhibition) and show a further increase after halving of the 3rd thoracic ganglion (removal of a reciprocal inhibition). The above results form part of a neurophysiological explanation for the change of locomotor activity in Orthoptera following the elimination of certain brain regions. The insects reacted with an increase in running and jumping activity which persisted for a long time[66,70]. The thoracic motor neurons responsible for the stridulatory movement seem also to be controlled by higher centres. Animals will not sing after removal of the brain and SEG; even though many of the reflexes including those involving the hind legs and the elytra are enhanced.

(*iii*) *Rhythmical discharge of motor neurons.* It has been found in vertebrates and invertebrates that a single pre-synaptic impulse, for instance from an interneuron, often releases in the motor neurons a repeated discharge, or modulates the resting discharge of the neurons for a short time[38–40, 121, 53, 14–16, 119]. One must, however, in each case determine whether the spontaneous discharge of the motor neurons is due to reverberating circuits, or whether it is a characteristic of the motor neurons themselves. At least the results on crickets suggest that the short elytral movements produced by stimulation of segmental sense organs are an expression of such a repetitive discharge of the motor neuron populations in the two halves of the ganglion.

(*iv*) *Possibilities of connections between thoracic neurons during the stridulatory movement.* The neural mechanism of the thoracic ganglia responsible for the stridulatory movement in gryllids and acridids must fulfill the following conditions: Impulses arriving in the ganglion via one of the connective tracts must be transferred in some way to functionally equivalent motor neurons of both sides. Moreover, the kind of circuit ensure synchronous activity. After sagittal sectioning of the ganglion the co-operation between the two halves is interrupted. These results and the anatomical findings concerned with the arrangement and connection of the thoracic neurons suggest the following two circuits:

Circuit 1 (Fig. 284(a)): Impulses transferred bilaterally from the brain to the thorax are sent directly via the ramified and intercrossing terminations of fibres to the dendrites of the motor neurons of both sides. This circuit allows the distribution of excitation even after eliminating one side. If this circuit is correct one might expect that halving the ganglion would not seriously impare the singing movements on each side. After such an operation the wings and legs could be moved either independently of each other, or, if the impulses arrived at the sane time or in the same rhythm, in a synchronous way. This, however, is not the case. Thus there seems to be a further connection between the co-ordinated motor neurons of both sides which is cut after halving the ganglion and so makes sound production impossible.

Circuit 2 (Fig. 284(b)): Impulses coming from the brain to the thorax are transferred to the motor neurons of both sides not directly, but by mediation of inter-

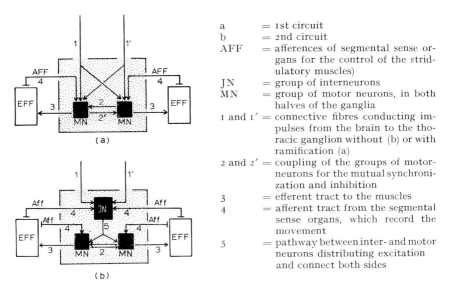

Fig. 284. Possibilities of connections in the thoracic neuron. Further explanation in the text. (Huber[72]).

neurons. These interneurons guarantee a symmetrical distribution of excitation and a synchronous discharge of the motor neurons in both halves. This circuit explains both the results after severance of one tract and the loss of stridulation after halving the ganglion. Such interneurons are known anatomically. The motor nerve cells also communicate directly with the connective fibres which could, for instance, control the muscle tonus. Whether the mutual adjustment of neuron populations is also accomplished via the interneurons or over a second direct coupling, is unknown. A direct connection between the dendrites of motor neurons of both sides has hitherto only been demonstrated in the larvae of Aeschna[133]. The transmission of excitation via interneurons is also likely in the co-ordination of the two drumming muscles of cicadas[94, 51–53]. Here the thoracic system obviously consists of an interneuron complex determining the rhythm of drumming activity, and of one motor neuron in each half which conducts the excitation to the muscle. In the first stage the arriving afferent or central excitation initiates activity in the pacemaker which discharges with a frequency of 200 impulses/sec. The pacemaker neuron or neurons then transmit the impulses to the motor neurons which fire with a frequency of 100 impulses/sec in strict alternation. Therefore in cicadas also one must assume a connection between two motor neurons which enables one motor neuron while discharging to inhibit the motor neuron of the opposite side. It should be stressed again that although in Orthoptera the proof of a rhythmically working system in the thoracic ganglion has been given, the functional connection between the participating neurons has still to be demonstrated electrophysiologically.

(g) Summary of the preceding results

Severance of connectives after halving-experiments, nervous isolation and local stimulation of the thoracic ganglia yield the following results:

(a) Ganglia which are isolated from the brain are unable to release a song pattern typical of the species. For this, excitation is necessary which arrives at the ganglion along the connective tracts. Impulses travelling in only one tract can still stimulate the thoracic system to produce stridulatory activity specific to the species.

(b) Severance of the connectives of both halves of the body and halving the ganglia located between the brain and the thoracic system have shown that brain impulses are transmitted to fibres which do not decussate in the anterior ganglia.

(c) The thoracic ganglion, however, when isolated from all central nervous regions, is still capable of co-ordinating a basic form of the stridulatory movement. It contains consequently all the nervous elements and circuits necessary for a distribution of excitation over the two halves of the ganglion and for rhythmical co-operation of the motor neurons. At the same time it has been shown that this system is not predetermined for one of the described chirp patterns. One can impose an entirely new rhythm upon it by stimulation of segmental sense organs or by electrical excitation of certain areas of the neuropile. In acridids the presence of such a mechanism in the 3rd thoracic ganglion has yet to be demonstrated.

(d) After halving the ganglion the stridulatory movement does not occur; the insects are unable to move the sound producing organs synchronously or even independently from each other. Two models are proposed which would explain the possible connections of the thoracic neurons.

4B. The Function of the Brain during Sound Production

(a) Anatomical remarks

The brain of Orthoptera consists of 3 parts, the proto-, deuto- and tritocerebrum. The cell bodies of the numerous neurons are situated immediately underneath the double layer of the brain sheath, partly in loose formation, partly concentrated into groups of cells (nuclei). Their fibres, the dendrites and neurites, enter the central neuropile and there synapse with each other (Figs. 285 and 286). The protocerebrum (PC) receives both laterally and dorsally sensory nerve fibres from the compound eyes and from the ocelli. The n.opticus (nop) after entering the brain subdivides into several bundles. One (1) of these terminates in the calyx of the ipsilateral mushroom body or (corpora pedunculata, mb, cp), the others (2 + 3) pass ventrally into the protocerebral neuropile. Between the 2 hemispheres run transverse connections, the largest of which is the optic commissure. The nerves of the ocelli (5) ramify as numerous dendrites and terminate near the pons cerebralis (b). The protocerebrum contains 3 systems of interneurons, (a) the paired and identically constructed mushroom bodies (cl), (b) the single central body (corpus centrale; cb) and (c) the protocerebral bridge (b). These systems are connected to afferent brain nerves, intercerebral tracts and fibres of the ventral nerve cord. The mb consist of numerous unipolar ganglion cells. On each side the cell bodies accumulate in two complexes in the dorsal region of the protocerebrum. Their fibres give off dendrites to the calyx; the neurites form the whole of the stalk system, which consists of the descending pendunculus, the "Balken" (β-lobe) and the ascending cauliculus (α-lobe)[54, 55, 117]. Afferent fibres from the area of the eye and the antennae, from the central body and the ventral nerve cord[117] terminate in the region of the calices, cauliculi and the pons. The central body in the middle of the brain consists of a dense net work of fibres. Its cell bodies are

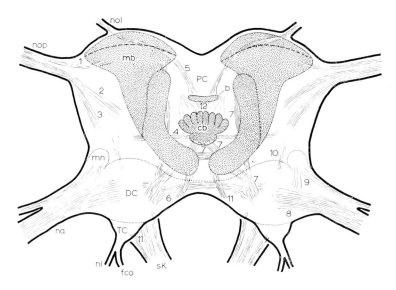

Fig. 285. Section through the brain of *G. campestris* as seen from the front.

PC = protocerebrum
DC = deutocerebrum
TC = tritocerebrum, brain nerves
nop = n.opticus
nol = n.ocellaris lateralis
na = n.antennalis
nl = n.labralis
fco = frontal connective
sk = circumoesophageal connective

The mushroom bodies are dotted (mb), the central body (cb) and the pons (b); the demarcations between the proto-deuto- and tritocerebrum are drawn in broken lines.

Tracts
1 = tr.optico-globularis
2/3 = tr.optico-protocerebellaris
4 = optic commissure
5 = tr.ocellaris-lateralis
6 = antennal commissure
7 = tr.olfactorio-globularis with crossing over to the opposite side
8 = sensory
9 = motoric antennal nerve
mn = motor nucleus of the antennal muscles
10 = fibres between proto- and deutocerebrum
11 = fibres descending to the nerve cord or ascending to the brain
12 = tract between pons and cb.

located in the lower margin of the pars intercerebralis, which lies on the dorsal midline of the protocerebrum. The cb is also connected with the optic ganglia, the deutocerebrum, the mb and the nerve cord. Right above the cb lies the bridge, a horseshoe-like network, whose cells are also distributed in the pars intercerebralis. The pons and cb communicate by numerous fibres which partly intercross.

The deutocerebrum receives laterally the sensory fibres of the antennal sense organs which end in numerous glomeruli. Dorsolaterally lie the nuclei of motor neurons whose fibres supply the antennal muscles. In the caudal region terminate fibres which originate from sense organs of the dorsal head capsule. Both halves of the second brain region communicate with each other by way of the antennal commissure which in gryllids is very strongly developped. From each deutocerebral half ascends a fibre tract, the tractus olfactorio-globularis, which connects the antennal region with the mb and the cb.

The tritocerebrum in Orthoptera is poorly developed. It innervates the labium and connects the brain with the stomatogastric nervous system. Further details

References p. 484

on the anatomy will be presented in the section "localisation of lesions and points of stimulation".

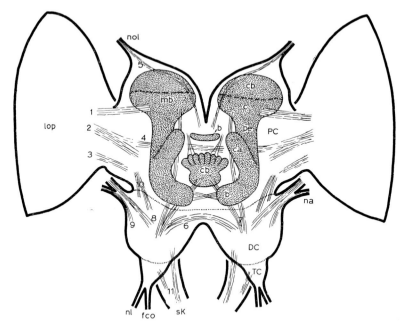

Fig. 286. Section through the brain of *G. rufus* as seen from the front. cb = cell body of the mushroom body, c = calyx, pe = pedunculus, ca = cauliculus, b = Balken, lop = region of the lobus opticus. For other symbols see Fig. 285.

(b) *Lesions in the brain*

Gryllus campestris: It has been found that male crickets can still sing after the elimination of important sense organs[66]. The occlusion of the compound eyes and ocelli or severance of their nerves did not change the acoustical behaviour. Amputation of the antennae which play a leading role in the recognition of sexes made orientation of the male during fighting and mating a little difficult, but the insects succeeded finally in copulating. Elimination of the anal cerci affected the female only while mounting the male and the copulatory position was not achieved in the normal way. Nevertheless, in this case also the male transferred the spermatophore successfully[66]. The function of the antennae as releasing organs for social behaviour can be partially replaced by sense organs on the body. Even the loss of the tympanic organs in the tibiae of the first pair of legs did not prevent the male from producing all types of songs.

In a further experiment parts of the brain were removed. The various operations are summarized in Table 35:

Males would not sing after large areas of the protocerebrum had been extirpated or after halving the brain. In both cases parts of the mb the bridge and the cb were damaged. Local operations on those systems showed that *one* undamaged mb was enough to release and to co-ordinate all types of gryllid songs. On the other hand,

TABLE 35

SONGS AND SEXUAL BEHAVIOUR OF THE MALE CRICKET AFTER OPERATIONS ON THE SENSE ORGANS AND THE BRAIN

Place of operation	N	gG	RG	Fight	WG	Courtship	Copulation
1. N.optici and N.ocellares severed	12	+	+	+	+	+	+
2. N.antennales severed	7	+	(+)	(+)	(+)	(+)	+
3. Tympanic organs destroyed	11	+	+	+	+	+	+
4. Cercal nerves cut	5	+	+	+	+	+	(+)
5. Brain halved	6	—	—	—	—	—	—
6. Bridge and cb destroyed	4	—	—	—	—	—	—
7. Right or left half of the protocerebrum removed	5	—	—	—	—	—	—
8. Right or left half of the calyx (mb) removed	6	+	+	(+)	+	+	+
9. Both mb removed	3	—	—	—	—	—	—

N = number of insects; gG = "calling"; RG = "rivalry"; WG = "courtship" song.
AG = responding song.

stridulation and acoustical behaviour were eliminated if the cb had been injured. Accordingly the brain structures described seem to be important centres for the stridulatory movement.

Gomphocerus rufus: In the male and female of this acridid similar operations on the sense organs, their nerves, and on the brain have been carried out. Table 36 shows the results.

TABLE 36

SONGS AND ACOUSTICAL BEHAVIOUR OF *G. rufus* AFTER EXTIRPATION OF VARIOUS SENSE ORGANS AND PARTS OF THE BRAIN
(For explanation of symbols see Table 35)

Place of operation	N	gG	WG	AG ♂ ♀	Courtship	Copulation
1. Extirpation of one compound eye	7	+	+	+ +	(+)	+
2. Extirpation of both compound eyes	5	+	—	+ +	—	—
3. Severance of the ocellar nerves	6	+	+	+ +	+	+
4. Abdominal tympanic organs destroyed	9	+	+	— —	+	+
5. Two antennae amputated	12	+	+	+ +	(+)	+
6. Brain halved	4	—	—	— —	—	—
7. 1 mb removed	6	+	+	+ —	+	+
8. 2 mb removed	6	—	—	— —	—	—
9. Cb destroyed	3	—	—	— —	—	—

The eyes and tympanic organs are necessary for the release of the "courtship" and "responding" song. But blinded males could still localize the responding female acoustically and jump upon it with accuracy. Here the chain of reactions broke down, as the visual stimulation necessary for the release of the "courtship" song was absent. Males with reduced thresholds, however, frequently produced the whole courtship in vacuo, which means that the movement pattern can also take place without visual control. Also in acridids brain lesions have shown that the mb and the cb are important nervous structures for sound production. As in gryllids one intact mb is sufficient for normal acoustical behaviour in the male. The females react to the extirpation of one mb by being distinctly more sensitive, but only further experiments will show whether the cerebral organisation for sound production of the female is similar to the male's or not. There may be several reasons for the loss of the "responding" song in the male and female after destruction of certain brain areas. If only the neural apparatus responsible for the co-ordination of the stridulatory muscles is eliminated, then the acoustical signal will still be perceived, but this information will not result in activation of the effectors. How does an insect behave under those circumstances? It is possible that the peripheral impulses arriving permanently in the brain would activate other motor systems whose pathways to the brain remained undamaged. This would cause substitute and displacement activities, which, however, have never been observed in acridids, although they do occur in gryllids[66]. A further reason for the lack of the "responding" song could be due to the destruction of perceptive centres. This would mean that information from the tympanic organs also arrives in the brain and is analysed there. One experimental result in G. rufus seems to support this idea. In a male a small dorsal area of the cb had been destroyed. The male sang, orientated, courted and mated, but did not respond to the female's song. Table 37 shows the result.

TABLE 37

SONGS AND COURTSHIP BEHAVIOUR OF A MALE AFTER DESTRUCTION OF A SMALL AREA IN THE DORSAL PROTOCEREBRUM

Releaser	Number of songs during 5 days of observation			
	gG	WG	Copulation	"Responding"
No releaser	17	1	—	—
Females in sight of the male	2	10	4	—
Female sings 14 times	1	—	—	—

It appears, therefore, that in G. rufus the central nervous analysis of acoustical signals and the transmission of the perception output to the stridulatory apparatus is localized in certain brain areas (see also [61,137]).

(c) Local injuries of the brain by punctures

Huber[66] describes a method for puncturing the brain. The head ganglion is exposed from the frontal side and damaged in various regions with a needle (50–80 μ in diameter). During or after puncturing, different reactions appear which last for variable lengths of time. Table 38 summarizes the various kinds of stridulation in Gryllus and Gomphocerus

TABLE 38

SONGS OF Gryllus AND Gomphocerus AFTER INJURY OF CERTAIN CEREBRAL PARTS

Centre of injury	Songs after puncture	
	Gryllus	Gomphocerus
1. Neuropile between the entry of the optical nerve and the mb	—	—
2. Region of the pars intercerebralis	—	—
3. Neuropile of the pons protocerebralis	—	—
4. Neuropile of the cb	Atypical song	Operation not performed
5. Mushroom bodies	gG, RG, WG	gG
6. Deutocerebrum	—	—

The songs of crickets were only released when the needle tip damaged parts of the calyx and stalk of the mb. Frequently the insects sang until completely exhausted suggesting that the needle had injured systems which under normal circumstances inhibited stridulation. If the neuropile of the cb was damaged, the males raised their wings and produced a more or less continuous whizzing song, a behaviour which normally never occurs. In *G. rufus* the mb have only been hit in a few cases; a single puncture released a continuous song.

Hitherto two groups of experiments have shown that for the release and co-ordination of a stridulatory movement two brain systems are necessary: the mb and the cb. Thus, for the first time, there is clear physiological evidence for Dujardin's[37] old hypothesis, that the higher psychical functions of insects take place in the mb. How the two brain structures co-operate has still not been explained. For further analysis, a method has been developed in which small areas of the brain can be stimulated electrically[69,89].

(d) *Electrical stimulation in the protocerebrum of crickets*

(i) *Method*: A holder is glued on the notum of a cricket anaesthesized with CO_2. The indifferent electrode (platinum wire) is soldered on this holder. The tip of the wire is bent in a loop which penetrates a prepared hole into the dorsal hemolymph cavity of the prothoracic segment. Then a window is cut out between the antennal bases of the head capsule, the tracheae which are in front of the brain and descending to the labrum removed, and the exposed brain area is cleant by absorbing the hemolymph with filter paper. After this operation the insect is fixed in the stimulatory apparatus (Fig. 287). The stimulating electrode (silver or tungsten wire) is 15–30 μ thick and insulated except for the tip, which can now be pushed through the neural lamella to a certain depth in the brain. After the cricket recovers, a cork ball is placed between its feet. The cricket turns this ball contrary to the running-direction, and drops it during jumping and flying. Details of this method are given in[89].

(ii) *Results*: By local electrical stimulation of the protocerebrum of gryllids, complex co-ordinations of movements could be released or inhibited. The impulses used differed in form (rectangular impulses and condensor discharges), length (0.1–1 msec), frequency (15–150 c/s) and amplitude (3–40 μamp). Each point of stimulation induced nearly the same behaviour if the stimulus values were kept constant. An increase of amperage or, to a minor extent, of the frequency shortened the latency of the reaction. Whole behavioural acts consisting of several successive phases could be released by stimuli of long duration. Short stimuli often released only the early phases of a behaviour. A strong stimulus usually released more of the successive phases of a behaviour than a weaker stimulus of the same duration.

(1) Song patterns of the "calling" songtype

During stimulation male crickets frequently produced a sound rhythm which in chirp rate and sequence was very similar to the "calling" song (Fig. 288). The insects put their antennae forward, the frequency of respiration increased, the elytra were raised 45° off the body and the insect began to sing. After cessation of the stimulus the song gradually weakened; the chirp interval remained constant but the pauses between the sequences became longer.

Chirp length and intensity sometimes showed irregularities, so that the song sounded somewhat distorted. This was due to the holder glued to the notum, which by its weight, changed the resonance properties of the stridulatory apparatus. The

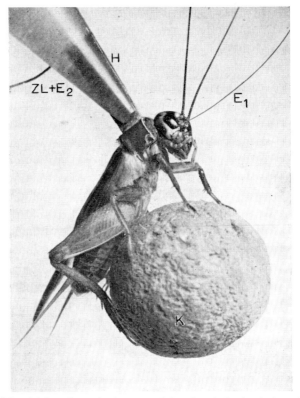

Fig. 287. Male cricket (*Acheta domesticus*) prepared for electrical stimulation of the brain. E_1 = stimulating electrode in the brain, E_2 = indifferent electrode with lead (ZL), H = holder, K = corkball for running on the spot. (Huber[72]).

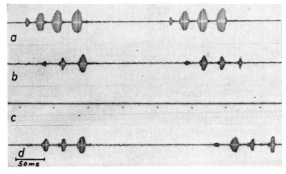

Fig. 288. Oscillograms of the song rhythm of *G. campestris* before (a), during (b) and after (d) electrical stimulation in the region of the cauliculus of the mb. Type calling song. Data: impulse frequency 15/sec (c), amperage 15 μamp.

released songs did not have the rhythm of the stimulation frequency. Contact therefore must have been made on a neural system which reacted with an innate rhythm. The long latency and the after-effect suggest a temporal summation of the activity in the neurons responsible at the beginning of the stimulus and a very gradual decrease of the induced excitation at the end of the stimulus, the neurons maintaining a rhythmical discharge over a long time. It is still not known whether this after-effect is an expression of an after-discharge from the neurons[15] or whether we have here a system with positive feed-back. Repeated stimulation of the same area at short intervals increased the stridulatory activity of the males considerably. One male, for instance, which was stimulated for the 4th time (20 sec stimulus duration and 15 sec between stimuli), only ceased to sing after 30 minutes. Thus within 140 sec a perfectly silent insect changed to a male ready for copulation. Furthermore, even after the stridulatory movement had faded, a central excitation still remained. Stimulation on the tactile sense organs on the abdomen or of the antennae caused the insect to sing again. It can be assumed, therefore, that the electrical stimulus affected a region of the brain which not only integrated peripheral afferent impulses, but is also related to some endogenous systems responsible for the readiness to copulate. In addition to the release of the stridulatory movement a stimulus also induces the insect to erect its body, to increase its respiratory activity and to lift the antennae; in short, it takes up that posture which accompanies the "calling" song. The neurons stimulated must, therefore, be part of a system which controls not one, but several motor mechanisms.

(2) Song pattern of the "rivalry" songtype

The "rivalry" song of a male cricket is an element of the fighting behaviour[134, 66, 60, 4]. Normally the song is released amongst males by whipping each others antennae, then stops during fight and reappears as a song of pursuit stridulated by the winner. This song, too, is correlated in time and space with a number of motor actions on the part of the insect whose state of excitation determines beginning, duration and intensity of the various parts of the fight[66].

During stimulation of certain protocerebral regions a sound pattern was released which corresponded exactly to the "rivalry" song. In the absence of an opponent the male, after a longer latent period, began with a distinct increase of respiratory activity, then raised its body, jerked its antennae upwards and beat into empty space. Then the "rivalry" song began accompanied as in normal fighting behaviour by a strong shaking of the body. The physical characteristics of such a released "rivalry" song, such as frequency, amplitude and modulation, are indistinguishable from the parameters of a natural "rivalry" song (Figs. 289 and 290). Stridulation stopped abruptly with the end of the stimulus, whereas antennabeating, shaking, and increased respiration decreased only gradually. But the male still remained in a state of general excitation for a long time afterwards and a vibration of the holder, or a light touch of the antennal tip or of the cerci immediately released the fighting behaviour and "rivalry" song. Here again the system stimulated must consist of a group of neurons which after a certain inertia respond with an innate rhythm independent of the stimulation frequency. Furthermore, this system not only controls the stridulatory movement, but also the motor systems of head and thorax which are responsible for the rivalry behaviour. One may suppose that this system of neurons plays a part in maintaining the readiness for fighting, although we still do not have any idea of the way in which such a preparedness could be expressed neurophysiologically.

The resemblance between the rivalry behaviour in nature and after electrical

References p. 484

468 F. HUBER

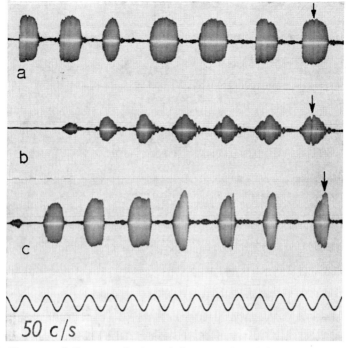

Fig. 289. Oscillograms of the song rhythm of *G. campestris*, (a) "rivalry" song without and (b) with fixed holder, (c) "rivalry" song during stimulation in the region of the tr.olfactorio-globularis. Impulse frequency 60/sec, amperage 24 μamp.

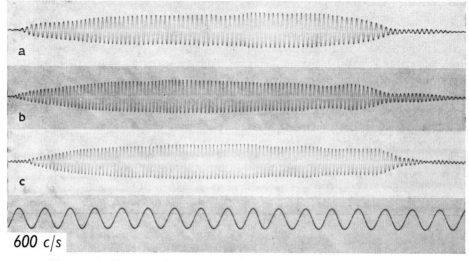

Fig. 290. Oscillograms of single chirps (↓) of the songs mentioned in Fig. 289.

stimulation is particularly interesting. In both cases the various components of fight follow each other rather flexibly[66]. The song may begin, for instance, after the shaking movements and after the mandibles are opened but before the insects run towards each other. The cessation of the song at the end of the stimulus is a further point of agreement. Under normal circumstances the "rivalry" song is always released by the beating of the antennae or by the song of a male of the same species. Stridulation continues until the opponent becomes silent, or flees and disappears out of the range of the winner's antennae. A new attack by the looser causes the winner to emit "rivalry" song readily over a long distance. Duration and intensity of the "rivalry" song in male crickets probably depends on the duration and intensity of the afferent excitation; in addition the threshold for the particular behaviour (fight or peace, victory or defeat) is co-determined by the previous history of the male. Similar rules seem to be valid for the song released by electrical stimulation. The song ceases with the end of the stimulus because the supply of excitation for the releasing system is interrupted. The threshold, however, still remains low for some time; this can be seen from Table 39.

TABLE 39

RIVALRY BEHAVIOUR OF A MALE CRICKET DURING STIMULATION IN THE PROTOCEREBRUM AND AFTER ANTENNAL STIMULATION 40 SEC LATER

No.	RD	La	Behaviour during electrical stimulation	AZ	Reaction after mechanical stimulation
1	30	18	"Rivalry" song, antennae beating, shaking	2	"Rivalry" song and antennae beating
3	30	11	"Rivalry" song, antennae beating, shaking	2	"Rivalry" song and antennae beating plus shaking
7	30	8	"Rivalry" song, antennae beating, shaking	4	"Rivalry" song and antennae beating plus shaking

No = serial number of stimulation; RD = duration of stimulation; La = latent period for the "rivalry" song; AZ = time of decrement for the "rivalry" song (time in sec).

The shortening of the latent period for the rivalry behaviour from 18 to 8 sec speaks for a change in the threshold; the reaction seems to be facilitated after repeated stimulation of the same centre. The cessation of the singing movement after stimulation suggests that afferent systems have been stimulated which are silent without excitation and do not possess after-discharges. The behaviour of the insect to other stimuli indicates a change in the nervous system which slowly decreases.

(3) Atypical song patterns

Abnormal song patterns are of great importance for the clarification of cerebral systems functioning during stridulation (Fig. 291). After electrical stimulation of a limited region of the protocerebrum the males only slightly raised their elytra and rubbed them irregularly but very quickly against each other without interruption (Flachvibrieren, Huber[66]). The continuous whizzing sounds are composed of short and highly modulated chirps. The repetition frequency oscillates between

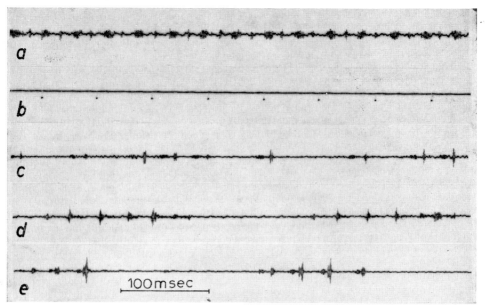

Fig. 291. Oscillogram of an atypical song of *G. campestris*, (a) during stimulation of the cb, (b) impulse frequency 15/sec, (c)–(d) sound patterns in different intervals after stimulation show the modification of the song to the rhythm of the "calling" song.

40 and 65 per second, considerably higher than in a normal cricket song pattern. Without photographic recordings of the elytral movement, however, it is not clear whether or not a single chirp corresponds to one movement of the elytra in either direction. Stimulation of various insects showed all transitions from a continuous song to irregularly composed songs.

Common to all of these atypical songs was the rhythm which did not show any relation to the stimulation frequency. Two groups can be distinguished according to the latent periods:

(a) Songs which began at full strength immediately after stimulation (0.5–1 sec) and stopped with the end of stimulation.

(b) Songs which began slowly and with great inertia (7–25 sec), but continued a long time after the end of stimulation.

In the second group the sound pattern was changed several times during the post-stimulatory phase (Figs. 291(c)–(e). The insect interrupted the stridulatory movement more and more frequently and sang at last with a rhythm comparable to the "calling" song, as far as the number of chirps in a sequence and the chirp interval are concerned (Fig. 291(e). The following conclusions can be drawn from the foregoing results: Atypical songs demonstrate that the mesothoracic apparatus in gryllids also responds to imposed atypical cerebral patterns of excitation. The thoracic mechanism is therefore not predestined for one rhythm. The frequency of elytral movement can be increased beyond the normal frequency, which shows that under normal conditions the motor system in the thorax does not work at the maximal stridulation frequency. It also follows that in the brain of a cricket two topographically different regions exist which influence sound production. These two systems must be responsible in an unknown way for the determination of chirp rate and sequence rate. By non-specific stimulation, however, their neurons can also be induced to produce different rhythmical discharges.

(4) Central inhibition of the stridulatory movement

So far we have only spoken about excitation of the stridulatory movement and the change of the sound pattern due to brain stimulation. Observations on cricket behaviour and the results of brain lesions indicate that the head ganglion plays an important part in the inhibition of sound production[66].

Here are some examples: Even with a fully developed spermatophore the cricket emits the "normal" song only during certain times of the day[134]. After copulation the male stops calling until the new spermatophore is pressed into the spermatophore sac[66]. The "courtship" song is also connected with the presence of the spermatophore, but the male at once stops singing when it is touched by a female. In acridids the song of the female occurs only at a certain time, during two ovipositions[79] (Huber, unpubl.). In the brain of the cricket certain areas, during stimulation and after a short latency, inhibit sound production, but the insect sings again as soon as the inhibition has faded. Table 40 shows some data.

Does the electrical stimulation cause a specific inhibition or a general damping of the whole motor activity? 8 out of 10 insects which ceased to sing during stimulation also stopped locomotion and the respiratory movements of the abdomen.

TABLE 40

INHIBITION OF STRIDULATION IN THE MALE OF THE FIELD CRICKET (G) AND THE HOUSE CRICKET (A) DURING STIMULATION OF THE PROTOCEREBRUM

Insect	Data of stimulation			Behaviour before stimulation	L_1	Stimulation with inhibitory effect	AZ	Behaviour after stimulation
	F	I	RD					
G 119	15	7	30	gG	12	—	25	gG
G 119	15	14	30	gG	3	—	47	gG
A 87	30	5	25	gG	22	—	12	gG
A 87	30	15	25	gG	5	—	77	gG

F = impulses/sec; I = amperage in μamp; RD = duration of stimulation in sec; L_1 = latency in sec; AZ = decrement of inhibition.

2 males reacted with stridulatory inhibition only. Thus it is shown that there are two systems in the protocerebrum (mb, cb) one which damps the general motor activity and another one which exerts a specific inhibition upon certain movements.

(5) Competitive inhibition of running and singing

Running and singing in gryllids are two expressions of behaviour which cannot take place at the same time. Morphological studies have shown that some muscles of the mesothorax raise and extend the elytra for stridulation and also move the coxa fore- and backwards while the insect runs. This result is in good accordance with observations of behaviour. Usually for the period of the "calling" song the male cricket remains in front of its earth-hole, sometimes cleaning one of its antennae and occasionally taking a few steps. Before the "rivalry" song the male stops running or walking, but body shaking and the rivalry song can occur simultaneously. During pursuit, rapid running and stopping to sing alternate frequently. The "courtship" song also ceases as soon as the male increases the locomotor activity beyond a certain measure. Single males also show competition of both movements which is demonstrated by the following records.

G121: 60 impulses/sec, 10 μamp, duration of stimulation 30 sec

Latency (in sec)	Reactions during stimulation	Behaviour after stimulation
2	Increase of respiration frequency, head and antennae are raised	
13	Beginning of running, slow raising of the elytra, singing	
21	Quick running, the elytra are laid back, singing stops	Running stops, the male begins to sing
3	Increase of respiration frequency head and antennae are raised	
9	Stridulation, beginning of running	
17	Running is accelerated, elytra are laid back on the thorax, singing stops	Running stops, the male begins to sing

Two examples of responses during and after stimulation showing inhibition between rapid running and singing.

The mutual competition can still be tested in another way. Male crickets which during brain stimulation stop to sing can be induced to defend themselves and to run by touching the cercal hairs. During this time they stop singing, but continue to hold their elytra in the stridulatory position. The question as to the region of the CNS involved and how the regulation and selection of the two movements is performed cannot yet be answered, nor is it known how specific movements are initiated. In general the behaviour of animals is an expression of a central competition between activation and inhibition. For the running and singing activity of crickets some of the same motor neurons are stimulated, although with different intensities and in different combinations. Weak running and singing can occur simultaneously. Therefore it seems unlikely that the reason that the two processes usually exclude each other is because the same descending fibers inhibit one activity while exciting the other. The simultaneous appearance and the superposition of the two activities rather suggests that competitive inhibition is connected with complex processes of the cerebral or thoracic systems and that a peripheral control may be involved.

(e) Localization of the points of stimulation releasing te stridulatory movement

(i) Method: After stimulation and electrocoagulation[89] the brains were removed under Ringer solution, put into the histological fixatives of Bouin and Bodian, embedded, and then sectioned in the sagittal plane. For staining, particularly of the points of stimulation, Azan (after Heidenhain) was used and for the investigation of the nervous tracts the method of impregnation (after Bodian[109]) proved to be convenient. The

histologically determined points of stimulation were then transferred to a brain map; the symbols indicate the spot reached by the tip of the electrode.

(ii) *Results*: Out of 143 histologically determined stimulation points in the protocerebrum of 143 male crickets, 38 are related to the acoustical behaviour. In Table 41 the points are put in topographical order. Figs. 292 (a and b) show their position in frontal and sagittal projection.

TABLE 41

DISTRIBUTION OF POINTS OF STIMULATION FOR SONGS AND STRIDULATORY INHIBITION IN THE PROTOCEREBRUM OF MALE CRICKETS

Regions of stimulation	NZ	B	NP	gG	RG	ATG	GH
Tractus olfactorio-globularis		+			2		1·
Range of cauliculus margin (mushroom body)		+	+	2			
Range of the "Balken" margin		+	+	3		1	
Cells and calyx system (mushroom body)	+		+	1			6
Pedunculus-"Balken"		+		2			2
Cauliculus		+		4	1		
Pons			+		1··		1···
Central body and marginal regions			+			7	
Caudal neuropile and descending tracts	+		+			4	

NZ = cell bodies; B = tracts; NP = neuropile; ATG = atypical song; GH = inhibition; 1· = area, where the tr.olfact.-globularis enters; 1·· = region between pons and tr.olfact.-globularis; 1··· = region between pons and neuropile of the calyx.

Whereas the points of stimulation regulating respiration[71] and locomotion[72] are not confined to certain limited regions of the protocerebrum, the points for the release and inhibition of the various types of songs and the atypical sound patterns are placed nearly entirely within the range of the two mb, the connective regions with afferent brain paths and the neuropile of the cb and its marginal regions. These results are in agreement with those of the previous brain lesions and the punctures in the protocerebrum[66]. The results of the local coagulation agree also with earlier findings. Coagulation points in one mb do not influence the acoustical behaviour, because one mb is enough to transfer information for the release, co-ordination and control of all songs[66]. Only large lesions in the cb and in its marginal regions, or the destruction of both ends of the "Balken" extinguish sound production. These results permit certain statements to be made on the structures releasing and inhibiting stridulation and the mechanisms which compose the sound patterns.

(f) *Discussion of the stimulation experiments on the brain*

It has been mentioned that in gryllids and acridids only certain sense organs are used for the reception of song-releasing stimuli (Table 42.) How are these organs connected to the brain structures and what is already known about path ways transmitting excitation? The following sections refer only to *Gryllus campestris*, as the experiments in *Gomphocerus rufus* are still in progress.

The antennae, cerci and tympanic organs have been found to play a significant role in acoustical behaviour.

References p. 484

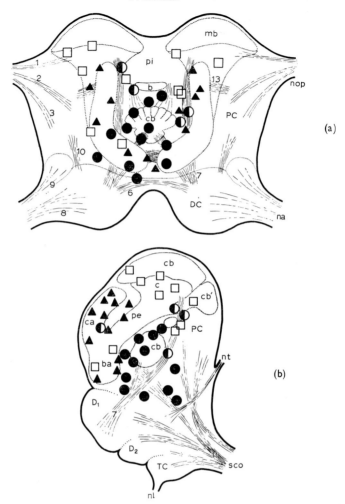

Fig. 292a and b. *G. campestris*. Diagram of the brain, with the stimulation points as seen from the front (292a) and sagitally (292b). ▲ = points of stimulation releasing the "calling" song, ◐ = points of stimulation releasing the "rivalry" song, ● = points of stimulation releasing atypical songs, □ = points of stimulation inhibiting the song (Huber[72]).

cb = cell body
pi = pars intercerebralis
D_1/D_2 = sensory lobe in the deutocerebrum
nt = n.tegumentarius-posterior

13-tract between the lateral regions of the neuropile of the protocerebrum and the pars intercerebralis.

(See Fig. 285 for further explanation of symbols.)

(*i*) *Afferent activities of the antennae.* The antennae, together with the anal cerci, are the most important organs of social orientation and bear amongst other structures the receptors for tactile stimuli. The nerve impulses originating from the antennal receptors can release "rivalry" song as well as "courtship" song[66,60]. In each half of the deutocerebrum branches of afferent antennal nerve fibres form numerous connections with 1. the dendrites of the motor neurons innervating the antennal muscles of the same side, 2. the interneurons of the antennal commissures and 3. fibres of the tractus olfactorio-globularis, the axons of which ascend to the mb and the cb.

TABLE 42

SENSE ORGANS KNOWN TO PARTICIPATE IN THE RELEASE AND INHIBITION OF SONGS

Song type	Eye	Ocellus	Antenna	Cercus	Tympanic organ	Copulatory apparatus
Gryllus						
gG	—	—	—	—	—	act., inh.
WG	—	—	act.	act.	—	act., inh.
RG	—	—	act.	act.	act.	—
Gomphocerus						
gG	—	—	—	—	—	—
WG	act.	—	act.	—	act.	—
AG	—	—	—	—	act.	—

— = not necessary; act. = stimulus leads to release; inh. = stimulus leads to inhibition; for other symbols see Table 33.

Anatomical considerations and lesion experiments make it seem likely that there is some direct transmission between antennae sensory and motor fibers. The antennae commissure appears necessary for coupled movements of the antennae. The tr.olfactorio-globularis belongs to song-releasing brain structures. This is shown by experiments using stimulation and localisation. 3 of 4 stimulation points for "rivalry" song lay along the tractus and near the region where it enters the mb (see Fig. 292 b).

In *Periplaneta americana*, Maynard[88] has stimulated the afferent antennal nerve and followed, with micro-electrodes, the propagation of excitation in the brain. After a latency of 40–70 msec impulses were recorded in the calyx of the equilateral mb, and were later recorded in the stalk system (pedunculus-Balken), but how the propagation progresses further is still unknown. In *Periplaneta* the tractus is very similar to that in the cricket([54,55]) and so we may expect a similar kind of transmission.

Thus, in the brain, one route taken by the impulses which release rivalry behaviour is as follows: after strong and repeated stimulation of the tactile receptors of the antennae the afferent impulses reach the first sensory lobe of each half of the deutocerebrum. Amongst other things the impulses are here switched over to fibres of the tr.olf.-glob. and arrive then in the mb and the cb. It is interesting to note that one can imitate the stimulus patterns of an opponents antenna beating by striking the antenna of a male with a vibrating brush. This stimulus leads males, which are prepared to fight, to show normal rivalry behaviour (see [65,66,4]).

(ii) Afferent activities of the cercus. The neurites of the numerous receptors of the cercus enter the last abdominal ganglion and form in part synapses with the giant fibres of the abdominal nerve cord[100,105]. It is still not known if the impulses released in the sensory cells by stimulation of the cercal sensory hairs are transmitted directly to the brain, *i.e.* without mediation of central neurons. Some transmission to the brain has to be assumed in view of observations on the behaviour; impulses from abdominal receptors are important for the release and inhibition of song (see[66]). In roaches transmission of impulses has been investigated with electrophysiological methods up to the 3rd thoracic ganglion[105]. Preliminary studies of crickets have shown

that after stimulation of the anal cerci (touch or air puffs) synchronized series of discharges in the nervous system can be recorded as far as the SEG (Huber, unpubl.).

(*iii*) *Acoustical afferent activities*. We don't discuss here the significance of hearing organs for the behaviour of females, but only their possible function in co-ordination of songs in males. According to Table 42 auditory excitations play a role in rivalry behaviour. But males with destroyed tympanic organs can still sing (p. 480). In crickets, impulses from the sensory cells of the tympanic organs travel along a special nerve to the 1st thoracic ganglion. According to [61,137] they are transmitted there to interneurons which may end in the head ganglia. A direct fibre connection between the tympanic organ and the brain is unknown. We are still ignorant of where and how the auditory excitation is analysed in the brain, although some results in acridids suggest that areas of the protocerebrum contribute to the analysis (Huber[67]).

(*iv*) *The role of the mushroom bodies during sound production*. Nearly all points of stimulation which release typical sound rhythms or inhibit stridulatory movements are placed in the mb or in their vicinity. It is this region where the fibres of the tr.olf.-glob. and the fibres of the central optic and antennal area enter the calyx and cauliculus. From lesions and brain stimulations we can state the following about the role of these systems during stridulation:

1. Afferent impulses which release or inhibit stridulation must be conducted to the mb. If these structures are eliminated, the conduction of excitation to the thoracic motor centres is interrupted at an important level. Without the mb the insect is unable to co-ordinate other movements coupled with stridulation, such as fight or courtship.

2. One mb with intact afferent tracts is sufficient to initiate and to maintain acoustical behaviour. It follows that the protocerebrum possesses two functionally equivalent systems between which there is now no evidence of interaction during stridulation. But the insect can then act correctly only if the activities of both systems are correlated. The synchronous excitation or inhibition can be obtained by simultaneous stimulation of the bilaterally placed sense organs which are in connection with the mb. In the case of the antennae one sense organ might be sufficient, as the fibres of one tr.olf.-glob. end in the calices of both mb.

3. The points of stimulation for inhibiting the song are mainly placed in the region of the cell bodies and the calyx of the mb. Releasing and inhibiting areas are topographically separated. There is no indication that a certain stimulation point can induce different song rhythms by changing the parameters of the stimulus; that is, the frequency, duration, form and intensity. Transitions from the "calling" to the "rivalry" song and *vice versa* have never been observed during stimulation of a distinct area with altered impulses. The afferent fibres which conduct the specific impulses are connected with certain neurons of the mb. As the course of fibres in the brain, and the number and arrangement in the mb are only incompletely known, it is impossible to delimit a certain area of the mb as being the locus for a certain type of song. Lesions and stimulations, however, have indicated that the mb of Orthopera contain mainly inhibitory neurons[66,70,71,72,104]. Out of 13 points of stimulation localized in the region of the cell bodies, where one is more certain of stimulation only the mb and not overlying structures, 11 inhibit stridulation and locomotion (see Fig. 292 a, b).

4. The afferent fibres form synaptic contact with the neurons of the mb mainly

in two places: in the region of the calyx and along the cauliculus[117]. Points of stimulation in those areas frequently release "calling" and "rivalry" songs. This is probably due to a local stimulation of afferent tracts entering the mb. It is not known where the impulses come from for the "calling" song.

5. Neither in the terminations of afferent tracts nor in the mb atypical songs can be released by stimulation. It is assumed, therefore, that these brain structures merely determine the beginning and duration of sound production, and do not carry out the command for translating song-releasing information into chirp- and sequence patterns. In the region of the "Balken" the mb can transmit their excitation to fibres of the cb and the ventral nerve cord. It seems that there the regulatory function of the mb comes into action.

(v) *The role of the central body during sound production.* Local injury and electrical stimulation in the cb and along its marginal areas release atypical songs. They have no relation to the stimulation frequency and are not "understood" by insects of the same species. The role of these brain structures for stridulation becomes clearer from the following results:

1. Lesions demonstrate that cerebral commands releasing stridulations are conducted via the cb or at least by mediation of its neurons to the thorax. Elimination of the cb suppresses the whole acoustical behaviour. After halving the brain the males cannot sing. There is histological evidence that in the cb there is at least a partial decussation of tracts from the mb to the thoracic nerve cord.

2. The cb is not only a passage way for excitation travelling from higher centres of the brain to the nerve cord, but also a second important cerebral apparatus for sound production. As the natural sound rhythm can only be disturbed in the cb and in the descending tracts, we may assume that its neurons participate in the translation of information from the mb into patterns of excitation for the chirp- and sequence rhythm. This system seems to be the highest motor centre for sound production. Apparently here the innate song rhythms are ready to be released, inhibited or changed by stimuli from the mb.

3. Normally the commands of the cb are conducted over both connective tracts to the thorax, but fibres of one tract alone are sufficient to transmit the information necessary for the song rhythm and to activate the motor rhythms.

5. Observations on a Reciprocal Action between the Brain and the Thoracic Motor System

So far we have found two regions which regulate and co-ordinate stridulatory movements: the brain and a thoracic ganglion. It has further been shown that the motor apparatus in the thoracic ganglion receives information from the brain which determines the beginning, duration and rhythm of the stridulatory movement. Now the question arises, whether stridulation in Orthoptera is controlled by a fixed program, in the brain or whether subordinated centres continuously report their state of excitation back to the brain so that acoustical behaviour is an expression of this interaction.

(a) *Relationship between the various elements of motion in the courtship of*
Gomphocerus rufus

The male has a very complicated courtship behaviour. In addition to the stridulatory movements of the hindlegs, the head with antennae and palpi and the anterior body

References p. 484

take part. The movements follow each other according to a fixed program. Faber[64] and Jacobs[79] distinguish between 3 phases:

Phase 1: after having taken a position diagonally in front of the female, the male begins to shake its head. The maxillary and labial palpi swing with the same frequency, but with opposite phases. The antennae are held forward and upwards. Near the end of this phase the hindfemora glide up and down with a small amplitude and almost soundlessly. Then the frequency of the head shaking increases, the palpi start to vibrate and the antennae are brought down and backwards describing semicircles.

Phase 2: during the last antennal movement the male suddenly raises the anterior part of its body, jerks both hindfemora upwards and flings the tibiae backwards once or twice. Then the antennae return to the initial position.

Phase 3: Now the "courtship" song follows, which may be interpreted as a prolonged "calling" song.

On the average one succession of movements lasts 10.5 sec and courtship consists of a more or less frequent repetition of all these movements. Immediately after the 3rd phase the males begin with the head-shaking, thus introducing the first phase again. If the metathoracic ganglion is eliminated, the male courts in silence. But the succession of the various movements is strictly observed and although, the song is omitted, its original duration is adhered to[67]. Blocking one or both antennae suppresses the antennal movements at the end of the first and the beginning of the second phase, but the other movements remain undisturbed. Therefore *G. rufus* possibly has a system within the brain which activates the various motor mechanisms of head and thorax in a temporarily fixed way, and it is not necessary that these movements be performed for the course of the courtship song. So the systems coupled to perform that song type are activated from a higher centre via several channels which apparently do not influence each other. (Further details in Huber and Loher[144].)

(b) Relations between the motor systems in complex situations of behaviour

During the fight and courtship of crickets the stridulatory movement is only one element of behaviour and is correlated with other processes in space and time. Is there an interplay between these activities?

The male cricket is able, even after section of connectives before the 2nd thoracic ganglion to recognize a female or an opponent. If such an operated male meets another male, it can be induced by the latter to perform all these fighting movements which are connected with the brain as co-ordinated systems. Only the song is missing. Now and then the insect begins to perform movements belonging to the fighting behaviour without a partner. It follows that feedback seems to be unnecessary between the thoracic stridulatory apparatus, the brain centres determining the fight in detail, and the motor systems of the other movements. In the case given the insect would have to stop or to change its rivalry behaviour as soon as one link of the functional chain was missing if feedback were important.

The situation during courtship is more complex. After the male with the sectioned thoracic nerve cord has touched a female, the male turns in the opposite direction. This movement corresponds to the taxis component of courtship and to the first phase of after-courtship. Earlier we have seen that after severance of the nerve cord the male is put in a conflict situation because the impulses from the receptors of the copulatory apparatus necessary for the release of the courtship song cannot be transmitted to the brain.

If, in this case, the song is eliminated and courtship is interrupted in spite of appropriate orientation, the cause does not necessarily lie in the paralysis of the stridulatory apparatus. The following observations, however, should be noted: operated male crickets, when meeting a female, frequently clean the antennae, the head and the forelegs, *i.e.* they respond with movements which do not fit into the given situation. These cleaning-movements can be compared to displacement activities[114, 115].

Little is know about the neurophysiological processes on which this change of behaviour is based. The system has been altered at two points: (a) the conduction of impulses from the abdominal proprioreceptors is arrested, and (b) the command coming from the brain cannot reach the thoracic stridulatory apparatus. We have seen that (a) by itself (achieved by cutting the abdominal nerve cord) causes a conflict situation between release and inhibition of courtship and after-courtship when the female is present. In addition it is possible that under the conditions imposed by having both (a) and (b), the afferent impulses activate other brain structures and motor systems which are normally inhibited.

(c) *Return of atypical songs to normal rhythms*

Atypical songs which go on after electrical stimulation can be changed in the post-stimulatory phase and their chirp rate re-adjusted to the "calling" and "rivalry" songs. Which mechanisms could be responsible for this process?

If we assume, with some reason, that chirp and sequence rate are based on the rhythmic properties of certain neuron groups, then one interpretation could be that during stimulation an alien rhythm of discharge is forced upon the neurons in the cb so that a changed pattern is sent to the thorax. In the post-stimulatory phase the neurons could return to their own former rhythm of discharge. In that case the frequently observed after-discharges would then be correlated with the re-establishment of the innate rhythm.

But there are other possibilities. Efferent signals from the cb could be compared with built-in standards and modified by feedback until they corresponded to one of these standards. The role of the imposed electrical stimulus might be to render this feedback ineffectual. There is no evidence, however, for built-in standards in the thoracic system. This system alone is unable to produce a normal song, even when electrically stimulated. A further possibility is that there are feedback loops between the cb and the mbs. Stimulation of the cb may secondarily activate also the mbs. The continued singing following stimulation might then be due to afterdischarge from the mb to the cb and return to normal song patterns due to return from the unspecific control of the imposed stimulus to the more natural control of the cb.

The examples in this section show the problems which neurophysiology has to face in the analysis of complex processes. Only more detailed results on the anatomy of the brain and the properties of central neurons will make definite statements possible.

6. The Peripheral Afferent Control of the Stridulatory Movement

So far we have discussed the activities of the CNS in Orthoptera during sound production. The way in which the sense organs participate in the co-ordination of the stridulatory movement is still unknown. Possible afferent organs of control are receptors

References p. 484

which receive excitation during the song, such as the acoustical system, sense organs in the region of the elytra and hindlegs.

(a) *The acoustical system*

The sound receptors in gryllids and acridids are the tympanic organs and mobile sensory hairs on the anal cerci and on the tergites of the abdominal segments. Their function is described by[8-12, 57,61,84,100-102,112,137]. The tympanic organs play the decisive role for the reception of sounds; if they are destroyed, a female cricket is unable to localize the sound source and acridids stop stridulating the "responding" song. But, after destruction of the tympanic organs, cricket males do stridulate the "calling" song as frequently and in the same rhythm as before. It can therefore be concluded that the stimulation of the insect's auditory receptors by its own song is not necessary for co-ordination of song rhythms. Although the insects have an acoustical feed-back system, it does not seem to play an important role in the co-ordination of their songs.

(b) *The system of proprioreceptors*

This system includes all sense organs, which are stimulated in the taking up of the stridulatory position and during stridulation itself. More and more studies point to the fact that the position and movement of arthropod legs is exactly recorded. The responsible sense organs are normally situated in the region of the articulations. They continuously report to the CNS information about the posture, the load upon the extremities, the direction of movement and its speed[92-97, 121-130, 138].

In gryllids we find directly in front of the elytral articulation, between the subcostal region and the lateral end of the tergal arm, a semicircular sensory cushion covered with short, mobile hairs. Stimulation of the cushion makes the cricket raise the elytrum or push the hindleg against the source of disturbance. The cushion is supplied by a branch of the n.anterior, which innervates with 3 further branches the sense organs on the elytral base[41,131]. Another sensory information comes from a stretch receptor at each wing hinge[135,139,143], but is not analysed in crickets.

From the position and innervation of the cushion one may conclude that it is stimulated during a change of position and during the movements of the wings in a similar way to the sense organs described on the wing articulations of other insects[120]. The experiments on the crickets which are summarized in Table 43 give so far only a few indications that the excitation of the proprioreceptors is necessary for the posture of the elytra during stridulation, but not for the co-ordination of the sound rhythm (see also[135]).

TABLE 43

OPERATION ON THE STRIDULATORY APPARATUS OF THE FIELD CRICKET AND THE EMITTED TYPES OF SONG

Type of operation	N	gG	RG	WG
1. Change of the resting position of the elytra by inversion	25	3	17	15
2. Fixation of 1 elytrum	4	2*	3*	4*
3. Amputation of 1 elytrum plus sensory cushion	5	3*	5*	5*
4. Thermic elimination of both the sensory cushions	7	5	7	4

N = number of insects; numbers followed by asterisks indicate soundless stridulatory movement.

Although crickets bear, on each elytrum, a complete stridulatory apparatus, it has been known for a long time that during stridulation they rub the file of the right elytrum against the scraper of the left. If the elytra are re-laid, i.e. the left over the right, the males try to re-establish the resting position before they stridulate. The males must therefore perceive the inverse elytral position in some way, even when the re-laying has taken place during deep anasthesia ([4,60,140]).

Males with an inverse elytral position can be produced experimentally, if immediately after ecdysis the still unpigmented and soft wings are re-laid before they harden (Keilbach[85] in *Gryllus campestris*, Stärk[111] in *Gryllus bimaculutus de Geer*). According to these authors the left elytrum then always remains over the right one and during stridulation the file of the left rubs against the scraper of the right elytrum. v. Hörmann[60] has repeated the experiments of Keilbach, but was unable to produce permanent "Linksgeiger" in *G. campestris*. The author's studies showed also that from 50 newly emerged adult males, 47 brought the inverted elytra back into the original position, usually before the first stridulatory movement. In *G. campestris* the sensitive period for re-laying the elytra seems to be very short, contrary to the case of *G. bimaculatus*[111]. But it is possible, by frequent re-laying, to alter the stridulatory apparatus to the extent that the male can sing in both positions for a short time. The differences between the songs stridulated in normal and inverted positions of the elytra consist of an important diminution of intensity (see also[111]), a shortening of the chirp length and a distinct modulation of the chirp amplitude. The song pattern remains unchanged and the members of the same species respond promptly to the signal (Huber, unpubl.). This is yet another proof that crickets responds especially to the rhythm.

According to the author's neurophysiological studies it was expected that re-laying the elytra would not change the song rhythm. Contrary to the theory of Golda and Ludwig[50], that the "right" and "left" songs are each controlled by a separate hereditory factor, the author's idea is that in both cases the same cerebral and thoracic mechanisms are active. Only the resting position of the elytra which is right over left in gryllids is determined and therefore during stridulation with inverted elytra only the angle of incidence in which the elytra are rubbed against each other, really changes. Consequently the intensity of contact between the two elytra and the resonance properties are altered. To our ear the song of a "Linksgeiger" differs from a normal stridulation, but a cricket does not seem to regulate its elytral position by hearing its own song. For even with the auditory apparatus (tympanic organs and cerci) eliminated it brings the elytra back into the correct resting position.

The uni- and bilateral extirpation of the tergal sensory cushions does not change the stridulatory position of the wings nor the rhythm. If one elytrum is blocked or amputated together with the sensory cushion the male raises the other elytrum and stridulates in silence according to the given situation. One can recognize the types of song by the different inclinations and movements of the single elytrum. Rhythm and position are thus undisturbed by the loss of one elytrum and the absence of a possible afferent activity of one side does not influence the efferent activity of the other side. This does not prove, however, that the cushion and other sense organs have no controlling effect upon position and motion of the same wing.

Receptors capable of recording the stridulatory movements are apparently absent from the thoracic muscles of insects[62,97]. On the other hand, it is conceivable

References p. 484

that the rhythmical vibration of the body caused by stridulation is reported and centrally analysed by the subgenual organs of the legs[8,9] and by other proprioreceptors[92,93] beside the stretch receptor which is not analysed[136,139,143].

Experiments on the stridulatory apparatus have been carried out in order to investigate the participation of the afferent activities in stridulation in *G. rufus* also. Table 44 summarizes the operations and results.

In *Periplaneta* Pringle[93] has proved electrophysiologically that the sensillae campaniformiae record the position of the legs and inform the CNS about the load upon each leg. Undoubtedly such sensillae are also stimulated during the stridulatory movements of the legs.

The short-termed phase shifting of the stridulatory movement of both legs due to an unilateral severance of the connectives is probably caused by the decrease

TABLE 44

OPERATIONS AND STRIDULATIONS IN *G. rufus*

Place and kind of operation	N	gG	WG	Remarks
1. Severance of one connective in front of the 3rd thoracic ganglion	7	+	+	The hindleg of the operated side stridulates after a delay of 0.5 to 1.0 sec
2. Amputation of the tarsi and tibiae of both hindlegs	8	+	+	Stridulatory rhythm undisturbed
3. The tarsi and tibiae are fixed against the hindfemora	4	+	+	No influence upon stridulation
4. The coxae of hindlegs are fixed without disturbing the coxatrochanteral articulation	6	+·	+·	Stridulatory movement maintained, although the sound is weak or absent, because the elytra are not touched anymore
5. One hindleg fixed in the coxal and coxatrochanteral articulation	10	+	+	The movement of the other hindleg is undisturbed
6. Severance of the connective in front of 3rd thoracic ganglion and fixing of the hindleg of intact side	3	+	+	Leg of operated side performs normal stridulatory movement

of the tonus (see p. 457). An alteration of the leg weight after extirpation of large femoral or tibial portions has no effect upon the stridulatory rhythm[79]. If one hindleg is locked in the coxotrochanteral joint, the other moves normally. The same is true after amputation. In each case the information from proprioreceptors about posture, intensity and movement of the leg can be different, but the thoracic apparatus behaves as if nothing has changed. In short, in acridids, as in gryllids, the loss of afferent activity from one side does not influence the efference of the other. Whether or not the peripheral control is necessary for the song rhythm of the same leg cannot be decided from these experiments. Although the amputation of the tarsus and the tibia changes the leg weight and its load, the rhythm remains constant. A further observation shows that a sudden raising of the femur in the second phase of courtship is unnecessary for the subsequent flinging out of the tibiae. The insect performs this movement even with the femur in a fixed position and within the same time. The last remarks make it probable that a peripheral control plays a role for the positioning of the legs, but not for the movement sequence even in one leg.

(c) Summary of Section 6

The acoustical system of gryllids and acridids is important for communication within the species. Undoubtedly the sound receptors are stimulated by the insect's own song, but their excitation does not seem to be essential for the control of the stridulatory process. One could object that an acoustical feed-back with a zero magnitude (elimination of the sound receptors) is without significance, contrary to a message which is not identical with the efferent activity (sound rhythm). But gryllids and acridids live in close proximity to insects of the same and other species, which may stridulate simultaneously and with different rhythms, and one never observes an adjustment to one type of song. In addition, an acoustical afferent activity of another kind which competes with the acoustical re-afference leaves the rhythm of movements undisturbed. Crickets possess a system which controls the resting position and the posture of the elytra during the song. It also causes the re-laid wings to return to the original position, but it does not affect the stridulatory rhythm of the "right" and "left" singers. Stimulation of sense organs on the wing and leg-joints are apparently important only when the relevant extremity takes up the stridulatory position. Their loss or change of excitation does not affect the movement of the extremity of the other side. The location of these control systems, their anatomy and excitatory conduction are the aim of further studies.

7. Summary

In this chapter the function of the CNS in Orthoptera during sound production was analysed and the following results were obtained:

1. The stridulatory movement is an expression of co-operation between certain thoracic muscles, whose activities are co-ordinated.

2. The central nervous structures concerned with sound production include 3 systems: (1) the mushroom bodies (2) the central body and (3) the thoracic apparatus in the 2nd or 3rd thoracic ganglion. The brain is connected with the thoracic apparatus by fibres which transmit the commands for stridulation. Fibres of one side alone are sufficient for that purpose.

3. Stimulation of the mushroom body-neurons or of afferent tracts terminating there release song rhythms which correspond to the usual pattern of the cricket's "calling" and the "rivalry" song. Stimulation of the calyx and the cell bodies of the mushroom bodies usually causes inhibition of song. The loss of one mushroom body does not interfere with acoustical behaviour. There is no evidence of interaction between the mushroom bodies during sound production but their activity must be synchronized. This might be done by a simultaneous stimulation of the paired sense organs or of one sense organ which can transmit excitation to both mushroom bodies.

A contra-lateral inhibition between the mushroom bodies seems to be absent in the acoustical behaviour, whereas it is present during regulation of locomotor activity[104,70].

4. The impulses releasing or inhibiting stridulation reach the mushroom bodies from the eyes, antennae, anal cerci, tympanic organs and from the receptors of the copulatory apparatus. The interactions of the impulses in these centres determine the beginning and duration of the stridulatory movement and communicate the specific pattern of song to the lower centres. A further important function of the

mushroom bodies consists in the co-ordination of several motor systems for instance during fight and courtship.

5. Electrical stimulation of the central body can release atypical songs. Elimination of this structure causes the loss of all songs. Impulses pass through the central body on the way from the mushroom bodies to the thoracic nerve cord, but its neurons also participate in the translation of the command "song type" into a pattern of excitation which determines the chirp and sequence rate.

6. The thoracic system controls and synchronizes the stridulatory muscles in both halves of the segment. When isolated from the brain, it is only able to start a basic form of the stridulatory movement and to maintain it for a short time. Stimulation of segmental sense organs or local electrical excitations of the respective ganglion causes atypical song rhythms. Therefore this apparatus seems to have no innate normal pattern. The brain impulses modulate the thoracic system with the rhythm of the chirp- and sequence rate, and it is assumed that the impulses are first transmitted to a system of interneurons and then distributed to motor neurons of both halves Halving the ganglion destroys co-ordination and stridulation is accordingly suppressed.

7. Observations on gryllids and acridids suggest that there is an unidirectional control from the brain to the thoracic ganglion during sound production.

8. Gryllids and acridids possess sense organs which are stimulated during the song and are likely to control the process of motion. Experiments showed that their stimulation or change of excitation can be used for the control of position of the relevant elytrum or leg. The activity of the opposite side remains undisturbed by the elimination of these organs. The stridulatory rhythm seems to be accomplished without the participation of controlling afferent activities.

9. The results given in this paper indicate that there are loci in the CNS of Orthoptera which analyse song-releasing and inhibiting stimuli, determine the rhythm, and co-ordinate the stridulatory muscles. Many questions are still unsolved. For instance: how do these centres or rather their neurons work; how are impulse patterns formed, transmitted and translated, and which nervous structure takes on the task of evaluating the sound signals?

The answers cannot come from lesions and stimulation experiments alone. The solution of the above problems is reserved for future electrophysiological research.

REFERENCES

[1] ALEXANDER, R. D., 1957a, Sound production and associated behavior in insects, *Ohio J. Sci.*, 57, (2) 101.
[2] ALEXANDER, R. D., 1957b, The song relationships of four species of ground crickets (Orthoptera: Gryllidae: Nemobius). *Ohio J. Sci.*, 57, 153.
[3] ALEXANDER, R. D., 1960, Sound communication in Orthoptera and Cicadidae, *Animal Sounds* (LANYON and TAVOLGA, ed.) The American Institute of Biological Sciences, 7, 38.
[4] ALEXANDER, R. D., 1961, Aggressiveness, territoriality, and sexual behavior in field crickets. *Behaviour*, 17, 130.
[5] ALEXANDER, R. D. and E. S. THOMAS, 1959, Systematic and behavioral studies on the crickets of the *Nemobius fasciatus* group (Orthoptera: Gryllidae: Nemobiinae) *Ann. Entom. Soc. America*, 52, 591.

[6] AUGER, D. and A. FESSARD, 1928, Recherches sur l'excitabilité du système nerveux des insectes, *C.R. Soc. Biol.*, *99*, 305.
[7] AUGER, D. and A. FESSARD, 1929, Observations complémentaires sur un phénomène de contractions rhythmées provoqués par excitation galvanique chez certains insectes, *C.R. Soc. Biol.*, *101*, 897.
[8] AUTRUM, H. J., 1940, Über Lautäusserung und Schallwahrnehmung bei Arthropoden. II. Das Richtungshören von Locusta und Versuch einer Hörtheorie für Tympanalorgane vom Locust-identyp. *Z. vergl. Physiol.*, *28*, 326.
[9] AUTRUM, H. J., 1941, Über Gehör und Erschütterungssinn bei Locustiden. *Z. vergl. Physiol.*, *28*, 580.
[10] AUTRUM, H. J., 1942, Schallempfang bei Tier und Mensch. *Naturwissenschaften*, *30*, 69.
[11] AUTRUM, H. J., 1955, Analyse physiologique de la réception des sons chez les Orthoptères. *L'Acoustique des Orthoptères*, I.N.R.A., Paris, pp. 338–355.
[12] AUTRUM, H. J., 1960, Phasische und tonische Antworten vom Tympanalorgan von *Tettigonia viridissima*. *Acustica*, *10*, 139.
[13] BETHE, A., 1897, Vergleichende Untersuchungen über die Funktion des Zentralnervensystems der Arthropoden. *Pflüg. Arch. ges. Physiol.*, *68*, 449.
[14] BULLOCK, TH. H., 1952, The invertebrate neuron junction. *Cold Spring Harbor Symp. Quant. Biol.*, *17*, 267.
[15] BULLOCK, TH. H., 1957, Neuronal integrative mechanism. *Recent Advances in Invertebrate Physiology*, Univ. Oregon Publ., Eugene, Oregon, pp. 1–20.
[16] BULLOCK, TH. H., 1959, Initiation of nerve impulses in receptor and central neurons. *Rev. mod. Physics*, *31*, 504.
[17] BUSNEL, M. C., 1953, Contribution à l'étude des émissions acoustiques des Orthoptères. I. Mém. Recherches sur les spectres de fréquence et sur les intensités. *Ann. Epiphytiés*, *III*, 333.
[18] BUSNEL, M. C., 1955, Etude des chants et du comportement acoustique d'*Oecanthus pellucens* Scop. *L'Acoustique des Orthoptères*, I.N.R.A., Paris, pp. 175–202.
[19] BUSNEL, M. C. and R. G. BUSNEL, 1954, La directivité acoustique des déplacements de la femelle d'*Oecanthus pellucens* Scop. *C.R. Soc. Biol.*, *148*, 830.
[20] BUSNEL, M. C. and R. G. BUSNEL, 1956, Sur une phonocinèse de certains acridiens à des signaux acoustiques synthétiques. *C.R. Acad. Sci.*, *242*, 292.
[21] BUSNEL, M. C., B. DUMORTIER AND R. G. BUSNEL, 1959, Recherches sur la phonocinèse de certains insectes. *Bull. Soc. Zool.*, *84*, 351.
[22] BUSNEL, R. G., 1955, Sur certains rapports entre le moyen d'information acoustique et le comportement acoustique des Orthoptères. *L'Acoustique des Orthoptères*, I.N.R.A., Paris, pp. 280–306.
[23] BUSNEL, R. G., 1956, Some new aspects of acoustical animal behaviour. *J. Sci. Industr. Research*, *15A*, 306.
[24] BUSNEL, R. G., B. DUMORTIER AND W. LOHER, 1953, Recherches sur le comportement des divers Acridoides mâles soumis à des stimuli acoustiques artificiels. *C.R. Acad. Sci.*, *237*, 1557.
[25] BUSNEL, R. G., B. DUMORTIER and W. LOHER, 1954, Sur l'étude du temps de la response au stimulus acoustique artificiel chez les Chorthippus et la rapidité de l'intégration du stimulus. *C.R. Soc. Biol.*, *148*, 862.
[26] BUSNEL, R. G and W. LOHER, 1954, Recherches sur le comportement de divers mâles d'acridiens des signaux acoustiques artificiels. *Ann. Sc. Nat., Zool.*, *11*, 271.
[27] BUSNEL, R. G., W. LOHER AND F. PASQUINELLY, 1954, Recherches sur les signaux acoustiques synthétiques réactogènes pour divers Acrididae ♂. *C.R. Soc. Biol.*, *148*, 1987.
[28] BUSNEL, R. G. and B. DUMORTIER, 1954, Observations sur le comportement acoustico-sexuel de la ♀ d'*Ephippiger bitterensis*. *C.R. Soc. Biol.*, *148*, 1589.
[29] BUSNEL, R. G. and W. LOHER, 1954, Mémoire acoustique directionelle du mâle de *Chorthippus biguttulus* L. *C.R. Soc. Biol.*, *148*, 993.
[30] BUSNEL, R. G and W. LOHER, 1955, Recherches sur les actions de signaux acoustiques artificiels sur le comportement de diverse Acrididae mâles. *L'Acoustique des Orthoptères*, I.N.R.A., Paris, pp. 365–394.
[31] BUSNEL, R. G. and B. DUMORTIER, 1955, Etude du cycle génital du mâle d'Ephippiger et son rapport avec le comportement acoustique. *Bull. Soc., Zool.*, *80*, 23.
[32] BUSNEL, R. G., B. DUMORTIER and F. PASQUINELLY, 1955, Phonotaxie de ♀ d'Ephippiger à des signaux acoustiques synthétiques. *C.R. Soc. Biol.*, *149*, 11.
[33] BUSNEL, R. G. and B. DUMORTIER, 1956, Rapport entre la vitesse de déplacements et l'intensité du stimulus dans le comportement acoustico-sexuel de la femelle de *Ephippiger bitterensis*. *C.R. Acad. Sci.*, *242*, 174.
[34] BUSNEL, R. G., M. C. BUSNEL and B. DUMORTIER, 1956, Relations acoustiques interspécifiques chez les Ephippigeres. *Ann. Epiphyties III*, 451.

[35] BUSNEL, R. G., B. DUMORTIER, M. C. BUSNEL, 1956, Recherches sur le comportement acoustiques des Ephippigeres. *Bull. Biol., Fasc. 3*, 15.
[36] BUSNEL, R. G. and W. LOHER, 1956, Etude des caractères physiques réactogènes de signaux acoustiques artificiels déclencheurs de phonotropismes chez les Acrididae. *Bull. Soc. Entomol., 61*, 52.
[37] DUJARDIN, F., 1850, Mémoires sur le système nerveux des insectes, *Ann. Sci. Nat. Zool., 14*.
[38] ECCLES, J. C., 1950, The responses of motoneurons, *Brit. Med. Bull., 6*, 304.
[39] ECCLES, J. C., 1953, *The Neurophysiological Basis of Mind*, Clarendon Press, Oxford.
[40] ECCLES, J. C., 1957, *The Physiology of Nerve Cells*, University Press, Oxford.
[41] ERHARDT, E., 1916, Zur Kenntnis der Innervierung und der Sinnesorgane der Flügel von Insekten, *Zool. Jb. Anat. u. Ontog., 39*, 293.
[42] FABER, A., 1929, Die Lautäusserungen der Orthopteren. Lauterzeugung, Lautabwandlung und deren biologische Bedeutung sowie Tonapparat der Geradflügler, *Z. Morphol. u. Ökol. Tiere, 13*, 745; 1932, Die Lautäusserungen der Orthopteren. Untersuchungen über die biozönotischen, tierpsychologischen und vergleichend-physiologischen Probleme der Orthopteren-Stridulation, *Ibid., 26*, 7.
[43] FABER, A., 1934, Neue Untersuchungen über die Lautäusserungen der Geradflügler (Orthopteren), *Der Biologe, III*, (10) 249.
[44] FABER, A., 1936, Die Laut- und Bewegungsäusserungen der Oedipodinen, *Z. wiss. Zool., 149*, 1.
[45] FABER, A., 1952, Ausdrucksbewegung und besonders Lautäusserung bei Insekten als Beispiel für eine vergleichend-morphologische Betrachtung der Zeitgestalten, *Verh. Zool. Ges. Freiburg, Zool. Anz., 46*, 106.
[46] FABER, A., 1953, Laut- und Gebärdensprache bei Insekten; Orthopteren, *Mitt. Staatl. Museum Stuttgart*.
[47] FABER, A., Über den Aufbau von Gesangsformen in der Gattung *Chorthippus* fieb, (Orthoptera), *Stuttgarter Beiträge zur Naturkunde, 1*, 1; 1957, Über parallele Abänderungen bei Lautäusserungen von Grylliden, *Ibid., 2*, 1.
[48] FRINGS, H. and M. FRINGS, 1957, The effects of temperature on the chirp-rate of male coneheaded grasshoppers, *Neoconocephalus ensiger*, *J. Exp. Biol., 134*, 411.
[50] GOLDA, H. und W. LUDWIG, 1958, Rechts-Links-Fragen bei der Flügellage und dem Zirpen von *Gryllotalpa vulgaris*, *Zool. Anz., 161*, 7.
[51] HAGIWARA, S., 1953, Neuro-muscular transmission in insects, *Japan J. Physiol., 3*, 284.
[52] HAGIWARA, 1956, Neuro-muscular mechanism of sound production in the cicada, *Physiol. Comparata et Oecol., 4*, 142.
[53] HAGIWARA, S. and A. WATANABE, 1956, Discharges in motoneurons of cicada, *J. Cell. Comp. Physiol., 47*, 415.
[54] HANSTRÖM, B., 1928, *Vergleichende Anatomie des Nervensystems der wirbellosen Tiere*, J. Springer-Verlag, Berlin.
[55] HANSTRÖM, B., 1940, *Inkretorische Organe, Sinnesorgane und Nervensystem des Kopfes einiger niederer Insektenordnungen*, J. Springer-Verlag, Berlin.
[56] HASKELL, P. T., 1953, The stridulation behaviour of the Domestic cricket, *Brit. J. Animal Beh., 1*, 120.
[57] HASKELL, P. T., 1956, Hearing in certain Orthoptera. 1. Physiology of sound receptors. 2. The nature of the response of certain receptors to natural and imitation stridulation, *J. Exp. Biol., 33*, 756.
[58] HASKELL, P. T., 1957, Stridulation and associated behaviour in certain Orthoptera. 1. Analyses of the stridulation of, and behaviour between, males, *Brit. J. Animal Beh., 5*, 139; 1958, 2. Stridulation of females and their behaviour with males, *Ibid., 6*, 27.
[59] HASKELL, P. T., 1960, Stridulation and associated behaviour in certain Orthoptera. 3. The influence of the gonads. *Brit. J. Animal Behav., 8*, 76.
[60] HÖRMANN, S. VON, 1957, Untersuchungen über den Erbgang einiger Verhaltensweisen bei Grillenbastarden, *Z. f. Tierpsychol., 14*, 137.
[61] HORRIDGE, G. A., 1960, Pitch discrimination in Orthoptera (Insecta) demonstrated by responses of central auditory neurones, *Nature, 185*, 623.
[62] HOYLE, G., 1957, *Comparative physiology of the nervous control of muscular contraction*, University Press, Cambridge.
[63] HOYLE, G., 1957, Nervous control of insect muslces, *Rec. Adv. Invertebrate Physiology*, Univ. Oregon Publ., Eugene, Oregon, pp. 73–99.
[64] HOLST, E. VON, 1935, Die Koordination der Bewegung bei den Arthropoden in Abhängigkeit von zentralen und peripheren Bedingungen, *Biol. Rev., 10*, 234.
[65] HUBER, F., 1952, Verhaltensstudien am Männchen der Feldgrille (*Gryllus campestris* L.) nach Eingriffen am Zentralnervensystem, *Verh. Zool. Ges. Freiburg, Zool. Anz., 46*, 138.
[66] HUBER, F., 1955a, Sitz und Bedeutung nervöser Zentren für Instinkthandlungen beim Männchen von *Gryllus campestris* L., *Z. f. Tierpsychol., 12*, 12.

[67] HUBER, F., 1955b, Über die Funktion der Pilzkörper (corpora pedunculata) beim Gesang der Keulenheuschrecke *Gomphocerus rufus* L. (Acridiidae), *Naturwiss.*, *42*, 566.
[68] HUBER, F., 1956, Heuschrecken- und Grillenlaute und ihre Bedeutung, *Naturwiss.*, *43*, 317.
[69] HUBER, F., 1957, Elektrische Reizung des Insektengehirnes mit einem Impuls- und Rechteckgenerator, *Industrie-Elektronik*, *2*, 17.
[70] HUBER, F., 1959, Auslösung von Bewegungsmustern durch elektrische Reizung des Oberschlundganglions bei Orthopteren (Saltatoria: Gryllidae, Acridiidae). *Verh. Deut. Zool. Ges. Münster, Zool. Anz., Suppl. 23*, 248.
[71] HUBER, F., 1960a, Experimentelle Untersuchungen zur nervösen Atmungsregulation der Orthopteren (Saltatoria: Gryllidae). *Z. vergl. Physiol.*, *43*, 359.
[72] HUBER, F., 1960b, Untersuchungen über die Funktion des Zentralnervensystems und insbesondere des Gehirnes bei der Fortbewegung und der Lauterzeugung der Grillen. *Z. vergl. Physiol.*, *44*, 60.
[73] HUGHES, G. M., 1952, Differential effects of direct current on insect ganglia, *J. Exp. Biol.*, *29*, 387.
[74] HUGHES, G. M., 1952, The co-ordination of insect movements. I. The walking movements of insects, *J. Exp. Biol.*, *29*, 267.
[75] HUGHES, G. M., 1957, The effect of limb amputation and the cutting of commissures in the cockroach (*Blatta orientalis*), *J. Exp. Biol.*, *34*, 306.
[76] HUGHES, G. M. and C. A. G. WIERSMA, 1960, Neuronal pathways and synaptic connexions in the abdominal cord of the crayfish. *J. exp. Biol.*, *37*, 291.
[77] JACOBS, W., 1944, Einige Beobachtungen über Lautäusserungen bei weiblichen Feldheuschrecken, *Z. f. Tierpsychol.*, *6*, 141.
[78] JACOBS, W., 1949, Über Lautäusserungen bei Insekten, insbesondere bei Heuschrecken, *Entomon*, *1*, 100.
[79] JACOBS, W., 1950, Vergleichende Verhaltensstudien an Feldheuschrecken, *Z. f. Tierpsychol.*, *7*, 169.
[80] JACOBS, W., 1952, Vergleichende Verhaltensstudien an Feldheuschrecken und einigen anderen Insekten, *Verh. Zool. Ges. Freiburg, Zool. Anz.*, *46*, 115.
[81] JACOBS, W., 1953, Verhaltensbiologische Studien an Feldheuschrecken, *Beiheft zur Z. f. Tierpsychol.*
[82] JACOBS, W., 1955, Problèmes relatifs à l'étude du comportement acoustique chez les Orthoptères, *L'acoustique des Orthoptères*, I.N.R.A., Paris, pp. 307–319.
[83] JACOBS, W., 1957, Einige Probleme der Verhaltensforschung bei Insekten, insbesondere bei Orthopteren, *Experientia*, *13*, 97.
[84] KATSUKI, Y. and N. SUGA, 1960, Neural mechanisms of hearing in insects. *J. exp. Biol.*, *37*, 279.
[85] KEILBACH, R., 1935, Über asymmetrische Flügellage bei Insekten und ihre Beziehungen zu anderen Asymmetrien, *Z. Morphol. u. Ökol. Tiere*, *29*, 1.
[86] LOHER, W., 1957, Untersuchungen über den Aufbau und die Entstehung der Gesänge einiger Feldheuschreckenarten und den Einfluss von Lautzeichen auf das akustische Verhalten, *Z. vergl. Physiol.*, *39*, 313.
[87] LOTTERMOSER, W., 1952, Aufnahme und Analyse von Insektenlauten. *Acustica*, *2*, 66.
[88] MAYNARD, D. M., 1956, Electrical activity in the cockroach cerebrum, *Nature*, *117*, 529.
[89] OBERHOLZER, R. J. H. and F. HUBER, 1957, Methodik der elektrischen Reizung und Ausschaltung im Oberschlundganglion (Gehirn) nicht narkotisierter Grillen (*Acheta domesticus* L. und *Gryllus campestris* L.), *Helv. Physiol. Acta*, *15*, 185.
[90] PASQUINELLY, F. and M. C. BUSNEL, 1955, Études préliminaires sur les mécanismes de la production des sons par les Orthoptères, *L'Acoustique des Orthoptères*, I.N.R.A., Paris, pp. 145–153.
[91] PIERCE, G. W., 1949, *The songs of insects*, Harvard University Press, Cambridge, U.S.A.
[91a] POWER, M. E., 1948, The thoracico-abdominal nervous system of an adult insect, *Drosophila melanogaster*. *J. comp. Neurol.*, *88*, 337.
[92] PRINGLE, J. W. S., 1938, Proprioception in insects. 1. A new Type of mechanical receptor from the palps of the cockroach. 2. The action of the campaniform sensilla on the legs, *J. Exp. Biol.*, *15*, 101; 1938, 3. The function of the air sensilla at the joints, *Ibid.*, *15*, 467.
[93] PRINGLE, J. W. S., 1940, The reflex mechanism of the insect leg, *J. Exp. Biol.*, *17*, 8.
[94] PRINGLE, J. W. S., 1954, A physiological analysis of cicada song, *J. Exp. Biol.*, *31*, 525.
[95] PRINGLE, J. W. S., 1956, Proprioception in Limulus, *J. Exp. Biol.*, *33*, 658.
[96] PRINGLE, J. W. S., 1957, Myogenic rhythms, *Rec. Adv. Invertebrate Physiology*, Univ. Oregon Publ., Eugene, Oregon, pp. 99–115.
[97] PRINGLE, J. W. S., 1957, *Insect Flight*, Cambridge Univ. Press.
[98] PROSSER, C. L., 1935, Central responses to proprioceptive and tactile stimulation, *J. Comp. Neurol.*, *62*, 495.

[100] PUMPHREY, R. J. and A. F. RAWDON-SMITH, 1936, Hearing in insects: The nature of the responses of certain receptors to auditory stimuli, *Proc. Roy. Soc. (London) B*, *121*, 18.
[101] PUMPHREY, R. J., 1936, Synchronized action potentials in the cereal nerve of the cockroach (*Periplanta americana*) in response to auditory stimuli, *J. Physiol.*, *87*, 4.
[102] PUMPHREY, R. J., 1939, Frequency discrimination in insects: a new theory, *Nature*, *143*, 806.
[103] PUMPHREY, R. J., 1940, Hearing in insects, *Biol. Rev.*, *15*, 107.
[104] ROEDER, K. D., 1937, The control of tonus and locomotor activity in the praying mantis (*Mantis reliogiosa* L.), *J. Exp. Zool.*, *76*, 353.
[105] ROEDER, K. D., 1948, Organisation of the ascending giant fiber system in the cockroach (*Periplaneta americana*), *J. Exp. Zool.*, *108*, 243.
[106] ROEDER, K. D., Electrical activity in nerves and ganglia, *Insect Physiology*, Vol. 17, New York, 423–462.
[107] ROEDER, K. D., 1953, Reflex activity and ganglion function, *Insect Physiology*, Vol. 18, New York, pp. 463–487.
[108] ROEDER, K. D., L. TOZIAN and E. A. WEIANT, 1960, Endogenous nerve activity and behaviour in the mantis and cockroach. *J. ins. Physiol.*, *4*, 45.
[109] ROMEIS, B., 1948, *Mikroskopische Technik*, Leibniz-Verlag, München.
[110] ROY-NOEL, J., 1954, Contribution à l'étude de l'appareil musical des Orthoptères ensifères, *Ann. Sci. Nat. Zool.*, *11*, 65.
[111] STÄRK, A. A., 1958, Untersuchungen am Lautorgan einiger Grillen- und Laubheuschrecken-Arten, zugleich ein Beitrag zum Rechts-Links-Problem, *Zool. Jb. Anat.*, *77*, 9.
[112] SUGA, N., 1960, Peripheral mechanism of hearing in locust. *Jap. J. Physiol.*, *10*, 533.
[113] TEN CATE, J., 1931, Physiologie der Gangliensysteme der Wirbellosen, *Ergeb. Physiol.*, *33*, 137.
[114] TINBERGEN, N., 1940, Die Übersprungsbewegung, *Z. f. Tierpsychol.*, *4*, 1.
[115] TINBERGEN, N., 1950, *The Study of Instinct*, Oxford Univ. Press.
[116] VOSS, FR., 1905, Über den Thorax von *Gryllus domesticus* L., mit besonderer Berücksichtigung des Flügelgelenkes und dessen Bewegung, *Z. wiss. Zool.*, *78*, 268; 645.
[117] VOWLES, D. M., 1955, The structure and connections of the corpora pedunculata in bees and ants, *Quart. J. Microscop. Sci.*, *96*, 239.
[118] WALKER, TH. J., JR., 1957, Specifity in the response of female tree crickets (Orthoptera, Gryllidae, Oecanthinae) to calling songs of the males, *Ann. Entom. Soc. Am.*, *50*, 626.
[119] WEIANT, E. A., 1958, Control of spontaneous activity in certain efferent nerve fibers from the metathoracic ganglion of the cockroach, *Periplaneta americana*, *Trans. 10th Int. Congr. Entomol.*, *2*, 81.
[120] WEIS-FOGH, T., 1956, Biology and physics of locust flight. IV. Notes on the sensory mechanism inlocust flight, *Phil. Trans. Roy. Soc. (London)*, *B*, *239*, 553.
[121] WIERSMA, C. A. G., 1952, Repetitive discharges of motor fibers caused by a single impulse in giant fibers of the crayfish, *J. Cell. Comp. Physiol.*, *40*, 399.
[122] WIERSMA, C. A. G., 1953, Neurons of Arthropods, *Cold Spring Harbor Symp. Quant. Biol.*, *17*, 155.
[123] WIERSMA, C. A. G., 1953, Neural transmission in invertebrates, *Physiol. Rev.*, *33*, 326.
[124] WIERSMA, C. A. G., 1955, The central representation of sensory stimulation in the crayfish, *J. Cell. Comp. Physiol.*, *46*, 307.
[125] WIERSMA, C. A. G., 1958, On the functional connections of single units in the central nervous system of the crayfish, *Procambarus clarkii girard*, *J. Comp. Neurol.*, *110*, 421.
[126] WIERSMA, C. A. G., 1959, Movement receptors in decapod crustacea, *J. Mar. Biol. Assoc. U.K.*, *38*, 143.
[127] WIERSMA, C. A. G. and W. SCHALLECK, 1947, Potentials from motor roots of the crustacean central nervous system, *J. Neurophysiol.*, *10*, 323.
[128] WIERSMA, C. A. G. and R. S. TURNER, 1950, The interaction between the synapses of a single motor fiber, *J. Gen. Physiol.*, *34*, 137.
[129] WIERSMA, C. A. G. and S. H. RIPLEY, 1952, Innervation patterns of crustacean limbs, *Physiol. Comparata et Oecol.*, *2*, 391.
[130] WIERSMA, C. A. G. and E. G. BOETTIGER, 1959, Unidirectional movement fibers from a proprioceptive organ of the crab, *Carcinus maenas*, *J. Exp. Biol.*, *36*, 102.
[131] ZACWILICHOWSKI, J., 1934a, Über die Innervierung und die Sinnesorgane der Flügel von der Schabe *Phyllodromia germanica* L., *Bull. int. Acad. Cracovie*, *B II*, p. 89; 1934b, Über die Innervierung und die Sinnesorgane der Flügel der Feldheuschrecke *Stauroderus biguttulus*, *Ibid.*, p. 187.
[132] ZAWARZIN, A., 1912, Histologische Studien über Insekten. II. Über das sensible Nervensystem der Äschna-Larven, *Z. wiss. Zool.*, *100*, 447.
[133] ZAWARZIN, A., 1924a, Histologische Studien über Insekten. V. Über die histologische Beschaffenheit des unpaaren ventralen Nerven der Insekten, *Z. wiss. Zool.*, *122*, 97; 1924b, VI. Zur Morphologie der Nervenzentren. Das Bauchmark der Insekten, *Ibid.*, *122*, 323.
[134] ZIPPELIUS, H. M., 1949, Die Paarungsbiologie einiger Orthopterenarten, *Z. f. Tierpsychol.*, *6*, 372.

(For References 135–144, see Addendum p. 802.)

CHAPTER 18

COMPARATIVE ANATOMY AND PERFORMANCE OF THE VOCAL ORGAN IN VERTEBRATES

by

G. KELEMEN*

1. Lower Vertebrates

(a) Fishes**

Among the vertebrates the fishes have no vocal organs and make sounds with the help of other anatomical elements. Vibrating the walls of the air bladder by muscle contraction results in a drumlike sound. A membranous window in the wall of the air bladder is beaten as if by drumsticks, by the rays of the fin. The bladder may vibrate like the strings of a violin. Or, sounds are produced by scraping two parts of the body against each other. Teeth, or toothless jaws are ground together.

Sounds made by fish, audible above the water surface, have been known for a long time. But subsurface sounds are by far the more important ones. Their study, initiated only recently, becomes a rapidly expanding specialized branch of comparative phonetics[9, 38]. The sounds show great diversity, between the frequencies of 20–4800 c/s with prevalence of the low tones between 75 and 300 c/s. Sounds, produced by activity in and around the air bladder as well as the sounds of stridulatory nature are, in their general characteristics, constant. Even single individuals can be identified by their particular sound patterns.

Although in the Dipnoi the gill and the pulmonary apparatus are equally important, the former has disappeared in the Tetrapods. Parts of the gill apparatus, claimed before for respiration, were now free to participate in the formation of a *larynx*. One is probably justified to speak of such an organ from the stage when, at its beginning, the *ductus pneumaticus* shows a narrowing, with muscles surrounding this orifice. In fishes, the most primitive forms carry only a sphincter mechanism, to prevent water from entering the lungs. This is a slitlike orifice, and its opening is effectuated by relaxation. An important step is development of a dilator group of muscle fibers offering a separate mechanism to pull apart the margins of the orifice.

In the Dipnoi no skeletal parts whatsoever are present; the sphincter of the larynx is made up by smooth muscle fibers. Orally from the aditus laryngis a strong plate is formed containing elastic elements; opposite this plate a voluminous fibrous cushion is located. Both are under the control of constrictors and dilators. But sounds in fishes, as shown are produced in different other ways; they do not originate in the larynx and consequently are not activated by the respiratory air current.

---

* Aided by a grant from the Wenner-Gren Foundation for Anthropological Research.
** See also J. M. Moulton, Chapter 22.

References p. 518

(b) Amphibians*

Arriving at the Amphibians, a simple new feature appears. This, however, presents a most important step on the way to transform the larynx into a vocal organ: two slender cartilaginous rods become embedded in the walls of the sphincter, the *cartilagines laterales*. Laryngeal movements are usually explained in a simplified manner, by describing the movements in the laryngeal cartilaginous skeleton. The appearance of these rods helps considerably in the understanding of laryngeal physiology.

From this primary form, the lateral rods, are derived the arytenoids, the cricoid and even the tracheal rings. They form the primitive laryngeal skeleton, enclosing the laryngeal box and dividing it into an arytenoid and a cricotracheal portion.

The larynx is an organ whose recent phylogenetic history shows marked developmental progress and its present state is that of considerable variation. The differentiation in the form of a larynx followed the increased importance of the lungs in land life and gave the entrance to them a more complex structure. Beyond the glottis, the entrance to the trachea enlarged with additional cartilages that tended to form a skeleton for this organ. The cartilages represent, partly, modified visceral arch structures. In the Tetrapod the posterior arches of the branchial skeleton become more and more specialized to support the tracheal entrance and are gradually converted into the laryngeal skeleton. But it is not merely the skeleton which furnishes evidence of the gradual modification of branchial arches; other illustrations are given by the vascular arches, the musculature, and the distribution of the original branchial branches of the vagus nerve.

Anura. With the Anura the tripartition of the skeleton of this part of the airway becomes definitive, to remain so until the highest forms are reached. The arytenoidal portion of the larynx becomes now the site of the vocal apparatus. *Rana esculenta* shows a fold, rising from the inner aspect of the arytenoid and covered with a mucous membrane. This is the vocal band. The pair protruding from both sides enclose the *rima glottidis*. Now a true *vestibulum laryngis* is clearly set off from a *cavum laryngotracheale*. The vocal band itself is broad, is divided by a longitudinal furrow in an upper and a lower half, and anticipates a much later developmental form.

Another important step is a newly won connection between the larynx and the hyobranchial apparatus. As a consequence the larynx has to follow the respiratory movements of the oral floor, as the latter is under the influence of the hyoid muscles. While the constrictors remain not much more than a simple sphincter, the dilatators undergo a more diversified evolution. Buccal airsacs (Fig. 293), inflated during vocalization, help in producing a loud and sustained voice, either as resonators or as reservoirs from which the widely inflatable lungs can be filled repeatedly, allegedly even under water.

(c) Reptiles

Division of the laryngo-tracheal skeleton in three parts: arytenoids, cricoid and tracheal rings had been now stabilized; arytenoid and cricoid, however, are sometimes in a cartilaginous connection, recalling their origin from the lateral cartilages.

---

* See also W. F. Blair, Chapter 23.

Fig. 293. *Rana* (Frog). Intraoral air sacs, inflated. (From F. Scheminzky, 1935, Die Welt des Schalles, *Das Bergland-Buch*, Wien–Leipzig–Berlin.)

On the other hand, defects, windows in the cricoid, point to the primitive independence of the two lateral cartilages from each other. In Lacerta the epiglottis has already taken the shape of an upturned collar. The cricoid becomes definitely the origin of a dilatator musculature.

Movable lips are absent, as in fishes and birds; although they are present in all mammals except the Monotremes.

Of outstanding importance is the tight connection established between the choanae and the aditus of the larynx. It is easy to see why safety of the airway through the nasal cavity and larynx must be guarded, *e.g.*, in the crocodile while the mouth is kept open under the water level. The connection between larynx and hyoid apparatus has become more intimate than in the Amphibia.

Inside the larynx one does not see any special equipment or arrangement to serve the production of sounds. The great majority are voiceless in the true sense. Snakes will hiss, but this sound is produced between the arytenoid edges by the expiratory air current. Some forms, however, have already developed membranous folds which can be considered as vocal cords, and a very few species are able to produce fairly loud tones, as some Gekkos. Laryngeal sacs are found in different locations. For species in possession of vocal power these sacs will serve as reinforcement; but they may be used, as in Chameleon, to lend a more formidable aspect to the animal, by inflation when irritated. It will be shown that inflation of air sacs with or without vocal production, when the animal is irritated, accompanies evolution as far as the Anthropoids.

Roaring, bellowing in Turtles and Crocodilia are signs that short, vocal cord-like structures are appearing. The long and rigid arytenoids make the reptilian larynx poorly adapted for sound production. Here, as in other forms, the dorsal connection of both arytenoids point to their origin from tracheal rings. In such cases the crestlike upper margins can be approximated for vocalization, but, except for short intervals, the arytenoids have to be kept sufficiently apart to secure the patency of the air passage.

2. Superior Vertebrates

(a) *Birds*

The inner wall of the laryngeal tube is smooth; otherwise, with its well developed elements, the larynx is roomy. Although located high in the pharynx, it can become, by different arrangements, a direct continuation of the oral passage. In this way the air current can reach the larynx through the nasal as well as through the oral cavity; this is an important factor in tone modulation. The mass of the larynx produces in the floor of the pharynx a considerable elevation, called the *torus pharyngeus*.

The constrictor musculature is represented by the *m. cricoarytenoideus lateralis*, corresponding to the laryngeus ventralis, and by the *m. interarytenoideus*, corresponding to the laryngei dorsales. Dilatation is effected through the cricoarytenoideus dorsalis. Because of this the larynx has a weak sphincter and a stronger dilatator, as well as an important levator for the larynx together with the trachea.

The larynx of birds lacks vocal cords. The separate parts of the larynx have but little mobility; it is, all in all, a well differentiated organ but no producer of sound. The latter originates in an apparatus which is the exclusive property of birds: the syrinx.

The *syrinx*[3,34] (Figs. 294–296) is located, in most cases, at the bifurcation of the trachea. Trachea and bronchi participate in its formation. The trachea itself forms loops which lie in a groove of the sternum. These loops elongate the trachea in such a way that the number of the tracheal rings can rise to 300.

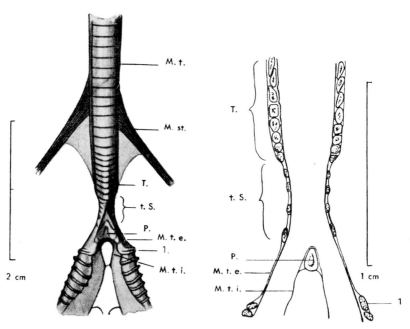

Fig. 294 and 295. *Gallus gallus domesticus* (Cock). Syrinx. (From W. Rueppel, 1933, Physiologie und Akustik der Vogelstimme, *J. Ornithol.*, 81, 433–542.)

Ventral view.   Frontal section.

M.t.	Musc. tracheolateralis	P.	Pessulus
M.st.	Musc. sternotrachealis	M.t.e.	Membrana tympaniformis externa
T.	Ossified lower tracheal ring (drum)	M.t.i.	Membrana tympaniformis interna
t.S.	Syrinx membrane, tracheal portion	1.	First bronchial ring

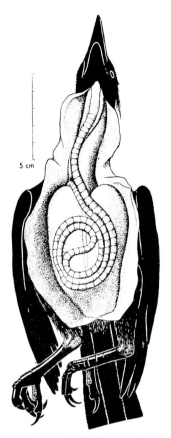

Fig. 296. *Phonygamus kerandreini* (Trumpet bird). Course of tracheal loops. (From W. Rueppel, 1933, same source as Fig. 294 and 295.)

Surrounding the syrinx is the *saccus clavicularis* (or interclavicularis). This is an air sac essential in the production of tones; together with the vibrating membranes and the musculature they make up the vocal system of the bird.

The vibrating membranes are the internal and the external tympaniform membrane. They are activated through the antagonism of the external pressure—inside the air sac—and the pressure of the air current running through the inner bronchotracheal tube. Rhythmic vibrations produce the sound in the column of air enclosed by the trachea.

The *internal tympaniform membrane* is located in the cranial extremity of the bronchi, which are here formed by rings open at their median side. The *external tympaniform membrane* is formed in the membrane connecting two bronchi. The caudal rings of the trachea are connected rigidly between themselves or are even grown together; they form the tympanum, divided in two halves by a ridge looking upward, the *pessulus*.

The system of the muscles is derived from the sternohyoid muscular apparatus. While a simple pair of antagonists, m. sternotrachealis and m. mylo-hyoideus, pulling the trachea up and down respectively, is the equipment of most birds, the singing

References p. 518

birds are equipped with a much more differentiated system. Here up to nine pairs of so-called syrinx muscles regulate the position of the uppermost bronchial rings and with it the tension of the tympaniform membranes. The regulation of the motility within the syrinx is centered in the hypoglossus nucleus.

The whole apparatus of the singing birds: bill, oral cavity, tongue, hyoid bone, larynx constitute an effective mechanism for modulation of the sounds produced in the syrinx. The heterogeneity of the sections composing the singing apparatus provides the bird with the ability to produce combinations resulting in a very large number of variations of tone. The vocal organs are placed in the anterior air chamber; this has partly resilient walls well suited to modify and amplify sounds. Step by step more parts of the surrounding regions are included in the vocal apparatus, resulting in the endless variety of bird song.

The glottis with its long and rigid arytenoids does not take part in the production of sound.

There are, allegedly, no absolutely mute birds. Some birds, as the vultures, are destitute of vocal organs; others, with poorly differentiated organs, are restricted to the utterance of a few notes.

The sounds of birds can be divided into two categories: call notes and true song. A call note is a brief sound with a relatively simple acoustic structure. It is used mainly as warning and is inherited, whereas the songs, although entirely or partly inherited, can be partly or entirely learned. Inborn and learned patterns are integrated with the maturing of the bird. A very involved anatomical arrangement is the basis for imitating human speech sounds or the utterances of other birds—mocking.

Double voices are sometimes produced, as by the Cock. They occur when the two voice-producing halves of the syrinx do not work in perfect tuning.

It was found, in birds[43] that some vocal organs function after the manner of a wind instrument having a resonating chamber with rigid walls while others function after manner of the human voice, coordinating syrinx and resonating chambers.

(b) Mammals

(i) General view on the mammalian larynx

Many vertebrates are voiceless, although certain of the lower Tetrapods can make hissing or roaring noises by a violent expulsion of air through the glottis; but in most mammals the larynx is a vocal organ.

In *Marsupials* a distinct cricoid cartilage, in addition to the arytenoids, appears already in Reptiles, but no true thyroid cartilage is found. In Marsupials, thyroid and cricoid are fused, and the arytenoids are grown together dorsally. Consequently, the musculature has remained at a more primitive level; there is *e.g.* no mechanism to stretch the vocal cords. Little use is made of a voice, although piercing cries or grunts are emitted occasionally. The laryngeal chamber is big; in spite of this the vocal production is poor, if not lacking altogether.

Monotremes with large arytenoids, but with thyroid and cricoid cartilages fused, are silent. The rigidity caused by the fusion of thyroid and cricoid is not favorable to laryngeal voice production. Silence, again, is the characteristic of the Edentates, although in Anteaters and Armadillos thyroid and cricoid are no longer fused and, consequently, the musculature shows a trend toward further differentiation.

Before turning to the Rodents a few remarks on the *mammalian larynx in general* are in order[8, 10, 21, 22, 28, 37, 40, 42].

The most important new acquisition of the mammalian larynx is the *thyroid cartilage*. It is the exclusive possession of the mammals. Division of the hyoid complex into hyoid and thyroid is already accomplished in the Marsupials. This newly liberated member of the group of laryngeal cartilages is captured again, in Marsupials and Monotremes, by its fusion to the cricoid, as if still looking for support. From here upward, however, a more or less free mobility between thyroid and cricoid becomes standard and is a most important factor in the production and modulation of voice.

To right and left *plicae aryepiglotticae* form the lateral boundary of the laryngeal aditus. Laterally of these folds the paired *recessus pharyngo-laryngei* transport food, appropriately prepared by chewing, from the mouth to the esophagus; a more bulky bolus will pass right over the larynx. Penetration into the latter is prevented partly by the epiglottis but with more effect by the sphincter action of the musculature.

From the viewpoint of phonation the *retrovelar or retropalatinal position of the epiglottis* (Fig. 297) is of importance. The rigid nasal chambers are directly connected with the larynx, and the separation of the food—from the air-passage is complete. The cuff-like *isthmus palato-pharyngeus (pharyngonasalis)*, at the level of the lower half of the nasopharynx, surrounds the larynx and makes the separation more stable.

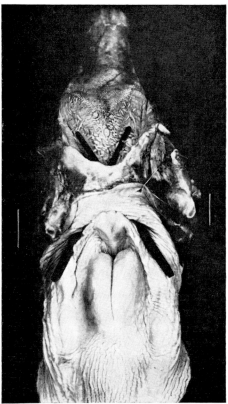

Fig. 297. *Otaria jubata* (Seal). Retrovelar position of the epiglottis. (From G. Kelemen and A. Hasskó, 1931, Das Stimmorgan des Seelöwen, *Z. Anat. Entw. gesch.*, 95, 497–511.)

References p. 518

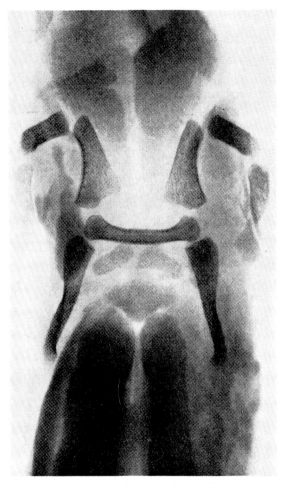

Fig. 298. *Phoca vitulina* (Seal). Protective "sheathing" around the "carrefour aéro-digestif" by the hyoid complex. (From G. Kelemen, 1929, Vergleichend-anatomische Röntgenbefunde am Zungenbeinapparate, *Arch. Ohren–Nasen–Kehlkopfheilk.*, *122*, 161–169.)

But this grip of the isthmus on the larynx gradually becomes looser. Horse and Swine *e.g.* are able to lower the laryngeal isthmus, temporarily creating a continuity between the latter and the oral cavity which can now be put to the service of vocal function. Carnivora, Primates show an especially loose connection between pharyngonasal isthmus and larynx. There are, however, uncounted modifications of the transition between larynx and the nasal or oral cavity, respectively. Odontoceti have developed a long tube between larynx and choanae[38]. In Anthropomorphs it is only the Orang which still shows its epiglottis in a retrovelar position. In all other Anthropomorphs the tip of the epiglottis hardly touches the lower edge of the velum. In Man, after the initial high position of the larynx in the embryo has been followed by a descensus, the larynx becomes completely liberated from the nasopharyngeal isthmus. The low suspension of the human larynx has resulted in participation of all parts of the pharynx, together with the nasal and oral cavities, in the production and modulation of sounds. With the descensus, the thyroid cartilage moved away from the hyoid, breaking this link, too, in the chain which hampered free mobility.

As a source of vocal production attention is to be given to the *arytenoids*[24, 25] (Fig. 299). Their upper edges, serving as a phonatory glottis, are thin and not capable of effectuating a strong closure of the aditus. Furthermore, their inner aspect is a concave one, forming a deep groove in the huge arytenoids of the Marsupials and the Monotremes. They show here, and in other forms, an elongation in the sagittal plane and themselves create the entire glottis. When closed, only the edges come in contact, below them the medial aspects of the two arytenoids cannot be approximated, because of their being excavated. This remaining opening, the *hiatus intervocalis* (Némai[24, 25]), represents a safety vent assuring the airway between the concave median surfaces. Moving to higher forms the arytenoids become smaller, and bound between themselves correspondingly smaller portions of the glottis. But even Man shows vestiges of the hiatus intervocalis, readily seen when the glottis is viewed from its tracheal aspect.

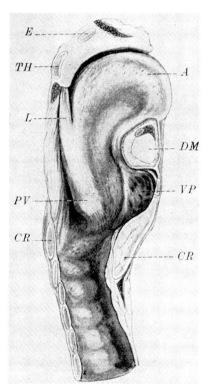

Fig. 299. *Otaria jubata* (Seal). Phonatory glottis formed between the upper arytenoid edges. (From G. Kelemen and A. Hasskó, same source as Fig. 297).

A: arytenoid	L: elastic bundle
CR: cricoid	PV: processus vocalis
DM: dorso-medial process	VP: posterior intralaryngeal sac
E: epiglottis	Th: thyroid

References p. 518

The *processus medialis* of the arytenoid, facing the thyroid, becomes a *processus vocalis* and serves as a basis for a fold which runs from this point to the inner surface of the thyroid. This fold did not develop in the Monotremes, but Marsupials and Edentates show it in a rudimentary form. In advancing evolution these folds take over the role of temporarily closing the laryngeal glottis, a function hitherto exercised by the arytenoid edges.

In the Placentalia the *cricoid* cartilage is characterized by development of the dorsal part of the ring to a broad plate. Sometimes the ring is interrupted in its ventral portion. The development of the massive plate occurred in relation to the dilatator laryngis, which moved in higher mammals from the thyroid to the cricoid.

The *epiglottis* shows considerable variety among the mammals. At its base, it can be divided in two halves, each one in connection with the plica ary-epiglottica. In Orang, it becomes broad and shovel-formed while the lateral food channels become less important, with regression of the ary-epiglottic folds.

(ii) The laryngeal musculature

A concise survey of the *laryngeal musculature* in mammals[10] can be given in the following way:

The *constrictors* show a primitive division into a ventral and a dorsal portion: *mm. laryngei ventrales* and *dorsales*. The laryngei ventrales are represented by the *mm. thyreo-crico-arytenoidei laterales*; the laryngei dorsales by the *mm. ary-proarytenoidei*, and after fusion of pro-arytenoid and arytenoid, by the *m. interarytenoideus*. Singularity of the *m. thyreoarytenoideus transversus* is observed in all Primates. In the Anthropomorphs a *m. interarytenoideus obliquus* will detach itself as a separate unit, becoming increasingly conspicuous along the line: Orang–Chimpanzee–Gorilla–Man.

In Placentals, the *m. crico-arytenoideus lateralis*, becomes sharply separated from the *m. thyreo-arytenoideus lateralis*. The latter is a massive plate resting on the lateral wall of the larynx. Its further differentiation is of basic importance for vocal production, because of the relation of this muscle to the vocal cord. At the level of the cord the muscle juts inward and is found in the basis of the cord. This part has been termed *m. vocalis*, although it never attained true independence. In case the cleft between the plica ventricularis and the plica vocalis is extended into a ventricle, the latter will divide the thyreoarytenoid muscle into an upper and a lower portion.

In a comparative study of the musculature directly related to the vocal cord the outstanding facts are a gradually increasing complexity of the *m. thyreoarytenoideus* and a closer approximation of it to the *ligamentum vocale*. Only by insertion of muscle fibers at the vocal cord itself does the glottis become a perfect vocal organ. Besides its use for vocalization, the closed glottis might help in fixation of the thorax during fore limb efforts (Negus)[22].

Among the *dilatators*, the *mm. cricoarytenoidei posteriores* developed from the *mm. kerato-crico-arytenoids* of the Marsupials and Cetacees. The part originating from the thyroid horn disappears. (This may serve as one example of simplification, which, as a counterpart to differentiation, dominates the further evolution of the larynx.)

The last acquisition in the muscle system of the Placentals is represented by the *mm. crico-thyreoideus internus* and *externus*, the latter again divided, ultimately, into a *pars recta* and a *pars obliqua*. It is of phonetical interest to recognize that in

Man one sees all varieties from a perfectly compact muscle to a clearly divided pars recta and pars obliqua. A *m. thyreoideus transversus*, connecting the edges of the inferior median incisura of the thyroid plate and originating from the *m. cricothyreoideus*, is found in widely divergent forms, as in Bears and Gibbons.

The *laryngeal cavity* claims the center of attention in a discussion whose primary aim is to demonstrate the arrangements in connection with vocal performance. In Monotremes, it is simply divided into an upper half between the arytenoids, and a lower half between the cricoids. In higher forms, with a few exceptions, the profile is complicated by the appearance of *vocal cords* between the processus vocales of the arytenoids and the inner aspect of the thyroid plate. These plicae are the reinforced upper edge of an elastic, submucous layer, the *conus elasticus*. The latter runs to the cricoid and includes the *ligamentum cricothyreoideum medium*. At this stage it became possible to add to the upper and lower portion of the cavity a *pars intermedia*, the space around the *plicae vocales* (*labia vocalia*, vocal cords). The latter bound the *rima glottidis*, divided into a *pars intermembranacea* between the vocal cords, and a *pars intercartilaginea* between the arytenoids. The intercartilaginous portion represents one half to one third of the glottis, or even less according to the species, and shows, in consequence of the concavity of the inner aspect of the arytenoids, the *hiatus intervocalis*.

The *plica ventricularis* represents the lower limitation of the laryngeal vestibulum, and is separated from the vocal cord by the entrance to the laryngeal ventricle.

In Marsupials the vocal cords[40] are marked by a very short fold. Even this is absent in the Cetacees. Production of sounds is achieved by a "glottis" formed by the sharp upper edges of the elongated arytenoids.

Stretching of the vocal cords—a basic factor in vocal production—presupposes an adequately firm structure in the cord itself and an appropriate muscular apparatus. Amphibians, Reptiles, Birds, Marsupials lack such mechanism. The folds may be too extensible to be stretched, as in Cats or some Primates.

In the Primates the vocal cords end often in a sharp edge directed orally. When sufficiently elongated this edge—the *vocal lip*—covers, from the medial side, the lower half of the entrance to the ventricle. The *m. thyreoarytenoideus* underlies, with its inner portion: the *m. vocalis*, the vocal cord but does not rise into the sharp, upward jutting lip[17].

Duplication, lengthwise, of the vocal cords occurs *e.g.* in Swine and Gibbon[27]. They are encountered, below the mammalian level, in Frogs. Production of double tones becomes possible with this mechanism, while double tones in Birds, as mentioned, originate by imperfect tuning between the two halves of the syrinx.

The evolution of vocal cords[40] is influenced by the presence of intralaryngeal air cushions or extralaryngeal air sacs.

Laryngeal air sacs, mostly absent in Marsupials and Monotremes, are developed in many Placentals. They appear in very different locations and sizes. Even nearly related species can show them in very different forms. All Primates, inclusive of Man, possess them in one form or other. A few instances are here enumerated:

Saccus thyreoideus: a ventral, median excavation of the thyroids, first seen in Marsupials. *Sinus laterales*: a continuation of the laryngeal ventricles. *Fovea centralis*:

situated ventral and medial in the plane of the ventricles, with the *saccus intercartilagineus anterior* as one variety. *Saccus dorsalis laryngo-trachealis*: excavation in the region of the oral edge of the cricoid. *Upper, median sinuses* extrude beyond the thyroid edges and develop, predominantly in Primates, to large extralaryngeal sacs. Some of these cavities find their way into the body of the hyoid bone and may become, with a bilateral origin, a single unit. In Orang the sacs penetrate under the platysma and reach under the mandible and to the sternum, with secondary spaces over the *m. pectoralis minor*; from here they continue into the fossae axillaris, supraspinata and subscapularis. These huge sacs proceed to confluence in the midline with frequently far reaching asymmetry. Mycetes (the Howler Monkey) carries a hyoid body blown up to a large bony sac with bilateral connecting ducts (*ductus pneumaticus*) to the laryngeal lumen. Besides this excavation and ballooning of the hyoid, Mycetes has developed paired *pharyngo-laryngeal sacs* in the lateral wall of the larynx.

The extraordinary variety in the development of the air sacs lead to numerous conjectures about their functional significance. Generally they are thought to be resonators, "loudspeakers" reinforcing the vocal production. They transform a very low voice, in Marsupials, to a more audible one; at the other extreme they reinforce the voice of the Howler Monkey to a fearful intensity.

The *mucous membrane* lining the greatest part of the larynx is a columnar, ciliated epithelium, while the vocal cords carry stratified squamous epithelium extending over the sharp vocal lips, where such are present. *Glands*, in conspicuous amount, are found in the basal parts of the epiglottis and around the opening of the laryngeal (Morgagni's) ventricles.

Innervation. The *superior and inferior* (or recurrent) *laryngeal nerves*, branches of the tenth cranial nerve, underlie the sensory and the motor functions. Their role is basically the same in all mammals. The *internal branch* of the superior laryngeal nerve is sensory, innervating the mucous membrane; its *external branch* innervates the cricothyroid muscle. The recurrent nerve serves all laryngeal muscles except the cricothyroid. To elucidate the role of the *sympathicus*, many experiments were carried out, predominantly on the Dog, without resulting in a concord of opinions.

The *arteries* of the larynx are branches of the *superior* and *inferior thyroid arteries*.

The innumerable variations of the *mammalian larynx*, and especially the arrangements serving the vocal production, may be illustrated by a number of characteristic instances.

The world of the lower mammals is a remarkably silent one. How is their larynx as an organ of voice organized? Taking *Atherura* (a Porcupine) (Fig. 300) as example, no vocal emission of this species was ever observed, either in free life or in captivity. All muscles around the glottis are present but are weakly developed: among the constrictors it is only the one *thyreoarytenoideus* which shows a somewhat more conspicuous size. The *m. cricoarytenoideus* is hardly capable of effectuating a closing of the glottis. Even so, in practically all its length the glottis is intercartilaginous with a very short fold marking the site of a vocal cord. Sounds could be formed nowhere except between the oral edges of the arytenoids.

This animal belongs to the *Rodents*, but we have to look back to a Marsupial, *Halmaturus* (Kangaroo) to find a somewhat longer fold at the site of the vocal cords; about one fifth of the glottis is enclosed by them. In the *Edentates*, one may see somewhat longer folds, enclosing about one third of the glottis.

Fig. 300. *Atherura macroura* (a Porcupine). Sagittal section of the larynx with short fold as "precursor" of a vocal cord. (From G. Kelemen, 1932, Das Stimmorgan der *Atherura macroura* (Erdstachelschwein), *Z. Anat. Entw. gesch.*, 99, 203–211.)

Rodents and *Lagomorphs* are either silent or produce a simple sound which can be intensified to loud shrieks in some species. Their epiglottis is generally large; the vocal folds are long and sharp-edged.

The higher the phylogenetic position of a mammal, the more one finds a reduction of the arytenoid and an elongation of the vocal cord. Gradually the formation of a real "vocal" glottis is reached, but initially formed to a considerable portion by approximation of cartilaginous parts. Horse, Deer, Swine exemplify–in this order–the gradual longitudinal growth of the vocal cords at the expense of the arytenoids.

Besides this material gain in length of the vocal cord, its relation to the musculature is of eminent importance from the viewpoint of vocalization. The very complicated pattern of muscle fibers in the human *m. thyreoarytenoideus*[11, 40, 41] sets it off against the simpler structure in all other mammals.

It is hardly possible to build up an ascending line among mammals solely by

References p. 518

proceeding from a more primitive to a more complex vocal utterance. The larynx of Felides, as in the Lion is formed as a more primitive type, as is in the Hare or the Antelope, but the vocal potential of the latter is minimal when compared with the Cats. A number of different details forbids it to construct a sequence of laryngeal development and still remain in step with the phylogenetic evolution of the entire organism. Only with this understanding can one proceed, beyond the mute lower mammals, with an analysis of the larynx as organ of voice.

Cetacees (Whales)[38] were the object of vivid interest in recent investigation (Fig. 301). Grating, groaning, creaking were observed with echo-sounding receivers and, more frequently, series of sharp clicks. Porpoises have been found to be taciturn when solitary but quite voluble in groups. Their larynxes (Fig. 302), varying among types with long or short arytenoids, will surely attract more anatomical attention now that the significance of their vocal production is well demonstrated.

Fig. 301. Registration of porpoise sounds with a hydrophone. (From F. S. Essapian, 1953, The Birth and Growth of a Porpoise, *Nat. Hist. Mag.*, Nov.)

Ungulata (Figs. 303, 305) present an endless variety of voices. In general, broad and flat thyreo-arytenoid folds are present. Air sacs play frequently a considerable role. The flat and broad vocal cord in *Bos taurus* (Fig. 304) shows an identical profile on its upper and its lower end making the use of the expiratory *and* the inspiratery air current equally possible in voice production.

Carnivora. In seals poor development or total absence of vocal cords is the most conspicuous feature. The arytenoids are grown together at their dorsal end, and are fixed in a position of moderate abduction. The adduction of the arytenoid edges, necessary to produce their extraordinary loud voice, is effectuated by the very strong m. thyreoarytenoideus. Voice is produced between the upper edges of this single united arytenoid by the inspiratory air current (Fig. 299). The roar of the Lion is produced with a comparatively simple vocal apparatus. Rounded vocal folds with small ventricles are present, in a very big laryngeal cavity. Dogs (Fig. 306) have sharp and long vocal folds, well developed ventricular bands, and a wide, deep ventricle; these show a tremendous difference from the larynx of the Cat (Fig. 307), in which a flatter vocal cord stands out in absence of a ventricle and with a hardly distinguishable ventricular band.

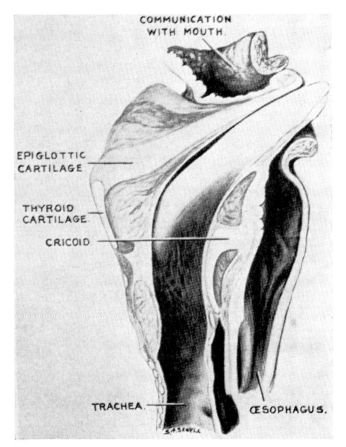

Fig. 302. Dolphin. Sagittal section of larynx. (From V. E. Negus, 1949, *Comparative Anatomy and Physiology of the Larynx*, Grune and Stratton, New York.)

The two latter forms show the way by which loud and richly modulated sounds can be produced by very different laryngeal mechanisms. The families of the Dogs and Cats belong to the same class, the Carnivora; they live under analogous conditions and show, generally a similar structure of their entire organism, *e.g.* regarding denture,

References p. 518

organs of the thorax, intestinal tract. But their larynx is so different that the separation must have occurred far down in the course of evolution. The *aditus laryngis* in the Felides remained the same as in Marsupials: pointing backward in evolution; Canides present a picture similar to the Prosimia: pointing forward in evolution. This is one of the examples showing that laryngeal evolution is in many respects out of step with other constituents of the surrounding organism.

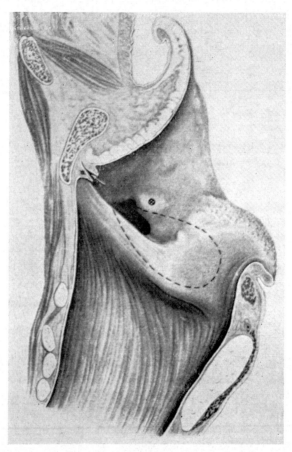

Fig. 303. *Equus caballus* (Horse). Sagittal section of larynx. (From F. Prodinger, 1940, Die Artmerkmale des Kehlkopfes der Haussäugetiere, *Z. Anat. Entw. gesch.*, *110*, 726–739.)

Chiroptera (Bats) were lately much discussed in connection with the phenomenon of echolocation and production of ultrasonic frequencies in addition to audible sounds[12] (Fig. 308). Two pairs of very thin membranes lying on the side walls of the larynx are stretched over a tiny, shallow cavity by contraction of the cricothyroid muscles. They are supposed to produce, by their vibrations, the high frequencies. Monostrosities occur, as in *Hypsognathus*, a tropical Bat, where the larynx occupies the greater part of the thoracal cavity.

(iii) Primates

The larynx of the *Primates*[13, 19, 24, 25, 28, 36, 44] is characterized by a high degree of development as against other mammals, the gap between the highest Anthropoid and Man remaining still very wide.

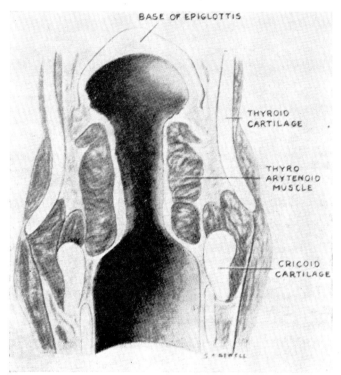

Fig. 304. *Bos taurus* (Ox). Sagittal section of larynx. (From V. E. Negus, 1949, same source as Fig. 302.)

Contrary to the condition in Lower Monkeys and in Man, the musculature, especially in Orang and Chimpanzee, shows a remarkable inconstancy. This variability becomes most conspicuous around the aditus of the larynx. As an interpretation one has to remember that parts showing the greatest inconstancy testify to an especially active phase of evolution.

Several species of *Lemurs* are noisy, without a reinforcement arrangement, as they lack air sacs. The only exception is an intralaryngeal median saccular outgrowth. Their vocal cords show a progressive evolution from an almost membranous condition in Tarsius to a structure more or less underlined by muscle. The thyreoarytenoid muscular complex shows many evidences of this tendency. In Marmoset the muscle is one compact bundle lying apart from the vocal cord, not incorporated in it. In Galago eight distinct notes were defined. The sounds of Lemurs are so characteristic that some of them gave rise to onomatopoetic native names.

Monkeys, in general, have an intralaryngeal reinforcement by resonance in lateral intralaryngeal air cushions. To this is added, in other forms, an extralaryngeal reinforcement by air sacs.

The *New World Monkeys*, Capuchins, Spider, have loud voices emitted by a big larynx; there are short, not valvular ventricular bands, and very long and sharp-edged vocal cords. The terrific roar, produced always in a chorus, by Mycetes (Howler Monkey)[6,26] (Fig. 310) is produced with very long, sharp vocal cords, supported by long and deep ventricles and reinforced by an osseous dilatation of the hyoid body.

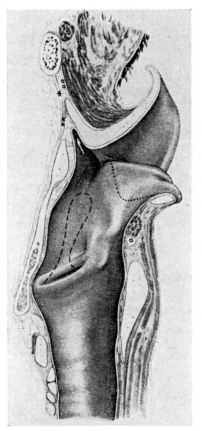
Fig. 305. *Sus scrofa* (Swine). Sagittal section of larynx. (From F. Prodinger, 1940, same source as Fig. 303.)

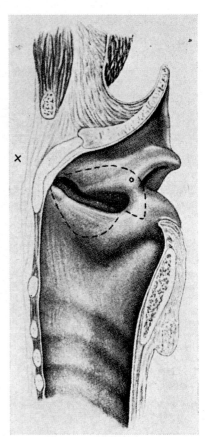
Fig. 306. *Canis familiaris* (Dog). Sagittal section of larynx. (From F. Prodinger, 1940, same source as Fig. 303.)

The *Old World Monkeys* (Fig. 309), Macaques, Baboons, are distinguished by air sacs, an outstanding characteristic. The middle portion of the hyoid body is inflated, and encloses the bony portion of the air sac which is continued in the membranous portion. The epiglottis is perforated, offering a communication between larynx and air sac. The well filled air sac serves as an effective resonator for this group, which includes, among others, the incessantly chattering Macaques. The skeleton of the larynx is built with a superior elasticity. The vocal cords jut only moderately into the lumen; they lie more flat along the lateral laryngeal wall. The free edge of the vocal cord, the vocal lip, is directed orally; this lip is, in strict sense, the vocal cord while the rest of the elastic fold remains embedded entirely into the lateral wall. On the other hand the lateral wall remains in more intimate relation to the vocal muscle, and can be adducted until the glottis is closed by this action. As the closing is effectuated not by fine edges but cushion to cushion, one cannot expect the more delicate modulatory capacity as the one connected with freely moving delicate lips. While the cushions close with the help of the expiratory air current, the free, upward-looking edges produce sound by an inspiratory mechanism.

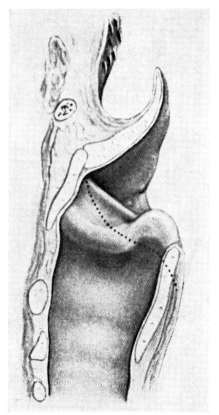

Fig. 307. *Felis domestica* (Cat.). Sagittal section of larynx. (From F. Prodinger, 1940, same source as Fig. 303.)

The voice of this group became more modulated, more colorful as the intercartilaginous part of the glottis can be closed to a very small hole. It is further affected by the ability to produce inspiratory sounds, by the elasticity of the cartilaginous skeleton, and by the pneumatic surrounding serving as a good resonator.

Finally, the connection of the larynx with the pharyngeal arch has become a very loose one. They do not embrace the epiglottis as in the Ungulates. The pharynx itself is further developed with differentiation of a uvula. However, this uvula hangs deep down, and as the oral aspect of the epiglottis is still overhung and covered by the soft palate a difference against human conditions remains.

References p. 518

In many respects, the *Anthropoids* reached the highest development of the laryngeal apparatus, but not without considerable differences between one another and not without features connecting conspicuously with lower forms (Fig. 318).

The larynx of Hylobates (Gibbon Fig. 311)[7, 26, 27] stands on a sideline on the road leading to human conditions. In spite of a very high degree of differentiation it is not on the straight line which leads from the low Mammals to the Primates.

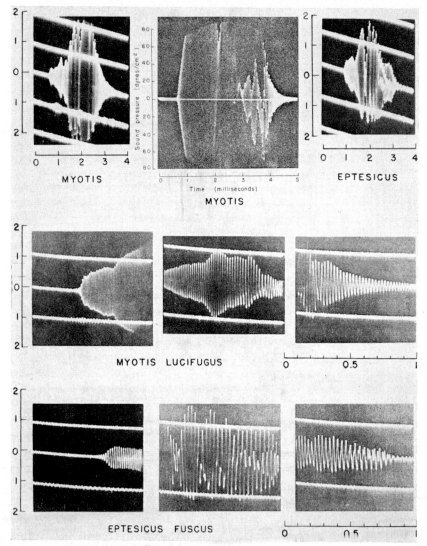

Fig. 308. *Myotis lucifugus* and *Eptesicus fuscus* (Bats). Cathode ray oscillograph records of sound pulses. (From D. R. Griffin, 1958, *Listening in the Dark*, Yale University Press, New Haven.)

It is the glottis itself which represents the highest degree in evolution: for the first time it can be closed along its whole extent. This makes it possible not merely to produce tones of a fixed frequency but to hold them at the same frequency for a certain time. The vocal cord is double, showing two parallel folds, although only the superior one can be considered as homologue to a vocal cord proper. The inferior fold, although jutting even farther out against the laryngeal lumen, does not find room for its insertion on the arytenoid, and runs backward to insert below the level of this cartilage. The closing of the glottis is very effective, as this is the only instance in the animal world, *not* excluding Man, where a *hiatus intervocalis* is totally absent. The complete closure is carried out with clear production of a single frequency without disturbing additions by the gap at the intervocal hiatus. On the other hand, as the accessory vocal cord does not insert on the vocal process of the arytenoid, finer modulation becomes questionable.

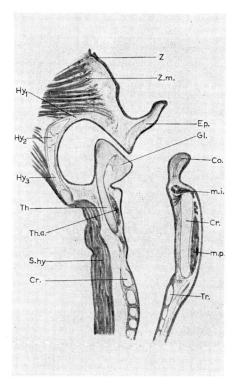

Fig. 309. *Cercopithecus sabeus* (Guenon). Sagittal section of larynx. (From J. Némai, 1920, Das Stimmorgan der Primaten, *Anat. Hefte*, 59, 259–292.)

Co.	Commissura interarytenoidea	Z.m.	Tongue, muscle
Gl.	Glands	Hy.1	Os. hyoideum
m.i.	Musc. interarytenoideus	Hy.2	Os. hyoideum, concavity
m.p.	Musc. cricoarytenoideus posticus	Hy.3	Os. hyoideum, inferior hook
Th.a.	Musc. thyreoarytenoideus	Ep.	Epiglottis
S.hy.	Musc. sternohyoideus	Th.	Thyroid
Z.	Tongue, mucous membrane	Cr.	Cricoid

There are many enthusiastic reports about the "musical" quality of the Gibbon voice or even about a "singing", which probably must be reduced to the fact that one hears clearly identifiable frequencies. It has to be added that double tones were

References p. 518

perceived by reliable observers; they may have been produced by the double vocal cord. This would be, besides the more complete closure of the glottis, another step leading farther than the point reached in human evolution. The double vocal cord is not lined by muscle in its free, upward pointing edge. To put the production of double tones on a firmer basis, the upper *and* the lower half of the vocal cord would have to come into a more intimate contact with the muscle. The mechanism to produce double tones, not only in Gibbon and Chimpanzee, but, as mentioned, in some lower forms, is not completely understood.

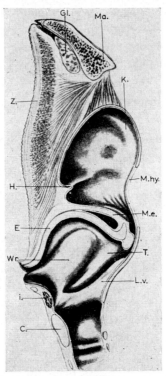

Fig. 310. *Mycetes seniculus* (Howler monkey). Sagittal section of the larynx. (From J. Némai, 1926, Das Stimmorgan der Primaten, *Z. Anat. Entw.gesch.*, *81*, 657–672.)

Ma.	Mandible	M.hy.	Membrana hyo-thyreoidea
Gl.	Glandula sublingualis	M.e.	Membrana hyo-epiglottica
Z.	Tongue	i.	Musc. interarytenoideus
H.	Hyoid	C.	Cricoid
K.	Bony vesicle		

Air sacs under the chin can be inflated from the throat, and aid in the production of the "most monumental ululations produced by any animal"[7]. The tremendous prolonged roar is led off by one male and is taken up by all others within hearing; it stops abruptly after a deafening crescendo.

The vocal cord of the *Orang* (Figs. 312–315) is embedded in the lateral laryngeal wall with a sharp edge pointing orally. Parallel to the vocal cord runs the ventricular band incorporating the lower corners of the broad epiglottic base and possibly serving as an effective damper to the vocal cord. The lateral laryngeal wall is pneumatized in its greater part. From the wide laryngeal ventricle the channel leads directly into the immense air sacs. Vocal production was rarely observed and was described as a roar or groan. The roaring is preceded by inflation of the air sacs, which are emptied visibly with the dying away of the vocal performance. The latter is uttered in conditions of excitement; if the cause for the stimulation stops, the air is emitted from the air sacs without having been used to support the voice. Air from the air sacs was considered to produce the deep humming sounds with the help of the ventricular bands. This would mean a third direction for the sound-producing air current besides the expiratory and the inspiratory.

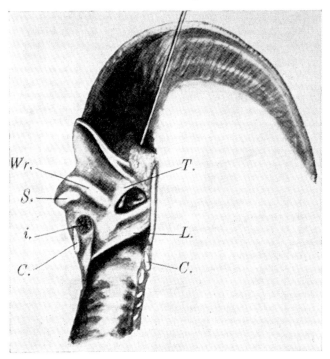

Fig. 311. *Hylobates syndactylus* (Gibbon). Probe in pneumatic space reaches the ventricle of Morgagni. (From J. Némai, 1926, Das Stimmorgan des Hylobates, *Z. Anat. Entw.gesch.*, *81*, 673–685.)

T.	Plica ventricularis	i.	Musc. interarytenoideus
Wr.	Cartilago Wrisbergi	C.	Cricoid
S.	Cartilago Santorini	L.	Chorda vocalis

In *Gorilla* the voice can take the character of a terrifice roar, but in captivity this animal is generally silent, a low murmur excepted. A *hiatus intervocalis* is present and is even wider than in Orang. Remarkable is the development of the inner laryngeal musculature. The *m. thyreoarytenoideus* shows the highest absolute and relative weight in comparison with Man and Orang—and the largest cross section. Here, as in other particulars of laryngeal elements, the Orang is placed between the Gorilla and Man.—An air sac develops unilaterally but extends beyond the midline and may end in the axillary fossa without, however, reaching dimensions as seen in Orang.

The *Chimpanzee* (Figs. 316, 317)[17, 29, 45] was widely studied regarding its vocal

References p. 518

production. In the free-living the amount and intensity of sound was described as overwhelming. Yerkes and Learned (1925) who compiled a list of 32 words or elements of speech for the Chimpanzee, characteristic of certain emotional situations, described calls which, with increased volume, burst into double tones.—The larynx is still in a high position, pressed against the lingual base, and, as in Orang and Gorilla there is still some difficulty in separating the soft palate and the epiglottis. This accounts for the absence of long drawn out sounds similar to those produced by Gibbon or Man; for similar sounds a constantly open oral resonant tube is necessary. The *hiatus intervocalis* is still a well-formed tunnel. The vocal lip, the sharp upward-turned edge of the vocal cord moves inward with displacement of the vocal process of the arytenoid, and is well suited to create inspiratory sounds which, in fact, were reported by various observers. Evacuation of the air sacs is accompanied by a grunting which is neither expiratory nor inspiratory, as the muscular mechanism regulating deflation of the air sacs is independent of the respiration.

(*iv*) *General considerations on mammalian phonation*

The organ of voice has to be conceived as a functional unit[15]. Phonetic achieve-

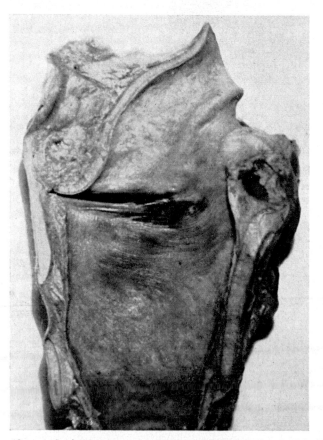

Fig. 312. *Pongo* (Orang). Sagittal section of larynx. (From J. Némai and G. L. Kelemen, 1929, Das Stimmorgan des Orang-Utan, *Z. Anat. Entw.gesch.*, *88*, 697–709.)

ment depends on cooperation of its component parts. Sites of junction where larynx, pharynx, oral and nasal cavities join each other, are of outstanding interest, as they influence the configuration of the resonant tube. These sections show considerable variety within the animal range. Continuity, characteristic of the human vocal organ, is attained in a rather roundabout way, by intercalation of regions without phonetic role in Man, or, in the opposite way, by elimination of parts, in Man, which were previously of decisive importance.

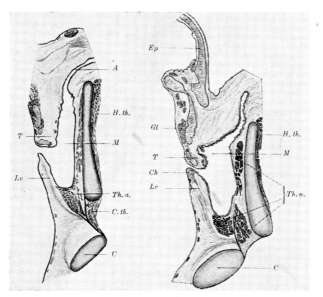

Fig. 313 and 314. *Pongo* (Orang). Frontal sections of one half of the larynx through anterior and middle third, respectively. (From J. Némai and G. L. Kelemen, 1929, same source as Fig. 312.)

Ep.	Epiglottis	Ch.	Chorda vocalis
T.	Plica ventricularis, with fat and cartilage	Lv.	Chorda vocalis, lip
		C.	Cricoid
Gl.	Glands	H.th.	Musc. hyothyreoideus
M.	Ventriculus Morgagni	Th.a.	Musc. thyreoarytenoideus
A.	Appendix ventriculi	C.th.	Musc. cricothyreoideus

Only in Man is the voice modifiable by muscular control of the upper air passages by palatal, lingual, nasal and other facial movements in a way to permit articular speech. While progressive *differentiation* produced the highest organized larynges in the Anthropoids, another factor, *simplification*, had to enter to create the human organ. The voluminous processi of the arytenoids in the laryngeal aditus are reduced in Man to the tiny corniculate (Santorini) cartilages. The steep rise of the glottis from dorsal to ventral, a more primitive condition, was eliminated.—Already in Orang, Gibbon and Chimpanzee the incorporation of muscular bundles had advanced gradually, with finer, more exact control of the vocal apparatus.

Taking the laryngeal structure as a basis, it is difficult to place any among the Anthropoids nearer to, or farther from the human level. Much depends upon the particular element chosen as a basis for comparison and upon the importance attributed

References p. 518

to one or the other of these elements by different observers[35]. The evolution of the larynx is frequently "out of step" amid the surrounding organism. Kangaroo, an aplacental, shows already a tiny vocal cord; *Atherura*, a rodent and consequently higher classified than the Kangaroo, shows merely a small, soft, not differentiated protuberance. In Felides the vocal cords rise high up to the aditus of the larynx, and in Seals the vocal cords step back completely. Double vocal cords appear, as mentioned, at different points of the phylogenetic sequence, to disappear through long intervals, as between Swine and Gibbon. The larynx of the Felides in general is more primitively constructed than the larynx of the Hare or of the Antelope, but in spite of this the vocal production of the latter is very much poorer. These and other examples testify to the difficulty of constructing a straight, ascending scale by tabulating the vocal capacity. Study of structure and of performance show many deviations from the straight evolutionary line.

Fig. 315. *Pongo* (Orang). Inflated air sac (plaster cast). (From R. Fick, 1929, Über die Körpermasse und den Kehlsack eines erwachsenen Orangs, *Z. f. Säugetierk., 4*, 65–80.)

Phonetical study of animal voices is still very much in its initial stage and only lately has due consideration been given to some chosen groups. The safest way to check bold theoretical construction is to direct full attention to anatomical

Fig. 316. *Troglodytes* (Chimpanzee). Sagittal section of larynx. (From G. Kelemen, 1948, The Anatomical Basis of Phonation in the Chimpanzee, *J. Morphol.*, *82*, 229–256.)

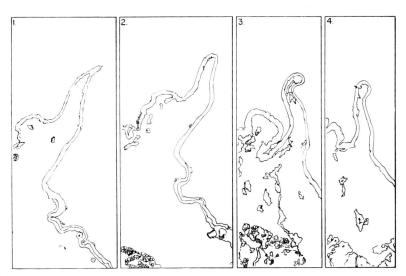

Fig. 317. *Troglodytes* (Chimpanzee). Sections of the vocal lip at different frontal levels. (From G. Kelemen, 1948, same source as Fig. 316.)

References p. 518

structure and to discard suggestions which disregard the structural-functional limits set by the apparatus itself[11,18].

A mechanism found in wide distribution among animals is the use of the inspiratory air current. As examples one may choose the Donkey and the Seal. The inspiratory I, followed by an expiratory A of the Donkey—reinforced by roomy lateral laryngeal air cushions—are known to everybody. Seal produces its voice during emergence and inhalation, between the free arytenoid edges; with this arrangement the resonating tube is situated caudad from the level of the voice production, with reversal of the entire resonating system. There exist a large number of arrangements in animals to produce inspiratory voice. The free vocal lip is one of these. In Monkeys, where the lip reaches its highest differentiation, the production of inspiratory sounds can be considered as an enrichment of the vocal performance. In Loris, inspiratory and expiratory notes seem to acquire equal importance, culminating in the "inspiratory chatter" in Galago. Inspiratory hissing is characteristic for so widely placed forms as Cobra and Loris. The tympaniform membranes in Birds[3,34] are placed between the air in the tracheo-bronchial system and the content of the air sac; expiration and inspiration are used continuously with quick reversals of the resonating system, and help to create the endless variety of the performance.

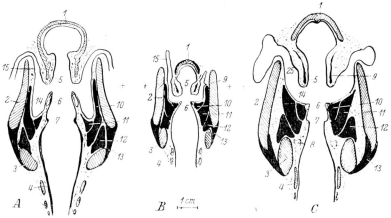

Fig. 318. (A) *Orang*; (B) *Homo*; (C) *Gorilla*. Frontal sections through the larynx. (From A. Kleinschmidt, 1938, Die Schlund- und Kehlorgane des Gorillas "Bobby", *Morphol. Jahrb.*, 81, 78–157.)

1. Epiglottis
2. Thyroid
3. Cricoid
4. First tracheal cartilage
5. Plica ventricularis
6. Plica vocalis with Ligamentum vocale
7. Plica glottidis
8. Protrusion of Conus elasticus
9. Musc. ventricularis
10. Musc. thyreoarytenoideus, Pars vocalis
11. Musc. thyreoarytenoideus, Pars lateralis
12. Musc. cricoarytenoideus lateralis
13. Musc. cricothyreoideus
14. Ventriculus Morgagni
15. Appendix ventriculi

One of the simplifications leading to the formation of the human vocal organ was elimination of the inspiratory voice, which now occurs in Man mainly as a pathological phenomenon.

Vowels are often heard in the vocal production of animals. In glyphic and graphic registration a surprising similarity with the human phonemes was found. Consonants are noted less frequently. A portamento-glissando is characteristic of the vocalization in many animals.

Intensity of the voice is much influenced by reinforcement through resonating pneumatic spaces; intralaryngeally, by air cushions located in the lateral laryngeal wall, or extralaryngeally, by air sacs. These structures have more to do with dynamics than with modulations of the voice. But even external air sacs may influence the quality of the voice by forming a glottis-like opening between the vocal cord and the ventricular band[17], where the air sac represents the bellow, connected with this "glottis" through the ventricle of Morgagni.

To produce loud sounds, the oral cavity has to serve in most cases as a resonating tube. Dog and Cat families, Monkeys are able to loosen easily the grip of the pharyngeal isthmus, and in this way to add the mouth to the resonating space. Where this is not possible the result is an inhibited vocal manifestation; this is the case with the majority of Mammals, particularly the lower forms, which are voiceless in any true sense. So are, in general, the majority of the Vertebrates.

Frequency of vocalization shows a very wide variety. Continuous "chorus" performance is characteristic of quite a number of groups. Sound production is infrequent in Gorilla, frequent and extremely varied in Chimpanzee, frequent with different degrees of modulations in Monkeys whose utterances are often described as "chattering". Against this continuous utterance other species accompany only vitally important situations with occasional sounds.

Numerous investigations have been conducted in the past few years into communication among animals[2, 5, 6, 7, 9, 33, 39, 45]. Sounds and sequences of sounds have a "meaning" which is "understood" by other members of the group; they have, consequently, a communicative function. It is evident that vocal communication is more highly developed in some infraprimate societies, as in birds, than among monkeys and apes. This may serve as an additional example of the difficulty in constructing a continuously ascending phylogenetic line on the basis of vocal performance.

Failure to teach Primates to imitate human sounds is caused by their entirely different laryngeal structure, which is the basis of a performance, highly evolved as to intensity and modulation but made up of entirely different phonetic elements[17, 18].

To associate sound with behavior or state of mind[14, 33, 39] is outside the scope of this presentation. Among the many intriguing problems falling in this line are the possible identification of group communications with language in the sense of the term as applied to man: sound production in choruses[6, 7], in some species to the exclusion of individual vocalization; complete cessation of a formerly frequently used voice when in captivity; and many others.

Rapid progress in application of physical methods of sound reproduction permits a hitherto unattainable precise analysis of animal sounds[1, 12, 15, 31]. The newly won knowledge, in its turn, asks for a more minute investigation of the anatomy of the sound-producing organs. Thorpe[43] emphasized that "the acoustic study of bird sounds has now far outdistanced the studies of the morphologist and... a most attractive field of investigation for the avian anatomist has been re-opened". This statement and postulate can be extended, more or less, to all vertebrates. Narrow anthropomorphic considerations, predominant in the past, have given way to treatment with full respect for the very considerable phonetic achievement encountered in the animal world. A newly awakening interest, resulting in a more penetrating study of structure and performance, is on its way toward gradually elucidating the numerous problems of comparative phonetics.

References p. 518

REFERENCES

For a comprehensive bibliography, see Kelemen, 1939[16].

[1] ANDERSON, J. W., 1954, The production of ultrasonic sounds by bats and other animals, *Science*, 119, 808–809.
[2] BIRCH, H. G., 1952, Communication between animals, *Cybernetics, 8th Conf.*, Josiah Macy Jr. Foundation, New York, pp. 134–172.
[3] BRAND, A. A., 1935, A method for the intensive study of bird song *Auk.*, 52, 40–52.
[4] BUSNEL, R. G., 1935, Psychophysiologie. Mise en évidence d'un caractère physique reactogène essentiel de signaux acoustiques synthétiques declenchent les phonotropismes dans le règne animal, *Compt. rend. Acad. Sci.*, 240, 1477–1479.
[5] BUSNEL, R. G., J. GIBAN, P. GRAMET, H. FRINGS, M. FRINGS and J. JUMBER, 1957, Interspecificité de signaux acoustiques ayant une valeur pour des Corvides européens et nord-americains, *Compt. rend. Acad. Sci.*, 245, 105–108.
[6] CARPENTER, C. R., 1934, A field study of the behavior and social relations of Howling monkeys, *Comp. Psychol. Monogr.*, 10, 1–158.
[7] CARPENTER, C. R., 1940, A field study in Siam of the behavior and social relations of the Gibbon (*Hylobates lar*), *Comp. Physiol. Monogr.*, 16, 169–182.
[8] DELIMA, E. E., 1954, Anatomia comparada da laringe, *Soc. oto-rhino-laryng. Latina*, 10, 55–72.
[9] FISH, M. P., 1954, The character and significance of sound production among fishes of the western North Atlantic, *Bull. Bingham Oceanogr. Coll.*, 14, 1–109.
[10] GOEPPERT, R., 1937, Kehlkopf und Trachea, in BOLK, GOEPPERT, LALLIUS and LUBOSCH, *Handb. d. vergl. Anat. d. Wirbeltiere*, Vol. 3, Urban and Schwarzenberg, Berlin–Wien, pp. 867–882.
[11] GOERTTLER, K., 1957, Le problème de la formation de la voix et du langage articulé du point de vue de l'anatomie, *Rev. Laryng.*, 78, Suppl., 487–493.
[12] GRIFFIN, D. R., 1958, *Listening in the Dark*, Yale University Press, New Haven.
[13] HILL, O., 1953, 1955, 1957, *Primates*, Comparative anatomy and taxonomy, Vol. 1, 2 and 3, University Press, Edinburgh.
[14] HUXLEY, J. and L. KOCH, 1938, *Animal Language*, Country Life Ltd., London.
[15] KATZENSTEIN, J., 1930, Methoden zur Erforschung der Taetigkeit des Kehlkopfes. sowie der Stimme und Sprache, in E. ABDERHALDEN, *Handb. d. biol. Arbeitsmethoden*, Abt. 5, Teil 7, 1. Haelfte, Urban and Schwarzenberg, Berlin–Wien, 261–418.
[16] KELEMEN, G., 1939, Vergleichende Anatomie und Physiologie der Stimmorgane, *Arch. Sprach- u. Stimmphysiol.*, 3, 213–237.
[17] KELEMEN, G., 1948, The anatomical basis of phonation in the Chimpanzee, *J. Morphol.*, 82, 229–256.
[18] KELEMEN, G., 1949, Structure and performance in animal language, *Arch. Otolaryng.*, 50, 740–744.
[19] KOLLMANN, M. and L. PAPIN, 1914, Études sur les Lemuriens, I. Le larynx et le pharynx, *Ann. sci. nat. zool.*, 88, ser. 9, t. 19, 227–318.
[20] LAWRENCE, B. and W. E. SCHEVILL, 1956, The functional anatomy of the Delphinid nose, *Bull. Mus. Comp. Zool.*, Harvard Coll., 114, 101–151.
[21] LULLIES, H., 1953, *Physiologie der Stimme und Sprache*, Springer Verlag, Berlin–Goettingen–Heidelberg.
[22] NEGUS, V. E., 1949, *The Comparative Anatomy and Physiology of the Larynx*, Grune and Stratton, New York.
[23] NEGUS, V. E., 1954–1955, The comparative anatomy of the larynx with particular reference to the function of the organ in man, *Lect. scient. basis of med.*, 4, 332–357.
[24] NÉMAI, J., 1920, Das Stimmorgan der Primaten, *Anat. Hefte*, 59, 259–291.
[25] NÉMAI, J., 1926, Das Stimmorgan der Primaten, *Z. Anat. Entw.-gesch.*, 81, 658–672.
[26] NÉMAI, J., 1926, Das Stimmorgan des Hylobates, *Z. Anat. Entw.-gesch.*, 81, 674–685.
[27] NÉMAI, J. and G. KELEMEN, 1933, Beitraege zur Kenntnis des Gibbonkehlkopfes. *Z. Anat. Entw.-gesch.*, 100, 512–520.
[28] NIERSTRASZ, H. F., 1927, Trachea und Larynx, in IHLE, VAN KAMPEN, NIERSTRASZ and VERSLUYS, *Vergleichende Anatomie der Wirbeltiere*, J. Springer, Berlin, 638–650.
[29] NISSEN, H. W., 1931, A field study of the Chimpanzees, *Comp. Psychol. Monogr.*, 8, 1–122.
[30] PRESSMAN, J. J. and G. KELEMEN, 1955, Physiology of the larynx, *Physiol. Reviews*, 35, 506–554.
[31] POTTER, R. K., G. A. KOPP and H. C. GREEN, 1947, *Visible Speech*, D. Van Nostrand Co., New York.
[32] PRODINGER, F., 1940, Die Artmerkmale des Kehlkopfes der Haussaeugetiere, *Z. Anat. Entw.-gesch.*, 110, 727–739.
[33] RÉVÉSZ, G., 1944, The language of animals, *J. Gen. Psychol.*, 30, 117–147.
[34] RUEPPELL, W., 1933, Physiologie und Akustik der Vogelstimme, *J. Ornith.*, 81, 433–542.
[35] ROMER, A. S., 1949, *The Vertebrate Body*, W. B. Saunders Co., Philadelphia–London.

[36] SANDERSON, I. T., 1957, *The Monkey Kingdom*, Doubleday and Co., Garden City, N.Y.
[37] SCHARRER, E., 1931, Stimm- und Musikapparate bei Tieren und ihre Funktionsweise, in BETHE, BERGMANN, EMBDEN and ELLINGER, *Handb. d. norm. u. pathol. Physiol.*, No. 15, 2. Haelfte, J. Springer, Berlin, pp. 1223–1254.
[38] SCHEVILL, W. E. and B. LAWRENCE, 1949, Underwater listening to the White Porpoise (*Delphinapterus leucas*), *Science*, *109*, 143–144.
[39] SCOTT, J. P., 1958, Language of animals, in *Animal Behavior*, University of Chicago Press, Chicago, 189–205.
[40] SEITER, GERTRUDE, 1955, Embryologie comparée des cordes vocales, *Ass. Franc. par l'Étude de Phonation et du Language*, *4*, 18–21.
[41] SIMONETTA, B., 1929, Studio comparativo sul m.tiro-aritenoideo e sul legamento vocale. *Ric. Morfol.*, *9*, 1–49.
[42] STRESEMANN, E., 1937, Syrinx, in BOLK, GOEPPERT, KALLIUS and LUBOSCH, *Handb. d. vergl. Anatomie d. Wirbeltiere*, Vol. 3, Urban and Schwarzenberg, Berlin–Wien, pp. 867–882.
[43] THORPE, W. H., 1959, Talking birds and the mode of action of the vocal apparatus of birds, *Proc. Zool. Soc. London*, *132*, 441–455.
[44] URBAIN, A. and P. RODE, 1946, *Les Singes Anthropoids*, Presses Universitaires de France, Paris, 111–115.
[45] YERKES, R. and B. LEARNED, 1925, *Chimpanzee Intelligence and its Vocal Expression*, The Williams and Wilkins Co., Baltimore.

For Addendum to Bibliography, see pp. 520, 521

Addendum to Bibliography Chapter 18

ALTMANN, S., 1959, Field observations on a Howling Monkey society, *J. Mammalogy*, *40*, 317–330.
BROWN, R., 1958, Animal languages, in: *Words and Things*, The Free Press, Glencoe, Ill.
BUSNEL, R.-G. et J.-C. BREMOND, 1961, Étude préliminaire du decodage des informations contenues dans le signal acoustique territorial du Rouge-Gorge (*Erithacus rubecula* L.), *Compt. rend. Acad. Sci.*, *252*, 608–610.
CAVE, A. J. E., 1960, The epipharynx, *J. Laryngol. and Otol.*, *74*, 713–717.
DuBRUL, E. L., 1958, *Evolution of the Speech Apparatus*, Charles C. Thomas, Springfield, Ill.
DuBRUL, E. L., 1960, Structural evidence in the brain for a theory of the evolution of behavior, *Perspectives in Biol. and Med.*, *4*, 40–57.
DuMENIL DuBUISSON, F. et R.-G. BUSNEL, 1960, Rôle d'un signal acoustique de verrat dans le comportement réactionnel de la truie en oestrus, *Compt. rend. Acad. Sci.*, *250*, 1355–1357.
ENZMANN, J., 1956, The structure and function of the laryngeal sac of hominoidea, *Anat. Rec.*, *124*, 286.
FESSARD, A. et B. VALLANCIEN, 1957, Données électrophysiologiques sur le fonctionnement de l'appareil phonataire du chien, *Folia phoniatr.*, *9*, 152–163.
FLOTTES, L., 1959, Le problème de la discrimination auditive en acoustique sous-marine, *J. Franc. oto-rhino-laryngol.*, *8*, 565–589.
FREEDMAN, M. L., 1955, The role of the cricothyroid muscle in tension of the vocal cord. An experimental study of dogs designed to release tension of the vocal cords in bilateral recurrent laryngeal nerve paralysis. *Arch. oto laryngol.*, *62*, 374–353.
FRINGS, H. and M. FRINGS, 1957, Recorded calls of the Eastern Crow as attractants and repellents, *J. Wildlife Management*, *21*, 91.
GOERTTLER, K., 1957, Le problème de la formation de la voix et du langage articulé du point de vue de l'anatomie., *Rev. laryngol.*, Suppl., 487–493.
HERRICK, E. H. and J. O. HARRIS, 1957, Singing female canaries, *Science*, *125*, 1299–1300.
HILL, O., 1960, *Primates*, Vol. 4, part 1, University Press, Edinburgh.
HILL, W. C. O. and A. H. BOOTH, 1957, Voice and larynx in African and Asiatic Colobidea, *J. Bombay Nat. Hist. Soc.*, *54*, 309–321.
HUECHTKER, T. und J. SCHWARTZKOPFF, 1958, Soziale Verhaltensweisen bei hörenden und gehörlosen Dompfaffen (*Pyrrhula pyrrhula* L.), *Experientia*, *14*, 106–111.
HUIZINGA, E., 1957, Larynx, voice, animals, *Annals otol., rhinol., laryngol.*, *66*, 679–690.
INGER, R. F., 1956, Morphology and development of the vocal apparatus in the African frog *Rana* (*Ptychadena*) *porosissima* Steunfachner, *J. Morphol.*, *99*, 57–72.
KELEMEN, G. 1961, Anatomy of the larynx as a vocal organ: evolutionary aspects, *Logos*, *4*, 46–55.
KELEMEN, G., 1962, Mimetic musculature and animal language, *Logos*, *5*, 18–21.
KELEMEN, G. and J. SADE, 1960, The vocal organ of the Howling Monkey, *J. Morphol.*, *107*, 123–140.
KELLOGG, W. N., 1958, Echo ranging in the porpoise, *Science*, *128*, 982–988.
KELLOGG, W. N., 1959, Auditory perception of submerged objects by porpoises, *J. Acoust. Soc. Amer.*, *31*, 1–6.
LANYON, W. E. and W. N. TAVOLGA, Eds., 1960, *Animal Sounds and Communication*, American Institute of Biological Sciences, Washington D.C.
MOULTON, J. M., 1956, Influencing the calling of sea robins (*Prionotus Spp.*) with sound, *Biol. Bull.*, *111*, 393–398.
MOULTON, J. M., 1958, The acoustical behavior of some fishes in the Bimini area. *Biol. Bull.*, *114*, 357–374.
MOULTON, J. M., 1960, Swimming sounds and the schooling of fishes, *Biol. Bull.*, *119*, 210–223.
NOVICK, A. and D. R. GRIFFIN, 1961, Laryngeal mechanism in Bats for the production of orientation sounds, *J. exp. Zool.*, *148*, 125–146.
ROE, A. and G. G. SIMPSON, Eds., 1958, *Behavior and Evolution*, Yale University Press, New Haven.
SCHEVILL, W. E., 1961, *Cetacea, Encycl. Biol. Sci.*, P. GRAY, Ed., New York, 205–209.
SCHLOSSHAUER, B., 1957, Sur quelques recherches experimentales relatives à la production de la voix chez le chien, *Rev. Laryngol.*, Suppl., 617–618.
SCOTT, J. P., 1958, *Animal Behavior*, The University of Chicago Press, Chicago, Ill.
STARCK, D. und R. SCHNEIDER, 1960, *Larynx*, in: H. HOFER, A. H. SCHULZ and D. STARCK, Eds., *Primatologia*, S. Karger, Basel and New York, Vol. 3, part 2, 423–587.
SMYTHE, R. H., 1961, *How Animals Talk*, Charles C. Thomas, Springfield, Ill.

THORPE, W. H., 1956, The language of birds, *Scient. Amer.*, *195*, 130–138.
THORPE, W. H., 1959, Talking birds and the mode of action of the vocal apparatus, *Proc. Zool. Soc. London*, *132*, 441–455.
VINCENT, F., 1960, Études préliminaires de certain émissions acoustiques de *Delphinus delphis* L. en captivité, *Bull. Inst. Océanogr.*, *57*, 1–23.
WORTHINGTON, L. V. and SCHEVILL, W. E., 1957, Underwater sounds heard from Sperm Whales, *Nature*, *180*, 291–292.

CHAPTER 19

COMPARATIVE ANATOMY AND PHYSIOLOGY OF THE AUDITORY ORGAN IN VERTEBRATES

by

B. VALLANCIEN

Studies in decriptive anatomy were followed by investigation in functional anatomy. Comparative physiology added a new feature to this evolutionary process. Consequently, it seems legitimate to place here all these disciplines under one and the same heading. Following this conception we had to start with an analysis of the elementary sensory structures in the most primitive forms. This way, the phylogenetic chain builds up in a natural way. It would be erroneous to attribute to these organs a determination of their own—valid, at any rate, to our own mind.

Every sensory organ capable of detecting stimuli set up by variations in pressure belongs to the terms of reference of a study of the comparative physiology of the auditory sense, the sonic stimulus being, when all is said, but a specific case of excitation by pressure, owing its specificity to its repetitive character and the rhythm of the repetitions. It is the branch of physics which we recognise as acoustics.

The primary purpose of all these receptors, whichever they be, is in fact to inform the organism of its position in space and of its displacements in a field of gravity; but they also keep it informed of its association with surrounding animate and inanimate objects. Hence, to amplify the information it receives from the world outside, the animal may well seek the echo of its own noise coming from inanimate obstacles or, better still, from living beings. It is thus that communication begins.

Perception of this kind presupposes the existence of an organ of reception placed in the best position for conducting the waves of pressure towards an organ of perception (transductor) which is capable of transforming the physical phenomena into nervous messages.

The whole vertebrate animal kingdom is provided with such organs. We may now fittingly proceed to compare the anatomical modifications they have undergone in the course of the evolution of the various species and to seek out the invariants among so many elements, the physiological role of which has not yet been conclusively established.

1. Primary Phenomena

(a) *Elementary structures*

From among the many works on the mechanism of the perception of vibrations we accord a special place to Dijkgraaf's observations[12] on the structure and functions of the lateral organs* and the labyrinth of the ear of fish. We shall place them within the context of more general phenomena as studied by Keidel[30] in relation to the potentials of the dorso-cutaneous nerve of the skin of the frog's back when subjected to vibrations of low frequency, and as studied by Fessard and Echlin[14] in association with the synchronised discharges of the receptors in the depth of the tissues in response to a vibratory stimulus in cat and frog.

The surprising anatomical feature is the invariable presence in all these receptors adapted to the vibratory sense of a rigid shell, capable of enclosing fluid, and cilia, the free extremity of which is directly or indirectly exposed to vibrations, the other end being attached to the nucleus of a sensory cell.

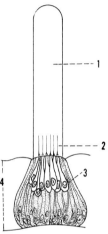

Fig. 319. Lateral organ. Sensory crista. (1) Cupula, (2) sensory cilia, (3) sensory cell, (4) epidermis. (After Dijkgraaf.[12])

The transductor is, in fact, always a ciliated cell whose cilia, held most of the time in a gelatinous mass (or cupula), bath in a fluid. The maximum stimulation is obtained by a displacement of this gelatinous mass, tangentially to the surface of the cell, from which the cilium emerges perpendicularly. It should be pointed out that this "shearing" of the cilia carried along by the displacement of the gelatinous mass operates in one direction only, producing an inhibition in the opposite direction.

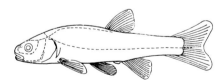

Fig. 320. Disposition of the lateral lines, on the fish body. (After Dijkgraaf.[12])

Let it be added that the movements of traction and pressure in the direction of the cilia also generate volleys of nervous influxes, but a different kind of excitation corresponds to this mode of stimulation.

---
* See the excellent article entitled: The Acoustics-lateralis system, by O. LOWENSTEIN in *Physiology of Fishes*, Vol. II, Academic Press, 1957.

References p. 554

If such structures are permanent features of the mammalian labyrinth, it is among fish that we find them in the pure state. In its original form, the system of the lateral organ appears like a sensory crest placed freely at the level of the epidermis. The elastic gelatinous mass enclosing the cilia (the cupula) is filled with water and is as transparent as glass. It is 10 μ and emerges from the surrounding water perpendicular to the surface of the skin (Fig. 319). These sensory crests, or ridges, are known to extend along three lines at the head and are prolonged down each side to the tail (Fig. 320). It is interesting to note that in certain permanently swimming fish the free ridges have a tendency to form what is called "canal organs" insensible to currents, which are thought to provide to some extent a reserve set of sheltered detectors. Dijkgraaf[12] has studied these "canal organs" thoroughly. He says: "One sees depressions like grooves forming in the epidermis right along the lateral lines, in such a way that one of the groups of free ridges is placed at the bottom of the groove, its edges reuniting above. The cupula changes shape and to a considerable extent keeps the light out of the canal thus formed" (Fig. 321).

Fig. 321. Formation of a lateral canal (cross-section). From left to right: free sensory crista, successive stages of progressive depression until the canal of the organ is formed. Bottom, centre: cross section through the canal between two sensory organs, at the level of a canal opening. (After Dijkgraaf.[12])

As can be seen in a longitudinal section, there is an opening leading outwards between two sensory organs of the canal. The function of these organs is to make the fish aware of the streaming water along its body and, at the same time, of its speed and position.

Anglo-Saxon authors have suggested that these were accessory organs of hearing, but their experiments[38] failed to gain full support, as it was shown[17] in *Phoxinus laevis* that the lateral organs do not participate in the perception of sound, not even in sound of low frequency. Pumphrey[40] nevertheless considers them to be organs of hearing, "because", he says, "sound is entirely mechanical agitation, of whatever nature, which emanates from some place exterior to the body of the animal. Hence an animal hears from the moment it shows itself capable of locating an object in movement without directly touching it".

Even if one cannot wholly subscribe to this broadened definition of sound, it remains true that stimulation by pressure, following its modality, calls for appropriate receptors and that the only difference between the vibratory sense, the static sense and the sense of hearing is one of receptive capacity.

Dijkgraaf[12] goes on to write: "When observing the free sensory ridges on a fish through binoculars, one notices that the movement of large objects makes the cupulae bend. In the same way, the cupulae of the "canal organs" are set in motion by a variation in close pressure which causes the fluid contained in the canal to move". Hence moving objects are a "biologically adequate" excitation (in the author's phrasing) for the lateral organs. And one could call the movements of fluid whose efficacy is immediate "physically adequate". It is, therefore, a reasonable proposition that the same sensory structure may possibly subserve the sense of balance just as well as that of hearing.

(b) *The nature of the excitation*

Electro-physiological experiments applied to certain specialised receptors have brought to light the frequent existence of permanent influxes, corresponding to what Hoagland[26] has called the spontaneous activity of the sensory crests. Excitation produces an intensification of those influxes, in the sense of increasing their frequency, *i.e.*, the number of discharges per unit time. We know from other sources that inhibition then results in their disappearance.

Thus this binary system enables us to make out three possible positions of electric nervous activity. In repose, the frequency of the spontaneous potentials is slow; it accelerates under excitation and is reduced to nil by inhibition. It should, however, be noted that, after excitation, there is a certain time-lag before the return to normal electric activity. After a period of acceleration there is invariably a short period of total inactivity. Conversely, after inhibition, a period of supernormal activity, or after-discharge, is observed to supervene (Fig. 322).

Fig. 322. Influx-discharge of a single receptor of the horizontal semicircular canal in *Raja clavata* reacting to an acceleration. (After Lowenstein and Sand, 1940, *J. Physiol.*, 99, 89.)

Several workers, Belgian and Dutch[13, 53, 54], tried to find out how this electric activity, which appears in one way only (traction of the cilia), acquires the form of alternate variations of the synchronous potentials of sonic vibrations when collected on the ciliated organs of the cochlea. So-called "microphonic" potentials are recorded when the cilia are pulled towards the cell, *i.e.*, perpendicular to its surface. That traction diminishes the potential of repose and compression has the reverse effect. Now, the cilia are pulled by a gliding movement of the cupula relative to the epithelium; therefore the frequency of the microphonic potentials should be double the frequency of stimulation, the cilia being pulled twice per period towards the stimulus, as is the case in the organs of the lateral line in fish (Fig. 323). If, however, the cilia are in an inclined position, they are pulled once only during the period of vibration and the frequency of the microphonic effect is found to be the same as that of the vibration. This fact is best exemplified by the cochlea, in which the external ciliated cells are inclined towards its axis.

As to the crests of the semicircular canals, Mygind[37] established the position of the cilia and a study of histological sections disclosed asymmetry in the distribution of the ciliated cells on the crest (Fig. 324). This explains why the displacement of the cupula produces a reaction only in the direction opposite to the site of the highest number of sensory organs. It is interesting to note[54], that this reasoning gives us exactly, for the three canals, the direction of excitation established by Ewald's second law. It is here that we find the link between the various organs of the labyrinth. Whether it be a matter of the microphonic potentials of the cochlea in phase with the displacement of the tectorial membrane, or of the movements of the cupula relative to the epithelium of the crest, we find a variation of potentials coinciding with the traction of the cilia.

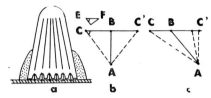

Fig. 323. Mecanism of the cilia. (a) Cupula, resting, (b) cilia, mobilized to and fro from AB, pulled twice during one period, (c) after being bent, traction is exerted only once. (After de Vries[53].)

By transmitting two sounds of different frequencies, it has thus been possible to demonstrate the effect of conjugated sounds upon the ciliated system experimentally in pigeons. During the negative phase of the low frequency the amplitude of the higher frequency increases, whereas it diminishes in the positive phase and is almost obliterated. This shall be an explanation of the hypothesis of ciliary asymmetry in pigeon.

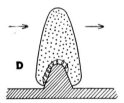

Fig. 324. Schematic sketch of a crista. (D) Side with predominating epithelium. Arrows: direction of the stimulation. (After J. R. Ewald, 1892.)

More recently, Trincker and Pratsch[50] have stated that the ciliated cells of all labyrinthine organs are apt to behave similarly in an electro-physiological sense. They *may* all be stimulated simultaneously by a displacement and by the whole field of vibrations of acoustic frequencies. But the results of their experiments cannot be accepted in normal physiological stimulation.

It remains for us to consider, therefore, how the sound-wave comes to exert its influence upon the cilia; that is, what is the mechanical transmitting agent and how does the nerve cell respond to this traction? *I.e.*, the nervous transmission of the stimulus and the mechanism of the transductor.

(c) *Mechanism of the transductor; nervous transmission of the stimulus*

The study of the mechanism contributing to the emission of volleys of influx under the influence of repeated stimulation has had to be confined to models. Many investigations have been facilitated by the fact that the elements of the skin which are sensitive to pressure are able to act like transformation organs in eliciting responses to superficial vibratory stimulation.

Intermittent repetitive excitation of the dorsal skin of frogs provokes a series of prolonged synchronised discharges with less adaptation than in excitation set up by prolonged continuous contact. All the same, the frequency of the synchronised discharges after a certain time. Keidel[30] has shown that these vibratory stimulations release rigorously synchronised discharges and that these appear in the growing

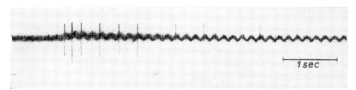

Fig. 325. Transitory phenomena in an isolated receptor; sudden change of stimulation. Short initial maximum, decrease of frequency of action currents to their value of limit (one action potential to 10 periods). The potentials are shown in the pression phase. (After Keidel.[30])

pressure phase of the period. If, on the other hand, the intensity of the stimulation is abruptly varied, the number of influxes per unit time is at first greatly increased, then declines to a final lower value, though retaining for minutes and even hours a high value which is proportional to the intensity of the stimulation. When the level of stimulation is low, a single fibre responds and its frequency of discharges declines with time until it reaches a final constant value synchronising with the stimulations (Fig. 325).

It has been shown[14] that when the tendon of a muscle in one of the digits of a cat, a rabbit or a frog is subjected to vibrations, this synchronisation can be improved by intensifying the stimulation or by contracting the muscle (Fig. 326). The tendon

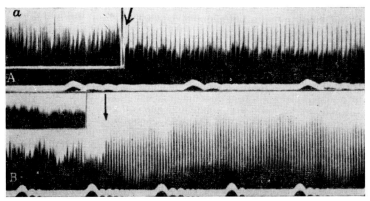

Fig. 326. M. gastrocnemius of the frog. (A,B). two examples of Afferent, asynchrone discharges at a strong tension: 100 g, after the arrow: synchronised discharge with application of a tuning fork on the tendon (After Echlin and Fessard.[14])

of the muscle being drawn back by traction upon the member, the discharges are more numerous and, if a tuning fork is placed on the course of the traction, it will be found that the discharges synchronise with the frequency of the tuning fork or a submultiple of it.

The transfer function of the *mechanical receptors-afferent fibres together* for different frequencies and for sudden variations in intensity is similar to that of an organ serving position and speed.

References p. 554

With a single receptor, the number of discharges for a given time depends first of all on the intensity of the stimulus, but also increases in proportion to the rate at which it is repeated. The informative element "frequency" is predominantly associated with synchronisation. At constant intensity, increased frequency of stimulation results in a larger number of discharges. With a preparation of several fibres, new fibres enter progressively into action until an *optimum rhythm* is attained, which varies, according to the given preparation, from one receptor to another. Beyond that the number of discharges decreases, while the frequency continues to increase. Seeking, for different frequencies, the threshold of intensity corresponding to the minimum of excitation required to evoke a discharge, one finds below the frequency of optimum sensitivity that the threshold values of intensity decrease linearly with the logarithm of the frequency. This produces a remarkable approximation to the configuration of the curve representing the subjective threshold of man's sensitivity to vibrations.

In 1939, Bekesy[3] showed that the curve of subjective intensity thresholds to vibrations can be represented by a linear law if the intensity of the excitation in the scale of low frequencies is represented, not by displacement, but by the speed of displacement as a function of the frequency; that is to say, the derivative of pressure in relation to time (if Hooke's law for low amplitudes is accepted). Hence to obtain the threshold, one reduces the intensity of stimulation in inverse proportion to the frequency.

Thus, when a preparation containing an assemblage of isolated *receptor elements and afferent fibres* is stimulated by vibrations, the frequency of the stimulations imposed is reflected by synchronised discharges, the intensity of which implies a certain integration of the total number of influxes within the time (when a single element, or, better still, a summation of several receptors is concerned) as a function of the number of fibres participating in the recorded final value. Abrupt variations in intensity call forth an initial transient maximum followed by a decline down to the final frequency of the discharges and, in the event of sudden negative variations, a period of silence (inhibition post) followed by an increase in the after-discharge up to an ultimate value proportional to the stimulation.

(d) The chemical mediator

The appearance of an initial maximum of discharges at the time of abrupt variations which subsequently tend towards an established stable order of low frequency is to be understood as a phenomenon of adaptation.

Many suggestions have been made to account for this fact, which appears to be common to all physiologico-sensory systems. Some authors, including Autrum[2], believe that different fibres exist for the whole sensory organ, one group reacting only to the derivatives (fibres on-off), other, fast ones reacting in proportion to the value of the stimulation, virtually without adaptation. (Tasaki's description[49] of the aortic vasomotor nerves.) Other authors, such as Hensel[24], for example, have it that the isolated system of receptor and afferent nerve exerts that function of characteristic transition. They account for it by supposing that there is an interaction between two substances, one inhibitory and the other stimulative, having different time constants.

Ranke[41] considers the receptor element-afferent fibres unit to be a "feeler" for internal regulations and a "releaser" of acquired and innate forms of behaviour. It is, he thinks, an organ of command, the properties of which can be determined quantitatively by means of the characteristics of its function of transition. For hearing and the vibratory sense this simple plan is complicated by the fact that not only does the frequency of the "feeler's" discharges depend upon the intensity and duration of the stimulus, but each individual discharge is in a phase relation with the particular period of stimulation by alternate pressures. The hypothesis therefore also involves the study of synchronisation.

The variation of the discharges with time can be accounted for by a difference in the concentration of ions in the two substances—inhibitory and stimulative—and that at once suggests the presence of a membrane which polarizes and depolarizes. This would confer plausibility upon the high value of the time constant in this phenomenon[24].

It has been possible to define the conditions of functioning by stating algebraically that the forces of diffusion, which alone could prevent the polarisation of the membrane, are proportional to the derivative of the difference between the concentrations. Hence it is not the variation in the concentration of the ions—*i.e.*, their speed of formation and displacement—but the polarising tension in the vicinity of the membrane which is the determining factor in the laws relating to stimulation. Thus the counter-polarisation, which may be regarded as the cause of the accommodation, subtracted from the polarizing tension, tends exponentially towards a limit value.

Bekesy's hypothesis[3], *viz.*, that it is not the differences in alternate potentials, but the continuous variations in tension of the cells of Corti's organ that determine the variation of the time of adaptation, corroborates the preceding assumptions.

2. The Development of the Stato-acoustic Organ

If we study the first stages of the development of the inner ear, we cannot but be struck by a remarkable constancy in the evolution of the stato-acoustic organ in the Vertebrata.

It is at the level of the first branchial cleft that a deep invagination appears, destined to become the birthplace of the auditory placode, the first element in the development of the stato-acoustic organ. The dorsal region of that invagination pushes out a projection, the future endolymphatic canal, while the base of the internal wall proliferates into a ganglionic formation. Soon the otocyst thus constituted will be constricted in its middle and divided into a dorsal vestibular section and a ventral auditory one. The dorsal section will give rise to the utricle and semi-circular canals; the ventral part will produce the sacculus, the lagena and the cochlear canal (Fig. 327).

(a) General aspect of sensory structures

The sacculus, the lagena and the cochlea are among the organs more especially to be involved in the detection of sound stimuli and their transformation into an auditory sensation. The sensory structures are the macula, the cristae and the papillae. They all have an epithelium and cells supporting the intercalated sensory cells. At their free end, these carry one cilium or several cilia embedded in an anhistic, gelatinous, transparent substance, the density of which is very much the same as that of the fluid in which they are immersed.

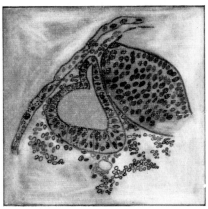

Fig. 327. Evolution of the stato-acoustic organ. Embryonic stage. (After H. W. Kaan, 1927, J. experim. Zool., 46, 13–63.)

(i) The maculae

The shapes and dimensions of the maculae vary. On account of its static role, only the saccular macula is invariably more or less vertical, placed on the internal surface of that organ. The macula of the lagena is located on the base of the cochlear canal in Reptiles and Birds. Among Mammalia, only the Monotremata have one; in all other Vertebrate the position of this macula is variable.

(ii) The papillae

They are always to be found in a thin part of the labyrinthine membrane and in the unpartitioned part of the perilymphatic space. E.g., the *papilla amphiborium* is situated near the orifice of the endolymphatic duct of the sacculus, at the base of a cul-de-sac, at a place where the wall is becoming progressively thinner.

The basilar papilla also occurs in Amphibia, in a recess adjacent to the lagena. It is covered by a membrane inserted here and there on the supporting cells which, according to McRoberts[44], are secreted by the internal epithelium.

In reptiles and birds it is highly developed and overlays the inner surface of the cochlea.

(iii) The auditory organ of Agnathes

In Petromyzon, the otocyst, where the only two semi-circular canals terminate is divided into two pockets, the anterior one assumed to be the sacculus and the posterior the *lagena*. In it are to be distinguished a saccular macula, a lagenar macula and another sensory formation said to be the *macula neglecta*. In addition, the epithelium of the inner wall is beset with vibratile cilia which, paradoxically, induce a rotary movement of the endolymph.

The ear is rudimentary in Myxines; it is a sac shaped like a disc, where the one and only semi-circular canal ends. The interior wall is covered with epithelium on which several sensory spots are to be seen, one being prolonged ventrally into the posterior section; this is said to be the lagenar macula.

In Elasmobranchii the organ is in direct communication with the outside world through the unclosed orifice of the otocyst. The similarity of the responses of sensory epithelium, whether of the otocyst or of the placodes of the lateral line, has emerged from a neuro-physiological study of ciliated organs.

In Teleostei, the presence of a communication between the right-hand and left-hand sacculi is an exception disposition which must be compared with the Weberian apparatus, the description of which will require an analysis (see p. 17).

Thus the auditory organ of Amphibia presents a very clear differentiation between the sacculus and the utricle, but, whereas the latter is exceedingly slender with its three well-differentiated semi-circular canals, the sacculus presents a wide pocket with two extensions, the basilar recess and the papilla amphiborium. There are also numerous anastomoses between the endolymphatic sacs.

In reptiles, birds and mammals the development of the perilymphatic and endolymphatic spaces begins to dominate the evolution of the various morphological stages which lead to the complex system in the higher vertebrates.

(b) *Evolution of the otic capsule*

If we consider the primitive formation of the otic capsule and the task devolved upon it of detecting sound-waves spatio-temporally, it will be evident that this membranous sac, which will be filled up with fluid, must be enclosed in a rigid envelope facing the outside only by one window, though the membrane of the sac must have scope to expand across another window to allow liquid displacement effected by acoustic waves; this is the secondary tympanum.

At the time of its invagination, the otocyst imprisoned within its wall a lamella of mesenchyma, thus making a chamber formed of wide-meshed tissue filled with perilymph. The exterior surface of that wall tends to ossify in contact with structures

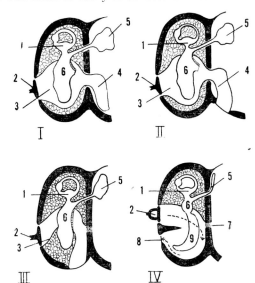

Fig. 328. Evolution of the otic capsule. Insertion of the round window. (1) Utricle, (2) columella, (3) perilymph, (4) diverticulum, (5) endolymphatic sac, (6) saccule, (7) perilymphatic aqueduct, (8) round window, (9) cochlear duct. (After H. M. de Burlet, modified by Cordier and Dalcq, in Grassé[19a].)

of nervous origin, encasing the nerves and vessels of the ear; the interior surface is overlaid with an epithelial and sensory membrane. Outside, its ossification continues until it comes into contact with the hyoid arch, from which the parts of the trans-

References p. 554

mitting apparatus will be formed. Inside and ventrally, it is obstructed by an evagination of perilymphatic tissue, *i.e.*, the perilymphatic sac, which is to become the expansion valve for the undulatory movements of the fluid; it forms the secondary tympanum (Fig. 328).

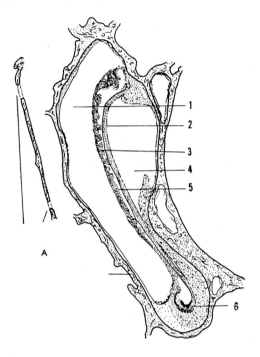

Fig. 329. Cochlear region of the pigeon. Location and size of the oval window which shown in another section came in relation to the figure A. (1) Vestibular scala, (2) basilar membrane, (3) vascular tegmen, (4) tympanic scala, (5) ductus cochlearis, (6) lagena. (After Schwartzkopff.[47])

In Amphibia, this perilymphatic evagination having passed the ossified shell through an internal orifice (future perilymphatic duct), extends to the external surface of the hind-brain. In Anura it curves in, emerging in the buccal cavity and, in Lacertilia, in the roof of the Eustachian tube; while in birds, owing to a coiling tendency of the cochlea, it comes close to the external wall and establishes itself in the vicinity of the primitive orifice of the otocyst, now become the *fenestra ovalis*. In this movement, the cochlear evagination stretches and the sensory epithelium of the basilar papilla extends on its inner surface, resting on an attenuated zone of the canal which becomes the basilar membrane. It is at the end of the tube that the *lagena* is found. The basilar papilla then assumes the name of Corti's organ, despite the fact that the arcade described by Corti as a typical morphological aspect of the sensory structures of the inner ear is lacking. It is replaced by an epithelial pad with alternate ciliated and supporting cells. The roof of the canal is covered with an epithelium rich in unicellular glands and small blood-vessels. This is the *vasculosum* (Fig. 329).

Outside this formation, the inner ear of birds is more like that of fish in the primitive stage than the highly differentiated structures of the mammalian organ of perception. In the latter, in fact, the cochlear duct has coiled into a spiral and Corti's organ has followed that trend, the cochlea now being situated in front and outside. It is a hollowed coiled tube 35 mm long with a bony axis called the *modiolus*.

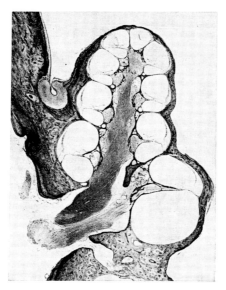

Fig. 330. Median section of the cochlea of a Guinea pig, showing the auditory nerve. The neural cells of the spiral ganglion and Corti's organ are degenerate. (After Davis[8], 1956.)

Its inner part is divided by a bony partition into compartments, this partition being an expansion of the modiolus, the *lamina spiralis*. This lamina does not reach the opposite wall and is connected to it by a fibrous expansion, the basilar membrane which fans out widely on the external wall to form the lamina of the contours. This partition, which divides the tube into two scalae, stops short of the apex, thus leaving a communicating aperture called the helicotrema. The width of its bony portion decreases from base to summit in line with the growth of its fibrous part (Fig. 330).

The upper—or more correctly anterior—scala, which corresponds to the *fenestra ovalis* and to the chain of transmission, is a diverticulum of the sacculus, hence its name, scala vestibuli. Through the helicotrema it communicates with the other scala, the scala tympani, which ends in the *fenestra rotunda*. Moreover, the scala vestibuli is itself separated into two parts by Reissner's membrane. The latter stretches from the insertion of the lamina spiralis on the modiolus to the lamina of the contours on which it bounds, with the basilar membrane, a cellular thickness known as the vascular stria (Fig. 331).

The duct of the cochlea thus formed (Scala Media) contains Corti's organ, which rests on the basilar membrane and the lamina spiralis. It has an arcade formed by two rods bounding the tunnel of Corti. On either side of this arcade Deiters' cells are juxtaposed, in three rows on the peripheral slope supporting the three external ciliated cells and in one row on the central slope supporting the internal ciliated cell. On each side of these formations there are the outer and inner grooves, the beds of which are beset with Claudius' cells. Further, these are separated from Deiters' cells by Hensen's cells.

The cochlear duct runs in the modiolus and provides ramifications to each row of ciliated cells, these ramifications passing through the thickness of the lamina spiralis before dividing into internal and external fibres. Between this stato-acoustic organ

References p. 554

—which, being deeply inserted in the skull, has become the inner ear—and the exterior environment there must be, anatomically, an organ of reception and one of transmission; these will be the external ear and the middle ear. *Physiologically*,

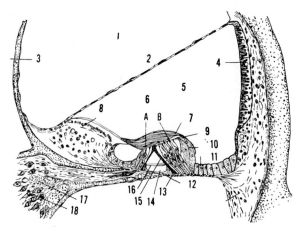

Fig. 331. Section of the cochlear duct of a mammal.

1. Vestibular scala
2. Reissner's membrane
3. Modiolus
4. Vascular stria
5. Cochlear duct
6A. Inner haircell
6B. Outer haircell
7. Tectorial membrane
8. Spiral limbus
9. Nuell's space
10. Hensen's cells
11. Claudius' cells
12. Basilar membrane
13. Deiter's cells
14. Tunnel of Corti
15. External pillar
16. Fibres of VIII.nerve
17. Spiral ganglion
18. Cells of the spiral ganglion

however, while perception includes only the function of the cell and of the nerve pathways, transmission comprises the mechanisms of all the structures, both solid and fluid, which help to propagate sound pressure to the sensory cell. A study of it therefore involves the functioning, not only of the middle ear, but of most of the inner ear as well.

(c) *The organ of reception*

This organ transmits sound by air conduction. It takes the form of a concha with the task of canalising sound-waves in the auditory passage; which is why it is found in birds and mammals.

In birds this concha of the ear is hidden in the head plumage. It is, in effect, a fold of the external auditory meatus, its size being adapted to its function. It is not developed at all in nocturnal birds. Its functions are to capture noises and to protect a *very* taut tympanum. Here the feathers are provided with looser barbs. Very often it is shaped like an ear trumpet open in front. Pigeon, duck, moor-hen and diving birds have no concha. The relationship between the tympanic surface and aperture of the concha has enabled us to evaluate the coefficient of sound "capture" by the external ear.

The vast majority of Birds have a muscular sphincter around the easily deformable auditory meatus which enables them to close or retract it at will.

In Mammals there is hypertrophy of the concha, which enjoys some degree of autonomy by virtue of its ability or tend towards the source of sound independently of movements of the head. These features are particularly striking in Chiroptera; the external ear of the bat group is of enormous proportions and has, moreover, an anterior eminence, the tragus*. The advantage to animals of finding their way in the dark by means of the reflection of the sound signals they send out from obstacles accounts for the very strong development of this receiving apparatus. Echo-location has begotten much research work. Perception of the return wave, or of the diminished intensity of the second sound, combined with the integration of the lapse of time between emission and reception, enables us to gauge the distance of the obstacle from the transmitter. If the sounds emitted by an animal sweep across a given surface, it will be able not only to gauge the distance of the obstacle but also, probably, its conformation within the sector thus swept.

Binaural hearing is used for finding the direction of the obstacle; this is determined by the difference in time and intensity and also in phase between the two ears. Some rats are able to make use of a similar process for finding their way in the dark.

A concha presenting a wide, mobile surface is, in fact, necessary to terrestrial animals, whose eye sight, moreover, is sometimes limited; hearing helps vision in warning them of danger. Where does the noise come from and what kind of noise is it? Binaural hearing locates it first, then mono-auricular hearing identifies it. A grazing horse can mobilise its auricles continually in every direction without ceasing to browse for an instant. This is because it possesses ten mobilising muscles against Man's three; in him the movements of the head and neck compensate for this lack.

Preyer's reflex

Dogs and guinea-pigs will jump and move their auricles in response to short sounds, while rats, mice and other rodents will also twitch their whiskers. This reflex, attributed to Preyer, is a part of phonokinesis. It is a non-orientated motor reflex provoked by an acoustic stimulus without any value as a signal in term of behaviour.

(d) *The transmitting apparatus*

(a) *Transmission in the middle ear*

It will be convenient to study the transforming process of the structures of the vertebrate middle ear as from the branchial arches to which they belong. Whereas each branchial arch is composed of four elements (the pharyngeal, epibranchial, ceratobranchial and hypobranchial), the mandibular arch and hyoid arch, placed one behind the other in front of the first branchial cleft, have only two (Fig. 332).

The cartilages of the mandibular arch are called the palatal quadrate and Meckel' cartilage. The ossified posterior part forms separate bones, the quadrate and the articular and, while the quadrate becomes the *incus* of mammals, the articular ossicle will form the *malleus*. The hyoid arch forms a dorsal, hyomandibular segment which attaches the lower jaw to the skull, providing the columella of the Reptilia and the stapes of Mammalia, while the ventral segment will remain isolated and later form the hyoid bone.

---

* See F. Vincent, Chapter 8.

These transformations can be verified by the connections of the branchial nerves with the branchiomere. The maxillo-mandibular nerve acts towards the mouth as though it were a branchial pouch and the facial spreads around the hyomandibular cleft. Hence its ramification, the tympanic cord, has the same relation to the quadrate and articular as it will have with the malleus and incus. Similarly, the nerve ramification of the tympanic tensor, a branch of the trigeminal, will have its replica in the innervation branch of the pterygoid which, in Reptilia, is inserted in the articular.

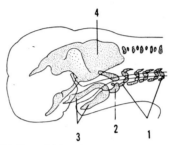

Fig. 332. Branchial cartilages. (1) Branchial arches, (2) hyomandibular, (3) hyoid arch, (4) mandibular arch. (After P. M. S. Watson in Grassé[19a].)

The relations of the vessels are even more suggestive. In its primitive disposition, the ventral aorta gives off an aortic arch at the level of each branchial arch; all the aortic arches collect in two dorsal aortae which unite posteriorly in a single trunk running along the rachis. In the course of evolution, the anterior arches have undergone radical changes. The first aortic arch, corresponding to the hyomandibular cleft, became atrophic from loss of respiratory function of that cleft, but the dorsal extremity of the second aortic arch gives off an artery which descends obliquely and anteriorly. In quadrupeds it enters into intimate relationship with the organs of the ear-drum and it is thus that it can sometimes perforate the stapes or columella after having enveloped the basal helix of the cochlea (Fig. 333), the stylostapedian artery.

This branchial evolution has given rise to some curious provisions in relation to the environment within which the organism moves.

Fig. 333. *Pachyuromis duprasi* Stapedial artery of the pericapsular arterial arch on the labyrinth wall of the tympanic cavity (original).

1. Aquatic Vertebrates

(a) *The bronchial columella.* As Witschi[58] has pointed out, a stem passing right through the dorsal aorta, thus connecting the fenestra rotunda with the bronchi, is commonly found in the larvae of Ranidae. It may be assumed that this arrangement serves the purpose of transmitting changes in pressure produced by acoustic waves at the level of the pulmonary sacs to the ear. This is the embryonic stage of the adult columella, but it is still called the bronchial columella. Foetally, therefore, it apparently forms part of the pulmonary apparatus (Fig. 334).

Fig. 334. Bronchial columella of the larva of *Rana sylvatica*. Cross section showing the columella connecting the bronchus to the round window crossing the dorsal aorta. (After Witschi.[58])

In the larval stage, the branchial sac communicates through the branchial clefts with the pharyngeal vestibule attached to the chondrocranium by thick, fibrous, supporting tissue.

The inner lamina of the otocyst, on the other hand, ends at the level of the orifice in a fibrous fascicular membrane.

At the level of the seventh branchial arch, this formation is produced up to the envelope of the branchial sac and the levator of the seventh arch thus becomes the homologue of the mammalian stapedius, seeing that it mobilises the columella when imposing traction upon the bronchial membrane.

When the branchial sacs have closed, the vessels of the aortic arches twist round the margin of the fenestra to reach the descending aorta.

These vessels are pushed back more and more upon the median line and then apply themselves directly to the membrane of the fenestra. "Meeting with the columellae, their walls first become indented, wrap themselves around the obstacles and finally completely enclose the upper part of each rod. A thin septum, which for a while still connects with the inner walls of the aortae, soon disintegrates and the columellae become fully engulfed by the aortic blood current." (Witschi[58]).

In the adult stage the organism adapts itself to the new amphibious function. The lungs and bronchi separate from the cephalic extremity which maintains the stato-acoustic organ; the columella carries a tubular bronchial process which is now united to the buccal pharynx; this is the Eustachian tube. The aorta is involved in this movement and the columella is freed from its vascular inclusion by a process in the reverse direction.

The inference to be drawn from this Odyssey is that the progression from the larval aquatic stage to the adult amphibious stage has exteriorized the tympanic membrane. The external part of the transmitting apparatus—here the peripheral extremity of the columella—found, at the level of the pulmonary sacs, the air cushion necessary to transmission which aquatic life otherwise denied to it. Once in the

References p. 554

open air, the respiratory portion migrates with the columella to the skin at the level of the pharynx.

A study of the auditory organ of certain Cetacea bears out this hypothesis. After studying the ear of the whale, Reysenbach de Haan[43] conjectured that in Devonian times a section of the submarine fauna was forced by a geological cataclysm to leave its aquatic environment. Some more highly developed fish were able to adapt themselves to this new amphibious life; functional necessity led to modification of the sense organs. The otic capsule encased in bone was virtually impermeable to this new environment, air. A bridge of bone (the columella) is formed from a portion of the hyoid arch (the hyomandibular), one end eroding the capsule to the point of creating the fenestra ovalis, the other end reaching out to the quadrate, thus forming a rudimentary transmitting apparatus. Later, an extra-columellar cartilage will develop on the side of the bone of the columella; it will be in contact with a section of the tegmentum which, at that place, forms the tympanic membrane.

For some reason unknown, certain animals which had already reached that stage evolved from air hearing returned to the aquatic life. The best known were the Cetacea, which were certainly predisposed to resume that kind of life.

Hanke[21] has shown that the auditory apparatus of the hippopotamous, the mammal most recently returning to an aquatic life, is already undergoing very marked changes. The aperture of the auditory meatus has been displaced upwards and distinctly reduced in volume; it can be closed by the active play of the muscles, while in seals this obturation has become passive.

In whales the auditory passage has atrophied so much that its meatus, which is difficult to find, sometimes has a diameter no more than 0.5 mm. From birth, the passage itself remains blocked by a large epidermal plug, but the umbilicated tympanum, inverted and facing outwards, is clearly recognisable (Fig. 335). The ossicular articulations are more or less welded together. This transmitting apparatus is attached to a large bone, the petromastoid, containing a highly developed cochlea. This bone is placed in an air pocket and is attached by a strong ligament only at the

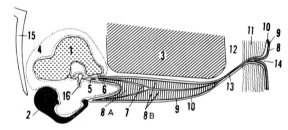

Fig. 335. Schematic frontal section of the tympanic region of a whale, modified according Hanke and Purves. (After Reysenbach de Haan.[43])

1. Petromastoid
2. Tympanicum
3. Squamosum
4. Tympanic cavity
5. Tympanic cone
6. Tympanic membrane protuding in the external auditory meatus
7. Section of ceruminal mass
8A and 8B. Ceruminal mass
9. Germinative stratum
10. Chorion
11. Fat
12. Connective tissue
13. True orifice of the external canal
14. External auditory meatus
15. Os occipitale
16. Ossicles (incus, malleus, stapes)

base of the skull, from which it is suspended. This curious arrangement is said to enable the auditory organ to collect ultrasonic signals (Fig. 336).

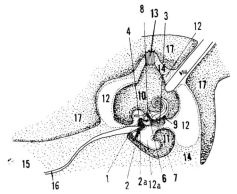

Fig. 336. Schematic section (dorso-ventral) of the external hearing apparatus of Odontoceti. (After Reysenbach de Haan.[43])

1. Tympanic cone
2. Malleus
2A. Processus gracilis
3. Tensor tympani
4. Incus
5. Stapes in the oval window
6. Perilymphatic space of the labyrinth
7. Endolymphatic space in the cochlear duct
8. Lateral semicircular canal
9. Round window, closed by its membrane
10. Petromastoid or perioticum
11. Tympanicum
12. Air pocket around the perioticum and
12A. In the tympanic cavity
13. Ligamentum suspensorium perioticum
14. Connective tissue
15. Fat
16. External auditory meatus
17. Cranial capsule
VII. Facial nerve
VIII Auditory nerve

(b) *Weber's organ.* This phenomenon should be seen in the light of the existence of the Weberian apparatus in Teleostomi, *i.e.*, the adaptation of the air-bladder to the inner ear.

In its simplest arrangement, the anterior portion of the two diverticula of the air-bladder is closely united to a fibrous membrane covering the fontanelle of the posterior wall of each otic capsule. The inner surface of this fibrous membrane, which is situated opposite the utriculus, bathes in the perilymph. Harden Jones and Marshall[22]. Wohlfahrt[59] adds that the perilymphatic duct in some species may be involved in this connection.

According to Weber's description, the right and left otocysts of the sacculus are attached to the transverse duct, the middle of which gives off a diverticulum, the endolymphatic sinus, which projects backwards into a central cavity filled with fluid, *viz.*, the impair sinus. From each side a chain of ossicles extends from the posterior section and from the lateral wall of this sinus to the air-bladder. A wide posterior ossicle is described, the tripus, which pivots around the centre of the third vertebra, attached at the back to the air-bladder and connected in front to the incus, or *intercalarium*, by a ligament. In its turn, the incus is in communication with a third ossicle, the stirrup or scaphium. The latter surrounds the ending of each lateral bifurcation of the impair sinus, completed by an inner bone, the *claustrum* (Fig. 337).

Although no electrophysiological experiment has hitherto succeeded in proving that the vibrations of the walls of the air-bladder generate potentials of action of the eighth pair, it is nevertheless true that fish possessing a Weberian apparatus respond to higher frequencies and lower intensities than fish not so equipped.

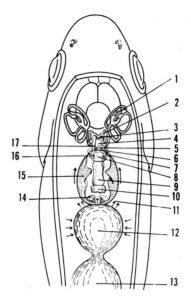

Fig. 337. Hearing apparatus of a fish with the system of Weber's ossicles (Ostariophysa). Animal opened from the dorsum. (After von Frisch[17], modified.)

1. Utricle
2. Saccule
3. Transversal canal
4. Sinus impar
5. Ligament of the stapes
6. Claustrum
7. Incus, or intercalarium

8, 9, 10. Vertebrae
11. Posterior ligament of the malleus
12, 13. Swimbladder
14. Suspensorium
15. Malleus
16. Ligament of the incus
17. Stapes, or scaphum.

Experiments performed by Von Frisch[17] tend to show that the Weberian ossicles have an auditory function and Evans[14] suggests that they may help to locate sound by virtue of the inter-communication between the endolymphatic fluids. The pressures of the sound-waves would come to bear unsymmetrically upon the air-bladder if the source of sound should be sideways relative to the axis of the fish.

Hence, in the case of aquatic Vertebrates, mechanical stimulation of the organ of transmission operates from an air pressure itself modified by the displacement of the organism in water.

Furthermore, discrimination between the sense of orientation and the sense of hearing resides in the nature of the process of excitation, a contingent function on the one hand and a sinusoidal one on the other.

What event will now mark the evolution of the auditory organ of an aquatic Vertebrate on dry land? The animal's new element failing to produce the necessary pressures for the stimulation of the sensory organs of hearing already in the cranial region, it is the lower jaw which, for a time, will now become the agent of sound transmission.

2. Reptiles

The modifications which have taken place in the acoustic apparatus of Vertebrates constitute a complex process which appears to be definitely associated with the animals' mode of life; *i.e.*, on the one hand with their environment and, on the other, with the capture of the prey they need for subsistence. Hence it is not surprising to find that the middle ear in the aquatic species remains in intimate contact with the respiratory tree or the air-bladder. Nor have we any greater reason to be amazed by the strong attachment of the transmitting apparatus to the masticatory organs in species living on land.

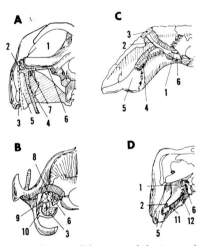

Fig. 338. Evolution of the mammalian ear. Diagrams of the transmission apparatus of 4 animals. (A) Reptile, avian type, (B) Young marsupial mammal, (C) Labyrinthodont mammal, (D) Mammal-type reptile. (After Watson.[54])

1. Dorsal process of the stapes
2. Tympanic process of the stapes
3. Tympanic membrane
4. Hyoid process of the stapes
5. Quadratum
6. Stapes
7. Internal process of the stapes
8. Incus
9. Hammer
10. Tympanic bone
11, 12. Ligament representing the hyoid process of the stapes of Labyrinthodont

May we not conclude that, as between Fishes and Reptiles, the difference in the pathways of the sound-waves reaching the otic capsule derives mainly from the way in which the acoustic pressure are propagated, having impinged upon the nearest sensory organs to the receptors. If the best way for sonic vibrations to travel towards the inner ear of the larvae of Batrachia is by the pulmonary sacs, or the air-bladder in Teleostei, it is highly probable that the ventral part of the head in Reptilia will be the organ of transmission best placed for the detection of vibrations emanating from the ground. It may possibly account for the deficiency of the tympanic cavity in Serpents, where the internal extremity of the columella is inserted in the quadrate whereas in Sauropsida, especially Crocodilia, the columella is upon a true tympanic membrane.

This arrangement recurs in Birds and is a sign that the mandible has withdrawn from the ground. There will be another transformation when the jaw, suspended

References p. 554

from the base of the skull, will need a more developed masticatory musculature to become mobile.

Watson[54] believes that the characteristic modifications of the apparatus for acoustic transmission in the mammalian ear are the unpredictable result of numerous changes in the ancestral cephalic structures, which appear, at least partly, to depend upon a reduction of the organs situated under the base of the skull, especially those representing an advance in the masticatory mechanism (Fig. 338).

In point of fact, the muscles attached to the skeleton of the mandibular arch formed by Meckel's cartilage and the palatoquadrate are the masticators. The hyoid arch, together with Reichert's cartilage, provides the suspensory system of the jaw.

While, according to the theory propounded by Reichert-Gaupp, the columella or stapes represents a vestigial hyoid arch, part of which has migrated to the cervical region to give insertion to the buccal muscles, the articulation of the jaw is formed by the articular bone, the posterior section of Meckel's cartilage and of the palatoquadrate. But in Mammals, the development of the masticatory muscles profoundly modifies the mandible which, upon the advent of the membrane bones (dentary, scale of the temporal), found appropriate materials with which to forge a new, more powerful articulation. This led to atrophy of the articular and palatoquadrate, which were destined to become the malleus and incus respectively, while their muscular insertions would account for the masticatory relations of the muscles of the ossicles.

3. Birds

Similarly, Denker[11] has shown that the ears of Birds were almost inaccessible to bone conduction. Admittedly, the pneumaticity of the cranium stands in the way of this form of propagation, while at most allowing for a reserve of air necessary for balancing in the drum of the ear those variations in pressure which are due to differences in altitude. It keeps the two middle ears in communication but, as Schwartzkopff[47] has proved, only slow variations in pressure can be established between one ear and the other. Maybe this mechanism enables birds also to perceive air currents. However that may be, it could not intervene in the process of sound location.

The amplifying power of the transmitting apparatus is important because the tympanum is strongly developed compared with the size of the base of the columella. This indicates exceptional acuity of hearing; nocturnal birds have it in the highest degree (Owls), their fenestra rotunda being approximately five times as large as the fenestra ovalis. This disproportion improves the mobility of the columellar basal plate; in Parrots, at all events, the fenestra rotunda is almost obturated by a fibrous plate[11].

Birds' hearing is attuned to higher pitches than those coming within Man's auditory range. Chicks, however, react very well indeed to the low cackling of their dam (lower than 400 c/s), while the clucking hen responds with the utmost sensitivity to its offsprings' shrillest cries of anguish (above 3000 c/s)[7]. Bird song can be accounted for by a fine capacity for differentiation. As a deaf bird continues to sing none the less, it is evidently not master of its own voice, during the first few months at any rate.

It is to be inferred from the small dimensions of the head that the difference in phase could not play an appreciable part in locating here. The bird must fortify it by mobilising the auricular folds, thus modifying the intensity and directional character of the sound-waves.

The columella, moreover, acts as an eccentric lever as well as a piston. The former produces a torsion couple which follows the rapidity of the sonic vibrations, reducing the pressure on the annular ligament to the same extent. This is made possible by the equilibration of the masse. Indeed, the inner surface of the columella's extremity is prolonged by a protuberance washed by the perilymph of the inner ear and is well designed to prevent the formation of eddies (Fig. 339).

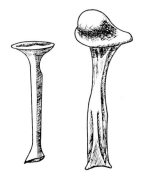

Fig. 339. Columella of *Tito alba*, and *Pyrrhula pyrrhula*. (After Schwartzkopff.[47])

The ear reaches its highest degree of perfection in Mammals. For this reason the anatomical description of the inner ear was given first to typify the most complete development of the stato-acoustic organ. The middle ear is similarly organised to adapt the message to a very subtle receptive function.

The tympanic membrane which obturates the external auditory passage is, roughly speaking, like a loud-speaker cone but, as it functions, it pivots about a

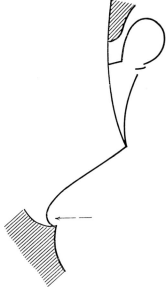

Fig. 340. Mobile fold of the lower part of the tympanic cavity permitting movements as a rigid cone. (After von Bekesy[3-4].)

References p. 554

hinge formed by its upper margin, in that attenuated portion which is called flaccid and which is subtended by two malleate tympanal ligaments.

It is the lower portion of the membrane, therefore, which has the maximum vibratory amplitude and here the pathway is limited only by the degree of elasticity of the fold which connects the membrane below to its bony insertion at the periphery of the *tympanic sulcus* (Fig. 340). This functional asymmetry, which is found throughout the chain of transmission, might well seem paradoxical to an acoustician if unaware of the ear's ability to produce combined sounds, the result of harmonic distortions. Tympanic vibrations have little amplitude, especially in sounds of high pitch and little intensity. This is why an amplifying system is necessary; it consists of the chain of ossicles, *i.e.*, malleus, incus and stapes. The purpose of this apparatus becomes clear when one considers the internal ear, enclosing the delicate structures of the sensory cells deeply inserted in the most compact bone of the skull, the otic bone, capable of responding equally to pressures as different as those which separate the threshold of auditory perception from that of pain, the variation of energy being in the proportion of 1 to 10^{12}. The apparatus must, therefore, amplify in order to make the inner ear sensitive to the most tenuous sounds; and it must attenuate to protect it from injury caused by very loud sounds. Thanks to the articulation of the ossicles, it is one and the same system which plays the two parts of amplifying and attenuating (Fig. 341).

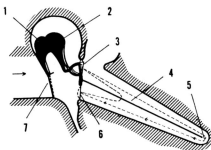

Fig. 341. Diagram of the vibratory movements of the tympanic membrane, of the ossicles and of the internal ear. Cochlea unrolled. (1) Hammer, (2) incus, (3) stapes, (4) vestibular scala, (5) helicotrema, (6) round window, (7) tympanic membrane. (After von Bekesy[3].)

In point of fact, the whole chain vibrates in the presence of noises of low intensity, producing a columellar effect. It is a mass disturbance coming nearer and nearer across a chain made rigid by the blockage of the articular capsules, the muscles of the stapes and of the malleus being relaxed. But, as soon as the amplitude of the vibrations exceeds the limits of laxity of the annular ligament which attaches the last link in the chain (the stapes) to the fenestra ovalis, the movements become fewer as the ossicular muscles begin to contract, stiffening the two extremities of the chain. The articulations absorb some of the energy which will be proportionally distributed among the various joints. The two muscles of the stapes and malleus are in fact synergic and antagonist, their rapid contraction standing in the way of transmission when this threatens to become injurious to Corti's organ. Comparative anatomy provides a weighty argument in favour of this synergy since in Seals the muscle fibres of the malleus are detached and insert themselves right into the head itself of the stapes[1].

Thus, the chain of transmission is able to pivot about two axes, an axis of rotation which allows the ossicles to be mobilised in one piece and which passes through the root of the long apophysis of the malleus and the horizontal branch of the incus; and an axis perpendicular to the other, passing through the anterior process of the malleus and the lenticular apophysis of the stapes; it is around this that the simplifying movements take place which absorb the energy of loud noises (Fig. 342).

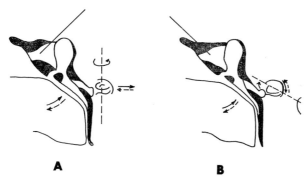

Fig. 342. Schema of the movements of the stapes: (A) at moderate intensity, (B) at high intensity (After von Bekesy[3]).

The articulation of incus and malleus is, however, less firmly fixed in some Rodents, in Monotremata and Cetacea. Furthermore, the malleus has a new apophysis, the *Folian process*, which is more or less concave. It offers an added point of support without impeding mobility, in the young animal at all events. This arrangement is said to favour the perception of shrill sounds, which is why it is found predominantly in Mammals having the most highly developed perception in the ultrasonic range (Bats, Whales)[43].

The amplification of sounds of low intensity is realised by the difference in diameter between the tympanum and fenestra ovalis. All the energy received by the tympanic membrane obtains in its entirety on the fenestra ovalis, because at that moment the chain of ossicles acts like a rigid columella.

A safety valve, the fenestra rotunda, makes the motion of the fluids of the inner ear possible and adequately proportional to the delicate structures of the sensory organ. Placed in a different plane from that of the fenestra ovalis, it vibrates in the opposite phase and the fluid waves coming from the stapes through the helicotrema reach it on its inner surface.

Bekesy's experiment, called the "nul method", has made the dephasage of the vibrations at the two fenestrae perceptible. A subject is taken, the transmitting apparatus of one of whose ears has been destroyed. Two sounds of very close frequency are transmitted to each ear through ear-phones. The subject then has the sensation as of the sound travelling in his head from one ear to the other; the insertion into the damaged ear of a plug of cotton-wool soaked in glycerine, creating as it does a columellar effect, causes an inversion in the direction of the displacement of the sound.

If the transmitting apparatus is to work steadily, the pressure on either side of the tympanum must be in equilibrium. In Batrachians, the canal which connects the pharynx with the otocyst maintains permanent communication. But the develop-

References p. 554

ment of the masticatory muscles in gnathostomous animals complicates the function by interposing muscular masses. Hence it is not surprising to find in the higher Vertebrates the belly of a muscle of the soft palate, the external peristaphyline, jutting out into the lumen of the Eustachian tube and intermittenly obturating it. In fact, through velar contraction, yawning and swallowing will circulate air in the drum by temporarily making the tubal canal permeable.

(ii) Transmission in the inner ear

The motion of the fluids of the inner ear is the sole stimulator of the sensory cell of Corti's organ. It may be brought about either by the transmitting apparatus just described, or by bone conduction.

The application of a vibrating system to any point of the skull produces an auditory sensation of the same frequency as that of the vibrator. Hence, just as a threshold of air conduction can be estimated, so can one evaluate a threshold of bone conduction.

If the auditory apparatus is defective, there may be a discrepancy between these two parallel thresholds, so that their divergence provides a means of locating the level of the lesion.

Many suggestions have been put forward to account for this phenomenon, but the following, advanced by Bekesy[3], is commonly accepted: When the vibrator emitting low frequencies is applied to any random region of the skull, it induces a synchronous disturbance in the whole brain-case. From 800 c/s the inertia of the opposite wall intervenes and the vibrations on either side are then in opposite phase. Above that frequency and within the whole high-pitch range, nodal lines are formed, the number and site of which vary with the frequency. They divide the brain-pan into sectors.

In its turn, the aggregate of ossicles, attached to the muscle and ligament walls, present sufficient inertia to these movements of low amplitude to annual the vibrations. Thus a movement of the stapes is realised relative to the fenestra ovalis. On account of its constitution, this fenestra does not exhibit the same elasticity towards alternate sound pressures as that of the fenestra rotunda, which appears to be far less rigid. The fluid of the scala vestibuli, therefore, is more involved than that of the tympanic scala and the basilar membrane naturally reflects this asymmetry; hence the possibility of auditory stimulation by bone conduction. More recently it was shown that phase was the main factor between the two ways of couduction.

It is again Bekesy whom we have to thank for a plausible explanation of the mechanism by which the fluids of the inner ear are set in motion. Using a chamber of plexiglass divided into two by a rubber partition, he studied the behaviour of eddies set up by a piston on one of the compartments. This *camera acustica* was intended to reproduce on a large scale the model of a cochlea, unrolled (Fig. 343). The liquid contents were carefully chosen to furnish a valid analogy with the viscosity of the perilymph. Metal filings in suspension showed the eddies induced by propagation and enabled low frequencies to be studied stroboscopically. With the reciprocating movements of the piston below 800 c/s, the whole of the membrane vibrates, exhibiting a specific deformation—"a pattern"—for each frequency. At frequencies above this level, the following is seen to occur: the outward displacement of the piston causes the membrane in the corresponding scala to rise to the maximum, while the

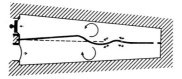

Fig. 343. Camera acustica, Bekesy's model. (After von Bekesy[3].)

reverse movement forces the membrane back. This undulation is propagated along the whole of its length, but abating very rapidly, and that portion of the membrane extending to the helicotrema remains immobile. The maximum point of amplitude corresponds to the formation of eddies in both compartments. These eddies are set up by a rotary movement in opposite directions and, according to Ranke[41], their angular velocity is proportional to the course of the piston in its fenestra (Fig. 344).

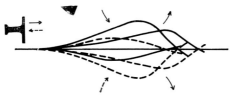

Fig. 344. Diagram of the disposition of the vibration of the basilar membrane during a complet cycle of a pure tone. (After von Bekesy[3].)

Above all, these eddies move about, following the frequency of the movements of the piston, *viz.*, towards the distal part of the membrane in response to low frequencies and towards its proximal section for high ones.

Satisfactory though this hypothesis of localisation be, there is one serious objection to it, *viz.*, the dimensions of the chamber with respect to the sound-waves stand in no relation to those of the cochlea. Ranke[41] pointed out in this regard that the length of a sound-wave of 20,000 c/s propagated in a fluid even exceeds the length of the basilar membrane. He added that no doubt the difference between the rigidity of the walls of the auditory canals and that of the chamber compensates sufficiently to make the comparison tenable. Actually, calculations show that, as soon as the wavelength is less than double the circumference of the scala, Bekesy's observation[3] is valid, as has been confirmed by Reboul[42] in a complete mathematical analysis of this phenomenon.

It now remains to reconcile this movement of the membrane with the requirements of a selective stimulation that must involve only a very few nerve cells simultaneously.

Two objections are at once obvious. One is that, if the membrane oscillates mechanically, the excitation of the cells must strike a by no means negligible field. Shower and Biddulph[46] have shown that there are 125 distinguishable gradations of frequency between 860 c/s and 1200 c/s and the result of calculations made from experimental data obtained upon the membrane, gives only two.

The other objection is that selectivity and damping are opponents. And, as the vibration of the membrane is damped, we must look elsewhere for the selectivity factor. Vibration is certainly effectuated as a whole. All that matters is the maximum point of amplitude.

References p.55

The basilar membrane, which Helmholtz, in his theory[23], wanted to make the resonator keyboard, cannot produce sufficient mobility to account for such delicate localisation. We shall have to look for other mechanisms. Gray's principle of maximum excitation[20], or Mach's law of contrasts[36], has been suggested, the sites having undergone less stimulation being apparently subject to all kinds of nerve inhibitions.

Two anatomical arguments that have been advanced, however, *viz.*, the presence of the *membrana tectoria* holding the cilia, which Bekesy[3] said was more rigid transversally than longitudinally; and the position of the internal and external ciliated cells inclined in opposite directions.

The authors are agreed that the whole of the basilar membrane is involved in stimulations by sounds of low pitch and that this widely curved deformation creates an image of excitation transmitted just as it is to the brain, which perceives a representation of it. The internal ciliated cells appear to be more implicated in this movement and their excitation is translated into a withdrawal of the cilia.

The problem becomes more complicated when stimulation is effected by high-pitched sounds above 800 or 1000 c/s. There is a partial and very localised rise—the site of which depending on the frequency—of the basilar membrane and the structures it supports; the higher the sound, the nearer is this rise to the base.

In Bekesy's view[3], this deformation which he studied in the fresh cadaver stroboscopically indicates that, near the site of maximum stimulation, a movement occurs of the ciliated cells under the membrana tectoria which, on the stapes side, takes place in a similar direction to that of the transverse fibres of the membrane, while towards the helicotrema this type of vibration ceases abruptly, when the ciliated cells take on a vertical movement. Further along, in the same direction, these vertical vibrations die out in longitudinal ones (Fig. 345).

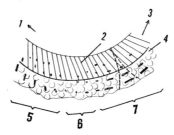

Fig. 345. Radial, vertical and longitudinal vibrations of the organ of Corti seen through Reissner's membrane. (1) In the direction of the stapes, (2) tectorial membrane ,(3) in the direction of the helicotrema, (4) Hensen's cells, (5) radial vibrations, (6) vertical vibrations, (7) longitudinal vibrations. (After von Bekesy[3].)

Proceeding from these minutiae, Huggins and Licklider[28] reason that the vertical movements of the basilar membrane involve to-and-fro movements of the membrana tectoria in view of the fact that its axis of rotation is above that of the basilar membrane. But, whereas the cilia of the external cells, being inclined, are shearing by this movement, the cilia of the internal cells, joined to the axial part of the membrane, are withdrawn and compressed (Fig. 346).

In the presence of low frequencies, the whole membrane vibrates in phase and all that matters is the degree of traction of the cilia, which, where the nervous system is concerned, thus interprets the sound image. From the moment the wavelength is registered within the length of the cochlea, the membrane shows the maximum stimulation. Thus this curvature implies a different up-and-down shearing movement of the cilia of the ciliated cells.

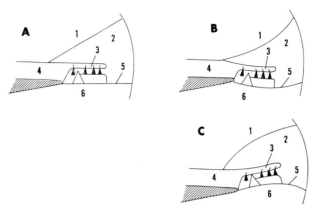

Fig. 346. Schema of the relative displacements of the basilar membrane and the elements supported by the same. A, B, C three phases of movement. (1) Vestibular scala, (2) cochlear duct, (3) tectorial membrane, (4) limbus, (5) basilar membrane, (6) tympanic scala. (After Huggins[27a].)

During the change in curvature of maximum stimulation, the cilia of the cells facing the stapes are stretched radially and those of the cells facing the helicotrema are stretched longitudinally, *i.e.*, in the direction in which the membrana tectoria is less rigid (Fig. 347). At their junction, therefore, the kind of stimulation of the cilia changes and the tractive force thus exercised will, of course, be proportional to the derivative of the amplitude of movement. The selectivity of the response is thus given scope for refinement. Moreover, by its filtering that the damping effect determines an angle of phase which gives additional precision.

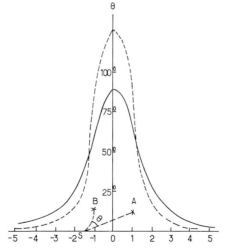

Fig. 347. Relation of the reaction of a superimposed wave S passing two filters A and B with intervention of the phase difference. (After Licklider.[33])

References p. 554

More recently, Bekesy[3] has been studying the vibratory sense of the skin of the fore-arm with the help of three different analogous models, in the light of the three principal theories of locating, resonance, telephone and propagating waves. The resonator and wave-propagating models provided the most exact locating. The most remarkable fact to emerge was the nervous inhibition which suppresses all sensations except that produced within the zone of maximum stimulation, as stated by Gray's principle[20] (Fig. 348).

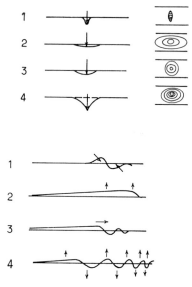

Fig. 348. Vibrations patterns in membranes for a continuous tone (with normal damping). Arrows indicating the direction of the movement at a given instant. According the theory. (1) of resonance, (2) telephone, (3) traveling waves, (4) standing waves. (After von Bekesy[3].)

Whereas physiologists think the subtle mechanism by which the mobilisation of the cochlear structures attains to so selective an excitation of the sensory cells is to be found in frequency perception, psychologists, by estimation of the complex stimuli, cast doubt upon the analytical capacity of the cochlea.

Fletcher[16], with the help of synthetics sounds, and Schouten[45] with periodical impulses of sound, have shown that it is possible to cut down certain frequencies of the stimulus without appreciably varying the sensation of pitch. Licklider[33] attributes this effect to relations of phase between the components. The facts are incompatible with any rigorous analytical distribution of tone by the cochlea and necessitate a reconsideration of the problem of discrimination.

Two recent theories seem to bear out the hypothesis of locating by maximum excitation, *viz.*, the volley theory propounded by Wever[56] and the duplex theory brought forward by Licklider[33]. It has been demonstrated that a group of nerve fibres is able to respond synchronously to a repeated stimulation following a far quicker cadence than a single fibre is able to do.

The principle of the volley theory is that the fibres in a group discharge in turn; hence the total response may be synchronous even if the individual fibres do not respond to each period of excitation (Fig. 349). The theory then recognises resonance as a means of distributing the stimulation on the basilar membrane following the frequency of the stimulus and at the same time invokes the volley principle to produce the diagram of excitation by nerve impulses. Wever[65] says that the frequency principle is practicable for low-pitched sounds and that the volley and locating principles come into play in the wide intermediate range. The volley effect does lead to a faithful representation of low and medium frequencies, but becomes inexact and inefficient for high frequencies; thus the spatial representation of locality has greater discriminative power in the higher and middle regions of the scale. The two variables, frequency and position, are partly complementary, therefore, and are mutually helpful in discriminating the level of sound.

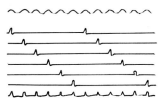

Fig. 349. Principle of the volley theory. Every fiber responds to one of the sound waves and their sum represents the total frequency of the stimulating wave. (After Wever.[55])

What, however, becomes of this representation when different synaptic transmissions conveying messages to the cortex slow down the rhythm of the influxes?

Licklider[33] considers the hypothesis of a nervous mechanism capable of converting the sound image into a representation of locality in the higher stages. He writes that the duplex theory of the perception of sound level depends ultimately upon the idea that the auditory system operates simultaneously by analysis into frequency and by analysis of autocorrelation. The analysis into frequencies is effected by the cochlea; that of autocorrelation by the nerve area of the auditory system. Therefore, this analysis is not that of the acoustic stimulus itself, but rather an analysis of nervous influxes resulting from the transformation of the stimulus by the cochlea (Fig. 350).

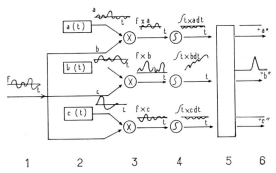

Fig. 350. Schema of the process of identification by correlation. (1) Function f(t) to be identified, (2) generator of patterns to compare a(t), b(t) and c(t), (3) multiplicator, (4) integrator, (5) comparator, (6) exit. (After Licklider[33].)

The stimulus, which is a function of time, is transformed into a space–time diagram as realised by a sonograph and this arrangement is transmitted thus by the neurons to the cortex. The author assumes that autocorrelation takes place in the course of this transmission. A chain of neurons certainly does provide an excellent delay line; the spatial aspect of a synaptic summation provides an approximate multiplication and, finally, the temporal aspect of that summation permits of integration.

Now, however, it is electrical phenomena, more even than mechanical movements, which appear to dominate the picture of nervous stimulation. Four kinds of phenomena have been identified in the cochlea. One is the potential of action, which is the expression of the nervous influx. The second is the potential of repose or of polarisation, positive in the endolymph and intracellularly negative. The other two are

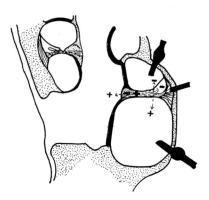
Fig. 351. The potentials of the cochlea of guinea-pig. (After Davis[10].)

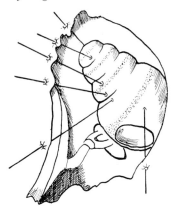
Fig. 352. Placement of the electrodes in the cochlea of guinea-pig. (After Davis[10].)

responses to acoustic stimuli, one, alternating, faithfully reproducing the wave-sound, which is the cochlear microphonic potential; the other is a positive or negative change in the value of the endolymphatic potential, which is the summating potential (H. Davis[8]; Fig. 351).

These potentials can be collected by electrodes placed in the various scalae (Fig. 352). The microphonic potential collected for the first time in 1932 by Wever and Bray[56] proves to be a considerable electrical phenomenon. It is proportional to the intensity of the stimulus, has neither a true threshold nor refractory period, depends on the integrity of the ciliated cells and changes phase at the same time as the stimulus. Though highly sensitive to anoxia, it can persist for four hours after death[15].

The intracellular and endolymphatic potentials are potentials of polarisation, negative for the former, positive for the latter. It may attain as much as 80 mV with reference to the perilymph. The source of the endocochlear potential has been found by Tasaki et al.[49] to reside at the level of the stria vasculoris. It is supposed that the endolymphatic potentials are dependent on the oxidation metabolism caused by the blood flow which would thus maintain a polarization, the importance of which is unique in electrophysiology since it contributes towards the extreme sensitiviness of the cochlea. This ion concentration would create in its vicinity a "battery effect" which could at any moment cause important electrical phenomena. The ciliated cell would include a receptor potential capable of creating a generator potential which would act as a starter for the action potential of the nerve.

The nerve's potential of action, which can be derived from the cochlea by eliminating the microphonic potential previously deducted in two scalae by opposition of phase, has the usual characteristics of nerve potentials.

Lastly, the summating potential represents the change in endolymphatic potential in response to a sound stimulus; so it is proportional to the square root of the acoustic pressure. It causes a unidirectional change of the base line of the microphonic potential, from which it apparently deviates with a delay of sometimes 1 msec. It can either increase or reduce the endolymphatic potential, for which reason it is qualified as a positive or negative summating potential (Fig. 353). The summating potential would rise from one basal turn alone. It would only appear for an intensity of 20 dB and would continue to rise above the microphonic, that is beyond 90 dB.

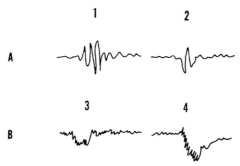

Fig. 353. Response to tone clicks in the basal turn of the guinea pig cochlea. (After Davis.[10]) A. Frequency: 2000 c/s; (1) response in the vestibular scala, (2) action potentials by annulment of the potentials of the vestibular and tympanic scalae. B. Frequency: 8000 c/s; (3) summation potential in the vestibular scala, (4) summation potential in the cohlear scala.

The foregoing study of the effect of traction upon the cilia of sensory cells makes it easier to understand the origin of these different potentials. We may assume with some confidence, explains Davis[8], that the curvature of the cilia of the ciliated cells must mobilise or modulate the biological source of energy.

Thus the endolymphatic potential could be identified with the cochlear microphonic potential of which it would be the source, at all events as far as the part of this potential which is sensitive to anoxia is concerned.

It would seem that the summating potential is likewise associated with both the endolymphatic and microphonic potentials, but its resistance to anoxia would bring it nearer to the fraction of the microphonic potential which persists after death. The summating and microphonic potentials are engendered by the traction of the cilia in their own direction; but, whereas the microphonic potential is produced by a symmetrical vibration, the summating potential is the effect of an asymmetrical displacement. According to Davis[8] it is the translation of the movement by which the propagating waves are travelling. Furthermore, the positive summating potential is, he says, brought about by the displacement of the tectorial membrane in a transverse direction. Bekesy[3] having observed that the external ciliated cells are mainly involved in transverse mobilisation, we may infer that anoxia increases the negativity of the summating potential, its positive fraction being more vulnerable than the corresponding fraction of the microphonic potential.

However that may be, the mechanical movements in different directions to which the cochlear membranes are subjected under the influence of propagating waves produced by sound vibration involve a ciliary traction which generates electric effects capable of triggering off volleys of influxes; also, however, of inducing inhibitory effects resulting in a more delicate analysis of the stimulus.

In the ultimate, the mechanistic arguments of Bekesy[3], Huggins[27] and Hilding[25] and the electrical data elaborated by Davis are only different ways of looking at the same phenomenon.

The volley theory had already shown that the synchronism of the influxes is the best way of translating frequency, but that its ingenuity failed to resolve the difficulties raised by psycho-physiologists.

The duplex theory passes the responsibility on to the nervous system and the coded representation in the potential of action is subjected to a succession of operations which, by autocorrelation, finally decode the message. This message is analysed peripherally and the nervous system transmits and interprets it in accordance with its own laws.

REFERENCES

[1] Ardouin, P., 1941, *L'Oreille Moyenne*, Masson et Cie., Paris.
[2] Autrum, H. J., 1952, Nerven und Sinnesphysiologie, *Fortschr. Zool.*, N.F., 9, 537–604.
[3] Bekesy, G. von, 1948, Elasticity of the cochlear partition, *J. Acoust. Soc. Am.*, 20, 227; 1949, Vibration of the cochlear partition in anatomical preparations and in models of the inner ear, *ibid.*, 21, 233; 1951, Microphonics produced by touching the cochlear partition with a vibrating electrode, *ibid.*, 23, 29; 1951, D.C. potentials and energy balance of the cochlear partition, *ibid.*, 23, 576; 1952, D.C. resting potentials inside the cochlear partition, *ibid.*, 24, 72; 1953, Shearing microphonics produced by vibrations near the inner and outer hair cells, *ibid.*, 25, 786; 1953, Description of some mechanical properties of the organ of Corti, *ibid.*, 25, 770; 1954, Some electromechanical properties of the organ of Corti, *Ann. Otol. Rhinol. Laryngol.*, 63, 448; 1955, Paradoxical direction of wave travel along the cochlear partition, *J. Acoust Soc. Am.*, 27, 137; 1956, Current status of theories of hearing, *Science*, 123, 779.
[4] Bekesy, G. von and W. A. Rosenblith, 1951, Mechanical properties of the ear, in *Handbook of Experimental Psychology*, Ed. S. S. Stevens, Wiley, New York.
[5] Borghesan, E., 1952, Tectorial membrane and organ of Corti considered as an unique anatomic and functional entity, *Acta Oto-Laryngol.*, 42, 473; 1954, Les récepteurs cochléaires et les théories modernes de l'audition, *J. franç. Oto-rhino-laryngol.*, 3, 213.
[6] Burlet, H. M. de, 1934, Vergleichende Anatomie des stato-akustischen Organs, *Handbuch vergl. Anat. Wirbelt.*, 2, 1293–1432.
[7] Collias, W. and M. Joos, 1953, The spectrographic analysis of sound signals of the domestic fowl, *Behaviour*, 5, 175.
[8] Davis, H., The four electric potentials of the cochlea, *Am. J. Physiol.*, in the press; 1953, The mechanism of hearing, in *Nerve Impulses* (Ed. D. Nachmansohn), Trans. Fourth Conf. Josiah Macy Jr Foundation, Caldwell (N.J.) Progr. Assoc.; 1956, Initiation of nerve impulses in the cochlea and other mechanoreceptors, in *Physiological Triggers* (Ed. T. H. Bullock), Am. Physiol. Soc., Washington, D.C.
[9] Davis, H., C. Fernandez and D. R. McAuliffe, 1950, Excitatory process in cochlea, *Proc. Natl. Acad. Sci.*, 36, 580–587.
[10] Davis, H., B. H. Deatherage, D. H. Eldridge and C. A. Smith, Summating potentials of cochlea, in the press.
[11] Denker, A., 1907, *Das Gehörorgan und die Sprechwerkzeuge der Papageien*, J. F. Bergmann, Wiesbaden.

[12] DIJKGRAAF, S. VON, 1952, Bau und Funktionen der Seitenorgane und des Ohrlabyrinths bei Fischen, *Z. vergleich. Physiol.*, *34*, 104–122; 1955, The physiological significance of the so-called proprioceptors, *Acta Physiol. et Pharmacol. Neerl.*, *4*, No. 1.

[13] ECHLIN, F. and A. FESSARD, 1938, Synchronized impulse discharges from receptors in the deep tissues in response to a vibrating stimulus, *J. Physiol. (London)*, *93*, 312–334.

[14] EVANS, H. M. and G. C. C. DAMANT, 1929, Physiology of the swimbladder in Cyprinoid fishes, *J. Exptl. Biol.*, *6*, 42.

[15] FERNANDEZ, C., 1955, Effect of oxygen lack on cochlear potentials, *Ann. Otol. Rhinol. Laryngol.*, *64*, 1193–1203.

[16] FLETCHER, H., 1951, Dynamics of the cochlea, *J. Acoust. Soc. Am.*, *23*, 637–645.

[17] FRISCH, K. VON, 1938, The sense of hearing in fish, *Nature*, *141*, 8.

[18] GALAMBOS, R. and H. DAVIS, 1948, Action potential from single auditory nerve fibers, *Science*, *108*, 513.

[19] GALAMBOS, G. and H. DAVIS, 1943, Response of single auditory nerve fibers to acoustic stimulation, *J. Neurophysiol.*, *6*, 39–57.

[19a] GRASSÉ (P. P.), 1954, *Traité de Zoologie*, Vol. XII, 1477 pp. Masson et Cie, Paris.

[20] GRAY, A. A., 1927, A re-statement of the resonance theory of hearing, *Acta Oto-Laryngol.*, *11*, 30–53.

[21] HANKE, H., 1914, Ein Beitrag zur Kenntnis der Anatomie des äusseren und mittleren Ohres der Bartenwale, Jena. *Z. Naturwiss.*, *51*.

[22] HARDEN-JONES, F. R. and N. B. MARSHALL, 1953, The structure and functions of the teleostan swimbladder, *Biol. Rev.*, *28*, 16.

[23] HELMHOLTZ, H., 1870, *Die Lehre von den Tonempfindungen als physiologische Grundlage für die Theorie der Musik*, third edition.

[24] HENSEL, H., 1953, Afferente Impulse aus des Kältererezeptoren der äusseren, *Arch. ges. Physiol.*, *Pflügers*, *256*, 195.

[25] HILDING, A. C., 1952, Origin and insertion of tectorial membrane, *Ann. Otol. Rhinol. Laryngol.*, *61*, 354–370.

[26] HILDING, A. C., 1952, A theory on the stimulation of the organ of Corti by sound vibration, *Ann. Otol. Rhinol. Laryngol.*, *61*, 371–383.

[27] HOAGLAND, H., 1933, Electrical responses from the lateral line nerves of catfish, *J. Gen. Physiol.*, *16*, 695.

[27a] HUGGINS, W. H. A., 1952, Phase principle for complex-frequency analysis and its implications in auditory theory, *J. Acoust. Soc. Am.*, *24*, 582–589.

[28] HUGGINS, W. H. A. and J. C. R. LICKLIDER, 1951, Place mechanisms of auditory frequency analysis, *J. Acoust. Soc. Am.*, *23*, 290–299.

[29] KATSUKI, J., S. YOSHINO and J. CHEW, 1950, *Japan. J. Physiol.*, 1–87.

[30] KEIDEL, W. D. VON, 1955, Aktions potentials des N. dorsocutaneus bei niederfrequenter Vibration der Froschrückenhaut, *Arch. ges. Physiol., Pflügers*, *260*, 416–436.

[31] KELLOGG, R., 1928, The history of whales. Their adaptation to life in the water, *Quart. Rev. Biol.*, *3*.

[32] LEGOUIX, J. P. and A. WISNER, 1955, Rôle fonctionnel des bulles tympaniques géantes de certains rongeurs (Mériones), *Acustica*, *5*, 208–216.

[33] LICKLIDER, J. C. R., 1951, A duplex theory of pitch perception, *Experientia*, *7*, 128–134; 1954, "Periodicity" pitch and "place" pitch, *J. Acoust. Soc. Am.*, *26*, 945.

[34] LOWENSTEIN, O. and T. D. M. ROBERTS, 1951, The localisation and analysis of the responses to vibration from the isolated elasmobranch labyrinth. A contribution to the problem of the evolution of hearing in vertebrates, *J. Physiol. (London)*, *114*, No. 4.

[35] LOWENSTEIN, O., 1957, The sense organs: the acoustico-lateralis system, in *The Physiology of Fishes*, Ed. MARGARET BROWN, Academic Press, New York, Vol. II, Chap. 2, pp. 155–186.

[36] MACH, E., 1865, Bemerkungen über die Akkomodation des Ohres, *Akad. Wiss. Wien*, *51*, 345–346.

[37] MYGIND, S. H., 1952, Function and diseases of the labyrinth, *Acta Oto-Laryngol.*, *41*, 5–6.

[38] PARKER, T. J., 1882, On the connection of the air bladder and the auditory organ in the red cod (*Lotella bacchus*), *Trans. N.Z. Inst.*, *15*, 234.

[39] PIERON, H., 1955, *La sensation, guide de vie*, Chap. III (Les qualités tonales), Gallimard Editions, Paris, pp. 224–225.

[40] PUMPHREY, R. J., 1950, *Hearing*, from the Symposia on Animal Behaviour of the Soc. Exptl. Biol. Cambridge, No 4.

[41] RANKE, O. F., 1950, Theory of operation of the cochlea: a contribution to the hydrodynamics of the cochlea, *J. Acoust. Soc. Am.*, *22*, 772–777.

[42] REBOUL, J., 1938, Théorie des phénomènes mécaniques se passant dans l'oreille interne, *J. phys. radium*, *9*, 185–194.

[43] REYSENBACH DE HAAN, M. D., Hearing in whales, *Acta Oto-laryngol.*, Stockholm, Suppl. 134.

[44] McROBERTS, D. D., 1934, Study of the development of the cochlea and cochlearis nerve in the fetal albino rat, *J. Morphol.*, *56*, 243–265.
[45] SCHOUTEN, J. F., 1940, The residue and mechanism of hearing, *Proc. Ned. Akad. Wetensch.*, *43*, 991.
[46] SHOWER, E. G. and R. BIDDULPH, 1931, Differential pitch sensitivity of the ear, *J. Acoust. Soc. Am.*, *3*, 275–287.
[47] SCHWARTZKOPFF, J., 1955, Schallsinnesorgane, ihre Funktion und Bedeutung bei Vögeln, *Acta XI Congr. Intern. Ornithol.*, 1954, Basel, pp. 189–208; 1958, Über nervenphysiologische Resonanz in Akustikussystem des Wellensittichs (*Melopsittacus undalatus* SHAW), *Z. Naturforsch.*, *13b*, 205–208; 1955, On the hearing of birds, *ibid.*, *72*, 340–347.
[48] TASAKI, I., H. DAVIS and J. P. LEGOUIX, 1952, Space-time pattern of cochlear microphonics (guinea pigs) as recorded by differential electrodes, *J. Acoust. Soc. Am.*, *24*, 502–519.
[49] TASAKI, I., H. DAVIS and D. H. ELDRIDGE, 1954, Exploration of cochlear potentials in guinea pig with microelectrodes, *J. Acoust. Soc. Am.*, *26*, 765.
[50] TRINCKER, D. and C. J. PARTSCH, 1959, The A.C. potentials (microphonics) from the vestibular apparatus, *Ann. Otol. Rhinol. Laryngol.*, *68*, 153.
[51] TUNTURI, A. R., 1950, Physiological determination of boundary of acoustic area in cerebral cortex of dog, *Am. J. Physiol.*, *160*, 395–401.
[52] VAN EYCK, M., 1951, Contribution à l'étude de l'électrophysiologie de l'appareil vestibulaire, *Acta med. belg.*, separata (82 pp.).
[53] VRIES, H. L. DE and J. M. VROLIJK, 1953, Phase relations between the microphonic crista effect of the three semi-circular canals, the cochlear microphonics and the motion of the stapes, *Acta Oto-Laryngol.*, *43*, 1.
[54] WATSON, D. M. S., 1952, The evolution of the mammalian ear, *Evolution*, *7*, No. 2.
[55] WEVER, E. G., 1938, Width of basilar membrane in man, *Ann. Otol. Rhinol. Laryngol.*, *47*, 37–47.
[56] WEVER, E. G., C. W. BRAY and M. LAWRENCE, 1941, Nature of cochlear activity after death, *Ann. Otol. Rhinol. Laryngol.*, *50*, 317–329.
[57] WEVER, E. G., M. LAWRENCE, R. W. HEMPHILL and C. B. STRAUT, 1949, Effects of oxygen deprivation upon cochlear potentials, *Am. J. Physiol.*, *159*, 199–208.
[58] WITSCHI, E., 1955, The branchial columella of the ear of larval ranidae, *J. Morphol.*, *96*, No. 3, 497–512.
[59] WOHLFAHRT, T. A., 1932, Anatomische Untersuchungen über das Labyrinth der Elritze (*Phoxinus laevis* L.), *Z. vergleich. Physiol.*, *17*, 659; 1950, Über die Beziehungen zwischen absolutem und relativem Tonunterscheidungsvermögen, sowie über Intervall, *Z. vergleich. Physiol.*, *32*, 151.

CHAPTER 20

EMISSION AND RECEPTION OF SOUNDS AT THE LEVEL OF THE CENTRAL NERVOUS SYSTEM IN VERTEBRATES

by

P. CHAUCHARD

1. Introduction: from Reflex to Mental

The objective character of sound, as a signal propagated in the external medium, leads comparative psychophysiology to be interested primarily in the organs of production and reception of sound and in the way in which they come into play in individual and social behaviour. Now what is really the most important thing is not merely to know how a sound is produced or how it affects the organism nor even to see the part the sound signal plays as a releaser of innate or acquired behaviour, nor the weight that attaches to sound emissions in such behaviour: what is important is to know the intimate mechanisms that make it possible for the sound to release a behaviour-pattern and thus give it a vital meaning; and to find out exactly how an animal is stimulated to utter sounds, its motivation proper to such behaviour.

In the framework of classical human psychology, using introspection and reporting, the sound appears from the reception point of view as a *conscious sensation*, and from the emission point of view as the *voluntary act* of speaking. The investigator is often tempted to avoid this aspect of subjectivity and self-consciousness, which to him seems a methodology divorced from his opportunities of objective analysis, and retrogressing to the domain of metaphysics. How indeed, without falling into the childish anthropomorphism of the ancient observers, *can* we know what the animal feels and wants, or even whether it is possessed of consciousness or will? So the zoopsychologist stays prudently on the outside of his subject and analyzes only its behaviour.

But if the idea of self-consciousness or interiority has a psychological connotation, has it not also a physiological aspect, that of the things that go on objectively in the interior of the subject and that specifically enable him to be a subject endowed with consciousness and spontaneity? One kind of elementary reflex neurophysiology, whether mere spinal reflexology or even the Pavlovian study of cerebral conditioned reflexes, has accustomed us to minimize this physiological self. The reflex concept entails the ideas of automatism and passivity: in certain conditions connected with the construction of the nervous system, sensory messages, either innately or by learning (*i.e.* association of signals), become motor messages responsible for behaviour, simply by passing through elective switchboards in the nerve centres. As mere switchboards, these centres in no sense create responses, and the subject appears—in Cartesian imagery—as a well-made automaton actuated by the outside environment. When it comes to Man, either he is radically separated from animals and endowed with a purely metaphysical self-consciousness that mysteriously actuates the nervous machinery, or his self-consciousness is asserted to be only an illusion, and he too is regarded as purely a machine actuated by sensory messages.

Both views appear equally lacking in real scientific objectivity: and they have had the deplorable result of rousing against reflex neurophysiology all the defenders of self-consciousness, subjectivity and awareness. Thus Gestalt philosophy has considered the functioning of the brain as an integrated whole, quite unrelated to the reflexes, the nerve impulses, the excitations and the inhibitions that are revealed by physiological analysis. Quite rightly, the psychophysiologists rejected this organicist "mythology", but they themselves were wrong in not realizing that it was in a better comprehension of cerebral mechanisms that the solution of the problem resided, as did also the just reply to the objections they had to face.

Luckily, scientific progress demands the rejection of elementary over-simplifications and the recognition of organic complexity. At the end of the 19th century it was the fashion to situate spiritual phenomena materially in the brain; did not the cerebral neurons of the temporal region contain peculiar chemical substances, the material support for all the sound images? Following on this, under the impetus of Pavlov, neurophysiology came into its domain of analysis of the cerebral clockwork, and no longer worried about a material explanation of psychology. It was through working in this objective way that progress was so considerable, particularly in the last 10 years since electrophysiology advanced to the study of the animal and human brain in normal function. We can now comprehend the neurophysiological aspect of self-consciousness and subjectivity as things not themselves localizable in the brain, but resulting from the sum total of functional activity, harmonized and integrated in that organ.

If the psychologist remains outside the animal he observes, the specialist in cerebral neurophysiology, *per contra*, penetrates inside it and analyzes the nervous processes responsible for its reactions. It was specifically the great pioneer merit of Pavlov that he used the age-old process of training to get to know, *from outside*, by sending signals and collecting the responses, *just what goes on inside*, in the cerebral cortex. Unfortunately, he stopped at a too analytical and too purely reflexological level, not taking account of all the possibilities of self-consciousness and spontaneity that could arise from cerebral integration.

Following the work of Sherrington and his school, after the demonstration by Lapicque of the processes of central autoregulation, after the more recent tracing of these processes to the reticular formation, the idea of reflex activity appears to have been profoundly transformed. A centre is no longer just an area of interconnections enabling variable switchings such that unchanged sensory messages become various motor messages. Much more independence appears between reception and execution, the more so as the centre becomes more complex. The central neurons make a synthesis of the manifold reflex and humoral influences that they receive, and from it, construct an adapted response which is no longer the passive transmission of the incident message, but an original and new message, reshaped by the central neurons. Instead of being automatic and obligatory, the response appears endowed with a *spontaneity* that is unpredictable for those who do not know the complexity of all the factors in play[5,6].

It is not a matter just of substituting this idea of the harmonious *integration* of multiple influences for that of the simple switching of an unchanged message; but of completely dropping the doctrine of a centre at rest, activated only during the course of the reflex. Ceaselessly receiving multiple sensory messages, subjected incessantly to humoral stimulation, a centre is *never* at rest; nerve impulses are propagated without intermission in its complex nervous structures. Granted, this actuation has *as its origin* the neuronic excitation which gives birth to it and maintains it, but it ends up as an uninterrupted *interior* activity at the centre, that no longer depends obligatorily on activation by a particular reflex message, nor of necessity manifests itself in a motor response. Every reflex message and every motor response, then, is intercalated into this central activity proper.

Even a centre as relatively simple as the spinal cord exhibits such an *autonomous* activity, well shown by the fact that the prolonged alternate movements of walking can be released by a cutaneous excitation of short duration. The incessant messages of muscular proprioception introduce elective reflex excitations and inhibitions into the cord, giving rise to fluctuating physiological *configurations* in time and space: *motor patterns*. In this way a reflex message comes to be switched into a framework which is self-differentiated by its own antecedent activity.

The harmonizing of behaviour components can no longer be thought of as an adaptation of the response to the stimulus by a simple reflex switching of the sensory message; it has a *primordially central motivation*. What is fundamental is not now the existence of a receptor organ which sends messages and an effector organ which performs acts, but the fact that these organs are in some way a part of the central nervous interiority, a part of the configurations that represent them, configurations that acquire a certain independence, a certain autonomy relative to the peripheral organs. A message is effective only through integration into the central sensory mechanisms that serves for its automatic recognition; the act is a central motor configuration before ever it can be an instruction to the peripheral effectors. The whole organism, whether in its sensory or in its motor aspect, is thus translated into neuron configurations internal to the nervous system, and the adapted response is in its origin a print of the central patterns, which are not otherwise externally manifested. What matters is what is inside, for it is there that the possibilities of integration, and hence of individual unity of behaviour, resides. The rôle of the reflex sensory message is no more than to adapt a pre-existing mechanism to the existing conditions; that of the motor message to put into practice the central motor pattern.

The importance of this *nervous interiority* depends on the degree of complexity, that is, on the number of interconnected neurons; there exists therefore a *hierarchy* of the afferent and efferent centres, which resides only partly in the fact that the elementary centres are in touch with but few peripheral elements whereas the higher ones integrate messages and orders relative to the whole organism: no, far more important—in the lower centres less rich in neurons, this interiority, this possibility of autonomous, spontaneous activity is slight and the classical reflex aspect predominates, whereas the complexity of the higher centres allows them to have their own permanent activity even in the absence of messages from or to the periphery; the neurophysiological interiority is great. This taking into account of nervous interiority thus enables us to *objectivize the subjective* and to give to psychological processes a physiological basis, it unable also not to make an absolute barrier between animal and Man, since the latter has a brain merely (from the point of view that interests us here) richer in neurons and therefore possessing a greater interiority and

References p. 580

a greater potential for autonomy. Objectively the great difference between the two is thus expressed in terms of nervous integration and we avoid the risks of error: both that which degrades the animal by denying it all interiority and awareness, and that which seeks to credit it with our own interiority and awareness. Thus we escape both evils: that of scientifically outlawing the specifically human psychological processes and minimizing them as a subjective illusion because of their dependence on human cerebral specificity; and that of overemphasizing these processes and idealizing them as independent spiritual phenomena.

There is no difference in kind between, on the one hand, the elementary configurations induced in the spinal cord of a vertebrate or the ventral ganglia of an invertebrate by the sensory messages, and, on the other hand, the configurations that these same messages provoke in the higher centres of the same animal—cerebral cortex or head ganglia as the case may be. The difference is merely in the degree of richness of the afferent neural network that permits better use of the messages; yet it is no less true that this quantitative complexification, when it attains such a degree, is indeed the source of new qualities. The elementary configurations only permit simple reflex-type control; the cerebral ones ensure an analysis so refined and informative as to produce a far better-adapted control and all the more so in that this control also benefits in quality from the richness of the higher networks of the motor command. It is such higher sensory configurations that permit feeling and percipience. In a sense, it is a case of the cerebral processes' *antedating* the sensation consciously grasped; this grasp creates nothing and can do no more than use the pre-existing cerebral patterns and configurations. What can become conscious as a sensation, then, is the cerebral configurations that are induced by the sensory messages; which, however, thanks to the cerebral autonomy, can be built up from memory *by the imagination* even in the absence of such messages. It is only the lesser degree of complexity and autonomy that distinguishes the nervous configurations of a lower centre from those of a higher centre or those of animal from those of human brains; it is also the fact that the brain alone permits *conscious apprehension*. But here again, this conscious apprehension, this peak of mentality and self-consciousness is not a disembodied metaphysical power peculiar to Man. It is the manifestation of the potentialities of *integration* that the brain possesses as an organ centralising and unifying the whole individual from the physiological and psychological points of view. Conscious apprehension is, in short, the taking over by the global cerebral integration, which brings the individual's own self into the unified patterns of his brain—those particular configurations which, if sensory, become sensations, if motor, voluntary commands.

Any *general neurophysiology of audition and speech* has therefore got to lift the *artificial* barrier which separates an objective neurophysiology of emission and reception of sound in animals, from the psychological aspects of sensation and deliberate speech that the study of Man involves. Sensation and the will are susceptible of objective study since they result from higher nervous activity. What creates their novelty is our reflective power, that makes us aware of them by objectifying them in our thought; and above all the fact that we can communicate that thought. But this is specifically because our human thought and power of communication alike are closely related to acoustic physiology, since we are concerned with speech and since a neurophysiology of sound production and reception is bound to culminate in the relationships between speech and thought in Man, demonstrating how the complexity of the human brain has enabled the sound signal to acquire such a *mental importance*. Having then come to understand that the whole of human self-consciousness is mediated by cerebral activity, the neurophysiologist would be totally lacking in objectivity if he did not allow that every animal must possess some sort of interiority, consciousness, possibility of sensation and of will, proportional to the complexity of its higher centres. The difficulty is that, due to the very insufficiency that arises from its cerebral inferiority, it has not the capacity to express itself in true speech; it is bound therefore to be difficult to assess this self-consciousness by analyzing its behaviour.

The head ganglion of a lower animal may be much less complex than the spinal cord of a mammal, but the animal may be no less rich for all that in interiority and consciousness; for the cord is only an elementary integrative organ of a part of the body, whereas, the cerebral ganglion may integrate and unify the whole organism, receiving messages from all the senses. Though in mammals the dominant organ is the cerebral cortex, and in Man in particular the prefrontal region, the zone of dominant integration and last to appear in evolution, yet it is true to say that in some lower vertebrates, interiority and mentality are not yet completely corticalized. The rich psychological possibilities of the bird are more bound up with the *hyperstriatum* than with the still incomplete cortex; man and ape deprived of cortex are totally blind and deaf, whereas dog and cat can still make use of these senses in their behaviour; it seems then legitimate to suppose that the sensory mechanisms in the basal ganglia are in animals still capable of giving rise to these sensations, but not now in Man. This is not surprising, for lower in the vertebrate scale, in fishes and amphibia, (where the cerebral cortex is lacking and the higher brain is limited to the rhinencephalon) sensory and motor integration, bases of mentality, are still exercised by the structures of the diencephalon and even the mescncephalon. It is at the level of the *optic tectum* that motor concentrations similar to those of the motor area of the brain of mammals can be demonstrated; it is at this level that the dominant integration permits such degree of consciousness and will as these animals have, and it is at this level that the conditioned reflex circuits are situated which in mammals are predominantly cortical. Right at the bottom of the zoological series, in unicels, below the level of specialized senses and a nervous system, it is in the protoplasmic configurations and in the integrations that make one whole of the cell, that lies the possibility of a very rudimentary mentality, self-consciousness and awareness—and here all likelihood of anthropomorphic fallacy will of course be excluded for those who link the degree of awareness with the objective development of the nervous system. In multicels, integration and individual mentality of course remain very deficient to the extent that integrating centres of the ganglionated nervous system type, with cerebral ganglia, have not appeared—which they do not until the level of the worms.

Thus before reviewing the main *principles* of central hierarchic configurations responsible for sound-signal reception and production in the animal series (including Man), it has seemed essential to make quite clear that present day trends in the neurophysiology of behaviour, consciousness and thought do permit to join in one scientific study both the objective and the subjective aspects of acoustic psychophysiology.

This objectivity of the subjective, at the level of neural self-consciousness, has

References p. 580

been postulated by another scientific school of thought, the so-called objectivist school of animal psychology which, under Lorenz and Tinbergen, has truly established the scientific study of the behaviour of the animal in its natural environment. It is very remarkable that this school, siding with the critics of reflexological conceptions and interesting itself only in behaviour, has on the one hand achieved an analysis of innate behaviour-sequences which interprets them as responses to simple sign-stimuli from the environment very comparable—apart from their innate character—to conditioned reflexes; whilst on the other, it has clearly perceived that these simple behaviour-reflexes do not cover the whole of animal behaviour and that what is important is the hierarchy of the centres in which originate the behaviour mechanisms that are selectively brought into play by the sign-stimuli; mechanisms which, if the motivation is strong, can be activated spontaneously or by means of a non-specific sign-stimulus. Thus this school has been led to objectively re-introduce the subjective, through showing the importance, to general behaviour, of the taking-over of all its elements by the individual integration machinery which utilizes or modifies them.

It is moreover of the utmost importance that progress in the embryology of behaviour[11] should enable us to understand how, in the process of formation of the organism, the autoregulations resulting from the interplay of heredity and environment permit the building up of all these neural configurations, as a species of "physiological organ" ensuring adaptive conduct. From this point of view, some especially interesting ideas have already been obtained about the primary configurations in the spinal cord.

It is not our function here to make a detailed study of the nerve centres implicated in audition or in speech in the various animal groups; indeed we have to admit the very insufficient state of knowledge in this field. Apart from the work of Huber[13] on the cricket and of Aranson and Noble[2] on the leopard frog, we are obliged to base our ideas mainly on the study of the audition and phonation centres in mammals and particularly in Man. And even here, our ignorance of the rôle and activity of the various levels of the centres is great. Fortunately it is to some extent easier to enumerate broad general principles; for audition is a particular case of sensory reception and phonation a particular case of neuro-motor action, and their central mechanisms can accordingly be readily explained within the general framework of activity of the sensory and motor centres. This is what we now propose to portray.

We shall study the various grades in succession: elementary centres at the level of the auditory sensory nuclei, proprioceptive sensory nuclei, and phonative motor nuclei of the medulla, with their potentialities of integration, of reflexes, the influences that emanate from these centres and those that operate on them; the importance for audition and phonation of the whole cerebellum-mesencephalon-ponsmedulla controlling and coordinating system, and conversely the influence of audition and phonation on these centres; audiophonative automatisms of behaviour at the level of the two systems of higher behaviour centres (motor and sensory cortex: corpora striata/thalamus and rhinencephalon/hypothalamus respectively); and we shall conclude by delineating the audiophonative peculiarities of the cerebral cortex in its relationships with sensation and consciousness demonstrating that in Man this is the foundation of *inward speech*, that peculiarl human mode of thought that gives to human language a significance quite distinct from that of animal language. We shall then have to question ourselves on the problem of the thinking of the dumb, since true human speech, being a *language*, is of cultural and social origin and therefore can only be achieved—despite any potentiality in the brain—through education in a human milieu (which is lacking in Kaspar Hauser children and deaf mutes; whilst aphasics betray, through cerebral lesions, serious disturbances in their cerebral control of speech, whether as a mode of communication or of thought).

Focussed then on the nervous system of mammals our study will bring out the analogies with other groups such as insects where our knowledge of the rôle of the various orders of centres is in general less complete.

2. The Elementary Centres of Audition and Speech

As with every sensory and motor structure the primary elements for reception and execution are respectively the *sensory relay neuron* which receives the sensory messages coming from the peripheral auditory receptor, and the *peripheral motor neuron* which controls the muscles operating the phonative organ. These are the afferent and efferent elements in the nerve centres whose presence is indispensable to reception and emission respectively, and whose elementary connections are the basis of the simple audiophonative reflexes (though it is often hard to pinpoint the part of the various orders of centres in the reflex influence of audition on phonation). In the classical words of Sherrington, the peripheral motor neuron is the *final common path* upon which converge all the manifold influences that can modify the utterance. The anatomical site of these elementary centres is very variable in the various invertebrate groups: the afferent centres are located in the ganglion corresponding to the segment of the receptor, *i.e.* one of the ventral ganglia for the numerous auditory receptors of insects, sometimes a head ganglion for head receptors; the efferent centres implicating in the same way the neurons of the ganglion that innervates the phonative organ, *e.g.* in the cricket, the second thoracic ganglion. In vertebrates[8], auditory afferents and phonative efferents are situated in the *pons* region. These include the sensory nuclei of the auditory nerve (the cochlear nucleus) where the axons of the peripheral auditory neurons end and where the cell-bodies of the primary relay neurons lie; their distribution in the nucleus virtually constitutes a medullary cochlea. Phonation is controlled from the motor nuclei of the numerous cranial nerves that innervate the phonator muscles, nuclei which therefore stretch over a considerable area: not just the vago-spinal nuclei which are motor centres to the larynx, but also the facial motor nuclei serving the articular muscles of the face, those of the trigeminal controlling the lower jaw, of the hypoglossal for the tongue, of the glossopharyngeal etc.; again, over and above the purely phonator motorization, there also comes in the respiratory motorization that controls the current of air in the larynx. Nor must we forget, side by side with these motor mechanisms, the importance of the sensory receptors of the complex feed-back sensitivities brought into play by the sound emission itself: the musculo-tendinous sensitivity of the phonator muscles, sensitivity of the buccal, pharyngeal and laryngeal mucosa to vibration, and again that of all the thoracic and abdominal proprioceptors. These mechanisms are sited in the sensory nuclei of the corresponding cranial nerves, encroaching on the dorsal horns of the cervical spinal cord, notably the trigeminal, the glossopharyngeal and the pneumogastric. The coordination of phonative movement is automatically taken care of by reflexes arising from all this sense-complex, acting on the motor nuclei and

References p. 580

controlling primarily the degree of tension of each muscle, that is, its *tonus*. The simplest monosynaptic primary reflex is the *myotatic* reflex, in which the stretching of a muscle makes it contract. Thus we know it is the state of tension of the muscle that, by acting in reflex manner on the excitability of its motor neurons, regulates its ability to receive and emit motor inputs; on these mechanisms rests the functional differentiation of the motor neurons of the antagonistic muscles that enables coordination of movement.

Although it is true that consciousness and volition in phonation are exercised primarily by virtue of the cortical auditory-afferent neurons and phonative-efferent neurons, we must still not forget that all this is only possible through the intermediation of the elementary centres. The cortex is connected not to the ear directly but to the (medullary) cochlear nucleus; it does not activate the phonator muscles, but sends its instructions to the phonative structures of the pons. Every phoneme, every sound, every word, every sentence is first of all in some sense imprinted on the medullary configurations of audition and of phonative motorization and sensitivity, where one day the progress of techniques of electrophysiological recording by multiple micro-electrodes will enable them to be recognized. These mechanisms have an anatomical existence which is in no sense specific, the same neurons subserving multiple functions, notably of the corresponding motor nerves. What *is* specific is the *physiological configuration* that is instantly built up, the spatiotemporal picture, the specific distribution of excitations and inhibitions that regulates the propagation of nervous impulses in the complex nervous network that unites the various sensory and motor neurons.

Nothing could be farther from the truth than to deny the complexity of functioning of these centres, however elementary, or to interpret them as no more than a passageway for messages going up or down. Local coordinations make the elementary centres modify the messages to a marked degree, introducing important regulating or perturbating factors of excitation or inhibition. The cortical centres can only put into gear, not create, a whole—pre-established—mechanism.

All the same, it is true that the elementary centres by themselves are insufficient; the distribution of the auditory inputs in the cochlear nucleus does not at this level permit all the tonal discrimination that is possible at the level of the cortex, which brings into play certain selective inhibitory processes operating to restrict the frequency-range to which each neuron reacts[16]. It has in fact been shown that although the various frequencies have receptors that are clearly localized in the auditory cortex, this is by no means the case at the lower levels, where the low frequencies have a more diffuse action; selectivity is not achieved by direct channels from ear to cerebrum, but establishes itself bit by bit in the course of its ascent through the centres by means of selective inhibitions. *Per contra*, from the phonational point of view, the lower centres, normal site of the sensoriomotor phonative patterns, are incapable on their own of initiating these patterns and of producing speech if they are not themselves motivated by the higher phonative centres. They can, however, either spontaneously or by direct or reflex excitation, bring about simple reactions such as the medullary yelp, described in dogs by Vulpian. Then again, in Huber's experiments on the cricket, isolation of the phonative ganglion from the cerebral ganglia blocks emission, allowing only the possibility of a phonative position that is acoustically ineffective.

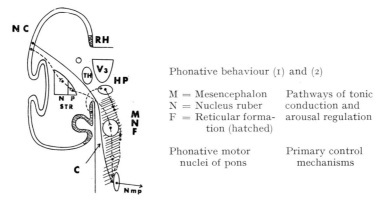

Fig. 354. Motor speech centres (frontal V.S. in Man). (1) Deliberate and automatic speech. NC = Neopallium; N and P str = neostriatum and palaeostriatum (corpora striata). Solid line = pyramidal neuron; broken line = extrapyramidal pathway. (2) Instinctive and affective speech. RH = Rhinencephalon; HP = Hypothalamus; V3 = 3rd ventricle; TH = Thalamus; Nmp = Peripheral motor neuron; C = Cerebellum.

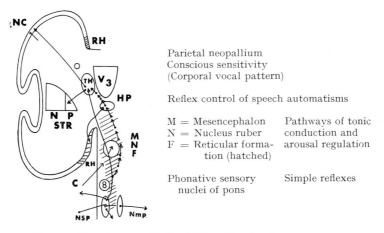

Fig. 355. Phonative sensitivity. NSP = Peripheral sensory neuron.

We are still very ignorant of this physiology of the elementary centres. Up till quite recent times the cochlear nucleus was thought of as a mere passive relay station; recent experiments, notably of Jouvet[14], show the importance of the excitation and inhibition reactions that may be located there. A psychophysiological process such as attention or inattention does not involve only the higher centres, but redounds on the functioning of the lower ones. Being preoccupied or attentive is not solely a matter of sensitivity or of utilization in the cerebral cortex of unchanged auditory messages, but also of a modification of the message itself by excitation and inhibition processes sited upwards of the cochlear nucleus; identical phenomena are produced in the field of muscular sensitivity, where the central mechanism introduces possibilities of modification upwards of the receptor apparatus.

Quite apart from all the influences of the higher centres, to complete the picture we should have to know all the interactions at medullary level, all the reflex influences

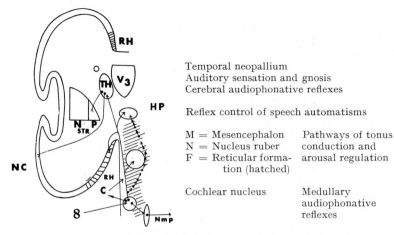

Fig. 356. Auditory sensitivity.

of whatever origin that could intrude, and the intrinsic sensitivity of these neurons to modifications of the internal medium, notably that produced by hormones. Though proprioceptive or audiophonative reflexes play a big part in the regulation of motor activity, there also exist audio-auditory reflexes that regulate ear function, for example the tonus of the muscles of the middle-ear ossicles. The pons region is such a centre (of sensory receptors and motor neurons more or less directly interconnected by white tracts or grey reticular substance) as to provide manifold possibilities for innumerable factors to influence auditory reception or phonative control; or for auditory messages, or proprioceptive phonative messages, or the indirect effect of phonative motor messages, to influence numerous central mechanisms of the region. It is in particular very likely that vestibular messages are not unique in influencing the medullary (sea-sickness) centres, but that intense auditory messages can also acts as perturbators, either of the whole sympathetic balance, or of tonic equilibrium through an action that may reach to the spinal cord.

Multiple interactions are possible at this level, between the somatic and the splanchnic spheres. The principal phonative nerve, the pneumogastric, is also after all the most important parasympathetic nerve. Vegetative balance thus influences speech; and we also have to take account of the action of vasomotor reflexes on auditory and phonative organs.

In some animals, sound emissions due to reflex excitation of certain cutaneous areas have been reported: for instance[21] the reflex croak of the frog whose back is stimulated, or the crow of cock or hen on stimulation of the region of the pubes, where the ischiopubic nerves spread out: in the cock, up to 64 consecutive crows have been achieved. We do not know the centres that are involved in these reflexes; in the frog the corpora bigemina appear to be necessary, but in the male, the courtship signal seems to depend on centres at the base of the 4th ventricle.

3. The Apparatus Controlling Nervous Activity

Between the elementary effector centres (whose complexity we have just seen as a source of considerable functional possibilities) and the higher centres of the cortical and basal ganglionic regions, whose presence is essential to a proper audiophonative behaviour, are located centres which, though not directly concerned like the other two in reception and effection, nevertheless have a fundamental bearing on the correct *interplay* of these functions, as also on the smooth running of all the nervous centres. To see the midbrain region as a mere corridor through which auditory messages ascend towards the thalamus relays, or pyramidal or extrapyramidal motor instructions descend to the medullary motor nuclei, would be a gross undervaluation of its part. Even at this level, this rôle of conduction by the collaterals of the responsible axons permits some switching of inputs to other formations. The corpora bigemina of lower vertebrates, like the posterior corpora quadrigemina of higher vertebrates, receiving auditory messages via collaterals, are thus an important reflex centre for such messages—which is not without influence on phonation. Though the pyramidal phonative neurons pass without relay through the geniculate tract, the plurineuronic extrapyramidal tract includes numerous relays at the level of the *nucleus ruber*, the *locus niger*, the corpora quadrigemina (*tectum*) and the *reticular formation*. At the level of these relays, multiple influences can act and there are possibilities of reflex control of the phonative neurons.

But the real importance of these centres appertains rather to their *non-specific regulatory function*, in which participate—in specific or non-specific fashion—the sensory messages of audition and phonation, and which in its turn operates upon the functioning of the peripheral or higher centres and of the auditory and phonative organs so as to adapt their activity automatically to requirements.

Historically, the recognition of these mechanisms originated with the discovery of the *regulatory centres for muscle tonus* of the whole organism that make posture and the correction of movement possible; in this way were delineated the rôle of the proprioceptive and vestibular sensory messages, that of the optic messages acting on the double system of the pontine tonic centres (which left to themselves cause decerebrate rigor) and the midbrain centres such as the nucleus ruber which, by modifying the activity of the tonus centres, enable proper regulation of tonus. Again, the delicate cerebellar mechanism makes possible a precision tonic regulation operating on these different centres. Nor is this system of the cerebellum and the nucleus ruber limited to a downward influence on the peripheral motor neurons via the extrapyramidal tracts such as the rubrospinal, tactospinal, reticulospinal, vestibulospinal, olivospinal... or via successive relays in the reticular formation; but it also has an upward influence on the motor neurons of the centres responsible for behaviour sequences, notably those of the cerebral cortex: which makes the neocerebellum indispensable to the effective carrying out of voluntary movement. Here again it is a question of reflex regulation: a nerve tract coming from the cerebral cortex carries information about its condition to the cerebellum, with collaterals and relays in the midbrain and pons regions. Though the sensory audiophonative messages contribute to this system of reflex tonus regulation, still more important is the fact that phonation, itself a particular case of motorization, is correct only in so far as the tonic regulation of the phonator muscles *under this system* can be properly carried out. The local control exerted by the proprioceptive reflexes in the primary centres would be inadequate in the absence of this general control. The modifications of the voice in disorders of this region are characteristic: scanning speech for example is one element of the cerebellar syndrome in Man.

References p. 580

For a long time the function of this region seemed only to be concerned with muscle tone, and it was not at once obvious that in fact it also embraced the control of *nervous tone*, of the excitability of the motor neurons responsible for this muscle tone. To Lapicque is due the credit of showing that it actually involves *centres regulating the excitability* of all the neurons, in other words centres ensuring the reflex self-regulation of the entire harmony of nervous activity. It is in effect in this region that he placed his "subordination centre" from 1928 onward, the initial observations of Mme Lapicque dating from 1923. At this time, apart from specialists in muscle tone, no neurophysiologist, whether of Sherrington's or Pavlov's schools, envisaged the necessity of such regulation, now universally recognized thanks to the recent discoveries of Magoun and Moruzzi which date only from 1949.

Thanks to the demonstration of this regulatory centre, it has also been possible to integrate the basilar sleep centre into the scheme of general control.

The *reticular formation* which extends from the medulla to the hypothalamus and to the diffuse thalamic system is in some degree a reserve of nervous input holding all the neurons under its influence and supplied in nonspecific fashion by all the sensory messages—somatic or otherwise. In it we can differentiate activator and inhibitor formations. Its influence is exerted on peripheral neurons basally and cerebral neurons higher up; hence it appears mainly as the *arousal centre* of the cerebral cortex, but also as the site of specific excitatory and inhibitory controls that enable harmonious functioning of the cortex, adjusting the periphery to its needs and it to the periphery's needs. The processes of excitation and inhibition that we have noted in conditions of attention and preoccupation at, for example, the level of the cochlear nucleus, depend on this formation. Recent studies show that the whole process of cortical functioning presupposes this preliminary control; this is the case for example at the various stages of reflex conditioning, where the new connection which is its basis is made, probably not in the cortex, but in these basilar control centres.

Cerebellum and tonus centres work in close association with the reticular formation, which turns out to be very sensitive to all the tensive and humoral disorders of the blood, notably variations of hormone balance. A study of the centres of audition and phonation has nowadays therefore to insist especially on the importance of the impact of audiophonative sensory messages on the control system whether by contributing to its correct functioning (alerting by sound) or by upsetting it (audiogenic seizures produced by intense sound); and it must again insist on the importance of this control over the activity of the audiophonative mechanisms in health or otherwise, whether for the specific problem of audiophonative control itself or for the repercussions on it of all the reflex humoral or cortical factors.

This control device exists in all vertebrates; in the invertebrate field we know less, but chronaximetric observations do attest the existence of central regulatory mechanisms.

4. Instinctive Centres of Audition and Speech

The centres located in front of the mesencephalon in higher vertebrates, like the head ganglia of invertebrates, are necessary for the execution of *adaptive behaviour reactions*. These are complex spatiotemporal configurations of the neuron networks of these centres, which are necessary for the activation of motor sequences present only potentially in the lower centres. Hence we find in this region auditory receptor centres and phonative effector centres which enable audition and phonation to play their part in animal behaviour.

It is not possible to make a systematization of the behaviour centres in the head ganglia of invertebrates, where our knowledge rests chiefly on the injections or localized excitations carried out by Huber on the brain of the cricket and which have enabled the localization of certain zones concerned with various types of song or posture. (An account of this work will be found in Chapter 17, p. 440.)

We again lack precise data on vocalization in lower vertebrates, where the corpora quadrigemina seem to be implicated along with the diencephalon. In birds, the higher motor control is not yet completely corticalized and depends mainly on the structures in the *hyperstriatum*. The neurophysiology of these is still in its infancy, which is a pity, for after all it is in birds that vocalization is most advanced, numerous species (especially the good singers) not only disposing of a vocabulary of inborn sounds but also having a learnt language resting on imitation of heard sounds, even those uttered by other species—as in the astonishing imitations of human speech by the parrot in virtue of the properties of articular control in its brain; these are infinitely more advanced than those of mammals and even the ape, which is only very poorly equipped to imitate human language, because of the inadequacy of its brain, even though the latter *as a whole* is more advanced and much more "intelligent" than that of the bird.

On the other hand, (although we are not yet very rich in detailed and precise information on the higher speech centres in mammals, apart from those of the human cerebral cortex) it remains possible, in the light of what we know about the rôle of the various grades and systems of behaviour centres, to systematically place the functions of audition and phonation there.

We can distinguish two main groups of centres enjoying, in spite of their interconnections, a certain independence[17]; one is the system ending in the cerebral cortex of the *neopallium* with auditory messages ascending to the temporal region via thalamic relays, and phonative messages of the frontal motor zones bringing into play the mechanisms of automatic motorization of the *neostriatum* and *paleostriatum*. Here we are dealing with mechanisms specific to heard and spoken language in both their automatic and voluntary aspects. But there exists another system, where audition and phonation come in more in relation to the great basic instinctive behaviour-complexes and the elementary affective complexes; this is the *hypothalamic* system of instinctive centres operating under the complicating influence of the primitive cortex or *rhinencephalon*. This system is phylogenetically the oldest, but no less complicated for that, even in Man, where it is relatively reduced by the formidable progress of the neopallium. This is the system that forms the subject of the greatest number of contemporary investigations. Of course the two systems are not independent and numerous interactions exist between the various ganglia of the basis, the rhinencephalon and the neopallium, notably at the level of the temporal region and the prefrontal zone.

But cerebrectomy experiments have shown that in mammals such as the dog and cat, complex audiophonative behaviour reflexes can still operate in an animal having nothing left by way of supramesencephalic centres except the hypothalamic nuclei. There exists a complex instinctive automatic activity no longer involving

References p. 580

the acquisitions of training, but in which the animal can exhibit an elementary adaptive behaviour seeking the agreeable and avoiding the disagreeable, guided in particular by audition and commonly including vocal emissions, *i.e.* calls. It is probable that the affective cry of the human suckling (whose cortex is not yet functional), though perhaps at first purely medullary, is not slow to become hypothalamic. It is only later that it is corticalized, when its motivation becomes more complex. In this very elementary type of affective speech common to Man and animals, the rôle of the rhinencephalon/hypothalamus system is considerable. We know that the configurations of reception and execution in this zone are quite *diffuse*, in contrast to the precise *somatotopic* localizations of the neopallium; complexly linked multiple centres intricate their excitatory and inhibitory influences. In a general way the cortical mechanisms often blocks the otherwise automatic release of the lower centres. It is thus that the decorticated individual can, under the influence of the unleashing of hypothalamic automatisms, manifest a stereotyped cry for long periods.

Although in ape and Man the individual deprived of the cerebral cortex cannot deploy the automatic activity of the decorticated dog, because their lower centres have seen the cortex take over part of their functions and are thus no longer in a condition to exert these functions autonomously, it is however true that, in association with the cortex, these primitive mechanisms still continue to play a part in reception and execution which is essential for elementary instinctive and affective behaviour. Man without a cortex is deaf and blind; but given his cortex he still uses cortical sensations to guide an elementary behaviour assured by the basal centres.

We are accustomed by introspection to allow an important rôle to consciousness and to the sensations in our *appetitive* states, such as hunger and thirst, and in our agreeable or disagreeable affective states. What neurophysiological experimentation has established beyond dispute is that these sensations are merely subsequent to the generalized tension set up in the cerebral cortex by the influence of disturbances situated specifically in the basal centres, especially the hypothalamus. The cortex moreover can be habituated by conditioning either to restrain or to trigger this basal reactivity. Elementary instinctive and affective reactions are automatic reflexes in which reflex or direct excitation of the hypothalamus triggers adapted reflex behaviour sequences. The modern studies of objectivist animal psychology on instinct, completed by neurophysiological observations on the centres of instinct, have clearly shown that instincts are no more than relatively simple chains of reflex reactions in which a sensory message, itself only one restricted element of the situation, becomes the *sign stimulus* or *releaser*. That is, its arrival in the basal centres triggers off, there, a spatiotemporal scheme, a specific neuronic configuration, in other words a certain coordination of executive neurons that automatically enables a certain type of response. This configuration has no existence as a permanent entity, but depends on a specific sensitization of the nervous mechanisms characteristic of the appetitive state, of which we now begin to grasp the psychological basis: sensitivity of certain hypothalamic centres to the composition of the blood, in eating and drinking behaviour; activation of adjacent centres by sex hormones for releasing complex sexual patterns. This specific activation, giving an elective switching character to a signal which in repose is indifferent, is preceded by a general indifferent activation of behaviour which puts the animal in a state of seeking, of *appetitive behaviour*, that brings it into the presence of the specific signal. It would seem that this part, common to all instinctive behaviour, depends on the non-specific activation of the reticular formation.

If the drive of the hypothalamic centres is very strong, it is possible to have *vacuum activity*, without any releaser, or through some external agency unrelated to instinct; in such conditions electrical excitation of the hypothalamus can activate neuronic instinct chains and release a complex stereotyped behaviour.

Among the sensory messages that reach the hypothalamus in this way, there are some which, in innate fashion, have the power to release configurations generating a specific affective behaviour, such as pain messages or more especially sensual messages[10]. At their strongest, such messages are capable of causing a violent shock at hypothalamus level; intense pain shock or manifestations of sexual orgasm. Through conditioning, the brain learns how to release such reactions: moral pain, sexual factors of psychological origin.

Auditory reception and manifestations by sound have their place in all such behaviour. This explains the relation between voice and emotionality—cries of joy and grief, emotive loss of the power of speech in conditions of stage fright.

It is very noteworthy that the hypothalamic centres are not only regulators of external behaviour but also coordinators of the whole sympathetic balance for good visceral functioning. Whether this be done through intra-organic physiological reflex controls or through extrinsic controls via behaviour, it is adjacent centres that mediate the maintenance of the constancy of the internal environment that is essential to the smooth running of all our cells. If the blood sugar happens to drop, the hypothalamic sensitivity proceeds at once to act on the glycogen reserves and to release an activation of behaviour, a nutritional sensitization which culminates in the intake of sugar and so remedies the deficiency in the internal regulation.

We cannot then too strongly stress the importance of these control centres, whether the instinctive behaviour and organic regulation centres of the hypothalamus or the regulatory centre of the reticular formation. These centres have got to function ceaselessly in response to all the demands of the organism and of the environment that entail adaptive reactions, not to mention all the affective shocks of psychological origin. Thus failure of these centres is the origin of the *disorders of adaptation* stressed by Selye, and which comprise at one and the same time a sympathetic imbalance (Reilly) and a hypophyso-suprarenal disturbance of the endocrine balance. These centres are responsible for the diverse organic and mental symptoms of *nervous fatigue*, a factor in insomnia. If this is specially noticed in the context of this paper it is because one of these principal factors, important in modern technical civilization, is *noise*. Intense auditory messages, whether or no they have an informative value (reflex at basal level or conscious at cortex level), are characterized by the upsetting influence they exert in their passage from medulla to hypothalamus. It is much too readily believed that noise fatigue is concerned with what is consciously disagreeable; but its harmfulness, which makes it a real poison, resides much more in all the reflex disturbances of the whole organic functioning at the level of the control centres. It is known that susceptible mice exposed to intense sound may suffer a fatal seizure. Chronaximetric analysis reveals that the excitation lies in the basal centres and is still manifested in an anaesthetized animal. We should here point out the possibility that an analogous but intrinsic factor may be at work in singers' fatigue, where the motor element is superadded to the auditory one.

References p. 580

It is hypothalamic disequilibrium that is to blame for those psychosomatic disorders that have a vocal aspect, often for example in *stammering*, a symptom with a very complex aetiology, in which nervous fatigue in relation to a conflict between the two hemispheres (especially suppressed lefthandedness) plays a notable part.

Though the study of various types of phonative behaviour in the animal scale is beginning to develop, neurophysiological studies are totally wanting. It is to be noted for example that in Orthoptera (Dumortier)[4] the existence of singing rhythms linked to optical messages, that can be upset by continuous lighting or obscuration, is comparable with the *nycthaemeral rhythms* reported in mammals and in Man, and which appear to be under control of biorhythmic control centres such as the sleep centre. The connection often discovered between a sound signal and hormones, notably in the sexual field[3], is in accord with the activation of instinctive centres, notably the hypothalamus, by these hormones in the vertebrates. Though the hormones can at any stage influence phonation by virtue of the general sensitivity of the neurons, it is at the hypothalamus and rhinencephalon level that a specific sensitivity is displayed.

5. Audiophonative Mechanisms of the Cerebral Cortex

Although for self-preservation and the propagation of the species, the animal can call upon the automatic instinct mechanisms of its lower centres, it would be a gross error to restrict its behaviour to such elementary components. Over and above those behaviour forms which conduce to unadaptive automatisms in which the releaser triggers an act outside its normal context, there is the whole range of learnt behaviour and even further, all the possibilities of ascendancy, of intelligent judgment, of the utilization of one experience to provide better adapted reactions to a new situation. The more developed the animal's higher nervous centres, the more it becomes capable of going beyond innate or even acquired automatisms. The head ganglia of the higher invertebrates and the cerebral cortex of mammals are alike the site of complex nervous configurations which, on the one hand, permit the acquisition of conditioned reflexes and, on the other, lie at the root of the possibility of ascendancy and conscious adjustment.

In its capacity as an organ of higher conscious and intelligent behaviour that reaches its peak in Man, the cerebral cortex possesses special mechanisms for the reception and linguistic utterance of sound, mechanisms which have come to permit conscious sensation and the deliberate expression of speech. But the fact of being conscious and deliberate is secondary with respect to the preadaptive cerebral mechanism. Granted that the cerebral mechanisms are indispensable, in Man and higher mammals, to consciousness and will, this is not to say that all cerebral activity is conscious and deliberate. Modern neurophysiology fully confirms from this point of view the indications of psychoanalysis, which teaches that attentive consciousness is no more than a thin skin incapable of even taking cognizance of the whole depth of the unconscious, even though this unconscious is itself of a psychological nature and takes its origin in cerebral mechanisms. All learnt reactions, at first conscious, become automatic routines that operate all the better for not being thought about.

When an auditory message arrives at the cerebral cortex, there is only a sensation in so far as we pay attention to it; but to the extent that such a message overcomes the reticular inhibition and impinges on even an inattentive brain, it is still registered unwittingly in the cerebral receptor mechanisms. In the same way, we can speak without paying attention and without intention: the cortical motor configurations are none the less activated. It is found, for example in human neurosurgery, that excitation of a sensory zone creates a hallucinatory sensation because it awakens the attention; on the other hand, though we are conscious of the act that an excitation of our motor cortex will provoke, we do not do it willingly, for it has been passively provoked.

We have seen that our ear is represented at the level of the medullary receptors; it is even more so at the level of the cerebral cortex in that temporal zone where the localized termination of the thalamocortical neurons creates now a true *cortical cochlea*[9], with different receptor neurons for different frequencies; each ear being connected to both hemispheres. At the level of this *primary area* of reception, the neurons pick up the patterns innately under the automatic stimulus of the arrival of the messages, these patterns being the neurophysiological aspect of the sound picture. It is the conscious integration of these patterns that produces the sensation; yet this latter is in a sense pre-recorded as a presentation in the cerebral configurations fluctuating in space and time as a function of the sensory messages. Modern research has also established that aside from these specific receptors there exist in broad cerebral zones certain nonspecific cerebral neurons on which converge inputs from all sources, including sound, and which appear to play a regulatory part. But above all the temporal zone, close to the receptors, becomes by education the seat of the *auditory gnoses*. In using them, that is, in creating associative conditioned reflexes out of them, the infant learns the perceptual meaning of sounds. It is in this zone that Penfield has pinpointed, in Man, a region where electrical excitation enables the recall of precise memories which are often auditory; the patient hears some tune that his mother sang to him when he was a child, believes some one is playing a record for him. It is not that the memories are materially present; memory relies on a cerebral tendency to recapture complex cerebral configurations that existed previously, and

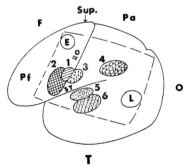

Fig. 357. The cerebrum of language (in Man). External face of left hemisphere. F = Frontal lobe; Pa = Parietal lobe; O = Occipetal lobe; T = Temporal lobe; Pf = Prefrontal integration centre; RO = Fissure of Rolando; SY = Fissure of Sylvius. (1) Pyramidal vocalization centre. (2) Extrapyramidal control centre of the corpora striata, and centre of Broca's motor aphasia. The arrow at the top indicates the position of the supplementary area on the internal face; (E) = writing centre. (3) Centre of phonative reception and sensitivity. (4) Gnosis of the corporal vocal pattern and the verbalized self-picture. (5) Reception centre for audition (cortical cochlea). (6) Auditory gnosis, verbal deafness centre; (L) = reading centre. The broken line indicates the approximate boundaries of the cerebrum of internal speech.

References p. 580

the electric current has shown itself able to provoke these configurations. The problem is in other ways complex and the very stimulation of the vicinity of the hippocampal zone itself prevents any immediate fixation of memories; a subject who counts during excitation has no recollection of it afterwards.

The direct excitation of the auditory zone gives rise to sound hallucinations which become more elaborate as the gnosis zone is more nearly approached.

This auditory zone which, in animals, simply gives information on sounds and on linguistic sound signals, comes in Man to acquire more importance in terms of the development of language. His greater wealth of neurons permits better analysis of sounds. We know that in human language the gnosis centre formerly described as the centre of *verbal deafness*, is found, like all the language centres, in the dominant hemisphere alone, that is—in the dextral individual—the left. There are specific centres for verbal gnoses, since certain patients only lose recognition of the sense of the words they hear, retaining sensations and other types of sound gnoses.

Language, from the emission viewpoint, comprises two other groups of closely associated centres in the cerebral cortex. There are first the motor centres. We have primarily the corticomotor neurons acting to control the peripheral phonative neurons, which are of two types. In the motor area of the *gyrus centralis anterior* are the *pyramidal neurons* whose axons directly stimulate the peripheral neurons through the geniculate tract; they are in their somatotopic position at the foot of the gyrus. Stimulation of them in Man or animals produces an inarticulate cry. These neurons are responsible for fine *precision* control. Close by, in the base of the premotor area, is the railhead of the extrapyramidal tract, which through successive relays, first in the corpora striata, then the mesencephalic and pons nuclei and via multiple parallel routes, also comes to activate the peripheral neurons. This is the zone of *global* overall control that allows the complex adaptations necessary to precision phonation. All vocal articulation presupposes coordination of these two zones. But the premotor zone is also a *praxis zone*, in other words the seat of those learnt conditionings that enable coordination at will of the activity of the various phonative neurons of the two zones so as to utter the various phonemes grouped as words and sentences. We have an inborn aptitude for such coordination, but we have to learn in infancy to employ it properly. This praxis zone is the *Broca's motor aphasia* or *anarthria* centre; it again is situated in the dominant hemisphere.

The difference between Man and animals resides in the neuronic richness that gives the former a greater wealth of verbal acts and hence of phonemes. It is because of the paucity of this control, due to its less numerous cortical neurons, that apes cannot be expected to speak, even though their phonative organs are as well developed as Man's. *The real phonative organ is the brain*; the periphery is no more than an effector mechanism.

Like Man, animals can at will pattern their cerebral motor zones to speak, but the sounds, poorer in content, are generally innate; *per contra*, Man, for all his conscious speech, also possesses a motor automatism that runs unconsciously. To the extent that speech is automatized and not deliberate, the neurons of the cerebral areas do not need to come into play. The motor configurations that arise at the second stage of the extrapyramidal tracts in the *corpora striata*—and in the case of Man, in the *neostriatum* especially—suffice to secure the automatism of language; in animals, innate language, like locomotion, is essentially a function of the corpora striata and only secondarily taken over by the cerebrum. This explains why in Broca's motor aphasia the lesions extend in depth and involve the corpus striatum, as Pierre Marie recognized. It is the language of the corpus striatum that must be implicated in sleep-talking. Conversely, song is a kind of speech in which will and consciousness play a considerable part, and which brings cortical mechanisms into play to a pre-eminent degree.

Though phonemes, sounds and words are first and foremost cortical motor configurations equivalent to corresponding auditory configurations, that is still only one aspect of cerebral language; in fact there are not only the configurations preparative to emission, but also each sound uttered demands a certain degree of contraction of the diverse phonator muscles, a certain type of vibration of the phonative apparatus and of the air-cavities of head, throat and chest. From this, result sensory messages of varied origin which come from all the receptors stimulated and which finally reach the lower part of the *ascending parietal* zone; here are located the receptors for the whole general sensitivity relative to the motor centres situated in front. All these messages are centralized by the postero-inferior part of the parietal zone, the *sensory gnosis centre* for speech, again unilateral in Man. This centre of sensitivity of the phonative apparatus has a considerable physiological importance, little though we may be aware of it (apart from singers, who know very well how to make use of these sensations.) The precision control of muscle tonus is not secured in the basal centres alone, but depends on interactions between cerebral sensory and motor zones. The motor configurations of speech have in some measure their counterpart in the parietal gnosis zone, constituting what Soulairac has called the *corporal vocal pattern*.

We know furthermore that this same parietal gnosis zone gives origin to the configurations of the *self-picture*, so vital for conscious apprehension. This is a real existence of the subject inside his own brain; in Man, this self-picture is verbalized and becomes the capital I, which is important for the expansion and humanization of conscious apprehension.

Thus there exist in the brain three zones connected with speech. In human neurosurgery Penfield[20] has shown that their electrical stimulation was incapable of producing the complex configurations of speech or of evoking articulated sounds; but if the subject is speaking already, such stimulations upset the configurations and cause a transitory aphasia, whether motor, sensory or auditory zones be involved; in this way these can be localized. In addition, excitation of the supplementary motor area situated on the inner face of the hemispheres in front of the pedal motor zone can provoke emission of a cry or of stammering.

It is not only the interrelations of sensory and motor zones that are close, but also those of auditory and motor zones; audiophonative reflexes are not purely medullary, and if there exists, as Tomatis has shown[22], a close relation between type of audition and type of emission, or if stammering can be produced by retarded audition of the subject's own voice or inadequacy of the controlling ear, then complex processes depending on relations inside the cortex must be involved.

It is at cortex level that the significant relationships between audition and phonation are set up, relationships which even in animals permit dialogue; but in animals sound always remains an *external signal* received or emitted, a means of communication alone and not as in Man a tool of thought.

In their capacity as a sensation and a voluntary act respectively, sound and speech

References p. 580

in animals as in Man are not phenomena of distinct kinds, but merely rest upon different cerebral mechanisms. To be conscious is to integrate the vocal pattern with the global cerebral activity, that is, with the self-picture; to will is to apply the configurations of this cerebral integration to the modulation of motor zone mechanisms. The cerebral cortex which, because of its complexity, has the power to retain the new pathways established by the coexistence of two signals (conditioned reflexes), which is the physiological basis of memory, has also the power to evoke memories through the imagination, that is to resurrect the corresponding cerebral configurations. Cerebral integration, when it functions properly—and the more so as the cerebrum becomes more complex—thus gives a possibility of mastery and decision that places the subject above the action; and this dominion is exercised not by a control outside the brain, but by an intrinsic cerebral integration mechanism that calls forth the subject and yet at the same time leaves it immanent in the brain patterns. The difference between animals and Man lies in the latter's greater complexity of cerebral integration.

6. Internal Speech and Human Thought

Up to this point nothing, apart from greater complexity, has seemed to differentiate Man from animals. The animal thinks with a quality of thought dependent on the complexity of its brain, and numerous psychological experiments show that it does enjoy a certain interiority, has something in its head which it can use to solve practical problems. The animal has a "language", a vocabulary, often with quite numerous meaningful signals.

As a means of communication, it is not only the greater articulatory richness that characterizes Man, in association with the progress of his cerebrum; it is above all the possibility of abstraction that his cerebral richness gives him, the greater possibilities of sensory analysis and synthesis. These things lead him to designate every object by a noun and to associate these nouns in an organized sentence with a grammatical structure. He thus becomes able to communicate abstract ideas. Animal speech on the other hand remains more signalling, intellectually poor and resting on an affective basis of warning signals notifying presence, analogous only to the cries and affective interjections of Man[7, 12, 15, 23].

In linguistic communication between animal and Man, we depend on the basis of animal signalling, a variety of social releaser. Animals do not attach to our words their dictionary meaning, but become trained to associate a sound signal with a situation. They are not skilled at imitating us, except the parrots; and even these, when they utter human phrases exactly pertinent to a situation, are not speaking humanly, but imitating a sound signal in the required conditions without understanding its true verbal meaning.

Animal and human speech both in their varying degrees permit dialogue, but animal language, whether innate or acquired, *is not susceptible* of progress; talented nightingales form schools of singing, but their innovations remain slight, and no infinite progress in expressive power of their language is observed. Therein lies an essential characteristic of human language, bound up with cerebral neuronic richness. We can ever increasingly perfect our speech—for example by learning another language.

Though the infant has an inborn aptitude for vocalizing (equally developed in the deaf who yet, through hearing nothing, become dumb); and though by playing at vocalization he learns to utter articulate sounds at will, it is above all with a utilitarian objective that he comes to strive to comprehend and imitate the language of his environment. This is a *social and cultural* mode of exploiting the phonative possibilities of the human brain. Were it not for this acquisition of social language, the child would remain at a very poor linguistic level, near to that of animals, even though he is more advanced from the articulatory point of view, and his imagination would hardly be able to make good the deficiency. Thus isolated children, for example infants reared by she-wolves, are profoundly dehumanized, in that they do not speak, they howl like wolves, and run four-footed with a blank face. Once past the normal age at which the infant learns to speak, it would seem that the cerebrum does not retain, from this point of view, all its possibilities; it is hard to re-educate the wolf-child, who can never acquire more than a few words. Language appears to be the instrument of humanity's intellectual progress. All men have at birth the same potentialities, which depend on the capacity of their brain; but they enjoy completely different possibilities of developing their intelligence, according as they learn a poor and primitive language or a cultural one, and according as they do or do not add the precision language of science or that variation of language which constitutes mathematics, and that other code allied to speech which is reading and writing. A little aboriginal reared less than the first five years in a primitive society can still by learning our language acquire all our culture; by contrast, after five, the primitive child has no longer the same cerebral potentialities.

From the time of the original mutation that endowed submen with a human brain, Man has had the aptitude for speech; but it is only slowly, by the striving of generations, that mankind has perfected language, learning in so doing to exploit the brain—a progress that is far from finished. The first men, despite their superior aptitude and intelligence, must have been very near the animal level, from the linguistic point of view.

But in order to understand the significance and value of human language and its specificity compared with animal language, it is no longer enough to study it by itself as a means of expression and communication; we have now to analyze from the physiological point of view the essential difference that arises as a result of cerebral complexity.

What is important is not just that man gives a name to everything, that he makes sentences, that his language progresses; but that this external code has become an *internal language*, that is, a specifically human mode of thinking. Like an animal but better, Man can think in images, can associate over his whole brain the cerebral configurations that any object imprints at the level of the various cerebral zones. This picture, the brain-image of the world, acquires autonomy as a cerebral configuration evocable by imagination even in the absence of its real counterpart in the outside world. Pictorial thinking, however, is a cumbrous mode of thought that lends itself but poorly to abstract ideation.

In Man this *primary system of signalization*, that of the conditioned reflex (to use Pavlov's terminology) comes to be replaced by a *second system*, that of linguistic signalization, the *verbalized image* of the world. In so far as a given situation rings a given bell, we develop the habit of associating a linguistic signal with each object; and then in thinking, we replace the pictorial brain-configurations with verbal ones.

References p. 580

From then on, the word is no longer a localized configuration in the motor zone, in the sensory zone, or in the auditory zone, narrowly tied to reception or emission; but becomes by confluence of the three configurations a *global* cerebral scheme that the brain can use in thinking. The articulate brain is no longer merely the brain of the execution and emission centres, but is *the brain of internal speech*, embracing complex neuronic circuits connecting these zones, multiway neuronic circuits and thalamic relays that a localized lesion cannot now interrupt. The cerebral inadequacy of animals has prevented the formation of this system that gives human thought all its scope. Internal speech is not just a mechanism in the service of thinking, but is that specific function of the human brain that gives to Man the power of thinking in peculiarly human fashion. There cannot be pure spiritual thought without speech: either we think in pictures or we think in words. We may not always take account of this, because language has become a mental automatism that functions without our heeding it; nevertheless the unexpressed thought is manifested in a certain electrical resonance in the phonator muscles.

To Pavlov must go the credit of throwing light on this human cerebral specificity, enabling objective study of the relations between the verbalized and the unverbalized, and considering psychotherapy in the objective aspect of the cerebral reconfigurations of speech.

However, he did not allow enough importance to the verbalization, no longer just of the outside world, but of the self-picture, which for the infant who speaks becomes the pronoun "I", subject of the sentence. It is when the child acquiring speech begins to say "I" that he makes intellectual progress which is staggering compared with that of the ape. The grammatical mechanisms (syntagnatic relations) common to the various languages are rooted in cerebral tendencies.

From the neurophysiological viewpoint, the human power of reflection, this consciousness of conscience which detaches the subject from his own action and enables judgment to be exercised, is associated with the greater possibilities of integration of the superbrain of Man; but this reflection would remain rudimentary without the intervention of speech and the possibility of saying "I" through verbalization of the self-picture, whose complex configuration—of which we have seen the value— is replaced by the simpler and more manageable configuration of the "I".

We cannot then too strongly insist on the importance, for human thought and awareness, of the cerebral mechanisms of speech. Nevertheless, it is also important not to forget that what makes Man is first and foremost his cerebrum. A man who cannot speak is still a man for all that. What about *his* intelligence?

The question is difficult to treat objectively, for we have to define the subject with which we are concerned, to know in what degree he is normal, and above all to find non-verbalized tests allowing us to assess his intelligence; which is rather ticklish. The first type is the Kaspar Hauser child or wolf child: he is not comparable to the savage because, desocialized, he is not a normal human being, the primitive man being social and possessing a rudimentary human dialogue. On the other hand, no truly scientific study has been made to assess the reflecting capacity of such (obviously) rare subjects; moreover we do not know to what degree they may have been actively dehumanized*. They show in any case the importance for human speech

---

* The sequestered infant, totally isolated from the world, does not develop his brain, and becomes an idiot.

of the social factor; but isn't this unnatural potentiality of adaptation, to wolf society—in the absence of anything in common with transformation of the animal by domestication—in itself a proof of intelligence?

In the case of deafmutes, the question is just as difficult, for we ought to distinguish between deafmutes without either language or social contact, deafmutes not re-educated but living in our society and striving to understand and communicate without the voice, and deafmutes re-educated with a sign-language or with true speech. Only the first two types should interest us, for the deafmute re-educated in a sign-language invented by the articulate to express his speech, receives normal language by a different sensoriomotor pathway, and thus finds himself provided with a ready-made normal internal language. Hence we find him of normal intelligence, with mere inadequacies in abstraction or grammar corresponding to the imperfections of sign-language. (It is to be noted that the electrical activity of their unexpressed thought involves those muscles of the hand that participate in the sign language.) Recently, Oleron[18] in comparing the intellectual evolution of normal and of deaf children has reconsidered, and weakened his idea that it is speech that enables the child to surpass the ape. In reality, the deaf child also makes rapid progress, and what is basic is not speech, but the cerebral development that normally permits it. In default of speech, intelligence develops, but with inadequacy of abstraction. We In default of speech, intelligence develops, but with inadequacy of abstraction. We must not confuse intelligence and speech; speech is normally the most suitable means of facilitating the rise and development of human intelligence, but in its absence Man has none the less a brain that gives him human potentialities superior to those of animals. Man's intelligence is not confined to knowledge, culture, verbalism; it is with reason that we emphasize the deep intelligence, the common sense of the simple and the primitive who do not possess a very rich language. They are no less capable of utilizing well their cerebral potentialities, and it must also be noted that these simple folk, for all their simplicity, have still got all the basic elements of a language, and are without doubt far removed from primeval Man, whose intelligence, practically unverbalized, was all the more remarkable.

There remains the case of the *aphasics*[19], brain cases who, having had the power of speech, have lost it. The interesting cases are not those of the loss of articular motorization in Broca's aphasia, but those of complex lesion troubles in *Wernicke's sensory aphasia*, where it is commonly stated that there is a loss of internal speech, from the point of view both of the unexpressed thought and of its utterance. The patient can articulate words, but without sequence or significance. Yet he is not deranged: tests show the retention of his intelligence, and his distress at not being able to speak is significant. Does this imply lack of connection between intelligence and speech? It would seem not, for today we have to amend the conception of what this illness means from the point of view of internal language. Alajouanine[1] has shown that in addition to internal speech, what disappears is the power to evoke words voluntarily, whether in thought or conversation; but automatisms fixed from infancy seem to be retained. The patient who cannot speak a given word says it without difficulty when it forms part of a proverbial saying, which he will blurt out all at once. Herein lie interesting possibilities for re-education. On the other hand, we know that in the adult, ablation of the speech hemisphere definitively suppresses speech, whereas in the infant there is the possibility of re-education in the other hemisphere.

References p. 580

This is not the place to dilate upon other aspects of human speech pathology. It was important to reintroduce it in its proper place in the evolution of animal language and thus to show how the complexification of the higher organ of reception and control of sound is reaching a peak in the uprising of novel qualities that confer upon the acoustic signal, as a means of communication, the privilege of becoming the basis of the most specifically human mode of thinking, origin in its turn of the spiritual progress of humanity.

REFERENCES

[1] ALAJOUANINE, TH., and P. MOZZICONACCI, 1948, *L'Aphasie*, Expansion Scientifique Française, Paris.
[2] ARANSON, L. R., and G. K. NOBLE, 1945, Sexual behaviour of Anoura, *Bull. Amer. Mus. Nat. Hist., 86*, No. 3, 89–139.
[3] BENOIT, J., 1956, États physiologiques et instinct de reproduction chez les oiseaux. In *L'Instinct dans le Comportement*, Colloque Singer-Polignac, Masson, Paris, 177–260.
[4] BUSNEL, R. G., 1955, L'Acoustique des Orthoptères, Colloque international Jouy 1954, *Suppl. Annales des Epiphyties*, I.N.R.A., Paris.
[5] CHAUCHARD, P., 1956, *Les Mécanismes Cérébraux de la Prise de Conscience*, Masson, Paris.
[6] CHAUCHARD, P., 1957, *Précis de Biologie Humaine*, Presses Universitaires de France, Paris.
[7] CHAUCHARD, P., 1958, *Le Langage et la Pensée*, 2ème Édition, Presses Universitaires de France, Paris.
[8] EYRIES, CH., and R. HUSSON, 1955, Physiologie de la phonation, *Encyclopédie médico-chirurgicale, Otorhinolaryngologie* 20.632 A (10) 1–5 and B (10) 1–11, Paris.
[9] FESSARD, Mme ALBE, and P. BUSER, 1957, Activités de projection et d'association du néocortex cérébral des Mammifères, *J. Physiologie, 49*, 521–656.
[10] GASTAUT, H., 1958, Données actuelles sur les mécanismes physiologiques centraux de l'émotion, *Psychol. franc., 3*, No. 1, 46–65.
[11] GESELL, A., and C. S. AMATRUDA, 1945, *The Embryology of Behavior*, Harper, New York.
[12] HALDANE, J. B., 1954, La signalisation animale, *L'année biologique, 30*, 3/4, 89–98.
[13] HUBER, F., 1955, Sitz und Bedeutung nervöser Zentren für Instinkthandlungen beim Männchen von *Gryllus campestris* L., *Z. Tierpsychol., 12*, 12–48.
[14] JOUVET, M., 1956, Aspects neurophysiologiques de quelques mécanismes du comportement, *J. Psychologie, 53*, 141.
[15] KOEHLER, O., 1956, Sprache und unbenanntes Denken, in *L'Instinct dans le Comportement*, Colloque Singer-Polignac, Masson, Paris, 647–674.
[16] LE MEE, J. M., and P. ABOULKER, 1953, Données actuelles de la physiologie de l'audition, *Congres Soc. Franç. d'O.R.L.*, Arnette, Paris.
[17] MORIN, G., 1958, *Physiologie du Système Nerveux Central*, 3ème Édition, Masson, Paris.
[18] OLERON, P., 1957, *Recherches sur le Développement Mental des Sourds-muets*, Centre National de Recherche Scientifique, Paris.
[19] OMBREDANE, A., 1951, *L'Aphasie et l'Élaboration de la Pensée Explicite*, Presses Universitaires de France. Paris.
[20] PENFIELD, W., 1953, A consideration of the neurophysiological mechanism of speech and some educational consequences, *Proc. Amer. Acad. Arts and Sci., 82*, 201–214.
[21] SCHWARTZKOPFF, J., 1954, Schallsinnesorgane, ihre Funktion und biologische Bedeutung bei Vögeln, *XIth Intern. Ornithologen Kongress*, Basel.
[22] TOMATIS, A., 1956, Relations entre l'audition et la phonation, *Ann. Télécommunications, 2*, 7–8; *Cahiers d'Acoustique*, No. 74, 1–8.
[23] VANDEL, A., 1947, *J. Psychologie, 44*, 129–153; *ibid.*, 1958, *55*, 151–171.

GENERAL BOOKS

Brain Mechanisms and Consciousness, Blackwell, Oxford, and Masson, Paris, 1954.
Larynx et Phonation, Symposium sur la Fonction Phonatoire du Larynx, Presses Universitaires de France, Paris, 1957.
La Voix, Cours International de Phonologie et de Phoniatrie, Maloine, Paris, 1953.
Cycle de cours sur Pavlov, La Raison, Paris, 1954, Vol. 8.
La Motivation, Congrès de Psychologie Scientifique de Florence, Presses Universitaires de France, Paris, 1958.

PART V

ACOUSTIC BEHAVIOUR

A – INVERTEBRATES (Chapter 21)
B – VERTEBRATES (Chapters 22–25)

CHAPTER 21

ETHOLOGICAL AND PHYSIOLOGICAL STUDY OF SOUND EMISSIONS IN ARTHROPODA

B. DUMORTIER

The ethological and physiological study of sound emissions in insects dates from the beginning of this century. The names of Regen, Fulton and Allard are among those of the pioneers in this branch of research which since that time has continued to develop and unfold.

Although today the descriptive study of acoustic behaviour is still—and rightly—being carried on, attention is also being given to the physiological factors which control this complex type of behaviour.

In this chapter we shall endeavour, by analysing sounds from the ethological, physiological and genetical point of view, to give a picture of the various aspects of modern bio-acoustic science.

1. Sound Emission and its Significance in Behaviour

(a) The place of the acoustic signal

The range of specific stimuli that cause particular behaviour is of such variety that each sense can be used as a receptor:

Osmo-receptors for the chemical emissions in Lepidoptera (Saturnidae and Bombycidae), and in Diptera (Trypetidae), etc.[73].

Chemo-receptors for the glandular secretion in various Gryllodea males (Hancock's gland in *Oecanthus*, the tibial gland of certain *Nemobius*).

Mechano-receptors for vibrations transmitted by the web (spiders), by the substratum (Orthoptera, Ephippigeridae and Oedipodinae), or for certain tactile stimulations (contact of the antennae) occurring in the prelude to mating (Orthoptera: Gryllidae).

Visual receptors for luminous stimuli (Coleoptera: Lampyridae and Elateridae), or for gestural messages (Arachnida: Salticidae; Crustacea: *Uca*).

Auditory receptors for acoustic emissions (see Chapter 15).

All the specific signals received by these various types of sensory apparatus are messages conveying particular information. In the very frequent cases where the signal is received by an individual in whom sufficient motivation (see p. 597) exists to induce it to seek the emitter (*e.g.* sexual motivation), the physical nature of the signal will have to be such as to provide the two individuals with the greatest chance of meeting. This chance depends on several factors, two of which are especially important: (i) the range, (ii) the resistance to lessening or interception by screens blocking the transmission channel.

If we omit vibratory or tactile stimuli which occur most often when the partners are near to one another, there still remain chemical, optic and acoustic messages.

References p. 649

Unquestionably chemical messages have the widest range, reaching several kilometres. In addition they are the only ones that are not instantaneous: their remanence enables them to impart lasting information. Their arrival at the receptor, however, is entirely dependent on the direction of the current transporting them. On the other hand there is practically no masking in this type of message.

Visual information, both luminous and gestural, is more limited in range because of the low acuity of the receptors (among arthropods, in any case). It is also restricted by the natural screening caused by the soil or by vegetation. Finally, its use is limited by the sensitivity of the eye: the gestural message is only perceptible in daytime*, the luminous only at night.

What position does the acoustic message occupy in this range of biological signalling?

As regards range, it is clearly greater than optical or gestural messages, but what must be taken into account is not the distance at which the signal is no longer physiologically perceptible, but the much shorter distance at which it ceases to have any value for releasing a reaction. Very few studies have been made in this direction. We must point out, however, that the female of the genus *Ephippiger* Berthold (Orthopt., Ephippigeridae) finds a stridulating male at a distance[45] of some 20 m. The female *Tettigonia cantans* (Fuessly) (Orthopt., Tettigoniidae) is similarly attracted up to 30 m or more and that of *Gryllus campestris* L. up to 10 m**. Among males *Pholidoptera* Wesmael (Orthopt., Tettigoniidae), meetings can take place between animals as far apart[24] as 13 m (see p. 588). The range is considerably less among the Acridinæ. It has been accepted[104] that the approximate limits of distance of audition of one male by another are the following: *Stenobrothrus lineatus*: (Panz.): 50 cm, *Chorthippus brunneus* Thunb.: 200 cm, *C. parallelus* (Zett.): 140 cm, *Omocestus viridulus* (L.): 140 cm. Although the data supplied by the human ear are of little interest in this sphere, a few maximum ranges for man can be given[3,24]:

Crioceris XII punctata L. (Coleopt., Chrysomelidae)	1.30 m
Lilioceris lilii Scop. (Coleopt., Chrysomelidae)	1.50 m
Nemobius sylvestris (Bosc) (Orthopt., Gryllidae)	8.00 m
Tettigonia viridissima L. (Orthopt., Tettigoniidae)	107 m
Gryllus campestris L. (Orthopt., Gryllidae)	134 m
Neoconocephalus nebrascensis (Bruner) (Orthopt., Conocephalidae)	250 m
Gryllotalpa hexadactyla Perty (Orthopt., Gryllotalpidae)	350 m
Neoconocephalus robustus crepitans (Scudder) (Orthopt., Conocephalidae)	500 m

Finally, it seems that although in natural conditions no screening can affect the reception of the sound emission, yet it can nevertheless be partially hindered by

---

* In fiddler-crabs the male beckons with his large cheliped during the day, but uses an acoustic and vibratory signal (vibration of the cheliped) during the night. Both have a sexual value[18-20,60,95]
** As regards the precision of the localisation of the transmitter by the receiver, it has generally been found that it was very exact.

climatic conditions (wind, differences in the density of the air near the ground). But these disturbances only cause a very limited reduction of the message intelligibility.

To sum up, the acoustic message undoubtedly represents the most complete and efficacious mode of imparting information, owing to its facility of diffusion, its resistance to disturbances and also to its possibility of creating a vocabulary by a variation of its different parameters*.

(b) *Principal types of sound emissions*

A rigid determinism governs, in most cases, sound production among arthropods. In a given environmental condition the signal is emitted; the animal will react to a certain situation by a well-defined type of sound response.

What these situations are, how the animal responds to them, what change in the situation is released by the "response", are the questions with which we shall now deal, leaving aside physiological conditions, which will be the subject of a special study (see pp. 639 ff.).

Faber defined a very precise terminology characterising the different types of song for the Insects, and especially for the Acrididae. The one which we are going to use here is directly inspired by it.

We think it possible to group sound phenomena under two headings according to the behaviour with which they are associated and the situation which conditions their appearance.

A. Emission ending with the creation of a situation which satisfies a need or a tendency.
   (a) Calling song
   (b) Congregational song
   (c) Premating songs (courtship, jumping, agreement songs).

B. Emission associated with a "hostile" or defensive attitude, often tending to put an end to a situation of constraint (not sought by the animal).
   (a) Rival's song
   (b) Disturbance sounds (fight, warning, contact sounds, etc.)
   (c) Protest sounds.

(b.1) *Emission ending with the creation of a situation which satisfies a need or a tendency*

(i) *Calling song* (German: *gewöhnlicher Gesang*, French: *chant d'appel*)

The various synonyms which designate this emission give its principal characteristics: ordinary, usual, wonted, spontaneous, indifferent, solitary, common song[77].

The calling song** is emitted by the male, sometimes for hours, and has perfectly defined physical characteristics (frequency, intensity, duration of the pulses and chirps, pulse rate) (see Chap. 12).

---

* Although by its very nature the acoustic signal has no remanence, it may appear at the level of the receiving individual under the form of "memory" (see p. 619).

** Insects are being referred to here, and especially Orthoptera because there is only very little information about other arthropods.

References p. 649

Though this song may be produced by an isolated male, its emission is, however, greatly facilitated by the animal's hearing the stridulation of other males of the same species and, in some cases, by artificial sound stimuli (see pp. 610 ff.).

In most examples known in the Orthoptera, the calling song of the male conditions the reaction of the female (attraction with or without sound response). Moreover, a taxic effect on other males (see p. 588) has also been established in some cases.

This acoustic behaviour has been recorded many times among the Orthoptera, since the experimental demonstration of its existence in the cricket made by Regen[153] in 1913* [40,64,65,178,185].

On the other hand, in the Homoptera the effect of this kind of song on the female is often quite indefinite.

The calling song is also used in the Heteroptera (Corixidae, Notonectidae, etc.) in which the attraction of the female *Buenoa limnocastris* Hung. by the male's stridulation has been recorded[110].

It is extremely probable that Coleoptera, with such varied apparatus of emission, also use stridulation in their sexual behaviour, but there have been only a few observations and without any experimental verification. We might, however, mention the attraction of male Anobiidae by the tapping of the female (see p. 326) and that of the male *Cerambyx scopolii* Fuessly by the sound emission of the female[181].

The significance of the stridulation made in flight by males of the Lepidoptera Arctiidae in the genera *Endrosa* Hb. and *Euprepia* Ol. is also uncertain, but it may be concerned with mating[29,141,142] (see p. 282).

Perhaps the vibration** which various species of fiddler-crabs (*Uca pugilator* Bosc, *Uca tangeri* Eydoux) make with their large cheliped should be included here (see p. 328). This emission is detected at night, and is given out by groups of crabs at the entrance of or near their burrows. It seems to have a sexual significance, just like beckoning gestures. This drumming may also be considered as a courtship display (see p. 597)[18-20,34,60,95].

Though it is not properly stridulation, we should also mention the flight noise of the female mosquito, which has an attractive effect for the male in many species (see p. 619).

In another Diptera of the genus *Dacus* (provided with a stridulatory apparatus), the emission of the male has been seen to attract the female (see p. 290).

Hence the calling song is the first stage in sexual behaviour: it leads up to the meeting of the male and female. When the partners are a short distance from each other, other song sequences occur in the more advanced groups. These are associated with other stimuli (visual, tactile, chemical) and set off copulation.

Though the role of the calling song in the meeting of the sexes is quite obvious in most cases, we still do not know how to interpret it as a neurophysiological phenom-

---

* However, the taxic effect of the cricket's calling song has been known for a long time. The French entomologist Goureau was already writing about this subject[91] in 1837: "(The male sings) to attract his female and to charm her". This same author had also recognised the courtship song: "When a female comes who has been attracted by this song, he goes up to her, touches her with his antennae and adapts his song, which becomes much sweeter and softer; it is interspersed with a short lively sound which recurs regularly at close intervals". Goureau had also noted now the female Ephippiger was attracted by the males stridulation: "The male songs are intended to call and charm the females".

** This vibration has been called; shivering, rapping, quivering, bouncing, drumming, tapping, *Trommelwirbel*.

enon. Some writers see it as resulting from a spontaneous and rhythmic activity of the central nervous system, similar to that frequently detected of the nerves in electrophysiology. Moreover, the work of F. Huber throws some new light on this aspect of the problem (see Chapter 17).

As regards behaviour, this type of sound manifestation should be considered not only at the level of the individual, but also at that of the group. So we may liken the calling song of the arthropod to the territory song of some birds (see Chapter 24). Actually, this kind of emission, besides inducing tropism in the female, is sometimes what we could in a figurative sense, call a "warning", usually given to other members of the same sex, so that they "know" not to come within a certain distance of the singer or they will set off the chain reaction of rivalry or combat. The fiddler crabs and the ocypods very clearly have this deliniation and preventive defence of their territory, and it is almost certain that the sound signal is part of this behaviour. These phenomena are not so clearly marked in the insects but it has been established that Orthoptera when singing are generally separated from one another. If this distance is reduced by the approach of another individual, the song and the behaviour are modified, setting off a "conflict" which one of the animals terminates by with-drawing.

As far as the sub-family Acridinae is concerned, the distance between the animals when the calling song gives way to the rival's song, may be considered as giving an idea of the extent of this territorial area.

This "respect of territory" may be of biological usefulness because rival's or disturbance songs have no attractive value for the female, and as they are emitted, any chance of mating (hence of activity propagating the species) is suppressed.

This defence of territory involves some sedentariness on the part of the animal, which can easily be checked. By marking, it has been established that many Orthoptera may stay within an area less than one m^2 for several days. In *Oecanthus pellucens* (Scop.) (Oecanthidae), the same insects have been found on the same leaves for fifteen nights consecutively; this was within an area of some dozen cm^2 [2,36]. Many Tettigoniids have also been reported sedentary: *Pterophylla camellifolia* (Fabr.) *Neoconocephalus ensiger* (Harris), *N. nebrascensis* (Bruner), *Orchelimum nigripes* (Scud.), *Amblycorypha rotundifolia* (Scud.), etc.[3]. In various *Chorthippus*, ten days after marking, 70% of the insects were in the same place and 95% had not moved more than 5 m[56]. Hence the individual territory seems to be integrated into a collective territory, which itself is very fixed. This may be evidence of a cohesion factor, (independently of ecological factors), which we shall discuss later (see p. 588).

Another problem to be considered on the group scale is that of the relations between various qualities of the signal (intensity, range, duration) and the density of the animal population. Busnel (1955)[37], produced some interesting concepts on this subject when considering the French Orthoptera. If the population is very dense, the signal may be faint in intensity (hence with a limited range), and yet at the same time able to reach a receiving animal. This is the general tendency in Acridids.

On the other hand, if the population is not dense, (a frequent case in Tettigoniids), the signal must have a range (hence an intensity) such as to reach a distant individual.

Another factor besides the intensity with which the message is emitted is important: its duration. The greater the distance between two individuals, the higher should be the repetition rate of the emission. The receiving animal can then direct itself towards the emitter and quickly correct any false step. On the other hand,

References p. 649

short or sporadic emissions are sufficient in the cases where many males and females live together in a confined space, since the distance to be travelled is always short.

Obviously these correlations have to be considered more as an attempt at rationalising insect communications, than as an expression of formal laws.

(ii) Congregational song (aggregating call)

The transition from the former song to this type is often quite indistinct, as regards the physical definition of the phenomenon and its ethological significance. Actually it is quite possible that the congregational song acts also as a calling song, and conversely the calling song seems definitely to have a gathering function in some cases (see below).

These two types of signal, however, may be distinguished by the fact that the congregational song does not only attract the opposite sex whereas the calling song does. The congregational song produces the grouping of males, females or larvae. It seems to be able to keep the group together in some cases, and in other cases to have a social character.

The congregational song is especially characteristic of the Homoptera Cicadidae in which it appears (see Chapter 14) in the form of a collective song, whether synchronised or not (see p. 617). The study on *Magicicada septemdecim* (L.) and *M. cassinii* (Fisher)[11] showed that it resulted in the gathering of both sexes into a confined area. Generally speaking, the males are more attracted by the congregational song of the other males than by the presence of females themselves. This chorusing behaviour which groups together closely the males and females of the same species, makes it easy for the sexes to meet and probably diminishes the risk of hybridisation.

One of the two songs of *Kleidocerys resedae* (Panzer) and *K. ericae* (Horvath) (Heteropt. Lygaeidae), which induces heterosexual gatherings, may also be put in the congregational song category. Moreover the stridulation is detected in both sexes[100].

Apart from the Hemipteroida the aggregating effect of the song has been recorded only a few times. Some cases have been reported in the Orthoptera in which the calling song of the male caused an attraction of other males. In this connection Baier (1930)[24] released thirty males of *Pholidoptera cinerea** (Tettigoniidae) 13 m from a cage of other males of this species, and found seventeen near the cage some time after. This same author also managed to establish the clear movement of a male *Gryllus campestris* L. towards another male which was stridulating.

Similar facts are recorded in the Acridinae[113]. After an acoustic isolation of one to three days, males *Stenobothrus lineatus* (Panz.), *Chorthippus brunneus* Thunb.**, *C. parallelus* (Zett) head for a loud-speaker relaying the calling song of the species. In this way they may travel a distance of 1.20 m. When they come near the apparatus, the insects sing in tune with the recorded song. In *C. parallelus*, it has been established that individuals, at first isolated, are attracted by a group of stridulating males[102].

Perhaps these cases should be regarded as a suggestion of the aggregating behaviour shown more fully and clearly by the Cicadidae. Though homospecific groups are also created as a result of this behaviour, the density of these groups will

---

* It is probably *P. griseoaptera* (De Geer).
** It will be noted that artificial acoustic stimuli very easily produce attraction in the males of this species, preparatory isolation not being necessary (see p. 620). The males moreover show a phonotaxis towards the females' stridulation.

certainly be rapidly limited by rivalry and defensive behaviour of a certain territorial perimeter.

The fact that the taxic reaction between males and females Acridinae only appears experimentally in animals previously isolated acoustically, suggests that these animals may seek a "specific sound atmosphere", the optimal level of which in the field is regulated by the distance of rivalry between the animals. This "need" may explain the tendency of subjects withdrawn from this "atmosphere" for some time or of the newly emerged adults, to head towards the source (loud-speaker or another male giving out the song of the species).

It is obvious that these gregarious displays, basically acoustic in the Orthoptera, have only been put under the heading "congregational song", because of the resemblance they have with those which we have previously reported in Hemipteroida. They are clearly distinguished from the latter, because the signal which produces them is the calling song, the fundamental role of which is heterosexual meeting*.

The Coleoptera of the family Passalidae give a quite remarkable example of the use of stridulation in the larva/parent relationship, which can be related to an aggregating behaviour. These insects live in worm-eaten wood with their larvae and it appears that they feed them[133]. An interesting observation has been made with a Brazilian species. After having extracted two adults and six larvae from a piece of wood, the observer found the two adults and four of the larvae together a few moments later under a new shelter, while the two other larvae hurried across the ground to meet the family which was stridulating continuously[136]. Even if olfaction is made to account for this regrouping, it is nevertheless true that a certain role can be attributed to stridulation (see pp. 289 and 331).

It may well be that in a few other cases the establishment of groups of the same species and the maintenance of their cohesion are also partially based on the sound signal. Here again it is a case of an aggregating effect of the signal, but not necessarily of an aggregating call, which can be differentiated as such from other types of emission which may possibly be used by the species. In this line of thought it does not seem improbable that sound emissions may come into play in certain gregarious displays such as the flight of mosquitoes or acridians. In the desert locust *Schistocerca gregaria* (Forsk.), the flight noise produces action potentials synchronous with the wing-beat frequency (17–20/sec), on the tympanic nerve of an individual situated at about 1 m from the flying insect. Although this flight noise (transmitted by the loudspeaker) does not seem to be able to cause a settled swarm to rise systematically[101], it is possible that it plays a part in maintaining individuals in flight in group formation.

Similarly, in the crabs of the genus *Dotilla*, the humming during flight from danger may very well be a means, along with other stimuli, of preventing the dispersion of group (see p. 312).

The ocypods' stridulation choruses, which have been described many times, are produced by animals standing at the entrance of their burrows and may keep the colony together. This grouping is, however, limited by the crab's highly developed instinct for defending his territory. The same supposition may be made for crabs living together in dense groups (as much as 80 to 1 m²), such as *Ilyoplax*, *Macrophthalmus*, *Uca*[25, 26, 94].

---

* We should remember that this role is not obvious in the Acridinae[113].

Some authors interpret the stridulation of ants (*Crematogaster scutellaris* Ol., *Megaponera, Messor*)[148,164] as a means of avoiding the dispersion of the group on the move, but it is also possible that it acts as a danger warning (a general agitation in a colony of *Pogonomyrmex* is caused by the stridulations of workers shut in a flask[180]), or again, as an appeal sent out to other members of the species by an individual in danger, wounded, or put out of action, (instance noted in *Megaponera*[165] and *Crematogaster scutellaris*[164]). There is evidence[165] that in some species (*Messor*), when a worker has found a source of food and has returned to her nest may "inform" her companions acoustically and lead them to her discovery. To the extent that the observation is accurate, this behaviour must be comparable to the bees' dance, as a means of communication.

Finally, a "nest music" has been postulated in the stridulating ants, fairly comparable with the "nest smell" which may act as a guide to animals when they are dispersed.

This collection of facts and suppositions furthers the idea that the sound emissions of ants may be closely related to the social life of these insects*.

In the termites, there is much evidence for denoting the tapping of the mandibles as a signal integrated in to the social life of the insect, (see p. 327)[33]. In the African species *Termes lilljeborgi* Sjost., for example, if a group engaged in gathering food for the colony is disturbed, the animals which are scattered across the vegetation emit their characteristic tapping and they all rush back to the nest[167]. Elsewhere soldiers have been seen to run to a breach made in a tunnel, beating the ground rapidly with their heads in order to gather the workers, which get to work repairing the damage[90].

There is obviously a need for experimental evidence, which cannot be replaced by chance observations from nature, to support the hypothesis that acoustic communications (or vibratory ones) exist in the termites.

To sum up, the congregational song is an attribute of the Hemipteroida, and more especially of the Cicadae. Though the song has an effect of gathering animals in some other cases, it is never of great importance, and very often, the role of the sound emission in the formation and furtherance of specific gatherings is still hypothetical

(iii) *Premating songs*

We shall put under this heading the various kinds of songs which are found especially in the Orthoptera Gryllidae, Oecanthidae and Acrididae. All of these songs lead up to mating more or less immediately. With the exception of the Orthoptera and the Hemipteroida (the latter are considered in Chapter 14), practically nothing is known of the existence of premating songs in the insects and other arthropods.

Among the Orthoptera, however, the song as a preliminary step to copulation does not have the same importance in all cases. For example, in the Tettigoniids, special stridulations which act as courting sounds when male and female meet have never been reported. Any individual attracted by the calling song of the partner

---

* The rapping of the hornet larvae (*Vespa crabro* L.) against the walls of their cell may be considered, in connection with a gestural signal, as a "food begging call", since it is only observed in hungry individuals and stops after feeding[89].

mates without any other type of emission being necessary*. In the Gryllids, on the other hand, the association is produced in two stages: a calling song which attracts the female and a courtship song when she comes near or in contact with the male. The Acridinae, however, show the most complex acoustico-sexual behaviour. Though the calling song does not always have a clearly taxic role with respect to the female[113], the different conducts of the male and female which result in copulation are regulated by a strict and developed sound ritual: the male's courtship, the response of the female (in some species), the alternating song of the male and female (in some species), the quest sound of the male (in some species), and jumping song of the male (see Fig. 361).

We shall consider these different kinds of emissions in order.

(1) Courtship song (Serenade, German: Werbegesang, French: chant de cour)**

This is always a display of the male, set off by a stimulus from the female. This stimulus is probably visual in the Acridinae and the Oecanthidae. It has been demonstrated with regard to the former that in the case of *Chorthippus brunneus* and *C. biguttulus*, the female did not produce any chemical emission with a stimulating effect on the male's song[140].

In the case of *Gryllus bimaculatus* De Geer (Gryllidae), however, an olfactory emission of the female seems to act as a stimulus on the male[106], but in most of the other Gryllidae, the triggering stimulus seems to be tactile. The courting behaviour of *Gryllus campestris* L. is caused by the partner's response to the blow from the male's antennae. If the second individual reacts with a whipping blow from his antennae, fight behaviour begins; if, on the other hand, the antennae move more gently, the male begins his courting activity, whether he is face to face with a male or a female[106, 108]. Sometimes, any meeting between animals of the same species provokes the emission of the fight song, which only turns into a courtship song if the other animal is recognised as a female (case of *Acheta veletis*)[4] (Fig. 358).

As a general rule, in the Gryllidae the courtship song is chiefly characterised by a considerable widening of the frequency spectrum (quite striking in insects which emit the calling song with an almost pure frequency) (Fig. 358). Moreover the elytra are hardly raised from the body, which frequently trembles, whereas they are clearly lifted up in the calling song. Finally, the number of pulses per chirp, the pulse-rate, the duration of the chirps, and the chirping rate, show some important and specific changes.

Acheta domesticus (L.) provides a good example of the courtship behaviour of the Gryllidae[3, 97]. In this species, the calling song is a chirp of two or three pulses (pulse-rate = 16 at 27°C), emitted at a rate of one per second. When the male is near a female, the courtship behaviour begins. The real courtship song, however, is only gradually emitted as the calling song is modified. Two stages have thus been identified: the pre-courtship, consisting of chirps from 1–2 pulses, with a spectrum already

---

* This does not necessarily mean that there is no "courting". Contact of antennae, chemical stimuli (Oecanthidae), vibrations, etc. may take place and play a role when the two partners are together. In *Ephippiger*, for example, a rapid vibration of the male's body has been noted when the female approaches. This vibration has a taxic effect on the female (see p. 59).

** The French entomologist Goureau was certainly of the first to describe the courtship song of the Acridids, since he was already writing in 1837[91]: "When the male sees a female, he runs to meet her and stops a little distance away; then he stridulates faintly with several cries, which you have to listen to carefully if you want to detect them. When the female stays still, he throws himself on her . . ."

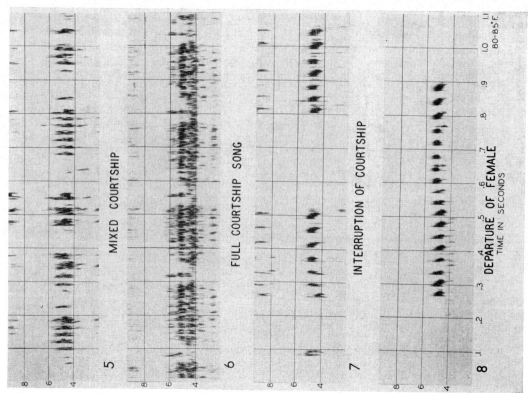
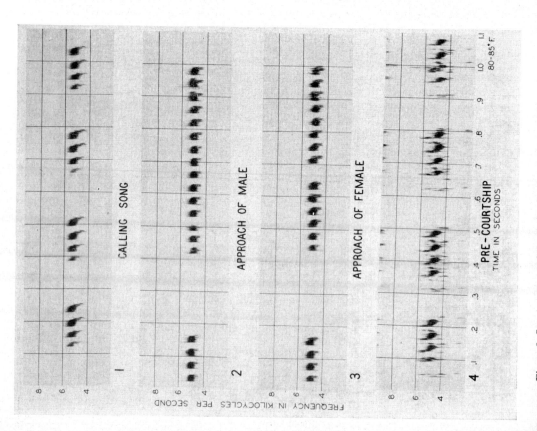

widened, and the mixed courtship similarly constructed to the courtship song, but interspersed with sharper pulses of the calling type. The mixed courtship is emitted as long as the female stays near the male but does not try to mount him. If the female tries to get away, the number of calling pulses increases*. It decreases when the female succumbs to the male's advances. When the calling pulses stop, real courtship begins. This has a wide spectrum with chirps of 8–12 pulses (2–3 chirps per second) with a pulse-rate beginning at 20 at the start of the chirp, decreasing to 12 at the end. The chirp ends with a sharp well detached pulse[3]. During the whole courtship song the male jerks his body, a behaviour pattern frequently observed in this family. In *Acheta assimilis* (Fabr.), the behaviour is very similar. The type of modification of the calling song leading to the courtship song is different in each species: in the case of *Anaxipha exigua* (Say), the continuous trill of the calling song is transformed into short trills; similarly the courtship song of *Nemobius fasciatus* (De Geer) is obtained by a shortening of the song; in *Miogryllus verticalis* (Serville), the chirping rate increases and the number of the pulses per chirp decreases (Fig. 359)[3].

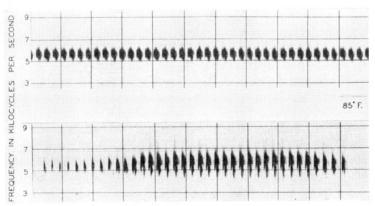

Fig. 359. Spectrograms of the calling (upper) and presumed courtship song (lower) of *Nemobius melodius* Thom. and Alex. (After Alexander, 1957[6].)

In the family Oecanthidae, the courtship behaviour and the courtship song are less well differentiated. In *Oecanthus argentinus* Saussure, the continuous trill of the calling song, is transformed into an irregular vibration of the elytra with or without sound emission. In *O. pellucens* (Scop.) the difference is slighter still, the frequency is not modified (contrary to what is seen in the Gryllidae), there is simply a decrease in the number of pulses per chirp and a grouping of the chirps in shorter series**[36].

Jacobs (1953)[113] has proposed a classification of the courtship for the Acridinae which corrobates the basic observations of Faber. He is led to distinguish four groups:

---

* The male reacts to the female's departure with the fight song (see p. 603).

** In the Nemobiinae and the Oecanthidae, the courtship song is generally speaking only slightly distinct. Without assuming a relation of cause and effect, we must note that in these two groups the male has a secretory organ, from which exudes a liquid which the female licks before or during copulation. In the Nemobiinae of America the secretory apparatus is situated in the proximal spine of the hind tibiae[80], and on the right elytron[159] in the French species. In the Oecanthidae it is Hancock's metanotal gland.

Group 1. No courtship. The male leaps on the female with or without jumping sound before (see p. 596):
Arcyptera fusca (Pall.), *Chorthippus pullus* (Phil.).

Group 2. Faintly expressed courtship; the courtship song is still very like the calling song:
Chrysochraon dispar (Germ.), *Euthystira brachyptera* (Ocsk.), *Chorthippus parallelus* (Azam.), *C. brunneus* Thunb., *C. vagans* (Eversm.).

Group 3. Long courtship; courtship song still like the calling song:
Chorthippus montanus (Charp.), *C. dorsatus* (Zett.), *Stenobothrus stigmaticus* (Ramb.), *Stauroderus scalaris* (Fisch. Waldh.).

Group 4. Production of a real and complex courtship song, of long duration; often broken with various short elements, whether sonic or not (rapid blows of the femur, for example):
Chorthippus biguttulus (L.), *C. mollis* (Charp.), *C. albomarginatus* (De Geer)[31], *C. apricarius* (L.), *Stenobothrus nigromaculatus* (H.S.), *S. lineatus* (Panz.), as well as various species of the genera *Omocestus* Bol. and *Gomphocerus* Thunb.

The Acridinae, then, provide within the sub-family an example of a very gradual development of courtship behaviour, starting with a complete lack of courtship song and ending with a long emission made up of various chirps resulting in variations on the theme of the calling song. In the species with very elaborate courtship, individual difference are noted. As a general rule the courtship behaviour ends with the jumping sound which is immediately followed by the male jumping on the female.

In the other Acridoidea families, the courtship displays are much less marked, and very often it is only possible to speak of courtship because the behaviour observed corresponds, in the progress of the premating activities, to the courtship song of the Acridinae. In this connection the lifting of the wing in *Bryodema tuberculata* (Fabr.) (Oedipodinae) and in *Tetrix tenuicornis* (Sahlb.) (Tetrigidae) is cited; or the extension of the tibiae in *Mecostethus grossus* L., *Aiolopus thalassinus* (Fabr.), *Bryodema tuberculata* (Oedipodinae).

Apart from the Orthoptera, manifestations of courtship are found among various Homoptera Cicadidae (see Chapter 14) and Heteroptera. Among the latter we may mention male *Sehirus bicolor* (L.) (Cydnidae) which, when he encounters a female produces a stridulation different from the calling song[100]. Among the auchenorrhynchous Homoptera, *Paropia scanica* (Fall) (Megophthalmidae) and *Eupelix depressa* (F.) (Eupelicidae), a courtship song has also been observed accompanied by dances near the female and attempts at copulation. A courtship song has similarly been described in *Doratura stylata* (Boh.) (Euscelidae)[137].

(2) Agreement song (Attraction song, call of invitation, German: Weibchengesang, French: chant d'acceptation)

This characteristic stridulation of the female of some species of Acridinae is considered as the expression of the willingness of the female to copulate[111, 113]. The stimulus which triggers this behaviour seems to be chiefly acoustic: the calling song of the male*. However, it has been shown[140] that in *Chorthippus biguttulus*, a specific olfactory stimulus from the male significantly increased the number of stridulations emitted by the female.

The agreement song is usually in a similar form to that of the calling song, but somewhat simplified: shortened (*Chorthippus apricarius, Omocestus viridulus*); lower pulse-rate (*C. parallelus, C. montanus*), etc. The signal sometimes has a similar intensity to that of the male (*C. dorsatus, brunneus, montanus*), but there are many individual differences. In addition a single insect shows great variations in its readiness to stridulate which are apparently related to the extent to which the ovocytes have matured[103] (see p. 640).

In most cases the female's agreement song stimulates a response in the male (usual calling song or slightly modified), followed by a new stridulation by the female, and the animals continue to alternate their song in this way (*Werbewechselgesang* of German authors). This is interspersed with short pauses during which the partners move towards one another. This exploratory behaviour also varies with the species. In *Chorthippus montanus*, male and female seek out each other[113]. In *Stenobothrus lineatus* and *Omocestus viridulus* the female goes to the male[103]. (We should note that in *C. parallelus*, the female is brachypterous and thus only makes stridulatory movements without sound emission[103].) On the other hand in *C. biguttulus* the male seeks out the female[47,179]. When the two partners come near to one another, the male, either begins his courtship song or assails the female directly after a jumping sound (see p. 596), according to the species.

Apart from the Acridinae, sound emissions of the female which demonstrate acceptance of copulation are apparently quite rare. However, a stridulation without taxic effect which is emitted by the female when the male is covering her, has been reported[35] in *Locusta migratoria* L. (*solitaria*).

There is good reason to think that the stridulation of some female Tettigoniidae (see p. 299), which is made in response to the calling song of the male, may be looked upon as an agreement song. Observations have been made in the American species *Scudderia furcata* Brunner, *S. curvicauda* (De Geer), *Microcentrum rhombifolium* (Sauss.), *Amblycorypha rotundifolia* (Scud.) and some others[82].

In *Scudderia texensis* Sauss. and Pict. it has been possible to note that the response of the female followed directly after the song of the male and that it was produced by a similar movement of the elytra. In this species as in *Microcentrum rhombifolium*, the male moves towards the female[16,82]. Hence it seems that the male induces by his song the female's response, which then produces the noted taxic reaction of the male. Here we have a complex behaviour pattern similar to the one we reported before in some Acridinae.

In the European forms, the stridulation of the female has been observed in *Tylopsis lilliifolia* (Fabr.) (Phaneroptidae)[87] and in various species of the genus *Ephippiger*[184]. In this latter case, stridulation only takes place in animals under keen sexual motivation and stops after copulation. The song is spontaneous, not produced in response to that of the male, and bears no relation to the protest sound which these insects produce when they are disturbed (see p. 605).

---

* A spontaneous emission of the agreement song has sometimes been noted (*Chorthippus dorsatus, C. brunneus*)[179].

There are very few examples of an agreement song in the other orders of insects. In the Homoptera, a stridulation of female *Doratura stylata* (Boh.) (Euscelidae), has been described. The acoustic behaviour of this species is singularly similar to that of the Acridinae, since the calling song of the male produces the emission of the female's agreement song which in its turn, determines the courtship song of the male followed by the approach towards the female[137,138]. Perhaps the emission of the Lepidoptera Arctiidae *Euprepia haroldi* Obth. female made in response to the stridulation in flight of the male[29,141,142] (see p. 282), should also be considered as an agreement song.

(3) Jumping sound (shout of triumph, German: Anspringlaut, French: cri de saut)

This is again a type of emission characteristic of the Acridinae, which comes before the leap of the male on the female. Generally it is a short stridulation. The trigger stimulus is probably the same as that of the courtship song, mainly visual. Actually it has been established in *C. biguttulus* and *C. brunneus* that the male jumped more frequently on a moving female rather than on a motionless one, whether of its own species or not[140]. In some species that have no courtship song (group 1 of Jacobs), the jumping sound is the only sound display in the sexual behaviour (*Arcyptera fusca, Chorthippus pullus*). It has also been noted that the males of some of the species which have only slightly specialized courtship (groups 2 and 3), sometimes throw themselves on the female without any courtship overtures, simply with the jumping sound (*Chrysochraon dispar, Eusthystira brachyptera, Chorthippus parallelus, C. vagans, C. montanus*, etc.).

Apart from the Acridinae, the phenomena which are comparable to the jumping sound are rarer or less well known. The mandible noises in *Calliptamus italicus* (L.), the wing movement in *Tetrix tenuicornis* (Sahlb.) or femur movement for *T. turki* (Krauss) and *T. subulata* (L.) (Tetrigidae)[113] have been likened to it as far as ethological significance goes.

In about a dozen species of Acridinae and Oedipodinae (*Stenobothrus stigmaticus, S. nigromaculatus, Omocestus haemorrhoidalis, O. viridulis, Chorthippus mollis*, etc.) a special emission has also been detected in the male which is called the "quest sound" (German: *Suchlaut*). This accompanies the movement of the male during the search for the female. These cries are always very simple and emitted irregularly[69,113].

(4) Meaning and role of the premating songs

In the males Acridinae, the courtship song seems to have a self-excitatory effect: as the courting progresses, so the song frequently gains in intensity; if it consists of isolated cries they become shorter and faster in rate (*Chorthippus dorsatus*). When the maximal excitation is attained the male utters his jumping sound and leaps on the female[113]. This is quite like the gradual charging of an electrostatic machine which produces a spark across the two terminals when the voltage is sufficiently high.

Actually, very often in the Acridinae, the male tries to "skip" the courtship song sequence by going straight for the female just with the jumping sound, and only after a series of fruitless attemps does he "decide" to engage in his courting behaviour pattern. Only in unusual circumstances does the female accept copulation without a courtship song as an introduction the duration of which varies according to her level of motivation (see below).

The male, which tries to mate with the female after each courtship song, may be repulsed several times and be forced to start his courting display from the beginning again before seeing the female give in. It is quite surprising to realise that even if the female is not ready to mate, both in the Acridinae and in the Gryllidae[97], she usually stays still until the end of the male's stridulation, without trying neither to repulse him nor to move away. She might be said to have been "fascinated"[113]. It has also been observed in the fiddler-crab *Uca tangeri*, that the vibration of the male's cheliped frequently prevented the escape of the female[95].

The courtship song, sometimes associated with gestures or vibrations, is thus apparently a stimulus which can have a double effect on the female: (i) to overcome the defense reflexes (especially in the Acrididae), (ii) to trigger the display of the mating motor patterns.

In most cases (see exceptions reported below), the song is apparently a *sine qua non* condition for the female's sexual behaviour. The female *Chorthippus brunneus*, for example, rejects the male with amputated elytra which no longer stridulates[103]. The sound signal of the male has an equally marked determining role in the species which have no courtship song, as in the *Ephippiger* for example. The female attracted by the calling song over a few cm from the male, stops and does not jump the tiny distance which separates her from him of he ceases stridulating[45] (or she takes much more time for it)*. Similarly in the mosquito *Aëdes aegypti* L. the males seem to ignore the motionless females even if they are in contact with them. But as soon as one of them flies off, her flight tone sets off the pursuit activity of the males[160]. Thus the song acts as a force pressing a trigger which releases an energy potentially present which is then spent in stereotyped activities leading to the union of the two partners.

By examining the female's behaviour we have a good basis for understanding her attitude towards the male and her reaction to the sound signal: the *level of motivation*, the reflection of the nervous and hormonal state of the insect. If this level, which can be imagined as moving according to an arbitrary scale of values, comes below a minimal threshold mark, the specific releasing signal (sound signal) has no effect. It becomes effective (*i.e.* it releases the reaction of the female) as soon as the level of motivation reaches this threshold, and it is more so as the level of motivation moves towards the high values on the scale (this is shown in the behaviour pattern by a shortening of the minimal duration of the releasing emission needed to gain the response of the female). For a still higher level of motivation, acceptance of the male by the female will be noted for instance after a simple jumping sound. Still higher up the scale the motivation will attain such a value that a trigger stimulus of the 2nd order, just the presence of the male (indicated by smell or sight), will be sufficient to incite the copulatory activity of the female. This is noted, for example, in *Ephippiger* where a female kept in isolation for a long time mates as soon as she is in the presence of a male, even though he does not sing (Fig. 360).

This is certainly a very general pattern which does not mean that we can always find all these developments in a single species, but some examples which correspond to each of these stages of motivation may be found in one group or another. It is possible, however, that in some forms with very ritualised behaviour (some Acridinae),

---

* Except if the level of motivation of the female is specially high.

all the chain of releasing stimuli is necessary to assure the response of the female, whatever her degree of motivation.

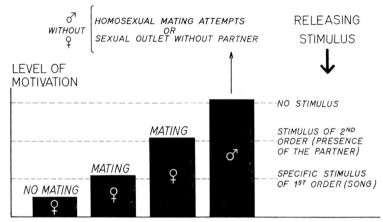

Fig. 360. Schematic representation of the level, of motivation. The black columns represents four different values of the level; the highest value is related to the male. The horizontal dotted lines indicate the minimal requisite stimulus for releasing the reaction at a given value of the level of motivation (see text).

This conception of a motivation level certainly does not apply only to insects; but rather it has been widely observed in the animal kingdom that the necessity for a very specific and special releasing stimulus gradually disappears as the motivation increases*. Obviously it would be useful to know what this term motivation actually covers in the physiological sense. We shall deal with this question later on (see pp. 639).

The premating songs, as we have seen, are not limited to the courtship song; the Acridinae especially show other acoustic display among which we shall consider the agreement song of the female and the jumping sound of the male.

The agreement song may be interpreted as the expression of a motivation state which incites the female to look for a mating partner[111,113]. This acoustic indication of a certain physiological condition is identical with that used by the male whose song is only heard if the animal is ready to mate. Like that of the male the song of the female has a taxic effect on members of the other sex, but only in some species. Moreover, as this emission of the female is made in alternation with that of the male, it seems to be quite a complex procedure of acoustic signalling: that of the two partners in motion specify or rectify the direction in which it is going, at each "response" given to its "call". From the evolution point of view, however, this display may be seen as a means of avoiding the "waste of energy" involved in the courting behaviour of the male in the presence of a female which is not ready to mate. The

---

* In the male the selectivity of the stimuli which release the copulating activity may also fade if the motivation is great (fig. 360). Often in the cases where the sexes are reared apart, males may be seen copulating among each other and trying to couple their genitalia; this implies the existence of motor patterns, which belong to the opposite sex, in the animal playing the role of female. The emission of the spermatophore may even occur without any specific stimulus (commonly observed in *Ephippiger*); the insect then rubs itself on the ground where it lays the sperm mass. This is onanism in the real Biblical sense of the word (Genesis 38:9).

agreement song does not always seem to be an indispensable point of the chain of stimuli which leads to mating. In fact this may be effected without such a behaviour pattern[113].

As far as the jumping sound is concerned (the only courting feature in some species), it has been suggested[146] that it might be a warning signal "telling" the female beforehand of the male's assault, and thus inhibiting the flight or defense reaction which she normally has towards any abrupt movement of an object or towards any unexpected contact. This may be similar to the behaviour of some mice which before moving give an ultra-sonic cry which indicates to the other mice that the noise which may follow is not to be taken as the announcement of an enemy's arrival.

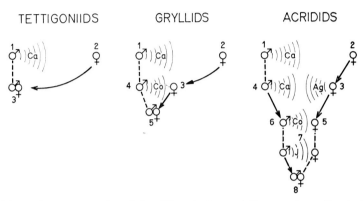

Fig. 361. Schematic representation of the different stages of the sexual meeting in Orthoptera. Ag: agreement song, Ca: calling song, Co: courtship song, J: jumping song. Dotted lines indicate the progress of the stages, arrows represent the movement of the partners.

The premating songs of insects should certainly be put in the same category as the sexual displays of which so many animals, vertebrates as well as invertebrates, provide examples. The movement of the plumose legs and the dance of some spiders (Salticidae), the beckoning with the big chelipeds in *Uca* crabs, the "wedding gifts" of Diptera (Empidinae), etc., are different behaviour patterns, but all contain the triggering stimulus which specifically conditions the female's copulatory activity. It is in this sense that the courtship song and the other sound displays connected with mating, may be compared with them.

(b.2) *Emission associated with a "hostile" or defensive attitude and tending to put an end to a situation of constraint*

This second category of emissions, quite unconnected with the meeting of the sexes, includes a certain number of sound manifestations, generally difficult to differentiate, and whose role and meaning are usually obscure.

From the point of view of behaviour, the main characteristic of these emissions is that they are not spontaneous, but made in reply to external stimuli, of a tactile, visual, (olfactory?), or acoustic nature.

(i) *Rival's song (German: Rivalengesang, French: chant de rivalité)*

An emission typical of a certain number of Acridinae peculiar to the males, which is produced, usually alternately, by two individuals a short distance from each other.

The rival's song generally differs from the calling song by: (i) its shortness, (ii) its changing intensity, (iii) the acceleration of the stridulatory movement.

In most species, the rival's song appears to be derived from the calling song, with one or more of the above modifications; in others, there seems to be no connection between the two, and the rival's song constitutes a completely original signal.

We give below some examples taken from Jacobs (1953)[113]:

1. Rival's song derived from calling song:
   Omocestus haemorrhoïdalis (Charp.)
   O. viridulus (L.)
   O. ventralis (Zett.)
   Chorthippus pullus (Phil.)
   C. vagans (Eversm.)
   C. biguttulus (L.)
   C. brunneus Thunb., etc.

2. Rival's song not derived from calling song:
   Stauroderus scalaris (Fisch. Waldh.)
   Chorthippus apricarius (L.)

Weih (1951)[179], on the other hand, gives the following classification:

1. Rival's song very different from calling song:
   Chorthippus brunneus Thunb.
   C. parallelus (Zett.)
   C. dorsatus (Zett.)

2. Rival's song with sound elements non-existant in calling song:
   C. biguttulus (L.)
   C. montanus (Charp.)
   Omocestus ventralis (Zett.)
   O. viridulus (L.)

3. Rival's song little different from calling song, except in length, intensity, pulse rate:
   Stenobothrus nigromaculatus (H.S.)
   Euthystira brachyptera (Osck.)
   Chrysochraon dispar (Germ.)

4. No rival's song:
   Gomphocerus rufus (L.)

The rival's song, as described above, denotes the so-called rivalry behaviour. The fact that this more or less characteristic signal is generally emitted in alternation by two individuals, creates a form of what Faber called *anaphony* (see p. 610). Certain writers, however, consider the anaphony resulting from the alternating emission of the calling song to be also a manifestation of rivalry (1st degree). This is apparently only a question of terminology and personal interpretation, as rivalry corresponds to no precise ethological criterion and can only be distinguished acoustically. All anaphony can be considered as rivalry, or rivalry can even be said to exist only when there is an emission of a particular type of song.*

Whatever the interpretation, the rival's song is the sign of a state of disturbance which increases in intensity as the distance between the individuals decreases. This

---

* Another possibility is to replace the expression "rival's song" by "answering song"[140]. The biological meaning of this sound behaviour is, of course, no clearer, but the word "answer" has at least the advantage of being the product of mere observation. The term "rival" on the other hand, already appears to be an interpretation—quite gratuitous—of the phenomenon.

growing reciprocal influence sometimes shows itself acoustically in the progressive variation of one or more parameters characterising this song (intensity, duration). In *Omocestus ventralis* (Zett.), for example, the length of the rival's song lessens at the distance between the two males decreases[179].

It is interesting to note that, on the exponential curve corresponding to this variation of length in relation to distance, a second curve can be almost superimposed which represents also in relation to distance, the variation in intensity (expressed in dB) of the stimulus received by the animal (Fig. 362). This means that the length of the stridulation varies in this particular case with the intensity of the stimulus provoking it*. As decibels are logarithmic units and the decrease in the length of the reply can be considered as expressing the intensity of the sensation, this is an illustration of the well-known Weber-Fechner law that the intensity of the sensation is proportional to the logarithm of the intensity of the stimulus: I sensation $\sim \log I$ stimulus. (The Fig. 365 shows another application of the same law.)

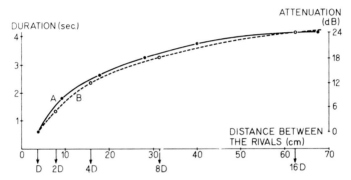

Fig. 362. Effect of the distance between rivals on the duration of the rival's song in *Omocestus ventralis* (curve A), and connection with the attenuation of the signal in terms of the distance (curve B). (Curve A, after the data of Weih, 1951[179].)*

In the case of *C. montanus*, a change in the length of the song can also be noted, passing from 0.8 sec for the contact cry** (maximum excitation, 0.5 cm) to 3.5 sec when the two animals sing separately (60 cm). The intensity of the song also gradually decreases as the two insects approach each other[179].

Factors other than duration or intensity can also intervene, such as the addition to the rival's song of various sound elements. *C. biguttulus*, for example, precedes the rivalry chirp (shortened calling song) by some short pulses, whose number is in inverse ratio to the intensity of the stimulus (in other words, the distance of the "rival")[179].

1.5 cm	isolated pulses (contact cry)**	
2 cm	3 pulses + chirp	
4.5 cm	2 pulses + chirp	rival's song
8 cm	1 pulse + chirp	
12 cm	chirp only	

* In the author's experiment, the intensity was not measured; it is known, however, that the initial intensity of a sound theoretically decreased by 6 db each time the distance between the source and the receiver is doubled. Thus, if I is the intensity at distance D, we would have $I - 6$ at $2D$, $I - 12$ at $4D$, $I - 18$ at $8D$, etc.
** This type of emission will be studied on p. 603.

References p. 649

Though such definite relations between the intensity of the stimulus and that of the reply cannot be found in all cases, the connection between the intensity of the rivalry displays and the distance from the "rival", is usually fairly clear.

In *Chorthippus biguttulus*, for instance, the following observations have been made:

0–2 cm	contact cry*
2–12 cm	alternating rival's song
12 to about 150 cm	alternating calling song
over 150 cm	the two animals no longer influence each other

In *C. montanus*, the rivalry sphere is different:

0–5 cm	contact cry*
5–15 cm	song halfway between contact cry and rival's song
15–30 cm	alternating rival's song
30–60 cm	alternating calling song
over 60 cm	the two animals no longer influence each other

The maximum interindividual distance at which the rival's song is observed therefore varies in different species:

C. brunneus	90 cm
C. dorsatus	75 cm
C. parallelus	17 cm

The ethological role of the rival's song is not yet completely clear and, in any case, varies according to the species. Thus, in *Chorthippus montanus* and *Omocestus ventralis*, the rival's song causes an attraction of individuals. On the other hand, in *C. dorsatus*, *C. brunneus* and *C. parallelus*, a repellent effect is observed[113,179]. It should be noted that in the first two species, the rival's song differs slightly from the calling song, whereas in the others the difference is very marked. We have already remarked on the aggregative effect of the calling song (see p. 588). It is therefore possible that this effect can also be produced with a faintly differentiated rival's song, the repellent effect only being observed at very short distance (at the moment of emission of the *contact cry** for instance).

In conclusion, we can say that the rival's song has two main roles:

(i) It seems to maintain a certain territorial area around each individual** (see p. 587).

(ii) A second function of the rival's song is, at least in certain species, that of sex recognition. In *Chorthippus pullus* and *C. montanus*, the male emitting the jumping sound (see p. 596) approaches any other member of the species. If the latter is a male it will react with the rival's song, and the "attacker" retreats and repeats the operation with another individual[113,179]. In this case, therefore, the rival's song acts as a signal device enabling the animal, which appears to be searching haphazardly, to recognise the sex of any met individual. The rival's song of the partner approached is then an inhibiting stimulus to the activities of copulation, which are only provoked in its absence (encounter with a female). This is a behaviour fairly similar to that of

---

* This type of emission will be studied on p. 603:
** The defense of this territory, however, is never violent in Acridinae, but seems to be carried on much more in a "symbolical" manner, by song; in Gryllidae, on the other hand, it is known that the untimely approach of one male to another gives rise to very aggressive behaviour.

some Gryllidae, in which the male that is looking for a mate waves its antennae at any other individual it meets; according to the response (waving or slow movement of the antennae), the partner is recognized as male or female (see p. 591).

Finally, it should be added that there has been another suggested explanation of the rival's song[140]. The "rivalry duets" might, in fact, be the result of an error, each of the males "thinking" it is in the presence of a female. If this theory (suggested by observation of *C. brunneus*) is valid, it cannot be considered to apply generally, for the rival's song is usually quite different from the agreement song of the female.

(ii) Disturbance songs

Under this heading we group the various emissions caused by the proximity of another member of the species, usually of the same sex. The names which have been given to these phenomena are quite varied: strife song (*Streitgesang*), cry of fear (*Schrecklaut*), alarm cry (*Störungslaut*), contact cry (*Kontaktlaut*), warning song, fight song (*Kampfgesang*), copulation song, angry cry, defensive stridulation, etc. However, without going as far as to suppose that a single denomination would have been sufficient, we have to agree that some authors show a tendency to subdivide too much and to clutter the vocabulary with terms which only describe some tiny features of a single phenomenon. This tendency combined with that of more or less indiscriminately renaming behaviour patterns formerly described under other names, complicates at the outset any attempt of analysing these various displays.

Of all the emissions which are caused by a specific disturbance in the creature, the fight song of the Gryllidae is one of the best defined*. Acoustically, it is characterized by an often great modification in the elements which make up the calling song (Fig. 358).

Generally speaking, an increase in the number of pulses in the chirp is noted: 25–30 as opposed to 2 or 3 in the calling song of *Acheta domesticus*; 30–35 as opposed to 15 in *Myogryllus verticalis* (Serv.)[3,5]. The frequency spectrum, generally very narrow in the calling song, often broadens a little in the fight song. Finally, during this emission the elytra form a greater angle with the axis of the body than in the two other types of song.

Many males Gryllidae, show a tendency to aggression towards any other member of the species, and the fight song is only the acoustic portion of a behaviour pattern in which some tactile stimuli also take part (antenna taps) and perhaps some visual and olfactory stimuli. In this way individuals recognize the other's sex and are thus incited to fight (with a male) or to court (with a female). The emission of the fight song is often accompanied by a rapid trembling of the body. In *G. campestris* the aggressive behaviour appears in the last two larval stages (the causative stimulus is not the whipping movement of the antennae, but just the contact), but obviously there is no acoustic sequence[106].

There are other conditions which also produce the emission of the fight song: the departure of the female, or the approach of another male during courtship[3,4,106]

---

* This signal is sometimes called "rival's song", but it seems inadequate to use the same expression for the Acridinae and the Gryllidae for the rivalry behaviour of the Acridinae is only slightly related to the fighting behaviour of the Gryllidae. We prefer to keep the designation "rival's song" for the Acridinae, and to use "fight song" for the Gryllidae.

References p. 649

(Fig. 358). It seems, therefore, that the fight song is triggered not by a well-defined stimulus, but, to be more precise, by any activity of a member of the species which produces a sudden change in prevailing conditions*.

The other kinds of emissions are obviously very varied, because they are recorded in different families, classes, or orders. They all originate, however, with the same kind of situation: the animal's perception of another member of the species, usually of the same sex.

Some males Acridinae (*Chorthippus biguttulus, C. montanus*), when they are a very short distance from each other, or in contact, produce a cry called "contact cry" (*Kontaktlaut*)[179]. It does not seem necessary to distinguish this emission from that which has been named "disturbance sound" (*Störungslaut*) in the same insects, and which has been reported in similar conditions. These signals are very brief and are emitted sometimes in alternation. They represent the highest level of the so-called rivalry behaviour.

Some Acridinae have also a special stridulation, called "copulation song", produced by the male during mating, when it is disturbed. In *Stenobothrus lineatus, Omocestus viridulus* and *Chorthippus brunneus*, it consists of series of clicks emitted at a rate 4 or 5 per second[102].

In Oedipodinae, *Locusta, migratoria* L. (*solitaria*) (Remaudiere), the male, copulating or not, emits a stridulation called "warning" when another male approaches to less than 12 cm; if the intruder comes still nearer, the song is accelerated and becomes a "contact cry" when the two animals touch[35].

The "warning song", the acoustic reaction of one member of the species to the arrival of another (usually of the same sex) before the animals come into contact, is also found in *Oecanthus pellucens* (Oecanthidae). This signal is quite similar to the "calling song", the greatest difference being in the grouping of the chirps in shorter series[36]. Apparently we may also place here the emission and behaviour noted when two males are a very short distance from one another, in the auchenorrhynchous Homoptera, *Achorotile albosignata* (Dahlb.) and *Criomorphus bicarinatus* (H.S.)[137].

The disturbance songs seem less common in Tettigonioidea. Some have a special stridulation, however, when males meet, such as the Phaneropteridae *Orphania denticauda* (Charp.)[72] and *Amblycorypha oblongifolia* (De Geer)[3]. A case like that of the genus *Ephippiger* is more debatable, since the same stridulation is employed by the male or the female, equally when meeting a member of the same sex during copulation or not, and when the insect is at grips with a predatory animal**.

The effect of the disturbance songs is usually quite clear; the disturbing animal moves away while the disturbed one often stays, at least if it belongs to a species with a territorial behaviour (Gryllidae, Oecanthidae). In *Locusta migratoria* (*solitaria*), it has been noted that in 90% of the cases, the warning song of the copulating male made

---

* Perhaps the "rasp" of the lobster *Palinurus vulgaris* Latr. noted during a combat between two individuals[61] may also be considered as a fight song.

** Thus a single acoustic phenomenon could be called contact cry, copulation song or protest sound according to the circumstances. This example gives us the opportunity of stressing that the classification of emissions which we suggest comprises (as all classifications do) a too rigid and narrow framework to easily fit the animal, with its elastic and polymorphic behaviour. The best way to look at this systematisation is only as an attempt at methodology, the aim being to define the general categories and to facilitate description. In no way does it claim to be the faithful image of reality in its minute details.

the approaching intruder withdraw[35]. In the Acridinae, the contact sound is followed (sometimes after a sequence of alternation) by a rapid separation of the two animals.

However, it would be hazardous to attribute everything in these examples to the acoustic stimulus, for some tactile, visual and olfactory stimuli occur at the same time, and they doubtless play a significant role.

(iii) Protest sound

Also called "call of distress, complaint cry, cry of agony"*, the protest sound occurs quite sporadically among the stridulating arthropods. It is produced in response to any excitation or situation of constraint, such as change in environment resulting from the imminence or presence of danger. Hence, the animal will emit the protest sound in a certain number of cases; at grips with an enemy (pursuit, fight, capture), following an unpleasant stimulation or a wound, etc.

In classifying emissions, we should emphasise that the term "protest sound" is kept for a sound reaction which is not caused by a member of the species. The acoustic phenomena set off by the presence of activity of a member of the species have been named disturbance songs (and in really definite cases rival's or fight songs). Some cases must be expected, however, where the emission in question will come in both categories (*Ephippiger*, for example)**.

Among insects, some examples are known in the order Orthoptera, where the protest sound only seems to occur in Tettigonioidea. The American species *Pterophylla camellifolia* (Fabr.) (Pseudophyllidae), *Amblycorypha oblongifolia* (De Geer) (Phaneropteridae), *Aglaothorax armiger* Rehn and Hebard (Tettigoniidae) and *Neoconocephalus exiliscanorus* (Davis) (Conocephalidae)[3, 52] make use of it. In Europa, the protest sound or *Orphania denticauda* (Charp.) (Phaneropteridae)[72] is well known as also that of species of the genus *Ephippiger* (Ephippigeridae), among which this cry is produced by the male and the female, either when the animal is captured or when it meets a member of the same sex**. This emission is still noted in animals which have been decapitated (Dumortier, unpubl.). Many Coleoptera stridulate just as readily when they are caught (Cerambycidae, Chrysomelidae for example) or when they are disturbed in any way: *Spercheus, Olocrates, Polyphylla, Passalus, Necrophorus*, etc.. The cicadas also emit a cry when they are attacked by a predator or held in the hand, and numerous other Homoptera males and females, in the genera *Aphrophora, Oncopsis, Macropsis*, etc.[137] are known to behave in the same way. Among the Heteroptera some protest sounds have been reported in the males, females, and larvae of *Coranus subapterus* (De Geer) (Reduviidae), and in various *Kleidocerys* males and females (Lygaeidae)[100].

In the Lepidoptera, the most famous case is that of the Death Head's Moth (see p. 323) which cries vigorously when it is held in the hand. Several other moths or butterflies also react in the same way to an excitation, either by a chirping accompanied by an emission of foam (as in *Rhodogastria bubo*, see p. 325), or by a grating sound produced by the rubbing of the wings, as in some species of the South American genus *Prepona*, or in *Charaxes sempronius* of Australia[67]. A crackling sound produced in the same way has also been reported in some species of the genus *Vanessa*[171] (see p. 329).

---

* All these names suggest too much and easily cause one to attribute to the phenomenon a meaning which the facts cannot support, as we shall be able to establish. The term "reflex-cry" would seem more suitable.
** See second footnote p. 604.

References p. 649

Apparently when caterpillars and pupae produce various sounds by the means described above, they are a result of an external excitation (see p. 331).

The Diptera and the Hymenoptera, when captured (for example in a spider's web), hum in a very special way. This sound has been attributed to the vibration of the thoracic muscles.

Finally, many termites react to shocks and vibrations with a rapid percussion of the head on the substratum. This tapping apparently acts as a communication between the various individuals (see p. 327).

Thus the protest sound is quite common among insects, but it is also found among other arthropods. In the same way stridulation in the scorpions (see p. 319) has only been noted in disturbed animals[2,145]. It is the myriapods, however, which definitely provide the most extraordinary example of the use of the protest sound. In some species the stridulation is not made by the animal itself, but by a leg autotomized as a result of an attack (see p. 322). The authors who have noted this behaviour[22,57,58,92] have put forward the hypothesis that this was perhaps a means of defence: the attention of the predator attracted by the movements and stridulation of the leg, giving the myriapod the opportunity to escape more safely. Clearly this would merit experimental verification.

Among the Crustaceans, some sound emissions have been recorded which are produced in conditions similar to those which set off the protest sound in insects. Among crabs, such sounds have been reported in the genera *Acanthocarpus*, *Matuta*, *Menippe*, *Ocypode*, and *Uca*, sometimes associated with the defence of their burrow (see pp. 312ff and 589). Rathbun (1937)[150] writes about *Acanthocarpus bispinosus* Milne Edwards: "When touched or taken in the fingers under water, the crab may set up such a vibratory grating of the hand (of the cheliped) against the suborbital tublercles as to make one's fingers literally tingle".

The emission of protest sounds has also been noted in mantis-shrimps and in lobsters (see p. 305 and 307).

All the classes in the sub-kingdom Arthropoda which include some stridulating species, thus have some forms which can react with a sound emission to an unpleasant stimulus.

Obviously writers have wondered what this display means and what use the animal might derive from it*. It was certainly tempting to think that the predator, surprised by the stridulation of its prey, might hesitate, thus giving the victim a chance to escape. Unfortunately it is easier to persuade oneself that the alarm squawk of the cicada captured by a wasp, or the violent stridulations of the grasshopper harpooned by a praying mantis, does not upset the predator at all. What value, then, can we think the minute chirping of the Cerambycidae or the Chrysomelidae has?

We certainly cannot come to a conclusion simply on the basis of chance observations, but the experiments made in this field have also turned out practically negative. In studying the effect of the stridulation of the scorpion *Opisthophthalmus latimanus* Koch (see p. 320), on various mammals, it has been established[2] that the rat, mongoose

---

* We should remember that a quite clear preliminary condition is that the predator should be able to distinguish between the animal which can make a protest sound and that which cannot. For this, it must have either an auditory apparatus or a mechanoreceptor. Actually, we should not forget that in many cases the sound emission is accompanied by a sometimes strong vibration of the body, and that if the aggressor is going to experience a sensation which will make it abandon its prey, this sensation may be equally tactile and vibratory as acoustic.

(*Suricata*) and shrewmouse (*Suncus*) ate without hesitation and indiscriminately the normal scorpion and the one in which the stridulatory apparatus had been destroyed. Faced, however, with the stridulation, the hedgehog (*Erinaceus*) showed some fear, whereas it pounced at once on a silent animal. On the second encounter with a stridulating scorpion, however, it hesitated only briefly and finally attacked the prey. Afterwards the hedgehog did not take any more notice of the stridulation.

Other animals such as the cat, dog, cercopithecus, galago are clearly afraid in the presence of *Opisthophthalmus*, but just as much when the scorpion does not stridulate.

The same type of experiment has been carried out on the Death's Head Moth with a similar result. When placed in the presence of the moth for the first time, a cercopithecus was alarmed by its squeaking. However, the monkey quickly overcame its fear and devoured the insect[129].

To sum up, we cannot note any definite proof of the protective value* of the protest sound. But in place of recorded facts we can develop some hypotheses. One of the most attractive is to correlate, in some instances, this type of sound emission with aposematic colourings as already suggested by Pumphrey (1955)[146]. Moreover, the parallel is easy to establish, on conditions, however, that the animal, besides emitting a protest sound, can also cause an unpleasant sensation in the predator (a nasty taste or smell, use of apparatus which can inflict a wound: claws, chelicerae, sting). We may then suppose that the bird or insectivore, after several experiences, will associate with this protest sound the "idea" of a painful feeling, which was linked with it in the first experiences. Hence, once conditioned, hearing the stridulation is enough to keep the animal at a distance.

Aposematic colouring or Protest sound
↓

Conditioned Stimulus { warning signal reminding the attacker that it has already been associated with:

Normal Stimulus   a painful sensation (unpleasant taste or wound)

In actual fact some insects are known which eject a foam with an unpleasant smell, at the same time chirping (see p. 324) and perhaps others have a hemolymph which holds a toxic or vesicatory substance (it has been definitely established that some insects were not eaten by birds or mammals). Again, arachnids or crabs are sufficiently armed to wound an assailant.

However, it should be noted that:

Only arthropods equipped with an efficient means of defence (taste, smell, weapons) can manage to condition the predator in this way.

Only animals with a sufficient developed brain (birds, mammals) can become conditioned in this way: hence they could only represents a part of the usual predators.

Therefore, it seems really doubtful whether there is an aposematic stridulation.

On the other hand, it seems much more probable (and more in line with the data) to regard the protest sound as a displacement activity. The great disturbance caused

---

* Nevertheless, it is possible that in animals living in groups, the protest sound may have a social value by warning other individuals of a danger (*cf.* alarm and distress calls of crows and starlings). Thus, the noise made by a captured spiny lobster (*Panulirus interruptus* (Randall)) makes all other individuals within a range of 1–2 m retire into their hiding places[125].

by capture or cornering must produce intense nervous discharges in the animal; there are no longer any readily available effectors for them and the stridulatory movements serve as an outlet.

That is why we feel it is more accurate to call this type of emission "reflex-cry". This would counteract the quite common tendency to think that the terms chosen to designate animal behaviour are intrinsic explanations of these behaviours.

Finally, it may very well be asked why the possible meaning of this acoustic display is being considered in the arthropods, whereas no consideration is being given to the role and efficiency which the cries of a mouse captured by a cat may have.

(iv) Sound emission and stridulatory movements in the pre-imago stages

We have described (pp. 330–334) the apparatuses and processes used by immature specimens (larvae, and pupae) in their acoustic manifestations. Usually it is a question of reactions to a disturbance and so they belong in this discussion.

As in most cases the only reasonably thorough studies have been made on the Orthoptera and, in particular, on the Acrididae[113, 179]. From the first instars they have some behaviour patterns which occur also in the adult. Some of these behaviour patterns are as efficient in the nymph as in the adult; for example the "protection behind a blade of grass" ("Sichern hinter dem Halme", of the German authors). On the other hand, some patterns are fruitlessly performed, because the organs or appendages involved are not developed enough. This is the case with the cleaning of the antenna by passing it under the tarsus of the front leg, which are resting on the ground. This is noted from the second instar, but the antennae are still too short, so the movement takes place in space. The stridulatory movements provide another example of this pointless activity (at all events an activity without the same result as in the adult). The wings and the elytra are still rudimentary and cannot act as a plectrum for the pars-stridens already outlined on the femur*; hence no sound emission is produced. In the cases noted (*Chorthippus brunneus, C. biguttulus, C. montanus, C. parallelus*), the stridulatory movement is made in response to an excitation which is caused by an adult of the species (contact or song). Similarly it has been noted that in the nymphs of *C. biguttulus* the stridulatory movements begin at the third instar. These movements correspond to those which produce the contact sound or the rival song in the adult. The movements of calling song become more frequent as the final moulting is approached[179].

In *Locusta pardalina* Walk, the nymphs when disturbed also react with stridulatory movements (without sound emission)[168].

In the family Catantopidae, some mandibular noises have been reported in the genera *Calliptamus* Serv. and *Podisma* Latr.[70,115].

It should be noted that stridulatory movements of the imago type are not possible in the Orthoptera *Ensifera*, since such movements obviously necessitate the movement of the elytra, which are not acquired until after the final moulting.

We shall not come back to stridulation in the larval forms of the other orders (Coleoptera, Lepidoptera especially), except to point out that they nearly always consists of a reaction to an excitation (usually tactile). The meaning of these emissions presents the same problems as those evoked in connection with the emissions made

---

* The number of teeth is fixed from the first nymphal instar, when the root follicles are visible. During the following instars the teeth develop gradually into their final form[113]

by the imagoes. There is this difference: apparently these sounds are always produced in response to a disturbing stimulus. Hence we can say the same about them as was said about the protest sounds. As far as pupal stridulation is concerned, the explanation based on the hypothesis of displacement activities fits especially well.

From the ontogenetic standpoint it will be noted that in the Holometabolae the stridulatory apparatus of the larva appears in a different place from that which it occupies in the adult (when the larva and the adult have such a device). Hence it is put into action by different processes, according to whether it is a larval or an imago apparatus. The differentiation and reconstitution which take place during metamorphosis thus have an simultaneous effect on the structures, the sensory-motor tracks, and the effector patterns. But the behaviour patterns remain constant: the larva, the adult, and sometimes even the pupa, respond with a sound emission to the same disturbing stimulus.

In the Heterometabolae, (especially Orthoptera Acridinae), on the other hand, the whole system is built up in progressive stages; the adult behaviour expressions are already adumbrated, or rather actually present, in the immature animal; they even appear before the structures which are used, have fully developed.

(c) The origin of the stridulatory movements

We know hardly any more about this point than about the origin of the stridulatory apparatus. Here again we have to resort to hypotheses.

There are only two kinds of data on which we may build a hypothesis:

(i) The different forms of stridulatory movements within a single family,

(ii) The song movements in the nymphs.

Thus we are forced to limit our attention to the Acrididae, the only family in which these two kinds of data are available.

(i) It is quite interesting to consider a case like that of *Mecosthetus grossus* (L.) and of some other Oedipodinae[113,114] since the stridulation is produced by a backward movement of the third tibiae; it is possibly a defensive movement. We have already recorded elsewhere (see p. 328) that in the Oedipodinae there is a change from a simple vibratory movement of the hind legs to a tapping on the substratum, then to a tapping accompanied by stridulation and finally to stridulation without tapping. Certainly nothing proves that this must be an orthogenetic evolution and the order in which these examples are given may be just an arbitrary creation of the mind bearing no relation to reality. Nevertheless, we may wonder whether the stridulation originates with an early defensive movement which gradually turns into a stridulatory or tapping movement[114], symbolizing of the defence. Moreover this ritualisation is common in the vertebrates displacement activities.

In the Orthoptera *Ensifera*[114] or the Coleoptera with elytro-abdominal apparatus, it has been suggested that stridulation would come from a ritualisation of the motions of taking flight.

(ii) From this aspect, do observations of the stridulatory movements of the larval instars justify us in deducing an ontogenic explanation for the song? As far as the Holometabola are concerned we cannot even put the question, and in the Heterometabola (we are thinking here of the Acridinae) nothing could be less certain. The conclusion, however, is possible that the unusual song movements which the larvae of *Chorthippus* show, may be the "neuter" form of stridulation, the differentiation into rival's, calling, or courtship song being later acquisitions*[179].

But let us say again that all of this is much more intuitive than deductive hypothesis.

2. Experimental Study of Acoustic Behaviour

In the first part of this Chapter we reviewed the various behaviour patterns associated with or set off by sound emission. Now we shall look more closely at two important aspects of these behaviour patterns: the acoustic reactions (Phonoresponse) and the taxic reactions (Phonotaxis).

(a) Phonoresponse

Although the calling song in the Orthoptera is the very type of the spontaneous stridulation, it is equally true that its production is generally facilitated, and often set off by the hearing of a sound stimulus which may be either natural (specific or not) or artificial. In general, the stimulus does not act as a straight forward catalyser which only has a part in starting a reaction potentially present. As long as the stimulus is produced, its effect is in evidence; this is shown by the changes which is causes in the song production rhythm.

Paradoxically, the acoustic stimulation may produce two opposite reactions, according to the species: (i) either the animal alternates its emission which that of the stimulus (alternation); (ii) or, in the case of species living closely packed together, the first emission sets off a collective and synchronous song of all the population (synchronism).

Faber (1932)[69] invented the term *Anaphony*** (*ana*: back, *phony*: voice) to describe this behaviour. This author distinguishes four types of anaphony:

1. *Amoebochronous anaphony*: single response to the stimulus.
2. *Metachronous anaphony*: alternating series of stimulus-response.
3. *Poïkilochronous anaphony*: the first stimulus causes several individuals' responses or several responses of the same individual.
4. *Synchronous anaphony*: collective emission made in a more or less regular way.

Here we shall only keep to the broad lines of this classification, not doing any more than distinguishing between *alternation* and *synchronism*.

(a.1) Alternation

(i) *Specific alternation*

In the Acridinae (see p. 600) we have already reported the alternation which is frequently displayed between two males at the emission of the calling song or rival's song. A phenomenon of the same kind is observed with a male and a female emitting the agreement song. The duration of these series of alternate songs is quite variable

---

* Any hypothesis based on the observation of the pre-image instars to explain the adult behaviour patterns, more or less tacitly implies an extension of Fritz Muller's famous aphorism "ontogeny recapitulates phylogeny", on the one hand at the level of the insect, and on the other on the ethological plane. This seems to be a hazardous extension to say the least.

** This name was replaced in his later works by *allelophony*. Weih (1951)[179] also uses the term: *Wechselsingen* (alternate song). Busnel in his turn, used the expression *Phonoresponse*, which we shall use here.

(approximately 12 sec); hence it is incomparably shorter than that which is often observed in the Tettigonioidea.

The rate, which has been shown experimentally (see below) to be one of the factors conditioning alternation, may take on different values according to the type of song considered. In *Chorthippus brunneus*, for example, an acceleration of the emission has been discovered in changing from the calling song of a single male to the alternate calling song of two males, and also in changing from the alternate calling song to the alternate rival's song[179].

An exact time study of alternation in the Tettigonioidea has been made in different species of the genus *Ephippiger*[45]. It has shown that the responses between two males might be exchanged for hours in three ways: (i) each insect keeps its own rate, and the responses are regularly interspersed in such a way that the final rhythm is doubled (Fig. 363, 2); (ii) the rate of each individual is halved and the two alternate so that the same rhythm is produced by two males singing in alternation as by a single male singing alone (Fig. 363, 3). This has been observed also in *Pterophylla camellifolia* (F.) (Pseudophyllidae)[8]; (iii) it has also been established that if one of the two insects stops stridulating, the other can to some extent respond in its place and so sing twice as much; this keeps the rate with or without alternation (Fig. 363, 4).

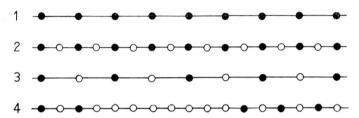

Fig. 363. The different ways of alternation in *Ephippiger*. 1, the rate of an individual alone; 2, 3, 4 see text. (After Busnel, Dumortier, Busnel, 1956[45].)

Regen's (1914)[154,155] experiments on the alternation of *Thamnotrizon apterus* Finot (= *Pholidoptera aptera* (F.) (Tettigoniidae) were undoubtedly the first in this field. They demonstrated that the alternate song only took place if the tympanic organs of the insect were not destroyed, and the stridulation went from one to the other with a sufficient intensity. Moreover Regen was able to establish that the stimulus was only transmitted through the air and not by the substratum, by putting his insects in nacelles suspended from balloons. An experiment of the same kind was performed by Baier (1930)[24] in *P. griseoaptera* (De Geer). It revealed that in this species an individual, in the absence of a real alternation, "avoids" starting its stridulation before another has finished its emission. An identical finding was made on *Orchelimum bradleyi* R. and H. (Conocephalidae)[83].

Until now we have only mentioned Orthoptera, the alternation of which has been the subject of research, but those in which it has simply been observed are far from few; let us quote, for example *Amblycorypha oblongifolia* (De Geer) (Phaneropteridae)[79,143]. In *Pterophylla camellifolia* (see above), collective alternation of a population has been observed. "The result is a great, pulsing song which fills the air for hours."[3] In the Acridian *Paratylotropidia bruneri* Scud. an alternate emission of mandibular noises has been reported[7].

References p. 649

Some other examples are also met among the Hemipteroïda. In the genus *Idiocerus*, in *Erythroneura hyperici* H.S. and *Lepyronia coleoptrata* L. (Homoptera, Auchenorrhynca), some degree of alternation has been established in the emission of the calling song[137]. It has also been reported that the Coleoptera Cerambycidae *Nothorrhina muricata* Dalm. frequently alternated its percussions (see p. 326) with those of another individual[71]. The same is the case in the Anobiidae[86].

In the crabs *Ocypode* and *Uca* the more or less regularly alternated stridulation of individuals in a colony has been reported several times[21,60,94].

Finally, by and large, even if it does not cause an exchange of alternate responses, the first individual's stridulation can often set off the song of other members of the same species which are nearby. Observation shows that it is often the same insect which takes the initiative of beginning the stridulation. Baier (1930)[24] gave it the name "leader". He defines it as "an individual of a species who singularizes himself by his more predominant chirping". The leader actually breaks the silence, but also sings more than the other individuals; this appears very clearly in *Ephippiger* the stridulation of which is alternated very regularly[45].

(ii) Interspecific alternation

Some reactions coming from the single response to the alternate song are observed among individuals belonging to different species, genera or families. This is the case with *Orchelimum militare* R. and H. and *O. bradleyi* R. and H.[83] and with the different combinations obtained, when *Ephippiger cunii* Bolivar, *E. provincialis* (Yersin), *E. bitterensis* Finot, *E. ephippiger* (Fiebig), etc.[45] are put together by twos. Alternation is easily obtained between an individual of one of the two former species and another belonging to one of the two latter, although the former has a chirp of three, four of five pulses and the second a one pulsed chirp (see Fig. 366).

Similar facts are met in the Acridinae[179], where, however, the percentage of interspecific responses is less. Table 45 gives the result of experiments carried out on *Chrysochraon dispar* (Germ.) and three species of the genus *Chorthippus*.

TABLE 45

PERCENTAGE OF INTERSPECIFIC OR INTERGENERIC RESPONSES IN VARIOUS ACRIDINAE (ACCORDING TO WEIH, 1951)[179]

Species present % reactions observed in	C. parellelus	C. montanus	C. dorsatus	Chrysochraon dispar
Chorthippus parellelus		71%(?)	29%	33%
C. montanus	21%		28%	25%
C. dorsatus	13%	2%		2%

It must, however, be noted that the responses are made up of rival's song or a disturbance song; hence they are only produced when the two insects are a little away from each other.

The interpretation of such results is quite difficult, for a certain percentage of responses to a non-specific stridulation may have several causes: a more or less swift summation effect due to the repetition of the stimulus under consideration, continuation of a state of excitation for some time caused by a series of phonoresponses with an individual of the species a little before experiment, etc.

At all events, even if it varies a lot from one species to another* the capacity of interspecific alternation, or more simply of response, is not uncommon in the arthropods. The innate releasing mechanism (I.R.M.) which causes the reaction "stridulation in response to a stridulation" seems to be switched on by stimuli which may have characteristics differing considerably. Moreover we shall find this same problem in connection with phonotaxis in the females (see pp. 618ff). Consequently two hypotheses are possible: (i) we may presume that a filtration of afferent messages which can have an effect on the I.R.M., occurs somewhere in the neuro-sensorial complex. The more or less wide range of selectivity in "filter", will condition the possibility that the animal has of reacting to stimuli different from those produced by the individuals of its species; (ii) in the second hypothesis, which from our viewpoint has more probability, the specific stimulus may be considered not as having an overall effect but only with one or more of its components being effective on their own. This means that it is sufficient for the real triggering element to be present in a foreign or artificial signal for it to take on the capacity of setting the I.R.M. going**. Experiments on alternation caused by artificial stimuli bring some light to bear on this subject.

(iii) Alternation with artificial stimuli

The use of artificial stimuli in the study of alternation or single responses, has thrown into relief the importance of several parameters. They may be considered as factors that set off the I.R.Ms which are involved in the triggering of the phono-responses.

TABLE 46

RESPONSE TO THE "MOUTH NOISE" ARTIFICIAL STIMULUS IN VARIOUS ACRIDINAE
(ACCORDING TO BUSNEL AND LOHER, 1954)[48]

Species	Number of emissions	Number of responses	Responses (%)
Chorthippus jucundus (Fisch.)	85	34	40
C. brunneus Thunb.	212	185	87
C. biguttulus (L.)	1000	950	95
C. longicornis (Latr.)	200	0	0
Omocestus ventralis (Zett.)	200	0	0

Chorthippus, for example, can easily be made to respond by artificial signals[69] such as sounds made with the mouth ("ffft"), or with various instruments: comb teeth, file, watch-winder, etc.[179] (Table 46).

The insect may even be attracted by particular types of signals such as the noise of a camera motor for *C. brunneus*, mouth noise ("prrrrr") for *C. biguttulus*[47–49, 51].

---

* In *Chorthippus brunneus* Thunb, some responses to the stridulation of *Pholidoptera griseoaptera* (De Geer), *Decticus verrucivorus* (L.) and *Metrioptera roeseli* (Hag.) (Tettigoniidae)[179] have been noted.

** Moreover, this is what happens in the european robin in which the aggressiveness of the male is set off merely by the sight of a tuft of red feathers, or again in the male three-spined stickleback where the different sequences of sexual behaviour (fight against the males, mating) are induced by the position the colouring and the volume of the belly[174]. Many other examples are known where a reaction is induced by a special stimulus extracted from a complicated group. Russel (1943)[162] called such triggering factors: "sign stimuli".

References p. 649

The common feature of all these stimuli is that they are more or less faithful imitations of real stridulation (rival's song or agreement song). The single response, however, or even alternation may be obtained in *C. biguttulus*, *C. brunneus*, *C. jucundus* and various Tettigonioidea with sound emissions without a resemblance either in modulation or frequency to the specific signal, like those made by Galton's whistle or by various simulated bird-calls, or again by pure frequencies or trains of impulses*.

These studies are intended to vary or remove certain parameters in order to show to what extent they are important in setting off the response. In this way it has been possible to reveal the role of the emission rate, the duration of the signal, and its form.

For example, in *C. brunneus* it was found that the best results were obtained with imitative signals of stridulation by not departing too far from the natural emission rate**[179]. It is evident that although the insect is always able to respond to a very slow rate of stimulation, it is not the same when the stimulatory rate is so rapid that the second stimulus arrives before the necessary time for the integration of the first has lapsed***. By keeping within the right limits, however, it has been established that, in the case of pure frequency signals, a normal rate produces a response with the calling song in *C. brunneus*, while a quick rate causes the emission of the rival's song[48]. Consequently, it may be asked whether it is not the repetition rate of the stridulations rather than their particular physical structure which involves the rivalry behaviour.

As far as the duration of the signal is concerned, it has been found, still in *C. brunneus*, that it cannot be less than about 20 msec (see Table 47).

TABLE 47

PERCENTAGE OF RESPONSES OF *C. brunneus* TO A SIGNAL OF PURE FREQUENCY ($I = 80$ db) PLOTTED AGAINST ITS DURATION (ACCORDING TO BUSNEL AND LOHER, 1956)[49]

Duration of signal in milliseconds	Number of signals emitted	Number of responses +	%
5	200	0	0
25	153	11	7
35	232	112	48
50	184	116	63
100	332	197	59

The upper duration threshold is more difficult to determine since, in this same species, a signal of 45 sec still causes a response. In *Ephippiger*, the study of long duration signals obtained with Galton's whistle, has shown on the one hand, that the insect never stridulated during the signal**** and, on the other hand, that, the longer

---

* Some muscular reactions (flight, fall, movements of the body or appendages) are also caused in a number of insects by sound. Special study of them has been made by Minnich, Frings, Treat Schaller and Busnel and we only mention them here as a reminder.
** Similar findings are also made in the tree frog *Hyla arborea* L. which can croak in alternation with the taps of a metronome. However, this result is only obtained within a quite narrow speed range of the instrument[41].
*** In the responses of the male to the female it is noticed that the male frequently increases his rate to such an extent that in the end the female cannot keep up the alternation[179]. Hence we may presume that the female's integration time is longer than the male's.
**** Moreover it has been frequently noticed in the Tettigoniids that an individual refrains from starting its song while another is stridulating (see p. 611).

was the stimulation time, the shorter was the integration time[45]. With *E. cunii*, this integration time goes on a average from 1.34 sec for a signal of 0.36 sec, to 0.77 sec for a signal of 5.5 sec.

Since such signals release reactions despite their considerable differences from those of the species itself, we have to admit that they have a *character* which can call forth the I.R.M. of the response behaviour*. This character, fundamentally "reactogenic", Busnel sees in the transitory phenomenon which arises at the start off and the cutting off of the signal[38, 39, 47–49]. This hypothesis is notably confirmed in the study of stimuli with variable build-up times made on *Chorthippus brunneus*[50]. For a signal of the same frequency (10 kc) and the same intensity (80 dB), the percentage of responses made by the insect is in proportion to the time which passed between the beginning of the signal (t_0) and the establishment of the constant intensity (t_1) (Fig. 364).

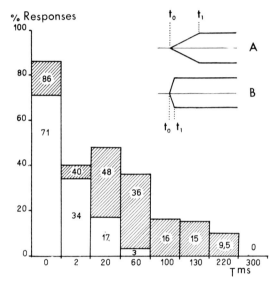

Fig. 364. Effect of the build-up time of the stimulus (10 kc) on the phonoresponse rate in *Chorthippus brunneus*. (After the data of Busnel and Loher, 1961[50].) The abcissa represents the build up time $(t_0 - t_1)$ in msec. Hatched parts: intensity 80 dB; white parts: 70 dB. The figures inside the columns give the percentage of responses. A and B are examples of two different build-up times.

When the phenomenon is instantaneous (rectangular shape on an oscillogram) the percentage of responses is maximal (86%). The build-up time only has to reach 0.2 sec for this percentage to be brought down to zero**, (see the detailed account of this research in Chapter 5, pp. 95 ff). Similar results have been obtained in phonotaxy studies in the females of the genus *Ephippiger*.

Other authors, however, think that the parameter which dominates the release of the reaction of the Acridinae (phonoresponse, alternance, sometimes also phono-

---

* Let us note that broadly speaking there is no preferred frequency value, and the intensity (which must reach at least 70 dB) is only a limiting factor.
** The analogy will be noticed between this requirement of a quick establishment of the sound stimulus and that, (which has been known for a long time), for the contraction of the muscle under the effect of an electric current.

References p. 649

taxis) lies in the pulse-rate. For example, in *Chorthippus parallelus* (Zett.) the percentage of responses at the diffusion of recorded calling songs, is almost the same (58% as opposed to 56%) when the signal is transmitted correctly or when distorted. On the other hand, the number of responses clearly decreases (20%) if the recording is played at half speed (halved pulse-rate)*[102].

Other experiments showed that the pulse-rate was an essential factor in the specific recognition of the song (see pp. 622ff); hence there is no question of minimising its value. We should take into account, however, that in the experiment made on *C. parallelus*, a recorded and more or less distorted natural signal was used, and no longer a synthetic signal unrelated to the stridulation of the insect. It would be hazardous to compare the results obtained under two very different sets of experimental conditions. The role of the signal's form (in the case of pure frequencies) seems to have been well demonstrated, but the pulse-rate plays a leading part in the case of natural signals or their imitation. Moreover, nothing proves that the reactions of these Acridinae to sound excitations are determined in the same way in the two cases.

In the preceding section we have limited our attention almost solely to the Orthoptera and in particular to the Acridinae. Some cases of phonoresponses or even of alternation, however, are known in other orders. Hence, the Coleoptera Anobiidae (see p. 326) respond to the tapping of a pencil on the table with their characteristic ticking, also to a clicking sound made with the tongue or even a quavering note produced by whistling[86]. In *Corixa striata* L. (Heteropt., Corixidae), males give a stridulation in response to pure tones frequency between 2 and 40 kc/s[166].

An example is even found in Crustacea where the fiddler crab *Uca pugilator* (Bosc) reacts with the sound vibration of its big cheliped (see p. 314) to a vibration made artificially near its burrow, from which it may even emerge under the effect of the stimulation[60].

(iv) The problem of learning

If song is an entirely innate behaviour (see p. 628), can we say that learning plays a role in the responses to song? In other words if insect sings spontaneously, does he learn to answer?

The old observations of Regen (1926)[156] on *Thamnotrizon apterus* (F.) (= *Pholidoptera aptera* (F.)) (Orth., Tettigoniidae) showed that responses to artificial signals were obtained with newly-moulted males kept in isolation, but not with insects which had already heard other males of the same species.

The experiments of Pierce (1948)[143] on *Pterophylla camellifolia* (F.) (Orth., Pseudophyllidae), proved that this insect was able to answer a two of five pulsed noise by a song in which the number of pulses was the same. This was also obtained with the noise of a type writer by Alexander (1960)[8].

Of course it would be convenient to explain both these results by "learning", but we may also speak of a "specific imprinting" in *Pholidoptera*, and of an "imitation capacity" in *Pterophylla*. All these "explanations" only mask our ignorance of the neural process involved in these reactions, and learning is anything but certain.

---

* It will also be noticed that the attacks have a more gentle slope, this amounts to saying that the transistory phenomena are less accentuated, and that the frequencies of the spectrum are halved.

(a.2) Synchronism

Fewer observations, and, in particular, experimental work, have been made on the synchronised song. Some cases are well know, however, in the Orthoptera. Colonies of *Oecanthus niveus* (De Geer) show distinct synchronisation, especially pronounced in the midst of their sound activity[3,14,63,83]. *Cyrtoxipha columbiana* Caudell (Gryllidae) has provided identical observation. It has been calculated that out of a total of 870 chirps emitted by a group of four of five insects, 808, that is 93%, were synchronous[14]. In the Acridid *Stenobothrus lineatus* (Panz.), synchronous song has been noted in groups of three of four individuals[102] *Neoconocephalus exiliscanorus* (Davis) (Tettigoniidae) stridulates in sequences of 15–20 chirps separated by a pause. When there are several insects together they do not synchronise their sequences, but rather the chirps within the sequences with a remarkable regularity. As the periods of an individual's silence considered in isolation fall during the stridulation sequences of its fellow members of the species, the impression is given of a perfectly synchronous, continuous emission[14].

Apart from the Orthoptera, synchronism has been reported in the Homoptera Cicadidae[17]. *Magicicada cassinii* (Fisher) shows a very clear synchronism "perhaps the most remarkable occurring anywhere in the animal kingdom". In this species the relay of different parts of the congregational song induces the population to emit collectively. Recordings played at half-speed and the song of another species, *M. septendecim* (L.), are both without effect. On the other hand, a recording played backwards gives good results[11].

A synchronism in the tapping of certain ants has also been reported[75], as in the sound emission of *Toxoptera aurantii* (Fonsc.) (Homoptera, Aphididae)[182] (see p. 284). None of these observations, however, throw any light on the mechanism of the synchronised song*. Moreover, this problem is not only posed for acoustic phenomena; it is well known that in certain fireflies, the flashing may be synchronised perfectly by a large number of individuals[32]. Similarly some simultaneous movements in the gathering of harvest-spiders and plant lice have been observed[13].

Synchronism is one of those cyclic activity manifestations of which living beings, animals and plants, provide so many examples. Life, from the physiological and ethological point of view, whether seen on the cell, organ, individual, or population scale, is partly the result of an infinite series of pulsations which are distinguished by their period and the nature of the phenomena on which they act: pulsation rhythm of vacuoles in Paramecium, nycthemeral, lunar or annual rhythms, migration rhythms, etc. without mentioning all the physiological rhythms. If the cause of some of these periodic activities is known to us, we are still looking for an explanation of the mechanism of many others. Acoustic synchronism is one of these.

---

* However, the following hypothesis may be suggested; the time gap between two chirps may be considered as a kind of refractory period characteristic of the species, during which the individual cannot emit another chirp. If the hearing of the first individual A's chirp by the individuals B, C, D ... n, which surround it, induces an inhibition of their stridulation (see p. 611) for a duration equal to the refractory period, one imagines that inhibition of the stridulation of B, C, D ... n and A's refractory period, will come to an end at the same time and, consequently, all the individuals under consideration will stridulate together.

(b) Phonotaxis

The term *phonotaxis* (= *phonotropism*)* was used by Busnel to designate the movements directed towards the source, caused by a natural or artificial acoustic signal. We have already mentioned the taxic reactions which were noted in connection with the calling, congregational, and agreement songs, either in the males or in the females (Orthoptera, mosquitoes, cicadas, fiddler-crabs, *Anobium*, etc.). Now we shall look at a more experimental aspect of the study of phonotaxis, which will bring to light some of the causative factors of the reaction, as in the case of phonoresponses.

(i) Specific phonotaxis

The first analytical study of the taxic value of the calling song with regard to the female, was made on 1913 by Regen[156]. He noted that the female of *Gryllus campestris* L. headed for a telephone receiver relaying the male's song. About fifteen years later, Baier (1930)[24] took up the work, with similar results, on the females *Tettigonia viridissima* L. and the males *Pholidoptera griseoaptera* (De Geer). Afterwards, some analogous experiments were conducted on various species (*Oecanthus, Ephippiger, Chorthippus*) by Busnel and coll. We shall take the work carried out on *Ephippiger* as an example of phonotaxis studies[40, 45].

As soon as a receptive female** hears the male's stridulation she stops her activity, quivers, then carries out some palp and antenna movements; this is the attention phase, which lasts several seconds. If she was not facing the male's direction at the beginning, she pivots to face the source; this is the orientation phase linked straightaway to the taxic phase proper, during which the female starts towards the male.

The orientation phase usually proceeds in jerks; the insect moves to turn itself in the direction of the partner only in response to the latter's stridulation. The taxic phase is also jerky with abrupt acceleration at each new stridulation. This is more marked the nearer the female gets to the male.

When the female begins to climb the plant at the top of which the male is sitting, the latter stops his stridulations and replaces them with a quick vibration of the body which also has an attractive effect on the female (see p. 59). The two animals meet, and mating begins.

Apparently no other stimulus apart from the sound stimulus has a part in the behaviour, since, on the one hand, a female with covered eyes finds the male without difficulty and, on the other hand, the female can carry out all of her movement in a tail wind (this excludes the effect of chemical signal coming from the male). Actually in *Chorthippus brunneus* it has been reported that the female when blinded, directs herself towards the male correctly but has great difficulty in placing him when she is no more than 5–6 cm from him. Some visual information must normally occur here[103].

The quantitative study of *Ephippiger*'s phonotaxis had made it possible to indicate accurately the connection between the releasing stimulus and the insect's reaction.

---

* This term was created by Fraenkel (1931), who seems to be the first to have considered this type of reaction in the general group of tropisms[76].
** See p. 640 for the physiological conditions which govern the receptivity of the females.

First of all, it can be established that the female when placed midway between two males heads for the one whose stridulation reaches her with the greatest intensity. The interdependence of the stimulus intensity and the reaction of the female, moreover, is accentuated as the distance between the two insects decreases. There is a linear relation between the intensity expressed in decibels and the speed of the female's movement[44]. An application of the Weber-Fechner's law (see also p. 601) may be seen in this (Fig. 365). The law is also found to be verified in some tropisms (phototropism of *Daphnia* and *Drosophila*, galvanotropism of *Paramecium*, etc.)[131,175].

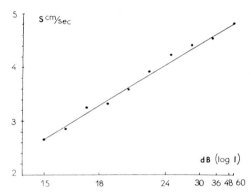

Fig. 365. Speed of the phonotaxic movement of the female *Ephippiger bitterensis* in relation to the intensity of the calling signal. Abcissae. log. intensity (decibels), ordinate: speed in cm/sec. (After Busnel and Dumortier, 1956[44].)

Another aspect of phonotaxis in *Ephippiger* is the existence of an "acoustic directional memory", which enables the female to continue to move in a straight line towards the sound source during 30 sec or more, even if the emission of signals has been interrupted. The insect can cover up to 1 or 2 m in this way. It can be said that this "memory" is actually nothing more than the temporary continuation of the state of excitation, which was created by the acoustic stimulus. As long as this state lasts, the reaction (locomotion) keeps going under its initial thrust[45].

This elementary "memory" is found in the Acridinae *Chorthippus biguttulus* (L.), the male of which shows a phonotaxis towards the female (see p. 595). Immediately following the stridulation of the female, the male moves in her direction, negotiating obstacles (grass) to keep by and large an almost rectilinear line of approach. The duration of this movement is much shorter than in *Ephippiger*, since it does not exceed 2–3 sec, at the end of which the male stops, stridulates again, and sets off anew guided by the female's response[48].

Very little thorough study has been done on phonotaxis except on some Orthoptera and mosquitoes. As far as the latter concerned, for a number of species it is known that the females' flight noise attracts the males and sets off their mating activity. In these insects, as opposed to the Orthoptera, the frequency is the factor which basically triggers the males' move[160, 183]. In *Aëdes aegypti* L., the frequency band which produces the reaction is quite narrow. Hence the sound produced by newly hatched females which is outside this band, has no effect on the males. On the other hand, the very young males, which have a flight noise similar to that of mature females, are often assaulted by adult males; this shows that sexual recognition is

References p. 649

based only on the sound signal and among the parameters of this signal, only on the frequency (see also p. 625). It has also been possible to show that the fundamental frequency was indispensable for the reaction of the male, the harmonics being unable to set it off on their own[183].

In the case of the Orthoptera, the signal (specific or artificial) only starts the first phase of sexual behaviour, that is, the move towards the partner of the opposite sex; the mating maneuvers themselves are released by other stimuli. In the case of *Aëdes aegypti*, on the other hand, the sound frequency which corresponds to the flight noise of the female also seems to cause the adoption of the mating posture (see p. 621).

(ii) Phonotaxis set off by artificial signals

In the Acridinae *Chorthippus biguttulus* and *C. brunneus*, the male can only be attracted by signals which can be considered as imitations of the agreement song of the female (mouth noise, camera motor noise)[47-49, 126]. In the female of *Ephippiger* which has been studied expressly on this subject, and which we shall again take as an example, phonotaxis may be released by totally artificial signals[43, 45], such as those obtained with Galton's whistle, various simulated bird calls or pure frequencies. The behaviour of the insect is on every point comparable with that noted in the natural phonotaxis. These experiments lead to the same conclusions as those to which the study of alternation and phonoresponses had led, *i.e.*:

(i) existence of a lower intensity threshold (somewhere between 50 and 65 dB according to the instrument used)*, below which the signal is ineffective,

(ii) necessity of establishing (or interrupting) the signal abruptly,

(iii) independence of the reaction with regard to the frequency (in the band 500–90,000 c/s in *E. bitterensis*). Hence we are led to wonder whether phonotaxis just like phonoresponse, when responding to artificial signals, is related to an I.R.M. really comparable to that which causes the response to the natural signal.

Whatever may be its exact mechanism, the phonotaxis shown under these conditions is spectacular, since the females may be attracted from a distance of several metres right on to the loud-speaker or on to the mouth of the man using Galton's whistle. The latter's attraction quality is, moreover, so great that, when the female is placed between the man who is using the whistle, and a stridulating male of the species, she heads for the former if the intensity of the sound is stronger than the male's.

In various American *Oecanthus*, attractions have also been obtained with signals of pure frequencies emitted at the pulse rate or chirp rate of the male[178].

Apart from these few Orthoptera, the only species the artificial phonotaxis of which has been really well studied, are the mosquitoes.

The first observation was made by chance in 1878 by Sir Hiram Maxim[130], who ascertained that a dynamo which hummed when running, attracted the male mosquitoes. He also noted that a tuning fork giving the same note as the female in flight, made a male turn towards the source while raising his antennae. In 1901, an electrical engineer, Weaver (quoted by Howard, 1901)[107] observed that a particular note given out by an electrical apparatus, made the mosquitoes in the room, and even those

---

* The lower threshold in the case of taxis obtained with the song of the male of the species is very distinctly lower, since the female is attracted by the stridulation of a partner placed at such a distance that no measurement is possible with decibelmeters at present available.

outside, flock to it. Laurence (1902)[123], reports that the human voice produces the same result with certain vowels.

In *Aëdes aegypti* L., a tuning fork or a loud speaker emitting a pure frequency within the limits of 300–800 c/s, may set off attraction and the adoption of the mating posture. The maximum number of responses is obtained between 400 and 600 c/s; this corresponds to the average spectrum of the flight noise of a female three or four days old (see p. 625).

Some comparable reactions have been obtained in *A. quadrimaculatus* Say and in *Culex pipiens* L. with frequencies within the range of 320–480 c/s for the first, and between 288 and 384 for the second[160].

In the same way that the fundamental of the female's emission rises in frequency as the insect ages, the maximum number of responses obtained in the male ascends with age towards the higher frequencies: 350 c/s at one day, 500 at three days, 600 at seven days. The intensity as well as the frequency has an optimal region: the highest percentage of responses is obtained for intensities within the range of 68–85 dB at 2.5 cm[183]. This is contrary to the results found in *Ephippiger* for which no upper threshold of intensity has been found, the strength of the reaction being, on the contrary, in proportion to the intensity*.

One of the most remarkable aspects of the effect of sound on mosquitoes, is certainly the triggering of preliminary mating movements noted in males of *Aëdes aegypti*[160]. We can actually determine three phases in this insect's phonotaxis:
(i) flight,
(ii) attraction towards the source of the stimulus,
(iii) mating movements carried out by the male clinging to the side of the cage in front of which is the tuning fork or the loud-speaker. In this third sequence, a vibration of the body and wings is noted, a behaviour pattern comparable to that which is seen when the male seizes the female in flight—that is the *"seizing response"*. Next there is a ventral flexion of the abdomen, bringing its end into contact with the cloth side of the cage, a movement which is analogous to the one which the male makes to fix his genitalia into the females, called *"clasping response"*.

Hence we are faced with a chain of reactions (flight, oriented taxis, mating movements) all of which are governed by the same stimulus. This is a much less well developed behaviour pattern than that which the Orthoptera Acridinae show, in which each sequence requires a special signal for its execution (calling, courtship, agreement, jumping songs with the possibility of visual, tactile or olfactory stimuli taking part).

Of course, the idea has been considered of using for practical ends this possibility of attracting mosquitoes by artificial stimuli or by the flight noise recorded and relayed by loudspeaker. The first attempt as such was made in 1901 by Weaver (quoted by Howard, 1901[107]). In front of a device generating the frequencies he was using, he placed a metal screen electrified with an alternating high tension current which electrocuted the insects which rushed on it. These experiments were taken up again in the field in 1949[117, 118], but the results obtained do not seem conclusive enough to allow a prediction of what the future of this new means of fighting mosquitoes will be. As far as laboratory technique is concerned, however, attraction by the sound of a tuning fork is a definite practical method of capturing and isolating male mosquitoes in laboratory colonies[161].

We should also mention here the numerous relations of attraction by very different noises in the cicadas[88, 134]. In Kansas, females of *Tibicen dorsata* (Say) have been seen to sit in great numbers on a tractor, the engine of which was making a noise resembling the male's song. In Malaya, the natives attract *Pomponia intemerata*

---

* Such a treshold, however, probably does exist, but in any case it is higher than 100 dB.

References p. 649

(Walk.) by clapping their hands. An identical observation has been made on a non-identified species in Chile[121,122]. Similarly a French species, (*Lyristes plebejus* Scop. probably), can be called by whistling[169].

Similar facts have been observed in *Gryllotalpa gryllotalpa* (L.) which was attracted by the noise of a tire pump[74].

In the ants, a positive taxis has been reported in a species of the genus *Crematogaster* (Formicidae), set off by a tuning fork vibrating at a frequency of 4,096 c/s[124].

The phonotaxis of female *Corixa striata* L. (Heteropt., Corixidae) was obtained with sonic or ultrasonic stimuli produced by a whistle or a frequency generator[166].

Finally, it is also appropriate to include the many reports (mostly dated) concerned with spiders "attracted" by the sounds of various musical instruments, harp, bagpipe, harpsichord, lute, violin[27,55]. It is possible, however, that mechanoreception is at the root of these taxic behaviour patterns, rather than audition proper*. Moreover, many authors since have obtained attractions of spiders toward a tuning fork put in contact with the web[30,93,132,139,147]. The sensitivity of these animals to the vibrations produced in this way seems to be restricted to a quite narrow band of low frequencies.

(iii) Interspecific phonotaxis and "elements of specific recognition"

Taxic reactions to artificial signals in many insects seem to show some flexibility of I.R.M.'s towards the stimuli which are presented to them. Consequently we may deduce, apparently, that non-specific natural signals, because of their relation to those of the species, may also be effective. Actually experiments have proved in many cases that interspecific attraction was possible. Nevertheless, some observations, which we have already emphasized, seem to show, in certain cases at least, that natural (whether specific or not) and artificial phonotaxis do not seem to be of exactly the same nature.

It seems that relatively few experiments have been made in the field of interspecific attractions. The only important results are concerned with the Orthoptera, Diptera (mosquitoes) and Homoptera (cicadas).

We shall take two examples from the Orthoptera, one from *Ephippiger* and one from *Oecanthus*.

The various French species of the genus *Ephippiger* have been put together in succession; in all cases the females showed a very clear, positive taxis towards the heterospecific males[46]. When the female had a choice between two groups of males, one from own species, the other from another species, she headed for the males of the second group less and less frequently in proportion to the difference of their songs from the species song; yet the number of heterospecific attractions is considerable. These results are summarized in Table 48.

Fig. 366 shows the oscillographic representation of the songs of these species, which differ especially in the number of pulses making up the chirp (case of *E. bitterensis*: 1 pulse, and *E. cunii*: 3-5 pulses) or in the length of the stridulatory movement (case of *Ephippigerida nigromarginata*). In comparing these analyses with the results laid out in Table 48, we are led to the opinion that specific recognition

---

* The reception of air-borne sounds, however, has been demonstrated by the electrophysiological study of the lyriform organ situated on the metatarsus of the spider's leg. This organ is also a vibration receiver[163,177].

TABLE 48

INTERSPECIFIC PHONOTAXIS IN *Ephippiger* (ORTHOPT. EPHIPPIGERIDAE)
(DUMORTIER, UNPUBL.)

Female	Males present	Respective attractions (%)	Total moves recorded	χ^2 Test $P = 0.05$
B_1	B_1 / E	43 / 57	207	S*
	B_1 / B_2	54 / 46	273	NS
	B_1 / B_3	60 / 40	182	S
	B_1 / C	65 / 35	143	S
	B_1 / N	70 / 30	208	S
B_2	B_2 / B_1	57 / 43	90	NS
C	C / B_1	71 / 29	145	S

The females were placed in a big cage with another cage at either end, one with males of the species, the other with those of a different species.
Key: B_1 = *E. bitterensis* Finot; B_2 = id., local form with a bipulsed chirp; B_3 = id., local form with a tripulsed chirp; E. = *E. ephippiger* (Fiebig); C = *E. cunii* Bolivar; N = *Ephippigerida nigromarginata* Lucas. S = significant; NS = non significant.

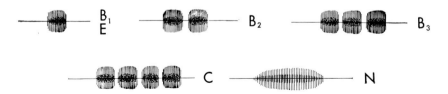

Fig. 366. Oscillographic representation of the songs of species used in experiments of interspecific phonotaxis in *Ephippiger* (diagrammatic). E, C, B_1, B_2, B_3, N, see key of the Table 48.

within the genus is established principally by judging the number of pulses making up the chirp**. This explains why the female is unable to recognise the male of her own species*** from a member of another species stridulating in the same way, yet this does not explain the marked preference of the females *E. bitterensis* for the males *E. ephippiger*. Besides, it is obvious that the number of pulses is an element of specific recognition only in as far as the pulses of two species under consideration are of the same structure. Hence the one-pulsed chirp of *Ephippigerida nigromarginata*

---

* Significant *in favour* of the heterospecific males.
** We propose the term *elements of specific recognition* to designate the parameters in the sound signal which enable an individual to recognise another individual of his own species. This recognition is shown by an appropriate behaviour: phonoresponse or phonotaxis.
*** Assuming, of course, that phonotaxis may be taken as a test of recognition.

has a very different structure from the one-pulsed chirp of *bitterensis* or *ephippiger* and the females of these two species distinguish them clearly. This seems to indicate that the toothstrike rate also plays a role.

As far as intensity is concerned, interspecific phonotaxis does not seem to necessitate levels higher than natural phonotaxis; the opposite is the case with artificial phonotaxis (see p. 620).

The second example in the Orthoptera relates to the North American *Oecanthus*, on which a very accurate study has been made[178]. In the first place it showed that the percentage of interspecific phonotaxis was very low and, secondly, that the *element of specific recognition* lay either in the pulse-rate or in the chirping-rate according to the species. Let us take as examples three species, the calling song of which is a continuous trill: *Oecanthus nigricornis* Walker, *O. quadripunctatus* Beuten., *O. argentinus* Sauss. At a given temperature, the pulse-rate of *nigricornis* is higher than that of *argentinus* or *quadripunctatus* (at 21°C, for example, we have the following values: 53, 42 and 33). The female of *nigricornis* does not react to the calling song of the other two species in these conditions. But, if a recording of one of these heterospecific songs made at a temperature* where the pulse-rate reaches that of *nigricornis* (26°C for *argentinus* and 31°C for *quadripunctatus*) is played to this same female placed at 21°C, she manifests a phonotaxis identical to that which she would have shown with a male of her own species. We should note, however, that the frequency of the signal is then appreciably higher than that of *nigricornis* (3.9 kc for *argentinus* and 4.4 kc for *quadripunctatus*, against 3.3 kc for *nigricornis* at the temperature under consideration). These results are confirmed by those obtained with artificial trills (produced by means of a frequency generator), which give positive responses in so far as the pulse-rate is close to that of the natural stridulation at the same temperature.

This same type of experimentation showed that in *O. niveus* (De Geer), the pulses of which are grouped in regular chirps, the *element of specific recognition* consisted of the chirping-rate.

The fact that *the element of specific recognition* modifies itself with the temperature (increase of the pulse or chirping rate with the rise in temperature) implies that the female's sensory-motor mechanisms which control the taxic behaviour, can respond to any values the *element* may take under the effect of external factors. This is achieved not by non-selectivity of the I.R.M., which would prevent any specific recognition, but by a permanent adjustment of the sensory-motor mechanisms with the variations of *the element of specific recognition*. At a given temperature the female is ready to respond to a pulse-rate (or a chirping rate) which corresponds exactly to that of the male of her species at the same temperature. Hence a female *quadripunctatus* at 22°C shows a taxic reaction to the song (natural or recorded) of the male kept at this same temperature (pulse-rate = 34), but not nearly so well to the broadcast of the song of a male recorded at 26.5°C for example (pulse-rate = 45).

Hence the temperature works symmetrically and simultaneously on the male and female, by modifying the pulse-rate of the one and tuning the sensitivity range of the other to the corresponding value.

The Diptera, with the mosquitoes, represent the second order in which interspecific phonotaxis or even intergeneric ones has been revealed. It has been shown in

---

* See p. 366 and Fig. 228.

Aëdes aegypti L. that any sound signal with a fundamental frequency around 500 c/s sets off the male's taxis. It is not surprising that the flight noise, whether of males or females, of several genera or species have been proved effective. Hence, the males *aegypti* are attracted by *Anopheles quadrimaculatus* Say males and females, *A. punctipennis* (Say) males and females, *Culex pipiens* L. males and engorged females*, *Aëdes trivitattus* (Coq.) females, *A. vexans* (Meig.) males and females, *Psorophora horrida* (D. and K.) females, *P. confinnis* (Lynch-Arr.) males**[160]. In most cases these attractions are followed by attempts at mating. Hence, as opposed to what we met in the Orthoptera, there is apparently not *element of specific recognition* in the mosquitoes, which we just have considered, other than frequency (see p. 621).

In this group of results connected with phonoresponse and phonotaxis, there are some important findings, though it must be remembered that they only have meaning within the limit of the examples quoted here.

(1) In a number of cases, phonotaxis and phonoresponse may be set off by non-specific and artificial signals.

(2) These behaviour patterns are obtained with artificial signals only if they are of a sufficient intensity, always appreciably higher than that needed to obtain reactions with natural signals. The intensity seems to be a limiting factor in every case; the frequency, within quite broad limits, apparently does not play an important role (except in the mosquitoes).

(3) Phonotaxis and phonoresponse do not seem to be caused by the acoustic signal as a whole, but only by some parameters*** (pulse, chirping-rate, frequency, transients, toothstrike rate, etc.).

(iv) *Psychological aspect: Phonotaxis and Tropism*

May natural phonotaxis be included under the many orienting and movement responses which have been called tropisms, or has it a special place? Obviously if one is referring to a general definition of tropisms like the following: "orientation and locomotion reaction set off and sustained by a source of energy, and leading the animal towards or away from this source"[144], then phonotaxis is a tropistic reaction. A more detailed comparative analysis of the characteristics of tropisms on the one hand, and of phonotaxis on the other, however, shows that the differences are actually more numerous than the analogies.

The characteristics of tropism of which we shall give an account are mostly borrowed from the authoritative work of Viaud[176].

	Natural phonotaxis	Tropism
Reaction:	1. Orientation	1. id.
	2. Move towards the source	2. id.
Stimulus:	3. Discontinuous stimulus****	3. Continuous stimulus
	4. Low intensity threshold	4. Often high intensity threshold

---

* Repletion raises the frequency of the flight noise,

** The age of the animals is important too, since it modifies, on the one hand, the emission frequency of the female, and, on the other hand, the area of the sound spectrum where the maximum sensitivity of the male is located. (See p. 621).

*** It is interesting to note that in the fiddler-crab *Uca tangeri*, the rate of beckoning gestures (day signal) and that of the drumming of the large cheliped (night signal) is about the same[95]. Both act as calling and courtship signals on the female (see p. 586). So it seems that the I.R.Ms which control the sexual behaviour of the female start by a certain specific rhythm in the production of the releasing stimulus either it is of a gestual or acoustic (or vibratory) nature.

**** Exception: the mosquitoes.

	5. Law of the maximum (= research for the maximum intensity of the stimulus)	5. id.
	6. Relation between the intensity of the stimulus and the intensity of the reaction (Weber-Fechner's law)	6. id.
	7. Total stimulus reducible to one or several trigger-parameters (sign-stimuli)	7. Stimulus acting as a whole, not reducible to trigger parameters
	8. Stimulus normally of an interindividual origin (coming from another member of the species or genus)	8. Stimulus with an origin outside the species of genus—either a normal part of the environment (geotropism rheotropism, sometimes phototropism)—or artificially imposed by man (galvanotropism, sometimes phototropism and chemotropism)
Behaviour:	9. Behaviour linked with motivation	9. No motivation
	10. Source and stimulus can be dissociated as far as the nature of released behaviour is concerned.	10. Source and stimulus cannot be dissociated
	11. Behaviour ending with the creation of a situation in which the motivation can be satisfied (mating, for instance)	11. Behaviour *apparently* representing an end in itself

As far as we can see, although the phenomenon has the same appearance in both cases (orientation and movement towards the source), and although the general laws governing the reaction are the same in phonotaxis and tropism (law of the maximum, law of Weber-Fechner), two levels of difference become clear: (i) one affecting the stimulus, (ii) the other related to the results of the analysis of the phonotaxic behaviour.

(i) In all known examples (mosquitoes excepted) the stimulus which releases a phonotaxis must have an element of discontinuity[*]. This is a natural part of the insect's stridulation: impact of the teeth against the plectrum, or silence separating the pulses or chirps. *Oecanthus, Chorthippus, Ephippiger*, do not react with a taxis to a continuous artificial signal of long duration. In tropism, on the other hand, although special reactions have been observed in response to stimuli with abrupt changes in intensity (case of the "shock-reactions" studied by Mast in the *Volvox*), such variations are not at all necessary in order to get the animal to move.

On the other hand, the stimulus generating a tropism acts as a whole; an analysis intended to reveal the fundamental triggering role of one or several of the constitutive parameters does not seem possible. Yet, this type of analysis conducted on the stimulus that causes a phonotaxis, shows that some elements can be extracted from its complex constituents, (pulse-rate, signal form, etc.) which are effective (sign-stimuli). The other components do not seem to play any role and may be considered as limiting factors. We have already shown (see footnote p. 613) that there is an analogy here between this and what is observed in the vertebrates where the trigger mechanisms of more developed behaviour patterns are only released by a limited number of stimuli "recognised" as signs and "extracted" from a context (stickleback, robin, gull, etc.).

Finally, when tropisms are responses to excitations coming from the inanimate environment, phonotaxis normally has a biological origin, coming from the species. Certainly we may wonder if this distinction is not artifical and only follows a methodological convention by which only those reactions which satisfy a certain condition concerning the origin of the stimulus would be considered as tropisms. In fact, the biological origin of the stimulus is not a special case or an exception to the rule; it goes hand in hand with other distinctive characteristics such as specific recognition, motivation, and aim of the behaviour. Therefore, this biological origin is one of the elements contributing to make phonotaxis a reaction distinct from orthodox tropisms.

(ii) From the aspect of behaviour we shall again find three of these elements of differentiation. The first is concerned with motivation (see p. 597). When tropism is a behaviour without motivation, phonotaxis is a reaction which can only be obtained in certain physiological conditions generally depending on the sexual state (see p. 639). The efficiency of the trigger stimulus is in

---

[*] This is a peculiarity of phonotaxis, but in other taxic behaviour patterns with similar characteritics (phototropism of some Coleoptera Lampyridae males to the luminous emissions of the females; chemotropism of many insects set off by the odoriferous emission of the partner), the factor of discontinuity need not exist.

proportion to the level of motivation, of the receiver. This level acts as a more or less thorough system of starting up the I.R.M.

The second characteristic of phonotaxis from a teleological point of view, is the fact that it is not an *end* in itself (which is the case with tropisms), but it seems to be a *means*, a link in a complex behaviour pattern, the climax of which is the meeting of the sexes and the result, the propagation of the species.

Finally, the third important point of comparison between taxis and tropism is related to the distinction between stimulus and source. Actually, as we have just said, the phonotaxis (response to stimulus) leads to the meeting with the source of the stimulus (individual of the opposite sex); this source then enters into the animal's field of perception by new sensory paths (olfactory, visual, tactil) and causes the second behaviour sequence (copulation). In tropism, on the other hand, the source only exists through the stimulus which comes from it, and to the animal it is nothing in itself (phototropism, galvanotropism); it may even be indefinable (rheotropism) or inaccessible (some cases of phototropism, geotropism).

On the whole, we feel that there are too many contradictions between phonotaxis and tropism for us to mix them. On the other hand, these elements: motivation, sign-stimuli within the stimulus, biological role of the phonotaxic behaviour (elements to which may be added the low value of the intensity threshold of the stimulus), enable the phonotaxis to be likened to what has been called in psychology "perceptive reactions". One could also put into this category the chemotropism of the males of the Lepidoptera Saturnidae to the odoriferous emissions of the female and the phototropism of the Coleoptera Lampyridae males to the light signals of the females. Nevertheless, for the psychologist, the conception of perceptive reaction implies a knowledge (innate or learned) of the meaning of the sign (which appears as a message of the future), and hence, a conduct adapted to the nature of this sign. It is understood that for man this supposes, on the one hand, the interpretation of the sign (decoding) and the realization of the phenomenon or event, of which the sign is the announcing symbol, and on the other hand, the choice of the conduct to be followed. All this takes for granted the participation of mental processes which are lacking in most vertebrates. Obviously this is a question of definition, but perceptive reaction has a definition too clearly "human" to fit unreservedly the phonotaxis of insects.

In addition, the problem is particularly complicated by the existence of the artificial phonotaxis which, by the high threshold of intensity they require, resemble tropisms.

To conclude, and as taking our basis the preceding examples, we shall say that phonotaxis is a *behavioural* reaction with a tropistic appearance, causing a locomotive activity in the individual, which leads it up to the source of the stimulus. This reaction is caused by *certain parameters* of the sound stimulus (*sign-stimuli*) acting as triggers on the one hand, and as *elements of specific recognition* on the other hand. This reaction, conditioned by a certain level of motivation (sexual), is only a link in a complex chain of behaviour, the result of which is the meeting of the sexes.

3. Song and Genetics

(a) The song as a genetic feature

Song is a specific feature in the same way as are the morphological features. It is passed on unchanged to each generation, but just as in other specific features, the parameters which caracterize it, are liable to some variation. One only has to hear a group of grasshoppers or crickets stridulating to ascertain the individual differences in frequency, intensity or pulse-rate. These differences become more marked when we look at forms with a wider "vocabulary", such as the Acridinae. The external factors, especially the temperature, may also considerably modify the song. These changes, however, are only simple responses to changes of environment, as changes in pigmentation or ornamentation.

In arthropods, the song seems to be altogether hereditary; no experiment has suggested that there may be the slightest process of learning, as opposed to what is

References p. 649

found in some birds for example*. In line with this, the behaviour set off by the signal of the species, corresponds to entirely innate patterns, which, however, show a quite broad selectivity (hence interspecific responses are possible). This plasticity, however, is preformed, and cannot be changed by learning. Hence by relaying to four groups of females *Chorthippus parallelus* of the third instar: a recording of *Omocestus viridulus*, (1st group), an imitation of the song of *parallelus* (2nd group), a recording slowed down (3rd group), a recording at the normal speed of the calling song of this same insect (4th group), three times a day for an hour, it was possible to establish that, once they become adults, only the females of groups 2 and 4 react to the corresponding recordings. Those of groups 1 and 3 showed no reaction to the recordings which had been relayed to them daily, but, on the other hand, responded to signals 2 and 4 which they had never heard before[103].

(b) The hybrids

Fulton (1933)[81] was the first to approach the problem of hereditary transmission of the song by the study of the hybrids. Since then this type of research has never been fully developed, and the sum of the results available to us today is rather slim. This is because the question has generally been studied somewhat superficially and on few species (belonging to the genera *Gryllus*, *Acheta*, *Nemobius* and *Chorthippus*), and sometimes without the use of thorough methods which distinguish genetic studies.

In the Gryllidae, crossing *Acheta rubens* (Scud.) ♂ × *A. assimilis* F.** ♀ gave in F_1 some hybrids, whose pulse-rate was midway between that of the parents. On the other hand, all the hybrids sang in chirps (*assimilis* feature) and not in trills (*rubens* feature)[28].

Crossing *A. rubens* × *A. veletis* Alex and Big. gave some interesting results which are quite like those obtained in the preceding case. Crossing *rubens* ♂ × *veletis* ♀ produced in F_1 some males (5 adults) with very slow development, the song of which was approximately of a length midway between the chirp of *veletis* and the trill of *rubens*. The number of teeth in the rasp was also intermediate. The only male hybrid of the reciprocal cross had an intermediate pulse-rate, but a song of the *veletis* type (chirp)[28]. It would be dangerous to conjecture the precise song-transmission mechanism from these few results. Nevertheless, we can say that the song does not depend on a single factor controlling all the parameters "en bloc", but several parameters (pulse-rate, chirp or trill, etc.) corresponds to different genetic actions. On the other hand, the intermediate values of the pulse-rate and number of teeth of the hybrid in comparison to those of the parents suggest to some extent a polygenic system of transmission.

The results have been less successful in connection with attempts at crossing the sub-species of *A. assimilis***. Mating occurs, nevertheless, in the laboratory, but there is no fertilisation and the eggs do not hatch[85]. In another experiments, however, an individual was obtained by the crossing *A. pennsylvanicus* (Burm.) ♂ × *A. vernalis*

---

* We must, however, point out that if the song is innate, some cases of response to song recall learning (see p. 616).
** The species *A. assimilis* was described in 1755 by Fabricius after a specimen from Jamaica. The same name was applied to some North American individuals, but a more precise study has shown some heterogeneity in their ecology and life cycle which has led to their division into different species, the distinction of which is difficult if only morphological features are considered (see p. 636).

(Blatchley) ♀, and two individuals by the crossing *A. pennsylvanicus* ♂ × *A. fultoni* Alexander ♀. The characteristics of the hybrids' songs are only known in a natural hybrid of these two species, found in a zone where they cohabit[5].

The study of the transmission of precourtship behaviour has been made on the hybrid resulting from crossing *Gryllus campestris* L. ♂ × *G. bimaculatus* De Geer ♀. In the first species, the precourtship consists of a single raising of the elytra without sound emission: in the second, the movement is repeated and accompanied by stridulation. Most F_1 individuals have a precourtship, which is intermediate in comparison with those of their parents, made up of a single sound[106].

The F_1 obtained by the crossing *N. allardi* Alex. and Thom.* × *N. tinnulus* Fulton, shows a chirping rate the value of which comes between that of the parents. In F_2 this value still remains intermediate, but the variability increases. The chirping rate of the back-crossed shows a slight tendency to be like that of the parent species used for the backcross[81] (Fig. 367). These results, the first obtained on the heredity of song, must really be considered as approximations, taking into account the small number of hybrids studied and the approximate methods of evaluating the chirping rate.

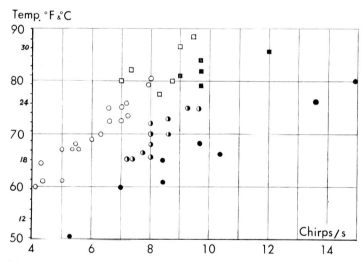

Fig. 367. Inheritance of song in F_1 hybrids of *Nemobius fasciatus fasciatus* (= *N. allardi*)* and *N. tinnulus* and progeny of backcrosses. Study of the chirping rate in relation to the temperature. ● = *N. allardi*, ○ = *N. tinnulus*, ◐ = F_1 *allardi* × *tinnulus*; ■ = backcross F_1 × *allardi*, □ = backcross F_1 × *tinnulus*. (After Fulton, 1933, simplified[81].)

As far as the Acrididae are concerned, the only studies made until now have dealt with the hybrids of *Chorthippus biguttulus* crossed with *C. brunneus*. The average duration of the parent's chirp is, for the first, 2.16 sec and for the second 0.18 sec. In F_1 the duration is intermediate: 0.50 sec (*biguttulus* ♂ × *brunneus* ♀), and 0.46 sec (*brunneus* ♂ × *biguttulus* ♀) (Fig. 97, p. 229). Some hybrids of this kind have been met in nature, but they are always special cases. Offspring of backcrosses have a a song which is like that of the parent species used for the backcross[140].

---

* Alexander and Thomas (1959)[12] designated this name for the species previously called *N.f. fasciatus* (De Geer); in line with this, and still according to the same authors, *N. fasciatus socius* Scud. becomes *N. fasciatus* (De Geer) and *N.f. tinnulus* Fulton, *N. tinnulus* Fulton.

References p. 649

TABLE 49

ISOLATING MECHANISMS

Hybrids can only appear if all the listed conditions (b) are fulfilled. The part (a) of each of the 9 alternatives represents the level where the barrier preventing hybridization may be situated. The asterisks show that the point under consideration may be illustrated by examples taken from *Acheta, Nemobius* or *Chorthippus*.

1
(a) Different geographical repartition*
 (geographical isolation)
(b) Identical geographical repartition*

2
(a) Different microhabitats (ecological isolation)*
(b) Common microhabitats or overlapping partially*

3
(a) Different annual cycle*
(b) Same annual cycle*

4
(a) Unphased daily rhythm of activity
(b) Phased daily rhythm*

5
(a) Sexual behaviour set off only by specific stimuli (ethological isolation)*
(b) Sexual behaviour set off by interspecific stimuli*

6
(a) Unsuited genitalia (morphological isolation)
(b) Suited genitalia

7
(a) No amphimixis or non viable zygote (cytological isolation)*
(b) Viable zygote*

8
(a) F_1 dies before maturity*
(b) F_1 reaches maturity*

9
(a) F_1 sterile*
(b) F_1 fertile*†

† (i) The hybrids may be inferior in comparison with the parental species (hybrid breakdown), and sterility is possible in F_2 or in the back-crossed.

(ii) The same group of conditions is again required for F_1 to give offspring. Another condition is that the song of the hybrid male can release the sexual behaviour of the hybrid female (see also footnote** on p. 635).

(c) Song and reproductive isolation among species

The study of phonotaxis has shown that—experimentally—it is not unusual to obtain interspecific attractions. The hybridization experiments prove that in some cases interspecific pairing can be fertile. However, hybrids are very seldom found in nature. Hence we must accept the existence of isolating mechanisms, which preserve the purity of the species[62]. Table 49 gives a plan of the various "levels" (geographical, ecological, ethological, etc.) where the isolating mechanism may appear. We shall try in particular to see what importance the song has among these mechanisms.

Let us say plainly that the very small number of species which have been observed or on which experiments have been done, does not allow for any generalisation. It is better to consider the results as only valid for the species on which they have been obtained and in so far as they have been produced by repeated experiments and not by more or less chance observation.

The only really thorough and valid work in this field is due to Perdeck (1957)[140], who showed point by point the hybridization mechanism between *Chorthippus biguttulus* and *C. brunneus*, by revealing the relative rôles of the different stimuli which are involved in pairing.

These species cohabit in the same biotope (sympatric species) and have a similar seasonal and daily rhythm; hence the first barrier can only be at the level 5 of the succession of isolating mechanisms, a level at which the triggering stimuli of the sexual behavior occur (Table 49).

It has been possible to establish that, taking the number of homogamic matings as equal to 100%, heterogamic matings for the same duration were of the order of 15% (Fig. 368). Already we can see a certain value in the sound emission as a preventing interbreeding system, since the song of the other species only sets off imperfectly the patterns of behaviour resulting in copulation (Table 50).

However, these behaviour patterns are better displayed if, in the heterogamic situation, the two partners can *hear* the specific song of an individual of the opposite sex (curve 2 and 3 of Fig. 368). This last observation shows, moreover, that the small

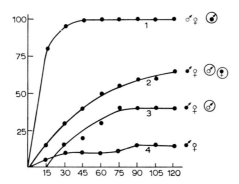

Fig. 368. Rate of copulation in homogamic et heterogamic situations (*Chorthippus biguttulus* and *C. brunneus*). Encircled signs: individuals enclosed in tubes, present during the mating of the partners. Ordinate: number of copulations (percentage) abcissae: time in min. Curve 1: homogamic matings. Curves 2, 3 and 4: heterogamic matings with or without presence of a conspecific individual of the other sex. In 1, the heterospecific male is used for giving a situation reciprocal to 3. (After Perdeck, 1957[140])

References p. 649

percentage of mating in the heterogamic situation (curve 4) is not caused by the absence of triggering stimuli of a non-acoustic kind*,**. Actually, if such stimuli were absent in situation 4, they were equally absent in situations 2 and 3 which only differ from 4 by the addition of the specific song. Hence we see once again that the specific song holds the power to set off sexual behaviour, and thereby that it is clearly an isolating mechanism (Table 51).

TABLE 50

ACTIVITIES OF THE MALE IN THE PRESENCE OF A SINGING FEMALE, REAL MATINGS BEING PREVENTED
6 INDIVIDUALS IN EACH EXPERIMENT (AFTER PERDECK 1957)[140]

♂ ♂ brunneus

	in the presence of ♀ brunneus	in the presence of ♀ biguttulus
Locomotion (big jumps included)	262†	78†
Encounters	19	4
Songs	918	235
Time (min)	90	90

♂ ♂ biguttulus

	in the presence of ♀ biguttulus	in the presence of ♀ brunneus
Locomotion (big jumps included)	535†	166†
Encounters	33	0
Songs	548	200
Time (min)	90	90

† These numbers express the result of a count carried out arbitrarily. Their value is only comparative.

However, in this experiment the song only has a partial isolating value, since in spite of everything there are quite a few heterogamic matings. This is probably due to the fact that the experimental conditions (high density of popylation, absence of specific partner, tests made on females with strong motivation), do not reproduce the natural ones***.

However, that may be, as soon as the female has agreed to mate, the genitalia come into contact just as quickly in the heterogamic situation as in the homogamic.

---

* In *biguttulus* female, however, the smell of the male increases the number of stridulations; this must normally facilitate the meeting.
** According to von Hörmann Heck (1957)[106] the courting behaviour of *Gryllus bimaculatus* male appears in response to stimuli coming from the female. In *G. campestris*, the mounting reaction of the female is caused by stimuli coming from the male. This author claims that the copulation *campestris* ♀ × *bimaculatus* ♂ is thus made very difficult "by the lack of necessary key-stimuli in both partners". In fact, this crossing was obtained by Cousin (1933)[59] a long time ago without any special difficulty.
*** This is also established for *Drosophila* in which some hybrids commonly obtained in the laboratory are practically never met in nature.

TABLE 51

HETEROSPECIFIC SOUND REACTIONS (AFTER PERDECK, 1957)[140]

(a) % responses of females to males
(b) % responses of males to females

(a)

	♂♂	
♀♀	brunneus	biguttulus
10 brunneus	36%	1%
10 biguttulus	1%	67%

(b)

	♀♀	
♂♂	brunneus	biguttulus
7 brunneus	31%	0%
7 biguttulus	0%	31%

Hence there is no morphological isolation.

The larval mortality of hybrids is not higher than that of members of the species. Hence it seems levels 7 and 8 of the isolation mechanisms are crossed without difficulty and there are no barriers at these embryonic or nymphal stages. Of course, if the F_1 reaches maturity, the same succession of barriers will have to be crossed after level 2 or 3, to ensure offspring. As far as we can see, ethological isolation will again be the most important factor, as in the case of the crossing between the two species. In actual fact, it has been established that the responses of females of the parental species to the song of F_1 males, are of as low a percentage (1 to 3%) as responses to heterospecific males. The males of the parental species seem to react to the hybrids of the opposite sex a little better than the females do (few observations). As for the reactions of the hybrids among themselves, there is a great deal of variety (Table 52).

TABLE 52

SONG REACTIONS OF THREE DIFFERENT F_1 FEMALES TO THE SONG OF F_1 MALES AND PARENTAL SPECIES MALES (AFTER PERDECK, 1957)[140]

♀ F_1	Reactions to the song of		
	♂ brunneus	♂ biguttulus	♂ F_1
♀ A	80%	57%	50%
♀ B	0%	72%	38%
♀ C	53%	0%	37%

As far as females issued from backcrosses are concerned, they seem to respond better to the male of the species used for the backcross than to the one of the other species.

References p. 649

To sum up, and in the case of these two species of *Chorthippus*, the principal, if not the only barrier to interspecific gene flow is of an ethological kind. The song of a species sets off, more or less with difficulty the behaviour patterns which result in mating with a partner of another species. The conditions of the experiment render this barrier more easily surmounted in the laboratory than in nature, but, though hybrids are produced in the F_1, they are partially isolated from one another and from the parental species.

We are now going to look at some other examples in Gryllidae, where the isolating mechanisms may consist of other factors than the song. *Acheta assimilis* (Jamaica) and *A. veletis* (Virginia) mate in the laboratory, but the eggs which are laid do not hatch; an isolating mechanism acts at level 7 (cytological isolation). It is obvious, however, that in this case the first barrier is geographical[9].

The crossing *A. veletis* ♂ × *A. rubens* ♀ gives in F_1 males with a very slow development, which in addition, seem sterile (isolation at level 9). On the other hand, it is at level 3 that the barrier for the hybrids between *A. assimilis* ♂ × *A. rubens* ♀ is situated, since the life cycle of the males F_1 is much longer than that of the females. All the females are dead before the males are adult. Moreover, the hybrid's mortality is very high. The same difference in annual rhythm is found again in *A. veletis* and *A. pennsylvanicus* (Burm.), sympatric species which have very similar songs. The females and males of *veletis* are dead before the males and females *pennsylvanicus* are adult. Experimental crossing has, in addition, not been successful. A difference of song is not needed, by all accounts, to avoid the gene flow between these two species[28].

These few examples are sufficient to show that it would be false to consider that in the singing Orthoptera, the song is in all cases the sole factor in isolating the species.

It is none the less true that examples of song being the barrier against interbreeding are many, and it must be realised in this connection that in general, sympatric species always differ in the structure of their songs. The case of *Chorthippus biguttulus* and *C. brunneus* is quite significant on this subject. Similarly in the *Nemobius fasciatus* group, the three species *N. fasciatus* (De Geer), *N. allardi* Alex. and Thom., *N. tinnulus* Fulton, live in different contiguous biotopes, and frequently the three forms are met together within a radius of a 100 m[12]. It is known, however, that the calling songs of these insects differ quite considerably (pulse-rate, trill or chirps), as do their courtships. There is no exaggeration in thinking that this specificity of song contributes completely or in part to the extreme rareness of natural hybridization.

In *Oecanthus*, the picture is the same. *O. quadripunctatus* and *O. nigricornis* are sometimes sympatric, as are *O. angustipennis*, *O. exclamationis* and *O. niveus*[78]. When the first two are trillers, their pulse-rates are very different, as is the case with *angustipennis* and *exclamationis*. Now, it has been shown[178] that the pulse-rate acts as the *element of specific recognition* for these species (see p. 624), hence the isolating mechanism.

Consequently, it seems that sympatry is coupled with some divergence in song characteristics. But if there is a relationship of cause and effect in this, it remains to be discovered what is the cause and what the effect. Let us say also that the contrary is not necessarily true: of course allopatry is not systematically coupled with an identity of song.

How can these facts help us to understand the mechanism of speciation? The results of which we have given an account are still too fragmentary to serve as a basis for theories. We shall just show how the song, in its double aspect of hereditary feature and isolating mechanism, can form one of the possible ways of speciation. We shall only refer to very classical data of Mendelian genetics to construct our theory, imagining a case of crossing between two species different by two pairs of alleles, located on two different chromosomes, causing two song parameters which make specific recognition possible. We make the formula of the first parent A A B B, and of the second C C D D*.

The F_1 individuals will be heterozygotes with the formula A C B D, the song of which will be either of an intermediate type between those of the parents or similar in whole or in part to the song of one of the two parents, according to whether a phenomenon of dominance occurs or not.

These hybrids between them will give rise to an F_2 in which nine kinds of combination will be possible: A A B D, A C B D, A C B B, A C D D, C C B D, A A B B, C C D D, A A D D, and C C B B. The last two correspond to homozygous individuals in which the parental traits in question are recombined in twos. These recombinants A A D D and C C B B will have a song of which one of the parameters will be of the first parental species' type, and the other of the second's. We acknowledged at the beginning that these two parameters represented the elements of specific recognition, but when the simultaneous possession of traits A and B for one species, C and D for the other, really makes up the key-stimulus, the reverse combination A D or C B is very likely not to be "recognised" by the parent species. With relation to the latter the two types of recombinants A A D D and C C B B will be isolated. It may be agreed, too, that A A D D and C C B B will be isolated in relation to each other for the same reasons. Hence each type of recombinant will only respond** to the stimuli of the individuals of its own kind; this causes the formation of two populations isolated ethologically from one another and from the parental species.

An important fact which distinguishes the heredity of the song from the heredity of morphological features usually studied, is that *the song is an isolating factor*. The appearance of new arrangements in its parameters is very likely *ipso facto* to produce the isolation of the individual from his species; this is a phenomenon which does not occur when traits like form, length or ornamentation of an appendage, for example, are in question. Now, the first condition of speciation is isolation, and this condition is fulfilled here. Certainly, the homozygotes of F_2 are not necessarily *species* distinct from the parental species, but, once the barrier has been placed between the two, progress towards speciation has already begun.

The objection will be raised that we have granted that the ethological barrier "song" must be broken down before the parent species can cross. The previous experiments showed that such a phenomenon could be seen in many experimental cases and its apparently extreme rareness in nature does not exclude the possibility of the situation where it is produced happening by chance. After all, do not all speciation and evolution lie in the fulfillment of improbable situations?

Certainly we do not claim that hybridization is *the* speciation process of the singing Orthoptera, but that it is a *theoretically* possible way. To sum up, it appears that:

(i) The song is a genetic trait essentially different from most morphological traits in that it is also an isolating factor.

(ii) Hence the homozygotes formed in F_2 may then evolve in isolation from the parental species.

(iii) Such homozygotes may give issue with at least a value of sub-species.

The hybrid AADD or CCBB will not perhaps be recognizable as such by its morphological characteristics but rather by its song. A systematist finding it in the

---

* The preceding evidence showed that the song was the manifestation of the action of genes or groups of genes, each having one of the parameters making up the sound signal dependent on it. These conditions, even if they are imaginary, are thus not altogether impossible.

** This involves the unwarranted assumption that the genetics structure which gives in the male a certain type of song, gives in the female, in a parallel way, the propensity to respond to *that* type of song.

References p. 649

field, will not hesitate to count it as a sub-species of one of the parent species, even, if the systematist is a "splitter", to raise it to the rank of *species nova*. We shall see, besides, that there are some examples of forms which are indistinguishable from known species by their morphology, but which are separated by their song.

(d) Song and systematics

We can approach the connections between song and classification from two angles: on the one hand by considering what help the song gives in the recognition of species, or at least sub-species, which are not distinguishable morphologically, and on the other hand by looking at the possibility of creating in some cases distinguishing standards based solely on acoustic features.

As far as the first point is concerned, the best example is provided by the American *Acheta* species. Various studies[128,157] had led to the conclusion that all the specimens found from Canada to Patagonia belonged to *A. assimilis* Fabr. In fact, study of life-cycle, song, and ecology showed that the name *A. assimilis* has actually covered many distinct *species*[5]. Morphological features are of practically no help in recognising the forms which are split up in this way, unless by a biometrical study, *i.e.* it is difficult to classify an individual simply by examining it. The biotope in which it was met is at once an indication, but the song seems to be the best criterion (Table 53).

TABLE 53

COMPARISON OF THE BIOLOGICAL PARAMETERS OF THE SONG OF FIVE EAST UNITED-STATES SPECIES RESULTING FROM THE SPLITTING OF THE SPECIES *A. assimilis* (ACCORDING TO ALEXANDER, 1957)[5]

Species		firma	pennsyl-vanicus	vernalis	fultoni	rubens
Number of individuals recorded		11	74	11	37	12
Biological song parameters (means)	Type of song	chirp	chirp	chirp	chirp	trill
	Pulse-rate	18	25	31	49	60
	Chirps/mn. (29°C)	100–120	150–240	180–200	300–360	—
	Pulses/chirp	4	4	3	2	—

The species *Nemobius melodius* Thomas and Alexander[173] has also been discovered by means of the song. It is morphologically confused with *N. carolinus* Scudder, but it is distinguished by its stridulation characteristics. Some infraspecific populations have been separated by examining their sound characteristics. This is the case in *Pterophylla camellifolia* (F.) (Pseudophyllidae) which has two different forms of song, one found northwest of the Appalachian Mountains, and one found southeast of them[10].

In the same way, two sibling and partially sympatric species are presently considered under the name *Amblycorypha rotundifolia* (De Geer) (Phaneropteridae). They are inseparable on a morphological basis, but have very distinct songs[8].

Hence we are naturally led to put song in the list of diagnostic features beside the usual morphological features, which still are the only means of distinguishing species. Several examples have shown that if we consider only the morphology we overlook sub-species, (or even, species), and at the same time collect under the same appelation, forms which are incontestably distinct, since very often they do not interbreed*.

Some authors have been able to propose standards based solely on the song characteristics detectable by ear[6,68,80,120] for determining the species of a given genus (Table 54).

TABLE 54

KEY FOR DETERMINING THE SPECIES OF THE GENUS *Nemobius* BY THE EXAMINATION OF THE CALLING SONG (FOR TEMPERATURES EQUAL TO OR HIGHER THAN 21°C((AFTER FULTON, 1931)[80]

I. Song made up of repeated short notes never much over a second in length and never so rapid but that the individual notes can be plainly heard.
  1. Sharp metallic chirps, with constant rhythm, 6–9 per sec.
    N. tinnulus Fulton**.
  2. Notes of rougher quality, observable variations in rhythm, 3–7 per sec.
    N. fasciatus (De Geer)**.
  3. Longer notes, one-half to one sec in length.
    (a) Non-rhythmical, occasional notes 2–3 sec long.
      N. sparsalsus Fulton.
    (b) Rhythmical but not very constant, notes of more uniform length.
      N. ambitiosus Scudder.
    (c) Each note begins as a low buzz with two undulations, audible only a few feet away, then increases in volume to the end, intervals between notes very brief; rhythmical, but not constant.
      N. confusus Blatchley.

II. Sound continuous for several seconds at a time.
  1. Notes and rests of about equal duration, usually 5–15 sec.
    N. cubensis Saussure.
    N. palustris Blatchley.
    N. palustris aurantius R. and H.
  2. Song interrupted at irregular intervals, usually 5–10 sec, by breaks of less than a second duration.
    N. griseus funeralis Hart.

III. Song usually continued for indefinite period.
  1. Of shrill tinkling quality without modulations.
    (a) With evident tremolo; at lower temperature song becomes a rapid series of sharp chirps, but at summer temperatures these are at a rate estimated at 12–15 per sec.
      N. allardi Alex. and Thom**.
    (b) A thin high pitched sound with a tremolo so rapid that it can scarcely be detected at summer temperature.
      N. bruneri Hebard.
  2. With modulations.
    (a) A weak tinkling sound with regular rhythmical undulations 5–7 per sec.
      N. maculatus Blatchley.
    (b) With a droning quality, variable in volume, part of the time louder and with rapid undulations, at other times weaker, lower pitched, without undulations.
      N. carolinus Scudder.

---

* What we have said about song is valid also for other ethological manifestations and for features of a physiological nature. The animal is certainly not made up solely of hairs, spines, segments of greater or lesser length, it also has behaviour. Though the specific or infraspecific individuality is generally expressed with sufficient clarity on the morphological plane, there are some cases where it is only shown in the animal's conduct. Hence it seems impossible today to base classification just on the appearance of the animal. Behaviour has a taxinomic value[127] that we cannot continue to overlook. Some of the most competent systematists, such as Chopard, recognise the value of using song to clarify the problems of species which are especially arduous, like that of the crickets[54]. Ethological features apart from song have actually been used already by some authors for systematics purposes[1,112].
** See footnote p. 629.

4. Factors Determining the Song and Associated Behaviour Patterns

(a) *External factors*

By external factors we mean any phenomenon coming from outside which can have an effect on the animal by setting off, stopping, or changing one of its sound activity manifestations. These factors are of two kinds: physical (light, heat, for example) or biological (the whole gamut of sign-stimuli).

If we leave out heat, which works in a relatively simple fashion because it only activates the biochemical processes involved in muscular activity, other factors have effects, the causation of which is much more complex. Actually the factor in question and the effect which it produces are far from being directly connected; their relation is only established according to physiological reactions which originate at the exteroceptor and end at the muscles controlling the stridulatory apparatus. The exact nature and importance of this chain of reactions are practically unknown.

(i) *Physical factors*

We have already mentioned the influence that variations in temperature can have on the various song parameters (see p. 366). We shall simply add that heat is only a limiting factor, *i.e.* a factor which must reach a particular value to enable the generator mechanism of the song to work, but it is not an element which directly sets off this mechanism. A temperature at the thermic optimum of the animal will enable its sound activity to be fully expressed (as also all the other forms of activity). Below the optimal value the activity will be slowed or even stopped. There seems no point in emphasizing these ideas which are familiar to everyone.

Light, at least in some cases, seems to have a much stronger effect. Actually it is a double action; that of regulating the daily activity rhythm (see p. 644), and, during periods of activity, that of disturbing or interrupting the song if the intensity happens to change suddenly. In the Cicadidae, *Magicicada septemdecim* (L.) and *M. cassinii* (Fish.), a cloud covering the sun makes the song's synchronism rate fall at once[11]. In *Chorthippus brunneus* Thunb. (Acrididae), the greater the intensity of the light the greater the stridulatory activity*[109]. There seems no reason to think that light is a limiting factor, however, for it is difficult, if not impossible, to define a light optimum as an absolute value as can be done with temperature**. Breeding by artificial light proves this, moreover, since most insects stridulate just as actively as in nature though the light intensity that they receive is much weaker. Sensitivity to light during the daily activity, seems especially a differential sensitivity, producing a reaction to any sudden change ΔI, whatever the value of I, within quite broad limits. However, that may be, light often acts in opposite ways according to whether the individuals under consideration have a day or night sound activity. In *Tettigonia viridissima* L. (Tettigoniidae) whose stridulation begins at the end of the day and lasts all night, artificial prolongation of the light has an inhibiting effect which delays the start of the song[135]. In *Ephippiger ephippiger* (Fiebig), (Ephippigeridae) on the other

---

* In *Orphania denticauda* (Charp.) (Orthopt. Phanaeropteridae), we noted that when lighting was extinguished during stridulation it made the insect stop immediately. In *Decticus*, the song is fully developed only in light. A sudden darkness makes the song irregular[35].

** It will be noted, on the other hand, that it is impossible to tell by observations made in nature what is due to the visible part of the spectrum (light) and what to the infra-red (heat).

hand, if an individual's daily rhythm of activity is disturbed by a prolonged stay in the darkness, a sudden stridulatory activity appears in response to several minutes light (Dumortier, unpubl.). In *Orchelimum agile* (De Geer) (Conocephalidae), which is a day singer, artificial lighting at night sets off stridulation at once[15].

However, it would not be right to generalise from these few examples, for susceptibility to light varies extremely from one family to another. Hence it has been possible to note that some night singing Tettigoniids stridulated normally in parks where there was strong lighting all night[53].

To conclude, let us say that it is not possible *a priori* to limit the physical factors which have an influence on the song or associated behaviour, to light and heat. We have to consider that all the parameters which make up the environment in which the insect lives can have a role. Atmospheric pressure, humidity, ionisation of the air for example, must also be considered as factors. Their effect, though as not well known, is not necessarily a negligible one*.

(ii) Biological factors

We shall only quote here as a reminder the different stimuli which can set off or keep going the stridulatory activity: natural or artificial acoustic stimuli, olfactory and tactile, which we have dealt with earlier.

(b) Internal factors

Though it is easier to divide the account into external and internal factors, this is necessarily a little artificial, since an external factor, like light, very probably works by the intermediary of a series of internal reactions, which it just sets off (see below). Hence we call a factor internal when the response it gives rise does not seem induced by any external phenomenon.

The problem which we shall tackle more precisely and which has been subject of some studies in Orthoptera, is that of the connections between the sexual state and the song in the male, and between the sexual state and receptivity in the female. Actually it is known that though the activity is regulated by a nycthemeral rhythm, (the study of which we shall review in the following pages), it is also subjected to some physiological conditions connected with sexuality.

(i) Observations on the male

The sexual state does not seem to have repercussions on the song according to a very consistent rule. For example, though it is common in the Acridinae to see the male resuming his stridulations again one hour after mating, for other Orthoptera, on the other hand, mating causes the stridulatory activity to stop for several days. In *Ephippiger*, the song is interrupted for a varying length of time from three to five days. The spermatophore in these insects shows a very sizeable loss of matter, equivalent to about 25% of the insect's weight. However, the curve of weight recovery does not seem to provide a solution to the problem of the causation of song resumption[42, 45] (see Table 55).

---

* In the mountains it has been noted that in hot but stormy weather no stridulation is heard[170]. However, the heat and light factors have not changed, whereas atmospheric pressure and ionisation of the air change at the approach of a storm.

References p. 649

TABLE 55

LENGTH OF SONG-INTERRUPTION AND WEIGHT RECOVERY (in g) IN *Ephippiger ephippiger* (MALE) AFTER MATING (ACCORDING TO BUSNEL, DUMORTIER AND BUSNEL, 1956)[45]

♂	Weight before mating	Weight of the sperm-atophore	Weight after mating	Weighed daily after mating						
				1st day	2nd day	3rd day	4th day	5th day	6th day	7th day
A	1.160	0.170	0.980	1.195	1.303	1.235	1.285 △ ☉			
cont'd	1.285	0.172	1.100	1.150	1.900	1.340	1.420 △ ☉			
B	1.685	0.353	1.328	1.430	1.566	1.625	1.660	1.700 △ ☉		
C	1.704	0.410	1.280	1.458	1.462	1.480	1.358	1.588	1.415 △ ☉	
D	1.300	0.350	0.930	1.342	1.258	1.283	△ ☉			
E	1.302	0.325	0.968	1.083	1.153	1.235	1.268	1.187	1.169 △ ☉	
F	1.740	0.398	1.331	1.738	1.540	1.492	1.563	1.780 △ ☉		

△ = reappearance of the song, ☉ = new mating.

In *Acheta domesticus* L. and other Gryllinae, the post-copulatory interruption of the song is not very marked. Moreover, these insects can form several spermatophores in the same day[119]. In *A. assimilis*, the artificial removal of the spermatophore has no effect on the song; the animals sometimes stridulate right at the end of the operation[149].

Many authors have tried to show a relation between the gonads and the song. There is quite good agreement among the few results. In the Acridinae *Chorthippus parallelus* (Zett.), castration, whether during the last nymphal instars or on the adult, does not disturb the song. The only modification is a lengthening of the duration of mating. Male and female may remain together from 2–10 h (against 30 min to 2 h in the case of non-castrated male)[105]. We should note, however, that this is perhaps only caused by a difficulty in forming a spermatophore after the operation; it indicates in no way that the motor patterns involved in this behaviour sequence are themselves disturbed.

In *Gryllus campestris*, castration has no effect on the song and behaviour, when the operation is done on the adult. With animals castrated before the final moulting (the next to last or last nymphal instar), Regen (1909, 1910)[151,152] observed that out of a total of eleven individuals nine stridulated normally and reacted in the presence of females (mating) like non-castrated males. One had only a weak and infrequent stridulation, and another only carried out weak elytral movements which did not produce any sound emission. This result, though a little disparate and obtained on a very small number of individuals, nevertheless encourages the thought that the gonads are without effect on the acoustico-sexual behaviour of this species. In *Ephippiger*, the removal of the testicles in the adult modifies neither the song nor the sexual behaviour. The castrated male mates just as often as a normal male, and the duration of the post-copulatory cessation of his song does not undergo any change[45].

To sum up, the removal of the gonads in the male does not seem to produce any particular changes of sexual behaviour, at least, as far as concerns the three species with which the experiments which we have quoted have dealt.

(ii) Observations on the female

(1) The cycle of sexual behaviour

Generally speaking, the females have a much more pronounced cyclic sexual behaviour than the males. In some periods, the female actively seeks a partner of the other sex; at other times, on the other hand, this reactive state disappears to give way to a passivity or even a defensive behaviour.

This succession of conditions connected on the one hand with fertilization, and on the other with oviposition, has been studied in particular in the Orthoptera Acridinae.

In *Chorthippus brunneus*, the first copulation is only accomplished a dozen days after the final moult; in *C. parallelus*, the newly emerged female only shows a responsive state at the end of two weeks. After mating, the female loses this responsive state, and goes about the process of repeated ovipositions without responding to other male songs. After a certain number of ovipositions the female is again ready for mating and the cycle repeats itself thus until death. In the twenty-four hours which precede oviposition, the female does not show any reactions towards the male. Hence there is a double mechanism for eliminating sexual behaviour in these species (to which we may add *Omocestus viridulus*) one appearing after copulation (long inhibition) the other appearing before oviposition (short inhibition)[105].

In *Euthystira brachyptera* (Ocsk.) the behaviour has the same general appearance, but several degrees have been recognisable in the cycle: active attitude (search for the male), or passive (male accepted if he appears), defence (male rejected), mating prevented (the male is not repulsed but the presence of remains of the spermatophore from a previous union prevents new coition)[158]. Fig. 369 gives the order of these different states.

Fig. 369. Cyclic evolution of the behaviour of the female *Euthystira brachyptera* towards the male in terms of copulation and oviposition. Each of the phases are only represented qualitatively, and not with their respective length. A = virgin females, B = mating ad libitum, I = imaginal moult, △ = defence, ⌐ = active attitude, + = passive attitude, ● = mating, ○ = mating prevented, ■ = oviposition. (After Renner, 1952[158].)

By and large we may say oviposition seems like a process for bringing things back to the start; it comes at the end of a cycle in progress and enables a new cycle to begin.

(2) *Factors determining the cycle*

The factors determining the female's receptivity cycle seem quite complex; they are certainly connected with the physiological maturation and migration cycle of the ovocytes, and various humoral phenomena (see below), but there is reason to think that other factors may also be concerned. In fact, in *Chorthippus brunneus, C. parallelus* and *Omocestus viridulus*, it has been possible to establish that the return to the responsive state after mating was quicker in females isolated from males than in those which stayed with them. Song (and perhaps other stimuli coming from the male, too) may, thus, have a slowing or even inhibiting effect[102]. This inhibiting action of the song—if it really exists—does not have its origin, at a peripheral level (*i.e.* at the level of the tympanic organs), for the action potentials recorded on the tympanal nerves

References p. 649

are the same, whether the female is virgin, is ready to oviposit, or has just copulated[98, 99*].

Dissections made on females *Euthystera*, at different stages of their sexual behaviour, demonstrated that acceptance of the male is connected with the position of the eggs in the genital tract. The male is accepted only when the eggs are still in the ovarioles (verified 40 times with 40 animals). The defence attitude, however, comes when the eggs are in the oviducti laterales (verified 37 times with 38 animals)[158]. The refusal phase which follows mating does not seem to have any connection with the quantity of sperm present in the receptaculum seminis. The extent of the stretching, of the receptaculum thus would not be a phenomenon transmitting along the nerves a stimulus inhibiting the sexual behaviour.

The injection of 2 mm³ of isotonic saline in the receptaculum of *Chorthippus parallelus* does not end the responsive state in which the females were before the experiment**[105]. It does not seem either that the introduction of the spermatophore in the genital passages of the female can in itself, mechanically, play a role. Actually, if the spermatophore is withdrawn, the female looks again for a mate, as is shown by observations made on *Euthystira*[158] and *Ephippiger*[45]. On the other hand, a chemical action by the seminal liquid is not to be ruled out, since in *Euthystira* females from which the receptaculum has been surgically removed, have a less pronounced post-copulatory refusal attitude[158]. Obviously it would be important to know what it is in the sperm that possesses this possible inhibitor role. The use of males made sterile by X-rays (destruction of the germinal cells) would certainly make possible the distinction between what is due to the spermatozoids, on the one hand, and, on the other, to liquids secreted by the accessory glands.

These first results suggest the possibility of a hormonal mechanism which eliminates the responsive state. Induced by a chemical (?) action of the sperm, it may have its origin in the female genital tract, and from there, the hormone carried by the circulation could reach the nerve centres responsible for the I.R.M.'s of the male acceptance behaviour and cause a blockage there. Moreover we shall look below at an experiment in which a hormonal action which reverses the behaviour seems to be demonstrated.

We should add that in the Tettigoniids with a voluminous spermatophore, the ingestion of the spermatophylax by the female perhaps plays a role in the establishing of the post-copulatory passivity***.

The role of the ovaries themselves has been studied by carrying out castrations. In *Euthystira*, the removal of the ovaries (and of the oviductus communis) performed on virgin females on the day of the imaginal moult, does not change the series of attitudes towards the male, but lengthens the time during which each one appears. If these ovariectomised females are left to mate, they keep a passive behaviour towards the male until the end of their lives. In addition, the spermatophore remains

---

* A cycle of sexual reaction with a peripheral origin has been outlined in the mosquito *Anopheles quadrimaculatus*. The males of this species react to the females' flight noise (stimulus triggering their sexual behaviour, see p. 619) only at certain periods; it was noted that this periodicity was connected with the spreading out and folding back of the antennal fibrillae (acoustic receivers). Hence it would be the cyclic functioning of the organ receiving the stimulus which governs and causes the rhythm—appearance and disappearance—of the behaviour[160].

** Observations made on 4 animals.

*** In *Ephippiger*, some preliminary experiments seem to show that if after each copulation the spermatophylax is removed before the female can begin to devour it (the sperm ampulla staying in position), mating is more frequent than in animals which have not undergone this treatment.

in the genital tract (*cf.* the phase "mating prevented" in normal females)[158]. In *Chorthippus parallelus*, castration performed in the responsive state puts an end to this condition for 24–48 h. Given to females which have lost their responsive state after castration, an injection of 2 mm³ of haemolymph taken from normal females in a responsive state, can apparently partially reinduce this state into the castrated females*[105].

The general impression which is drawn from this first group of results (some of which have to be accepted with reservations because of the small number of observations on which they are based) is that:

1. Castration in the male (performed on the adult or the nymph) does not modify the sound emission or the associated behaviour patterns. The only change is the lengthening of the mating-time (*Chorthippus*).

2. In the female, the effect of castration is a little more pronounced: the gonad partially control the cyclic character of sexual behaviour; the migration of the ovocytes plays an especially important role (*Euthystira*).

3. The post-copulatory interruption of the responsive state seems to occur only after the penetration of the sperm into the genital tract.

4. The effect of the seminal liquid seems to work by chemical rather than neural or mechanical means.

5. Return to the responsive state could be delayed by the repeated hearing of the males' song (*Chorthippus*).

There is no doubt that these observations, which have a definite importance, nevertheless only give a very partial picture of the phenomenon, in particular revealing the terminal portion of the chain of physiological reactions which govern the sexual cycle. A fuller view will have to include endocrine organs such as the corpora allata, which have an important role in the maturation of the eggs. Finally, even when the complete chain of hormonal or other processes have been explained, it will still have to be discovered how a physiological state can block or liberate an I.R.M. It is at the juncture of physiology and ethology that, for the biologist, the great problem of all animal behaviour patterns and equally human ones, is placed.

(c) *Action of pharmacodynamic drugs*

The action of chemical compounds on insects has hardly been studied except for practical purposes to obtain toxics or to do pure physiological research. The few experiments which have been made with non-fatal doses have not given very clear results as far as behaviour is concerned. Scopolamine (ester of *l*-tropic acid and scopine), a substance with a parasympatholytic effect on mammals, in *Gryllus bimaculatus* causes a raising of the elytra. Pervitine (N-methylamphetamine), sympathomimetic product, and quinine, slow down and shorten the song of *Chorthippus montanus*[23].

Research of this kind, so advanced in vertebrates will certainly only be really fruitful in the insects when the biochemistry of the nervous system is well known, and especially that of the transmitter substances.

(d) *Neurophysiology of acoustic behaviour*

We shall not deal with this very important question here, since it is the subject

---

* Observation made on 3 animals.

References p. 649

of a special chapter in this textbook (see Chapter 17).

(e) *Combined action of external and internal factors: daily rhythms of activity*

(i) *Origin of the daily rhythm*

Very many studies have shown that animal activity is usually not distributed evenly in time, but that periods of activity and rest alternate regularly. The concept of daily rhythm of activity may be perfectly well extended to the insects' song. It has actually been commonly observed that stridulation only occurs for a certain definite time which generally begins and ends with great regularity one day to another. There are very few studies at present which aim at revealing the causation of this phenomenon, at least if only those which deal solely with stridulatory activity are considered. We shall take as an example two pieces of research made on two European Tettigoniids: *Tettigonia viridissima* L. and *Ephippiger ephippiger* (Fiebig).

Daily rhythms can have two origins: either they are caused by an external factor, or they seem independent of the external world and they are regulated by the play of internal phenomena (internal clock). In the two examples which we shall look at here, it seems that the cyclic activity is governed by an external factor: light. However, even if the external factor is the *primum movens* of the reaction, it never works alone; its action involves the concomitant participation of internal mechanisms. Moreover, it can only work if an appropriate physiological state is in existence at the start.

Hence daily rhythm, even when it is caused by an external factor, may always be considered as the result of a double action, external and internal. That is why we place this study of periodic activity here.

The normal rhythm of stridulatory activity of *Tettigonia viridissima* can be kept up in the laboratory, with artifical lighting (glow-lamp) from 8 a.m. to 8 p.m. (12 h light, 12 h darkness) (Fig. 370). It is found then that the song starts about the same time as in nature (7 p.m.), *i.e.* before the light is put out [135]. *Ephippiger ephippiger*, like the

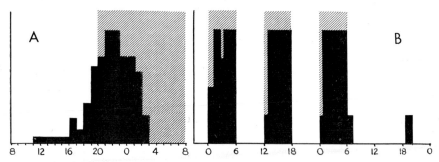

Fig. 370. Stridulatory activity of *Tettigonia viridissima*. (A) 24-h nycthemeral cycle (light from 8 a.m. to 8 p.m.), (B) 12-h nycthemeral cycle (6 h light, 6 h darkness), three cycles and continuous lighting. Ordinate: relative activity, abscissae time in hours. Hatching represents dark periods. (After Nielsen, 1938[135].)

other species of the genus, is a day and night singer (Fig. 371). In the laboratory, with artifical lighting (fluorescent tubes) and at a constant temperature (25°C), its activity begins during the night (about midnight) and ends about 2 p.m.[66]. As in the case of *Tettigonia*, the beginning of the stridulatory activity is not connected with the reversal of the lighting conditions (change from light to darkness for *Tettigonia* or darkness to light for *Ephippiger*). This finding is particularly important for the understanding of the phenomenon.

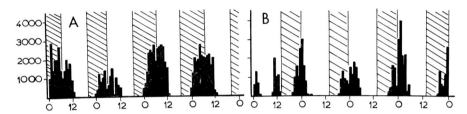

Fig. 371. Stridulatory activity of *Ephippiger ephippiger*. (A) natural nycthemeral cycle, (B) reversed cycle. Ordinate: total number of stridulations per hour (two insects), abscissae: time in h. Hatching represents dark periods. (After Dumortier, Brieu and Pasquinelly, 1957[66].)

The role of light, or more precisely of the nycthermeral cycle light/darkness, appears clearly in the experiments where this cycle is shortened or displaced in time by comparison with the natural cycle. By submitting *Tettigonia viridissima* to a quicker light cycle (6 h lighting and 6 h darkness), a correlative acceleration of the stridulatory activity rhythm of the insect is obtained (Fig. 370)[135]. In *Ephippiger*, by reversing the light cycle, *i.e.* by putting the insect in the light during the night, and in the darkness during the day, a reversal of the daily rhythm of stridulation is noted (Fig. 371)[66].

The fact that, in the two examples quoted here, the song is not set off by the establishment of either light or darkness, seems to indicate that there is no direct connection between sound activity and light (or darkness), but that this connection is established by the intermediary of a chain of internal reactions creating and ending periodically the physiological state necessary for the expression of the song.

The dephasing which is noted between the establishment or the cutting out of the light factor and the appearance or cessation of the song, is quite suggestive of an endocrine mechanism which gives rise to a substance effecting the stridulatory activity after a threshold. It is conceivable that if light, for example, starts off the secretion mechanism, the substance formed will reach the concentration at which it becomes active only after a certain time, hence the dephasing between the inductor physical factor and the manifestation of the song.

We should add that the theory of a hormonal action controlling an activity rhythm is not altogether unfounded, since it has been verified on the cockroach. In this insect, for the locomotor activity, Harker (1956)[96] showed that the 24-h rhythm was caused by a periodic neurosecretion, connected with the nycthemeral cycle, of the sub-oesophageal ganglia.

As far as the two species which we have taken as examples here are concerned, we can apparently take it as demonstrated that, in natural conditions, the alternation rhythm light/darkness is the inductor of the stridulatory activity. However, this does not justify the thinking that this activity has a purely exogenous causation. There are many cases where an endogenous mechanism ("internal clock") is found phased with

References p. 649

an exogenous phenomenon having cyclic variations, the second "reinforcing" the functioning of the first*.

In concluding let us mention that a quite original study made on the Coleoptera *Cerambyx cerdo* L., showed that this insect's ability to produce a protest sound (insect held between the fingers) follows a daily rhythm. The percentage of reactions is at its minimum at 8 a.m. (27%) and increases until reaching its maximum between 8 and 11 p.m. (83%). Animals kept in the dark for a week show a similar rhythm**[172].

(ii) Daily and annual rhythm

Though practically nothing is known of daily stridulatory activity in other species, it is nevertheless interesting to ascertain that the distribution of the song over 24h is quite characteristic of each family. Table 56 shows clearly, for example, that though night stridulation is very common in Tettigoniids, it is, on the other hand, much more rare in the Acridids. It will be noted besides that as well as day or night species, some species are encountered which have both a day and a night stridulatory activity. Often in this latter case, the night is different from the day song. Some information connected with the seasonal succession of orthopteran stridulation, will be found in Table 57.

ACKNOWLEDGEMENTS

The task which we have striven to handle successfully over the Chapters 11, 12 and 21, has been made much easier by the suggestions and comments which have been made and by the documents and information which have been kindly passed on to us.

Our thanks, which go first to Professor Richard D. Alexander (U.S.A.), are extended also to all those without whose help the imperfections which are present in this work would have been many more: Professors and Doctors Berland, Boesiger, Busnel, Chopard, Demange, Dresco, Guinot, Vachon, Viette (France), Cloudsley-Thompson (England), Altevogt, von Hagen, Jacobs, Loher and Roewer (Germany).

---

* The "internal clock" can however act alone in *Ephippiger* since (i) insects reared in constant light from hatching exhibit a 24-h rhythm of stridulatory activity when adults; (ii) the 24-h rhythm persists in insects kept in constant darkness. In these two examples, the "clock" may be only internal because of the lack of external change (Dumortier, unpubl.).

** A daily rhythm has also been reported in the snapping of the Alpheidae[116].

TABLE 56

DAILY RHYTHM OF STRIDULATORY ACTIVITY OF SOME ORTHOPTERA AND CICADIDAE. UPPER: INSECTS OF THE EASTERN UNITED STATES. |(AFTER ALEXANDER, 1956[3].) LOWER: INSECTS OF FRANCE (AFTER BUSNEL, MODIFIED, 1955[37].) FOR (1) SEE THE SYNONYMY IN THE LEGEND OF THE TABLE 57.

SPECIES	DAY	NIGHT	DAY
Amblycorypha uhleri		▬▬▬▬▬▬▬▬▬▬	
Pterophylla camellifolia		▬▬▬▬▬▬▬▬	
Neoconocephalus ensiger		▬▬▬▬▬▬▬▬	
Orchelimum vulgare	----	▬▬▬▬▬▬▬▬	----
Conocephalus fasciatus	----	▬▬▬▬▬▬▬▬	----
C. brevipennis	----	▬▬▬▬▬▬▬▬	----
Atlanticus testaceous	-- -- --	▬▬▬▬▬▬▬▬	
Nemobius F. fasciatus N.F. socius N.F. tinnulus[(1)]	▬▬▬▬▬▬▬▬▬▬▬▬▬▬▬▬▬▬▬▬▬▬	----
Gryllus assimilis (woods)[(1)]	-- -- ▬▬▬▬▬▬▬▬▬▬▬▬▬▬▬▬▬▬ -- --	
Œcanthus 4-punctatus	▬▬▬▬▬▬▬▬▬▬▬▬▬▬▬▬▬▬▬▬▬▬	
O. latipennis		▬▬▬▬▬▬▬ -- --	
O. pini		▬▬▬▬▬▬▬▬	
Phyllopalpus pulchellus	▬▬▬▬▬▬▬▬▬▬▬▬▬▬▬▬▬▬▬▬▬▬	
Orocharis saltator		▬▬▬▬▬▬▬ -- --	
Gryllotalpa hexadactyla	-- -- ▬▬▬▬▬▬▬▬	
Tibicen canicularis	▬▬▬		▬▬▬
T. linnei	▬▬▬▬▬ ...		▬▬ -- --
T. auletes	▬▬		
Chorthippus biguttulus	▬▬▬▬		▬▬▬▬
C. mollis	▬▬▬		▬▬▬
C. jucundus	▬▬▬		▬▬
Ephippiger bitterensis	▬	-- -- --	
E. ephippiger	▬	-- --	
E. cunii	▬	-- --	
Tettigonia viridissima		-- ▬▬▬▬▬▬	
Homorocoryphus nitidulus		▬▬▬▬▬▬▬	
Barbitistes fischeri		-- ▬▬▬▬	

TABLE 57

SONG PERIODS OF SOME ORTHOPTERA OF THE EASTERN UNITED STATES (NORTH CAROLINA). (AFTER FULTON, 1951[84].) 1 = *N. allardi* ALEX. AND THOM., 2 = *N. fasciatus* (DE GEER), 3 = *N. tinnulus* FULTON, 4 = *Acheta rubens* (SCUD.), 5 = *A. fultoni* ALEX.

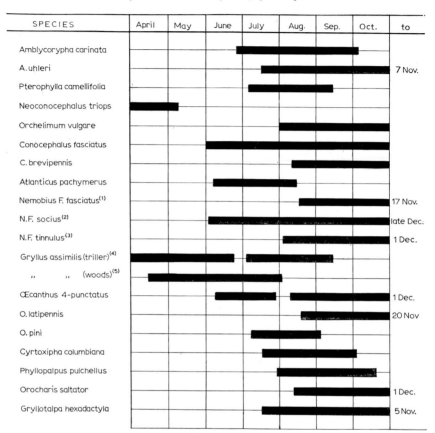

REFERENCES

[1] ADRIAANSEE, A., 1947, *Ammophila campestris* Latr. und *A. adriaansei* Wilcke. Ein Beitrag zur vergleichenden Verhaltensforschung, *Behaviour*, *1*, 1.

[2] ALEXANDER, A., 1958, On the stridulation of Scorpions, *Behaviour*, *12*, 339.

[3] ALEXANDER, R. D., 1956, A comparative study of sound production in Insects, with special reference to the singing Orthoptera and Cicadidae of the eastern United States, *Thesis*, Ohio State University, 529 pp.

[4] ALEXANDER, R. D., 1957, Sound production and associated behaviour in Insects, *Ohio J. Sci.*, *57*, 101.

[5] ALEXANDER, R. D., 1957, The taxonomy of the field crickets of the eastern United States, *Ann. Entom. Soc. Am.*, *50*, 584.

[6] ALEXANDER, R. D., 1957, The song relationships of four species of ground Crickets (Orthopt., Gryllidae, *Nemobius*), *Ohio J. Sci.*, *57*, 153.

[7] ALEXANDER, R. D., 1960, Communicative mandible-snapping in Acrididae, *Science*, *132*, 152.

[8] ALEXANDER, R. D., 1960, Sound communication in Orthoptera and Cicadidae, in Animal Sounds and Communication, *Proc. of the A.I.B.S. Meetings Bloomington (Indiana)*, W. E. LANYON and W. N. TAVOLGA (Ed.), Am. Inst. Biol. Sci., Washington, 1958, p. 38.

[9] ALEXANDER, R. D. and R. S. BIGELOW, 1960, Allochronic speciation in field crickets, and a new species *Acheta veletis*, *Evolution*, *14*, 334.

[10] ALEXANDER, R. D. and D. J. BORROR, 1955, The use of insect sounds in systematics. *Communication at Symposium of Entom. Soc. Am. on New Approaches to Systematics*, Cincinnati (Ohio), 1955.

[11] ALEXANDER, R. D. and T. E. MOORE, 1958, Studies on the acoustical behaviour of 17-years Cicadas, *Ohio J. Sci.*, *58*, 107.

[12] ALEXANDER, R. D. and E. S. THOMAS, 1959, Systematic and behavioural studies on the crickets of the *Nemobius fasciatus* group, *Ann. Entom. Soc. Am.*, *52* 591.

[13] ALLARD, H. A., 1917, Synchronism and synchronic rhythm in the behavior of certain creatures, *Am. Naturalist*, *51*, 438.

[14] ALLARD, H. A., 1918, Rhythmic synchronism in the chirping of certain Crickets and Locusts, *Am. Naturalist*, *52*, 548.

[15] ALLARD, H. A., 1929, The last meadow Katydid; a study of its musical reactions to light and temperature, *Trans. Am. Entom. Soc.*, *55*, 155.

[16] ALLARD, H. A., 1929, Our insect instrumentalists and their musical technique, *Annual Report Smithsonian Inst. for 1928*, 563.

[17] ALLARD, H. A., 1946, Synchronous singing of 17-year Cicadas, *Proc. Entom. Soc. Washington*, *48*, 93.

[18] ALTEVOGT, R., 1955, Beobachtungen und Untersuchungen an indischen Winkerkrabben, *Z. Morphol. u. Okol. Tiere*, *43*, 501.

[19] ALTEVOGT, R., 1957, Untersuchungen zur Biologie, Ökologie und Physiologie indischer Winkerkrabben, *Z. Morphol. u. Okol. Tiere*, *46*, 1.

[20] ALTEVOGT, R., 1958, Ökologische und ethologische Studien an Europas einziger Winkerkrabbe *Uca tangeri* Eydoux, *Z. Morphol. u. Okol. Tiere*, *48*, 123.

[21] ANDERSON, A. R., 1894, Note on the sound produced by the Ocypode Crab, *Ocypoda ceratophthalma*, *J. Asiat. Soc. Bengal, (Calcutta)*, *63*, 138.

[22] ANNANDALE, N., J. COGGIN BROWN and F. H. GRAVELY, 1913, The Limestone Caves of Burma and the Malay Peninsula; Myriapoda, *J. Asiat. Soc. Bengal, (Calcutta)* [N.S.], *9*, 415.

[23] AUMILLER, W., 1958, Die Beeinflussung von *Gryllus bimaculatus* De Geer und *Chorthippus montanus* (Charp.) durch einige Pharmaka, *Wissenschaftliche Prüfung fur das Lehramt an Höheren Schulen im Jahre 1958 in München*, 41 pp.

[24] BAIER, L. J., 1930, Contribution to the physiology of the stridulation and hearing of Insects, *Zool. Jb.*, *47*, 151.

[25] BALSS, H., 1921, Über Stridulationsorgane bei Dekapoden Crustaceen, *Naturwiss. Wochschr.*, *20*, 697.

[26] BALSS H., 1956, Decapoda, in BRONNS, *Klassen und Ordnungen des Tierreichs*, Band 5, I. Abt. Leipzig, p. 1369.

[27] BERLAND L., 1932, *Les Arachnides*, Lechevalier Ed., Paris, 485 pp.

[28] BIGELOW R. S., 1960, Interspecific hybrids and speciation in the genus *Acheta*, *Can. J. Zool.*, *38*, 509.

[29] BOURGOGNE J., 1951, Lépidoptères, in GRASSÉ, *Traité de Zoologie*, Vol. 10 (1), Masson Ed., Paris, p. 174.

[30] BOYS, C. V., 1881, The influence of a tuning fork on the Garden Spider, *Nature*, *28*, 149.

[31] BROWN, E. S., 1955, Mécanismes du comportement dans les émissions sonores chez les Orthoptères, *Colloque sur l'Acoustique des Orthoptères*, Fascicule hors série des *Annales des Epiphyties*, INRA, Paris, p. 168.

[32] Buck, J. B., 1938, Synchronous rhythmic flashing of fireflies, *Quart. Rev. Biol.*, *13*, 301.
[33] Bugnion, E., 1910–1917, Le bruissement des Termites, *Bull. Soc. Entom. Suisse*, *12*, 125.
[34] Burkenroad, M. D., 1947, Production of sound by the fiddler crab *Uca pugilator* (Bosc), with remarks on its nocturnal and mating behavior, *Ecology*, *28*, 458.
[35] Busnel, M. C., 1953, Contribution à l'étude des émissions acoustiques des Orthoptères 1er mémoire: recherches sur les spectres de fréquence et sur les intensités, *Ann. des Epiphyties*, *3*, 333.
[36] Busnel, M. C., 1955, Etude des chants et du comportement acoustique du mâle *d'Oecanthus pellucens* (Scop.), *Colloque sur l'Acoustique des Orthoptères*, Fascicule hors série des *Annales des Epiphyties*, INRA, Paris, p. 175.
[37] Busnel, R. G., 1955, Sur certains rapports entre le moyen d'information acoustique et le comportement acoustique des Orthoptères, *Colloque sur l'Acoustique des Orthoptères*, Fascicule hors série des *Annales des Epiphyties*, INRA, Paris, p. 281.
[38] Busnel, R. G., 1955, Mise en évidence d'un caractère physique réactogène essentiel des signaux acoustiques synthétiques déclenchant les phonotropismes dans les règne animal, *Compt. rend. Acad. Sci.*, *240*, 1477.
[39] Busnel, R. G., 1956, Etude de l'un des caractères physiques essentiels des signaux acoustiques artificials réactogènes sur les Orthoptères et d'autres groupes d'Insectes, *Insectes Sociaux*, *3*, 11.
[40] Busnel, R. G. and B. Dumortier, 1954, Observations sur le comportement acoustico-sexuel de la femelle *d'Ephippiger bitterensis*, *Compt. rend. Soc. Biol.*, *148*, 1589.
[41] Busnel, R. G. and B. Dumortier, 1955, Phonoréactions du mâle *d'Hyla arborea* à des signaux acoustiquqs artificiels, *Bull. Soc. Zool. France*, *80*, 66.
[42] Busnel, R. G. and B. Dumortier, 1955, Etude du cycle génital du mâle *d'Ephippiger*, et son rapport avec le comportement acoustique, *Bull. Soc. Zool. France*, *80*, 23.
[43] Busnel, R. G. and B. Dumortier, 1955, Phonotaxie de la femelle *d'Ephippiger* (Orthoptères) à des signaux acoustiques synthétiques *Compt. rend. Soc. Biol.*, *149*, 11.
[44] Busnel, R. G. and B. Dumortier, 1956, Rapport entre la vitesse de déplacement et l'intensité du stimulus dans le comportement acoustico-sexuel de la femelle *d'Ephippiger bitterensis*, *Compt. rend. Acad. Sci.*, *242*, 174.
[45] Busnel, R. G., B. Dumortier and M. C. Busnel, 1956, Recherches sur le comportement acoustique des Ephippigères, *Bull. Biol. France et Belg.*, *90*, 219.
[46] Busnel, R. G., M. C. Busnel and B. Dumortier, 1956, Relations acoustiques interspécifiques chez les Ephippigères, *Ann. des Epiphyties*, *3*, 451.
[47] Busnel, R. G. and W. Loher, 1953, Recherches sur le comportement de divers mâles d'Acridoidea soumis à des stimuli acoustiques artificiels, *Compt. rend. Acad. Sci.*, *237*, 1557.
[48] Busnel, R. G. and W. Loher, 1954, Recherches sur le comportement de divers mâles d'Acridiens à des signaux acoustiques artificiels *Ann. Sci. Nat. Zool.*, *16*, 271.
[49] Busnel, R. G. and W. Loher, 1956, Etude des caractères physiques réactogènes de signaux acoustiques artificiels déclencheurs de phonotropismes chez les Acrididae, *Bull. Soc. Entom. France*, *61*, 52.
[50] Busnel, R. G. and W. Loher, 1961, Déclenchement de phonoréponses chez *Chorthippus brunneus* Thunb. (Acridinae), *Acoustica*, *11*, 65.
[51] Busnel, R. G., W. Loher and F. Pasquinelly, 1954, Recherches sur les signaux acoustiques réactogènes pour divers mâles d'Acrididia *Compt. Rend. Soc. Biol.*, *148*, 1987.
[52] Caudell, A. N., 1906, The Cyrtophylli of the United States, *J.N.Y. Entom.*, *14*, 32.
[53] Chopard, L., 1938, *La Biologie des Orthoptères*, Lechevalier Ed., Paris, 541 pp.
[54] Chopard, L., 1954, Le difficile problème de systématique posé par les Grillons, *Ann. Mus. Congo Tervuren (Zool.)*, *1*, 326.
[55] Chrysanthus, F., 1953, Hearing and stridulation in Spiders, *Tijdschr. Entom.* *96*, 57.
[56] Clark, E. J., 1948, Studies on the ecology of British grasshoppers, *Trans. Roy. Entom. Soc, (London)*, *99*, 173.
[57] Cloudsley-Thompson, J. L., 1958, *Spiders, Scorpions, Centipedes and Mites*, Pergamon Press, 228 pp.
[58] Cloudsley-Thompson, J. L., 1961, A new sound-producing mechanism in Centipedes, *Entom. Month. Mag.*, *96*, 110.
[59] Cousin, G., 1933, Sur l'hybridation de deux espèces de Gryllidae (*Acheta campestris* et *bimaculatus*), *Bull. Soc. Entom. France*, *38*, 189.
[60] Dembowski, J., 1925, On the "speech" of the fiddler crab *Uca pugilator*, *Trav. Inst. Nenckiego*, *3*, 1.
[61] Dijkgraaf, S., 1955, Lauterzeugung und Schallwahrnehmung bei der Languste (*Palinurus vulgaris*), *Experientia*, *11*, 330.
[62] Dobzhansky, T., 1937, Genetic nature of species differences, *Am. Naturalist*, *71*, 404.
[63] Dolbear, A. E., 1897, The Cricket as a thermometer, *Am. Naturalist*, *31*, 970.
[64] Duijm, M. and T. Van Oyen, 1948, Sprinkhanen, *De Levende Natuur*, *51*, 1.
[65] Duijm, M. and T. Van Oyen, 1948, Het sjirpen van de zadelsprinkhaan, *De Levende Natuur*, *51*, 81.

[66] DUMORTIER, B., S. BRIEU and F. PASQUINELLY, 1957, Facteurs externes controiant le ryhtme des périodes de chant chez le mâle d'*Ephippiger ephippiger* (Fieb.), *Compt. rend. Acad. Sci.*, *244*, 2315.
[67] EDWARDS, H., 1889, Notes on noises made by Lepidoptera, *Insect Life*, *2*, 11.
[68] FABER A., 1928, Die Bestimmung der deutschen Geradflügler nach ihren Lautäusserungen, *Z. Wiss. Ins. Biol.*, *23*, 209.
[69] FABER, A., 1932, Die Lautäusserungen der Orthopteren, II, *Z. Morphol. Okol. Tiere*, *26*, 1.
[70] FABER A., 1949, Eine bisher unbekannte Art der Lauterzeugung europäischer Orthopteren: Mandibellaut von *Calliptamus italicus* (L.), *Z. Naturforsch. Tübingen*, *4b*, 367.
[71] FABER, A., 1953, Eine unbekannte Art der Lauterzeugung bei Käfern: das Trommeln von *Nothorrhina muricata*, *Jb. Ver. Vaterl. Naturk. Württemberg*, *108*, 71.
[72] FABER, A., 1953, *Laut- und Gebärdensprache bei Insekten: Orthoptera*. Teil I, Vergleichende Darstellung von Ausdrucksformen als Zeitgestalten und ihre Funktionen, Stuttgart, 198 pp.
[73] FERON, M., 1959, Attraction chimique du mâle de *Ceratitis capitata* Wied. (Dipt. Trypetidae) par la femelle, *Compt. rend. Acad. Sci.*, *248*, 2403.
[74] FEYTAUD, J., 1933, La Courtilière et les moyens de la combattre, *Rev. Zool. Agr.*, 5.
[75] FORBES, H. O., 1881, Sound producing Ants, *Nature*, *24*, 102.
[76] FRAENKEL, G., 1931, Die Mechanik der Orientierung der Tiere im Raum, *Biol. Rev.*, *6*, 36.
[77] FRINGS, H. and M. FRINGS, 1958, Uses of sounds by Insects, *Ann. Rev. Entom.*, *3*, 87.
[78] FULTON, B. B., 1915, The tree Crickets of New-York: life history and bionomics, *N.Y. Agr. Experiment Station, Techn. Bull.*, *42*, 1.
[79] FULTON, B. B., 1928, Sound perception by insects, *Sci. Monthly*, *27*, 552.
[80] FULTON, B. B., 1931, A study of the genus *Nemobius*, *Ann. Entom. Soc. Am.*, *24*, 205.
[81] FULTON, B. B., 1933, Inheritance of song in hybrids of two subspecies of *Nemobius fasciatus*, *Ann. Entom. Soc. Am.*, *26*, 368.
[82] FULTON, B. B., 1933, Stridulating organs of female *Tettigoniidae* (Orthopt.), *Entom. News*, *44*, 270.
[83] FULTON, B. B., 1934, Rhythm, synchronism and alternation in the stridulation of Orthoptera. *J. Elisha Mitchell Sci. Soc.*, *50*, 263.
[84] FULTON, B. B., 1951, The seasonal succession of Orthopteran stridulation near Raleigh, North Carolina, *J. Elisha Mitchell Sci. Soc.*, *67*, 87.
[85] FULTON, B. B., 1952, Speciation in the field cricket, *Evolution*, *6*, 283.
[86] GAHAN, C. J., 1918, The death-watch: notes and observations, *Entomologist*, *51*, 153.
[87] GERHARDT, U., 1913, Copulation und Spermatophoren von Grylliden und Locustiden, I, *Zool. Jb. Syst.*, *35*, 415.
[88] GIARD, A., 1895, La Cigale et d'autres insectes attirés par certains bruits, *Actes Soc. Sci. Chili*, 5, LXV.
[89] GONTARSKI, H., 1941, Lautäusserungen bei Larven der Hornisse (*Vespa crabro*), *Natur u. Volk*, *71*, 291.
[90] GOUNELLE, E., 1900, Sur les bruits produits par deux espèces américaines de Fourmis et de Termites, *Bull. Soc. Entom. France*, 168.
[91] GOUREAU, 1837, Essai sur la stridulation des insectes, *Ann. Soc. Entom. France*, *6*, 31.
[92] GRAVELY, F. H., 1915, Notes on the habits of Indian Insects, Myriapods and Arachnids, *Rec. Indian Mus.*, *11*, 483.
[93] GRÜNBAUM, A. A., 1927, Über das Verhalten der Spinne (*Epeira diademata*) besonders gegenüber vibratorischen Reizen, *Psychol. Forsch.*, *9*, 275.
[94] GUINOT-DUMORTIER, D. and B. DUMORTIER, 1960, La stridulation chez les Crabes, *Crustaceana*, *1*, 117.
[95] HAGEN, H. O. VON, 1961, Personnal communication of *Thesis* abstracts (in press.)
[96] HARKER, J. E., 1956, Factors controlling the diurnal rhythm of activity in *Periplaneta americana* L., *J. Exp. Biol.*, *33*, 224.
[97] HASKELL, P. T., 1953, The stridulation behaviour of the domestric cricket, *Brit. J. Animal Beh.* *1*, 120.
[98] HASKELL, P. T., 1956, Hearing in certain Orthoptera; I. Physiology of sound receptors, *J. Exp. Biol.* *33*, 756.
[99] HASKELL, 1956, Hearing in certain Orthoptera; II. The nature of the response of certain receptors to natural and imitation stridulation, *J. Exp. Biol.*, *33*, 767.
[100] HASKELL, P. T., 1957, Stridulation and its analysis in certain Geocorisae (Hemipt,. Heteropt.), *Proc. Zool. Soc. (London)*, *129*, 351.
[101] HASKELL, P. T., 1957, The influence of flight noise on behaviour in the desert locust *Schistocerca gregaria* (Forsk.), *J. Ins. Physiol.*, *1*, 52.
[102] HASKELL, P. T., 1957, Stridulation and associated behaviour in certain Orthoptera, 1. Analysis of the stridulation of, and behaviour between, males, *Brit. J. Animal Beh.*, *5*, 139.

[103] HASKELL, P. T., 1958, Stridulation and associated behaviour in certain Orthoptera. 2. Stridulation of females, and their behaviour with males, *Brit. J. Animal Beh., 6*, 27.
[104] HASKELL, P. T., 1958, The relation of stridulation behaviour to ecology in certain grasshoppers, *Insectes Sociaux, 5*, 287.
[105] HASKELL, P. T., 1960, Stridulation and associated behaviour in certain Orthoptera. 3. The influence of gonads, *Brit. J. Animal Beh., 8*, 76.
[106] HÖRMANN-HECK, S. VON, 1957, Untersuchungen über den Erbgang einiger Verhaltensweisen bei Grillenbastarden (*G. campestris* L., *G. bimaculatus* De Geer), *Z. f. Tierpsychol., 14*, 137.
[107] HOWARD, L. O., 1901, *Mosquitoes. How they live; how they carry disease; how they are classified; how they may be destroyed*, Mc. Clure, Philips & Co., N.Y., p. 15.
[108] HUBER, F., 1955, Sitz und Bedeutung nervöser Zentren für Instinkthandlungen beim Männchen von *Gryllus campestris* L., *Z. f. Tierpsychol., 12*, 12.
[109] HUKUSIMA, S., 1948, On the diurnal rhythm in the stridulating activity of *Chorthippus bicolor* (Charp.), *Sevro Seital (Physiol., Ecol.) Kyoto, 2*, 94.
[110] HUNGERFORD, H. B., 1924, Stridulation of *Buenoa limnocastris* Hung. and systematic notes on the *Buenoa* of the Douglas Lake region of Michigan, with the description of a new form, *Ann. Entom. Soc. Am., 17*, 223.
[111] JACOBS, W., 1944, Einige Beobachtungen über Lautäusserungen bei weiblichen Feldheuschrecken, *Z. f. Tierpsychol., 6*, 141.
[112] JACOBS, W., 1952, Vergleichende Verhaltenstudien an Feldheuschrecken und einigen anderen Insekten, *Verhandl. Deut. Zool. Ges.*, 115.
[113] JACOBS, W., 1953, Verhaltensbiologische Studien an Feldheuschrecken, *Z. f. Tierpsychol.*, Beih. 1, 228 pp.
[114] JACOBS, W., 1955, Problèmes relatifs à l'étude du comportement acoustique chez les Orthoptères *Colloque sur l'Acoustique des Orthoptères*, Fascicule hors-série des *Annales des Epiphyties*, INRA, Paris, p. 307.
[115] JACOBS, W., 1957, Einige Probleme der Verhaltensforschung bei Insekten, insbesondere Orthopteren, *Experientia, 13*, 97.
[116] JOHNSON, M. W., E. A. EVEREST and R. W. YOUNG, 1947, The role of snapping shrimps *Crangon* and *Synalpheus* in the production of underwater noise in the sea, *Biol. Bull. 93*, 122.
[117] KAHN, M. C. and W. OFFENHAUSER JR., 1949, The identification of certain West African Mosquitoes by sound, *Am. J. Trop. Med., 29*, 827.
[118] KAHN, M. C. and W. OFFENHAUSER JR., 1949, First field tests of recorded Mosquito sound used for Mosquito destruction, *Am. J. Trop. Med., 29*, 811.
[119] KHALIFA, A., 1950, Sexual behaviour in *Gryllus domesticus* L., *Behaviour, 2*, 264.
[120] KNEISSL, L., 1900, Die Lautäserungen der Heuschrecken Bayerns, *Natur Offenbarung, 46*, 41.
[121] LATASTE, F., 1894-95, Un procédé pour capturer les Cigales, *Feuille des Jeunes Naturalistes, 25*, 157.
[122] LATASTE, F., 1895, Un procédé pour capturer les Cigales, *Bull. Soc. Entom. France*, 182.
[123] LAURENCE, S. M., 1902, Mosquitoes attracted by sounds, *Brit. Med. J., 1*, 64.
[124] LE ROY WELD, D., 1899, The sens of hearing in Ants, *Science, 10*, 766.
[125] LINDBERG, R. G., 1955, Growth, population dynamics and field behavior in the spiny lobster *Panulirus interruptus* (Randall), *Univ. Calif. Publ. Zool., 59*, 157.
[126] LOHER, W., 1957, Untersuchungen über den Aufbau und die Entstehung der Gesänge einiger Feldheuschreckenarten und den Einfluss von Lautzeichen auf das akustische Verhalten, *Z. f. vergl. Physiol., 39*, 313.
[127] LORENZ, K., 1939, Vergleichende Verhaltensforschung, *Verhandl. deuts. Zool. Ges. 41*, 69.
[128] LUTZ, F. E., 1908, The variation and correlations of certain taxonomic characters of *Gryllus*, *Carnegie Inst. of Washington, 101*, 1.
[129] MARSHALL, G. A. K., 1902, Experimental evidence of terror caused by the squeak of *Acherontia atropos*, *Trans. Entom. Soc. (London)*, 402.
[130] MAXIM, SIR H., 1901, On mosquito responding to sounds. Letter to London Times, Oct. 29, quoted in *J. Hyg., 2*, 79; *Nature, 64*, 655.
[131] MEDIONI, J., 1956, Analyse expérimentale du phototropisme de la Drosophile sauvage, in *Union Internationale des Sciences Biologiques*, section de Psychologie expérimentale et de Comportement animal, p. 26 (Strasbourg).
[132] MEYER, E., 1928, Neue sinnesbiologische Beobachtungen an Spinnen, *Z. Morphol. u. Okol. Tiere, 12*, 1.
[133] MILLER, W. C., 1931, Some observations on the habits of *Passalus cornutus*, *Ohio J. Sci., 31*, 266.
[134] MYERS, J. G. and T. H. MYERS, 1928, The significance of Cicada song. A problem in insect communication, *Psyche* (Brit.), *8*, 40.
[135] NIELSEN, E. T., 1938, Zur Oekologie der Laubheuschrecken. *Entom. Meddel. (Copenhagen), 20* 121.

136 OHAUS, F., 1900, Bericht über eine entomologische Reise nach Zentrabrasilien (Fortsetzung), Berlingska Boktryckerier, VI + 145 pp.
137 OSSIANNILSSON, F., 1949, Insect Drummers, *Opuscula Entomologica* supplementum X, Lund, Berlingska Boktryckerier, VI + 145 pp.
138 OSSIANNILSSON, F., 1953, On the music of some european leaf-hoppers and its relation to courtship, *Trans. 9th Intern. Congr. Entom.*, 2, 139.
139 PECKHAM, G. W. and E. G. PECKHAM, 1887, Some observations on the mental powers of Spiders, *J. Morphol.*, 1, 383.
140 PERDECK, A. C., 1957, The isolating value of specific song patterns in two sibling species of Grasshoppers (*Chorthippus brunneus* Thunb. and *C. biguttulus* L.), *Behaviour*, 12, 1.
141 PETER, K., 1910, Über einen Schmetterling mit Schallapparat, *Endrosa (Setina) aurita* var. *ramosa*, *Mitt. Naturwiss. Ver. Neuvorpommern u. Rügen*, 42, 24.
142 PETER, K., 1912, Versuche über das Hörvermögen eines Schmetterlings (*Endrosa* var. *ramosa*), *Biol. Zentr.*, 32, 724.
143 PIERCE, G. W., 1948, *The Songs of Insects*, with related material on the production, detection and measurement of sonic and supersonic vibrations, Harvard, Univ. Press., *Cambridge (Mass)*, VII + 329 pp.
144 PIERON, H., 1951, *Vocabulaire de la psychologie*, Presses Universitaires de France, Paris, IX + 356 pp.
145 POCOCK, R. I., 1896, How and why scorpions hiss, *Nat. Sci.*, 9, 17.
146 PUMPHREY, R. J., 1955, Rapports entre la réception des sons et le comportement, *Colloque sur l'Acoustique des Orthoptères*, Fascicule hors série des *Annales des Epiphyties*, INRA, Paris, p. 320.
147 RABAUD, E., 1921, Recherches expérimentales sur le comportement des Araignées, *Année psychol.*, 22, 21.
148 RAIGNIER, A., 1933, Introduction critique à l'étude phonique et psychologique de la stridulation des fourmis, *Brotéria, Revista de Sciencias Naturaes (Lisbon)*, 2, 51.
149 RAKSHPAL, R., 1960, Sound producing organs and mechanism of song production in field crickets of the genus *Acheta* Fabr. (Orthop., Gryllidae), *Can. J. Zool.*, 38, 499.
150 RATHBUN, M. J., 1937, The Oxystomatous and allied Crabs of America, *U.S. Nat. Mus. Bull.* 166, VI + 272 pp.
151 REGEN, J., 1909, Kastration und ihre Folgeerscheinungen bei dem Männchen von *Gryllus campestris* L., *Zool. Anz.*, 34, 477.
152 REGEN, J., 1910, Kastration und ihre Folgeerscheinungen bei dem Männchen von *Gryllus campestris* L., *Zool. Anz.*, 35, 427.
153 REGEN, J., 1913, Über die Anlockung des Weibchens von *Gryllus campestris* L. durch telephonisch übertragene Stridulationslaute des Männchens. Ein Beitrag zur Frage der Orientierung bei den Insekten, *Pflügers Arch. Ges. Physiol.*, 155, 193.
154 REGEN, J., 1914, Untersuchungen über die Stridulation und das Gehör von *Thamnotrizon apterus* (Männchen), *Anz. Akad. Wiss. Wien, Math-Nat. Kl.*, 344.
155 REGEN, J., 1914, Untersuchungen über die Stridulation und das Gehör von *Thamnotrizon apterus* (Männchen), *Sitz ber. Akad. Wiss. Wien, Math-Nat. Kl. 123*, 853.
156 REGEN, J., 1926, Über die Beeinflussung der Stridulation von *Thamnotrizon apterus* F. (Männchen) durch kunstlich erzeugte Töne und verschiedenartige Geräusche, *Sitz ber. Akad. Wiss. Wien, Math-Nat. Kl.*, 135, 329.
157 REHN, J. A. G. and M. HEBARD, 1915, The genus *Gryllus* (Orthop.) as found in America, *Proc. Acad. Natl. Sci. Philadelphia*, 67, 293.
158 RENNER, M., 1952, Analyse der Kopulationsbereitschaft des Weibchens der Feldheuschrecke *Euthystira brachyptera* (Osck.) in ihrer Abhängigkeit vom Zustand des Geschlechtsapparatus, *Z. f. Tierpsychol.*, 9, 122.
159 RICHARDS, I. J., 1952, *Nemobius sylvestris* in S. E. Devon, *Entomologist*, 85, 83, 108, 136, 161.
160 ROTH, L. M., 1948, A study of Mosquito behavior. An experimental laboratory study of the sexual behavior of *Aëdes aegypti* L., *Am. Midland Naturalist*, 40, 265.
161 ROTH, L. M. and E. R. WILLIS, 1952, Method for isolating males and females in laboratory colonies of *Aëdes aegypti*, *J. Econ. Entom.*, 45, 344.
162 RUSSEL, E. S., 1943, Perceptual and sensory signs in instinctive behaviour, *Proc. Linn. Soc. (London)*, 154, 195.
163 SALPETER, M. M. and C. WALCOTT, 1960, An electron microscopical study of a vibration receptor in the Spider, *Exp. Neurol.*, 2, 232.
164 SANTSCHI, F., 1923, Les différentes orientations chez les Fourmis, *Rev, Zool. Afr.*, 12, 124.
165 SANTSCHI, F., 1926, Deux notices sur les Fourmis, *Bull. et Ann. Soc. Entom. Belg.*, 66, 327.
166 SCHALLER, F., 1951, Lauterzeugung und Hörvermögen von *Corixa striata* L., *Z. vergl. Physiol.* 33, 476.
167 SJÖSTEDT, Y., 1900, Monographie der Termiten Afrikas, *K. sv. vet Acad. Handl.*, 34, suppl. 1904, 38.

[168] SMIT, C. J. B. and A. L. REYNEKE, 1940, Do nymphs of Acrididae stridulate ?, *J. Entom. Soc. South, Afr.*, *3*, 72.
[169] SOLIER, A. J., 1837, Observation sur quelques particularités de la stridulation des Insectes, et en particulier sur le chant de la Cigale, *Ann. Soc. Entom. France*, *6*, 199.
[170] STAEGER, R., 1930, Beiträge zur Biologie einiger einheimischer Heuschreckenarten. *Z. Wiss. Insektenbiol.*, *25*, 36, 53.
[171] SWINTON, A. H., 1876-7, On stridulation in the genus *Vanessa*, *Entom. Month. Mag.*, *13*, 169.
[172] TEMBROCK, G., 1960, Stridulation und Tagesperiodik bei *Cerambyx cerdo* L., *Zool. Beitr.* *5*, [N.F.] 419.
[173] THOMAS, E. S. and L. D. ALEXANDER, 1957, *Nemobius melodius*, a new species of cricket from Ohio, *Ohio. J. Sci.*, *57*, 148.
[174] TINBERGEN, N., 1951, *The Study of Instinct*, Clarendon Press, London, XII + 228 pp.
[175] VIAUD G., 1938, *Recherches expérimentales sur le phototropisme des Daphnies*, Publications de la Faculté des Lettres de Strasbourg, 2e série, fasc. 84, 196 pp.
[176] VIAUD, G., 1956, Taxies et tropismes dans le comportement instinctif, in *L'instinct dans le comportement des animaux et de l'homme*, Masson Ed., Paris, p. 5.
[177] WALCOTT, C. and W. G. VAN DER KLOOT, 1959, The physiology of the spider vibration receptor, *J. Exp. Zool.*, *141*, 191.
[178] WALKER, T. J., 1958, Specificity in the response of female Tree Cricket (Orthop. Gryllidae Oecanthinae) to calling songs of males, *Ann. Entom. Soc. Am.*, *50*, 626.
[179] WEIH, A. S., 1951, Untersuchungen über das Wechselsingen (Anaphonie) und über das angeborene Lautschema einiger Feldheuschrecken, *Z. f. Tierpsychol.*, *8*, 1.
[180] WHEELER, W. M., 1926, *Ants*, Columbia University Press, 663 pp.
[181] WILL, F., 1885, Das Geschmackorgan des Insekten, *Z. Wiss. Zool.*, *42*, 674.
[182] WILLIAMS, C. B., 1922, Co-ordinated rhythm in insects; with a record of sound production in an Aphid, *Entomologist*, *55*, 173.
[183] WISHART, G. and D. F. RIORDAN, 1959, Flight responses to various sounds by adult males of *Aëdes aegypti* L. (Diptera, Culicidae), *Can. Entom.*, *91*, 181.
[184] YERSIN, A., 1856, Mémoire sur quelques faits relatifs à la stridulation des Orthoptères et à leur distribution géographique en Europe, *Bull. Soc. Vaud. Sci. Nat.*, *4*, 108.
[185] ZIPPELIUS, H. M., 1948, Die Paarungsbiolgie einiger Orthopterenarten, *Z. f. Tierpsychol.*, *6*, 372.

For a more detailed bibliography on Insect sounds, see M. FRINGS and H. FRINGS, *Sound Production and Sound Perception by Insects*, A bibliography, The Pennsylvania State University Press, 108 pp.

CHAPTER 22

ACOUSTIC BEHAVIOUR OF FISHES

by

J. M. MOULTON*

1. Introduction

For many centuries, the business of fishing in the oceans was based on experience of the fishermen concerned. The nets were lowered where the catch had been good in the past; methods of fishing were relatively slow to change. Within quite recent years, a number of technological advances have been applied to the fisheries which promise to increase the yield of the seas. Television cameras have made it possible to examine the action of nets in the water[162,189], telemetering devices have made possible accurate depth adjustment of the nets[57,172], and electrical fishing devices have been designed to control the movements of fishes in salt[99] and fresh[8,121] water.

Perhaps most useful of the tools added to the fisherman's equipment has been the echo-sounder, which enables fishermen to locate schools of fish at sea and to measure their depth through reflection from the schools of high frequency sound pulses. The echo-sounder, first adapted to British fisheries about twenty-five years ago[96], reveals the presence of fish schools without scaring them away[114,131]. More recently, a high frequency sound transducer has been attached to fishes themselves in order to track their movements over short ranges[182,183]. The echo-sounder and miniature transducer are sonic devices which transmit frequencies higher than fishes are able to hear; their use suggests the significance of adequate information on the acoustical biology of fishes to improvement of the fisheries.

People have been interested in observing and studying fishes throughout recorded history. The Egyptians practiced fish culture of Tilapia 3000 years ago[36], and the ruins of ancient Pompeii include what appear to have been fish breeding pools[125]. It is not surprising that the literature of antiquity contains numerous allusions to the acoustical behavior of fishes.

Many fishes produce sounds within the hearing ranges of various species which have been studied. The attention these sounds attracted during the Second World War gave a sharp stimulus to the study of fish acoustical biology. Fish calls were considered in the design of several instruments of war[55]. The Chesapeake Bay hydrophone net alerted personnel to possible submarine menace when the croakers, *Micropogon undulatus* (Linnaeus), were calling during the spring and early summer[108,119a,124], until these sounds were filtered out. The Japanese experienced similar difficulties with listening devices[14,65], as did our submariners listening for enemy ships in the Pacific Ocean[65]. With the information accumulated during the

---

* Some of the research upon which part of this chapter is based was performed by the author under grants of the Woods Hole Oceanographic Institution, of the Bowdoin College Faculty Research Fund established by the Class of 1928, and of the National Science Foundation, Research Grant NSF-G4403.

References p. 687

1940's, it became apparent that the oceans are very noisy places biologically, and fish acoustical biology became a very active field of study.

The sounds of fishes are of interest to biologists for a number of reasons. They vary annually, seasonally, and even diurnally with changes in behavior, distribution and perhaps with individual growth within fish populations[55, 119a]. They provide a means for studying the distribution of sonic species in nature. It is probable that specific sounds can be related to rather definite patterns of behavior[137]. They certainly suggest that sound is important in the biology of fishes, and that it may in the future be used to influence the behavior of wild fishes in ways beneficial to mankind.

Sound travels over four times as rapidly in sea water as it does in air at comparable temperature. Due to reflection of sound waves from the surface and bottom interfaces of a body of water, and to temperature gradients which serve to mediate this reflection, water acts as a sound guide, funneling sound waves which may be detected with appropriate equipment over surprising distances. Marshall[124], in a discussion of this matter, cites an experiment in which the explosion of a 6-lb. T.N.T. bomb in Dakar, West Africa, was heard with appropriate listening equipment 3,100 miles away in the Bahama Islands. The adaptations of fishes to the diverse features of their habitats are so numerous, whichever of their systems one considers, that it would seem more surprising if fishes were deaf in the excellent sound conductor which is their environment, than is the fact that they hear.

This chapter attempts to summarize the acoustical biology of fishes—their hearing and sound production, their related anatomy, and the behavior of fishes in relation to sound production.

2. The Hearing of Fishes

While it has been recognized for a long time that fishes are sensitive to sounds, many early investigators of the problem denied a power of hearing in fishes. Even as late as the early 1930's, investigators had been about equally divided as to whether or not fishes hear[186]. Early observations were those of owners of private pools who thought fishes were attracted to bells at feeding time[19]. Among the earliest experiments on the matter were those of Kreidl[109] who after removing the inner ears from goldfish found that the fish still responded to sounds. He concluded that fishes perceive sound vibrations through the skin, and thus cannot be said to hear.

The experiments of G. H. Parker and his student, H. B. Bigelow[19], Professor emeritus of Harvard University and noted pioneer in oceanography, led them to disagree with Kreidl's conclusions. Parker[146, 147] demonstrated that 96% of normal *Fundulus heteroclitus* (Linnaeus) responded to sounds engendered by plucking a bass-viol string attached to the aquarium. He further showed that rendering of the skin essentially insensitive did not prevent this response, but cutting of the auditory nerves eliminated the response in 78% of the trials.

Bigelow[19] repeated Parker's experiments with goldfish and obtained similar results. He then repeated Kreidl's operations and experiments, obtaining results similar to those of Kreidl. But through careful dissection, Bigelow found that by Kreidl's method, only the semi-circular canals and utriculus of the ear were removed. The sacculus-lagena portion, or pars inferior—the portion of the ear most probably responsible for hearing in the goldfish—was left intact.

In refuting Kreidl's conclusion, Bigelow established the facts, since thoroughly confirmed, that fishes receive sound stimuli through the inner ear, and that these stimuli are conveyed to the brain over the auditory nerve—in other words, that fishes hear. An important criterion of an auditory function of the fish inner ear was satisfied by Piper[157] when he demonstrated action potentials under sound stimuli in the auditory nerves of the pike, *Esox lucius* Linnaeus, and the eel, *Anguilla vulgaris* Linnaeus; the question then became one of what fishes hear, rather than of whether or not they hear[61]. Parker subsequently demonstrated what he interpreted as responses to sound in a variety of fishes under various conditions[148, 149, 150, 151, 152, 153].

Kreidl's pioneering research into the matter of fish hearing must not, however, be underestimated. Prior to his time, dicussion of fish hearing had been based mainly on the behavior of fishes in nature—for example, Aristotle noted that fishes fled from loud noises such as those of galley oars[143], Hunter in 1762 that fishes fled from the sound of a gun[154]—and on purely anatomical evidence. To most early anatomists and naturalists, the presence of well-defined inner ears in fishes was strongly suggestive of the power of hearing.

But even these observers differed in interpretation of their observations, and those differences of opinion are reflected in the texts of their times. Thus Carus[34] concluded from a review of ear anatomy in teleosts and elasmobranchs that the fish ear "must be very well fitted for participating in even the slightest vibrations conveyed to the internal ear by the bones of the cranium, and for thereby exciting the auditory impressions in the branches of the (eighth) nerves, which are distributed in great numbers on the sacculus and vestibule, and are even attached by their extremities to the bony connections." Home[98] on the other hand, from a similar review, concluded that "fishes, from the structure of the (ear), can only hear sounds which agitate the water immediately in contact with the head of the fish; so that the impulse is conveyed without interruption from the liquid in which they live, to the organ of hearing." T. R. Jones[103] felt it probable that noise produces in fish a "powerful sensation", while Sanborn Tenney[177], forty-five years after Weber[185] had correctly postulated an auditory function for the Weberian ossicles of ostariophysan fishes, asserted that all vibrations reaching the ears of fishes must come through the hard covering of the head, and that fishes could scarcely hear more than the loudest sounds. A footnote added that some naturalists accredited fishes with a "less obtuse" sense of hearing than he proposed. Even a quite recent work in comparative anatomy[141] has questioned the hearing ability of fishes, as well as the ability of fishes to produce sounds.

The physiology of hearing in fishes has been well-reviewed in recent years by Kleerekoper and Chagnon[107], Griffin[83], and Lowenstein[119], and for complete discussions of the problem, one should refer to these reviews.

The sensory areas of the fish inner ear differ from those of tetrapod vertebrates. There is, for example, no homologue of the basilar papilla which appears only on the tetrapod line, first among amphibians, and from which the Organ of Corti, the sensory area for hearing of some reptiles, of birds and mammals, has derived[161]. Some fish ears possess sensory areas for which homologues are lacking in the ears of other vertebrates—for example, the macula neglecta and lacinia which are sensitive to sound vibrations[117, 118, 184].

In all fishes, the inner ear is divided into pars superior and pars inferior (Figs.

372 and 373). The former consists of the semicircular canals (one or two in cyclostomes[161]; three in other fishes roughly arranged in the three planes of space) and a utriculus into which the canals open at either end. The pars inferior is comprised of the sacculus and a more posterior lagena. The utriculus, sacculus and lagena together comprise the vestibulum of the inner ear.

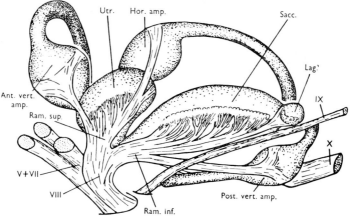

Fig. 372. Ventro-lateral view of the left membranous labyrinth of the ray (*Raja clavata*). (From Lowenstein and Roberts, 1951[118].)

Ant. vert. amp. = Anterior vertical ampulla
Hor. amp. = Horizontal ampulla
Lag. = Lagena
Post. vert. amp. = Posterior vertical ampulla
Ram. inf. = Ramus inferior
Ram. sup. = Ramus superior
Sacc. = Sacculus
Utr. = Utriculus
V-X = 5th–10th cranial nerves

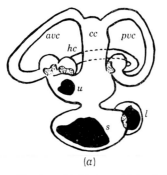 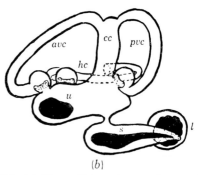

Fig. 373. The two chief types of labyrinths in fish. (a) 'Normal" type (trout); (b) ostariophysan type (minnow). Right labyrinths, median aspect. (Modified from von Frisch, 1936[70]; from Lowenstein, 1957[119].)

avc = Anterior vertical canal
cc = Crus commune
hc = Horizontal canal
l = Lagena with astericus
pvc = Posterior vertical canal
s = Sacculus with sagitta
u = Utriculus with lapillus

Each semicircular canal possesses an enlargement or ampulla containing a crista, a sensory area the impulses from which lend information as to position in space, but which may also be sensitive to sound under special conditions[118]. The utriculus, sacculus and lagena each contains a characteristic macula, a sensory area, of which the conditions for response appear to vary even among fishes[117,118]. Whereas sound perception is probably normally limited to the pars inferior of higher vertebrates, through the Organ of Corti, in fishes both divisions may participate, as in elasmobranchs and non-ostariophysan teleosts[50,118]. The lagena, although not its macula, is the homologue of the cochlea of the mammalian ear; in elasmobranchs its macula does not respond to vibrational stimuli[118], although it is sensitive to sounds in ray-finned fishes[50,71].

Carus'[34] generalizations concerning the anatomy of fish ears might in several regards be carried forward into present thinking. Although the point is certainly made by other authors, it is often overlooked in accounts of fish ear anatomy and function that there is a dichotomy between ear structure in relation to the skull in elasmobranchs and teleosts. The elasmobranch ear, which most students of anatomy first become familiar with, is essentially completely enclosed within a cartilaginous capsule—the otic capsule[184]. This enclosure is incomplete in elasmobranchs at the fenestra ovalis, where a thin membrane in close association with the endolymphatic duct, which rises to the top of the head, might conceivably provide for passage of vibrational stimuli to the inner ear[34,118].

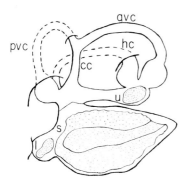

Fig. 374. Left labyrinth of *Prionotus carolinus* (Triglidae), median aspect. The posterior vertical and horizontal canals pass through bony canals of the skull. Abbreviations as in Fig. 373.

In teleosts, the form of the ear, particularly of the perilabyrinthine system surrounding the membranous ear, shows considerable variation. A utriculo-saccular canal, consistently present in elasmobranchs, may or may not be present in teleosts[154]. It is lacking, for example, in *Prionotus* (Fig. 374). The lagena may be divorced from the sacculus, as in *Balistes*. In most teleosts, only the posterior vertical and horizontal semicircular canals are enclosed by skeleton (Fig. 374), or as in *Cyprinus* and *Carassius*, only the horizontal canal[19]. In some teleosts, *Palinurichthys* for example, a loop of chondrocranium is thrown also round the anterior vertical canal.

Whatever the relationships of the canals, the utriculus, sacculus and lagena of teleosts generally lie immediately adjacent to the brain, embedded in soft tissue, protected to a greater or lesser degree by skull convolutions. It is to sensory areas adjacent to soft tissue that the branches of the auditory nerve make connection to the maculae and cristae of the inner ear, in a remarkable uniformity of gross pattern. As Griffin[83] has suggested in another connection, the soft tissues of the fish are nearly enough like water so that no barrier to sound vibrations coming through the water into the body is presented—"a 'pure' fish is relatively 'transparent' to underwater sound." The anatomy of most teleosts provides some degree of soft

References p. 687

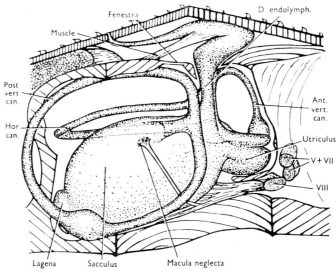

Fig. 375. Inside view of the left membranous labyrinth of the ray (*Raja clavata*), showing the position and innervation of the macula neglecta and the spatial relationship between the sacculus and the fenestra of Scarpa, and the ductus endolymphaticus. (From Lowenstein and Roberts, 1951[118].)

Ant. vert. can. = Anterior vertical canal
D. endolymph. = Ductus endolymphaticus
Hor. can. = Horizontal canal
Post. vert. can. = Posterior vertical canal
V–VII, VIII = 5th to 7th and 8th cranial nerves

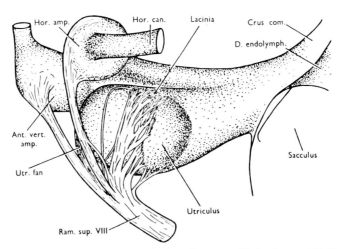

Fig. 376. Postero-lateral view of the left utriculus of the ray (*Raja clavata*), showing the lacinia utriculi and its innervation. (From Lowenstein and Roberts, 1951[118].)

Ant. vert. amp. = Anterior vertical ampulla
Crus com. = Crus commune
D. endolymph. = Ductus endolymphaticus
Hor. amp. = Horizontal ampulla
Hor. can. = Horizontal canal
Ram. sup. VIII = Ramus superior of eighth nerve
Utr. fan = Main utricular nerve fan

tissue continuity between water and inner ears, if only through the central nervous system and its membranes, and adjacent soft tissues of the head.

The recording of action potentials from the isolated elasmobranch labyrinth and the results of operations on the labyrinth of the intact fish have yielded somewhat different results as to localization of sound perception in elasmobranchs. Vilstrup[184] denervated various portions of the labyrinth of *Squalus acanthias* Linnaeus and concluded from conditioning experiments that reactions to sound are probably elicited from vibratory stimulation of the macula of the sacculus. Lowenstein and Roberts[118] recorded action potentials in the isolated labyrinth of *Raja clavata* Linnaeus, not only from the macula neglecta (Fig. 375) and macula of the sacculus, but also from the lacinia (Fig. 376) of the utriculus during sound stimulation.

The upper frequency response of the macula of the sacculus in the isolated inner ear of *Raja clavata* is rarely above 120 c/s, as indicated by action potentials[118]. On the other hand, the upper limits of frequency perception of all teleosts thus far tested[70,106,119], usually by conditioning experiments, are considerably above this level, whether or not accessory hearing organs are present (Table 58). In *Amiurus nebulosus* (LeSueur), the upper limit approaches that of man[13,159].

TABLE 58

A LIST OF TELEOSTS IN WHICH HEARING HAS BEEN RELIABLY DEMONSTRATED, WITH DATA FOR FREQUENCY RANGES WHERE AVAILABLE
(No similarly reliable demonstration of hearing is available for non-teleostean fishes*)

Family	Species	Frequency range (c/s) Low	High	Reference
NON-OSTARIOPHYSI				
Anabantidae	*Anabas scandens* (= *A. testudineus*)		over 659	Diesselhorst (1938)
	Betta splendens			Schneider (1941)
	Colisa lalia			Schneider (1941)
	Macropodus cupanus			Schneider (1941)
	Macropodus opercularis		2637–4699	Schneider (1941)
	Trichogaster leeri			Schneider (1941)
	Trichogaster trichopterus			Schneider (1941)
Anguillidae	*Anguilla anguilla*			Bull (1918)
	Anguilla anguilla	36	488–650	Diesselhorst (1938)
Cottidae	*Cottus gobio*			Stetter (1929)
	Cottus scorpius			Froloff (1925)
Cyprinodontidae	*Fundulus heteroclitus*			Parker (1903)
	Lebistes reticulatus			Farkas (1935)
	Lebistes reticulatus	44	1200–2068	Farkas (1936)
Embiotocidae	*Cymatogaster aggregatus*			Moorhouse (1933)
Esocidae	*Umbra limi*			Westerfield (1921)
	Umbra pygmaea			von Frisch and Stetter (1932)
Gadidae	*Gadus aeglefinus*			Froloff (1925)
	Gadus callarias			Froloff (1925)

continued on page 662

TABLE 58 *(continued)*

Family	Species	Frequency range (c/s) Low	High	Reference
Gobiidae	*Gobius niger*		800	Dijkgraaf (1952)
	Gobius paganellus		600–800	Dijkgraaf (1950)
	Periophthalmus koelreuteri		up to 651	Diesselhorst (1938)
Labridae	*Crenilabrus griseus*			Froloff (1925)
	Crenilabrus melops			Bull (1928)
	Crenilabrus pavo			Froloff (1925)
Mormyridae	*Gnathonemus* sp.		2794–3136	Stipetic (1939)
	Marcusenius isodori		2069–3100	Diesselhorst (1938)
Percidae	*Acerina cernua*			Froloff (1925)
	Perca fluviatilis			Froloff (1925)
Sciaenidae	*Corvina nigra*		1024	Dijkgraaf (1950)
	Corvina nigra		1000	Dijkgraaf (1952)
	Corvina nigra			Froloff (1925)
Sparidae	*Sargus annularis*		1250	Dijkgraaf (1952)
OSTARIOPHYSI				
Characinidae	*Hemigrammus caudovittatus*			von Boutteville (1935)
	Hyphessobrycon flammeus		6960	von Boutteville (1935)
	Pyrrhulina rachoviana			von Boutteville (1935)
Cobitidae	*Nemacheilus barbatula*		1740–3480	Stetter (1929)
Cyprinidae	*Alburnus lucidus* (= *A. alburnus*)			Zenneck (1903)
	Carassius auratus			Bigelow (1904); Manning (1924); Denker (1931)
	Carassius auratus		3480	Stetter (1929)
	Carassius carassius			Froloff (1925)
	Idus melanotus (= *Leuciscus idus*)			Denker (1931)
	Idus melanotus (= *Leuciscus idus*)		5524	Stetter (1929)
	Leuciscus dobula			Zenneck (1903)
	Rutilus rutilus			Zenneck (1903)
	Phoxinus phoxinus	25	5000–7000	von Frisch (1938)
	Phoxinus phoxinus	16	5000–6000	von Frisch and Stetter (1932)
	Phoxinus phoxinus			Stetter (1929); Denker (1931); Benjamins (1934); von Boutteville (1935); Hafen (1935)
	Pimephales notatus			McDonald (1921)
	Tinca tinca			Froloff (1925, 1928)
Gymnotidae	*Electrophorus electricus*		870–1035	von Boutteville (1935)
Siluridae	*Ameiurus nebulosus*		over 13,000	Maier (1909); Haempel (1911); Parker and Van Heusen (1917); Krausse (1918); von Frisch (1923); Stetter (1929); von Frisch (1938); Autrum and Poggendorf (1951); Poggendorf (1952)

* From Lowenstein (1957); Autrum and Poggendorf (1951); Poggendorf (1952).

Intact elasmobranchs seem to respond only to rather intense sounds[148,151,184]. In my experiments[135] in conditioning intact smooth dogfish, *Mustelus canis* (Mitchill), to sounds at Woods Hole during 1957, swimming responses developed to sound signals of mixed frequencies and high gain using mild shock as the unconditioned stimulus, a method adapted from Froloff[72], Bull[30], and Griffin[80]. However, responses were inconsistent even after prolonged training. Anecdotal reports that sharks are repelled from underwater human screams and attracted by the fin flutters of dying fish[54] and to the grunts of speared grouper, are unconfirmed by experimental evidence. (See later comments on an Indonesian sound fishery.) There is little concrete evidence that sharks use acoustical stimuli in their ordinary behavior[118], although underwater swimmers have reported hearing sounds from passing sharks[35], and professional Indonesian fish listeners attribute to some sharks a sound like a sharply indrawn breath[92].

That the presence of an action potential in the auditory nerve during sound stimulus is not sufficient to establish the power of hearing in fishes has been underlined by Lowenstein and Roberts[118]. Hearing implies that the animal uses the information coming to the central nervous system in some concrete way. From the point of view of the experimenter, a bothersome feature of fish biology is the rapidity with which fishes adapt to a variety of stimuli. Wild fishes may initially react to a number of stimuli to which, within a short time they adapt, showing no further obvious reactions. Such stimuli include not only sounds[33,127,131], but also odors[26] and even electric shock[106]. Yet it is likely that both sounds and odors play important roles in the biology of at least some fishes[94,132,187]. In elasmobranchs, in which behavioral responses to sound in nature still remain questionable, even sensory areas for vibrational stimuli show an unusually high rate of adaptation[118].

The application of conditioning techniques to problems of fish hearing by von Frisch and his school[69,70,71,173,191], largely alleviated this problem in laboratory investigation. With combinations of feeding and mild punishment as the unconditioned stimuli, various kinds of fishes, especially *Amiurus nebulosus* and *Phoxinus laevis* Agassiz, were trained to respond to sounds and to discriminate between sounds of different frequencies. After training, the intensity thresholds could also be measured by lowering sound intensities until responses were extinguished.

That sounds may be perceived by fishes in a variety of ways has impeded resolution of questions anent their acoustical physiology. While the inner ears are responsible for perception of most frequencies to which fishes tested are sensitive, it is possible that sense organs in the lateral line canals and possibly isolated skin receptors are at least partly responsible for receiving the lowest frequencies of sound which fishes perceive[71,106]. The closely related embryological origins of inner ears and lateral line organs, itself suggests that their functions may be closely related.

3. The Definition of Hearing in Fishes

The possible multiplicity of sensory areas responsible for vibration perception in fishes raises a problem of defining hearing within this group which Griffin[80], and Lowenstein and Roberts[118] have quite recently reviewed. To follow the summary of the latter authors, von Frisch considers "hearing" a valid term if sound perception is localized within the inner ear, Pumphrey that difference in intensity levels may be

References p. 687

a useful criterion for distinguishing between auditory and tactile sound perception, and Autrum that it is often impossible to distinguish between hearing and tactile sound perception, and that such distinction is not especially valuable to advancing the general problem of sound perception.

The point of view of Lowenstein and Roberts[118] is a reasonable one, and implies that behavioral evidence must be an important criterion of hearing capacity. These authors state: "To define our attitude in the matter, we would say that, given reliable evidence for a utilization by the animal of the sensory input from the vibration sensitive areas of the labyrinth, we should be well content to use the term hearing. As this evidence is at present still absent in the case of the *elasmobranchs* we want to confine ourselves strictly to the description of the peripheral sensory responses to vibration. This need not, however, preclude us from speculating on the *potential* significance of vibration sensitivity in well-defined sensory areas of the labyrinth."

Unfortunately the absence of information on utilization of sound which Lowenstein and Roberts note for the elasmobranchs exists for nearly all kinds of fishes in nature. Elasmobranchs are not easily trained in any event, but the behavior of conditioned fishes generally, while certainly revealing their ability to perceive stimuli and to respond to them in predictable ways, lends little information as to how fishes utilize sound perception and sound production in nature[131]. In many ways, the behavior of captive fishes differs from the behavior of the same species in nature. For example, nearly all known sound-producing species tend to become silent in captivity, and some sounds known to be produced by certain species in nature have never been recorded from captive fishes no matter how closely natural conditions are simulated[65,132]. In addition to laboratory investigation, there remains to be done a tremendous amount of field observation in the relation of sound to the biology of wild fishes before the significance of sound to fish behavior is fully appreciated.

Another problem in generalizing on the acoustical biology of fishes lies in the fact that hearing has developed more or less independently in various groups of fishes. In clupeids sound perception seems to be centered in the utriculus[49], in ostariophysans in the pars inferior (sacculus and lagena)[71]. In elasmobranchs, areas sensitive to sound vibrations, as indicated by action potentials, are located both in the pars superior and pars inferior of the inner ear. Part of the macula of the sacculus of *Raja clavata* responds to sound stimuli, part to positional changes[118]. To Lowenstein and Roberts, this has indicated that "phylogenetically the elasmobranch vestibulum may thus be considered to represent an extremely generalized state comprising the morphological and functional raw materials from which, during the evolution of the vertebrates, the one or the other potentiality has found its full functional realization." That most living elasmobranchs must be regarded as specialized rather than as primitive forms, does not remove the probable truth from the implication that the various divisions between hearing and positional sensory functions of the teleost inner ears may have arisen from a situation like that existing in living elasmobranchs. In all actinopterygians studied, except ostariophysans, there is some acoustical sensitivity in the utriculus; sound is perceived in the sacculus and lagena of all actinopterygians so far as is known, and in the sacculus but not the lagena of elasmobranchs[50,118].

Strong evidence of a significance of sound to the biology of fishes lies in the relationships of varied accessory hearing organs which serve to establish a close connection between the inner ear and the resonating air bladder, which have recently been reviewed by Jones and Marshall[102]. In the Holocentridae, as in others of the berycoid fishes, and similarly within some other teleost fishes, the anterior end of the air bladder is in direct apposition to a thin membrane of the auditory region of the skull[17,171]. In the Holocentridae, this apposition is either through paired diverticula, each attached to a cartilaginous window between the proötic, exoccipital and basioccipital (Myripristinae), or through simpler anterior bulges (*Holocentrus* sp.) of the air bladder, the membranous wall of which forms a window or "tympanum" between the same three bones. The arrangement varies with the orientation of the membranous portion of the auditory region of the skull[142,171]. A similar arrangement occurs in *Sargus annularis* (Sparidae)[51].

In the mormyrids, or elephant fishes of Africa, a somewhat more elaborate connection prevails in that anterior diverticula of the air bladder are constricted from it during development, and remain as small air-filled resonating chambers in direct contact with the sacculus wall[50,71,102]. Filling these chambers with water depresses the hearing sensitivity of these fishes[50]. In the Clupeidae, anterior tubular extensions of the air bladder run forward into the skull, each terminating in an enlargement within the proötic bulla adjacent to the perilymph surrounding much of the inner ears[122,144,180,181,190].

In many of the ostariophysans, the catfishes and their allies distributed among some 29 families, the most noted of these accessory hearing devices occurs[185]. Although the mechanism varies, it typically consists of four small bones called the Weberian ossicles, modifications of anterior vertebrae and ribs, which lie between the air bladder and a fluid-filled chamber which is part of the perilymphatic system surrounding the inner ears. Because of the terminology used by Weber[185], in anteroposterior sequence the claustrum, stapes, incus and malleus, in his original descriptions of these ossicles, three of them have been erroneously considered as homologues of the malleus, incus and stapes of mammals[101]. Most authors have, however, emphasized their correct relationship[161], and many have adopted a terminology preventing a confusion of terms and homologies[111], in anteroposterior sequence the claustrum, scaphium, intercalarium and tripus of Bridge and Haddon[28].

As in the case of the ear, the air bladder serves a number of functions, and there is considerable evidence that in the ostariophysans at least, the accessory organs of hearing as represented by the Weberian ossicles have more than a single purpose. Thus there is good evidence that they are concerned in perception of changes in hydrostatic pressure as such changes influence the air bladder[102], and it has been suggested that very slight pressure changes may be of great importance in the orientation of fishes to various objects[187]. (There is good evidence that this is the case in clupeids, for example[139].) But the important point here is the clear evidence[13,69,71,159] that the Weberian ossicles of ostariophysans lend to this group a higher frequency sensitivity and a lower sound pressure threshold than is possessed by fishes lacking connections between the air bladder and inner ear[71], indeed than by any other fishes thus far tested. Fishes with the inner ear coupled to the air bladder or a derivative thereof possess keener hearing—lower thresholds and wider frequency ranges—than other fishes[83]. Kleerekoper has shown for four cyprinids that the most acute hearing lies at the resonant frequencies of the respective air bladders experimentally reduced. It seems likely that connections between air bladders and inner ears are an adaptation for improved hearing in many fishes. Weber[185] accredited the first expression of this idea to Aristotle[178].

Accurate information on the sensitivity to sound of clupeids, with the tubular air bladder extensions to the inner ear, is lacking, largely due to the difficulty of maintaining these fishes during transport and captivity. Menhaden are acutely sensitive to high intensity sounds of mixed frequencies at close range[134]. Pounding on the decks of herring fishing boats at night, with wooden mallets, startles nearby herring schools and their quickened swimming creates a bioluminesence in the water by means of which the schools may be located[131]. The bioluminesence may itself serve to help bunch the fishes together, as it is believed to do in a seine net fishery of Malaya[155]. During the closing of a net about a pocket of menhaden, Maine fishermen create a loud noise by rapping on the boat, by splashing the water and by any other means at hand, to prevent the fish from escaping through the still open gap.

Certain aspects of ear anatomy in clupeids, aside from the air bladder extensions, are further suggestive of adaptations for sound perception. The relationship of the proötic expansion of the air bladder extensions to the utriculus suggests that the pars superior, rather than the sacculus or lagena, of the clupeid ear is primarily responsible for sound perception[50, 144, 190]. In adult menhaden, *Brevoortia tyrannus* Latrobe, the thin-walled proötic bulla containing the enlarged terminal vesicle of the air bladder diverticulum places only a thin wall between perilymph and the environment; in menhaden larvae, the developing thin-walled bulla projects for a short distance into the water of the opercular cavity in a way one might liken to the suspension of a hydrophone into the sea (Moulton, unpublished).

There are other specialized features of the clupeid skull which would appear to establish an intimacy between the environment and the perilymphatic system surrounding the inner ears[180]: the auditory foramen, the temporal foramen, and communication of the lateral recess with the exterior through the lateral line canals. While as Tracy[180, 181] emphasized, these arrangements may serve to facilitate detection of pressure changes, they would also seem to facilitate passage of sound vibrations to the inner ears. Although Tracy rejected an auditory function for these arrangements, the results of more recent work with clupeids and with other fishes would suggest that the anatomy of clupeids abets sound perception.

Although various accounts of sound production or possible sound production by clupeids exist[9, 96, 100, 102, 140], it appears probable that, despite their anatomical adaptations for sound perception, the clupeids are largely silent fishes. The sound of large numbers of them moving through the water may, however, be of considerable intensity[9, 139]. Such sound has been recorded and has been described by underwater swimmers. A sudden dive by a net-encircled school of menhaden is referred to as "thundering" by Maine fishermen, due to the loud sound produced. The possibility of sound production during release of air through the post-anal aperture of the herring air bladder in the form of a bubble stream has been suggested[62, 140]; sound production due to this cause has been observed in the European minnow, *P. laevis*[16], and in a mullet or chub sucker, *Moxostoma* sp. Several ostariophysans, on the other hand, with equally or more remarkable accessory hearing organs, are rather noisy fishes[41, 46, 47, 65], as are many other kinds of fishes without accessory hearing organs, such as sea robins or gurnards (*Triglidae*) and toadfishes (*Batrachoididae*). The distribution of fish sound production does not show any clear correlation with the distribution of anatomical specializations for hearing in fishes.

Dijkgraaf[47] concluded, after an acute study of the matter, that the most noticeable sound components produced by the ostariophysan *Phoxinus laevis* were due to gas release from the air bladder during quickened swimming and diving, and that there was no apparent biological significance of the resulting sounds to this species. He agreed with von Frisch[70] that sound production in a species with good hearing is not a necessary corollary of that fish's hearing ability, but recognized that many sounds exist in the water, the perception of which might enhance a fish's survival. Whether or not the sound of gas release by *Phoxinus laevis* has a significance to other fishes was not examined. It is conceivable that predators might be attracted to such noises. At Bermuda during the summer of 1958, a barracuda was apparently attracted to amplified stridulatory sounds of small jacks, played back into the water[139,215].

Available information on frequency sensitivity, sound pressure sensitivity, and tone discrimination of fishes is based chiefly on experiments with conditioned specimens, and is available for relatively few species. The recent review tables of Kleerekoper and Chagnon[106] and of Lowenstein[119] are most complete (Tables 58 and 59). The most definitive data available on sound pressure sensitivity are those

TABLE 59

TONE DISTINCTION IN VARIOUS SPECIES*

Species	Range of Tone Distinction (c/s)	Reference
NON-OSTARIOPHYSI		
Umbra sp.	288 and 426	Westerfield (1921)
Anguilla anguilla	Less than one octave**	Diesselhorst (1938)
Gobius niger	$3/4$ up to $1^{1}/_{4}$ tone** (9–15% frequency distinction)	Dijkgraaf (1952)
Corvina nigra	$3/4$ up to $1^{1}/_{4}$ tone** (9–15% frequency distinction)	Dijkgraaf (1952)
Sargus annularis	$3/4$ up to $1^{1}/_{4}$ tone**	Dijkgraaf (1952)
Sargus annularis	$3/4$ up to $1^{1}/_{4}$ tone** 9–15% frequency distinction)	Dijkgraaf (1952)
Marcusenius isodori (Mormyridae)	$1/4$ Octave**	Diesselhorst (1938)
Gnathonemus sp. (Mormyridae)	One tone**	Stipetic (1939)
Macropodus opercularis (Anabantidae)	Varying between one tone and $1^{1}/_{3}$ octaves**	Schneider (1941)
OSTARIOPHYSI		
Phoxinus phoxinus	A major third in 821–651 range; a minor third in 290–345 range	Stetter (1929)
Phoxinus phoxinus	$1/2$ tone in 987.7–1046.5 range (6% frequency distinction)	Wohlfahrt (1939)
Phoxinus phoxinus	$1/4$ tone in 400–800 range	Dijkgraaf and Verheijen (1949)

* From Kleerekoper and Chagnon (1954).
** The frequency range in which distinction of tones was examined was not mentioned by these authors.

of Autrum and Poggendorf[13] and Poggendorf[159], and of Kleerekoper and Chagnon[106], for two ostariophysans, the silurid *Amiurus nebulosus* and the cyprinid *Semotilus a. atromaculatus*, respectively. These authors have erected threshold sensitivity curves over a range of frequencies for these species, with different methods and with differing results. In *Amiurus* (Fig. 377) hearing sensitivity is approximately constant from 60–1600 c/s, with an optimum at 800 c/s; sensitivity falls steadily to approximately 13,200 c/s, the upper frequency limit of sensitivity, approximating a value which had previously been determined by Stetter[173]. In Semotilus (Fig. 378), Kleere-

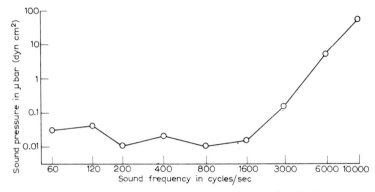

Fig. 377. *Amiurus nebulosus*. Absolute hearing threshold curve in relation to sound pressure. Average values of the measurements in the best animal. (From Poggendorf, 1952[159].)

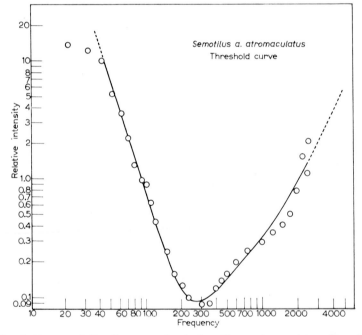

Fig. 378. Threshold curve of vibration perception in *Semotilus a. atromaculatus*. (From Kleerekoper and Chagnon, 1954[106].)

koper and Chagnon found, the greatest sensitivity is at 300 c/s, the threshold rising on either side to upper and lower frequency limits not actually determined. The curve resembles the form for that of human hearing. The differences between the results of Autrum and Poggendorf and of Kleerekoper and Chagnon may be related to differences in methods employed, as well as to the use of different species. *Semotilus* with the ears removed responds to frequencies of 20–200 c/s, so far as has been determined. Von Frisch and his students[71] concluded that *Phoxinus* perceives frequencies from 25–130 c/s with ears as well as skin. Of the most important sound producers among fishes, hearing sensitivity has been studied only in sea robins (*Triglidae*) by Griffin, who did not attempt to determine threshold sensitivity over the whole frequency range.

Two interesting aspects of the results of Kleerekoper and Chagnon, suggestive of a utilization of sound in orientation of the creek chub, were the following. Measurements of sound pressures within the experimental tank indicated that the conditioned fishes moved to the feeding station along lines approximating crests of highest intensity, and when sound sources of two different intensities were used simultaneously, the fish moved to the source of greater intensity. Menhaden, in the confining limits of a live car, were exposed to high intensity sounds at Woods Hole during the summer of 1955 and behaved in an analogous fashion, their rather frenzied swimming concentrating the fish before the face of a semi-directional transducer[130], the position of which was changed in the course of the experiments. Trained fishes, at least, appear to have no problem in determining the location of sound sources used to produce conditioned stimuli, as Moorhouse observed in the case of a surf perch, *Cymatogaster aggregatus* Gibbons, twenty-five years ago[128].

4. Uses of Sound in the Fisheries

Before turning to a discussion of the production of sound by fishes, we may briefly examine the utilization of sound by fisheries. Despite the absence of clear evidence that fishes orient to sound in their normal behavior, some fisheries have, probably for centuries, utilized sounds to increase the catch. These uses themselves would seem to indicate that sound perception is of significance to the behavior of several kinds of fishes in nature.

Uses of sound in the herring and menhaden fisheries have already been mentioned. To them may be added the use of a small explosive charge in scaring herring schools at night to momentarily increased activity in the California herring fishery[131], the fishermen sighting the school by the glowing of phosphorescent organisms in the water. A similar method is used in a Malayan mackerel fishery, in which the sound source is a powered boat[105]. Recently streams of air bubbles have been introduced into the water in Maine to guide herring schools.

Anecdotal reports of freshwater fishermen, always to be treated with a certain amount of reservation, indicate that at times creating noise in the water seems to improve the fishing. This belief is reflected in the relatively recent appearance on the American market of various patented "sonic lures" which purport to introduce into the water sounds which fishes might encounter in nature—the whine of a mosquito, the grunt of a certain kind of bait fish or a more random noise which the manufacturer avers attracts fishes to it. Available evidence would seem to indicate that such devices may be more certain of luring fishermen than fish. Griffin[83] has noted that artificial sounds tend to repel fishes rather than to attract them.

References p. 687

Yet in European and Asiatic fisheries, sound has been used for centuries. Ryukyuan fishermen use two sound sources in conjunction with gill nets[7, 131]. One of these is a sound tub, "a wooden barrel with a wooden handle fastened across the open top. Weights are hung from the handle. When the waves move the floating tub, the weights hit the sides of the barrel and make considerable noise. Sound tubs are effective in driving fishes out of their hiding places into the net." The other source is the scare rope, used in gill net and drive-in-net fisheries. Swimming fishermen drag metal rings or stones over the reefs by means of ropes with white cloth and/or leaves tied along them at intervals. These ropes with their sound-producing devices are thought to scare fishes from crevices into the nets.

Indonesian fishermen employ sound in a hook and line fishery. The instrument used, variously referred to as a "kuruck-kuruck"[92] and "uruck-uruck"[187] is a triangular bamboo frame with coconut half-shells strung along one side. This is shaken under water by means of a short handle at one angle. The fishermen are convinced that the sound created attracts certain kinds of fishes to hooks hung from the boats, and that the species caught will depend on the depth of shaking. It is said to be particularly effective in fishing for snapper and shark in Bontang, East Borneo[92].

Another method with acoustical implications, analogous to the last, is that of mackerel fishermen of the Madeira Islands who chop their bait in open boats at night after arrival on the fishing grounds. The sound of the chopping is thought to attract fish to the area[2].

The sound of the human voice has been described occasionally as a fishing adjunct. The most interesting is one described by Westenberg[187] in an ancient Javanese fishery for *Stromateus niger*. "When a free-swimming school... is seen near the surface, a boy jumps overboard with a bamboo, about as long as his body, and fitted with a cross-bar near its upper end. As he floats with his arms resting over the cross-bar, he sings in a monotonous voice, uttering a prolonged "ooh". This makes the fish flock around the singer, and sometimes even jump in his face from excitement. The school is then easily encircled with a net." Traditional singing is practiced while fishing for morays (*Muraenidae*) among the rocks of the Madeira Islands[2], and singing was described by Pliny as having been practiced by mullet fishermen of Nismes centuries ago in calling dolphins to scare the mullet into shallows so the fishermen might net them[158]. The attraction of dolphins by music, singing, and calling is also mentioned elsewhere by Pliny, and the phenomenon is mentioned in connection with another fish (*Trichochos*) by Athenaeus[193], giving Aristotle as an authority. This fish is not indexed in Thompson[178].

Sound is employed in scaring fishes into nets or into areas where they may be netted, in widely separated areas. Its use in the Maine menhaden fishery has already been mentioned. Another example is seen in "blashing", a restricted salmon and sea trout fishery of Robin Hood's Bay, Yorkshire, England[38]. Here fishermen create a loud noise while rowing parallel to the shore at night, afterward placing a seine around the area between shore and their course, the ends of the seine being hauled in from shore. "The point is," writes Dr. John S. Colman, Director of the Isle of Man Marine Biological Station, "that the men believe that the noise of blashing scares the fish and that any fish between the boat's course and the sound rush into the shallow water away from the noise, where they are rounded up in the subsequent haul of the seine net."

Other uses of techniques engendering sounds are employed in various Indonesian[187] and Malayan[105] fisheries. This is particularly notable in tuna fisheries of Indonesia and Japan[187] in which baiting and chumming is accompanied by splashing the water, at times with a special instrument designed for the purpose. The splashing is believed to attract tuna to the bait. In such methods, however, the possibility that visual stimuli are primarily involved cannot be discounted. The same may be said of a number of methods which in daylight employ or involve splashed water in scaring fish into nets, for example, the beating of stones on the water[105] and the thrashing of a pole ornamented with shells or tin[155]. The splashing may also serve to obscure the sight of the net, as waves are utilized in a push-net fishery of Malaya[155].

Westenberg[187] suggests that the rustling sounds of attached fronds and leaves are responsible for the grouping of many small fishes under floating debris in the ocean, and describes a raft fishery which takes advantage of this habit of several small fishes of tropical waters.

If the sounds of bait-chopping of the Madeiran Islanders and of the "kuruck-kuruck" of Indonesian fishermen serve to attract fishes to bait, and the fishermen have used these methods for many, many years under such an impression, it cannot be true that all artificially engendered sounds are repelling to fishes. Yet reports of the consistent influencing of unconditioned fish behavior through artificial sounds under experimental conditions are rare[131,132,171]. The main difficulty is one of observing fish behavior in relation to sound in most oceanic waters. We shall return to this latter problem following a consideration of sound production by fishes.

5. Sound Production by Fishes

Perhaps the most striking evidence of significance of sound to the biology of fishes lies in the fact that many fishes produce species specific sounds, and in anatomical arrangements which appear to be special adaptations for sound production. This area of fish acoustical biology has attracted considerable attention within the last decade and a half, attention generated by publication of a number of studies attempting to define these sounds.

The popular notion of the sea as a silent place does not hold true for near-shore water, especially in warmer seas[137], nor even for greater depths[83,124]. Recent intensive studies of fish sound production have, like studies in many other areas of investigation, raised more questions than they have answered. It is difficult to reconcile the production of sound by many species of fishes with the existence of so little concrete evidence that fishes orient to sound, or that underwater sound is of significance to fish behavior generally. Initial attempts to correlate sound production with fish behavior have been those describing conditions under which sound is produced in nature. That sound production in most sonic species seem to occur under rather circumscribed conditions has indicated that sound production is a significant part of certain behavioral patterns[137]. Still, evidence that fishes move in relation to sound sources, either toward the sounds or away from them, is scarce, and is largely of an indirect nature in untrained fishes, although the behavior of conditioned fishes in relation to sound sources can be predicted.

Interest in the sounds produced by fishes is reflected in writings since the time of Aristotle, who compared the sounds of fishes to those of other animals[11]. Various fishes of ancient times were named by Aristotle and others according to the sounds they produced[178]. Aristotle[11, 12], Athenaeus[19], Pliny[158], and the Greek Anthology[56], refer directly or indirectly to the sounds and hearing of fishes, as well as of other marine organisms. Dufossé[58, 59, 60], an early and thorough French investigator of the mechanisms and circumstances of fish sound production, and Day[43], cite other classical references dealing with underwater biological sounds. The sound-producing powers of certain *Siluridae, Sciaenidae* and *Triglidae*, for example, were well-known to fishermen of Aristotle's time[178].

Early authoritative modern accounts of the sounds of fishes were those of Darwin[41] and Alcock of the *Investigator*[3]; other accounts were those of mariners who heard the sounds resonated by their ship's hulls. (See review of Fish[65].)

The earliest American accounts of sound-producing fishes were all of sciaenids— the croakers, drums, and their relatives. William Penn, the founder of Pennsylvania, observed a valuable food source in the drum, *Pogonias* sp. "Tis so called," he wrote in 1685[156], "from a noise it makes in its Belly, when it is taken, resembling a Drum." The French explorer, de Bienville, founder of New Orleans and later Governor of Louisiana[75], reported hearing the drum during exploration of the mouths of the Mississippi River in 1699[77]. The drum is produced as the fish root in the mud[183a]. John Lawson, the British Surveyor-General of North Carolina in the early 1700's, a superior natural history observer for his time[4], described the sounds of the croaker, *Micropogon undulatus* in 1714[112]. Of even earlier date for the western hemisphere (1623) is Captain John Smith's comment on the Bermuda "Growper" (*Serranidae*), so-called "from his odd and strange grunting"[113].

The sounds of fishes have worked themselves into the folklore of Japanese fishermen. A Japanese proverb, cited by Fish[65], is: "Therefore, joy or sadness, without a croaking of guchi, there can be no good catch for fishermen." "Guchi" is a common name for various Japanese sciaenids[145].

Definitive studies of the sounds produced by aquatic animals awaited the development of adequate equipment for reliable recording and analysis, and it was the rapid development of such equipment during and after the Second World War that intensified the study of underwater biological sounds. As early as 1909, David Starr Jordan, the distinguished student of fishes and president of Stanford University, and C. F. Holder, biographer of Charles Darwin and Louis Agassiz, predicted that "doubtless one of these days, a scientist will be born... (whom) we shall all see... going down in a diving bell and taking the language of fishes into a phonograph for posterity"[97].

That the underwater sounds of the sea have been familiar to fishermen for ages is indicated by the use of listening techniques in various fisheries, techniques which are still employed today in some parts of the world. Both ambient sound, the sound of surface waves and of water moving over the bottom, and biological sounds have been used as indicators of good fishing grounds. Syrian fishing boat crews station a crew member deep in the boat's hull to detect the resonance of a certain category of ambient noise which they have learned to associate with a particular bottom conformation over which fishes feed in large numbers[39]. It is possible that Leonardo da Vinci experimented with listening to the sounds of fishes by a method still used in India, holding the end of an oar to the ear, the blade being in the water[124]. A similar method is used in Indonesia by fishermen trained for years and is thought to be most effective at night[92].

The art of fish listening as an adjunct to the fisheries reaches its highest refinement on the east coast of Malaya[105,155] where specially trained members of fishing crews submerge their whole bodies in the water to listen for sounds which will guide the placing of nets. The method is used especially in fishing for sciaenids and carangids in 15 to 20 fathoms of water in the method known as the *Pukat Payang*. Parry[155] gives the following interesting account: "The (fishing boat) carries abaft the main mast a small 8-ft. boat... When the fishing ground is reached, this is launched and the (fish listener net leader or *juru sĕlam*) gets into it. He is then paddled about while he dives overboard and listens at intervals, until he locates a shoal of fish. Location is done by sound, although increasing numbers of net leaders now carry goggles as well.

"The art of detecting fish by ear reaches a very useful refinement among the most skilled *juru sĕlam*. Not only is it possible for them to detect the presence of the fish, but many other details beside. The fish are said to "croak or chuckle". When a few voices are heard fairly regularly, it is said that there is a fair-sized shoal, with only the leaders "talking". When, however, irregular but frequent croaks are heard, then the shoal is assumed to be scattered, with no single leading group, but many little units making the noise. The presence of rocks and water snags shows itself in a change in the note of the croaks. Not all fish croak, apparently, but (Jewfish—*Sciaenidae*, and Scabbardfish), the kinds most frequently caught by this method, are said to have clear voices. All fish probably make a noise while swimming, from the beat of their tails. Even prawns are said to make a faint "bleat".

"The *juru sĕlam* relies on the exploitation of this faculty for his prestige and ability to get a crew together and control them. Anyone with any unusual qualification receives a respect tinged with awe. It is said that no shark will attack a *juru sĕlam*. Sharks of sufficient size to do so are very rare in these waters, but nonetheless the reason given is interesting. If a shark approaches a *juru sĕlam*, it is said to realize that the man is "clever". It then becomes afraid and ashamed, and will not open its mouth (to bite), until it has swum off to its home."

A further interesting comment on this method occurs in a letter to the author from Mr. E. R. A. de Zylva, Deputy Director of Fisheries in Ceylon, telling of a conversation with a Malayan fish listener[133]. "My conversation with the fish listener (in Malaya) is... still very fresh in my mind. He described to me the different sounds which he is able to distinguish under water, *e.g.*, kris-kris, groh-groh, and others such as the rustling of silk... Each distinct sound is made by one species of fish according to him. He also stated that it was possible for him to say in what direction the fish were proceeding and how far away they were from him. One of the explanations he gave me of this ability was as follows: "When you are walking in the sun and pass into the shade of the forest you feel a difference on your skin."

"This seemed very incredible to me and I asked him how he picked up his knowledge. He told me that it was the result of many years of study and that it had taken him seven years to achieve the power of independent fish listening..."

When during World War II biological sounds of the oceans interfered with listening for enemy ships, the sounds of fishes, as well as of other marine organisms, became of interest for other than economic reasons, and intensive studies of their sources and distributions were carried out, especially by United States and Japanese scientists. The refinement of suitable listening equipment, hydrophones, amplifiers,

References p. 687

and recorders, has made it possible to obtain recordings of these sounds with a high degree of fidelity. In early descriptions of marine noises, the sounds were likened to grunts, groans, whistles and other familiar sounds, and attempts were made to write the sounds in words. It has now become possible to define these sounds on a quantitative basis, and to prepare sound spectrograms which can be used by comparison in identifications of other sounds recorded and analyzed on similar equipment. For the interested investigator, lacking specialized high fidelity equipment, it is possible to use inexpensive, water-proofed microphones with appropriate amplification and recording equipment to listen to, and to record, underwater noises connected with various behavioral problems. Such recordings, while lacking in fidelity, may be useful in studying some aspects of acoustical behavior. An "aquatic stethoscope" has been designed for simple underwater listening[23].

For the most reliable descriptions of underwater sounds, however, use has been made of equipment which provides accurate information on intensity and frequency levels of the sounds recorded. At the Woods Hole Oceanographic Institution, for example, where I have done most of my work, this equipment includes various crystal hydrophones which can be lowered to great depths in the ocean from a quiet ship; the sounds coming to the hydrophone are recorded on board. These same hydrophones as well as others can be used in shallow waters and in tanks, where the behavior of fishes and other marine animals under observation can be correlated with the sounds they produce.

Following tape recording of the sounds, the recordings can be played through analysis equipment of various types, the oscilloscope and the Sonagraph and Vibralyzer of the Kay Electric Co., for example, so that actual patterns of frequency and intensity distributions of each kind of sound can be obtained by photograph or on a special paper (teledeltos) with the Sonagraph and Vibralyzer. The duration of the teledeltos record varies with the frequency range scanned. Different intensities are shown in the teledeltos recordings by varying darknesses of the record, the darkest areas being those of greatest intensity. When data from these records are combined with anatomical and physiological studies, it is possible to correlate, in some cases, the mechanism of sound production with the type of sound produced. Each method of sound production has its own pattern of frequency distributions, in relation to time and intensity, and even similar methods, when used by different species or by adults of the same species but of different sizes, will show characteristic shifts in the sound spectrogram.

A typical example of a marine sound analyzed on the Vibralyzer is that of a portion of a staccato call of a sea robin (*Triglidae*) recorded at the Woods Hole Oceanographic Institution (Fig. 379). The sound of the sea robin is produced by muscles in the walls of a bilobed air bladder, the left lobe (rarely the right lobe instead) of which is sub-divided by a partition perforated in its center. The muscles of each lobe of the air bladder are discrete from those of the other lobe; they are innervated by the vagus nerves. The two lobes differ somewhat in size.

When the staccato call is produced, the call is of something over 2 seconds in duration. On vibration frequency analysis, it is seen that intensity peaks center at two different frequency levels, each series of peaks being produced by contraction of the muscles on one of the two lobes. At times in a given call, one lobe may drop out for a portion of the call, so that only the higher or the lower row of pulses will be seen for a portion of the record. No difference is, however, detectable to the ear. Similar records from a variety of fishes have recently been published[137]. By measuring such records horizontally and vertically, one can define the location of the sound pulses and thus their characteristics in terms of time duration and of frequency.

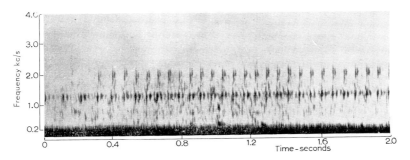

Fig. 379. A sonagram of the staccato call of a sea robin (*Prionotus* sp.). The fish makes the sound by vibrating the bilobed swimbladder. Note that pairs of 2-kc/s pulses alternate with pulses at 1.3 kc/s. It has been suggested that each series of pulses corresponds to vibrations of one of the two lobes of the swimbladder[132]. The right lobe of the swimbladder in *Prionotus carolinus* is a simple chamber (1.3-kc/s pulses?) communicating by a small opening with the left lobe which has a perforated transverse partition (2-kc/s pulses?)[179]. Alternate stimulation of the lobes would explain the alternation of the individual sets of pulses. (From Backus, 1958[14].)

The Vibralyzer includes another device, a sectioner, which makes it possible to analyze any point of the recording in terms of the relative intensities of the frequency components at that point. Another type of sound intensity analysis, the contour "map", may be obtained by gradually lowering the attenuation of the recording through several analyses so that various frequency components appear at successive gain settings as less and less intense components are brought in to successive records. In this way, contours of the sounds may be drawn, similar to topographic maps, and these too, are found to vary from one kind of marine noise to another. This method has been developed and much information obtained from it at the Woods Hole Oceanographic Institution. It is described by Backus[14] and an example has been published by Hersey[95].

Such methods and analyses, combined with behavioral observations and anatomical studies, have yielded considerable information on the acoustical biology of marine fishes and have attracted the attention of many biologists, and of oceanographers generally, to the problems concerned in marine animal sounds. A major obstacle is still that of the difficulty of observation of fishes at sea. Many sounds which have been recorded are ascribed to fishes only because it seems likely they represent the sources, and these unseen fishes have been tentatively ascribed common names simply on the basis of the sounds they produce, for example, the echo fish, the whooping fish, and the carpenter fish. Studies on shallow water fishes which can be seen during recording, have been most profitable in correlating sound production with behavioral patterns[137]. Studies on sound production by fishes in aquaria have been of somewhat doubtful value, since it is clear that acoustical behavior of most captive fishes differs markedly from that of fishes in nature. Most sound-producing species become silent in captivity, or produce only a portion of the sounds which have been attributed to them on the basis of other evidence. Many sounds which

References p. 687

have been described from captive fishes have been described on the basis of sounds produced during shock, handling or other stimuli of an artificial nature. While informing us, perhaps, of the frequencies and mechanics of sound production, such studies can do little toward clarifying the circumstances under which sounds are produced in nature, nor toward appreciating the full vocabulary of any species.

Some of the sound of fishes are contingent on the movements and feeding activity of the animals concerned. Dr. Marie Fish of the Narragansett Marine Laboratory of the University of Rhode Island calls these "mechanical sounds"[68]. They are primarily due to friction between skeletal parts, between teeth and food, or between body and the bottom. Such sounds may possess biological significance for some species; for example, puffers, *Spheroides maculatus*, may be attracted to the chewing of other members of the species[25]. There is another category of sound in the sea which is more apparently purposeful and which current research in some areas indicates may be of importance to the behavior and life histories of the source species. Sounds of this latter category have been termed by Fish[68] "biological sounds", although these sounds too may be produced by skeletal parts. In many fishes, these sounds stem from organs apparently specially adapted for sound production. Thus in diverse groups of fishes, there are skeletal stridulatory mechanisms or there are specialized muscles on the walls of or adjacent to the air bladder. When these muscles contract, they make of the air bladder a percussion instrument and various kinds of throbbing, drumming, grunting or whining sounds result. Such sounds may be created through muscles in the wall of the air bladder, through fluttering of body wall musculature, or through fin drums against areas where the air bladder closely approaches the body surface. Air bladders specialized for sound production Bertin[18] calls "organes de phonation" since the sounds emitted can be placed in the musical scale.

The air bladder may also be used to resonate sounds created by other parts. Thus as in the grunts and carangids, pharyngeal teeth are stridulated adjacent to the air bladder so that the rather faint stridulatory sounds are magnified to respectable noises clearly audible to the human ear. Many sounds of fishes may be heard by the human ear, even underwater.

Of those fish sounds commonly heard at sea or in shore waters, perhaps the most notable is booming of the drum, *Pogonias cromis*, which is one of the loudest calls of fishes known[37,194]. This sound, reported by early travellers along various parts of the North American coast, is frequently heard from boat decks and by shore fishermen. It has been confused by at least one sea captain with "insects in the spirits room"[65]. It was thought by Gunther[88] to be due either to the clapping of pharyngeal teeth together or, more likely since the sounds sometimes accompanied vibrations of ships, to the fishes' beating their tails against the bottoms of vessels to rid them of parasites.

On the other hand, the sounds of fishes range to the very faint, such as those of the release of gas bubbles from the air bladders of various fishes, or the tooth snaps of the desmoiselle, *Pomacentrus leucostictus*, in the West Indies[137]. The latter sound was not audible to the unaided ear in air close to the aquarium, but was clearly audible with a hydrophone and amplifier. Gas bubble release in the cyprinid *Phoxinus laevis* may be heard up to a meter from the aquarium[64,47]. Thus fish sounds range from very faint to very loud.

More information is available on the sound production of salt water fishes than of fresh-water fishes. Of the latter, sound production has been described in *Cyprinidae*, *Cobitidae*, and *Ophiocephalidae* of Japan and the western Pacific islands[65], in the *Bagridae*, a family of Japanese catfishes[145], in *Amiurus (Siluridae)*[1,110] as well as other siluroids[60], in *Anguilla (Anguillidae)*[1,60], and in a number of European cyprinids[42,46,47,60,93], either through chewing or gas release.

Sound production is much more widespread and more highly evolved among marine fishes than among fresh-water ones. Even among the catfishes, marine species are more important sound producers than are fresh-water ones[65]. The most common sounds of fresh-water fishes are the "bruits de souffle" of Dufossé[60], caused by gas release from the air bladder or anus[29,58,59]. Such sounds may also originate from a number of salt water fishes, especially when first drawn from the water. This is true, for example, of the conger eel, *Leptocephalus conger* Linnaeus, which, when first caught, produces a grunting sound during gas release through the pharynx.

Most fresh-water sound producers which have been described among fishes are cyprinids and Siluroidea, both ostariophysan groups possessing Weberian ossicles and, to judge from those species studied, forms with acute hearing. This has given rise to the belief that sound production and good hearing seem to go together, a matter which von Frisch and Dijkgraaf have discussed and rejected. The clupeids with the air bladder connected to the inner ears have no special arrangements for sound production; the noisy oceanic sciaenids have no accessory organs for hearing. Sound production appears to be distributed independently of accessory hearing organs, and does not necessarily imply excellence of hearing. In fact, as von Frisch[70] and Griffin[80] have pointed out, it may be that the predominant sound production of toadfishes, sea robins and sciaenids may occur because members of these groups have relatively poor hearing. On the other hand, the predominant sound producer of the western edge of the Great Bahama Bank[137] and of the Bermuda Bank[139] is the squirrelfish, *Holocentrus ascensionis* (Osbeck) with the air bladder connected to the inner ear[142].

Among salt water fishes, sound production is of wide occurrence. Fish[65] described it as occurring in 41 families of Pacific fishes beside various siluroids, and later[66,67,68] described the sounds of 28 fishes distributed among 21 families of the western North Atlantic. Moulton[137] has described the sounds of 12 species in 9 families of the western edge of the Great Bahama Bank. Marshall[124] has presented convincing evidence that sound production is widespread among deep-sea macrourids. Oceanographic vessels of the Woods Hole Oceanographic Institution, for example, have recorded many sounds at sea which have been ascribed to fishes as the most probable sources, the fishes remaining in many cases unidentified. Judging from available evidence, hundreds of species of marine fishes are capable of producing sounds, either as by-products of feeding and other activity, or through the use of organs apparently especially adapted to the production of sound. It is important to recognize, in considering the acoustical biology of fishes, that both types of sound may possess biological significance.

Examples of some of the more interesting methods of sound production among various fishes, described by Bridge[29] and others, will demonstrate the variety of acoustical adaptations occurring among fishes.

Several different methods of sound production have been described in triggerfishes (*Balistidae*)[65], although they are not common to all genera of the family[137]. In *Balistes* and *Melichthys*, for example, the air bladder is intimately connected with the skin, forming a thin window behind the gill opening and above the base of the pectoral fin[24, 29, 79, 137, 167]. In young specimens, this area is translucent to light shown through the fish; in adults it becomes pigmented and opaque. Such triggerfishes when caught or handled elevate the pectoral fins to an abnormally high position and flutter them rapidly against this membrane; a drumming sound is simultaneously created[137, 167] which Schultz[167] and Moulton[137] attribute to the fin drum and which Cunningham[40] attributes either to fin action or to pectoral girdle movements. Vibragrams of the drumming are like those of a repeated percussive sound rather than of a stridulatory sound which would be anticipated were skeletal friction the sole source[136, 137].

A second method of sound production among balistids is through grinding or clicking of the toothplates, especially during feeding, as well as during handling[65, 129, 137, 167]. Norman[143] described pharyngeal and tooth stridulation which is also implicated by Fish[65].

Bridge[29] described stridulation due to movement of dorsal fin spines. The most frequently described mechanism of sound production among balistids, however, has been that of stridulation between the postclavicles and the cleithrum, upon the latter of which Mobius[126] observed a longitudinally grooved area[29, 40, 65, 170]. Stridulation between these skeletal elements occurs adjacent to the air bladder which resonates the sound[102]. Breder and Clark[24] have recently reviewed sound production in plectognaths generally, for which they found balistids most specialized. They observed that holding the finger against the "tympanum" above the pectoral fin suppressed a "grunting sound" sometimes hear from balistids during handling. At such times, vibrations of the tympanum could be felt. Many sound-producing fishes possess a close connection between the air bladder and the skin which may serve to introduce sounds produced into the water[29].

Siluroid fishes show perhaps the greatest diversity of elaborate sound-producing mechanisms. A mechanism attributed by Fish[65] to siluroids generally, but actually described by Haddon[89], and by Bridge and Haddon[27] for a unique Indian form[29] is the stridulatory apparatus of the Indian siluroid, *Callomystax gagata*. Here, interlocking spines of the third, fourth, and fifth vertebrae form a plate divided in the mid-line posteriorly. The latter division holds the first interspinous of the dorsal fin; the inner walls of the division are grooved and ridged, as is the first interspinous. Vertical flexions of the body produce a "harsh, grating sound", due to movements of the ridged surfaces over each other.

Another elaborate device producing sounds in siluroids is the "elastic-spring" device, so-called, of a number of catfishes of South America, Africa, and India[29, 65, 170]. The transverse processes of the fourth vertebra are bent down and backward, and are broadened distally to form bony plates facing ventrally and embedded in the antero-dorsal wall of the air bladder. Muscles originating from the occipital region of the skull insert on the spring-like transverse processes and their rapid contractions vibrate the air bladder wall. This vibration of the gas-filled chamber creates a sound audible out of water up to 100 ft. from a catfish with this device[29]. Other methods of sound production by siluroids are described by Fish[65].

While it is possible that the sound-producing devices of balistids, triglids, silurids, and many other families of fishes, serve other functions, their adaptation for sound production can hardly be doubted.

Common sound-producing mechanisms among marine fishes involve the stridulation of skeletal parts, the stridulation being resonated by the air bladder, and the rapid contraction and relaxation of muscles attached to or adjacent to the air bladder. Devices of the latter category are often as elaborate as adaptations of skeletal parts for sound production, and a variety of such devices are demonstrated by gadids, sciaenids, triglids, batrachoidids, chaetodontids, holocentrids and serranids.

In some of the angelfishes (*Chaetodontitae*), for example in *Pomacanthus* and *Angelichthys*, the thick-walled air bladder is embedded completely, except in its mid-ventral line, within the heavy musculature surrounding the ribs. Many muscle fibers insert on the wall of the air bladder and cutting of these fibers in a freshly killed fish creates a resonance within the air bladder.

In the serranids (*Epinephelus striatus*) the air bladder is thin-walled and lies entirely within the coelom. The peritoneum is, however, unusually heavy and tough. In the intact fish it is stretched over the air bladder and other visceral organs. In *E. striatus* considerable sound is produced even in the opened fish by strong contraction of axial fibers attached to the heavy peritoneum. In these fishes nearly any quick movement of the body results in brief sounds. Spear fishermen of Bermuda and the Bahamas frequently hear similar percussive sounds from fishes fleeing suddenly under attack.

In addition to the attachment of muscle fibers to specially modified parts of the skeleton creating sound within the air bladder, and to creation of sound through simple contraction of axial musculature, there are two other generalized relationships of muscle to air bladder remaining to be discussed.

In one of these, the musculature lies entirely within the bladder wall, on the lateral or antero-lateral surfaces. Such arrangements are found in the Triglidae, Batrachoididae, Macrouridae, and Zeidae[102,124]. The loudest sound producer of the east coast of the United States is the toadfish, *Opsanus tau*[55,86,167,179]. In *Prionotus* (Triglidae), this air bladder musculature is innervated by a branch of the vagus nerve which emerges from the skull through a foramen, courses lateral to the roots of the bachial plexus, and emerges from a retroperitoneal course to perforate the dorsal connective tissue of the air bladder, subsequently running ventrally to enter the drumming muscles. Stimulation of the nerve causes contraction of the musculature, and sounds occurring in nature can be roughly reproduced by appropriate stimulation of the nerve or of the musculature directly[132].

A second type of arrangement of musculature developed in close association with the air bladder is seen in some Gadidae and Sciaenidae. In these cases, the musculature originates from a tendon dorsal to the air bladder and inserts ventrally on axial musculature, or the muscles arise from the wall of the bladder itself, and extend laterally to attach to axial musculature. In either case, rapid contractions of the musculature against the air bladder are responsible for sound production. Within both of these groups, as also in siluroids, occur subdivisions and diverticula of the air bladder, sometimes with constricted openings and separating, perforated diaphragms. The passage of gas between the various chambers adds to the sounds produced. In some sciaenids only the males possess the drumming muscles[179], in *Pogonias* the females produce a softer sound than the males[77], and in gadids there is a seasonal differentiation of the drumming muscles which are larger in males than in females[78].

The most notable sounds of fishes are summarized by Jones and Marshall[102] from Knudsen, Alford and Emling[108]. Sciaenids produce the loudest sounds which have been described in open water[167], and in these fishes, most elaborately in *Pogonias*, subdivided air bladders are the rule. The sounds of the croaker, *Micropogon undulatus*, may attain an intensity of 110 dB above 0.0002 dynes/cm^2 and the sounds of the toadfish, *Opsanus tau*, approach that level[55,102]. Also, sciaenids show the broadest frequency range of sound production among fishes—for *Micropogon* from 100–10,000 c/s, although there is a seasonal shift in the principal frequency from 550 in early June to 250 in early July.

The frequency components of most fish sounds lie below 8 kc/s and in most cases the principal frequencies are below 4 kc/s; results vary somewhat with methods used in analysis. These characteristics, as well as others, can help to distinguish the sounds of fishes recorded at sea from those of cetaceans.

A number of fishes employ skeletal stridulation in less direct relationship to an air bladder than in the cases already described. For example, in the carangids[43,65,137], diodontids[32,65,137], tetradontids[65,137] and haemulids[31,65,137,167], the predominant form of sound production is through stridulation of maxillary and mandibular or of pharyngeal teeth. In both cases, these sounds are produced in proximity to an air-filled chamber, whether the fish is held above or below the water. In haemulids, for example, the dorsal pharyngeal teeth lie just below and against the anterior end of the air bladder, and posterior extensions of the lower pharyngeal tooth plates lie against the air bladder; the rubbing of the ventral teeth over the dorsal is resonated by this relation to the air bladder[31,102,137,167]. Some works have implicated the air bladder directly in haemulid sound production[87,169]. Tooth stridulation is also a secondary means of sound production in angelfishes and triggerfishes, in addition to the other methods already described.

In view of the variety of mechanisms producing sounds among fishes, perhaps it is not surprising that each species produces sounds of characteristic frequency and intensity patterns. Although predominant intensities may vary from one part of the year to another[55,108], as has been shown among sciaenids, presumably due to age difference within a population, the pictures obtained on vibration frequency analysis of specific sounds are characteristic of the species studied, and comparisons of such sound spectrograms can assist in the identification of sounds heard at sea. A prime interest of the International Committee on Biological Acoustics is the establishment of a library of identifiable sounds with recording data carefully preserved, to allow for such comparisons and identifications.

6. The Relations of Sound to Fish Behaviour

What is the usefulness of sound production and perception to fishes? Under what conditions do fishes produce sounds? Is there a directional sensitivity to sound among fishes so that they can orient to sounds? We have already mentioned the rapid adaptation of fishes to many stimuli introduced into the water by man, and here sound stimuli are included. Sounds initially causing startled swimming, after a short time have no notable effect upon the fishes concerned.

Early investigations of fish sound production concerned mostly the sounds produced by fishes during handling in air. Burkenroad[31, 32], during a year's collection of a variety of sound-producing marine fishes of Louisiana, studied sounds produced when fishes were taken from the hook, trawl or seine. He heard no vocal sounds from free, undisturbed fishes during his observations, although numerous early accounts of travellers and reviews of fisheries included numerous references to fish sounds heard from fishing and other vessels. He was sure, however, that sound must be produced under other circumstances than during capture by man and pointed out that human handling seemed to cause sound production in most fishes capable of sound production.

However, Day[43], for example, an early observer of fish sound production, had noted sound production by fishes under a variety of circumstances and observed that sounds were produced under circumstances other than those suggestive of fear or anger.

In a recent article on marine biological sounds of crustaceans, fishes, and mammals, Backus[14] summarizes the purposes thus far suggested for the sounds of fishes under five categories. These are attraction between the sexes[43,68,124,132,167], orientation[80,83], defence against enemies[32,68,136], communication generally, and intimidation. It is important to bear in mind that accidental sounds, too, may have their significance to the biology of fishes[131], and thus may fall within one or more of these categories. Most of the sounds apparently purposefully produced by fishes in nature can probably be included under one or more of Backus' groups.

To define purposeful sounds of fishes, we might say that these sounds are those produced by organs especially adapted for sound production, or by actions of muscles and skeletal parts which create vibrations of the gas of the air bladder. A useful purpose of such sounds by fishes has been shown in very few instances. The vocabulary of most fishes is probably not so varied as that of birds and mammals, and the rapid adaptation of unconditioned fishes to sounds introduced from other sources into the water is a rather serious obstacle to experimental work.

The opportunities to observe fish sound production and reactions during sound production in nature have been limited, and most experiments have been carried on in aquaria where the behavior of fishes in easily seen. It is self-evident, however, that a fish once confined is a modified animal, no longer behaving as he does at liberty, and among sound-producing species, nearly all tend to become quiet in captivity. This problem has been partially relieved by Fish through the application of various stimuli to many kinds of fishes contained in tanks to see if any sounds may be elicited[65,68]. Such experiments, while important in determining possible sources of marine sounds, reveal little information as to the conditions under which sounds are produced by the same species in nature, nor is it possible in aquarium experiments to specify the reactions of other members of a given species to the sounds produced by one of its members. With smaller fishes this problem is somewhat reduced, in cases where the normal habitat can be rather accurately duplicated. The better but more difficult approach is that of visiting the fishes at home. While this present difficulties, advances in diving techniques and the use of glass panels over fish habitats in shallow water have made it possible to observe considerable fish behavior in nature during underwater listening and to specify at least some of the conditions under which fishes use sound production in their normal associations

References p. 687

even though the diver or the conveyance of the observer interjects an abnormal factor into the situation[100].

It would be convenient to generalize that the noisiest fishes are non-schooling species, tending to live solitarily. This is true to some extent of the majority of sound-producers among fishes, but not of all. Some schooling sciaenids, such as *Cynoscion*[20,76], are among the noisiest of fishes, while other schooling sciaenids, such as *Menticirrhus*[68], may be almost completely silent. (The latter genus lacks an air bladder[20].) Scombrids, many of which are schooling fishes, are of questionable sonic importance with the exception of isolated species[65]. And the schooling clupeids are vocally silent, although they may produce considerable sound through swimming movements[9,139] and gas release. Other than the sciaenids, the Western North Atlantic fishes producing the loudest sounds are non-schooling fishes, serranids, holocentrids, triglids, batrachoidids, balistids, chaetodontids, haemulids, all fishes that tend to disperse over the bottom or to inhabit isolated crevices; in such cases, sound production conceivably plays a part in communication between members of the species[14,62,100].

Drumming is so characteristic of most sciaenids, that those of the Atlantic Coast of America have been classified according to the drumming function, or the lack thereof[168]. It was this group of fishes which early provided evidence for sexual attraction through sound production in fishes[77]. Both males and females of the black drum, *Pogonias chromis* of the southern United States, produce a drumming sound[76,77], although the females are said to produce a softer note[104]. Goode[77] quotes A. W. Roberts, then a curator of the New York Aquarium: "During the spring months the males constantly pursued the females, and on such occasions, both the males and females gave out a series of very musical and liquid drum-like sounds, which could be distinctly heard in any part of the Aquarium."

Similarly the male of the weakfish, *Cynoscion regalis*, is the more important sound producer[20], although the female will on occasion produce sounds[68]. Specimens of this species as small as two inches in length can produce audible sounds[179]. Sound production is said to increase during the offshore spawning migration[6].

Fish[68] finds the early part of the breeding season of the toadfish, *Opsanus tau*, in Rhode Island waters, to be characterized by production of a "boat-whistle" sound, characteristic of pre-spawning activity, and that males guarding the nest of eggs are easily stimulated to production of other sounds of a grunting nature. When males are silent while guarding the nest, they are completely spent. Different females apparently add their eggs to a nest guarded by a single male. There is no sexual difference in the sound-producing air bladder of the toadfish with its intrinsic musculature[68,179], and Fish has suggested the possibility that certain pairs of sounds of differing intensities bear the relationship of one fish calling in response to another, as sea robins (*Triglidae*) respond to imitations or recordings of their calls[132].

Sea robins, *Prionotus evolans* and *P. carolinus* of the Woods Hole region are the source of at least two different sounds[132], and they have been accorded a wider vocabulary[05,08,76]. They produce a grunting sound by means of an air bladder with intrinsic muscles similar to that of the toadfish[68,74,179]. This sound, frequently described and endowing triglids in America and Europe with a number of common names relatable to this sound, occurs when the fish are removed from hooks or handled out of water. It can be heard above the water when the fish is submerged, and is easily heard in air over a considerable distance. It is produced spontaneously by captive sea robins and by free specimens. Dorsal fin erection usually accompanies the grunt.

The two species of triglids at Woods Hole are responsible also for another kind of sound—a staccato call of approximately 2.5 sec duration, which is heard from these species only during the breeding season[132], to judge from listening experiments conducted at one station from mid-June to late August of 1954 and 1955. (Both calling and fish were scarce in 1957[138].) The waters of the Woods Hole area where this staccato call has been heard are murky, and visibility to a human diver does not generally exceed 6 ft. in the summer. The production of a call to attract members of the opposite sex would seem a useful thing. But the calls of sea robins are probably the same in the two sexes. There is no obvious difference in their acoustical anatomy, nor are the intrinsic muscles of the air bladder of different sizes during the calling period. Stimulation of the sound production in physiological preparations elicits similar sounds in the two sexes. These considerations would suggest that the staccato call is a more generalized mechanism of communication than a breeding call. However, that it is a communicative device used in the breeding season seems to be the case in Great Harbor at Woods Hole[132]. If recordings of the staccato call are made and played back into the water, answers can be obtained from fishes in the water. Very crude imitations of the call and even explosive sounds will also elicit the staccato call. Other sounds played back into the water seem to suppress the calling. It is thus possible to exert some influence with man-made sounds over the acoustical behavior of these species, an exception to the general rule[131]. The staccato call has not been heard from sea robins under direct observation in captivity, and it is not known how movements of the fishes in nature may relate to it, but the incidence of its production reached its peak in Great Harbor during two seasons at the height of breeding activity[132]. A scarcity of triglids is accompanied by an absence of the staccato calling[138].

Tavolga's recent studies on the reproductive behavior of the goby, *Bathygobius soporator*, which makes its characteristic sounds in aquaria, furnishes another interesting case of a species apparently orienting to sound during the breeding season[10, 175, 176]. Grunts are produced by the male only during courtship behavior, their rate of production varying with the intensity of courtship behavior[175a]. When recordings are played back to an isolated female, the female darts about the tank momentarily, then becomes quiet. Continuing trials extinguish the response, which is not displayed by non-gravid females. Some of the darting movements are oriented toward a male in an adjoining flask, but the sight of the male does not alone elicit the response, although vision appears to play a role in the acoustical behavior of the male, and the female follows a flask containing a male during playing of the recorded sounds. In the absence of a male, the movements of the female are unoriented.

A possible predominance of sound production by the black drum during the breeding migration has already been mentioned. Similar predominance of sound production during the breeding season of other sciaenids[55, 108], of triglids[132], and of gobies[176] has been cited. The drumming muscles of the haddock, *Melanogramus angelifinis*[91], are larger in males than in females, and the difference is more apparent during the extensive winter spawning season than during the summer months[78]. (Observations of sound production by cod have suggested that some gadid sounds

References p. 687

are probably produced under threat or in fear[22].) Although seasonal variation in sound production recorded at sea might be explainable on other grounds than changes in behavior, for example that sound sources move out of the listening area, especially in relation to fixed hydrophone stations, many evidences indicate that natural sounds play a role in the reproductive behavior of several kinds of fishes.

There is some evidence that sound production and detection are useful to fishes as an orientation device as they are to some mammals[5,82,84,85,163,164] and birds[81]. Evidence already reviewed that the lateral line and skin receptors of fishes are sensitive to low frequency vibrations is suggestive that their sensitivity may provide information among schooling fishes which results in the spacing and coordination of members of the school. The possibility that blind fishes may orient by these mechanisms is suggested by evidence recently considered by Backus[14].

Griffin[80,83] has studied a series of paired sounds recorded in deep water below the level of light penetration about 170 miles north of Porto Rico by a ship of the Woods Hole Oceanographic Institution. These paired sounds, each similar to a drawn-out whistle, Griffin proposed after careful study, were the calls of a fish moving along below the recording vessel, each call followed by its bottom echo. Each echo, fainter than the corresponding call, followed the call by an interval consistent throughout the recording studied. On the basis of careful calculations, Griffin suggests that an unknown animal, presumably a fish, was swimming through dark water orienting to the echoes of its own call in maintaining a constant height above the bottom. The echoes were of sufficient intensity at the calculated depth of the fish, and of such a frequency range as to be audible to fishes in which hearing sensitivity has been studied[80].

Echoes have been detected from other sounds recorded at sea, and presumed to stem from fishes. The raps of the "carpenterfish" for example, recorded by the Woods Hole Oceanographic Institution on the edge of the Continental Shelf, and each pulse of the staccato call of Woods Hole sea robins, are followed by probable echo components.

The acoustical behavior of some other fishes is suggestive of echolocation, although none has been demonstrated. The Nassau grouper or hamlet, *Epinephelus striatus*, and the squirrelfish, *Holocentrus ascensionis*, are the predominant sound sources among fishes of the Bahama and Bermuda Banks. The acoustical behavior of these species is similar during the daytime[137], although the former is primarily a daylight feeder, the latter crepuscular or nocturnal[15,160]. During the summer of 1956, recordings of their calls were made along the eastern edge of the Great Bahama Bank in the vicinity of Bimini, from the Lerner Marine Laboratory of the American Museum of Natural History[137]. Although Bardach[15] found through tagging and recapture that this grouper moves rather freely over its home reefs in Bermuda, both of these species are usually observed near underwater crevices during the day when viewed in relatively shallow water through glass panels from a boat, or during underwater swimming. Upon approach of a hydrophone, the grouper emits vibrant, prolonged grunts and moves into hiding; the squirrelfish emits a volley of thump-like sounds before entering small crevices, usually higher in reefs and ledges than those of the grouper. Various considerations—clarity of the water, spontaneous calling of individuals from considerable distances, lower incidence of the calling at night, the obvious role of vision in this acoustical behavior, rough contours of the habitat—would seem to reject echolocation of the approaching hydrophone as a function of these sounds. Both species erect their spiny fins during sound production, and the calling seems to be an audible component of defensive behavior[137]. At Bermuda during the summer of 1958, squirrelfish on the edges of reefs often produced their volleyed thumps upon approach of other fishes to their crevices and while in pursuit of one another, both on the reefs, and in the Government Aquarium. Yet no obvious reactions of other fishes to these sounds have been observed.

A complicating factor in correlating behavior with sound production is the fact that many fishes with well-developed air bladders create some sound during any vigorous movement. This is true, for example, of some chaetodontids, serranids, holocentrids, and haemulids. But the sounds produced during such activity are briefer, less intense, and otherwise unlike sounds most characteristic of the species concerned. The black angelfish of the Bahamas, *Pomacanthus arcuatus*, a fish which nibbles at a suspended hydrophone, produces short, rather percussive grunts during darting at the rubber surface, or during feeding activity. But when encountering another of its species at sea, the same species emits considerably longer outbursts of sound of quite a different pattern[137]. Thus sound production appears to be an expression of communication in recognition behavior in this species which, like other species of *Pomacanthus* and species of *Angelichthys*, tends to occur in pairs or larger groups in the Bahamas and at Bermuda.

Perhaps the most frequently suggested function of the sounds of fishes is that of defence, offence or intimidation. The conditions under which many sounds are produced are suggestive of the importance of such a function. The sounds accompanied by spiny fin erection of triglids, serranids and holocentrids, tooth stridulation of carangids, balistids, ostraciids, diodontids, tetradontids, and chaetodontids, the fin drumming of balistids, the snaps of a sciaenid[48] and a pomacentrid[137] during protection of a hiding place, all are produced under circumstances suggestive of intimidative functions. The significance of these sounds to the survival of the species concerned is entirely unknown; as Burkenroad[32] has pointed out, it is hardly likely that they are adaptations to handling by man, yet it is only under such conditions that several of these sounds have been recorded. It is conceivable that the production of these sounds during attack by predators might so startle the attacking animal that it would abandon the chase. Evidence is lacking.

The most frequently described response of fishes to sound is that of quickened swimming[33,127,131,134], and this seems to be a generalized response to several sound sources. It is possible that this effect is of more importance than hitherto accorded it in the behavior of fishes, in determining their distributions, and in assuring other breeding phenomena. The rapidity with which this type of response is extinguished during repeated trials under experimental conditions obscures an understanding of its significance, if any, to the biology of fishes. Reactions of fishes to sounds they are likely to meet in nature are essentially limited to such responses as those of sea robins to recordings of their staccato call, the behavior of female gobies in relation to recorded calling of males, movements of clupeids away from recorded sounds of predators, the quickened swimming of a sciaenid, *Corvina nigra*[53], during stridulation by a spiny lobster, *Palinurus vulgaris*[52]. Reaction of the California spiny lobster, *Panuliris interruptus*, to a fish's approach may include stridulation[115].

Fishes are adapted to hear sounds and to produce sounds. Their calls are being

References p. 687

recorded and analyzed and libraries of the sounds of fishes, as of other animal sounds, are being rapidly accumulated. Much remains to be done in the study of the production by fishes of sound in their natural environment. With continuing observation, it is not unlikely that information may come to light which will allow a greater control over a most important food source than has been possible until now.

7. Addendum

Some publications dealing with the acoustical biology of fishes which have come to the author's attention since completion of the manuscript of this chapter are mentioned in this brief addendum. References are numbered consecutively with those previously listed.

The hearing of fishes has been furthur elucidated. The role of the air bladder in hearing of *Ameiurus* has been examined by Kleerekoper and Roggenkamp[210]; mutilation of the air bladder depressed sensitivity to all frequencies between 330 and 1840 c/s. Sensitivity increased between 210 and 330 c/s, and this increase was abolished by elimination of the lateral line system. Clark[202], through conditioning experiments, demonstrated sensitivity of lemon shark, *Negaprion brevirostris*, to ringing of a "submerged bell".

Contributions to the acoustical anatomy of fishes include publications of Bougis and Ruivo[200], Harrington[207], Robertson[217], and Walters[227]. Bougis and Ruivo describe muscles probably possessing a drumming function in a deep-sea brotulid, *Benthocometes robustus*, and Walters similar muscles combined with a specialized tendon in the Indo-Pacific veliferid, *Velifer hypselopterus*. Harrington presents a valuable study of the skull of the American cyprinid, *Notropis bifrenatus*, including a good description of the otic region; Robertson reviews the entire skeleton of *Brevoortia spp.* (Clupeidae).

Sound production by fishes has received considerable attention. Tavolga[223] has compared the sounds of two species of *Opsanus*, *O. tau* and *O. beta*, Gray and Winn[205] have correlated sound production with behavior of the male *O. tau*, and Fish and Mowbray[204] have identified a new species of *Opsanus* from the Bahamas, initially on the basis of its sound production. Tavolga[224] has further clarified the relation between sound production and reproduction in *Bathygobius soporator* and has suggested a mechanism of sound production. Stout and Winn[222] and Stout[221] have examined sound production in relation to reproduction in the freshwater satinfin shiner, *Notropis analostanus*. Tavolga[225] has described courtship sounds of a blenniid, *Chasmodes bosquianus* of the Gulf of Mexico. Hubbs and Miller[209] note sound production in an ariid catfish of American fresh waters. Shemanskiy[218] has commented on the sounds of fishes. Van Bergeijk[226] discusses ably the physical characteristics of a sea robin (Triglidae) call.

Klausewitz[211] has described the acoustical behavior of the cobitid *Botia hymenophysa* and describes the anatomy of sound production, and Benl[199] the acoustical behavior of *Botia* spp. and of the African knife-fish, *Xenomystus nigri* (Notopteridae). Moulton[213] has conducted a study of marine biological sound sources of the Bermuda area.

Swimming sounds stemming from movements of a carangid, *Trachurus trachurus*, in shoals of the Black Sea have been examined by Shishkova[219, 220], who describes also other marine biological sounds, by Hashimoto et al.[208] who in a very recent paper discuss detection of fish by sonabuoy, and by Moulton[215] who examined the relation between swimming sounds and schooling of some Bermuda fishes. Sounds which Azhazha[197] describes as stemming from sharks the translator, R. H. Backus[198], and myself concur more probably came from cetaceans.

Gunter and McChaughan[206] and Packard[216] refer to sound production in physiological studies of sound-producing fishes. Considerable data has been summarized by Fish[203] which deals with fish sound production, and Moulton[214] has prepared a reference list of use to students of underwater sound and of acoustical biology.

Busnel[201, 196] has discussed sound producing instruments used in African river fisheries and compares these with a Yugoslavian fishery using a similar method. Meschkat[212] has examined several acoustical fisheries, especially recently in the Amazon River valley. Moulton[215] has described, through information supplied by David Hammond, Fisheries Officer of Accra, a fish listening method used in a Ghana sea fishery.

The author wishes to thank Mrs. Robert H. Glover for her help in preparation of the main manuscript of this chapter.

REFERENCES

[1] Abbott, C. P., 1877, Traces of a voice is fishes, *Am. Nat.*, *11*, 147–156.
[2] Aflalo, F. G., 1905, *The Salt of My Life*, Sir Isaac Pitman and Sons Ltd., London.
[3] Alcock, A., 1902, *A Naturalist in Indian Seas*, John Murray, London.
[4] Allen, E. G., 1951, The history of American ornithology before Audubon, *Trans. Am. Phil. Soc.*, *41*, 387–591.
[5] Anderson, J. W., 1954, The production of ultrasonic sounds by laboratory rats and other mammals, *Science*, *119*, 808–809.
[6] Anon., 1944, *Fish and Shellfish of the south Atlantic and Gulf coasts*, Conservation Bull. No. 37, U. S. Dept. of the Interior.
[7] Anon., 1948, *Aquatic resources of the Ryukyu area*. Report No. 117, National Resources Section, General Headquarters, Supreme Commander for the Allied Powers, Tokyo. Reproduced as *U. S. Fish and Wildlife Serv.*, Leaflet No. 333.
[8] Anon., 1953, Electrical devices for controlling the movements of anadramous fishes, *Nature*, *171*, 591–592.
[9] Anon., 1953, Underwater listening experiments near schools of menhaden and little tuna, *Comm. Fisheries Review*, *15*, (11) 32.
[10] Anon., 1958, *Research Reviews*, ONR. Jan.
[11] Aristotle's *Historia Animalium*, *4*, lib. 55, 9, trans. by D'A. W. Thompson, Clarendon Press, Oxford, 1910.
[12] Aristotle's *De Anima*, in the version of William of Moerbeke and the commentary of St. Thomas Aquinas, trans. by K. Foster and S. Humphries, Yale Univ. Press, 1951.
[13] Autrum, H. and D. Poggendorf, 1951, Messung der absoluten Hörshwelle bei Fischen (*Amiurus nebulosus*), *Naturwissenschaften*, *38*, 434–435.
[14] Backus, R. H., 1958, Sound production in marine animals, *U. S. Navy J. Underwater Acoustics*, *8*, (2).
[15] Bardach, J. E., 1958, On the movements of certain Bermuda reef fishes, *Ecology*, *39*, 139–146.
[16] Benjamins, C. E., 1934, La fonction du Saccule, *Rev. laryngol. oto. rhinol.*, *55*, 1233–1242.
[17] Berg, L. S., 1947, *Classification of Fishes both Recent and Fossil*, J. W. Edwards, Ann Arbor, Michigan.
[18] Bertin, L., 1958, Organes sonores, in *Traité de Zoologie*, Masson et Cie, Paris, *13*, 1239–1247.
[19] Bigelow, H. B., 1904, The sense of hearing in the goldfish *Carassius auratus* L., *Am. Nat.*, *38*, 275–284.

[20] BIGELOW, H. B. and W. C. SCHROEDER, 1953, Fishes of the Gulf of Maine, *U. S. Fish and Wildlife Serv.*, Bull. No. 874.
[21] BOUTTEVILLE, K. F. VON, 1935, Untersuchungen über den Gehörsinn bei Characiniden und Gymnotiden und den Bau ihres Labyrinthes, *Z. vergl. Physiol.*, 22, 162–191.
[22] BRAWN, V., personal communication.
[23] BREDER, C. M., JR. and P. RASQUIN, 1943, Chemical sensory reactions in the Mexican blind characins, *Zoologica*, 28, 169–200.
[24] BREDER, C. M., JR. and E. CLARK, 1947, A contribution to the visceral anatomy, development, and relationships of the Plectognathi, *Bull. Am. Museum of Nat. Hist.*, 88, 291–319.
[25] BREDER, C. M., JR., personal communication.
[26] BRETT, J. R. and D. MACKINNON, 1954, Some aspects of olfactory perception in migrating adult coho and spring salmon, *J. Fish. Res. Bd. Canada*, 11, 310–318.
[27] BRIDGE, T. W. and A. C. HADDON, 1883, Contributions to the anatomy of fishes II. Air-bladder and Weberian ossicles in the siluroid fishes, *Phil. Trans. Roy. Soc. London*, 184, 65–333.
[28] BRIDGE, T. W. and A. C. HADDON, 1889, Contributions to the anatomy of fishes I. The Air-bladder and Weberian ossicles in the Siluridae. *Proc. Roy. Soc. London*, 46, 309–328.
[29] BRIDGE, T. W., 1910, Fishes, *Cambridge Natural History*, 7, Macmillan and Co.Ltd., London.
[30] BULL, H. O., 1928, Studies on conditioned responses in fishes, *I. J. Mar. Biol. Ass.*, 15, 485–533.
[31] BURKENROAD, M. D., 1930, Sound production in the Haemulidae, *Copeia*, (1) 17–18.
[32] BURKENROAD, M. D., 1931, Notes on the sound producing marine fishes of Louisiana, *Copeia*, (1) 20–28.
[33] BURNER, C. J. and H. L. MOORE, 1953, Attempts to guide small fish with underwater sound, *U. S. Fish and Wildlife Serv.*, Spec. Sci. Report, Fisheries No. 111.
[34] CARUS, C. G., 1827, *An Introduction to Comparative Anatomy of Animals*, trans. by R. T. GORE, Longman, Rees, Orme, Brown and Green, London.
[35] CASTAGNA, M., personal communication.
[36] CHIMITS, P., 1957, Tilapia in ancient Egypt, *FAO Fisheries Bull.*, 10, 211–215.
[37] COATES, C. W. and J. W. ATZ, 1954, *The Animal Kingdom*, Book 4, Fishes of the World, Doubleday and Co. Inc., Garden City, New York.
[38] COLMAN, J. S., personal communication.
[39] COUSTEAU, J. Y., 1953, *The Silent World*, Harper and Brothers, New York.
[40] CUNNINGHAM, J. J., 1910, On the marine fishes and invertebrates of St.Helena, *Proc. Zool. Soc. London*, 86.
[41] DARWIN, C., 1878, *Journal of Researches into the Natural History and Geology of the Countries visited during the Voyage of H. M. S. Beagle around the World, under the command of Capt. Fitzroy*, R. N. D. Appleton and Co., New York, new ed.
[42] DAVID, K., 1941, Lautäusserungen beim Gründling, *Zool. Anz.*, 134, 90–91.
[43] DAY, F., 1878, Remarks on Mr. Whittmer's paper on the manifestations of fear and anger in fishes, *Proc. Zool. Soc. London*, 214–221.
[44] DENKER, A., 1931, Über das Hörvermögen der Fische, *Oto-laryng.*, 15, 247–260.
[45] DIESSELHORST, G., 1938, Hörversuche on Fischen ohne Weberschen Apparat, *Z. vergl. Physiol.*, 25, 748–783.
[46] DIJKGRAAF, S., 1932, Über Lautäusserungen der Elritze, *Z. vergl. Physiol.*, 17, 802–805.
[47] DIJKGRAAF, S., 1941, Haben die Lautäusserungen der Elritze eine biologische Bedeutung, *Zool. Anz.*, 136, 103–106.
[48] DIJKGRAAF, S., 1947, Ein Töne erzeugender Fisch im Neapler Aquarium, *Experientia*, 3, 493.
[49] DIJKGRAAF, S., 1949, Untersuchungen über die Funktionen des Ohrlabyrinths bei Meerfischen, *Physiol. Comp. et Oecol.*, 2, 81–106.
[50] DIJKGRAAF, S., 1952, Bau und Funktionen der Seitenorgane und des Ohrlabyrinths bei Fischen, *Experientia*, 8, 205–216.
[51] DIJKGRAAF, S., 1952, Über die Schallwahrnehmung bei Meeresfischen, *Z. vergl. Physiol.*, 34, 104–122.
[52] DIJKGRAAF, S., 1955, Lauterzeugung und Schallwahrnehmung bei der Languste (*Palinurus vulgaris*), *Experientia*, 11, 330–331.
[53] DIJKGRAAF, S., personal communication.
[54] DIOLE, P., 1953, *The Undersea Adventure*, Julian Messner Inc. Cited in SPECTORCKY, A. C., *1 he Book of the Sea*, Appleton–Century–Crofts Inc., New York, 1954, pp. 244–246.
[55] DOBRIN, M. B., 1947, Measurements of underwater noise produced by marine life, *Science*, 105, 19–23.
[56] DOUGLAS, N., 1927, *Birds and Beasts of the Greek Anthology*, privately printed, Florence, Italy.
[57] DOW, W., 1954, Underwater telemetering, A telemetering depth meter, *Deep-Sea Res.*, 2, 145–151.
[58] DUFOSSÉ, A., 1858–1862, Des différents phenomenes physiologiques nommés voix des poissons, *Compt. rend. Acad. Sci.*, Paris, 46, 352–356; 47, 916; 54, 393–395.

59 Dufossé, A., 1866, De l'ichthyopsophie ou des différents phenomenes physiologiques nommés "voix des poissons", *Compt. rend. Acad. Sci.*, Paris, *62*, 978–980.
60 Dufossé, A., 1874, Recherches sur les bruits et les sons expressifs que font entendu les poissons d'Europe, *Ann. des Sci. Naturales*, 5ème ses., *19*, Art. 3 and 5.
61 Edinger, L., 1908, Ueber das Hören der Fische und anderer mederer Vertebraten, *Zentr. Physiol.*, *22*, 1–4.
62 Evans, H. M., 1940, *Brain and Body of Fish*, Blakiston, Philadelphia.
63 Farkas, B., 1935, Untersuchungen über das Hörvermögen bei Fischen, *Allattani Közlem.*, *32*, 1–20.
64 Farkas, B., 1936, Zur Kenntnis des Hörvermögens und des Gehörorgans der Fische, *Acta Oto-laryng.*, *23*, 499–532.
65 Fish, M. P., 1948, *Sonic fishes of the Pacific*, Proj. NR 083-003, Contr. N6 ori-195, t.o.i. between ONR and Woods Hole Oceanographic Institution, Techn. Report No. 2.
66 Fish, M. P., A. S. Kelsey Jr. and W. H. Mowbray, 1952, Studies on the production of underwater sound by North Atlantic coastal fishes, *J. Mar. Res.*, *11*, (2) 180–193.
67 Fish, M. P., 1953, The production of underwater sound by the northern seahorse, *Hippocampus hudsonius*, *Copeia*, 98–99.
68 Fish, M. P., 1954, The character and significance of sound production among fishes of the western North Atlantic, *Bull. Bingham Oceanogr. Coll.*, *14*, Art. 3, 1–109.
69 Frisch, K. von and H. Stetter, 1932, Untersuchungen über den Sitz des Gehörsinnes bei der Elritze, *Z. vergl. Physiol.*, *17*, 686–801.
70 Frisch, K. von, 1936, Ueber den Gehörsinn der Fische, *Biol. Rev.*, *11*, 210–246.
71 Frisch, K. von, 1938, The sense of hearing in fish, *Nature*, *141*, 8–11.
72 Froloff, J. P., 1925, Bedingte Reflexe bei Fischen I, *Pflüger's Archiv.*, *208*, 261–271.
73 Froloff, J. P., 1928, Bedingte Reflexe bei Fischen II, *Pflüger's Archiv.*, *220*, 339–349.
74 Galton, J. C., 1874, The song of fishes, *Popular Sci. Rev.*, *13*, 337–349.
75 Gayarre, C., 1851, *Louisiana, its Colonial History and Romance*, Harper and Brothers, New York.
76 Goode, G. B., 1884, The fisheries and fishery of the United States, Section I, *Natural History of useful aquatic animals*, U. S. Gov't Printing Off., Washington, D. C.
77 Goode, G. B., 1888, *American Fishes*, Standard Book Co., New York.
78 Graham, H. W., personal communication.
79 Gregory, W. K., 1933, Fish skulls: A study of the evolution of natural mechanisms, *Trans. Am. Phil. Soc.*, *23*, 75–481.
80 Griffin, D. R., 1950, *Underwater sounds and the orientation of marine animals*, a preliminary survey, Proj. NR 162-429, t.o. 9, between ONR and Cornell University, Techn. Report No. 3.
81 Griffin, D. R., 1953, Acoustic orientation in the oil bird, *Steatornis*, *Proc. Natl. Acad. Sci.*, *39*, 884–893.
82 Griffin, D. R., 1953, Bat sounds under natural conditions, with evidence for echolocation of insect prey, *J. Exp. Zool.*, *123*, 435–465.
83 Griffin, D. R., 1955, Hearing and acoustic orientation in marine animals, *Papers Mar. Biol. and Oceanogr.*, Deep-Sea Research, suppl. to Vol .3, pp. 406–417.
84 Griffin, D. R., 1958, *Listening in the Dark*, Yale Univ. Press, New Haven, Chap. 10, 13.
85 Grinnell, A. D. and D. R. Griffin, 1958, The sensitivity of echolocation in bats, *Biol. Bull.*, *114*, 10–22.
86 Gudger, E. W., 1908, Habits and life history of the toadfish (*Opsanus tau*), *Bull. U. S. Bur. Fisheries*, *28*, Part 2, 1071–1109.
87 Gudger, E. W., 1927, Teleostean fishes of Tortugas, *Pap. Lab. Carn. Inst. Wash.*, 26.
88 Gunther, A. C. L. G., 1880, *An Introduction to the Study of Fishes*, Adam and Charles Black, Edinburgh.
89 Haddon, A. C., 1881, On the stridulating apparatus of Callomystax gagata, *J. Anat. Physiol.*, *15*, 322.
90 Hafen, G., 1935, Zur Physiologie der Dressurversuche, *Z. vergl. Physiol.*, *22*, 192–220.
91 Hagman, N., 1921, *Studien über die Schwimmblase beniger Gadiden und Macruriden*, Doctoral dissertation, Univ. of Lund. Carl Blum, Lund.
92 Hakim, H. and J. R. Pattinasarany, personal communication.
93 Hase, F., 1941, Lautäusserungen bei einheimischen Fischen, *Zool. Anz.*, *135*, 176.
94 Hasler, A. D., 1956, Perception of pathways by fishes, *Quart. Rev. Biol.*, *31*, 200–209.
95 Hersey, J. B., 1957, Electronics in Oceanography, *Advances in Electronics and Electron Physics*, *9*, 239–295. Academic Press Inc., New York.
96 Hodgson, W. C., 1957, *The herring and its fishery*, Routledge and Kegan Paul, London, Chap. 5, Science and fishing.
97 Holder, C. F. and D. S. Jordan, 1909, *Fish Stories*, Henry Holt and Co., New York.

[98] Home, E., 1823, *Lectures on Comparative Anatomy*, Vol. 3, Longman, Hurst, Rees, Orme, and Brown, London.
[99] Houston, R. B., Jr., 1949, German commercial electrical fishing device, *U. S. Fish and Wildlife Serv.*, Fishery Leaflet No. 348, 1–4, with supplement.
[100] Huxley, T. H., 1881, The herring, *Nature*, 23, 607.
[101] Jahn, T. L. and V. J. Wulff, 1950, Phonoreception, Chapter 13 of *Comparative Animal Physiology*, ed. by C. L. Prosser, W. B. Saunders Co., Philadelphia.
[102] Jones, F. R. H. and N. B. Marshall, 1953, The structure and functions of the teleostean swimbladder, *Biol. Rev.*, 28, 16–83.
[103] Jones, T. R., 1855, *General Outline of the Organisation of the Animal Kingdom and Manual of Comparative Anatomy*, John van Voorst, London, 2nd ed.
[104] Jordan, D. S. and B. W. Evermann, 1902, *American Food and Game Fishes*, Doubleday, Page and Co., New York.
[105] Kesteven, G. L., 1949, *Malayan Fisheries*, Malaya Publishing House Ltd., Singapore.
[106] Kleerekoper, H. and E. C. Chagnon, 1954, Hearing in fish, with special reference to *Semotilus atromaculatus atromaculatus* (Mitchill), *J. Fish. Res. Bd. Can.*, 11, 130–152.
[107] Kleerekoper, H., 1955, Threshold curves for hearing in four species of Cyprinidae, *Anat. Rec.*, 122, 460.
[108] Knudsen, V. O., R. S. Alford and J. W. Emling, 1948, Underwater ambient noise, *J. Mar. Res.*, 7, 410–429.
[109] Kreidl, A., 1895, Ueber Die Perception der Schallwellen bei den Fischen, *Arch. f. ges. Physiol.*, 61, 450–464.
[110] Kreutzig, K., personal communication.
[111] Krumholz, L. A., 1943, A comparative study of the Weberian ossicles in North American ostariophysine fishes, *Copeia*, 33–40.
[112] Lawson, J., 1714, *The History of Carolina*, W. Taylor and T. Baker, London.
[113] Lefroy, J. H., 1876, *Memorials of the discovery and early settlement of the Bermuda or Somers Islands*, Bermuda Gov't Library, reprinted Jan. 1932.
[114] Leim, A. H., 1952, Do echo sounders frighten fish? *Canadian Fisherman*, June, 17–18.
[115] Lindberg, R. G., 1955, Growth, population dynamics, and field behavior in the spiny lobster, *Panuluris interruptus* (Randall), *Univ. Calif. Publ. Zool.*, 59, 157–247.
[116] Loukashkin, A. and N. Grant, 1954, Further studies of the behavior of the Pacific sardine (*Sardinops caerulea*) in an electrical field, *Proc. Calif. Acad. Sci.*, Ser. 4, 28, 323–337.
[117] Lowenstein, O. and T. D. M. Roberts, 1949, The equilibrium function of the otolith organs of the thornback ray (*Raja clavata*), *J. Physiol.*, 110, 392–415.
[118] Lowenstein, O. and T. D. M. Roberts, 1951, The localization and analysis of the responses to vibration from the isolated elasmobranch labyrinth. A contribution to the problem of the evolution of hearing in vertebrates, *J. Physiol.*, 114, 471–489.
[119] Lowenstein, O., 1957, The Sense Organs: The Acousticolateralis System, Chap. 2, Part 2, of *The Physiology of Fishes*, Vol. 2, ed. by M. E. Brown, Academic Press., New York.
[119a] Loye, D. P. and D. A. Proudfoot, 1946, Underwater noise due to marine life, *J. Acous. Soc. Am.*, 18, 446–449.
[120] McDonald, H. E., 1921, Ability of *Pimephales notatus* to form associations with sound vibrations, *J. Comp. Psychol.*, 2, 191–193.
[121] McLain, A. L. and W. L. Nielsen, 1953, Directing the movement of fish with electricity, *U. S. Fish and Wildlife Serv.*, Spec. Sci. Report, Fisheries No. 93, 1–24.
[122] Maier, H. N. and L. Scheuring, 1923, Entwicklung der Schwimmblase und ihre Beziehungen zum statischen Organ und der Kloake bei Clupeiden, *Wiss. Meeresuntersuch.* (Abt. Helgoland), 15, 1–22.
[123] Manning, F. B., 1924, Hearing in the goldfish in relation to the structure of its ear, *J. Exp. Zool.*, 41, 5–20.
[124] Marshall, N. B., 1954, *Aspects of Deep Sea Biology*, Philosophical Library, New York.
[125] Mauiri, A. D., 1958, Pompeii, *Scientific Am.*, 198, (4) 68–78.
[126] Mobius, K. A., 1889, *Balistes aculeatus*, ein trommelender Fisch, *Sitz. Ber. preuss. Akad. Wiss.*, 999.
[127] Moore, H. L. and H. W. Newman, 1956, Effects of sound waves on young salmon, *U. S. Fish and Wildlife Serv.*, Spec. Sci. Report, Fisheries No. 172.
[128] Moorhouse, V. H. K., 1933, Reactions of fish to noise, contr. to *Can. Biol. and Fish.*, N. S. 7, 465–475.
[129] Moseley, H. N., 1879, *Notes by a naturalist on the "Challenger", being an account of various observations made during the voyage of H. M. S. "Challenger" round the world 1872–1876*, Macmillan and Co., London.
[130] Moulton, J. M., 1955, *Report to the Director of the Woods Hole Oceanographic Institution*, unpublished.

[131] MOULTON, J. M. and R. H. BACKUS, 1955, Annotated references concerning the effects of man-made sounds on the movements of fishes, *Maine Dept. of Sea and Shore Fisheries*, Fisheries Circ. No. 17, 1–7.
[132] MOULTON, J. M., 1956, Influencing the calling of sea robins (*Prionotus* spp.) with sound, *Biol. Bull.*, *111*, 393–398.
[133] MOULTON, J. M., 1956, Fishes and sound in the sea, *Bowdoin Alumnus*, *30*, (2) 5–7.
[134] MOULTON, J. M., 1956, The movements of menhaden and butterfish in a sound field, *Anat. Rec.*, *125*, 592.
[135] MOULTON, J. M., 1957, *Report to the Director of the Woods Hole Oceanographic Institution*, unpublished.
[136] MOULTON, J. M., 1957, Sound production in the spiny lobster *Panulirus argus* (Latreille), *Biol. Bull.*, *113*, 286–295.
[137] MOULTON, J. M., 1958, The acoustical behavior of some fishes in the Bimini area, *Biol. Bull.*, *114*, 357–374.
[138] MOULTON, J. M., 1958, A summer silence of sea robins, *Copeia*, 234–235.
[139] MOULTON, J. M., 1958, *Report to the Director of the Woods Hole Oceanographic Institution*, unpublished.
[140] MURRAY, J., 1831, Herring making sounds, *Mag. Nat. Hist.*, 148.
[141] NEAL, H. V. and H. W. RAND, 1936, *Comparative Anatomy*, P. Blakiston's Son and Co. Inc., Philadelphia.
[142] NELSON, E. M., 1955, The morphology of the swimbladder and auditory bulla in the Holocentridae, *Fieldiana, Zoology*, *37*, 121–130.
[143] NORMAN, J. R., 1951, *A History of Fishes*, Ernest Benn Ltd., London, 4th ed.
[144] O'CONNELL, C. P., 1955, The gas bladder and its relation to the inner ear in *Sardinops caerulea* and *Engraulis mordax*, *U. S. Fish and Wildlife Serv.*, Fishery Bull. No. 104.
[145] OKADA, Y., 1955, *Fishes of Japan*, Maruzin Co. Ltd., Tokyo.
[146] PARKER, G. H., 1903, Hearing and allied senses in fishes, *Bull. U. S. Bur. Fish. Comm.*, *22*, 45–64.
[147] PARKER, G. H., 1903, The sense of hearing in fishes, *Am. Nat.*, *37*, 185–204.
[148] PARKER, G. H., 1909, The sense of hearing in the dogfish, *Science*, *29*, 428.
[149] PARKER, G. H., 1910, The function of the ear in cyclostomes, *Science*, *31*, 470.
[150] PARKER, G. H., 1910, The structure and function of the ear of the squeteague, *Bull. U. S. Bur. Fish.*, *28*, 1211–1224.
[151] PARKER, G. H., 1911, Influence of the eyes, ears, and other allied sense organs on the movements of the dogfish, *Mustelus canis* (Mitchell), *Bull. U. S. Bur. Fish.*, *29*, 43–57.
[152] PARKER, G. H., 1911, Effects of explosive sounds, such as those produced by motor boats and guns, upon fishes, *Bur. Fish. Doc.*, No. 752, 1–9.
[153] PARKER, G. H., 1912, Sound as a directive influence in the movements of fishes, *Bull. U. S. Bur. Fish.*, *30*, 97–104.
[154] PARKER, G. H., 1918, A critical survey of the sense of hearing in fishes, *Proc. Am. Phil. Soc.*, *57*, 69–98.
[155] PARRY, M. L., 1954, The fishing methods of Kelantan and Trengganu, in Malayan Fishing Methods, by T. W. BURDEN and M. L. PARRY, *J. Malayan Br. Royal Asiatic Soc.*, *27*, Part 2.
[156] PENN, W., 1685, A further account of the Province of Pennsylvania, in *Narratives of Early Pennsylvania, West New Jersey and Delaware*, ed. by ALBERT COOK, Charles Scribner's Sons, New York, 1912, 259–278.
[157] PIPER, H., 1906, Aktionsströme vom Gehörorgane der Fische bei Schallreizung, *Zentr. Physiol.*, *20*, 293–297.
[158] PLINY's *Naturalis Historia*, *3*, Lib. 9, Cap. 8, trans. by H. RACKHAM, Harvard U. Press, Cambridge, 1940.
[159] POGGENDORF, D., 1952, Die absoluten Hörschwellen des Zwergwelses (*Amiurus nebulosus*) und Beiträge zur Physik des Weberschen Apparates der Ostariophysen, *Z. f. vergl. Physiol.*, *34*, 222–257.
[160] RAY, C. and E. CIAMPI, 1956, *The underwater guide to marine life*, A. S. Barnes and Co., New York.
[161] ROMER, A. S., 1955, *The Vertebrate Body*, W. B. Saunders Co., Philadelphia, 2nd ed.
[162] SAND, R. F., 1957, Application of underwater television to fishing gear research, *Trans. Am. Fish. Soc.*, (1956) 158–160.
[163] SCHEVILL, W. E., 1956, Evidence for echolocation by cetaceans, *Deep-Sea Res.*, *3*, 153–154.
[164] SCHEVILL, W. E. and B. LAWRENCE, 1956, Food finding by a captive porpoise *(Tursiops truncatus)*, *Breviora*, *53*, 1–15.
[165] SCHNEIDER, H., 1941, Die Bedeutung der Atemhöhle der Labyrinth fische für ihr Hörvermögen, *Z. vergl. Physiol.*, *29*, 172–194.
[166] SCHMITT, W., 1931, Crustaceans, *Smithsonian Scientific Series*, *10*, Part 2, 192–199.
[167] SCHULTZ, L. P. and E. M. STERN, 1948, *The ways of fishes*, D. VanNostrand Co. Inc., New York.

[168] Smith, H. M., 1905, The drumming of the drum-fishes (Sciaenidae), Science, 22, 376–378.
[169] Smith, H. M., 1907, Fishes of North Carolina.
[170] Sorenson, W., 1884, Om lydorganer hosfiske, en physiologisk og comparativ-anatomisk undersoguse, Copenhagen.
[171] Starks, E. C., 1908, On a communication between the air-bladder and the ear in certain spiny-rayed fishes, Science, 28, 613–614.
[172] Stephens, F. H., Jr. and F. J. Shea, 1956, Underwater telemeter for depth and temperature, U. S. Fish and Wildlife Serv., Spec. Sci. Report, Fisheries No. 181, 1–22.
[173] Stetter, H., 1929, Untersuchungen über den Gehörsinn der Fische, besonders von Phoxinus laevis L. und Ameiurus nebulosus Raf., Z. vergl. Physiol., 9, 339–447.
[174] Stipetic, E., 1939, Ueber das Gehörorgan der Mormyriben, Z. vergl. Physiol., 26, 740–752.
[175] Tavolga, W. N. and R. W. Neelands, 1954, Life History of a goby, Bathygobius soporator. AIBS Bull., 4 (4) 28, (title only).
[175a] Tavolga, W. N., 1954, Reproductive Behavior of the gobiid fish, Bathygobius soporator, Bull. Am. Museum of Nat. Hist., 104, Art. 5.
[176] Tavolga, W. N., 1956, Visual, chemical and sound stimuli as cues in the sex discrimination behavior of the gobiid fish, Bathygobius soporator, Zoologica, 41, Part 2, 49–64.
[177] Tenney, S., 1865, A Manual of Zoology, Charles Scribner and Co., New York.
[178] Thompson, D'A. W., 1947, A Glossary of Greek Fishes, Oxford University Press, London, p. 291.
[179] Tower, R. W., 1908, The production of sound in the drum-fishes, the sea-robin, and the toadfish, Ann. N. Y. Acad. Sci., 18 (5) Part 2, 149–180.
[180] Tracy, H. C., 1920, The clupeoid cranium in its relation to the swimbladder diverticulum and the membranous labyrinth, J. Morph., 33, 439–483.
[181] Tracy, H. C., 1920, The membranous labyrinth and its relation to the precoelomic diverticulum in clupeoids, J. Comp. Neurol., 31, 219–257.
[182] Trefethen, P. S., 1956, Sonic equipment for tracking individual fish, U. S. Fish and Wildlife Serv., Spec. Sci. Report, Fisheries No. 179, 1–11.
[183] Trefethen, P. S., J. W. Dudley and M. R. Smith, 1957, Ultrasonic tracer follows tagged fish, Electronics, April.
[183a] Turner, H. A., personal communication.
[184] Vilstrup, T., 1951, Structure and Function of the Membranous Sacs of the Labyrinth, Ejnar Munksgaard, Copenhagen.
[185] Weber, E. H., 1820, De aure et auditu hominis et animalium. Pars I, De aure animalium aquatilium, Lipsiae.
[186] Werner, L. H., 1932, The sensitivity of fishes to sound and to other mechanical stimulation, Quart. Rev. Biol., 7, 326–339.
[187] Westenberg, J., 1953, Acoustical aspects of some Indonesian fisheries, J. du Conseil pour l'exploration de la Mer, 18, 311–325.
[188] Westerfield, F., 1921, The ability of mud-minnows to form associations with sounds, J. Comp. Psychol., 2, 187–190.
[189] Whiteleather, R. T., 1957, Some interesting aspects of fisheries exploration, 1956, Trans. Am. Fish. Soc., (1956) 195–198.
[190] Wohlfahrt, T. A., 1936, Das Ohrlabyrinth der Sardine (Clupea pilchardus Walb.) und seine Beziehungen zur Schwimmblase und Seitenlinie, Z. Morphol. und Ökol. d. Tiere, 31, 371–410.
[191] Wohlfahrt, T. A., 1939, Untersuchungen über das Tonunterscheidungsvermögen der Elritze (Phoxinus laevis Agass.), Z. vergl. Physiol., 26, 570–604.
[192] Woods Hole Oceanographic Institution, 1953, Revised Collection of Sea Noises.
[193] Yonge, C. D., 1854, Translation of The Deipnosophists or Banquet of the Learned of Athenaeus, Vol. 2, Henry G. Bohn, London.
[194] Young, J. Z., 1950, The Life of Vertebrates, Clarendon Press, Oxford.
[195] Zenneck, J., 1903, Reagieren die Fische auf Tone? Pflüger's Arch. ges. Physiol., 95, 346–356.
[196] Anon., 1959, Appeaux de pêche acoustique utilisés au Sénégal et au Niger, La Nature, 87, 486–487.
[197] Azhazha, V. G., 1958, Sharks receive and emit ultrasounds (translated by R. H. Backus), Rybnoye Khoziaistvo, 34, (3) 30–32.
[198] Backus, R. H., personal communication.
[199] Benl, G., 1959, Lautäusserungen beim Afrikanischen Messerfish und bei Doticn, Die Aquar.- und Terrar.-Zeitschr. (Datz), 12, (4) 108–111.
[200] Bougis, P., and M. Ruivo, 1954, Recherches sur le poisson de profondeur Benthocometes robustus (Goode et Bean), Vie et Milieu, (suppl. 3) 155–209.
[201] Busnel, R. G., 1959, Étude d'un appeau acoustique pour la pêche, utilisé au Sénégal et au Niger, Bull. de l'I.F.A.N., 21, sér. A, (1) 346–360.
[202] Clark, E. C., 1959, Instrumental conditioning of lemon sharks, Science, 130, 217–218.

203 FISH, M. P., 1958, An outline of sounds produced by fishes in Atlantic coastal waters, sound measurements and ecological notes, *ONR Contr.*, *Nonr-396(02)*, *NR 083-054*, *Techn. Report No. 12*.
204 FISH, M. P., and W. H. MOWBRAY, 1959, The production of underwater sound by *Opsanus* sp., a new toadfish from Bimini, Bahamas, *Zoologica*, *44*, Part 2, 71–76.
205 GRAY, G-A., and H. E. WINN, 1958, The sound production and behavior of the guarding male toadfish (*Opsanus tau*), *Anat. Rec.*, *132*, 446.
206 GUNTER, G., and D. MCCAUGHAN, 1959, Catalepsy in two common marine animals, *Science*, *130*, 1194–1195.
207 HARRINGTON, R. W., JR., 1955, The osteocranium of the American cyprinid fish, *Notropis bifrenatus*, with an annotated synonymy of teleost skull bones, *Copeia*, (4) 267–290.
208 HASHIMOTO, T., M. NISHIMURA and Y. MANIWA, 1960, Detection of fish by sonabuoy, *Bull. Jap. Soc. Sci. Fish.*, *26*, 245–249.
209 HUBBS, C. L. and R. R. MILLER, 1960, *Potamarius*, a new genus of ariid catfishes from the fresh waters of Middle America, *Copeia*, (2) 101–112.
210 KLEEREKOPER, H., and P. A. ROGGENKAMP, 1959, An experimental study on the effect of the swimbladder on hearing sensitivity in *Ameiurus nebulosus nebulosus* (LeSueur), *Can. J. Zool.*, *37*, 1–8.
211 KLAUSEWITZ, W., 1958, Lauterzeugung als abwehrwaffe bei der hinterindischen Tigerschmerle (*Botia hymenophysa*), *Natur und Volk*, *88*, 343–349.
212 MESCHKAT, A., personal communication.
213 MOULTON, J. M., 1959, Sound production by some marine animals at Bermuda, *Anat. Rec.*, *134*, 612.
214 MOULTON, J. M., 1960, References dealing with animal acoustics, particularly of marine forms. Mimeographed.
215 MOULTON, J. M., 1960, Swimming sounds and the schooling of fishes, *Biol. Bull.*, *119*, 210–223.
216 PACKARD, A., 1960, Electrophysiological observations on a sound-producing fish, *Nature*, *186*, (July 2) 63.
217 ROBERTSON, W. V., 1959, The osteology of *Brevoortia patronus* and *Brevoortia gunteri* (Pisces; Clupeidae), *Thesis*, Texas A. and M. College, unpublished.
218 SHEMANSKIY, Y. A., 1958, The "language" of fish (in Russian), *Nauka i Zhizn*, (8) 42.
219 SHISHKOVA, E. V., 1958, Concerning the reactions of fish to sounds and the spectrum of trawler noise (translated by J. M. MOULTON), *Rybnoye Khoziaistvo*, *34*, (3) 33–39.
220 SHISHKOVA, E. V., 1958, Notes on research on sounds made by fish (translated by the Bell Telephone Laboratories, Inc.), *Ryb. Khoziai. i Okeano.* (*VNIRO*), *Trudy 36*, 280–294.
221 STOUT, J. F., 1959, The reproductive behavior and sound production of the satinfin shiner, *Anat. Rec.*, *130*, 643–644.
222 STOUT, J. F., and H. E. WINN, 1958, The reproductive behavior and sound production of the satinfin shiner, *Anat. Rec.*, *132*, 511.
223 TAVOLGA, W. N., 1958, Underwater sounds produced by two species of toadfish, *Opsanus tau* and *Opsanus beta*, *Bull. Mar. Sci. of the Gulf and Carribean.*, *8*, 278–284.
224 TAVOLGA, W. N., 1958, The significance of underwater sounds produced by males of the gobiid fish, *Bathygobius soporator*, *Physiol. Zool.*, *31*, 259–271.
225 TAVOLGA, W. N., 1958, Underwater sounds produced by males of the blenniid fish, *Chasmodes bosquianus*, *Ecology*, *39*, 759–760.
226 VAN BERGEIJK, W. A., 1960, *Waves and the Ear*, Doubleday and Co. Inc., Garden City, N.Y.
227 WALTERS, V., 1960, The swimbladder of *Velifer hypselopterus*, *Copeia*, (2) 144–145.

CHAPTER 23

ACOUSTIC BEHAVIOUR OF AMPHIBIA

by

W. F. BLAIR

1. Introduction

Vocalization has evolved to a high degree in the acaudate amphibians, where it serves primarily to aggregate the sexes for breeding and in species identification. Knowledge of the vocalizations in this group has increased greatly since about 1953. The perfection of portable magnetic tape recorders has permitted the "capture" of sounds in the field, and the sound spectrograph and cathode ray oscilloscope have permitted objective measurement and comparison of these sounds. Most of the following discussion will concern vocalizations in the acaudate amphibians as they have been studied in reference to problems of vertebrate speciation*.

In the caudate amphibians, vocalization is of relatively little significance, and we will treat this group briefly before going on to a more detailed discussion of the Salientia.

2. The Caudata

Vocalization in the caudate amphibians has been recently reviewed[43] and the following discussion is based largely on this review. Various salamanders produce sounds more or less incidentical to the breathing process. The simplest type of sound production is found in the families Salamandridae and Ambystomidae (*Salamandra salamandra, Triturus alpestris, Taricha torosa, Taricha similans, Ambystoma tigrinum*). In these a simple clicking, clucking or kissing sound is produced by opening the mouth and thus allowing incoming air to break the moist seal of the adpressed lips.

Another salamander, *Amphiuma means*, is said[3] to produce a whistling sound by forcing air through the branchial fissures. The membranous flaps of the branchial fissure act as vibratory cords to produce this sound[2], which is most often uttered when the animals are disturbed.

Some of the lungless salamanders (family Plethodontidae) including *Desmognathus*, *Aneides lugubris* and *Eurycea bislineata* have been reported to utter a squeaking sound, presumably by use of a vibrating valve through which air may be forced under pressure, although the method of sound production in these salamanders has never been established.

Vocal cords are present in the larynx of the ambystomid *Dicamptodon ensatus*. The giant salamander, *Megalobatrachus japonicus* and another ambystomid, *Ambystoma maculatum*, are also suspected of using this mechanism to produce sounds.

---

* This work has centered at the University of Texas and the University of Western Australia, and additional work has developed at the American Museum of Natural History in New York City.

Many of the sounds reported in the caudate amphibians are apparently incidental to the breathing process and without significance. Some are thought to have survival value, however, and in some species (*Aneides lugubris, Dicamptodon ensatus*) acts of defensive or escape behavior accompany the sound production. It has been suggested, but not proved, that the sounds have sexual significance in some species (*Eurycea bislineata, Ambystoma maculatum, Salamandra salamandra*).

3. The Acaudata

The voice is highly developed in the males of most species of salientian amphibians in relation to its functions of attracting a female and of species identification. The voice has been secondarily lost or reduced in magnitude, however, in various evolutionary lines. Among North American salientians, this has occurred in species of the genera *Rana* and *Bufo* of the western part of the continent, where there are few species per genus and where the identification function of voice consequently is of relatively little significance[17]. Weak voices have been attributed to the females of several species of Salientia, and the female of the midwife toad, *Alytes obstetricans*, has been reported as being louder than that of the male[47]. The females of most species appear to be voiceless, and confirmation of the voice in even those species for which it has been reported would be desirable. An exception to this is found in the leptodactylid genus *Tomodactylus*, in which the eggs are layed on land and undergo direct development, and in which there is consequently no aggregation into breeding swarms. The voices of the males and females are reported to be distinguishable, and males have been seen orienting on calling females and moving to them after the females had responded to their own calls by calling themselves[27].

(a) Emission of sounds

(i) Morphology of organs and mechanisms of function

The vocal structures may be classified as primary or secondary. The tissues lining the laryngeal cartilages, which may be set in motion by expiration of air from the lungs and the trunk musculature by which air is forced from the lungs into the mouth, comprise the primary vocal structures[30]. The mouth, openings from the mouth into the vocal sac, vocal sac, and associated gular muscles comprise the secondary vocal structures[40]. The secondary vocal structures are not essential to the production of sound but serve to modify the sounds produced by vibrations of the laryngeal tissues.

The mating call is produced in a closed system, since the mouth and nostrils are closed while the animal is calling. Generally speaking, air forced from the lungs by contraction of the trunk muscles sets the vocal cords in motion as it passes through the larynx. The character of the sound emitted is dependent on the nature of the vocal cord vibrations and to a large extent on the acoustical properties of the secondary vocal structures. Beyond this general statement, very little is known about the mechanics of sound production in the Salientia. The only exception is the pelobatid genus *Scaphiopus*, which has recently been the subject of a comparative study of the primary and secondary vocal structures and the mechanics of sound production[40]. It may be surmised that many of the features of sound production in *Scaphiopus* characterize the Salientia generally, but this remains to be determined.

References p. 707

The larynx of *Scaphiopus* is described by McAlister[40] as follows: "a bean shaped structure set between the pharynx and lungs with the convex portion facing forward and opening directly into the pharynx through the slit-like glottis. The glottis is flanged by a pair of fleshy, muscular lips and the whole larynx is bound with muscle." The laryngeal structures show interspecific variation in the shape of the depression in the laryngeal wall anterior to the vocal cords and in the structure of the latter. In *Scaphiopus hammondi* and *S. bombifrons* of the subgenus *Spea*, the cavity is relatively deep and steep walled. In *S. holbrooki* of the subgenus *Scaphiopus* the cavity is quite shallow and broader. In *S. couchi* (subgenus *Scaphiopus*) and *S. multiplicatus* (subgenus *Spea*) the condition is intermediate. The vocal cords are a pair of thickened, rather muscular membranes stretching from the walls of the larynx and abutting in the center of the laryngeal cavity. The thin, free edge of each cord is recurved downward (posteriorly) and is often curled. The cords are relatively thick and fleshy in the subgenus *Spea*; they are less bulky in *S. couchi*. An accessory fold anterior to the vocal cords is present in *S. holbrooki*.

The vocal sacs are the most important of the secondary vocal structures. No sacs are found in the primitive families Ascaphidae and Pipidae, and in the Discoglossidae poorly developed sacs are found in only *Bombina bombina*[39]. In the more advanced families, the vocal sacs are sometimes lost in connection with loss of the mating call in various species. In the Colorado River Toad *(Bufo alvarius)* of southwestern North America, the vocal sacs are vestigial and the call is weak and probably of little significance in attracting the female[21, 31]. A Philippine rhacophorid, *Rhacophorus leucomystax*, shows geographic variation in the presence or absence of vocal sacs[29].

The vocal sacs are diverticulations from the lining of the oral cavity, with which they remain connected by one or two slits. They are inflated by air passing into the oral cavity and from it into the sacs. The sac is covered with striated muscle, mostly from the subhyoid. There is considerable variation in the location and orientation of the vocal sacs in relation to other structures. The following classification[39] has been used:

1. *Median subgular.* An unpaired vocal sac beneath the throat. This may be either "internal", with no modification (thinning or folding) of the throat skin in correlation with the vocal sac, or it may be "external", with the throat skin obviously modified. The terms "external" and "internal" in this context are unfortunate ones, but they are in general use. Classification of a particular sac as external or internal is sometimes highly subjective.

2. *Paired subgular.* Paired sacs located beneath the throat and may be either internal or external.

3. *Paired lateral.* Vocal sacs located behind and below the angles of the jaws and may be either internal or external.

It has been hypothesized that the median subgular type of sac is ancestral[39], but recent evidence tends to discount this. In the ontogeny of an African toad, *Bufo regularis*, the median vocal sac develops from a lateral slit or slits and passes through a stage in which it is a paired structure[32] thus strongly suggesting that the ancestral condition was paired. In *Scaphiopus* of the primitive family Pelobatidae, species of the subgenus *Spea* have paired subgular sacs, while *S. holbrooki* of the subgenus *Scaphiopus* has remnants of a median membrane and *S. couchi* has a groove on the lower margin of the inflated median pouch[40].

The mechanics of sound production have received adequate study only in *Scaphiopus*[40]. During emission of the call there is a continual, smooth contraction of the body wall muscles, a continual expansion of the vocal pouch, and a steady decrease in the size of the body cavity. There is no reverse flow of air in the system during the production of a call, but most of the air in the vocal sac is returned to the lungs after termination of a call, and, as the nostrils are opened, there is probably some air exchange. Elasticity of the gular skin appears to limit initial inflation of the vocal sac and its final volume and also to limit the duration of the call, since the call must cease once the vocal sac reaches its limit of expansion.

The production of a single call note (trill) begins with the forcing open of the adpressed vocal cords by the column of air from the lungs. The cords will begin to vibrate at their natural frequency and will continue to vibrate as long as the proper pressure level is maintained. The effect of the cord vibration is to chop the DC air flow into a series of separate air parcels. In the period that the vocal cords burst apart, their thin, curved edges are free to vibrate, giving forth eddy currents which are impressed on the air parcel. This second, AC, component is in the range of audible frequencies and is the primary source of the sound produced. The above applies in the subgenus *Spea*, where the vocal cords as a whole oscillate at an inaudible frequency. In the subgenus *Scaphiopus* the vocal cords vibrate at an audible frequency (126 c/s or higher), and this vibration is audible as the fundamental frequency. In this subgenus, then, there are two sources of sound, the whole vocal cords and their edges In The subgenus *Spea* the only source of sound is the vocal cord edges In the subgenus *Scaphiopus*, the call is the relatively unmodified noise of the fundamental and its harmonics. There is some emphasis on a few harmonics, but there is relatively little damping of others. The call in the subgenus *Spea* is well modulated, with some harmonics emphasized and others completely damped, giving it a musical quality. The specific vocal structures responsible for the modulation in *Spea* are not identified, but they may include the cavity in the laryngeal wall anterior to the vocal cords, the accessory fold on the vocal cords of *S. holbrooki*, change in larynx shape, the edges of the glottis, mouth lining, mouth slits, or vocal sac. Interspecific differences in call are attributed more to nervous and muscular control of the vocal structures than to morphological differences.

(ii) Physical characters of emissions

The structure of the mating call has been determined by sound spectrography in most Nearctic Salientia, and extensive studies are in progress on Australian, Mexican and Central American groups In general the dominant frequency lies in the range of 2000–4000 c/s. The principal variables involved in the interspecies differentiation of call are duration, dominant frequency, pulse (trill) rate, and degree of modulation. The presently available information about call structure will be discussed by families.

Pelobatidae. The two subgenera of *Scaphiopus* differ in call structure as previously stated. In the subgenus *Scaphiopus*, the call is a relatively unmodulated sound (Fig. 380), although some harmonics are slightly emphasized and some tend to be damped. The chief difference between the calls of the two species, *S. couchi* and *S. holbrooki*, is in duration of the call[9]. The geographically disjunct subspecies (incipient species, probably isolated since early Wisconsin or Wurm glaciation) *S. hol-*

brooki holbrooki and *S. h. hurteri* show no evident differentiation in call structure[18].

In the subgenus *Spea* the call is well modulated and audibly pulsed (Fig. 380). The sympatric species *S. hammondi* and *S. bombifrons* have the call differentiated principally in pulse rate and to a lesser extent in duration[9]. A third, apparently allopatric, species *S. intermontanus* has a short, weak, rapidly pulsed call[12].

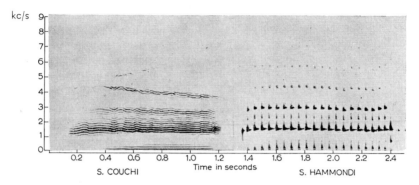

Fig. 380. Sonagrams of a call of *Scaphiopus couchi* (subgenus *Scaphiopus*) and one of *S. hammondi* (subgenus *Spea*). Note suggeneric difference in degree of modulation.

Bufonidae. The call structure has been determined in 14 of the 16 species of *Bufo* that occur in the United States. One species, *B. boreas*, appears to have lost the mating call. For another, *B. canorus*, information about call structure is not available. Six of the 14 species belong to a phylogenetically close group. One of these, *B. woodhousei*, with two secondarily interbreeding subspecies, occurs sympatrically with all of the 5 remaining species, all of which are allopatric in respect to one another. The call of *B. woodhousei* is relatively short (usually about one to three seconds in duration), inaudibly pulsed, and relatively unmodulated, although a series of emphasized harmonics forms a dominant frequency band in the range of about 1500–2000 c/s[11]*. The five remaining species of this group have the call audibly pulsed and highly modulated. *B. microscaphus* of the southwestern United States[15] and *B. hemiophrys* of north-central United States and central Canada[14] have relatively short, rapidly pulsed calls. Three eastern U.S. species, *B. terrestris*, *B. americanus*** and *B. houstonensis* have relatively longer calls. The slowest pulse rates are found in *B. americanus* and *B. houstonensis*, and the greatest duration is found in the former. All 5 species have finely tuned, musical calls, with the greatest modification in this direction in *B. houstonensis*. Two small toads, *B. punctatus* and B. *debilis*, of southwestern North America are possibly more closely related to this group than to any other. Both have rapidly pulsed calls; the call of *B. punctatus* is well modulated and musical, while that of *B. debilis* is poorly modulated.

Two morphologically related species, *B. cognatus* and *B. compactilis*, have the call strikingly differentiated in duration and in pulse rate[11]. The call of *B. cognatus* averages about 20 sec in duration and may extend to 54 sec, while that of *B. compactilis* averages about 0.4 sec in length. The pulse rate averages about 13/sec in *B. cognatus* and is about three times greater in *B. compactilis*. The call is well modulated in

---

* and **, see P. Marler, Chapter 9, with the sonagram of this species.

both species. Each pulse note in *B. cognatus* shows a marked frequency shift, rising sharply to a peak and then dropping sharply off.

The Mexican toad, *B. valliceps*, has a pulsed, well modulated call. The largest U.S. toad, *B. marinus*, has the lowest dominant frequency (about 640 c/s) and the slowest pulse rate (about 10/sec). Another large toad, *B. alvarius*, has a weak, vestigial call of about 0.7 sec duration, with most of its energy centering in the dominant frequency at about 1135 c/s. The smallest U.S. toad, *B. quercicus*, has the highest dominant frequency (up to 5200 c/s) and differs from other species in showing a sharp rise and fall in frequency in the course of each call.

Interspecific hybrids between closely related species have calls intermediate between those of the parent species; hybrids between the distantly related *B. valliceps* and *B. woodhousei* have imperfect calls of the *valliceps* type[13].

Leptodactylidae. The large leptodactylid family remains largely unknown as to call structure in its many species and genera, but work is presently in progress with Australian and Middle American representatives. In Australian *Crinia* of the superspecies, *signifera* and *insignifera*[37] the call shows strong differentiation among species that are little differentiated in external morphology. Differentiation has occurred in duration and modulation. In some, the fundamental and its harmonics are strongly expressed; in others certain bands of harmonics are damped and others are emphasized. In species with pulsed calls, the number and rate of the pulses are differentiated.

The three species, *C. signifera*, *C. sloanei* and *C. parinsignifera*, are strongly differentiated in duration of call (0.15, 0.06 and 0.48 sec, respectively), number of pulses (5.0, 13.0 and 78.0), pulse repetition frequency in c/s (31.0, 214.0, 167.0) and in call repetition rate in sec (6.0, 9.7 and 61.5)[36]. Calls of four western Australian species of *Neobatrachus* have a simple pulse structure with broad tuning and are strongly differentiated interspecifically in pulse rate[38]. In the genus *Heleioporus*, on the other hand, the calls are finely tuned and unpulsed and differ principally in duration and in frequency shifts.

The few calls of American leptodactylids available at present indicate that there is great diversification of call in this family (Fig. 381). The call of *Eleutherodactylus augusti*, recorded in central Texas is noise-like, with little modulation of the lower harmonics. The call of *Tomodactylus nitidus* from the state of Oaxaca, is very finely tuned. Another leptodactylid, *Syrrhophus marnocki*, of central Texas, has two distinct calls. One is a simple chirp; the other is pulsed. The second type is reported to be given only when in amplexus or when a male and female are near one another[33]. A pronounced frequency shift in the course of the call occurs in some leptodactylids; *Leptodactylus labialis*, *Eupemphix pustulosus*). The latter achieves a complicated call by the addition of one to three notes of quite different structure at the end of the main call.

Hylidae. The characters of the call in this primarily New World family are known mostly from work with U.S. representatives, although work is currently being done with Central American and Mexican groups. The two species of *Acris* (small, terrestrial, littoral hylids) have a pulsed call. Where the ranges of the two overlap on the coastal plain of the southeastern United States, the calls differ in pulse structure and timing[17]. In *A. crepitans* there is an audible pulsing within each pulse note; in *A. gryllus* there is none. In the former the interval between pulse notes varies little in the course of the call but does increase slightly in the terminal part. In the latter, there is a relatively long interval between pulses at the start of the call, and in the last part the pulse repetition rate is greatly increased. Outside of the zone of overlap (Texas and Florida) the calls of the two species are essentially indistinguishable.

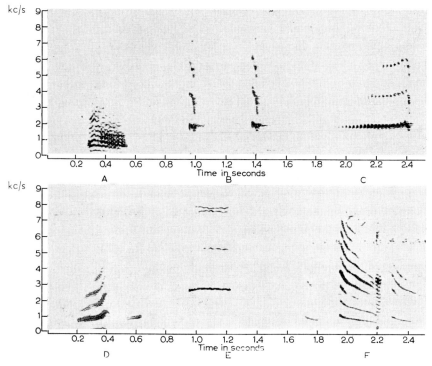

Fig. 381. Sonagrams of calls of some North American Leptodactylidae. (A) noise-like "bark" of *Eleutherodactylus augusti*; (B) two whistled calls of *Syrrhophus marnocki*; (C) pulsed "climax" call of *Syrrhophus marnocki*; (D) call of *Leptodactylus labialis* showing rising frequency, followed by dominant frequency band of a background individual and with extraneous noise between 7 and 8 kc/s, (E) finely tuned call of *Tomodactylus nitidus*; (F) call of *Eupemphix pustulosus* showing descending frequency and ending in a noise-like note, preceded and followed by fragments of calls of background individuals and with extraneous noise between 5 and 6 kc/s.

The calls of 12 U.S. species of *Hyla* and five related species of the nominal genus *Pseudacris* have been studied[16, 17, 20, 26]. Two of the five members of the *versicolor* groups have pulsed, highly modulated calls (Fig. 382). They differ principally in the shorter call of *H. versicolor* and longer call, with a marked frequency shift in each pulse note, of the sympatric *H. phaeocrypta* (= *H. avivoca*). The other three, *H. femoralis*, *H. arenicolor* and *H. baudini*, have the call notes relatively unmodulated and harsh. *H. femoralis* and *H. arenicolor* differ principally in that the former has the second harmonic at about 200 c/s and harmonics at about 2600 and 4800 c/s relatively emphasized, while in the latter the fourth harmonic at about 500 c/s is emphasized and others between 1800 and 2200 are less emphasized. *H. baudini* has a call remarkably similar to that of *H. femoralis* except that the dominant frequency is at about 2500 c/s.

The calls of the three species of the *cinerea* group, *H. cinerea*, sympatric *H. grati*-

osa, and allopatric *H. andersoni*, are structurally similar and unique among U.S. species of *Hyla* in the relatively even distribution of energy in their numerous harmonics. They differ principally in the specific harmonics emphasized (Fig. 383). In another group, which may or may not be a phylogenetic grouping, *H. squirella* (Fig. 383), *H. wrightorum*, and *H. regilla*, have calls characterized by the superimposition of a dominant frequency band and its harmonics on the pattern of poorly modulated (some harmonics more or less damped) harmonics of the fundamental. At least one additional type of call has been heard in *H. regilla*, but its structure has not been analyzed.

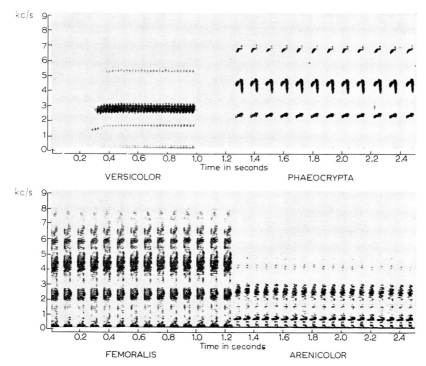

Fig. 382. Sonagrams of one complete call of *Hyla versicolor* and segments of the call of three other members of the *versicolor* species group. Note musical calls of *H. versicolor* and *H. phaeocrypta* and noise-like calls of *H. femoralis* and *H. arenicolor*. (From Blair, 1959[20].)

Hyla crucifer has a finely tuned, relatively short call that more closely resembles the calls of *Pseudacris ornata* and *P. streckeri* than it does the calls of other U.S. *Hyla* as the genus is presently limited. Its repertoire also includes a pulsed call that is not known in the species of *Pseudacris*. In the *Pseudacris nigrita* group, change in pulse rate has been the principal means of interspecific differentiation. Two species, *P. clarki* and *P. nigrita* which are narrowly sympatric in eastern Texas and Oklahoma differ in the rapidly pulsed call of the former and slowly pulsed call of the latter. A similar relationship exists between *P. feriarum* (rapidly pulsed) and *P. nigrita* in the southeastern United States.

Hyla septentrionalis, a representative of a tropical group that barely reaches the United States, has a complex call in which a noise-like segment is followed by an audibly pulsed segment.

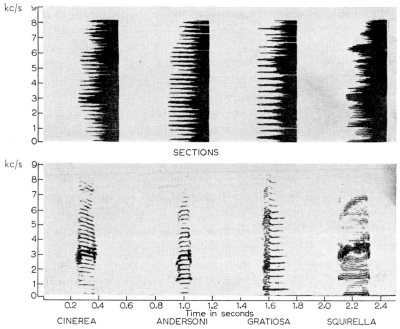

Fig. 383. "Sections" and sonagrams of the calls of the three members of the *Hyla cinerea* species group and of *H. squirella*. In sections, the length of the horizontal lines represents amplitude at a selected moment in time. In the three members of the *cinerea* group, note relatively even distribution of energy in the various harmonics by contrast with emphasis on a dominant frequency band in *H. squirella*. (From Blair, 1959[20].)

Microhylidae. The call structure in this large and widely distributed family is known only from a few North American species. Three species, *Microhyla olivacea*, *M. carolinensis*, and *Hypopachus cuneus*, have noise-like calls of about one to three seconds duration with relatively little emphasis on some harmonics and moderate damping of others. The two species of *Microhyla* show reinforcement of their differences in duration and dominant frequency where they occur sympatrically[10]. A Mexican species, *Microhyla usta*, has a more modulated call than the above, with an emphasized band between three and four kc/s and its harmonics. The pulse rate (about 80 c/s) is resolvable by the sound spectrograph, while that of the others is faster and not resolvable.

Ranidae. Structure of the call in this large, primarily African family is known only from a few of the North American species of *Rana*. The most commonly heard call of *Rana pipiens* is an audibly pulsed one of usually less than one second duration (up to 1.4) and in which the higher harmonics tend to be damped, particularly in the initial part of the call. At the height of the breeding season, a "chuckling" call is heard which is structurally similar to the regular cell except that each pulse note rises and falls in frequency. A related species, *R. areolata*, differs in having a faster pulse rate and an emphasized frequency band at about 1000–1400 c/s, with its harmonics expressed. Another, more distantly related, species, *R. clamitans* has a short call of about 0.15–0.30 sec duration with a long series of expressed harmonics, of which the third and fourth are usually emphasized. The higher harmonics are progressively damped in the closing part of the call.

(iii) Physiological characters

Some generalizations may be made concerning internal and external effects on the call of Salientia. The pulse rate and dominant frequency usually vary inversely with body size[17]. The large *Bufo marinus* has a dominant frequency of about 640 c/s, while the smallest North American toad, *Bufo quercicus*, has one of about 5000 c/s. The pulse rate in *B. marinus* is about 10 notes/sec, while that in *B. debilis* the second smallest North American toad is about 120.

Temperature is the one environmental factor that has been clearly shown to affect the call. It has been claimed[48] that frequency decreases with increased temperature in *Pseudacris feriarum*, but this is contrary to all other evidence for this and other groups of salientians. The dominant frequency increases with temperature in *Microhyla olivacea* and *M. carolinensis*, with the regression coefficient differing significantly between the two species[10]. The calls of a single individual *Hyla versicolor* recorded over a range of 9°C ambient air temperatures showed an average increase in frequency of 60 c/s per 1°C rise in temperature and an average increase of 0.44 pulses/sec in repetition rate[17]. A significant positive correlation between frequency of calling and water temperature has been reported in *Pseudacris nigrita* and *P. clarki*, and a negative correlation between call length and temperature has been reported in *P. nigrita* and *Bufo americanus*[4]. In the range of 15–25°C, the Q_{10} value for *P. nigrita* was estimated at 2.21 and that for *P. clarki* at 2.25.

It should be pointed out that temperature effects on pulse rate and dominant frequency would not destroy the effectiveness of such differences as isolating mechanisms, as the calls of both species would be affected, thus maintaining the relative difference, although the difference would be slightly increased or decreased where the Q^{10} was different for the two.

(iv) Behavior and vocabulary

Most of the vocalizations of salientians appear to be solely sex calls, although a few possible exceptions will be discussed below. Calling is initiated by whatever endocrine and external factors stimulate the reproductive effort. In aquatic breeders this typically starts as the building of an aggregation in the breeding pool; in terrestrial breeders this involves the finding or attraction of a mate without preliminary aggregating of the breeding population. In some species, such as the semi-aquatic and littoral *Rana pipiens* in the southern United States, the sex calls may be heard throughout most of the year, although there are peaks of calling correlated with peaks of reproductive activity in the Spring and Fall. Species adapted to life in xeric regions may have a brief period of calling correlated with brief reproductive activity after heavy rainfall. An example of this type is *Scaphiopus holbrooki hurteri* of Texas and Oklahoma in which the entire period of calling and of reproductive activity may include no more than one or two nights after flooding rain in the Spring. A related species, *Scaphiopus couchi*, may call several times during a season if there are several periods of heavy rainfall. Various species that have succeeded in surviving in the very xeric deserts of the Great Basin of western North America call and breed in permanent water holes without the stimulus of rain. In most of these, such as *Bufo woodhousei, B. microscaphus, Scaphiopus intermontanus, Hyla arenicolor*, the calling and breeding extend over a considerable period of time. An exception to this was seen on a night in central Arizona when thousands of *Bufo punctatus* moved in to a desert pool on a rainless night in what was obviously a peak of reproductive activity.

It is generally believed that the calls of the initial males that reach a breeding pool serve to attract other males until an aggregation of calling males is built up[23]. Evidence suggesting that this does occur is seen in the segregation of different species in different parts of the same breeding pool. One such example was noted in a mixed population of *Bufo woodhousei* and *B. americanus*[7]. Another was noted in a mixed population of *Bufo cognatus, B. debilis, B. compactilis* and *B. woodhousei*[24]. Each species was segregated in different parts of the breeding pool.

The call of most salientians with which we are familiar is a very stereotyped sound. There is little variation from time to time in the call of an individual except as the call is influenced by changes in temperature mentioned earlier. Evidence from the calls of hybrids[13] indicates that the character of the call is genetically determined. Geographic variation in call may occur in widely distributed species[17].

There is general evidence of some social interrelations in a chorus of calling males. It has been suggested that the calls of *Rana clamitans* serve as cues in maintaining the orientation of males in respect to one another in a simple social organization[41]. The phenomenon of a "chorus leader" who calls first and stimulates other males in the pool to follow with their calls is widespread. This phenomenon has been reported as being particularly pronounced in *Hyla versicolor, Pseudacris streckeri*, and *Microhyla olivacea*[4]. The habit of calling in trios has been claimed for *Hyla crucifer*[28] and for *Hyla gratiosa*[45], and choral groups have been indicated for various other salientians[34]. The postulation of different notes in the individuals of the trios appears to result from the directional nature of the human ear rather than from any actual difference in calls. The calls of the three *H. gratiosa* comprising a chorus sounded different to our ears in the field, but no significant differences were seen in sound spectrograms of these calls. Social interrelations in breeding choruses provide a profitable field for extensive and critical observation.

While the stereotyped vocalizations of salientians appear to be exclusively sex calls, a few interesting variations from these fixed patterns are known. Various treefrogs are known to have a "tree" call that differs or appears to the human ear to differ from the "water" call that is emitted at or in the breeding pool. The tree call of *Hyla gratiosa* has been said to sound very different from the breeding call and has been attributed a possible function in maintaining orientation of individuals in respect to one another and to the breeding pool[45]. The tree call of *Hyla squirella*, as shown by sound spectrograms, differs from the water call in the same general way that a whispered (unvoiced) human word differs from a voiced one. The two calls show resemblances in general pattern but differ in the absence of a harmonic pattern or frequency shift in the tree call[17].

Some species are known to have what might be termed a "climax" call that is added to the repertoire at the peak of the reproductive season. In this category are the pulsed calls of *Hyla crucifer* and *Syrrhophus marnocki* and the "chuckle" of *Rana pipiens*, in which a frequency shift is incorporated in each call note. Most North American species of salientians apparently lack any such climax call.

A male release call or warning vibration is generally distributed among salientian species. The significance of this call is obvious in view of the known tendency of sexually excited males to attempt amplexus with virtually any object of approximately appropriate size and shape. Release of the clasped male by the clasping male in response to the vibration serves to conserve reproductive effort. The warning vibration has been most thoroughly investigated in North American *Bufo*[1]. Males have been shown to release males of their own species group in response to the warning vibration but to show variable response with males of other groups[8].

Finally, the so-called alarm and distress calls of various salientian species are of debated significance. Some littoral species (*e.g.*, *Rana pipiens*) utter a short, sharp sound as they leap into the water upon being disturbed. This might be regarded as an alarm call, alerting other members of the population, but it has yet to be demonstrated that this call is of any social significance. It could result from the inadvertent release of air from the lungs as the startled animal makes its leap. A distress call is frequently heard when a salientian is grasped by a predator, but this call may also be inadvertent. It has been suggested that such calls serve to alert other members of the population[47]; however, calling frogs gave no apparent response to a "screaming" *Rana pipiens* that had been caught by a raccoon[23].

(b) *Reception of sounds*

The ability of salientians to hear sounds has been long assumed by naturalists because they were highly vocal animals and because of the prominent tympanum present in most kinds, and as early as 1905[50] it was demonstrated that *Rana clamitans*, *R. pipiens* and *R. catesbeiana*, are affected by vibrations in the range of 50–10,000 c/s. Response to these vibrations disappeared after cutting of the eighth cranial nerve, thus indicating that the responses were truly auditory.

(i) *Morphology of receptor apparatus*

An external tympanum of thinned integument is present on the sides of the head in most salientians, although various burrowing or aquatic types lack both tympanum and middle ear[47], and in one Asiatic *Rana* the tympanum is internal within an external ear canal. Sound waves impinging on the tympanum are transmitted through an ear ossicle (columella) via the oval window to the perilymph and in turn to the tectorial membranes overlying the sensory areas that are believed sensitive to sound waves. These sensory areas include the basilar papilla and the amphibian papilla[6]. Recent studies of the basilar papilla[5,6] suggest that this structure has the capacity of discriminating different sound frequencies. The evidence comes partly from experiments with models of the basilar papilla which are interpreted as indicating that this structure can perform frequency analysis according to the place principle and partly from the morphological evidence that the papillary nerve is divided into five bundles, which supports the belief that the sensory cells are differentiated in regard to their frequency specificity.

The relative roles of the central nervous system and the sensory structures of the inner ear in discriminating frequency and other variations in sounds remain to be determined.

(ii) Behavioral responses

It has generally been assumed that the vocalizations of salientians serve to accumulate breeding aggregations or attract a mate in solitary breeders and as a means of species discrimination. Until recently, the evidence has been derived mostly from indirect sources or from field observations rather than from precise experimentation. One deterrent to the experimental approach has been the fact that responsiveness of the females to the call of the males is limited to the brief period in which the female is physiologically ripe.

Observations by various workers on a variety of species in nature have shown the female coming to a calling male of her species[17, 19, 23, 25, 36, 44, 46, 49]. A common feature of these observations is the notation that the choice lies with the female. She approaches the male and attracts his attention by moving in front of him or by body contact. In a recent laboratory experiment[42] it has been shown that female *Pseudacris nigrita* respond positively to the calls of hidden males and to tape recorded male calls. Other experiments[38a] have shown conclusively that female *Pseudacris streckeri* respond preferentially to the calls of their own males when presented a choice of their calls or the calls of *P. clarki*. Presently unpublished data from tests of other combinations, including tests between allopatric species with relatively minor differentiation in call, indicate high discriminatory ability in some U.S. hylids. Experiments with *Bufo terrestris* and *Hyla gratiosa* also indicate a positive response of females, and of males as well in the toad, to calling males[23]. Inconclusive results in some field experiments with toads[7, 23] are attributable to the fact that the animals tested were not in a high state of reproductive excitement. In a field experiment with *Bufo terrestris*[22] the use of auditory cues offered the best explanation of the homing of toads which had been marked and moved distances up to 700 yards.

In addition to the clear-cut experimental evidence now available as to the ability of salientians to discriminate interspecific differences in call, there is indirect evidence from the observed facts that sympatric, closely related species differ in at least one, and usually more, of the major attributes of the mating call and that there may be reinforcement of call differences between sibling species in the zone of overlap (*e.g.*, *Microhyla*, *Acris*)[10, 17].

In closing, it may be emphasized that difference in mating call is only one of a complex of isolation mechanisms, including ethological and ecological ones, that restrict cross-mating of natural populations. Natural hybridization may result from the occasional break-down of one or more of these mechanisms. Call difference could fail as an isolation mechanism when a female going to the call of a male of her own species in a mixed population passed too closely to a male of another species or inadvertently bumped into him.

[For Addendum to this Chapter, see p. 803]

REFERENCES

[1] ARONSON, LESTER R., 1944, The sexual behavior of Anura. 6. The mating pattern of *Bufo americanus, Bufo fowleri*, and *Bufo terrestris*. *Am. Museum Novitates*, 1250, 1–15.
[2] BAKER, C. L., 1945, The natural history and morphology of Amphiuma, *Report Reelfoot Lake Biol. Station*, 9, 55–91.
[3] BAKER, LOUISE C., 1937, Mating habits and life history of *Amphiuma tridactylum* Cuvier and effect of pituitary injections, *J. Tenn. Acad. Sci.*, 12, 206–218.
[4] BELLIS, EDWARD D., 1957, The effects of temperature on salientian breeding calls, *Copeia*, 85–89.
[5] BERGEIJK, WILLEM A. VAN, 1957, Observations on models of the basilar papilla of the frog's ear, *J. Acous. Soc. Am.*, 29, 1159–1162.
[6] BERGEIJK, WILLEM A. VAN and EMIL WITSCHI, 1957, The basilar papilla of the anuran ear, *Acta Anatomica*, 30, 81–91.
[7] BLAIR, ALBERT P., 1942, Isolating mechanisms in a complex of four species of toads, *Biol. Symposia*, 6, 235–249.
[8] BLAIR, ALBERT P., 1947, The male warning vibration in *Bufo*, *Am. Museum Novitates*, No. 1344, 1–7.
[9] BLAIR, W. FRANK, 1955, Differentiation of mating call in spadefoots, genus *Scaphiopus*, *Tex. J. Sci.*, 7, 183–188.
[10] BLAIR, W. FRANK, 1955, Mating call and stage of speciation in the *Microhyla olivacea–M. carolinensis* complex, *Evolution*, 9, 469–480.
[11] BLAIR, W. FRANK, 1956, Call difference as an isolation mechanism in southwestern toads (genus *Bufo*), *Tex. J. Sci.*, 8, 87–106.
[12] BLAIR, W. FRANK, 1956, Mating call and possible stage of speciation in the Great Basin spadefoot, *Tex. J. Sci.*, 8, 236–238.
[13] BLAIR, W. FRANK, 1956, The mating calls of hybrid toads, *Tex. J. Sci.*, 8, 350–355.
[14] BLAIR, W. FRANK, 1957, Mating call and relationships of *Bufo hemiophrys* Cope, *Tex. J. Sci.*, 9, 99–108.
[15] BLAIR, W. FRANK, 1957, Structure of the call and relationships of *Bufo microscaphus* Cope, *Copeia*, 208–212.
[16] BLAIR, W. FRANK, 1958, Call difference as an isolation mechanism in Florida species of hylid frogs, *Quart. J. Fla. Acad. Sci.*, 21, 32–48.
[17] BLAIR, W. FRANK, 1958, Mating call in the speciation of anuran amphibians, *Am. Nat.*, 92, 27–51.
[18] BLAIR, W. FRANK, 1958, Mating call and stage of speciation of two allopatric populations of spadefoots (*Scaphiopus*), *Tex. J. Sci.*, 11, 484–488.
[19] BLAIR, W. FRANK, 1958, Response of a green treefrog (*Hyla cinerea*) to the call of the male, *Copeia*, 333–334.
[20] BLAIR, W. FRANK, 1959, Call structure and species groups in U. S. treefrogs (*Hyla*), *Southwestern Naturalist*, 3, 77–89.
[21] BLAIR, W. FRANK and DAVID PETTUS, 1954, The mating call and its significance in the Colorado River toad (*Bufo alvarius* Girard), *Tex. J. Sci.*, 6, 72–77.
[22] BOGERT, CHARLES M., 1947, A field study of homing in the Carolina toad, *Am. Museum Novitates*, No. 1355, 1–24.
[23] BOGERT, CHARLES M., 1958, The biological significance of voice in frogs, Booklet accompanying *Folkways Records Album*, No. FX 6166, 18 pp.
[24] BRAGG, ARTHUR N., 1940, Habits, habitat and breeding of *Bufo woodhousii woodhousii* (Girard) in Oklahoma, *Am. Mid. Nat.*, 24, 306–335.
[25] COURTIS, S. A., 1907, Response of toads to sound stimuli, *Am. Nat.*, 41, 677–682.
[26] CRENSHAW, JOHN W., JR. and W. FRANK BLAIR, 1959, Relationships in the *Pseudacris nigrita* complex in southwestern Georgia, *Copeia*, 215–222.
[27] DIXON, JAMES RAY, 1957, Geographic variation and distribution of the genus Tomodactylus in Mexico, *Tex. J. Sci.*, 9, 379–409.
[28] GOIN, COLEMAN, J., 1948, The peep order in peepers; a swamp water serenade. *Quart. J. Fla. Acad. Sci.*, 11, 59–61.
[29] INGER, ROBERT F., 1954, Systematics and zoogeography of Philippine amphibia, *Fieldiana, Zoology*, 33, 183–531.
[30] INGER, ROBERT F., 1956, Morphology and development of the vocal sac apparatus in the African frog *Rana (Ptychadena) porosissima* Steindachner, *J. Morphol.*, 99, 57–72.
[31] INGER, ROBERT F., 1958, The vocal sac of the Colorado River toad (*Bufo alvarius* Girard), *Tex. J. Sci.*, 10, 319–324.
[32] INGER, ROBERT F. and BERNARD GREENBERG, 1956, Morphology and seasonal development of sex characters in two sympatric African toads, *J. Morphol.*, 99, 549–574.
[33] JAMESON, DAVID L., 1954, Social patterns in the leptodactylid frogs, *Syrrhophus* and *Eleutherodactylus*, *Copeia*, 36–38.

[34] JAMESON, DAVID L., 1955, Evolutionary trends in the courtship and mating behavior of Salientia, *Systematic Zool.*, *4*, 105–119.
[35] LITTLEJOHN, MURRAY J., 1958, A new species of frog of the genus *Crinia* Tschudi from south-eastern Australia. *Proc. Linn. Soc. New South Wales*, *83*, 222–226.
[36] LITTLEJOHN, MURRAY J., 1958, Mating behavior in the treefrog, *Hyla versicolor. Copeia*, 222–223.
[37] LITTLEJOHN, MURRAY J., 1959, Call differentiation in a complex of seven species of *Crinia* (Anura, Leptodactylidae), *Evolution*, *13*, 452–468.
[38] LITTLEJOHN, M. J. and A. R. MAIN, 1959, Call structure in two leptodactylid genera of Australian burrowing frogs, *Copeia*, 266–270.
[38a] LITTLEJOHN, MURRAY J. and TED C. MICHAUD, 1959, Mating call discrimination by females of Strecker's chorus frog (*Pseudacris streckeri*), *Tex. J. Sci.*, *11*, 86–92.
[39] LIU, CH'ENG CHAO, 1935, Types of vocal sac in the Salientia, *Proc. Boston Soc. Nat. Hist.*, *41*, 19–40.
[40] MCALISTER, WAYNE H., 1959, The vocal structures and method of call production in the genus *Scaphiopus* Holbrook. *Tex. J. Sci.*, *11*, 60–77.
[41] MARTOF, BERNARD S., 1953, Territoriality in the green frog, Rana clamitans, *Ecology*, *34*, 165–174.
[42] MARTOF, BERNARD S. and ERIC F. THOMPSON, JR., 1958, Reproductive behavior of the chorus frog, *Pseudacris nigrita*, *Behaviour*, *13*, 243–258.
[43] MASLIN, T. PAUL, 1950, The production of sound in caudate amphibia, *Univ. Colorado Studies Series in Biology*, *1*, 29–45.
[44] MILLER, NEWTON, 1909, The American toad (*Bufo lentiginosus americanus*, LeConte), *Am. Nat.*, *43*, 641–668.
[45] NEILL, WILFRED T., 1958, The varied calls of the barking treefrog, *Hyla gratiosa* LeConte, *Copeia*, 44–46.
[46] NOBLE, G. KINGSLEY, 1923, Voice as a factor in the mating of batrachians, *Science*, *58*, 270–271.
[47] NOBLE, G. KINGSLEY, 1931, *The Biology of the Amphibia*, McGraw-Hill, New York.
[48] THOMPSON, ERIC F. and BERNARD S. MARTOF, 1957, A comparison of the physical characteristics of frog calls (Pseudacris), *Physiol. Zool.*, *30*, 328–341.
[49] WELLMAN, GORDON B., 1917, Notes on breeding of the American toad, *Copeia*, No. 51, 107–108.
[50] YERKES, ROBERT M., 1905, The sense of hearing in frogs. *J. Comp. Neurol. and Psychol.*, *15*, 279–304.

CHAPTER 24

ACOUSTIC BEHAVIOUR OF BIRDS

by

J. C. BREMOND

1. Introduction

Birds have without doubt been the subject of most of the observations that have been made on the vocalization and acoustic behavior of animals. Naturalists have gathered a vast general knowledge of the subject in a large number of books and articles that have been extended by an almost equally wide knowledge of cynegetic origin.

In the limited space of an article it is, therefore, impossible to make a synthesis of all the results obtained. This review is concerned mainly with developments extracted from experimental studies made during the last ten to fifteen years, with an attempt to stress certain aspects of the studies which are related to physical techniques necessary for a psychophysiological study of the various types of acoustic behavior of birds.

Very important modern work on innate and acquired behavior, such as that by Thorpe, Messmer, Sauer and Marler has been discussed in more detail in other articles of this book. The subject has therefore only been lightly touched here in order to situate it in the overall picture.

The bibliography, which has been limited to about 150 relatively recent references, contains works which appear to be necessary for a knowledge of this vast subject.

2. Classification of Sonic Emissions

(a) *The problem of description*

There is, as yet, no adequate solution to the problem of describing the acoustic signals of birds, and, in a more general manner, that of their visual reproduction. The sounds produced by each species of bird are particularly constant and specific and may be learned and immediately recognized and identified by the listener, but it is impossible to describe them in a satisfactory manner, *i.e.* in a way in which the uninitiated may visualize an actual representation[25]. Evidence for this is given by the many graphic transcriptions proposed as solutions by many authors, from simple onomatopoeia to comparisons with other sounds, compositions derived from musical representation and even the creation of whole codes of symbols. All have little use except for the author himself. The older methods of description are gradually being abandoned and replaced by diagrams given by apparatus used for analysing the signals (oscillograms, sonagrams) which define a complex sound precisely, in a manner similar to that of a chemical formula, and give information on the constituent elements[156].

References p. 746

In addition to these processes, the most practical method of illustrating these signals is to compare them with sounds or phrases taken from outside the musical domaine. Modern methods of analysis have confirmed the dissimilarity to notes or chords.

A classification of sonic emissions involves the integration of all the elements which characterize sounds. Consideration of the multiplicity and intricacy of physical factors renders this task difficult, even impossible, unless generalisations are made which may themselves be falsified in many individual cases due to the present inadequacy of observations. Since each worker has attached importance to a certain particularity it will be necessary to regroup or subdivide existing classifications. Whatever the type of classification adopted the problems of designating the signals remain. Two principal tendencies are, however, evident from the trials that have been made.

(a) A signal may be described in relation to the state of the donor during emission. Thus expressions such as "warning call", "anxiety call", "excited call", etc., have been defined.

(b) A signal may also be described in relation to its action on the receptor, *i.e.* song of territorial defence, courtship song, feeding calls, etc.

The second method of classification appears preferable as it is more objective. The two methods are actually used simultaneously even though they may sometimes both be untrue, since a particular call may correspond to different behaviour elements and several calls to the same situation.

The problem is further complicated by the fact that the tone of a cry may vary with the circumstances and the individual[132]. These variations of tone are sometimes accompanied by a change in a typical call which may assume an intermediate form between that corresponding to the psychological situation that the bird is leaving as it passes gradually to one representing another psychological state. For example, the bullfinch (*Pyrrhula pyrrhula*) gives a series of gradually changing short calls before singing and then gives an intermediate sound between this signal and the first notes of the song.

The most practical classification of acoustic signals seems to be one which is related to factors which either motivate or play a part in changing behaviour. Based principally on experimental evidence, this method is less susceptible to criticism of interpretation than the others and is particularly suitable for integrating the results of future work[145].

It is possible to outline schematically two groups of motivating factors amongst those inducing the bird to give signals. The first is made up of the total of information originating from the channel; this for example reflects conditions specific to the environment in which the animal is living. The second group contains internal factors specific to the individual nervous system, endocrine system, metabolism etc. Experimental work has shown that acts of behavior produced by this latter group of factors are not produced spontaneously in the majority of cases; the physiological state of the bird only produces a tendency towards a certain action which is generally only brought about under the influence of a signal.

(b) The bird's repertoire of acoustic signals

Estimates of the richness of vocabulary have been made through systematic studies of some species. Some of these results are summarized in Table 60. The various categories of signals are given in the first column, *i.e.* the different types of motivation producing signals, *e.g.* summoning or warning calls, each category possibly containing several distinct signals. The total number of acoustic signals used by the bird are given in the second column.

TABLE 60

RANGE OF VOCABULARY IN CERTAIN SPECIES

Species	Number of categories ofs acoustic signals	Number of acoustic signals	Authors	Ref.
Cettia cetti	6		Trouche	132
Corvus monedula		9	Lorenz	148
Domestic chick (adult)		20	Schwelderup-Ebbe	103–104
Domestic chick (young)		5	Schwelderup-Ebbe	103–104
Fringilla coelebs		20	Marler	79
Garrulus glandarius	6		Goodwin	49–50
Hylocichla mustelina	6		Brackbill	21–23
Larus argentatus	6		Goethe	51
Melospiza melodia		24	Nice	91–92
Parus atricapillus		17	Odum	150
Sylvia communis		25	Sauer	101
Turdus merula		13–15	Messmer	87

The repertoire of a bird is studied under three headings: (a) song; (b) "calls" or "cries"; and (c) noises.

(a) Song may be defined as a series of notes, generally of more than one type, emitted in succession, thus producing a recognizable sequence or scheme. It shows traces of rhythm and length which are difficult to distinguish from those found in the group of signals designated by the general term "calls". Song-production is essentially under the influence of sexual hormones.

The song of young birds (the sub-song of British workers)—or juvenile song—should be mentioned in addition to that of the adults. Although similar to the adult song it shows greater variation in many species. It nevertheless contains the majority of the elements of the actual song.

(b) Calls are, in most cases, mono- or disyllabic sounds which do not contain more than four or five notes. They are not organized into sequences or phrases of a definite length. They are generally produced in a variety of situations, not necessarily connected with sexuality.

(c) Noises are produced by the bird with various parts of the body (feathers, beak, feet), quite apart from the songs and calls emanating from the syrinx.

3. Signals Emitted under the Influence of Internal Inducing Factors and External Factors Instigating Behavior

(a) General aspects

In the annual biological cycle, much acoustic behaviour occurs in the same order and with the same details, thus demonstrating that preregulated nervous mechanisms

References p. 746

exist. These may come into play after they have been rendered more sensitive, through the effect of sex hormones, to the external stimuli which set in motion the various acts of behavior. Another proof of the permanence of these nervous mechanisms is given by the work of Tinbergen[129] on the Cormorant (*Phalacrocorax carbo sinensis*). Even though sexual behaviour is only slightly or not at all in evidence in autumn and in early winter in this species, changes in posture, derived from sexual behavior but motivated by the aggressive tendency occur throughout the year, irrespective of fluctuations in frequency and intensity depending on changes in the sexual instinct. This observation shows that the nervous mechanism is present throughout the year and that variations in the "autochthonous" sexual behaviour depend on hormonal fluctuations. As Benoit[10] has said, "a hormone has a motor effect but is not itself the motor agent, it only starts the action".

These remarks on behaviour mechanisms apply in general to acoustic signals and in particular to the signal that has been studied most—the song.

(*i*) *Observations on the hormonal state as an internal motivating factor*

The number and intensity of songs heard in spring, the normal breeding season of birds in our climate, suggests that there is a liaison between the hormonal state and sound production. The song changes during the season; it starts as infrequent and often incomplete phrases and then increases to a maximum in volume, duration and frequency of repetitions, then dies down, changes and disappears. In many species of passerines it reappears in autumn in a more or less normal form after a period of regression, even though there is no fresh clutch of eggs; it is related at this period to the incomplete regression of the gonads[85]. Many birds rear two broods a year; for example during the first brood in an american species of thrush (*Hylocichla mustelina*), the male actively looks after the female and helps in the rearing of the young and its song is very frequent during this period. However, during the second brood its behaviour shows that it takes less care of the family and at the same time its song decreases in frequency and changes[21, 23]. Similar observations have been made with Cetti's warbler (*Cettia cetti*)[132] and the majority of song birds. In the common snipe (*Capella gallinago*) there is a correlation between the occupation of a territory, the appearance of song and development of the gonads. The non-migratory starling (*Sturnus vulgaris*) of Great Britain which starts singing at the beginning of winter, shows a more premature development of the testes than do individuals in migratory populations which do not start to sing until just before their departure[27], (that is, even before they have chosen their territory). A relationship between the development of song and the age of the bird has also been shown in the wren (*Troglodytes troglodytes*)[62]. This bird has no territory when one year old and gives infrequent and abnormal song. At two years of age it sings almost normally and decides on a territory although it does not mate until the following year, when it sings and reproduces on the same territory.

(*ii*) *Experimental proof*

The majority of manifestations of sexual behaviour are governed by the gonadal hormones.

It has been known for a long time that castration results in the disappearance of sexual behavior in the cock. The first experiment was carried out in 1918. When a capon is injected with extracts of pig's testees it shows the normal behavior of the cock, including a normal vocalization[97]. This has also been shown and confirmed several times in a number of species. Sexual manifestations reappear in castrated females when the oestrogenic hormone is injected. Song appears, in the majority of cases, to be under the influence of the male hormone. It reappears in the capon 48 hours after injection[37, 38, 40]. Analogous results have been obtained in experiments carried out on the gulls: *Larus atricilla* and *Larus argentatus*. In the case of the domestic hen and the canary, (*Serinus canaria*), injection of male hormone into females results in the production of the male song and stops all female sexual behaviour[106].

Even though hormones play a necessary role in the manifestation of song, in certain cases, the nervous system alone is capable of determining its production. A cock, castrated as an adult, still sings for several months[8]. The thyroid hormone also plays a role in the manifestation of various acts of behavior. In the cock, thyroidectomy results in a loss of aggressiveness, song and signs of courting. This behavior courting. This behaviour is reversed by injection of thyroxin[13]. Complete removal of the suprarenal glands is difficult to achieve without causing a serious operational trauma and so conclusions from such experiments are not very significant, but it is undeniable that injection of cortisone stimulates growth of the testes and causes the premature appearance of song and the aggressive tendency.

(b) Influence of external factors on the development of internal motivating factors

Psychological factors influence secretion of the hypophysis by way of the hypothalamic centres. The appearance of a particular behaviour element may be accelerated by seeing or hearing other individuals. This explains the particular importance of the often long and complex "parades" in certain species. Since the female is less advanced, physiologically, than the male in her sexual development she needs to receive psychological stimulation from the male to produce a normal sequence of behaviour. The secretion of the hypophysis may itself be activated by external factors, in particular by light. This has been shown in the domestic duck[9], starling[11,12], house sparrow (*Passer domesticus*)[136,137], pheasant and grouse[32], robin (*Erithacus rubecula*)[102], serin[134,135], and other species, showing that the phenomenon is a general one. The hypophysis may, however, do without this stimulation for its control over secretion of the sexual hormones as has been shown by experiments on immature ducks reared in the dark, which, despite a definite retardation, finally developed very large testees[10].

It often happens in gregarious species which perform certain activities in large numbers, that the birds take up a call after one of their number has emitted this call. This type of behavior is designated by the term "sonic panurgism". In this case, where birds in the group are leading similar lives, it may be supposed that most of them are in a similar physical and physiological state at any given moment. A signal circulating in the channel aids a tendency to emit the same signal. This phenomenon of panurgism is reflected in certain signals in the repertoire of the bird. The ease in which a bird may emit a signal depends on its physical and physiological state. "Gobbling" may be produced in the domestic turkey in reply to the transmission of artificial sounds of pure frequency, but the threshold of this reaction, a subject studied in detail by Schleidt[105], varies with the season and the physiological state of the bird.

References p. 746

While some external factors play a stimulating role, others, on the contrary, have an inhibiting action. It has been shown that low temperatures have an inhibiting effect on the song of the american song-sparrow (*Melopiza melodia*)[91] although the temperature has to be progressively lowered to obtain the same effect as spring advances (Fig. 384).

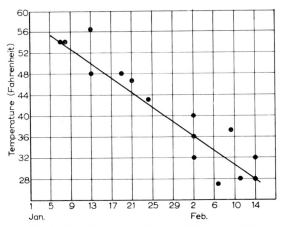

Fig. 384. Changes in the threshold of thermal inhibition of the song of the american song sparrow (*Melospiza melodia*) during January and February. Ordinate: temperature in °F (from Nice, 1937)[91].

(c) *The song and its role in instigating acts of behaviour*

Song and reproduction appear to be closely associated in the bird. Song has a preliminary role which is considered to be indispensable for the successful conduct of reproduction. It is associated with various optical signals:—postures, movements, exhibition of coloured plumage, etc. However, some preliminary phases have been obtained without it and certain songless species exist. In these species, calls or even sounds then have the role that is played by song in the other species in instigating behavior. For example, storks (*Ciconiidae*) give mandibular noises as signals. Song is not the only type of acoustic signal employed during the series of actions related to reproduction. It may be accompanied by calls, or even noises such as those made by the wings during "parades" and in the flight of many birds.

Song is the attribute of the male in many species, at least during the period of reproduction. Many cases of females singing in autumn have been reported, even though they do not do so in spring. This is due to the bipotentiality of the ovaries of the bird since only the left ovary is well-developed and functional. The right ovary, which is responsible for this type of song and whose histological structure is similar to that of the testees, is no longer inhibited by the left ovary during the period of sexual latency and so gives a ♂ character even in the ♀[85]. Autumnal songs seem to have little role as signals, except perhaps in relation to the defence of territory. This does not apply to species in which song is used by the two sexes as a means of reciprocal stimulation.

Song has many potentialities with respect to the information that it is able to transmit. It may indicate:

(1) The identity of a bird to another of the same species.
(2) Situation of a bird in the canal and the limits of its territory.
   (a) Three dimensional position of the emittor.
   (b) Two dimensional position of the emittor (situation of the nest, the approximate place chosen for building).
   (c) Situation of a bird in, near, or looking for a collective perch.
(3) Stage which the bird has reached in the sexual cycle.
(4) Potential dominance of a bird.
(5) Behaviour of the emittor.

Songs are very often associated with territorial behaviour and ceremonies of greeting and invitation. Due to the frequency in which they occur during these activities they have been the subject the most detailed studies.

(d) *Patterns of song*

Each species has its individual song and so the number of songs is considerable when one realizes the diversity of birds. Some of these songs are monotonous or very poor, others, on the contrary, are complex and rich in motifs, and some even possess elements that are musical to the human ear. It is possible, in addition to this diversity, to distinguish a certain number of general rules concerning their constitution[56]. Each species possesses a certain number of elementary sounds in its repertoire, which may average 200 to 300 in species which have very diverse songs. These sounds, or notes as they are termed by analogy with music, follow one another in a rigid order, thus constituting definite patterns that are reproducible by the bird. They last for 3 to 4 seconds on the average, rarely longer in the majority of species. Parrots have difficulty in learning motifs that last for more than 3 seconds[55]. The European wren seems to have attained the maximum of the possibilities in this field since it is capable of emitting a pattern lasting for 10 seconds.

Only a small number of the possible combinations of sounds are used by the bird. A phrase is made up of a succession of patterns and a song by a connection of phrases. That is to say, that the bird specialises in a certain style of singing, which eliminates a large number of theoretically possible combinations in the same repertoire. This phenomenon is also found in imitative species which have a tendency to deform some of the copied motifs to a greater or lesser extent, if these are too different from their own song. This restriction in variability is illustrated by the following examples: The meadow-lark (*Sturnella magna*) has a repertoire of about 300 notes. The patterns, each composed of three to six notes, only give about 50 types of song. In spite of the fact that the repertoire of the nightingale (*Luscinia megarhynchos*) is as rich, it only has 24 different types of song with patterns composed on the average of 11 notes. Imitative birds have a larger number of songs, but their repertoire does not appear to contain more than 500 to 600 notes.

There is a minimum of variability which cannot be surpassed. This is shown by contrasts in the various phrases. The greater the variation in the song, the shorter are the silences it contains. The sonic elements are particularly susceptible to variation although sometimes only the pauses and silences separating the various songs and their constituent phrases may be affected. This occurs in species which have a monotonous and repetitive song where the pauses are longer, and in this case the length of these pauses varies. The relative disposition of the various elements of the song in general does not seem to be the result of chance but is governed by a certain number of factors. Although these factors are not precisely known their existence has been shown by counts made on the song of the chaffinch[60]. In this bird, the various motifs of spontaneous song are not emitted with repetitions of equal frequencies. The rate of the song (number of songs per unit time) also has an effect on its composition. The emission of previously recorded songs stimulated the finch. It replies by a song in which the motifs that it had heard attain an abnormally high proportion. This should not be considered as an imitation phenomenon but as a facilitation in favour of a choice between several possibilities. The distribution and length of the silences are also governed by definite rules and do not occur just by chance. An interval of less than three seconds between two songs is rare. This suggests that each song is followed by an inhibitory effect which disappears with time. The fact that the interval separating an incomplete (abbreviated or interrupted) song from a normal (complete) song of the same type tends to be shorter than that between two normal songs, confirms the hypothesis that the inhibitory effect is a consequence of the song that has just been emitted.

These rules, comparable to those of music make bird-song resemble a relatively poor, very primitive music. This construction of signals is much more evolved than that of batrachians or insects.

What is the semantic value of the phrase thus constituted? The robin gives various different songs in succession while defending its territory. By altering the natural order of these songs experimentally it has been shown that their disposition does not have any information content and that a single one of them, given repeatedly, has the same information content as the normal sequence[144]. The constituent elements, or notes, of the song do not, in themselves, have any information content. They are grouped into motifs and a series of motifs constitute a song. Artificially isolated motifs have only very little information content. If their order is altered experimentally, the resulting territorial defence behaviour varies with the particular arrangement. It therefore appears that, for this species, information is based on a syntax of notes constituting the motifs. Does a syntax necessary for transmitting information exist in other species that have a less flexible song than the robin? This problem can only be solved by further experimental work. The various dialects of individuals of the same species which are separated geographically by quite large distances perhaps depend on syntactic variations of this type. In addition it has also been proved that there are slightly different elementary notes in certain species (see p. 734).

(e) Song and territory

Territory in song-birds may be defined as being especially the area chosen for breeding, and secondly, the space where food necessary for rearing the young may be found (the limits of these two types of territory are not the same). The extent of the territory can vary with the number of individuals, even in one species. It is defended during the breeding season (sometimes even during the whole year) against intruders of the same species, particularly if they are sexually mature. Attacks between species are uncommon. A territory is defended with increasing vigour the smaller its size and the nearer the intruders approach its center. Acoustic signals, mainly songs, are the principal means used in this defence. Visual signals are often associated with them but they are of less importance since the reaction in question can occur when it is impossible for the bird to see them, as is very often the case in biotopes where vision is limited. The observation below, taken from Lack[63], demonstrates the preponderance of acoustic signals over visual signals in the robin.

"...On May 27, 1937, an unringed newcomer started to sing on the territory owned by ♂ 44. The ♂ 44 started to sing in the opposite corner of the territory, the newcomer (which had no means of knowing that he was on territory that was already occupied) sang again and the owner replied nearer at hand, after which the newcomer again sang. This happened twice, the owner, who was hidden all the time in the thick bushes, finally replying from a distance of about 15 meters. At this, the newcomer flew away even though he had never seen and never would see his rival."

We have been able to verify experimentally, as a confirmation of this observation, that the reaction of the owner is immediate and very definite, by broadcasting a previously recorded song in the territory of a robin. The bird approaches and sings near the loudspeaker. The intensity of the bird's defensive reaction varies with the season, with a maximum in springtime. In the middle of the summer some phonotaxes towards the loudspeaker may still be obtained, but the robin attracted in this way uses a few short sequences of song and very soon shows no reaction to the transmitted signal.

There is, therefore, a relation between the sexual state of the bird and the defence of its territory during the breeding season. However, during experimental broadcasts carried out between July and August, accompanied by captures in a net, the signal transmitted by the loudspeaker drew more immature birds hatched in the preceding spring than adults. These observations do not minimize the value of the relation that exists between the song, behaviour and the hormonal state of the bird, but demonstrate the role the song has in instigating the process of attraction between individuals in relation to behaviour other than that connected with reproduction. It is very likely that adults which are attracted at this season to the sonic source broadcasting the signal, are induced to show this behaviour by the same stimulus as that acting on the immature birds. Additional experiments on castrated animals would clarify this point.

The song is very often the first manifestation of a whole chain of actions which result in reproduction, since its hormonal threshold of motivation is lower than that of other acts of behaviour. It can even appear before the bird has started to look for a territory, *e.g. Sturnus vulgaris* whose departure for migration at the end of winter is preceded by song.

In addition to the use of song for defence of territory it is also a signal affirming its possession. After having chosen a territory, birds returning from migration sing there with great frequency. A newcomer can find out whether the territory is free by the reply that is given to its song. The intensity of reply provoked in this manner will also give it information of the sexual state of the occupant.

Territorial defence reactions have been known to bird shooters for a long time and used by them to attract birds by imitating their signals by means of birdcalls. During the breeding season these phonotaxes are easy to obtain because the threshold of the birds' sensitivity to warning or escape signals is higher at this time.

Many species of birds which do not possess a song in the sense in which it has been defined here, defend their territory none the less. A common tern (*Sterna hirundo*) in possession of a territory reacts to intrusion by a neighbour by giving special calls

References p. 746

of increasing intensity before adopting a threat and then a fighting posture[96]. In the case of the woodpeckers (*Dendrocopos major* and *Dendrocopos minor*) a noise, "drumming", is used as a signal[39,119]. This noise, produced by a series of rapid blows with the beak (10 to 20 times per second) on a tree trunk specially chosen for its acoustic properties, has a role in defence or advertising of possession of the territory. It is possible to obtain drumming from the woodpecker experimentally in reply to an analogous sound produced by knocking a tree trunk with the same cadence and rhythm as those used by the bird. The woodpecker approaches near to the source of the sound if the signal is continued in spite of its replies. These drummings only have signal value during the breeding season. If they are given experimentally outside this season they have no effect.

Therefore, it does not necessarily mean that a bird must have a song in its repertoire, for it to show a territorial behaviour. On the other hand, species that possess a song may have little territorial behaviour. This, for example, is the case in the bullfinch and many passerines which have a song composed of many learned motifs.

(f) The song and its role during greeting and invitation ceremonies

Pair formation is the result of a whole chain of reactions. There are numerous internal, hormonal and external motivation factors. These vary progressively during the various stages of the breeding season. Through variation of the threshold of sensitivity to signals coming from the channel these factors lead to a behaviour that is adapted to the physiological state of the bird. Mechanisms of interaction between signals that are perceived and those that are emitted (feedback) which lead to a gradual modification of behaviour, are comparable with induction between the various organs concerned in the development of an embryo, a condition that may be only achieved after having been initiated by preceding conditions. Compensating mechanisms can eventually intervene in the absence of a behaviour element. An example of this occurs during the mating of birds that have been artificially deafened. The occurrence of various behaviour elements is more or less modified depending on the importance of acoustic factors. It stops at a certain stage, which varies with the species, thus showing the indispensable role that acoustic signals play in the sequence of events. The sources of impulses during these ceremonies are mainly optic and acoustic and the two are often associated. Song does not appear to play the essential role among them which it has in territorial defence. However, it has an attractive function; for example if two previously paired bullfinches are placed in two separate cages at a distance one from another, the number of songs from the ♂ increases. In the robin the number of songs tends to diminish in the ♂ after pairing. If the ♀ is taken away, the song of the male becomes more frequent until he has found her again or until she is replaced by another ♀[63].

The role of song in maintaining contact between two birds, in addition to all sexual phenomena, is suggested by the large number of species giving vocal duets. A second bird starts to sing immediately after, and sometimes even before the first has finished its song. Thorpe[124] cites the two following observations on the subject:

A musician possessed a bullfinch trained to whistle "God save the King". An untrained canary put into contact with the bullfinch also learned to repeat this air after a year. The bullfinch often paused after the third line for a little longer than the melody required. The canary then took up where the bullfinch left off and ended it correctly. The other observation concerned two Australian magpies. The first bird had learned a flute melody of 19 notes in two distinct phrases. The second bird afterwards learned the melody from the first. They then took up the habit of always singing together, in counter melody, the first bird singing the first phrase and the second only the second. Later, the second, younger bird, died and then the first managed to sing the complete melody.

The first case mentioned shows that the synchronizing action of song may become artificially interspecific.

(g) *Acoustic signals other than song used during greeting and invitation ceremonies*

In order that future pairs may meet it is necessary for one of them, which up till then is considered as stranger, to penetrate into a territory which is particularly well defended at that time of the year. The first exchanges of signals can only be made at a distance. Presence of a newcomer immediately initiates a defensive reaction if appeasing signals have not already been given. Observations carried out on the night heron (*Nycticorax nycticorax*)[73,95], shows that the ♂ calls the ♀ from the position it has chosen for building a nest. If the ♀ moves too rapidly towards it a hostile reaction is produced in the ♂, which is all the stronger the nearer the two birds are together. It is probable that there is a synchronization in the sexual states of the two individuals during this period which may spread over several days. The ♂ heron gives a clicking of the beak in addition to vocal phenomena. This behaviour is similar to that of the stork. Some species use noises as acoustic signals during their "parade". During certain phases of the flight of the wood pigeon (*Columba palumbus*) it claps it wings to give a characteristic sound. The nightjar (*Caprimulgus europaeus*) behaves in a similar manner. The woodcock (*Scolopax rusticola*) and many species of snipe produce a sound during gliding that is caused by vibration of wing or tail feathers which are specially adapted morphologically to play this role.

Quite often the two paired birds give each other food or materials which may be used for building the nest. This behaviour is accompanied by acoustic signals characteristic of young birds, and is more often exhibited by the ♀ than by the ♂. Thus the ♀ robin invites[63] the ♂ to feed her by vigorously agitating her wings and producing the call of the fledgling. She then starts to behave like the young when they ask for food. The red-backed shrike (*Lanius collurio*) behaves in a similar way when she is fed by the ♂ while she is sitting on the nest. The ♂ tern gives a call which is very similar to that of the fledgling in his relations with the female. This return to a juvenile behavior seems to be quite widespread in birds. The "parades" therefore form a series of reactions differing with the species: they are silent in the ruff (*Philomachus pugnax*), while in the robin they are accompanied by some acoustic signals, and in the heron they become very complex and experimental deafness inhibits their development. Pairing is fairly normal in the case of artificially deafened bullfinches but the young produced by these couples (even if only one of the two partners is deaf) are abandoned when one or two weeks old[61]. Acoustic signals therefore play an important and generally indispensable role for the complete development of the numerous behaviour elements which culminate in reproduction, although the position of these signals in the sequence of the various acts varies with the species.

References p. 746

(h) Individual recognition

(i) Between adults

In the majority of cases amongst species defending a territory the individuals of the neighbourhood are generally known. Defence reactions are produced less easily by intrusion of neighbours than those made by strangers. The members of a colony of jackdaws (*Corvus monedula*)[71,72] recognize one another. Acoustics play an important role in the means by which pairs identify eachother. Mediterranean gulls (*Larus melanocephalus*) only recognize the other individuals in the colony which are living in their immediate neighbourhood. The possibilities of acoustic memory and subtlety in perception of details may be observed in many species. A great crested grebe (*Podiceps cristatus*) can discern the noise made by a particular motor-boat from others on the same lake. A ♀ allied woodhewer (*Lepidocolaptes affinis*) can distinguish the noise made by its mate alighting on a trunk, from the noise of other species alighting on the same trunk[115]. A herring gull (*Larus argentatus*) will only react to a call of its mate, in spite of a large number of similar calls in the channel[124].

(ii) Between parents and their young

Recognition between the parents and their young is an acquired phenomenon which appears at a more or less early stage. The parents generally only recognize their young after several days. This delay varies with the species. In experiments on jackdaw fledglings of about 15 days old it is possible to make changes from one nest to an other or to replace them by very young individuals. In the blackbird (*Turdus merula*) the parents oscillate and vibrate the nest support by alighting nearby and the young are conditioned to this signal very early and rapidly. It may be replaced experimentally by another signal. Young birds, reared in individual nests isolated from all vibration, reply to stimuli of pure frequencies produced by a generator, (300 to 10,000 c/s), by opening their beaks wide just as they do for a parent that is coming to feed them[87]. It has been shown in further trials that the signals has been learned after the third day (conditioning) and the birds stop replying to another signal. Stimuli produced by the parent or from another source that the bird considers as a partner are therefore learned quite rapidly. This non reversible phenomenon of "rapid learning" (imprinting) only occurs during a relatively short period in the life of an individual. It is designated under the name "Prägung" by german workers. The concept of a partner is also shown by this experiment.

Young geese[148] (*Anser anser*) that have been separated from their parents before hatching become attached to another bird or to a human being if this is the first creature that they see. It is impossible, afterwards, to make them follow individuals of their own species, even though they may be their parents. This type of conditioning does not take more than a minute, perhaps even less. Young mallards thus become attached to a human as soon as they have hatched if at this moment he imitates the mother's call[148]. It is often difficult to prove the existence of such phenomena in the acoustic field, even though they probably exist in many species.

Even though the young of some species are able to recognize the parent-partner in other birds, this identification does not appear, judging from feeding behaviour, to be specific to one or both of the parents, because young rooks (*Corvus frugilegus*) that have just begun to fly and are still being fed by their parents jump towards every adult passing nearby which provokes the demand for food. The same birds run after many moving objects when in captivity. A robin feeds any young bird which gives the food begging call, even though it is much older than its own[63]. Thus, in this case, acoustical signals are stronger than all other recognition stimuli, but there is as yet no definite proof that the latter do not exist. The behaviour of the ♀ mallard (*Anas platyrhynchos*) is an example. She comes to the help of every duckling that cries for help and after having chased away the aggressor she kills the young bird if it is not one of her own. The domestic hen also reacts to the distress signals of a strange chick. Turkey hens, which had been deafened as poults bred like normal hens, but behaved as if they could not differentiate poults and predators. Thus they even killed their own young immediately after hatching, as they would do with predators approaching the nest[153].

(iii) Recognition of the sex of a partner

The mechanism of acoustic signals may be responsible for this recognition, as, for example, in *Thryomanes bewicki*, or optical means may be used, for example in *Colaptes auratus*, in which only the ♂ has moustaches. If a female has moustaches painted on her, she is treated as a male[85]. In the majority of cases both acoustic behaviour and postures together reveal the sex of a bird, even in species in which the sexes have differing plumage. A male wren reacts to a stuffed wren placed on its territory by giving intimidating postures, songs and calls, which under normal conditions would chase the intruder away if it was a male. Since the stuffed bird does not reply it is therefore considered to be a ♀ and copulation is attempted[4]. In the American song sparrow the sexes, which are similar in appearance, recognize one another by the behaviour of the partner[91].

4. Emission of Acoustic Signals Induced by External Stimulation Factors

These signals constitute the group known as "call notes"—a term that has already been defined (see p. 711).

(a) Call notes for position and coordination of activities

These are given by the young when they are looking for their parents and by the parents when they are gathering their young together. Paired birds also use localization calls when they are separated. The bullfinch possesses two types of signal, one that is given when the partner is thought to be far away and the other, a "conversational" call, that is used when the birds are close together. The cohesion of winter flocks of bullfinches and chaffinches is ensured by sonic signals of this type. Many birds that carry out certain actions in common, gather together and coordinate their activities by means of acoustic signals. At nightfall, mallards exchange acoustic signals in order to gather together before leaving the place where they have spent the day. The calls they use are at first spaced out and then become more frequent, the rhythm of their emission becoming faster until they have almost finished grouping together and flight is imminent. If, at this precise moment, the experimenter gives a flight call the whole flock flies away, each bird repeating the call. An analogous behaviour in the coordination of movement has been observed in the grey-lag goose (*Anser anser*). When a group of these birds is at rest, or when they are feeding peacefully, they give a very slow chattering sound in which 6–7 syllables may be distinguished by the human ear. The syllables are reduced in number in relation to an increase in the tendency to move on, and at 5 syllables the geese already "think" more about moving on than of feeding. 3 syllables correspond to an accelerated walking pace and to intention flight, unless these 3 syllables are emitted at a higher tone than normal; in that case the flock only moves off at a walking or swimming pace. Finally, 2 syllables signify that the geese are ready to fly, and the monosyllabic call is the alarm signal to which all the adults reply by abruptly flying off without any preliminary call and at which the young run towards their mother[153].

Many flights of small Passerines are accompanied by acoustic signals. During their journey, migrating birds are particularly sensitive to the calls of their own species. Hunts organized in south-east France to catch wood-pigeons are based on this phenomena. The birds reply to visual signals (beating of wings) and to noises (wing-beats made by a decoy which is invisible to the bird, cooing noises[34]) and they go towards the source of the sound. A similar experiment has been carried out on the crane (*Grus grus*). During migration these birds have the habit of giving a characteristic call when they are in flight. If this call is given when a group of cranes is seen looking for a place to alight, thus simulating other birds on migration, the cranes start to turn above the experimenter and will even come down and alight if it is suitable ground. This signal has a strong effect as birds of this species are very cautious concerning halts they make during migration.

(b) *Calls related to environment*

The alarm call is one of the signals that has been most studied in this field as the bird may be persuaded to give it without much difficulty. It has an additional interest in that some cases of interspecific reaction have been observed. Most species possess a signal that has this function. The domestic hen warns her chicks of danger by means of two types of signal, one that corresponds to a potential danger and the other to an imminent danger. The corresponding behaviour produced in the young is, therefore, different, depending on the type of signal used. The herring gull has two types of alarm call, one that is used in its relations with other adult members of the group, and one that is specially destined for young birds[51]. In the pigeon, noises such as the beating of wings during an escape flight assume the role of an alarm signal. If an individual bird produces this noise, it causes panic amongst the others and induces flight in the whole group.

The alarm signal quite often has an interspecific action among small Passerines. For example tits give an alarm signal from their normal perch in the higher branches of trees. This alarm signal is interpolated into the calls of other birds living in the lower branches (other species of tits and finches). The alarm signal of the jay and the green woodpecker also have an interspecific action.

Distress calls have also been the object of experimental study. These calls are generally given by a bird that is caught by a predator but they may be obtained artificially by holding the bird in the hand. From experiments carried out on starlings[46,48], herring gulls[44] and rooks[31,34,43,47] it has been demonstrated that this signal has an effect on the whole group. When a recording of a distress call is transmitted by a loudspeaker in broad daylight a phonotaxis towards the point of emission is provoked in a group of rooks, after which the birds disperse. When this recording is broadcast at the night near the roosts of starlings or crows (in which several thousand birds may be gathered) it causes the birds to abandon the roosts almost immediately in a general escape flight. It has been observed, during different experiments, that starlings exhibit specific reactions. This specificity includes certain geographical races or sub-races. An alarm call of *Larus argentatus* recorded in America produces a reaction in birds in that country but when it is broadcast in Holland it has only a very slight stimulatory value. On the other hand, the same signal recorded from a herring gull captured in Holland and transmitted in the same country produces a normal escape reaction. In both cases, however, the same zoological species, *Larus argentatus*, is concerned. An analogous phenomenon of narrow specifity has been shown in the distress signals of jackdaws during experiments carried out in the north and south of France[34].

As a contrast to the preceding example, it has been shown in other species of birds that signals of one species can be interpreted by another species and will incite phonoreactions of the same type. The distress call of the jackdaw (*Corvus monedula*) stimulates reaction from the rook (*Corvus frugilegus*), the carrion crow (*Corvus corone*) and the magpie (*Pica pica*). All these species have been used in these experiments. When the distress signal of the jackdaw is transmitted over a loudspeaker it has at best only a very slight effect on the jay (*Garrulus glandarius*). The distress call of the jay, on the other hand, produces attraction and then escape behaviour in birds of all the other species cited above. This interspecificity, which will be discussed later, seems to be due to a mutual conditioning by geographic cohabitation or to the presence of physical characteristics of the reaction common to several species[34].

The assembling of individuals incited by an acoustic signal in order to frighten away a predator is a particular type of alarm behavior. The signal that is used differs from the alarm signal in that it is easily localized, lasts quite a long time and is repeated many times, its rhythm of emission increasing with the degree of excitation of the caller. The behavior it produces differs from that stimulated by the alarm call in that:

(i) the emittor "voluntarily" remains visible to the predator and draws its attention;

(ii) the emittor can, in certain cases, prevent the aggressive behavior of the predator into which it tries to instil fear.

This behavior has been observed in the house sparrow, particularly during the breeding season when, for example a cat kills a young bird. By giving this signal, the parents attract a large number of individuals which all stay at a certain distance from the predator and try to frighten it away. The postures that the cat then adopts show fear (ears laid back, walking with bent legs, search for cover). The mere sight of a cat prowling on territory near its young, produces this behavior in the blackbird. Jays and magpies often gather in order to chase an enemy.

This type of call may also be emitted by an isolated bird. In this way the ♂ pheasant (*Phasianus colchicus*) tries to frighten away a dog that has followed it and forced it to perch.

(c) Calls accompanying feeding behavior

The ♀ bullfinch has priority over the ♂ in feeding. Under these circumstances, in addition to other signals, she uses a particular call which is also used by the ♂

when he wishes to dominate other males and chase them away from food they desire.

The food begging call and posture of young birds are signals that have a strong influence on the behaviour of the adult (see p. 720–721).

Feeding behaviour may be produced solely by an acoustic signal in the herring gull. This signal, which corresponds to feeding, attracts other members of the species which are within hearing distance around the caller[44,45]. When a record of this signal is broadcast it attracts a group of birds that happen to be in the environment. They wait, immobile, for 10–15 minutes around the loud-speaker and then fly away again. Seagulls may be attracted from a radius of 3–5 km over the sea towards a powerful loud-speaker.

One of the most curious cases of acoustic signals related to feeding behavior is that of the African Honey-guides (*Indicator sp.*), which is reported in the detailed study of this bird published by Friedmann[42]. The bird attracts the attention and a phonotaxis of an animal of another species, a mammal—*Melivora capensis* or man and then it moves away and guides the animal towards a bee's nest by means of a signal. If the human does not react, this signal increases in intensity, and if the receptor does not advance the bird stops and then goes away. The method used for guiding the receptor may be considered as almost completely acoustic since the bird is for the most of the time out of sight. When it comes near the nest, the Honey-guide perches and stops calling. It stays waiting nearby and does not approach the nest to feed on the debris until it has just been destroyed by the follower. The reason why the bird guides towards the nest is still a problem.

It is highly probable that Melivora is interested from the food point of view. But, as far as the Honey-guide is concerned, it has been shown that birds may guide when their crop is full of food. It has also been shown that, in certain regions of South Africa where guiding used to be frequent, it has gradually disappeared during the last 20 years as a result of the increasing development of civilization which has caused the disappearance of possible associates (*Melivora*) or man's loss of interest.

The fact that the bird does not guide towards a bee's nest whose combs are exposed, and does not eat it, complicates the problem of finding motivating stimuli.

It may therefore be concluded that a temporary symbiotic behaviour is concerned and that the signal has an interspecific value which results from training, *i.e.* a mutual conditioning of the receptor and the emittor.

(d) "Irrelevant" behaviour

It sometimes happens that, in the middle of performing one action, an animal introduces an element of an entirely irrelevant piece of behaviour, without any apparent change in the external situation. Such irrelevant elements are called "displacement activities". They have often been described in relation to the postures of birds, but they are difficult to identify in their acoustic behaviour, particularly where the normal acoustic behaviour is imperfectly known. The following instances of song occurring under abnormal circumstances could be interpreted as examples of displacement in the acoustic field:

(a) A wounded chestnut shouldered hangnest (*Icterus pyrrhopterus*) that has fallen into the middle of a river started to sing while she was washed away by the current and continued to sing when she was saved.

(b) A robin caught in a trap sings for several seconds when it is taken up into the hand[63].

(c) A warbler that is hurt with a stone from a catapult starts to sing.

(d) A male wren released after ringing sings at the moment it comes to rest.

(e) Blue tits (*Parus caeruleus*) caught in a net often emit phrases of song in addition to the distress calls that are normal in such circumstances.

(f) Song after copulation of cock and pigeon.

Such behaviour appears to be "irrelevant", although further data might contradict this conclusion.

Under the heading of "irrelevant behaviour" one can also consider another phenomenon—that of "vacuum" activities. For instance, a starling kept in captivity was observed to go through the whole fly-catching behaviour when no fly was present[129]. It is possible that something similar accounts for the extraordinary amount of song in many cage-birds kept in captivity. It might be that there is a surplus of energy which appears in the form of song. Certainly, this opinion has long been held by birdfanciers, and is the basis of techniques which were employed in "song schools" where canaries were trained for the song competitions which used to be held in Western Europe and elsewhere. Thus Adams, a nineteenth-century bird fancier, as cited by Armstrong[4], writes that, to a certain extent, the less a bird uses its normal energy the better it will be as a pupil. It used to be the practice to keep the birds in dark cages, or only rarely bring them into the light, in order to reduce their activity. This complete or partial darkness did not interfere with the development of the testes and their hormonal secretions.

Manifestations of the conflicts of impulses in the bird may sometimes be shown by abnormal calls[3]. When there is a tendency to give two types of call at one and the same time the form of one of these may be imposed on the other. For example, a flight call which is normally produced after repetitions, may be given alone, or an alarm call may be repeated at a rapid tempo in the same way as food begging call.

5. Development of the Bird's Vocabulary with Relation to Age

(a) *Appearance of the first sounds*

The bird gives sounds from the earliest stage. An extreme case is that of the young herring gull which is capable of emitting feeble cries in reply to sounds reaching it even before it has hatched[51].

In nesting species, the first signals are generally manifestations of discomfort, then cries are produced in relation to feeding very soon afterwards. Calls of localization do not appear until the bird is ready to leave the nest. On the other hand, in species that do not build nests, such as the goose, these signals appear very early from the very first day[74].

There is always a change from simple to complex signals. At the beginning they are given alone, then together, until finally they are repeated several times, some of them being replaced by others that have just appeared. In the majority of cases it is the rule that a new sound appears suddenly, at a definite constant age, at least under rearing conditions.

The song appears in its juvenile form at an early age, when the young are just beginning to fly, at the age of 19 days in the blackbird[87] and 38 days in the white-

TABLE 61

TIME OF APPEARANCE OF THE SIGNALS OF YOUNG CHICKS (FROM SCHWELDERUP-EBBE[104])

A — Group reared in isolation; B — Group reared in the presence of adults; C — Group reared in isolation, but able to see and hear the adults.

Type of signal	Time of appearance of the signals in days					
	A		B		C	
	♂	♀	♂	♀	♂	♀
Laying call	—	255	—	263	—	258
Rhythmic call	208	262	213	264	210	259
Warning call	162	236	161	239	160	232
Threatening call	109	220	117	224	109	221
Nesting call	—	269	—	274	—	272
Summoning call	146	239	149	255	143	249
Song of the cock	232	—	209	—	200	—

throat (*Sylvia c. communis*)[101]. The signals may appear at different ages in the different sexes, for example, in the domestic hen (*cf.* Table 61)[103, 104].

(b) *Distinction between innate and acquired signals**

All the individuals of a birds species possess signals derived from a common genetic origin, which are, therefore, innate, in contrast to other signals which are acquired at different ages as a result of parental or social contacts. The idea of learning was already known by bird breeders several centuries ago, particularly with respect to song.

It is not easy to distinguish experimentally between innate and acquired signals. A fairly rigorous technique is used consisting of placing isolated individuals in soundproof chambers as soon before hatching as possible, and rearing them under these conditions. The individuals obtained by this method are named "Kaspar Hausers" by german workers from the name given to a child who was brought up isolated in a cave. Three different groups are recognized, depending on whether they are hatched in a soundproof chamber after artificial incubation or have been isolated at a more or less early age. Another method reported by Schwartzkopff[109] consist of the surgical removal of the cochlea of young animals, thus rendering them deaf. This method allows the birds to be left in contact with normal individuals avoids the effects resulting from isolation. Unfortunately it is not possible to carry out this operation on very young individuals. Another method evolved particularly by Nicolai[93] on the bullfinch and canary, consists in obtaining batches of interspecific eggs and rearing the young birds.

Differences between the signals of birds submitted to experiments and those of controls reared under normal conditions may be distinguished in recordings or with a very practised ear.

A systematic study of the development of the vocabulary has been obtained in some species by using these methods:

---

* See the chapter on learning (Marler), page 228.

In the blackbird (Turdus merula)[87]

Parts of the song of the young male are entirely innate: because songs of young wild birds and those of Kasper Hausers or birds rendered deaf are all similar. From the first spring, the young bird is able, by imitating the adults, to incorporate partial or entire motifs of their song into its own song. It is immediately able to imitate the exact tone, but is has to acquire a respiratory technique in order to imitate the rhythm and intensity. Parental training of the young ends in the middle of summer when the song of the adults ceases. The motifs that the young birds has learned are the only ones that may be distinguished in its song during the winter. The song the young bird has learned is the one that reappears the following spring, the juvenile components having disappeared. The young that have hatched from late nests and which have not therefore had the time to learn, as parental education stops in mid-summer, retain their learning faculties, and incorporate the songs of species living in the neighbourhood at the beginning of spring. Eight motifs which are all used in the song in varying succession originating from other species or variants of the individual signal, have been distinguished in the blackbird in this way.

In the domestic hen[104]

The signals of female chicks are entirely innate and those of the male are partially so. Vocal development may even be retarded by contact with the adults, probably due to the highly hierarchical despotism of a group of hens. The juvenile song appears more rapidly in the male if it is in contact with the adult as an example (*cf.* Table 61).

In the chaffinch (Fringilla coelebs)[122, 125]

Studies have been carried out mainly on the male song.

First type of experiment: The young birds are reared by their parents and isolated from the adults from the beginning of September. In the following spring all the birds, even those which were associated experimentally with other species during the winter, have the normal song of their species. Signals of other species are therefore not acquired but those that were learned during contact with the parents are retained.

Second type of experiment: The young birds are isolated as soon as they are hatched and put in small groups in a soundproof chamber so that they have no contact with song-birds of their species. In this experiment, the songs that are obtained in spring are different from normal songs, such as those obtained in the first type of experiment. The birds in each group have their individual songs which are related to one another physically. Although analysis has not allowed an appreciable discrimination between the different groups, it is, however, possible to recognize slight differences between them.

Differences between the songs of birds reared in isolation immediately after hatching and those of birds reared in isolation after September (having thus had a normal early youth), suggest that the very young birds acquired a certain number of characteristics of the song even before they were able to produce the full song. It has been verified experimentally that the song is fixed as soon as it has reached its full intensity and stays more or less unchanged for the whole life of the individual.

Songs that are induced by injection of testosterone propionate or by lighting do not differ from normal songs even though they are emitted during unnatural periods.

References p. 746

The song of the male chaffinch is thus almost completely innate; the bird only incorporates minor modifications in detail that it has learned.

In the whitethroat (Sylvia c. communis)[101]

This warbler possesses 25 different signals. These are emitted in the same way and at the same phases of the biological cycle by both the isolated and the wild individuals, and so are therefore innate. However, the bird reared in isolation lacks the trills of the courting song.

In the herring gull (Larus argentatus)[51]

Comparable results have been obtained in studies on the vocabulary of the herring gull. The major part of this is innate although two signals, at least, are learned from the adults.

In the bullfinch (Pyrrhula pyrrhula)[61,93]

Although the songs vary from one geographical region to another, they possess characteristics in common, for instance, certain qualities of sound and the disposition of several elements of the phrase. These constant characteristics constitute the innate portion of the song, as they are still present in birds that are reared without acoustic contact with their species. Learning starts at about the seventh week and is completed in one year for the male and in two years for the female. At the end of the first year the female uses the song that she has learned from the father and does not retain the acoustic signals of the mother. She completes her vocabulary during the first mating by incorporating the song of her mate. After this she loses the ability to assimilate new signals, even those of another male during a later mating. These facts have been shown by experiments of the following type. If a female bullfinch, reared by a father that possesses the normal song of the species, is mated with a male that has been reared by canaries, she will use both their signals for her song, which will be composed of alternating motifs borrowed from the two species, bullfinch and canary. There are two signals amongst those related to a social life which serve to maintain acoustic contact between individuals. The first, which is used when the birds are close, is completely innate. The second, which is emitted at a greater distance, is partially acquired. Only its monosyllabic character and certain qualities of timbre are innate. The call emitted by young birds that are reared in isolation is slightly different from that of wild, normal individuals. If they are reared by parents that do not belong to the species but whose signal largely corresponds to the natural characteristics of that of the bullfinch, they adopt this foreign signal and retain it even if, later on, they are put into contact with their own species.

Therefore, in conclusion, the proportion of innate and acquired signals and the length of time taken for learning vary with the species.

(c) Sub-song

The term sub-song was first used by Nicholson in 1927. He used the term to designate songs of weak sonic volume. It is now only applied to the individual who has not yet acquired its full song. Sub-song becomes progressively modified, and new motifs randomly dispersed and similar to those of the full song appear. The song is at this point named rehearsed song[147]. The development of the full song from the rehearsed song consists of an increase in the number of motifs belonging to the full song. In young birds this transition typically occurs during their first winter and early spring. In many species, (for instance: *Fringilla coelebs*[122], *Sylvia communis*[101], *Turdus merula*[87]) a similar process recurs each spring as the adult bird passes from a sexually quiescent phase into the reproductive cycle. The inverse phenomenon has been observed on some sick or aged birds (whitethroat)[152].

The sub-song has been particularly well studied in the chaffinch[125, 126] where it is emitted until the birds are about eleven months old. In some cases it is emitted by older birds, when it is related to the appearance of a weak motivation due to an increasing amount of sexual hormone.

Thorpe[126] recognizes seven characteristics:

1. The main fundamental frequency or pitch of the notes is apt to be lower than in the full song.

2. The frequency (or pitch) range of the sub-song as a whole and of the individual notes of which it is composed tends to be greater.

3. The sub-song is much quieter.

4. The overall pattern of notes is entirely different.

5. The length of the phrases of the song bursts is different, tending to be longer.

6. The sub-song is characteristic of lower sexual motivation, being generally produced earlier in the breeding season.

7. There is some evidence that, especially in young birds, the sub-song is in the nature of practice for the full song.

In many of the Turdidae, the juvenile song has these seven characteristics. The following species possess only four or five: mistle thrush (*Turdus viscivorus*), brambling (*Fringilla montifringilla*), canary (*Serinus canaria*), American goldfinch (*Spinus tristis*), dunnock (*Prunella modularis*). The sub-song is difficult to recognize in the Emberizidae, and many cases are still doubtful.

Fig. 385 illustrates of the sub-song towards the full song in *Fringilla coelebs*. The progressive diminution in the range of frequencies used and the appearance of the pattern of the adult song may be remarked.

Since the sub-song has a weak sonic level, is variable in composition and is different from the full song, it has not yet been recognized in many species. This is a field of research that has hardly been touched.

(d) *Imitation*

Imitation is proverbial in certain birds. Those that are "good singers" have songs that are composed of a genetic origin that is common to all individuals of the species, into which motifs are incorporated by learning and memory at certain ages during parental, social or biogeographical contacts. These motifs can be:

(a) Integral or partial copies of sequences heard in adults of the same or another species.

(b) Inventions of the individual, but arising from alteration of sequences heard in adults of the same or another species. This ability to incorporate signals appears to be quite widespread, but to a more or less developed degree. The chaffinch, for example, can learn to improve its adult song up to the age of 13 months but it never incorporates signals that are foreign to the species[125], even though these motifs may be present in its subsong.

References p. 746

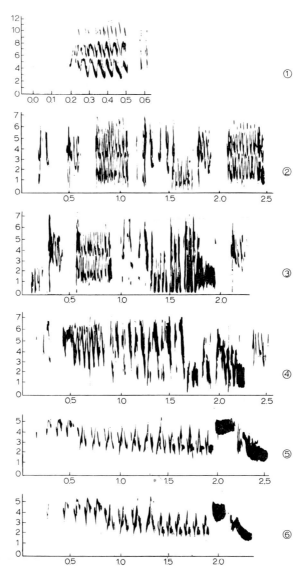

Fig. 385. Gradual evolution of the subsong to the full song in the chaffinch (*Fringilla coelebs*). 1–2: first elements of the subsong. 3–4: appearance of the pattern of the full song. 5–6: appearance of the full song. Ordinate, frequency in c/s; Abscissa, time in 0.1 seconds (from Thorpe, 1958)[125].

The best known example of an imitating bird is that of the canary, which is able to improve its song by imitating. In order to achieve this, breeders train the birds simply by isolating them after the first moult, or they may make several of them learn from a common "tutor". Nowadays records are used in the musical teaching of birds. Another case of imitation, which has been studied by Borror[20] in particular, is that of the mocking bird (*Mimus polyglottos*) which imitates the carolina wren

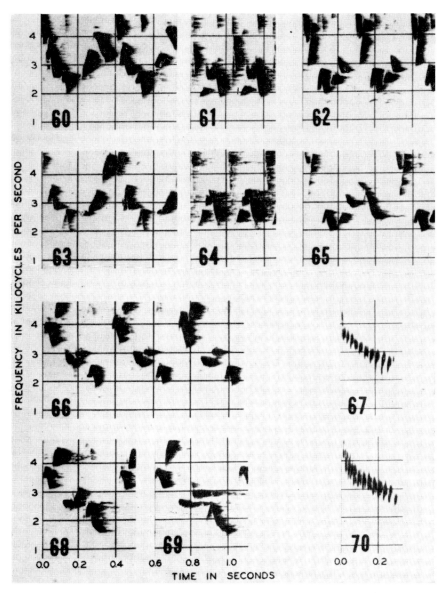

Fig. 386. Imitations produced by the mocking bird (*Mimus polyglottos*) of sounds borrowed from the repertory of the Carolina wren (*Thryothorus ludovicianus*). 63, 64, 65, 68, 69: elements of the wren's repertory imitated by the mocking bird 60, 61, 62, 66. Trill of the wren: 70, as imitated by the mocking bird, 67 (from Borror and Reese, 1956)[20].

(*Thryothorus ludovicianus*). The imitations are very faithfully reproduced (Fig. 386) and vary with the model song of birds from different geographical regions. The main difference is in the way in which the mocking bird sings, as the phrases last longer than those of the wren, and in addition it interprets other motifs which do not belong to its model. It is able to imitate the trill exactly.

Rearing in captivity increases the possibilities of imitation in many species. This

References p. 746

seems to be due to the fact that since the bird is able to accomplish the acts necessary for subsistence with ease and rapidly, a certain amount of motor energy is left over which is shown as song. Observations made on wild individuals have shown that imitation is practised mainly in spring during the increase in sexual activity[82]. This suggests a relationship to the sub-song. Gonadectomy suppresses it, and, reciprocally, injection of testosterone into the castrated bird results in its reappearance. It is generally more developed in the male than the female. It is difficult to say at the present state of knowledge whether this ability is beneficial for the bird. Has the signal any value? In any case, in Australia, where the greatest proportion of imitating birds are found, they generally live in dense forests. Does environment have an influence on imitation? At least 30 of the 160 species of passerines that live in Great Britain are imitators[83].

Imitation may be more or less developed in the different species. The robin, for example, imitates very little. On the other hand in the song of the yellow-hammer (*Emberiza citrinella*) "snatches" of song borrowed from the following species and lasting only a few minutes have been observed: greenfinch, chaffinch, sparrow, bullfinch, skylark, crested lark, wagtail, blue tit and coal tit, swallow, gold-crest, redstart, nightingale, thrush and blackbird, *i.e.* 15 species[116]. Imitation is not only confined to the song but may also be found in the calls and even the noises. The jay reproduces the sounds made by the woodcock with its wings during courting cere-

TABLE 62

EXAMPLES OF BIRD SPECIES THAT ARE ABLE TO LEARN AND INCORPORATE IN THEIR OWN SONGS THOSE OF OTHER SPECIES

Acrocephalus palustris	Marsh warbler
Acrocephalus schoenobaenus	Sedge warbler
Alauda a. arvensis	Skylark
Ammaspiza henslowii	Henslow's sparrow
Anthus pratensis	Meadow pipit
Carduelis cannabina	Linnet
Chloris chloris	Greenfinch
Coccothraustes coccothraustes	Hawfinch
Cyanosylvia svecica	Bluethroat
Emberiza citrinella	Yellowhammer
Emberiza schoeniclus	Reed bunting
Erithacus rubecula	Robin (European)
Fringilla coelebs	Chaffinch
Garrulus glandarius	Jay
Hippolais icterina	Icterine warbler
Hippolais polyglotta	Melodius warbler
Icterus galbula	Baltimore oriole
Lanius collurio	Red-backed shrike
Mimus polyglottos	Mocking bird
Phoenicurus phoenicurus	Redstart
Pyrrhula pyrrhula	Bullfinch
Saxicola rubetra	Whinchat
Serinus canaria	Canary
Sturnus vulgaris	Starling
Sylvia communis	Whitethroat
Turdus ericetorum	Song thrush
Turdus merula	Blackbird
Turdus migratorius	Robin (USA)
Zonotrichia leucophris	White-crowned sparrow

(e) Psittacism

In contrast to the natural learning abilities, psittacism only develops in captivity and consists of incorporating certain sounds borrowed from the human language into the birds' repertoire. Birds are the only vertebrates that have this faculty.

The parrot, jay and blackbird and certain crows can talk, or rather imitate human words (Fig. 387) when in captivity, but a long time is generally required to teach them a word or a phrase. In exceptional cases they can learn it after only having heard it once and it is then associated with an extra-ordinary emotion. Lorenz[148] cites a case of a jackdow that learned to say "He's caught in the trap" and of his parrot that was able to repeat "here is the chimney-sweep". We had a jay that was able to repeat a bird-call that it had only heard on the day that it was captured. Parrots have an excellent memory: von Lukanus had one that lived with a tame hoopoe that he had named Höpfchen, and the parrot quickly learned its name. When the hoopoe died the parrot no longer repeated its name. Nine years later von Lukanus acquired

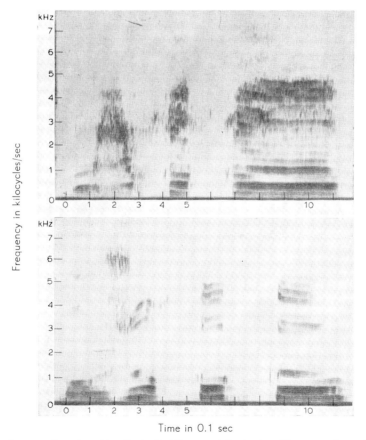

Fig. 387. Comparative phonetics: "voice" of the parrot (on top) and human voice saying "Bonjour Coco" (below). (Busnel, unpublished.)

References p. 746

another hoopoe and the parrot immediately cried "Höpfchen, Höpfchen", several times. Talking parrots do not often only say "hello" once, *i.e.* in an intelligent manner. The bird does not, in fact, understand what it says. It is able to associate a word with a well-defined fact (conditioning) but it can only in very exceptional cases use this ability to obtain a simple result. In no case has a parrot ever been taught to say that it wants to eat when it is hungry or that it wants to drink when it is thirsty.

(f) Dialects

In many birds where the song is only partially hereditary, variations in pitch, rhythm and the disposition of the sequences are added. This is the result of learning. Since the adult song is remarkably stable, slight geographical variations result from the fact that the birds have learned together and have adopted the same variations, thus giving rise to dialects[116]. The term dialect as it is used in modern literature only applies to adult birds, as the young do not include in their full song many of the passing imitations that they used the year before.

Knowledge of these dialects is still based on too large a number of subjective interpretations of many workers who encountered difficulties in describing or communicating the signals that they recognized. Systematic modern analyses by magnetic recording are still too few, and it is difficult to come to a conclusion on the extent and importance of dialects. It appears that many birds are able to use and conserve local variations and analyses would help in gaining a knowledge of their exact importance. Fig. 388 shows the range common to two songs of an american meadow-lark

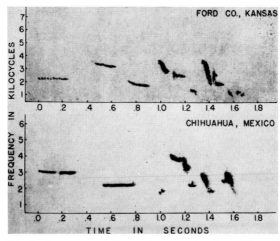

Fig. 388. Study of dialects: two songs of the western meadow lark (*Sturnella neglecta*) recorded from two different populations (from Lanyon and Fish, 1958)[68]

(*Sturnella neglecta*) in two different regions that have been studied and the variations in detail that occur due to the dialect. The formation of dialects appears to result from a very strong tendency of the bird to stereotype its signals. Is this phenomenon a reaction to the conditions of living by those species demonstrating it so as to produce a mechanism whereby they may be isolated from the numerous signals circulating in the canal? A greater variation in songs is observed among the fauna of small islands, where there are fewer species, than on the continent. Is it necessary for continental species to "standardize" and fix their repertoire in order to avoid confusion between species? Here again a solution may only be provided by further experiments.

6. Acoustic Signals – Their Role in Relation to Their Physical Properties

(a) Range of frequencies emitted and comparison with their auditory possibilities

Some examples of the frequencies used and perceived by the bird are given in Table 63. Even though the values given here may only be considered as indicators, several remarks may be made. The auditory possibilities are greater at both ends of the spectrum than the frequencies used. The maximum points of auditory sensitivity occur near the usual frequencies. The bird is barely able to hear sound below 50 c/s and above 15 to 20,000 c/s. Perception and emission of high frequencies would be only of slight value as their range in air is limited; in addition the signals are likely to be confused after numerous diffractions on intervening obstacles.

TABLE 63

COMPARISON OF AUDIBLE FREQUENCIES AND FREQUENCIES USED IN THE SONGS BY CERTAIN BIRDS SPECIES

Species	Range of sensitivity c/s			Range used in the song	Method	Authors
	min.	optimum	max.			
Anas platyrhynchos	< 300	2,000–3,000	> 8,000		T	Trainer[131]
Asio otus	< 100	6,000	18,000		T	Schwartzkopff[113]
Bubo bubo	60	1,000	> 8,000		T	Trainer[121]
Chloris chloris			20,000		T	Granit[52]
Columba livia	< 300	1,000–2,000			T	Trainer[131]
Corvus brachyrhynchos	< 300	1,000–2,000	> 8,000		T	Trainer[131]
Erithacus rubecala			21,000		T	Granit[52]
				2,000–13,000	P	Bremond (unpublished)
Falco sparverius	< 300	2,000	> 10,000		T	Trainer[131]
Fringilla coelebs	< 200	3,200	29,000		E	Schwartzkopff[112]
Loxia curvirostra			20,000		T	Knecht[146]
Melopsittacus undulatus	40	2,000	14,000		T	Knecht[146]
Passer domesticus			18,000		T	Granit[52]
Passerherbus henslowii				3,100 10,200	P	Borror[142]
Pica pica	< 100	800–1,600	21,000		E	Schwartzkopff[112]
Pyrrhula pyrrhula	< 100	3,200			T	Schwartzkopff[109]
	200	3,200	20–25,000		E	Schwartzkopff[111]
Strix aluco	< 100	3,000–6,000	21,000		E	Schwartzkopff[112]
Sturnus vulgaris	< 100	2,000	15,000		T	Granit[52]
Thryothorus ludovicianus				1,300–7,000	P	Borror[143]

P — Physical analysis of the song; T — Training; E — Electrophysiology.

It has been difficult to establish whether sound is perceived by the auditory pathway or by the vibratory sense (Herbst corpuscles) during work carried out on the perception of lower frequencies[107]. It is possible to make this discrimination by experimenting on subjects that have been rendered totally deaf by removing the inner ear. Even though the majority of birds are not able to hear deep sounds so well as man, Schwartzkopff has been able to show that the bullfinch has a better perception of vibrations than man (Fig. 389)[108], although this vibratory sense has no greater sensitivity than that of the ear. In fact it was shown during training that this bird perceives sounds and vibrations in the same way. Training of one sense is very easily carried over to the other[112].

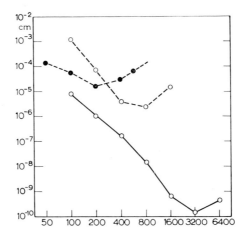

Fig. 389. Threshold of vibration in the bullfinch (*Pyrrhula*) --- O --- and man --- ● ---. Auditory threshold in the Bullfinch ——— O ———. Ordinate-amplitude of vibration. Abscissa = frequency of vibration (from Schwartzkopff, 1954)[112].

Audiograms may be established in two ways. The first depends on conditioning to a sound. The second, electrophysiological method, is based on the microphone potentials of the cochlea.

This latter method, which gives an electrical curve of the response of the ear, does not, however, produce information on the coincidence of this response with the cortical sensation. On the other hand, using the first method of conditioning to sounds, information on the auditory frequencies and the minimum thresholds of perception may be obtained (Figs. 390, 391). Sensitivity is high in the optimal zone of frequencies, near to the physical limits.

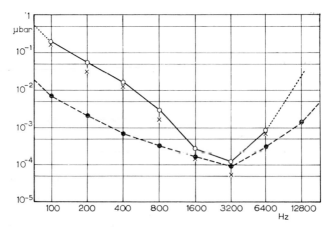

Fig. 390. Auditory threshold in the bullfinch (*Pyrrhula pyrrhula*) ——— O ——— and man --- ● ---. Abscissa = sonic pressure (from Schwartzkopff, 1954)[112].

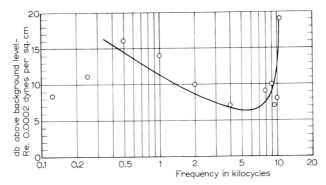

Fig. 391. Auditory threshold in the pheasant (*Phasianus colchicus*) (from Stewart, 1955)[118].

It is sometimes possible to obtain an approximate result by using the spontaneous reactions of the bird, without having to turn to training methods, for example by using phenomena such as those described previously on the subject of "gobbling" in the turkey. This response may be provoked by pure sounds of diverse strength, although each frequency has a minimum sonic level. Arrangement of these thresholds gives a curve which is very similar, although not exactly corresponding to the thresholds of minimum perception[105].

Auditory performances vary with the species, but the Strigidae may be distinguished from other groups[112] since, in relation to their nocturnal way of life, the majority of these birds have a particularly well-developed sense of hearing which may be confirmed by the following anatomical details:

The ratio of the surfaces between the orifice of the external ear and the tympanum are 3 or 6 in the Passerines and between 15 and 50 in the Strigidae.

The position and surfaces of insertion of the columella are specific to this group and modify the sensitivity of the ear. The surface of the fenestra rotunda is almost double than that of the fenestra ovalis in the majority of birds. However, in the Strigidae it is five times greater, thus increasing the mobility of the basal plate.

The middle ear plays an important role in the bird—its destruction leads to a considerable lowering in auditory sensitivity. It is, therefore, probable, that the improvements found in the Strigidae give this group a superior auditory sensitivity to other birds.

The differential tonal sensitivity in the optimal zone of perception appears to vary with the species (Table 64), but is, of necessity, of a high value in birds that

TABLE 64

DIFFERENTIAL TONAL SENSITIVITY IN MAN AND CERTAIN BIRDS (FROM SCHWARTZKOPFF[113])

Species	Zone of optimal sensitivity	Threshold of differential tonal sensitivity in %	Authors
Homo sapiens	1,000–3,000	0.3	Ranke[151]
Melopsittacus undulatus	1,000–3,000	0.3–0.7	Knecht[146]
Loxia curvirostra	1,000–3,000	0.3–0.7	Knecht[146]
Columba livia	1,000–2,000	6	Wassiljew[157]

References p. 746

are able to give very faithful imitations. Problems concerning the mechanisms of sound control have not yet been solved. Deafness in man causes modification in the voice, but auditory control in the bird does not appear to play an important role since the greater part of its repertoire remains unchanged after the inner ear has been destroyed.

(b) Recognition of the direction of a sonic source

Due to the vibratory nature of sound, the bird has three means whereby it may localize the direction of a sonic source. These require a binaural comparison by the two ears between the phase difference, time and intensity.

Small species find it difficult to make a precise comparison between phase difference, particularly in the lower frequencies, since the distance separating the two ears is small.

The difference between the time in which the song reaches the two ears is independent of the frequency, as is also the case for the speed of sound. This time difference is also used very little by small species.

On the other hand, perception of differences in intensity due to the acoustic shadow thrown by the head is not incompatible with the small distance separating the two ears. This shadow, which becomes more definite the higher the frequency, has been shown by measuring the microphone potentials of the cochlea while an intensity and frequency source is moved round an angle of 360°[110].

The bird probably uses these three means at the same time, especially when the physical characteristics of the signal may be so adapted. In any case, as experimental work has shown, the direction of a sound may only be recognized by using the two ears. In some species it has been possible, after training, to show the birds' ability for angular separation.

With two artificial signals it has been shown to be 20–30° in the Pine grosbeak (*Pinicola enucleator*)[52] and 20–25° in the Bullfinch (*Pyrrhula pyrrhula*)[110]. As a result of experiments on the behaviour of the domestic hen produced by the call of a chicken, this separating ability was found to be approximately 4° in this species.

(c) Physical characteristics of sounds emitted during behaviour

It is necessary to review the principal physical characteristics of acoustic signals that are used by birds since they are so diverse and have an importance in behaviour.

(a) *Frequency*. The frequencies emitted may either be pure and constant during the whole of the signal (Fig. 392).

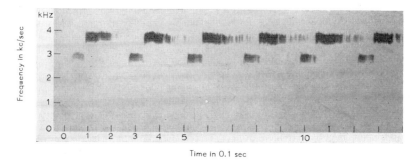

Fig. 392. Song of the great tit (*Parus major*) (original)

or vary slowly at the two ends of the signal (Fig. 393)
or vary sharply at the ends of the signal (Fig. 394) or during it (Fig. 395)
or be accompanied by numrous harmonies.

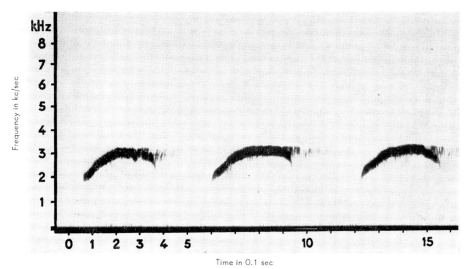

Fig. 393. Song of the nuthatch (*Sitta europaea*) (original)

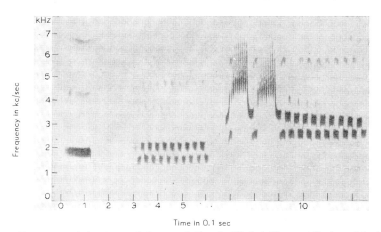

Fig. 394. Fragment of the song of the wood thrush (*Hylocichla mustelina*) (original analysis)[*].

(b) *Amplitude*. The amplitude varies during a signal, in most cases very rapidly. A very sharp slope where the maximum sonic energy is concentrated at the beginning or the end of a signal is termed a transient. This occurs frequently.

(c) *Segmentation, rhythm*. A signal, particularly when it is fairly long, is composed of brief but numerous interruptions which follow one another in a certain defined rhythm depending on the species. This rhythm may also be given by the repetition of the same note or phrase (Figs. 392, 393). The sounds, which cover a wide frequency

---

[*] Recording made by Prof. Kellog and published by the Cornell University.

References p. 746

band, are often composed of very short impulses which follow one another at intervals which may be less than 10^{-2} sec (Fig. 394).

(d) *Form*. The form of a signal embraces all the characteristics given above. It is a general term which is often employed in the description of signals. A characteristic frequency rarely has any value in itself as information. It has to be associated with characteristics of form. Sometimes, even, the form of a signal is more important than its frequency range.

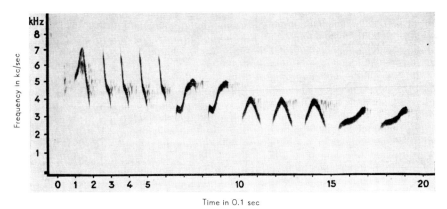

Fig. 395. Song of the willow warbler (*Phylloscopus trochilus*) (original).

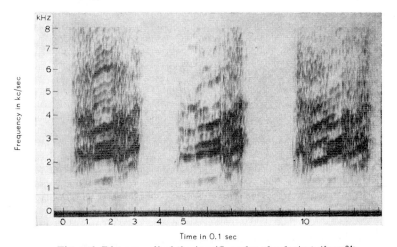

Fig. 396. Distress call of the jay (*Garrulus glandarius*) (from[34]).

(d) *Easily localized acoustic signals*

Certain signals, in order to achieve their aim, should indicate the position of the emittor. Songs have a role in indicating position and the limits of a territory, whereas many calls serve particularly for gathering individuals together (Fig. 397). For example, the signals used by the hen for gathering her chicks together are segmented and composed of numerous repetitions of relatively short notes with the maximum concentration of energy situated at the ends of the signal. The juxtaposition of these characteristics permits the simultaneous use of processes of adjustment of the sonic source and gives the signal maximum efficiency.

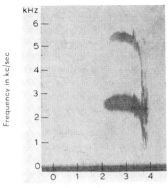

Fig. 397. Call emitted by the bullfinch (*Pyrrhula pyrrhula*) during regrouping behaviour (original).

Signals having these properties have a limited number of ranges, thus explaining the phenomenon of convergence that is illustrated in Fig. 398. A particular case of acoustic auto-localization is found in the cave species *Steatornis*[53] and *Collocalia*[86] which employ echolocation. Here again, the signal, which is of audible frequency, is composed of a succession of straight-front waves*.

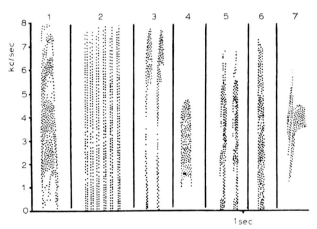

Fig. 398. Signals emitted by various species of birds at the sight of an Owl on a perch: (*Turdus merula*) (1); *T. viscivorus* (2); *Erithacus rubecula* (3); *Sylvia borin* (4); *Troglodytes troglodytes* (5); *Saxicola torquata* (6); *Fringilla coelebs* (7) (from Marler, 1957)[81].

(e) *Acoustic signals that may be situated with difficulty*

Sometimes birds are found in situations where it is useful for them to exchange signals without, however, revealing their position. The chaffinch, for example, emits

---

* See chapter on echolocation, p. 218.

two types of signals at the sight of a predator[78]. If the latter is perched, the signals are of a localizable type and attract the attention of other individuals towards a precise position. The signal is completely different if the predator is hunting; warning of danger is signalled by means of a call which gives the minimum of information on the emitters' situation (Fig. 399). Such signals are characterized by their long duration and stable frequency. They start and finish imperceptibly. Here again there is a difference in the form of these signals with the species (Fig. 399). Many "conversational" signals emitted by birds that are not far from one another are of this type. The young begging for food also emit calls of this type, since they are particularly exposed to the action of predators (Fig. 400).

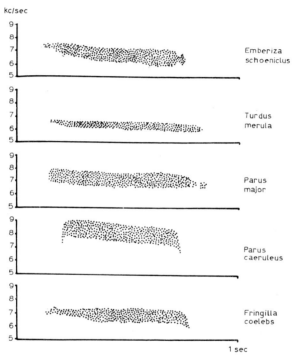

Fig. 399. Alarm signals emitted by various species at the sight of a hunting predator: *Emberiza schoeniclus*; *Turdus merula*; *Parus major*; *P. caeruleus*; *Fringilla coelebs* (from Marler, 1957)[81].

(f) Acoustic signals with an intraspecific stimulus value

This phenomenon of specificity is a necessary attribute of signals that are perceived at a distance and risk being received by species for which they are not destined. The songs of many zoological species occupying the same or neighboring territories are very different and play a large role in their isolation during reproduction[81]. This is the case for example in the warblers, *Phylloscopus collybita*, *P. trochilus*, *P. sibilatrix* and also *Fringilla coelebs* and *F. montifringilla*, amongst whom hybridization is possible. The closely related species of fly-catchers, *Muscicapa hypoleuca* and *M. albicollis*, have practically identical vocabularies but their songs are very different. The separation of these songs appears to have resulted from some external action,

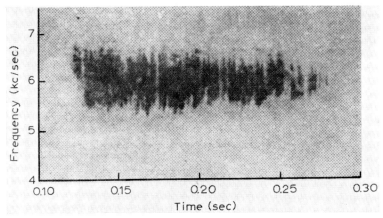

Fig. 400. Food begging call of a young wren (from Marler, 1955)[78].

presumably through selection due to environmental influences. It has been observed that *Fringilla coelebs* and *Regulus regulus* living on small island where the fauna is restricted, have much more variable and less well-fixed songs than birds of the same species living on the continent. A narrower divergence in physical characteristics is found in the signals related to behaviour of birds close to one another. Since these signals are of a weak sonic level and are associated with optical signals, it may be assumed that, in this case, risks of confusion are lessened. It is not possible to obtain these conditions in the case of signals that are exchanged at a great distance by individuals that are unable to see one another.

(g) *Acoustic signals with an interspecific stimulus value*

Signals that have an interspecific action are always related to situations connected with the environment (presence of food, danger, etc.).

In addition to these instances of interspecificity resulting from training to a signal foreign to the species, signals exist which can have an interspecific action due to their similar physical constitution. For example, in winter, the signals of tits often cause escape reactions in groups of chaffinches (Fig. 399). Interspecificity is due in this case to a phenomenon of convergence of all the characteristics of the signal[81]. In other cases, signals may be quite different but may possess one or a small number of these characteristics in common. If these are of an instigating type, the signal will have an interspecific action. This is noticeable in the jay and the carrion crow. The distress call of the jay covers a wide frequency range (Fig. 396) with a slight maximum of amplitude in the region of 4,000 c/s, where a modulation of amplitudes may also be observed. The signal has a very abrupt start and end. Even though the distress call of the crow is not fundamentally different, it has a different aspect (Fig. 401). The crow reacts to this call of the jay by a positive phonotaxy followed by flight. An artificial bird-call of the jay (whose sound is represented in Fig. 402) is shown to have an instigating action on the jay and the crow, in an equal manner. The form of this signal is the only attribute that it has in common with the two others. The range of frequency is markedly reduced and there is no

References p. 746

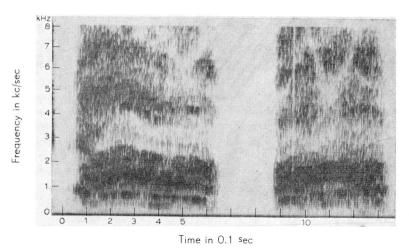

Fig. 401. Distress call emitted by the carrion crow (*Corvus corone*) (from[34]).

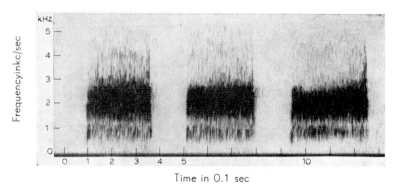

Fig. 402. Noise of an artificial bird-call used to attract the jay (original).

internal modulation. It may be concluded that the character "presence of transient" is the principal instigating element which gives it an interspecific value. If the distress signal of the jay is altered by successive copies across an apparatus where the bands pass too near to one another, thus changing the transient but not modifying the fundamental frequencies and their modulation, it will have almost no action on the jay but will still retain its effect on the crow. These experiments show that the transients at the start of the signal are an important character for the jay and the crow, as they, and they only, can provoke the reaction, but that the crow has a less specific reaction, as it is not indispensable for this species that they should be complete. Therefore, the whole of the signal does not have an interspecific value but only some of its characteristics.

In conclusion, the same phenomenon that has been shown for optical signals is again found in the acoustic field, that only some of the many characteristics of a signal are taken into consideration by the bird and are capable of stimulating behaviour.

(h) *Does the concept of sonic environment exist in the bird?*

Man is sensitive to the manner in which sounds reach or come back to him. This is called sonic environment. The perception of environment is the result of complex physical phenomena in addition to an analysis at the level of the cortex. Time and intensity of reverberation certainly predominate amongst the physical phenomena. In order to realize this it is sufficient to compare the acoustic properties of the same place before and after a fall of snow. The distance of a sonic source may be judged from its sonic level and, also, by comparing the frequency spectrum which is perceived with that produced at the point of emission. Is the bird able to perceive these phenomena? Even though no systematic experimental work has been undertaken on this subject there are certain pointers which favour a positive answer. Perhaps it takes advantage of this in certain situations. The great spotted woodpecker (*Dendrocopos major*) appears to choose a branch for "drumming" whose sonic qualities are such that they assure that the signal will reach the maximum possible distance. Among the different types of territory that birds choose is one that is called the song territory. It is a zone, sometimes even a precise point, where the bird sings. Has this localization acoustic reasons? It is essential that territorial songs should be widely diffused. Woodland species generally sing from a high perch. Species inhabiting open land have to sing in flight to obtain the same result. There is a correlated change in the form of the song; those species that sing from a perch tend to have a short song that is often repeated whereas the others tend to have a long song that is less often repeated. These modifications are found independently in different families[81]. The following observations which we have made on a hooded crow (*Corvus corone cornix*) lead to the conclusion that this bird is sensitive to reverberative properties. It stayed obstinately silent when it was held by its feet and carried from one room to another in the laboratory, in spite of the shaking-up it received. It immediately gave distress calls when it was taken into a sound-proof room, but did not repeat these cries once it was taken outside this room again.

The signals used by *Steatornis* and *Collocalia* for echolocation are composed of a series of short transient with straight slopes, a form which appears to be particularly well adapted to the analysis of echos. In this case, it is the emittor itself which judges the distance, but perhaps species exist where the receptor may make this measurement from a signal produced by another bird. The bullfinch has an attraction signal that it only gives when it is at a distance from other birds of its species. This call (Fig. 397) has a form which is analogous to that of signals used by other birds which use echo location. Starting at a weak sonic level, it increases progressively in amplitude and finishes by a very abrupt transient covering a wide range of frequencies. Does it give the receptor the chance of judging the distance of the emittor in this way? This is what remains to be shown and which would open up a new and interesting aspect of acoustic signals.

7. Conclusion

This chapter has been particularly designed to show the importance of acoustic signals in the behavior of the birds. Without pretending to be a complete resume of all the problems that are being investigated at the present moment, it was of particular advantage, within the limits of this book, to emphasize certain modern aspects of experimental work in this field which are based on the electro-acoustic processes of recording, reproduction and physical analysis of sounds, that are in the process of development. It is certainly impossible to isolate one line of behavior without always referring to immediately preceding and succeeding accompanying events, which are invariable and should always be recognized.

In any case, we think that these studies, limited to isolated acoustic signals, which can be carried out with greater accuracy than others, will provide the ethologist with a precise knowledge of one of the links of the whole process.

We have also intended to emphasize that the bird has been good material for this type of study, both in the laboratory and in the field, allowing the general and classical problems of acoustic psycho-physiology to be attacked and developed by means of new methods.

REFERENCES

[1] ALLEY, R. and H. BOYD, 1950, Parent-young recognition in the coot (*Fulica atra*), *Ibis*, 92, 46-51.
[2] ANDREW, R. J., 1956, Intention movements of flight in certain Passerines and their use in systematics, *Behaviour*, 10, 179-203.
[3] ANDREW, R. J., 1957, Comparative study of the calls of *Emberiza* sp. (buntings), *Ibis*, 1, 27-42.
[4] ARMSTRONG, E. A., 1947, *Bird display and behaviour*, L. DRUMMON, Ed., London.
[5] ASH, J., 1952, Habituated fear response in blue tits, *Brit. Birds*, 45, 288-289.
[6] AUTRUM, H., 1942, Schallempfang bei Tier und Mensch, *Naturwiss.*, 30, 69-85.
[7] BARBER, D. R., 1959, Singing pattern of the common chaffinch, *Fringilla coelebs* L., *Nature*, 183, No. 4654, 129.
[8] BENOIT, J., 1929, Le déterminisme des caractères sexuels secondaires du coq domestique. Etude physiologique et histophysiologique, *Arch. Zool. Exptl.*, 69, 219-499.
[9] BENOIT, J., 1937, Facteurs externes et internes de l'activité sexuelle. II. Etude du mécanisme de la stimulation par la lumière de l'activité testiculaire chez le canard domestique. Rôle de l'hypophyse, *Bull. Biol. France et Belg.*, 71, 393-437.
[10] BENOIT, J., 1956, L'instinct dans le comportement des animaux et de l'homme. VI. Etats physiologiques et instinct de reproduction chez les oiseaux, Masson, Paris, 177-260.
[11] BISSONNETTE, T. H., 1930, III. The hormonal regressive changes in the testis of the European starling (*Sturnus vulgaris*) from May to November, *Am. J. Anat.*, 46, 477-497.
[12] BISSONNETTE, T. H., 1931, V. Effects of lights of different intensities upon the testis activity of the European starling (*Sturnus vulgaris*), *Physiol. Zool.*, 4, 542-574.
[13] BLIVAISS, B. B., 1947, Interrelations of thyroid and gonad in the developments of plumage and other sex characters in brown leghorn roosters, *Phys. Zool.*, 20, 67-107.
[14] Boss, W. R., 1943, Hormonal determination of adult character and sex behavior in the herring gull (*Larus argentatus*), *J. Exptl. Zool.*, 94, No. 2, 181-206.
[15] BOYCOT, B. B. and J. Z. YOUNG, 1950, The comparative study of learning, *Symposia Soc. Exptl. Biol.*, 4, 432-453.
[16] BORROR, D. J., 1956, Variation in Carolina wren songs, *Auk*, 73, 211-229.
[17] BORROR, D. J., 1959, Variation in the song of the rufous-sided towhee, *Wilson Bull.*, 71, 54-72.
[18] BORROR, D. J. and C. R. REESE, 1953, The analysis of bird sounds by means of a vibralyzer, *Wilson Bull.*, 65, 271-276.
[19] BORROR, D. J. and C. R. REESE, 1954, Analytical studies of Henslow's sparrow songs, *Wilson Bull.*, 66, 243-252.
[20] BORROR, D. J. and C. R. REESE, 1956, Mockingbird imitation of Carolina wren, *Bull. Mass. Audub. Soc.*, 40, 309-318.
[21] BRACKBILL, H., 1943, A nesting study of the wood thrush (*Hylocichla mustelina*), *Wilson Bull.*, 55, No. 2, 73-87.
[22] BRACKBILL, H., 1948, A singing female wood thrush, *Wilson Bull.*, 60, No. 2, 98-102.

[23] BRACKBIL, H., 1958, Nesting behavior of the wood thrush, *Wilson Bull.*, *70*, No. 1, 70–89.
[24] BRAND, A. R., 1935, A method for the intensive study of bird song, *Auk*, *52*, 40–52.
[25] BRAND, A. R., 1937, Why bird song cannot be described adequately, *Wilson Bull.*, *64*, 11–14.
[26] BRAND, A. R., 1938, Vibration frequencies of passerine bird song, *Auk*, *55*, 263–268.
[27] BULLOUGH, W. S., 1942, The reproductive cycles of the British and continental races of the starling, *Phil. Trans. Roy. Soc., (London), Series B*, *231*, 165–246.
[28] BURGER, J. W., 1941, Experimental modification of the plumage cycle of the male European starling, *Bird Banding*, *12*, 27–29.
[29] BURGER, J. W., 1947, On the relation of daylength to the phases of testicular involution and inactivity of the spermatogenic cycle of the starling, *J. Exptl. Zool.*, *105*, 259–267.
[30] BURGER, J. W., 1949, A review of experimental investigations on seasonal reproduction in birds, *Wilson Bull.*, *61*, 211–230.
[31] BUSNEL, R. G., J. GIBAN, P. GRAMET, H. FRINGS, M. FRINGS and J. JUMBER, 1957, Interspécificité de signaux acoustiques ayant une valeur sémantique pour des corvidés européens et nord-américains, *Compt. rend. acad. sci.*, *245*, 105–108.
[32] CLARCK, L. B., S. L. LEONARD and G. BUMP, 1937, Light and the sexual cycle of game birds, *Science*, *85*, 339–340.
[33] COLLIAS, N. and M. JOOS, 1953, The spectrographic analysis of sound signals of the domestic fowl, *Behaviour*, *5*, 175–188.
[34] Colloque (1960) sur la protection acoustique des cultures et autres moyens d'effarouchement des oiseaux, *Ann. Epiphyties, I.N.R.A., Hors Sér.*
[35] DAVIS, J., 1958, Singing behavior and the gonad cycle of the rufous-sided towhee, *Condor*, *60*, No. 5, 308–336.
[36] DAVIS, L. I., 1958, Acoustic evidence of relationship in North American crows, *Wilson Bull.*, *70*, No. 2, 151–167.
[37] DAVIS, E. D. and L. V. DOMM, 1941, The sexual behavior of hormonally treated domestic fowl, *Proc. Soc. Exptl. Biol. Med.*, *48*, 667–669.
[38] DAVIS, E. D. and L. V. DOMM, 1943, The influence of hormones on the sexual behavior of domestic fowl, *Essays in Biology*, 171–181.
[39] DELAMAIN, J. and M. DELAMAIN, 1937, Le tambourinage des pics. *Alauda*, *IX*, 46–63.
[40] DOMM, L. V., D. E. DAVIS and B. B. BLIVAISS, 1942, Observations on the sexual behavior of hormonally treated brown leghorn fowl, *Anat. Rec.*, *84*, 481–482.
[41] FALCONER, D. S., 1941, Observations on the singing of chaffinches, *Brit. Birds*, *35*, 98–104.
[42] FRIEDMANN, H., 1955, The honey-guides, *U.S. Nat. Mus. Bull.*, *208*, 1–279.
[43] FRINGS, H. and M. FRINGS, 1957, Recorded calls of the eastern crow as attractants and repellents, *J. Wildlife Management*, *21*, No. 1, 91.
[44] FRINGS, H., M. FRINGS, B. COX and L. PEISSNER, 1955, Recorded calls of herring gulls (*Larus argentatus*), as repellants and attractants, *Science*, *121*, No. 3140, 340–341.
[45] FRINGS, H., M. FRINGS, B. COX and L. PEISSNER, 1955, Auditory and visual mechanisms in food-finding behavior of the herring gull, *Wilson Bull.*, *67*, 155–170.
[46] FRINGS, H. and J. JUMBER, 1954, Preliminary studies on the use of a specific sound to repel starlings (*Sturnus vulgaris*) from objectionable roosts, *Science*, *119*, No. 3088, 318–319.
[47] FRINGS, H., M. FRINGS, J. JUMBER, R. G. BUSNEL, J. GIBAN and P. GRAMET, 1958, Reactions of American and French species of *Corvus* and *Larus* to recorded communications signals tested reciprocally, *Ecology*, *39*, No. 1, 126–132.
[48] FRINGS, H., J. JUMBER and M. FRINGS, 1955, Studies on the repellent properties of the distress call of the European starling (*Sturnus vulgaris*), Occasional Papers Dept. Zool. Entomol., *55*, 1, University Park, Pennsylvania, May, 1955.
[49] GOODWIN, D., 1952, A comparative study of the voice and some aspects of behaviour in two Old World jays, *Behaviour*, *4*, 293–316.
[50] GOODWIN, D., 1956, Further observations on the behaviour of the jay (*Garrulus glandarius*), *Ibis*, *98*, 186–219.
[51] GOETHE, F., 1955, Beobachtungen bei der Aufzucht junger Silbermöwen, *Z. Tierpsychol.*, *12*, 3, 402–433.
[52] GRANIT, O., 1941, Beiträge zur Kenntnis des Gehörsinns der Vögel, *Ornis Fennica*, *18*, 49–71.
[53] GRIFFIN, D. R., 1957, Acoustic orientation in the oil bird, *Steatornis*, *Proc. U.S. Nat. Acad. Sci.*, *39*, 884–893.
[54] HARTSHORNE, C., 1956, The monotony threshold in singing birds, *Auk*, *73*, 176–192.
[55] HARTSHORNE, C., 1958, The relation of bird song to music, *Ibis*, *100*, 421–445.
[56] HARTSHORNE, C., 1958, Some biological principles applicable to song behaviour, *Wilson Bull.*, *70*, 41–56.
[57] HINDE, R. A., 1952, The behaviour of the great tit and some other related species, *Behaviour, Supplement*, *2*, 1–201.

[58] HINDE, R. A., 1959, The conflict between drives in the courtship and copulation of the chaffinch (*Fringilla coelebs*), *Behaviour*, 5, 1–31.
[59] HINDE, R. A., 1954, Factors governing the changes in strength of a partially inborn response, as shown by the mobbing behaviour of the chaffinch (*Fringilla coelebs*), *Proc. Roy. Soc.*, 142, 306–331.
[60] HINDE, R. A., 1958, Alternative motor patterns in chaffinch song, *Animal Behaviour*, VI, No. 3–4, 211–218.
[61] HÜCHTKER, R. and J. SCHWARTZKOPFF, 1958, Soziale Verhaltensweisen bei hörenden und gehörlosen Dompfaffen (*Pyrrhula pyrrhula*), *Experientia*, XIV, No. 3, 106–111.
[62] KLUIJVER, H. N., J. LIGTVOET, C. VAN DEN OUWELANT and F. ZEGWAARD, 1940, De levenswijze van den winterkoning *Troglodytes tr. troglodytes* (L.), *Limosa*, 13, 1–51.
[63] LACK, D., 1939, The behaviour of the robin. I. The life history with special reference to aggressive behaviour, sexual behaviour and territory. II. A partial analysis of aggressive and recognitional behaviour, *Proc. Zool. Soc. Lond.*, A, 109, 169–178.
[64] LACK, D., 1940, Observations on captive robin, *Brit. Birds*, 38, 133.
[65] LACK, D., 1940, Pair formation in birds, *Condor*, 42, 269–286.
[66] LACK, D., 1943, The life of the robin, *London H.F.A.G.*, Witherby.
[67] LACK, D. and H. N. SOUTHERN, 1949, Birds on Tenerife, *Ibis*, 91, 607–626.
[68] LANYON, N. E., E. WESLEY and W. R. FISH, 1958, Geographical variations in the vocalisation of the western meadowlark, *Condor*, 60, No. 5, 339–341.
[69] LEROY, P., 1952, Effets de la cortisone sur le développement et les glandes endocrines des rats et des hamsters, Thèse de Science, Paris, pp. 1–164.
[70] LÖHRL, H., 1958, Das Verhalten des Kleibers (*Sitta europaea caesia* Wolf), *Z. Tierpsychol.*, 15, 2, 191–252.
[71] LORENZ, K., 1931, Beiträge zur Ethologie sozialer Corviden, *J. Ornithol*, 79, 67–127.
[72] LORENZ, K., 1935, Der Kumpan in der Umwelt des Vogels: Der Artgenosse als auslösendes Moment sozialer Verhaltungsweisen, *J. Ornithol*, 83, 137–213; 289–413.
[73] LORENZ, K., 1938, A contribution to the comparative sociology of colonial nesting birds, *Proc. 8th Intern. Orn. Congr. Oxford (1934)*, 207–218.
[74] LORENZ, K., 1941, Vergleichende Bewegungsstudien an Anatinen, *J. Ornithol.*, 89, 19–29.
[75] LORENZ, K., 1950, The comparative method in studying innate behaviour patterns, *Symp. Soc. Exptl. Biol.*, 4, 221–268.
[76] LYNES, H., 1913, Early "drumming" of the snipe and its significance, *Brit. Birds*, 6, 354–359.
[77] MARLER, P. R., 1952, Variation in the song of the chaffinch (*Fringilla coelebs*), *Ibis*, 94, 458–472.
[78] MARLER, P., 1955, The characteristics of some animal calls, *Nature*, 176, 6–8.
[79] MARLER, P., 1955, Vocal communication in the chaffinch (*Fringilla coelebs*), *Proc. Assoc. for Study of Animal Behaviour*, 3, No. 2, 35.
[80] MARLER, P., 1956, The voice of the chaffinch and its function as a language, *Ibis*, 98, 231–261
[81] MARLER, P., 1957, Specific distinctiveness in the communication signals of birds, *Behaviour*, 11, 13–39.
[82] MARSHALL, A. J., 1950, The function of the bower of the satin bowerbird in the light of experimental modifications of the breeding cycle, *Nature*, 165, 388.
[83] MARSHALL, A. J., 1950, The function of vocal mimicry of birds, *Emu*, 50, 5–16.
[84] MARSHALL, A. J., 1951, Food availability as a timing factor in the sexual cycle of birds, *Emu*, 50, 267–282.
[85] MAYAUD, N., 1950, in GRASSÉ, *Traité de Zoologie*, Oiseaux, Vol. XV, Masson, Paris, pp. 539–653.
[86] MEDWAY, L., 1959, Echo location among *Collocalia*, *Nature*, 184, 1352–1353.
[87] MESMER. E. and I. MESMER 1956, Die Entwicklung der Lautäusserungen und einiger Verhaltensweisen der Amsel (*Turdus merula merula* L.) unter natürlichen Bedingungen und nach Einzelaufzucht in schalldichten Räumen, *Z. Tierpsychol.*, 13, 3, 341–441.
[88] MORLEY, A., 1953, Field observations on the biology of the marsh tit: *Parus palustris*, *Brit. Birds*, 46, No. 7, 233–238.
[89] MORRIS, D., 1956, L'instinct dans le comportement des animaux et de l'homme. VIII. The function and causation of courtship ceremonies, *Colloque Singer Polignac, Paris*, Masson Editeur, Paris, pp. 261–286.
[90] MOYNIHAN, M., 1953, Some displacement activities of the black headed gull (*Larus ridibundus*), *Behaviour*, 5, 1, 58–80.
[91] NICE, M., 1937, Studies in the life history of the song sparrow (I), *Trans. Linnaeus Soc.*, N.Y.4.
[92] NICE, M. M., 1943, Studies in the life history of the song sparrow (II), *Trans. Linnaeus Soc.*, N.Y.6, 1–328.
[93] NICOLAI, J., 1959, Familientradition in der Gesangentwicklung des Gimpels (*Pyrrhula pyrrhula*), *J. Ornithol.*, 100, 1, 39–46.
[94] NOBLE, G. K. and M. WURM, 1940, The effect of testosterone propionate on the black-crowned night heron, *Endocrinol.*, 26, 837–850.

[95] NOBLE, G. K. and M. WURM, 1942, Further analysis of the social behavior of the black-crowned night heron, *Auk*, *59*, 205–224.
[96] PALMER, R. S., 1941, A behavior study of the common tern, *Proc. Boston Soc. Nat. Hist.*, *42*, 1–119.
[97] PEZARD, A., 1918, Le conditionnement physiologique des caractères sexuels secondaires chez les oiseaux, *Bull. Biol. France et Belg.*, *52*, 1–176.
[98] POULSEN, H., 1951, Inheritance and learning in the song of the chaffinch (*Fringilla coelebs* L.), *Behaviouc*, *3*, 216–228.
[99] POULSEN, H., 1951, Inheritance and learning in the song of the chaffinch (*Fringilla coelebs* L.), *Behaviour*, *3*, 216–228.
[100] POULSEN, H., 1954, On the song of the linnet (*Carduelis cannabina*), *Dansk. Ornithot. Foren. Tidskr.*, *48*, 32–37.
[101] SAUER, F., 1954, Die Entwicklung der Lautäusserungen vom Ei ab schalldicht gehaltener Dorngrasmücken (*Sylvia c. communis* Lath.) im Vergleich mit später isolierten und mit wildlebenden Artgenossen, *Z. Tierpsychol.*, *11*, No. 1, 10–93.
[102] SCHILDMACHER, H., 1939, Über die künstliche Aktivierung der Hoden einigerVogelarten im Herbst durch Belichtung und Vorderlappenhormone, *Biol. Zentr.*, *59*, 653–657.
[103] SCHJELDERUP-EBBE, T., 1922, Beiträge zur Sozialpsychologie des Haushuhns, *Z. Psychol.*, *88*, 225–252.
[104] SCHJELDERUP-EBBE, T., 1923, Weitere Beiträge zur Sozial- und Individualpsychologie des Haushuhns, *Z. Psychol.*, *92*, 60–87.
[105] SCHLEIDT, M., 1955, Untersuchungen über die Auslösung des Kollerns beim Truthahn (*Meleagris gallopavo*), *Z. Tierpsychol.*, *11*, 3, 417–435.
[106] SCHOEMAKER, H. H., 1939, Effect of testosterone propionate on behaviour of the female canary, *Proc. Soc. Exptl. Biol. Med.*, *41*, 299–302.
[107] SCHWARTZKOPFF, J., 1948, Die Hörschwellen des Dompfaffen (*Pyrrhula pyrrhula minor* Brehm), *Naturwissenschaften*, *9*, 287.
[108] SCHWARTZKOPFF, J., 1948, Der Vibrationssinn der Vögel, *Naturwissenschaften*, *10*, 318.
[109] SCHWARTZKOPFF, J., 1949, Über Sitz und Leistung von Gehör- und Vibrationssinn bei Vögeln, *Z. Vergleich Physiol.*, *31*, 527–608.
[110] SCHWARTZKOPFF, J., 1950, Beitrag zum Problem des Richtungshörens bei Vögeln, *Z. Vergleich Physiol.*, *32*, 319–327.
[111] SCHWARTZKOPFF, J., 1952, Untersuchungen über die Arbeitsweise des Mittelohres und das Richtungshören der Singvögel unter Verwendung von Cochlea-Potentialen, *Z. Vergleich Physiol.*, *34*, 46–68.
[112] SCHWARTZKOPFF, J., 1954, Schallsinnesorgane, ihre Funktion und biologische Bedeutung bei Vögeln, *Acta XI. Congr. Intern. Ornithol.*, 189–208, Basel, 1955.
[113] SCHWARTZKOPFF, J., 1955, On the hearing of birds, *Auk*, *72*, 340–347.
[114] SIMMONS, K. E. L., 1951, Inter-specific territorialism, *Ibis*, *93*, 407–413.
[115] SKUTCH, A. F., 1945, Life history of the allied woodhewer, *Condor*, *47*, 85–94.
[116] STADLER, H., 1930, Vogeldialekt, *Alauda*, Suppl. to No. 1, 1–66.
[117] STADLER, H., 1934, Der Vogel kann transponieren, *Ornithol. Monatsschr.*, *59*, 1–9.
[118] STEWART, P. A., 1955, An audibility curve for two ring-necked pheasants, *Ohio J. Sci.*, *LV*, No. 2, 122–125.
[119] THIBAUT DE MAISIERES, C., 1940, Observations sur les picidés du Mont Bükk (Nord de la Hongrie), *Alauda*, Series III, *XII*, 17–65.
[120] THORPE, W. H., 1950, The concepts of learning and their relation to those of instinct, *Symposia Soc. Exptl. Biol.*, *4*, 387–408.
[121] THORPE, W. H., 1951, The learning abilities of birds, *Ibis*, *93*, 1–52 and 252–296.
[122] THORPE, W. H., 1954, The process of song learning in the chaffinch as studied by means of the sound spectograph, *Nature*, *173*, 465–469.
[123] THORPE, W. H., 1955, Comments on the bird fancier's delight together with notes on imitation in the subsong of the chaffinch, *Ibis*, *97*, No. 2, 247–251.
[124] THORPE, W. H., 1956, *Learning and Instinct in Animals*, Methuen, London.
[125] THORPE, W. H., 1958, The learning of song patterns by birds, with especial reference to the song of the chaffinch (*Fringilla coelebs*), *Ibis.*, *100*, 535–570.
[126] THORPE, W. H. and P. M. PILCHER, 1956, The nature and characteristics of subsong, *Brit. Birds*, *51*, 509–514.
[127] TINBERGEN, N., 1936, Zur Soziologie der Silbermöve (*Larus a. argentatus*, Pontopp.), *Beitr. Fortpflanzungsbiol. Vogel.*, *12*, 89–96.
[128] TINBERGEN, N., 1948, Social releasers and the experimental method required for their study, *Wilson Bull.*, *60*, 6–51.
[129] TINBERGEN, N., 1952, *The study of instinct*, Clarendon Press, Oxford.
[130] TINBERGEN, N., 1953, *The Herring Gull's World*, Collins, London.

[131] TRAINER, J. R., 1946, The auditory acuity of certain birds, *Cornell Univ., Abstracts of Theses*, pp. 246–251.
[132] TROUCHE, L., 1939, Nouvelles observations sur les manifestations vocales de la Bouscarle de cetti (Cettia cetti) *Alauda, III*, No. 2–4, 181–210.
[133] VALLENCIEN, B., W. LOHER and R. G. BUSNEL, 1955, Recherches sur l'audition et le comportement acoustique de quelques oiseaux d'intérêt agricole. I. Etude sur l'audation des Corbeaux à l'aide de la méthode du potentiel microphonique cochléaire, *Ann. Epiphyties, 2,* 185–200.
[134] VAUGIEN, L., 1946, Influence de l'éclairement artificiel sur la reproduction et le métabolisme du Serin: *Serinus canaria, Compt. rend. Acad. Sci., 222,* 926.
[135] VAUGIEN, L., 1948, Recherches biologiques et expérimentales sur le cycle reproducteur et la mue des oiseaux passeriformes, *Bull. Biol. France et Belg., 82,* 166–213.
[136] VAUGIEN, L., 1951, Sur le conditionnement, par la lumière et la chaleur, du cycle testiculaire du moineau domestique, *Bull. Soc. Zool. Franç., 76,* 335–339.
[137] VAUGIEN, L., 1952, Sur l'activité testiculaire, la teinte du bec et la mue du moineau domestique soumis, en hiver, à l'éclairement artificiel continu, *Bull. Soc. Zool. Franç., 77,* 385–407.
[138] WASSILJEW, PH., 1933, Über das Unterscheidungsvermögen der Vögel für die hohen Töne, *Z. Vergleich. Physiol., 19,* 424–438.
[139] WEIDMANN, U., 1958, Verhaltensstudien an der Stockente (*Anas platyrhynchos* L.), *Z. Tierpsychol., 15,* 3, 277–300.
[140] WEIDMANN, R. and U. WEIDMANN, 1958, An analysis of the stimulus situation releasing foodbegging in the black-headed gull, *Animal Behaviour, VI,* No. 1–2, 114.
[141] WEVER, E. G. and C. W. BRAY, 1936, Hearing in pigeon as studied by the electrical responses in the inner ear, *J. Comp. Psychol., 22,* 353–363.
[142] BORROR, D. J. and C. R. REESE, 1954, Analytical studies of Henslow Sparrow's songs, *Wilson Bull., 66,* 243–252.
[143] BORROR, D. J., 1956, Variation in Carolina Wren songs, *Auk, 76,* 211–229.
[144] BUSNEL, R. G. and J. C. BREMOND, 1961, Etude préliminaire du décodage des informations contenues dans le signal acoustique territorial du Rouge gorge (*Erithacus nubecula*), *C.R. Acad. Sci., 252,* 608–610.
[145] COLLIAS, N. E., 1960, An ecological and fonctional classification of animal sounds, in *Animal Sounds and Communication*, American Institute of Biological Sciences, Washington, Publication No. 7, 368–391.
[146] KNECHT, S., 1940, Über den Gehörsinn und die Musikalität der Vögel, *Z. vergl. Physiol., 27,* 169–232.
[147] LANYON, W. E., 1960, The ontogeny of vocalisations in Birds, in *Animal Sounds and Communication*, American Institute of Biological Sciences, Washington, Publication No. 7, 321–347.
[148] LORENZ, K. Z., 1952, *King Solomon's ring*, Methuen, London.
[149] MARLER, P. and D. ISAAC, 1960, Analysis of syllable structure in songs of the Brown Towhee, *Auk, 77,* 433–444.
[150] ODUM, E. O., 1942, Annual cycle of the black-capped Chickadee, *Auk, 59,* 499–531.
[151] RANKE, O. F. 1953, *Physiologie des Gehörs; Lehrbuch der Physiologie* (Trendelburg und Schütz), Vol. *Gehör-Stimme-Sprache*, Berlin, Springer Verlag.
[152] SAUER, F., 1955, Entwicklung und Regression angeborenen Verhaltens bei der Dorngrasmücke (*Sylvia c. communic*), *Proc. XIth International. Ornithol. Congr.*, (1954), 218–226.
[153] SCHLEIDT, W. M., M. SCHLEIDT and M. MAGG, 1960, Störung der Mutter–Kind-Beziehung bei Truthühnern durch Gehörverlust, *Behaviour, 16,* 255–260.
[154] THIELCKE, G., 1960, Mischgesang der Baumläufer, *Journ. f. Ornithol., 101,* 286–290.
[155] TINBERGEN, N., 1959, Comparative studies of the behaviour of Gulls (Laridae): a progress report, *Behaviour, 15,* 1–70.
[156] THORPE, W. H. and B. I. LADE, 1961, The song of some families of the Passeriformes. I. Introduction: The analysis of bird songs and their expression in graphic notation, *Ibis, 103a,* 231–245.
[157] WASSILJEW, PH., 1933, Über das Unterscheidungsvermögen der Vögel für die hohen Töne, *Z. vergl. Physiol., 19,* 424–438.

CHAPTER 25

ACOUSTIC BEHAVIOUR OF MAMMALS

by

G. TEMBROCK

1. Introduction

Although mammals have a very well developed vocal organ and many of them produce fairly differentiated forms of sound, they have hardly been dealt with in modern animal acoustics. Only in very narrowly circumscribed fields, especially within the framework of ultrasonics, have illuminating and interesting results been obtained. The classic laboratory animals, such as the white rat, have been tested for their sound responses in connection with neurological problems. There have been also excellent investigations on the functions of the middle and inner ear. So far, however, there is no comprehensive study of the sounds produced by the dog, our oldest domestic animal. A large number of investigations have been carried out in which recordings of the sounds of insects and birds, on ordinary records or magnetic tape, have been analyzed. Yet, corresponding research for the case of mammals is very much in its beginnings. One reason for this may be found in the fact that it is far more difficult to obtain useful recordings with mammals than with insects and birds and that large animals cannot be kept in laboratories. Recordings in the open air suffer from the difficulties which quite generally beset the systematic observation of mammals under natural conditions. Moreover, many of the sounds produced by them are connected with definite situations. The different forms of sound which are at the disposal of most species, are produced only under specific biological conditions and serve much more rarely for general communication than is the case with birds and insects. Olfactory stimuli determine the behaviour of many mammals to a much greater extent than acoustical ones.

All these factors lead to an almost frightening lack of knowledge in this field, and yet, we humans should have the greatest interest in learning more about the utterings of mammals: after all we belong to the *Hominoidea* and the genealogy of our sound forms (not of language itself!) must have its roots within the *Mammalia*. There are general laws in the formation of sounds and sound sequences in mammals which may be illuminating for the rhythm of our own language.

A systematic survey of the sounds of mammals must start with the erection of a framework which can be filled in later on. The aim must be to find the broad lines and regularities which would lead to an understanding of the wealth and diversity of the observations.

The production of sounds is not a fancy of Nature, but an expression of biological needs. Sounds are subjected to and modulated by the pressure of the environment, even though they are regarded primarily as epiphenomena of neurological processes

References p. 783

in the organism. In addition there are internal regulations which result from physiological functions. Finally genetic principles come into play which modify existing sound forms, create phylogenetic sequences and thereby offer a fruitful field for homology research.

Thus, we have to sift the available material, order and classify the diverse forms of sound so as to be able to define and characterize them objectively in the same way as was done in morphology 150 years ago. The fact that we can now preserve sounds (as we have been able to do it with organs and tissues for a long time) forms the decisive foundation for our task. At the same time we have to get away from the hardly usable phonetic transcriptions which are only very imperfect means for representing animal sounds. One could doubtless find numerous references in the literature on mammals in which animal sounds are transcribed by the symbols of human speech, but modern animal acoustics has no use for them. They could do no more than give us indications where to look for sounds. In the following survey we have, therefore, not attempted to give an exhaustive collection of all references on mammalian sounds. The aim is rather to survey and order the whole field and, using recordings, to evaluate some analyses of specific sound forms. Much would be gained if this attempt would show up the wealth of unsolved problems and provide a certain system for the somewhat haphazard research.

So far we have put the comparative-systematic and hence phylogenetic problems in the foreground. In addition to and in connection with this we have the ontogenesis with its special problems. We have also carried out preliminary investigations on the stimulating mechanisms which are designed to throw light on the functions of the individual components of the sound forms. All this is hardly more than a beginning and demands modesty and the confession of our ignorance. We shall have to collect a large amount of material if we want to catch up with the physiologists and anatomists. On the other hand we have the advantage of being able to use now and in all later stages of our own researches, their methods and findings. The bio-acoustician needs a good (and expensive) laboratory. Then he will get at the bottom of the "speech" of mammals.

2. The Physical Characteristics of Sounds and the Physiological Factors in the Production of Sounds

The physical characteristics of the sounds of mammals depend largely on the manner of the production of the sounds. The mechanical sounds are noises, while those formed with the larynx are predominantly harmonic sounds. In some isolated cases almost pure tones of single frequency can be detected (*Rousettus*[44]). (See Chapter of Kelemen p. 489.)

Harmonic sounds are combinations of frequencies. The vibrating system, the *ligamenta vocalica*, generates a fundamental frequency which is joined by harmonics. In addition there are vibrations within the "mouthpiece", the cavity of the mouth and its fittings (organs contained in it). Here velar and other frictional noises can superimpose themselves on the sounds proper. These supra-glottal cavities differ markedly from one mammal to the other.

The amplitude of the sound waves comes next. Besides the frequency, the amplitude is an essential physical characteristic of sound waves. Frequency and intensity of sound show certain relations to the size of the animal body which follow the physical laws underlying their generation. Small bodies can produce only relatively high and weak sounds. However, forms of sound with marked differences in pitch and intensity can exist within a given species. *Dasyprocta* can produce sounds which can be heard over distances beyond 100 m[31]. The voice of *Vulpes vulpes* L. extends well over 5 octaves! So far, complete surveys of the possible forms of sound are available only for very few species of mammals. In *Felis catus* L. 21 groups of sound can be distinguished[54], in *Vulpes vulpes* L. about 40 forms of sound (in roughly 28 groups)[85], in *Pan troglodytes* Blumenb. there are roughly 32 forms of sound[97]. This applies only to adult animals. About 23 types of sound can be distinguished in the domestic pig and 11 in the cattle of Camargue. In *Potos flavus* 7 different vocal expressions are known: (1) twitter, infantile sound of displeasure and for begging, at high intensity shrill and ending in a whistle; (2) barking, begging and contact call; (3) puffing, mainly used as an introduction to the barking sound; (4) chirping, soothing sound in the maternal behaviour; greeting sound; (5) spitting, fright and warning sound; (6) hissing, expression of strong excitement; (7) screaming, sign of highest excitation, associated with attack.

Physical analyses have been carried out so far only for very few species of mammals and principally in connection with the generation of ultrasound (*Rodentia, Chiroptera, Cetacea*). Table 65 gives the frequencies of sounds of various mammals based on original investigations by the author.

TABLE 65

FREQUENCIES OF SOUNDS OF VARIOUS MAMMALS
(Original)

Species	Type of sound	Behaviour	Principal frequencies (c/s)
MARSUPIALIA			
Phascolarctos cinereus Goldf.		not known	380–538
RODENTIA			
Microtus oeconomus Pall.	mono-syllab.	when catching	6080–7232
Microtus brandti Radd.	mono-syllab.	warning	3600–4300
Cynomys ludovicianus Ord	short sound	excitement	5120–7232
Citellus citellus L.		warning	4304–6080
Cricetus cricetus L.		defence	2560–3040 (3616)
		in the den	6080
Mesocricetus auratus Waterh.		defence	(904–1076) 2152–3616
Glis glis L.			3040–3616
Rattus norvegicus Berk.	multi-syllab.	2 days old (at touch)	6080–7232
Sciurus sciurus L.		excitement	3040–3535
Myocastor coypus Mol.	mono-syllab.	defence	1280–1520
Myocastor coypus Mol.	single sound	voice cont.	640 (–1280)
Dolichotus patagona Zimm.		defence	1520, 1808 (–3040)
Lagostomus sp.			(640) 2560
Erethizon dorsatus L.	sound sequence	voice cont.	269–538
	short sound	defence	2560–4304
Hystrix cristata L.	short sound	voice cont.	452, 904–1280
INSECTIVORA			
Erinaceus europaeus L.		defence	225–380 and 1808

[continued]

754 G. TEMBROCK

TABLE 65 (continued)

Species	Type of sound	Behaviour	Principal frequencies (c/s)
Aethechinus frontalis Smith		defence	2560, 6080, 7232, 8608
CHIROPTERA			
Pteropus medius Temm.		defence	3616, 4304, 6080–10240
PRIMATES			
Lemuroidea			
Lemur mongoz L.		voice contact	380
Lemur catta L.	mewing	voice contact	904, 1076
Ceboidea			
Alouatta sp.		voice contact	452, 538, 640, princ. frequ. 760, 1076
Cacajao (Uakari) sp.			1080, 2150, 2560–6080
Cebus albifrons Humb.	sound sequence	voice cont.	a. 6080, 7232, 10250, 12150 b. 1280, 1520
Cebus apella Kuhl			1080, 1280 and 1770–3335
Cebus capucinus L.	2-syllab. trilling	unspecif. encounter with other ind. of the species	907 and 2152
Cebus nigrivittatus Wagner			3616
Cebus sp. (Zimt-Kapuz.)		voice cont.	6080–8608
Saimiri sciurea L.	trilling		7080
		excitement	2560–4300 and 5120–7080
Lagothrix lagotricha Humb.		voice contact	640 and 905
Leontocebus rosalia Wied.			3040–3535
Oedipomidas oedipus Wgn.			1770–4300
Oedipomidas spixi			2560–3535 and incr. to 6080–8625
Callimico goeldii Thomas		excitement	8608–10240
Hapale jacchus L.		excitement	2560–4304 (3616)
Ateles geoffroy v. Hasselt & Kuhl		voice cont.	3040–3616
Ateles sp.	"barking"	excitement	1076–3040 (2152)
	sound sequence	voice cont.	5120–6080 (10240)
Cercopithecoidea			
Simia (Macaca) speciosa Temm.	grunting	excitement	(453) 538
Macaca maura Cuv.		excitement	(3616) 4304
Macaca mulatta Zimm.	hissing sound	defence	640–4304
	grunting sound	excitement	538–640
	whimpering	unspecif.	538–3616
Macaca silenus L.	hissing sound	defence	320–1520 and 760
Macaca memestrina L.	whimpering	unspecif.	340–452
Macaca radiata Geoff.	screaming ("ke")	excitement	2152–2560
Macaca speciosa Temm.	2-syllab. call		1076–1520
Papio sphinx L.	short vowel sound	excitement	380–538
	cry	voice contact	600–1200
Papio leucophaeus Cuv.	grunting	excitement	452 (95–1076)
	screaming ("ke-ke")	excitement	600–1200
	cry	voice contact	300–4800
Papio doguera Puch.	screaming	excitement	538 (320–1280)
	bleating sound		452–904 (3616–4304)
	short vowel sound	excitement	640
	barking sound	warning	269–1808

TABLE 65 (continued)

Species	Type of sound	Behaviour	Principal frequencies (c/s)
Papio hamadryas L.	screaming	excitement	670–1808
	short vowel sound	excitement	640 (320–1280)
	barking sound	warning	640–1520
Theropithecus sp.	barking sound		538–640
Cercocebus torquatus Kerr.	call	excitement	300–4800 (beginning 1200–2400)
	trilling	unspecif.	2152–3616 (1808–4304)
Cercopithecus aethiops L.	hissing	threatening	320–538 and 640–1076
Cercopithecus nicticans L. and *petaurista* Schreb.	hissing	threatening	640 (538–904)
	grunting	excitement	640
	trilling	voice contact	3040–3616
	chirping	warning	3040–3616 (2560–7232)
Cercopithecus talapoin Schreb.	hissing	threatening	1280–5120
	screaming	excitement	2125 (1280–5120)
	trilling-chirping	excitement	3040–3616
Cercopithecus diana L.		voice cont.	1808–3616
Cercopithecus mitis Wolf		voice cont.	4304 (8608)
Presbytes entellus Dufr.		defence	4304–7232
Pongidae			
Symphalangus syndactylus Desm.	call	excitement	760–1080
Hylobates hoolock Harl.	calling-sequence (howling)	excitement	640, 760
Hylobates lar L.	calling	excitement	438–904
Pan troglodytes Blum.	oo-howling	excitement	380–452
	oo-sounds (staccato)	excitement	269–320 rising to 904 at the end
	oh-sounds (staccato)	excitement	538–904
	clear sounds	feeding	904–1076
Gorilla gorilla Sav. e. Wym. (2 years old)	sound sequence (staccato)	play	(134) 229–640
CARNIVORA: FISSIPEDIA			
Canis familiaris L. (Alsatian)	barking	excitement	452–(904)
(Eskimo-dog)	howling	excitement	538 (904 and 1080)
(Dachshund)	growling ("yum-yum")	pairing	452
Canis dingo Blum.	barking	excitement	538 (904, 1520)
	howling		538–760 end: 452
Canis lupus L.	barking	excitement	320–904 (452)
	howling		380 (760)
	growling ("yum-yum")	pairing	380 and 450 (whining component: 760)
Canis aureus L.	barking	excitement	1076–1280
	howling		904 (1808) falling to 760–640
Canis anthus Cuv.	barking	excitement	904–1076
	howling		760–1076
Canis latrans Say	barking	excitement	1076–1280
	howling		1076–2152
Cuon alpinus Pall.	whimpering	(male)	8608–10240
Lycaon pictus Temm.	sound sequence	female (mother)	(640) 760
	whimpering		3616

[continued]

TABLE 65 (continued)

Species	Type of sound	Behaviour	Principal frequencies (c/s)
Lycaon pictus Temm.	barking	excitement	640
Nyctereutes procyonides Gray	whining	excitement	1080
Alopex lagopus L.	barking	excitement	904 (3216)
	barking sequence	mating contact	904–1076
	growling ("yum-yum")	mating contact	160–380
Vulpes corsac L.	whimpering	defence	(1220) 2152–3616
	yelping	fighting	1808–2152
Vulpes vulpes L.	barking	excitement	640–(5120)
	barking sequence	mating contact	640–1076
	growling ("yum-yum")		113–380
	whimpering	greeting	1808–4304
	whimpering	inferiority	2560
	whimpering	threatening	640 incr. to 1280
	yelping ("ke-ke")	defence	740–904
	whimpering growl	contact	113 (134) (2560–4304)
Megalotis chama Smith	growling	mating contact	380–538
Urocyon cinereo-argentatus	barking	excitement	904–1076
Ailuropoda melanoleucos Edw. (2 years old)	2-syllab.	expecting food	640–1808 (–3616)
	mono-syllab.	defence	1280–1520
Ursus arctos L.		defence	380–538
Tremarctos ornatus Cuv.		defence	320, 640
Thalarctos maritimus Desm.	call	voice contact	160–640
Selenarctos tibetanus Cuv.	sound sequence	male (mating)	640–760
	call	female	(269) 640 (–904)
Helarctos malayanus Raffl.	call	female	904 (1280–3040)
Bassariscus astutus Licht.			640–904
Lutreola lutreola L.	mono-syllab.	defence	1808–2152
Mustela nivalis L.		defence	2560–3040
Mustela putorius L.	sound sequence	mother with young	640
Martes martes L.	sound sequence	defence	452–1220
	yelping	play	538, 640
Grison vittatus Schreb.	sound sequence	while feeding	2152 (1808–2560)
Gulo gulo L.		defence	538, 760 (2152–2560)
Mellivora capensis Schreb.	sound sequence	voice cont.	2152–4304 (3040)
Amblonyx sp.		voice cont.	2152–4304 (3040)
Pteronura brasiliensis Forst.	purring		380–640 (452)
	sound sequence	voice cont.	(760, 904) 1520–3040
	bi-syllabic		452–760
Suricata tetradactyla Schreb.	mono-syllab.	voice contact	452–1026
Crossarchus fasciatus Desm.	mono-syllab.	voice contact	(538)–640
	trilling	person. contact	905–1020
	yelping ("ke")	defence	905–3040
Hyaena hyaena L.	sound sequence	while feeding	1808–2152 (904)
		defence	905, 1080–1520
Hyaena brunnea Thunb.	growling	defence	452–640
			452–904
Crocuta crocuta Erxl.	"laughing"		538–904 (–2560)
	growling		160–380
	sound sequence	voice contact	from 226 incr. to 452–904
Lynx lynx L.	sound sequence	(male) mating	640–904 (760)
Lynx caracal Schreb.	mewing	male	1280–1808

TABLE 65 (continued)

Species	Type of sound	Behaviour	Principal frequencies (c/s)
Felis chaus Güld.		female (mating)	640–904 (760)
	mewing	male (mat.)	904–1076 (1808)
	call	male (mat.)	640–904
Felis pardalis L.	mewing	male	904–1520 (1076)
Felis catus L. (Siam)	mewing		760–1520
Felis sylvestris griselda Thomas	mewing		600–1200
Felis wiedi Schinz	mewing		1808
Felis serval Schreb.	mewing		2150, 2560 (1808)
Felis chaus Güld.	mewing		538–1520
Puma concolor L.	mewing		452, 905–1520
	purring		269 (226)
Panthera leo L.	roaring		160, 190–1076
	(princip. sound)		(226–905)
	roaring (last sounds)		134–160
	rumble	defence	134–160
	cooing	mating beh.	113–134
Panthera tigris L.	call		320–538(160,190)
	snorting		113
		defence	640–1080
Panthera pardus L.	call sequence		190–226
	cooing		190–269
Panthera onca L.	mewing		640–1080
	call sequence		135–640
Panthera nebulosa Griff.	mewing		600–2400
Panthera uncia Schreb.		defence	1076–3616
Acinonyx jubatus Erxl.	"barking"(mewing)		760, 905 (1770–2150)
CARNIVORA: PINNIPEDIA			
Zalophus californianus Less.	call sequence	awaiting food	380–1280 (640)
	call sequence	defending territory	380–1080 (640)
Otaria byronica Blv.	sound sequence	mating	226–905 (380)
Arctocephalus pusillus Schreb.	a-sounds	territ. beh.	640–1080 (760)
Callorhinus ursinus L.	sound sequence	voice cont.	452–904
Odobenus rosmarus L.	sound sequence	voice cont. (young)	380–538
Halichoerus grypus F.		defence	380–904
Phoca vitulina L.	"grunting"	mating (male)	640–1280
Mirounga leonina L.	sound cont.	voice cont.	640–904
Macrorhinus angustirostris Gill.	howling		320–380(226–905)
PROBOSCIDEA			
Elephas maximus L.	rumble	excitement	95–380 (–640)
	trumpeting (long)	excitement	640, 760 (–1520)
	trumpeting (short)	awaiting food (begging)	640
Loxodonta africana Blumb.	rumble	excitement	80–905
SIRENIA			
Trichechus manatus L.			4304
PERISSODACTYLA			
Tapirus terrestris L.	call	excitement	2152–3616
	2-syllab. sound	excitement (female)	3040, 3535
Diceros bicornis L.	call	voice contact	904
	rumble	threatening	269–452
Rhinoceros unicornis L.	call	voice contact	320, 380
Equus asinus L.	call sequence "I"		1770–3040
	Call sequence "A"		640–905
Equus grevyi Oust.	call sequence "I"		640–1520
	Call sequence "A"		190–452
Equus quagga antiquorum Sm.	short sounds	defence (female)	905–1076 (–1280)
Equus quagga granti Wint.	neighing	(female)	640–1808
	neighing	(male)	640–905

[continued]

758   G. TEMBROCK

TABLE 65 (continued)

Species	Type of sound	Behaviour	Principal frequencies (c/s)
Equus caballus L.	neighing		1076–2152
	neighing (the last sounds)		380–538
(Isabell)	neighing		2150–3040 (2560)
(Konik)		threatening	320, 380, 538
Equus zebra hartmannae Matsch.	neighing	(male)	1520, 1808 (2152)
Hemionus h. hemionus Pall.	call sequence	(male)	a. 640, 760
			b. 1520 (1280–3040)
Hemionus h. onager Bodd.	call sequence	(male)	a. 538, 640
			b. 1520–2152
Equus przewalskii Pol.	neighing		1520–2152 (2560)
Asinus somaliensis Sclat.	call sequence	(male)	a. 380 (–904)
			b. 1520–2152
ARTIODACTYLA			
Sus scrofa L.	grunting		113–760 (905)
Mangalica pig	grunting	(female)	160, 190
Domestic pig	grunting	(female)	269–320 (95)
Dicotyles pecari Fisch.	sound sequence	excitement	160–538
Dicotyles tajacu L.	short grunting	unspecif.	80, 95 (–538)
Potamochoerus porcus L. (= *penicillatus* Schinz)	grunting		135, 160–538
Potamochoerus larvatus Cuv.	grunting		80–460 (538)
Phacochoerus aethiopicus Pall.	sequences	mating (male)	380–1770
Hippopotamus amphibius L.	call sequence	voice contact	95–113 at the end: 380–640
Choeropsis liberiensis Mort.	grunting		190–270 (640)
Camelus dromedarius L.	roaring		226–269
		mother with young	126, 190 (–1520)
Camelus bactrianus L.		mother with young	190
Lama glama L.	short sound	defence	320–452
Lama huanacus Molina	sound sequence	excitement	1520–1808
Muntiacus muntjac Zimm.	sound sequence	(male)	7232–8608
Capreolus c. pygargus Pall.	barking		640–760, 904–1076
Alces alces L.	sound sequence	(female)	80, 95, 640–1520
Rangifer tarandus L.	sound sequence	(male) rut	160–226
Cervus elaphus bactrianus Lyd.	call sequence	rut	a. 760–1076
			b. 1280–3040
Cervus elaphus L.	"bellow"	rut	320–452
Cervus elaphus canadensis Erxl.	call		1280–2560 decr. to 640–905
Axis axis Erxl.	call	(male)	538–1770 (640–760)
Dama dama L.	call	(female)	452
	call	rut (male)	380–538
Pseudaxis sika Temm.	call		1520–3535
Rucervus duvauceli Cuv.	call	rut	453 and 538
Elaphurus davidianus M.Edw.	call	rut	226–538 (–1080)
Capreolus capreolus L.	call	(young)	3535–4300 decr. to 760–905
Taurotragus oryx Pall.	long call		861, 904, 1076–3616
Taurotragus oryx Pall.	call		135–1080
Oryx algazel Oken	call	mother with young	380, 452–760
Connochaetes taurinus albojubatus Thom.	call sequence		95–160 (226)
Connochaetes gnu Zimm.	call sequence		452–538
Antilope cervicapra Pall.	call sequence	male (mating)	552–640
	short sound	female	760–1280

TABLE 65 (continued)

Species	Type of sound	Behaviour	Principal frequencies (c/s)
Addax nasomaculatus Blainv.	call		(760) 1808–3040
Gazella subgutturosa Güld.	call	male	380–538
Gazella soemmeringi Cretzschm.	call		1280–2560 (–3616)
Antidorcas marsupialis Zimm.	call	male	1076, 1280 (–3616)
Rupicapra rupicapra L.		male	640–904 (2540–3616)
Ovis aries musimon Pall.	call	awaiting food	620, 760, 1520–1808 (–3616)
Capra hircus L.	call		226, 270–3040
Ovis aries L.			
(Moorland sheep)	call	awaiting food	760–1808 (380)
(Soay sheep)	call	awaiting food	760–1520 (380)
(Eastfrisian sheep)	call	awaiting food	640–1520 (380)
Bos taurus L. (Aurochs)	call	awaiting food	190–904 (380)
(Watussi)	call	awaiting food	226–1520 (640)
Bos indicus (Noelle-zebu)	call	awaiting food	270–320
(Dwarf zebu)	call	cow with young	320–538
Bubalis bubalis L.	short sound		904, 1076–1280
Bibos banteng Raffl.	long call	female (awaiting food)	904–2560
Poëphagus grunniens L.	call	cow with young	113–190
Bison bison L.	call	cow with young	48–95 (160)
Bison bonasus L.	call		57–95 (–270)

The data presented in Table 65 serve merely as preliminary information. They have been obtained from recordings of a few (or only a single) specimens of the species in question. The frequencies were determined with the sound spectrometer SSP-10 (RFT-Köpenick) which filters out 36 quarter octaves (from 40 to 17,216 c/s). The numbers given in the table are the mean values of the filter ranges. The recording instruments have an upper frequency limit of about 10–14 kc/s. Some examples of recordings are shown in Fig. 403.

Apart from the physiology of the production of the sounds which will not be dealt with in this chapter, there are numerous other physiological factors which influence the formation of the sounds, and are often agents communicating certain environmental conditions. The hormone-controlled sexual cycle gives rise to the periodicity of certain sound forms. Mono-oestric species have therefore a specific annual cycle. *Vulpes*, *Alopex* and *Canis* give out particular sound forms, heat-growling ("um-um") exclusively during the propagation period; *Vulpes* starts the luring-growling ("yum-yum") together with bringing of food about three to four weeks after mating. At this time the barking sequence recedes as voice contact, becomes incomplete, and finally ceases soon after the birth of the young[85]. The *Cervidae* too having mating sounds tied to the seasons.

Besides the annual periodicity there is periodicity of many physiological processes caused by the changes from day to night over the 24 hours. The general activity too is subjected to this rhythm[86]. Animal sounds are related to the diurnal periodicity which is superimposed on the annual rhythm. Special investigations on this problem are not available.

References p. 783

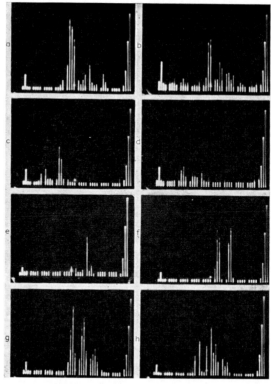

Fig. 403. Sound frequency spectrum of *Vulpes vulpes* L. and *Alopex lagopus* L. (a) *Vulpes* ♂, barking stanza; (b) *Alopex* ♂, barking stanza; (c) *Vulpes* ♂, growling; (d) *Alopex* ♂, growling; (e) *Vulpes* ♂, whimpering of inferiority; (f) *Vulpes* ♀, barking; (g) *Vulpes* ♀, greetings whimpering; (h) *Vulpes* ♂, threatening whimpering.

MEAN VALUES OF THE FILTER

Octave	I	II	III	IV	V	VI	VII	VIII	IX
Filter 1	40	80	160	320	640	1,280	2,560	5,120	10,240
Filter 2	48	95	190	380	760	1,520	3,040	6,080	12,160
Filter 3	57	113	226	452	904	1,808	3,616	7,232	14,464
Filter 4	67	134	269	538	1,076	2,152	4,304	8,608	17,216

Light and temperature are the most important meteorological factors which influence the production of sounds, directly or through other mechanisms. The influence of temperature is particularly marked with nestlings, since at first their thermal regulation is very incomplete. In *Vulpes* this starts only on the 20th day (first meat meal), when the vocalizations become much more infrequent, and are stimulated only by special situations. New born dogs (*Canis*) having a body temperature of 38.5–38.8° were put on a plate at 12.7–18°, with either the lower part or the whole body in contact. The sounds given out were twice as frequent when the whole body was on the plate[22].

Rattus norvegicus Berk. males and females can be distinguished by their voices.

It could be shown that these differences were caused by the hormone of the male as well as by a non-hormonal sexual factor. Both have an effect on the quality of the voice. Females and castrated males become hoarse after an injection of testosterone propionate. The change in the voice is reversible[25].

3. Types of Vocalizations and Their Functions

(a) *The types of sound*

In mammals there is a large variety of types of sounds depending largely on the construction and use of the vocal and subsidiary organs. In view of the present imperfect state of our knowledge a classification must be regarded as provisional. Schneider[77] has pointed to some regularities.

The following attempt is meant to provide a survey of the field:

(*i*) Unvoiced noises: without participation of the vocal organ.
    1. Mechanically generated noises,
    2. Noises generated by air flow stream current.

(*ii*) Voiced sounds: with participation of vocal organ (and subsidiary organs).
    1. Single sounds,
        (a) mono-syllabic
            (α) short sounds
            (β) long sounds
        (b) multi-syllabic
    2. Sound sequences,
        (a) homotypical
        (b) heterotypical
            (α) short sequences (phrase)
            (β) long sequences (strophical)
    3. Mixed sounds.

(*i*) *Unvoiced noises* (very common in mammals)

1. *Mechanically generated noises.* *Hystrix* on excitation through the prickles, rattles the tail by vibrating tail motions similar to those of other *Rodentia*[21]. Tail rattling is also known in *Neotoma fuscipes* Baird: where the vibrating tail hits the nest material. *Castor* too claps alarm signals with its tail[20]. Many of the species of *Glires* hit the ground with their hindleg when excited (*Lepus*?, *Oryctolagus*, *Meriones shawi* Roz., *Dipus aegypticus* Hasselq., *Pachyuromys duprasi* Lat.[3, 20, 21]). In marsupialians the same type of hitting sound has been known for a long time[33]. *Macropus* and other species with biped motion have the same warning signal[30]. *Dasyprocta aguti* L. males drum with the front legs on the ground during courtship[66].

Many Ungulates stamp with their front legs (*Cervus*, *Ovis*, *Rupicapra* and other *Bovidae*[30] (and other refs.).

Many of the species of the Primates, when excited drum on the ground, scme with the flat of the hand, others with the knuckles. The males of *Pan troglodytes* Blum. stamp the ground rhythmically in their dance[95] while hitting with the knuckles on resonating objects[74]. *Gorilla* when excited drums with the flat of the hand, or, more

References p. 783

rarely with the fist on his breast, *Pan* drums sometimes rhythmically on the belly with the flat of the hand[27,36,74].

In open hunting it drums mainly on hollow trees and tree trunks[60]. Further mechanical sounds are the gnashing of teeth in *Rodentia* (*e.g. Cricetus*[21], *Hydrochoerus, Cuniculus*), the clicking of the jaws in canids (*Canis lupus* L., *C. familiaris* L. and subgenera *Thos*) and in *Papio*. Audible snapping of the jaws was described also in dolphins[83], chattering of the teeth in *Ailuropoda* and *Muntiacus, Cervus, Phacochoerus* (Fig. 404).

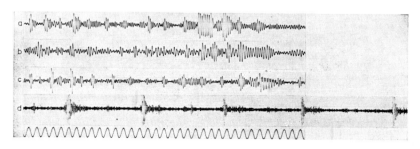

Fig. 404. Noises in mammals, partly produced mechanically. (a) *Panthera pardus* L. ♀, gargling, oscillogram using filter passing 150–300 c/s; (b) *Panthera leo* L., gargling, using filter passing 150–300 c/s; (c) *Puma concolor* L. ♂, purring, filtered 150–300 c/s; (d) *Phacochoerus aethiopicus* Pall., rhythmic grinding of teeth in heat, filtered 1200–2400 c/s. Frequency markers in all figures: 50 c/s.

2. *Noises generated by air flow stream current*. Numerous un-voiced sounds can be produced by means of the flow of breath. Some of these are certainly not only subsidiary effects of the breathing process, but are genuine means of expression. Such sounds can arise during exhalation (hackling) or during inhalation. Characteristic sounds are produced particularly when breathing-in air through the nose for scenting. Clicking, hissing and spitting sounds belong to the class of unvoiced sounds which may contain ultrasonic components. Young *Ornithorhynchus, Caluromys philander* L. and *Sminthopsis crassicaudata* Gould are reported to give out hissing noises. The same holds for *Vombatus ursinus* Shaw[29]. Spitting noises are widespread, for example in *Dendrolagus leucogenys* Mtsch.[74], *Rodentia* (*Glis, Cricetus* etc.), *Carnivora* (especially *Felidae*). Clicking noises were reported for *Didelphys virginiana* Kerr and for tylopods[63]. A whistling sound occurs in canids. Further unvoiced sounds are the snorting of *Equidae*, of *Diceros* and of *Panthera tigris* L. Another sound produced in the nose is the trumpeting of the elephants. *Cystophora cristata* Erxl. one of the Pinnipeds, produces an unvoiced sound by blowing up the nostrils[12,56]. Labial sounds are known in various mammals. Among the *Rodentia* they occur in *Capromys pilorides* Say[31]. They also occur frequently in the Primates.

(*ii*) *Voiced sounds*

1. *Single sounds*. Voiced sounds are produced by most mammals. The quality of the sound depends largely on the construction of the larynx, on the resonant cavities etc., and on the size of the species. The frequency range, in particular, is determined by the mass of the vibrating system. Thus, we can expect in smaller species

(*Insectivora, Chiroptera,* many *Rodentia* and *Prosimii*) only relatively high-pitched sounds which might even reach into the ultrasonic range. Investigations with *Petaurus, Prosimii* and *Rodentia* show that their sounds contain harmonics which go up to 40 kc/s. In *Rattus* sounds of 23–28 kc/s have been found which last for 1–2 sec[2]. Young animals of *Mus musculus* L., *Apodemus flavicollis* Melchior and *Microtus arvalis* Pallas give out sounds with single components up to 100 kc/s[100].

(a) The *mono-syllabic* sounds represent the phonetically simplest element of voiced sounds. They show (especially as short sounds) no significant variation of pitch and are uniform also as regards the sound quality. They require only a setting adjustment of the sound producing system.

(α) Within this group short sounds are certainly the original form of phonetic expression. They are given out with a short breath as the expression of a state of excitement. As an upper limit one could define the length of a syllable of 1 sec, but the decisive criterion is the fact that they are individual, isolated sounds which are not repeated regularly. The sounds of voice contact of *Crossarchus* and *Suricata* belong to this category, as does the well-known single barking sound of the canids (*Canis, Vulpes, Alopex* etc.) (Fig. 405). This is a short sound too, although its length is influenced by the degree of excitement, since it can be given out in different steps of intensity (its duration in *Vulpes* is $1/2$–$1/3$ sec). The principal frequencies lie between 700 and 900 c/s, but harmonics range up to 4,300 c/s (Fig. 412).

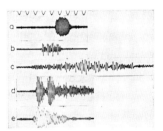

Fig. 405. Mono-syllabic short sounds. (a) *Crossarchus fasciatus* Desm., sound of voice contact filtered 800–1600 c/s; (b) *Dicoty'es tajacu* L., short grunting sound, filtered 150–300 c/s; (c) Maskenschwein (Domestic pig) grunting, filtered 150–300 c/s; (d) *Vulpes vulpes* L. staccato yelping in defence, filtered 800–1600 c/s; (e) *Camelus dromedarius* L. ♀, defence sound, filtered 300–600 c/s.

Similar barking sounds have been reported for Marsupialia: *Phascolarctos, Sarcophilus, Thylacinus*[29]. The warning sounds of the cervids are of a similar character. Among the bovids, lowing single sounds are very widespread. However, there is a tendency to prolong them so that we find all intermediate stages to long sounds. Short sounds are also frequent in the *Suidae* (sounds ranging from grunting to barking). Whistling sounds too can be regarded as short sounds. They are frequent in *Glires* and can also be found in *Tapirus* and *Diceros*.

(β) We may consider genuine long sounds as single sounds if their (single) syllable lasts longer than 1 sec. They are extended single sounds. They have to be distinguished strictly from the extended sound forms which arise from sound sequences closely fused (see below), such as the howling of the canids. An exact analysis is often required to decide which of the two classes a particular sound belongs to. Long sounds are probably derived (phylogenetically) from short sounds. Examples of this can be found in various groups, of which we mention only some characteristic ones: the

References p. 783

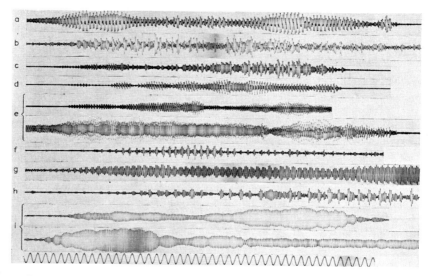

Fig. 406. Long sounds. (a) *Cervus elaphus* L. ♂, rutting call, filtered 300–600 c/s; (b) *Oryx algazel* Oken ♀, mothers call with young, filtered 300–600 c/s; (c) *Ovis aries* L. (Moorland sheep), bleating, filtered 800–1600 c/s; (d) *Ovis aries* L. (Eastfriesian sheep), bleating while waiting for food, filtered 300–600 c/s; (e) *Bos taurus* L., roaring while waiting for food, filtered 600–1200 c/s (next line follows immediately); (f) *Poëphagus grunniens* L. ♀, mother sound, filtered 300–600 c/s; (g) *Camelus dromedarius* L., roaring, filtered 200–400 c/s; (h) *Diceros bicornis* L. ♂, rumbling in defensive threat, filtered 200–400 s/s; (i) *Macrorhinus angustirostris* Gill., howling, filtered 150–300 c/s. Next line joins on; total duration of sound about 8.8 sec.

excited crying of *Diceros*, the heat cries of some bovids and cervids, the barking cry in *Vulpes* (a prolonged bark)[89], possibly also the curious howling sound of *Macrorhinus*, though this appears to be bi-syllabic (in connection with breathing). Threatening sounds of felids can also become long sounds. In certain circumstances *Pan* gives out a long-drawn "hoo" sound.

(b) Single sounds may become *multi-syllabic* if during sound formation the position of phonation is varied regularly (see Fig. 407). In this way the quality of sound is changed and this is often accompanied by a change of pitch. A bi-syllabic "shehek" sound has been reported for *Dendrolagus leucogenys* Matsch. one of the *Marsupialia*[74]. The bi-syllabic "meow" sound of the felids are well-known. They occur above all particularly in *Felinae*, but also in *Pantherinae*. The principal call of the tiger (*Panthera tigris* L.) also appears to be bi-syllabic, just as the sound of voice contact of *Cebus*. In the *Cervidae* sounds of this group occur also (*Axis, Cervus candensis* Erxl.).

2. *Sound sequences.* Sound sequences can arise in two ways: by a close succession of single sounds (short sounds) or by a rhythmic subdivision of long sounds.

(a) *Homotypical* sound sequences (without rhythm) occur when single sounds are produced one after the other without a definite time interval between them. The characteristic of these sounds lies in their repetition, not in a rhythm. Canids can join barking sounds in irregular intervals, felids do the same with their mewing, *Vulpes* join their yelping ("ke-ke") in defence. The pinnipeds (especially *Otaridae*) often join their sounds in irregular fashion.

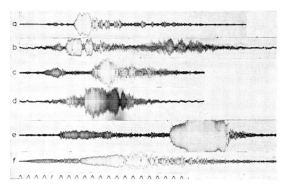

Fig. 407. Multi-syllabic single sounds. (a) *Tapirus terrestris* L. ♂, clear call ("whistle"), first syllable stressed, filtered 1200–1400 c/s; (b) *Diceros bicornis* L. ♂, clear call, filtered 600–1200 c/s; (c) *Cervus canadensis* Erxl. ♂, clear short call, first syllable higher pitch than second, filtered 2400–4800 c/s; (d) *Axis axis* Erxl. ♂, (e) *Ailuropoda melanoleucus* E.M.Edw., short call, second syllable stressed (when waiting for food), filtered 600–1200 c/s; (f) *Crocuta crocuta* Erxl., single call out of the call sequence (voice contact), three-syllabic with strong variation in pitch, filtered 400–800 c/s.
(a) and (b) could be homologous, also (c) and (d).

In the simplest case of rhythmic sound forms the stanzas have no definite lengths. They are characterized solely by a fixed time interval between the single sounds, which gives the whole sound sequence its own rhythm. These sequences are derived sound sequences without rhythm and can be observed very strikingly in the pinnipeds. *Arctocephalus pusillus* Schreb. has a rhythmic sound sequence; that of *Zalophus californianus* Less. is well-known. The luring sounds of *Vulpes* have a typical rhythm, but often no fixed, definite division into stanzas. A special case is represented by rhythmic sound sequences which are produced by regular bursts of breath, such as the hackling (in canids, an expressive sound of greeting) or the laughing (*Hominoidea*). The transition into group (b) is gradual. These heterotypic sound sequences are divided into stanzas and can even have regular internal subdivisions in their structure. They represent the most highly developed sound production of mammals (the most highly developed phylogenetically as well). The luring sounds of *Vulpes* are an intermediate case. When luring the young for feeding, these sounds are usually not divided into stanzas. However, when they occur as contact sounds between a pair in heat, they are sharply delimited in length, the stanza consisting of about four sounds and lasting about 0.6 sec (the single sounds 0.06 sec).

The corresponding stanza of *Alopex lagopus* L. consists of about 7 sounds and lasts 0.8 sec. The frequencies in *Vulpes* lie mainly between 113 and 380 c/s, in *Alopex* between 160 and 380 c/s. The domestic dog (Dachshund) was found to produce 4–5 sounds per stanza in 0.8–1.0 sec[34]. *Capreolus* L. gives out similar sounds in heat. Many mustelids give out stanzas of grumbling sounds[26]. A barking stanza which in its rhythm is similar to the luring sound (growling, "Muffen") occurs in the canids, especially in *Vulpes* and *Alopex*, also in *Cerdocyon* and modified in *Nyctereutes*. Phonetically these sounds are similar to a short barking sound, but the stanza is based on a specific rhythm. In *Vulpes* it consists of 4–7 sounds, the stanza lasting about 1.5 sec. The principal frequencies lie at 640–1080 c/s. For *Alopex* whose stanza is shorter, they lie between 900 and 1080 c/s. The howling stanzas (where, in a secondary development, the sharp temporary delimination has been lost) are derived from these barking stanzas (*Canis lupus* L., *C. familiaris* L., *C. latrans* Lay., *C. dingo* Blumenb., *C. aureus* L., *C. anthus* Cuv., *C. mesomelas* Schreb., *C. adustus* Sund.). Barking sounds introduce the stanza and also occur in it[34].

Call sequences in the form of stanzas appear to be frequent in *Insectivora*[32]. *Crocidura leucodon* Herm. gives out several variants forms[32]. As examples for the *Glires* be mentioned *Citellus beecheyi* Richards, *Castor fiber* L. and *Lepus timidus* L.

(b) Heterotypical sound sequences (call series[77]) with more pronounced internal sub-division are found in various more highly specialized mammals. Among the carnivores this holds particularly for representatives of the felids. While the *Felinae* (e.g. *Felis serval* Schreb.) form simple stanzas from a few sounds (about 5 sounds per stanza in 3.5 sec), *Pantherinae* give out highly subdivided call sequences (*Panthera onca, pardus, leo*). *Panthera pardus* L. gives out a rhythmic sound sequence apparently consisting of two parts (interval between sounds about 0.5 sec), the first part louder than the second. In *Panthera leo* L. the (rarely absent) starting call is followed by the main series of roaring calls with typical rhythm, the sound sequence is concluded by calls with decreasing intensity and increasing intervals. At the maximum intensity (main call) the rhythm is most marked and least variable. It was disputed for a long time whether *Puma concolor* L. gives out comparable sequences[99]. Recently roaring was heard which sounded slightly higher and shorter than in *Panthera leo* L.[52]. The hyenas also have highly subdivided sound sequences. *Crocuta* begins usually with starting calls, these are followed by the main sequence and end-calls which gradullay become slower[77]. Besides this *Crocuta* and *Hyaena* have rhythmic sound sequences, which are known as "laughing" in *Crocuta* (Fig. 408).

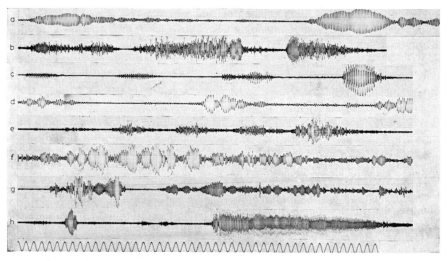

Fig. 408 Sound sequences with rhythm and partial internal subdivision. (a) *Pan troglodytes* Blumenb. ♀, hoo-calls, filtered 150–300 c/s; (b) *Equus grevyi* Oust. ♂, call sequence with internal subdivision, here the final staccato filtered 800–1600 c/s; (c) *Arctocephalus pusillus* Schreb., call series (uniform spacing), filtered 400–800 c/s; (d) *Panthera pardus* L. ♀, call sequence, filtered 200–400 c/s; (e) *Hyaena hyaena* L., call series (sharply defined sound series), filtered 600–1200 c/s; (f) *Hippopotamus amphibius* L. ♂, Voice contact call series, internal subdivision, fixed form of stanza, filtered 400–800 c/s; (g) *Dicotyles pecari* Fisch., subdivided call series, filtered 400–800 c/s. (h) *Equus asinus* L., call sequence (heterotypical), First "I" ("E"), then "A" (long), filtered 1600–3200 c/s.

In the Ungulata also rhythmical sound sequences occur in stanza form with sometimes strong internal subdivision. The sequences happen to be similar to the ones just mentioned. The "neighing" of the equids is an example. The few starting calls are followed by a main stanza which, in turn, is followed by a concluding call of single sounds with increasing intervals. In the domestic horse (*Equus caballus* L.) several variants can be distinguished. Comparable sequences occur in *Equus quagga*. The groups *Equus asinus* L. and *Equus grevyi* Dust are modified. In *Connochaetes gnu* Zimm. the barking calls are also built into a grouped series consisting on the average of 9–10 single sounds[77] during which changes of pitch and intensity occur. Specific call series are developed also in the higher Primates.

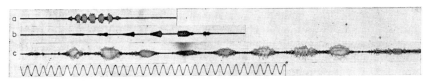

Fig. 409. Trills and subdivided sounds. (a) *Suricata tetradactyla* Schreb., trill of voice contact, filtered 300–600 c/s; (b) *Cebus capucinus* L. (*gracilis*), trill of excitement, filtered 2400–4800 c/s; (c) *Capra hircus* L. (dwarf goat), bleating (when waiting for food), filtered 1200–2400 c/s.

In *Hylobates lar* L. there are sequences of calls with increasing modulation: speed and pitch rise, then follow two or three lower-pitched sounds[15]. The "song" of *Pongo pygmaeus* Hopp is possibly comparable[10, 11]. (Further remarks on the sound formation in Primates under Section 3d.)

There is a way in which rhythmic sound sequences in stanza form can arise, namely by subdivision of single sounds into trilling sequences. This occurs especially in Insectivora, and is based on long sounds. In the *Rodentia* some subspecies of *Proechimys* trill like birds[31]. The very extended begging sounds of young nesting mammals may occur subdivided, ranging from a twittering to a trilling form. Trilling sounds occur also in *Crossarchus*, and frequently in Primates (*e.g. Cercopithecus*). They have also been described in the *Chiroptera* (*Nyctalus noctula* Schreb.)[67]. Rhythmic fluctuations have also been found in the long sounds of *Ruminantia* (*Ovis, Capra* and others) which makes them sound like a cackle.

3. *Mixed sounds*. In mixed sounds as understood here, at least two different detectable sound forms are superimposed. Very few observations are available so far. In *Vulpes* a few such sound forms could be detected. Thus growling and barking become "growl-barking", or whining and cackle can be superimposed whereby the primarily activated sound form (yelping) can be modified in sound by the secondarily activated one (whining). The lure-growling too can may be superimposed on whining to become a whimpering-growl. This combination is particularly interesting since the growling contains very low frequencies, the whimpering very high ones[88]. In *Vulpes* there is an intermediate phase between the various seasonal moods (pairing, rearing the young) when the corresponding sounds may be superimposed. If they are in a state of equilibrium, the fox utters neither the one nor the other, but an infantile whimpering.

In the higher Primates especially the *Hominoidea* several sound variants can be assumed to arise from superimposition.

References p. 783

(b) *Sounds in the service of pairing*

The often sexually differentiated vocal organs of mammals point to a connection between voice and reproduction. Various species produce sounds only during the reproductive period and then the sounds are often produced only by the male. In some cases it can be assumed that these sounds serve for marking territory. However, detailed investigations do not exist. The strong rutting cries of the bovids and cervids, in particular, are directed against rivals. On the whole two groups of sounds can be distinguished in connection with heat:

(*i*) Rival sounds: sounds for marking out of the territory and/or warding off competitors.
  1. Warning sounds.
  2. Threatening sounds.
(*ii*) Pairing sounds: sounds creating and maintaining contact between mating partners.
  1. Sounds for communication and contact over a distance.
  2. Sounds during actual contact.

(*i*) *Rival sounds*

1. *Warning sounds.* Rival sounds have the function of keeping rivals away. They are very strong and can be heard from afar. They have a wide frequency spectrum and can therefore be easily located. The well-known rutting cries of the cervids and the roaring cries of the bovids (*e.g. Bibos*) belong to this category. *Giraffa* bulls utter a hoarse cry against rivals[84]. Roaring cries of *Zalophus wollebaeki* Siev. have a definite territorial function[19].

2. *Threatening sounds* are produced when the rival is sighted, and are usually also very strong, mostly extended and with a certain modulation of pitch. They often lead to the withdrawal of the rival without fight. Well-known examples are the "caterwauling" in *Felis catus* L. and a "threatening song" of lower intensity. The latter lasts about 2–10 sec with strong variations in pitch between 262 and 528 c/s. The caterwauling proper is much louder, lasts up to 30 sec, and has a frequency of 396–792 c/s with appreciable contributions beyond this.

Vulpes vulpes L., against a rival gives out a very strong whining threat which may rise to a threatening cry. The principal frequencies lie between 1600 and 3200 c/s, but the variations in pitch are very large so that the starting sound can lie between 800 and 1600 c/s with strong harmonic sounds between 3200 and 6400 c/s. The sound is long and extended, increasing like a siren ot its full intensity. The mustelids too have shrill threatening cries.

(*ii*) *Pairing sounds*

Paring sounds have an entirely different phonetic character. They are usually sound sequences, often rhythmic and subdivided into stanzas. The degree of complexity of the basic structure of these sounds allows individual variations which facilitate recognition between the pairing partners.

1. *Sounds for creating contact* over a distance are strong. The intermittant character facilitates the location of the partner. In *Vulpes* and *Alopex* the sounds promoting distant "contact" are barking stanzas. In *Vulpes*, individuals are distinguished by

the different mean length of their barking stanzas as well as by differences in the quality of the sounds. A similar call sequence has been described for *Castor. Trichosurus vulpecula* Kerr produces a sound sequence which is answered by the partner[29]. For *Crocidura leucodon* Herm. a sound sequence "tji-tji-tji-tji-tji" was described[32]. The calls of *Hydrochoerus* attract besides the mating partner, also the yaguar![31]. A specific luring sound is the chirruping of *Capreolus*. The call sequences of the equids also belong here. The caterwauling of *Felis catus* L. is a mating sound too. Caterwauling consists of: a rhythmic sound sequence: one two-syllabic sound follows several monosyllabic ones, the frequencies lying between 600 and 3200 c/s, the principal components between 800 and 1600 c/s; this sound lasts for 1–2 sec; (b) a sound that lasts mostly 0.5–1.5 sec. It contains a "rrr"-component and has principal frequencies between 800 and 1600 c/s.

Other *Felinae*, such as *Felis serval* Schreb. give out in the mating season a sequence of mewing sounds.

2. *Sounds of "close" contact* (*i.e.* actual contact) are very widespread, particularly in species where the mating partners remain together for a shorter or longer time. These sounds are relatively weak and therefore in many species not known in detail. In *Ornithorhynchus* the female first gives out squeaking sounds which are answered by the male before pairing takes place. In Insectivora, they are mostly sequences of clear whispering sounds which serve for the voice contact between the partners. In many cases where the partners get together merely for copulation, the intraspecific attack and defence behaviour must be suppressed. Thus sound forms characteristic of these groups of behaviour become mixed into the mating relationship. In this, one of the partners may use infantile sound forms which curbs the attacking behaviour. The male of *Sciurus* approaches the female in this way. This is also known in *Cricetus*. The males of *Evotomys glareolus* Schreb. follow the females with ultrasonic sounds, which are supposed to "neutralise" the rustling of the leaves which would normally lead to flight[68]. Series of growling, whispering or "wailing" faint calls, mostly rhythmic, are often uttered between the partners of the *Rodentia* and accompany the pairing.

Sequences of growling sounds are widespread in these circumstances in mustelids and canids: *Canis lupus* L., *C. familiaris* L., *Vulpes vulpes* L., *Megalotis chama* A. Smith, *Alopex lagopus* L., *Urocyon*, *Fennecus*.

The principal mating sounds of the felids are purring and mewing. The purring is found in the *Felinae* and *Pantherinae*. Lionesses kept in isolation (*Panthera leo* Cl.) rub along objects in the cage and take up the mating position. *Panthera tigris* L. utters a snort of greeting[45].

(c) *Sounds in the service of rearing the young*

(i) *Sounds of the adult*

Adults use various sound forms in rearing their young, namely warning, enticing, luring, to keep voice contact.

(a) *Warning sounds* often show certain physical characteristics which make the location more difficult for the enemy. Thus they often have a relatively narrow frequency range, no interruption, and hardly any variation in pitch[51]. The warning sound of *Cercopithecus talapoin* Schreb. has its fundamental frequencies between

References p. 783

3040 and 3615 c/s (and harmonics in the next higher octave) without interruptions or fluctuations of intensity which would facilitate the location in this frequency range. The warning sound of *Papio*, *Cercocebus* and other *Cercopithecus* species are also difficult to locate (for man)[94]. *Vulpes vulpes* L. warns, near its den, with a faint barking sound (with closed mouth). This bark is short and has a frequency range of 150–800 c/s. The location is difficult. If the adult fox is far from its den, it barks very loudly, and the frequencies of this sound lie between 640 and 5120 c/s (harmonics). The location is easy and the young near the den (as well as the partner) can localize the enemy[85], *Alopex lagopus* L. behaves very similarly. The *Cervidae* have corresponding warning sounds, while the *Rodentia* have clear cries ("whistles") which have similar functions (*Marmota*[58]). The *Suidae* often warn with short grunting or barking sounds which are difficult to locate.

(b) *Luring sounds* must be easy to locate. They are therefore often repeated (call series), and lie usually in the lower frequency region. Here phase differences make the location possible[51]. Sound sequences (often rhythmic) serve not only to call the young back, but are used also in pairing (see above). The relationship between male and female resembles often that of the young to adult. Thus, the male uses the same sounds with the female as the parents use with the young. In *Vulpes* this is rhythmic growling[79]. It serves as a call for feeding. The puppies can stimulate the production of this sound in adult foxes which have no young ones themselves. *Panthera pardus* L. and *Ursus arctos* L. have similar sounds (short sequences).

(c) *Sounds of voice contact* occur in mammals living together for some time as a family unit. Such units can be regarded as intermediate to the higher sociological order (see Section 3d). Sounds of this kind are essential for young strays. The relatively early independence of the young necessitates corresponding sound forms. While *Oryx algazel* Oken normally very rarely produces sounds, a mother with offspring is often heard, without external cause, to produce a bleating sound. Similar sounds occur in *Bos*, *Poëphagus*, *Bison* and other bovids. The purring of *Felis catus* L. has probably a similar function. It lasts on the average 0.5 sec, the main frequencies lie between 200 and 400 c/s. The females of *Lama glama* L. make a sound for voice contact even before giving birth, a "clacking"[63]. In *Camelus* the sound ranges from growling to roaring. A mother whose foal had been taken away, continued to howl for 15 days. The young as well as the mother of *Lama* produce specific sounds during the first contact[41].

(ii) Sounds of the young

Functionally two main groups of sounds can be distinguished in the young: calls for care, and voice contact sounds.

(a) *The "care-calls"* are sounds calling the mother or parents to look after the needs of the young. They are used mainly by nestlings. Low temperature, humidity (dampness), hunger, and other factors stimulate such calls from the young. Normally the parent will come and attend to the needs of the young until the call stops. Frequently newborn nestlings can be distinguished by their voices (*e.g. Vulpes vulpes* L.[87]). The "care-calls" are often long (length of breath) and are repeated. Young *Myotis* give out "calls of abandonment" when parted from their mother. With increasing age these sounds change over into the ultrasonic range. Apparently many mothers of the *Chiroptera* can recognize their young by their calls. Young *Erinaceus* when abandoned, give out a "twittering whistle"[32]. Even in the first hours after birth, one can hear a faint squeaking. Young *Sciurus* which have fallen from their nests, give out a loud whistle; with *Cricetus* it is a peeping in the same situation. In young *Meriones persicus* Blanf.; it sounds like the scraping of a pencil on paper[21]. The young of *Apodemus flavicollis* Melch. call with frequencies of 70–80 kc/s[100]. The young of *Bradypus* has a peeping and squeaking sound. The mother reacts only to this call. Even if it is only half a tone higher or lower, she does not react[5]. The playing of recorded puppy sounds stimulated the male fox of *Vulpes vulpes* L. to collect and carry food to the (empty) brood box even before the birth of the young[85].

The squalling sounds of the newborn *Vulpes* have here their main frequencies between 1200 and 2400 c/s, with harmonics up to 9000 c/s. With increasing age these sounds soon become differentiated. From the 8th day one can hear a rhythmic swing which arises mainly from amplitude variations of harmonics between 1600 and 3200 c/s. This has developed into a "rhythmic squalling" by the 18th day. The fundamental frequency lies at 720 c/s, strong harmonics at 1520–2150 c/s with about 8 pulses/sec. By the 21st day a sound sequence is formed with this rhythm which goes over into the typical barking stanza. The barking sound proper can be heard on the 19th day at the earliest. The rhythm of the barking stanza is maintained to old age and serves now, as later, for the voice contact (see below)[87,88].

Fig. 410. Oscillograms of squeaking (croaking) sounds of new-born *Vulpes vulpes* L. (a) sound of ♂, (b) sound of a ♀ with high-pitched end sounding like "ee".

The newborn of *Thalarctos maritimus* Desm. give out four different sound forms: buzzing (when sucking), crying (especially when looking for teat), and a kind of "grunting", which frequently changes into squeaking (in situations of discomfort). The buzzing becomes later a purring. Polar bears, about 14 weeks old can be heard to growl for the first time. Later snorting and labial sounds are added as expressions of excitement[71].

The newborn of *Felis silvestris* L. use hunger sound-sequences sounding like "eee". At the age of three weeks, two sound types can be distinguished: the "EE-ee-yee-aa" sequence as a sign of hunger and the "eeee-üüü" form when abandoned. The hunger sound is further changed phonetically. The sound of abandonment becomes later the voice contact sound and a "begging" sound which are distinguished phonetically[47].

The whimpering sounds of nestling mammals stimulate the mother to carry them back if they have strayed[21]. However, experiments with deaf rats have shown that other additional factors may come into operation. Observations in felids have shown, moreover, that the reaction can show "fatigue"[45]. The young of *Felis catus* L. have a call with a mean duration of 0.5–1.0 sec and frequencies between 528 and 2112 c/s. It is very frequent during the first 20 days, and shows sexual-dimorphic differences[54].

Isolated young of *Phoca vitulina* L. are called "howlers" on account of their calls.

In *Ovis aries musimon* auditive stimuli (voices of lambs) from a recorder caused the mother-sheep to look for their young, although the mothers did not recognise these voices as those of their own young.

(b) *Voice contact sounds.* The walking and the nesting suckling of corresponding age give out voice contact sounds if the family unit keeps together for a longer period. The genesis of the barking stanza of *Vulpes vulpes* L. has already been referred to. In this case the care-call becomes the voice contact as soon as the young start to walk about independently. At this time the rhythmic "growling" develops; this is a contact sound in adult foxes (the pair) as well.

In other species too, these sound forms seem to correspond mostly with those of the adult animals. This will be discussed in more detail in connection with sociological phenomena.

(d) Sounds and sociology

The contribution which sounds make to the cohesion of social units in mammals varies from case to case. In species living in the open field, optical contact predominates, in inhabitants of the woods, acoustical contact. There are four main groups of sounds with social functions: (i) alarm calls, (ii) voice contact sounds, (iii) group sounds and (iv) sounds expressing special moods.

(i) Alarm calls

The alarm calls used in societies have probably developed from the corresponding sounds used in the rearing of the young. While in the latter case they are confined to a definite season, in societies they are permanently in readiness, even if sometimes certain individuals take over the task of calling the alarm. Prairie dogs (*Cynomys*) are a good example. They have well "organized" societies in sub-divided territories. When enemies are near, a short nasal warning call is made, which causes general attentiveness and spreads rapidly through the whole colony. In the case of enemies in the air, a piercing cry drives them back into their dens[37]. In *Cervus canadensis* Erxl. (*nelsoni*) some animals are always in places with a wide view, whence, in case of danger, they make a bleating sound of warning. *Alces alces columbae* Lyd. also has a warning call[17]. After a short grunting or barking call of warning the *Suidae* stop their usually very noisy activities immediately and the whole group leaves the danger area almost noiselessly.

Amongst the Primates with their highly differentiated social groups, the sound forms are very diversified. In *Ateles*, the warning calls sound like "ŏ" and "hö"[93]. *Alouatta palliata* Gray has various warning sounds. A "trouble sound" of the male leads to the stopping of all activities. A warning sound rouses only the males. Specific grunting alarm calls cause the young to be quiet. Finally a "signal-grunting" rouses the whole group[13]. In *Semnopithecus entellus* Dufr. all animals flee to the trees as soon as the warning call, which sounds like "hoon" is made[61]. *Ateles geoffroyi* Kuhl has a doglike warning bark[13].

In *Gorilla*, the alarm cry is loud and high, a mixture of barking and roaring[7]. In *Pan*, vibrations produced by stamping on the ground or by shaking of the branches on trees, appear to act as alarm signals. In any case, the animals are immediately quiet in cases of danger[60].

(*ii*) *Voice contact sounds*

Voice contact sounds firstly keep the family unit together and are then transferred to social groupings of higher order. They are mostly call sequences and are very widespread in the societies of *Pinnipedia*. In *Odobaenus* they seem to consist of two-syllable calls which are produced in sequence[57]. Two-syllable calls have given *Ctenomys* the name of "tucotuco". The animals make these calls almost continuously in their digging way of life[31].

The *Hippopotamus* bull replies to the call of the cow with a characteristic voice contact call, a sound sequence consisting of 4–5 impulses lasting about 1 sec and with fundamental frequencies (Fig. 408) between 80 and 538 c/s. In *Ruminantia*, repeated single sounds are used for voice contact (*Ovis*, *Capra* etc.); *Ocapia* too has a voice contact sound[46]. Certain variants of the well-known neighing of the equids have the same function. Among the *Herpestinae* there are species with strongly social characteristics, like *Crossarchus*, which emit almost continuous short voice contact sounds (Fig. 405). It seems to be a general fact that the voice contact sound are either subdivided call sequences, and therefore unmistakable, or they are repeated short sounds. In the former case they are used infrequently in the latter, very often. The voice contact call of *Crossarchus fasciatus* Desm. is at 640 c/s and lasts about 0.05 sec. The subdivided call series of *Panthera* and *Crocuta* serve evidently also as voice contact sounds (Fig. 408).

The social species of the *Canidae* have the well-known howling stanzas as their voice contact sounds (Fig. 411). These keep hunting packs in contact[98]. The howling sounds deep and guttural. It seems to have certain variants, possibly with different functions (see under (*iii*)). The howling hunting calls of the *Simocyoninae* seem to serve the same purpose. Isolation stimulates yelping in puppies.

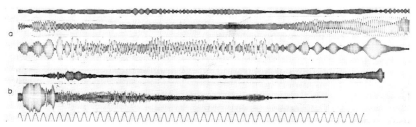

Fig. 411. Howling stanza of Canids. (a) *Canis lupus* L. (the first three lines belong together), filtered 300–600 c/s; (b) *Canis familiaris* L. (Eskimo dog) (two lines), filtered 1200–2400 c/s.

Our knowledge of the sounds of the Primates is very imperfect*. Especially in the higher species there are a large variety of sounds, serving social functions, which so far cannot be identified in detail. In fact this may not be possible in every case (see under (*iv*)). In *Ateles*, specific calls have been heard in a "dialogue". Here the voice contact sounds sound like a and ee[93]. In *Alouatta* also, sounds have been detected which effect coordination. In *Papio porcarius* Br. the group seems to be kept together by grunting sounds[8]. In *Cercopithecus* trilling sounds can produce voice contact. In *Zati* these calls sound whimpering, like "heee", and are mono-syllabic. The voice

---

* See also Chapter 7 (Zhinkin), p. 132 ff.

References p. 783

contact sounds of *Ateles* are reported to be similar to the neighing of horses[13]. Gorillas often have only a "staccato bark"[7] probably serving for voice contact.

In *Pan* the most important forms of communication are[60]:
1. Optical signals by posture, gesticulation, and mime.
2. Contact signals through direct bodily contact.
3. Vibrations by stamping the ground or shaking of branches.

Apart from this, the production of sound as such, must be supposed to serve here—as in several other Primates—simply for voice contact. In danger, chimpanzees turn silent immediately.

(iii) Group sounds

Group sounds are collective utterings within a society. These are always call sequences, mostly having a complicated, subdivided, rhythmical structure. Well-known is the "howling in unison" of the canids. Widespread within the *Caninae*, it is "infectious" even amongst individuals of different species. The same holds for the roar of the lions or the "singing" of the *Hylobates* and *Pongo*. *Pan*, too, has sounds expressing excitement which are transferred to other individuals, which in turn join in these sound sequences "oo-oo-oo-oo". Very little is known about the function of these group calls. They serve possibly as voice contact with other groups to whom they might possibly indicate the present occupation of a territory. This may explain the roaring in unison of the lions after a meal. This occurs in both sexes, often while lying down, or even when lying on their sides, since often remainders of the meal are lying about. It may serve for frightening off other groups. Apparently such group sounds occur mostly in nomadic species, for they do not thus risk giving away a permanently occupied territory. The vocal chorus can, at the same time, call back lost members, which can be particularly important in hunting. This factor could also be relevant in the choral roaring of lions after getting their prey.

(iv) Sounds expressing a special mood

Sounds expressing a mood occur especially in highly differentiated societies which generally have a rich and well-developed range of sound forms. This holds particularly for the Primates. Many of the sound forms of mammals stem from a specific mood. In the social groups this characteristic can become much more differentiated if the protection by the society allows a strong vocal activity. Young animals especially tend to use numerous forms of expression, since the environmental factors have not yet increased the thresholds of stimulation. They express every mood as long as a negative experience has not provided a check.

In this way, a very rich variety of sound forms can be developed. The well-organised warning system of the social group makes this vocal activity possible. It comes to an immediate stop only in case of a general alarm. These social factors permit certain infantile habits to be preserved even in the adult. In animals not forming part of a social group such habits reappear only during pairing, again irrespective of environmental factors. This demonstration of a mood can in fact help the society since it is important for everyone in the group to be informed of the mood of individual members. This often leads to complicated intra-social relationships.

The general expression of mood is, however, not as strongly "functional" as sound forms which are related either to specific environmental situations or to elementary

intra-specific mutual relationships. The expressions of mood can remain much more variable and show many shades, as for instance in the sounds of *Pan* which can be subdivided into four groups:
   1. Agressive sounds (barking).
   2. Sounds of ease and comfort (soft, nearly soundless barking).
   3. Sounds of fear and pain (howling).
   4. Sounds of excitement (very variable with oo-oo sound)[39].

The loose functional connection of the sounds of mood in members of social units may have been the essential basis for the development of human speech. Three functional systems have to be considered for the evolution of sounds in *Hominoidea*: Brain, larynx, and organs of the mouth[38]. Progress in this branch of anthropology will require a coordination of these three fields.

In the Primates the oral cavity has closer functional relationships with nutrition; in Man its evolution is related to that of speech. The variety of the sounds of the Primates has not yet been classified phonetically, apart from the difficulties of tracing the functional relationships already mentioned.

(e) *Sound and protection*

Various mammalian species give out defensive sounds when threatened. Observations on individual species indicate that there are intra- and inter-specific sounds of defence. *Vulpes vulpes* L. seems to produce the "push" yelp, only against animals outside his species. It is an explosive sound with simultaneous pushing forward of the body and opening of the mouth, the ears stay upright and turned forward. The sound is made for long intervals. Very similar, but also intra-specific, is the yelping (Fig. 405) and explosive call which is emitted in quick succession. Both sound forms have their main frequencies between 600 and 1600 c/s. The explosive start is fundamentally different from the positive social calls which start softly and increase gradually. While this type of sound accompanies the actual defence behaviour, there are also sounds of defensive threatening. They signal the defensive attitude and can often cause the enemy to retire. The snarling of the canids and the hissing of the felids belong to this category.

Sounds of pain, which are often very intense, represent a further group of defence sounds. They seem to have a great restraining effect in intra-specific disputes. They can often be heard among young animals and determine the limitations for the bitting games, while serving also as important checks for the actions of the grown animals.

4. Sounds and Behaviour

In general, sounds are symptoms of certain sequences of behaviour. If they are specialized for remote action, they can become the principal medium of expression. In other cases, they accompany conspicuous sequences of behaviour and support acoustically their optical effect. In addition, definite positions are taken up for certain sounds, ("compulsory positions"). It is not always easy to decide whether a position is conditioned physiologically by the sound, or whether it is a genuine expression of behaviour. Both can coincide. A "compulsory position" during sound production can acquire a special character of expression by "mimic exaggeration".

References p. 783

(a) Sounds and forms of behaviour

Many mammals assume specific positions for certain sounds. Raising of the head is very widespread. In this, physiological factors play a part: the position of the larynx and the connection between nose and cavity of the mouth (soft pallet, etc.) can determine the quality of the sound. Such positions are found particularly in vocalized forms of sound (harmonic sounds), while sounds with noise character (hissing, whispering) which are widespread among the Insectivora and Rodentia, rarely require special positions. Almost all hooved animals (Ungulata) lift their head and stretch their neck, thereby forming an almost straight larynx for the sound (*Bovidae*). The equids, too, lift their head when making their call sequences. The same applies for the higher beasts of prey. In the canids this tendency is further developed. *Canis lupus* L., *C. latrans* Say., *Thos*, lift their heads high during howling. In the barking stanzas of *Canis familiaris* L., which can go over into howling, the same tendency is noticeable. *C. dingo* Blum., too, lifts its head. Although *Alopex* does not howl, it lifts the head slightly during short "multi-syllabic" barking stanzas, while *Vulpes*, in the corresponding sound sequence, lifts the head hardly above the horizontal line. In *Pantherinae* there are typical positions of phonation, especially in the sound sequences of the lion (*Panthera leo* L.) and other species of *Panthera*. In the *Felinae* such positions of phonation are not highly developed. They are, however, very marked in the *Pinnipedia*: most seals lift the head clearly when they make their calls and sound sequences.

Macrorhinus lifts only its head high out of the water, opens the mouth wide and makes a long-drawn peculiar howling sound. Elephants lift up their trunks when trumpeting. *Diceros* lifts its head horizontally when it makes its peculiar sounds. *Tapirus terrestris* L. too lifts its head when whistling. Occasionally there is also the opposite movement of the head; thus *Crocuta* lowers its head to the ground when making its call sequence.

In *Monotremata*, *Insectivora* and many *Glires*, the sounds accompany different modes of behaviour without being specially stressed by particular positions. However, this support by a definite position becomes more distinct with increasing differentiation of the type of behaviour. Even in *Rodentia*, there are strong expressive motions which are accompanied by certain sounds and thus become more effective, (*e.g. Sciurus*). This holds to a high degree for the Primates. We find here all stages of specialization. The voice contact sounds of *Cebus* accompany the general modes of motion, but the strong defensive sounds of excitement are coupled with strong expressive gestures. Excitement is connected with vegetative symptoms[20]. In the acoustic field, sounds belonging to this category are the many hissing, snorting blowing, spitting and squeaking noises, which represent breathing sounds without participation of the vocal organ. In situations of uncertainty, canids give out a squeaking sound, equids snort and many carnivores spit. Such sounds occur also in *Rodentia* (*Cricetus* and others). The excited panting in canids can be put in this group. It may be that the growling sounds are derived from these vegetative accompanying noises.

The forms of expression controlled by the central nervous system include instinctive, intentional and displacement activities. They give rise to mechanical noises (hitting of the ground with the hindlegs or buttocks in *Marsupialia* and *Rodentia* or with the hands in the Primates). The body itself can provide resonant cavities in this case (chest of *Gorilla* and *Pan*[74]). Sounds can occur as displacement activities if other forms of reaction are blocked[98].

Play can also lead to sound production: an Indian elephant (*Elephas maximus* L.), in the Berlin Zoological Gardens, had the habit of hitting with the tail against the wooden fence around a tree in rhythmic sequence.

(b) Sounds as stimuli

According to Lorenz (1952), an instinctive motion is released by an "innate releasing mechanism" (IRM). The "releasing agent" is composed of "key stimuli" which in general are additive in their effect: "law of heterogenous summation"[91]. Busnel calls this reaction to particular sound forms "phonotaxis".

For the mammals there is an almost complete lack of special analyses of the releasing effect of particular sound forms. The warning sounds of *Microtus brandti* Radd. can be evoked by high-pitched sounds, such as the ringing of a telephone. In this case, the frequency seems to be decisive. The frequencies of the telephone used were between 4300 and 8600 c/s and the voice of *Microtus* contained frequencies of 3600–4300 c/s. Apart from this, the quick succession of individual impulses forming the syllable could be relevant. There are of course numerous observations on the intra-specific effect of particular forms of sound. The lion's roar is transferred to members of the species and occasionally also to tigers and leopards. Other *Canis* species often join in the howling of *Canis lupus* L. The effect of many luring sounds is unmistakable, although it is not always certain whether it is based on innate or acquired reactions. The latter can help in the identification of individual sounds: a male fox (*Vulpes vulpes* L.) kept by us tested with tape recordings reacted only on the screaming sound of greeting of his "own" vixen (he was monogamous).

5. Problems of Homology Research

There are three criteria defining homology: (1) the criterion of position; (2) the criterion of the special quality of the structures; (3) the criterion of interconnection by intermediate forms (criterion of continuity). Besides these principal ones, auxiliary criteria can be used for simple structures: (a) simple structures can be regarded as homologous if they occur in a large number of nearly similar species; (b) the probability of a homology in simple structures increases with the existence of further similarities of equal distribution in nearly-similar species; and (c) the probably of homology of a characteristic decreases with the frequency of occurrence of this characteristic in species which are definitely not related[64]. These homology criteria were developed from morphological structures and tested on numerous examples.

The sounds of most mammals prove to be characteristic for the species. Their variability is no greater than that of the morphological characteristics. There are individual variants which show up even at birth, as in the croaking sounds of newborn foxes. The length of the barking stanzas of *Vulpes* differs from one individual to another and stays the same throughout life.

Even accepting that in principle the foundations of homology research are secure, there are still difficulties in the application of the criteria. Sounds are not always at hand, as are morphological structures, but occur only under certain conditions. Thus the "criterion of position" could be referred to the situation that exists at the time when the sound is made. This concept of "position" would have to include all available external and internal factors. Characteristics of position would be age, sexual state, presence of a mating partner or of young, enemies, prey etc. as well as climatic factors, time of day, and time of year. The other two principal criteria can be transferred directly to the case of sound forms. With modern technics, structural analysis of sounds can be carried very far and is almost equal to that possible in morphological studies. The problem of intermediate forms is soluble in principle, but it requires comprehensive data which, in the case of mammals, are so far not available. The auxiliary criteria can be transferred without difficulty to the sound forms. They are essential for single sounds, which often represent very simple "structures".

As sounds are produced, coordinated and controlled by the animal itself, they are not subject to the pressure of environmental factors to the same extent as are morphological characteristics. The determining factors are predominantly of an intra-specific nature. For this reason, sounds as a whole could be regarded as more conservative than many of the morphological characteristics. The systematic study of sound forms can therefore provide very useful criteria for the analysis of inter-specific relationships. Systematic investigations in mammals are, however, very incomplete. Preliminary studies in canids have given us a few pointers which allow a test of the value of homology research in the field of sounds[34]. In this, one must start by correlating phenomenologically similar characteristics. The application of the homology criteria thus allows us to separate convergencies. The best-known canid sound is barking, which is certainly very widespread within this group. The phonetic transcriptions in the literature do not permit an application of the homology criteria. However, acoustic recordings make it likely that these sounds in *Canis*, *Thos*, *Vulpes*, *Alopex*, and *Urocyon* are homologous (Fig. 412). In *Urocyon*, *Vulpes* and *Alopex* the barking syllable lasts longer than in *Canis* and *Lyacon*. The steps of phylogenetic differentiation in barking and howling in the Canidae seems to be the following: barking stanza (voice contact, pairing)—pair-barking—pair-howling—choris howling (howling in unison, sometimes mixed with barking). The sound can be a general unspecific expression of excitement or, in connection with the rearing of the young, it is specialized as a warning sound. The house dog tends to use this sound

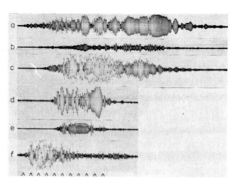

Fig. 412. Barking sounds of Canids. (a) *Alopex lagopus* L. ♂, filtered 600–1200 c/s; (b) *Urocyon cinereo-argentatus* Schreb., filtered 600–1200 c/s; (c) *Vulpes vulpes* (L.) ♀, filtered 600–1200 c/s; (d) *Canis familiaris* L. (Dachshund), filtered 300–600 c/s; (e) *Canis familiaris* L. (Eskimo dog), filtered 600–1200 c/s; (f) *Canis lupus* L. ♂, filtered 600–1200 c/s.

very much more, since domestication has done away with inhibiting external factors.

Yelping ("ke-ke-ke") does not seem to be as widespread as barking. It was definitely detected in *Canis mesomelas* Schreb., *Vulpes vulpes* L., *Alopex lagopus* L., *Urocyon cinereo-argentatus* Schreb., *Fennecus zerda* Zimm., *Nyctereutes procyonoides* Gray, *Speothos venaticus* Lund and *Lycaon pictus* Temm.[53,79,85,65,35].

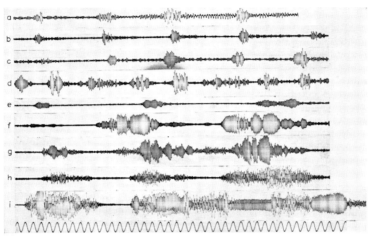

Fig. 413. Rhythmic sound sequences of Canids. (a) *Vulpes vulpes* L. ♂, growling in heat, filtered 200–400 c/s; (b) *Megalotis chama* A. Smith, growling ("Muffen") while pairing, filtered 400–800 c/s; (c) *Canis familiaris* L. (Dachshund) growling while pairing, filtered 300–600 c/s; (d) *Alopex lagopus* L. growling while pairing, filtered 300–600 c/s; (e) *Canis lupus* L., whimpering growling before howling stanza, filtered 300–600 c/s; (f) *Alopex lagopus* L., barking stanza, filtered 600–1200 c/s; (g) *Vulpes vulpes* L. barking stanza, filtered 800–1600 c/s; (h) *Alopex lagopus* L. ♂, rhythmic whimpering of greeting, filtered 600–1200 c/s; (i) *Canis familiaris* L. (Eskimo dog) barking while howling, filtered 600–1200 c/s.

(a)–(c) Show the same rhythm and a very similar sound structure. (d) Has faster rhythm but a sound structure which is similar to (a)–(c). (e) Has the typical rhythm of the barking stanza of the Canids, see (f) and (g) and introduces the howling which is homologous to these barking stanzas. Also within the howling barking sounds may occur which show the old rhythm (i). The whimper of greeting has a different sound structure but the same rhythm (h) correspondingly in *Vulpes*.

The peculiar howling is widespread amongst the canids, without homology having been tested in every case. It can be regarded as established generally in the species *Canis* (including *Thos*). *Canis familiaris* starts howling on the 43rd day at the earliest[4]. Some breeds bark very little and howl a great deal (dogs of the Bantu-negroes, eskimo hounds[29]). *Canis dingo* Blch. howls a great deal. The howling of *Canis lupus* L. is certainly homologous[9,28,29,55,59,62,69,79,90,98]. As in *Canis familiaris* L., it occurs in several variants (Fig. 411). *Canis pallipes* Sykes which is probably genetically connected with the dog, is also reported to howl[50,55].

There are also several reports on howling for the sub-genus *Thos*[29,42,53,55,80], especially for *Canis (Thos) aureus* L., *Canis anthus* Cuv., *Canis adustus* Sund. and *Canis mesomelas* Schreb. The literature also mentions the howling of *Canis latrans* Say.[1,6,55,81]. All data on these sounds forms outside the species *Canis* are problematical. There are references to *Vulpes vulpes* L. and *Alopex lagopus* L.[82,73], but the homology is most doubtful. *Nyctereutes* has a howl-like sound[80]. *Chrysocyon jubatus* Desm. has

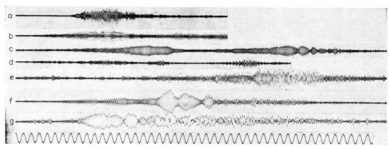

Fig. 414. Mewing sounds of Felids. (a) *Felis serval* Schreb. ♂, nearly mono-syllabic short sound, filtered 3200–6400 c/s; (b) *Felis wiedi* Schinz. ♂, two-syllabic sound (second syllabic here not complete), filtered 1200–2400 c/s; (c) *Felis sylvestris griselda* Thomas ♀, typically two-syllabic, filtered 600–1200 c/s; (d) *Lynx chaus* Güld. ♂, two-syllabic, filtered 600–1200 c/s; (e) *Puma concolor* L. ♀, stretched with crescendo, filtered 800–1600 c/s; (f) *Panthera tigris* L. ♀, filtered 300–600 c/s; (g) *Panthera leo* ♂, single sound out of the voice contact calls, filtered 200–400 c/s.

a sound which has a certain similarity with howling, but which is unlikely to be homologous[43]. Thus all definite findings are confined to *Canis*. All the same, there are evidently homologous sound forms in other canid species, as tape recordings have shown. This holds particularly for the species *Vulpes* and *Alopex*, and probably for *Megalotus* and *Cercocyon azarae* Wied., but definite proofs of howling are missing here. Under similar circumstances (first homology criterion) the barking stanza occurs in its place, which seems to be shortest in *Alopex*. The structure of the howling stanza of *Canis* has distinctly recognizable barking components; often howling is preceded by barking sounds, or barking sounds are mixed with howling. Moreover, doubtful reports on the howling in *Vulpes* state that the last sound of the barking stanza may be drawn out. On one occasion, we have heard something similar. In order to reconstruct, on the basis of these results, a relationship between these sound forms, one would have to start with the single barking sound. The next stage would be their combination into a barking stanza, which then becomes longer, and eventually turns to howling. Accordingly *Alopex* would phylogenetically precede *Vulpes*, while *Canis* would represent the latest stage of phylogenetic development.

In this connection, it is interesting to note that a rhythmically similar sound sequences, growling ("Muffen"): "ium-ium-ium", has been found so far in the same species which produce howling or barking stanzas. Both in *Canis* and *Vulpes*, growling can precede howling or barking stanzas. This can be taken as an indication of homology, in which the peculiar growling, as far as it is known, can be regarded as an homologous sound form in the various species (Fig. 413).

In the *Felinae*, there are numerous homologous sound forms. The best-known is mewing, a sound which is formed in very many variants, so that only investigations based on extensive material could lead to a fruitful analysis of the homology problem. *Felis catus* L. has three principal forms of mewing which are sounded only during expiration (in contrast to earlier views!). Purring, however, is produced both during inspiration and expiration and is characteristic of *Felinae*[54]. The *Pantherinae* produce a homologous sound ("gargling") (Fig. 404).

In the *Felinae* an increasing differentiation of sound sequences is found: *Panthera tigris—P. leo—P. onca—P. pardus*. A first attempt at homology research in *Cervidae* is given by Donath.

The *Ungulata*, too, provide a wide field for homology research. The peculiar clear sound of *Tapirus terrestris* L. is evidently homologous with certain short sounds of *Diceros bicornis* L. (Fig. 407).

It has yet to be shown how far the rumbling sounds of the elephant are comparable with those of the rhinoceros. The sounds of *Ovis* show frequently the vibrations which are characteristic for *Capra*. In the cervids, various sound forms are probably comparable (Fig. 406); they may possibly also be compared with those of the bovids.

In the case of strictly rhythmical sound sequences, one could think of similarities within the *Ferungulata*, particularly for sounds which are produced in mating contact.

The rhythmic voice contact calls of *Hippopotamus* could be compared with those of *Suidae* (suborder *Suina*) (Fig. 408).

In the *Perissodactyla*, neighing sound sequences offer themselves for homology analysis. Special relationships can be established easily between *Equus asinus* L. and *Equus grévyi*. It is not yet possible to extend comparisons beyond the orders. Such comparisons may one day throw light on the genetics of certain sound forms.

Homology research is very difficult in the primates. Within the *Cercopithecinae*, of which *Macaca mulatta* Zimm. has so far been investigated in most detail[16,23,24], some sound types occurring in various species in similar form can be provisionally delimited[94]:

1. Spitting sounds: these start as a noise and may change into a vocal sound (*Macaca mulatta* Zimm., *M. radiata* Geoff., *M. silenus* L., *Papio sphinx* L., *P. leucophaeus* Cuv., *P. hamadryas* L., *Cercopithecus aethiops* L., *C. nicticans* L. and *C. talapoin* Schreb.).

2. Grunting sounds: nasal, with vowel-like sound "O" ("chrong") (*M. mulatta* Zimm., *Papio leucophaeus* Cuv., *C. nicticans* L.).

3. Cackling, "kay-kay-kay" etc. (*Macacus memestrina* L., *M. radiata* Geoff., *Papio sphinx* L., *P. leucophaeus* Cuv., *P. hamadryas* L., *P. doguera* Puch., *Cercopithecus aethiops* L., *C. nicticans* L. and *C. petaurista* Schreb., *C. talapoin* Schreb.).

4. Whimpering sounds: short sounds, with strong modulation of pitch, mostly in two parts (*Macaca mulatta* Zimm., *M. memestrina* L., *M. radiata* Geoff.).

5. Bleating sounds: in *Papio doguera* Puch., similar to the bleating of sheep.

6. Vocal short sounds (similar to 5): sounds like "a", "ö" (diphthong) or "o" (*Papio hamadryas* L., *P. sphinx* L., *P. doguera* Puch).

7. Barking sounds: pulsative sounds similar to the barking of dogs (*Papio hamadryas* L., often in series; in *P. doguera* Puch. bi-syllabic.)

8. Chirping sounds: sounding like "tshe(a)k" (*Cercopithecus nicticans* L. and *C. petaurista* Schreb.).

9. Trilling sounds: *Cercocebus torquatus* Kerr., *Cercopithecus nicticans* L. and *C. petaurista* Schreb. In *C. talapoin* Schreb. 7 and 8 may be combined to form a trilling-chirping sound.

10. Cries: in *Papio sphinx* L. and *P. leucophaeus* Cuv. the sound has three phases, the middle one being rhythmic and the final one more strongly fluctuating in frequency and amplitude.

In addition special calls have been heard in individual species. "Babbling" should also be mentioned, produced by fast movements of the lips (*e.g. M. radiata* Geoff. and *Cercopithecus* species).

Trilling sounds are frequent also in *Plathyrrhini*, usually as expression of excite-

References p. 783

ment. Much more material is required for a systematic and successful homology research in Primates. The structure of the larynx and the sound forms lead to the conclusion that the African *Procolobus* has a special position, as opposed to the asiatic *Colobus* species[34a].

In a systematic, comparative study of sounds, the phonetic characteristics have to be distinguished from the structural ones. Particular sound sequences often have a specific rhythm which is completely independent of the phonation. Thus canids can make the rhythm of hackling (fast breathing) acoustically more effective by using phonetic elements. In related species, sound sequences are often found to have similar structure (subdivision) but different acoustic character. For example, *Pan troglodytes* Blum. and *Pan paniscus* Schwarz differ in the manner of vocalization of certain sound sequences. In *P. paniscus* Schwarz, the vowels "ah" and "a" (as in "calm") predominate; in *P. troglodytes* Blum. "o" ("lot") and "oo" ("moon") (Fig. 408)[92].

The sound sequences of the *Pinnipedia* also seem to be based on certain rhythms, which are independent of their sound character. Thus, *Arctocephalus pusillus* Schreb. produces a very typical sequence of "a" sounds (like "take"). The duration of the sound is about 0.13 sec (with certain variations), the mean interval between the sounds is about 0.15 sec. In the rutting season of *Otaria byronica* Bev., a rather similar sound sequence is heard, but with a substantially different sound character.

The analysis of the ontogenesis of particular sound forms will give us important information on the genetic relationships between these sounds. The generic evolution will in some cases pass through stages which illuminate structural relationships much more clearly than the fully developed sound systems. Moreover, frequently infantile sound forms are taken over into adult life by acquiring a renewed functional character. In this way, the phylogenesis of the formation of sounds may provide a pointer for a "juvenile preponderance" in the sense of Eimer, *i.e.* for an anticipation in the infantile stage of phylogenetically future characteristics.

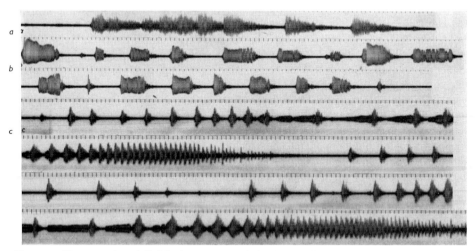

Fig. 415. Three species of *Hylobatinae* compared. (a) *Symphalangus syndactylus* Desm, strophic call sequence; (b) *Hylobates lar* L., unequal sequence of different modulated sounds; (c) *Hylobates hoolock* Harl., heterotypical strophic call-sequences, two strophes. Each strophe contains 10 introduction sounds, followed by 4 crescendo sounds and ends with a trilling sequence. 9 short calls are between the two strophes. They are similar to the short calls of *Hylobates lar*.

From this point of view, mammalian sounds offer a promising field for research, particularly fruitful in relation to *Hominoidea*.

Our still rather imperfect survey of the whole field of the sounds of mammals hardly justifies general conclusions based on homology research. The line of thought presented here is meant merely to convey general impressions derived from existing data. It appears that the development of sounds can proceed in two different directions:

(a) Extension of sounds, that is to say formation of long sounds.

(b) Abbreviation of sounds and formation of sound sequences.

A tendency for the extension of sounds occurs in *Artiodactyla*. In *Camelus* roaring sounds can last up to 3.7 sec. In this group, many sounds have these characteristics, although there are also some sequences of short sounds (presumably genetically old ones). Among the *Cervidae* and *Bovidae* also long sounds are conspicuous. In the *Carnivora* this tendency shows up in two ways: in the *Pinnipedia* by true sound extension, and here *Macrorhinus* holds a "record" with its sound of 8.8 sec duration. On the other hand, in the canids, the sound extension tendency appears as a slurring of sound sequences. Within the *Carnivora*, the short sounds are doubtless the original ones. The *Perissodactyla* have strong tendencies for the formation of sounds sequences.

The length of sounds is a function of the breathing rhythm. Sounds sequences are certainly related to physiological factors. Excitement leads to shortness of breath and promotes the formation of sound sequences. Young mammals can hardly produce long sounds, so that short sounds and sound sequences built up from them, are dominant. Phylogenetically, the sound sequences develop into subdivided sound series with a fixed basic rhythm of the same kind as has been known in birds for a long time (*e.g. Fringilla*[60]).

The degree of definiteness of a sound form is the result of the phylogenetic mechanisms which led to its development. The definition grows with the extent of functional character which it attains in the framework of vocal and other expressional behaviour.

Thus, a rational homology research has to be just as much research on the functional character of sounds since structural analysis is meaningless without a knowledge of the factors determining the development and production of the detailed forms of sound.

REFERENCES

[1] ALCORN, J. R., 1946, On the decoying of coyotes, *J. Mammal.*, 27, 122.
[2] ANDERSON, J. W., 1954, The production of ultrasonic sounds by laboratory rats and other mammals, *Science*, *119*, 808.
[3] ANONYMUS, 1926, Stimmlaute des Schneehasen, *Deut. Jägerztg.*, 87, 344.
[4] BAEGE, B., 1933, Zur Entwicklung der Verhaltensweisen junger Hunde in den ersten drei Lebensmonaten, *Z. Hundeforsch.*, *3*.
[5] BEEBE, W., 1940, *Dschungelleben*, Leipzig.
[6] BENSON, S. B., 1948, Decoying coyotes and deer, *J. Mammal.*, 29, 406.
[7] BLOWER, J., 1956 The mountain gorilla and its habitat in the Birunga volcanoes, *Oryx*, 3, 6.
[8] BOLWIG, N., 1957, Some observations on the habits of the Chacma baboon, *Papio ursinus*, *S. African J. Sci.*, *53*, (10) 255.

[9] Botezat, E., 1931, Neues aus dem Leben des Wolfes und des Wildschweins, *Bull. Fac. Sti. Cernauti*, 5, 158.
[10] Brandes, G., 1939, *Buschi, Vom Orang-Säugling zum Backenwülster*, Leipzig.
[11] Brandes, G., 1938/39, Das Singen der Orangmänner, *Zool. Garten*, 10/11.
[12] Brown, R., 1868, Notes on the history and geographical relations of the Pinnipeds frequenting the Spitzbergen and Greenland Seas, *Proc. Zool. Soc. (London)*, 405.
[13] Carpenter, C. R., 1934, A field study of the behavior and social relations of the howling monkeys (*Alouatta palliata*), *Comp. Psychol. Monographies*, 10.
[14] Carpenter, C. R., 1935, Behavior of red spider monkeys in Panama (*Ateles geoffroyi* Kühl.), *J. Mammal.*, 16, 171.
[15] Carpenter, C. R., 1940, A field study in Siam of the behavior and social relations of the gibbon (*Hylobates lar*.), *Comp. Psychol. Monographies*, 16.
[16] Chance, M. R. A., 1956, Social structure of a colony of *Macaca mulatta*, *Brit. J. Animal Beh.* 4,1.
[17] De Vos, A., 1958, Summer observations on moose behavior in Ontario, *J. Mammal.*, 39, 128.
[18] Eibl-Eibesfeldt, I., 1951, Beobachtungen zur Fortpflanzungsbiologie und Jugendentwicklung des Eichhörnchens, *Z. f. Tierpsychol.*, 8, 370.
[19] Eibl-Eibesfeldt, I., 1955, Ethologische Studien am Galápagos-Seelöwen (*Zalophus wollebaeki* Sivertsen), *Z. f. Tierpsychol.*, 12, 286.
[20] Eibl-Eibesfeldt, I., 1957, Ausdrucksformen der Säugetiere, *Kükenthal Handbuch der Zoologie*, Vol. 8, Part 10, Berlin.
[21] Eibl-Eibesfeldt, I., 1958, Das Verhalten der Nagetiere, *Kükenthal Handbuch der Zoologie*, Vol. 8, Berlin.
[22] Fredericson, E., N. Gunrey and E. Dubois, 1956, The relationship between environmental temperature and behavior in neonatal puppies, *J. Comp. Physiol. Psychol.*, 49, 278.
[23] Garner, R. L., 1900, *Apes and Monkeys, their Life and Language*, Boston.
[24] Garner, R. L., 1900, *Die Sprache der Affen*, Leipzig.
[25] Gilse, P. H. G. v., R. T. de Jongh, H. F. de Nooy and M. Wijnans, 1952, Sexual influences on the voice of rats, *Acta Physiol. et Pharmacol. Neerl.*, 2, 237.
[26] Goethe, E., 1940, Beiträge zur Biologie des Iltis, *Z. f. Säugetierk.*, 15, 180.
[27] Grabowsky, F., 1905, Mitteilungen über den Gorilla des Breslauer Zoologischen Gartens, *Verhandl. Deut. Naturforsch. u. Ärzte*, 255.
[28] Grzimek, B., 1945, *Wolf Dschingis*, Stuttgart.
[29] Heck, L., 1912, Säugetiere, in *Brehms Tierleben*, Leipzig.
[30] Hediger, H., 1958, Verhalten der Beuteltiere (*Marsupialia*), *Kükenthal Handbuch der Zoologie*, Vol. 8, Part 10, Berlin.
[31] Hershkovitz, P., 1955, On the cheek pouches of the tropical American paca, *Agouti paca* L., *Säugetierk. Mitt.*, 3, 67.
[32] Herter, K., 1957, Das Verhalten der Insektivoren, *Kükenthal Handbuch der Zoologie*, Vol. 8, Berlin.
[33] Heubel, F., 1939, Beobachtungen und Versuche über das Sinnesleben und die Intelligenz eines *Macropus giganteus* Zimm., *Bijdr. tot de Dierk.*, 417.
[34] Heydecke, R., Vergleichende Untersuchungen zur Lautgebung der Caniden, Thesis, Berlin.
[34a] Hill, W. C. and A. H. Booth, 1957, Voice and larynx in African and Asiatic *Colobidae*, *J. Bombay Nat. Hist. Soc.*, 54, 309.
[35] Jobaert, A. J., 1951, Les lycaons ou chiens chasseurs, *Zool. N.S.*, 12, 133.
[36] Johnson, M., 1933, *Congorilla*, Leipzig, p. 91.
[37] King, A., 1955, Social behavior, social organization, and population dynamics in a black-tailed prairiedog town in the black hills of South Dakota, *Contrib. Lab. Vertebr. Biol. Univ. Michigan*, 67, 1.
[38] Kipp, F.A., 1955, Die Entstehung der menschlichen Lautbildungsfähigkeit als Evolutionsproblem, *Experientia*, 11, 89.
[39] Kohts, N., 1930, Infant ape and human child, *J. Psychol.*, 27, 412.
[40] Koppers, W., 1942, Haushund und Wildhund in Zentralindien, *Bull. soc. fribourg. sci. nat.*, 36, 113.
[41] Kraft, H., 1957, Das Verhalten von Muttertier und Neugeborenen bei Cameliden, *Säugetierk. Mitt.*, 5, 174.
[42] Kühn, W., 1935, Die Dalmatinischen Schakale, *Z. f. Säugetierk.*, 10, 144.
[43] Krieg, H., 1940, Im Lande des Mähnenwolfes, *Deut. Zool. Gart.*, N.F., 12, 157.
[44] Kulzer, E., 1958, Untersuchungen über die Biologie von Flughunden der Gattung *Rousettus* Gray., *Z. vergl. Physiol.*, 47, 374.
[45] Leyhausen, P., 1956, Das Verhalten der Katzen (*Felidae*) *Kükenthal Handbuch der Zoologie*, Vol. 8, Berlin.
[46] Lang, E. M., 1957, Stimmlicher Kontakt von Okapis untereinander, *Säugetierkundl. Mitt.*, 5, (4) 171.

[47] LINDEMANN, W. and W. RIECK, 1953, Beobachtungen bei der Aufzucht von Wildkatzen, *Z. f. Tierpsychol.*, *10*, 92.
[48] LORENZ, K., 1953, Die Entwicklung der vergleichenden Verhaltensforschung, *Verhandl. Deut. Zool. Ges. Freiburg*, 36.
[49] LOVERIDGE, A., 1946, Kip, an interesting and unusual pet of East Africa, *Nat. Hist. New York*, *55*, 172.
[50] LYDEKKER, R., 1926, *Jackals, The game animals of Africa*, London, 2nd ed., p. 456.
[51] MARLER, P., 1956, Über die Eigenschaften einiger tierlicher Rufe, *J. f. Ornithol.*, *97*, 220.
[52] MCCABE, R. A., 1949, The scream of the Mountain lion, *J. Mammal.*, *30*, 305.
[53] MERVE, N. J. VAN DER, 1953, *The Jackal, Fauna and Flora*, An official publication of the Transvaal Provincial Administration No. 4, Pretoria.
[54] METZE, M., 1958, Die Lautgebung der Hauskatze, *Thesis*, Berlin.
[55] MIVART, ST. G., 1890, *Monograph of the Canidae*, London.
[56] MOHR, E., 1951, Die Stimme der Robben in den europäischen Gewässern. *Schr. Naturwiss. Ver. Schlesw.-Holst.*, *25*, (Karl-Gripp-Festschr.), 29.
[57] MOHR, E., 1956, Das Verhalten der Pinnipedier, *Kükenthal Handbuch der Zoologie*, Vol. 8, Part 10, Berlin.
[58] MÜLLER-USING, R. and D., 1955, Vom "Pfeifen" des Murmeltieres, *Z. Jagdwiss.*, *1*, (1) 32.
[59] MURIE, A., 1944, The Wolves of Mount Mac Kinley, *Fauna of the Nat. Parks of U.S.*, 20.
[60] NISSEN, H. W., 1931, A field study of the chimpanzee, *Comp. Psychol. Monographies*, 8.
[61] NOLTE, A., 1955, Freilandbeobachtungen über das Verhalten von *Macaca radiata* in Südindien, *Z. f. Tierpsychol.*, *12*, 77.
[62] PFUNGST, O., 1914, Versuche und Beobachtungen an jungen Wölfen, *Ber. 6. Kongr. exp. Psychol. Göttingen*.
[63] PILTERS, H., 1955, Das Verhalten der Tylopoden, *Kükenthal Handbuch der Zoologie*, Vol. 8, Part 10, Berlin.
[64] REMANE, A., 1952, *Die Grundlagen des natürlichen Systems, der vergleichenden Anatomie und der Phylogenetik*, Leipzig.
[65] RENSCH, B., 1950, Beobachtungen an einem Fennek, *Megalotis zerda* Zimm., *Der Zool. Gart., N.F.*, *17*, 30.
[66] ROTH-KOLAR, H., 1957, Beiträge zu einem Aktionssystem des Aguti (*Dasyprocta aguti aguti* L.), *Z. f. Tierpsychol.*, *14*.
[67] RYBERG, O., 1947, *Studies on Bats and Bat Parasites*, Stockholm.
[68] SCHLEIDT, W., 1948, Töne hoher Frequenz bei Mäusen, *Experientia*, *4*, 145.
[69] SCHMID, B., 1923, *Die Sprache und andere Ausdrucksformen der Tiere*, München.
[70] SCHMID, B., 1928, Über die Phonetik der Tiersprache (Sichtbarmachung tierischer Laute), *Zool. Anz.*, 89.
[71] SCHMID, B., 1930, Tierphonetik, *Z. vergl. Physiol.*, *12*, 760.
[72] SCHMID, B., 1936, Zur Psychologie der Caniden, *Kleintier und Pelztier*, *12*, 11.
[73] SCHMOOK, A., 1949, *Der Fuchs*, Heidelberg.
[74] SCHNEIDER, K. M., 1933, Zur Jugendentwicklung eines Eisbären, *Deut. Zool. Gart., N.F.*, *6*, 156; 224.
[75] SCHNEIDER, K. M., 1937, Kann der Schimpanse "brusttrommeln"?, *Deut. Zool. Gart., N.F.*, *9*, 161.
[76] SCHNEIDER, K. M., 1954, Vom Baumkänguruh (*Dendrolagus leucogenys* Matsch.), *Deut. Zool. Gart., N.F.*, *21*, 63.
[77] SCHNEIDER, K. M., 1956, Über gegliederte Rufweisen bei Tieren, *Beitr. z. Vogelk.*, *5*, Part 2, (2) 168.
[78] SEITZ, A., 1940, Die Paarbildung bei einigen Cichliden, I. Die Paarbildung bei *Astatotilapia strigigena*, *Z. f. Tierpsychol.*, *4*, 40.
[79] SEITZ, A., 1949, Lautäusserungen beim Fuchs, *Natur u. Volk*, *79*, 252.
[80] SEITZ, A., 1955, Untersuchungen über angeborene Verhaltensweisen bei Caniden, III. Beobachtungen an Marderhunden, *Z. f. Tierpsychol.*, *12*, 463.
[81] SETON, E. TH., 1926, *Animals*, selected from *Life Histories of Northern Animals*, Nelson Doubleday.
[82] SIEDEL, F., 1951, *Wildtiere unter Menschen*, Jena.
[83] SLIJPER, E. J., 1955, Geluiden van walvissen en dolfijnen, *Vakbl. Biol.*, *35*, 193.
[84] STANTON, J., 1955, Is the Giraffe mute?, *Nat. Hist. New York*, *64*, 128.
[85] TEMBROCK, G., 1957, Zur Ethologie des Rotfuchses (*Vulpes vulpes* L.), unter besonderer Berücksichtigung der Fortpflanzung, *Deut. Zool. Gart., N.F.*, *23*, 289.
[86] TEMBROCK, G., 1958, Zur Aktivitätsperiodik bei *Vulpes* und *Alopex*, *Zool. Jb., Abt. Allgem. Zool. Physiol. Tiere*, *68*, 297.
[87] TEMBROCK, G., 1958, Lautentwicklung beim Fuchs; sichtbar gemacht, *Umschau*, *58*, 566.

[88] TEMBROCK, G., 1959, Zur Entwicklung rhythmischer Lautformen bei *Vulpes, Zool. Anz.*, 22, suppl.
[89] TEMBROCK, G., 1959, Beobachtungen zur Fuchsranz unter besonderer Berücksichtigung der Lautgebung, *Z. f. Tierpsychol.*, 16, 351.
[90] THORPE, W. H., 1954, The process of song-learning in the chaffinch as studied by means of the sound spectrograph, *Nature*, 173, 465.
[91] TINBERGEN, N., 1952, *The Study of Instinct*, Oxford.
[92] TRATZ, E. and H. HECK, 1954, Der Bonobo—eine neue Menschenaffen-Gattung, *Säugetierk. Mitt.*, 2, 97.
[93] WAGNER, H. O., 1956, Freilandbeobachtungen an Klammeraffen, *Z. f. Tierpsychol.*, 13, 302.
[94] WAPPLER, E., 1958, Vergleichende Untersuchung zur Lautgebung der Cercopitheciden, *Thesis*, Berlin.
[95] WEINERT, H., 1940, *Der geistige Aufstieg der Menschheit*, Stuttgart.
[96] WOLBURG, I., 1956, Über die Entwicklung der Temperaturregulation bei Jungsäugern, *Verhandl. Zool. Ges. Hamburg*, 29.
[97] YERKES, R. M. and B. W. LEARNED, 1925, *Chimpanzee Intelligence and its Vocal Expressions*, Baltimore.
[98] YOUNG, S. P. and E. A. GOLDMANN, 1944, *The Wolves of North America*, Washington.
[99] YOUNG, S. P. and E. A. GOLDMANN, 1946, The Puma, *Am. Wildlife Inst.*
[100] ZIPPELIUS, H. M. and W. SCHLEIDT, 1956, Ultraschall-Laute bei jungen Mäusen, *Naturwiss.*, 43, 502.

References added in proof

GRAUVOGL, A., 1958, Über das Verhalten des Hausschweines unter besonderer Berücksichtigung des Fortpflanzungsverhaltens, *Thesis*, Berlin (F.U.).
DONATH, P., 1960, Untersuchungen zur Lautgebung der Cerviden, *Thesis*, Berlin.
RESCHKE, B., 1960, Untersuchungen zur Lautgebung der Feliden, *Thesis*, Berlin.
ROSS, S., J. P. SCOTT, M. CHERNER and V. DENENBERG, 1960, Effects of restraint and isolation on yelping in puppies, *Anim. Behav.* 8, 1.
POGLAYEN-NEUWALL, I., 1962, Beiträge zu einem Ethogramm des Wickelbären (*Potos flavus* Schreber), *Z. Säugetierk.*, 27, 1.
SCHLOETH, R., 1961, Das Sozialleben des Camargue-Rindes, *Z. f. Tierpsych.*, 18, 574.
TEMBROCK, G., 1959, Stimmliche Verständigung unter Tieren, *Wiss. Fortschr.*, 9, 302.
TEMBROCK, G., 1959, *Tierstimmen, eine Einführung in die Bio-Akustik*, Wittenberg.
TEMBROCK, G., 1960, Probleme der Bio-Akustik, *Wiss. Z. Humb. Univ., Math. Nat. R.*, 8, 573.
TEMBROCK, G., 1960, Struktur- und Homologie-Probleme bei Lauten höherer Wirbeltiere, *Festschr. d. Humb. Univ.*, II, 329.
TEMBROCK, G., 1960, Spezifische Lautformen bei *Vulpes* und ihre Beziehungen zum Verhalten, *Säugetierkdl. Mitt.*, 8, 150.
TEMBROCK, G., 1960, Homologie-Forschung an Caniden-Lauten, *Zool. Anz. Suppl.*, 23, 320.
TEMBROCK, G., 1961, Lautforschung an *Vulpes* und anderen Caniden, *Zool. Anz. Suppl.*, 24, 482.
TEMBROCK, G., 1962, Methoden der vergleichenden Lautforschung, *Symp. Theriol. Brno*, 329.
TSCHANZ, B., 1962, Über die Beziehung zwischen Muttertier und Jungen beim Mufflon (*Ovis aries musimon*, Pall.), *Experientia*, 18, 187.

PART VI
APPENDIX

Addendum to Chapter 8

(ACOUSTIC SIGNALS FOR AUTO-INFORMATION OR ECHOLOCATION)

F. VINCENT

(1) Bats

(a) The larynx of Microchiroptera

A study of "general morphology" has been performed on *Myotis*[4]. Its architecture is comparable to that of the human laryngeal tracts. The way the motor fibres end in the striated muscles has been examined in *Myotis* and *Rhinolophus*[17]. An elaborate differentiation has been revealed and it could be related to the emission of ultrasounds. So far such a differentiation had only been found in Man where it seems connected also to the extraordinary possibilities of tonal modulation[18]. The phoniatric physiology of *Myotis* has been the subject of a theory similar to one of those put forward for Man[4,9].

(b) Cochlear potentials

Further experiments were carried out by recording cochlear potentials in *Myotis lucifugus*[29]. The sensitivity of its ear is inferior to that of other Mammals tested in this way, as far as the low frequencies are concerned. Above 20 kc/s, the sensitivity is much better in certain narrow frequency ranges (between 10 and 60 kc/s, and most often at 40 kc/s). Some results have been obtained on the effect of several kinds of interruptions of the ossicular chain.

(c) Plecotus

Griffin and Grinnell[7] observed the masking-effect of continuous noise covering the whole frequency range of echolocation sounds in *Plecotus rafinesquii*. This bat can hear echoes that are as inferior as 35 dB to the level of the noise. This involves an acute recognition of the echo which must not be shared by the thermal noise (*cf.* paragraph on theories of echolocation in the addendum). By switching on and off the latter at different rates up to 1000/s, it was impossible to decrease the animal's ability.

(2) Crustacean

A freshwater Cladocera, *Mixodiaptomus laciniatus*, is able to detect an obstacle in front of it at a distance of 1 mm[25] and to orientate properly in the narrow passages of a plastic maze (75 to 90 % of trials without hits). The females, bigger and more slowly moving than males, are more clever. A phototropotactism can inhibit this ability. When the maze is made of a non transparent, thermo- and phono-isolated substance, the percentage of misses drops to 33.7. Besides, if the *Mixodiaptomus* is going perpendicularly to the wall of the maze, it collides with it most of the time. The reception seems to be located in the first pair of antenna. These results brought Schröder to

References p. 792

conclude in a sound orientation; the sounds might be generated by the swimming movements of the animal.

This behavior puts *Mixodiaptomus laciniatus* completely apart from other Cladocera which are unable to avoid an obstacle.

(3) Cetacea Odontoceti

Referring to the publications of numerous XIXth and XXth centuries authors, Tomilin[28] quotes several Odontoceti which emit different categories of sounds. Those whose description seems to correspond to echolocation are found in *Hyperoodon ampullatus, H. planifrons, Mesoplodon bidens, Delphinapterus leucas, Monodon monoceros*, as well as in a dolphin of the Chinese rivers, *Lipotes vexillifer*. Besides, Fish and Mowbray[5] have studied the acoustic emissions of the captive Beluga, *Delphinapterus leucas*, and describe those used by this animal for echolocation. Moreover, Busnel, Vincent, Van Heel and Dziedzic have been able to record on the high sea, then analyse echolocation clicks from *Delphinus delphis, Globicephala melaena* and *Grampus griseus* (1962, not yet published). It would be very surprising if *Platanista*, which lives in the muddy waters of India and is almost completely blind, was not endowed with a very good echolocation system.

Norris and his co-workers[21] have tested the echolocation system in *Tursiops truncatus* entirely deprived of the use of its eyesight*.

Cruising sounds and searching sounds were thus recorded, as well as in the case of bats (see Chap. 8). The sound beam was found to have directionality, the transmission of which seems to be connected with the upper jaw and melon region. Several tests on the echolocation ability of this animal were successfully performed: passing between two poles, discriminating between objects of the same size but of different sound-reflecting characteristics. In two months the animal hit only once a pole. Although the circumstances and the animal were different from those of Kellogg's previous experiment[12] one is tempted to think that the information obtained through the eyes can slightly inhibit that which is obtained through hearing, as Griffin proved it to be for bats (see p. 185).

Lilly and Miller[16] on the same species showed that echolocation clicks or squawks can be emitted simultaneously with whistles. The authors conclude that "such observations demonstrate that the bottlenosed dolphin has at least two separately controllable sonic emitters, one for the production of clicks and one for the production of whistles". Besides, they think that "some of the slow trains of clicks may be ... a form of communication"**, when "rapid trains of clicks (squawks, quacks, and so on) occur most frequently ... during intense emotional situations ...".

Dudok van Heel[2] presented an interesting hypothesis for the numerous cetacean strandings which remained surprisingly unexplained since we know that the animals are equipped with a very good "sonar". He thinks that some gently shelving coasts cannot, according to their slope, send any echo in the direction of the animal or the herd***. Therefore the animal goes on swimming and gets stranded without being able to detect the coast in these cases. This gives a solution to an old enigma.

---

* These experiments have been performed again by Wood (cited by Dudok van Heel[2]).
** See note p. 220.
*** This has been found in bats too by Möhres and Dijkgraaf (cited by Dudok van Heel[2]).

However as we noticed several times at sea, dolphins often swim very fast without emitting any noise. It is possible that they do that only in places they know very well (dolphins territories may exist). Besides, Gray[6ter] reports about one *Globicephala*'s stranding, that "some kindhearted people had thought to rescue them by tying ropes around their tails and towing them out to sea, but it was very disheartening to see that the whales turned right around and headed back for the shore", and further "For it seems that the whale must in some way be unhealthy before he will run ashore". It may be possible that Dudok van Heel's theory is not the only cause for whales stranding.

(4) Blind men

It has been well known for many years that some blind men, if not all of them, can very cleverly detect obstacles at a small distance. The first time hearing has been mentioned as being partly involved in such a detection, was probably in the early XIXth century ([14] and [8] cited by Griffin[26]). Some experiments were performed by Heller[8] who concludes that blind subjects detect a change of sound in their own footsteps at 3 or 4 meters from the obstacle used, and that a tactile sensation is involved at 60 or 70 centimeters.

Then an interesting series of tests has been carried out by a group of blind and blindfolded psychologists[27,1] who detected a standardized obstacle by listening to the noise of their own footsteps, fingersnaps, and so on; these experiments have been performed on subjects having their skin covered or uncovered, their ears plugged or unplugged, walking on hard or soft floor, so as to show the importance of the only sense of hearing. Going further, they made the subject stand still, listen through earphones to the noise of another walking subject, or of a moving apparatus emitting noises. So they could test different kinds of noises: white noises and pure tones of 10,000 c/s were found to be the most useful ones for echolocation.

Recently Kellogg[13] tested the ability of blind men to discriminate between objects of different sizes or sound reflecting characteristics (wood, painted wood, velvet, and so on).

These results are highly interesting, scientifically speaking, and we hope, as Griffin[26] does, that they will soon be used practically to help more blind men in their every day life.

(5) Theories of echolocation

Several theories have been recently presented, in an attempt to explain how animals can use echolocation without getting disturbed by the emission of their companions, the differences of intensity between the pulses and their echoes, and so on.

Nordmark[19] brings to our mind that two pulse trains of identical repetition, but slightly separated in time (pulses emitted by a non-moving animal, and their echoes coming back from a non moving obstacle for instance), give rise to a tone the frequency of which depends on the time interval. If this time decreases (when the relative distance between the animal and the obstacle decreases) the tone frequency is raised, and reciprocally. Differences of intensity have no importance within certain limits. Man's ears are very sensitive to such tones. If this is how echolocative animals find their way, distances are translated into frequencies, and the shorter the distance to the obstacle, the higher the frequency sweep. The author thinks the lowest audible

References p. 792

tone of this kind to be 100 c/s; this means that a bat cannot detect an object further than 2 or 3 meters (which is comparable to the distance of which a bat increases its emission rate), and a dolphin not further than 9–15 meters (which seems small compared to the speed at which such animals can run). This hypothesis has the advantage of explaining how the animals can orientate themselves clearly when the echoes are very faint and the emitted pulses overlap. Kay[10, 11] and Pye[23] made several criticisms of this theory. Testing blind guidance devices they found that frequency modulation is much more important than the effect described by Nordmark for a human ear. As bats emit frequency modulated pulses, they are thought to locate obstacles more easily in this way. However bats' and men's ears are not identical. Besides, the time difference notes may be useful in some circumstances, in cases such as reinforcement for the information, so complicated is the phenomenon of echolocation. When one cuts the cricothyroid muscles of *Eptesicus*, there is an end to the frequency modulation of its pulses; nevertheless, it can perfectly well avoid obstacles of medium size. In this case, for instance, Nordmark[20] shows there is evidence of the non-necessity of the frequency modulation to locate obstacles.

Strother[26] elaborated a new interesting theory, based on the principle of ultrasonic pulse compression, which is applied in some radars.

The inner ear of the animal is supposed to be provided with a kind of filter that introduces a linear time delay proportional to the received frequency; such a filter will make the pulses shorter and of greater amplitude. Overlapping of pulses by their echoes will occur no more; the resolution range of the animal will be much better, and jamming noises will be of little effect since the amplitude is increased by some 35 dB (see above in addendum). Not only does this conform to the radar experiments, but it may give us the solution to numerous puzzling characteristics of echolocative animal behaviour.

(6) *Conclusion*

The further research advances, the more animals are found using echolocation. However, they can be divided into two main groups. In the first one we have several very different kinds of animals (including Man occasionally) which listen to noises directly connected with a normal activity (walking, swimming, sneezing, and so on). In the second group we see animals which listen to sounds especially emitted for echolocation: Bats, Cetacean and Birds; these sounds are not normally used for or connected with another activity, although Lilly and Miller[16] think that some clicks are used for communication between dolphins. In other words, animals of the former groups locate the obstacles by listening only, while those of the latter do the same thing by emitting and listening. Respectively, the terms of passive and active echolocation must characterize the possibilities of these two groups.

REFERENCES

[1] COTZIN, M. and K. M. DALLENBACH, 1950, Facial vision: the role of pitch and loudness in the perception of obstacles by the blind, *Am. J. Psychol.*, *63*, 485–515.
[2] DUDOK VAN HEEL, W. H., 1962, *Sound and Cetacea*, Thesis, Utrecht (published by J. B. Wolters, Groningen).
[3] EDINGER, T., 1955, Hearing and smell in cetacean history, *Monatsschr. Psychiatr. Neurol.*, *129*, 37–58.

[4] FISCHER, H. and H. GERKEN, 1961, Le larynx de la Chauve-Souris (*Myotis myotis*) et le larynx humain, *Annales Otolaryngol.*, *78*, 577–585.
[5] FISH, M. P. and W. H. MOWBRAY, 1962, Production of underwater sound by the white whale or Beluga, *Delphinapterus leucas.*, *20*, 149.
[6] FRASER, F. C. and P. E. PURVES, 1959, Hearing in whales, *Endeavour*, *18*, 93–98.
[6bis] GILMORE, R. M., 1961, Hvalene of den amerikanske Marine, *Norsk Hvalfangst-Tidende*, *50*, 89–117.
[6ter] GRAY, W. B., 1960, *Creatures of the Sea*, American Book Stratford Press, Inc., 215 pp.
[7] GRIFFIN, D. R. and A. D. GRINELL, 1958, Ability of bats to discriminate echoes from louder noise, *Science U.S.A.*, *128*, 145–146.
[8] HELLER, T., 1904, *Studien zur Blindenpsychologie*, Leipzig, W. Engelmann, 136 pp.
[9] HUSSON, R., 1962, Physiologie de la phonation, *C.R. Acad. Sci.*, *254*, 3250–3252.
[10] KAY, L., 1961, Perception of distance in animal echolocation, *Nature*, *190*, 361–362.
[11] KAY, L., 1961, Orientation of bats and men by ultrasonic echolocation, *Brit. Communic. Electron.*, *8*, 582–586.
[12] KELLOGG, W. N., 1961, *Porpoises and Sonar*, The University of Chicago Press, 177 pp.
[13] KELLOGG, W. N., 1962, Personal communication to the author.
[14] KUHNAU, J. C. W., 1810, *Die blinden Tonkünstler*, Berlin.
[15] LILLY, J. C., 1961, *Man and Dolphin*, Doubleday, New-York, 312 pp.
[16] LILLY, J. C. and A. M. MILLER, 1961, Sounds emitted by the bottlenosed dolphin, *Science*, *133*, 1689–1693.
[17] MICHEL, VAN C., 1961, Différenciation dans les cordes vocales de la Chauve-Souris en rapport avec l'émission d'ultrasons, *C.R. Soc. Biol.*, *CLV*, 1143–1145.
[18] MICHEL, VAN C. and M. A. GEREBTZOFF, 1960, *1° Congrès Européen d'Anatomie*, Strasbourg (in the press).
[19] NORDMARK, J., 1960, Perception of distance in animal echolocation, *Nature*, *188*, 1009–1010.
[20] NORDMARK, J., 1961, Perception of distance in animal echolocation, *Nature*, *190*, 363–364.
[21] NORRIS, K. S. et al., 1961, An experimental demonstration of echolocation behavior in the porpoise, *Tursiops truncatus* (Montagu), *Biol. bull.*, *120*, 163–176.
[22] PYE, J. D., 1960, A theory of echolocation by bats, *J. Laryngol. Otol.*, *74*, 718–729.
[23] PYE, J. D., 1961, Perception of distance in animal echolocation, *Nature*, *190*, 362.
[24] ROEDER, K. D. and A. E. TREAT, 1961, The detection and evasion of bats by moths, *Amer. Sci.*, *49*, 135–148.
[25] SCHRÖDER, R., 1960, Echoorientierung bei *Mixodiaptomus laciniatus*, *Naturwiss.*, *47*, 548–549.
[26] STROTHER, G. K., 1961, Note on the possible use of ultrasonic pulse compression by bats, *J.A.S.A.*, *33*, 696–697.
[27] SUPA, M., M. COTZIN and K. M. DALLENBACH, 1944, "Facial vision: the perception of obstacles by the blind", *Am. J. Psychol.*, *57*, 133–183.
[28] TOMILIN, A. G., 1955, On the behaviour and sonic signalling of whales. Translation made by the Fisheries Research Board of Canada, no. 377, from *Trud. Inst. Okeanol. Akad. Nauk SSSR*, *18*, 28–47 (in Russian).
[29] WEVER, E. G. and J. A. VERNON, 1961, Hearing in the bat, *Myotis lucifugus*, as shown by the cochlear potentials, *J. audit. Res.*, *1*, 158–175.

Addendum to Chapter 9

(INHERITANCE AND LEARNING IN THE DEVELOPMENT OF ANIMAL VOCALIZATIONS)

P. MARLER

The number of papers which has appeared since preparation of Chapter 9 is evidence of the rapidly expanding interest in the field of bioacoustics, and only a few can be mentioned here. In Orthoptera, the taxonomic significance of singing behavior has been extensively reviewed by Alexander[1], who discusses the various kinds of selective influence to which cricket song is subject. Both he and Bigelow[3] add to the information upon sound production by hybrid crickets. Alexander reminds us again of the significance of Huber's demonstration[7] of the role of central nervous patterns of excitation in the control of singing patterns, supplementary to the dictates of the structure of the stridulating organs. Furthermore Huber finds that auditory feedback is not necessary for species-specific sound production. Here, as in physiological studies of the basis of sound production in cicadas[6] and various other behavior patterns of arthropods (e.g. Wilson[17]; Wiersma[16]) there is an increasing emphasis upon direct nervous control of behavior which, although often normally modulated by various kinds of sensory feedback, is often capable of maintaining species-specific patterns on its own. It seems probable that the inheritance of arthropod sound patterns is achieved by concerted genetic control of both the form of the sound producing organs and the pattern with which the nervous system excites them, with responsiveness to particular patterns of sensory feedback imposed as a further factor in some cases.

Among the Anura, Bogert[5] has written an extensive review of all aspects of sound production. Included are new data on the vocal behavior of hybrids within the genera *Scaphiopus*, *Bufo* and *Hyla*, showing that the calls are sometimes intermediate between those of the parent species in some respects while conforming to one of the parents in other respects. Here too the assumption that physical structure of the sound producing organs dictates all of the properties of the vocal pattern has been questioned[11] and this promises to be a fruitful line for further work.

Birds continue to be the major focus for investigations of vocal development with much of the recent material reviewed by Tembrock[12], Lanyon[8], Thielcke[13] and Thorpe[15]. In an important paper Thielcke-Poltz and Thielcke[14] have continued and extended Messmer and Messmer's study of song development in *Turdus merula*, tracing in detail the process by which sounds of sibling and adult blackbirds, as well as alien sounds, such as cricket calls and the sound of a tape-recorder switch, are incorporated into the full song. The ability to learn appears at about the 28th day of age, and seems to persist from then on, whenever the bird is in the appropriate mood. Sometimes the learning may coincide with singing, so that practice begins immediately, but in some cases at least this is not necessary, so that an interval of up to two months may intervene between hearing a sound and producing a complete imitation of it. In tests of the minimum numbers of presentations of a model necessary for imitation to result, some birds learned with as few as 15–50 presentations, all given on one day, although there was wide variation in ability in individual birds, apparently depending upon the physiological state at the time of the experiment. Once again Thielcke demonstrates how individuality is introduced into the song, by imperfections in imitations, by the invention of new themes, by temporal rearrangement of phrases within the song, and by transposition from one frequency range to another. In a bird raised in isolation, the song lacks many of the variations introduced in these various ways, and since they are consistently present in the song of wild birds, Thielcke concludes that exposure to adult song is necessary for completely normal song development to occur in *Turdus merula*. In *Lanius collurio*, which also incorporates imitations of other species into its song, the situation is somewhat similar. Blase[4] has shown how the song of birds raised in isolation resembles only the juvenile song of wild birds. The calls develop normally, but the opportunity to imitate other species is necessary for normal development of song to occur.

In *Junco oreganus*, whose species specific song is a simple trill of similar, repeated syllables, auditory experience seems to impinge upon the development process in a somewhat different way[9]. Birds raised in isolated groups developed a variety of song types, many of which conform to the wild type in overall pattern, but differed in the simpler structure of the syllables from which they were composed. Birds which were allowed to hear singing of adult juncos and other species produced songs with more elaborate syllable types. Some of these arose as approximate imitations. Many, however, bore no resemblance to the sounds to which they were exposed, and seem to arise through an intensification of the process of vocal invention or improvisation, elicited by exposure to a richer auditory environment. As in *Lanius collurio* and *Turdus merula*, song development is not restricted to one phase of the life history, since in captive juncos at least, new song patterns may also arise in the second year. It begins to appear as though in species in which vocal inventiveness is a major factor, the process of song development is generally more attenuated than in species in which more or less precise learning from adult conspecifics is a dominant factor such as *Fringilla coelebs*, *Sturnella magna* and *S. neglecta*.

There are some consistent changes during the development of particular themes in *Junco oreganus*. The syllable repetitions are usually not identical in the first renderings of a theme, and it is only later that the characteristic stereotypy appears. The variability is partly a result of wavering in the frequency of the notes, also characteristic of the early renderings of themes in *Turdus merula*[14] and *Zonotrichia leucophrys*[10]. The process by which this wavering is eliminated remains to be determined.

Studies in progress on song development in *Zonotrichia leucophrys*[10] suggest many parallels with *Fringilla coelebs*. Here also there are song dialects, clearly differentiating the song patterns in separate populations. Within a given population there is considerable homogeneity, especially in the terminal part of the song (Marler and Tamura, in the press). This confers a great advantage for developmental studies since it makes it possible to predict in detail some of the characteristics of the song patterns which experimental birds would have produced if left to grow up in the natural state. Birds taken as nestlings and raised with siblings in a sound proof room developed songs

References p. 796

which showed no sign of the details of the local song type from the area where they were born, and lacked the overall pattern of whistles and a trill. However, some of the notes corresponded in temporal structure and tonal quality with the whistles occurring in the first part of the natural song. Attempts to train such birds with normal song at an age of ten months and older failed. Then it was found that a male captured two or three weeks after fledging and placed in a sound-proof box with one other male of the same age, developed a detailed rendering of the local song type, having learned it in the first month or so of life. Attempts to train such experienced young birds with a song type from another area were unsuccessful after an age of four months. Training earlier than this produced some effect, although the effects of experience of the parental song type were still dominant. Further work in progress appears to confirm that learning of the local song pattern normally occurs between the ages of about 10–100 days, while the young birds are still in the area of birth.

This differs from the situation in *Fringilla coelebs*, in which most of the learning of song patterns from adults takes place at an age of about 9 months, after dispersal and possible exchange of birds between populations has been completed. In *Zonotrichia leucophrys* most of the learning seems to occur before young birds have dispersed from the place of birth. The stereotypy of the song dialects, and their stability from year to year suggests that little exchange of individuals takes place between separate populations, after the song patterns have been learned. If either males or females prove to be attracted to breed in areas where they hear the song type learned in their youth, the learned song tradition may have direct repercussions upon the genetic constitution of the populations to which they belong, raising once more the possibility that song "dialects" are signs of incipient speciation.

Further work on mammals continues to indicate that learning from conspecifics plays a minimal role, if any, in the development of species-specific vocalizations. Both in foxes[12] and in the elephant seal, *Mirounga angustirostris*[2] it is possible to trace the continuous development of some adult sounds from calls of the young, but there is no evidence that the change involves imitation of others. Perhaps the expanding studies of cetacean and primate vocal behavior are the most likely to reveal vocal traditions, if indeed they exist in sub-human mammals at all.

REFERENCES

[1] ALEXANDER, R. D., 1962, The role of behavioral study in cricket classification, *Syst. Zool. 11*, 53–72.
[2] BARTHOLOMEW, G. A. and N. E. COLLIAS, 1962, The role of vocalizations in the social behavior of the Northern Elephant Seal, *Animal Behaviour, 10*, 7–14.
[3] BIGELOW, R. S., 1960, Interspecific hybrids and speciation in the genus *Acheta* (Orthoperta, Gryllidae), *Can. J. Zool., 38*, 509–524.
[4] BLASE, B., 1960, Die Lautäusserungen des Neuntöters (*Lanius c. collurio* L.), Freilandbeobachtungen und Kaspar-Hauser-Versuche, *Z. Tierpsychol., 17*, 293–344.
[5] BOGERT, C. M., 1960, Effects of sound on the behavior of amphibians and reptiles, In, W. F. LANYON and W. N. TAVOLGA (Eds.), *Animal Sounds and Communication*, Publ. No. 7, AIBS Washington, D.C.
[6] HAGIWARA, S. and A. WATANABE, 1956, Discharges in motoneurones of cicada, *J. Cell. and Comp. Physiol., 47*, 415–428.
[7] HUBER, F., 1960, Untersuchungen über die Funktion des Zentralnervensystems und insbesondere des Gehirnes bei der Fortbewegung und der Lauterzeugung der Grillen, *Z. vergl. Physiol., 44*, 60–132.
[8] LANYON, N. E., 1960. The ontogeny of vocalizations in birds. In W. E. LANYON and W. N. TAVOLGA (Eds.), *Animal Sounds and Communication*, Publ. No. 7, AIBS Washington, D.C.

[9] MARLER, P., M. KREITH and M. TAMURA, 1962, Song development in hand-raised Oregon Juncos, *Auk*, 79, 12–30.
[10] MARLER, P. and M. TAMURA. Song "dialects" in three populations of White-crowned Sparrow, *Condor* (in the press).
[11] MCALISTER, W. H., 1959, The vocal structures and method of call production in the genus *Scaphiopus*, Holbrook, *Texas J. Sci.*, *11*, 60–77.
[12] TEMBROCK, G., 1959, *Tierstimmen. Eine Einführung in die Bioakustik*, A. Ziemsen, Wittenberg Lutherstadt.
[13] THIELCKE, G., 1961, Ergebnisse der Vogelstimmen-Analyse, *J. Ornithol.*, *102*, 285–300.
[14] THIELCKE-POLTZ, H. and G. THIELCKE, 1960, Akustisches Lernen verschieden alter isolierter Amseln (*Turdus merula* L.) und die Entwicklung erlernter Motive ohne und mit künstlichem Einfluss von Testosteron, *Z. Tierpsychol.*, *17*, 211–244.
[15] THORPE, W. H., 1961, *Bird Song. The Biology of Vocal Communication and Expression in Birds*, Cambridge Univ. Press.
[16] WIERSMA, C. A. G., 1962, The organization of the arthropod central nervous system, *Am. Zoologist*, *2*, 67–78.
[17] WILSON, D. M., 1961, The central nervous control of flight in a locust, *J. Exp. Biol.*, *38*, 471–490.

Addendum to Chapter 14

(ACOUSTICAL BEHAVIOUR OF HEMIPTERA)

D. LESTON AND J. W. S. PRINGLE

Stridulation is now shown to be of widespread occurrence in Psyllidae[3] (see page 401). Apparently parts of the metathorax are set in motion to produce sound but the possibility of a wing strigil being present has not been excluded. Elucidation of the problem might throw light on the question of the origin of the family and the relationship of Sternorrhyncha to Auchenorrhyncha.

Recent work confirms the contention that Pentatomomorpha are able to evolve intraspecific stridulating structures and behaviour with some facility. The genus *Strombosoma*, left unplaced in Pentatomoidea for many years, may be placed within Cydnidae although fitting none of the three established subfamilies. Females of *S. unipunctatum* have a typically cydnid strigil and plectrum, the former with about 50 teeth[4]. The pentatomomorphan *Thaumastella aradoides* has paired limae dorsally between abdominal terga I and II and a strigil, apparently on the cubitus, on each hindwing[7]. The bug is placed within a monotypic subfamily in Lygaeidae[7] but general morphology scarcely supports this: the positions of strigil and plectrum differ from any so far described and suggest ranking the bug as an independent family (on trichobothrial evidence close to or within Pentatomoidea).

The stridulatory apparatus of certain coreids (see page 403) has been found[8]. A lima on a process of the modified forewing articular region plays on a strigil situated on the prothorax ventro-laterally: this again is a new form of strigil and plectrum. Sounds have been elicited by touch from both sexes but known mating behaviour suggests that the genera concerned, *Centrocoris*, *Phyllomorpha*, etc., employ intraspecific sound communication.

A structural modification which must considerably re-enforce the efficiency of pentatomid tymbals in communication is the presence of air sacs[4]. In *Eysarcoris* (= *Stollia*) males there are two large air sacs occupying, during the spring mating period, over half the volume of the abdomen and in more or less immediate contact with the tymbal area. The sacs are analogous to those of Cicadidae.

Sound spectrographs of the calls of a number of North American Homoptera and Heteroptera are now available[5]. The author restricts the term stridulation to that which is correctly strigilation, mis-interprets extraspecific communication and lays no stress on whether a sound is pulsed or not. But comparison of the spectrographs of, say, the reduviids *Stenopoda*, *Melanolestes* and *Sinea* and the phymatid *Phymata wolffi* with the cydnid *Corimelaena lateralis* or the cercopid *Lepyronia* shows an irregular pulse repetition rate in the first group and a regular one in the second. In the case of cydnids—*C. lateralis*[5], *Sehirus melanopterus*[4] and *Amaurocoris* sp[4]— the call elicited by touching (defence call or disturbance sound) is sometimes, but not always, distinctly pulsed: it is suggested that in these pentatomomorphans extra-specific calling has evolved in what are basically intraspecific-calling bugs.

Further proof, if any is now required, of the reality of intraspecific acoustic communication, comes from study of closely allied species. There are differences in pulse repetition frequency between the lygaeids *Kleidocerys ericae* and *K. resedae*[2], between *Sehirus bicolor* and *S. melanopterus*, between species of the periodical cicada complex[6]. The differences become meaningful only if regarded as functioning as behavioural barriers to cross mating in the field.

REFERENCES

[1] FRINGS, H. and M. FRINGS, 1960, *Sound Production and Sound Reception by Insects. A Bibliography*, Penn. State Univ. Press, State College.
[2] HASKELL, P. T., 1961, *Insect Sounds*, Witherby, London.
[3] HESLOP-HARRISON, G., 1961, Sound production in the Homoptera with special reference to sound producing mechanisms in the Psyllidae, *Ann. Mag. nat. Hist.*, (13) *3*, 633–640.
[4] LESTON, D., unpublished.
[5] MOORE, T. E., 1961, Audiospectrographic analysis of sounds of Hemiptera and Homoptera, *Ann. ent. Soc. Amer.*, *54*, 273–291.
[6] MOORE, T. E. and R. D. ALEXANDER, 1958, The periodical cicada complex (Homoptera: Cicadidae), *Proc. 10th int. Congr. Ent.*, *1*, 349–355.
[7] SEIDENSTÜCKER, G., 1960, Heteropteren aus Iran 1956, III; Thaumastella aradoides Horv., eine Lygaeide ohne Ovipositor, *Stuttgart. Beitr. Naturk.*, *38*, 1–4.
[8] STYS, P., 1961, The stridulatory mechanism in Centrocoris spiniger (F.) and some other Coreidae (Heteroptera), *Acta ent. Mus. nat. Prag.*, *34*, 427–431.

Addendum to Chapter 16

(SOUND RECEPTION IN LEPIDOPTERA)

ASHER E. TREAT

In the three years that have passed since Chapter 16 was written, several advances have been made, and much new work, which can be reviewed only tentatively, has been initiated. Evidence has accumulated that acoustic sensitivity has positive survival value for tympanate moths in the avoidance of capture by bats[4,5,6,8]. On the other hand, the suggestion (Roeder and Treat, 1957) that the wing sounds of noctuids contain ultrasonic components audible to other moths has not been supported by subsequent experiments.

Although the wingbeats may be inaudible as such, it is nevertheless likely that certain ultrasounds produced by other means in moths of some groups are detectable by tympanate moths as well as by bats and perhaps by other predators. The British investigators, Blest, Collett, and Pye, have recently studied the so-called "striated band" of the arctiid metepisternum[2]. They find that this structure is actually a multiple microtymbal, and that at least in some species it is the source of characteristic ultrasounds as well as of faintly audible clicks. The ecological significance of the organ remains to be determined.

Knowledge of the noctuid tympana has been extended by the work of Suga[7] in Japan. He finds that the two acoustic neurones have the same frequency of maximum sensitivity and therefore provide no basis for frequency discrimination. The more sensitive and more slowly adapting neurone, however, shows curious hysteretic effects, both augmentative and inhibitory, when stimulated simultaneously over a period of several seconds by both tone bursts and continuous pure tones. He suggests that this faculty may aid in the detection of pulsed sounds against a background of continuous noise, and compares the two noctuid neurones with tonic and phasic mechanoreceptors in other animals.

Roeder and Treat[5] observed that a differential (right *vs.* left) tympanic nerve response is present in noctuids only when the stimulating ultrasonic pulses are of low intensity. At high intensities, because of the complete saturation of the acoustic receptors, this differential is absent. This may account for a tendency noted by Roeder[3] for free-flying moths to turn away from a source of faint ultrasonic pulses, while they show seemingly non-directional responses to ultrasonic pulse sequences of high intensity.

Belton, in Ontario, has succeeded in obtaining the first records from the tympanic nerves of pyralid moths[1]. He finds evidence of at least three acoustic cells, sensitive only in the range above 18 kcps. No non-acoustic fiber (B cell) appeared to be active in his preparations. Belton and others at the Belleville laboratory are also experimenting with ultrasound as a possible factor in the behavioral control of crop pests.

REFERENCES

[1] BELTON, P., Electrical response to sound in pyralid moths (in the press).
[2] BLEST, A. D., T. S. COLLETT and J. D. PYE, The generation of ultrasonic signals by a New World arctiid moth (MS submitted for publication, 1962).
[3] ROEDER, K. D., The behavior of free-flying moths in the presence of artificial ultrasonic pulses, *Animal Behavior* (in the press).
[4] ROEDER, K. D. and A. E. TREAT, 1961, The reception of bat cries by the tympanic organ of noctuid moths, Chapter 28 in *Sensory Communication* (W. ROSENBLITH, ed.), Massachusetts Inst. Technol. Press.
[5] ROEDER, K. D. and A. E. TREAT, 1961, The detection and evasion of bats by moths, *American Scientist*, *49*, 135–148.
[6] ROEDER, K. D. and A. E. TREAT, 1960, The acoustic detection of bats by moths, *Proc. XI Int. Entom. Congr., Vienna* (in the press).
[7] SUGA, N., 1961, Functional organization of the two tympanic neurones in noctuid moths, *Japanese J. Physiol.*, *11*, 666–677.
[8] TREAT, A. E., Comparative moth catches by an ultrasonic and a silent light trap, *Ann. Entom. Soc. America* (in the press).

Additional References to Chapter 17

(THE ROLE OF THE CENTRAL NERVOUS SYSTEM IN ORTHOPTERA DURING THE CO-ORDINATION AND CONTROL OF STRIDULATION)

F. HUBER

[135] WILSON, D. M., 1961, The central nervous control of flight in a locust, *J. Exp. Biol.*, *38*, 471.
[136] WILSON, D. M. and T. WEIS-FOGH, 1962, Patterned activity of co-ordinated motor units studied in flying locusts, *J. Exp. Biol.* (in the press).
[137] SUGA, N. and Y. KATSUKI, 1961, Central mechanism of hearing in insects, *J. Exp. Biol.*, *38*, 545.
[138] WENDLER, G., 1961, Die Regelung der Körperhaltung bei Stabheuschrecken (*Carausius morosus*), *Naturwissenschaften*, *48*, 676.
[139] GETTRUP, E., 1962, Thoracic proprioceptors in the flight system of locusts, *Nature*, *193*, 498.
[140] HUBER, F., 1962a, Central nervous control of sound production and some speculations on its evolution, *Evolution* (in the press).
[141] HUBER, F., 1962b, Vergleichende Physiologie der Nervensysteme von Evertebraten, *Fortschr. Zool.*, *15/2*, 166.
[142] AUTRUM, H., J. SCHWARTZKOPFF und H. SWOBODA, 1961, Der Einfluss der Schallrichtung auf die Tympanal-Potentiale von *Locusta migratoria* L., *Biol. Zbl.*, *80*, 385.
[143] PABST, H. und J. SCHWARTZKOPFF, 1962, Zur Leistung der Flügelgelenk-Rezeptoren von *Locusta migratoria*, *Z. vergl. Physiol.*, *45*, 396.
[144] HUBER, F. und W. LOHER, 1962, Verhaltensphysiologische Studien an der Feldheuschrecke *Gomphocerus rufus* L.; I. Zentrale und periphere Kontrolle der Lautäusserungen des Männchens (in preparation).

Addendum to Chapter 23

(ACOUSTIC BEHAVIOUR OF AMPHIBIA)

W. F. BLAIR

Much progress in the field of salientian bioacoustics in the time since Chapter 23 was written makes necessary some elaboration on our discussion and a few corrections. A major review paper has appeared dealing with the influence of sound on the behavior of amphibians and reptiles[4], and we will refer mostly here only to work that has appeared subsequent to publication of that article. Another major review paper deals with vocalization as one of the most important of the complex of isolating mechanisms in salientian amphibians[15]. Another review paper discusses structure of the mating call as one of various classes of non-morphological data of use in classification of salientian amphibians[2].

An important paper dealing with comparative morphology of vocal structures and with the mechanisms of vocalization in 16 species of North American toads (*Bufo*) has appeared[13]. One significant point is that vocalization is clearly demonstrated to occur in a closed system; air is not alternately pumped between lungs and vocal sac during the course of the call.

Increased use of the mating call as a clue to taxonomic relationships has made some of the names used in Chapter 23 incorrect in application to the populations referred to. All references to *Bufo compactilis* are to a species now known as *B. speciosus*. It has been shown[4] that two allopatric species that are morphologically rather similar have strikingly differentiated calls in duration (about 0.4 to 0.7 second in *speciosus*; 12 to 37 seconds in *compactilis*). The name *speciosus* applies to all U.S. records of this complex of toads, with *compactilis* restricted to central and southern Mexico. The name *Hyla versicolor* as used in Chapter 23 encompasses a pair of sibling species that are sharply differentiated in pulse rate of the mating call[10,11], and the sonagram labeled *versicolor* in Fig. 382 is actually that of the rapidly pulsing population to which the name *chrysoscelis* applies. Mating call discrimination experiments show that the females of these two species respond preferentially when presented with a choice of the calls representative of the two species[12]. Our statement (p. 698) that *Bufo punctatus* is possibly more closely related to the *americanus* group of toads than to any other must be amended in the light of more recent evidence. Our unpublished data from hybridization indicate that this species is not close to the *americanus* group, and the similarity in call structure seems to be the result of convergent evolution.

A recent study[7] of call structure in 13 species belonging to 5 genera of the Leptodactylidae adds materially to our knowledge of this large family. It is common for the fundamental frequency to be also the dominant frequency. In general the genus *Eleutherodactylus* shows relatively greater interspecies variation in call structure, while the genus *Leptodactylus* shows relatively less. Two studies have treated differentiation of call structure in species complexes of *Hyla*. The *H. microcephala–phlebodes–ebraccata* complex[8] in the Canal Zone shows strong differentiation. One interesting feature of these Tropical treefrogs is the increased complexity of the call. Whereas the hylas of temperate North America typically have a call that is formed by the repetition of the same note, these have differentiated the notes. It might be theorized that the large number of sympatric species of *Hyla* in the Tropics has pressured the evolution of the more complex calls as isolating mechanisms. The *Hyla eximia* group has been defined after comparative analysis of call structure in included and related species[1]. The group centers in Mexico with outliers in both the eastern and western United States.

One hopeful development of the immediate past is the tendency to include structural analyses of the mating call in the descriptions of new species of frogs[5,6,9]. It is to be hoped that this trend continues, since the structure of the call is often the most distinctive feature of a species population.

Attention is turning increasingly to the testing of female frogs for their responses to the recorded calls of their own and other species. Female *Pseudacris streckeri* are able to discriminate between recorded calls of their own species and those of the allopatric, closely related species *P. ornata*[3]. The only noticeable difference in the calls is a difference of about 300–400 c/s in the dominant frequency. One research group[14] has concluded that some species with highly developed vocalization depend on call as an isolation mechanism, while others do not. Their data are unconvincing at this stage because of questions about methodology and experimental design.

REFERENCES

[1] BLAIR, W. F., 1960, Mating call as evidence of relations in the *Hyla eximia* group, *Southwestern Naturalist*, 5, 129–135.
[2] BLAIR, W. F., 1962, Non-morphological data in anuran classification, *Syst. Zool.*, 11, 72–84.
[3] BLAIR, W. F. and M. J. LITTLEJOHN, 1960, Stage of speciation of two allopatric populations of chorus frogs (*Pseudacris*), *Evolution*, 14, 82–87.
[4] BOGERT, CH. M., 1960, The influence of sound on the behavior of amphibians and reptiles. In W. E. LANYON and W. N. TAVOLGA (eds.), *Animal sounds and communication*, Am. Inst. Biol. Sci. Publ. No. 7, Washington, D.C.
[5] DUELLMAN, W. E., 1961, Description of a new species of *Hyla* from Chiapas, Mexico, *Copeia*, 414–417.
[6] FOUQUETTE, M. J., 1958, A new tree frog, genus *Hyla*, from the Canal Zone, *Herpetologica*, 14, 125–128.
[7] FOUQUETTE, M. J., 1960, Call structure in frogs of the family Leptodactylidae, *Tex. J. Sci.*, 12, 201–215.
[8] FOUQUETTE, M. J., 1960, Isolating mechanisms in three sympatric treefrogs in the Canal Zone, *Evolution*, 14, 484–497.
[9] FOUQUETTE, M. J., 1961, Status of the frog *Hyla albomarginata* in Central America, *Fieldiana, Zoology*, 39, 595–601.
[10] JOHNSON, F. C., 1959, Genetic incompatibility in the call races of *Hyla versicolor* Le Conte in Texas, *Copeia*, 327–335.
[11] JOHNSON, F. C., 1961, Cryptic speciation in the *Hyla versicolor* complex. Unpub. doctoral dissertation, Univ. Texas.
[12] LITTLEJOHN, M. J., M. J. FOUQUETTE and C. JOHNSON, 1960, Call discrimination by female frogs of the *Hyla versicolor* complex, *Copeia*, 47–49.
[13] MCALISTER, W. H., 1961, The mechanics of sound production in North American *Bufo*, *Copeia*, 86–95.
[14] MARTOF, B. S., 1961, Vocalization as an isolating mechanism in frogs, *Am. Mid. Nat.*, 65, 118–126.
[15] MECHAM, J. S., 1961, Isolating mechanisms in anuran amphibians, in W. F. BLAIR (ed.), *Vertebrate Speciation*, Univ. Texas Press, Austin.

SYSTEMATIC INDEX

(The page numbers with an asterisk refer to illustrations)

ARTHROPODA–CRUSTACEA

BRANCHIOPODA

DIPLOSTRACA

Cladocera

Mixodiaptomus laciniatus 789

CIRRIPEDIA

THORACICA

Balanus 315

MALACOSTRACA

ISOPODA

Androniscus 314

DECAPODA

Macrura

Alpheidae 73, 306, 646 (footnote)
Alpheus 306*, 349, 354
– *malleodigitus* 307
– *strenuus* 307
Amphibetaeus 306 (footnote)
Coralliocaris 306
Crangon (see *Alpheus* 306)
Linuparus 308
Metapenaeopsis acclivis 306
– *barbatus* 305, 306*
– *durus* 306
Palaemonidae 306
Palinuridae 307–309
Palinurus argus 308, 349
– *vulgaris* 307, 308*, 428, 604 (footnote), 685
Panulirus, 308
– *interruptus* 607 (footnote), 685
Parapenaeus akayebi (see *Metapenaeopsis barbatus* 305, 306)
– *acclivis* (see *Metapenaepsis acclivis* 306)
Penaeidae 305
Penaeopsis stridulans 305, 318
Pontonia 306
Synalpheus 62, 306
– *lockingtoni* 307
Typton 306

Anomura

Aniculus (see *Trizopagurus* 310)
Clibanarius (see *Trizopagurus* 310)
Coenobita perlatus 309
– *rugosus* 309*
Gebia issaeffi 309*
Paguridea
Pagurus (see *Trizopagurus* 310)
Thalassina anomala 309
Thalassinidea 309
Trizopagurus 310
– *strigimanus* 310
– *tenebrarum* 310*

Brachyura

Acanthocarpus 312, 606
– *bispinosus* 606
Acmaeopleura 311
– *parvula* 311*
Brachygnatha 336
Dotilla 312, 589
– *fenestrata* 313*
Globopilumnus 314
– *stridulans* 315*
Helice tridens 311*
Hexaplax 312
Hexapus 312
Ilyoplax 590
Lambdophallus 312
Macropathalmus 311, 590
Matuta 311, 606

[Matuta]
– lunaris 312*
Menippe 312, 606
– obtusa 312*
Metaplax 311
Ocypode 313, 334, 335, 587, 589, 606, 612
– cordimana 313 (footnote)
– cursor 314*
– nobilii 314*
– platytarsis 313*
Ommatocarcinus macgillivrayi 310
Ovalipes ocellatus 310, 311*, 314, 336, 337 (footnote)
– punctatus 314, 315*, 336, 337 (footnote)
Oxystomata 336
Potamon 312
– africanum 312, 313*

Pseudozius bouvieri 311
Sesarma 314
Uca 74, 314, 583, 584 (footnote), 587, 590, 599, 606, 612, 618
– musica 314*
– pugilator 328, 586, 616
– tangeri 586, 597, 625 (footnote)
– terpsichores 314

STOMATOPODA

Gonodactylus chiragra 305
– demani 305
– glabrous 305
– oerstedti 305
Lysiosquilla excavatrix 305
Squilla empusa 305
– mantis 305

ARTHROPODA–MYRIAPODA

CHILOPODA

Alipes 321
Rhysida nuda togoensis 322
Scutigera decipiens 321

DIPLOPODA

Arthrosphera aurocineta 322
Borneopoeus costatus 322*

Bournellum retusum 322*
Sphaeropoeus musicus 322*
– volzi 321
Sphaerotheridae 321
Sphaerotherium acteon 323*
– anomalum 322*
– campanulatum 322*
– coquerelianum 322*

ARTHROPODA–INSECTA

ODONATA

Aeschna (larva) 453, 459
Epiophlebia superstes 333, 334* (larva)

DICTYOPTERA

Archiblatta hoeveni 289
Periplaneta 420, 423, 424*, 645
– americana 412*, 423, 475, 482

ISOPTERA

Armitermes 327
Bellicositermes 327
Capritermes 327
Cornitermes 327
Rhinotermes 327
Syntermes 327
Termes lilljeborgi 590

PLECOPTERA

Chloroperla grammatica 328
Dinoceras cephalotes 328
Perlidae 328

PHASMOPTERA

Palophinae 329
Phyllium athanysus 279
Pulchriphyllium crurifolium 279*

ORTHOPTERA

Acheta assimilis 303, 348, 360, 361, 369*, 422, 443, 591, 593, 634, 636, 640
– domesticus 302*, 303, 348, 353, 360, 361, 443, 447, 466*, 591, 603, 640
– firma 636
– fultoni 636, 647, 648
– pennsylvanicus 359, 360, 362, 636
– pennsylvanicus × fultoni 628
– pennsylvanicus × vernalis 628
– rubens, 360, 636, 648
– rubens × veletis 628, 634
– rubens × assimilis 628, 634
– veletis 362, 592*, 634
– veletis × pennsylvanicus 634
– vernalis 636

Acrididae 76, 296–298, 347*, 351*, 359, 365–366, 415, 421–423, 427, 440, 445, 454, 464, 484, 585, 590, 608
Acridinae 365 (footnote), 587, 589, 591, 593, 594, 596, 597, 599–602, 603 (footnote), 604, 605, 609, 610, 612
Acridoidea 335, 350, 352*, 357*, 599*
Aglaothorax armiger 605
Aiolopus strepens 328
– thalassinus 594
Amblycorypha carinata 648
– oblongifolia 348, 359, 421, 604, 605, 611
– rotundifolia 300, 303, 587, 595, 636
– uhleri 359, 647, 648
Ametroides kibonotensis 285*
Anacridium aegyptum 421
Anaxipha exigua 348, 361, 593
Anoedopoda lamellata 305
Arcyptera fusca 352*, 594, 596
Arphia sulfurea 304, 421
Atlanticus pachymerus 648
– testaceous 348, 360, 368*, 647
Barbitistes fischeri 61*, 348, 647
Batrachotetriginae 286
Bryodema tuberculata 327, 328, 594
Bullacris 286
Calliptamus 333 (nymph), 608 (nymph)
– italicus 329, 596
Celes variabilis 328
Charora 284
– crassinervosa 285*
Chloealtis conspersa 303, 349, 361
Chorthippus 71, 74, 79, 91, 366, 587, 610 (nymph), 618, 626
– albomarginatus 357*, 594
– apricarius 303, 594, 595, 600
– bicolor (see – brunneus)
– biguttulus 92, 228, 229*, 303, 349, 351*, 353, 360, 445, 591, 594, 595, 596, 600, 601, 602, 604, 608 (nymph), 613, 614, 619, 620, 631*, 632, 633, 647
– biguttulus × brunneus 229*, 629, 631–633,
– brunneus 20, 21*, 64*, 79, 92, 95–100, 228, 229*, 297*, 303, 351*, 353, 358*, 360, 361, 365, 584, 588, 591, 594, 595, 596, 597, 600, 602, 603, 604, 608 (nymph), 611, 613, 614, 615*, 618, 620, 631*, 632, 633, 638, 641
– curtipennis 303, 349
– dorsatus 594, 595, 596, 600, 602, 612
– jucundus 296*, 353, 360, 613, 614, 647
– longicornis (see Chorthippus parallelus)
– mollis 594, 596, 647
– montanus 297*, 594, 595, 596, 600, 601, 602, 604, 608 (nymph), 612, 634
– parallelus 20, 21*, 297, 303, 353, 584, 588, 594, 595, 596, 600, 602, 608 (nymph), 612, 613, 616, 628, 640, 641, 642, 643

– pullus 594, 596, 600, 602
– vagans 594, 596, 600
Chrysochraon dispar 594, 596, 600, 612
Cirtophyllus perspicillatus 367*
Conocephalidae 348, 359
Conocephalus 74, 301
– brevipennis 303, 348, 359, 647, 648
– dorsalis 8*
– faciatus 359, 647, 648
– maculatus 298*
– spartinae 348
– strictus 421*
Cophus 419*
Cylindracheta arenivaga 279*, 296 (footnote)
Cyphoderris monstrosus 282*, 299, 336
Cyrtaspis scutata 300*
Cyrtoxipha columbiana 617, 648
Decticus 301, 359, 648, 422, 429, 638 (footnote)
– albifrons 357
– verrucivorus 348, 350, 351*, 419*, 421, 613 (footnote)
Deinacrida megacephala 285
Dictyophorus 324
Egnatioides 284
Encoptolophus sordidus 328
Endacusta 419*
Ephippiger 59*, 60*, 62*, 64, 71, 72, 76, 77, 79, 91, 100, 104*, 105, 301, 360, 364, 365, 366, 584, 586 (footnote), 591 (footnote), 595, 597, 598 (footnote), 604, 605, 611*, 612, 614, 615, 618, 619, 620, 622, 626, 639, 640, 642, 646 (footnote)
– bitterensis 44*, 301*, 303, 359, 365, 443, 612, 619*, 620, 623, 647
– cunii 612, 623, 647
– ephippiger 299*, 348, 352*, 358*, 359, 360, 364*, 612, 623, 638, 640, 644, 645*, 647
– provincialis 361, 367, 369*, 612
– terrestris 353
Ephippigerida nigromarginata 622, 623
Ephippigeridae 299, 359, 583
Eremogryllus hamadae 366
Euthystira brachyptera 594, 596, 600, 641*, 642, 643
Gampsocleis buergeri 426
– glabra 357, 366
Gomphocerus 594
– rufus 297*, 303, 600, chapter 17
Griotettrix verruculatus 304
Gryllacrididae 338 (footnote)
Gryllacridoidea 285, 299
Gryllidae 348, 351*, 355 (footnote), 359, 392, 440, 464, 484, 583, 590, 591, 597, 603, 604
Grylloidea 302–303, 335, 347*, 350, 352*, 360, 362–364, 418, 419*, 422, 583, 599*
Gryllotalpa 419*
– gryllotalpa 348, 359, 360, 361, 622

[*Gryllotalpa*]
- *hexadactyla* 348, 361, 584, 647, 648
Gryllotalpidae 302, 303, 348, 359
Gryllus 233, 424*
- *abbreviatus* 422*
- *assimilis* (see *Acheta assimilis*)
- *bimaculatus* 302*, 303, 360, 481, 591, 643
- *campestris* 71, 76, 302*, 348, 352*, 358*, 359, 360, 366, 369, 419*, 422, 423, 431, 584, 588, 591, 603, 618, 640, chap. 17
- *campestris* × *bimaculatus* 229, 629, 632 (footnote)
- *toltecus* 419*
Haaria 419*
Henicus monstrosus 329
Homorocoryphus nitidulus 348, 359, 647
Lamarckiana 298
Locusta 74
- *migratoria* 71, 298*, 329, 348, 422, 429*, 447
- *migratoria* (*solitaria*) 328, 353, 595, 604
- *migratoria gallica* (*gregaria*) 304; (*solitaria*) 304
- *migratoria migratorioides* 421
- *pardalina* 333 (nymph), 608 (nymph)
Locustidae (see Acrididae)
Meconema thalassina 328, 329
Mecosthetus grossus 416*, 594, 609
Mesembria dubia 329
Metrioptera bicolor 351*
- *brachyptera* 351*
- *roëseli* 613 (footnote)
Microcentrum 299
- *rhombifolium* 300, 357, 595
Miogryllus verticalis 348, 593, 603
Myrmeleotettix maculatus 297*
Nemobiinae 583, 593 (footnote)
Nemobius allardi 228, 348, 367, 634, 637, 647, 648
- *allardi* × *tinnulus* 629*
- *ambitiosus* 637
- *aurantius* 637
- *bruneri* 637
- *carolinus* 348, 363, 369*, 370*, 636, 637
- *confusus* 361, 369*, 370*, 637
- *cubensis* 637
- *fasciatus* 303, 351*, 359, 369*, 593, 634, 637, 647, 648
- *fasciatus fasciatus* (see *Nemobius allardi*)
- *fasciatus socius* (see *Nemobius fasciatus*)
- *fasciatus tinnulus* (see *Nemobius tinnulus*)
- *griseus funeralis* 637
- *maculatus* 637
- *melodius* 359, 593*, 636
- *palustris* 348, 637
- *sparsalsus* 637
- *sylvestris* 584
- *tinnulus* 228, 358*, 359, 361, 369*, 634, 637, 647
Neoconocephalus 299

- *ensiger* 348, 367, 368*, 443, 587, 647
- *exiliscanorus* 605, 617
- *nebrascensis* 368*, 584, 587
- *robustus* 359, 584
- *triops* 300*
Neoxabea bipunctata 348, 359
Nomadacris septemfasciata 103*, 298
Oecanthidae 348, 355 (footnote), 359, 583, 590, 591, 593 (footnote), 604
Oecanthus 74, 350, 354, 419*, 447, 618, 620, 624, 626
- *angustipennis* 358*, 634
- *argentinus* 361, 593, 624
- *exclamationis* 348, 634
- *latipennis* 348, 647, 648
- *nigricornis* 348, 359, 369*, 370*, 624, 634
- *niveus* 303, 348, 358*, 359, 361, 363, 366, 368, 369*, 370*, 617, 624, 634
- *pellucens* 71, 348, 351*, 353, 359, 361, 363*, 369, 444, 587, 593, 604
- *pini* 647, 648
- *quadripunctatus* 303, 351*, 359, 361, 369*, 624, 634, 647, 648
Oedaleonotus fuscipes 329
Oedipoda coerulescens 327, 417*
Oedipodinae 297, 298*, 327, 328, 330, 583, 594, 596, 609
Omocestus 594
- *haemorrhoidalis* 596, 600
- *rufipes* (see *Omocestus ventralis*)
- *ventralis* 303, 600, 601*, 602, 613
- *viridulus* 297, 353, 361, 584, 595, 596, 600, 604, 628, 641
Orchelimum 299, 360
- *agile* 367 (footnote), 639
- *bradleyi* 611, 612
- *concinnum* 359
- *militare* 612
- *nigripes* 587
- *vulgare* 348, 359, 368*, 647, 648
Orocharis 419*
- *saltator* 348, 363, 369*, 370*, 647, 648
Orphania denticauda 604, 605, 638 (footnote)
Orphulella speciosa 330
Pamphagidae 286, 298
Paratylotropidia brunneri 329, 611
Paroxya atlantica 421*
Phaneroptera 419*
Phanaeropteridae 348
Pholidoptera aptera 421, 611, 616
- *cinerea* (see *Pholidoptera griseoaptera*)
- *griseoaptera* 584, 588, 611, 613 (footnote), 618
Phonogaster cariniventris 286
Phyllopalpus pulchellus 647, 648
Phyllophoridae 291, 300
Phylloptera 419*

Phymateus 329
Physemacris 286
– *variolosa* 285*
Platydactylus 419*
Pneumora 286
Pneumoridae 286
Podisma 333 (nymph), 608 (nymph)
Porthetini 298
Prophalangopsidae 282*, 299, 337, 338 (footnote)
Psectrocnemus longiceps 298*
Psophus stridulus 327, 328
Pterophylla camellifolia 299, 300*, 303, 348, 358*, 366, 421, 587, 605, 611, 616, 636, 647
Schistocerca gregaria 59, 299 (footnote), 329, 427, 447, 589
Scudderia 299
– *curvicauda* 300*, 348, 595
– *furcata* 300, 595
– *septentrionalis* 359
– *texensis* 300, 595
Sphingonotus coerulans 421
Stauroderus scalaris 594, 600
Stenobothrus lineatus 297, 303, 349, 353, 358*, 359, 361, 584, 588, 594, 595, 604, 617
– *nigromaculatus* 594, 596, 600
– *rubicundus* 330
– *stigmaticus* 303, 594, 596
Stetophyma grossum 298
Tachycines asinamorus 423
Taeniopoda picticornis 324
Tetrix subulata 596
– *tenuicornis* 594, 596
– *turki* 596
Tettigonia 301, 420, 422, 423, 427, 430*
– *cantans* 351*, 421, 423, 431*, 584
– *viridissima* 100 (footnote), 348, 353, 421, 429, 584, 618, 638, 644*, 645, 647
Tettigoniidae 348, 351*, 418, 419*, 421, 425*, 440, 595
Tettigonioidea 299–302, 335, 347*, 350, 352*, 355, 357*, 360, 364–365, 422, 587, 599*, 605 639, 642
Thamnotrizon (see *Pholidoptera*)
Trachypterella andersoni 286
Tridactylidae 279, 305
Tylopsis liliifolia 595
Uromenus rugosicollis 357, 358*, 361
Xyronotus aztecus 286*

COLEOPTERA

Acalles 293
Aegidinus 291
Aegidium 291
Amphisternus 302,
Anobium 618
– *pertinax* 326

– *striatum* 326
Anoxia villosa 288
Aptinus 325
Aromia moschata 351*
Berosus 287
Blethisa 288, 289*
Bolbelasmus bocchus 289
– *gallicus* 289
– *unicornis* 289
Bolitotherus cornutus 326
Brachynus 325
Cacicus americanus 295*, 296
Camelonotus 288
Camptorrhinus 288
Cantharis 379, 380
Carabidae 287, 336
Carabus 424*
– *hortensis* 423
Cerambycidae 281, 282*, 335 (footnote), 605
Cerambyx cerdo 304, 349, 358*, 646
– *scopolii* 586
Cetonia aurata 376*
Cetoninae 330
Ceutorrhynchini 304
Chiasognathus granthi 296
Chrysomelidae 281, 287, 336, 605
Cicindelidae 336
Clythra 281, 287
Copris 293, 337
Crioceris duodecimpunctata 287*, 584
– *merdigera* 304
Cryptorrhynchus 288
– *lapathi* 293
Ctenoscelis 296
Curculionidae 287, 288, 293, 336
Cybister aegyptiacus 304
– *confusus* 325, 331 (larva)
Dorcus (larva) 330
Doritomus longimanus 293*
Dynastinae 288
Dytiscus 386, 413
Ectatorrhinus 288
Elaphrus 288
– *riparius* 304
Elateridae 583
Endomychidae 280, 304
Enoplopus velikensis 280*, 304
Ergastes faber 282*
Erotylidae 304
Frickius 291
Gasterocercus 288
Geniates 296
Geotrupes 288, 291, 349
– *stercorarius* 304, 376*
– *sylvaticus* 331* (larva)
– *vernalis* 290*
Geotrupidae 330

Graphopterus 334
— *serrator* 296
— *variegatus* 296*
Gyrinus 219, 413
Heliocopris 291
Heliophilus cribratostriatus 287
Heteroceridae 286
Hispa testacea 304
Hispidae 280
Hornia minutipennis 387
Hybalus 291
Hydrophylidae 287, 336
Hydrophilus piceus 287
Hygrobia hermanni 293
Idiostoma 291
Lagochile 292
Lampyridae 583
Languriidae 280
Lema trilineata 287*
Lepyrus 293
Ligyrus 293
Lilioceris lilii 360, 584
Lomaptera 286
Lucanidae 331, 336
Lucanus (larva) 330
Macraspis 292
Masoreidae 336
Meloidae 387
Melolontha 376*
Melolonthinae 330, 331
Necrophorus 74, 288, 334, 349, 605
— *vespillo* 289*, 304
— *vespilloides* 349, 351*
Nitidulidae 280
Notorrhina muricata 326, 612
Ochodaeus 293
— *maculipennis* 293*
Olocrates 605
— *abbreviatus* 287, 288*
— *gibbus* 287
Orphnus 291
Oryctes 334
— *nasicornis* 288*
— *rhinoceros* 288, 330* (larva), 331 (pupa)
Oxychila 296
Passalidae 330, 331* (larva), 589
Passalus 605
— *cornutus* 289*, 349
Phileurus 290
Phonapate nitidipennis 281*
Pissodes 293
Plagithmysus 291, 296
Platycerus 330 (larva)
Polyphylla 605
— *fullo* 288
Priobium castaneum 280*
Prionus coriarius 296

Proculus 290
Rutelinae 330, 331
Scarabaeidae 287, 336
Scarabaeoidea 291, 330 (larva)
Scolytidae 281
Scolytus destructor 280*
Serica brunnea 281, 304
Siagona 292
— *fuscipes* 291*
Sibinia 293
Silphidae 287, 336
Sinodendron 330 (larva)
Spercheus 605
— *emarginatus* 287
Synapsis 291
Taurocerastes 291
Tenebrionidae 281, 287, 336
Termitogaster emersoni 327
Trox 293
Xestobium tesselatum 326*

NEUROPTERA

Myrmeleontidae 428

LEPIDOPTERA

Acherontia atropos 73, 106, 323, 324*, 332 (pupa), 349, 351*, 353, 605, 607
Aegeriidae 388
Aegocera mahdi 294
— *tripartita* 294*
Ageronia arethusa 295*
Aglaïs polychloros 329
— *urticae* 329
Agrotis 423
Amatidae 437
Arcte caerulea 283, 284*
Arctia caja 323 (footnote)
Arctiidae 420, 435, 437, 800
Arctiinae 415
Blenina metascia 332 (pupa)
Catocala 415*, 416*, 434
Cerura 332 (larva)
Charaxes sempronius 605
Chelonia pudica 434
Cimelia 435 (footnote)
Composia fidelissima 325
Cossidae 436 (footnote)
Dicranura 332 (larva)
Drepana 332 (larva)
Dudgeonea 436 (footnote)
Dysschema tiresias 324
Eligma hypsoides 333* (pupa)
— *narcissus* 333 (pupa)
Endrosa 71, 282, 586
— *aurita ramosa* 434
Erasmia sanguiflua 325
Euprepia 282, 586

– *haroldi* 596
Euvanessa antiopa 329
Gangara thyrsis 332 (pupa)
Geometridae 415, 428, 434
Glyphipterygidae 436 (footnote)
Hecatesia fenestrata 294
– *thyridion* 294
Heliochelus paradoxus 294
Hemaris fuciformis 379
Hesperiidae 332 (pupa)
Hylophila prasinana 323
Lycaenidae 332 (pupa)
Lymantria monacha 283
– *viola* 332* (pupa)
Lymantriidae 332 (pupa)
Lymantriinae 415
Microlepidoptera 435
Musurgina laeta 294*
Noctuidae 294, 332 (pupa), 415, 427, 428, 436, 437, 800
Notodontidae 415, 420, 435
Papilionidae 332 (pupa)
Parnassius mnemosyne 330
Pemphigostola synemonistis 294, 295*
Pergesa elpenor 376*
Platagarista tetraplura 294
Plotheia decrescens 333 (pupa)
Plusia gamma 376*
Prepona 605
Prodenia evidania 218, 421, 428
Psychidae 436 (footnote)
Pyralididae 415, 436, 800
Pyrameis 424*
– *atalanta* 423
Rhodia fugax 331 (larva)
Rhodogastria bubo 282, 325, 605
– *lupia* 282*, 325
Rhopalocera 435
Saturnidae 74, 331 (larva), 332 (pupa), 435, 436 (footnote)
Satyridae 415, 423
Smerinthus populi 323 (footnote)
Sphecodina abbotti 331 (larva)
Sphingidae 331, 435
Sphinx ligustri 376*
Stilnoptia salicis 283
Symitha molalella 332 (pupa)
Thecophora fovea 294, 295*
Tetracis crocallata 437
Vanessa io 329

DIPTERA

Aëdes aegypti 349, 354, 387, 420, 423, 424, 597, 619, 620, 621, 625
– *trivittatus* 625
– *vexans* 625
Anopheles 426

– *gambiae* 349
– *punctipennis* 625
– *quadrimaculatus* 621, 625, 642 (footnote)
– *subpictus* 420
Boreellus 379
– *terraenovae* 379
Calliphora erythrocephala 430*
Ceratitis capitata 73
Ceratopogonidae 428
Chironomidae 413
Chironomus 376*, 382, 383*
Culex 71, 351*, 415
– *pipiens* 621, 625
Culicidae 376*, 379, 413, 414*, 426, 428
Dacus 586
– *oleae* 290*
– *tryoni* 290*, 349
Drosophila 54, 55*, 379, 380, 453
– *funebris* 379, 385*
– *melanogaster* 379
– *repleta* 379
Empidinae 599
Eohelea stridulans 302
Eristalis 423
– *tenax* 378, 388
Forcipomyia 382, 383*
Musca autumnalis 379
– *domestica* 376*, 379
Pentaneura aspersa 420
Psorophora confinnis 625
– *horrida* 625
Syrphidae 386–388
Tipula 379

HYMENOPTERA

Absyrtus luteus 379
Andrena nitida 423
Anthidium 387
Anthophora 387
Apis mellifica 40*, 58, 59, 71, 325, 349, 376*, 379, 380, 381, 382, 383*, 386, 423, 583, 724
Bombus 380, 381, 383, 386, 388, 424*
– *agrorum* 379
– *lucorum* 376*, 382, 383*
– *scoroeensis* 423
Braconidae 387
Camponotus 423
Crematogaster 622
– *scutellaris* 304, 590
Dendromyrmex 328
Dorylidae 283
Dorymyrmex emmaericellus 283
Formica sanguinea 413*
Megachile 387
Megaponera 590
Messor 350, 590
Monomorium salomonis 350

Mutilla barbata brutia 283
– *europaea* 283
Mutillidae 283
Myrmica rubra laevinodis 283*
– *ruginodis* 350
– *sulcinodis* 304
Myrmicidae 283
Neomyrma rubida 304
Ophion luteus 376*
Phytodyctus polyzonias 334 (pupa)
Pogonomyrmex 509
– *marcusi* 283*
Polistes 380, 381
Poneridae 283
Solenopsis fugax 304
Tetramorium caespitum 350
Vespa 380, 424*
– *crabo* 334 (larva), 423, 590 (larva, footnote)
– *vulgaris* 376, 379

PSOCOPTERA

Atropos pulsatorium 326*, 327
Lepinotus inquilinus 327

THYSANOPTERA

Anactinothrips 291
Diceratothrips 291
– *princeps* 291
Sporothrips 291

HOMOPTERA

Achorotile albosignata 604
Aetalion reticulatum 396
Aetalionidae 393, 400 (larva)
Aphididae 284*, 401, 402
Aphis 284
Aphrodes bicinctus 393*
Aphrophora 605
Auchenorrhyncha 335, 392–401, 798
Biturridae 400
Cercopidae 393, 394, 397, 400
Cercopoidea 394
Cicada 74, 398*
Cicadellidae 71, 393, 397, 400
Cicadelloidea 394
Cicadetta coriaria 417*
Cicadidae 334, 347, 393, 396, 400, 415, 418, 422, 427, 618
Coleorrhyncha 393
Criomorphus bicarinatus 604
Cryptotympana exalbida 399*
Doratura stylata 77, 370, 594, 596
Erythroneura hyperici 612
Eupelix depressa 594
Eurymelidae 393, 400
Fulgoroidea 393, 394, 400
Graptopsaltria nigrofuscata 398

Huechys incarnata 418*
Idiocerus 612
Lepyronia 798
– *coleoptrata* 612
Livia juncorum 401
Longiunguis 284
Lyristes plebejus 622
Macropsis 605
Magicicada cassinii 354, 392, 588, 617, 638
– *septemdecim* 354, 392, 398, 588, 617, 638
Meimuna 398
Membracidae 393, 400
Neophilaenus campestris 393*
Octotympana 398
Okanagana rimosa 349
Oncopsis 605
Paropia scanica 396, 594
Peloridiidae 393
Platypleura 396*, 398
– *capitata* 395*, 397*, 399*
– *octoguttata* 349, 397*, 399*, 400
– *westwoodii* 399*
Pomponia intemerata 621
Psyllidae 393, 401, 402, 798
Purana campanula 399*
Rihana mixta 399*
Sternorrhyncha 392, 393, 401, 402, 798
Tanna 398
– *japonensis* 422
Terpnosia ransonetti 397*, 399*
– *ridens* 399*
– *stipata* 399*
Tettigadinae 397
Tettigarcta tomentosa 393, 394*, 396*, 400, 401*
Tettigarctidae 393
Tibicen auletes 647
– *canicularis* 349, 647
– *dorsata* 621
– *linnei* 647
Toxoptera aurantii 284*, 401, 617
– *citricidus* 401
– *coffeae* (see *Toxoptera aurantii*)
– *odinae* 401
– *taversi* (see *Toxoptera citridus*)
Trioza acutipennis 401
– *nigricornis* 401
Typhlocybidae 400

HETEROPTERA

Acanthosomidae 403
Aelia acuminata 407*
Aethus flavicornis 402
Agonoscelis 404
Amaurocoris, 798
Amphibicorisae 402, 405, 408
Amyoteinae 403, 404, 407

SYSTEMATIC INDEX

Anisops 292, 405
Aradacanthia 403
Aradidae 286, 403, 408, 409
Arctocorisa 405
Arhaphe 403, 404
Artabanus 403
Buenoa 292*, 405
– *limnocastris* 405, 407, 586
Calisiinae 403
Callicorixa distincta 405
Carpocoris 404, 407
– *pudicus* 408
Centrocoris 403, 798
Cimicomorpha 402, 404, 405, 409
Coranus subapterus 281*, 406, 407, 605
Coreidae 403
Coreinae 403
Coreoidea 402
Corimelaena lateralis 798
Corimelaeninae 403
Corixa 280, 292, 405, 408, 415, 420
– *striata* 280*, 610 (footnote), 616, 622
Corixidae 405, 408, 586
Corixinae 405
Cryptocerata 415
Cydnidae 392, 402, 403, 408
Cydninae 403
Elasmodemidae 404
Elasmucha 403
Eysarcoris 798
Geocorisae 402, 408
Glypsus 404
Hesperocorixa 405
Hydrocorisae 402, 405, 406
Illibius 403
Kleidocerys 403, 406, 408, 605
– *ericae* 588, 799
– *resedae* 588, 799
Largidae 403
Larginae 403
Lygaeidae 392, 403
Lygaeoidea 402
Mecideinae 403
Mecistorhinus 404
Melanolestes 798
Mezirinae 403, 404
Micronecta poweri 407
Micronectinae 405
Microvelia diluta 405
Nabidae 405
Nabis flavomarginatus 284
Naucoridae 406
Nepidae 406
Nezara viridula 404
Notonecta 405
Notonectidae 405, 408, 586

Notonectinae 405
Orsilinae 403
Pachychoris 284
Palmacorixa 405
Palomena 404
Pentatomidae 392, 403, 404, 408
Pentatominae 403, 404, 407, 409
Pentatomoidea 402, 403
Pentatomomorpha, 402, 403, 404, 406, 408, 409
Phyllomorpha 328, 403, 798
Phymata wolffi 798
Phymatidae 281, 404
Pictinus 286*, 403
Piesma quadrata 402*, 406
Piesmidae 392, 403, 408
Piezosterninae 403
Pirates 405
Plea 415
Pleidae 406
Psectrocoris 403
Ptenidiophyes 405
Ranatra quadridentata 290
Reduviidae 281, 404, 407
Rhodnius prolixus 423
Rossius 403
Saldidae 405
Schizopteridae 405
Sciocoris cursitans 404*
Scutelleridae 403
Scutellerinae 403
Sehirinae 403
Sehirus bicolor, 406*, 408, 594, 799
– *melanopterus* 798, 799
Sigara 280, 292, 405
– *distincta*, 405
– *dorsalis* 407, 408
– *striata* 407
Sinea, 798
Spathocera 403
Sphaerocorini 403
Stelgidocoris 403
Stenopoda 798
Stiretrus 404
Stollia (see *Eysarcoris* 798)
Stridulovelia 292, 405
Strigicoris 403
Strombosoma unipunctatum 798
Tessaratoma papillosa 289*, 290
Tessarotomidae 403
Tessaratominae 403
Tetyra 284
Tetyrinae 403
Thaumastella aradoides 798
Triatominae 405
Velia 292
Veliidae 405

SYSTEMATIC INDEX

ARACHNIDA

SCORPIONIDEA
Androctonus 320
Heterometrus costimanus 319*
– *swammerdami* 319
Opisthophthalmus 607
– *glabrifrons* 320
– *latimanus* 320*, 354, 359, 606
– *wahlbergi* 320
Pandinus imperator 319
Parabuthus brachystylus 320
– *flavidus* 320
– *laevifrons* 321*
– *planicauda* 320
Rhopalurus borelli 320, 321*

AMBLYPYGI
Musicodamon atlanteus 318*

SOLIFUGAE
Galeodes araneoides 318*

OPILIONIDEA
Biacumontia 318
Cryptobunus 318
– *sylvicolus* 319*
Lawrencella 318
Lispomontia 318
Nemastoma argenteo lunulatum 318, 319*
– *dentipalpes* 318, 319*

ARANEIDA
Achaearanea tepidariorum 423
Argyrodes 318
Asagena 318
Cambridgea 318
Chilobrachys samarae 317*
– *stridulans* 316
Diguetia 316
Entelecara broccha 317
Erigone 317
Euphrictus 316
Gongylidiellum mursidum 317
– *vivum* 317*
Hysterocrates 316
Leptiphantes 316
Lithyphantes 318
Lycosa chelata 328
Musagetes 316
Pardosa lugubris 328
Pisaura mirabilis 328
Salticidae 583, 599
Selenocosmia crassipes 316, 317*
Selenogyrus 316
– *aureus* 316*
Sicarius hahni 316*
Steatoda bipunctata 318, 349
– *castanea* 317, 349
Stridulattus stridulans 316
Tarentula pulverulenta 328
Theridiidae 317, 336, 349
Theridium ovatum 317*
Zygiella x-notata 423

CHORDATA
(VERTEBRATA)

AGNATHA

CYCLOSTOMATA

PETROMYZONTIA
Petromyzon 530

MYXINOIDEA
Myxine 530

GNATHOSTOMA

CHONDRICHTHYES (ELASMOBRANCHII)

EUSELACHI
Mustelus canis 663
Negaprion brevirostris 686

Raja clavata 525*, 658*, 660*, 661, 664
Squalus acanthias 661

OSTEICHTHYES (ACTINOPTERYGII)

TELEOSTEI
Acerina cernua 662
Actinopterygii 664
Alburnus lucidus 662
Amblyopsidae 224
Amblyopsis spelaea 224
Ameiurus 677, 686
– *nebulosus* 661, 662, 663, 667, 668*
Anabantidae 661
Anabas scandens 661
Angelichtys 679, 685
Anguilla 677
– *anguilla* 661, 667
– *rostrata* 71
– *vulgaris* 657
Anguillidae 661
Aplodinotus grunniens 78
Bagridae 677
Balistes 659, 678
Balistidae 678, 680, 682, 685
Bathygobius soporator 78, 92, 101, 683, 686
Batrachoididae 666, 677, 679, 682, 686
Benthocometes robustus 686
Betta splendens 661
Bleniidae 686
Botia horae 224
– *hymenophysa* 686
Brevoortia 686
– *tyrannus* 666, 669, 670
Brotulidae 224, 686
Callomystax gagata 678
Carangidae 673, 676, 680, 685, 686
Carassius 659
– *auratus* 662
– *carassius* 662
Chaetodontidae 679, 682, 685
Characinidae 662
Chasmodes bosquianus 92, 686
Clupeidae 664, 665, 666, 669, 677, 682
Cobitidae 662, 677, 686
Colisa lalia 661
Corvina nigra 662, 667, 685
Cottidae 661
Cottus gobio 661
– *scorpius* 661
Crenilabrus griseus 662
– *melops* 662
– *pavo* 662
Cymatogaster aggregatus 661, 669
Cynoscion 682
– *regalis* 670, 682
Cyprinidae 662, 665, 677, 686
Cyprinodontidae 661
Cyprinus 656, 659
Diodontidae 680, 685

Electrophorus electricus 662
Embiotocidae 661
Epinephelus striatus 679, 684
Esocidae 661
Esox lucius 657
Fundulus heteroclitus 656, 661
Gadidae 661, 679, 683
Gadus aeglefinus 661
– *callarias* 661
Gnathonemus 662, 667
Gobiidae 662
Gobius niger 662, 667
– *paganellus* 662
Gymnotidae 662
Haemulidae 680, 682, 685
H. nigrammus caudovittatus 662
Hippocampus 71
Holocentridae 665, 679, 682, 685
Holocentrus 665
– *ascensionis* 677, 684, 685
Hyphessobrycon flammeus 662
Idus melanotus 662
Labridae 662
Lebistes reticulatus 661
Leiostomus xanthurus 222
Leptocephalus conger (= *Conger conger*), 677
Leuciscus dobula 662
– *idus* (*Idus melanotus*) 662
Macropodus cupanus 661
– *opercularis* 661, 667
Macrouridae 677, 679
Marcusenius isidori 662, 667
Melanogramus angelifinis 683
Melichthys 678
Menticirrhus 682
Micropogon undulatus 655, 672, 680
Mormyridae 662, 665
Moxostoma 666
Mugil cephalus 222
Muraenidae 670
Myripristinae 665
Noemacheilus barbatulus 662
Notopteridae 224
Notropis analostanus 686
– *bifrenatus* 686
Ophiocephalidae 677
Opsanus beta 686
– *tau* 679, 680, 682, 686
Ostariophysi 540*, 662, 665–667, 677
Ostraciidae 685
Palinurichthys 659
Perca fluviatilis 662
Percidae 662
Periophthalmus koelreuteri 662
Phoxinus laevis 102, 524, 658*, 663, 666, 667, 676

[Phoxinus]
– *phoxinus* 662, 667, 669
Pimephales notatus 662
Pogonias 672, 679, 680
– *chromis* 676, 682
Pomacanthus 679
– *arcuatus* 685
Pomacentridae 685
Pomacentrus leucostictus 676
Prionotus 71, 674, 675*, 679
– *carolinus* 659*, 675, 682
– *evolans*, 682
Pyrrhulina rachoviana 662
Rhombus maximus 58*
Rutilus rutilus 662
Sargus annularis 662, 665, 667
Sciaenidae 655, 662, 672, 673, 677, 679, 680, 682
Scombridae 682
Semotilus a. atromaculatus 667, 668*, 669
Serranidae 672, 679, 682, 685
Siluridae 662, 672
Siluroidea 677, 678, 679

Solea solea 58*
Sparidae 662
Spheroides maculatus 676
Stigicola dentatus 224
Stromateus niger 670
Teleostei 531, 539, 541, 657, 659, 664, 665
Tetraodontidae 680, 685
Tilapia 655
Tinca tinca 662
Trachurus trachurus 686
Trichochos 670
Trichogaster leeri 661
– *trichopterus* 661
Triglidae 666, 669, 672, 677, 678, 679, 682, 683, 685, 686
Umbra 667
– *limi* 661
– *pygmaea* 661
Velifer hypselopterus 686
Veliferidae 686
Xenomystus nigri 224, 686
Zeidae 679

CROSSOPTERYGII (CHOANICHTHYES)

DIPNOI
Dipnoi 489

Protopterus 105

AMPHIBIA

URODELA = CAUDATA
Ambystoma maculata 694, 695
– *tigrinum* 694
Ambystomidae 694
Amphiuma means 694
Aneides lugubris 694, 695
Desmognathus 694
Dicamptodon ensatus 694, 695
Eurycea bislineata 694, 695
Megalobatrachus japonicus 694
Plethodontidae 694
Salamandra salamandra 694, 695
Salamandridae 694
Taricha similans 694
– *torosa* 694
Triturus alpestris 694

ANURA = SALIENTIA
Acris 706
– *crepitans* 699
– *gryllus* 699
Alytes obstetricans 695
Ascaphidae 696

Bombina bombina 696
– *variegata variegata* 370
Bufo 695, 705, 794, 803
– *alvarius* 696, 699
– *americanus* 230*, 698, 703, 704, 803
– *americanus* × *woodhousei* 230*
– *boreas* 698
– *canorus* 698
– *cognatus* 698, 699, 704
– *compactilis* 698, 704, 803
– *debilis* 698, 703, 704
– *hemiophrys* 698
– *houstonensis* 698
– *marinus* 699, 703
– *microscaphus* 698, 704
– *punctatus* 698, 704, 803
– *quercicus* 699, 703
– *regularis* 696
– *speciosus* 803
– *terrestris* 698, 706
– *valliceps* 230, 699
– *woodhousei* 230*, 698, 699, 704
Bufonidae 698–699

SYSTEMATIC INDEX

Crinia insignifera 699
– *parinsignifera* 699
– *signifera* 699
– *sloanei* 699
Discoglossidae 696
Eleutherodactylus augusti 699, 700*, 803
Eupemphix pustulosus 699, 700*
Gastrophryne (see *Microhyla* 706)
Heleioporus 699
Hyla 700, 794, 804
– *andersoni* 701, 702*
– *arborea* 104, 614 (footnote)
– *arenicolor* 700, 701*, 704
– *baudini* 700
– *chrysoscelis* 803
– *cinerea* 700, 702*
– *crucifer* 701, 704
– *ebraccata* 804
– *eximia* 804
– *femoralis* 700, 701*
– *gratiosa* 701, 702*, 704, 706
– *microcephala* 804
– *phaeocrypta* 700, 701*
– *phlebodes* 804
– *regilla* 701
– *septentrionalis* 701
– *squirella* 701, 702*, 704
– *versicolor* 700, 701*, 703, 704, 803
– *wrightorum* 701
Hylidae 699–702
Hypopachus cuneus 702
Leptodactylidae 699, 700*, 803
Leptodactylus labialis 699, 700*, 803
Microhyla 706
– *carolinensis* 702, 703

– *olivacea* 702, 703, 704
– – × *carolinensis* 229
– *usta* 702
Microhylidae 702
Neobatrachus 699
Pelobatidae 695, 697
Pipidae 696
Pseudacris 700, 701
– *clarki* 701, 703, 706
– *feriarum* 701, 703
– *nigrita* 701, 703, 706
– *ornata* 701, 804
– *streckeri* 701, 704, 706, 804
Rana 695
– *areolata* 702
– *catesbeiana* 705
– *clamitans* 703, 704, 705
– *esculenta* 490
– *pipiens* 71, 702, 703, 705
– *sylvatica* 537*
Ranidae 702, 703
Rhacophorus leucomystax 696
Scaphiopus 695, 696, 697, 794
– *bombifrons* 696, 698
– *couchi* 696, 697, 698*, 703
– *hammondi* 696, 698*
– *holbrooki* 696, 697, 703
– *hurteri* 698, 703
– *intermontanus* 698, 704
– *multiplicatus* 696
Spea 696, 697, 698
Syrrhophus marnocki 699, 700*, 704
Tomodactylus 695
– *nitidus* 699, 700*

REPTILIA

SQUAMATA
Ancistrodon mokasen 85
Lacerta 491
– *viridis* 103
Lacertilia 532

CROCODILIA
Alligator mississippiensis 71, 104
Crocodilia 541

AVES

PODICIPITIFORMES
Podiceps cristatus 720

PELECANIFORMES
Phalacrocorax carbo sinensis 88, 105, 712

CICONIIFORMES
Ciconiidae 78, 82, 714, 719

Nycticorax nycticorax 719

ANSERIFORMES
Anas platyrhynchos 105, 713, 720, 721, 735
Anser 725
– *anser* 71, 81, 83, 720, 721

ANHIMIFORMES
Chauna torquata 82

LARIFORMES
Larus argentatus 44, 71, 85–87, 90, 107, 711, 713, 720, 722, 723, 724, 725, 728
– *atricilla* 713
– *melanocephalus* 720
Sterna hirundo 80, 717

CHARADRIIFORMES
Belonopterus chilensis 83
Capella gallinago 76, 712, 719
Haematopus ostralegus 83
Philomachus pugnax 719
Scolopax rusticola 719, 732

RALLIFORMES
Aramides cajanea 82
Choriotis australis 78
Grus grus 722
Otis tarda 78

GALLIFORMES
Gallus gallus 57*, 70, 71, 72*, 81, 82, 93, 231, 233, 492*, 542, 566, 711, 712, 713, 721, 722, 725, 726, 727, 738, 740
Odontophorus guyanensis marmoratus 82
Phasianus colchicus 81, 713, 723, 737*
Tetrao 80

COLUMBIFORMES
Columba livia 735, 737
– *palumbus* 719, 722
Geopelia 231
Streptopelia risoria 231

FALCONIFORMES
Falco sparverius 82, 735

STRIGIFORMES
Asio otus 735
Bubo bubo 735
Strigidae 78, 83, 86, 214, 216, 542, 737
Strix aluco 64*, 735
Tyto alba 216, 543*

PSITTACIFORMES
Melopsittacus undulatus 735, 737

SPHENISCIFORMES
Aptenodytes 88
Pygoscelis 88
Spheniscus 88

CUCULIFORMES
Crotophaga ani 85
Cuculus canorus 231

PICIFORMES
Colaptes auratus 721
Dendrocopus major 74, 718, 745
– *minor* 718
Indicator 91, 106, 724

CAPRIMULGIFORMES
Caprimulgidae 216
Caprimulgus europaeus 76, 719
Steatornis caripensis 12, 95 (footnote), 216, 217, 741, 745

APODIFORMES
Apodidae 217, 218
Collocalia brevirostris 95 (footnote), 217, 218*, 741–745

CORACIADIFORMES
Upupa epops 733

PASSERIFORMES
Acrocephalus palustris 238, 239, 732
– *schoenobaenus* 732
Agelaius phoeniceus 237, 238
Alauda arvensis 235, 732
Ammospiza henslowii 732, 735
Anthus pratensis 235, 238, 732
– *trivialis* 231, 240
Carduelis cannabina 237, 240, 732
Carpodacus mexicanus 231
Certhia brachydactyla 231
Cettia cetti 711, 712
Chlamydera maculata 238
Chloris chloris 732, 735
Coccothraustes coccothraustes 732
Contopus virens 237
Corvidae 44, 57*, 74, 81, 85, 86, 87, 89, 90, 93, 94, 95, 104, 105, 566, 723, 733, 744
Corvus brachyrhynchos 88, 89, 90, 735
– *corone* 88, 723, 743, 744*
– – *cornix* 745
– *frugilegus* 88, 93, 720, 722, 723
– *monedula* 71, 88, 89, 93, 711, 720, 723
Cyanosylvia svecica 732
Dolychonyx oryzivorus 238
Dumetella carolinensis 238
Emberiza calandra 231
– *citrinella* 237, 732
– *schoeniclus* 231, 233, 732, 742*
Emberizidae 729
Erithacus rubecula 79, 86, 613 (footnote), 713, 716, 717, 718, 719, 721, 725, 732, 735, 741*
Fringilla coelebs 71, 231, 235, 236*, 237, 240, 241, 711, 716, 721, 727, 728, 729, 730*, 732, 735, 741*, 742*, 743, 781, 795, 796
– *montifringilla* 729, 742
Furnarius rufus 82

Galerida cristata 732
Garrulus glandarius 74, 88, 90 (footnote), 91, 93, 94*, 105, 711, 722, 723, 732, 733, 740*, 743, 744*
Geothyplis trichas 237
Hippolais icterina 732*
- *polyglotta* 732
Hirundo rustica 231, 732
Hylocichla mustelina 238, 711, 712, 739*
Icteria virens 238
Icterus galbula 238, 732
- *pyrrhopterus* 724
- *spurius* 238
Junco oreganus 795
Lanius collurio 719, 732, 795
Lepidocolaptes affinis 720
Locustella naevia 231
Loxia curvirostra 735, 737
Luscinia megarhynchos 74, 79, 80, 238, 715, 732
Melospiza melodia 71, , 231, 237, 711, 714*, 721
Mimus polyglottos 85, 233, 238, 239, 240, 730, 731*, 732
Molothrus ater 238
Muscicapa albicollis 231, 237, 742
- *hypoleuca* 231, 237, 742
Oriolus oriolus 231
Parus ater 732
- *atricapillus* 71, 711
- *caeruleus* 725, 732, 742*
- *carolinensis* 85
- *major* 235, 738*, 742*
Passer domesticus 39*, 41*, 42*, 713, 723, 732, 735
Passerherbulus (see *Ammospiza*)
Passerines 722, 737
Pheucticus ludovicianus 238
Phoenicurus phoenicurus 732
Phonygamus kerandrenii 493*
Phylloscopus collybita 231, 742
- *sibilatrix* 231, 742

- *trochilus* 740*, 742
Pica pica 39*, 88, 105, 723, 735
Pinicola enucleator 738
Prunella modularis 729
Ptilonorhynchidae 238
Pyrrhula pyrrhula 231, 233, 234, 235, 239, 240, 241, 543*, 710, 718, 719, 721, 723, 726, 728, 732, 735, 736*, 738, 741*, 745
Quelea quelea 86 (footnote)
Quiscalus quiscula 231, 238
Regulus regulus 732, 743
Richmondena cardinalis 238
Riparia riparia 231
Saxicola rubetra 732
- *torquata* 741*
Serinus canaria 71, 231, 232, 234, 713, 719, 726, 729, 732
Sialia sialis 238
Sitta europaea 739*
Spinus tristis 729
Sturnella magna 231, 237, 715, 795
- *neglecta* 231, 237, 240, 734*, 795
Sturnus vulgaris 44, 85, 87, 90 (footnote), 91, 93, 107, 238, 239, 712, 713, 717, 722, 723, 725, 732, 735
Sylvia borin 741*
- *communis* 71, 231–233, 711, 725, 728, 729, 732
- *curruca* 238
Thryomanes bewicki 77, 721
Thryothorus ludovicianus 730, 731*, 735
Toxostoma rufum 238
Troglodytes aëdon 77
- *troglodytes* 231, 712, 715, 721, 725, 741*, 743*
Turdus ericetorum 231, 234, 238, 732
- *merula* 71, 81, 103, 231, 233, 234, 240, 711, 720, 723, 725, 726, 729, 732, 733, 741*, 742*, 794, 795
- *migratorius* 238, 732
- *viscivorus* 231, 729, 741*
Zonotrichia leucophris 732, 795, 796

MAMMALIA

MONOTREMATA
Ornithorhynchus 762, 769

MARSUPIALIA
Caluromys philander 762
Dendrolagus leucogenys 762, 764
Didelphys virginiana 762
Halmaturus 501
Macropus 516, 761
- *benetti* 84
Petaurus 763
Phascolarctos cinereus 753, 763

Sarcophilus 763
Sminthopsis crassicaudata 762
Thylacinus 763
Trichosurus vulpecula 769
Vombatus ursinus 762

INSECTIVORA
Aetechinus frontalis, 754
Crocidura leucodon 766, 769
Erinaceus 607, 770
- *europaeus* 753

CHIROPTERA

Artibeus jamaicensis palmarum 209, 210
Asellia tridens 204, 208, 214
Carollia perspicillata azteca 200, 209, 209–211
Chilonycterinae 207, 208 (footnote)
Cynopterus 214
Desmodontidae 208, 210–211
Dirias albiventer minor 205
Eidolon 214
Emballonuridae 207
Eonycteris 214
Eptesicus fuscus 189–190*, 193*, 198*, 206, 508*, 792
Glossophaga soricina leachii 209, 210
Hypsognathus 504
Lasiurus borealis 212
Lissonycteris 214
Lonchophylla robusta 209, 210
Lonchorhina aurita 210
Macroglossus 214
Macrophyllum macrophyllum 210
Megachiroptera 211–215
Megadermidae 211
Microchiroptera 185–211
Molossidae 204
Myotis 185–205, 197*, 209, 210, 213, 770, 789
– koenii 193*
– lucifugus 185–204, 189*, 193*, 213, 508*, 789
– macrotarsus 205
– natteri 194
– (Rickettia) pilosa 205
Noctilio leporinus 205, 206
Noctilionidae 205, 206
Nyctalus noctula 189, 212, 767
Nycteridae 208, 211 (footnote)
Nycteris thebaica 208
Phyllostomidae 207, 208–210
Phyllostomus hastatus panamensis 210
Pipistrellus subflavus 193*
Pizonyx vivesi 205
Plecotus 190, 192*, 204–205, 208
– auritus 190, 192*, 194
– rafinesquei 205, 789
Ptenochirus 214
Pteropus 211, 214
– giganteus 211
– medius 754
– poliocephalus 211
Rhinolophidae 71, 199–208, 215
Rhinolophus 199–204, 200*, 201*, 203 (footnote), 789
– ferrum-equinum 199, 200*, 201
– hipposideros 199, 200
Rhinopoma microphyllum 207, 208
Rhinopomatidae 207, 208
Rhynchiscus naso 207, 208, 210
Rousettus 185, 214, 752

– aegyptiacus 211, 212*, 213*, 214
– amplexicaudatus 212
– seminudus 212
Saccopteryx bilineata 207, 208
Tadarida 204
Taphozous 207, 208
Uroderma bilobatum 209–210
Vespertilionidae 71, 185–208, 210, 212, 213, 215

EDENTATA

Bradypus 771
Edentata 494, 501

PRIMATES

Alouatta 500, 506, 754, 773
– palliata 772
– seniculus 510*
Ateles 506, 754, 772, 773, 774
– geoffroyi 754, 772
Cacajao 754
Callimico goeldii 754
Ceboidea 754
Cebus 754, 764, 776
– albifrons 754
– apella 754
– capucinus 506, 754, 767*
– nigrivittatus 754
Cercocebus 770
– torquatus 755, 781
Cercopithecinae 781
Cercopithecoidea 754
Cercopithecus 607, 767, 770, 773, 781
– aethiops 755, 781
– diana 755
– mitis 755
– nicticans 755, 781
– petaurista 755, 781
– sabeus 509*
– talapoin 755, 769, 781
Colobus 782
Cynocephalus (see Papio)
Galago 505, 518
Gorilla gorilla 498, 511, 512, 518*, 519, 755, 761, 772, 777
Hapale jacchus 754
Hominoidea 765, 767, 775, 783
Hylobates 232, 499, 508, 511*, 513, 516, 774
– hoolock 755, 782*
– lar 71, 755, 767, 782*
– leucogenys 232
Hylobatinae 782*
Lagothrix lagotricha 754
Lemur 505
– catta 754
– mongoz 754
Lemuroidea 754
Leontocebus 505

– oedipus 754
– rosalia 754
– spixi 754
Loris 518
Macaca 151, 157*, 506, (see also Zati)
– maura 754
– mulatta 754, 781
– nemestrina 754, 781
– radiata 754, 781
– silenus 754, 781
– speciosa 754
Mycetes (see Alouatta)
Oedipomidas (see Leontocebus)
Pan troglodytes 71, 132, 232, 498, 505, 510–513, 517*, 519, 753, 755, 761, 764, 766*, 772, 774, 775, 777, 782.
– paniscus 782
Papio 83, 762, 770
– anubis 132
– doguera 754, 781
– hamadryas 136, 150 → 755, 781
– leucophaeus 754, 781
– porcarius 773
– sphinx 754, 781
Plathyrrhini 781
Pongidae 755
Pongo pygmaeus 496, 498, 500, 505, 510, 511, 512*, 513*, 516*, 518*, 767, 774
Presbytes entellus 755
Procolobus 782
Prosimii 763
Saimiri sciurea 754
Semnopithecus entellus 772
Simia (Macaca) speciosa 754
Symphalangus syndactylus 511*, 755, 782*
Tarsius 505
Theropithecus 755
Zati 773

RODENTIA

Apodemus flavicollis 763, 771
Atherurus 516
– marcrourus 500, 501*
Capromys pilorides 762
Castor 761, 769
– fiber 766
Cavia cobaya 215
Citellus beecheri 765
– citellus 753
Clethrionomys glareolus 769
Cricetus 216, 762, 771, 776
– cricetus 753
Ctenomys 773
Cuniculus 762
Cynomys ludovicianus 71, 83, 126, 753, 772
Dasyprocta 753
– agouti 761

Dipus aegypticus 761
Dolichotis patagona 753
Erithizon dorsatus 753
Evotomys (see Clethrionomys)
Glis 216, 762
– glis 753
Hydrochoerus 762, 769
Hystrix 761
– cristata 753
Lagostomus 753
Marmota 74, 84, 770
– sibirica 84
Meriones persicus 771
– shawi 761
Mesocricetus auratus 753
Microtus arvalis 763
– brandti 753, 777
– oeconomus 753
Mus musculus 84, 103, 244 → 535, 563
Myocastor coypus 753
Neotoma fuscipes 761
Pachyuromys duprasi 536*, 761
Peromyscus 248
Proechimys 767
Rattus 74, 103, 247 →, 535, 763, 771
Rattus norvegicus 215, 753, 760
Sciurus 769, 771, 776
– sciurus 753

LAGOMORPHA

Lagomorpha 501
Lepus 502, 516, 761
– timidus 766
Ochotona princeps 232
Oryctolagus 90, 91, 104, 247, 761

CETACEA

Balaenidae 223
Balaenoptera musculus 223
– physalus 223
Delphinapterus leucas 71, 220, 790
Delphinidae 81, 220–223, 503*, 670, 762
Delphinus delphis 220–221*, 790
Eubalena 223
Globicephala 220
– melaena 71, 790
Grampus griseus 790
Hyperoodon ampullatus 790
– planifrons 790
Lipotes vexillifer 790
Megaptera boops 223
Mesoplodon bidens 790
Monodon monoceros 790
Mysticeti 223
Odontoceti 220–223, 496, 539*
Phocaena phocaena 223, 502*
Physeter catodon 74, 81, 220

Platanista 790
Stenella 220
Tursiops truncatus 71, 220, 223, 790

CARNIVORA-FISSIPEDIA
Acinonyx jubatus 757
Ailuropoda melanoleucus 756, 762, 765*
Alopex lagopus 756, 759, 760*, 763, 765, 768, 769, 770, 776, 778*, 779, 780
Amblonyx 756
Bassariscus astutus 756
Canidae 504, 764, 765, 773
Caninae 774, 779*
Canis 759, 760, 763, 778, 779, 780
 – *adustus* 766, 779
 – *anthus* 755, 766, 779
 – *aureus* 755, 766, 779
 – *dingo* 755, 766, 776, 779
 – *familiaris* 74, 90, 91, 103, 133, 500, 503, 506*, 519, 535, 564, 755, 762, 765, 766, 769, 773*, 776, 778*, 779*
 – *latrans* 755, 766, 776, 779
 – *lupus* 83, 755, 762, 766, 769, 773*, 776, 777, 778*, 779*
 – *mesomelas* 766, 779
 – *pallipes* 779
Cerdocyon 765
 – *azarae* 780
Chrysocyon jubatus 779
Crocuta 766, 773, 776
 – *crocuta* 756, 765*
Crossarchus 763, 767
 – *fasciatus* 756, 763*, 773
Cuon alpinus 755
Felidae 502, 504, 516, 762, 780*
Felinae 764, 766, 769, 780
Felis catus 232, 499, 502, 503, 507*, 519, 753, 757, 768, 769, 770, 771, 780
 – *chaus* 757
 – *pardalis* 757
 – *puma concolor* (see *Puma concolor*)
 – *serval* 757, 766, 769, 780*
 – *sylvestris* 771
 – *sylvestris griselda* 757, 780*
 – *wiedi* 757, 780*
Fennecus 769
 – *zerda* 779
Grison vittatus 756
Gulo gulo 756
Helarctos malayanus 756
Herpestinae 773
Hyaena 766
 – *brunnea* 756
 – *hyaena* 756, 766*
Lutreola lutreola 756
Lycaon pictus 756, 778, 779
Lynx caracal 756

 – *chaus* 780*
 – *lynx* 756
Martes martes 756
Megalotis 780
 – (*Vulpes*) *chama* 756, 769, 779*
Mellivora capensis 724, 756
Mephitis 74
Mustela nivalis 756
 – *putorius* 756
Nyctereutes procyonides 232, 756, 765, 779
Panthera 773
 – *leo* 74, 502, 503, 757, 762*, 766, 769, 774, 776, 777, 780*
 – *nebulosa* 757
 – *onca* 757, 766, 780
 – *pardus* 757, 762*, 766*, 769, 770, 777, 780
 – *tigris* 757, 762, 764, 777, 780*
 – *uncia* 757
Pantherinae 757, 764, 769, 776, 780
Potos flavus 753
Pteronura brasiliensis 756
Puma concolor 757, 762*, 766, 780*
Selenarctos tibetanus 756
Simocyoninae 773
Speothos venaticus 779
Suricata 607, 663
 – *tetradactyla* 756, 767*
Thalarctos maritimus 756, 771
Thos 762, 776, 778, 779
Tremarctos ornatus 756
Urocyon 769, 778
 – *cinereo-argentatus* 756, 778*, 779
Ursus arctos 756, 770
Vulpes 756, 759, 763, 764, 765, 767, 768, 770, 771, 776, 778, 779
 – *corsac* 756
 – *vulpes* 71, 753, 760*, 763*, 765, 769, 770, 771*, 772, 775, 777, 778*, 779

CARNIVORA-PINNIPEDIA
Arctocephalus pusillus 757, 765, 766*, 782
Callorhinus ursinus 757
Cystophora cristata 762
Halichoerus grypus 757
Macrorhinus 764, 776, 783
 – *angustirostris* 757, 764*
Mirounga angustirostris 796
 – *leonina* 757
Odobenus 773
 – *rosmarus* 757
Otaria byronica 757, 782
 – *jubata* 495*, 497*
Otariidae 764
Phoca vitulina 496*, 757, 771
Pinnipedia 756, 774, 779, 780
Zalophus californianus 757, 765
 – *wollebaeki* 768

SYSTEMATIC INDEX

PROBOSCOIDEA
Elephas maximus 757, 777
Loxodonta africana 757

SIRENIA
Trichechus manatus 757

PERISSODACTYLA
Asinus somaliensis 758
Diceros 762, 763, 764, 776
– *bicornis* 757, 764*, 765*, 780
Equidae 762, 767, 769
Equus asinus 757, 766*, 767, 781
– *caballus* 74, 83, 90 (footnote), 496, 501, 504*, 535, 758, 767
– *grevyi* 757, 766*, 767, 781
– *przewalskii* 758
– *quagga antiquorum* 757
– – *granti* 757
– *zebra hartmannae* 758
Hemionus hemionus hemionus 758
– – *onager* 758
Rhinoceros unicornis 757
Tapirus terrestris 757, 763, 765*, 776, 780

ARTIODACTYLA
Addax nasomaculatus 759
Alces 80
– *alces* 758
– – *columbae* 772
Antidorcas marsupialis 759
Antilope cervicapra 758
Axis axis 758, 764
Bibos 768
– *banteng* 759
Bison 770
– *bison* 759
– *bonasus* 759
Bos 770
– *indicus* 759
– *taurus* 56*, 71, 90 (footnote), 502, 505*, 759, 764*
Bovidae 761, 770, 776, 789
Bubalis bubalis 759
Camelus 770, 783
– *bactrianus* 758
– *dromedarius* 758, 763*, 764*

Capra 74, 767, 773, 780
– *hircus* 759, 767*
Capreolus 765, 769
– *capreolus* 758
– – *pygargus* 758
Cervidae 759, 764, 770, 780, 783
Cervus 80, 761, 762
– *elaphus* 758, 764*
– – *bactrianus* 758
– – *canadensis* 758, 764, 765*, 772
Choeropsis liberiensis 758
Connochaetes taurinus albojubatus 758
– *gnu* 758, 767
Dama dama 501, 758
Dicotyles pecari 758, 766*
– *tajacu* 758, 763*
Elaphurus davidianus 758
Gazella soemmeringi 759
– *subgutturosa* 759
Giraffa 74, 768
Hippopotamus 538, 773, 781
– *amphibius* 758, 766*
Lama glama 758, 770
– *huanacus* 758
Muntiacus 762
– *muntjac* 758
Okapia 773
Oryx algazel 758, 764*
Ovis 761, 767, 773, 780
– *aries* 759, 764*
– – *musimon* 759, 772
Phacochoerus 762
– *aethiopicus* 758, 762*
Poëphagus grunniens 759, 764*, 770
Potamochoerus larvatus 758
– *penicillatus* (see *porcus*)
– *porcus* 758
Pseudaxis sika 758
Rangifer tarandus 758
Rucervus duvauceli 758
Ruminantia 767, 773
Rupicapra 761
– *rupicapra* 759
Suidae 763, 770, 772, 781
Suinae 781
Sus scrofa 71, 78, 102 (footnote), 506*, 758
Taurotragus oryx 758

GLOSSARIAL INDEX
by
W. B. BROUGHTON

(Chapter I was written before I had access to the other contributions; what follows was necessarily written after, since it incorporates the subject index. Its compilation, both in its glossarial and its simple index aspects, revealed further points that might profitably have been discussed in Chapter I; these are reviewed now or under appropriate entries in the glossary.)

"Les biologistes," our toiling editor has often remarked, "ont, vous le savez, une attitude toujours personnelle face à tous les problèmes, et ils entendent assumer leur propre responsabilité, *sans autre considération que leur point de vue propre*" (*in litt.*, but italics mine). It is the more remarkable that in a book such as this, the work of no less than 24 of these anarchists, there should be so little more than incidental and sporadic discordance with a terminological framework set up in the first chapter and unseen by nearly all of them.

Where such differences exist, I have adopted the expedient that seems most logical and convenient from the point of view of the reader accepting or using Chapter I (especially the Key, pp. 22–3) as a general guide: this is, to index in general under the terms there proposed, adding the author's equivalent term in brackets and quotation-marks against the relevant page reference(s). This has seemed to me a useful, indeed essential, means of reducing varied usages to an at least provisional common denominator. At the same time, of course, I have also entered such terms in their own alphabetical place in the index, and with a similar concordance.

I must particularly crave the indulgence of those who have adopted my former usage (1952, 1954) of the word *chirp*, and who may now find my abandonment of it vexatious. However, if I argue that others should give up cherished usages, I must be prepared, if sufficient reason be shown, to do the same myself; reason enough to yield my point is given in Chapter I. Another comment worth making about *chirp* is that it may seem at first sight rather odd to describe a bark, a yelp or a pop as a variety of chirp; on the other hand, the only alternative is yet another latin or greek neologism, and after all, it is common enough to speak of human beings as being chirpy, so why not sea-lions, dogs or crustaceans?

Unifications additional to those of Chapter I have also been attempted in this Glossary. I am sure these are very imperfect (though the quasi-dogmatic style imposed by the need for brevity may appear at times to contradict this belief); if they form some sort of adequate foundation for further and better schemes, I shall be satisfied. These remarks apply particularly to the long entry on *Song, calls and callnotes:* here I have adopted the same expedient of laying down an at least provisional common terminology as a yard-stick for all groups, gleaned from the best elements in them and stripped of terms which for one reason or another are unsatisfactory; incorporating all index references under this main scheme, with cross-references from the more individual usages elsewhere in the glossary as before.

Where I have had to deprecate terms, I have given brief reasons; which will, I hope, prove convincing. A few obvious translation errors have been given the correct word in the Index, usually without comment.

One neologism I have been driven to add to the jargon of scientific literature. This is discussed under "mechanism of sound production".

Key to the signs, arrangements, etc., in the glossarial index

(*A* **Synoptic Key to Terminology** *appears in pp. 22–3 of Chapter I; the* **Ethological Terminology of Song** *is discussed under the glossary entry* song, etc., *categories.*)

f following a page number indicates that the entry refers to that page and one or more following it;

GLOSSARIAL INDEX

* following a page number relating to a *term*, indicates that the term is defined, lengthily discussed, or illustrated on that page;

‖ the double-bar, is used where necessary to delimit the end of a definition from ensuing discussion or other material.

In cross-references, the entries referred *to* are separated by semi-colons, not commas, which, with hyphens, are used only to separate subdivisions of a single entry. Thus, in the cross-reference: "*see* song, calls and call-notes, categories – courtship; sound production", there are just two entries, one beginning *song*, and one *sound*.

A

a!, 157*, 757
A.C.T.H. and the audiogenic seizure, 252
A.D.P. transducers, 51
abbreviation of sounds in evolution, 783
abdomen, *see also* opisthosoma:
   in percussive sound production, insects, 326f, 332
                                 Meconema, disputed, 328
   in stridulation, Crustacea, 305
                      insects, 282f, 292f, 332f, 403, 405
abdominal resonator, Hemiptera, 392, 396, 798
abstract letter, 174*
       symbolism absent from animal sound, 69, 576
                present in human language, 576
       word, 177*
abused terms, 6–13*, *see also* song, calls and call-notes.
Academy of Medical Sciences, U.S.S.R., 150
acceptance song,
acceptation, chant d' *(F.)*, *see* song, calls and call-notes, categories – courtship.
accidental sounds, *see* sound, biological.
accuracy of prise-de-son, 25
acetazolamide and the audiogenic seizure, 252, 262
acetylcholine and the audiogenic seizure, 255
acoustic anatomy, *see* ear; sound production apparatus.
acoustic message, 113*, 583f
acoustic releasers, sign-stimuli, 18*, 91, 570, 614f, 622f, 703, 777
acoustico-lateralis, *see* acustico-lateralis.
acquired signals, birds *see* learning.
acridid-type stridulation 295f
actogram, actography, 54*f, 245, 247
acustico-lateralis, 223, 523f, 663, 684
   canal organs, 524
   crista, 523f, 526
   nature of excitation, 525
   transductor always a rigid shell and sensory-hair cells 523
acute and grave sounds, 166
adaptation, disorders of (Selve), 571
adaptation of existing structures for sound production, 334
adaptation, sensory, in fish, 663, 680
                     Lepidoptera, 436
ad hoc definitions, dangers of, 3, 5
aditus laryngis, 491, 495f, 504

adjacent word 138*, 177*
adjective, 141
adrenals, adrenalectomy and the audiogenic seizure, 252
advantages of acoustic signals, 73
adverb, 141
affective nature of animal sound, 69
affectors, 112*
after-courtship, significance in *Gryllus*, 450
after-discharge, 428, 436
age and the audiogenic seizure, *see* audiogenic.
      song in birds, 712
aggregation, sound and, 386, 391, 398f, 694, 703f, 706, 717, *see also* cohesion.
aggregative song, *see* song, calls and call-notes, categories.
agonistic song behaviour, *see* song, etc, categories – rivalry, protection, raptorial.
agreement song (female receptivity song), *see* song, etc, categories – courtship.
air-bladder of fish, 224, 489, 665f, 686, *see also* phonation; sound-production apparatus.
air current, effect on stridulation, 368
air cushions, *see* air sacs.
air flow and sound propagation, distinction, 385
air pressure and flight sound, 373
air sacs and tymbals, 396, 798
air sacs and tympana, insects, 415f, 436
air sacs, vocal, birds, 493, 499
          bustard, 78
          chimpanzee, 512
          frogs, 490f, 696
          gibbon, 510
          reptiles, 491
ak ak ak, 151f, 155f
alarm behaviour, 83, 85
alarm reactions and Johnston's organ, 420
alarm signals, *see* song, calls and call-notes, categories – protection.
alary stridulation, insects, 289f, 294f, 303f, 329f, 401f, *see also* elytra; flight sound; stridulation; tegmina.
albumin, amino-acids in, as final words, 138
alerting by sound, 568, *see also* song, etc, categories – protection.
algebra, constructive, and language, 133
algorithm: thirteenth-century name of what is now known as arithmetic; now used with a different connotation, defined on pp. 132f, 138, 173
  for change of formant frequencies, 167
  for exchange of words; 149
  for selection of concrete words, 145f
  for transformation of sounds into a syllabic line, absent from monkeys, 161
  governing structure of human speech, 166*
  logical, 150*
  of complete syntagma, 143*
  of correctness and incorrectness, 150
  of exchange of a uniform spectrum of random noise for a non-uniform selective spectrum, 155
  of objective vertical substitution regulating baboon calls, 151
  supplementary, of syllabic modifications, 168

call-notes, loosely used concepts
asynchronous *(neurol.)* sound production, 377
attack quality, 156, 164, *see also* transient
attack song (Angrifflaut), *see* song, etc, categories – rivalry.
attacking birds of prey, *see* mobbing.
attention, 565, 572
attraction by sound, 85f, 89, 107, 386f, 667, 670, 681
au au au, 151f, 155f
audio-auditive reflexes, 566, *see also* feedback.
audioepileptic seizure, 247
audiogenic seizure, 244*f, 568, 571
   in various animals:
      chicken, 245
      dog, 245
      goat, 245
      mouse, 244f, 250f
      rabbit, 245, 247, 250
      rat, 245, 247f, 250f, 254f
   general aspects
      actogram, 245, 247
      and age, 248, 250, 256, 260, 262
      alkalosis, 252, 262
      behaviour, 244, 246f, 251, 254, 258
      biochemistry, 248, 254f, 260f
      and blood, 249, 251f, 257, 261
      and brain, 248f, 254f
      and central nervous system, 244, 254, 256, 261
      chemical mediator, 261f
      coagulation, 252
      death, 245, 248, 254, 256, 258f
      drugs, 257f, 261
      ear, 244, 250f, 257
      electroencephalogram, 254, 258
      electrophoresis, 252f
      enzymes, 250, 256, 261f
      epilepsy, 247f, 254f, 262
      exophthalmia, 244
      eyes, 244, 254
      genetics, 248, 250, 256, 261
      glands, 252, 256
      haemorrhage, cerebral, 248f, 252
      hormones, 252, 256
      intensity of, 248, 254f, 259f
      latent period, 244, 246f, 250, 254, 260f
      light cycle, 251
      micturition, 244, 254
      myoclonus, 247f
      nystagmus, 248
      oscillograms, 245
      parabiotic, 248
      pH, 252, 261f

audiogenic seizure, *general aspects, continued:*
    phase, 244f, 254, 262
    record, 245, 247
    "recovery", 244
    resistance, 247, 250, 252f, 255f, 259f
    respiratory rhythm, 244
    running phase, 244, 246f, 254, 259, 262
    senescence, *see* age, *above.*
    severity, 260f
    sex, 250
    strain, 245, 247f, 250, 252f, 256f
    surgery, 254
    susceptibility, 248, 250f, 254f, 262
    temperature, 251
    tone, pure, 248f
    tranquillizers, 247f, 258, 262
    violence, *see* intensity, *above.*
audiogram, B.S. 661/3021 : a chart or table relating hearing loss for pure tones to frequency. ‖
    736f, *see also* sensitivity.
audio-mimicry of aculeate hymenoptera, 387f
audio-phonative reflexes, 566, *see also* feedback.
        cortical mechanisms, 572f
audiospectrum, *see* sound spectrum.
audiospectrogram, correctly used not for the above, but for the graph of its variation
    with time; *see, however,* sound spectrum; sound spectrogram.
audition, *see also* hearing:
  subjective human, 3, 375
auditory and emission apparatus, relationship, 337, *see also* echolocation.
       feedback, *see* feedback.
       gnosis, 566, 573f
       meatus, 534f
       observation and musical training, 375
       organs, receptors, see ear; hearing; kinaesthesis; receptors, tuned; tympanal
         organs.
       sensation, 566
       sensitivity curves, *see* hearing; sensitivity.
augmented song (congested song), *see* song, calls and call-notes, categories – courtship
aural analysis, 16f, 136, 375
autocorrelation in messages, 117
auto-information, 183
automatism, 557
autotomized leg stridulating, Chilopoda, 321
autumnal song of female birds, 714

B

B cells, 435, 437, 800
background noise, 32*, 50, 73, 387
Balz *(G.)*, courtship display, serenade, *see* song, calls and call-notes, categories.
banality, 118*
band oscillograph, 154
barium titanate transducers, 27, 50

bark, barking, 84, 700, 753f, 763, 768f, 775
  homology of, 778, 781
basal ganglia, 561
basal plate of ear, 737
basic signal, 22f*, *see also* elementary waveform.
basic sound, 17*
basic word, 177*
basilar membrane, in bats, 192
                 insufficiency of Helmholtz's theory, 548
                 vibration of, 547
basilar papilla, 532, 657, 705
    recess, 531
    sleep centre, 568
Batesian mimicry, acoustic, 85
battery recorders, 36
beak noises, 711, 714, 719
beat, (1), *music:* rhythmic pulsation, as e.g. three beats to the bar, etc; Sotavalta *extends for bird song as follows:* the smallest rhythmic unit; a single rhythmic accent that can be repeated after successive intervals, regular or irregular. Between beats may be sub-beats of a different order.
  (2), *physics* (beats). B.S. 661/1032: the periodic variations of amplitude resulting from the addition of two periodic quantities of the same kind but of slightly different frequency.
beats, used in echolocation (Pye's theory), 193
beehive, *see also* queen bee.
  actography, 58
  noise spectrum, 40
begging behaviour, begging response, 81, 103, 334, 720, 724, 753, 770
  in adult female, 719
  blackbird responding to pure tone or human whistle, 81
behaviour, *see also* audiogenic seizure.
  acoustic, of particular groups, Part V, 581f
    amphibians, 694f, 703, 706
    arthropods, 391f, 583f
    birds, 709f
    fish, 655f, 680f
    mammals, 751f
  acoustic, primordially central motivation, 559
behaviourist psychology, 119
Békésy's camera acustica, 546f
        null experiment, 545
Bel, *see* decibel; level.
bell, provoking audiogenic seizure, 248
bellow, bellowing, 491, 759
benzoquinolizine and the audiogenic seizure, 259
bi-directional microphone, 33
binary messages, 115*
binaural hearing; hearing with two ears. The counterpart is uniaural, not the hybrid monaural.
biological determinants of sound form, *see* phonomorphosis.
        parameters of arthropod emissions, 361

biological and physical events, 11
bioluminescence in fisheries, 666
biophysical relation between signal and behaviour, 73
bird calls, artificial, 15, 77
    as stimuli to insects, 614
              birds, 717, 743
bird's nest soup bird, 217
bi-syllabic: hybrid word; the correct form is disyllabic, q.v.
bit, 115*
blashing, fish, 670
bleat, bleating, in mammals, 754f, 764, 770f
              in prawns, 673
              homology, 781
blinding bats, 183f, 196, 209
blood, *see* audiogenic seizure.
boat-whistle, 682
booming of drum (fish), 676
borers, actography, 56
brain, *see* audiogenic seizure; central nervous system.
brain-image of the world, 577
branchial arches, 535f
branchial fissures, whistling, 694
braying, 516
breeding and sound production, *see* inheritance; song, calls and call-notes, categories; *and* sound production.
broadcast song, *see* song, calls and call notes, categories – proclamation.
Broca's motor aphasia, 573f, 579
bronchial columella, 536f
buccal stridulation, insects, 279, 292, 329f, *see also relevant appendages*.
building-up time, build-up time. B.S. 204/1312: the interval of time that elapses between the instant when the envelope of a transmitted wave of specified frequency is first received, and the instant when it first attains a specified fraction of its steady-state magnitude. ‖
    Hence, for the animal emission: lapse of time from the beginning of a given wave-train to the instant when it attains its maximum amplitude (see figure).

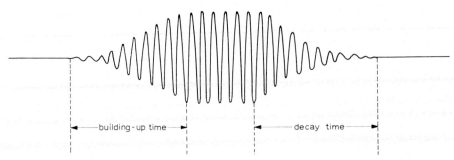

burrowers, actography, 56
burst, 5, 18, 217
buzz, 187, 200, 211, 771

C

C.N.E.T. frequency analyzer, 41, 107
C.N.S., *see* central nervous system
cackling, homology of, 781
calcium and the audiogenic seizure, 256f, 262
calculus, grammatical, 178
call, *see* song, calls and call-notes.
call territory, 75*
calling song, *see* song, calls and call-notes, categories – proclamation.
camera acustica, Békésy's, 546f
campaniform sensilla, 482
cancrizans *(music)*: imitation *al rovescio*, i.e. repetition of a phrase or figure in reverse in another part. Especially *canon cancrizans* or "crayfish counterpoint" (contrepoint de l'écrevisse). On the locomotion of crab and crayfish, the French are clearly the better observers.
    By extension, applied to the playback to experimental animals of their own recorded song *in reverse*, 93f, 240
canto (= Lied), *plural* canti, *adj.* cantical, 14*, *see also* song.
cantus (= Gesang), *plural* cantus, *adj.* cantual, *see* song.
  fortior, 444, 448
  mitior, 444, 448, 769
  submitior, 448
  See also song, calls and call-notes, categories – courtship.
capsule, head, in stridulation, insects, 280
captivity, conditioning rapid in, 105, *see also* domestication.
    and imitation, 731
capture clicks, 197
carapace in stridulation, Crustacea, 309
carbon dioxide and the audiogenic seizure, 252, 262
cardiac rhythm, *see* audiogenic seizure
cardioid microphone, 33
Caripe, 216
carpenter fish, 675, 684
carrier-wave, elementary wave-form as, 392, 447
Cartesian imagery, 557
    man-animal barrier, 559f
castration and song, 450, 640f, 712f, 717, 761
catatonic states and the audiogenic seizure, 248
caterwauling, 768f
Caucasus, 150
cave birds, 216f
  fish, 224f
cavum laryngis, *see* larynx.
celeste, 17*
cello and alligator, 104
central body, 460f
central nervous system, *see also* audiogenic seizure; *and under the various anatomical elements*.
  in birds, 561
  in cicada, 398
  in Lepidoptera, pathways, 437

central nervous system, *continued:*
  in Orthoptera, 440f, 483f
    antennae, rôle of, 473f
    auditory, visual, antennal connections, 452
    brain function, 460f
      lesions, 462f
      maps, 472f
      puncturation, 464f
      -thorax reciprocity, 477f
    central analysis located in brain, 464
      body, rôle of, 477
      coordination, 445
      coupling of bilateral symphasic movement 449
      inhibition, 471
    cerci, rôle of, 475f
    competitive inhibition, 471
    decussation, absence of, 455
    electric stimulation of protocerebrum, 465f
                              thoracic neuropile, 455f
    final central control of stridulatory movement, 450
    fusion frequency, 456
    motor interrelations, 478f
    mushroom bodies, rôle of, 476
    nerves of stridulatory muscles, acridid, 445f
                                    gryllid, 441f
    nervous coupling of halves of thoracic ganglion, 455
    operations on nerve cord, 450f
    parts necessary for whole repertoire, 450f
    peripheral afferent control, 479
    post-operative readjustment, 479f
    rhythmic neural control mechanisms, 449f
    song pattern as a tool, 445
    thoracic circuits, 458f
    thoracic ganglia, rôle of, 441f, 445f, 453f, 459, 563
           mechanism not stereotyped to one rhythm, 470
    tonus control 457f
    tympanal organ, rôle of, 476
  in vertebrates:
    adaptive behaviour reactions, 568f
    apprehension, conscious, 560
    arousal centre for cerebral cortex 568
    central integration, 558f
           motor configuration precedes motor message, 559
           neurons, rôle of, 558
    as complex of fluctuating, physiological configurations, 559
    corridor fallacy, 558, 564, 567
    doctrine of root between reflexes, 559
           autonomous activity, 559f
    dominance, regions of, 561
    elementary centres of audition and speech, 563f
    fatigue, "nervous", 571

feedback, pontine, 566
frequency discrimination, central, 564
general function, 577f
hierarchy principle, 559, 562
hormones, interaction, 568
intermediate centres, 566
   non-specific regulatory function, 567
   compared with invertebrates, 568
instinctive centres, upper, 568f
language, cerebrum of, 573f
Lapicque's subordination centre, 568
lower centres, 559
neural configurations as a sort of "physiological" organ, 562
noise, a poison even below disagreeable level, 571
percipience antedated by cerebral processes, 560
peripheral motor neuron is final common path, 563
phonation, coordination, 563
        nuclei, 563f
praxis zone, 574
precision control, 574
protozoan analogue of central configurations, 561
response a print of central patterns, 559
seasickness centre, 566
somato-splanchnic interaction, 566
speech hemisphere ablation, 579f
suckling-cry, 570
supramesencephalic instinctive centres, 568f
tonus regulation centres, 567f
visceral functioning and nervous fatigue, 571
cephalic percussion, insects, 326f, 334
      stridulation, insects, 277f, 405f
cephalothorax in stridulation, *see also* carapace; prosoma.
  Chelicerata, *see* prosoma.
  Crustacea, 305, 310f
cercal hairs of cockroach, 409, 412
  frequency compass, 420
cerebellum, 562, 567
cerebellar syndrome, 567
cerebral clockwork, 558
      conditioned reflexes, 557
      cortex, 561f, 572f
      hemispheres, *see* audiogenic seizure; central nervous system.
      integration, 558
      mapping, crickets, 107, 472f
cerebrectomy and sound behaviour, 569f
ceremonies, 715, *see also* greeting; invitation
channel, 114*, 130
  sound, 121*
chant *(F.)*, *see* song.
characteristics of a sound, 156*f, 175*f
         particular emissions, *see* physical characteristics.

chattering, monkeys, 517
    teeth, 762
checked and unchecked sounds, 166
chela, cheliped, in stridulation, Crustacea, 306, 309f, 314
chelicera in stridulation, Chelicerata, 316, 318f
chemical mediator, 261f, 528f
    message, 113, 583f
        remanence of, 584
chemoreceptors, 583
chest-drumming, 762
China, 217
chirp, *as equated to one movement* (= Silbe, G.), Broughton, *now abandoned and deprecated as over-restricted (see p. 12, and introduction to Glossary)*, 445f, 447.
    Preferred unitary dictionary sense, *as first empirical parameter of analysis:* the shortest unitary rhythm-element of a sound emission that can readily be distinguished as such by the human ear, 12*, 16*, 22f*, 133, 591, 699, 762 (single sound), *see also* song, calls and call-notes, categories.
    figured, 20f*
        as purely arbitrary collection of motor-correlated units, 12*, 356f *(deprecated)*
chirping, 324, 753f
    homology, 781
chirrup, 5
choanae, 491, 496
chord, 8
chordal pulse, 9f*
    utterance, double-voice, *see* phonation.
chordotonal organs, 400f, 408, 413, 422
chorus leader, 704
chorus performance, 83, 517, 589, 704, 774, *see also* song, calls and call-notes, categories – aggregative responding; *and* sonic panurgism.
chromophotography, 375
chuckle, chuckling, 673, 702, 704
cinematography, 374
clacking, 770
clap, shrimps, 307
clarinet, waveform, 10
classification of sound and song: *see* song.
Claudius cells, 533f
claustrum, 665
claws in stridulation, Crustacea, 305f
cleithrum in stridulation, fish, 678
click, clicking, 12*, 95, 187f, 205, 207, 211f, 216f, 397f, 407, 694, 762
    and tympanal organ, 420
climax call, frogs, 704, *see also* song, calls and call-notes, categories – courtship
clocks, internal, 644f
clonic convulsions, seizures, 244f, 250, 253f, 258f
clucking, 694
clypeus in stridulation, insects, 292
coactant, 14
    spider and bug, 407
    bat and moth, 87, 428, 434

coagulation (audiogenic seizure), 252
coaptation of stridulatory apparatus, 335
cochlea, 192, 199, 525, 529, 532f, 659
   cortical, 573
   medullary, 563
cochlear microphonic potentials, 102, see also sensitivity, auditory.
        nucleus, 563f
cocoon in sound production, 332
code, 112
coded transitions in translation, 134
codification, 73
coding of biological parameters into sound forms, see phonomorphosis.
cohesion, 587, 717, 721, see also aggregation.
   of a pair, 718
coin-tossing and information theory, 117
collective life and sound, 83f, 132f, 150, 385, 589, see also song, calls and call-notes, categories – aggregative responding.
   baboons, 132f, 150
   swarms of insects, 385, 589
columella, 535f, 705, 737
combination sound, see tone, combination.
"combination of sounds", 137
common chord, 8, 9*f.
common song, see song, calls and call-notes, categories – proclamation.
communal singing, see chorus.
communication (see also information), 114*, 337, 517, 681, 683, 685
   extraspecific, 391*, 407, 409, 798, see also interspecific.
   intraspecific, 391*, 407, 409, 798, see also specificity, stenospecificity.
   theory, 112*, 130
communicative systems, 133f, 173*
   baboon, 168f
   ultimate aim of, 150
communicative value of stridulation, 440
compact and diffuse sounds, 166
compass (of an emission apparatus): the range between highest and lowest frequencies emissible, see sound spectrum. Preferable to range, because of the latter's alternative use in the topographic sense.
   (of a receptor): the range of frequencies receivable, from the highest to the lowest. See above; and range; sensitivity.
compatibility of words, 143
compatible and incompatible words, 145, 149
competition between signals, 105
complete syntagma, 143*, 147, 177*
complete and incomplete words, 142*, 149, 166, 177f
complex free vibrations: free vibrations which are not simple sine-waves of steady amplitude. See also normal modes.
complex sound, 165, see also tone, complex.
complexity, 115*f
composition of a word, 138*
compression horn speakers, 45
compulsory positions for sound emission, exaggeration of, 775

concept, 148*, 171, 173*
conceptual thought, 132, 171, 561, 575
concha, 534, see also ear; tragus; antitragus.
   hypertrophy in mammals, 534f
concrete letter, 174*
       word, 134, 178*
           unequal, 148
condenser microphone, 28
conditioned reflex, 102
conditioning in echolocation, 195f
congested song, see song, calls and call-notes, categories – courtship
congregative song, see song, etc, categories – aggregative responding
consciousness, 576
   of consciousness, 578
conservatism of sound compared with morphology, 778
consonant, 5, 122, 156, 162f; see also under phoneme *for distinction between* written consonant *and* consonantal sound.
   animals, 516
   genetic precursor, 163
   sonorous, 163
consonantal and non-consonantal sounds, 166
continuant and interrupted sounds, 166
continuous waves B.S.204/4118: waves in which the successive oscillations are identical as soon as a steady state is reached. ‖
   See damped waves.
control, pest, by sound, 387
       of phonation, 136
controlling ear, 575
conus elasticus, 499
conventional signs in algorithm theory, 179*
convergence, see evolution
conversation song, see song, calls and call-notes, categories.
conversing, 133
convulsions, audiogenic, 244, 247, 255f
cooing, 757
copulation song, see song, calls and call-notes, loosely used concepts.
        timing, in grasshoppers, 641
corniculate cartilages, 513
corpora bigemina, 567
       quadrigemina, 567, 569
corporal vocal pattern, 565, 573, 575
corpus callosum, 254
      centrale, 460f, 477
        pedunculatum, 460f, 476
          evidence for higher psychic function, 465
        striatum, 573f, see also hyper-, neo-, and palaeostriatum.
correctness and incorrectness, 150
cortex, cerebral, 561f, 572f, see also audiogenic seizure.
cortical projection, 103
Corti, organ of, 532f, 657, 659
cotio-cotio, 86

coupler, acoustic, Orthoptera, 366
cour, chant de *(F.)*, *see* song.
courtship song, *see* song, calls and call-notes, categories.
courtship, 77*
  in various animals:
    acridid, classification of, 594
    birds, 78, 718
    braconids, 387
    fish, 78
    Hemiptera, 391
    Lepidoptera, 437
    mammals, 78
    mosquitoes, 386f
    Orthoptera, 8, 21, 78, 477f
    syrphids, 387
  general aspects, 77
    defence reactions, overcoming by courtship song, 597, 599
    evolution, 594
    female appetitive behaviour, 598
    tremulation in Tettigoniidae, 591, 618
coxal stridulation, Chelicerata, 317
                Crustacea, 312
                insects, 290f, 326, 406
crackling, 191, 221, 233
cranial nerves, cricket, 461f
cranium in stridulation, insects, 280
creaking, centipedes, 322
      whales, 220
cricoid, cricothyroid, *see* larynx skeleton.
crista acustica, *see* acustico-lateralis; ear.
critical periods in vocal development, 240f
croak, 673
cruising clicks, 197, 217, 790
cry, crying, 172, 754f, 771
  homology, 781
crystal microphone, 27, 50
      transducer, 50
cud-chewing, actography, 56
curare and the audiogenic seizure, 254
cut signals, 94
cutaneously produced sound, 566
cybernetics, 112*, 171f
cycle, B.S.661/1006: of a periodic quantity – a complete repetition of the series of changes that takes place during the period of a recurring variable quantity.
    1007, note 2. By extension, *cycle* is applied to recurring quantities even when successive cycles are not identical.
cycles per second, *see* frequency.
cynegetic: connected with hunting, 709

D

daily rhythms, *see also* nycthemeral rhythm.

daily rhythm, *continued:*
  actography, 61
  in insects, 647
  in mammals, 759
  origin of, 644f
damped waves; waves whose amplitudes progressively decrease with respect to time. *See also* building-up time.
damping, B.S.661/1026: the effect of the dissipation of energy of an oscillating system on its operation.
  in cicada song, 397
  in seawater, 49
dance-language of bees, 69
danger call, *see* song, calls and call-notes, categories – protection.
dazzle, acoustic, 213
dB, *see* decibel.
deaf mutes, 563, 577, 579
deafening, as acoustic Kaspar-Hauserization, 726
      effect on mating behaviour, 718
          parental behaviour, 719, 721
            vocal behaviour, 231, 233
        experimental, of bats, 184f
deafness, verbal, 573f
death, *see* audiogenic seizure.
decay-time: lapse of time from the instant maximum amplitude of a wave-form begins to decrease until it has fallen to zero or a specified level near zero. *See diagram under* building-up time.
decerebrate rigor, 567
decibel (dB), B.S.661/2022: one tenth of a bel. Two powers P_1 and P_2 are said to be separated by an interval of n bels (or $10n$ decibels) when

$$n = \log_{10}\left(\frac{P_1}{P_2}\right)$$

    The unit is dimensionless. ‖ *For reference levels, see* pp. 37f*. *See also* level.
  difficulties with transients, 38
  recorder, 44
  in water, reference level, 49
declension, 140
decoding, 145
decorticated dog and *canon cancrizans*, 102
decoy, response to, 15, 64, 77, 80, 86, 119, 614, 717, 722, 743
decoying birds, an ancient practice, 106, 717
defence, collective, 83
defence as possible origin of stridulation, 609
defence sounds, *see* song, calls and call-notes, categories – protection.
defining *ad hoc*, dangers of, 3, 5
Deiter's cells, 533f
delayed speech, 125
density of information, 115*, 126
descent of larynx, 496
description of sounds, the problem, 5f, 709f

desperation-call, more precise term for one category of distress call, *see* song, calls and call-notes, loosely used concepts.
detection clicks, 197
detector, grid, 55
determination of form of sound by generating organ, *see* phonomorphosis.
determinism of sound production in Arthropoda, 585
deutocerebrum, 461f
development of sound behaviour, 171, 228f, 725f, 782
diagnostic, behaviour as, 637, *see also* systematics.
dialects in animals, 87
   bird song, genetic implications, 237, 795
     validity of data, 734
dialogue, 773, *see also* song, calls and call-notes, categories – responding.
 of queen bees, 386
diatonic scale, 8
dictionary, 115*
diel, *Webster, 1961:* involving a 24-hour period. *O.E.D. and Shorter Oxford*, no entry. The word is an illegitimate neologism, coined in the absence of need; where *daily* will not do, owing to suggestion of daytime occurrence, *nycthemeral* will, and is of respectable antiquity and regularly formed.
diencephalon, 561, 569
differential thresholds, *cybernetics*, 122*
differentiation of sound-production apparatus, 334
diffuse and compact sounds, 166
diffusion, sociometric, 125
diminished song, *see* song, calls and call-notes, categories – courtship.
diphthongs, 159f
diplosyllable (Doppelsilbe) 8*, 19*
 Orthoptera ('double pulse') 360
 Tettigoniid ('double pulse') 357*
directed behaviour, 62, *see also* orientation; phonotaxis; phonotropism; *and discussion under* taxis and tropism.
directional acoustic memory, 75*, 105, 619
   sensitivity, *see* hearing.
directivity of microphones, 33, 52
disc recording 34, 516 ('glyphic registration')
disjunct subspecies and sound, 698
disorder, 144
disorders of adaptation (Selye), 571
dispersal response to recorded desperation (distress) calls, 85
disphony, 65
displacement activity, 607, 724f, 776f
display, *see also* song, calls and call-notes, categories.
 actography of, 64
 arena, 80
distress call, *see* song, calls and call-notes, loosely used concepts.
disturbance sounds, *see* song, calls and call-notes, categories – protection.
disyllabic sound, 19*, 711, 754f, 764
diversity, maximal, 124
   of sound-production apparatus in some groups, 336, 402
 dogs and lamp-posts, 128

domestication and the transformation of bird song, 231
dominance, potential, of singing bird, 715
dominant frequency, 6*, 349, 697f, 803, *see also* physical characteristics.
Doppelsilbe *(G.)* = diplosyllable, not disyllabic sound: 8*, 19*, 357*, 360
Doppler effect, B.S.661/2019: the change in the observed frequency of a wave caused by the time rate of change in the length of the path of travel between the source and the observer.
    and echolocation, 203
dormitories, bird, actography, 57
          desertion after playback of desperation (distress) call, 86
double tone, double voice, *see* phonation.
drugs, *see* audiogenic seizure.
    and song, 643
drum, *see* ear; mirror; tymbal; tympanum; *and species index (fish)*
drumming, 676, 682, 685f, *see also* chest.
    snipe, woodcock, 719
    woodpecker, 718, 745
      actography, 60
dual sounds, fish, 682f, 685
ductus pneumaticus, 490
duet as pair-link, 718
dummies, 119, *see also* decoys.
duration as a parameter, *see* physical characteristics of emissions.
    and range of signal, 74
    and population density, 587
    and temperature, 369
dynamic level, baboon, 160

E

ear, *see also* acustico-lateralis; audiogenic seizure; echolocation; hearing; lateral line; tympanal organ; *and under the various anatomical elements.*
    in amniotes, 531f, 541f
    amphibia, 531f, 536f, 705f
    aquatic vertebrates, 536f
    bats, 192f, 202, 204, 208f, 535
    birds, 542f, 737f, *see also* amniotes, *above.*
    Cetacea, 538
    cyclostomes, 530
    elasmobranchs, 531, 659
    fishes, 656f, 686
    mammals, 543f, *see also* amniotes, *above.*
    reptiles, 541f, *see also* amniotes, *above.*
    teleosts, 531, 659
    vertebrates, 522f
      aquatic, 536f
      Dèhèuy's work, 543f
      blood vessels, 536
      "ciliated" organs of cochlea, 525
      cristae, 529f, 658
      development, recapitulatory aspects, 529

directionality, 535
drum, *see* tympanum, *below*
elementary structures for vertebrate hearing, 522
external, 192, 202, 204, 208f, 212, 534f
inner, 534, 657, 751
labyrinth, 523, 656f
maculae, 529f, 657
middle, 534f, 705, 737, 751
otic capsule, 531f, 659, 666, 686
papillae, 529f, 657, 705
partes inferior and superior, 656f
rhythm analysis by 8
sensory-hair cells, 529f
sensory structures, 529f
sphinctrate meatus, birds, 534
tympanum, 534, 705
   exteriorization in development, 537f
   and pulmonary sacs, 537
   secondary, 531*
tympanic membrane, vibration, 544f
        sulcus, 544
ease and comfort calls, 153, 774
echo, *see* reverberation; reflection.
echo fish, 224, 657, 684
echolocation, 12, 95, 106, 183f, 185*, 535, 791
  in certain groups:
    bats, 185f, 789
      audible sounds unconnected with, 187f
      breathing problem, 191
      dazzling with white noise, 213
      different types, 186f, 204f
      fish-bats, special features, 205f
      flight-chamber tests, 194f, 202f, 209, 213
      Griffin's Doppler effect theory, 203
      habituation, 195
      hearing aspects, 191f, 202f, 208f
      homing after blinding, 196
      Megachiroptera, summary, 214
      Microchiroptera, summary, 208
      Möhres' intensity discrimination theory, 203
      not universal, 214
      outstanding problems, 215
      phylogenetic aspects, 214
      posture, 189
      preying posture, 197
      Pye's theory of beat detection, 193
      speeding of emissions during approach, 199, 211
      summary, 214
      whispering bats, 208
    birds, 216f
    blind men, 791

echolocation, *in certain groups, continued*:
    Cetacea, 219f, 790f
    fish, 223f, 675, 684
    insects, 218f, 437
        compared with fish, 219
    oil-bird, 216f, 745
        flight-chamber tests, 217
    owls, 216
    rodents, 215f
    swifts, 217f
  general aspects:
    antennae, 219, 789
    brain specialization, 193
    cruising clicks, 197, 217, 790
    detection of prey, 218f
    emitter-receptor systems, 185, 199f, 207, 209, 212
    intensity, use of, 203, 217
    interference mechanism of Rhinolophidae, 201
    mazes, 222
    not automatic, 196
    passive and active, 792
    physical characteristics of emissions, 187f, 199f, 205, 207, 209, 212, 217f
    in pursuit of prey, 196, 221
    range of beam, 202
    rates of capture of prey, 196
    reversing receptors, effect of, 185
    of static food, 208f
    and stranded whales problem, 790
    theories of, 193, 203, 791
echosounder, 655
écrevisse, contrepoint de l' *(F.)*: *see* cancrizans.
education and speech, 563
ee ee ee, 152f
efficiency of wing motor, 379
Egyptians, ancient, use of desperation (distress) call, 88
élan vital as source of sound production, 337
electric signal a transformed letter, 145
electro-acoustic apparatus, 107, 346
electrodynamic microphone, 26
              transducer, 50
electro-encephalogram, *see* audiogenic seizure.
electromechanical recording, 34
electrophoresis, *see* audiogenic seizure.
electrophysiological methods, 102
    and insect hearing, 106, 435f, 800
elementary wave-form, vibration: the hypothetical continuous wave-form to which any given wave-train emitted by the animal approximates; the basic signal, 6*, 22*f
    as carrier-wave, 392, 447
elytra in stridulation, 286f, 293, 295f, 329, *see also* tegmen.
embedding (algorithm theory), 178*

embedding alphabet, 146
embryology of behaviour, 562
emitter *(cybernetics)*, 114*, 117, 118 ("transmitter"), 147, 337 ("transmitter"), 710 ("donor")
emotion and psittacism, 733
emotional states and animal sounds, 151
empirical first description of a sound, 14, 16*
encephalic centres, *see* cerebral hemispheres.
endolymphatic duct, 659f
energy expenditure in sound production, 359
English language, 159
entropy in sound, 133
entry of a word, 138*, 174*
envelope of a wave-form, 9*, 43*, 123*, 156f
environment, sonic, 744*f
enzymes, *see* audiogenic seizure.
eosinopenia (audiogenic seizure) 251f
eosinophiles (,,   ,,   ) 251f
epiglottis, 491, 495, 498, 512
   of bat, 191
   and transients, 135
epilaryngeal tube, baboon, 136, 162
              child, 162
epilepsy, audiogenic, 247f, 254f, 262
equal temperament, scale of, 9f
equation relating wing beat to structure, 377
escape reactions, 87, 428, 695, 723, *see also* defence; song, calls and call-notes, categories – protection.
estrus, *see* oestrus.
ethology, Lorenz-Tinbergen, 119, 562, 570
      statistical, 130
ethological terminology, 18f, *see also* song, calls and call notes, categories.
Eustachian tube, 532, 537
events, biological and physical, as criteria for definitions, 11
evolution: of air-sacs and vocal production, 491f
      of brain, 561
      convergent, in amphibian courtship calls, 803
            concealment calls of birds, 742
            Hemiptera, 402
            mobbing calls of birds, 741
      of courtship behaviour in Acrididae, 594
      effect of island life on song variability, 734
      homology research on sounds, 752, 777f
      independent, of hearing in fish stocks, 664
            of wing strigils, 404
      juvenile preponderance, 782
      neotenic sounds, 782
      of otic capsule, 531f
      of sound production in Hemiptera, 409
      of speech, 171f
      of stridulatory apparatus, 336f

evolution: of stridulatory movements, 609f
    of strigil-plectrum method, 409
    of tymbal from strigillar mechanism, 409
evolutionary aspects of echolocation, 214
    potential of wing-strigil, 402
exact words, importance, 5
excitement, generalized, sounds of, 710, 753, 774
exophthalmia, audiogenic, 244
expanding messages, 134f, 149*, 171
experimental study of behaviour, 610f
explosive sounds, reactions to, fish, 683
extensions of meaning, when legitimate and when not, 3, 5f, 8, 13
extension of sounds in evolution, 783
external ear, *see* ear.
extinction of response to sound, 685
extrapyramidal tract, 567, 574
    centre, 573
extraspecific communication, 391*, 407, 409, 798, *see also* interspecific.
eyes and the audiogenic seizure, 244, 254

F

facilitation as explanation of apparent imitation, 716
family signals, 82f, *see also* song, calls and call-notes, categories.
fatigue in flight, 380, 382
feather noises, 78, 711
Fechner, *see* Weber-Fechner.
feedback, 125, 563, 718, 738, 794, *see also* audio-auditive reflexes; audiophonative reflexes.
    auditory, in vocal development, 233, 794
    positive, 467
feeding actography, cow, 56
    behaviour, 86, 723
feet noises, birds, 711
female, reduced pars stridens, in Orthoptera, 365
female song, *see also* song, calls and call-notes, categories – courtship.
    birds, 714
    Hemiptera, 394, 396, 404, 406, 408
    proclamation in Orthoptera, 595
femur in stridulation, insects, 281, 285, 291f, 295f, 333, 403, 405f
fenestra ovalis, 192, 532f, 545, 659, 705, 737
    rotunda, 192, 531f, 536f, 545f, 737
field playback apparatus, 44f
field, sonic, 25
fighting song, *see* song, calls and call-notes, categories – rivalry.
figure, melodic, 17*, 22*f: the smallest melodic unit which has a pattern; consists of one
    or more beats (and/or sub-beats). (Sotavalta).
    in birds, 18*, 715 ("patterns")
    in lemurs, 10*
figure, rhythmic (adaptation of the term *figure* to rhythmic concepts) 18*f
figured chirp, 20*f
    sequence, 18*, 20*f
    syllable, 20*f

file, stridulatory, 278*, 441*
filming speed, 34
filter analyzers, drawbacks, 40
fin drumming, 676, 682, 685f
fin spines, dorsal, in stridulation, 678
final common path (Sherrington) 563
    length of a word, 142
    message, 164
    word, 138*, 143, 145, 149, 177*
fisheries, listening techniques, 672f, 687
      technological advances, 655
      use of sound in, 666, 669f, 687
flat and natural sounds, 166
flight chamber, *see* echolocation.
    muscles, direct and indirect, 377
        physiology, 377
    sound, birds, 719
        insects, 330, 374f, 589
  analysis, 380f
  biological significance, 386, 428
  cinematography, 374
  a complex tone, 382
  difficulties in recording, 375
  and Johnston's organ, 420
  modification of 385
  parameters of, 382f
  physiology, 377
  recording methods, 374f
  species specificity, 375
  swarms, 385, 589
  ultrasonics lacking from, 375
  and wing coupling, 377
flight-tone signals, 73
focussing sound, insects, 353
Foliar process, 545
food call, *see* song, calls and call-notes, categories – family and aggregative.
foot *(poetry)*, 14
forced vibration, oscillation, B.S.661/1016: a vibration directly maintained in a system by a periodic force, and having the frequency of the force. ‖ 362
forewings, *see* alary stridulation; elytral stridulation; flight sound; tegmen.
form of signal, *see* wave-form.
formal word, 141*, 147*, 169, 178*
formalized approach to animal language, 136
formants, formant frequencies: principal frequencies in the sound spectrum of a human (or analogous) utterance, which, for a particular individual, are diagnostic of particular speech-elements, especially vowels, in which some three characteristic formants (different for each vowel and for each individual) are present. They are resonance frequencies originating in the vocal tract, and in some but not all cases can be attributed to particular parts of it. The individual differences, especially between ♂ and ♀, in the formant complement appropriate to a given vowel might be expected to lead to confusion by a hearer, and indeed Broadbent (Proc. Roy. Inst.

G. Brit. 38: 430, 1961) succeeded in making an experimental audience identify one and the same synthetic sound as *bit* in one context and *bet* in another; the diagnostic properties of the formant-set of a particular sound are thus *relative* to those of the rest of the utterance, which is taken into account, computer-wise, by the hearer; 101, 154*f, 158f, 167

formant, formant-word *(algorithm theory,)* 174*

forme, 130

formula, the sonogram or oscillogram as a, 709

foster parents, 81

founded alphabets, 144*f

Fourier analysis, 6, 355

fragile message, 119*

frame *(frequency analysis)*, 154*

free vibration, B.S. 661/1017: a vibration resulting from a disturbance of a system and having a period depending solely on the properties of the system. ‖ *See also* forced vibration.

frequency, *see also* phonomorphosis; physical characteristics; sound production; temperature; etc.

B.S. 661/1007: the rate of repetition of the cycles of a periodic quantity. The reciprocal of the period.

The unit is the cycle per second (in Europe, the *hertz:* 1 c/s = 1 Hz.)

By extension, applied to recurring quantities even when successive cycles are not identical. ‖

In bio-acoustics, there is some tendency to reserve the term for the spectral components of the elementary wave-form alone, a useful expedient for avoiding the all-too-easy confusion of these with all the other periodicities (such as tooth-impact rate); *periodicity* and *rate* themselves are convenient terms for these latter.

analyzers, 39f, 107

carrier, *see* elementary wave-form.

discrimination, *see* echolocation; hearing; Johnston's organ.

in cortex and cochlear nucleus, 564

dominant, 6*, 349, 697f, 803

formant, *see* formants.

fundamental, 6*, 136, 158, 392, 752, 803

high, in playback, 46

low, reception by skin of fish, 663

modulation, 17*, 354*, 392

natural, B.S. 661/1021: the frequency of a free vibration, 392, *cf.* resonance frequency, below.

principal, 6*, 349, 697f, 803

resonance, resonant, B.S. 661/1022: a frequency at which resonance *(q.v.)* occurs in a system. ‖

N.B., this and *natural frequency* are not the same thing, though often used as if synonymous by bio-acousticians.

The natural frequency of free vibration, at which a system will oscillate if merely triggered (by, e.g. a blow), and then left to itself, is given by

$$F_{nat} = \frac{1}{2\pi}\sqrt{\frac{1}{LC} - \frac{R^2}{4L^2}}$$

In the more familiar electrical case, L is inductance, C capacitance and R resistance; in the acoustic case, L is inertia or mass, C elasticity and R made up of e.g. air resistance and similar quantities. Resonance effectively eliminates this, so that

$$F_{res} = \frac{1}{2\pi} \sqrt{\frac{1}{LC}}$$

and this is the frequency an applied periodic force of minimal value has to have in order to keep the system vibrating. It is clearly higher than the natural frequency.

shifts in sound production, 199, 699f, 704, 739, 767
spectrum, *see* sound spectrum.
and temperature, insects, 368f, 379
  compared with vertebrates, 367, 370
and tooth-impact rate, 362, 444
of wing beat, 375f, *see also* flight-sound.
  equation relating morphological characters to, 377
  factors affecting, 377
  and humidity, 380
  myogenic, 377
  neurogenic, 377
  and temperature, 379f
  volitional control, 382
fret call, more precise term for one type of distress call, *see* song, calls and call-notes, loosely used concepts.
friction, emission by, 278f, 332, *see also* stridulation; strigilation.
frontal lobe, 573
full song: fully developed adult song under active sex-hormone influence, in contradistinction to subsong, juvenile song, 235, *see also* song, calls and call-notes, categories.
  evolution from juvenile song, 730
function of songs, Hemiptera, 400, 407, *see also* song, etc, categories.
fundamental, *see* frequency.
funding alphabet, 146

G

Galton whistle, 104, 248, 421, 437, 614, 620
  preferred to natural signal, 620
genealogy of language, 751
generating apparatus, determination of sound form by, *see* phonomorphosis.
genetic aspects of the audiogenic seizure, 248, 250, 256, 261
  control of vocal behaviour, 232f, 627f, 704, 726
  signals, 63
geniculate bodies, *see* audiogenic seizure; central nervous system
  tract, 567
genitalia in stridulation, Diplopoda, 321
geographic conditioning of interspecific response, 88
  factors, *see* island; isolation
  variation of calls, 698f, 704, 706
    in vocal-sac development, 696
Gestalt, 130
  philosophy, psychology, 95, 119, 558
Gewöhnlichegesang *(G.)*, *see* song, calls and call-notes, categories – proclamation
gin-trap stridulation, pupal, 331
glands *(audiogenic seizure)*, 252

glands *(sound production)*, 324f
glass-bottomed boats, 681
global message, 115*
glossopharyngeal nerve, *see* audiogenic seizure; central nervous system
glottis, 490, 493, 497f, 508f, *see also* larynx.
    false, 517
    fully closable in gibbon, not in man, 509
glucose and the audiogenic seizure, 252, 255
glutamic acid and the audiogenic seizure, 250, 256, 261
glycogen and the audiogenic seizure, 255
glyphic registration 516 (? translator's *lapsus* for *groove recording*, i.e. disc recording).
gnashing 762, *see also* stridulatory apparatus.
gnosis, *see* auditory gnosis; parietal gnosis; sensory gnosis.
gobbling, turkey, 713
gonads, relation to song, 450, 640f, 712f
                imitation, 732
grammar, genesis of, 169
         use of, 576, 578
grammatical calculus, 178
gramophone in frequency determination, 346
graphic registration, 516 (? oscillographic recording)
grating, fish, 678
       whales, 220
grave and acute sounds, 166
greeting ceremonies, 715, 718f, 769, *see also* song, calls and call-notes, categories – courtship.
gregarious behaviour of baboons, 150
         signal sounds, *see* song, calls and call-notes, categories – aggregative responding.
grid detector, 55
grinding, palinurids, 308
groaning, fish, 674
        monkey, 132, 172
        whale, 220
groh-groh, 673
ground rhythm, audible, 16*
  independent of melody, 18
ground vibration, perception of, *see* vibration.
growling, 755f
  homology, 780
grunting, fish, 101, 224, 674, 676, 683f
        homology, 781
        mammals, 220, 754f, 770f
        whales, 220
gryllid-type stridulation, 302
gyrus centralis anterior, 574

H

habituation, in echolocation, 195
hackling, 762, 765
haemorrhage, audiogenic, 248f
hair sensilla, *see* hearing; vibration receptors.
halteres, as sound generators, 386

hand rearing, effect on bird song, 235
haplosyllable, 19*, 360, 357 ("pulse")
harmonic, 6*, 739f, 752
   analysis, 8*f, 18, 383*f
Hartshorne's anti-monotony principle, 715
head in percussive sound production, insects, 326f, 334
       stridulation, insects, 277f, 405f
health of experimental animals, importance to reliable results, 235
hearer, relevant, concept of, 3, 14*, 407
hearing:

   Moulton, *supra*, pp. 663f, discusses definitions of hearing, primarily in the vertebrate context. Of Pumphrey's definitions, there mentioned, that of 1950 (Symp. S.E.B. No. IV; "an animal *hears* when it *behaves as if* it has located a moving object (a sound source) not in contact with it.") is rejected as making hearing a definitively directional phenomenon.

   His earlier definition (1940, Biol. Rev. 15: 107: "demonstrable responsiveness to *sound*", *q.v.*) is on the other hand too broad for some workers, and Dijkgraaf incorporates it as only the first of *two* mandatory criteria, returning to von Buddenbrock (1937) for his second, though removing the undue restrictiveness of von Buddenbrock's "resonant membranes":

   (1) demonstrable sensitivity to air- or waterborne sound; together with
   (2) detection of these stimuli with special receptors primarily used for this purpose
      (1960, Proc. Roy. Soc. B, 152: 51–54).

He terms (1) by itself, without (2), merely *sound reception;* and responsiveness to sound or vibration reaching the animal through the solid substratum, *vibration perception.*

   Though these distinctions between vibration perception, sound perception, and "echter Gehörsinn" remain as artificial as they were in 1937, they are related more realistically to everyday experience than the attempt to treat all that shudders as sound and every reaction to it as a hearing response – and all that glisters as gold. Zoology, indeed, science itself, is founded on the arbitrary subdivision of intergrading series into finite classes, simply as working units for study; in the present context we can recognize the infinite series without having to reject the useful working distinctions between its parts.

   See also acustico-lateralis; audiogenic seizure; ear; echolocation; kinaesthesis; tympanal organs; vibration receptors.

in various animal groups:
  amphibia, 705f
  arthropods, 412f, 583
  bats, 193f, 202f, 204, 789
  birds, 735f
  cetaceans, 220f, 790f
  crustacean antennae, 789
  Diptera, 413f, 420, 426
  elasmobranchs, 663
  fish, 656f, 686
    first demonstrated, 106
  Hemiptera, 392, 400, 408f, 415f, 420f, 427
  Lepidoptera, 415, 428, 434f
  Orthoptera, 416f, 420f, 426f
  Strigids *versus* passerines, 737

hearing, *continued:*
  vertebrates, 522f, *see also* acustico-lateralis; ear.
  general aspects:
    absolute sensitivity, 423f
    after-discharge, 428, 436
    behaviour responses, 428f, 435, 437
    binaural, 535
    chemical mediator, 528f
    chordotonal organs, 400, 408, 413, 422, *see also* scolopophorous organs, *below.*
    conditioning fish, 663
    dephasage, fenestral, 545
    directional, 425f, 428f, 680f, 738
    during flight, insects, 386
    electrical response, 420f, 435f
    escape by, 218f, 387, *see also* ultrasonics.
    frequency compass, *see also* echolocation.
      arthropods, 420f
      birds, 735f
      fish, 661f
      frogs, 705
    frequency discrimination, *see also* sensitivity, auditory.
      birds, 737f
      fish, 667
      frogs, 705
      insects, 424, 426, 436, 800
      spiders, 423f, 447
    hair sensilla, 409, 412, 422, 434, 436f
    human, theories of, 548f
    initial transient, 528
    lyriform organs, spiders, 412, 420, 423
    mechanism of transductor in nervous transmission, 526
    optimal rhythm, 528
    organs in water, 48
    orientation by, *see also* echolocation.
      bats, 184
      fish, 669
      rats, 216
    parallel evolution in fishes, 664
    perception *(contra* reception), 522f
    predation by, 387, *see also* echolocation
    primary phenomena, vertebrates, 522f
    reception *(contra* perception), 534f
    recruitment, 528
    scolopophorous organs, 412f, 422, 435f, *see also* chordotonal organs, *supra*
    sensory adaptation, 428, 436, 663
    theories of human, 548f
    transfer function, 527
    transmission, intra-aural, 535f, 546f
    tuning to carrier frequency, insects, 392
heart beat, 7*f, 11
heat-growling, 759

heat production by flight muscles, 380
hee-haw, 516
helicotrema, 533f
Helmholtz, resonator theory, 548f
hemisyllable: either of the elements of a diplosyllable, 21*f
    N.B., not a haplosyllable.
Hensen's cells, 533f
Herbst corpuscles, 735f
herd coordination by sound, see song, calls and call-notes, categories – aggregative responding
herd leader, baboons, 150
    relations, baboons, 150, 152
Hertz = cycles per second, see frequency
heterogeneous summation, 777
heterospecific sound reactions, see extraspecific; interspecific
hiatus intervocalis, 497, 512
  uniquely absent from gibbon, 509
hiccup, 132
hierarchy of afferent and efferent centres, 559, 562
        animal signals 70f
        founded alphabets 144*f
        repertoires, 130
high-energy phosphate and the audiogenic seizure, 256, 262
hippocampus, 574
hiss, inspiratory, 516
    loris, 516
    mammals, 753f, 762
    snake, 491
homing of toads, 706
homology, criteria of, 777
      research and sound forms, 752, 765, 777f
          primates, 781
homophony, (1) *music:* opposite of polyphony, i.e. without contrapuntal interplay, melody and harmony moving by-and-large in the same direction
    (2) *linguistics:* the quality of being a homophone (a word looking or sounding like another, but with a different meaning, as *sun* and *son*), 65
hon..., 153
honey-guide bird, 86, 91, 724
horizontal substitution, 138*, 176*
hormonal rhythms, actography, 63
    threshold of song, 717
hormones, 252, 256, 566, 568, 572, 639f, 711f, see also castration; gonads
horn, French, and alligator, 104
horseshoe, of Rhinolophids, 199f
howling, 755f, 764, 773f
  homology of, 779
hugging, 152
human audition, subjective, as first tool of analysis, 3
    language, 118, 133
      speech, first origin, 135
humidity, effect on signal, 380, see also physical characteristics; song.

humidity, stimulating care-calls 770, *see also* song.
hunger, stimulating care-calls, 770f, *see also* song.
hybrid species, song characters, 92, 228, 628f, 704
hydrophone, 25, 51*, 674
   for actography, 58
   net, 655
   sensitivity, 51*
hydrostatic perception, 408, 665
5-hydroxytryptamine and the audiogenic seizure, 258f
5-hydroxytryptophane and the audiogenic seizure, 260f
hymns and psalms, rhythm and melody relations, 17
hyoid apparatus, 490, 496
     arch, 535f
     bone, 538
     muscles, 490, 493
hyperstriatum, 561, 569
hypoglossal nerve, rôle in phonation, bats, 191, 212
        nucleus, 494
hypopharynx, man, 162
hypophysis, 713
hypothalamus, 568f, 713
hysteresis effect in noctuid tympana, 800
Hz. = Hertz = cycles per second, *see* frequency

I

I, capital, 575
I.C.B.A., International Committee for Biological Acoustics, Introduction and p. 680
I.R.M., innate releasing mechanism, 570, 615, 711f, 777
imagination, 560
imitation, apparent, due to facilitation, 716
      controls upon, 240, 731
      as a factor in vocal development, 234, 729f
      of human sounds, ape and parrot compared, 569
         failure to teach, 517
      human, of monkey, 151, 154f
imitations, artificial, as stimuli, 614, *see also* decoy
imitative species, 732f
imitativeness in birds, inherited, 239
        grasshoppers, alleged, 616
immature stages, *see* under stage concerned
   behaviour, 608f
impact pulses, 9*f
impedance, acoustic, 48*
implication, 142*
imprinting, 240, 614, 720
improvisation in bird-song development, 233, 795
Impuls *(G.)* = pulse, 8, *see also entry under* pulse
impulse, B.S 204/1114: a *brief* change of current produced in a circuit ‖ (my italics). By extension, it could be used for a train of waves (electrical or acoustic) of which it constituted the envelope; but this train would have to be short relative to the interval between it and the next.

Example of this usage, pp. 364, 365; though the wave-trains are strictly speaking too long to qualify as *impulses*, and should properly be called *pulses*, according to the British and American Standards – a procedure made impossible, however, by the context. *See* Chapter 1, and entry under *pulse*.

impulse counter, 55, 61

impulsion *(F.)*, strictly = *impulse* as defined above; in my experience, *not* used in the sense in which *pulse* (q.v.) is used in English, until recently, and then only in partial equivalence, as translated below from Grand Larousse, 1962:

Impulsion, *radiotech:* train of very high frequency waves, used notably in radar, which succeed one another periodically in time at intervals of some decades of microseconds, and whose duration varies from some hundreds of microseconds to a fraction of a microsecond.

television: sudden variation of amplitude of the modulation, used for transmitting time-base sync. signals. ‖

The second of these definitions is much the same as the B.S. for *impulse;* the first equates to certain limited kinds of *pulse,* (as well as to *impulses*), namely

(1) of high-frequency waves; and at the same time

(2) of short pulse duration (e.g., less than a millisecond)

It does not, however, specify that the pulse duration shall be short *relative to* interval or period, and this is where this new definition appears to me to depart from time-honoured practice in the French language, which, when I was working in France 10 years ago, appeared to have no equivalent whatever to the everyday concept of *pulse* in the English-speaking world; at that time, *impulsion*, at least among the physicists I worked with, equated rigorously to *impulse*, not *pulse*.

Thus the present situation is that in the electrical context, *impulsion* can be used for anything from *impulse, s.s.* to *pulse*, provided the latter be of inframillisecond duration and of h.f. current. *Pulse* in acoustics still has no equivalent.

inattention, 565
incidental sounds, *see* sound, biological
incompatible and compatible words, 145, 149
incomplete song and post-cantical inhibition, 716
incomplete sound (= phoneme), 167
incomplete syntagma, 142*f, 177*
incomplete words (= morphemes) and complete words, 142*, 149, 166, 177*f
incus, 535f, 665
indifferent song, *see* song, calls and call-notes, categories – proclamation
individual recognition, 82f, 720f
Indonesia, 671
inferiority, *see* appeasement
information, 69, 73, 92*f, 112, 115*f, 130*, 147, 337, 714
  inherited, 137, 228f
  intrinsic 69
  "other", 15
  pulse, 392
  theory, 100, 112f
  value, 100, 112, 716
informational machines, 133
infra-red visual sensitivity, owls, 216
infrasonic: *for definition, see* sound; *see also* hearing
  signals, in frogs, 697
  orientation, bats, 184

inheritance of vocal behaviour, 228f, 494, 627f, 704, 794f, *see also* innate signals
inherited information, 137, 228f
inhibition periods after song, 716
initial word, 138*, 142, 145, 177*
innate releasing mechanism, 570, 615, 711f, 777
innate signals, birds, 107, 494, 726
input, 145, 149, 157
insect cidal intoxication, 54, 55
instinctive movements, 776
integration-time (phonoresponses), 104
intelligence *(information theory)*, 15
intelligent judgment, 572
intelligibility, 118*
intensity of convulsions *(audiogenic seizure)*, 248, 254f, 258f
    sound, *see* sound intensity *for definition and discussion*
    baboon signal not controlled, 153
intensity and echolocation, 203
    factors affecting, 350, 353, *see also* phonomorphosis
    and population density, 73, 587
    of voice, 517
intention, 131
  movements, 776
interaction of behaviour, 63
intercalarium, 665
interchangeable signs, 138
interchirp rhythm, 448
intergeneric response, 612
interindividual messages, 114*, 116
interiority, psychological and physiological, 557f
interjections, no algorithm for, 141, 170
intermedial organ, 419f
internal speech, 562, 576f
International Committee for Biological Acoustics, Introduction *and* p. 680
interphyletic phonoresponse, 321
interrupted and continuant sounds, 166
interspecific alarm calls, 722
    attraction, 769
    hybrids and vocal inheritance, 228f, 726f
    imitation, 237, 719, 730f
    response, 63, 84, 86f, 105, 612f, 633, 722
        nil, between macaque and baboon "ak ak", 151
    signals, 114*, 743
interspecificity, *see also* extraspecific; stenospecific
intersyllabic (Zwischensilbe) rhythm, 448
interval of recurrent animal sounds: lapse of time between the *end* of one member of a series of more or less irregularly periodic sounds and the *beginning* of the next
    For the concept to have any real meaning in respect of irregular sounds, it must obviously be a mean value of a fair sample.
intervals between calls, semantic value, 95
    in bird song, 716
intimidation, 681, 684f, *see also* song, calls and call-notes, loosely used concepts

intraspecific communication, 391*, 407, 409, 798
intrasyllabic transition *(linguistics)*, 162
intrinsic information, 69
introspection, 557
invariants, 139*, 158, 174*
   of an incomplete sound, 167
inventiveness, vocal, 238, 729f
inversion of signals, birds – little reduction of response, 93
inverted tegmina, effect on stridulation, 360
invitation ceremonies, 715, 718f, *see also* song, etc., categories – courtship
inward speech, 562, 576f
ionophone, 45f
iproniazid and the audiogenic seizure, 259f
island birds, enhanced song variation, 734
isolating factors in hybridization, 630
isolation, *see also* Kaspar Hauser
   and song learning, 228f, 237, 726f
   and responsiveness, 588f, 641f
   reproductive, by song, 631f, 706, 799, 804
isthmus palatopharyngeus, 495f
        pharyngonasalis, 495f

J

Japan, 671
jaw-clicking, 762
Johnston's organ, 413f, 420
   directional characteristic, 426
   frequency discrimination, 429
   response to female flight tone, 420
judgment, intelligent, 572
juru selam, fish listener, 673
juvenile song, *see* song, calls and call-notes, categories – subsong
juvenilism, 719, 769

K

Kaspar Hauser individuals, 228f, 237, 563, 577, 726f
kc/s = kilocycles per second, *see* frequency
key to American *Nemobius* species by song, 637
keys and the audiogenic seizure, 247f
keratis, 248
kHz, kilohertz = kilocycles per second, *see* frequency
kinaesthesis: primarily, perception of movement of any part of the body, by musculo-
        skeletal proprioceptors; by extension, including posture information *via*
        semicircular canals, lateral line, etc.; by further extension, including
        vibration, *via* various vibration receptors, 69, 223, 408
kinesis: movement, or increase of movement (not necessarily locomotory), whose di-
       rection, if any, is unrelated to that of the source of stimulation, but whose amount
       is related to the intensity of stimulation, 428
kissing sounds, 694
Klein ionophone, 45f

knocks, whales, 220
kris-kris, 673
kuruck-kuruck, 670
kymograph in actography, 435
 and flight-sound, 375

L

labia vocalia, vocal cords, *see* larynx
labial sounds, 762, 771, *see also* lips
laboratory playback apparatus, 44f
labyrinth, 523, 656f
lacinia utriculi, 657, 660f
lagena, 529, 532, 656, 658f
lamina spiralis, 533
lamp-posts and dogs, 128
language, use of term, 69, 114*, 132f, 170
    animal, 106, 112, 132, 576
    human, 118, 563, 572
    internal, inward, 562, 576f
    a mental automatism, 578
    social tendency in, 129, 577
    statistical structure of, 127, 129
    technology of, 120
larvae, dumbness of, 76
larval (*see also* nymphal) sound production, 330f, 404
laryngeal nerves, rôle in phonation,
  bats, 190, 209, 212
  man, 190
laryngo-nasal passage, bats, 200, 212
laryngophone, 29*, 56
larynx, *of various vertebrates:*
  amphibians, 490, 694, 696
  apes, 508f
  bats, 189f, 199f, 212, 504, 508, 789
  birds, 491f
  carnivores, 502f, 506f
  cetaceans, 499, 502f
  Dipnoi, 489
  lemurs, 505
  mammals, 494f, 752
  man, 162, 190, 516f
  marsupials, 494
  monkeys, 162, 505f, 509f
  monotremes, 494
  primates, 505f
  Procolobus, 782
  rodents, 500f
  seals, 502f
  ungulates, 502, 505f

larynx, *general aspects* (*see also* phonation; sound production)
    air-sacs, 499f, *see also general entry* air sacs
    arteries, 500
    cavity, 499
    choanae and, 496
    descent, 496
    muscles, 489f, 494, 498
    nerves, 190, 209, 212, 500
    phylogenetic aspects, 501f, 513f, 696
    skeleton, 188f, 498f, 494f, 695
    variations, 500f, 696
    ventricle, 499, 517
    vestibulum, 499
    vocal cords, 491f, 494, 499, 502, 509, 694f
        double, 514
        seals without, 502f
latch-trap, 116
latent period *(audiogenic seizure)*, 244, 246f, 250, 254, 260f
lateral line, *see* acustico-lateralis
    as accessory hearing organs, 524
laugh, 172, 756, 765f
lax and tense sounds, 166
leader of herd, 150
        influence, 152
        of song, 704
leaf, nose, 209
learned signals, 107, 726
learning, 63, 572, 616f
    of calls, birds, monkeys, 170; insects, 229, 234f
    disposition for, inherited, 239f
    and emotional bonds, 240
    as factor in vocal development, 234f, 494, 628, 794f
    latent, 241
    of vocal behaviour, characteristics of process, 239f
    vocalization a self-rewarding phenomenon, 241
least effort *(linguistics)*, 129
leaves, rustling, response of mice, 84
Lee effect, 125*, 575
lefthandedness in crickets, 360f, 443, 480f
legs in stridulation, *see also under various podomeres*,
    Chelicerata, 317, 319
    Chilopoda, 321f
    Crustacea, 309, 312, 314f
    Diplopoda, 321f
    insects, 290f, 330f
legs, autotomized, stridulating (Chilopoda), 321
letter, 174*
    abstract, 174*
    concrete, 174*
    justification for use to denote a sound, 165*
Lettish language, 159

level, *see also* sound level,
>B.S.661/2025: of a quantity related to power: the ratio, expressed in decibels, of the magnitude of the quantity to a specified reference magnitude.
>*Note 1:* the word *level* is frequently used in compound terms such as "intensity level" and "sound pressure level" to relate a magnitude to a common reference magnitude or "zero level".
>*Note 2:* in the absence of any statement to the contrary, the reference magnitude for stating sound pressure levels in air shall be 0.0002 dyne/sq.cm, and that for sound intensity levels 10^{-16} watt/sq.cm. ‖. *See further* pages 37*, 49*

level recorder, 44
levels of analysis of animal sounds, 3, 14*
lexical meaning, 148*, 174*
lexicon, 148
libraries of animal sound, 686
Licklider's duplex theory, 550f
Lied *(G.)* = song, 5, *see also* song, calls and call-notes – canto
ligamentum vocale, 498, 752
light cycle and the audiogenic seizure, 251
light, effect on song behaviour, 638, 760
lima, 403, 798
limen: name for the minimum detectable difference in pitch, *q.v.*
limited nature of animal sounds, 172
line, 5
line of formal words, 141*
    a complete syntagma, 149*
linear equivalence *(algorithm theory)* 144*
lingual nerve, 212
linguistics, 127, 133
    animal, 129
Linksgeiger *(G.)* = lefthander, *see* lefthandedness; stammering
lips, *see also* labial sounds,
    amphibians, 694
    monkey calls, 160, 164
    vertebrates, 491
locatability of acoustic signals, 73, *see also* physical characteristics
location of sound apparatus, diversity in Arthropoda, 335
locus niger, 567
logical algorithm, 150*
logon, 122*
long chirps (single sounds) 763, *see also* song, calls and call-notes, categories.
loosely used concepts in song, *see* song, calls, call-notes, loosely used concepts.
loudness, *for definition, see* sound
    factors in, 382
    no transition in baboon, 153
loudspeaker, underwater, 53, *see also* transducers
lure, response to, 64, 77, 80, 86, 105, 119, 123*, *see also* bird call; decoy
    supranormal, 92
lyriform organs, 412, 420, 423
lysergic acid diethylamide and the audiogenic seizure, 260f

M

machinal translation, 146
macula neglecta, 530f, 657, 660f
maculae, *see* ear
Mader analysis, 383f
magnesium and the audiogenic seizure, 256
magnetic recording, 34f
magnetostrictive transducers, 50*
Malaya, 671, 673
malleus, 535f, 665
man-animal barrier, 559f
mandible in stridulation, insects, 279, 329f, 333
mandibular arch, 535f
mandibular noise, stork, 714
marine sound, 106, 219f, 656f, 681f, *see also* underwater sound
   acoustic damping of, 49*
masking, aural, B.S. 661/3023: the increase expressed in decibels in the threshold of hearing
   of the masked sound due to the presence of the masking sound
mastery, 576
matching, *see* feedback
maternal sounds, *see* song, calls and call-notes, categories – family
mating sounds, *see* song, etc., categories – courtship
maxilla in stridulation, insects, 279, 339
maximal diversity, 124
maximum information, 116*
maze, 116
Mc/s = megacycles per second, *see* frequency
meaning, 133, 517
   lexical, 148*
   objective, 151
   of a word, 134, 137
measurement accuracy and microphone 25
         of signals, 37f
mechanical sounds, *see* sound, biological
"mechanism" of sound production:
   (1) the *apparatus* by which the sound is produced, e.g. strigil, plectrum, etc.;
   (2) the *mechanics* of its utilization, e.g. friction, buckling, air-streams, etc. without refer-
      ence to any relationship between this and the character of the sound;
   (3) the *process of translating* the physical and biological properties of the apparatus and
      its mechanics into the physical parameters of the sound (e.g. tooth impacts per second
      into pulses per second).
  These three meanings are quite distinct, and usually "mechanism" only means one of them; but at different times it is used in any or all of the three senses. This appalling confusion can only be cleared up if we say *apparatus* when we mean apparatus *(see therefore*, phonative apparatus, sound-production apparatus); *mechanics* when we mean mechanics – or possibly just *phonation, sound production*, as they stand *(which therefore see)*; and for sense (3), to avoid repeated circumlocution, use a neologism, for no single word already exists. The best candidate seems to be *phonomorphosis*, defined as in (3); it is a word neither ugly, complicated, hybrid, nor irregular, and has a clear etymological relationship to its utilization (phonomorphogenesis might indeed better suit the pedant, but its length eliminates it). *See, therefore*, phonomorphosis.

If these recommendations be accepted, *mechanism* itself will remain as an invaluable umbrella term for those contexts where more than one of the three aspects are to be encompassed by a single word (*as e.g. in the discussion under the entry*, phonation) "mechanism" of stridulation: the same criticisms apply. *See therefore the above references, plus* stridulation; stridulatory apparatus; strigil; strigilation

mechano-receptors, 583
  rôle in audition, 338
Meckel's cartilage, 535
medulla, 562, 566
medullary yelp, 564
mel, unit of pitch, *q.v.*
mellow and strident sounds, 166
melodic figure, *see* figure
melody, acquired, 237
membranes, vibrating, *see* phonation; sound production
memory, 63, 105, 573, 576, 720, 733
  Penfield's work, 573
mental importance of sound signal, 561
mentality *(cybernetics)*, 126
meow, 764
mesencephalon, 561f, 574
mesonotum, *see* mesothorax
mesopharynx, man, 162
mesosoma in stridulation, Chelicerata, 320f
mesosternum, *see* mesothorax
mesothorax in stridulation, insects, 281, 397
message, 122*f, 134f, 147*, 174*, 557
metachronous anaphony, 610*
meta-language, 150*
  objective, 151
metanotum, *see* metathorax
metaphysics, 557
metasternum, *see* metathorax
metathorax in stridulation, insects, 282, 291, 403
metronome, response of alligator to, 104
               frog to, 65
mewing, 220, 754f, 769
  dolphins, 220
  homology, 780
MHz, Megahertz, megacycles per second, *see* frequency
miaow, 764
microclimate, effect on amplitude of signal, 350
microphones, 25f, 50
  characteristics and conditions of use, 29f
  sensitivity, 51*
microphone-source distance, 32
microtymbal, multiple, of arctiid moths, 800
micturition in the audiogenic seizure, 244, 254
mimic exaggeration of emitting posture, 775
mimicry, Batesian acoustic, 85
       in insect flight sound, 387f

mimicry, mocking, *see* psittacism
    pulse, 9*f
    species-specific, in bird song, 238, *see also* psittacism
"mirror", tettigonioid, minor acoustic rôle, 366
misunderstanding, mutual, 149
mobbing behaviour, 723, 741
mocking, *see* psittacism
models, theory of, 112
modiolus, *see* cochlea
modulation, various types, 17*, 354*, *see also* physical characteristics
    of vocal apparatus, 136, 494, 697
moment, thickness of the, 102*, 122
"monaural", *deprecated hybrid for* one-eared *(q.v.)* or single-eared; *properly*, either uniaural *(in contradistinction to* binaural, *for hearing); or* platyphonic – *or at worst,* monophonic – *(in contradistinction to* stereophonic, *for reproduction)*.
mono-amino-oxidase inhibitors and the audiogenic seizure, 259f
monocyclic, 19*, 22*f
  pulse, response to, 427
monofactorial inheritance, 229
monopulsate, *illegitimate, see* unipulsate
monosyllabic communication system, 161
    sounds, 711, 753f
        (motor-defined syllable) 763* (*see also* song, calls and call-notes, categories), 359 ("one-pulsed")
mooing of cow, 56
Morgagni, ventricle of, 511, 517
morpheme, 142*, 167*
morphological determinants of sound form, *see* phonomorphosis
morphology of animal sound, Part IV, pp. 275f, *see also* phonative apparatus; sound-production apparatus; stridulatory apparatus
  and vocal behaviour, 229, 232
Morse code, 115
motivation, stimulus-specific, 72
motive (*music, preferable to* motif): a figure, group of figures, or a phrase which has significance or is otherwise characteristic of a section, phrase, sentence or period. During the succession of the latter, a motive recurs, varied or unvaried, and therefore can give the whole succession – *song* – a cyclic impression (Sotavalta).
    Armstrong uses the term to describe the acoustic individuality of one, or perhaps two or more phrases. Thorpe uses it but little.
    The differences are of degree, and not substantial at that.
motive, 715f
motor-correlated terminology, 20*, 22*f
motor messages, 557
mounting song, *see* song, calls and call-notes, categories – courtship
mouth imitations of animal sounds, 15
  as artificial stimulus, 613f
mouth in phonation, baboon, 162
        bat, 190f, 207
        mammals, 752
        man, 162
        moth, 323f

See also phonation
moving-coil microphone, 26
     transducer, 50
multiple message, 113*
multipulsate, 22*f
multipulse, *deprecated for above*
multisyllabic, *hybrid for* polysyllabic, *q.v.*
munching sounds, 132
musculi laryngis, 489f, 494, 498
mushroom bodies, 460f, 476
 physiological evidence for higher psychic function, 465
music, 5, 136, 716
musical form, 5f
   instruments, 9f, 104
 confusing by mutilation of sound, 100f
 imitating by animals, 239
musical notation, 23, 346, 709
    sounds, 698f
    training essential to auditory observation, 375
muted sound, baboon, 153
muteness, birds, 494, 714
 edentates, 494
 marsupials, 494
 monotremes 494
 rodents, 500f
mutilation of messages, effects, 94, 100f, 123, 716
    sound-production apparatus, 365, 378
mutual conditioning, interspecific responses, 88
    misunderstanding, 149
    satisfaction signal, baboons, 153
myoclonus, 247f
myogenic vibration, 377, 398
myotatic reflex, 564

N

nasal appendage, bats, 199, 208f
nasal and oral sounds, 166, 185f
nasal resonance, 161
nasopharynx, 495f
natural, flat and sharp sounds, 166
natural frequency, not the same thing as resonance frequency, *see* frequency.
   period, B.S 661/1020: period of a free vibration.
neighing, 757f, 767
neocerebellum, 567
neopallium, 565f, 569f
neostriatum, 569f, 574
neotenic sound forms, 782
nervi laryngis, 190, 209, 212, 500
"nervous" fatigue *(med.)*, 571
nervous system as unique determinant of bird song, 713

neurochronaxy, 191
neurogenic vibration, 377
neurological problems and sound, 571, 751, *see also* audiogenic seizure
neuromuscular coordination inherited, 229
neurophysiological basis of sound production, 450f
neurophysiology of audition and speech, 560
        and information theory, 124
        reflex, 557
        of self-consciousness, 558
        of sound picture, 573
        suitability of Orthoptera for, 440
neurosis, 132
new methods of study, 102, 655
nociceptive action, 100
node, B.S. 661/2016: a point, line or surface of an interference pattern at which the amplitude of the sound pressure *(pressure node)* or particle velocity *(velocity node)* is zero (or, in practice, a minimum)
noise *(acoustics)*:
  (1) B.S. 661/1003: sound which is undesired by the recipient
      Undesired electrical disturbances in a transmission channel or device may also be termed *noise*, in which case the qualification *electrical* should be included unless it is self-evident. ‖
  (2) Sound without regularities, confused sum of many unrelated sounds heard together. ‖ (*supra*, p. 18*, and e.g. van Bergeijk *et al.*, Waves and the Ear)
  (3) "white" noise, the extreme case of (2), in which the sound spectrum contains all frequencies in more or less equal measure. ‖
noise *(cybernetics)*, 114*, 119*
  background, 32*, 50, 73, 387
  mechanical, *see discussion under entry* sound, biological
    birds, 711
    mammals, 752
  and nervous fatigue, 571
  thermodynamic threshold, 49
  white, *see* (3) *above*
    dazzling bat, 213
non-conditioned interspecific response, 90
non-existence of an algorithm, 141
non-expanding messages, 134f, 148*, 169, 171
non-null word, 144
non-phonetic transcription, 136
noradrenaline and the audiogenic seizure, 258f
normal mode of vibration, B.S 661/1018: a characteristic distribution of vibration amplitudes, among the parts of a system each part of which is vibrating freely at the same frequency. Complex free vibrations are combinations of these simple normal modes.
normal song, *see* song, calls and call-notes, categories – proclamation
nostrils in phonation, bats, 190, 199, 205, 208, 212
note, B.S. 661/9021: *a.* a conventional sign for a tone which indicates its pitch and duration; *b.* the tone itself (*see* tone). ‖
    N.B. The tone may be, and usually is, complex.
    5, 16, 17*, 22*f, 715f

noun, 140f
nucleus ruber, 567
null-word, 144, 148, 178*
numerals, 141
nuptial display, *see* courtship
nursery for monkeys, U.S.S.R., 150
nycthemeral rhythm: rhythm with a 24-hour repetition period. *The best term of several, see discussion under* diel.
   572, 639, 645, 759, *see also* daily rhythm
nymphal *(see also* larval) sound production, 330f, 396, 400
nystagmus, 248

O

objective meaning of animal calls, 151
      series, 148*
objectivist psychology = ethology, *q.v.*
objectivization of empirical subjective analysis, 3
      the subjective, 559
obstacles, *see* echolocation
oestrus, 78
offence sounds, *see* song calls and call-notes, categories – rivalry
oh, 755
olfaction, bats, 201, 212
olfactory channel, 128
      message, 113*
      sentence, 128f
omnidirectional microphone, 33
one-eared orientation, 202
onomatopoeisms, 5, 18, 673, 709
ontogeny of human speech, 171
      sound production 228f, 725f 782
oo, 755, 775
Opien, 106
opisthosoma, percussion sound production, Chelicerata, 328
      stridulation, Chelicerata 317f, 320
optical message, 583
oral and nasal sounds, 166
oral resonance, *see* phonation
ordinary song, *see* song, calls and all-notes, categories – proclamation
organic sounds, 132
organicism, 558
orientation by sound, 681, 684, 695, 789, *see also* echolocation; hearing; homing
orientative signals, *see also* echolocation; physical characteristics; song, calls and call-notes
  baboon, 153
  bugs, 408
oriented reaction, *see* taxis and tropism
originality *(information theory)* 117*f
oropharyngeal chamber, 135
Orpheus, 106

oscillogram, oscillograph, oscilloscope, 38f, 136, 375, 516, 694, *see also* audiogenic seizure
osmoreceptors, 583
otic capsule, *see* ear
otitis, *see* audiogenic seizure
otocysts, 530f, *see also* otic capsule, *under* ear
ou ou ou, 152f
output *(cybernetics)*, 145, 149, 157
oval window, 192, 532f, 545, 659, 705, 737
ovary, right, of bird, and sound production, 714
overflow activity, = vacuum activity, *q.v.*
overtone, B.S. 661/9003: a partial whose frequency is higher than, but not necessarily an integral multiple of that of the pure tone to which the pitch would normally be ascribed
oxygen and the audiogenic seizure, 244

P

pairing, rôle of song in, 718f, *see also* song, calls and call-notes, categories – courtship
palae, corixid, in stridulation, 405
palaeostriatum, 569
palate, soft, 512, 752
palatoquadrate, 535
panoramic frequency analyzers, 107
panurgism, sonic, 88*, 713, *see also* chorus performance
papilla amphibiorum, 531, 705
parabiotic state (Pavlov), 248
parades, bird, 713, 719f
parasite, host-recognition by sound, 387
parathyroidectomy and the audiogenic seizure, 252, 256
parent and young, *see* song, calls and call-notes, categories – family
parietal gnosis zone, 575
pars stridens, 278, *see also* strigil
  abdominal, 282
  cephalic, 279f
  on legs, 290
  quantitative data, 303f
  thoracic, 281
partial, B.S. 661/9002: a pure-tone component of a complex tone. ||, 101
passivity, central nervous, 557
Pavlovian cerebral conditioned reflexes, 557
pecten in stridulation, chelicerates, 320f
pedal motor zone, 575
pedipalp in stridulation, chelicerates, 316f, 319
"peep-bo behaviour" of grasshoppers: reaction especially of Acrididae, in which the insect takes up a position facing upwards on the opposite side of a haulm to the human observer, and maintains this position relative to the observer as the latter tries by sideways head movements to get a view of the insect, 608
peeping sounds, 771
pen oscillograph, 39
Penfield's recall experiments, 573
perceptual meaning of sounds, 573

percipience: conscious grasp, 560
percussion sound production, arthropods, 325f, 332
perfumiers and the language of scents, 128
perilymph, 705
periodic quantity, B.S. 661/1004: an oscillatory quantity whose values recur for certain equal increments of time ‖, 8
period (1) *acoustics*, B.S. 661/1005: of a periodic quantity: the smallest time-unit for which the quantity repeats itself.

By extension, the term *period* is applied even when successive cycles of oscillation are not identical. ‖

The extension means that the term is applied to the time elapsing between corresponding points of two successive members of a series of more or less irregularly periodic sounds. In the last resort, as irregularities increase, the only corresponding points will be the beginnings, hence the ultimate permissible extension of the term for bio-acoustic purposes:

lapse of time between two successive homologous members of a series of emissions measured from the beginning of one to the beginning of the other.

Clearly, however, the concept of period has no meaning outside a repetitive context, and for the definition just given to have any meaning it is necessary to take a mean of the time-lapses between successive members in the whole series, or in a fair sample of it.

(2) *music, applied to ornithology* by Sotavalta: the smallest independent unit of expression: a unit of higher order than the sentence, consisting of one or several sentences which describe the whole expression. Periods are generally coordinated, and separated from each other by a silent interval. A period can be an exact repetition of the preceding period, or be different from it; though it often has a similar structure.

6, 16, 22*f

period, natural: period of a free vibration, *q.v.*
periodicity: neutral term for any count per unit time of a recurrent sound. If frequency be limited in use to the spectral components only, periodicity becomes a useful term for all other repetition rates.
pervitine and song, 643
pessulus, 493
pest control, 387
  by ultrasound, 800
pH, *see* audiogenic seizure
pharmacodynamic drugs and song, 643
pharyngeal modulation, man, 160
pharynx, birds, 492
        as feedback control system, 163
        human, as preadaptive to conceptual thought 171
        rôle in phonation, 162f, 169, 496f, 507
        transients in, 135
phase: the particular point in the cycle that a sinusoidal quantity has reached at the instant of investigation. *See also* audiogenic seizure; *and* phase difference, *infra*.
phase difference, B.S. 661/1013:
    (1) between two instantaneous values of the same sinusoidal quantity: the fraction of the whole period that elapses between their occurrence;
    (2) between two sinusoidal quantities having the same frequency: the fraction of the whole period that elapses between the occurrence of an instantaneous value of one

and the instantaneous value of the other at the corresponding point of the cycle. ‖

By natural extension, the concepts both of phase and phase difference are applied to non-sinusoidal periodic quantities, even such as the left and right leg movements in locomotion or stridulation. These are *symphasic*, or *in phase*, when there is no phase difference, and the legs are moving in perfect unison; and *antiphasic*, or *fully out of phase*, when the phase difference is half a period and the legs are moving in exact antithesis.

phased learning of chaffinch's song, 235f

phon: a psycho-acoustic unit of equivalent loudness, the need for which arises from the different sensitivity of the ear to different frequencies.

The loudness level in phons of a pure tone is the intensity (in dB above reference level) of a 1000 c/s pure tone assessed (as the modal value of judgments by "normal" observers) as being equally loud.

phonation: a common definition is *the production of sound by voice;* which presents few difficulties if discussion is limited to the higher vertebrates. It will be clear, however, from study of this treatise, that there are many other mechanisms in the animal kingdom where vibrating membranes cooperate with resonance cavities of some sort to produce what are essentially vocalized sounds in a manner essentially analogous to that of the vertebrate larynx and its accessories. There is therefore no real justification for limiting the use of the term to mechanisms of the higher vertebrate type. Nevertheless, in deference to common practice, and perhaps common sense, references relating to what is commonly accepted as vocal production have been entered under *phonation*, while other mechanisms have on the whole been entered under *sound production, stridulation,* or *strigilation*. Some overlapping is inevitable.

See also central nervous system; resonance; song, calls and call-notes.

in various vertebrates, Chapter 18, p. 489f:
amphibians, 490, 694f, 803f
baboons, 133, 136, 162, 169
bats, 189f, 207, 212
birds, 491f, 710f
donkey, 516
fish, 489
frog, 697
gibbon, more advanced than man in some respects, 509f
mammals, 494f, 512f, 762f
man, 133, 162, 513
reptiles, 490f
seals, 516

general aspects (mechanics), *see also* sound production; stridulation; strigilation
articular speech, 513
double voice, 494, 499, 509f, 512
factors affecting, *see* song, calls and call-notes
feedback, *see general entry thereunder*
auditory, unimportant to bird phonation, 738
female voice, 695
inspiratory air current, 516
modulation, 697
motor template, 233
neurochronaxic theory, 191
neuromuscular theory, 191
parallel reduction in amphibian lines, 695f

phonation, *general aspects, continued:*
    phonative motorization centre, 565
            sensitivity centre, 565, 573
    and phonetics, independence, 783
    phylogenetic aspects, 765, 777f
    respiratory motorization, 563
    "singing" of gibbon, 509f
phonative act, 3
phonative apparatus, *see also* air sacs; epiglottis; larynx; pharynx; sound-production apparatus; stridulatory apparatus
    air bladder, 224, 489
    laryngo-pharyngo-oro-nasal junctions, importance of 513
    mouthpiece, 752, 775
    muscles, 563
        lingual, palatal and nasal in man, 513
    mutilation of, bats, 190f, 212
    primary vocal structures, 695*
    resonant tube of mammal, 513, 752
    resonators, air sacs as, 517
        oral cavity as, 135, 517, 752
        pharyngeal cavity as, 135, 752
    respiratory motorization, 563
    secondary vocal structures, 695*f
phone: umbrella term for a simple vowel or consonantal sound, 5, 14
phoneme: (1) Prague school: functional unit of language realizable as a sound.
    (2) van Bergeijk *et al., op. cit.:* the smallest unit in a language that distinguishes one utterance from another.

    Some of the phones in a language are also phonemes; for example, *t* and *sh* in the word *tush* (note that the phone *sh* is rendered in English by two *written* consonants, though it is itself a single consonantal sound). But the phone *ch* in *church* is not a phoneme, since it can be divided into the two already considered: *t* and *sh*.

    A phoneme is however a theoretical concept, which in practice is spoken in different ways by different speakers, and by the same speaker in different verbal contexts. For example the *t* sounds in *ate* and *eight* are not distinguishable, but when the *th* sound is added to the latter (a written *t* is lost in the process, but the *t* sound is not) to make eighth (ate-th), the *t* sound changes a little, losing some of its crispness; and the Cockney, saying: "'e didn 't oughter 'ave ate it", makes all the *t* sounds by means of a glottal stop instead of a dental stop: yet all these sounds are identifiable to an Englishman as belonging to the class *t* sound. In this sense, *phoneme* is the name for the class, and all its variants are called *allophones* of the phoneme
    (3) an incomplete speech-sound (*supra*, pp. 167, 174*). This connotation is a little difficult to integrate with the more classical concepts outlined above.
    5, 14f, 124, 133, 167, 174*
phonemes, human, in animals, 516
            synthesized in the cerebral cortex, 574
phonetics, the science of spoken sounds, 5, 14f, 20, 63, 136
            comparison of human and parrot voice, 733
            independence of phonation, 782
phonokinesis, 103*f, 122, 535, *see also* kinesis
phonomorphosis: *neologism for* the process of translating the physical and biological properties of the sound-production apparatus, and its mechanics, into the physical

GLOSSARIAL INDEX 871

parameters of the sound produced; the *shaping* of the sound (including its spectrum) by its generative machinery. E.g., determination of frequency by properties of resonating cavities etc., of rhythm pattern by rates of movement and dimensions of moving structures etc.
  For discussion and justification, see entry: mechanism of sound production.
acoustic couplers, rôle of, 366
body size, rôle of, 703
damping in, 397
detailed, in stridulating acridid, 445f
             cricket, 441f
        calling frog, 696
frequency generation, 362f, 375f, 397f, 407, 444, 697
inversion of tegmina, effect of, 360, 480f
mirror, tettigonioid, rôle of 366
primary generator, rôle of, 19, 362f
pulse generation, 364f ("impulse"), 397f, 407, 697
resonance, 19, 364f, 397, 407
resonant membranes, rôle of, 364f, 397
         air-columns, rôle of, 19, 397
of stridulation in arthropods, 360f
vocalization, 19
phonoreaction, any form of reaction to an acoustic stimulus, 64, 89f, 97, 104, 122, 535, 610, 680, 685
phonoresponse, 64, 97f, 104*, 610. If Busnel's definition (p. 104) as a response invariably involving *emission* of sound is to be respected, the rather natural tendency to use the term for *any* response to sound is to be avoided (*see* phonoreaction)
  interphyletic, 321
phonotaxis, 89f, 105*, 122, 618f, 777, *see also* taxis and tropism
  artificially stimulated, 620f
  factors in, 619f, 777
  intergeneric, 624f
  interspecific, 622f
  in mosquitoes, 386, 619
  sexual, 76f, 386
  specific, 618
  stages, 618, 621
phonotropism, 618, *see also* taxis and tropism
phosphorus and the audiogenic seizure, 252
photocell in frequency determination, 375
photoelectric recording (photographic recording), 34
photography, 374f
phrase, 3, 5, 12*f, 22*f, 359, 715*, 761: an entity formed by one figure or by several figures or groups of figures belonging together. Can frequently be subdivided into *sections*, which consist of one or several figures, and each of which expresses an essential part of the phrase. Even a phrase as such can still be incomplete and unbalanced.
phylogeny of human speech, 171   [(Sotavalta)
physical characteristics of emissions, *see also* amplitude; echolocation; frequency; intensity; modulation; phonomorphosis; pulse; song, calls and call-notes, categories; systematics; transient.
  in various groups:
    Amphibia, 697f

physical characteristics of emissions, *in various groups, continued:*
    arthropods, 346f, 350f, 354f, 373
    bats, 187f, 199f, 205, 207, 209, 212
    birds, 217f, 715f, 735f, 738f
    fish, 489, 680t
    Hemiptera, 391f, 397f, 407, *and under* insects
    insects, 346f
        flight sound, 382f, *see also* frequency, wing-beat
    mammals, 752f, 768
    Orthoptera, 447f, 449f, *and under* insects
  general aspects:
    atypical patterns on electrostimulation, 465f
    based on biological events, 11f, 356f, *see also* pulse
    for concealment, 741f, 769
    correlation with morphology, *see* phonomorphosis
    of directional, locatable signals, 740f, 768f
    for echolocation *(which also see)*, 745
    for effective decoying, 743f, *see also insect references above*
    for extraspecific communication, 391
    factors affecting, 350, 353, 377f, *see also* phonomorphosis
    geographical variation, 698f
    highly interspecific signals, 743f
    for intraspecific communication, 391
    of locatable signals, 740f, 768f
    musical transcription, 22f, 346
    notation of, 346
    pulsed character in Hemiptera, 406
    for self-concealment, 741f, 769
    of subsong, 729, 769
    typical patterns on electrostimulation, 465f
physical events and biological events as criteria of analysis, 11f, 356f, *see also* pulse
physico-chemical criteria of signals, 69
physiological determinants of sound form, *see* phonomorphosis
        state and sound production, 448f
physiology of animal sound, Part IV, pp. 275f, *see also* phonation; sound production; stridulation
piano scale, 9f
piezo-electric microphones and transducers, 26*, 50*
Pimonow analyzer, 41
pinna, focussing in bats, 192, 202
piping of queen bee, 325, 374, 386
pitch, B.S. 661/9004a: that subjective quality of a sound which determines its position in a musical scale. ‖

    For *pure tones* heard by human beings, the relationship of pitch to frequency is found to vary with frequency in two complementary ways:
        (1) high in the auditory range, an octave rise in frequency appears more than an octave higher in pitch, and low in the auditory range, an octave rise in frequency appears less than an octave higher in pitch;
        (2) increasing the loudness of a high tone raises its pitch, increasing the loudness of a low tone lowers its pitch.
    By using the modal values of subjective judgments by "normal" subjects, of

equal pitch intervals in varying parts of the scale, a pitch/frequency graph can be constructed *(see below)* analogous to the loudness/intensity graph on which the *phon* is based. The unit here is the *mel*: a pure tone of 1000 c/s at 40 dB above threshold has a pitch of 1000 mels; and the pitch of any sound judged to be n times that of 1 mel is n mels. More comfort will be got from reading off mels in terms of cycles per second from the graph, than will be got by attempting to make anything from the definition.

(I am indebted to the Librarian of the Science Library, South Kensington, for helping me with information relating to the above discussion.)

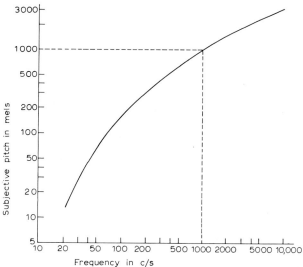

B.S. 661/9004b: the general pitch level of a passage of music
pitch, relative: the capacity to judge musical intervals correctly
    absolute: the capacity to judge the position of a note on the musical scale without any previous standard having been produced
    discrimination, 8
    innate, 237
    of flight sounds, 375f
plastic behaviour, arthropods, 362
plasticity, evolutionary, Hemiptera, 391
play, 777
play calls, 755
playback to experimental animals, apparatus for, 44f
    experimental, attracting birds, 724
        fish, 667
        frogs, 706
        discrimination, 804
        groups studied, 76f
        response, 683, 685, 716f, 771, *see also* attracting, *above*
pleasure calls, 153, 774
plectrum, 278*, 391*, 397, 401f, 798
plica ventricularis, 498f
plicae aryepiglotticae, 495, 498

plicae vocales, vocal cords, *see* larynx
pneumatic duct, 499
   spaces, 517, *see also* air sacs
poetry, 5, 14
poikilochronous anaphony, 610*
polycyclic, consisting of many cycles, *cf.* monocyclic, 22*f
polygenic control of vocal behaviour, birds, 233
polynucleotide chain code, 137
polypulsate, *inadmissible hybrid for* multipulsate
polysyllabic sounds, 357, 753f, 761f, 764f, *see also* song, calls and call-notes, categories
pons cerebralis, 460f
     Varolii, 562f, 574f, *see also* central nervous system
         as feedback centre, 566
pop, Crustacea, 309
population density and signal properties, 73f, 587
portamento-glissando, 516:
   portamento is defined as a transition from one note to another higher or lower without a break. It is only possible on such instruments as voice and violin, where there are no fixed stopping places on the scale;
   glissando is a rapid run up or down the scale of an instrument with fixed keys or positions, such as piano or trombone
postclavicle in stridulation, fish, 678
potassium in the audiogenic seizure, 252
"potential danger" signals, *see* song, calls and call-notes, categories – protection
pouncing insects, actography, 59
practice song, *see* song, etc. categories – subsong
Prägung *(G.)* = imprinting, *q.v.*
predator, mobbing, 723, 741
predator-prey acoustics, 90, *see also* prey; song, etc. categories – protection
prefrontal region, 561, 569, 573
pre-mating song, *see* song, etc. categories – courtship
premotor area, 574
present, thickness of, 102*, 122
"presentation" (rump) 153
pre-spawning sounds, 682
prey, pursuit of audible, 196
    rates of capture by bats, 196
Preyer's reflex, 103*, 535
preying by echolocation, 196
   posture, 197
principal frequency, 6*, 349, 697f, 803, *see also* physical characteristics
prise-de-son *(F.)*. The nearest English rendering is "recording of the sound", literally "capture of the sound", that is the act of take-up by the microphone. No accepted English expression means exactly this and nothing else, a serious deficiency; "recording of the sound" (p. 25, 32) has a much wider connotation and would lead to confusion if generally used; "act of recording" is little better.
    One might propose *sound-uptake;* or we might do better to use the French expression itself.
    25, 32
probabilities of occurrence *(information theory)*, 120
proboscis in stridulation, pupal, 332

procesuss medialis, 498
vocalis, 498
prolegs, anal, in sound production, 332
pronotum, *see* prothorax
pronoun, 141
propagation of sound in water, 48f
prosencephalon, 568f
prosoma in stridulation, Chelicerata, 317f, 320
prosternum, *see* prothorax
protection against birds, 85f
prothorax in stridulation, insects, 280f, 290, 292, 404
protocerebral bridge, 460f
protocerebrum, 460f
prototympana, 436
psalms and hymns, relation of melody to rhythm, 17
pseudolanguage, monkeys, 106
psittacism: parrot-language, imitation without apprehension, 569, 733*
    success after single hearing, associated with extraordinary emotion, 733
psycho-analysis, 572
psychology, classical, and sound, 557
               and cybernetics, 119
psychophysiological receptivity, 105
pt pt, 153
puffing, 753
pukat payang (fish listening), 673
pulsatile signal transformed to harmonic signal, 10
pulse, 7*f, 18*
  American Standard, 11*
  as a biologically-defined parameter:
        It is shown in Chapter I (p. 7–12) that there is no justification for defining a pulse in terms of a biological event (such as wing-stroke or leg-stroke) since the sound thus produced may, and usually does, contain many pulses in the correct sense in which the word is used in physical acoustics (see p. 18).
        The error becomes peculiarly apparent when the oscillogram of one of these "biological pulses" is examined; the author is then obliged to invent some new term, or misuse another (such as *impulse*, *q.v.*) for those constituent elements that the physicist would call *pulses*.
        A particular difficulty arises in translating from French, which apparently contains no word equivalent to our pulse (*see* Impulsion).
    7*f, 18*, 356f, 360f
  in its correct sense as a physically-defined parameter (which may, of course, *happen* to correspond to a single biological event)
    by bat and marine workers, Möhres, Pringle, Thorpe, 8*, 18*
    by Alexander, 357*, footnote
    by Blair, 698f
    by Dumortier, 591
    by Leston and Pringle, 392, 406*f, 798
    by Moulton, 674f, 684
    by Vincent, 187f
  chordal, 9*f
  "false", 10*

pulse, *continued:*
  impact, 9*f
  information, 392
  mimicry, 9*f
  modulation, 17*, 406
  myogenic, 398
  neurogenic, 398
  pattern, 391
  repetition rate, 6*, 356*, 391f, *see also* physical characteristics
    as releaser, 616
    change of, during echolocation, 199, 211
  secondary, 10*
  square, 7*, 11
  transient, 398, 400
pulse-train, 8, 11, 22*f
  figured, 22*f
pupal sound production, 330f, 434
pure tone, *see* tone
purines as letters of an alphabet, 137
purposeful sounds, *see* sound, biological
purring, 757, 769f
pygidium in stridulation, *see also* abdomen
  Diplopoda, 321
pyramidal vocalization centre, 573
pyridoxine and the audiogenic seizure, 256
pyrimidines as letters of an alphabet, 137

Q

quadrate, 535
quantification, 112
quantifier of existence, 148*
           generality, 148*
quantity of information, 116*
quantum *(information theory)*, 115*
queen bee, dialogue, 386
           piping, 325, 374, 386
questionnaire on terminology, 4, 11, 24
quinine and song, 643

R

r.m.s. value, B.S. 661/1014: root mean square value of a varying quantity: the square root of the mean value of the squares of the instantaneous values of the quantity. In the case of a periodic quantity, the mean is taken over one period.
rain as releaser, 703
range, auditory, *see* sensitivity. Compass, *q.v.* is a more precise term
           frequency, of emissions, *see* sound spectrum. Compass, here too, is a more precise term than range
           topographic, of signals: the distance at which a signal suffers a given loss of intensity relative to its initial intensity (Busnel), 73f, 407, 584, 678, 753
ranking messages, 130
rapping, 761

rarity of signals *(information theory)*, 127
rasping, Crustacea, 308
        whales, 220
rate, repetition: a count per unit time of sound rhythm components; used instead of repetition *frequency*, the term serves to emphasize the distinction between these modulation phenomena and the elementary waves themselves, whose frequencies form the sound spectrum. *See also* frequency, pulse repetition rate
Ratterlaut, 187
rattle, Crustacea, 308
reaction to sound, *see* releaser; phonokinesis; phonoreaction; phonoresponse; phonotaxis
reactogenic signal factors, 91
        value, 105
reading centre, 573
real series, 147*
reassurance call, *see* song, calls and call-notes, loosely used concepts
receiver, recipient *(cybernetics)*, 114*, 117, 147, 337 ("receptor"), 710
reception of sound, *see* hearing; kinaesthesis; vibration
receptors, tuned, 392, 400
recessus pharyngo-laryngeus, 495
recognition, individual, 82f, 720f
        parent-young, 720f
        rivalry song functioning for, 602
        of sex, 721
        signals, 82, *see also* song, calls and call-notes, categories.
        of sound of a particular boat by grebe, 720
        species, 622f, 706
        specific, elements of, 685, 715, *see also* releaser
recording, act of, 25, *see also* prise-de-son
        conditions and technique, 25, 153
        as means of storage, 33
rectangular pulse or wave, 7*
rectification, 9*, 55*
rectrices of humming-bird snapping, 78
red nucleus, 567
reduced song, diminished song, *see* song, calls and call-notes, categories – courtship
redundancy, 113*, 116*f, 119*, 130*, 147
reflection, dangers of, 10, 32, 353, 397
        internal, in small emitters, 10
reflectors, parabolic and other, 32
        tracheal air-sacs as, 436
reflex, classical, 559
        epilepsy, 247
refraction, effect on amplitude, 350
regional variation, 107, *see also* geographic variation
regrouping call, 741
Reissner's membrane, 533
release call, frogs, 705, *see also* song, calls and call-notes, loosely used concepts
releasers, acoustic, 18*, 91, 570, 614f, 622f, 703, 777
relevant hearer concept, 3, 14*, 407. [Note that this can include the emitter itself.]
repertoire, 70f, 112, 114, 116, 117*, 120*, 130, *see also* vocabulary
   birds, 711

repertoire, *continued*:
  rank (Zipf's Law), 128*
  size difference inherited, 237
reporting, 557
reserpine and the audiogenic seizure, 258f
resistance, *see* audiogenic seizure
  air, *see* frequency, natural and resonant
resonance, B.S. 661/1022: a condition resulting from the combinations of the reactances of a system, in which a response to a sinusoidal stimulus of constant magnitude reaches a maximum at a particular frequency. ‖
  chamber, 10, 135, 158f, *see also* air-bladder; air sacs; phonation; sound production
  frequency, *see* frequency
resonating underwater sound, 672
resonator, abdominal, of Cicada, 392, 396
          Heteroptera, 798
    air sacs as, 500, 517, 665f, 676f
    tracheal air sacs – or reflectors?, 436
    tegmen as, Orthoptera, 363f
respiratory rate, bat, 191
      rhythm and the audiogenic seizure, 244
      tract, bat, 189
      vocal movements, Anura, 103
responding song, *see* song, calls and call-notes, categories
retarded audition (Lee effect) 125*, 575
reticular formation, 558, 565f
  and appetitive behaviour, 571
reticulated substance and the audiogenic seizure, 258, 261
retropalatine epiglottis, 495
retrovelar epiglottis, 495
reveille, bumble-bees, 388
reverberation, 353, 397
  chamber, 10, 19, 135
reverberation, sensitivity to, 745
reversal, song, *see* cancrizans
rhinencephalon, 568f
rhythm, acoustic:
  acquired, 237
  audible ground, 16*
  and melody, independence, 17
  patterns and different ears, 8
      as releaser, 616
rhythm, biological, *see also* annual; daily; diel; nycthemeral; sexual
  hormonal, actography of, 61f
rhythmic figures, 18*f
ribbon microphone, 26
ripple: term proposed (p. 8*, 16*f, 22*f) as umbrella for both *trill* and *tremolo*, especially useful when it is not known whether or not frequency modulation is present
ritual, acoustic, 16
ritualization of defence movements as stridulation, 609f
ritualized behaviour in Orthoptera, 597f
rivalry, 79f, *see also* song, calls and call-notes, categories

rivalry tremulation in Gryllids, 603
roaring, 491, 757f, 764, 768f
Rochelle salt transducers, 27, 50
roll call in baboons, 152
roosts, actography, 57
        abandonment after playback of desperation (distress) call, 86
rostrum in stridulation, Crustacea, 308
                     insects, 280f, 292, 404
round window, 192, 531f, 536f, 545f, 737
rovescio, al, *see* cancrizans
rumble, 757, 764
   elephant and rhinoceros homology?, 781
running phase, fit, *see* audiogenic seizure
Russian nouns, 140f

S

saccule, sacculus, 529f, 656f
saccus clavicularis, 493
saccus pneumaticus, *see* air sacs; larynx
safety call, reassurance call, before moving, mice, *see* song, calls and call-notes, loosely used
Santorini cartilages, 513   [concepts
satisfaction signal, baboons, 153, 774
saut, cri de *(F.)* = mounting call, *see* song, etc. categories – courtship
scala media, 533f
       vestibuli, 533f
scales, musical, 8f
scanning speech, 567
scaphium, 665
scare rope in fisheries, 670
schooling, fish, and sound, 682, 687
scoloparia, scolopidia, scolopophorous organs, 408, 412f, 416, 420, 422, 435f, *see also* chor-
scolops, 412   [dotonal organs
scopolamine and song, 643
scraper, 278*, 441*, *see also* plectrum
scraping sounds, 332, *see also* stridulation
   aphids, 401
   mammals, 771
screaming, 753f
sea, *see* marine; underwater
search for lost conspecific, baboons, 152
"searching" (cleaning) in monkeys, 151
searching sounds, *see* song, etc, categories
seasonal rhythms 90, *see also* annual *and* sexual rhythms
   actography, 61
seawater, *see* marine; underwater
secret language, 114*
section, amplitude, 43 (fig. 27)*, 675, 702
       musical, *see* phrase
sedentariness of Orthoptera, 587
segmentation of sounds in evolution, 783
selective actography, 57, 60

880 GLOSSARIAL INDEX

self-awareness, self-consciousness, 557f
self-control, 100, 576
self-image, self-picture, 573, 575f
semantic, semantics, 14, 69, 716
semantic interspecificity, 87f, see also extraspecific; interspecific; and song, calls and call-
    notes, categories
semicircular canals, 656f, see also labyrinth
sender, see emitter
Senegalese fishermen, sound decoys, 86
senescence, see audiogenic seizure
sensation, conscious, 557
sensitive periods in vocal development, 240f
sensitivity, auditory, see also hearing,
    bats, 193, 789
    birds, 736f
    fish, 667
sensory gnosis centre for speech, 575
        messages, 557
sentence *(linguistics)*, 15f, 125
        (music, and birds), the smallest unit of form that is independent of its context.
        May consist of one or several phrases. (Sotavalta). 15f, 22*f
    stereotyped: such songs as the Wren's and the Chaffinch's, consisting of several phrases
        olfactory, 128f   [(Armstrong)
separate sound, 137
sequence, 16*, 359f, 447f, 753f, 764f
    arising by addition of short, or subdivision of long, sounds, 761, 767
    homotypic and heterotypic, 764f, see also song, etc. categories.
    long and short, 761
    rate, 447
series, 5
serotonin and the audiogenic seizure, see 5-hydroxytryptamine
serum proteins and the audiogenic seizure, 252
sex, see also audiogenic seizure; song, calls and call-notes, categories – courtship
    recognition, 602
sexual behaviour, autochthonous, variations, 712
            and stage in cycle, 715
            and territory, 717
        dimorphism in sound and sound production, 77, 396, 402, 404, 768, 771
        rhythm, factors determining, 641f
            and song, 639f
        signals, see song, etc. categories – courtship
shake *(music)*, 16*f, 13 ("trill")
sharp and natural sounds, 166
shehek, 764
"short" sound, 753f, 763*
    homology, 781
shouting, 172
sideband frequencies, 355
Siemens analyzer, 40
sign, 116*
signs, conventional, in algorithm theory, 179*
        interchangeable, 138

sign-set, 114*, *see also* alphabet; repertoire; vocabulary
sign-stimulus, 562, 570, 626, 777
sign-stimuli, acoustic, 18*, *see also* releaser; social releaser
signal, *see also* song, calls and call-notes
   animal, justification for the term, 69, 72
   basic, 22*f, *see also* elementary wave-form
   -to-noise ratio, 32
   physico-chemical criteria of a, 69
   of signals, the *word* as a, 171*
   synthesis, 15, *see also* artificial
   -type dependent on population density, 75
   value, 73, *see also* song, calls and call-notes
Silbe *(G.)* = syllable, *q.v.*, 19*f, 22*f
silence of spentness, 639f, 682
silent animals, *see* muteness
"silent" stridulation, 333
simple message, 113*
sine-waves, 6, 101
singer's fatigue, 571
single sound, 753f, 762*f
sinusoidal component, 6
sirens and the audiogenic seizure, 248
      for bird scaring, 86
skin as sound receptor in fish, 656, 663, 684
skull of fish, 666, 686
slapping, 761
sleep centre, basilar, 568, 572
sleep-talking, 575
slope of signal, 96, *see also* attack; transient
small emitters, special features, 10
snapping, shrimps, 306f
snaps, fish, 685
snorting, 757, 762, 769f
social character of song, 588
      interrelations, 704
      releasers, 76, 84, 576, *see also* sign-stimulus; releaser
      tendency in language, 129
sociometric diffusion, 125
soft palate, 512, 752
solitary song, *see* song, calls and call-notes, categories – proclamation
Solomon, 106
somatotopic localization in neopallium, 570
sonabuoy, 687
sonagram, sonagraph, *see* sound spectrograph
sone *(psycho-acoustics)*, B.S. 661/3011: unit of loudness on a scale designed to give scale
      numbers proportional to loudness. ‖
        1 sone = 40 phons *(q.v.)* and a sound of loudness 2 sones is subjectively twice as
      loud (modal values of many judgments)
        Amphibia, 694
        Arthropoda, 323f, 331, 374
        fish, 676
      appendages vibrating, 328
      differentiation of, 334

SONG, CALLS AND CALL-NOTES: (alphabet continued on p. 891)

W. H. Thorpe, in his recent monograph, Bird Song (1961), grants that the title *Bird Vocalizations* would be both more accurate and more objective, because of the anthropocentric artistic connotations of the word *song*. Wisely, however, he avoids the *reductio ad absurdum* of the ultimate unattainable in objectivity, and sticks to the familiar word despite its disadvantages, justifying his choice with some success. In passing, he designates *Bird Vocalizations* as "almost the only practical alternative."

The learned contributions of my colleagues in this book show clearly that, so far as animal sound in its entirety is concerned, even this alternative must also be rejected as being no alternative at all: for there are many sounds of animal origin which can by no stretch of the imagination be called vocalizations *(see relevant entry)*, and yet are certainly as entitled to be called song as are the rasping vocalizations of crow or corncrake, or the squirts, squeaks and sneezes of some of the Estrildid finches' serenades.

To serve its purpose, an umbrella term for animal utterances must have implicit within it the biological origin of the sound. This rules out such a term as *sound* itself, unless cumbrously qualified every time by *animal*. Even *signal*, in these word-corrupting days of star-signals from outer space, has no particular connotations of biological origin, and has the added disadvantage, as Busnel shows (p. 73) that until an emission is proved to have what he calls *signal value*, either to its emitter or to some other *relevant hearer* (as I have called it, p. 14), there is no real excuse for calling it a signal. Furthermore, the sound of approaching thunder is certainly a signal to the man without an overcoat or umbrella – a signal coming from a source unequivocally inanimate. Therefore we must reject *signal*, and are left with *song* as the only candidate that, in the global field of animal acoustics, comes anywhere near fulfilling our requirements for an umbrella term which will cover calls, cries, creaks, clicks, rasps, hisses and so on, as well as song in the more familiar sense. Song, *thus used, may be defined as:* Sound of animal origin which is not *both* accidental *and* meaningless.

Such an extension of the conventional meaning is properly inclusive *of* it, and does not transgress the tenets by which, in Chapter I, I have suggested the bio-acoustician should be guided in his search for terms. The negativeness of the definition allows inclusion both of an apparently accidental sound which has meaning of some sort in the life of its utterer, and of an apparently meaningless sound which is nevertheless clearly produced by a "deliberate" act. Thus those many "warning sounds" which seem to be merely incidental to other activities but which do appear to have a protective value, are given equal status as "song" with the – plainly directed – alarm and distress calls of birds and other animals; whilst the "ordinary song" of grasshoppers, manifestly not an accidental emission, yet whose biological function remains still in most cases in the realms of speculation, continues to qualify for the term song that has indeed been used for it over many decades.

Within his broad category of bird song, Thorpe recognizes a distinction between a narrower category of *song* ("a series of notes, generally of more than one type, uttered in succession and so related as to form a recognizable sequence or pattern in time"; *see also* Brémond, p. 711, *supra*), on the one hand; and the mostly single or duple *call-notes* (sometimes continued into longer series, but then rarely if ever tailored into any organized sequence-patterns), on the other. This is a useful distinction, provided we recognize, as these authors do, that the two forms represent end-points of a graded series. This in turn means that in the still wider field of all animal sound, we are going to get a certain overlapping at the edges between some author's usages of *call* and others' usages of *song*. We need not get many grey hairs over this; we have, however, to delete the *-note* from *call-note* in this wider field, embracing as it does many calls which are in no sense notes.

It might be worth using a special term for *song* in this still collective but restricted sense, in contradistinction to *calls:* the word *cantus* is ready to hand, has the right connotations, and is already used in this sense by German entomologists (*cantus fortior*, etc. q.v.)

German workers are, indeed, lucky in having two distinct native words for the collective and finite senses, respectively, of song: *Gesang* for *song*, and *Lied* for *a song*. This latter is a third sense in which song is used, defined by Sotavalta as "the whole succession of periods" (see p. 22–3); Thorpe, *op. cit.* and Brémond, *supra* p. 715, treat it as a "succession of phrases". The difference is but one of analytical refinement. In Chapter I, I rendered *Lied* of the German orthopterists by *canto*, but doubted anyway the value of retaining for entomology anything more than the word *syllable* from that set of terms. *Canto* could, however, be very useful in the general field, in this sense of a finite song or *Lied*; and has the right connotations from its Elizabethan naturalization

Synopsis of suggested definitions

Song *(sensu latissimo)*: sound of animal origin which is not *both* accidental *and* meaningless. *Includes* calls and call-notes.

Cantus *(song, sensu stricto)*: a series of notes, generally of more than one type, uttered in succession and so related as to form a recognizable sequence or pattern in time.
(plural, *cantus*
adjective, *cantual*)
Contrasts *with* calls and call-notes.

Canto (a song, *sensu strictissimo*): a complete succession of periods (or of phrases).
(plural, *canti*   A finite piece of *cantus*.
adjective, *cantical*)

Index references

Song *(s.l.)*, *see also* dialects; flight sound; imitation; psittacism

absence from certain forms, *see* muteness.
acquired (partly) in some birds, 238
alien, learning of, birds, 234
autumnal, birds, 714
behaviour, insects, 585f
bird, as primitive music, 716
chaffinch, learning and inheritance, 235
  and display, birds, 714, 716
divergence, without divergence of vocabulary, 742f
and drugs, 643
factors affecting, 638f, 703f, 711f, 721f, 759f, *see also* temperature
factors initiating, 703, 710f
imitation and facilitation, 716
information transmitted, 714f, *see also* general entry, information.
inheritance, birds, 231
island life, effect on variability, 734
an isolating factor, 631f
minimum of variability law, 715f
as mood expression:

consummation calls, *see appropriate categories, infra.*
ease and comfort, pleasure, satisfaction, 153, 774
displeasure, aggressiveness, 753, 774
generalized anxiety, 710, 774
excitement, 710, 753f, 774
play, 755
mutilation of, 94, 100f, 123, 716
non-vocal in some birds, 714
and pairing, birds, 718
post-cantical inhibition, birds, 716
rehearsed, 728*
social character, 588, 717
species-specificity
  birds, 238, 715
  fish, 680
  frogs, 697
stereotyping tendency, 734

Cantus (song, *s.s.*) 444, 448, 494, 711
human, 172, 575

884 [SONG]   GLOSSARIAL INDEX

Canto (a song) 14, 715*
post-cantical inhibition, birds, 716
Calls and call-notes, *general aspects (for categories, see next pages)* 494*, 711*, 721f, 753f
abnormal, 725
centres for, 570

concealment calls, 742
deafening, effect on social call, 233
environment, effect, 722
inheritance of, 231, 494
learning, effect on social call, 235
localizable, 721, 740
migration and, 722

Song, calls and call-notes, loosely used concepts

"Warning" calls

Most of the succeeding sections contain so-called warning sounds of one kind or another. *Warning* is a useful lay term which covers a multitude of shades of meaning; it is thereby the less fit to use as a precision term, and should probably be avoided wherever the state of knowledge allows a more exact word. It has been used by one author or another in all the following senses:

Description	Suggested general term	Indexed in Section: (below)
warning of impending movement, to reassure conspecifics who would otherwise react by alarm behaviour (e.g. ultrasonic cry of woodmice, 88f, 769)	**Reassurance call**	Courtship
warning of ♂ to another ♂ attempting copulation with him, or disturbing him *in cop.* (♂ release call of frogs, cop. disturbance song of grasshoppers, 705, 604)	**Disabusing call**	Courtship
warning to potential rivals, territorial or otherwise	**Rivalry-proclamation**	Rivalry
warning to sighted or heard rival (threat)	**Intimidation call** (Moulton, 681, 684f)	Rivalry
warning of imminent intent to attack (some categories of contact and disturbance sounds)	**Attack sound** (Angrifflaut)	Rivalry
warning of potential danger (e.g. 84; and "trouble-sound" of ♂ *Alouatta*, 772)	**Alerting call**	Protection
warning to defence, arousing only ♂♂	**Mobilizing call**	Protection
warning to the young to be quiet	**Quelling call**	Protection
warning of pressing danger (danger calls)	**General alarm call**	Protection
"warning" to predator, aposematic sense (with or without social appeal function to conspecifics or others for help) (one category of "distress" call, v. infra)	**Desperation call**	Protection

"Distress" calls

(1) the term has been long used for the piping of offspring (or fellows) lost or in other adverse conditions (cold, hunger etc.);

(2) it has been more recently used (Frings, Busnel) for the cry of terror of a captured individual, in the abundant literature that has grown up round the attempts to control bird-roosts by playback of such sounds.

The word is a perfectly apt one for either connotation, but quite different in each. Both Busnel *(supra,* p. 81, 85) and Thorpe (Bird Song p. 19, 21) use it in both senses, though Thorpe tends to use *fear trill,* or *fear squeal,* in preference to *Distress call,* for (2).

Clearly two separate terms, each more precise than "distress", are needed, to express the *kind* of distress whilst retaining the sense of extremity. The following are put forward:
 (1) **Fret call** or **fretting call**   *indexed* family; *or* social
 (2) **Desperation call**   *indexed* protection

"Contact" calls
    Contact has greatly changed its meaning in the last 2–3 decades. From meaning actual touch, it has come to mean any process of establishing relationship; we contact people by telephone, by radio, even by spiritualistic medium, if we happen to be spiritualists. Thus Tembrock's usages, *distant contact* and *close contact* (e.g. p. 768f), are essentially modern, though in view of the ambiguity of the word, perhaps better replaced, in English, by *proclamation* and *proximity* respectively.
    Contact calls in grasshoppers (Kontaktlaute, e.g. p. 603) may mean anything from mild protest at slight "accidental" disturbance by another individual, through *disabusing calls (supra)* to highly intimidating attack calls (Angrifflaute) at the peak of rivalry sequences. It would seem better, as Dumortier suggests (p. 604) to abandon the term; and to replace it in each context by the precise term relating to that context (see succeeding sections).

Mating song, copulation song:
    These terms ought in all justice to be reserved for calls given *during* copulation without external disturbance, such as the continuous stridulation of *Omocestus haemorrhoidalis* during copulation (Jacobs, Verhaltensbiologische Studien an Feldheuschrecken, 1953, p. 105).
Disturbance sounds during copulation *(supra,* p. 604) would best be otherwise described *(see* Warning calls, *supra)*. Index references under courtship, *infra.*

Song, calls and call-notes – Categories

On p. 14 we reviewed the modes and levels in which sound can be analyzed, and on pp. 16f outlined a suggested procedure
 (1) beginning, in classical manner, with empirical (aural) study,
 (2) objectivizing this by correlating the dissected empirical elements with measured physical data, and motor data where known; and
 (3) adding meaning to it by studying the biological context, electrophysiologically or ethologically or both.
We there followed up (1) and (2) in some detail *(indexed:* physical characteristics of emissions) and no further discussion is necessary, except perhaps to record with lively interest that an essentially similar but independent approach in Chapter 25 (p. 761, *supra*) has given an essentially similar framework for the particular field of mammalian sound, differing only in the actual words used. The *single sound* of Tembrock's classification would seem to equate to the *chirp* of the Synoptic Key (pp. 22–3); his *syllables* seem to be unitarily motor-correlated units of the type on p. 23, since he speaks of a monosyllabic single sound as only requiring a "setting adjustment" of the apparatus (p. 763); he clearly makes use of the *audible ground rhythm*, when present, in the same way; his *homotypic sequences*, though not so much defined as described, do seem to equate to *1st-order sequences*, and his *heterotypic sequences* to *2nd-order sequences* in my Key – and I felicitate him on having found more informative names for them; and finally, his *mixed sounds* would appear to include, if not to be *in toto*, what I have called *figured* sounds. This independently achieved harmony of ideas in two groups as far apart as mammals and grasshoppers, is a source of some satisfaction to at least one of us, and I hope to the other.

A terminology of ethologically classified sound was not pursued in Chapter 1 because at the time of writing it did not appear controversial enough to warrant special discussion there. While this remains roughly true, the study of this book and of other material in the interim, shows a substantial need for some sort of concordance between the terminologies used in different animal groups. We have just seen for example ("loosely used concepts") that no fewer than ten radically distinct types of sound have all at one time or another been labelled simply "warning" sounds; and some other similar cases.

On the other hand, despite this superficial appearance of striking variety in terminology, there is already a large mass of common usage upon which to build an integrated system at least provisionally; enough, it seems to me, to make the effort worth while and timely. I therefore attempt, not primarily the ethological classification itself (ably done already in Chapters 5 and 14 and the whole of Part V), but the extraction of the best terminological usages from each of these and others. This does, however, demand some sort of classification: but an overall animalian one which is wide and inclusive enough to allow the plugging of all the particular group classifications into it with a maximum of successful socketry; this in turn has required more subdivision than in Chapter 5, if only in the interests of easy index reference. Otherwise, I hold no particular brief for the classification which follows, *qua* classification.

Proclamation Song

In birds, there seems little doubt that most song in the sense of **cantus** is sexually motivated and thus intimately bound up with reproductive behaviour (p. 714). This seems to be the case in mammals too, whose proclamation calls (distance-contact, p. 768) are in many cases designated rutting or heat calls (759, 764, 768f). Again, in Amphibia, nearly all the calls are freely labelled mating calls (p. 695f).

The only other group well worked in this sense of functional significance is the insects. Here, though there are many examples of song labelled *courtship* (corresponding broadly to close-contact pairing calls of mammals, p. 769) or *rivalry*, with abundant evidence for these interpretations, the uncertainty of workers concerning the interpretation of the large category of proclamation song is reflected in the number of synonyms by which it has been, and still is, known – mostly synonymic renderings of the original German in which it was defined by Faber and later workers: Gewöhnlichegesang, for which Cassell gives *usual, ordinary, customary, trite, commonplace, inferior, common, vulgar, familiar, everyday*. Of these and others, *ordinary, common* and *normal* are in frequent use as a name. From French schools come *calling song* (chant d'appel) and *indifferent song* (chant indifférent). We now discuss these.

Normal seems undesirable because it implies that the categories of courtship and rivalry are in some way *abnormal; ordinary* and *extraordinary* make a similar pair.

Common implies that this form of song occurs more often than others, which may not be the case.

Indifferent, a gallant attempt at objectivity, goes too far, allowing the song no function whatever.

Calling song, on the other hand, in English at least, has connotations far too positively implying *summons hither* to be acceptable before such a summons is proven in a given case – particularly as it may function equally importantly in keeping individuals of the same sex at a distance (as Dumortier points out, p. 587).

Jacobs tried to solve the problem with the new term *spontaneous song*, but this again has undesirable implications of a wholly internal motivation, making it unusable in cases where this has not yet been proved.

When I started this discussion, and before I put a title to it, I had every intention of suggesting as definitive choice one term, out of those already proposed, which should seem to have, on assessment, the least disadvantages (since none were without); and at the same time be a reasonably good fit in the rest of the animal kingdom. But to describe them, I used a term, inadvertently as I thought: proclamation song; and this is a better candidate than any. It makes no assumptions about motivation, about hither-calling or thither–sending (it is in fact the same as *calling*, without the summons connotation – perhaps in this a better equivalent of *appel*? I don't know); the songs in fact *are* proclaimed and differ in this from both definitive courtship and definitive alternation and rivalry behaviour, which are at closer quarters and involve orientation contrasting suitably with the concept of proclamation; and there seems nothing against including, in such a category of proclamation song, the territorial and pre-courtship (distance-contact) emissions of amphibians, birds and mammals – a useful piece of unification.

At this stage, I found my choice was less inadvertent than I thought; Thorpe, indeed, had used the term – qualified by *territorial* – for that particular type of proclamation, and this is presumably where I had picked it up. It is already, then, in use for birds; unqualified, it is clearly quite neutral and therefore suitable for all unmapped sectors of song behaviour, *except alarm and its analogues*. Alarm signals may be *broadcast*, but they are not *proclaimed*; this is however useful, since we set out in the first place to seek a term that should exclude that category *(Gewöhnlichegesang* does not include alarm either). (If we ever need an umbrella term to cover both, *broadcast song* will do).

Index references – Proclamation Song

(ordinary, common, spontaneous, normal, indifferent, calling song) 400, 406f, 447f, 465f,
    585f, 704 *(Hyla)*, 710, 716f, 721, 740, 757f, 764, 768f

Proclamation song, female, 595
               motivation, 587

Responding in general

Responding behaviour of the type designated by Faber as *anaphonic* (p. 610) is written into much of insect ritual as the basis from which more complex reciprocity stems. The same basic type (triggered, perhaps, even, by simple neutral proclamation song) may issue in simple aggregation, in rivalry, or in courtship, even homosexual courtship; it cannot therefore be studied, or regarded, as a mere part of one of these, but as a kind of pool from which they all arise – grading, however, of course, into each of them. If birds (or mammals) ever had a common pool like this, it seems to have gone, and their *antiphony* (p. 82, and glossary entry) seems to be tailored to the specific branch it belongs to (courtship), resembling in this only the more differentiated parts of the insect rituals. Consequently, though *anaphony* itself is a term thus far limited to insects, the terms *alternation*, *conversation*, *exchange-calls*, and *responding* all seem to have wider currencies; all seem to be used without restriction for anything from the most generalized exchanges to the most differentiated.

Index references – Responding

anaphony, 600, 610*
alternation (conversation, exchange), 63, 591, 602, 610, 721, 773f
responding, 407f, 449, 600, 602, 682
synchronism, 610*, 617f

Sexual responding and courtship ritual

The early stages often, at least, operate over a distance (Tembrock's distance voice-

888 [SONG]   GLOSSARIAL INDEX

contact, e.g.), triggered perhaps by a proclamation song that may not even be sexually motivated (in some cases); these stages should perhaps be designated *precourtship* rather than *courtship*, this latter more aptly applied to relations at much closer quarters. *Mating sounds* should surely be restricted to sounds emitted during the act of mating itself, as a natural accompaniment of the act. The very general term *pre-mating*, covering as it does everything before the act itself, has, where used, been difficult to index; I have tried to sort it out adequately by reference to the context, like other terms of analogous generality. I hope I have done this satisfactorily. *See also general entry*, courtship.

Index references – Courtship Responding

(Proclamation, *q.v.* grading into)
Pre-courtship song (pair-formation song, distance mating calls, etc.) 76, 391, 681f, 695, 706, 718, 768f, 803
Courtship responding (premating, proximity calls), 77, 391, 595, 710, 718, 755f, 768f
    greeting (pair cohesion), both sexes: 718f, 753f
    parades, both sexes: 713, 719
    antiphony, both sexes: 82
    male courtship song (often leading to highly ritualized serenade)
      (Werbegesang→ Balz) 400, 406f, 440, 444, 448f, 585, 590f, 686, 695, 704
      meaning of, 596f
      motivation of 597f
        Tettigoniidae, 590, 595
    female receptivity (acceptance or agreement) song, 365, 408, 594f
    (for anonymous and other female songs, see general entry under female)
    special variants or sub-variants in the course of courtship:
      cantus mitior (Verminderte Gesang, Faber; diminished song, reduced song): phase often present in Orthopteran courtship song, reduced in volume and/or other respects, characteristically occurring at the opening of a sequence, 444, 448*
        The same term might well be applied to a similar type of emission in mammals, 769
      cantus fortior: full normal-strength courtship song of grasshopper 444, 448*
      cantus submitior: intermediate type, 448*
      congested song: bird song under the influence of high sexual stimulation, the songs [canti] being repeated rapidly and completely without intervals (Thorpe)
        This is almost exactly paralleled in grasshoppers; in *Chorthippus biguttulus*, the songs come tumbling out in a mad rush of high excitement, almost tripping over one another. Jacobs calls it, in the final phase of high excitement, immediately preceding mounting song and copulation, Verstärkte Gesang (augmented song, or cantus ultrafortior?). There seems no reason why Thorpe's term should not be used here, with, however, a shade wider latitude, possibly also for the climax call of frogs (704) though this is for workers in that field to consider.
      searching song, of insect for partner lost during ritual (Suchlaut) 591, 596
Mounting song or call (Anspringlaut or cri-de-saut): call emitted by ♂ at the moment of intended copulation. Not adequately translated by *jumping-song*, which is highly ambiguous and unspecific to courtship, 400, 440, 591, 596, 599
Mating song, copulation song:
      As already remarked in this section, and under *loosely used concepts*, these terms should be reserved for sounds emitted as a normal accompaniment of the act of mating, during the course of it.
      Though such songs do occur *(see* loosely used concepts, *above)* I find no references here that I can with certainty index under this head; all references to mating calls

have therefore been indexed under the nearest appropriate more general heading of courtship, *above*.

Disabusing calls: (copulation disturbance, grasshoppers; male release, frogs; *see* loosely used concepts, *above*,) 604, 705

Reassurance calls (warning of impending movement)
　　male mouse before approaching female (*see* loosely used concepts), 88f, 769

Territorial responding and rivalry ritual
　　Here again, early stages may operate over a distance, triggered by a proclamation song that may have been territorially neutral (i.e. not "intended" as a challenge); or the sequence may be triggered by other releasers, such as trespass, evoking a specifically territorial challenge at closer quarters. See therefore, *proclamation song* as well as below.

territorial song, territorial rutting calls: 408, 587, 602, 710, 714, 716f, 753, 757f, 764, 768f
rivalry responding, rivalry exchange calls: 21, 63f, 79, 408, 440, 444, 448, 467f, 585, 599f, 603f
　　classifications of, 600
　　relation to spacing of rivals, 601
　　significance, 602
intimidation call: quite the best term; used by Moulton, above, for fish, and applicable to the more advanced stages of rivalry, in which real threat can be recognized. *See also* loosely used concepts, *above*. 604 ("warning" in *Oecanthus*), 681, 684f, 755f, 768
attack call (Angrifflaut): warning of imminent attack, or uttered during attack. Some types of "contact" or "disturbance" calls, *see* loosely used concepts; 602, 604, 753
fighting, quarrelling, combat songs: sounds emitted during more or less prolonged physical combat, usually by both participants; in contradistinction to the attack sounds above, usually emitted by but one participant as the accompaniment to a single attack, which drives off the other. But the rivalry sequence may end either way, depending on the reaction of the attacked animal.
　　360, 448, 603f, 756f
appeasement (inferiority, subordination, submission): the antithesis of rivalry, inhibiting or terminating attack. Usually gestural, but sometimes using sounds. 78, 82, 756

Family responding sounds, 81, 769f
maternal sounds, 753, 764
parent and young, infant-contact, 81, 753f
enticing, coaxing calls, 759, 770
food summons, 82
begging, care-calls, invitation, 81, 103, 334, 718f, 724, 743, 753f, 770f
family cohesion, 770
fret call (distress) 70f, 81f, 152, 770f
play calls, 755

Aggregative responding and social calls, 233, 235
　　Aggregational is not in the O E.D., nor in Webster, 1961. *Congregational* is defined as "pertaining to a congregation". The correct adjectives for "tending towards aggregation, congregation" are *aggregative, congregative*
　　See responding in general, *supra*, for references not specifically aggregative

aggregative, congregative responding 132, 400, 408, 585, 588f, 721, 741
   assembly, 89, 721
   cohesion, 587, 717, 721
   conversation, 721, 773f
   community chorus (sonic panurgism), 88*, 713, 774
   coordination, 83, 721
food summons, 86, 710, 725
flight summons, 721
fret call, 152 (lost individual)
mobbing calls, see next section

Protection, self and social
   The only communicative behaviour that from its very nature cannot include alternation; but it does include one form of anaphony: synchronous (p. 610*) – during mobbing behaviour.
   I include here (as well as elsewhere, to ensure ease of reference), as early stages in the protection sequence, calls of unknown provocation which might be of a protective nature.

simple displeasure calls (reason unknown to observer) 753
generalized anxiety calls (reason unknown to observer) 710, 774
generalized perturbation calls (reason a mild disturbance):
   e.g. some "contact" sounds *(see* loosely-used concepts), disturbance sounds, sounds of mild protest, 331, 585, 753, 798
   Grading at higher levels of excitation into rivalry and fighting, q.v.
danger, defence and defiance calls, various grades, 82f, 151, 157, 391, 494, 681, 685, 695, 710, 753f, 764, 772, 798
   alerting call (potential danger), 84, 769f, 772 ("trouble sound")
   mobilizing call (♂♂ arousal to potential danger) 772
   quelling call (to young to be quiet) 772
   general alarm call (pressing danger) 722
   desperation call (of terrified captive, "distress" call sense (2), *see* loosely used concepts):
      terror, defiance at bay, "warning" to predator: 39f, 85f, 89f, 605f, 685, 722f, 740, 743
      significance, 606
         aposematic, or displacement?, 607
reassurance call, warning of "no danger", *see* courtship above

Raptor's sounds
   Capture whoop, 753f

Cantus incertae sedis – Subsong, juvenile song
   The term subsong, for birds, was originally proposed by E. M. Nicholson, (1927), in the simple sense of quiet song, not reaching far and presumed to have little communicatory function. See Thorpe (Bird Song), and Brémond, *supra*, p. 728, for discussions.
   In some species, subsong is characterized also by indefiniteness of acoustic properties and the semblance of being a practice routine out of which the full song "crystallizes" as development proceeds; and this has led to a tendency to *define* it in terms of its acoustic properties alone. E. A. Armstrong, *in litt.*, deprecates restriction of the word to but a single sub-category of an originally extensive term, in this way; he would retain subsong for the

whole, and in the absence of his preferred criterion – communicatory function – by which to define the part, would rather *qualify* the main term, e.g. developmental subsong. Though Thorpe gives full data on acoustic properties, he seems now to lean more than formerly towards Armstrong's view. Brémond *(supra,* p. 728) frankly equates subsong to juvenile song (developmental subsong of Armstrong). Marler *(supra,* p. 234f) seems to use the term less restrictively and to use *juvenile song* itself as definitive term for the developmental sub-category.

See also physical characteristics of emissions

subsong, 18, 235, 728*f

juvenile song, 234, 711, 725

in many species, not yet recognized, 729

evolution to full song, 730

diminished song, cantus mitior, reduced song (insects). These were carelessly referred to as synonyms of subsong on page 18. They are of course names for a highly differentiated part of courtship ritual, and therefore in no sense synonymous.

---

sonic field and microphone, 25

sonic level, *see* sound level

sonic panurgism, 88*, 713, *see also* chorus performance

sonogenic convulsions, 247

sonograph, sonogram, *see* sound spectrograph

soprano–tenor error, 375*

in Landor's work, 375

sound, *see also* physical characteristics, B.S. 661/1001:

a. Mechanical disturbance, propagated in an elastic medium, of such a character as to be capable of exciting the sensation of hearing. By extension, the term "sound" is sometimes applied to any disturbance irrespective of frequency, which may be propagated as a wave motion in an elastic medium.

Disturbances of frequency too high to be capable of exciting the sensation of hearing are described as *ultrasonic*.

Disturbances of frequency too low to be capable of exciting the sensation of hearing are described as *infrasonic*.

b. The sensation of hearing excited by mechanical disturbance. ‖

These two definitions, designed as working definitions for physical acousticians, have not satisfied bio-acousticians, for whom their anthropocentricity is a drawback. Pumphrey (1940, Biol. Rev. 15· 107) defined it as any disturbance of *low intensity,* in contradistinction to high-intensity disturbances which would be tactile in their effect; but he foresaw the later discovery of unclassifiable intermediate phenomena. (He also limited his discussion to *air,* but only for the purposes of that paper). By 1950 his forecast was materializing, and he then defined sound as any mechanical disturbance whatever that is potentially referable to an external and localized source (p. 524, *supra)*. This, like his definition of hearing, has come under criticism for excessive generality *(see discussion under hearing)*; and Dijkgraaf (1960, Proc. Roy. Soc. B, 152: 51–4) returns to the distinction between *sound* (carried in a *medium)* and *ground vibration* (carried through a *substratum)*, with his definition: **a succession of pressure waves propagated with a characteristic velocity through the medium involved (air or water).** The distinction is again (as in his definition of hearing) arbitrary but practical; the reference to velocity is a useful way of excluding non-acoustic mechanical disturbances.

sound, accidental, *see next entry*

[SOUND]

sound, biological and mechanical, purposeful and accidental (incidental):

Moulton uses *biological sound* for sound of animal origin *(supra,* p. 673). M. P. Fish restricts the term to sounds of importance to the behaviour and life histories of the source species (cited p. 676); for these, Moulton uses the term *purposeful* sounds. For their counterpart, sounds of no known utility to the species, merely contingent on movement and feeding, he uses *accidental* (p. 681), Fish, *mechanical* (cited p. 676). Whilst most such sounds are in fact mechanically produced, there are, as Moulton points out, plenty of other sounds equally mechanically produced, but of undoubted biological significance. Tembrock (pp. 752, 761) thus uses *mechanical* purely in relation to mode of production, and includes in it many sounds of biological significance.

There seems no doubt that the virtual relegation by Fish of all mechanically produced sounds to the category of biological insignificance is unfortunate and confusing; and her restriction of biological sound is equally unfortunate, since all animal sound is *ipso facto* biological, whether it is significant or not.

In the face of these conflicting usages, the safest practice would seem to be to use terms such as *mechanically produced* and *vocally produced (not* voiced – *salve* Tembrock! – because that would relegate the unvoiced whisper to the category of mechanical sound); together with *biologically significant* (or *purposeful* – though the anti-teleologists will rostrospect at this term) and *incidental* (this latter expressing, better than *accidental*, the notion of contingency upon other activities). The distinctions are of course arbitrary: the larynx is itself a mechanical structure, and some apparently incidental sounds may prove to have significance; but they are useful, and to reject them is a counsel of despair.

Index references to mechanically-produced sound *(for vocally-, see* phonation):

Biologically significant	Incidental
676, 681, 752, 761f	676, 685, 695, 752

See also sound production; stridulation; strigilation

---

sound channel, 121*
    significance relative to other channels, 583
  chart, 122
  contour map, 675*
  damping in seawater, 49
  distinctive features of a, 165*, 175*f
  event, 122
  form, determination by generating apparatus, *see* phonomorphosis
    genealogy of human, 751
  harmonic, 752*
  incidental, *see* sound, biological
  intensity, *see also* level; sound level; sound loudness.
    B.S. 661/2031: at a point in a progressive wave: the mean rate of flow of sound energy per unit area normal to the direction of propagation. This is also the normal sound energy flux per unit area in the wave. The unit is the erg per second per sq. cm. [= dyne/sq. cm.], but sound intensity may also be expressed in watts per sq. cm.
    Sometimes used more generally as some quantity proportional to the sound power or energy. ‖ 48*f
  level, B.S. 661/2032: a weighted value of the sound pressure level as determined by a sound level meter ‖ [which is an objective noise meter designed to measure a frequency-weighted value of the sound pressure level in accordance with an American Standard specification].

In ordinary parlance, however, "sound level" is often taken with a less restricted meaning, more or less synonymous with the simple term *level, q.v.*
level recorder, 44*
loudness, B.S. 661/3010: an observer's auditory impression of the strength of a sound
  See also general entries, loudness; phon; sone.
mechanical, *see* sound, biological
picture neurophysiological aspect, 573
production:
  in certain groups:
    amphibians, 694f, 803f
    birds, 711f
    cetaceans, 687, *see also* echolocation
    fish, 657, 671f, 677f, 686f
    mammals, 751f
    sharks, 687
  general aspects (mechanics), *see also* evolution; flight sound; phonation; phonomorphosis; physical characteristics; sound-production apparatus; stridulation; strigilation.
    by airstream, 761, *see also* sound- production apparatus
    asynchronous *(neurol.)*, 377
    breeding and, 682f, 714f, 717
    central nervous organization, 398, 440f, *see also* central nervous system
    chorussing, *see* chorus performance; sonic panurgism
    distribution in fish, 677
    factors affecting, *see* hormones; song, calls and call-notes
    form of sound, determination by apparatus, *see* phonomorphosis
    frequency compass, *see* sound spectrum
    functional significance, 681f, 686, 714f, *see also* song, etc., categories
    gesturing and, 776, *see also* song, calls and call-notes *and* display
    due to handling in air, fish, 681f
    and hearing, correlation, 677, 735
    loudest fish, 680
    marine and freshwater compared, 677
    mechanical, 761, *see also* sound-production apparatus
    myogenic, 377
    natural conditions, 681
    nervous system unique determinant in birds, 713
    neurogenic, 377, 398
    neurophysiological basis, 450f
    noisiest fish, 680
    non-schooling fish, 682
    posture exaggeration in, 775f
    due to shock etc. in aquaria, 676
    synchronous *(neurol.)* 377
sound-production apparatus of animals, Part IV, p. 275f, *see also* larynx; phonation; stridulation; stridulatory apparatus; syrinx; systematics; *and under the various organs*.
  in certain groups:
    Amphibia, 694f, 803f
    Arthropoda, 277f
    Auchenorhyncha, 393f
    birds, 491f, 710f
    cave fish, 224

sound production apparatus, *in certain groups, continued:*
    fish, 224, 489, 676, 686
  general aspects:
    air or liquid stream:
      Amphibia, 694
      Arthropoda, 323f, 331, 374
      fish, 676
    appendages, vibrating, 328
    differentiation of, 334
    dimorphism, sexual, 396, 402, 404, 768
    diversity in Coleoptera, 336
                Pentatomorpha, 402
    evolutionary origins, arthropods, 336
    and form of sound, relationship, *see* phonomorphosis
    gin-trap, in pupae, 331
    homogeneity, Auchenorhyncha, 335
                 Reduvioidea, 404
    of immature stages, 330f
    "internal airstream", insects, 325
                           disputed, 374
    membranes, vibrating, amphibians, 694
                    fish, 676, 678f
                    insects, 323, 391f
    mirror, tettigonioid, rôle of, 366
    mutilation of, in bats, *see* echolocation
               in Orthoptera, 365
    parallelism, Hemiptera, 392f
    and perception apparatus, relationship, 337
    problems, biological, connected with, 334f
    sexual dimorphism, 396, 402, 404, 768
    using substratum, chelicerates, 328
                   insects, 325f, 332, 334
    tymbal, *see general entry,* tymbal
sound
  propagation in water, 48
  purposeful, *see* sound, biological
  significance to fish, 664f, 667f, 671, 685
  spectogram, B.S. 661/1037 (actually "speech spectrogram" in the Standard):
        A graphical representation of the variations with time of the spectrum of a sound. ‖
        For discussion, *see* sound spectrum, *infra*
  spectrogram, spectrograph: 42*, 43*, 107, 122, 375, 674f, 694f, 698, 700f, 709f, 798
  spectrometer, 153f, 759f
  spectrum, B.S. 661/1034: of a complex sound: a representation of the amplitudes of the components arranged as a function of frequency. Depending upon the nature of the sound its spectrum may be a line spectrum or a continuous spectrum or a combination of the two types. ‖
    Line spectrum, B.S. 661/1035: a spectrum of a sound whose components are confined to a number of discrete frequencies. ‖
    Continuous spectrum, B.S. 661/1036: a spectrum of a sound whose energy is continuously distributed over a frequency range. ‖

The sound spectrum is an instantaneous picture of the frequencies emitted at one particular moment (fig. 23, e.g.); the spectrogram shows how this picture varies from instant to instant over a finite time (fig. 24)
  39, 40*, 43*, 122, 136, 176*, 346. *For spectra of particular groups, see* physical characteristics of emissions.
 binary analysis, 165*f
 breadth in arthropods, 350, 392
    birds, 735
    fish, 680
    mammals, 753
   see also echolocation
  change of, in bats, during pursuit of prey, 199
  human and monkey compared, 154f, 165
  influence of generating apparatus, *see* phonomorphosis
  monkey and human compared, 154f, 165
  non-random, 176*
  of random noise, 164, 176*
  seasonal shifts, 680
  sidebands, 355
  superfamily specificity in arthropods, 350
  tonal, 165
  ultrasonic content in arthropods, 350
 speed in water, 49
 storage, 33
 tub in fisheries, 670
 uptake, *see* prise-de-son
 velocity in water, 49
sounding-board, tegmen as, 363
source *(cybernetics)*, *see* emitter
source–microphone distance, 32
spatial message, 114*
    signal, 114*
speciation, 634f, 694
species recognition, 622f, 706
specific recognition, elements of, 685, 715, *see also* releaser
specificity (stenospecificity), 63, 87, 105, 742f
spectral substitution, 166
spectrogram, *see* sound spectrogram
spectrum, *see* sound spectrum
speech, 15f, *see also* central nervous system; language; neurophysiology; singer's; song;
    stammering; thought
  affective, 565
  animal, 15f, 106, 132f
  articular, 513
  automatic, 565f
  deliberate, 565, 572
  education, 563
  gnosis centre for, 575
  human, 15f, 100, 106, 562
  instinctive, 565
  internal, inward, 562, 576f

speech, *continued:*
  meaningless, 134
  ontogeny and phylogeny, 162, 172
  scanning, 567
  sensory gnosis centre for, 575
  sound, complete (= syllable, *algorithm theory*), 175*
  spectrogram: the sound spectrogram of connected speech. Sometimes called *visible speech*
  spectrum frame, 154
  in stage fright, 571
  synthetic, 19
  and thought, important relation, 576f
spentness, silence of, 639f, 682
spermatophore, relation of mass to stridulation, 639f
spermatophylax ingestion and post-copulatory passivity, 642
spikes, 11, 101
spinal cord, 561f
spinal haemorrhage in audiogenic seizure, 256f
spinal nerves and the   ,,   ,,   254
spinal reflexology, 557
spiracles in sound production, 324f, 331, 374, 386
spitting, 753f, 762
    homology, 781
spontaneity, 558
spontaneous song (Spontangesang), *see* song, calls and call-notes, categories – proclamation
"sportsmen", use of bird calls by, 717
Sprenglaute, 191*
squalling, 771
square waves, 11, 95, 101
squeaking, amphibia, 694
            chilopods, 322
            mammals, 769f
squeal, 152f
  duration, 161
  bumble-bees, mosquitoes, syrphids, 386
staccato, 5, 674, 683, 755
stage fright, 571
stammering, 572, 575
stanza, 14, 765, 768f
stapedius, 537
stapes, 535f, 665
statistical behaviour study, 112, 130
            structure of language, 127, 129
stato-acoustic organ of vertebrates, *see* acustico-lateralis; ear; hearing
steady state, B.S. 661/1030: the state which a system ultimately attains under persisting
      conditions
stenospecificity, 63, 87, 105, 742f
  of desperation (distress) call response, 723
stereotyping tendency in bird song, 734
sternohyoid, 493
stethoscope, aquatic, 674
sting in stridulation, scorpions, 320f

storage of sounds, 33
    tape, 37
Störungslaut *(g)*, disturbance sound, *see* song, calls and call-notes, categories
strain, *see* audiogenic seizure
straying and care calls, 770, *see also* song, calls and call-notes, categories
streptomycin and the audiogenic seizure, 251
stretch-sensitive elements (B cells), 435, 437, 800
striated band, arctiid microtymbal, 800
strident and mellow sounds, 166
stridulation: the term means literally the *production of a creaking sound* (L. *stridulus)* and nothing more.

    O.E.D.: stridulate: to make a harsh, grating, shrill noise; said specifically of certain insects. ‖

    It is frequently used with a specific connotation of *friction*, and the Concise Oxford Dictionary defines in this sense, though it ruins its own definition by including cicadas as examples! The O.E.D. itself makes no reference to friction and gives as one example: 1854, Badham: Women are obliged to stridulate louder at each other as the wind rises and threatens to drown their voices.

    Leston and Pringle (p. 391*f) are guided by the inclusiveness of the Oxford definition, and of the etymology, and use *stridulation* accordingly without connotations of friction, using *strigilation* (*q.v.*) as a precision term for the frictional type. This practice clearly merits general adoption, leaving *stridulation* as an empirical term descriptive only of the sound itself, and *strigilation* as a motor-correlated term descriptive of the mode of production of certain types.

    The distinction could not however be applied here in indexing all authors, because of their different usages.

    5, 278*, 391*, 798*, *see also* physical characteristics; stridulatory apparatus; *and under the various organs concerned.*
apparatus, *see* stridulatory apparatus
auditory feedback unimportant in cricket, 480, 483
and castration, 450
communicative value, 440
fish, 489, 676, 678f, 685
genital proprioceptors and, 450f
and limb operations in *G. rufus*, 482
by lobster on approach of fish, 685
mechanism, *see* phonomorphosis
mechanisms, *see* stridulatory apparatus
as mood expression, 441, 448f
and muscle tonus, 457f
origin of movements, 609f
peripheral afferent control, 479f
proprioceptors, rôle of, 480f
as releaser, 441
releasing stimuli for, 450
and sense organs, 462f
"silent", 333
and strigilation, 391*, 798*
systematics of, *see* systematics of sound production
stridulators, artificial, 15
stridulatory apparatus, 19, *see also* sound-production apparatus

stridulatory apparatus, *of various groups, continued:*
    acridid type, 365f, 444f
    arthropods, 278*
    chelicerates (arachnids) 316
    chilopods, 321f
    Crustacea, 305f
    diplopods, 320f
    ephippigerid type, 364f
    fish (teeth, jaws, skeleton), 489, 676, 678f
    gryllid type, 362f, 392f, 441f
    Hemiptera, 329f, 798
    insects, 279f, 391f
  general aspects, see also general entries under the various structures used
    factors affecting emission behaviour, *see* song
                  emitted sounds, *see* phonomorphosis
    final central control, 450
    left-right inversion of (tegmina), 360f, 443, 480f
    mutilation of, 445
    programme of stridulatory movements, detailed, 443, 445
    wide distribution of, evolutionary aspects, 409
strigil: literally, a scraper
    O.E.D. (1) sweat scraper; (3) comb.
      At first used in entomology without connotations of sound production (e.g. F. B. White, Ent. mon. mag; 1873, X: 60, quoted by O.E.D. (2): "certain ♂♂ of *Corixa* with a curious structure resembling a curry comb", which he named *strigil;* and Tillyard, 1926, Insects of Australia and New Zealand, p. 399, a cleaning function suggested for the strigil of the lepidopteran fore tibia), it seems to have been brought into substantive use as a name for a stridulatory file (or pars stridens, *q.v.*) by Leston (1954, Ent. mon. mag. XC: 49–56).
        278*, 391*, 397, 401f, 798
strigilation: O.E.D. cites Cockeram, 1623: to currie a horse
        Term used by Leston and Pringle (p. 391) as a specific term for *frictional stridu-*
           lation (which see, and strigil) 278*, 391*, 401, 798
string of sounds, 18
stroboscope for frequency determination, 346, 375
Strophe *(G.)* = verse, stanza, 18*
strophic sequences, 761
structure of emission apparatus, *see* phonative apparatus; sound-production apparatus;
    stridulatory apparatus
structure *(psychology)*, 119*
strychnine and the audiogenic seizure, 244
subgenual organ, 413, 419f, 422f
subjective human audition as tool of analysis, 3, 14f
subjective, objectivization of the, 559
subjectivity, 136
subjectivity of animal language, 69
submission, *see* appeasement
subordination centre (Lapicque), 568
substitution, vertical and horizontal *(algorithm theory)*, 137*f, 176*
substratum in sound production, 325f, 332
subtlety of perception *(cybernetics)*, 120
Suchlaut *(G.)* = search sound, *see* song, calls and call-notes, categories – courtship

Sukhumi, 150
summation counter, 62
        heterogeneous, 777
superimposition of sounds, 767
supersign, 116*
supraglottal cavities, 752
supranormal response to artificial signals, 92
surrealism in grasshopper, 126
swarms of insects, flight sound, 385
swim-bladder, 224, 489, 665f, *see also* air bladder; phonation; sound-production apparatus
swimming, quickened, as response to sound, 685
        experimental extinction, 685
        sounds, 685f
syllabic conjunction, 161f
syllable (1) as a biological parameter; a sound equivalent to one specified motor movement in certain groups (= Silbe, *G*.) 19*f, 23*, 356f ("pulse"), 360f ("pulse"), 444, 447
  figured, 20*f
    (2) *(linguistics, poetry)*: unit of pronunciation forming a word or part of a word and containing one vowel sound with or without consonants either or both sides of it. 5, 15, 30, 137, 175*
    (3) *(ornithology)*, 19, 233 (this latter may indeed be in sense (1) – it is not possible to tell from the context)
symbiosis, acoustic, of Honey-guide bird, 724
symbol, 115*
symbolic codes for sound representation, 709
symbolism, absence from animal sound, 69
sympatric species, hybrids, 629f
        prevention of hybridization, 631f
sympatry and divergence of song, 634, 698, 706, 742f, 804
        interspecific alarm response, 723
symphasis: relationship between two oscillations or oscillatory movements which reach corresponding positions simultaneously; the in-phase condition, *see* phase; syndromism
synapses and information theory, 123
synchronous anaphony, 610*
    (neurol.) sound production, 377
syndromism (Jaçobs): bilateral symphasis of stridulatory movement. *Compare* antidromism
syntagma, 134, 142*f, 177*
syntagmatic algorithm, 143*
        relation, 578
syntax of bird song, 716
synthesis of animal sound, 15
synthetic languages, 144
syrinx, 232, 492f, 711f
syringeal muscles, 493f
systematics and sound, 350, 402, 636f, 686, 694f, 803
        of sound-production apparatus, 335f, 360

T

t- t-, 15, 19
tabac, 15

tactile message, 113*
tail-rattling, 761
talk, 172
tanks, experiments on fish in, 681
tape pre-amplifier, 37
    recorder characteristics, 35f
        precautions, 36
    recording, 34, 107, 694
    storage of magnetic, 37
tarsal inhibition, insects, 103
tarsus in percussive sound production, 327f
  grading to strigilation, 328
tarsus in stridulation, 280, 292
tattoo, 18
taxis and tropism:

    It is an interesting fact that whereas "tropism" has practically disappeared from zoological textbooks (except in relation to sedentary animals such as colonial coelenterates showing growth curvatures analogous to those universally defined as tropisms in plants), in psychological texts it may still be equated, or even preferred, to "taxis". This cannot be better illustrated than in two dictionaries from the same publishers (Penguin Books):–

	Abercrombie et al., 1951 Dictionary of Biology	Drever, 1952 Dictionary of Psychology
Taxis:	Locomotory movement of an organism or cell, e.g. gamete, in response to a stimulus, e.g. light or temperature, the direction of movement being towards or away from the place where the stimulus originates.	An orienting response of organisms to physical forces, having direction; usually included under *tropism*.
Tropism	(1) response to stimulus, e.g. gravity or light, in plants and sedentary animals by growth curvature, the direction of curvature being determined by the direction from which the stimulus originates. (2) (Zool.) equivalent to *taxis*, but this usage is becoming obsolete.	An orienting (and movement) response to physical and chemical agencies in cells, organs and organisms explicable (Loeb) in purely physico-chemical terms; may be either positive or negative, with reference to the determining agency's direction, and may result, in the case of organisms, in actual movement, or in the assumption of a definite axial position.

    The statement of Abercrombie et al., that equivalence between the terms was in 1951 becoming obsolete seems now more valid than ever, in the ranks of psychologists as well as zoologists. There can be no question that the distinction is a useful one contributing powerfully to precision of expression, and more so as later workers add further refinements to either term (e.g. Lorenz and Tinbergen, distinguishing, in complex behaviour sequences, a *kinetic* component *(see kinesis, supra) released* by sign-stimuli; and a *taxic* component, *directed* by stimuli or sign-stimuli).

In this context, the discussion of "phonotaxis" by Dumortier (p. 625, *supra*) is interesting. Using the terminology of Loeb, and Drever, he shows that "phonotaxis" is not a *tropism* – and therefore not a *taxis* either, since in that terminology the two are the same. If *taxis* does differ from *tropism*, it does so through being defined in the sense of Kühn, Fraenkel and Gunn, and Abercrombie *et al.* above, and phono*taxis* will be legitimate to the extent that it conforms to such a specification. In fact, as Dumortier shows, it is far too complex for that simple concept, and observation shows that it has many of the properties of the dual mechanism proposed by Lorenz and Tinbergen in other contexts. If their well-founded restriction of *taxis* to but the taxic part, in such complex reactions, is to be respected, a term is badly needed to characterize briefly the whole reaction; we cannot use a long circumlocution every time, such as: partly kinetic, partly taxic reaction triggered by some elements and directed by others. Yet no term seems to have been suggested. Perhaps qualification by "complex", though scarcely accurate, would meet the case (e.g. *complex phonotaxis*, in the present context); or one might propose *mixotaxis* as the general term – but with a good deal of hesitation.

taxonomic value of behaviour, 637, *foot*
tectorial membrane, 526, 548, 705
tectum, optic, 561, 567
teeth- chattering, 762
        gnashing, 762
tegmen *(entomology)*: term originally specific to the Orthopteromorph orders of insects, for the characteristically hardened forewings of some of them, in contradistinction to the differently hardened *elytra* of beetles and *hemelytra* of Heteroptera. Now used also by Leston and Pringle (p. 397, *supra*) for analogous hardening in Homoptera.
    A useful term, very current in English, but which does not yet seem to have caught on in non-English-speaking countries. *See also therefore*, elytra.
tegmen in stridulation, acridids, 10, 295f
        cicadidae, 397
        gryllids, 302f
            an acoustic coupler, 366
            effect of left-right inversion, 360
        Meconema, 329
        tettigoniids, 297f
teledeltos paper, 674
telemetering *(marine zoology)*, 655
telephone transmission *(Regen)*, 76, 586
television, 655
telson in stridulation, crustacea, 305
temperament, scale of equal, 9f
temperature and the audiogenic seizure, 251
    care-calls stimulated by, 770
    effects on signal, 62, 366f, 379, 447, 449, 627, 638, 703, 714, 760
    internal, insects, 380
    of language, 128*
template, auditory, for monitoring emission, 233
temporal message, signal, 114*
temporal quantum, 115*
    summation, Orthoptera, 466
tense and lax sounds, 166

territory, 587, 716
    Call-, 75*
    female, 77, 714
territorial song, *see* song, calls and call-notes, categories
testiform right ovary and autumn song in the female, 714
testosterone, 732, *see also* hormone
tettigoniid-type stridulation, 299f
theme, *see* motive
thermal inhibition, *see* temperature
    message, 113*
thermodynamic noise threshold, 49*
thermometer crickets, 62, 366
thickness of the moment, of the present, 102*, 122
thought, 132, 171, 561, 575
thorax in stridulation, insects, 280f, 290f
threat, *see* song, calls and call-notes, loosely used concepts
threshold, auditory, B.S. 661/3014: measured under specified conditions and at a specified frequency: the minimum r.m.s. value of the sound pressure of a sinusoidal sound wave of that frequency which excites the sensation of hearing.
    By extension the term is also used to cover complex waves such as speech or music. ‖ *See also* sensitivity, auditory
thresholds of excitation, 100*
throat, *see* pharynx
    microphone, 29
throbbing, 676
thumps, 684
thundering of menhaden, 666
thyroid cartilage, 190, 494f, *see also* larynx skeleton
thyroidectomy, 713
thyroxin, 713
tibia in stridulation, insects, 284, 292, 298, 403, 406
tics, 247
Ticklaut, 187, 207, 209
timbre, B.S. 661/3002: that subjective quality of a sound which enables a listener to judge that two sounds having the same loudness and pitch are dissimilar; 122
    innate, 237
time-quantum, 115*
tji tji tji tji tji, 769
tonal perception, 101, *see also* frequency discrimination
tone, *see also* frequency
    B.S. 661/3003 *a.* a sound giving a definite pitch sensation
        b. sometimes also the physical stimulus giving rise to the sensation
    combination, B.S. 661/3007: a tone produced by a non-linear relationship between response and stimulus in the ear (i.e. as a subjective tone, *q.v.*), or in a transducer, and derived from two sinusoidal components in the stimulus
    complex: a combination of sine waves of more than one frequency which may be of constant or varying amplitude; a tone which is not a pure tone *(q.v. infra)*, 382
    difference, B.S. 661/3008: the combination tone whose frequency is the difference between the frequencies of two generating tones
    fundamental, 136, *see also* frequency, fundamental
    modulation, birds, 492

muscle and nerve, *see* tonus; tonic
pure, B.S. 661/3004: A tone in which the sound pressure varies sinusoidally with time. ∥
    8*, 351, 358, 392, 443, 509, 752, *see also* audiogenic seizure.
  in gibbon, 509
  stimulating begging response of birds, 81, 720
                response of *Chorthippus brunneus*, 614
                turkey, 713
  underwater, 53
subjective, B.S. 661/3006: an aural sensation of a tone of a particular frequency in the absence of stimulation by sound of that frequency
summation, B.S. 661/3009: the combination tone whose frequency is the sum of the frequencies of two generating tones
variation with circumstances, 710
warble, B.S. 661/3005: a tone whose frequency is continuously varying in a regular manner between fixed limits
tongue in phonation, bats, 212
tonic convulsions, seizures, 244f, 250, 253f, 258f
tonus control in Orthoptera, 457f
            vertebrates, 567f
tooth impact rate and frequency, relationship, arthropods, 362f, *see also* phonomorphosis
toothplates in stridulation, fish, 678
torque in insect flight, 378
torus pharyngeus, 492
tossing coins and information theory, 117
touch, bats, 183
trachea, birds, 492f
tragus, 192, 199, 208, 535
tranquillizers, *see* audiogenic seizure
transcription, musical, 23, 346, 709,
          phonetic, 752
transducer, 48, 50*, *see also* microphone; loudspeaker
  attached to a fish, 655
  biological, *see* sound-production apparatus; transductor
transductor, 522*
  in nerve transmission, 526
transfer function, mechanical receptor to afferent fibre 527
transformation, algorithmic, 173*
  of a word, 134
transient, B.S. 661/1031: a phenomenon which occurs during the change of a system from one steady state to another as a result of a sudden change of stimulus
    21*, 92, 95*, 122, 135, 355*
distortion, B.S. 661/4035: a type of distortion which occurs only as the result of rapid fluctuation in amplitude and/or frequency of the stimulus. Transient distortion is frequently manifested as a duration of frequency components in the response in excess of the duration of the stimulus
transients in arthropods, 21, 355, 365, 427
  Fourier analysis, 355, 356*
transients, initial, in receptor-nerve preparation, 528
        measuring level, 38
        and playback, 46
        pulse, 397, 400

transients, *continued:*
    as releasers, 615, 744
        for phonotaxis, 620
    and sound spectra, 41, 355
    and tympanal organs, 427
    and ultrasonics, 355
transitions, acoustic, 135, 139
    coded, 134
    intrasyllabic, 162
translating machine, 134
translation of one code into another, 138*
translator's chain of coded transitions, 134
transmission channel, 114*, 121*
    level, 49
    of message, 116*, 147
transmitter, *see* emitter
tree call, *Hyla*, 704
tremolo, 13*, 16*f
tremulation, insects, 59, 591, 603, 618
tricarboxylic cycle and the audiogenic seizure, 256, 261
trichobothria, 408f, 412, 420, 798
trill, 13*, 16*f, 359, 361, 447, 593, 697, 754f, 767
  capuchin, 163
  homology, 781
tripus, 665
tritocerebrum, 461f
triumph, shout of (originally "shriek of") picturesque expression used (almost certainly without intent to perpetuate) by Pumphrey on a somewhat informal occasion, for the mounting call of Acrididae, *see* song, calls and call-notes, categories – courtship.
triungulin recognizing host by sound, 387
trochanter in stridulation, 330f
tropism, *see* taxis and tropism
trumpeting, 757f, 762
  bumble-bees, 388
tuco-tuco, 773
Tüllgren's trichobothria, 408f
tuned sounds, finely, 698f
tuning fork and alligator, 104
tuning fork and mosquito, 105
tuning, sharp, to carrier frequency (insect hearing), 392, 400
tut-tut, 15, 19
twitter, 753, 767, 771
tymbal, 377, 392f, 400, 403f, 407, 409, 798, 800
  evolution of, 401, 404, 409
  homogeneity in Auchenorhyncha, 335, 393f
  muscles, 393
  parallel evolution in Hemiptera, 392f, 798
tympanal bodies, 416
tympanal organs, *see also* chordotonal organs; hearing; scoloparia etc.
  anatomy, comparative, 415f
    B-cells, 435, 437, 800

and bats, 218f
and clicks (reaction to), 420
"counter-tympanum", 418
detensor tympani, muscle, 400
directional characteristics, 425f, 428f
frequency discrimination, 427, 447
       range, 420
hemipteran, 400f
hysteresis effect, 800
lepidopteran, 415f, 434f, 800
orthopteran, 415f, 421f, 426f, 429f
properties responded to by, 427
prototympana, 436
and phylogeny in moths, 435
reaction-time, 435
resting discharge, 436
sensitivity, absolute, 423f
       maximum, region of, 421f
and sound-production apparatus, relationship, 337
stretch-sensitive elements (B-cells), 435, 437, 800
tensor tympani muscle, 418
tonic-phasic discrimination, 427
ultrasound, response to, 420
wing sounds, response to, 435, 437
tympaniform membranes, birds, 493, 516
tympanum,
  bird phonation, 493
  insect audition, *see* tympanal organs
       sound production, *see* mirror
  vertebrate audition, *see* ear

U

ultrasonic, ultrasound, 133; *for definition, see* sound
  analysis, 37f
  care calls, 770
  clicking, 762, *see also* echolocation
  emission in bats, 106f, 184f, *see also* echolocation
       guinea-pig, 215
       intersyllable of *Gryllus*, 449
       mammals, 751f
       mouse (reassurance call before moving), 84, 769
       rats, 215
  hissing, 762
  microphones, 29
  orientation, 48
    in bats, 184f, *see also* echolocation
  perception, *see also* echolocation
    in dogs escaping vampires, 210
    grasshoppers, 106
    moths escaping bats, 87, 428, 434f
  pest control, 800

ultrasonic, *continued:*
  spitting, 762
  in transients, 355
  and tympanal organs, 420
  in varied sounds, 763
um-um, 759
Umwelt, 112*f, 116, 127
unconscious, the, 572
understanding, 118*
underwater acoustics, 48, 656f, 665, 672f, 687, *see also* echolocation
  range, *Micronecta*, 407
underwater stridulation, Hemiptera, 405f
        transducers, 48
        "transparency to sound" (diaphony) of fish, 659
unequal concrete words, 148
unicyclic, *deprecated hybrid for* monocyclic, *q.v.*
unipulsate, 19
unipulse, *deprecated for above*
unit of information (bit), 115*
unoriented reaction, *see* kinesis
unvoiced sounds, 761*
uropods in stridulation, Crustacea, 305
uruck-uruck, 670
usual song, *see* song, calls and call-notes, categories – proclamation
utilization of sound as a criterion of hearing, 664
utriculus, 531, 656, 658f
uvula, 507

V

vacuum activity, 725
  driven by hypothalamic centres, 571
vacuum copulation, insects, 598
van 't Hoff's Law and stridulation, 367
variable expressivity, in genetics of vocal behaviour, 233
variation, regional, 107, *see also* geographic
vascular stria, 533f
vasculosum of ear, 532
vegetation, effect on amplitude, 350
velar noise, 752
velum, 512
Venezuela, 216
ventilation flight tone, bumble-bee, 388
       rate, bats, 191
ventricle, *see* larynx
ventricular pulse, 7*, 11
verb, 141
verbalized self-image, 573
      world-image, 577
Vers *(G.)*, 5*, 18*
verse: better to use stanza to avoid confusion with German Vers, *q.v.*
vertebral spines in stridulation, 678

vertical substitution, 138*, 146*, 155, 166, 176*
   final, 148*, 176*
vestibule of ear, 657f
vestibulum laryngis, 499
vibralyzer, vibrogram, *see* sound spectrograph
vibration, free, *see* free vibration
vibration perception, birds, 735f
               fish, 663f
     pick-up, 55
     receptors, 413, 422f
     species-specific, 59
vibrissae, rodent, 103
violoncello and alligator, 104
visible speech, 107, *see also* sound spectrogram
vision, bats, 212, 214
     owls, 216
visual message, 113*, 583f
     observation of flight, 375
     receptors, 583
     and sound signals, interplay, 78
vitamins and the audiogenic seizure, 256
vocabulary, 70f, 124, *see also* alphabet; repertoire.
  amphibians, 703f
  chimpanzee, 512, 576, 627
  fish, 676, 681f
  mammals, 753
  and age, birds, 725
  inheritance, 228f
  variations, 62, 70
  volume, 148*, 232, 711
vocal apparatus, *see also* larynx; phonative apparatus; sound-production apparatus; voice.
  absence from some birds, 494
  experimental mutilation, 190f, 212
vocal band, 490
     behaviour, *see* inheritance; innate signals; learning.
     cords, *see* larynx
     psychosomatic disorders and the hypothalamus, 572
     sacs, *see* air sacs
vocalic and non-vocalic sounds, 166
vocalization (1): any utterance made with the voice. *Used by, e.g., Marler, as a more objective quasi-synonym of* song, *to avoid the musical connotations of the latter.*
     (2): the *process* of rendering an utterance *sonant* by the use of the voice, e.g., the conversion of *p* to *b*.
     See discussion under song, *and see* larynx; phonation; vowel.
  and inheritance, 228f
  of whistle, 172
voice, *see* phonation
  development, no data on Cetacea, Chiroptera, Insectivora, Pinnipedia, 232
  intensity, 517
  synthetic, 15, 19
voiced and voiceless sounds, 166, 761, *see also* phonation; sound, biological

voiceless sound, baboon, 153
volition in phonation, 564, *see also* will
volley, 5, 18:
> (1) a discrete series of nerve impulses, *see* impulse
> (2) by extension, a series of *pulses* (q.v.), and thus further for a series of unitary sounds.
>
> The term is a useful empirical one, and there seems no reason why it should not be further extended to such uses as a volley of syllables, a volley of phrases and so on.

voluntary act, 557, 567, *see also* will
vowel, 5, 122, 135, 156, 162f *see also* phone; phoneme.
> in animals, 516, 753.

W

wailing, 769
warming up before flight, insects, 380, 386
warning song, *see* song, calls and call-notes, loosely used concepts.
water call, *Hyla*, 704
water, sound and ultrasound in, *see* underwater
wave, B.S. 661/2001: a disturbance which travels through a medium by virtue of the inertia and elasticity or analogous properties of the medium. Usually, the passage of a wave involves only temporary departure of the state of the medium from its equilibrium state. ‖ 6, *see also* damped waves.
> and airflow, distinction, 383

wave-form, B.S. 661/1033:
> a. the shape of the graph representing the successive values of a varying quantity, usually plotted in a rectangular coordinate system;
> b. by extension, the characteristics of a varying quantity which can be represented by a graph as in a. ‖
>
> 6, 740, *see also* elementary wave-form.
> and direction, relationship, 383f
> determination by generator, *see* phonomorphosis.

wave-front, B.S. 661/2005: of a progressive periodic wave; a continuous surface at all points of which the disturbance has the same phase at a given instant. ‖
> A continuous isophasic surface.

wavelength: the distance in space corresponding to one complete period of a wave.
wave, rectangular, 7*
wave slippage reflector, 32
wave-train, B.S. 661/4116: an unbroken group of waves. ‖
> 7*, *see also* building-up time; pulse

Weber-Fechner Law, 105, 385, 601, 619
Weber's organ, Weberian apparatus, Weberian ossicles, 539f, 657f, 665, 677
> influence of, 665

Wechselgesang *(G.)* = responding song ⎫
Weibchengesang ,, = female song ⎪ *see* song, calls and call-notes, categories
Werbegesang ,, = courtship song ⎬
Werbewechselgesang = courtship responding ⎭
Wernicke's sensory aphasia, 579
Wever's volley theory, 550f
whimper, 754f, 771f
> homology, 781

whining, 676, 768
whisper, baboon, 153
    bat, 209
    frog, 704
    mammal, 769
whistling sounds, 324, 674, 694, 700, 753, 762f, 770f.
whistling speech, Canary Is., 172
white noise, *see* noise
  dazzling bats, 213
whizzing song of operated *Gryllus*, 465, 469f
whooping fish, 675
will, 576, *see also* volition; voluntary.
wind instruments, analogy of bird voice, 494.
wing inertia, 377f
    loading, alteration experiments, 377
    receptors for orientation in bats, 186f
    sounds 435, 437, 800
    in stridulation, *see* alary stridulation; elytral stridulation; flight sound; tegmen.
wing-beat amplitude, 377
        frequency, *see* flight sound parameters
wing-clap of nightjar, woodpigeon, 719
wonted song, *archaic for* ordinary song, *see* song, calls and call-notes, categories – proclamation.
Woods Hole, 674
word, 15f
  (1) *(information theory)*, an element of a repertoire of communication-symbols, 115*
  (2) *(algorithm theory)*, 138*, 145*, (includes words without meaning)
  (3) *(linguistics):* any sound or combination of sounds recognized as a part of speech, conveying an idea or alternative ideas, and capable of serving as a member of, the whole of, or a substitute for, a sentence.
  abstract, 177*
  adjacent, 138*, 177*
  basic, 177*
  compatible, incompatible, 145, 149
  complete, incomplete, 142*, 149, 166, 177*f
  composition of 138*
  concrete, 134, 178*
  entry of, 138*, 174*
  exact, 5
  final, 138*, 143, 145, 149, 177*
  formal, 141*, 147*, 169, 178*
  initial, 138*, 142, 145, 177*
  meaning of, 134, 137
  null, 144, 148, 178*
  as a signal *of* signals, 171*
  simple: a word in sense (3) above conveying a single undivided idea, 16.
  structure of, 134
  transformation of, 134
  unequal concrete, 148
writing centre, 573

X

X-rays, epilaryngeal, 162f

Y

yelp, yelping, 757, 764
    homology, 779
    medullary, 564
    stimulated by isolation, 773
yum-yum, 755

Z

Zipf's Law, 127*f
zirconate, lead, in transducers, 52.

ALPHABETIC INDEX

Page numbers in parentheses refer to taxonomic units in the systematic index

Absyrtus luteus 379
Acalles 293
Acanthocarpus 312, 606
– *bispinosus* 606
Acanthosomidae 403
Acerina cernua 662
Achaearanea tepidariorum 423
Acherontia atropos 73, 106, 323, 324*, 332
  (pupa), 349, 351*, 353, 605, 607
Acheta assimilis 303, 348, 360, 361, 369*, 422,
  443, 591, 593, 634, 636, 640
– *domestica* 302, 303, 348, 353, 360, 361, 443,
  447, 466*, 591, 603, 640
– *firma* 636
– *fultoni* 636, 647, 648
– *pennsylvanicus* 359, 360, 362, 636
– – × *fultoni* 628
– – × *vernalis* 628
– *rubens* 360, 636, 648
– – × *assimilis* 628, 634
– – × *veletis* 628, 634
– *veletis* 362, 592*, 634
– – × *pennsylvanicus* 634
– *vernalis* 636
Achorotile albosignata 604
Acinonyx jubatus 757
Acmaeopleura 311
– *parvula* 311*
Acrididae 76, 296–298, 347*, 351*, 359, 365–366,
  415, 421–423, 427, 440, 445, 454, 464, 484,
  585, 590, 608
Acridinae 365 (footnote), 587, 589, 591, 593,
  594, 596, 597, 599–602, 603 (footnote), 604,
  605, 609, 610, 612
Acridoidea 335, 350, 352*, 357*, 599*
Acris 706
– *crepitans* 699
– *gryllus* 699
Acrocephalus palustris 238, 239, 732
– *schoenobaenus* 237, 238
Actinopterygii 664, (815)

Addax nasomaculatus 759
Aëdes aegypti 349, 354, 387, 420, 423, 424, 597,
  619, 620, 621, 625
– *trivittatus* 625
– *vexans* 625
Aegeriidae 388
Aegidinus 291
Aegidium 291
Aegocera mahdi 294
– *tripartita* 294*
Aelia acuminata 407*
Aeschna (larva) 453, 459
Aetalion reticulatum 396
Aetalionidae 393, 400 (larva)
Aetechinus frontalis 754
Aethus flavicornis 402
Agelaius phoenicus 237, 238
Ageronia arethusa 295*
Aglaïs polychloros 329
– *urticae* 329
Aglaothorax armiger 605
Agnatha (814)
Agonoscelis 404
Agouti *(Dasyprocta agouti)* 761
Agrotis 423
Ailuropoda melanoleucus 756, 762, 765*
Aiolopus strepens 328
– *thalassinus* 594
Alauda arvensis 235, 732
Alburnus lucidus 662
Alces 80
– *alces* 758
– – *columbae* 772
Alipes 321
Allied woodhewer *(Lepidocolaptes affinis)* 720
Alligator mississippiensis 71, 104
Alopex lagopus 756, 759, 760*, 763, 765,
  768–770, 776, 778*–780
Alouatta 500, 506, 754, 773
– *palliata* 772
– *seniculus* 510*

Alpheidae 73, 306, 646 (footnote)
Alpheus 306, 349, 354
— *malleodigitus* 307
— *strenuus* 307
Alytes obstetricans 695
Amatidae 437
Amaurocoris 798
Amblonyx 756
Amblycorypha carinata 648
— *oblongifolia* 348, 359, 421, 604, 605, 611
— *rotundifolia* 300, 303, 587, 595, 636
— *uhleri* 359, 647, 648
Amblyopsidae 224
Amblyopsis spelaea 224
Amblypygi (814)
Ambystoma maculata 694, 695
— *tigrinum* 694
Ambystomidae 694
Ameiurus 677, 686
— *nebulosus* 661, 662, 663, 667, 668*
Ametroides kibonotensis 285*
Ammospiza henslowii 732, 735
Amphibetaeus 306 (footnote)
Amphibia (816)
Amphibicorisae 402, 405, 408
Amphisternus 302*
Amphiuma means 694
Amyoteinae 403, 404, 407
Anabantidae 661
Anabas scandens (= *testudineus*) 661
Anacridium aegyptum 421
Anactinothrips 291
Anas platyrhynchos 105, 713, 720, 721, 735
Anaxipha exigua 348, 361, 593
Ancistrodon mokasen 85
Andrena nitida 423
Androctonus 320
Androniscus 314
Aneides lugubris 694, 695
Angelfish 679
—, Black (see *Pomacanthus arcuatus*)
Angelichthys 679, 685
Anguilla 677
— *anguilla* 661, 667
— *rostrata* 71
— *vulgaris* (= *anguilla*) 657
Anguillidae 661
Angular-winged katydid *(Microcentrum)* 299
Anhimiformes (818)
Ani, Smooth-billed *(Crotophaga ani)* 85
Aniculus, (Trizopagurus) 310
Anisops 292, 405
Anobium 618
— *pertinax* 326
— *striatum* 326
Anoedopoda lamellata 305
Anomura (805)

Anopheles 426
— *gambiae* 349
— *punctipennis* 625
— *quadrimaculatus* 621, 625, 642 (footnote)
— *subpictus* 420
Anoxia villosa 288
Anser 725
— *anser* 71, 81, 83, 720, 721
Anseriformes (817)
Ant *(Camponotus)* 423
— *(Crematogaster)* 622
— *(Dendromyrmex)* 328
— *(Dorylidae)* 283
— *(Dorymyrmex)* 283
— *(Formica)* 413
— *(Megaponera)* 590
— *(Messor)* 350, 590
— *(Monomorium)* 350
— *(Myrmica)* 283, 304, 350
— (Myrmicidae) 283
— *(Neomyrma)* 304
— *(Pogonomyrmex)* 590
— (Poneridae) 283
— *(Solenopsis)* 304
— *(Tetramorium)* 350
Anteater 494
Antelope 502, 516
Anthidium 387
Anthophora 387
Anthus pratensis 235, 238, 732
— *trivialis* 231, 240
Antidorcas marsupialis 759
Antilope cervicapra 758
Ant-lion (see Myrmeleontidae)
Anura (816)
Aphididae 284*, 401, 402
Aphis 284
Aphrodes bicinctus 393*
Aphrophora 605
Apis mellifica 40*, 58, 59, 71, 325, 349, 376*,
   379–383*, 386, 423, 583, 724
Aplodinotus grunniens 78
Apodemus flavicollis 763, 771
Apodidae 217, 218
Apodiformes (818)
Aptenodytes 88
Aptinus 325
Arachnida (814)
Aradacanthia 403
Aradidae 286, 403, 408, 409
Aramides cajanea 82
Araneida (814)
Archer 74
Archiblatta hoeveni 289
Arcte caerulea 283, 284*
Arctia caja 323 (footnote)
Arctiidae 420, 435, 437, 800

Arctiinae 415
Arctocephalus pusillus 757, 765, 766*, 782
Arctocorisa 405
Arcyptera fusca 352*, 594, 596
Argus 74, 80
Argyrodes 318
Arhaphe 403, 404
Armadillo 494
Armitermes 327
Aromia moschata 351*
Arphia sulfurea 304, 421
Artabanus 403
Arthropoda (805)
Arthrosphera aurocineta 322
Artibeus jamaicensis palmarum 209, 210
Artiodactyla (823)
Asagena 318
Ascaphidae 696
Asellia tridens 204, 208, 214
Asinus somaliensis 758
Asio otus 735
Ateles 506, 754, 772, 773, 774
– *geoffroyi* 754, 772
Atherurus 516
– *marcrourus* 500, 501*
Atlanticus pachymerus 648
– *testaceous* 348, 360, 368*, 647
Atropos pulsatorium 326*, 327
Auchenorrhyncha 335, 392–401, 798
Aves (817)
Axis axis 758, 764

Baboon 132, 136, 150 →, 162*, 506
– *(Papio)* 83, 762, 770
– *(Theropithecus)* 755
Bagridae 677
Balaenidae 223
Balaenoptera musculus 223
– *physalus* 223
Balanus 315
Balistes 659, 678
Balistidae 678, 680, 682, 685
Barbitistes fischeri 61*, 348, 647
Barnacle *(Balanus)* 315
Barracuda 667
Bassariscus astutus 756
Bat 12, 36, 71, 83, 87, 106, 183–215, 223, 435, 437, 504, 535, 545, 800
–, Free-tailed (Molossidae) 204
–, Horseshoe (see Rhinolophidae)
–, Long-eared (see *Plecotus auritus*)
–, Mouse-tailed (Rhinopomatidae) 207, 208
–, Piscivorous 205
–, Sac-winged (Emballonuridae) 207
–, Romb (see *Rousettus aegypiacus*)
Bathygobius soporator 78, 92, 101, 683, 686
Batrachoididae 666, 677, 679, 682, 686

Batrachotetriginae 286
Bear 82
– *(Helarctos)* 756
– *(Selenarctos)* 756
– *(Tremarctos)* 756
– *(Ursus)* 756, 770
–, Polar *(Thalarctos maritimus)* 756, 771
Beaver *(Castor)* 761, 766, 769
Bee *(Apis mellifica)* 40, 71, 325, 349, 376*, 379–383*, 386, 423, 583, 724
Beetle (see Scarabaeidae, Scarabaeoidea), Coleoptera (809)
Bellicositermes 327
Belonopterus chilensis 83
Beluga (see *Delphinapterus leucas*)
Benthocometes robustus 686
Bermuda grouper 672
Berosus 287
Berycoid fishes 665
Betta splendens 661
Biacumontia 318
Bibos 768
– *banteng* 759
Bison 770
– *bison* 759
– *bonasus* 759
Biturridae 400
Blackbird (see *Turdus merula*)
Black drum (see *Pogonias chromis*)
Blenina metascia 332 (pupa)
Blenniidae 686
Blenny 74
Blethisa 288, 289*
Bluethroat *(Cyanosylvia svecica)* 732
Bolbelasmus bocchus 289
– *gallicus* 289
– *unicornis* 289
Bolitotherus cornutus 326
Bombina bombina 696
– *variegata variegata* 370
Bombus 380, 381, 383, 386, 388, 424*
– *agrorum* 379
– *lucorum* 376*, 382, 383*
– *scoroeensis* 423
Boreellus 379
Borneopoeus costatus 322*
Bos 770
– *indicus* 759
– *taurus* 56*, 71, 90 (footnote), 502, 505*, 759, 764*
Bot-fly 387
Botia horae 224
– *hymenophysa* 686
Bournellum retusum 322*
Bovidae 761, 770, 776, 789
Brachygnatha 336
Brachyura (805)

Brachynus 325
Bradypus 771
Brambling *(Fringilla montifringilla)* 729, 742
Branchiopoda (805)
Brevoortia 686
— *tyrannus* 666, 669, 670
Brotulidae 224, 686
Bryodema tuberculata 327, 328, 594
Bubalis bubalis 759
Bubo bubo 735
Buenoa 292*, 405
— *limnocastris* 405, 407, 586
Bufo 695, 705, 794, 803
— *alvarius* 696, 699
— *americanus* 230, 698, 703, 704, 803
— — × *woodhousei* 230
— *boreas* 698
— *canorus* 698
— *cognatus* 698, 699, 704
— *compactilis* 698, 704, 803
— *debilis* 698, 703, 704
— *hemiophrys* 698
— *houstonensis* 698
— *marinus* 699, 703
— *microscaphus* 698, 704
— *punctatus* 698, 704, 803
— *quercicus* 699, 703
— *regularis* 696
— *speciosus* 803
— *terrestris* 698, 706
— *valliceps* 230, 699
— *woodhousei* 230*, 698, 704, 699
Bufonidae 698–699
Bullacris 286
Bullfinch (see *Pyrrhula pyrrhula*)
Bullfrog *(Rana catesbeiana)* 705
Bumble bee *(Bombus)* 376, 380–383, 386, 388, 423, 424
Bunting
—, Corn *(Emberiza calandra)* 231
—, Reed (see *Emberiza schoeniclus*)
Bush cricket, Handsome (see *Phyllopalpus pulchellus*) 647
— —, Jumping (see *Orocharis saltator*)
— -dog *(Speothos)* 779
— katydid (see *Scudderia* 299)
Bustard, Great *(Otis tarda)* 78
Buzzard 82, 85

Cacajao 754
Cacicus americanus 295*, 296
Calisiinae 403
Callicorixa distincta 405
Callimico goeldii 754
Calliphora erythrocephala 430*
Calliptamus 333 (nymph), 608 (nymph)
— *italicus* 329, 596

Callomystax gagata 678
Callorhinus ursinus 757
Caluromys philander 762
Cambridgea 318
Camel (see *Camelus*)
Camelonotus 288
Camelus 770, 783
— *bactrianus* 758
— *dromedarius* 758, 763*, 764*
Camponotus 423
Camptorrhinus 288
Canary (see *Serinus canaria*)
Canidae 504, 764, 765, 773
Caninae 764, 769*
Canis 759, 760, 763, 778–780
— *adustus* 766, 779
— *anthus* 755, 766, 779
— *aureus* 755, 766, 779
— *dingo* 755, 766, 776, 779
— *familiaris* 74, 90, 91, 103, 133, 500, 503, 506*, 519, 535, 564, 755, 762, 765, 766, 769, 773*, 776, 778*, 779*
— *latrans* 755, 766, 776, 779
— *lupus* 83, 755, 762, 766, 769, 773, 766–779*
— *mesomelas* 766, 779
— *pallipes* 779
Cantharis 379, 380
Capella gallinago 76, 712–719
Capra 74, 767, 773, 780
— *hircus* 759, 767*
Capreolus 765, 769
— *capreolus* 758
— — *pygargus* 758
Caprimulgidae 216
Caprimulgiformes (818)
Caprimulgus europaeus 76, 719
Capritermes 327
Capromys pilorides 762
Capuchin *(Cebus capucinus)* 506, 754, 767*
Capybara *(Hydrochoerus)* 762, 760
Carabidae 287, 336
Carabus 424*
— *hortensis* 423
Carangidae 673, 676, 680, 685, 686
Carassius 659
— *auratus* 662
— *carassius* 662
Cardinal *(Richmondia cardinalis)* 238
Carduelis cannabina 237, 240, 732
Caribou (see *Rangifer*)
Carnivora (822)
Carollia perspicillata azteca 200, 209–211
Carpenterfish 675, 684
Carpocoris 404, 407
— *pudicus* 408
Carpodacus mexicanus 231
Castor 761, 769

– *fiber* 766
Cat (see *Felis catus*)
–, Ring tailed *(Bassariscus)* 756
Caudata (816)
Catbird *(Dumetella carolinensis)* 238
Caterpillar 103, 331, 387
Catfishes (Siluridae) 665, 677, 678
Catocala 415*, 416*, 434
Cavia cobaya 215
Cavy, Patagonian *(Dolichotis)* 753
Ceboidea 754
Cebus 754, 764, 776
– *albifrons* 754
– *apella* 754
– *capucinus* 506, 754, 767*
– *nigrivittatus* 754
Celes variabilis 328
Centipede 321, 322, 606
Centrocoris 403, 798
Cerambycidae 281, 282*, 335 (footnote), 605
Cerambyx cerdo 304, 349, 358*, 646
– *scopolii* 586
Ceratitis capitata 73
Ceratopogonidae 428
Cercocebus 770
– *torquatus* 755, 781
Cercopidae 393, 394, 397, 400
Cercopithecinae 781
Cercopithecoidea 754
Cercopithecus 607, 767, 770, 773, 781
– *aethiops* 755, 781
– *diana* 755
– *mitis* 755
– *nicticans* 755, 781
– *petaurista* 755, 781
– *sabeus* 509*
– *talapoin* 755, 769, 781
Cercopoidea 394
Cerdocyon 765
– *azarae* 780
Certhia brachydactyla 231
Cerura 332 (larva)
Cervidae 759, 764, 770, 780, 783
Cervus 80, 761, 762
– *elaphus* 758, 764*
– – *bactrianus* 758
– – *canadiensis* 758, 764, 765*, 772
Cetacea (821)
Cetonia aurata 376*
Cetoninae 330
Cettia cetti 711, 712
Ceutorrhynchini 304
Chaetodontidae 679, 682, 685
Chaffinch (see *Fringilla coelebs*)
Chameleon 491
Characidae (Characinidae) 662
Charadriiformes (818)

Charaxes sempronius 506
Charora 284
– *crassinervosa* 285*
Chasmodes bosquianus 92, 686
Chauna torquata 82
Cheetah *(Acinonyx)* 757
Chelonia *(Turtles)* 491
Chelonia pudica 434
Chiasognathus granthi 296
Chickadee
–, Black capped *(Parus atricapillus)* 71, 711
–, Carolina *(Parus carolinensis)* 81
Chiffchaff *(Phylloscopus collybita)* 231, 742
Chilobrachys samarae 317*
– *stridulans* 316
Chilonycterinae 207, 208 (footnote)
Chilopoda (806)
Chimpanzee (see *Pan troglodytes*)
Chironomidae 413
Chironomus 376*, 382, 383*
Chiroptera (820)
Chlamydera maculata 238
Chloealtis conspersa 303, 349, 361
Chloris chloris 732, 735
Chloroperla grammatica 328
Choanichthyes (816)
Choeropsis liberiensis 756
Chondrichthyes (814)
Chordata (814)
Choriotis australis 78
Chorthippus 71, 74, 79, 91, 366, 587, 610 (nymph), 618, 626
– *albomarginatus* 357*, 594
– *apricarius* 303, 594, 595, 600
– *bicolor* (see *Chorthippus brunneus*)
– *biguttulus* 92, 228, 229*, 303, 349, 351*, 353, 360, 445, 591, 594, 595, 596, 600, 601, 602, 604, 608 (nymph), 613, 614, 619, 620, 631*, 632, 633, 634, 647
– – × *brunneus* 229*, 629, 631, 632, 633
– *brunneus* 20, 21*, 64*, 79, 92, 95–100, 228, 229*, 297*, 303, 351*, 353, 358*, 360, 361, 365, 584, 588, 591, 594, 595, 596, 597, 600, 602, 603, 604, 608 (nymph), 611, 613, 614, 615*, 618, 620, 631*, 632, 633, 638, 641
– *curtipennis* 303, 349
– *dorsatus* 594, 595, 596, 600, 602, 612
– *jucundus* 296*, 353, 360, 613, 614, 647
– *longicornis* (see *Chorthippus parallelus*)
– *mollis* 594, 596, 647
– *montanus* 297*, 594, 595, 596, 600, 601, 602, 604, 608 (nymph), 612, 643
– *parallelus* 20, 21*, 297, 303, 353, 584, 588, 594, 595, 596, 600, 602, 608 (nymph), 612, 613, 616, 628, 640, 641, 642, 643
– *pullus* 594, 596, 600, 602
– *vagans* 594, 596, 600

Chorus frog *(Pseudacris nigrita)* 701, 703, 706
– –, Ornated *(Pseudacris ornata)* 701, 804
– –, Spotted *(Pseudacris clarki)* 701, 703, 706
– –, Strecker's *(Pseudacris streckeri)* 701, 704, 706, 804
Chrysochraon dispar 594, 596, 600, 612
Chrysocyon jubatus 779
Chrysomelidae 281, 287, 336, 605
Chub sucker (see *Moxostoma*) 666
Cicada 74, 398*
Cicadellidae 71, 393, 397, 400
Cicadelloidea 394
Cicadetta coriaria 417*
Cicadidae 334, 347, 359, 393, 396, 400, 415, 418, 422, 427, 618
Cicindelidae 336
Ciconiidae 78, 82, 714, 719
Ciconiiformes (817)
Cimelia 435 (footnote)
Cimicomorpha 402, 404, 405, 409
Cirripedia (805)
Cirtophyllus perspicillatus 367*
Citellus beecheri 765
– *citellus* 753
Cladocera (805)
Clethrionomys glareolus 769
Clibanarius (see *Trizopagurus*)
Clupeidae 664–666, 669, 677, 682
Clythra 281, 287
Cobitidae 662, 677, 686
Cobra 518
Coccidae 393
Coccothraustes coccothraustes 732
Cock (see *Gallus gallus*)
Cockroach (see Dictyoptera)
Codfish (see *Gadus*, Gadidae)
Coenobita perlatus 309
– *rugosus* 309*
Colaptes auratus 721
Coleoptera (809)
Coleorrhyncha 393
Colisa lalia 661
Collocalia brevirostris 95 (footnote), 217, 218*, 741–745
Colobus 782
Columba livia 735, 737
– *palumbus* 719, 722
Columbiformes (818)
Composia fidelissima 325
Conehead (see *Neoconocephalus* 299)
Conger eel *(Conger conger;* see *Leptocephalus conger)*
Connochaetes taurinus albojubatus 758
– *gnu* 758, 767
Conocephalidae 348, 359
Conocephalus 74, 301
– *brevipennis* 303, 348, 359, 647, 648

– *dorsalis* 8
– *fasciatus* 359, 647, 648
– *maculatus* 298
– *spartinae* 348
– *strictus* 421*
Contopus virens 237
Cophus 419
Copris 293, 337
Coraciadiformes (818)
Coralliocaris 306
Coranus subapterus 281*, 406, 407, 605
Coreidae 403
Coreinae 403
Coreoidea 402
Corimelaena lateralis 798
Corimelaeninae 403
Corixa 280, 292, 405, 408, 415, 420
– *striata* 280*, 610 (footnote), 616, 622
Corixidae 405, 408, 586
Corixinae 405
Cormorant *(Phalacrocorax)* 88, 105, 712
Cornitermes 327
Corvidae 44, 57*, 74, 81, 85–87, 89, 90, 93, 94, 95, 104, 105, 566, 723, 733, 744
Corvina nigra 662, 667, 685
Corvus brachyrhynchos 88–90, 735
– *corone* 88, 723, 743, 744
– – *cornix* 745
– *frugilegus* 88, 93, 720, 722, 723
– *monedula* 71, 88, 89, 93, 711, 720, 723
Cossidae 436 (footnote)
Cottidae 661
Cottus gobio 661
– *scorpius* 661
Cow (see *Bos taurus)*
Cowbird, Brown-headed *(Molothrus ater)* 238
Coyote *(Canis latrans)* 755, 766, 776, 779
Coypu *(Myocastor coypus)* 753
Crab 310–315
–, Hermit 309–310
Crane *(Grus grus)* 722
Crangon (see *Alpheus)*
Creeper, Short-toed tree *(Certhia brachydactyla)* 231
Crematogaster 622
– *scutellaris* 304, 590
Crenilabrus griseus 662
– *melops* 662
– *pavo* 662
Cricetus 216, 762, 771, 776
– *cricetus* 753
Cricket (see Grylloidea, Gryllidae, *Gryllus, Acheta)*
–, Ground *(Nemobius)*
–, House *(Acheta domestica)*
–, Tree *(Oecanthus)*
Crinia insignifera 699

ALPHABETIC INDEX 917

– *parinsignifera* 699
– *signifera* 699
– *sloanei* 699
Crioceris duodecimpunctata 287*, 584
– *merdigera* 304
Criomorphus bicarinatus 604
Croakers (see Sciaenidae)
Crocidura leucodon 766, 769
Crocodile 74, 81, 491
Crocodilia 541, (817)
Crocuta 766, 773, 776
– *crocuta* 756, 765*
Crossarchus 763, 767
– *fasciatus* 756, 763*, 773
Crossbill *(Loxia curvirostra)* 735, 737
Crossopterygii (816)
Crotophaga ani 85
Crow (see Corvidae)
–, American *(Corvus brachyrhynchos)* 88–90, 735
–, Carrion (see *Corvus corone*)
–, Hooded (see *Corvus corone cornix*)
Crustacea (805)
Cryptobunus 318
– *sylvicolus* 319*
Cryptocerata 415
Cryptorrhynchus 288
– *lapathi* 293
Cryptotympana exalbida 399*
Ctenomys 773
Ctenoscelis 296
Cuckoo *(Cuculus canorus)* 231
Cuculiformes (818)
Cuculus canorus 231
Culex 71, 351*, 415
– *pipiens* 621, 625
Culicidae 376*, 379, 413, 414*, 426, 428
Cuniculus 762
Cuon alpinus 755
Curculionidae 287, 288, 293, 336
Cyanosylvia svecica 732
Cybister aegyptiacus 304
– *confusus* 325, 331 (larva)
Cyclostomata (814)
Cydnidae 392, 402, 403, 408
Cylindracheta arenivaga 279*, 296 (footnote)
Cymatogaster aggregatus 661, 669
Cynocephalus (see *Papio*)
Cynomys ludovicianus 71, 83, 126, 753, 772
Cynopterus 214
Cynoscion 682
– *regalis* 670, 682
Cyphoderris monstrosus 282*, 299, 336
Cyprinidae 662, 665, 677, 686
Cyprinodontidae 661
Cyprinus 656, 659
Cyrtaspis scutata 300*
Cyrtoxipha columbiana 617, 648

Cystophora cristata 762

Dacus 486
– *oleae* 290*
– *tryoni* 290*, 349
Dama dama 501, 758
Dasyprocta 753
– *agouti* 761
Death head's moth *(Acherontia atropos)* 73, 106, 323, 324*, 332 (pupa), 349, 351*, 353, 605, 607
Death-watch 326
Decapoda (805)
Decticus 301, 359, 422, 429, 638 (footnote)
– *albifrons* 357
– *verrucivorus* 348, 350, 351*, 419*, 421, 613 (footnote)
Deer
–, Fallow (see *Dama dama*)
–, Red (see *Cervus elaphus*)
Deinacrida megacephala 285
Delphinapterus leucas 71, 220, 790
Delphinidae 81, 220–223, 503*, 670, 762
Delphinus delphis 220–221*, 790
Dendrocopus major 74, 718, 745
– *minor* 718
Dendrolagus leucogenys 762, 764
Dendromyrmex 328
Desmodontidae 208, 210, 211
Desmognathus ensatus 694, 695
Diceratothrips 291
– *princeps* 291*
Diceros 762, 763, 764, 776
– *bicornis* 757, 764*, 765*, 780
Dicotyles pecari 758, 766*
– *tajacu* 758, 763*
Dicranura 332 (larva)
Dictyophorus 324
Dictyoptera (806)
Didelphys virginiana 762
Diguetia 316
Dingo *(Canis dingo)* 755, 766, 776, 779
Dinoceras cephalotes 328
Diodontidae 680, 685
Diplopoda (806)
Diplostraca (805)
Dipnoi 489, (816)
Diptera (811)
Dipus aegypticus 761
Dirias albiventer minor 205
Discoglossidae 696
Dog (see *Canis familiaris*)
–, Bush *(Speothos)* 779
–, Prairie *(Cynomys)* 71, 83, 126, 753, 772
–, Raccoon-like *(Nyctereutes)* 232, 756, 765, 779
Dogfish (see *Mustelus canis*)
Dolichotis patagona 753

Dolphin (see Delphinidae)
–, Bottlenosed *(Tursiops truncatus)* 71, 220, 223, 790
–, Common *(Delphinus delphis)* 220–221*, 790
–, Risso's *(Grampus griseus)* 790
–, Spotted *(Stenella)* 220
Dolychonyx oryzivorus 238
Donkey 518
Doratura stylata 77, 370, 594, 596
Dorcus (larva) 330
Doritomus longimanus 293*
Dormouse (see *Glis*)
Dorylidae 283
Dorymyrmex emmaericellus 283
Dotilla 312, 589
– *fenestrata* 313*
Dove, Ringed turtle- *(Stretopelia risoria)* 231
–, Rock *(Columba livia)* 735, 737
Dragonfly (see *Aeschna* 453, 459, *Epiophlebia* 333, 334*)
– (larva) 74
Drepana 332 (larva)
Drosophila 54, 55*, 379, 380, 453
– *funebris* 379, 385*
– *melanogaster* 379
– *repleta* 379
Drumfish 672 (see also *Pogonias*)
Duck 86 (footnote)
–, Domestic *(Anas platyrhynchos)* 105, 713, 720, 721, 735
–, Pintail 88
Dudgeona 436 (footnote)
Dumetella carolinensis 238
Dunnock *(Prunella modularis)* 729
Dynastinae 288, 330, 331
Dysschema tiresias 324
Dytiscus 386, 413

Echo fish 224, 675
Ectatorrhinus 288
Edentates 494, 501, (820)
–, Conger (see *Conger conger, Leptocephalus conger*)
–, Moray (see Muraenidae) 670
Eel *(Anguilla)* 677
Egnatioides 284
Eidolon 214
Eland (see *Taurotragus oryx*)
Elaphrus 288
– *riparius* 304
Elaphurus davidianus 758
Elasmobranchs 531, 657, 659, 661, 663, 664, (814)
Flasmodemidae 404
Elateridae 583
Electrophorus electricus 662
Elephant 83, 762, 776
– fish (see Mormyridae) 662, 665

Elephas maximus 757, 777
Eleutherodactylus augusti 699, 700*, 803
Eligma hypsoides 333* (pupa)
– *narcissus* 333* (pupa)
Emballonuridae 207
Emberiza calandra 231
– *citrinella* 237, 732
– *schoeniclus* 231, 233, 732, 742
Emberizidae 729
Embiotocidae 661
Empidinae 599
Encoptolophys sordidus 328
Endacusta 419*
Endomychidae 280, 304
Endrosa 71, 282, 586
– *aurita ramosa* 434
Enoplopus velikensis 280*, 304
Entelecara broccha 317
Eohelea stridulans 302
Eonycteris 214
Ephigger 59*, 60*, 62*, 64, 71, 72, 76, 77, 79, 91, 100, 104*, 105, 301, 360, 364, 365, 366, 584, 586 (footnote), 591, (footnote), 595, 597, 598 (footnote), 604, 605, 611*, 612, 614, 615, 618, 619, 620, 622, 626, 639, 640, 642, 646 (footnote)
– *bitterensis* 44*, 60*, 63*, 301*, 303, 359, 365, 443, 612, 619*, 620, 622, 623, 647
– *cunii* 612, 615, 622, 623, 647
– *ephippiger* 299*, 348, 352*, 358*, 359, 360, 364*, 612, 623, 638, 640, 644, 645*, 647
– *provincialis* 63*, 361, 367, 369*, 612
– *terrestris* 353
Ephippigerida nigromarginata 622, 623
Ephippigeridae 299, 359, 583
Epinephelus striatus 679, 684
Epiophlebia superstes (larva) 333, 334
Eptesicus fuscus 189–190*, 193*, 198*, 206, 508*, 792
Equidae 762, 767, 769
Equus asinus 757, 766*, 767, 781
– *caballus* 74, 83, 90 (footnote), 496, 501, 504*, 535, 758, 767
– *grevyi* 757, 766*, 767, 781
– *przewalskii* 758
– *quagga antiquorum* 757
– – *granti* 757
– *zebra hartmannae* 758
Erasmia sanguiflua 325
Eremogryllus hamadae 366
Erethizon dorsatus 753
Ergastes faber 282*
Erigone 317
Erinaceus 607, 770
– *europaeus* 753
Eristalis 423
– *tenax* 378, 388

Erithacus rubecula 79, 80, 613 (footnote), 713, 716–719, 721, 725, 732, 735, 741
Erotylidae 304
Erythroneura hyperici 612
Esocidae 661
Esox lucius 657
Eubalena 223
Eupelix depressa 594
Eupemphix pustulosus 699, 700*
Euphrictus 316
Euprepia 282, 586
– *haroldi* 596
Eurycea bislineata 694, 695
Eurymelidae 393, 400
Euselachil (814)
Euthystira brachyptera 594, 596, 600, 641*, 642, 643
Euvanessa antiopa 329
Evotomys (see *Clethrionomys* 769)
Eysarcoris 798

Falco sparverius 82, 735
Falconiformes (818)
Felidae 502, 504, 516, 762, 780*
Felinae 764, 766, 769, 780
Felis catus 232, 499, 502, 503, 507*, 519, 753, 757, 768, 769, 770, 771, 780
– *chaus* 757
– *pardalis* 757
– *(Puma) concolor (Puma concolor)* 757, 762*, 766, 780*
– *serval* 757, 766, 769, 780*
– *sylvestris* 771
– – *griselda* 757, 780*
– *wiedi* 757, 780*
Fennecus 769
– *zerda* 779
Fiddler crab (see *Uca* 74, 314, 583, 590, 599, 606, 612, 618)
Field cricket (see *Acheta assimilis, Gryllus campestris*)
Finch 722
–, House *(Carpodacus mexicanus)* 231
Firefly 74, 617
Fissipedia (822)
Fly (see *Calliphora, Musca*)
–, Bot 387
–, Fruit (see *Ceratitis, Dacus, Drosophila*)
–, Gad 387
–, Horse 387
–, Warble 387
Flycatcher
–, Collard (see *Muscicapa albicollis*)
–, Pied (see *Muscicapa hypoleuca*)
Flying fox *(Megachiroptera)* 211–215
Forcipomyia 382, 383*
Formica sanguinea 413*

Fox (see *Vulpes vulpes*)
–, Arctic (see *Alopex lagopus*)
–, Gray (see *Urocyon*)
Frickius 291
Fringilla coelebs 71, 231, 235–237, 240, 241, 711, 716, 721, 727–730*, 732, 735, 741*, 742, 743, 783, 795, 796
– *montifringilla* 729, 742
Frog 76, 103, 104, 491*, 499, 527*, 537, 566
–, Chorus (see *Pseudacris nigrita*) 701, 703, 706
–, Cliff *(Syrrhophus marnocki)* 699, 700*, 704
–, Crawfish *(Rana areolata)* 702
–, Cricket *(Acris)* 706
–, Edible *(Rana esculenta)* 490
–, Green *(Rana clamitans)* 703–705
–, Leopard *(Rana pipiens)* 71, 702–705
–, Mexican white-lipped *(Leptodactylus labialis)* 700*, 803
–, Texan barking *(Eleutherodactylus augusti)* 699, 700*, 803
–, Tree, see Treefrog
–, Wood *(Rana sylvatica)* 537*
Fulgoroidea 393, 394, 400
Fundulus heteroclitus 656, 661
Furnarius rufus 82

Gad-fly 387
Gadidae 661, 679, 683
Gadus aeglefinis 661
– *callarias* 661
Galago 505, 518
Galerida cristata 732
Galeodes araneoides 318*
Galliformes (818)
Gallus gallus 57*, 70, 71, 72*, 81, 82, 93, 231, 233, 492*, 542, 566, 711–713, 721, 722, 725–726, 738, 740
Gampsocleis buergeri 426
– *glabra* 357, 366
Gangara thyrsis 332 (pupa)
Garrulus glandarius 74, 88, 90 (footnote), 91, 93, 91*, 105, 711, 722, 723, 732, 733, 740*, 743, 744
Gasterocercus 288
Gastrophryne (see *Microhyla*)
Gazella soemmeringi 759
– *subgutturosa* 759
Gebia issaeffi 309*
Gekko 491
Geniates 296
Geocorisae 402, 408
Geometridae 415, 428, 434
Geopelia 231
Geothyplis trichas 237
Geotrupes 288, 291, 349
– *stercorarius* 304, 376*
– *sylvaticus* 331* (larva)

– *vernalis* 290*
Geotrupidae 330
Gibbon (see *Hylobates, Symphalangus*)
Giraffa 74, 768
Glis 216, 762
– *glis* 753
Globicephala 220, 791
– *melaena* 71, 790
– *sylvaticus* 331* (larva)
– *vernalis* 290*
Globopilumnus 314
– *stridulans* 315*
Glossophaga soricina leachi 209, 210
Glow worm 74
Glutton (see *Gulo gulo*)
Glyphipterygidae 436 (footnote)
Glypsus 404
Gnathonemus 662, 667
Gnu (see *Connochaetes*)
Goat (see *Capra hircus*)
Gobiidae 662
Gobius niger 662, 667
– *paganellus* 662
Goldcrest *(Regulus regulus)* 732, 734
Goldfinch
–, American (see *Spinus tristis*)
Goldfish (see *Carassius, Cyprinus*) 656, 659
Gomphocerus 594
– *rufus* 297*, 303, 600, chapter 17
Gongylidiellum mursidum 317
– *vivum* 317*
Gonodactylus chiragra 305
– *demani* 305
– *glabrous* 305
– *oerstedti* 305
Goose, grey-leg *(Anser anser)* 71, 81, 83, 720, 721
Gorilla gorilla 498, 511, 512, 518*, 519, 755, 761, 772, 777
Gnathostoma (814)
Grackle, Purple *(Quiscalus quiscula)* 231, 238
Grampus griseus 790
Graphopterus 334
– *serrator* 296
– *variegatus* 296*
Graptopsaltria nigrofuscata 398
Grasshopper 126, 127 (see Tettigoniidae, Tettigonioidea)
–, Meadow (see *Conocephalus, Orchelimum*)
Great crested grebe *(Podiceps cristatus)* 720
Greenfinch *(Chloris chloris)* 732, 735
Grey-leg goose *(Anser anser)* 71, 81, 83, 720, 721
Griotettrix verruculatus 304
Grison vittatus 756
Grosbeak, Pine *(Pinicola enucleator)* 738
–, Rose-breasted *(Pheucticus ludovicianus)* 238
Ground cricket (see *Nemobius*)
Grouper *(Epinephelus)* 679, 684

Grouse 80, 713
–, Long-tailed 74
–, Wood 74
Grunt 676
Grus grus 722
Gryllacrididae 338 (footnote)
Gryllacridoidea 285, 299
Gryllidae 348, 351*, 355 (footnote), 359, 392, 440, 464, 484, 583, 590, 591, 597, 603, 604
Grulloidea 299, 302, 303, 335, 347*, 350, 352*, 360, 362–364, 418, 419*, 422, 583, 599*
Gryllotalpa 419
– *gryllotalpa* 348, 359, 360, 361, 622
– *hexadactyla* 348, 361, 584, 647, 648
Gryllotalpidae 302, 303, 348, 359
Gryllus 233, 424*
– *abbreviatus* 422*
– *assimilis* (see *Acheta assimilis*)
– *bimaculatus* 302*, 303, 360, 481, 591, 643
– *campestris* 71, 76, 302*, 348, 352*, 358*, 359, 360, 366, 369, 419*, 422, 423, 431, 584, 588, 591, 603, 618, 640, chapter 17
– – × *bimaculatus* 229, 629, 632 (footnote)
– *toltecus* 419*
Guacharo *(Steatornis caripensis)* 12, 95 (footnote), 216, 217, 741, 745
Guinea pig 103, 533*, 535, 552*, 553*
Gull
–, Black-headed 85
–, Herring *(Larus argentatus)*
–, Laughing *(Larus atricilla)*
–, Mediterranean *(Larus melanocephalus)* 720
Gulo gulo 756
Gurnard (see *Triglidae*)
Gymnotidae 662
Gyrinus 218, 219, 413

Haaria 419*
Haddock *(Gadus aeglefinus)* 661, (see also *Melanogramus angelifinus* 683)
Haematopus ostralegus 83
Haemulidae 680, 682, 685
Halichoerus grypus 757
Halmaturus 501
Hamlet (see *Epinephelus*) 679, 684
Hamster (see *Cricetus*)
Hapale jacchus 754
Hare *(Lepus)* 766
Harvest spider *(Cryptobunus)* 318
Hawfinch *(Coccothraustes coccothraustes)* 732
Hecatesia fenestrata 294
– *thyridion* 294
Hedgehog *(Erinaceus)* 607, 770
Helarctos malayanus 756
Heleioporus 699
Helice tridens 311*
Heliochelus paradoxus 294

Heliocopris 291
Heliophilus cribratostriatus 287
Hemaris fuciformis 379
Hemigrammus caudovittatus 662
Hemionus hemionus hemionus 758
– – *onager* 758
Hen (see *Gallus gallus*)
Henicus monstrosus 329
Hermit crab 309–310
Heron 78
–, Night *(Nycticorax nycticorax)* 719
Herring (see Clupeidae)
Herpestinae 773
Hesperiidae 332 (pupa)
Hesperocorixa 405
Heteroceridae 286
Heterometrus costimanus 319*
– *swammerdami* 319
Heteroptera (812)
Hexaplax 312
Hexapoda (see Insects)
Hexapus 312
Hippocampus 71
Hippolais icterina 732*
– *polyglotta* 732
Hippopotamus 538, 773, 781
– *amphibius* 758, 766*
–, Pigmy (see *Choeropsis*)
Hirundo rustica 231, 732
Hispa testacea 304
Hispidae 280
Holocentridae 665, 679, 682, 685
Holocentrus 665
– *ascensionis* 677, 684, 685
Hominoidea 765, 767, 775, 783
Homo see Man
Homoptera (812)
Homorocoryphus nitidulus 348, 359, 647
Honey bee *(Apis mellifica)* 40, 71, 325, 349, 376*, 379–383*, 386, 423, 583, 724
Honey guide *(Indicator)* 91, 106, 724
Hoopoe *(Upupa epops)* 733
Hornia minutipennis 387
Horse (see *Equus caballus*)
Horse-fly 387
House cricket (see *Acheta domestica*)
Huechys incarnata 418*
Hyaena 766
– *brunnea* 756
– *hyaena* 756, 766*
Hyena (see *Crocuta, Hyaena*)
–, Painted (see *Lycaon pictus*)
Hybalus 291
Hydrochoerus 762, 769
Hydrocorisae 402, 405, 406
Hydrophylidae 297, 336
Hydrophilus piceus 287

Hygrobia hermanni 293
Hyla 700, 794, 804
– *andersoni* 701, 702*
– *arborea* 104, 614 (footnote)
– *arenicolor* 700, 701*, 704
– *baudini* 700
– *chrysoscelis* 803
– *cinerea* 700, 702*
– *crucifer* 701, 704
– *ebraccata* 804
– *eximia* 804
– *femoralis* 700, 701*
– *gratiosa* 701, 702*, 704, 706
– *microcephala* 804
– *phaeocrypta* 700, 701*
– *phlebodes* 804
– *regilla* 701
– *septentrionalis* 701
– *squirella* 701, 702*, 704
– *versicolor* 700, 701*, 703, 704, 803
– *wrightorum* 701
Hylidae 699–702
Hylobates 232, 499, 508, 511*, 513, 516, 774
– *hoolock* 755, 782*
– *lar* 71, 755, 767, 782*
– *leucogenys* 232
Hylobatinae 782*
Hylocichla mustelina 238, 711, 712, 739
Hylophila prasinana 323
Hymenoptera (811)
Hyperoodon ampullatus 790
– *planifrons* 790
Hyphessobrycon flammeus 662
Hypopachus cuneus 702
Hypsognathus 504
Hysterocrates 316
Hystrix 761
– *cristata* 753

Icteria virens 238
Icterus galbula 238, 732
– *pyrrhopterus* 724
– *spurius* 238
Idiocerus 612
Idiostoma 291
Idus melanotus 662
Illibius 403
Ilyoplax 590
Indicator 91, 106, 724
Insecta (806)
Insectivora (819)
Isopoda (805)
Isoptera (806)

Jackal (see *Canis adustus, aureus* and *mesomelas*)
Jackdaw (see *Corvus monedula*)

Jaguar (see *Panthera onca*)
Jay (see *Garrulus glandarius*)
Jerboa *(Dipus)* 761
Jewfish 673
Junco oreganus 795

Kangaroo *(Macropus)* 84, 516, 761
Katydid
-, Angular-winged (see *Microcentrum*)
-, Bush (see *Scudderia*)
-, Northern true (see *Pterophylla camellifolia*)
-, Oblong-winged (see *Amblycorypha oblongifolia*)
-, Round-winged (see *Amblycorypha rotundifolia*)
Kinkajou (see *Potos flavus*)
Kleidocerys 403, 406, 408, 605
– *ericae* 508, 799
– *resedae* 588, 799
Knife fish *(Xenomystus nigri)* 224, 686
Koala *(Phascolarctos cinereus)* 753, 763

Labridae 662
Lacerta 491
– *viridis* 103
Lacertilia 532
Lagochile 292
Lagomorphs 501, (821)
Lagostomus 753
Lagotricha 754
Lama glama 758, 770
– *huanacus* 758
Lamarckiana 298
Lambdophallus 312
Lamprey (see *Petromyzon*)
Lampyridae 583
Languriidae 280
Lanius collurio 719, 732, 795
Lapwing
-, Spur winged *(Belonopterus chilensis)* 83
Largidae 403
Larginae 403
Lariformes (818)
Lark
-, Crested *(Galerida cristata)* 732
-, Sky *(Alauda arvensis)* 235, 732
Larus argentatus 44, 71, 85-87, 90, 107, 711, 713, 720-725, 728
– *atricilla* 713
– *melanocephalus* 720
Lasiurus borealis 212
Lawrencella 318
Leaf-insect (see *Phyllium, Pulchriphyllium*)
Lebistes reticulatus 661
Leiostomus xanthurus 222
Lema trilineata 287*

Lemur 505
– *catta* 754
– *mongoz* 754
Lemuroidea 754
Leontocebus 505
– *oedipus* 754
– *rosalia* 754
– *spixi* 754
Leopard (see *Panthera pardus*)
-, Clouded (see *Panthera nebulosa*)
Lepidocolaptes affinis 720
Lepidoptera (810)
Lepinotus inquilinus 327
Leptiphantes 316
Leptodactylidae 699, 700*, 803
Leptodactylus labialis 699, 700*, 803
Lepus 502, 516, 761
– *timidus* 766
Lepyronia 798
– *coleoptrata* 612
Lepyrus 293
Leuciscus dobula 662
– *idus (Idus melanotus)* 662
Ligyrus 293
Lilioceris lilii 360, 584
Linnet (see *Carduelis cannabina*)
Linuparus 308
Lion (see *Panthera leo*)
-, Ant (see Myrmeleontidae)
-, Mountain (see *Puma concolor*)
-, Sea *(Zalophus)* 757, 765, 768
Lipotes vexillifer 790
Lispomontia 318
Lissonycteris 214
Lithyphantes 318
Livia juncorum 401
Lizard (Lacertilia: *Lacerta*) 103, 491, 532
Lobster, Spiny (see Palinuridae)
Locust (see *Locusta, Nomadacris*)
-, Sprinkled (see *Chloealtis conspersa*)
Locusta 74
– *migratoria* 71, 298*, 329, 348, 422, 429*, 447
– – *(solitaria)* 328, 353, 595, 604
– – *gallica (gregaria)* 304; *(solitaria)* 304
– – *migratorioides* 421
– *pardalina* 333 (nymph), 608 (nymph)
Locustella naevia 231
Locustidae (see Acrididae)
Lomaptera 286
Longiphylla robusta 209, 210
Lonchorhina aurita 210
Longiunguis 284
Loris 518
Louse, Book (see *Psocus*)
Loxia curvirostra 735, 737
Loxodonta africana 757
Lucanidae 331, 336

Lucanus (larva) 330
Luscinia megarhynchos 74, 79, 80, 238, 715, 732
Lutrecla lutreola 756
Lycaenidae 232 (pupa)
Lycaon pictus 756, 778, 779
Lycosa chelata 328
Lygaeidae 392, 403
Lygaeoidea 402
Lymantria monacha 283
– *viola* 332* (pupa)
Lymantriidae 332 (pupa)
Lymantriinae 415
Lynx caracal 756
– *chaus* 780*
– *lynx* 756
Lyre bird 74
Lyristes plebejus 622
Lysiosquilla excavatrix 305

Macaca 151, 157*, 506 (see also *Zati* 773)
– *maura* 754
– *mulatta* 754, 781
– *nemestrina* 754, 781
– *radiata* 754, 781
– *silenus* 754, 781
– *speciosa* 754
Macaque (see *Macaca*)
Mackerel 669, 670
Macraspis 292
Macropodus cupanus 661
– *opercularis* 677, 679
Macropsis 605
Macrouridae 677, 679
Macrura (805)
Magicicada cassinii 354, 392, 588, 617, 638
– *septemdecim* 354, 392, 398, 588, 617, 638
Macroglossus 214
Macrophthalmus 311, 590
Macrophyllum macrophyllum 210
Macropus 516, 761
– *benetti* 84
Macrorhinus 764, 776, 783
– *angustirostris* 757, 764*
Magpie (see *Pica pica*)
Malacostraca (805)
Mallard *(Anas platyrhynchos)* 105, 713, 720, 721, 735
Mammalia (819)
Man *(Homo)* 80, 155*, 156*, 162, 423, 496, 498, 499, 505, 509, 513, 518*, 565*, 566*, 571*, 736*, 738, 775, 791, chapter 20
Manatee *(Trichechus manatus)* 757
Mangalica pig 758
Mantis shrimp (see *Gonodactylus*, *Squilla*)
Marcusenius isidori 662, 667
Margay (see *Felis wiedi*)
Marmoset *(Leontocebus)* 505

Marmot (see *Marmota*)
Marmota 74, 84, 770
– *sibirica* 84
Marsupials 494, 497–501, 504, (819)
Martes martes 756
Martin, Sand (see *Riparia riparia*) 231
Masoreidae 336
Matuta 311, 606
– *lunaris* 312*
Meadow grasshopper (see *Conocephalus*, *Orchelimum*)
– lark
– –, Eastern (see *Sturnella magna*)
– –, Western (see *Sturnella neglecta*)
Mecideinae 403
Mecistorhinus 404
Meconema thalassina 328, 329
Mecosthetus grossus 416*, 594, 609
Megachile 387
Megachiroptera 211–215
Megadermidae 211
Megalobatrachus japonicus 694
Megalotis 780
– *(Vulpes) chama* 756, 769, 779*
Megaponera 590
Megaptera boops 223
Meimuna 398
Melanogramus angelifinis (Gadus aeglefinis) 683
Melanolestes 798
Melichthys 678
Mellivora capensis 724, 756
Meloidae 387
Melolontha 376*
Melolonthinae 330, 331
Melopsittacus undulatus 735, 737
Melospiza melodia 71, 231, 237, 711, 714, 721
Membracidae 393, 400
Menhaden (see *Brevoortia tyranus*) 686
Menippe 312, 606
– *obtusa* 312*
Menthicirrhus 682
Mephitis 74
Meriones persicus 771
– *shawi* 761
Mesembria dubia 329
Mesocricetus auratus 753
Mesoplodon bidens 790
Messor 350, 590
Metapenaeopsis acclivis 306
– *barbatus* 305, 306
– *durus* 306
Metaplax 311
Metrioptera bicolor 351*
– *brachyptera* 351*
– *roëseli* 613 (footnote)
Mezirinae 403, 404

Microcentrum 299
– *rhombifolium* 300, 357, 595
Microchiroptera 185–211
Microhyla 706
– *carolinensis* 702, 703
– *olivacea* 702–704
– – × *carolinensis* 229
– *usta* 702
Microhylidae 699
Microlepidoptera 435
Micronecta poweri 407
Micropogon undulatus 655, 672, 680
Microtus arvalis 763
– *brandti* 753, 777
– *oeconomus* 753
Microvelia diluta 405
Millipede 321–323
Mimus polyglottos 85, 233, 238–240, 730–732
Mink (see *Lutreola*)
Minnow (see *Phoxinus laevis*)
Miogryllus verticalis 348, 593, 603
Mirounga angustirostris 796
– *leonina* 757
Mixodiaptomus laciniatus 789
Mockingbird (see *Mimus polyglottos*)
Molossidae 204
Molothrus ater 238
Mongoose (see *Crossarchus, Suricata*)
Monkey 106, 518, 519
–, Howler (see *Alouatta*)
–, Spider (see *Ateles*)
–, Wolly (see *Lagothrix*)
Monodon monoceros 790
Monomorium salomonis 350
Monotremes 494, 497–499, 530, (819)
Moose (see *Alces*)
Moray eel (see Muraenidae) 670
Mormyridae 662, 665
Mosquito 74, 76, 107, 379, 380, 386, 618, 620, 621, 625 (see also *Aëdes, Anopheles, Chironomus, Culex,* Culicidae, *Psorophora*)
Moth 87
Mouse (see *Mus*)
–, Field (see *Apodemus*)
–, Pouched (*Sminthopsis crassicaudata*) 762
Moxostoma 666
Mugil cephalus 222
Mullet 670
Muntiacus 762
– *muntjac* 758
Muraenidae 670
Mus musculus 84, 103, 244 → , 535, 763
Musagetes 316
Musca autumnalis 379
– *domestica* 376*, 379
Muscicapa albicollis 231, 237, 742
– *hypoleuca* 231, 237, 742

Musicodamon atlanteus 318*
Mustela nivalis 756
– *putorius* 756
Mustelus canis 663
Musurgina laeta 294*
Mutilla barbata brutia 283
– *europaea* 283
Mutillidae 283
Mycetes (see *Alouatta*)
Myocastor coypus 753
Myotis 185–204, 209, 210, 213, 770, 789
– *koenii* 193*
– *lucifugus* 185–204, 189*, 193*, 213, 508*, 789
– *macrotarsus* 205
– *natteri* 194
– *(Rickettia) pilosa* 205
Myriapoda (806)
Myripristinae 665
Myrmeleontidae 428
Myrmeleotettix maculatus 297*
Myrmica rubra laevinodis 283*
– *ruginodis* 350
– *sulcinodis* 304
Myrmicidae 283
Mysticeti 223
Myxine 530
Myxinoidea (814)
Nabidae 405
Nabis flavomarginatus 284
Narwhal *(Monodon)* 790
Naucoridae 406
Necrophorus 74, 288, 334, 349, 605
– *vespillo* 289*, 304
– *vespilloïdes* 349, 351*
Negaprion brevirostris 686
Nemacheilus (see *Noemacheilus*)
Nemastoma argenteo lunulatum 318, 319*
– *dentipalpes* 318, 319*
Nemobiinae 583, 593 (footnote)
Nemobius allardi 228, 348, 367, 634, 637, 647, 648
– *allardi* × *tinnulus* 629*
– *ambitiosus* 637
– *aurantius* 637
– *bruneri* 637
– *carolinus* 348, 363, 369*, 370*, 636, 637
– *confusus* 361, 369*, 370*, 637
– *cubensis* 637
– *fasciatus* 303, 351*, 369*, 593, 634, 637, 647, 648
– – *fasciatus* (see *allardi*)
– – *socius* (see *fasciatus*)
– – *tinnulus* (see *tinnulus*)
– *griseus funeralis* 637
– *maculatus* 637
– *melodius* 359, 593*, 636
– *palustris* 348, 637

– *sparsalsus* 637
– *sylvestris* 584
– *tinnulus* 228, 358*, 359, 361, 369*, 634, 637 647, 648
Neoconocephalus 299
– *ensiger* 348, 367, 368*, 443, 587, 647
– *exiliscanorus* 605, 617
– *nebrascensis* 368*, 584, 587
– *robustus* 359, 584
– *triops* 300*, 648
Neomyrma rubida 304
Neophilaenus campestris 393*
Neotoma fuscipes 761
Neoxabea bipunctata 348, 359
Nepidae 406
Neuroptera (810)
Newt *(Taricha, Triturus)* 694
Nezara viridula 404
Nighthawk 218
Nightingale (see *Luscinia megarhynchos*)
Nightjar *(Caprimulgus europaeus)* 76, 719
Nitidulidae 280
Noctilio leporinus 205, 206
Noctilionidae 205, 206
Noctuidae 294, 332 (pupa), 415, 427, 428, 436, 437, 800
Noemacheilus barbatulus 662
Nomadacris septemfasciata 103*, 298
Northern true katydid (see *Pterophylla camellifolia*)
Notodontidae 415, 420, 435
Notonecta 405
Notonectidae 405, 408, 586
Notonectinae 405
Notopteridae 224
Notorrhina muricata 326, 612
Notropis analostanus 686
– *bifrenatus* 686
Nuthatch (see *Sitta europaea*)
Nyctalus noctula 189, 212, 767
Nycteridae 208, 211 (footnote)
Nycteris thebaica 208
Nyctereutes procyonides 232, 756, 765, 779
Nycticorax nycticorax 719

Oblong winged katydid (see *Amblycorypha oblongifolia*)
Ochodaeus 293
– *maculipennis* 293*
Ochotona princeps 232
Octotympana 398
Ocypode 313, 334, 335, 587, 589, 606, 612
– *cordimana* 313 (footnote)
– *cursor* 314*
– *nobilii* 314*
– *platytarsis* 313*
Odobenus 773

– *rosmarus* 757
Odonata (806)
Odontoceti 220–223, 496, 539*
Odontophorus guyanensis marmoratus 82
Oecanthidae 348, 355 (footnote), 359, 583, 590, 591, 593 (footnote), 604
Oecanthus 74, 350, 354, 419*, 447, 618, 620, 624, 626
– *angustipennis* 358*, 634
– *argentinus* 361, 593, 624
– *exclamationis* 348, 634
– *latipennis* 348, 647, 648
– *nigricornis* 348, 359, 369*, 370*, 624, 634
– *niveus* 303, 348, 358*, 359, 361, 363, 366, 368, 369*, 370*, 617, 624, 634
– *pellucens* 71, 348, 351*, 353, 359, 361, 363*, 369, 444, 587, 593, 604
– *pini* 647, 648
– *quadripunctatus* 303, 351*, 359, 361, 369*, 624, 634, 647, 648
Oedaleonotus fuscipes 329
Oedipoda coerulescens 327, 417*
Oedipodinae 297, 298*, 327, 328, 330, 583, 594, 596, 609
Oedipomidas (see *Leontocebus*)
Oilbird *(Steatornis caripensis)* 12, 95 (footnote), 216, 217, 741, 745
Okanagana rimosa 349
Okapia 773
Olocrates 605
– *abbreviatus* 287, 288*
– *gibbus* 287
Ommatocarcinus macgillivrayi 310
Omocestus 594
Omocestus haemorrhoidalis 596, 600
– *rufipes* (see – *ventralis*)
– *ventralis* 303, 600, 601*, 602, 613
– *viridulus* 297, 353, 361, 584, 595, 596, 600, 604, 628, 641
Oncopsis 605
Ophiocephalidae 677
Ophion luteus 376*
Opilionidea (814)
Opisthophthalmus 607
– *glabrifrons* 320
– *latimanus* 320*, 354, 359, 606
– *wahlbergi* 320
Opossum
–, Australian *(Trichosurus vulspecula)* 769
–, Virginian *(Didelphys virginiana)* 762
Opsanus beta 686
– *tau* 679, 680, 682, 686
Orang-outang (see *Pongo pygmaeus*)
Orchelimum 299, 360
– *agile* 367 (footnote), 639
– *bradleyi* 611, 612
– *concinnum* 359

[Orchelimum]
- militare 612
- nigripes 587
- vulgare 348, 359, 368*, 647, 648
Oriole, Baltimore (Icterus galbula) 238, 732
-, Golden (Oriolus oriolus) 231
-, Orchard (Icterus spurius) 238
Oriolus oriolus 231
Ornithorhynchus 762, 769
Orocharis 419
- saltator 348, 363, 369*, 370*, 647, 648
Orphania denticauda 604, 605, 638 (footnote)
Orphnus 291
Orphulella speciosa 330
Orsilinae 403
Orthoptera (806)
Oryctes 334
- nasicornis 288*
- rhinoceros 288, 330* (larva), 331 (pupa)
Oryctolagus 90, 91, 104, 247, 761
Oryx algazel 758, 764*, 770
Ostariophysi 540*, 662, 665–667
Osteichthyes (815)
Ostraciidae (Ostraciontidae) 685
Otaria byronica 757, 782
- jubata 495*, 497*
Otariidae 764
Otis tarda 78
Ovalipes ocellatus 310, 311, 314, 336, 337 (footnote)
- punctatus 314, 315*, 336, 337 (footnote)
Ovis 761, 767, 773, 780
- aries 759, 764*
- - musimon 759, 772
Owl (see Strigidae)
-, Barn (Tyto alba) 216, 543*
-, Eagle (Bubo bubo) 735
-, Long-eared (Asio otus) 735
-, Tawny (Strix aluco) 735
Oxychila 296
Oxystomata 336
Oystercatcher (Haematopus ostralegus) 83

Pachychoris 284
Pachyuromys duprasi 536*, 761
Packrat, Dusky footed (Neotoma) 761
Paguridea
Pagurus (see Trizopagurus)
Palaemonidae 306
Palinurichthys 659
Palinuridae 307–309
Palinurus argus 308, 349
- vulgaris 307, 308, 410, 604 (footnote), 685
Palophinae 329
Palmacorixa 405
Palomena 404
Pamphagidae 286, 298

Pan troglodytes 71, 132, 232, 498, 505, 510–513, 517*, 519, 753, 755, 761, 764, 766*, 772, 774, 775, 777, 782
- paniscus 782
Panda (see Ailuropoda melanoleucus)
Pandinus imperator 319
Panthera 773
- leo 74, 502, 503, 757, 762*, 766, 774, 780*
- nebulosa 757
- onca 757, 766, 780
- pardus 757, 762*, 766*, 769, 770, 777, 780
- tigris 757, 762, 764, 777, 780*
- uncia 757
Pantherinae 757, 764, 769, 776, 780
Panulirus 308
- interruptus 607 (footnote), 685
Papilionidae 332 (pupa)
Papio 83, 762, 770
- anubis 132
- doguera 754, 781
- hamadryas 136, 150→, 755, 781
- leucophaeus 754, 781
- porcarius 773
- sphinx 754, 781
Parabuthus brachystylus 320
- flavidus 320
- laevifrons 321*
- planicauda 320
Parapenaeus akayebi (see Metapenaeopsis barbatus)
- acclivis (see Metapenaeopsis acclivis)
Paratylotropidia brunneri 329, 611
Pardosa lugubris 328
Parnassius mnemosyne 330
Paropia scanica 396, 594
Paroxya atlantica 421*
Parrot 239, 241, 542, 715, 733*
Parus ater 732
- atricapillus 71, 711 ,
- caeruleus 725, 732, 742
- carolinensis 85
- major 235, 738*, 742*
Passalidae 330, 331* (larva), 589
Passalus 605
- cornutus 289*, 349
Passer domesticus 39*, 41*, 42*, 713, 723, 732, 735
Passerherbulus (see Ammospiza)
Passeriformes (818)
Passerines 722, 737
Patagonian cavy (Dolichotis) 753
Peccary (see Dicotyles)
Pelecaniformes (817)
Pelobatidae 695, 697
Peloridiidae 393
Pemphigostola synemonistis 294, 295*
Penaeidae 305

Penaeopsis stridulans 305, 318, 336
Pentaneura aspersa 420
Pentatomidae 392, 403, 404, 408
Pentatominae 403, 404, 407, 409
Pentatomoidea 402, 403
Pentatomomorpha 402, 404, 406, 408, 409
Perca fluviatilis 662
Percidae 662
Perch *(Perca)* 662
Pergesa elpenor 376*
Periophthalmus koelreuteri 662
Periplaneta 420, 423, 424*, 645
– *americana* 412*, 423, 475, 482
Perissodactyla (823)
Perlidae 328
Peromyscus 248
Petaurus 763
Petromyzon 530
Petromyzontes (814)
Phacochoerus 762
– *aethiopicus* 758, 762*
Phalacrocorax carbo sinensis 88, 105, 712
Phaneroptera 419*
Phanaeropteridae 348
Phascolarctos cinereus 753, 763
Phasianus colchicus 81, 713, 723, 737*
Phasmoptera (806)
Pheasant (see *Phasianus colchicus*)
Pheuticus ludovicianus 238
Phileurus 290
Philomachus pugnax 719
Phoca vitulina 496*, 757, 771
Phocaena phocaena 223, 502*
Phoenicurus phoenicurus 732
Pholidoptera aptera 421, 611, 616
– *cinerea* (see *griseoaptera*)
– *griseoaptera* 584, 588, 611, 613 (footnote), 618
Phonapate nitidipennis 281*
Phonogaster cariniventris 286
Phonygammus kerandrenii 493*
Phoxinus laevis 102, 524, 658*, 663, 666, 667, 676
 phoxinus 662, 667, 669
Phyllium athanysus 279
Phyllomorpha 328, 403, 798
Phyllopalpus pulchellus 647, 648
Phyllophoridae 291, 300
Phylloptera 419*
Phylloscopus collybita 231, 742
– *sibilatrix* 231, 742
– *trochilus* 740, 742
Phyllostomidae 207, 208, 210
Phyllostomus hastatus panamensis 210
Phymata wolffi 798
Phymateus 329
Phymatidae 281, 404
Physemacris 286
– *variolosa* 285*

Physeter catodon 74, 81, 220
Phytodyctus polyzonias 334 (pupa)
Pica pica 39*, 88, 105, 723, 735
Piciformes (818)
Pictinus 286*, 403
Piesma quadrata 402*, 406
Piesmidae 392, 403, 408
Piezosterninae 403
Pig 496, 499, 501, 516, 758, 763*, (see also *Sus scropha*)
Pigeon 74, 78, 532, 725
–, Trumpeter 231
–, Wood *(Columba palumbus)* 719, 722
Pika *(Ochotona)* 232
Pike *(Esox lucius)* 657
Pimephalus notatus 662
Pine marten *(Martes martes)* 756
Pinicola enucleator 738
Pinnipedia 756, 774, 779, 780, (822)
Pipidae 696
Pipistrellus subflavus 193*
Pipit, Meadow *(Anthus pratensis)* 235, 238, 732
Pirates 405
Pisaura mirabilis 328
Pissodes 293
Pistol shrimp 306–307
Pizonyx vivesi 205
Plagithmysus 291, 296
Platagarista tetraplura 294
Platanista 790
Platycerus 330 (larva)
Platydactylus 419*
Platypleura 396*, 398
– *capitata* 395*, 397*, 399*
– *octoguttata* 349, 397*, 399*, 400
– *westwoodii* 399*
Platyrrhini 781
Plea 415
Pleidae 406
Plecoptera (806)
Plecotus 190, 192*, 204, 205, 208
– *auritus* 190, 192*, 194
– *rafinesquei* 205, 789
Plectognaths 678
Plethodontidae 694
Plotheia decrescens 333 (pupa)
Plusia gamma 376*
Pneumora 286
Pneumoridae 286
Podicepitiformes (817)
Podiceps cristatus 720
Podisma 333 (nymph), 608 (nymph)
Poëphagus grunniens 759, 764*, 770
Pogonias 672, 679, 680
– *chromis* 676, 682
Pogonomyrmex 590
– *marcusi* 283*

Polecat *(Mustella putorius)* 756
Polistes 380, 381
Polyphylla 605
– *fullo* 288
Pomacanthus 679
– *arcuatus* 685
Pomacentridae 685
Pomacentrus leucostictus 676
Pomponia intemerata 621
Poneridae 283
Pongidae 755
Pongo pygmaeus 496, 498, 500, 505, 510, 511, 512*, 513*, 516*, 518*, 767, 774
Pontonia 306
Porcupine *(Atherurus)* 500, 516
– *(Erethizon)* 753
– *(Hystrix)* 761
Porpoise
–, Common *(Phocaena)* 223, 502*
–, White *(Delphinapterus)* 71, 220, 790
Porthetini 298
Potamochoerus larvatus 758
– *penicillatus* (see – *porcus*)
– *porcus* 758
Potamon 312
– *africanum* 312, 313*
Potos flavus 753
Pouched mouse *(Sminthopsis crassicaudata)* 762
Prairie dog *(Cynomys)* 71, 83, 126, 753, 772
Prepona 605
Presbytes entellus 755
Primates 496, 499, 500, 505, 519, 762, 767, 773, 774, 776, (820)
Priobium castaneum 280*
Prionotus 71, 674, 675*, 679
– *carolinus* 659*, 675, 682
– *evolans* 682
Prionus coriarius 296
Proboscoidea (823)
Procolobus 782
Proculus 290
Prodenia evidania 218, 421, 428
Proechimys 767
Prophalangopsidae 282*, 299, 337, 338 (footnote)
Prosimii 763
Protopterus 105
Prunella modularis 729
Psectrocnemus longiceps 298*
Psectrocoris 403
Pseudacris 700, 701
– *clarki* 701, 703, 706
– *feriarum* 701, 703
– *nigrita* 701, 703, 706
– *ornata* 701, 804
– *streckeri* 701, 704, 706, 804

Pseudaxis sika 758
Pseudozius bouvieri 311
Psittaciformes (818)
Psocoptera (812)
Psocus 290 (footnote), 326
Psophus stridulus 327, 328
Psorophora confinnis 625
– *horrida* 625
Psychidae 436 (footnote)
Psyllidae 393, 401, 402, 798
Ptenidiophyes 405
Ptenochirus 214
Pteronura brasiliensis 756
Pterophylla camellifolia 299, 300*, 303, 348, 358*, 366, 421, 587, 605, 611, 616, 636, 647
Pteropus 211, 214
– *giganteus* 211
– *medius* 754
– *poliocephalus* 211
Ptilonorhynchidae 238
Puffer fish *(Spheroides maculatus)* 676
Pulchriphyllium crucifolium 279*
Puma concolor 757, 762*, 766, 780*
Purana campanula 399*
Pygoscelis 88
Pyralididae 415, 436, 800
Pyrameis 424*
– *atalanta* 423
Pyrrhula pyrrhula 231, 233–235, 239–241, 543, 710, 718, 719, 721, 723, 726, 728, 732, 735, 736, 738, 741*, 745
Pyrrhulina rachoviana 662

Quail, Wood *(Odontophorus guyanensis marmoratus)* 82
Quelea quelea 86 (footnote)
Quiscalus quiscula 231, 238

Rabbit *(Oryctolagus)* 90, 104, 761
Raccoon-like dog (see *Nyctereutes*)
Raja clavata 525*, 658*, 660*, 661, 664
Ralliformes (818)
Rana 695
– *areolata* 702
– *catesbeiana* 705
– *clamitans* 703–705
– *esculenta* 490
– *pipiens* 71, 702, 703, 705
– *sylvatica* 537
Ranidae 702, 703
Ranatra quadridentata 290
Rangifer tarandus 758
Rat (see *Rattus*)
Ratel *(Mellivora)* 724, 756
Rattus 74, 103, 247 →, 535, 763, 771
– *norvegicus* 215, 753, 760

ALPHABETIC INDEX

Redstart *(Phoenicurus phoenicurus)* 732
Reduviidae 281, 404, 407
Regulus regulus 732, 743
Rhacophorus leucomystax 696
Rhinoceros unicornis 757
Rhinolophidae 71, 199–208, 215
Rhinolophus 199, 200*, 201*, 203 (footnote), 204, 789
– *ferrum-equinum* 199, 200*, 201
– *hipposideros* 199, 200
Rhinopoma microphyllum 207, 208
Rhinopomatidae 207, 208
Rhinotermes 327
Rhodia fugax 331 (larva)
Rhodnius prolixus 423
Rhodogastria bubo 282, 325, 605
– *lupia* 282*, 325
Rhombus maximus 58*
Rhopalocera 435
Rhopalurus borelli 320, 321*
Rhynchiscus naso 207, 208, 210
Rhysida nuda-togoensis 322
Richmondena cardinalis 238
Rihana mixta 399*
Ringtail (see *Bassariscus*)
Riparia riparia 231
Robin
–, American *(Turdus migratorius)* 238, 732
–, European (see *Erithacus rubecula*)
–, Sea (Triglidae: *Prionotus*)
Rodents 215, 216, 501, (821)
Roe (see *Capreolus*)
Rook *(Corvus frugilegus)* 88, 93, 720–723
Rossius 403
Round winged katydid (see *Amblycorypha rotundifolia*)
Rousettus 185, 214, 752
– *aegyptiacus* 211–214
– *amplexicaudatus* 212
– *seminudus* 212
Rucervus duvauceli 758
Ruff *(Philomachus pugnax)* 719
Ruminantia 767, 773
Rupricapra 761
– *rupricapra* 759
Rutelinae 330, 331
Rutilus rutilus 662

Saccopteryx bilineata 207, 208
Saimiri sciurea 754
Salamander *(Desmognathus, Salamandra)* 694, 685
–, Arboreal *(Aneides lugubris)* 694, 695
–, Japanese giant *(Megalobatrachus japonicus)* 694
–, Pacific giant *(Dicamptodon ensatus)* 694, 695
–, Spotted *(Ambystoma maculata)* 694, 695
–, Tiger *(Ambystoma tigrinum)* 694
–, Two lined *(Eurycea bislineata)* 694, 695
Salamandra salamandra 694, 695
Salamandridae 694
Saldidae 405
Salientia (816)
Salmon 670
Salticidae 583, 599
Sand martin *(Riparia riparia)* 231
Sarcophilus 761
Sargus annularis 662, 665, 667
Sarigue *(Caluromys philander)* 762
Saturniidae 74, 331 (larva), 322 (pupa), 435, 436 (footnote)
Satyridae 415, 423
Saxicola rubetra 732
– *torquata* 741*
Scabbard fish 673
Scaphiopus 695–697, 794
– *bombifrons* 696, 698
– *couchi* 696–698*, 703
– *hammondi* 696, 698
– *holbrooki* 696, 697, 703
– *hurteri* 698, 703
– *intermontanus* 698, 704
– *multiplicatus* 696
Scarabaeidae 287, 336
Scarabaeoidea 291, 330 (larva)
Schistocerca gregaria 59, 299 (footnote), 329, 427, 447, 589
Schizopteridae 405
Sciaenidae 655, 662, 672, 673, 677, 679, 680, 682
Sciocoris cursitans 404*
Sciurus 769, 771, 776
– *sciurus* 753
Scolopax rusticola 719, 732
Scolytidae 281
Scolytus destructor 280*
Scombridae 682
Scorpion 318–320
Scorpionidea (814)
Screamer, Common *(Chauna torquata)* 82
Scudderia 299
– *curvicauda* 300*, 348, 595
– *furcata* 300, 595
– *septentrionalis* 359
– *texensis* 300, 595
Scutelleridae 403
Scutellerinae 403
Scutigera decipiens 321
Seal 502, 516, 518, 544
–, Eared (see *Arctocephalus, Callorhinus, Otaria,* and *Zalophus*)
–, Elephant (see *Mirounga*)
–, Fur (see *Arctocephalus*)
–, Harbour (see *Phoca vitulina*)
–, Hooded (see *Cystophora cristata*)

Sea robin (Triglidae: *Prionotus*)
Sea trout *(Cynoscion regalis)* 670, 682
Sehirinae 403
Sehirus bicolor 406*, 408, 594, 799
– *melanopterus* 798, 799
Selenarctos tibetanus 756
Selenocosmia crassipes 316, 317*
Selenogyrus 316
– *aureus* 316*
Semnopithecus entellus 772
Semotilus atromaculatus 667, 668*, 669
Serica brunnea 281, 304
Serinus canaria 71, 231, 232, 234, 713, 719, 726, 729, 732
Serranidae 672, 679, 682, 685
Serval (see *Felis serval*)
Sesarma 314
Shark 663, 670, 673, 687
–, Lemon (see *Negaprion*)
Sheep (see *Ovis aries*)
Shield bearer (see *Atlanticus*)
Shieldbugs 391, 392, 402–404
Shrew 607 (see also *Crocidura*)
Shrike 85
–, Red backed (see *Lanius collurio*)
Shrimp (see Penaeidae, Palaemonidae)
–, Mantis (see *Gonodactylus, Squilla*)
–, Pistol (*Alpheus*) 306, 349, 354
–, Snapping (see Pistol shrimp)
Siagona 292
– *fuscipes* 291*
Sialia sialis 238
Sibinia 293
Sicarius hahni 316*
Sigara 280, 292, 405
– *distincta* 405
– *dorsalis* 407, 408
– *striata* 407
Silphidae 287, 336
Siluridae 662, 672
Siluroidea 677, 678, 679
Simia (Macaca) speciosa 754
Simocyoninae 773
Sinea 798
Sinodendron 333 (larva)
Sirenia (823)
Sitta europaea 739
Skate (see *Raja clavata*)
Skunk (see *Mephitis*)
Sloth *(Bradypus)* 771
Smerinthus populi 323 (footnote)
Sminthopsis crassicaudata 762
Snapper 670
Snipe *(Capella)* 76, 712, 719
Sole (*Solea solea*)
Solea solea 58*
Solenopsis fugax 304

Solifugae (814)
Spadefoot, Couch's *(Scaphiopus couchi)* 696–698, 703
–, Eastern *(Scaphiopus holbrooki)* 696, 697, 703
–, Great basin *(Scaphiopus intermontanus)* 698, 704
–, Hammond's *(Scaphiopus hammondi)* 696, 698*
–, Hurter's *(Scaphiopus hurteri)* 698, 703
–, Plains *(Scaphiopus bombifrons)* 696, 698
Sparidae 662
Sparrow 85, 88
–, American song (see *Melospiza melodia*)
–, Henslow's *(Ammospiza henslowii)* 732, 735
–, House (see *Passer domesticus*)
–, White crowned *(Zonotrichia leucophris)* 732, 795, 796
Sparrow hawk *(Falco sparverius)* 82, 735
Spathocera 403
Spea 696–698
Spearfish 679
Speothos venaticus 779
Spercheus 605
– *emarginatus* 287
Spermophile (see *Citellus*)
Sphaerocorini 403
Sphaeropoeus musicus 322*
Sphaeropoeus volzi 321
Sphaerotheriidae 321
Sphaerotherium acteon 323*
– *anomalum* 322*
– *campanulatum* 322*
– *coquerelianum* 322*
Sphecodina abbotti 331 (larva)
Spheniscus 88
Sphenisciformes (818)
Spheroides maculatus 676
Sphingidae 331, 435
Sphingonotus coerulans 421
Sphinx ligustri 376*
Spider 74, 315–318, 328, 336, 420, 423
–, Harvest 318
Spinus tristis 729
Sporothrips 291
Springbok *(Antidorcas marsupialis)* 759
Spring peeper *(Hyla crucifer)* 701, 704
Sprinkled locust (see *Chloealtis conspera*)
Squamata (817)
Squilla empusa 305
– *mantis* 305
Squirrel (see *Sciurus*)
–, Ground (see *Citellus*)
Squirrelfish *(Holocentrus ascensionis)* 677, 684, 685
Stag (see *Cervus*)
Starling (see *Sturnus vulgaris*)

Stauroderus scalaris 594, 600
Steatoda bipunctata 318, 349
— *castanea* 317, 349
Steatornis caripensis 12, 95 (footnote), 216, 217, 741, 745
Stelgidocoris 403
Stenella 220
Stenobothrus lineatus 297, 303, 349, 353, 358*, 359, 361, 584, 588, 594, 595, 604, 617
— *nigromaculatus* 594, 596, 600
— *rubicundus* 330
— *stigmaticus* 303, 594, 596
Stenopoda 798
Sterna hirundo 80, 717
Sternorrhyncha 392, 393, 401, 402, 798
Stetophyma grossum 298
Stick insect (see Palophinae)
Stickleback, Threespined 613 (footnote)
Stigicola dentatus 224
Stilnoptia salicis 283
Stiretrus 404
Stollia (see *Eysarcoris* 798)
Stomatopoda (806)
Stonefly (see Plecoptera)
Stork (Ciconiidae) 78, 82, 714, 719
Streptopelia risoria 231
Stridulattus stridulans 316
Stridulovelia 292, 405
Strigicoris 403
Strigidae 78, 83, 86, 214, 216, 542, 737
Strigiformes (818)
Strix aluco 735
Stromateus niger 670
Strombosoma unipunctatum 798
Sturnella magna 231, 237, 715, 795
— *neglecta* 231, 237, 240, 734
Sturnus vulgaris 44, 85, 87, 90 (footnote), 91, 93, 107, 238, 239, 712, 713, 717, 722, 723, 725, 732, 735
Suidae 763, 770, 772, 781
Suinae 781
Suricata 607, 663
— *tetradactyla* 756, 767*
Sus scrofa 71, 68, 102 (footnote) 506*, 758
Swallow *(Hirundo rustica)* 231, 732
Swift 218
Sylvia borin 741
— *communis* 238
— *curruca* 238
Symitha molalella 332 (pupa)
Symphalangus syndactylus 511, 755, 782*
Synalpheus 62, 306
— *lockingtoni* 307
Synapsis 291
Syntermes 327
Syrphidae 386–388
Syrrhophus marnocki 699, 700*, 704

Tachycines asinamorus 423
Tadarida 204
Taeniopoda picticornis 324
Tanna 398
— *japonensis* 422
Taphozous 207, 208
Tapirus terrestris 757, 763, 765*, 776, 780
Tarbagan (see *Marmota sibirica*)
Tarentula pulverulenta 328
Taricha similans 694
— *torosa* 694
Tarsius 505
Tasmanian devil *(Sarcophilus)* 761
Taurocerastes 291
Taurotragus oryx 758
Teleosts 531, 539, 541, 657, 659, 664, 665
Tench *(Tinca tinca)* 662
Tenebrionidae 281, 287, 336
Termes lilljeborgi 590
Termites (see Isoptera)
Termitogaster emersoni 327
Tern, 85, 719
—, Common *(Sterna hirundo)* 80, 717
Terpnosia ransonetti 397*, 399*
— *ridens* 399*
— *stipata* 399*
Tessaratoma papillosa 289*, 290
Tessarotomidae 403
Tessaratominae 403
Tetracis crocallata 437
Tetramorium caespitum 350
Tetrao 80
Tetraodontidae 680, 685
Tetrix subulata 596
— *tenuicornis* 594, 596
— *turki* 596
Tettigadinae 397
Tettigarcta tomentosa 393, 394*, 396*, 400, 401*
Tettigarctidae 393
Tettigonia 301, 420, 422, 423, 427, 430*
— *cantans* 351*, 421, 423, 431*, 584
— *viridissima* 100 (footnote), 348, 353, 421, 429, 584, 618, 638, 644*, 645, 647
Tettigoniidae 348, 351*, 418, 419*, 421, 425*, 440, 595
Tettigonioidea 299, 302, 335, 347*, 350, 352*, 355, 357*, 360, 364, 365, 422, 587, 599*, 605, 639, 642
Tetyra 284
Tetyrinae 403
Thalarctos maritimus 756, 771
Thalassina anomala 309
Thalassinidea 309
Thamnotrizon (see *Pholidoptera*)
Thaumastella aradoides 798
Thecophora fovea 294, 295*
Theridiidae 317, 336, 349

Theridium ovatum 317*
Theropithecus 755
Thoracica (805)
Thos 762, 776, 778, 779
Thrasher, Bown *(Toxostoma rufum)* 238
Threespined stickleback 613 (footnote)
Thrush 732
–, Mistle (see *Turdus viscivorus*)
–, Song (see *Turdus ericetorum*)
–, Wood (see *Hylocichla mustelina*)
Thryomanes bewicki 77, 721
Thryothorus ludovicianus 730, 731*, 735
Thylacinus 763
Thysanoptera (812)
Tibicen auletes 647
– *canicularis* 349, 647
– *dorsata* 621
– *linnei* 647
Tiger (see *Panthera tigris*)
Tilapia 655
Tinca tinca 662
Tipula 379
Tit 85, 722, 743
–, Blue (see *Parus caeruleus*)
–, Coal (see *Parus ater*)
–, Great (see *Parus major*)
–, Willow (see *Parus atricapillus*)
Toad 76
–, American *(Bufo americanus)* 230*, 698, 703, 704, 803
–, Canyon *(Bufo punctatus)* 698, 704, 803
–, Colorado *(Bufo alvarius)* 696, 699
–, Dakota *(Bufo hemiophrys)* 698
–, European fire-bellied *(Bombina bombina)* 696
–, Giant *(Bufo marinus)* 699, 703
–, Great plains *(Bufo cognatus)* 698, 699, 704
–, Green *(Bufo debilis)* 698, 703, 704
–, Gulf coast *(Bufo valliceps)* 230, 699
–, Mexican *(Bufo valliceps)* 230, 699
–, – narrow-mouthed *(Hypopachus cuneus)* 702
–, Midwife *(Alytes obstetricans)* 695
–, Narrow-mouthed *(Microhyla)* 706
–, Oak *(Bufo quercicus)* 699, 703
–, Sonoran *(Bufo compactilis)* 698, 704, 803
–, Southern *(Bufo terrestris)* 698, 706
–, Southwestern *(Bufo microscaphus)* 698, 704
–, Western *(Bufo boreas)* 698
–, Woodhouse's *(Bufo woodhousei)* 230*, 698, 704
–, Yosemite *(Bufo canorus)* 698
Toadfish (Batrachoididae: *Opsanus)* 679–682, 686
Tomodactylus 695
– *nitidus* 699, 700*
Toxoptera aurantii 284*, 401, 617
– *citricidus* 401

– *coffeae* (see *aurantii*)
– *odinae* 401
– *taversi* (see *citridus*)
Toxostoma rufum 238
Trachurus trachurus 670
Trachypterella andersoni 286
Tree cricket (see *Oecanthus*)
Tree-creeper, Short-toed *(Certhia brachydactyla)* 231
Treefrog
–, Arizona *(Hyla wrightorum)* 701
–, Barking *(Hyla gratiosa)* 701, 702*, 704, 706
–, Canyon *(Hyla arenicolor)* 700, 701*, 704
–, Common *(Hyla versicolor)* 700, 701*, 703, 704, 803
–, Cuban *(Hyla septentrionalis)* 701
–, Green *(Hyla cinerea)* 700, 702*
–, Mexican *(Hyla baudini)* 700
–, Pacific *(Hyla regilla)* 701
–, Pine barrens *(Hyla andersoni)* 701, 702*
–, Piney woods *(Hyla femoralis)* 700, 701*
Tremarctos ornatus 756
Triatominae 405
Trichechus manatus 757
Trichochos 670
Trichogaster leeri 661
– *trichopodus* 661
Trichosurus vulpecula 769
Tridactylidae 279, 305
Triggerfish (Balistidae: *Balistes*) 659, 678, 682
Triglidae 666, 669, 672, 677–679, 682, 683, 685, 686
Trioza acutipennis 401
– *nigricornis* 401
Triturus alpestris 694
Trizopagurus 310
– *strigimanus* 310
– *tenebrarum* 310
Troglodytes aëdon 77
– *troglodytes* 231, 712, 715, 725, 741*, 743*
Trox 293
Trumpet bird *(Phonygammus kerandrenii)* 493*
Tuco-tuco (see *Ctenomys*)
Tuna 671
Turbot (see *Rhombus maximus*)
Turdus ericetorum 231, 234, 238, 732
– *merula* 71, 81, 103, 231, 233, 234, 240, 711, 720, 723, 725, 726, 729, 732, 733, 741*
– *migratorius* 238, 732
– *viscivorus* 231, 729, 741*
Turkey, Domestic 81, 104, 713, 721, 737
Tursiops truncatus 71, 220, 223, 790
Turtle 491
Tylopsis liliifolia 595
Typhlocybidae 400
Typton 306
Tyto alba 216, 543*

Uakari (see *Cacajao*)
Uca 74, 314, 583, 584 (footnote), 587, 590, 599 606, 612, 618
— *musica* 314*
— *pugilator* 328, 586, 616
— *tangeri* 586, 597, 625 (footnote)
— *terpsichores* 314
Umbra 667
— *limi* 661
— *pygmaea* 661
Upupa epops 733
Urocyon 769, 778
— *cinereo-argentatus* 756, 778*, 779
Urodela (816)
Uroderma bilobatum 209, 210
Uromenus rugosicollis 357, 358*, 361
Ursus arctos 756, 770

Vampire (Desmodontidae) 208, 210
Vanessa io 329
Velia 292
Velifer hypselopterus 686
Veliferidae 686
Veliidae 405
Vertebrata (814)
Vespa 380, 424*
— *crabro* 334 (larva), 423, 590 (larva; footnote)
— *vulgaris* 376, 379
Vespertilionidae 71, 185–215
Vole (see *Microtus*)
—, Bank (see *Clethrionomys*)
—, Continental field (see *Microtus arvalis*)
Vombatus ursinus 762
Vulpes 756, 759, 763–765, 767, 768, 770, 771, 776, 778
— *corsac* 756
— *vulpes* 71, 753, 760*, 763*, 765, 769–772, 775, 777–779*
Vulture 74

Wagtail 732
Walrus (see *Odobenus*)
Wapiti (see *Cervus canadensis*)
Warble-fly 387
Warbler
—, Cetti's *(Cettia cetti)* 711, 712
—, Garden *(Sylvia borin)* 741
—, Grasshopper *(Locustella naevia)* 231
—, Icterine *(Hippolais icterina)* 732*

—, Marsh (see *Acrocephalus palustris*)
—, Melodious *(Hippolais polyglotta)* 732
—, Sedge *(Acrocephalus schoenobaenus)* 732
Wart-hog *(Phacochoerus)* 762
Waterbug 391, 402, 405–409
Weakfish *(Cynoscion regalis)* 670, 682
Weasel (see *Mustela nivalis*)
Whale 538*, 545
—, Beaked *(Mesoplodon)* 790
—, Bottlenose *(Hyperoodon)* 790
—, Blue *(Balaenoptera)* 223
—, Fin *(Balaenoptera)* 223
—, Humpback *(Megaptera)* 223
—, Pilot *(Globicephala)* 790
—, Sperm *(Physeter)* 74, 81, 220
Whinchat (see *Saxicola rubetra*)
White-ants (see Isoptera)
Whitethroat (see *Sylvia communis*)
—, Lesser (see *Sylvia corruca*)
Whooping fish 675
Wolf (see *Canis lupus*)
Woodcock *(Scolopax rusticola)* 719, 732
Woodpecker 61*, 76
—, Black 74
—, Great spotted *(Dendrocopus major)* 74, 718, 745
—, Green 722
Wood rail *(Aramides cajanea)* 82
Wren, Carolina (see *Thryothorus ludovicianus*)
—, European (see *Troglodytes troglodytes*)
—, House *(Troglodytes aëdon)* 77
Wryneck 85

Xenomystus nigri 224, 686
Xestobium tesselatum 326*
Xyronotus aztecus 286*

Yak (see *Poëphagus grunniens*)
Yellowhammer (see *Emberiza citrinella*)
Yellowthroat
—, Maryland (see *Geothlypis trichas*)

Zalophus californianus 757, 765
— *wollebaeki* 768
Zati 773
Zebu (see *Bos indicus*)
Zeidae 679
Zonothrichia leucophris 732, 795, 796
Zygiella x-notata 423

PRINTED IN THE NETHERLANDS BY
DRUKKERIJ MEIJER, N.V. WORMERVEER